Handbook of
BIOENERGY
CROP
PLANTS

Handbook of
BIOENERGY CROP PLANTS

EDITED BY

CHITTARANJAN KOLE
CHANDRASHEKHAR P. JOSHI
DAVID R. SHONNARD

CRC Press
Taylor & Francis Group
Boca Raton London New York

CRC Press is an imprint of the
Taylor & Francis Group, an **informa** business

CRC Press
Taylor & Francis Group
6000 Broken Sound Parkway NW, Suite 300
Boca Raton, FL 33487-2742

First issued in paperback 2019

ISBN-13: 978-1-4398-1684-4 (hbk)
ISBN-13: 978-0-367-38159-2 (pbk)

This book contains information obtained from authentic and highly regarded sources. Reasonable efforts have been made to publish reliable data and information, but the author and publisher cannot assume responsibility for the validity of all materials or the consequences of their use. The authors and publishers have attempted to trace the copyright holders of all material reproduced in this publication and apologize to copyright holders if permission to publish in this form has not been obtained. If any copyright material has not been acknowledged please write and let us know so we may rectify in any future reprint.

Library of Congress Cataloging-in-Publication Data

Handbook of bioenergy crop plants / edited by Chittaranjan Kole, Chandrashekhar P. Joshi, and David R. Shonnard.
    p. cm.
Includes bibliographical references and index.
ISBN 978-1-4398-1684-4
1. Energy crops. I. Kole, Chittaranjan. II. Joshi, Chandrashekhar P. III. Shonnard, David.

SB288.H36 2012
662'.88--dc23
    2011034452

Visit the Taylor & Francis Web site at
http://www.taylorandfrancis.com

and the CRC Press Web site at
http://www.crcpress.com

Dedicated to

Prof. M. S. Swaminathan
"A world scientist of rare distinction"

For his invaluable contributions to agricultural science and society

Contents

SECTION I

SECTION II

SECTION III

Preface

All living organisms need energy for their survival, and all societies need energy for development. However, the energy requirements of modern humans have become far more complex than that of just food for survival. We need energy to heat our homes, light up the night, undertake unprecedented feats of art and architecture, manufacture the goods on which our health and prosperity depend, and, most importantly, move ourselves and our goods to and fro. More than a century of past consumption and projections of population growth to 10 billion by 2100 will lead to future centuries of fossil fuel shortages unless renewable alternatives are developed. Without immediate and innovative measures, we face a progressive scarcity of the liquid petroleum products on which we depend. Bioenergy derived from biomass provides a glimmer of hope of at least partially resolving some of the most vexing and pressing issues of supplying humanity with transportation fuels.

The immediacy of the search for alternative and renewable sources of energy overshadows the work of nearly everyone, from scientists to politicians, farmers to soldiers. Liquid fossil fuel reserves, ominously, have been predicted to soon decline globally. Additional concerns about climate change, the energy demands of burgeoning populations, particularly in developing economies, and the political instability of petroleum availability and pricing have also called for attention and resources on this search. The real potential of domestic job creation for nations with plentiful biomass resources is also important. Bioenergy, in the form of cellulosic biomass, starch from crops, sugar from cane, and oils from plants, has emerged as a potential cost-effective and environmentally sustainable alternative feedstock. Dedicated energy crops whose growth does not compete with food production and that have a minimal ecosystem footprint have been developed or are in development to provide plant biomass for the production of bioenergy.

According to the U.S. Energy Information Administration (EIA), an independent agency that provides statistics and analysis of global energy needs and supplies, 2008 worldwide petroleum production and consumption was approximately 85,000,000 barrels every single day. However, biofuels accounted for only a tiny percentage of this, providing approximately 2% of only the liquid transportation fuel markets. The United States, the major importer and consumer of petroleum products, has recently mandated that approximately 18% of the liquid fuels used in the United States should be replaced with biofuels by the year 2022. Other countries have set similar goals. However, biofuel initiatives have been dogged by criticisms: they could change the current land usage to the detriment of its biodiversity and the economic status of its inhabitants; they could compete with food crops, resulting in food price increases; and, contrary to the general outlook, they could even increase greenhouse gases.

The engineering of biofuel production is dogged by its own issues and limitations. All biofuels are not created equal. Ethanol yields only 25–50% more energy than the fossil energy invested in its production, whereas biodiesel prodigiously gives 90%. As compared with fossil fuels, greenhouse gas emissions are reduced at least 12% by the production and combustion of corn ethanol and 41% by biodiesel. Starch and sugars are commonly used for making ethanol, but cellulosic ethanol could be derived from the cell walls of plants more plentifully and with less pollution than corn or sugar ethanol. However, difficult issues linger about producing sufficient biomass for future biofuel needs: the efficiency of bioconversion of recalcitrant cellulosic biomass to bioethanol and the cost competitiveness of biofuels over fossil fuels. As always, to properly evaluate and answer these questions, more research must be done.

The choice of plant species for biofuel production also requires additional thought and evaluation. In general, perennial plants will be preferred over annuals with unconventional sources such as algae. An ideal bioenergy crop would grow fast, prodigiously, and densely; use water and nutrients efficiently; tolerate stress well; have a low invasive potential; contain cell walls predisposed for high conversion to biofuels; and be cheaply harvestable—A very full wish-list indeed! It is unlikely that an existing plant possesses all of these characteristics, and, in any case, a crop's geographic region will dictate at least some of its necessary characteristics.

This three-section book is the first compilation of its kind. It provides the most current thinking in bioenergy production from crop plants with highlights of special requirements, major achievements, and unresolved concerns. We believe that this overview of currently used and emerging bioenergy crops is timely and necessary for academia, government agencies, and industry, and we hope that policy-makers and the general public may turn to it as a useful resource.

The first section covers general concepts of and concerns about bioenergy production and includes 12 chapters dealing with topics across a broad range of issues in the biofuel value chain, including several chapters on concepts and strategies of genetic improvement of bioenergy crops using conventional and molecular breeding, genetic engineering, and genomics; ecological issues and biodiversity of dedicated bioenergy crops; feedstock logistsics and enzymatic cell wall degradation to produce biofuels; a comprehensive review of current and future process technologies of liquid transportation fuels production; international standards for fuel quality; unique issues of biofuel-powered engines; life-cycle environmental impacts of biofuels compared with fossil fuels; and social concerns in areas of public policy, economics, and perceptions of biofuels. The second section deals with commercialized bioenergy crops and includes ten chapters covering potentially leading bioenergy crops or their groups such as cassava, jatropha, forest trees, maize, oil palm, oilseed Brassicas, sorghum, soybean, sugarcane, and switchgrass. The third section includes 11 chapters, 8 of which deliberate on emerging crops or their groups for bioenergy production, such as brachypodium, diesel trees, minor oilseeds, lower plants, paulowniea, shrub willow, sugarbeet, sunflower, and sweet potato. In addition, two chapters have been devoted to unconventional biomass resources, such as vegetable oils, organic waste, and municipal sludge.

The 33 chapters of this handbook have been contributed by 97 globally reputed experts from 12 countries, including Australia, Brazil, Canada, India, Malaysia, Mexico, New Zealand, Palestine, Slovenia, South Africa, Switzerland, and the United States. We extend our thanks to them for their useful contributions and for their cooperation throughout the period of preparation of this volume.

We are also thankful to Ms. Randy Brehm and Ms. Kari Budyk of CRC Press/Taylor & Francis Group for their constant cooperation since inception until completion of this book. We extend our thanks to Ms. Phullara Kole and Ms. Sandra Hubscher for their highly useful and timely editorial assistance.

With ongoing research and development efforts leading to discovery and production of additional bioenergy crops, we are certain that not all species were included in this work. However, we believe that this unique and exhaustive compilation about bioenergy crops will lead to future discoveries related to the use of plants for bioenergy production and will assist in developing novel ways of alleviating serious energy problems that we are facing in today's world. These are indeed challenging times, and we feel privileged to be part of this collaborative book project involving a community of highly talented scientists from all over the world. We are hopeful that this handbook will be a valuable resource for students, faculty, educational institutions, private companies, and government agencies interested in bioenergy production.

Editors

Chittaranjan Kole is an internationally renowned scientist with approximately 27 years of experience in teaching and research on plant genetics, genomics, and biotechnology. During this period, he has guided 30 students and published more than 140 research articles, most of which appear in the leading peer-reviewed journals and proceedings of international meetings. Several of his pioneering contributions, particularly on molecular evolution of flowering stress-related and highly repetitive genes, have paved the way for several new avenues for fellow scientists. His excellence in research and editing has been generously appreciated by many leading scientists of the world, including seven Nobel Laureates—Arthur Kornberg, Normal Borlaug, Werner Arber, Phillip Sharp, Gunter Blobel, Lee Hartwell, and Roger Kornberg.

Professor Kole has already edited 40 books and is presently editing another 40+ volumes for leading publishers such as CRC Press/Taylor & Francis, Springer, Wiley-Blackwell, and Science Publishers. He is an associate editor, editorial board member, regional editor, and member of the international advisory board of a number of leading journals, scientific networks, and publication houses. He has earned several awards and recognitions for his service to science and society.

Chandrashekhar P. Joshi is a professor of plant molecular genetics in the School of Forest Resources and Environmental Science (SFRES) at Michigan Technological University. His current research focuses on wood cell wall development in bioenergy trees, with particular emphasis on unraveling the process of cellulose synthesis for improved bioenergy production. Dr. Joshi has recently served as the director of the Biotechnology Research Center and also as the director of graduate programs at SFRES at Michigan Technological University. He has coauthored 100 presentations, published 67 papers in peer-reviewed journals/books, and served on 48 graduate student committees. Dr. Joshi teaches courses in genomics, molecular genetics, bioinformatics, and grantsmanship. He is a recipient of the National Science Foundation's CAREER award and Michigan Technological University's 2011 Research Award. He is a fellow of the Sustainable Futures Institute and an inductee of the Academy of Teaching Excellence at Michigan Technological University. Dr. Joshi is a coauthor of three approved U.S. patents related to cellulose synthesis for improved cellulose and reduced lignin production in bioenergy trees, and he has recently coedited *Genetics, Genomics and Breeding of Poplars*. Since 2010, he has participated as a distinguished visiting professor in the world class university–supported program in bioenergy science and technology at Chonnam National University in South Korea.

David R. Shonnard received a BS in chemical/metallurgical engineering from the University of Nevada, Reno in 1983, an MS in chemical engineering from the University of California, Davis in 1985, a PhD from the University of California, Davis in 1991, and postdoctoral training in bioengineering at the Lawrence Livermore National Laboratory from 1990–1993. He has been on the faculty of the Department of Chemical Engineering at Michigan Technological University since 1993. Dr. Shonnard has over 20 years of academic experience in sustainability issues in the chemical industry and green engineering. He is coauthor of the textbook *Green Engineering: Environmentally-Conscious Design of Chemical Processes*, published by Prentice Hall in 2002. His current research interests focus on investigations of new forest-based biorefinery processes for production of transportation fuels (such as cellulosic ethanol) from woody biomass using recombinant DNA approaches. Another active research area is life-cycle assessment (LCA) of biofuels and other biorefinery products to determine greenhouse gas emissions and net energy balances. His experiences in LCA methods and applications include a one-year sabbatical at the Eco-Efficiency Analysis Group at BASF AG in Ludwigshafen, Germany, and contributions to National Academy of Sciences publications on green chemistry/engineering/sustainability in the chemical industry. Dr. Shonnard has coauthored 70 peer-reviewed publications and has received numerous honors and awards for teaching and research into environmental issues of the chemical industry.

Contributors

Hani Al-Ahmad
Department of Plant Sciences
University of Tennessee
Knoxville, Tennessee

and

Department of Biology and Biotechnology
An-Najah National University
Nablus, Palestine

Alphonse Anderson
School of Natural Resources and Environment
University of Michigan
Ann Arbor, Michigan

Utku Avci
Complex Carbohydrate Research Center
University of Georgia
Athens, Georgia

Ajay K. Badhan
Department of Biological Sciences
University of Alberta
Edmonton, Alberta, Canada

Anjanabha Bhattacharya
National Environmentally Sound Production
    Agriculture Laboratory
University of Georgia
Tifton, Georgia

Ajaya Kumar Biswal
Complex Carbohydrate Research Center
University of Georgia
Athens, Georgia

Conrad Bonsi
College of Agricultural, Environmental, and
    Natural Resources
Tuskegee University
Tuskegee, Alabama

Michael A. Borowitzka
School of Biological Sciences and Biotechnology
Murdoch University
Murdoch, Australia

Jennifer N. Bragg
Western Regional Research Center,
    Agricultural Research Service
U.S. Department of Agriculture
Albany, California

Eckehard Brockerhoff
Scion
New Zealand Forest Research
    Institute
Christchurch, New Zealand

Michael J. Brodeur-Campbell
Department of Chemical Engineering
Michigan Technological University
Houghton, Michigan

Doug Brownell
Department of Agricultural and
    Biological Engineering
Pennsylvania State University
University Park, Pennsylvania

Marcos Silveira Buckeridge
Departament of Botany
University of São Paulo
São Paulo, Brazil

Kimberly D. Cameron
New York State Agricultural
    Experiment Station
Cornell University
Geneva, New York

Heitor Cantarella
Centre for Soil and Environmental
    Resources
Agronomic Institute
Campinas, Brazil

Michael D. Casler
U.S. Dairy Forage Research Center
    Agricultural Research Service
U.S. Department of Agriculture
Madison, Wisconsin

Feng Chen
Department of Plant Sciences
University of Tennessee
Knoxville, Tennessee

Yuen May Choo
Research & Development
Malaysian Palm Oil Board
Kajang, Malaysia

Sarah Collinson
Crop Genetics Research & Development
Pioneer Hi-Bred International, Inc.
Woodland, California

Carlos Henrique de Brito Cruz
Gleb Wataghin Institute of Physics
State University of Campinas
Campinas, Brazil

and

São Paulo Research Foundation (FAPESP)
São Paulo, Brazil

Anete Pereira de Souza
Center for Molecular Biology and Genetic
   Engineering
State University of Campinas
Campinas, Brazil

Kanwarpal S. Dhugga
DuPont Agricultural Biotechnology
Pioneer Hi-Bred International, Inc.
Johnston, Iowa

Ismail Dweikat
Agronomy and Horticulture Department
University of Nebraska–Lincoln
Lincoln, Nebraska

Marceline Egnin
College of Agricultural, Environmental, and
   Natural Resources
Tuskegee University
Tuskegee, Alabama

Rubens Maciel Filho
School of Chemical Engineering
University of Campinas
Campinas, Brazil

David Flaspohler
School of Forest Resources and Environmental
   Science
Michigan Technological University
Houghton, Michigan

Fatima Fofanah
Pennsylvania State University–Harrisburg
Middletown, Pennsylvania

Antonio Augusto Franco Garcia
Department of Genetics
University of São Paulo
Piracicaba, Brazil

V. Edwin Geo
Department of Mechanical Engineering
Anna University
Chennai, India

A. Gopinath
Department of Mechanical Engineering
Anna University
Chennai, India

Rajeev Gupta
DuPont Agricultural Biotechnology
Pioneer Hi-Bred International, Inc.
Johnston, Iowa

Robert J. Henry
Queensland Alliance for Agriculture and Food
   Innovation
University of Queensland
Brisbane Australia

Judson G. Isebrands
Environmental Forestry Consultants
New London, Wisconsin

Jaclyn E. Johnson
Mechanical Engineering/Engineering Mechanics
Michigan Technological University
Houghton, Michigan

Nicholas H. Johnson
Department of Energy and Mineral Engineering
Pennsylvania State University
University Park, Pennsylvania

Puthiyaparambil Josekutty
Pennsylvania State University–Harrisburg
Middletown, Pennsylvania

Nirmal Joshee
Agricultural Research Station
Fort Valley State University
Fort Valley, Georgia

Blake Lee Joyce
Department of Plant Sciences
University of Tennessee
Knoxville, Tennessee

Tom N. Kalnes
Renewable Energy and Chemicals
    Research
UOP LLC
Des Plaines, Illinois

Krishan L. Kalra
Department of Microbiology
Punjab Agricultural University
Ludhiana, India

Deepkamal Karelia
Pennsylvania State University–Harrisburg
Middletown, Pennsylvania

Nilkamal Karelia
Cheyney University of Pennsylvania
Cheyney, Pennsylvania

Gregory Keoleian
School of Natural Resources and
    Environment
University of Michigan
Ann Arbor, Michigan

Matias Kirst
School of Forest Resources and
    Conservation
University of Florida
Gainesville, Florida

Joseph E. Knoll
Crop Genetics and Breeding
    Research Unit, Agricultural
    Research Service
U.S. Department of Agriculture
Tifton, Georgia

Gerhard Knothe
National Center for Agricultural Utilization
    Research, Agricultural Research Service
U.S. Department of Agriculture
Peoria, Illinois

Gurvinder S. Kocher
Department of Microbiology
Punjab Agricultural University
Ludhiana, India

Pawan Kumar
National Environmental Sound Production
    Agriculture Laboratory
University of Georgia
Tifton, Georgia

Jeong-Dong Lee
Division of Plant Biosciences
Kyungpook National University
Republic of Korea

Jude Liu
Department of Agricultural and Biological
    Engineering
Pennsylvania State University
University Park, Pennsylvania

Ah-Ngan Ma
Engineering & Processing
    Research Division
Malaysian Palm Oil Board
Kajang, Malaysia

Abraham R. Martin-Garcia
Department of Chemical Engineering
University of Sonora
Hermosillo, México

Karen A. McDonald
Department of Chemical Engineering and
    Materials Science
University of California, Davis
Davis, California

Peter B.E. McVetty
Department of Plant Science
University of Manitoba
Winnipeg, Manitoba, Canada

Scott Miers
Mechanical Engineering/Engineering
  Mechanics
Michigan Technological University
Houghton, Michigan

Robert B. Mitchell
U.S. Department of Agriculture, Agricultural
  Research Service
University of Nebraska–Lincoln
Lincoln, Nebraska

Diego Morales
Pennsylvania State University–Harrisburg
Middletown, Pennsylvania

Desmond Mortley
College of Agricultural, Environmental,
  and Natural Resources
Tuskegee University
Tuskegee, Alabama

Jeffrey Naber
Mechanical Engineering/Engineering
  Mechanics
Michigan Technological University
Houghton, Michigan

G. Nagarajan
Department of Mechanical
  Engineering
Anna University
Chennai, India

Satya S. Narina
Department of Biology
Virginia State University
Petersburg, Virginia

Milton Yutaka Nishiyama, Jr.
Department of Biochemistry
University of São Paulo
São Paulo, Brazil

Henry T. Nguyen
National Center for Soybean
  Biotechnology and Division of
  Plant Sciences
University of Missouri
Columbia, Missouri

Karine Zinkeng Nyiawung
College of Agricultural, Environmental, and
  Natural Resources
Tuskegee University
Tuskegee, Alabama

Damaris Odeny
Biotechnology Platform
Agricultural Research Council
Onderstepoort, South Africa

Sivakumar Pattathil
Complex Carbohydrate Research Center
University of Georgia
Athens, Georgia

Steve Pawson
Scion
New Zealand Forest Research Institute
Christchurch, New Zealand

Shobha Devi Potlakayala
Pennsylvania State University–Harrisburg
Middletown, Pennsylvania

Sukumar Puhan
Department of Mechanical Engineering
GKM College of Engineering & Technology
Chennai, India

Hifjur Raheman
Agricultural and Food Engineering Department
Indian Institute of Technology
Kharagpur, India

Vaman Rao
Nitte University & NMAM Institute
  of Technology
Nitte, India

Tom L. Richard
Department of Agricultural and Biological
  Engineering
Pennsylvania State University
University Park, Pennsylvania

Donald L. Rockwood
School of Forest Resources
  and Conservation
University of Florida
Gainesville, Florida

Milenko Roš
Institute for Environmental Protection
    and Sensors
Maribor, Slovenia

Kusumal Ruamsook
Department of Supply Chain and
    Information Systems
Pennsylvania State University
University Park, Pennsylvania

Sairam Rudrabhatla
Pennsylvania State University–Harrisburg
Middletown, Pennsylvania

Sarah Ryan
Pennsylvania State University–Harrisburg
Middletown, Pennsylvania

Georgios Sarantakos
Swiss Federal Institute of Technology in
    Lausanne
Lausanne, Switzerland

Grover J. Shannon
National Center for Soybean
    Biotechnology and Division of
    Plant Sciences
University of Missouri
Columbia, Missouri

Sanjeev K. Sharma
DGM-Biotechnology, Biosciences Centre
Unichem Laboratories Ltd.
Pilerne, India

David R. Shonnard
Department of Chemical Engineering
Michigan Technological University
Houghton, Michigan

Rippy Singh
National Environmental Sound Production
    Agriculture Laboratory
University of Georgia
Tifton, Georgia

Shoba Sivasankar
DuPont Agricultural Biotechnology
Pioneer Hi-Bred International, Inc.
Johnston, Iowa

Lawrence B. Smart
New York State Agricultural Experiment Station
Cornell University
Geneva, New York

Barry D. Solomon
Department of Social Sciences
Michigan Technological University
Houghton, Michigan

Glaucia Mendes Souza
Department of Biochemistry
University of São Paulo
São Paulo, Brazil

C. Neal Stewart, Jr.
Department of Plant Sciences
University of Tennessee
Knoxville, Tennessee

Behnam Tabatabai
Pennsylvania State University–Harrisburg
Middletown, Pennsylvania

Muhammad Tahir
Department of Plant Science
University of Manitoba
Winnipeg, Manitoba, Canada

Rebekah Templin
Pennsylvania State University–Harrisburg
Middletown, Pennsylvania

Evelyn Thomchick
Department of Supply Chain and Information
    Systems
Pennsylvania State University
University Park, Pennsylvania

Vilas Tonapi
National Research Center for Sorghum
Hyderabad, India

Ludmila Tyler
University of California, Berkeley
Berkeley, California

and

Western Regional Research Center, Agricultural
    Research Service
U.S. Department of Agriculture
Albany, California

Alankar Vaidya
Matrix—The Innovation Center
Praj Industries Limited
Pune, India

Babu Valliyodan
National Center for Soybean Biotechnology
    and Division of Plant Sciences
University of Missouri
Columbia, Missouri

Marie-Anne Van Sluys
Department of Botany
University of São Paulo
São Paulo, Brazil

Barrett Vaughan
College of Agricultural, Environmental and
    Natural Resources
Tuskegee University
Tuskegee, Alabama

N. Vedaraman
Department of Chemical Engineering
Central Leather Research Institute
Chennai, India

K. C. Velappan
Department of Chemical Engineering
Central Leather Research Institute
Chennai, India

John P. Vogel
Western Regional Research
    Center, Agricultural Research
    Service
U.S. Department of Agriculture
Albany, California

Kenneth P. Vogel
Agricultural Research Service
U.S. Department of Agriculture
University of Nebraska–Lincoln
Lincoln, Nebraska

Christopher R. Webster
School of Forest Resources and
    Environmental Science
Michigan Technological University
Houghton, Michigan

Jeremy Worm
Mechanical Engineering/Engineering
    Mechanics
Michigan Technological University
Houghton, Michigan

Chee-Liang Yung
Engineering & Processing Research
    Division
Malaysian Palm Oil Board
Kajang, Malaysia

Carla D. Zelmer
Department of Plant Science
University of Manitoba
Winnipeg, Manitoba, Canada

J. Y. Zhu
Forest Products Laboratory
U.S. Department of Agriculture
Madison, Wisconsin

Gregor D. Zupančič
Institute for Environmental Protection
    and Sensors
Maribor, Slovenia

Abbreviations

2,4-D	2,4-dichlorophenoxyacetic acid
3-PGA	3-phosphoglyceric acid
4CL	4-coumarate CoA ligase
ACGT	Asiatic Centre for Genome Technology
ABNT	Brazilian Association for Technical Standards
ABRACAVE	Associação Brasileira de Florestas Renováveis
ACF	Agricultural Consultative Forum
ACP	acyl-carrier protein
ADC	analog-to-digital converter
ADH	alcohol dehydrogenase
ADP	adenosine 5′ diphosphate
ADPGlc	adenosine 5′ diphosphate glucose
AFEX	ammonia fiber explosion
AFLP	amplified fragment length polymorphism
AGPase	ADP-glucose pyrophosphorylase
AGP	arabinogalactan protein
ai	active ingredient
AMF	arbuscular mycorrhizal fungi
ANP	Agency of Petroleum, Natural Gas and Biofuels
AOSA	Association of Official Seed Analysts
APEC	Asia-Pacific Economic Cooperation
ARAD1	arabinan deficient 1
arad1	*ARABINAN DEFICIENT 1* (gene)
ARDD	Alberta Renewable Diesel Demonstration
ARP	ammonia recycle percolation
ARS	Agricultural Research Service of USDA
ASBR	anaerobic sludge blanket reactor
ASGR	apospory-specific genomic region
ASTM	American Society for Testing and Materials
ATP	adenosine 5′ triphosphate
B20	biodiesel blend (biodiesel 20% and diesel 80% on volume basis)
B40	biodiesel blend (biodiesel 40% and diesel 60% on volume basis)
BAC	bacterial artificial chromosome
Bc-1	brittle culm-1 (gene)
BDC	bottom dead center
BFDP	Bioenergy Feedstock Development Program
BglB	β-glucosidase
BIS	Bureau of Indian Standards
Bk2	brittle stalk-2 (gene)
bmr	brown midrib (gene)
BNF	biological nitrogen fixation
BOD	biochemical/biological oxygen demand
BPR	biogas production rate
BSEC	brake-specific energy consumption
BSFC	brake-specific fuel consumption
BSU	Bosch smoke unit

Bt	*Bacillus thuringiensis*
BTE	brake thermal efficiency
BTL	biomass-to-liquid
C_1	exo-1,4-β-glucanase
C10:0	capric (decanoic) acid
C12:0	lauric (dodecanoic) acid
C14:0	myristic (tetradecanoic) acid
C16:0	palmitic (hexadecanoic) acid
C16:1	palmitoleic (9(Z)-octadecenoic) acid
C18:0	stearic (octadecanoic) acid
C18:1	oleic (9(Z)-octadecenoic) acid
C18:2	linoleic (9(Z),12(Z)-octadecadienoic) acid
C18:3	linolenic (9(Z),12(Z),15(Z)-octadecadienoic) acid
C20:0	eicosanoic (arachidic) acid
C20:1	eicosenoic acid
C22:0	behenic (docosanoic) acid
C22:1	erucic (13(Z)-docosenoic) acid
C24:0	lignoceric (tetracosanoic) acid
C3H	*p*-coumarate 3-hydroxylase
C4H	cinnamate-4-hydroxylase
C6:0	caproic (hexanoic) acid
C8:0	octanoic (caprylic) acid
CA	crank angle
CAA	Clean Air Act
CAD	cinnamyl alcohol dehydrogenase
CAFÉ	corporate average fuel economy
CaMV	cauliflower mosaic virus
CAPS	cleaved amplified polymorphic sequences
CBD	cellulose-binding domain
CBH	cellobiohydrolase
CBM	carbohydrate binding modules
CBP	consolidated bioprocessing
cc	cubic capacity
CCR	cinnamoyl CoA-reductase
CDM	clean development mechanism
cDNA	complementary DNA
CED	cumulative energy demand
Cel1	*Arabidopsis* endo-(1-4)-β-glucanase gene
CEN	European Committee for Standardization
CesA	cellulose synthase
CesA	cellulose synthase-*A* (gene)
CesAs	cellulose synthase genes (*CesAs*)
Cesl	cellulose synthase like
CFPP	cold filter plugging point
CGSB	Canadian General Standards Board
CHP	combined heat and power (unit)
CI	compression ignition
CIAT	International Center for Tropical Agriculture
CIP	Centro Internacional de la Papa (International Potato Center)
CMB	cassava mealybug
CMC	carboxymethyl cellulose

CMD	cassava mosaic disease
CMV	cucumber mosaic virus
CN	cetane number
CO	carbon monoxide
CO_2	carbon dioxide
CO_2e	carbon dioxide equivalence
CobL4	cobra-like-4 (gene)
COC	copper oxychloride
COD	chemical oxygen demand
CP	constant pressure
cpDNA	chloroplast DNA
CP	cloud point
CPO	crude palm oil
CR	compression ratio
CRP	Conservation Reserve Program
CRW	corn root worm
Cry1Ab	cryptochrome 1
cSt	centistoke
CT	computer tomography
CTMP	chemical-thermomechanical pulp
CV	combustion value/constant volume
Cyt P450	cytochrome P450
DAG	diacyl glycerol
DArT	diversity array technology
DBH	diameter at breast height
DDGS	dried distiller's (distiller's dried) grains with solubles
DEB	Department of Energy business
deg	degree
DGAT1	diacylglycerol transferase 1
DGS	distiller's grains and solubles
DIN	Deutsches Institut für Normung (German Institute for Standardization)
DM	dry matter
DME	dimethyl ether
DOE	U.S. Department of Energy
DOY	day of year
DP	degree of polymerization
DW	dry weight
Dw	dwarfing
E10	fuel blend of 10% ethanol and 90% gasoline
E85	fuel blend of 85% ethanol and 15% gasoline
EA	*E. amplifolia*
EBAMM	ERG biofuel analysis metamodel
ECB	European corn borer
EFB	empty fruit bunches
EG	*Eucalyptus grandis*
EGR	exhaust gas recirculation
EGSB	expanded granular sludge bed digester
EGT	exhaust gas temperature
EISA	Energy Independence and Security Act
EMS	ethyl methanesulfonate
EPA	U.S. Environmental Protection Agency

EPAct	Energy Policy Act of 2005 (United States)
EPRI	Electric Power Research Institute (Palo Alto, CA)
ER	endoplasmic reticulum
EST	expressed sequence tag
EU	European Union
EU	*E. urophylla*
EUCAGEN	International *Eucalyptus* Genome Network
F5H	ferulate-5-hydroxylase
FA	fatty acid
FAD	fatty acid desaturase
FAE1	fatty acid elongase 1
FAEE	fatty acid ethyl esters
FAME	fatty acid methyl ester
FAO	Food and Agriculture Organization of the United Nations
FED	fossil energy demand
FFA	free fatty acid
FFB	fresh fruit bunches
FFV	flex fuel vehicles
FIBEX	fiber extrusion
FISH	fluorescence in situ hybridization
FNR	fast neutron radiation
FPP	farnesyl pyrophosphate
FPU	filter paper units
fra	fragile fiber (gene)
FT	Fischer–Tropsch
FTS	Fischer–Tropsch synthesis
FW	fresh weight
G	guaiacyl
GA	gibberellin
GalA	galacturonic acid
GATL	galacturonosyltransferase-like (gene)
GAUT1	galacturonosyltransferase 1
GAX	glucuronoarabinoxylan
GBSS	granule bound starch synthase
GD	green diesel
GDD	growing degree days
GDP	guanosine 5'-diphosphate
GEM	gene expression microarrays
GFP	green fluorescent protein
GGPP	geranyl geranyl pyrophosphate
GHG	greenhouse gas
GISH	genomic in situ hybridization
GL	gigaliter
GLA	gamma-linolenic acid
Glc	glucose
Glc-1-P	glucose-1-phosphate
Glc-6-P	glucose 6-phosphate
GM	genetically modified
GPI	glyco(syl)phosphatidylinositol
GPP	geranyl pyrophosphate
GREET	Greenhouse Gases, Regulated Emissions, and Energy Use in Transportation model

GRIN	Germplasm Resources Information Network (USDA)
Gt	gigaton
GT	glycosyltransferase
gt1	grassy tiller 1
GUS	beta-glucuronidase
GWh	gigawatt hour
GWP	global warming potential
H	p-hydroxyphenyl
ha	hectare
HC	hydrocarbon
HCl	hydrochloric acid
HDV	heavy duty vehicle
HEAR	high erucic acid rapeseed cultivar
HEV	hybrid electric vehicles
HFD	hydrotreated renewable diesel
HG	homogalacturonan
HG:α1,4GalAT	homogalactronanan α-1,4-galacturonosyltransferase
HI	harvest index
HMF	hydroxymethylfurfural
HOME	honge oil methyl ester
hptII	hygromycin-phosphotransferase
HR	herbicide resistance
HRGP	hydroxyproline-rich glycoprotein
HRD	hydrotreated renewable diesel
HRJ	hydrotreated renewable jet
HRR	heat release rate
HRT	hydraulic retention time
HVO	hydrogenated vegetable oil
IBP	International Biological Program (National Academy of Science, United States)
IC	internal combustion
ICRISAT	International Crops Research Institute for the Semi-Arid Tropics
IEA	International Energy Agency
IGCC	integrated gasification combined-cycle
IITA	International Institute of Tropical Agriculture
ILUC	indirect land-use change
indels	insertions and deletions
INMETRO	National Institute of Metrology, Standardization and Industrial Quality
IOP	injector opening pressure
IPC	International Poplar Commission
IPCC	Intergovernmental Panel on Climate Change
irx	irregular xylem (gene)
Irx4	irregular xylem4 (gene)
ISO	International Organization for Standardization
ISS	Institute of Science in the Society
ISSR	inter-simple sequence repeat
ITS	internal transcribed spacer
IU	international unit
IV	iodine value
IVDMD	in vitro dry matter digestibility
JB	jatropha biodiesel
JBB	jatropha biodiesel blend

JGI	Joint Genome Institute (USA)
JISC	Japanese Industrial Standards Committee
JO	jatropha oil
JO/D	jatropha oil/diesel
JOME	jatropha oil methyl ester
JSA	Japanese Standards Association
KAS	acyl-acyl carrier protein
KCS	beta ketoacyl-CoA synthase
kg	kilogram
KOH	potassium hydroxide
KOME	karanja oil methyl ester
kW	kilowatt
LB	Luria–Bertani
LCA	life-cycle assessment
LCI	life-cycle inventory
LD	linkage disequilibrium
LDV	light-duty vehicles
LG	linkage group
LOME	linseed oil methyl ester
LPG	liquid petroleum gas
LSU	livestock unit
LT	low temperature
LUC	land-use change
Ma	*Maturity*
ManS	mannan synthase (gene)
MARS	marker-assisted recurrent selection
MAS	marker-assisted selection
MBRC	miles between road calls
MDF	medium-density fiberboard
MDV	medium duty vehicle
MEP	2-C-methylerythritol-4-phosphate pathway
MESP	minimum ethanol selling price
MFO	medium fuel oil
Mg ha^{-1}	megagram per hectare
Mha	million hectare
MJ	megajoule
MLG	mixed-linkage glucan
mmBtu	million British thermal unit
MOME	mahua oil methyl ester
mpg	miles per gallon
MPOB	Malaysian Palm Oil Board
MSW	municipal solid waste
Mt	metric ton
MT/year	metric ton per year
MTBE	methyl tertbutylether
Mtoe	million ton of oil equivalent
MUFA	monounsaturated fatty acid
MVA	mevalonic acid pathway
MYR	Malaysian ringgit
N	nitrogen
NAA	napthalene acetic acid

NaOH	sodium hydroxide
NAS	National Academy of Sciences (United States)
NER	net energy ratio
NEV	net energy value
NEY	net energy yield
NGCC	natural gas combined-cycle
NMR	nuclear magnetic resonance
NOME	neem oil methyl ester
$NO_{x/x}$	oxides of nitrogen
NPGS	National Plant Germplasm System (USDA)
NPK	nitrogen phosphorus potash
NRC	National Research Council (Canada)
NSF	National Science Foundation (United States)
OBD	on-board-diagnostic
OCT	order cycle time
OEM	original equipment manufacturer
OLR	organic loading rate
OM	organic matter
OP	Oxford Paper
OPC	Oglethorpe Power Corporation/Open-pollinated cultivars
P450	P450 oxygenase of cytochrome
PAL	phenyl ammonia lyase
PCA	principal component analysis
PCR	polymerase chain reaction
PDC	pyruvate decarboxylase
PE	population equivalent
PER	petroleum energy ratio
PFAD	palm fatty acid distillate
pH	potenz hydrogen
Pi	inorganic phosphate, HPO_4^{2-}
PM	particulate matter
Pol	polarization value (a measure of sucrose content)
POME	palm oil mill effluent
PORIM	Palm Oil Research Institute of Malaysia
PPi	pyrophosphate anion, $P_2O_7^{4-}$
ppm	parts per million
PRP	proline-rich protein
PtAO	*Pinus taeda* P450 abietadienol/abietadienal oxidase
PTGS	post-transcriptional gene silencing
PtTPS-LAS	*Pinus taeda* abietadienol/levopimaradiene synthase
PUFA	poly-unsaturated fatty acid
PWB	*Paulownia* witch's broom
pyMBMS	pyrolysis molecular beam mass spectrometry
PZEV	partial zero emissions vehicle
QTL	quantitative trait loci
Qua2	Quasimodo2
R&D	research and development
R:S	root:shoot
R1	hydrodeoxygenation reactor
R2	hydroisomerization reactor
RAE	rapeseed methyl ester

RAPD	random amplified polymorphic DNA
RBDPO	refined, bleached and deodorized palm oil
RBDPOo	refined, bleached and deodorized palm olein
REE	rapeseed ethyl ester
RFLP	restriction fragment length polymorphism
RFS	Renewable Fuel Standards
RG-I	rhamnogalacturonan I
RG-II	rhamnogalacturonan II
RME	rapeseed methyl ester
RMP	refiner mechanical pulp
RNAi	RNA interference
ROME	rubberseed oil methyl ester
ROS	reactive oxygen species
rpm	revolutions per minute
RRPS	restricted recurrent phenotypic selection
RUBISCO	ribulose-1,5-bisphosphate carboxylase
S	syringyl
SAA	soaking in aqueous ammonia
SABS	South African Bureau of Standards
SBCR	slurry bubble column reactors
SBE	starch branching enzyme
SBP	specific biogas productivity
SBR	sequencing batch reactor
SCAR	sequence characterized amplified region
sCOD	soluble chemical oxygen demand
SCR	selective catalytic reduction
SFA	saturated fatty acid
SFP	single-feature polymorphism
SHF	separate hydrolysis and fermentation
SGI	Synthetic Genomics, Inc.
SI	spark ignition
SMD	Sauter mean diameter
SmF	submerged fermentation
SNP	single nucleotide polymorphism
SOC	soil organic carbon
SOC	soluble organic compound
SOME	sesame oil methyl ester
SP	starch phosphorylase
SPORL	sulfite pretreatment to overcome recalcitrance of lignocellulose
SRAP	sequence-related amplified polymorphism
SRWC	short-rotation woody crop
SS	starch synthase
SSCF	simultaneous saccharification and co-fermentation
SSF	solid-state fermentation
SSF	simultaneous saccharification and fermentation
SSR	simple sequence repeat
SUCEST	project that has generated the largest collection of ESTs
SUNY	State University of New York
SVO	straight vegetable oil
TAG	triacylglycerol
tb1	teosinte branched1 (gene)

TDC	top dead center
TDF	transcript-derived fragment
T-DNA	transfer(red)-DNA
TE	transposable element
TFBR	tubular fixed-bed reactors
tga1	teosinte glume architecture1 (gene)
Ti	tumor-inducing
TILLING	target-induced local lesion in genome
TISI	Thai Industrial Standards Institute
TMD	transmembrane domain
TMP	thermomechanical pulp
TMV	tobacco mosaic virus
TNB	Tenaga Malaysia Berhad
TNC	total nonstructural carbohydrates
TPP	thiamine pyrophosphate
TPS	terpene synthase
TRAP	target region amplification polymorphism
TRBO	tobacco mosaic virus RNA-based overexpression
TS	total solids
TSP	total soluble protein
TSS	total suspended solids
tVFA	total volatile fatty acids
TW	terawatts
UASB	upflow anaerobic sludge blanket reactor
UBHC/HC	unburned hydrocarbons
ULSD	ultra-low sulfur diesel
ULSF	ultra-low sulfur fuel
USDA	U.S. Department of Agriculture
USFS	USDA Forest Service
VBETC	Volumetric Biodiesel Excise Tax Credit
VCR	variable compression ratio
VEETC	Volumetric Ethanol Excise Tax Credit
VFA	volatile fatty acid
VLCFA	very-long-chain fatty acid
VOC	volatile organic compound
VS	volatile solids
VSS	volatile suspended solids
WA	Western Australia
WPI	World *Paulownia* Institute
WTP	willingness to pay
WUE	water-use efficiency
WVO	waste vegetable oil
WWFC	Worldwide Fuel Charter
WWTP	wastewater treatment plant
XGA	xylogalacturonan
xgd1	xylogalacturonan deficient 1 (gene)
XTR	XET-related (gene)
XyG, XG	xyloglucan
ZM 4	*Zymomonas mobilis* 4
ZSM-5	zeolite sieve of molecular porosity-5

Section I

1 Conventional and Molecular Breeding for Improvement of Biofuel Crops
Past, Present, and Future

Anjanabha Bhattacharya
University of Georgia

Joseph E. Knoll
U.S. Department of Agriculture

CONTENTS

1.1 INTRODUCTION

Initially when the potential of plants to produce liquid fuels was realized, the immediate focus was to use food and feed crops such as sugarcane (sugar), maize (starch), or soybean (oil) for biofuel production. Sugar can be directly fermented by yeast to produce ethanol, and starch can be easily converted into fermentable sugars. Many types of fats and oils can be chemically converted into biodiesel. Collectively, these fuels are known as first-generation biofuels. However, it was soon realized that these crops alone cannot meet the current demand for fuel because converting these crops into fuels is relatively inefficient with respect to the ratio energy output to input, mostly because only a fraction of the plant is utilized. Also, increased production of crops for fuel use has raised concerns about the environmental impacts of intensive agriculture and changing land usage. In addition, with the rapidly growing human population and increasing demand for animal products, the diversion of food and feed to fuel has become controversial. Biofuel production from food crops such as sugarcane, sunflower, soybean, sugarbeet, rapeseed, and maize has been blamed for triggering a food crisis in recent years. Whether or not increased biofuel production has really displaced a significant quantity of food is still highly debated. Nonetheless, this does raise a question: which should be given a priority when it comes to making a choice between energy and food resources? The answer to this paradox lies in utilization of crop residues and the many potential dedicated nonfood biofuel crops, including perennial grasses, such as switchgrass (*Panicum virgatum* L.) and *Miscanthus* spp.; fast-growing trees including poplar (*Populus* spp.) and willow (*Salix* spp.); fiber crops such as kenaf (*Hibiscus cannabinus* L.); and oil-rich nonedible crops such as *Jatropha curcas* L. and *Millettia pinnata* (L.) Panigrahi. Production of biofuels from food crop residues or from dedicated nonfood lignocellulosic crops utilizes the whole plant, thus capturing more energy per unit of land area.

Liquid biofuel production from plant cell-wall material is almost a half-century-old practice (Himmel and Bayer 2009), but its potential is only beginning to be realized. One of the greatest obstacles to producing liquid fuels from these materials is that conversion of cellulosic matter into fermentable sugars is much more difficult than conversion of starch. This "cellulosic" ethanol is known as a second-generation biofuel. This chapter addresses the progress made in the development of crop cultivars for first- and second-generation biofuels, with a major emphasis on perennial grasses for second-generation biofuels, and considers further improvements that could be made in the future. This chapter will focus on traditional plant breeding approaches and molecular tools available to the plant breeder. Transgenic approaches, which will certainly be important in the future, will be considered in detail in Chapter 3 of Section 1 of this handbook. Finally, this chapter will conclude with a brief look toward the future possibilities of other alternative biofuel sources.

1.2   THE PAST AND PRESENT—ETHANOL AND BIODIESEL: THE FIRST-GENERATION BIOFUELS

1.2.1   Ethanol from Sugarcane

Brazil has a long history of producing ethanol fuel from fermentation of sugar from sugarcane. However, large-scale production did not begin until the late 1970s with the government-mandated ProAlcohol program, which made it compulsory to blend ethanol with gasoline. Even then, success of this program was mixed, and the popularity of ethanol among consumers in Brazil tended to vary with the price of oil. Consumer acceptance of fuel ethanol increased dramatically with the introduction of flex-fuel vehicles, those able to operate on any blend of gasoline and ethanol (Matsuoka et al. 2009). In addition to liquid fuel, electricity is produced by burning the leftover sugarcane bagasse, which increases the energy efficiency of the whole process. To date, the Brazilian sugarcane-based ethanol industry is the most successful example of biofuel production in the world. Many sources now consider Brazil to be "fuel independent." Although technological advances in mechanization and processing were critical for this industry to thrive, genetic improvement of the feedstock undoubtedly also played a significant role.

Modern sugarcane is a complex hybrid derived primarily from *Saccharum officinarum* and *Saccharum spontaneum*. *S. officinarum* is believed to contribute the sweet stalk trait, whereas *S. spontaneum* contributes genes for stress tolerance and disease resistance. After the initial crosses were made, *S. officinarum* was used as the recurrent female parent in multiple backcrosses, resulting in several hybrids, which became the foundation stock for modern sugarcane cultivar development (Jannoo et al. 1999; Lakshmanan et al. 2005). Because of the large size, high ploidy level, and complexity of the sugarcane genome, molecular tools available to sugarcane breeders are somewhat lagging behind those for crops with simpler genomes such as rice. *S. officinarum* is octaploid ($2n = 80$), and cultivated sugarcane is even more complex ($2n = 100$–130). Still, because of its tremendous economic importance for food and fuel, progress is being made to develop these technologies for sugarcane. Molecular markers have been developed for sugarcane, and several studies have been conducted to assess the genetic diversity of the crop. Jannoo et al. (1999) surveyed a large collection of 109 sugarcane cultivars, mostly from Barbados and Mauritius, and 53 *S. officinarum* germplasm clones using low-copy restriction fragment length polymorphism (RFLP) markers. Aitken et al. (2006) surveyed 270 *S. officinarum* clones and 151 Australian cultivars and breeding lines using amplified fragment length polymorphism (AFLP) markers. The results of both of these studies were surprisingly similar. All clones tended to produce multiple markers, demonstrating the heterozygous nature of the complex polyploids. Although there was more diversity in the sugarcane cultivars because of hybridization, the most diversity within *S. officinarum* was found among accessions from New Guinea by both of the studies, supporting the hypothesis that the island of New Guinea is the center of origin for *S. officinarum*. Jannoo et al. (1999) identified a set of *S. officinarum* clones from New Caledonia as a distinct group, whereas Aitken et al. (2006) also identified a distinct group of *S. officinarum* clones from the South Pacific (Hawaii and Fiji). Both studies concluded that hybridization must have occurred after *S. officinarum* was disseminated away from its center of origin. Jannoo et al. (1999) noted that most of these clones had more than 80 chromosomes. A surprising result of both studies was that most of the diversity found within *S. officinarum* (85%, Jannoo et al. 1999; 90%, Aitken et al. 2006) was retained in the sugarcane cultivars, although relatively few clones formed the basis for modern sugarcane breeding.

The greatest challenge to molecular mapping in sugarcane is the high ploidy of its genome. Because of a high degree of homology between subgenomes, a given marker may represent more than one locus, which complicates mapping. Marker dosage, the number of loci associated with a particular marker, must be determined to map it. Chi-square tests for Mendelian segregation ratios are generally accepted as a means to determine marker dosage in polyploids. However, this approach has limitations because of segregation distortion (departures from expected ratios) and overdispersion (greater than expected variance in the distribution of marker data), often resulting in markers that cannot be assigned a dosage. To address these limitations, Baker et al. (2010) advocate a Bayesian mixture model for assigning marker dosage in the complex polyploids like sugarcane. Simplex markers, those representing only one locus, are the most informative and are used to construct the framework maps. The positions of duplex markers can be estimated to increase marker density and consolidate broken linkage groups (Aitken et al. 2007). Ambiguous markers, such as AFLP, can be used for genetic mapping in sugarcane, but a higher number of simplex markers can be obtained using functional genomics information such as expressed sequence tag (EST) data. A large database of sugarcane EST data, named SUCEST, has been developed, which contains over 237,000 ESTs, representing about 43,000 putative transcripts (Vettore et al. 2003). Using RFLP and simple sequence repeat (SSR) markers derived from sequence information in the SUCEST database, one of the most complete genetic maps for sugarcane was reported by Oliveira et al. (2007). This map spanned over 6,000 centi-Morgans (cM) and contained 664 markers in 192 cosegregation groups, which is more than the expected number of chromosomes, indicating that even this map is not sufficiently saturated.

Despite the low resolution of sugarcane genetic maps, quantitative trait loci (QTL) have been identified in this crop, which could be useful in marker-assisted selection (MAS). Pinto et al. (2010) used single marker analysis to identify single-dose markers associated with important traits including fiber content, cane yield, Pol (polarization value; a measure of sucrose content), and total sugar yield. An advantage of single marker analysis is that a linkage map is not required, and associations between unlinked markers can be detected. However, the probability of identifying false-positive associations is increased compared with interval mapping approaches that utilize a genetic map. Alwala et al. (2009) developed genetic maps from an interspecific cross (*S. officinarum* "Louisiana Striped" × *S. spontaneum* "SES 147B") using AFLP, sequence-related amplified polymorphism (SRAP), and target region amplification polymorphism (TRAP) markers. SRAPs are semiambiguous markers designed to amplify within genes, and TRAPs are targeted to specific genes; in this case sucrose metabolism-related genes were targeted. QTLs for sugar content (Brix and Pol) were identified using composite interval mapping (CIM; Zeng 1994) and discriminant analysis.

Sequence information from the SUCEST database has been used to create a complementary DNA (cDNA) microarray for sugarcane (Rocha et al. 2007). This technology allows for the direct identification of the genes underlying important traits. Genes associated with hormone response and environmental stress were identified using the microarray (Rocha et al. 2007). Sucrose metabolism-associated transcripts were also identified in a later study using the same microarray (Papini-Terzi et al. 2009). Molecular markers for use in breeding could be developed using this genomic information.

1.2.2 ETHANOL FROM GRAIN

In the United States, ethanol is produced primarily from maize (*Zea mays* L.) grain, although some is also produced from sorghum [*Sorghum bicolor* (L.) Moench] and other grains. Large-scale production of grain ethanol for biofuel has increased dramatically in the last 2 decades, mostly because of increased processing capacity as more ethanol plants were constructed in the Corn Belt. High-yielding maize varieties have allowed grain producers to meet the demands of the ethanol industry. The United States is the top maize producer in the world (FAO 2010), and much of that is due to a long and successful history of maize breeding. The greatest leap in maize production came with the introduction of hybrids in the 1930s. The first hybrids were double crosses, produced by crossing two F_1 hybrids, resulting in a mixed population of mostly heterozygous individuals. So-called three-way hybrids, produced by crossing an F_1 female parent to an elite inbred male parent, gave slightly better uniformity and performance. Continued improvement of inbred lines eventually allowed for the production of single cross hybrid (F_1) seed on a larger scale. Single cross hybrids offered better vigor and uniformity than double cross or three-way hybrids. The discovery of cytoplasmic-genetic male sterility in sorghum also allowed for production of high-yielding grain sorghum hybrids.

In addition to yield, grain composition and starch quality are important factors to consider in selecting grain for ethanol production. For example, in sorghum, protein digestibility was found to affect starch conversion because the proteins are believed to protect the starch granules from amylase enzymes (Wu et al. 2007). Starch is a polymer of glucose molecules linked by α-glycosidic bonds that consists of two forms: the mostly unbranched amylose and the highly branched amylopectin. Wu et al. (2006) found that higher percentages of amylose in maize starch mixtures and in other grains resulted in decreased conversion efficiency. A significant amount of breeding has been conducted by the major seed companies (Pioneer and Monsanto) to produce maize hybrids with grain qualities specifically designed for the fuel ethanol industry (Bothast and Schlicher 2005).

1.2.2.1 Sweet Sorghum

Some research has been done to investigate the use of sweet sorghum as a source of fermentable sugars in places where sugarcane cannot be grown or to supplement the sugarcane crop in warm temperate areas. Like sugarcane, sweet sorghums accumulate free sugars in their stems, which can be easily fermented into ethanol. In the United States, most sweet sorghum cultivars in use today are pure lines, developed by the U.S. Department of Agriculture (USDA) in Meridian, MS (Murray et al. 2009). This is in contrast to grain sorghums, which are almost exclusively hybrids. In general, sweet sorghums are very tall plants with limited seed production. Some research has been conducted to investigate the potential of producing hybrid sweet sorghum seed using short-statured, easily harvested seed parents (Corn and Rooney 2008; Makanda et al. 2009). Many genetic markers have been developed for sorghum, including RFLP markers, some of which are derived from other grasses (Bowers et al. 2003), several collections of SSR markers (Kong et al. 2000; Li et al. 2009; Yonemaru et al. 2009), and recently chip-based Diversity Array Technology (DArT) markers (Mace et al. 2008). Many linkage maps of sorghum have been published as well (Bhattramakki et al. 2000;

Haussmann et al. 2002; Bowers et al. 2003; Mace et al. 2009). Recently QTLs have been identified in sorghum for sugar-related traits using traditional linkage mapping (Natoli et al. 2002; Murray et al. 2008) as well as association mapping approaches (Murray et al. 2009). These QTLs may be useful in marker-assisted breeding to improve the productivity of sweet sorghum. The sorghum genome has also recently been sequenced (Paterson et al. 2009), which provides a tremendous opportunity to further analyze genes in this species and to compare similar genes in other grass species. The sorghum genome is relatively small in size (750 Mbp) and can act as a connecting link between small cereal genomes such as rice (480 Mbp) and very large complex genomes like sugarcane, which is related to sorghum. Comparative genomics using sorghum as a model C_4 grass will also help us to understand the evolutionary aspect of genes involved in biofuel traits and aid in designing strategies for enhancing biomass amenable to biofuel production.

1.2.3 BIODIESEL FROM CONVENTIONAL OILSEED CROPS

Biodiesel is produced from fats and oils (triglycerides) by transesterification with alcohol. Usually methanol is used to produce fatty acid methyl esters (Fukuda et al. 2001) and glycerol as a byproduct. Almost any source of naturally occurring triglycerides can be used, but the composition of the various fatty acids in the fuel affects its properties, including viscosity and crystallization temperature. These properties are especially important at lower temperatures. Poor performance in cold weather is currently a major limitation to acceptance of biodiesel. Thus, in addition to increased oil content, modification of fatty acid components of oilseed crops is a primary goal to produce high-quality biodiesel. Davis et al. (2008) observed that saturated long-chain fatty acid methyl esters were primarily responsible for crystallization in peanut biodiesel at higher temperatures than those observed for soybean or canola biodiesels. They suggested that decreasing the concentration of long-chain fatty acids, thus increasing the proportion of short-chain fatty acids by processing or by plant breeding, will be necessary to improve the properties of peanut biodiesel at low temperatures. Similarly, Krahl et al. (2007) advocated the use of plant breeding to improve the fuel qualities of rapeseed methyl ester (RME) biodiesel by selecting for short-chain (C_{12}–C_{16}) fatty acids, although Friedt and Luhs (1998) suggested that such progress may be limited by traditional plant breeding methods and advocated transgenic approaches for oil modification in rapeseed.

In addition to fatty acid chain length, the degree of unsaturation in fatty acids will also affect biodiesel properties. Peanut and soybean cultivars with high oleic (18:1) to linoleic (18:2) acid ratios have been developed through conventional breeding and marker-assisted approaches (Takagi and Rahman 1996; Chu et al. 2009). In peanut, a 1-bp insertion mutation was found to cause a frameshift in the *ahFAD2B* gene, leading to a loss of function of the enzyme that catalyzes the production of linoleic acid from oleic acid. A polymerase chain reaction (PCR)-based cleaved amplified polymorphic sequences (CAPS) marker was developed to screen for the presence of the mutant allele (Chu et al. 2009). Because oleic acid is monounsaturated, it has good properties for biodiesel, including lower crystallization temperature than saturated fatty acids and better oxidative stability than linoleic acid (Knothe 2009). It is possible that these cultivars could be used to produce superior biodiesel, although Davis et al. (2009) did not observe any significant differences in fuel viscosity between the high- and low-oleic peanut cultivars in their study. However, Tat et al. (2007) observed a reduction in the emissions of oxides of nitrogen (NO_x) from burning high-oleic soybean biodiesel compared with conventional soy biodiesel.

1.2.4 JATROPHA—A POSSIBLE DEDICATED SOURCE OF BIODIESEL

Jatropha (*Jatropha curcas* L.) is gaining importance as an emerging biofuel crop. Jatropha is a relatively long-lived tropical shrub or tree that produces oil-rich, usually inedible, nuts. The oil can be used to produce biodiesel as with edible oils. Jatropha is reported to grow on marginal lands not suitable for many other crops, so it should compete less for prime farmland. The major disadvantage of this crop is that it is relatively undomesticated, so yields are typically low. Improvements in seed yield and oil percentage and quality are needed for a jatropha-based fuel industry to be successful. At current yield levels, Lam et al. (2009) found that biodiesel production from Jatropha would not be economically or environmentally competitive with oil palm (*Elaeis guineensis* Jacq.) biodiesel in Malaysia. As a relatively new crop, limited knowledge of genetic potential pertaining to specific traits is a major hurdle in genetic improvement. Additionally, little is known of the naturally occurring genetic diversity of the crop. Several studies using molecular markers have been conducted to assess this diversity. Because little a priori knowledge of genomic sequences exists for this crop, markers such as random amplified polymorphic DNA (RAPD; Kumar et al. 2009b; Ikbal et al. 2010), intersimple sequence repeat (ISSR; Kumar et al. 2009a), and AFLP (Tatikonda et al. 2009) are well suited to these analyses. Ikbal et al. (2010) found moderate diversity among 40 Indian *J. curcas* accessions (average similarity coefficient of 0.73, with a range of 0.44–0.92) using RAPD markers. Kumar et al. (2009a) used ISSR markers to estimate genetic variation between *J. curcas* accessions and related *Jatropha* species; genetic similarity ranged from 0.346 to 0.807. Tatikonda et al. (2009) genotyped 48 accessions collected from various locations in India using AFLP markers. Using principal component analysis (PCA) based on 680 polymorphic markers, they identified five major groups. The genetic diversity did not appear to correlate with oil content or seed weight, indicating multiple sources of variation for desirable traits.

Some SSR markers have been developed recently for jatropha (Pamidimarri et al. 2009), and these will be useful for further diversity analyses, for creating a genetic map for this species, and eventually for marker-assisted breeding and selection. Although jatropha is being developed as a fuel crop, reduced toxicity is desirable for the purpose of utilizing the leftover seed meal as high-protein animal feed. By screening a collection of 72 accessions representing germplasm from 13 countries with RAPD and ISSR markers, Basha et al. (2009) identified several markers that appeared to be associated with a reduced toxicity trait found in some Mexican *J. curcas* accessions. Some of these markers were converted to more robust sequence-characterized amplified region (SCAR) markers for use in MAS, although more detailed linkage mapping analysis will be needed to verify their utility. This study also revealed limited genetic variability among most of the cultivated jatropha accessions from around the world, and identified Mexico as a possible center of origin.

Given the relatively narrow genetic base of cultivated *J. curcas*, improvement of the crop may be facilitated by crossing with other *Jatropha* species. Desirable traits such as photoperiod insensitivity, stress tolerance, and oil quality, to name just a few, have been identified in related *Jatropha* species (Basha and Sujatha 2009). Parthiban et al. (2009) reported high levels of incompatibility between *Jatropha* species, whereas Basha and Sujatha (2009) were able to create F_1 hybrids between *J. curcas* and several other *Jatropha* species by simple cross-pollination. Although self-compatible, *J. curcas* is monoecious (Divakara et al. 2010), which facilitates cross-pollination. Successful crosses were verified by genotyping the progeny with parent-specific markers. Pollen fertility in the hybrids ranged from 42 to 69%, suggesting that further crossing would be possible. Additionally, a hybrid between *J. curcas* and *J. integerrima* was successfully advanced to the F_2 and reciprocal backcross generations, and significant phenotypic variation was observed. Recently completion of the first genomic sequence for *J. curcas* was reported (Synthetic Genomics, Inc., 2009), resulting from a private venture between Synthetic Genomics, Inc. and the Asiatic Center for Genome Technology. This should accelerate the development of genetic markers and the directed integration of desirable genes for improvement of jatropha as a biofuel crop.

1.3 THE PRESENT AND NEAR FUTURE—CELLULOSIC ETHANOL FROM CROP RESIDUES AND SECOND-GENERATION BIOFUEL CROPS

Use of food crops for producing fuel is relatively inefficient because only a small portion of the total plant, and thus of the total captured solar energy, is utilized. The term "second-generation biofuels" usually refers to ethanol made from the breakdown and fermentation of whole-plant biomass. Most of the dry biomass of a plant consists of the cell walls, which are composed mainly of cellulose, hemicellulose, lignin, and a few other minor components, collectively known as lignocellulosic biomass. Cellulose is composed of long chains of glucose molecules bound by β-glycosidic bonds, which comprise between 30% and 40% of the dry mass of the cell wall (Vermerris 2008). Other structural carbohydrates are collectively referred to as hemicelluloses, which are highly varied among the plant kingdom. In grasses most of the hemicellulose consists of glucuronarabinoxylans, complex polymers of primarily xylose molecules, which also contain glucuronic acid and arabinose residues (Vermerris 2008). Lignin is a complex polymer of phenolic molecules derived from the phenylpropanoid pathway, which has an important function in providing strength and flexibility to plant tissues and plays a role in plant reactions to diseases and insect attacks.

The glucose subunits in the cellulose are of primary interest in the production of ethanol from biomass. In cellulosic ethanol production, enzymes such as cellulase break down the cellulose chains into glucose molecules that can then be fermented into ethanol by yeast. However, cellulose is difficult to extract from the cell wall because it is embedded in a cross-linked matrix with the other components, hemicellulose, lignin, pectins, and proteins. It is thought that lignin physically limits the access of cellulase enzymes to interact with the cellulose chains and can thus decrease the overall conversion efficiency. Using a transgenic approach, Chen and Dixon (2007) down-regulated six different genes in the lignin biosynthesis pathway in alfalfa (*Medicago sativa* L.). They observed that enzymatic saccharification efficiency of acid-pretreated stem biomass was directly related to lignin content. Composition of the lignin subunits also appears to influence conversion efficiency, but in general lignin is very resistant to chemical breakdown and is thus difficult to eliminate in the conversion process. One of the goals in developing biofuel crops is to select species or cultivars with reduced lignin content, or to modify the structural components of the lignin, without rendering plants susceptible to lodging, diseases, or insect pests.

1.3.1　Crop Residues as a Source of Biofuels

After crops such as wheat and maize are harvested, a significant amount of crop residue is usually left in the field. With the advent of cellulosic ethanol technology, it has become possible to harvest and utilize this material to make fuel, although how much residue can be removed without negative environmental and agronomic effects is still being debated. In addition to grain yield and quality, it has now become increasingly important for maize breeders to consider the properties of leaves and stems. Fortunately, the maize plant is also used as forage and silage, so some knowledge of its fiber properties already exists. Many of the characteristics important in forage digestibility are also important in conversion of lignocellulosic biomass into fuel.

Four low-lignin mutants of maize have been identified, *bm1* (Halpin et al. 1998; Vermerris et al. 2002), *bm2* (Vermerris and Boon 2001), *bm3* (Miller et al. 1983; Vignols et al. 1995), and *bm4*, all of which are associated with a visual phenotypic marker—brown pigmentation in the stalks and midribs. These mutations—blocks in the lignin biosynthesis pathway—are recessive, and the brown coloration is caused by the accumulation of phenylpropanoid lignin precursors. In maize, these low-lignin mutants generally have very good forage digestibility, but the severe decrease in lignin tends to be associated with lower grain yields, lodging, and disease susceptibility (reviewed by Pedersen et al. 2005). Although some brown-midrib (*bm3*) maize hybrids are available for forage use (de Leon and Coors 2008), it is not likely that a dual-purpose grain and biomass type will be developed using this mutation. The other three *bm* loci have not been studied as extensively and could potentially be useful in biomass improvement of maize stover. In addition to lignin content, the structure and composition of the lignin is also known to affect cell wall digestibility (reviewed by Barrière et al. 2009). Thus, altering lignin structure, rather than severely reducing the lignin content, appears to be the best strategy for improving grain maize stover as a biomass source.

Similar brown-midrib (*bmr*) mutations have been reported in sorghum and are also associated with reduced lignin content. Oliver et al. (2005) tested isolines of sorghum carrying the *bmr6* or *bmr12* mutation. Interestingly, in one specific genetic background (a hybrid of A Wheatland × RTx430), the *bmr12* mutation did not affect grain yield compared with the wild-type hybrid. Thus, in sorghum it appears possible to maintain grain yield and reduce lignin in the stover within the same hybrid. It has also been observed that in sorghum, *bmr* mutations did not increase susceptibility to *Alternaria*, and in some lines the *bmr* genes were actually associated with increased resistance to disease caused by *Fusarium* (Funnell and Pedersen 2006). A possible explanation for these observations is that the phenylpropanoid lignin precursor molecules are redirected into biochemical resistance pathways.

Several studies have been conducted to identify QTL associated with digestibility and other forage properties of leaves and stems in maize (Cardinal et al. 2003; Wei et al. 2009; Lorenzana et al. 2010). Perhaps because of varying environmental effects, differences in phenotypic evaluation, and differing genetic backgrounds, only a few common QTL are apparent when these studies are compared. Because of the tremendous genomic resources available for maize (Guillaumie et al. 2007; Penning et al. 2009), including a fully sequenced genome (Schnable et al. 2009), some of these QTL can be putatively assigned to a causative underlying gene on the basis of their location in the genome (Wei et al. 2009). Still, the biosynthesis of plant cell walls is a very complex process, and very few major QTLs have been identified. For example, Lorenzana et al. (2010) reported 152 QTLs for various stover composition traits in maize, each with relatively small effects.

In addition to utilizing crop residues, dedicated photoperiod-sensitive maize and sorghum varieties are being developed for the cellulosic ethanol industry. Under the long day lengths of temperate summers, these plants will not flower and will continue to grow vegetatively, producing large quantities of biomass.

1.3.2 PERENNIAL GRASSES AS DEDICATED BIOFUEL CROPS

Plant-based products (like agricultural and forestry residues) and paper and fiber wastes provide various feedstocks for the emerging cellulosic biofuels industry, but these materials alone cannot meet the increasing demand. Perennial grasses grown specifically for biomass production have the potential to meet a large portion of this demand for energy feedstock. Some of the species currently being studied for biomass production in warmer climates include sugarcane and energycanes (*Saccharum* hybrids), napiergrass (*Pennisetum purpureum* Schum.), bermudagrass [*Cynodon dactylon* (L) Pers.], and giant reed (*Arundo donax* L.). In cooler temperate locations, native prairie grasses including switchgrass (*P. virgatum* L.), big bluestem (*Andropogon gerardii* Vitman), prairie cordgrass (*Spartina pectinata*), and little bluestem (*Schizachyrium scoparium*) are being studied for their potential as biomass crops (Gonzalez-Hernandez et al. 2009). Several species and hybrids in the genus *Miscanthus* are also promising candidates for bioenergy production in temperate regions. This is by no means an exhaustive list of plant species with biomass production potential, although these are among the most studied of the perennial grasses. Most of these crops are undomesticated or only a few generations removed from their wild progenitors, and so there should be room for improvement to develop superior cultivars specifically suited to production of large quantities of high-quality biomass. In the development of these cultivars, a combination of traditional breeding, molecular approaches, and transgenic plant technologies should be considered.

In the early 1990s, the U.S. Department of Energy identified switchgrass as a highly promising source of biomass for energy and fuels (Sanderson et al. 1996). Since then, a considerable amount of research has been conducted on the utilization and improvement of switchgrass for biofuel purposes. Switchgrass had been used for years as a forage crop, and several named cultivars have been widely planted. Most sources identify two major ecotypes: lowland and upland. The lowland ecotypes tend to be thicker stemmed, taller, and higher yielding plants adapted to wetter sites, whereas the upland ecotypes, although thinner stemmed and lower yielding, are adapted to drier conditions (Bouton 2008). 'Alamo' and 'Kanlow' are two commonly planted lowland cultivars, whereas the upland cultivars include 'Cave-in-Rock,' 'Blackwell,' and 'Trailblazer,' among many others (Sanderson et al. 1996). Ploidy in switchgrass is quite variable, ranging from diploid to dodecaploid, with most of the lowland types being tetraploid and most of the uplands being octaploid (Bouton 2008). One of the first studies utilizing molecular markers in switchgrass was a diversity assessment of a representative sample of the germplasm. Because no species-specific markers had been developed at the time, sorghum chloroplast DNA probes were used to identify RFLPs. Of 80 probe/enzyme combinations, only one polymorphism was found, located within the ribulose-1,5-bisphosphate carboxylase large subunit gene (*pLD 5*). This indicates that there is relatively little genetic diversity within the switchgrass chloroplast genome, but the surprising result was that this one marker clearly differentiated the two ecotypes (Hultquist et al. 1996). Around the same time, Gunter et al. (1996) analyzed genetic diversity in switchgrass using RAPD markers. Although many more polymorphisms were observed in this study, cluster analysis of the marker data separated the accessions into upland and lowland groups, lending additional support to the genetic distinction between the two ecotypes. Recently, EST sequences were used to develop SSR markers for switchgrass (Tobias et al. 2006). Markers such as these will be useful tools for creating a genetic map for this species.

Switchgrass is propagated by seed, although it is highly self-incompatible. Thus, switchgrass cultivars are actually populations or synthetic varieties maintained by generations of random mating. The possibility of producing F_1 hybrid seed in switchgrass has been proposed. Martinez-Reyna and Vogel (2008) created reciprocal F_1 hybrids between the cultivars 'Kanlow' and 'Summer' by bagging a panicle of each parent together in the same bag, taking advantage of self-incompatibility. Twelve such crosses were made, and the hybrid seed bulked to capture the variation within each parent. These are contrasting ecotypes, but both are tetraploid, thus allowing crossing to take place. Although the cultivars themselves are heterogeneous, significant high-parent heterosis was observed for biomass yield and plant height in the F_1 generation when the plants were planted in swards (Vogel and Mitchell 2008), but only moderate midparent heterosis for second season yield was observed in individually spaced plants (Martinez-Reyna and Vogel 2008). When the hybrids were advanced to the F_2 and F_3 generations (actually syn_2 and syn_3 generations), the observed heterosis disappeared (Vogel and Mitchell 2008). These studies identified the tetraploid upland and tetraploid lowland ecotypes as heterotic groups for switchgrass breeding. However, the heterogeneity within switchgrass accessions makes the establishment of seed production fields difficult if the goal is to produce true F_1 seed. To produce true F_1 hybrids, clonal propagation of parent lines would be required to establish seed production fields. Traditional vegetative propagation in switchgrass is slow and cumbersome, although tissue culture techniques for rapid cloning have been developed (Bouton 2008). An alternative is to produce semihybrids (Brummer 1999). In this scheme, two parent populations from contrasting heterotic groups are planted together in isolation and allowed to randomly mate. The resulting seed will be approximately 50% hybrid and 50% nonhybrid.

Vegetatively propagated perennials offer the advantage of perpetuated hybrid vigor and genetic uniformity, although initial planting is more labor-intensive than for seeded crops. Energycanes, which are propagated from cane cuttings, were developed by sugarcane breeders with the USDA in Florida and Louisiana. Although the main product of sugarcane is sucrose for food or fuel, energycanes were developed specifically for the purpose of producing energy and biofuels. Whereas sugarcanes are usually highly advanced backcrosses, most energycanes are F_1 or BC_1 hybrids between cultivated sugarcane and closely related species, often *S. spontaneum*. Legendre and Burner (1995) showed that these early-generation hybrids tended to be superior to sugarcane cultivars in biomass yield and ratooning ability. They also have greater cold tolerance and can be overwintered in mild temperate locations where sugarcane will not survive. In addition to sugarcanes grown for sucrose production, Tew and Cobill (2008) define two types of energycanes. Type I energycanes were developed as dual-purpose plants, from which the sugars can be extracted for direct fermentation to

ethanol, and the remaining biomass, or bagasse, can be used for cellulosic ethanol or direct combustion to power the ethanol refinery. Type II energycanes do not produce as much sugar and are
primarily grown for cellulosic biomass. The genera *Saccharum*, *Erianthus*, and *Miscanthus* are
all very closely related, and taxonomists still do not agree on many species classifications within
these genera. Crosses between *Erianthus arundinaceum* and *Saccharum* (Cai et al. 2005) and
crosses between *Saccharum* and *Miscanthus* (Burner et al. 2009), often called "miscane," have
been reported. Because of their highly heterozygous nature and generally high ploidy levels, the F_1
generation of a cross will be very heterogeneous. Preliminary selections are made from single-plant
nurseries and can then be propagated into larger, replicated plots for further testing.

Several species and interspecific hybrids in the genus *Miscanthus* are being developed
as cellulosic energy crops for colder temperate regions. The species of primary interest are
M. sinensis, *M. sacchariflorus*, and *M. giganteus*, a naturally occurring hybrid of the former two
(Linde-Laursen 1993), often called "giant miscanthus." In some parts of Europe, miscanthus is
already being produced commercially for combustion to generate heat and power (Clifton-Brown
et al. 2008), but miscanthus production in the United States is lagging. "Freedom" is a named selection of miscanthus developed by Mississippi State University that has recently been commercially
licensed (Franco 2011). It is expected to be produced primarily in the Southeast for the cellulosic
ethanol industry. Other private ventures are also developing miscanthus varieties for biomass
production.

Miscanthus, unlike its relatives described above, must be propagated by rhizomes rather than cane
cuttings. Logistically, this is more difficult because propagation material must be dug from nurseries
and then planted manually or using special equipment. Thus seed propagation would be a desirable
trait in miscanthus, although this is not possible in the high-yielding *M. giganteus*, which is a sterile
triploid. Glowacka et al. (2009) subjected *M. sinensis* (diploids and triploids) and *M. giganteus* to
colchicine treatments in attempts to double their chromosome numbers. Tetraploids and hexaploids
of *M. sinensis* were recovered, but no chromosome doubling was observed for *M. giganteus*. This
actually supports Linde-Laursen's (1993) assertion that on the basis of observations of meiotic chromosomes in *M. giganteus*, recovery of a stable fertile plant by chromosome doubling would be near
impossible. Still, if one could be produced, a hexaploid *M. giganteus* might allow for seed production. Additional benefits of increased ploidy, such as enhanced vegetative growth, are also possible.
Miscanthus tends to establish slowly, taking up to three years to reach maximum yields; faster
establishment is another trait that needs to be improved in this crop.

Because of its status as a relatively new crop and the fact that genomes tend to be quite large,
genomics resources for miscanthus are rather limited (Armstead et al. 2009). Ambiguous markers
such as AFLP and RAPD are quite useful in such circumstances. Greef et al. (1997) used AFLP
markers to assess the genetic diversity in European *Miscanthus* accessions. They observed considerable variation among *M. sinensis* clones, but very little among clones of *M. giganteus*, suggesting
that all clones of the hybrid are probably derived from a single cross. RAPD markers have been used
to create a genetic linkage map in *M. sinensis*, which was then used for identification of QTL for
plant height, flag leaf height, and stem diameter, traits that are important biomass yield components.
Because *M. sinensis* is highly heterozygous and self-incompatible, the mapping strategy (offspring
cross mapping) was a bit different than for model crop plants like sorghum or rice. The mapping
population consisted of 89 F_1 plants, originating from a cross between two sibling F_1 plants, which
were produced by crossing two *M. sinensis* accessions with contrasting traits of interest (Atienza et al.
2003). The identified QTL could be used in MAS. Although genomic information for *Miscanthus*
is still being developed, the genus is closely related to several other C_4 grasses for which significant
genomics resources already exist. These crops include maize, sorghum, and sugarcane. Even rice
(*Oryza sativa* L), although more distantly related, is expected to share considerable synteny with
Miscanthus. This allows for comparative genomic studies and the transferability of molecular markers and other genetic information. For example, Hernandez et al. (2001) demonstrated that RFLP
probes from maize hybridized to miscanthus DNA (100% of probes tested), and SSR markers were

also transferrable (75% of primer pairs tested). Polymorphisms were detected among 11 miscanthus accessions using some of the maize-derived SSR markers. Such markers could be added to the *M. sinensis* map, and it may be possible to correlate known QTL positions in other grass genomes with those in *M. sinensis*.

Napiergrass, also known as elephantgrass, has been developed primarily as a forage species for the tropics. Because forage is harvested while the growth is still quite young and digestible, it has high protein content and has a greater proportion of biomass in the leaves. Dwarf leafy types, such as the cultivar 'Mott' (Sollenberger et al. 1989), have been developed specifically as high-quality forage varieties. For bioenergy production, the opposite plant type is desired: tall plants with most of the biomass in the stems. The cultivar 'Merkeron' was also developed initially as a high-yielding forage cultivar, but it is capable of reaching a height of 4 m, with a considerable percentage of stem biomass. Even under low fertility, substantial yields are possible. With no added fertilizer, 'Merkeron' yielded over 25 Mg ha^{-1} yr^{-1} dry matter for the first two seasons at Tifton, GA, although yields declined without additional fertilization in subsequent seasons (Knoll et al. 2011). 'Merkeron' was selected from a cross between a leafy dwarf type and a tall type (Burton 1989), and selfed populations of progeny from 'Merkeron' segregate for the dwarf characteristic. In addition to 'Merkeron,' several other promising napiergrass selections are being evaluated at Tifton, GA, and other locations in the southeastern United States for their potential as biomass feedstocks. Important traits include biomass yield, cold tolerance and overwintering ability, nitrogen use efficiency, and improved biomass digestibility for cellulosic ethanol conversion.

A few investigations of genetic diversity in napiergrass have been conducted. Isozymes, iso-electric or molecular size variations in specific proteins, have been used successfully to fingerprint napiergrass accessions and evaluate their diversity (Daher et al. 1997; Bhandari et al. 2006). More recently, Pereira et al. (2008) surveyed the diversity in a Brazilian collection of 30 napiergrass accessions using RAPD markers. Twenty primers were tested, which yielded 88 scorable bands, of which 64 showed polymorphism among the accessions. Moderate genetic diversity (average genetic distance 0.21, maximum 0.34) was revealed, along with several possible duplicated accessions. Babu et al. (2009) also used DNA markers (RAPD and ISSR) to survey genetic diversity among a large collection of napiergrass accessions. Interestingly, the data from the two different marker systems generated different dendrograms, although both data sets showed correlation with geographic origins of the accessions. AFLP markers have also been applied to assess diversity among napiergrass accessions (Anderson et al. 2008). SSR markers have been developed for the related species pearl millet [*Pennisetum glaucum* (L) R. Br.; Senthilvel et al. 2008], and it is likely that some of these could be used in diversity assessments, genetic mapping, and eventually marker-assisted breeding in napiergrass.

Napiergrass is protogynous, so crossing is facilitated by pollination before the anthers emerge, although selfing is also possible (Hanna et al. 2004). The breeding scheme for napiergrass is similar to that for other clonally propagated species. F$_1$ hybrids are generally heterogeneous, and single-plant selections are increased by stem cuttings, in a similar manner as for sugarcane, for further testing. Napiergrass can also be successfully crossed with pearl millet; the resulting plant was called "elephantmillet" by Woodard and Prine (1993). Napiergrass is allotetraploid, and crosses with pearl millet result in triploid offspring which are sterile, although fertile hexaploids can be recovered through chromosome doubling with colchicine (Hanna 1981; Anderson et al. 2008). Napiergrass has also been successfully crossed with an apomictic relative *P. squamulatum*; the hybrid has some fertility (Hanna et al. 2004; Anderson et al. 2008). The apomictic trait in *P. squamulatum* is carried on a hemizygous chromosome segment, termed the apospory-specific genomic region (ASGR; Goel et al. 2003). Although the underlying genetic mechanism is still unclear, molecular markers located on the ASGR should facilitate selection for the trait. Several backcrosses could result in an apomictic napiergrass, which would have the advantages of seed propagation while maintaining hybrid vigor and the uniformity of vegetatively propagated material. Seed production would be limited to tropical and subtropical areas because flowering in napiergrass is short-day sensitive.

1.3.3 TREES AS A BIOFUEL SOURCE

Wood is perhaps the oldest sources of bioenergy in the world. As with perennial grasses, the potential now exists to use wood from fast-growing tree species as a feedstock to produce liquid fuels. Candidate trees include several species of pine (*Pinus* spp.), *Eucalyptus* spp., willows (*Salix* spp.), and poplar (*Populus* spp.). Because of their importance in the timber and paper industries, considerable genetic resources are already in place for many tree species. *Populus trichocarpa*, also known as black cottonwood, has a genome size of 480 Mbp and a haploid chromosome number of 19 (Tuskan et al. 2006). This is perhaps the smallest genome size among tree species, and so it was selected for a genomic sequencing project. A consortium of researchers from governmental organizations and universities from the United States, Canada, and several European countries was formed to conduct the project (International *Populus* Genome Consortium 2010). The strategy used was whole genome shotgun sequencing, in which thousands of randomly selected genomic clones are sequenced and are then assembled in silico based on overlapping sequences. The assembled sequences were then anchored to specific linkage groups on the basis of known locations of SSR markers identified in the earlier stages of the project (Tuskan et al. 2004). Completion of the assembled sequence was reported in 2006 by Tuskan et al. An international collaborative effort has also been formed to sequence the *Eucalyptus* genome. The International *Eucalyptus* Genome Network (EUCAGEN) includes researchers from both public and private sectors from every continent. The first draft sequence of *Eucalyptus grandis* was recently reported (EUCAGEN 2010). A critical next phase for these projects is to annotate the genome sequences; that is, to assign putative functions to the genes. Such information should be useful to tree breeders to assist in selecting plants with desirable traits. This is especially important given the long generation times for woody plants. Genomics resources such as these should help to hasten the development of new cultivars with improved biofuel traits.

1.4 CONCLUSIONS AND FUTURE PROSPECTS FOR BIOFUEL CROPS

There are many plant species that can be used for production of biofuels, from existing food and feed crops to dedicated biomass crops like trees and perennial grasses. Existing crops have, in the past, been selected mostly for yield of food and feed, and many of the emerging biomass crops are barely domesticated at all. These crops will require specific improvements to make them more amenable to biofuel production. Fortunately, considerable natural variation is present among most cultivated crop species, and many of these crops also have a repository of genes in their wild relatives that can be exploited for crop improvement. However, there are many challenges for plant breeders in development of biofuel crops. Some of these species have very long generation times and are highly outcrossing, preventing the creation of inbred lines for classical genetic studies. In addition, the genomes of many biomass crops (e.g., sugarcane) are very large and complex. Still, significant progress has been made in genetic improvement of biofuel feedstocks.

A combination of traditional and molecular tools has been used in biofuel crop improvement, and the molecular tools are becoming increasingly important. Molecular markers are now routinely used to assess the genetic diversity in germplasm collections, to map important quantitative traits (QTL), and to select for desirable traits linked to those markers. Although progress has been made by utilizing QTL analysis and MAS, it has been met with some limitations, especially for highly polygenic traits. However, it is now possible to quickly genotype many markers, essentially covering the entire genome, at relatively little cost. Thus, whole-genome selection has been proposed as the next phase in marker-assisted plant breeding (Heffner et al. 2009). The procedure begins with collecting phenotypic and genotypic data from a large collection of breeding material and germplasm of interest. This "training population" is used to create a complex prediction model of trait-marker associations. This information can then used to predict the breeding value of any individual on the basis of all of its available genetic marker information. Data from further evaluations in the field can also be used to refine the prediction model. The various statistical methodologies that can be used in this process are reviewed by Heffner et al. (2009).

With the advent of genome sequencing, it is now possible to associate each gene, or cluster of genes, with a particular trait of importance. The genome sequence of *Arabidopsis thaliana* var. Columbia was completed in the year 2000 (*Arabidopsis* Genome Initiative 2000), and most of the genes were annotated by 2005. This opened new vistas for genome sequencing and mapping in other plant species. However, *Arabidopsis* is a dicot, and not a cultivated crop, so there was a need to sequence an important monocotyledonous crop plant. Thus, rice was sequenced to act as a model crop for cereal genomics. The rice sequencing project was completed in the year 2005 (International Rice Genome Sequencing Project 2005), opening the door for detailed comparative genomic studies among grasses. Two other grasses with relatively small genomes have also been sequenced: sorghum (Paterson et al. 2009) and the model species *Brachypodium distachyon*. Together, these three genome sequences provide references for the three major branches of the grass family (International *Brachypodium* Initiative 2010). Comparative genomics studies have revealed that even distantly related species can show considerable homology between genes controlling specific traits. Thus, the information gleaned from the smaller model species genomes can be applied to the more complex biofuel crop species.

As demand for DNA sequencing increased, new technologies were invented to reduce costs, time, and labor involved and to increase reliability. Typical capillary electrophoresis sequencers utilize fluorescently labeled dideoxynucleotide terminators (often referred to as "Sanger" sequencing; Sanger et al. 1977). These systems can read several hundred base pairs in one fragment. Depending on the system, 96 or 384 samples can be sequenced in parallel. Each sample must be an individual clone, thus requiring some preparation time. The next generation of sequencers use microscopic beads to capture individual DNA molecules. DNA to be sequenced is fragmented and then ligated to beads. Each bead captures a single fragment, and then each fragment is amplified and sequenced in parallel. Examples include the SOLiD™ System (Applied Biosystems 2010) and the 454 Sequencing System (Roche Diagnostics Corporation 2011). The Illumina sequencing systems (Illumina, Inc. 2010) use a similar approach, but the DNA is captured by oligonucleotides bound to a slide. These types of sequencers generate shorter reads, usually around 100 bp, but they generate so many reads that the overall throughput greatly exceeds that of capillary electrophoresis systems. One run can generate several giga-base pairs of sequence information. Powerful computer software is required to assemble these short reads into contigs. These short sequences can also be combined with known longer sequences to generate complete assemblies. Rapid sequencing technologies are increasing the pace at which more and more plant genomes are being sequenced. Also, high-throughput resequencing of genomes allows for the discovery of thousands of single nucleotide polymorphisms in natural and mutagenized populations. As these technologies improve, read lengths and throughput will undoubtedly continue to increase. The greatest limitation, at least in the near future, will not be in sequencing and genotyping, but in deciphering all of this genomic information to pinpoint the sequences responsible for particular phenotypes.

Currently, much of the focus in biofuel crop improvement is on selecting for biomass traits that allow for easier conversion of lignocellulosic material into sugars by enzymatic or microbial saccharification for subsequent fermentation into alcohol. However, other technologies for producing liquid fuels from biomass are being developed, such as pyrolysis (Elliott 2007; Balat et al. 2009a) and gasification (Balat et al. 2009b). There are likely to be other technologies in the future as well. All of these conversion procedures will likely require specific properties in their feedstocks. Future biofuel crops will need to be tailored to specific conversion technologies, and this will guide future breeding efforts. In addition, biomass production will need to occur close to processing facilities, and those facilities will require a consistent supply of feedstock throughout most of the year. Thus, there is no single biofuel crop that will fulfil the need for fuel everywhere. Rather, a combination of crops with different growth cycles and habits, each suited to a particular geography and end use, will need to be developed. There may even be other species with potential as biofuel crops that have yet to be discovered. For example, research is currently being conducted to determine if aquatic weeds like water hyacinth (*Eichhornia crassipes*; Aswathy et al. 2010) or even microalgae (Williams and Laurens 2010) can be utilized to produce biofuels.

As the human population continues to expand and the standard of living in much of the world continues to improve, the demand for transportation fuels will continue to increase as a result. Even with advances in fuel efficiency and improvements in mass-transit systems to offset a portion of this demand, the need for fuel will continue to grow. The environmental consequences of our continued usage of fossil fuels are becoming all too apparent, as demonstrated by the recent disaster in the Gulf of Mexico. In addition, the release of carbon dioxide and other greenhouse gases into the atmosphere from burning these fuels is suspected of causing global climate change. Plant-based biofuels hold immense promise for meeting our future fuel needs in a more sustainable way. The challenge to develop new, dedicated biofuel crops is great, but plant breeders now have an incredible array of genetic and genomic resources to help them meet this challenge.

DISCLAIMER

The mention of trade names or commercial products is solely for the purpose of providing specific information and does not imply recommendation or endorsement by the U.S. Department of Agriculture.

REFERENCES

Aitken KS, Jackson PA, McIntyre CL (2007) Construction of a genetic linkage map for *Saccharum officinarum* incorporating both simplex and duplex markers to increase genome coverage. *Genome* 50:742–756

Aitken KS, Li J-C, Jackson P., Piperidis G, McIntyre CL (2006) AFLP analysis of genetic diversity within *Saccharum officinarum* and comparison with sugarcane cultivars. *Aust J Agric Res* 57:1167–1184

Alwala S, Kimbeng CA, Veremis JC, Gravois KA (2009) Identification of molecular markers associated with sugar-related traits in a *Saccharum* interspecific cross. *Euphytica* 167:127–142

Anderson WF, Casler MD, Baldwin BS (2008) Improvement of perennial forage species as feedstock for bioenergy. In: Vermerris W (ed), *Genetic Improvement of Bioenergy Crops*. Springer, New York

Applied Biosystems (2010) The SOLiD™ System: Next generation sequencing, available at http://www.appliedbiosystems.com/absite/us/en/home/applications-technologies/solid-next-generation-sequencing.html (accessed June, 2010)

Arabidopsis Genome Initiative (2000) Analysis of the genome sequence of the flowering plant *Arabidopsis thaliana*. *Nature* 408:796–815

Armstead I, Huang L, Ravagnani A, Robson P, Ougham H (2009) Bioinformatics in the orphan crops. *Briefings Bioinform* 10:645–653

Aswathy US, Sukumaran RK, Devi GL, Rajasree KP, Singhania RR, Pandey A (2010) Bio-ethanol from water hyacinth biomass: An evaluation of enzymatic saccharification strategy. *Bioresour Technol* 101:925–930

Atienza SG, Satovic Z, Petersen KK, Dolstra O, Martin A (2003) Identification of QTLs influencing agronomic traits in *Miscanthus sinensis* Anderss. I. Total height, flag-leaf height and stem diameter. *Theor Appl Genet* 107:123–129

Babu C, Sundaramoorthi J, Vijayakumar G, Ram SG (2009) Analysis of genetic diversity in Napier grass (*Pennisetum purpureum* Schum) as detected by RAPID and ISSR markers. *J Plant Biochem Biotechnol* 18:181–187

Baker P, Jackson P, Aitken K (2010) Bayesian estimation of marker dosage in sugarcane and other autopolyploids. *Theor Appl Genet* 120:1653–1672

Balat M, Balat M, Kirtay E, Balat H (2009a) Main routes for the thermo-conversion of biomass into fuels and chemicals. Part 1: Pyrolysis systems. *Energy Convers Manag* 50:3147–3157

Balat M, Balat M, Kirtay E, Balat H (2009b) Main routes for the thermo-conversion of biomass into fuels and chemicals. Part 2: Gasification systems. *Energy Convers Manag* 50:3158–3168

Barrière Y, Méchin V, Riboulet C, Guillaumie S, Thomas J, Bosio M, Fabre F, Goffner D, Pichon M, Lapierre C, Martinant J-P (2009) Genetic and genomic approaches for improving biofuel production from maize. *Euphytica* 170:183–202

Basha SD, Francis G, Makkar HPS, Becker K, Sujatha M (2009) A comparative study of biochemical traits and molecular markers for assessment of genetic relationships between *Jatropha curcas* L. germplasm from different countries. *Plant Sci* 176: 812–823

Bhandari AP, Sukanya DH, Ramesh CR (2006) Application of isozyme data in fingerprinting Napier grass (*Pennisetum purpureum* Schum.) for germplasm management. *Genet Resour Crop Evol* 53:253–264

Bhattramakki D, Dong J, Chhabra AK, Hart GE (2000) An integrated SSR and RFLP linkage map of Sorghum bicolor (L.) Moench. *Genome* 43:988–1002

Bothast RJ, Schlicher MA (2005) Biotechnological processes for conversion of corn into ethanol. *Appl Microbiol Biotechnol* 67:19–25

Bouton J (2008) Improvement of switchgrass as a bioenergy crop. In: Vermerris W (ed), *Genetic Improvement of Bioenergy Crops*. Springer, New York

Bowers JE, Abbey C, Anderson S, Chang C, Draye X, Hoppe AH, Jessup R, Lemke C, Lennington J, Li ZK, et al. (2003) A high-density genetic recombination map of sequence-tagged sites for Sorghum, as a framework for comparative structural and evolutionary genomics of tropical grains and grasses. *Genetics* 165:367–386

Brummer EC (1999) Capturing heterosis in forage crop cultivar development. *Crop Sci* 39:943–954

Burner DM, Tew TL, Harvey JJ, Belsky DP (2009) Dry matter partitioning and quality of *Miscanthus*, *Panicum*, and *Saccharum* genotypes in Arkansas, USA. *Biomass Bioenergy* 33:610–619

Burton GW (1989) Registration of 'Merkeron' napiergrass. *Crop Sci* 29:1327

Cai Q, Aitken K, Deng HH, Chen XW, Fu C, Jackson PA, McIntyre CL (2005) Verification of the introgression of *Erianthus arundinaceus* germplasm into sugarcane using molecular markers. *Plant Breed* 124:322–328

Cardinal AJ, Lee M, Moore KJ (2003) Genetic mapping and analysis of quantitative trait loci affecting fiber and lignin content in maize. *Theor Appl Genet* 106:866–874

Chen F, Dixon RA (2007) Lignin modification improves fermentable sugar yields for biofuel production. *Nat Biotechnol* 25:759–761

Chu Y, Holbrook CC, Ozias-Akins P (2009) Two alleles of ahFAD2B control the high oleic acid trait in cultivated peanut. *Crop Sci* 49:2029–2036

Clifton-Brown J, Chiang Y-C, Hodkinson TR (2008) *Miscanthus*: Genetic resources and breeding potential to enhance bioenergy production. In: Vermerris W (ed), *Genetic Improvement of Bioenergy Crops*. Springer, New York

Corn R, Rooney WL (2008) Sweet sorghum heterosis. Paper presented at the Joint Annual Meeting, October 5–9, 2008, Houston, TX, available at http://a-c-s.confex.com/crops/2008am/webprogram/Paper42644.html (accessed June 2010)

Daher RF, deMoraes CF, Cruz CD, VanderPereira A, Xavier DF (1997) Morphological and isozymatic diversity in elephantgrass (*Pennisetum purpureum* Schum). *Roy Soc Bras Zootec* 26:255–264

Davis JP, Geller ED, Faircloth WH, Sanders TH (2009) Comparisons of biodiesel produced from unrefined oils of different peanut cultivars. *J Amer Oil Chem Soc* 86:353–361

de Leon N, Coors JG (2008) Genetic improvement of corn for lignocellulosic feedstock production. In: Vermerris W (ed), *Genetic Improvement of Bioenergy Crops*. Springer, New York

Divakara BN, Upadhyaya HD, Wani SP, Gowda CLL (2010) Biology and genetic improvement of *Jatropha curcas* L.: A review. *Appl Energy* 87:732–742

Elliott DC (2007) Historical developments in hydroprocessing bio-oils. *Energy Fuels* 21:1792–1815

EUCAGEN (2010) The International *Eucalyptus* Genome Network, available at http://www.eucagen.org/ (accessed June 2010)

FAO (2010) Food and Agriculture Organization of the United Nations, FAOSTAT, available at http://faostat.fao.org/ (accessed June 2010)

Franco C (2011) Field day promotes giant miscanthus. Mississippi State University, available at http://msucares.com/news/print/agnews/an11/110203.html (accessed Sept. 2011)

Friedt W, Luhs W (1998) Recent developments and perspectives of industrial rapeseed breeding. *Fett-Lipid* 100:219–226

Fukuda H, Kondo A, Noda H (2001) Biodiesel fuel production by transesterification of oils. *J Biosci Bioeng* 92:405–416

Funnell DL, Pedersen JF (2006) Reaction of sorghum lines genetically modified for reduced lignin content to infection by *Fusarium* and *Alternaria* spp. *Plant Dis* 90:331–338

Glowacka K, Jezowski S, Kaczmarek Z (2009) Polyploidization of *Miscanthus sinensis* and *Miscanthus* × *giganteus* by plant colchicine treatment. *Indust Crops Prod* 30:444–446

Goel S, Chen ZB, Conner JA, Akiyama Y, Hanna WW, Ozias-Akins P (2003) Delineation by fluorescence in situ hybridization of a single hemizygous chromosomal region associated with aposporous embryo sac formation in *Pennisetum squamulatum* and *Cenchrus ciliaris*. *Genetics* 163:1069–1082

Gonzalez-Hernandez JL, Sarath G, Stein JM, Owens V, Gedye K, Boe A (2009) A multiple species approach to biomass production from native herbaceous perennial feedstocks. *In Vitro Cell Dev Biol-Plant* 45:267–281

Greef JM, Deuter M, Jung C, Schondelmaier J (1997) Genetic diversity of European *Miscanthus* species revealed by AFLP fingerprinting. *Genet Resour Crop Evol* 44:185–195

Guillaumie S, San-Clemente H, Deswarte C, Martinez Y, Lapierre C, Murigneux A, Barriere Y, Pichon M, Goffner D (2007) MAIZEWALL. Database and developmental gene expression profiling of cell wall biosynthesis and assembly in maize. *Plant Physiol* 143:339–363

Gunter LE, Tuskan GA, Wullschleger SD (1996) Diversity among populations of switchgrass based on RAPD markers. *Crop Sci* 36:1017–1022

Halpin C, Holt K, Chojecki J, Oliver D, Chabbert B, Monties B, Edwards K, Barakate A, Foxon GA (1998) Brown-midrib maize (*bm1*)—A mutation affecting the cinnamyl alcohol dehydrogenase gene. *Plant J* 14:545–553

Hanna WW (1981) Method of reproduction in Napiergrass and in the 3 × and 6 × alloploid hybrids with pearl millet. *Crop Sci* 21:123–126

Hanna WW, Chaparro CJ, Mathews BW, Burns JC, Sollenberger LE, Carpenter JR (2004) Perennial *Pennisetums*. In: Moser LE, Burson BL, Sollenberger LE (eds), *Warm-Season (C4) Grasses*. ASA/CSSA/SSSA, Madison, WI

Haussmann BIG, Hess DE, Seetharama N, Welz HG, Geiger HH (2002) Construction of a combined sorghum linkage map from two recombinant inbred populations using AFLP, SSR, RFLP, and RAPD markers, and comparison with other sorghum maps. *Theor Appl Genet* 105:629–637

Heffner EL, Sorrells ME, Jannink JL (2009) Genomic selection for crop improvement. *Crop Sci* 49:1–12

Hernandez P, Dorado G, Laurie DA, Martin A, Snape JW (2001) Microsatellites and RFLP probes from maize are efficient sources of molecular markers for the biomass energy crop *Miscanthus*. *Theor Appl Genet* 102:616–622

Himmel ME, Bayer EA (2009) Lignocellulose conversion to biofuels: current challenges, global perspectives. *Curr Opin Biotechnol* 20:316–317

Hultquist SJ, Vogel KP, Lee DJ, Arumuganathan K, Kaeppler S (1996) Chloroplast DNA and nuclear DNA content variations among cultivars of switchgrass, *Panicum virgatum* L. *Crop Sci* 36:1049–1052

Ikbal, Boora KS, Dhillon RS (2010) Evaluation of genetic diversity in *Jatropha curcas* L. using RAPD markers. *Indian J Biotechnol* 9:50–57

Illumina, Inc (2010) Illumina—Sequencing technology, available at http://www.illumina.com/technology/sequencing_technology.ilmn (accessed June 2010)

International *Brachypodium* Initiative (2010) Genome sequencing and analysis of the model grass *Brachypodium distachyon*. *Nature* 463:763–768

International *Populus* Genome Consortium (2010), available at http://www.ornl.gov/sci/ipgc/ (accessed June 2010)

International Rice Genome Sequencing Project (2005) The map-based sequence of the rice genome. *Nature* 436:793–799

Jannoo N, Grivet L, Seguin M, Paulet F, Domaingue R, Rao PS, Dookun A, D'Hont A, Glaszmann JC (1999) Molecular investigation of the genetic base of sugarcane cultivars. *Theor Appl Genet* 99:171–184

Knoll JE, Anderson WF, Strickland TC, Hubbard RK, Malik R (2011) Low0input production of biomass from perennial grasses in the Coastal Plain of Georgia, USA. Bioenerg Res. DOI 10.1007/s12155-011-9122-x

Knothe G (2005) Dependence of biodiesel fuel properties on the structure of fatty acid alkyl esters. *Fuel Process Technol* 86:1059–1070

Kong L, Dong J, Hart GE (2000) Characteristics, linkage-map positions, and allelic differentiation of *Sorghum bicolor* (L.) Moench DNA simple-sequence repeats (SSRs). *Theor Appl Genet* 101:438–448

Krahl J, Munack A, Bockey D (2007) Property demands on future biodiesel. *Landbauf Volkenrode* 57:415–418

Kumar RS, Parthiban KT, Rao MG (2009a) Molecular characterization of *Jatropha* genetic resources through inter-simple sequence repeat (ISSR) markers. *Mol Biol Rep* 36:1951–1956

Kumar RV, Tripathi YK, Shukla P, Ahlawat SP, Gupta VK (2009b) Genetic diversity and relationships among germplasm of *Jatropha curcas* L. revealed by RAPDs. *Trees Struct Funct* 23:1075–1079

Lakshmanan P, Geijskes RJ, Aitken KS, Grof CLP, Bonnett GD, Smith GR (2005) Sugarcane biotechnology: The challenges and opportunities. *In Vitro Cell Dev Biol-Plant* 41:345–363

Lam MK, Lee KT, Mohamed AR (2009) Life cycle assessment for the production of biodiesel: A case study in Malaysia for palm oil versus jatropha oil. *Biofuels Bioprod Biorefin* 3:601–612

Legendre BL, Burner DM (1995) Biomass production of sugarcane cultivars and early-generation hybrids. *Biomass Bioenergy* 8:55–61

Li ML, Yuyama NN, Luo L, Hirata M, Cai HW (2009) In silico mapping of 1758 new SSR markers developed from public genomic sequences for sorghum. *Mol Breed* 24:41–47

Linde-Laursen I (1993) Cytogenetic analysis of *Miscanthus* 'Giganteus', an interspecific hybrid. *Hereditas* 119:297–300

Lorenzana RE, Lewis MF, Jung HJG, Bernardo R. (2010) Quantitative trait loci and trait correlations for maize stover cell wall composition and glucose release for cellulosic ethanol. *Crop Sci* 50:541–555

Mace ES, Xia L, Jordan DR, Halloran K, Parh DK, Huttner E, Wenzl P, Kilian A (2008) DArT markers: Diversity analyses and mapping in *Sorghum bicolor*. *BMC Genom* 9:26

Mace ES, Rami JF, Bouchet S, Klein PE, Klein RR, Kilian A, Wenzl P, Xia L, Halloran K, Jordan DR (2009) A consensus genetic map of sorghum that integrates multiple component maps and high-throughput Diversity Array Technology (DArT) markers. *BMC Plant Biol* 9:13

Makanda I, Tongoona P, Derera J (2009) Combining ability and heterosis of sorghum germplasm for stem sugar traits under off-season conditions in tropical lowland environments. *Field Crops Res* 114:272–279

Martinez-Reyna JM, Vogel KP (2008) Heterosis in switchgrass: Spaced plants. *Crop Sci* 48:1312–1320

Matsuoka S, Ferro J, Arruda P (2009) The Brazilian experience of sugarcane ethanol industry. *In Vitro Cell Dev Biol-Plant* 45:372–381

Miller JE, Geadelmann JL, Marten GC (1983) Effect of the brown midrib-allele on maize silage quality and yield. *Crop Sci* 2:493–496

Murray SC, Rooney WL, Hamblin MT, Mitchell SE, Kresovich S (2009) Sweet sorghum genetic diversity and association mapping for Brix and height. *Plant Genome* 2:48–62

Murray SC, Sharma A, Rooney WL, Klein PE, Mullet JE, Mitchell SE, Kresovich S (2008) Genetic improvement of sorghum as a biofuel feedstock: I. QTL for stem sugar and grain nonstructural carbohydrates. *Crop Sci* 48:2165–2179

Natoli A, Gorni C, Chegdani F, Marsan PA, Colombi C, Lorenzoni C, Marocco A (2002) Identification of QTLs associated with sweet sorghum quality. *Maydica* 47:311–322

Oliveira KM, Pinto LR, Marconi TG, Margarido GRA, Pastina MM, Teixeira LHM, Figueira AV, Ulian EC, Garcia AAF, Souza AF (2007) Functional integrated genetic linkage map based on EST-markers for a sugarcane (Saccharum spp.) commercial cross. *Mol Breed* 20:189–208

Oliver AL, Pedersen JF, Grant RJ, Klopfenstein TJ, Jose HD (2005) Comparative effects of the sorghum *bmr-6* and *bmr-12* genes: II. Grain yield, stover yield, and stover quality in grain sorghum. *Crop Sci* 45:2240–2245

Pamidimarri DVNS, Sinha R, Kothari P, Reddy MP (2009) Isolation of novel microsatellites from *Jatropha curcas* L. and their cross-species amplification. *Mol Ecol Resour* 9:431–433

Papini-Terzi FS, Rocha FR, Vencio RZN, Felix JM, Branco DS, Waclawovsky AJ, Del Bem LEV, Lembke CG, Costa MDL, Nishiyama Jr MY, et al. (2009) Sugarcane genes associated with sucrose content. *BMC Genom* 10:120

Parthiban KT, Kumar RS, Thiyagarajan P, Subbulakshmi V, Vennila S, Rao MG (2009) Hybrid progenies in Jatropha—a new development. *Curr Sci* 96:815–823

Paterson AH, Bowers JE, Bruggmann R, Dubchak I, Grimwood J, Gundlach H, Haberer G, Hellsten U, Mitros T, Poliakov A, et al. (2009) The *Sorghum bicolor* genome and the diversification of grasses. *Nature* 457:551–556

Pedersen JF, Vogel KP, Funnell DL (2005) Impact of reduced lignin on plant fitness. *Crop Sci* 45:812–819

Penning BW, Hunter CT, Tayengwa R, Eveland AL, Dugard CK, Olek AT, Vermerris W, Koch KE, McCarty DR, Davis MF, et al. (2009) Genetic resources for maize cell wall biology. *Plant Physiol* 151:1703–1728

Pereira AV, Machado MA, Azevedo ALS, do Nascimento CS, Campos AL, Ledo FJD (2008) Genetic diversity among elephantgrass accessions estimated by molecular markers. *Roy Soc Bras Zootec* 37:1216–1221

Pinto LR, Garcia AAF, Pastina MM, Teixeira LHM, Bressiani JA, Ulian EC, Bidoia MAP, Souza AP (2010) Analysis of genomic and functional RFLP derived markers associated with sucrose content, fiber and yield QTLs in a sugarcane (*Saccharum* spp.) commercial cross. *Euphytica* 172:313–327

Rocha FR, Papini-Terzi FS, Nishiyama Jr MY, Vencio RZN, Vicentini R, Duarte RDC, de Rosa Jr VE, Vinagre F, Barsalobres C, Medeiros AH, et al. (2007) Signal transduction-related responses to phytohormones and environmental challenges in sugarcane. *BMC Genom* 8:71

Roche Diagnostics Corporation (2011) 454 Sequencing, available at http://my454.com/ (accessed Sept. 2011)

Sanderson MA, Reed RL, McLaughlin SB, Wullschleger SD, Conger BV, Parrish DJ, Wolf DD, Taliaferro C, Hopkins AA, Ocumpaugh WR, et al. (1996) Switchgrass as a sustainable bioenergy crop. *Bioresour Technol* 56:83–93

Sanger F, Nicklen S, Coulson AR (1977) DNA sequencing with chain-terminating inhibitors. *Proc Natl Acad Sci USA* 74:5463–5467

Schnable PS, Ware D, Fulton RS, Stein JC, Wei FS, Pasternak S, Liang CZ, Zhang JW, Fulton L, Graves TA, et al. (2009) The B73 maize genome: Complexity, diversity, and dynamics. *Science* 326:1112–1115

Senthilvel S, Jayashree B, Mahalakshmi V, Kumar PS, Nakka S, Nepolean T, Hash CT (2008) Development and mapping of simple sequence repeat markers for pearl millet from data mining of expressed sequence tags. *BMC Plant Biol* 8:119

Sollenberger LE, Prine GM, Ocumpaugh WR, Hanna WW, Jones, Jr CS, Schank SC, Kalmbacher RS (1989) Registration of 'Mott' dwarf elephantgrass. *Crop Sci* 29:827–828

Synthetic Genomics Inc. (2009) First jatropha genome completed by Synthetic Genomics Inc. and Asian Centre for Genome Technology. Press Release, available at http://www.syntheticgenomics.com/media/press/52009.html (accessed Sept. 2011)

Takagi Y, Rahman SM (1996) Inheritance of high oleic acid content in the seed oil of soybean mutant M23. *Theor Appl Genet* 92:179–182

Tat ME, Wang PS, Van Gerpen JH, Clemente TE (2007) Exhaust emissions from an engine fueled with biodiesel from high-oleic soybeans. *J Amer Chem Soc* 84:865–869

Tatikonda L, Wani SP, Kannan S, Beerelli N, Sreedevi TK, Hoisington DA, Devi P, Varshney RK (2009) AFLP-based molecular characterization of an elite germplasm collection of *Jatropha curcas* L., a biofuel plant. *Plant Sci* 176:505–513

Tew TL, Cobill RM (2008) Genetic improvement of sugarcane (*Saccharum* spp.) as an energy crop. In: Vermerris W (ed), *Genetic Improvement of Bioenergy Crops*. Springer, New York

Tobias CM, Hayden DM, Twigg P, Sarath G (2006) Genic microsatellite markers derived from EST sequences of switchgrass (*Panicum virgatum* L.). *Mol Ecol Notes* 6:185–187

Tuskan GA, DiFazio S, Jansson S, Bohlmann J, Grigoriev L, Hellsten U, Putnam N, Ralph S, Rombauts S, Salamov A, et al. (2006) The genome of black cottonwood, *Populus trichocarpa* (Torr. & Gray). *Science* 313:1596–1604

Tuskan GA, Gunter LE, Yang ZK, Yin TM, Sewell MM, DiFazio SP (2004) Characterization of microsatellites revealed by genomic sequencing of *Populus trichocarpa*. *Can J For Res* 34:85–93

Vermerris W (2008) Composition and biosynthesis of lignocellulosic biomass. In: Vermerris W (ed), *Genetic Improvement of Bioenergy Crops*. Springer, New York

Vermerris W, Boon JJ (2001) Tissue-specific patterns of lignification are disturbed in the brown midrib2 mutant of maize (Zea mays L.) *J Agric Food Chem* 49:721–728

Vettore AL, da Silva FR, Kemper EL, Souza GM, da Silva AM, Ferro MIT, Henrique-Silva F, Giglioti EA, Lemos MVF, Coutinho LL, et al. (2003) Analysis and functional annotation of an expressed sequence tag collection for tropical crop sugarcane. *Genome Res* 13:2725–2735

Vignols F, Rigau J, Torres MA, Capellades M, Puigdomenech P (1995) The brown midrib3 (*bm3*) mutation in maize occurs in the gene encoding caffeic acid *O*-methyltransferase. *Plant Cell* 4:407–416

Vogel KP, Mitchell RB (2008) Heterosis in switchgrass: biomass yield in swards. *Crop Sci* 48:2159–2164

Wei MG, Li XH, Li JZ, Fu JF, Wang YZ, Li YL (2009) QTL detection for stover yield and quality traits using two connected populations in high-oil maize. *Plant Physiol Biochem* 47: 886–894

Williams PJL, Laurens LML (2010) Microalgae as biodiesel & biomass feedstocks: Review & analysis of the biochemistry, energetics & economics. *Energy Environ Sci* 3:554–590

Woodard KR, Prine GM (1993) Dry-matter accumulation of elephantgrass, energycane, and elephantmillet in a subtropical climate. *Crop Sci* 33:818–824

Wu X, Zhao R, Bean SR, Seib PA, McLaren JS, Madl RL, Tuinstra M, Lenz MC, Wang D (2007) Factors impacting ethanol production from grain sorghum in the dry-grind process. *Cereal Chem* 84:130–136

Wu X, Zhao R, Wang D, Bean SR, Seib PA, Tuinstra MR, Campbell M, O'Brien A (2006) Effects of amylose, corn protein, and corn fiber contents on production of ethanol from starch-rich media. *Cereal Chem* 83:569–575

Yonemaru J, Ando T, Mizubayashi T, Kasuga S, Matsumoto T, Yano M (2009) Development of genome-wide simple sequence repeat markers using whole-genome shotgun sequences of sorghum (*Sorghum bicolor* (L.) Moench). *DNA Res* 16:187–193

Zeng Z-B (1994) Precision mapping of quantitative trait loci. *Genetics* 136:1457–1468

2 Genomics for Bioenergy Production

Robert J. Henry
University of Queensland

CONTENTS

2.1 INTRODUCTION

Genomics has become a key tool for the analysis of plants and their performance for different end uses (Henry 2010a and b). Genomics allows for the simultaneous examination of the genes within a given organism, rather than the single-gene approach traditionally used. This paradigm has provided a tool that has greatly increased our knowledge of biological systems. Applications of genomics to food and feed uses of plants have previously dominated research efforts, but the application to energy use is now beginning. Plants have always been the basis of human food and have been burnt for heating, but they have only more recently been considered as potential sources of energy more widely. Humans have come to rely on energy from fossil plants in the form of oil or coal. The direct use of plants to satisfy energy requirements has become a key alternative as these fossil resources become limiting and their consumption threatens to cause global warming along with all of its well-known consequences. Humans domesticated plants by selecting genotypes that perform well under cultivation but may not be well equipped to survive in the wild (Purugganan and Fuller 2009). Crop plants were also mostly domesticated by selection of the genome for food use. The selection of plants for energy production will require isolating and selecting for previously ignored traits in already domesticated species or selecting previously undomesticated but promising species (Simmons et al. 2008; Henry 2010b). Domestication of plants for food has produced many domesticated plants with nonstructural carbohydrates (e.g., sugars and starches) that are sources of energy for animals and humans. These carbohydrates have been exploited in the first generation of bioenergy production from domesticated plants. Second-generation bioenergy crops are widely recognized as those that are available to be utilized for their more abundant structural (cell wall) carbohydrate with the potential to utilize much more of the carbon in these plants for energy. The ancient process of plant domestication was not always a deliberate one, but the opportunity now exists to domesticate with specific objectives and using the tools of modern science to achieve in a few generations what previously required hundreds. Genomics provides tools for a comprehensive analysis of the plant and the potential for selection of genotypes better suited to any given human use. Thus, genomics is a powerful tool for use in the accelerated domestication and improvement of plants as energy crops. Genomics may also be applied to the engineering of plant and nonplant organisms for the conversion of plant biomass to bioenergy. The application of genomics at these two levels may be complementary or potentially even synergistic in achieving the goal of energy- and cost-efficient bioenergy production.

2.2 APPLICATIONS OF GENOMICS IN THE DEVELOPMENT OF ENERGY CROPS

Genomics can be applied at many different levels in the development of plant species as bioenergy crops:

- In identifying higher plants for which genomes show potential as bioenergy crops,
- In identifying genes for desirable bioenergy traits,
- In screening and selecting superior bioenergy genotypes, and
- In supporting efforts to modify plant genomes, making them better bioenergy crops.

2.3 EVOLUTIONARY RELATIONSHIPS IN HIGHER PLANTS AND THEIR GENOMES

Evolutionary or phylogenetic approaches (Henry 2005) can be applied to the search for suitable species or traits for bioenergy production. The composition of a given plant's biomass is a major determinant of the suitability of that plant for use in specific bioenergy production processes. Cell wall composition is a major factor, and this varies by plant group. A molecular phylogenetic approach will ultimately allow the plants with the appropriate sets of genes for the desired composition to be found efficiently.

2.4 GENOME SEQUENCING

The last few years have seen the emergence of radically improved technology for DNA sequencing (Schuste 2008), making large-scale plant genome analysis much more feasible. Genome sequencing of potential bioenergy crop plants provides a platform for analysis of the genetic potential of these species and targets for their genetic improvement.

The sequence of the sorghum genome (Paterson et al. 2009) provides a reference genome for not only sorghum but also the many closely related grass species that are potential bioenergy crops (e.g., sugarcane and *Miscanthus*). Despite the polyploidy complexity of the sugarcane genome, significant efforts are now being made to obtain a reference genomic sequence for this species because of its existing importance as an energy crop. The eucalypt, the most widely planted forest species in tropical and subtropical parts of the world, is being sequenced because of its energy potential. This, combined with the available poplar genome sequence, will provide a foundation for genomics approaches in woody biomass crops.

2.5   ANALYSIS OF GENOME VARIATION

Genotyping an entire genome allows association genetics to be applied to the discovery of the genetic basis of important traits. Analysis of variation in plant genomes can now be considered an analysis of all of the single-feature polymorphisms. This includes all of the insets and deletions (indels) and the single nucleotide polymorphisms (SNPs). Analytically indels can usually be detected or assayed as SNPs. SNP discovery and analysis in plants has recently been reviewed (Henry and Edwards 2009). SNP discovery even in complex plant genomes is now facilitated by advancing DNA sequencing technologies (Bundock et al. 2009; Trick et al. 2009). Efficient targeted mutagenesis techniques have been developed for the discovery of naturally occurring or induced mutations (Cross et al. 2008). Analysis of variation in known SNPs in a plant population is now possible with very high-throughput techniques (Edwards et al. 2009; Masouleh et al. 2009) and is being advanced by the application of nanotechnology (Pattemore et al. 2008). Association genetics has often considered only a few candidate genes. The identification of the genetic basis of gelatinization temperature in plant starches is a good example of this because this trait may be important in determining the energy required to gelatinize starch for conversion to sugars and subsequently into a biofuel. Starch biosynthesis genes were the obvious candidates, and analysis revealed that SNPs altering key amino acid residues in the soluble starch synthases of rice could individually be responsible for an 8°C reduction in gelatinization temperature (Waters et al. 2007). This alters the structure of the amylopectin, allowing water to penetrate the starch granule more easily. This approach would complement the expression of thermostable amylases in plants to digest the starch during processing to biofuels (Wolt and Kraman 2007). Genomics tools are making the consideration of all options possible, and these developments are important when candidate genes are not readily identifiable for the trait of interest.

2.6   TRANSCRIPTOME ANALYSIS

The same tools that have revolutionized DNA sequencing have the potential to allow detailed analyses of genes expressed in different plants or tissues at different times. Transcriptome analysis allows the identification of genes controlling key traits for selection in plant domestication and improvement for human use. This approach involves the identification of candidate genes by determining their differential expression in association with the trait of interest. Candidate genes are then usually screened to determine function and to confirm association with the trait. Microarray analysis has been widely applied to transcript analysis in gene discovery. Early complementary DNA (cDNA)-based arrays were replaced by oligonucleotide arrays as the availability of more gene sequence data for more species made this possible. Advances in DNA sequencing technology offer techniques for comprehensive transcript profiling by sequencing, even in species without well-characterized genomes.

2.7   PROTEOMICS OF BIOENERGY CROPS

In some cases, the plant proteome will be the key area for analysis, although for bioenergy traits associated with carbohydrate composition this is probably less likely. However, proteins (for animal feed or other higher value applications) may be an important co-product in the energy crop. Proteomics is likely to make important contributions to understanding of the regulation of genes determining bioenergy traits.

2.8   METABOLOMICS IN RELATION TO BIOENERGY PRODUCTION

Metabolomics will be important in understanding and improving bioenergy crops, especially in establishing maximal metabolic efficiency. Metabolomics will also guide metabolic engineering to ensure carbon is stored in the most accessible form for use in bioenergy production. Metabolomics may also be important if high-value co-product molecules are being produced in the energy crop. For example, the production of plastics in plants has been approached by developing the ability to produce compounds such as polyhydroxybutyrate (Somleva et al. 2008). The production of high-value co-products may be necessary to ensure that the processing of the biofuel crop is economically attractive.

2.9   TARGET TRAITS FOR BIOENERGY PLANT IMPROVEMENT

Identifying traits for improvement in potential bioenergy crops remains a major challenge. A general list of these desirable traits is provided in Table 2.1. Many of these are complex traits that are important targets for dissection using genomic approaches. Additional species-specific traits and traits suiting specific bioenergy applications need to be added to this list to target specific species at desired energy outcomes.

The development of second generation bioenergy crops that utilize plant cell walls will be greatly advanced by an increased understanding of the genomics of cell wall biosynthesis and degradation (Fincher 2009a, 2009b).

2.10   GENETIC MODIFICATION TO ADD NOVEL BIOENERGY TRAITS

Transgenic plants with novel attributes may have superior performance as bioenergy crops. The expression of enzymes in plants to aid their processing into fuel molecules is a good example of this. Expression of carbohydrate degrading enzymes (Taylor et al. 2008) that target structural or nonstructural carbohydrates has the potential to greatly increase the efficiency of biofuel production and reduce the cost of conversion of biomass to biofuel. Genes improving lignin degradation may also be useful in increasing the efficiency of biofuel production (Liang et al. 2008). Genetic engineering of plants for biofuel production will be aided by genomic knowledge, especially of the

TABLE 2.1
Traits That May Be Considered in Bioenergy Crop Development

High biomass accumulation
High harvest index
High fraction of biofuel in harvested biomass
Nutrients partition to nonharvested parts
Able to be grown on marginal lands
Harvested material able to be stored in the field
High bulk density
High water use efficiency
High nitrogen use efficiency
Low potential as a weed
High co-product potential
Optimal biomass composition
Large scale of potential production
Low cost of harvest
High suitability for genetic improvement

species being transformed. Metabolic engineering requires knowledge of the existing pathways within the plant, which are best defined at the genomic level. The use of techniques such as zinc-finger nucleases may allow specific modification of plants to produce superior biofuel crops by the addition of new genes (Shukla et al. 2009) or modification of existing ones (Townsend et al. 2009). Thus, genomics has accelerated the discovery of genes that might be manipulated to produce superior bioenergy crops.

2.11 PROMOTERS AND CONTROL OF EXPRESSION

Knowledge of regulatory systems at the genome level will be important for breeding and selecting superior crop varieties and especially for the precise control of important transgenes. The manipulation of regulatory genes may be an important path to significant gene improvements in energy crops. The production of transgenic bioenergy crops with useful transgenes will require the use of gene promoters that direct transgene expression, at appropriate levels, in the required tissue and at the necessary time during plant development. The species specificity of these processes (Furtado et al. 2008) will require detailed analysis and understanding of regulatory processes controlling gene expression in target species. Exploiting our growing knowledge (Held et al. 2008) of the role of small RNA molecules in the regulation of plant performance and composition may also be important in the development of bioenergy crops.

2.12 MODEL BIOENERGY CROPS

Several crops for which genomics tools are available are model systems for analyses of the processes required to adapt plants to be better energy sources. Although not an energy crop, *Arabidopsis* (a general plant model) and rice (a grass species model) models are very useful for research on energy crops. The rice genome is now very well documented and characterized. Sequencing of many genotypes and of related species in the genus is increasing genomic knowledge of rice. However, the sorghum plant is a much better model for other C_4 bioenergy species such as sugarcane and maize than are C_3 plants like rice. Maize has been proposed as a useful model genome for energy crops in the grasses (Lawrence and Walbort 2007).

2.13 GENOMICS OF SPECIFIC BIOENERGY SPECIES

Plant genomes are generally large compared with other organisms (Table 2.2). However, recent advances in DNA sequencing technology have greatly increased the rate at which plant genomes are being characterized.

2.13.1 SORGHUM

Sorghum is a very efficient C_4 plant with good biomass potential and tolerance to hot and dry environments. Sorghum was domesticated as a food crop and is used today mainly as a source of animal feed. The development of sorghum as an energy crop will require selection for increased plant biomass rather than grain yield. This could lead to a significant change in plant architecture. The genome sequence of sorghum has been reported by Paterson et al. (2009). Sweet sorghum has been identified as a source of biomass for first-generation biofuel production. However, the development of high-biomass sorghum genotypes may provide an important source of biomass for biofuel production in the environments to which sorghum is adapted. Sorghum can be produced in locations without the water that might be required for other energy crops such as maize or sugarcane. The availability of the genome sequence and other genomic resources in sorghum should facilitate the rapid development of these improved types.

TABLE 2.2
Genomes of Some Potential Energy Crop Species

Species	Genome Size (pg)[a]	Genome Sequence
Sorghum	0.75	Patterson et al. 2009
Maize	2.73	www.maizegdb.org
Sugarcane	7.0 (Edme et al. 2005)	
Saccharum officinarum	4.05	
Saccharum spontaneaum	3.15	
Poplar (*Populus trichocarpa*)	0.5	Tuskan et al. 2006
Eucalyptus (*Eucalyptus globulus*)	0.58	www.eucagen.org
Willow (*Salix fragilis*)	0.86	

[a] From http://data.kew.org/cvalues.

2.13.2 SUGARCANE (AND CLOSE RELATIVES)

Sugarcane is a very efficient C_4 plant grown in tropical and subtropical regions (Henry 2010). Sugarcane (a *Saccharum* hybrid) is probably the leading industrial crop internationally with significant amounts being used for ethanol production and electricity generation. Sugarcane is currently a first-generation biofuel crop with the sugar being used to produce ethanol on a large scale. Sugarcane was first domesticated for the high sugar content of the canes, but development of sugarcane as an energy crop may result in the selection of plants for biomass yield rather than sugar content, resulting in a reduced sugar content.

Close relatives of sugarcane such as *Miscanthus* (Yamada and Henry 2011) may also have potential as energy crops. Hybrids between sugarcane and other species within the *Saccharum* complex (e.g., *Miscanthus* and *Erianthus* species) may provide important options for the development of energy crops. Molecular tools will be essential to the efficient introgression of genes from these other species.

2.13.3 MAIZE

Maize is a major food crop with most maize being used to feed farmed animals. The development of maize as an energy crop or as a dual-purpose food and energy crop will be greatly advanced by applications of genomics. The protein residue remaining after converting the grain carbohydrates to ethanol has become a key animal feed ingredient. Genomics approaches may be used to improve the nutritional value of this protein. Maize may be a useful model genome for grass improvement for bioenergy (Carpita and McCann 2008).

2.13.4 OTHER GRASSES

Most grasses have been domesticated and used for food and pasture production. The many grass species (more than 10,000 worldwide) may include species overlooked for these uses but that have potential as energy crops. Species such as switchgrass (*Panicum virgatum*), a native grass from North America, are being intensely evaluated as potential bioenergy crops (McLaughlin et al. 1999). Analysis of grasses that might suit bioenergy production in specific environments may identify many new options for domestication.

2.13.5 POPLAR

The poplar is a fast-growing woody biomass option for many temperate and cooler climates. The poplar genome has been sequenced by Tuskan et al. (2006). Willows (*Salix* species), a closely related species, are also potential bioenergy crops. The poplar genome has provided a model genome for research on wood plant species and their use for bioenergy and other applications.

2.13.6 EUCALYPTUS

Eucalypts are the most planted hardwood tress in tropical and subtropical parts of the world (Henry 2011). Members of this group of more than 700 species are adapted to a wide range of environments. Eucalypts have high biomass accumulation rates and a tolerance of marginal growing conditions, making them good candidates for the production of wood biomass. Many genomics tools for eucalypts, including genetic maps and gene and genome sequences have been developed in recent research (Henry 2011).

2.14 FUTURE PROSPECTS FOR USE OF GENOMICS IN BIOENERGY PRODUCTION FROM PLANTS

The future of energy crops should see the application of genomics to the discovery and manipulation of genes to create optimally designed energy plants. The efficiency of photosynthesis may be enhanced by selecting or engineering plants with an optimal metabolism for specific environments. For example, C_4 pathways of photosynthesis may improve the efficiency of carbon fixation, especially at high light intensities in warm and dry environments. Plant architecture may be optimized to capture solar energy and use it to fix and store the carbon at a high density in tissues that can then be easily harvested and converted to useful bioenergy molecules. Biomass composition may be selected to best suit available biofuel conversion technologies. For example, altered lignin content and modified carbohydrate composition are major options. Also, the proportion of cellulose and noncellulosic polysaccharides may be altered. Most importantly, the linkages between different carbohydrate polymers and between carbohydrate and lignin may be reduced or altered. Genomics will provide the tools to characterize the genetic control of these key biofuel traits. Enzymes required in any conversion process could all be produced in the plant to reduce costs and improve energy efficiency.

Future biomass crops need to be designed for sustainable production with all nutrients being recycled to the soil. Nitrogen needs to be supplied by the crop itself or by other crops in the rotation. Plants also need to be designed to maximize the value of co-products remaining after energy extraction. This may be an essential feature of the economics of growing bioenergy crops. Some of these co-products may be industrial or pharmaceutical products. However, high-value animal feeds and human food co-products have the advantage of reducing the competition between food and energy production. Protein residues remaining after conversion of plant carbohydrates to energy may be significant co-product options. Genomics approaches may allow for analyses of bioenergy and co-product potential and allow selection or breeding for both characteristics.

Genomics also has the potential to make a major contribution to future bioenergy production on the basis of the novel production of bioenergy from algae. Algae also have potential as a direct source of high-value fuel molecules (e.g., alkanes). Thus, algae may also represent an alternative source of biomass.

REFERENCES

Bundock PC, Eliott F, Ablett G, Benson AD, Casu RE, Aitken KS, Henry RJ (2009) Targeted SNP discovery in sugarcane using 454 sequencing. *Plant Biotechnol J* 7:347–354
Carpita NC, McCann MC (2008) Maize and sorghum: Genetic resources for bioenergy grasses. *Trends Plant Sci* 13:415–420
Cross M, Waters D, Lee LS, Henry RJ (2008) Endonucleolytic mutation analysis by internal labeling (EMAIL). *Electrophoresis* 29:1291–1301

Edme SJ, Comstock JC, Miller JD, Tai PYP (2005) Determination of DNA content and genome size in sugarcane. *J Amer Soc Sugarcane Technol* 25:1–15

Edwards KJ, Reid AL, Coghill JA, Berry ST, Barker GL (2009) Multiplex single nucleotide polymorphism (SNP)–based genotyping in allohexaploid wheat using padlock probes. *Plant Biotechnol J* 7:375–390.

Fincher GB (2009a) Exploring the evolution of (1,3;1,4)-beta-D-glucans in plant cell walls: Comparative genomics can help! *Curr Opin Plant Biol* 12:140–147

Fincher GB (2009b) Revolutionary times in our understanding of cell wall biosynthesis and remodeling in the grasses. *Plant Physiol* 149:27–37

Furtado A, Henry RJ, Takaiwa F (2008) Comparison of promoters in transgenic rice. *Plant Biotechnol J* 6:679–693

Held MA, Penning B, Brandt AS, Kessans SA, Yong W, Scofield SR, Carpita NC (2008) Small-interfering RNAs from natural antisense transcripts derived from a cellulose synthase gene modulate cell wall biosynthesis in barley. *Proc Natl Acad Sci USA* 105:20534–20539

Henry RJ, Edwards K (2009) New tools for single nucleotide polymorphism (SNP) discovery and analysis accelerating plant biotechnology. *Plant Biotechnol J* 7:311

Henry RJ (2005) *Plant Diversity and Evolution: Genotypic and Phenotypic Variation in Higher Plants*. CABI Publishers, Oxon, United Kingdom

Henry RJ (2010a) An overview of advances in genomics in the new millennium. In: Kole C, Abbott AG (eds), *Principles and Practices of Plant Genomics. Volume 3: Advanced Genomics*. Science Publishers Enfield, NH and CRC Press, Boca Raton, FL

Henry RJ (2010b) *Plant Resources for Food, Fuel and Conservation*. Earthscan, London, pp 200

Henry RJ (2010) Basic information on the sugarcane plant. In: Henry RJ, Kole C (eds), *Genetics, Genomics and Breeding of Sugarcane*. Science Publishers, Enfield, NH and CRC Press, Boca Raton, FL, pp 1–7

Henry RJ (2011) Eucalyptus. In: Kole C (ed), *Wild Crop Relatives: Genomic and Breeding Resources. Vol 10: Forest Trees*. Springer, New York, in press

Lawrence CJ, Walbot V (2007) Translational genomics for bioenergy production from grasses: Maize as the model species. *Plant Cell* 19:2091–2094

Liang HY, Frost CJ, Wei XP, Brown NR, Carlson JE, Tien M (2008) Improved sugar release from lignocellulosic material by introducing a tyrosine-rich cell wall peptide gene in poplar. *Clean Soil Air Water* 36:662–668

Masouleh A, Waters D, Reinke R, Henry RJ (2009) A high-throughput assay for rapid and simultaneous analysis of perfect markers for important quality and agronomic traits in rice using multiplexed MALDI-TOF mass spectrometry. *Plant Biotechnol J* 7:355–363

McLaughlin S, Bouton JD, Bransby D, Conger B, Ocumpaugh W, Parrish D, Taliaferro C, Vogel K, and Wullschleger S (1999) Developing switchgrass as a bioenergy crop. In: Janick J (ed), *Perspectives on New Crops and New Uses*. ASHS Press, Alexandria, VA pp 282–299

Paterson AH, Bowers JE, Bruggmann R, Dubchak I, Grimwood J, Gundlach H, Haberer G, Hellsten U, Mitros T, Poliakov A, et al. (2009) The Sorghum bicolor genome and the diversification of grasses. *Nature* 457:551–556

Pattemore JA, Trau M, Henry RJ (2008) Nanotechnology: The future of cost-effective plant genotyping. In: Henry RJ (ed), *Plant Genotyping II: SNP Technology*. CABI Publishers, Wallingford, United Kingdom, pp 133–153

Purugganan MD, Fuller DQ (2009) The nature of selection during domestication *Nature* 457:843–848

Schuste SC (2008) Next-generation sequencing transforms today's biology. *Nature Meth* 5:16–18

Shukla VK, Doyon Y, Miller JC, DeKelver RC, Moehle EA, Worden SE, Mitchell JC, Arnold NL, Gopalan S, Meng X, et al. (2009) Precise genome modification in the crop species *Zea mays* using zinc-finger nucleases. *Nature* 459:437–441

Simmons BA, Loque D, Blanch HW (2008) Next-generation biomass feedstocks for biofuel. *Genome Biol* 9:242

Somleva MN, Snell KD, Beaulieu JJ, Peoples OP, Garrison BR, Patterson NA (2008) Production of polyhydroxybytyrate in switchgrass, a value-added co-product in an important lignocellulosic biomass crop. *Plant Biotechnol J* 6:663–678

Taylor LE, Dai ZY, Decker SR, Brunecky R, Adney WS, Ding S-Y, Himmel ME (2008) Heterologous expression of glycosyl hydrolases in planta: A new departure for biofuels. *Trends Biotechnol* 26:413–424

Townsend JA, Wright DA, Winfrey RJ, Fu F, Maeder ML, Joung JK, Voytas DF (2009) High-frequency modification of plant genes using engineered zinc-finger nucleases. *Nature* 459:442–445

Trick M, Long Y, Meng L, Bancroft I (2009) Single nucleotide polymorphism (SNP) discovery in the polyploidy *Brassica napus* using Solexa transcriptome sequencing. *Plant Biotechnol J* 7:334–346

Tuskan GA, DiFazio S, Jansson S, Bohlmann J, Grigoriev I, Hellsten U, Putnam N, Ralph S, Rombauts S, Salamov A, et al. (2006) The genome of black cottonwood, *Populus trichocarpa* Torr. & Gray. *Science* 313:1596–1604

Waters DLE, Henry RJ, Reinke RF, et al. (2006) Gelatinization temperature of rice explained by polymorphisms in starch synthase. *Plant Biotechnol J* 4:115–122

Wolt JD, Karaman S (2007) Estimated environmental loads of alpha-amylase from transgenic high-amylase maize. *Biomass Bioenergy* 31:831–835

Yamada T, Henry RJ (2011) *Miscanthus*. In: Kole C (ed), *Wild Crop Relatives: Genomic and Breeding Resources. Vol 8: Industrial Crops*. Springer, Heidelberg, Dordrecht, London, New York, in press

3 Genetic Engineering for Bioenergy Crops

Puthiyaparambil Josekutty, Shobha Devi Potlakayala, and Rebekah Templin
Pennsylvania State University–Harrisburg

Alankar Vaidya
Praj Industries Limited

Sarah Ryan and Deepkamal Karelia
Pennsylvania State University–Harrisburg

Nilkamal Karelia
Cheyney University of Pennsylvania

Vaman Rao
NITTE University & NMAM Institute of Technology

Vilas Tonapi
National Research Center for Sorghum

Behnam Tabatabai, Fatima Fofanah, Diego Morales, and Sairam Rudrabhatla
Pennsylvania State University–Harrisburg

CONTENTS

3.1 INTRODUCTION

Fossil fuels (coal, petroleum, and natural gas) are neither sustainable nor ecofriendly because the source is finite and their use cause considerable pollution. (Naik et al. 2010). There has been a dramatic increase in the price of oil in 2008, and more drastic price increases nicknamed "peak oil" in the coming years are predicted (Goldemberg 2007; Potters et al. 2010). Dwindling oil reserves, less than adequate investments into oil exploration and production, and rising demand for oil are major reasons for the anticipated oil price increase (Lloyd's 2011). Most greenhouse gas emissions are the result of electricity production and heating (27%) using fossil fuels. Other causes of greenhouse gas emissions include land use/change and forestry (18%), agriculture (13%), other energy sectors (13%), transportation (12%), manufacturing and construction (11%), and industrial process (3%) (World Resource Institute 2011). The concerns about dwindling fossil fuel reserves and oil price increases, and the relationship between fossil fuels and global climate change have generated great interest in bioenergy/biofuels.

Unlike fossil fuels, biofuels are renewable because they can be grown repeatedly. Also, biofuels are carbon neutral and essentially reduce the carbon emissions (Naik et al. 2010). Using biofuels to produce bioenergy is one part of the solution to curb global climate change and allow the United States to become energy independent.

Bioenergy is broadly defined as renewable energy derived from biological materials that are used to produce heat, generate electricity, and provide energy for transportation (Yuan et al. 2008). Generally, biofuels are broadly categorized into four categories—first-generation, second-generation, third-generation, and fourth-generation biofuels—depending on the type of feedstock being used. First-generation biofuels are based primarily on corn and soybean and other edible food crops that compete for agricultural croplands, natural fresh water resources, and fertilizers. These fuels are used primarily as small blends, and the energy input and output ratios do not meet the large-scale commercial use. Second-generation biofuels are derived from cellulosic biomass (e.g., miscanthus, switchgrass, sweet sorghum). Biofuels derived from nonedible plant resources are also considered to be second-generation biofuels. Algae are the source of third-generation biofuels, but fourth-generation biofuels are chemically created with the help of petroleum-like hydroprocessing and revolutionary processes such as Joule's "solar-to-fuel" method (Zarrilli 2007; Yuan et al. 2008). Biodiesel is derived from animal fat and vegetable oil (Jena et al. 2010).

Biomass is a general term used to describe materials of biological origin, including all living matter derived from plants and animals. Energy from biomass is gained from several sources, such as wood, grasses, and animal materials (Babu 2008). Traditionally, biomass is similar to fossil fuels in that it is burned to heat water or produce steam and generate electricity. There are four categories of biomass conversion: direct combustion, thermochemical, biochemical, and agrochemical. These aspects have been described in detail in Rooney et al. (2007).

In 2007, the United States used roughly 542×10^9 L of gasoline, approximately one quarter of the global oil consumption (Vermerris 2008). The United States has been importing more fossil fuels than it produces within the country (EIA 2011). This dependence on imported oil, combined with the political instability and ongoing conflicts of the major oil-producing nations has made the energy crisis a political priority. As a result, the Energy Independence and Security Act (EISA) was enacted in 2007 to help stimulate biofuel production in the United States. According to the EISA, by 2022, a minimum of 16 billion gal of cellulosic ethanol per year must be produced (Leistritz and Hodur 2008). Biofuels will play a major role in the future economics of energy production (McLaughlin et al. 2011).

Not only will the U.S. economy benefit from an increase in biofuels production, but there are also large economic benefits that developing countries can gain by joining the biofuel industry. Conventional biomass products (e.g., wood) are used to provide household energy for people of the developing world. Here, biomass fuels meet the energy needs of households; however, the combustion of biofuels pollutes the air, causing serious health problems. Reducing these emissions using improved stoves and better fuels can reduce respiratory illnesses and greenhouse gas emissions (Kammen 2006). Developing countries would be able to replace a higher percentage of their oil use because of their smaller consumption level, thus decreasing these emissions. Additionally, the biofuel industry is labor-intensive and has the potential to create many new jobs. For example, the bioethanol industry offers 4.2 million jobs in Brazil. In the short term, until higher grain prices stimulate a renewed emphasis on agricultural development, areas that face food shortages or import most of their food could experience higher food insecurity challenges (Cassman and Liska 2007).

However, bioenergy is not without drawbacks. First-generation biofuels received criticism because they are produced from food crops. Crops with dual use as food and fuel get rated by their comparative value as food and biofuel feedstock, leading to an increase in food costs (Cassman and Liska 2007). For example, corn prices have seen an abrupt increase because of corn's use in ethanol production. Another criticism is about the replacement of arable land with biofuel crops in place of food crops, which threatens the sustainability of food production. Other criticisms surrounding first-generation biofuels include poor water use efficiency of the feedstocks, inability to meet large volume requirements (except for sugarcane), and the large carbon footprint of ethanol (Fargione et al. 2008; Searchinger et al. 2008; Stoeglehner and Narodoslawsky 2009; Rathmann et al. 2010). Because, at the moment, ethanol is only profitable or competitive around $50 per barrel, subsidies offered by the government play a great role in the expansion of the biofuel market and its increased production (Ruth 2008).

Second- through fourth-generation biofuels attract less criticism regarding water use efficiency of the crops and the carbon footprint of the fuel. With the aid of other sources of renewable energy, problems associated with first-generation biofuels can be reduced in second-generation biofuels. Using the cellulosic biomass crops of second-generation biofuels instead of food crops of the first generation would alleviate the food versus fuel competition. Biofuel crops such as sweet sorghum and jatropha could be grown on marginal lands and contaminated croplands that are deemed unsuitable for crop production. This approach will offer the greater environmental benefits of reduced soil erosion, carbon sequestration, and better utilization of land resources. Former wastelands could then be used in a profitable way, thus leading to true economic growth and development.

3.2 BIOFUEL CONVERSION

A first-generation biofuel, corn grain, is the most important biomass for bioethanol production in the United States (~97%) followed by sorghum (2%) and 1% from other crops, beverage/juice waste, and food processing waste (Nichols and Bothast 2008). Corn and sorghum used for ethanol production account for approximately 20% of the U.S. corn crop and 15% of the sorghum crop. The fermentation process of starch to ethanol is similar for all grains. Essentially, starch (a polymer of glucose) is enzymatically converted to sugar, which is then fermented to produce ethanol. There are two methods of ethanol production from corn, namely, wet mill and dry grind processes (Nichols and Bothast 2008). In the dry grind process, saccharification and fermentation occur simultaneously after the dry grinding of corn. In the wet mill process, saccharification and fermentation are carried out in separate steps. Starch in the wet mill process is fairly pure, allowing for the separation of other components such as protein, lipids, vitamins, and fiber. Although ethanol production from starchy grains works well, it could result in competition between energy and food production and may not be sustainable in the long run (Yuan et al. 2008; Potters et al. 2010), hence, the drive to produce second-generation biofuels.

Plant cell walls are mostly composed of cellulose, lignin, hemicelluloses, and pectin. These compounds are integrated together to form a strong backbone of the cell wall that maintains the structural and physiological integrity of the cell. Because cellulose and hemicellulose are polysaccharides, they can be broken down into simple sugars and used for the fermentation of alcohol. However, cellulose microfibrils are embedded in a matrix where lignin is a part, which resists degradation. Lignin is composed of different subunits, namely, p-hydroxyphenyl (H), guaiacyl (G), and syringyl (S). The bonds among the polymers are less reactive; therefore, a single enzyme cannot degrade them all. Lignification is correlated with secondary wall thickening (Weng et al. 2008). Energy from this source is appealing in that it is portable and compatible with the current fuel infrastructure (Rubin 2008). Pretreatment of the complex cell wall is essential to achieve enzymatic removal of sugars from the cell wall.

The general process of cellulosic ethanol production includes pretreatment, enzymatic hydrolysis, fermentation, and distillation and has been described in detail in Mosier et al. (2004).

3.3 BIOETHANOL

The steps in the process of cellulosic ethanol production are presented in Figure 3.1. There are various pretreatment methods, including steam explosion, liquid hot water, pH, dilute acid, concentrated acid, alkaline-based treatments of lime, and ammonia fiber expansion used in cellulosic ethanol production (Yang and Wyman 2008; Zhu and Pan 2010). In the steam explosion method, superheated steam approximately 160–260°C kept under high pressure (100–700 psi) is exposed to lignocellulose for a short period of time followed by a flashing process to release the steam in an explosion. This causes the lignocellulose to open and expose the cellulose, allowing for increased digestibility. The sugars are found in the liquid stream, but because of the nature of this process, several compounds (furfural, 5-hydroxymethylfurfural) that inhibit fermentation are formed (Abogbo and Coward-Kelly 2008; Lu and Mosier 2008; Brethauer and Wyman 2010).

In a hot water pretreatment, the explosive decompression of the steam explosion is replaced by controlled cooling to keep the water in the liquid phase throughout (Lu and Mosier 2008; Yang and Wyman 2008). Advantages of this method include making complete hydrolysis of hemicelluloses possible and making treated material highly digestible during enzymatic saccharification. The controlled pH liquid hot water treatment is a modified version of the hot water (140–220°C, for 10–30 min) pretreatment and provides greater control of the chemical reactions that occur during pretreatment. An advantage of this procedure is that it minimizes the formation of degradation products (Lu and Mosier 2008; Yang and Wyman 2008).

Dilute mineral acids such as hydrochloric acid, phosphoric acid, nitric acid, and sulfuric acid have been studied for their efficacy as a pretreatment in cellulosic ethanol production. Sulfuric acid hydrolysis seems promising because of its low cost and effectiveness (Lu and Mosier 2008; Yang and Wyman 2008). For effective acid hydrolysis, ideal operation conditions may be 0.5–1.4% (w/w) sulfuric acid treatment, 100–250 g/L biomass solid loading, and a residence time between 3 and 12 min at 165–195°C. A downside of this pretreatment is that it produces compounds that can inhibit

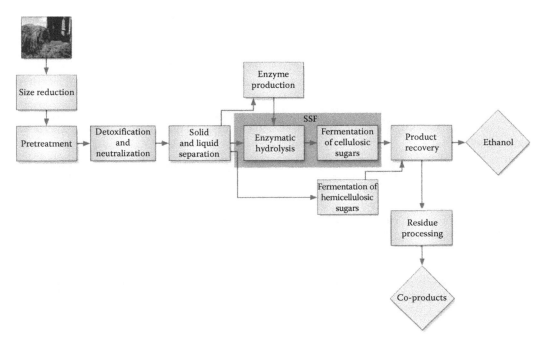

FIGURE 3.1 Traditional cellulosic biomass conversion to ethanol on the basis of concentrated acid pretreatment followed by hydrolysis and fermentation. (From U.S. DOE, *Breaking the Biological Barriers to Cellulosic Ethanol: A Joint Research Agenda*, DOE/SC/EE-0095, U.S. Department of Energy Office of Science and Office of Energy Efficiency and Renewable Energy, 2006, available at http://genomics.energy.gov/gallery/.)

the fermentation process. Recycling and disposing of the large quantity of sulfuric acid used in this process also make it expensive (Lu and Mosier 2008; Yang and Wyman 2008).

Lime removes lignin under various conditions depending on the type of the lignocellulosic feedstock used (e.g., 100°C for 13 h for corn stover, 100°C for 2 h for switchgrass and 150°C for 6 h at 14 atm for poplar wood). When pretreating woody biomass containing a lot of lignin (e.g., poplar), adding oxygen can improve the process (Lu and Mosier 2008; Yang and Wyman 2008).

There are two pretreatments that use ammonia: ammonia fiber expansion (AFEX) pretreatment and ammonia recycle percolation (ARP) pretreatment. The AFEX process is widely applicable to pretreating biomass from grasses including sugarcane bagasse. This process permits nearly a complete conversion of cellulose and hemicelluloses to fermentable sugars at very low enzyme loadings. Being an efficient procedure, it is also less expensive than other methods (Holtzapple et al. 1991; Delarosa et al. 1994; Reshamwalla et al. 1995; Dale et al. 1996; Moniruzzaman et al. 1997). In the ARP process, aqueous ammonia solution, 15% of weight, is passed through the lignocellulosic biomass in a reactor at 80–180°C and then the ammonia is separated and recycled. This process is very efficient in separating fermentable sugars from the lignin component and also in recycling the ammonia (Iyer et al. 1996; Lee et al. 1996; Wu and Lee 1997; Lu and Mosier 2008).

Enzymatic hydrolysis is performed by cellulase and hemicellulase enzymes, either individually or in tandem (cellulosomes). These enzymes break down polymeric substrates into glucose. The best understood system is from *Trichoderma reesei*, a fungus that contains several enzymes (Lu and Mosier 2008; Nakagame et al. 2010). These enzymes include endoglucanases that "decrease the degree of polymerization of macromolecular cellulose by attacking accessible sites and breaking the linear cellulose chain" (Mousdale 2008), cellobiohydrolases that "attack the chain ends of the cellulose polymers, liberating the disaccharide cellobiose" (Mousdale 2008), and β-glycosidases that "hydrolyze soluble cellodextrins and cellobiose to glucose" (Mousdale 2008). Additionally, there is a fourth enzyme, cellodextrinase, which "attacks the chain ends of the cellulose polymers, liberating glucose" (Mousdale 2008). For the best productivity (converting all of the sugars at high rates), microorganisms that bring about fermentation should be able to withstand ethanol and inhibitory compounds as well as be resistant to contamination. The following microbes were improved to impart these properties: yeast (*Saccharomyces cerevisiae*), *Zymomonas mobilis*, and *Escherichia coli* (Lu and Mosier 2008; Kim et al. 2010a,b; Sun and Cheng 2002).

Yeast can tolerate ethanol and exists at a low pH, reducing the risk of contamination of other microorganisms. Various strains of yeast can ferment an array of sugars such as monosaccharides (glucose, mannose) and also disaccharides such as sucrose and maltose to ethanol. *Z. mobilis* is a Gram-negative bacterium that uses the Entner–Doudoroff pathway anaerobically to produce 5–10% more ethanol than yeast. This strain also cannot ferment pentoses effectively. *E. coli* can generate ethanol in small quantities because it can ferment sugars into lactic acid, formic acid, and acetic acid (Lu and Mosier 2008). However, most strains of yeast are unable to ferment pentose sugars such as xylose and arabinose that represent up to 40% of total biomass carbohydrates. *S. cerevisiae*, *E. coli*, and *Z. mobilis* have been genetically modified to enable them to ferment pentose sugars (Lu and Mosier 2008; Kim et al. 2010a, 2010b; Weber et al. 2010). Current trends indicate that corn grain ethanol will be displaced by cellulosic ethanol, ethanol production from sugarcane will include production from sugar and cellulosic biomass, considerable quantities of biodiesel will be produced from nonedible oil and cellulosic feedstock, and ethanol may be replaced by higher energy (more reduced) compounds, as biofuel discovery and practical applications of synthetic catalysts become game-changers in the biofuel production scenario.

The future of genomics-based biotechnology research for bioenergy crops has been detailed in a recent review by Yuan et al. (2008). Genetic improvement of biofuel crops through biotechnology will play an important role in improving biofuel production and making biofuels more sustainable (Gressel 2008; Vega-Sánchez and Ronald 2010; Harfouche et al. 2011). Incorporating new genes into plants uses various techniques for delivery. These genes are then made part of the plant's chromosomal DNA through recombination. Particle bombardment (gene gun) forces the genes into the cell through pressure. To gain specificity in the plant cell requires the use of *Agrobacterium tumefaciens*, which allows the genes to enter the nucleus and combine with the host DNA. Although *A. tumefaciens* is only found naturally among dicot species, certain strains are able to infect monocot plants, such as corn, sorghum, and switchgrass (Sticklen 2008). The gene of interest is placed under the control of a promoter, allowing for tissue-specific inducible expression. Adding a selectable or screenable marker to the vector simplifies "the identification of transformed plants and increases the efficiency of recovery of transgenic plants" (Skinner et al. 2004). The selectable markers typically used are *gus* and *gfp* or those that confer antibiotic resistance (screenable): kanamycin, hygromycin, glyphosate, etc. Being able to observe the level of gene expression of a species is very useful in determining phenotypes that are desirable, leading to a better understanding of the specific genes that affect important features (Heaton et al. 2008). For the confirmation of a successful transformation, PCR, Southern blotting, and progeny analysis can be used (Skinner et al. 2004). Using the techniques of genomics allows for faster selection of desirable characteristics than traditional cross-breeding. Biotechnology using tissue culture techniques is also an important area of research. One method useful to researchers is somaclonal variation (spontaneous changes in plants), in which the opportunity arises to "develop new germplasm, better adapted to end-user demands" (Schroder et al. 2008). These variations could lead to better adaptations of plants to unfavorable conditions.

In addition to the use of genetics to improve bioenergy crops, it is equally important to use genetically improved organisms that will lower the operating costs and allow for faster and more efficient ethanol production. A recombinant strain of *S. cerevisiae* allows for the co-fermentation of glucose and xylose. The transgenic *Z. mobilis* strain (*Z. mobilis* CP4), producing up to 95% ethanol, can grow on a mixture of glucose and xylose to produce 95% ethanol. In *E. coli*, overexpressing pyruvate decarboxylase (*PDC*) and alcohol dehydrogenase (*ADH*) genes can produce a high percentage of alcohol (Lu and Mosier 2008). The ATCC11303 (*E. coli*) strain B seems to be the best host for incorporating the PET vector, "producing more than 1000 mM ethanol from hemicellulose hydrolysate sugars" (Lu and Mosier 2008).

In addition to having improved microorganisms, incorporating these enzymes (namely cellulases) directly into plants gives them a distinct advantage over microorganisms, namely, a lower energy input than microbial production (Sticklen 2008). These hydrolytic enzymes are mainly microbial in nature and require codon alterations to become suitable for plant expression. Also, for the correct expression of these enzymes, proper protein folding is necessary. This problem can be fixed if the enzymes accrue in subcellular components rather than in the cytosol. Extraction from the plants will enable researchers to apply these enzymes as "part of the plant total soluble protein (TSP)" (Sticklen 2008) to biomass for the conversion into sugars. This process is relatively simple and can become incorporated into traditional cellulosic ethanol production. More research is needed to increase the levels of production and the biological activity of the heterologous enzymes (Sticklen 2008). An additional area of research is incorporating heat-induced enzymes into the plant to help increase its biomass conversion efficiency. One such enzyme is the *Acidothermus celluloyticus* cellulase E1 (Yuan et al. 2008).

Lignin plays an important role in structural support, so the crop does not lodge, making harvesting difficult (Tew and Cobill 2008). Also, lignin is physically important in protecting against pathogens and helps water transport through the xylem (Torney et al. 2007). Downregulating the biosynthesis of lignin is an important part of reducing the amount of lignin present in plant cell walls, thereby reducing the cost of pretreatment (Sticklen 2008; Yuan et al. 2008). Cinnamyl alcohol dehydrogenase (CAD) downregulation in poplar resulted in improved lignin solubility in an alkaline medium, leading to more efficient delignification (Abramson et al. 2010; Harfouche et al. 2011). Having a cell wall more uniform in nature could also help in lowering the amount of pretreatment needed. Another method is to redirect carbon flux from lignin biosynthesis to overall biomass built up, helping to increase sugar release during enzymatic hydrolysis. This was seen in aspen when 4-coumarate CoA ligase (4CL) was downregulated .Although more research is needed in bioenergy crops and the downregulation of lignin biosynthesis, this is an important alternative in reducing pretreatment costs (Sticklen 2008).

Use of genetically modified (GM) feedstocks for the production of bioenergy has been complicated by public acceptance of these crops. The public (environmental organizations, consumer advocacy groups, and scientific community) scrutinizes these crops because of their concerns over "health and environmental safety and socioeconomic considerations" (Chapotin and Wolt 2007). Because biofuels are considered a greener technology and are intended to replace petroleum usage, this technology could be held to a higher environmental standard. Concerns regarding the use of genetic engineering have the potential to slow down the adoption of GM crops. General public perception of GM crops has been affected based on incidents of agricultural biotechnology crops such as the StarLink™ corn, co-mingling of pharmaceutical plants and food crops, unapproved GM rice, and *Bt* corn effects on monarch butterflies.

3.4 BIOENERGY CROPS

3.4.1 CORN

Corn is an important cereal grain and a staple food. Corn is more productive as a biofuel crop than wheat because of its more efficient C_4 photosynthetic pathway compared with the C_3 pathway of wheat. C_4 plants are generally more efficient users of water and nitrogen compared with C_3 plants (Karp and Shield 2008). Corn is adapted to various soils and environmental conditions. Because corn is a widely available starchy feedstock, it is the principal source of ethanol production in the United States. Corn is in high demand as a food and feed, therefore criticism for its use as a biofuel feedstock is mounting (Fargione et al. 2008; Stoeglehner and Narodoslawsky 2009; Ajanovic 2010). One way to help reduce competition between food and fuel is the use of corn stover, the remaining vegetation after grain harvesting, for cellulosic ethanol. Traditionally, corn stover is left remaining on the fields to help reduce soil erosion and replenish nutrients. It is estimated that in 2030, 256 million dry tons per year of corn stover (de Leon and Coors 2008) will become available in the United States. The use of corn stover depends on its yield potential and carbohydrate composition of the cell wall. Carbohydrate composition of corn stover is 37% cellulose, 28% hemicellulose, and 18% lignin. Corn will be very important in the immediate production of cellulosic ethanol (de Leon and Coors 2008).

Cell wall composition in corn stover is being manipulated using various traits. The *brown midrib* mutations (*bm1, bm2, bm3, bm4*) are naturally occurring mutations found in corn that are known to alter the "lignin concentration and/or composition of the plant" (de Leon and Coors 2008). Maize *bm1* is found to affect the expression of *CAD*, and *bm2* plants have lower lignin contents and significantly diminished levels of Ferulic acid (FA) ether (Barriere et al. 2004). The *bm3* allele is the most efficient at enhancing cell wall digestibility. Expressing the gene trait of *Lfy1* allows corn hybrids to generate more forage yields than other normal corn hybrids. These plants tend to have more nodes and leaves on their main stalk. It is also thought that, through altered lateral branch formation, corn could have increased biomass. The *grassy tiller 1 (gt1)* and *teosinte branched1 (tb1)* genes are connected with activation of "lateral meristems and reduced apical dominance" (de Leon and Coors 2008). Expansins (proteins) play a part in relaxing cell walls for growth and expansion. Although these proteins have been found in corn stover, more research needs to be done to determine their ability to reduce pretreatment cost through modifying cell walls (Sticklen 2008).

Enabling plants to survive in adverse environments would allow for the use of a broader range of land, resulting in more biofuel production. Resistance to biotic and abiotic stresses is a key part to making it possible. It has been suggested that reactive oxygen species (ROS) "act as intermediate signaling molecules to regulate gene expression ... a central component of plant adaption" (Schroder et al. 2008). ROS are very toxic in that they react with several cell components, such as lipids, proteins, and/or nucleic acids. Higher production of ROS is induced by stresses, but plants have some control by expressing different mechanisms, mainly "enzymatic and non-enzymatic reactions" (Schroder et al. 2008). Such enzymes/proteins include "SOD, APOD, CAT, GST, GPOD, enzymes of ascorbate-glutathione pathway dehydrin, actin, histone" (Schroder et al. 2008).

3.4.2 SUGARCANE

Sugarcane is a large perennial C_4 grass found in tropical and subtropical regions. Mostly table sugar (~70%) is produced from sugarcane (Matsuoka et al. 2009). Because sugarcane is efficient at converting solar energy into chemical energy, this crop is important in ethanol fuel. The three types of energy canes include sugarcane primarily grown for sugar production; type I energy cane, which is grown for sugar and fiber production; and type II energy cane, which primarily yields lingo cellulosic fiber (Tew and Cobill 2008). Bagasse, the residue left over from sugar mills, can be used to generate heat and electricity. The separation of juice from the fiber is done through milling or diffusion. Brazil has the most advanced sugarcane-based ethanol industry, and ethanol replaces a significant proportion of transportation fuel in Brazil (Goldemberg et al. 2008; Balat and Balat 2009; Matsuoka et al. 2009).

Genetic engineering of sugarcane focuses on several agronomic traits (e.g., resistance to disease and pests, herbicide resistance, increased sucrose content) because it is difficult and time-consuming to develop sugarcane exhibiting such qualities through conventional breeding. Recent reviews (Hotta et al. 2010; Watt et al. 2010) have discussed various methods used for transformation of sugarcane and applications of genetic engineering for improvement of sugarcane to make it a better biomass crop. Biolistic transformation using microprojectiles dominates sugarcane transformation studies, but there are reports of *Agrobacterium*-mediated and electroporation-based transformation of sugarcane (Arencibia et al. 1995, 1999; Enriquez et al. 2000; Manickavasagam et al. 2004; Lakshmanan et al. 2005; Jain et al. 2007; Molinari et al. 2007; Wu and Birch 2007;

Hotta et al. 2010). As of 2005, sugarcane was genetically transformed with three genes conferring resistance to herbicides, six genes offering resistance to diseases, and five genes each for imparting resistance to pests and modifying the metabolomics of sugarcane. However, only a few of the transgenic sugarcane were field tested (Lakshmanan et al. 2005). In South Africa, herbicide-resistant sugarcanes (glufosinate ammonium and glyphosate resistant) have reached field trial stage. Transgenic sugarcane resistant to Sugarcane Mosaic Potty virus with heterologous expression in antisense and untranslatable form has reached field trial stage, whereas insect-resistant sugarcane [Cry1A (c)] heterologous expression is still in the pot bioassay stage. Several transgenic crops with modifications to sucrose metabolism have completed glasshouse trials or are being studied in the field trial stage (Watt et al. 2010). Recent trends in the genetic engineering of sugarcane are directed at the modification of sucrose metabolism to enhance sucrose production and accumulation, as well as recovery of high-value products from sugarcane (McQualter et al. 2004; Petrasovits et al. 2007; Wu and Birch 2007, 2010). Also, the range of transgenic sugarcane, especially those with modification of sugar metabolism, has increased in recent years to enhance the sucrose content of sugarcane (Ma et al. 2000; Wu and Birch 2007).

3.4.3 SWEET SORGHUM

Sweet sorghum is the fifth most important cereal crop worldwide. Although it can grow in harsh environmental areas, sweet sorghum is mainly found in hot/dry tropical and subtropical areas. Several factors make sorghum a good choice for the biofuel industry, including "yield potential and composition, water-use efficiency and drought tolerance, established production systems, and potential for genetic improvement" (Rooney et al. 2007). Several traits have been found to be connected with drought tolerance, which have been enhanced by breeders to make sorghum variants highly drought tolerant. These traits include "heat tolerance, osmotic adjustment, transpiration efficiency, rooting depth, epicuticular wax, and stay green" (Rooney et al. 2007). Grain sorghum provides starch and sweet sorghum produces sugar, and both types provide cellulose for biofuel conversion. The average yields from grain sorghum are only slightly lower than those of corn, if not the same as corn grain, with the potential of genetic improvement to increase yield. Hybrids are being cultivated for regions where sugarcane production is limited. Biomass sorghum has the potential to become a dedicated energy crop on the basis of high yield capability and broad growth range.

One characteristic feature of sorghum useful in improving its yield is its height. There are four genes known to influence this characteristic, the *Dwarfing* (*dw*)1-4 genes. These genes impart partial dominance of the tallness trait to plants, the effects of which are additive in nature (meaning a plant with *dw*1, 2, and 3 would be taller than a plant with *dw*1 only). Adapting sorghum to long days, as found in temperate regions, led to the discovery of *Maturity* (*Ma*) genes. Specifically *Ma*1 gene has been shown to be involved in controlling the rate of maturity, making the plant carrying this gene in its recessive form react to long days as it normally would to short days (photoperiod insensitive). Increasing yield is highly dependent on changing the source/sink balance. Drought-resistant sorghum keeps a higher photosynthetic rate during conditions of low water. One gene locus (Alt_{sb}) is known to provide aluminum tolerance. Aluminum is found in acidic soils. These soils make up about half of the possible arable land available. However, iron is the main problem for alkaline soils, and some iron-tolerant lines have been found in sorghum (Saballos 2008).

Sweet sorghum is highly recalcitrant to in vitro manipulations, such as tissue culture and transformation. Reports on tissue culture of sweet sorghum are limited to MacKinnon et al. (1986), Rao et al. (1995), and Raghuwanshi and Birch (2010). There is only one published report on transformation of sweet sorghum. Raghuwanshi and Birch (2010) reported transformation of sweet sorghum variety Ramada using microprojectile bombardment. They reported development of a transformation system for sweet sorghum and demonstrated production of transgenic sweet sorghum resistant to the antibiotic hygromycin. Luciferase was used as the reporter gene in this case. Although the reported transformation frequency is low (0.09%) for hygromycin resistance, the coexpression frequency of the reporter gene was 62.5%. With the development of this transformation system, a platform is laid to make further improvements for transformation of sweet sorghum. This development may pave the way for metabolic engineering of sugar metabolism in sweet sorghum. Genetic modifications that will impart sweet sorghum resistance to biotic and abiotic stresses may also be achieved with further refining of this protocol.

3.4.4 SWITCHGRASS

Switchgrass is a perennial high biomass grass native to the prairies of North America. It is identified by the U.S. Department of Energy (DOE) as one of the United States' promising energy crops for many reasons. It is drought tolerant, high yielding, perennial, enhances soil and wildlife, can be established from seed, and is adaptable to marginal lands (Bouton 2008). Several soil types are tolerated, ranging from sands to heavy clays and pH levels of 5–7. Yields are approximately 10–25 Mg/ha per year (Yuan et al. 2008). Switchgrass is taxonomically divided into two groups: lowland and upland. Lowland plants are considered coarse and tall with large biomass yields and are found in wet regions with milder winters; upland plants are shorter, have a lower biomass yield, and are found in drier and colder regions. Studies have shown two types of chloroplast DNA, U and L, which show differences that can determine if the plant is an upland or lowland plant (Bouton 2008).

There are a few published reports on genetic engineering of switchgrass (Richards et al. 2001; Somleva et al. 2002, 2008; Xi et al. 2009; Li and Qu 2010). Richards et al. (2001) reported production of transgenic switchgrass expressing *GFP* reporter gene. Somleva et al. (2002) generated transgenic switchgrass expressing the *bar* gene, which confers resistance to the herbicide Basta. More recently, Somleva et al. (2008) developed transgenic switchgrass producing polyhydroxybutyrate, a value-added co-product. Xi et al. (2009) demonstrated transformation of switchgrass with a chimeric hygromycin phosphor-transferage gene. Li and Qu (2010) have developed a high-throughput, *Agrobacterium*-mediated transformation system for switchgrass cv. Alamo. Using this modified protocol, they have been able to transform switchgrass cv. Performer with 90% efficiency and cv. Alamo and Colony with 50% efficiency. With the advancement in transformation protocols, it is anticipated that other genes of interest to cellulosic biofuel production (e.g., genes controlling cellulose metabolism, lignin metabolism) may be introduced to this important biofuel feedstock in the near future.

3.4.5 MISCANTHUS

Miscanthus is a high biomass, cold-tolerant, perennial grass easily propagated by rhizomes and has the potential to be a contributor to ethanol production (Clifton-Brown et al. 2008). Originating in East Asia, its natural range stretches from northeastern Siberia to the temperate area of Polynesia, and toward central India. Needless to say, miscanthus thrives in a wide range of climates. Currently *Miscanthus* spp., specifically *Miscanthus* × *giganteus*, is being used in Europe as the main feedstock for biofuel. This hybrid species is sterile and requires vegetative propagation. Miscanthus has shown better cold tolerance than switchgrass, possibly allowing for growth at high latitudes; it can possibly use nitrogen better than switchgrass; and has a yield between 7 and 38 Mg/ha per year (Yuan et al. 2008).

Although there are a couple of published reports on tissue culture of miscanthus (Holmes and Petersen 1996; Kim et al. 2010b), there is only one report of genetic engineering of miscanthus. Callus initiated from immature spikelets or germinating seeds was transformed using *A. tumefaciens*. Selection was carried out using antibiotic G-418 with a *npt II* selectable marker and plants were regenerated from callus (Engler and Chen 2009). It is anticipated that, with further refining of transformation protocols, this important cellulosic ethanol biomass will be genetically engineered to improve agronomic traits (growth, yield, and resistance to biotic and abiotic stress) and to make it more amenable to pretreatment and saccharification through modification of cell wall properties.

3.4.6 POPLAR

Poplar is discussed as the representative tree biofuel crop because it is a model crop. Poplar is the model tree biofuel feedstock because of its wide range of adaptation, available genome, fast growth, clonal propagation ease, sexual compatibility with other species, and available transformation techniques (Davis 2008; Yuan et al. 2008). These trees are found throughout the northern hemisphere, are shade intolerant (grow best with complete weed control), grow in moist areas, and have medium to short life spans. Woody biomass plants have several advantages, including flexible range of harvesting time and good environmental impacts (Davis 2008). Yields for poplar are between 5 and 20 Mg/yr or 10 and 30 Mg/yr based on genotype, site, region, etc. Poplar is used for cellulosic ethanol or for co-firing with coal, but the costs for harvesting and chipping are the major disadvantages to using poplar for bioenergy needs. Areas of genetic improvement would include genes influencing growth, branching, stem thickness, light response competition, plant height, and cell wall makeup (Rubin 2008).

Poplar is highly amenable to genetic transformation (Cseke et al. 2007). Therefore, poplar has been variously genetically engineered to improve its growth and wood properties (Park et al. 2004; Cseke et al. 2007; Baba et al. 2009). Expression of carbohydrate-binding modules (CBMs) has been shown to increase cell growth in poplars transformed with cell-wall-targeted *Clostridium cellulovorans* CBM (Shani et al. 1999). Xyloglucanse was overexpressed in poplar to enhance the growth and yield in poplar. The overexpression of xyloglucanse reduced crosslinks in their cell wall, which, in turn, increased the plasticity under turgor pressure during growth (Park et al. 2004). Similarly, the expression of *Arabidopsis* endo-(1-4)-ß-glucanase gene (*Cel1*) in poplar trees resulted in longer internodes compared with their wild-type control (Shani et al. 2004). Accelerated growth of poplar expressing expansin has been reported recently (Gray-Mitsumune et al. 2008). Hu et al. (1999) produced lines of transgenic poplar (*Populus tremuloides* Michx), which exhibited a 45% reduction of lignin and an increase of 15% in cellulose. This altered cell wall composition enhanced leaf, root, and stem growth without affecting the structural integrity of the transgenic plant, one of the key requirements of successful modification. Introduction of tyrosine-rich peptides to poplar trees through genetic transformation resulted in the formation of wood that is more susceptible to protease digestion than wild-type plants. This genetic modification caused greater release of sugar from lingocellulose complex during saccharification (Liang et al. 2008). Transgenic poplar resistant to heavy metals such as mercury, cadmium, arsenic, and lead have been developed. Poplars were engineered with *merA*, *merB*, *gsh1*, *CYP2E1*, *MnP*, *cys1*, and *PsMT$_A$1* genes. The introduced bacterial genes *merA* and *merB* conferred resistance to organomercurial pollutants (Yadav et al. 2010). Genetic engineering of trees in relation to their application for forestry was recently reviewed (Harfouche et al. 2011). This review reported the issue of 25 permits for field trials of GM poplars in Europe since 1991. These GM poplars are being evaluated for altered wood composition, altered wood properties, sterility, lignin modification, herbicide tolerance, faster growth, and phytoremediation.

3.4.7 EUCALYPTUS

Eucalyptus is a large genus of trees and large shrubs of more than 700 species. Although many species of eucalyptus are naturalized across the globe, they originated in Australia. Eucalyptus comprises multipurpose trees used for timber and firewood or for the extraction of high-value chemicals (Eldridge 1993; Ladiges et al. 2003; Domingues et al. 2011). Many species of *Eucalyptus* grow fast and generate large biomass suitable for bioenergy production (Stricker et al. 2000). Eucalyptus hybrid (*E. grandis* × *E. urophylla*) has proved to be ideal for forestry because of their growth characteristics and wood properties. This hybrid is also amenable to in vitro manipulations including genetic transformation.

Recent reviews (Hinchee et al. 2009; Abramson et al. 2010; Harfouche et al. 2011) have considered genetic engineering of *Eucalyptus* along with other tree crops. Eucalyptus genome sequencing was also reviewed recently (Grattapaglia and Kirst 2008). Several microsatellite markers of *Eucalyptus* were developed and characterized by Brondani et al. (1998). Complete nucleotide sequence of the chloroplast genome from the Tasmanian *E. globulus* was reported by Steane in 2005. Harcourt et al. (2000) reported development of transgenic *Eucalyptus*. They used *Agrobacterium*-mediated transformation to transfer insecticidal *cry3A* gene to *E. camaldulensis*. Recently, the forestry biotechnology company has patented transgenic Eucalyptus hybrid (*E. grandis* × *E. urophylla*). These freeze-tolerant transgenic plants harbor transcription factor *CBF2* from *Arabidopsis* driven by the *Arabidopsis rd29a* stress-inducible promoter (Hinchee et al. 2009). According to a recent report from the Institute of Science in the Society (ISS), Arborgen's transgenic *Eucalyptus* were modified with genes conferring resistance to antibiotics and altering the lignin pathway in addition to the CBF transcription factor (http://www.i-sis.org.uk/FTGEEEASI.php).

3.5 BIODIESEL

The diesel engine, made by Dr. Rudolph Diesel, was originally run on peanut oil at the Paris Exposition of 1900. Various vegetable oils were used to run these first engines, and Dr. Diesel stated "the diesel engine can be fed with vegetable oils and will help considerably in the development of the agriculture of the countries which use it" (Demirbas 2008). Transesterification is the process of making biodiesel from triglycerides (fatty acids). During this process, triglycerides are reacted with methanol or ethanol in the presence of a catalyst (that speeds up the reaction rate) to produce "biodiesel, (m)ethylesters, and glycerin" (Demirbas 2008) (Figure 3.2). Some of the benefits of biodiesel are availability, renewability, biodegradability, lower exhaust emissions, and it is nonflammable and nontoxic (Demirbas 2008).

Although the use of biodiesels can help stem the use of foreign oil, the present cost of commercial production outweighs its benefits, with a gallon of biodiesel (100%) costing from $2.00 to $2.50 plus taxes (Demirbas 2008). Also, "the competitiveness of biodiesel relies on the prices of biomass feedstock and costs, linked to the conversion technology" (Demirbas 2009). When blended with petrol (up to ~20%), there seems to be no problem incorporating the new fuel in engines or equipment, and little modification is required with higher percentage blends (Demirbas 2009). Production of biodiesel can lead to useful by-products that may help reduce the total cost of seed cake, fruit husks, and glycerin (Achten and Mathijs 2007). The use of vegetable oils is likely an inexhaustible source of renewable energy with several different sources: cottonseed, rapeseed, sunflower seed, soybean, jatropha, and palm. In the United States, soybean oil is the primary biodiesel source, palm oil is the source in Indonesia and Malaysia, rapeseed oil in Europe, and jatropha in India and Southeast Asia (Demirbas 2008).

3.5.1 BIODIESEL PRODUCTION METHODS

A general process for biodiesel production is represented in Figure 3.2. Various methods of biodiesel production from different sources were reviewed recently (Andrade et al. 2011). This review described the transesterification of vegetable oils, animal fats, and oil from algae to produce biodiesel using homogeneous, heterogeneous, and enzyme catalysts along with ultrasound, microwave, and supercritical alcohol techniques. Transesterification (base, acid, or enzyme catalyzed) and noncatalytic transesterification are two general methods used for biodiesel production. During transesterification, methanol is preferred over the use of ethanol because of its relatively low cost, lower moisture sensitivity, miscibility with biodiesel, and ability to reduce the viscosity of biodiesel (Demirbas 2008).

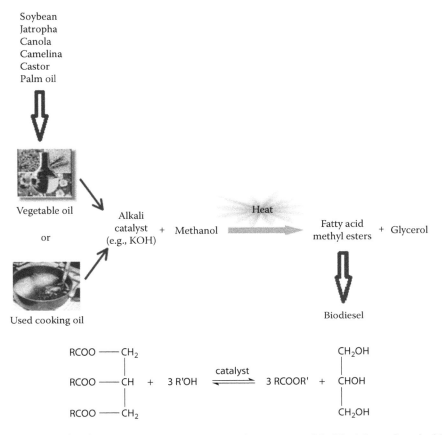

FIGURE 3.2 (See color insert) The biodiesel production process. (Modified from Sustainable Green Technologies, sgth2.com/bio-diesel_faq)

3.5.1.1 Catalytic Transesterfication

Base-catalyzed reactions involve sodium hydroxide or potassium hydroxide, with potassium hydroxide being preferred because of the fast reaction rates, cheap catalyst, and less corrosive reaction. One disadvantage of using a base catalyst is its reaction with free fatty acids to form soaps, producing alkaline water that requires energy-intensive waste treatment. Although acid catalyst transesterification using sulfuric acid and/or phosphoric acid can handle larger amounts of free fatty acids and water, its disadvantages are a longer reaction time and require higher temperatures (>100°C) (Demirbas 2008; Nag 2008; Andrade et al. 2011).

Enzymatic transesterification can be used instead of chemical catalysts for several reasons. The enzymes are generally more selective, allow for easier glycerol removal, convert free fatty acids, perform at lower temperatures, and have increased reusability of the catalyst. Lipases are derived from microbes or fungi. According to Al-Zuhair (2007), although the general lipase used is from *Candida antarctica* B, the *Pseudomonas fluorescens* lipase had the better enzymatic activity. Other factors such as water content, type of alcohol used, type of lipase, and temperature all affect the usefulness of the lipase in biodiesel production.

3.5.1.2 Noncatalytic Transesterfication

Noncatalytic methods are supercritical methanol and BIOX. Supercritical methanol has a simpler procedure and shorter reaction time than catalytic methods as well as being environmentally friendly, but under cost-intensive higher temperature and pressure (Al-Zuhair 2007; Demirbas 2008; Andrade et al. 2011). BIOX is a new method that "converts both triglycerides and free fatty acids in a two step, single phase, continuous process at atmospheric pressures and near-ambient temperatures, all in less than 90 minutes" (http://www.bioxcorp.com/). This process can use grain, cooking grease, and animal fats as feedstocks.

3.5.2 Major Biodiesel Crops

3.5.2.1 Soybean

Soybean is a biodiesel plant that has garnered a lot of interest. It is thought to have originated from China and is cultivated worldwide. It is a bushy, green legume with seeds typically consisting of approximately 20% oil and is the world's main supply of vegetable oil. The oil has five fatty acids that make up its composition: palmitic, stearic, oleic, linoleic, and linolenic. There are many uses of soybean in areas such as livestock/poultry feed, plastics, paints, cosmetics, pharmaceuticals, and diesel fuel. In general, soybean is planted in spring/early summer and harvested in fall. Tropical areas allow for soybean growth year round. Lately, genetic improvement of this crop has been focused on several traits such as yield, pest resistance, and seed oil composition (Lee et al. 2007), but emphasis is geared more toward modifying oil content (fatty acid composition). There are numerous genetic maps of soybean that help in identifying potential target genes to modify, such as those affecting growth/flowering (*E1-5*), leaf form (*lf1, Lf2, Ln, Lo, Lnr, lw1, lw2, lb1 lb2*), disease (*Hb, Hm, hs1-hs 3, CP4, Als1*), sterility (*st1-st8, ms1, ms6, fs1, fs2, msp*), and fatty acid composition (*fap2, Fas, St1, St2, Ol, fan1-fan3*). Currently, soybean is transformed through *Agrobacterium* or particle bombardment using one of three typical explants: cotyledonary node, embryonic axis, or somatic embryos (Lee et al. 2007). *Agrobacterium* relies on a biological method of transformation, whereas particle bombardment uses physical induction (Finer and Larkin 2008). One advantage of using soybean is its ability to grow without the addition of nitrogen as fertilizer. Yields in the United States for soybean are 2668 kg/ha, with an overall net negative energy return when producing biodiesel (Pimentel and Patzek 2005) with costs of $0.70/L ($2.64/gal) (Demirbas 2009). One of the main disadvantages to using soybean oil is that it is a major food crop in the United States, with similar problems to that of incorporating corn for bioethanol.

3.5.2.2 Jatropha

Jatropha is an important nonfood crop adapted to environments of tropical and subtropical climates, semi-arid, and marginal lands. Jatropha can grow in various soils, but it is sensitive to frost and waterlogging (Achten and Mathijs 2007). Other attractive features of jatropha include drought hardiness, rapid growth, easy propagation, low cost of seeds, and small gestation period (Sujatha et al. 2008). Jatropha oil is odor and colorless, with its seed oil content variously reported to be between 30% and 50%. The main disadvantage of using jatropha oil for biodiesel is its high viscosity (Pramanik 2003). The high viscocity of jatropha oil reduces the efficiency of fuel injectors in the diesel engines (Demirbas 2008). Reducing the viscosity was done through dilution with diesel blends, with 70–80% diesel showing the greatest improvement (Pramanik 2003). In addition to being cultivated for its oil content, jatropha plants are being used in water and wind erosion control as a living fence to protect food crops and to make soap (Achten and Mathijs 2007; FAO 2010). Citing Bayer Crop Science AG as source, CSA News (2008) reported that Jatropha may provide up to 2270 L/ha oil and boasted that Jatropha oil is a high octane oil/Octane 60 oil (CSA News 2008). Because of this, jatropha is one of the very efficient biofuels and can be used with slight modifications to the diesel engine. Also, this oil is environmentally clean (Bayer Crop Science AG 2008). Overall, jatropha is expected to have a positive effect on land currently thought to be wasteland. Because jatropha harvesting is currently not mechanized, manual labor will be needed, potentially increasing the number of jobs available in rural environments (FAO 2010).

Genetic transformation of jatropha is still in the developmental stage because not many genes of significant value have been incorporated through genetic engineering. More research is needed in jatropha to make this crop more reliable as a bioenergy source. Currently, the germplasm available has many setbacks, including the lack of genetic information, poor yields, vulnerability to pests, and low genetic diversity. Areas of improvement should focus on improving yields, higher oil content, and achieving faster maturity and enhanced fuel properties (Sujatha et al. 2008). *Agrobacterium*-mediated transformation has been performed with the *SaDREB1* gene, with the *bar* gene for selection on phosphinothricin and *gus* as a reporter gene (Sujatha et al. 2008). Genes such as curcin and stearoyl-acyl carrier protein denaturase and *JcERF* (demonstrating salt and frost tolerance) have been identified. Most recently, the jatropha genome (~400 million base pairs) has been sequenced by a collaborative effort between Synthetic Genomics, Inc. (SGI) and the Asiatic Centre for Genome Technology (ACGT) (http://checkbiotech.org/node/26008). A high-quality normalized cDNA library using developing jatropha seeds has been developed by Natarajan et al. (2010). Li et al. (2008) have demonstrated *Agrobacterium*-mediated transformation of jatropha, and Purkayastha et al. (2010) have demonstrated genetic transformation of jatropha using a particle bombardment method. These technologies will be useful in speeding up the genetic engineering process of jatropha.

3.5.2.3 Canola

The genus *Brassica* consists of approximately 100 species including *Brassica napus* (canola), which is believed to have originated in the Mediterranean region or in northern Europe. Through breeding, excellent varieties of canola lines have been developed in the Organisation for Economic Co-operation and Development (OECD) countries (OECD Paris 1997). Edible oil, low in erucic acid, was first extracted in Canada in 1956 (Colton and Potter 1999). Canola is currently grown for its seeds, which yield from 35% to more than 45% oil. Canola oil is an excellent cooking oil and can be used to manufacture biodiesel through enzymatic and chemical processes (Dizge and Keskinler 2008; Issariyakul et al. 2008; Cheng et al. 2010). The remaining by-product after seed oil extraction, canola seed meal, is used as a high-protein animal feed.

Canola is highly amenable to in vitro manipulations, including tissue culture and genetic engineering. Transformation efficiency of canola was improved to make a working protocol that suits multiple cultivars (Cardoza and Stewart 2004, 2007; Bhalla and Singh 2008). Canola has been genetically engineered using *Agrobacterium* to impart herbicide-resistance to imidazoline, glufosinate, glyphosate, sulfonylurea, and bromoxynil (Blackshaw et al. 1994; Zhong et al. 1997; Cardoza and Stewart 2007). Canola became the number one crop of Canada because of the use of GM canola. The Canola Council of Canada provides a wide variety of information on the significance of GM canola (http://www.canolacouncil.org/facts_gmo.aspx). Oleic acid level in *B. napus* was increased by silencing the endogenous oleate desaturase gene (Stoutjesdijk et al. 2000). Canola that can produce high levels of γ-linolenic acid was achieved by introducing δ12-desaturase genes from the fungus *Mortierella alpina* (Liu et al. 2001). Transgenic canola with elevated levels of stearate content was obtained by the overexpression of the *Garm FatA1*, an acyl-carrier protein (ACP) thioesterase, isolated from *Garcinia mangostana* (Hawkins and Kridl 1998). The seed-specific mutants derived from engineering *Garm FatA1* gene resulted in transgenic plants that can accumulate 55–58% more stearate than the wild-type plants (Facciotti et al. 1999). In addition, these lines also showed an increase in laurate at the sn-2 position (Knutzon et al. 1999). Because there are increasing pest problems in canola cultivation, insect resistance is a target trait for genetic improvement. For example, canola is very susceptible to the diamond back moth. Halfhill et al. (2000) introduced *Bt* toxin through the *Bt cry1A(c)* gene to *B. napus* to develop insect-resistant canola. The *B. napus* genome has been sequenced recently, and a sequence-level comparative analysis at the scale of the complete bacterial artificial chromosome (BAC) clones was conducted (Cho et al. 2010).

3.5.2.4 Camelina

Camelina was used for oil production long before World War II and grown all over Europe, but this practice declined after the war. The oil is mostly unsaturated, making it a good source of omega-3 fatty acids. Unfortunately, a large portion of the fats are polyunsaturated, making this oil difficult to work with for fuel production (Lu and Kang 2008). Camelina is adaptable to harsh environments such as "semiarid regions and in low-fertility or saline soils" (Budin et al. 1995), is resistant to pests, and is high in nutrient efficiency with a short vegetation period. The high amount C_{18} fatty acids makes camelina a renewable source of oleochemicals, which are used in varnishes and in drying oils for paint. Camelina yields range from 2 to 3 t/ha, with its oil content between 28% and 42% (Gehringer et al. 2006).

There are a few published reports of transformation of Camelina using *Agrobacterium* (Lu and Kang 2008; Kuvshinov et al. 2009; Nguyen et al. 2009). Also, there are three patent applications for transformation of Camelina and transgenic plants produced. This indicates the potential of Camelina as a dedicated biodiesel crop. Lu and Kang (2008) used DsRed as a fluorescent protein marker and also transformed Camelina with castor fatty acid hydroxylase, resulting in a change in the fatty acid profile of Camelina. With the availability of these protocols, it is anticipated that more value-added genes that will confer resistance to biotic and abiotic stresses, herbicide resistance, enhance oil yield, and increase quality of oil can be generated in the near future.

3.5.2.5 Castor

Castor is a monotypic genus belonging to the family *Euphorbiaceae*. The center of origin of castor is believed to be Africa and India (Ramaprasad and Bandopadhyay 2010). Castor is widely used as a lubricant and has many other medicinal properties. Castor oil is widely used in India and several countries as a source of biodiesel (Ogunniyi 2006; Sujatha et al. 2008). Such diversity of use has led to a steady increase in the demand for castor oil on the world market.

Malathi et al. (2006) developed semilooper-resistant transgenic castor using *Agrobacterium*-mediated transformation. The transgenics contained *bar* and *cryIAb* genes, and 88% of the semilooper larvae that fed on these castor plants did not survive. Sujatha et al. (2008) generated stable tranformants of castor using particle bombardment, although this study only used reporter gene *UidA* encoding ß-*galacturonidase* (GUS) and selectable marker hygromycin-phosphotransferase (*hptII*). They standardized the conditions for transformation by particle bombardment, including the helium pressure (1100 psi), target distance (6.0 cm), and size of the gold particles (0.6 μm). *Agrobacterium*-mediated transformation of castor with the *cryIEC* gene offered field resistance to tobacco caterpillar larva and semilooper larva (Sujatha et al. 2009). *Ricinus communis* genome has been sequenced quite recently and is estimated to be approximately 320 Mb in size (Chan et al. 2010). Approximately, 50% of this genome is made up of repetitive DNA. With these advances in genetic engineering and genomics, GM castor with many favorable traits could be developed in the near future, which would enhance the value of castor as a biodiesel crop.

3.5.2.6 Oil Palm

Oil palm is known to have originated from West Africa (Hardon et al. 1985). It is currently a significant cash crop in Malaysia and Indonesia (Sambanthamurthi et al. 2000, 2002, 2009). Palm oil contributes to approximately 20% of world oil and fat production (Oil World Annual 2001). Current and rising demand for biodiesel will increase the demand for palm oil. Therefore, conventional breeding, genomics, and genetic engineering are being applied to achieve genetic improvement of oil palm (Parveez 1998).

Genetic engineering is applied to oil palm to reduce the time needed to develop improved varieties by accelerated breeding approaches, achieve precision gene transfer, and widen the genetic base of oil palm (Sambanthamurthi et al. 2009). Abdulla et al. (2005) reported genetic transformation of immature embryos of oil palm using gene gun and *Agrobacterium*-mediated transformation.

An immature embryo system enhanced the amenability of oil palm to in vitro manipulations. Oil palm was recently transformed with polyhydroxyvalerate gene using *Agrobacterium*-mediated gene transfer (Faud et al. 2008). The giant gene construct used in this experiment included pFA2 (binary vector, ~24.4 kb) and pFA3 (super binary vector, ~39.4 kb) that contained several genes (*bkt*B, *pha*B, *pha*C, and *tdc*B) essential for the synthesis of polyhydroxyvalerate. This construct was effectively transferred to the immature embryos using *Agrobacterium*-mediated transformation.

3.6 CONCLUSIONS

Biofuel is the future fuel because it is a renewable, less polluting, and environment friendly sustainable fuel. Major challenges for the biofuel industry include but are not limited to (1) production of biofuel feedstocks at a reasonable cost and in sufficient quantities, (2) cost-effective transportation of the biomass to processing facilities, and (3) cost-effective processing of biomass into biofuels and developing methods for cost-effective, environment friendly utilization of by-products of the processing. A combination of biotechnologies such as genomics, marker-assisted breeding, and genetic engineering has the potential to accelerate breeding to develop biofuel crops that are more productive and highly adapted to abiotic and biotic stress. Availability of such novel biofuel crop cultivars may increase acreage and production of biofuel crops to meet the ever-increasing demand for biofuel feedstocks without affecting the global food security. However, it is not clear at this moment as to which crop will be the "winner" among several crops in terms of being cost-effective for large-scale production of biofuels.

REFERENCES

Abdulla R, Zainal A, Heng WY, Li LC, Beng YC, Phing LM, Sirajuddin SA, Ping WPS, Joseph JL (2005) Immature embryo: A useful tool for oil palm (*Elaeis guineensis* Jacq.) genetic transformation studies. *Electron J Biotechnol* 8(1):24–34

Abogbo FK, Coward-Kelly G (2008) Cellulosic ethanol production using the naturally occurring xylose-fermenting yeast. *Pichia stipitis Biotechnol Lett* 30:1515–1524

Abramson M, Shoseyov O, Shani Z (2010) Plant cell wall reconstruction toward improved lignocellulosic production and processability. *Plant Sci* 178:61–72

Achten WMJ, Mathijs E (2007) Jatropha biodiesel fueling sustainability? *Biofuels Bioproducts Biorefining* 1:283–291

Ajanovic A (2010) Biofuels versus food production: Does biofuels production increase food prices? *Energy*, doi:10.1016/j.energy.2010.05.019

Al-Zuhair S (2007) Production of biodiesel: Possibilities and challenges. *Biofuel Bioprod Biorefin* 1:57–66

Andrade JE, Pérez A, Sebastian PJ, Eapen D (2011) A review of bio-diesel production processes. *Biomass Bioenergy* 35:1008–1020

Arencibia A, Carmona E, Cornide M, Oramas P, Sala F (1999) Somaclonal variation in insect resistant transgenic sugarcane (*Saccharum* hybrids) plants produced by cell electroporation. *Transgen Res* 8:349–360

Arencibia A, Monila P, de la Riva G, Selman-Houssein G (1995) Production of transgenic sugarcane (*Saccharum officinarum* L) plants by intact cell electroporation. *Biotechnol Appl* 9:156–165

Baba K, Park YW, Kaku T, Kaida R, Takeuchi M, Yoshida M, Hosoo Y, Ojio Y, Okuyama T, Taniguchi T, Ohmiya Y, Kondo T, Shani Z, Shoseyov O, Awano T, Serada S, Norioka N, Norioka S, Hayashi T (2009) Xyloglucan for Generating Tensile Stress to Bend Tree Stem. *Molecular Plant* 2(5):893–903

Babu BV (2008) Biomass pyrolysis: A state-of-the-art review. *Biofuel Bioprod Biorefin* 2:393–414

Balat M, Balat H (2009) Recent trends in global production and utilization of bio-ethanol fuel. *Appl Energy* 86:2273–2282

Barriere Y, Ralph J, Mechin V, Guillaumie S, Grabber JH, Argillier O, Chabbert B, Lapierre C (2004) Genetic and molecular basis of grass cell wall biosynthesis and degradability. II. Lessons from brown-midrib mutants. *Crit Rev Biol* 327:847–860

Bhalla PL, Singh MB (2008) *Agrobacterium*-mediated transformation of *Brassica napus* and *Brassica oleracea*. *Nat Protocols* 3:181–189

Blackshaw RE, Kanashiro D, Moloney MM, Crosby WL (1994) Growth, yield and quality of canola expressing resistance to acetolactate synthase inhibiting herbicides. *Can J Plant Sci* 74:745–751

Bouton J (2008) Improvement of switchgrass as a bioenergy crop. In: Vermeris W (ed), *Genetic Improvement of Bioenergy Crops* (Part 2). Springer, New York, pp 295–308

Brethauer S, Wyman CE (2010) Review: Continuous hydrolysis and fermentation for cellulosic ethanol production. *Bioresour Technol* 101:4862–4874

Brondani RPV, Brondani C, Tarchini R, Grattapaglia D (1998) Development, characterization and mapping of microsatellite markers in *Eucalyptus grandis* and *E. urophylla*. *Theor Appl Genet* 97:816–827

Budin JT, Breene WM, Putnam DH (1995) Some compositional properties of Camelina (*Camelina sativa* L. Crantz) seeds and oils. *JAOCS* 72:309–315

Cardoza V, Stewart CN (2004) *Agrobacterium*-mediated transformation of canola. In: Curtis IS (ed), *Transgenic Crops of the World—Essential Protocols*. Kluwer, Dordrecht, The Netherlands, pp 379–387

Cardoza V, Stewart Jr CN (2007) Canola. In: Pua EC, Davey MR (eds), *Biotechnology in Agriculture and Forestry, Vol 61: Transgenic Crops VI*. Springer, Berlin Heidelberg, Germany, pp 29–37

Cassman KG, Liska AJ (2007) Food and fuel for all: Realistic or foolish? *Biofuel Bioprod Biorefin* 1:18–23

Chan AP, Crabtree J, Zhao Q, Lorenzi H, Orvis J, Puiu D, Melake-Berhan A, Jones KM, Redman J, Chen G, Cahoon EB, Gedil M, Stanke M, Haas BJ, Wortman JR, Fraser-Liggett M, Ravel J, Rabinowicz PD (2010) Draft genome sequence of the oilseed species. *Ricinus communis*. *Nat Biotechnol* 28:951–956

Chapotin SM, Wolt JD (2007) Genetically modified crops for the bioeconomy: Meeting public and regulatory expectations. *Transgen Res* 16:675–688

Cheng L, Yen S, Su L, Chen J (2010) Study on membrane reactors for biodiesel production by phase behaviors of canola oil methanolysis in batch reactors. *Bioresour Technol* 101:6663–6668

Cho K, O'Neill M, Kwon SJ, Yang TJ, Smooker A, Fraser F, Bancroft I (2010) Sequence-level comparative analysis of the *Brassica napus* genome around two stearoyl-ACP desaturase loci. *Plant J* 61:591–599

Clifton-Brown J, Chiang Y, Hodkinson TR (2008) Miscanthus: Genetic resources and breeding potential to enhance bioenergy production. In: Vermeris W (ed), *Genetic Improvement of Bioenergy Crops*. Springer, New York, pp 273–294

Colton B, Potter T (1999) History. In: Salisbury PA, Potter T, McDonald G, Green AG (eds), *Canola in Australia: The First Thirty Years*. Organising Committee of the 10th International Rapeseed Congress, Canbera, Australia, pp 1–4

CSA News (2008) Is Jatropha the next big thing in biofuels? *CSA News* 53:2–4, available at www.agronomy.org/csa-news. (accessed February 2, 2011).

Cseke LJ, Cseke SB, Podila GK (2007) High efficiency poplar transformation. *Plant Cell Rep* 26:1529–1538

Dale BE, Leong CK, Pham TK, Esquivel VM, Rios I, Latimer VM (1996) Hydrolysis of lignocellulosics at low enzyme levels: Application of the AFEX process. *Bioresour Technol* 56:111–116

Davis JM (2008) Genetic improvement of poplar (*Populus* spp.) as a bioenergy crop. In: Vermerris W (ed), *Genetic Improvement of Bioenergy Crops*. Springer, New York, pp 377–396

de Leon N, Coors JG (2008) Genetic improvement of corn for lignocellulosic feedstock production. In: Vermerris W (ed), *Genetic Improvement of Bioenergy Crops*. Springer, New York, pp 185–210

Delarosa LB, Reshamwalla S, Latimer VM, Shawky BT, Dale BE, Stuart ED (1994) Integrated production of ethanol fuel and protein from coastal Barmuda grass. *Appl Biochem Biotechnol* 45–46:483–497

Demirbas A (2008) *Biodiesel a Realistic Fuel Alternative for Diesel Engines*. Springer, London, p 214

Demirbas A (2009) *Biofuels Securing the Planet's Future Energy Needs*. Springer, London, p 336

Dizge N, Keskinler B (2008) Enzymatic production of biodiesel from canola oil using immobilized lipase. *Biomass Bioenergy* 32:1274–1278

Domingues RMA, Sousa GDA, Silva CM, Freire CSR, Silvestre AJD, Pascoal Neto P (2011) High value triterpenic compounds from the outer barks of several *Eucalyptus* species cultivated in Brazil and in Portugal. *Indust Crops Prod* 33:158–164

EIA (2011) http://www.eia.doe.gov. (accessed January 4, 2011)

Eldridge K, Davidson J, Harwood C, van Wyk G (1993) *Eucalypt Domestication and Breeding*. Claredon, Oxford, United Kingdom, pp 60–72

Engler D, Chen J (2009) *Transformation and Engineered Trait Modification in Miscanthus Species*. World Intellectual Property Organization, New York

Enriquez GA, Trujillo IE, Menendz C (2000) Sugarcane (*Saccharum* hybrid) genetic transformation mediated by *Agrobacterium tumefaciens*; production of transgenic plants expressing proteins with agronomic and industrial value. In: Arencibia AD (ed), *Plant Genetic Engineering towards the Third Millennium*. Elsevier Science, Amsterdam, The Netherlands, pp 76–81

Facciotti MT, Bertain PB, Yuan L (1999) Improved stearate phenotype in transgenic canola expressing a modified acyl–acyl carrier protein thioesterase. *Nat Biotechnol* 17:593–597

FAO (2010) Jatropha: A Smallholder Bioenergy Crop. The Potential for Pro-Poor Development, available at http://www.fao.org/docrep/012/i1219e/i1219e00.htm (accessed December 20, 2010)

Fargione J, Hill J, Tilman D, Polasky S, Hawthorne P (2008) Land clearing and the biofuel carbon debt. *Science* 319:1235–1238

Faud FAA, Ismail I, Sidik SI, Zain CRCN, Abdullah R (2008). Super binary vector system enhanced transformation frequency and expression level of polyhydroxyvalerate gene in oil palm immature embryo. *Asian J Plant Sci* 7:526–535

Finer JJ, Larkin KM (2008) Genetic transformation of soybean using particle bombardment and SAAT approaches. In: Kirti P (ed), *Handbook of New Technologies for Genetic Improvement of Legumes*. CRC Press, Boca Raton, FL, pp 103–123

Gehringer A, Friedt W, Luhs W, Snowdon RJ (2006) Genetic mapping of agronomic traits in false flax (*Camelina sativa* subsp. *sativa*). *Genome* 49:1555–1563

Goldemberg J, Coelho ST, Guardabassi P (2008) The sustainability of ethanol production from sugarcane. *Energy Policy* 36:2086–2097

Grattapaglia D, Kirst M (2008) Eucalyptus applied genomics: From gene sequences to breeding tools. *New Phytol* 179:911–929

Gray-Mitsumune M, Blomquist K, McQueen-Mason S, Teeri T, Sundberg B, Mellerowicz E (2008) Ectopic expression of a wood-abundant expansin *PttEXPA1* promotes cell expansion in primary and secondary tissues in aspen. *Plant Biotechnol J* 6:62–72

Gressel J (2008) Transgenics are imperative for biofuel crops. *Plant Sci* 147:246–263

Halfhill MD, Richards HA, Mabon SA, Stewart Jr CN (2001) Expression of *GFP* and *Bt* transgenes in *Brassica napus* and hybridization with *Brassica rapa*. *Theor Appl Genet* 1003:659–667

Harcourt RL, Kyozuka J, Floyd RB (2000) Insect- and herbicide-resistant transgenic eucalypts. *Mol Breed* 6:307–315

Hardon JJ, Rao V, Rajanaidu N (1985) A review of oil palm breeding. In: Rusell GE (ed), *Progress in Plant Breeding*. Butterworths, London, pp 139–163

Harfouche A, Meilan R, Altman A (2011) Tree genetic engineering and applications to sustainable forestry and biomass production. *Trends Biotechnol* 29:9–17

Hawkins D, Kridl L (1998) Characerization of acyl-ACP thioesterase of mangosteen (Garciniamangosteena) seed and high levels of state productionin transgenic canola. *Plant J* 13:743–752

Heaton EA, Flavell RB, Mascia PN, Thomas SR, Dohleman FG, Long SP (2008) Herbaceous energy crop development: Recent progress and future prospects. *Curr Opin Biotechnol* 19:202–209

Hinchee M, Rottmann W, Mullinax L, Zhang C, Chang S, Cunningham M, Pearson L, Nehra N (2009) Short-rotation woody crops for bioenergy and biofuels applications. *In Vitro Cell Dev Biol-Plant* 45:619–629

Holme IB, Petersen KK (1996) Callus induction and plant regeneration from different explant types of *Miscanthus* x *ogiformis* Honda 'Giganteus'. *Plant Cell Tiss Org Cult* 45:43–52

Holtzapple MT, Jun J, Ashok G, Patibhandla SL, Dale BE (1991) The ammonia freeze explosion (AFEX) process: A practical lignocellulose pretreatment. *Appl Biochem Biotechnol* 28–29:59–74

Hotta CT, Lembke CG, Domingues DS, Ochoa EA, MQ Cruz GMQ, DM Melotto-Passarin, Marconi TG, Santos MO, Mollinari M, Margarido GRA et al. (2010) The biotechnology roadmap for sugarcane improvement. *Trop Plant Biol* 3:75–87

Hu W, Harding SA, Lung JL, Popko JL, Ralph J, Stokke DD, Tsai CJ, Chiang VL (1999) Repression of lignin biosynthesis promotes cellulose accumulation and growth in transgenic trees. *Nat Biotechnol* 17:808–812

Issariyakul T, Kulkarni MG, Meher MC, Dalai AK, Bakhshi NN (2008) Biodiesel production from mixtures of canola oil and used cooking oil. *Chem Eng J* 140:77–85

Iyer PV, Wu ZW, Kim SB, Lee YY (1996) Ammonia recycled percolation process for pretreatment of herbaceous biomass. *Appl Biochem Biotechnol* 57–58:121–132

Jain M, Chengalrayan K, Abouzid A, Gallo M (2007) Prospecting the utility of a PMI/mannose selection system for the recovery of transgenic sugarcane (*Saccharum* spp. hybrid) plants. *Plant Cell Rep* 26:581–590

Jena PC, Raheman H, Kumar GVP, Machavaram R (2010) Biodiesel production from mixture of mahua and simarouba oils with high free fatty acids. *Biomass Bioenergy* 34:1108–1116

Kammen D (2006) Bioenergy in developing countries: Experiences and prospects. In: Hazell P, Pachauri RK (eds), *Bioenergy and Agriculture: Promises and Challenges*. IFPRI, Washington DC, pp 21–22

Karp A, Shield I (2008) Bioenergy from plants and the sustainable yield challenge. *New Phytol* 179:15–32

Kim HS, Zhang G, Juvic JA, Widholm JM (2010a) *Miscanthus* x *giganteus* plant regeneration: Effect of callus types, ages and culture methods on regeneration competence. *Global Change Biol Bioenergy* 2:192–200

Kim J, Block DE, Mills DA (2010b) Simultaneous consumption of pentose and hexose sugars: An optimal microbial phenotype for efficient fermentation of lignocellulosic biomass. *Appl Microbiol Biotechnol* 88:1077–1085

Knutzon DS, Hayes TR, Wyrick A, Xiong H, Davies HM, Voelker TA (1999) Lysophosphatidic acid acyltransferase from coconut endosperm mediates the insertion of laurate at the sn-2 position of triacylglycerols in lauric rapeseed oil and can increase total laurate levels. *Plant Physiol* 120:739–746

Kuvshinov V, Anne K, Kimmo K, Svetlana K, Eija P (2009) U.S. Patent Application 12/288791, publication date June 11, 2009

Ladiges PY, Udovicic F and Nelson G (2003) Australian biogeographical connections and the phylogeny of large genera in the plant family Myrtaceae. *Journal of Biogeography* 30:989–998

Lakshmanan P, Geiskes RJ, Aitken KS, Grof CLP, Bonnett GD, Smith GR (2005) Sugarcane biotechnology: The challenges and opportunities. *In Vitro Cell Devl Biol-Plant* 41:345–361

Leistritz FL, Hodur NM (2008) Biofuels: A major rural economic development opportunity. *Biofuel Bioprod Biorefin* 2:501–504

Lee G, Wu X, Groover SJ, David SA, Henry NT (2007) Soybean. In: Kole C (ed), *Genome Mapping and Molecular Breeding in Plants. Vol 2: Oilseeds.* Springer, Berlin, Heidelberg, Germany, pp 1–53

Lee YY, Iyer P, Wu Z, Kim SB (1996) Ammonia recycled percolation process for pretreatment of herbaceous biomass. *Appl Biochem Biotechnol* 57–58:121–132

Li MR, Li HQ, Jiang HW, Pan XP, Wu GJ (2008) Establishment of an *Agrobacteriuim*-mediated cotyledon disc transformation method for *Jatropha curcas*. *Plant Cell Tiss Org Cult* 92:173–181

Li R, Qu R (2010) High throughput Agrobacterium-mediated switchgrass transformation, *Biomass Bioenergy* 35:1046–1054

Liang H, Frost CJ, Wei X, Brown NR, Carlson JR, Tien M (2008) Improved sugar release from lignocellulosic material by introducing a tyrosine-rich cell wall peptide gene in poplar. *Clean Soil Air Water* 36:662–668

Liu JW, DeMichele S, Bergana M, Bobik Jr E, Hastilow C, Chuang L-T, Mukerji P, Huang Y-S (2001) Characterization of oil exhibiting high gamma-linolenic acid froma genetically transformed canola strain. *J Amer Oil Chem Soc* 78:489–493

Lloyd's (2011) White Paper: Sustainable Energy Security Report, available at http://www.lloyds.com /~/media/ Lloyds/Reports/360%20Energy%20Security/7238_Lloyds_360_Energy_Pages.pdf (accessed January 3, 2011)

Lu C, Kang J (2008) Generation of transgenic plants of a potential oilseed crop *Camelina sativa* by *Agrobacterium*-mediated transformation. *Plant Cell Rep* 27:273–278

Lu Y, Mosier NS (2008) Current technologies for fuel ethanol production from lignocellulosic plant biomass. In: Vermerris W (ed), *Genetic Improvement of Bioenergy Crops*, Springer, New York, pp 161–182

Ma H, Albert HH, Paul R, Moore PH (2000) Metabolic engineering of invertase activities in different subcellular compartments affects sucrose accumulation in sugarcane cells. *Aust J Plant Physiol* 27:1021–1030

MacKinnon C, Gunderson G, Nabros MW (1986) Plant regeneration by somatic embryogenesis from callus cultures of sweet sorghum. *Plant Cell Rep* 5:349–351

Malathi B, Ramesh S, Venkateswara RK, Dashavantha RV (2006) *Agrobacterium*-mediated genetic transformation and production of semilooper resistant transgenic castor (*Ricinus communis* L.). *Euphytica* 147:441–449

Manickavasagam M, Ganapathi A, Anbazhagan VR, Sudhakar B, Selvaraj N, Vasudevan A, Kasthurirangan S (2004) *Agrobacterium*-mediated transformation and development of herbicide resistant sugarcane (*Saccharum* species hybrids) using axillary buds. *Plant Cell Rep* 23:134–143

Matsuoka S, Ferro J, Arruda P (2009) The Brazilian experience of sugarcane ethanol industry. *In Vitro Cell Dev Biol-Plant* 45:372–381

McLaughlin W, Falconer LL, Conrad A, Lacewell RD, Falconer LL, Blumenthal JM, Roone WL, Sturdivant AW, McCorkle DA (2011) The Economic and Financial Implications of Supplying a Bioenergy Conversion Facility with Cellulosic Biomass Feedstocks, available at http://ageconsearch.umn.edu/ bitstream/98809/2/McLaughlinSAEA0114.pdf (accessed February 2, 2011)

McQualter RB, Dale JL, Harding RM, McMahon JA (2004) Production and evaluation of transgenic sugarcane containing a Fiji disease virus (FDV) genome segment S9-derived synthetic resistance gene. *Aust J Agric Res* 55:139–145

Molinari HBC, Marur CJ, Daros E, Campos MKF, Carvalho JFRP, Filho BJC, Perrira LFP, Viera LGE (2007) Evaluation of a stress-inducible production of oroline in transgenic sugarcane (*Saccharum* spp.): Osmotic adjustment, chlorophyll fluorescence and oxidative stress. *Physiol Plant* 130:218–229

Moniruzzaman M, Dale BE, Hespell RB, Bothast RJ (1997) Enzymatic hydrolysis of high moisture corn fibre pretreated by AFEX and recovery and recycling of the enzyme complex. *Appl Biochem Biotechnol* 67:113–126

Mosier N, Wyman C, Dale B, Elander R, Lee YY, Holtzapple M, Ladisch M (2004) Features of promising technologies for pre-treatment of lignocellulosic biomass. *Bioresour Technol* 96:673–686

Mousdale DM (2008) *Biofuels Biotechnology, Chemistry, and Sustainable Development*. CRC Press, Boca Raton, FL

Nag A (2008) Processing of vegetable oils as biodiesel and engine performance. In: *Biofuels Refining and Performance*. McGraw Hill, New York, pp 165–189

Naik SN, Goud VV, Rout PK, Dalai AK (2010) Production of first and second generation biofuels: A comprehensive review. *Renewab Sustainab Energy Rev* 14:578–597

Nakagame S, Chandra RP, Kadla JF, Saddler JN (2010) Enhancing the enzymatic hydrolysis of lignocellulosic biomass by increasing the carboxylic acid content of the associated lignin. *Biotechnol Bioeng* 108:538–548

Natarajan P, Kanagasabapathy D, Gunadayalan G, Panchalingam J, Shree N, Sugantham PA, Singh KK, Madasamy P (2010) Gene discovery from *Jatropha curcas* by sequencing of ESTs from normalized and full-length enriched cDNA library from developing seeds. *BMC Genom* 11:1–7

Nguyen T, Xunjia L, Derocher J (2009) Floral Dip Method for Transformation of *Camelina*. US Patent application # US2009/037627 Patent Publ (WO/2009/117555)

Nichols NN, Bothast RJ (2008) Production of ethanol from grain. In: Vermerris W (ed), *Genetic Improvement of Bioenergy Crops*. Springer Science & Business Media, Philadelphia

OECD Paris (1997) Consensus document on the biology of *Brassica napus* L. (Oilseed rape), available at http://bch.cbd.int/database/record-v4.shtml?documentid = 101193

Ogunniyi DS (2006) Castor oil: A vital industrial raw material. *Bioresour Technol* 97:1086–1091

Oil World Annual (2001) ISTA Meilke, Germany, available at http://www.oilworld.biz/app.php (accessed February 2, 2011).

Park YW, Baba K, Furuta Y, Iida I, Sameshima K, Arai M, Hayashi T (2004) Enhancement of growth and cellulose accumulation by overexpression of xyloglucanase in poplar. *FEBS Lett* 564:183–187

Parveez GKA (1998) Optimization of parameters involved in the transformation of oil palm using the biolistic method. PhD Thesis, University of Putra, Malaysia

Petrasovits LA, Purnell MP, Nielsen LK, Brumbley SM (2007) Production of polyhydroxybutyrate in sugarcane. *Plant Biotechnol J* 5:162–172

Pimentel D, Patzek TW (2005) Ethanol production using corn, switchgrass, and wood : Biodiesel production using soybean and sunflower. *Nat Resour Res* 14:65–76

Potters G, Van Goethem D, Schutte F (2010) Promising biofuel resources: Lignocellulose and algae. *Nat Edn* 3:14

Pramanik K (2003) Properties and use of *Jatropha curcas* oil and diesel fuel blends in compression ignition engine. *Renewab Energy* 28:239–248

Purkayastha J, Sugla T, Paul A, Solleti SK, Mazumdar P, Basu A, Mohommad A, Ahmed Z, Sahoo L (2010) Efficient in vitro plant regeneration from shoot apices and gene transfer by particle bombardment in *Jatropha curcas*. *Biol Plant* 54:13–20

Raghuwanshi A, Birch RG (2010) Genetic transformation of sweet sorghum. *Plant Cell Rep* 29:997–1005

Ramaprasad R, Bandopadhyay R (2010) Future of *Ricinus communis* after completion of the draft genome sequence. *Curr Sci* 99:1316–1318

Rao AM, Padma Sree K, Kavi Kishor PB (1995) Enhanced plant regeneration in grain and sweet sorghum by asparagines, proline and cefotaxime. *Plant Cell Rep* 5:72–75

Rathmann R, Szklo A, Schaeffer R (2010) Land use competition for production of food and liquid biofuels: An analysis of the arguments in the current debate. *Renewab Energy* 35:14–22

Reshamwalla S, Shwanky BT, Dale BE (1995) Ethanol production from enzymatic hydrolysates of AFEX-treated coastal Burmuda grass and switchgrass. *Appl Biochem Biotechnol* 51–52:43–55

Richards HA, Rudas VA, Sun JK, McDaniel Z, Tomaszewski Z, Conger BV (2001) Construction of a GFP-BAR plasmid and its use for switchgrass transformation. *Plant Cell Rep* 20:48–54

Rooney WL, Blumenthal J, Bean B, Mullet JE (2007) Designing sorghum as a dedicated bioenergy feedstock. *Biofuels Bioprod Biorefining* 1:147–157

Rubin EM (2008) Genomics of cellulosic biofuels. *Nature* 454:841–845

Ruth L (2008) Bio or bust? The economic and ecological costs of biofuels. *EMBO Rep* 9:130–133

Saballos A (2008) Development and utilization of sorghum as a bioenergy crop. In: Vermerris W (ed), *Genetic Improvement of Bioenergy Crops*. Springer, New York, pp 211–248

Sambanthamurthi R, Parveez GKA, Cheah SC (2000) Genetic engineering of the oil palm. In: Yusof B, Jalani BS, Chan KW (eds), *Advances in Oil Palm Research*. MPOB, Kuala Lumpur, Malaysia, pp 284–331

Sambanthamurthi R, Singh R, Kadir APG, Abdullah MO, Kushairi A (2009) Opportunities for the oil palm via breeding and biotechnology. In: Jain SM, Priyadarshan PM (eds), *Breeding Plantation Tree Crops: Tropical Species*. Springer, New York, pp 377–421

Sambanthamurthi R, Siti Nor Akmar A, Parveez GKA (2002) Genetic manipulation of the oil palm—Challenges and prospects. *Planter* 78:547–562

Schroder P, Herzig R, Bojinov B, Ruttens A, Nehnevajova E, Stamatiadis S, Memon A, Vassilev A, Caviezel M, Vangronsveld J (2008) Bioenergy to save the world. Producing novel energy plants for growth on abandoned land. *Environ Sci Pollut Res* 15:196–204

Searchinger T, Heimlich R, Houghton RA, Dong F, Elobeid A, Fabiosa J, Tokgoz S, Hayes D, Yu T-H (2008) Use of US croplands for biofuels increases greenhouse gases through emissions from land-use change. *Science* 319:1238–1240

Shani Z, Dekel M, Tsabary G, Goren R, Shoseyov O (2004) Growth enhancement of transgenic poplar plants by overexpression of *Arabidopsis thaliana* endo-1,4-bglucanase (cel1). *Mol Breed* 14:321–330

Shani Z, Shpigel E, Roiz L, Goren R, Vinocur B, Tzfira T, Altman A, Shoseyov O (1999) Cellulose-binding domain increases cellulose synthase activity in *Acetobacter xylinum* and biomass of transgenic plants. In: Altman A, Ziv M, Izhar S (ed.), *Plant Biotechnology and In Vitro Biology in the 21st Century*. Kluwer Academic Publishers, Dordrecht, The Netherlands, pp 213–218

Skinner DZ, Muthukrishnan S, Liang GH (2004) Transformation: A powerful tool for crop improvement. In: Skinner DZ, Liang GH (eds), *Genetically Modified Crops: Their Development, Uses, and Risks*. Food Products Press, New York, pp 1–16

Somleva M, Snell K, Beaulieu J, Peoples O, Patterson N (2008) Production of polyhydroxyburate in switchgrass, a value-added co-product in an important lignocellulosic biomass crop. *Plant Biotechnol J* 6:663–678

Somleva MN, Tomazewski Z, Conger BV (2002) *Agrobacterium*-mediated genetic transformation of switchgrass. *Crop Sci* 42:2080–2087

Steane DA (2005) Complete nucleotide sequence of the chloroplast genome from the Tasmanian blue gum, *Eucalyptus globulus* (Myrtaceae). *DNA Res* 12:215–220

Stoeglehner G, Narodoslawsky M (2009) How sustainable are biofuels? Answers and further questions arising from an ecological footprint perspective. *Bioresour Technol* 100:3825–3830

Stoutjesdijk PA, Hurlestone C, Singh SP, Green AG (2000) High-oleic acid Australian *Brassica napus* and B. *juncea* varieties produced by co-suppression of endogenous delta 12-desaturases. *Biochem Soc Trans* 28:938–940

Stricker JA, Rockwood DL, Segrest SA, Alker GR, Prine GM, Carter DR (2000) Short Rotation Woody Crops for Florida, available at http://www.treepower.org/ papers/strickernydoc (accessed January 21, 2011)

Sticklen MB (2008) Plant genetic engineering for biofuel production: Towards affordable cellulosic ethanol. *Nat Rev Genet* 9:433–443

Sujatha M, Lakshminarayana M, Tarakeswari M, Singh PK, Tuli R (2009) Expression of the *cry1EC* gene in castor (*Ricinus communis* L.) confers field resistance to tobacco caterpillar (*Spodoptera lituraFabr*) and castor semilooper (*Achoea janata* L.). *Plant Cell Rep* 28:935–946

Sujatha M, Reddy TP, Mahasi MJ (2008) Role of biotechnological interventions in the improvement of castor (*Ricinus communis* L.) and *Jatropha curcas* L. *Biotechnol Adv* 26:424–435

Sun Y, Cheng J (2002) Hydrolysis of lignocellulosic matrials for ethanol production: A review. *Bioresour Technol* 83:1–11

Tew TL, Cobill RM (2008) Genetic improvement of sugarcane (*Saccharum* spp.) as an energy crop. In: Vermerris W (ed), *Genetic Improvement of Bioenergy Crops*. Springer, New York, pp 249–272

Torney F, Moeller L, Scarpa A, Wang K (2007) Genetic engineering approaches to improve bioethanol production from maize. *Curr Opin Biotechnol* 18:193–199

U.S. DOE (2006) *Breaking the Biological Barriers to Cellulosic Ethanol: A Joint Research Agenda*, DOE/SC/ EE-0095, U.S. Department of Energy Office of Science and Office of Energy Efficiency and Renewable Energy, available at http://genomics.energy.gov/gallery/. (accessed January 29, 2011).

Vega-Sánchez ME, Ronald PC (2010) Genetic and biotechnological approaches for biofuel crop improvement. *Curr Opin Biol* 21:218–224

Vermerris W (2008) Why bioenergy makes sense. In: Vermerris W (ed), *Genetic Improvement of Bioenergy Crops*. Springer, New York, pp 3–42

Watt D A, Sweby DL, Potier BAM, Snyman SJ (2010) Sugarcane genetic engineering research in South Africa: From gene discovery to transgene expression. *Sugar Technol* 12:85–90

Weber C, Farwick A, Benisch F, Brat D, Dietz H, Subtil T, Boles E (2010) Trends and challenges in the microbial production of lignocellulosic bioalcohol fuels. *Appl Microbiol Biotechnol* 87:1303–1315

Weng J, Li X, Bonawitz ND, Chapple C (2008) Emerging strategies of lignin engineering and degradation for cellulosic biofuel production. *Curr Opin Biotechnol* 19:166–172

World Resource Institute (2011) Expanding Agriculture and Protecting Ecosystems: Can Payments to Farmers Accomplish Both? available at http://earthtrends.wri.org/updates/node/296 (accessed January 4, 2011)

Wu L, Birch R (2007) Doubled sugar content in sugarcane plants modified to produce a sucrose isomer. *Plant Biotechnol J* 5:109–117

Wu L, Birch RG (2010) Physiological basis for enhanced sucrose accumulation in an engineered sugarcane cell line. *Funct Plant Biol* 37:1161–1174

Wu ZW, Lee YY (1997) Ammonia recycled percolation as a complemantary pretreatment to the dilute acid process. *Appl Biochem Biotechnol* 63–65:21–34

Xi Y, Fu C, Ge Y, Nandakumar R, Hisano H, Bouton J, Wang Z (2009) *Agrobacterium*-mediated transformation of switchgrass. *Bioenergy Res* 2:275–283

Yadav R, Arora P, Kumar S, Chaudhury A (2010) Perspectives for genetic engineering of poplars for enhanced phytoremediation abilities. *Ecotoxicology* 19:1574–1588

Yang B, Wyman CB (2008) Pretreatment: The key to unlocking low-cost cellulosic ethanol. *Biofuels Bioproducts Biorefining* 2:26–40

Yuan JS, Tiller KH, Al-Almad H, Stewart NR, Stewart Jr. NS (2008) Plants to power: Bioenergy to fuel the future. *Trends Plant Sci* 13:421–429

Zarrilli S (2007) Global Perspective on Production of Biotechnology-based Bioenergy and Major Trends, available at http://www.fao.org/biotech/seminaroct2007.htm (accessed January 30, 2011)

Zhong R, Zhu F, Liu YL, Li SG, Kang LY, Luo P (1997) Oilseed rape transformation and the establishment of a bromoxynil-resistant transgenic oilseed rape. *Acta Bot Sin* 39:22–27

Zhu JY, Pan XJ (2010) Woody biomass pretreatment for cellulosic ethanol production: technology and energy consumption evaluation. *Bioresour Technol* 101:4992–5002

4 *In Planta* Production of Cell Wall Degrading Enzymes

Karen A. McDonald
University of California, Davis

CONTENTS

4.1 INTRODUCTION

As a result of requirements of the Energy Independence and Security Act of 2007, by the year 2022, 36 billion gal of biofuels will need to be produced to meet liquid transportation fuel demand, with at least 21 billion gal of "advanced biofuels," defined as renewable fuels derived from non-cornstarch sources achieving greater than 50% reduction in greenhouse gas (GHG) emission. In meeting this challenge, cellulosic biofuels will likely be a major contributor because of the resource potential of cellulosic feedstocks, which are estimated to be over 1 billion dry tons per year in the United States, sufficient to produce enough biofuels to replace 30% of current demand for transportation fuels (Perlack et al. 2005). The process of breaking down a complex polysaccharide carbohydrate (such as starch, cellulose, or hemicellulose) into monosaccharide components that can be fermented into biofuels is called saccharification. Saccharification of corn starch, alpha-linked glucose polymers, is relatively easy compared with breaking down the beta-linked glucose polymers that make up the structurally aligned and hydrogen-bonded cellulose polymers in cellulose microfibrils. In addition, the biological decomposition of cellulosic biomass presents a formidable challenge because of the recalcitrance of cellulose microfibrils embedded within the complex, heterogeneous structure of the plant cell wall composed of cellulose, hemicellulose, and lignan (Figure 4.1). This necessitates not only costly, energy-intensive, and environmentally detrimental biomass pretreatment steps (usually involving high temperatures, acids, and/or enzymes) to increase the accessibility and effectiveness of cellulase enzymes, but also high cellulase enzyme loadings (ratio of enzyme mass to biomass), currently approximately 100 times the loadings used for corn starch saccharification.

For the enzymatic conversion routes, enzyme preparations (referred to as cellulases), composed of mixtures of endoglucanses, exoglucanases, hemicellulases, and/or beta-glucosidases, are produced

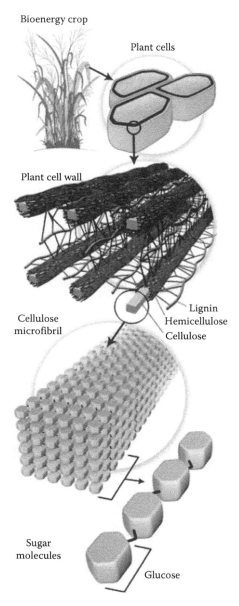

FIGURE 4.1 (See color insert)   Plant wall recalcitrance: A key scientific challenge. (From U.S. DOE, Bioenergy research centers: An overview of the science, Genome Management Information Systems, Oak Ridge National Laboratory, 2009.)

in separate microbial fermentations on-site or at commercial enzyme manufacturing plants, typically utilizing genetically modified fungal hosts (*Trichoderma reesei, Aspergillus niger, Penicillium funiculosum, Humicola insolens,* etc.) grown in expensive stainless steel fermenters under aerobic and aseptic conditions as submerged cultures. Commercial enzyme manufacturers produce these enzymes during large-scale (up to ~300,000 L) aerobic fermentation, which requires energy for fermenter and media sterilization, agitation for mixing viscous fungal suspensions, gas compression for air/oxygen sparging, fluid heating for temperature control, pumps for fluid transfer, etc. This is followed by downstream processing, which may include host cell deactivation and/or removal, protein concentration, and stabilization/formulation.

The scale required for production of enzymes to meet the needs of the cellulosic biofuel industry will likely be unprecedented in the industrial enzyme industry. For example, to meet the requirements for annual production of 21 billion gal of cellulosic biofuels using current cellulase enzyme loadings for steam-pretreated corn stover [~1 kg enzyme per dry ton biomass (K. McCall, personal comm)] and assuming 72 gal of ethanol produced per dry ton of corn stover, it would require an annual production of close to 300,000 t of cellulase enzyme per year, assuming that the cellulase enzymes are not reused/recycled. If this amount of enzyme was produced using traditional fungal fermentation at a relatively high titer of 100 g enzyme/L with a 200 h batch turnaround time, it would require a total fermenter capacity of over 74 million L. The current worldwide industrial enzyme fermenter capacity is estimated at approximately 20 million L. It is of course expected that protein engineering will result in significant reductions in enzyme loadings and increased stability enabling reuse of enzymes and thereby reducing the required fermenter capacity; however, the capital investment for this will still be substantial.

Further, the amount of energy consumption, CO_2 generation, and organic carbon (cellulosic) nutrient source consumption for a fungal fermentation-based cellulase enzyme production is often neglected in cellulosic biofuel life-cycle analyses. Saez and coworkers showed that for cellulase production in an aerobic, submerged bioreactor containing *T. reesei* growing on an insoluble cellulosic carbon source (5% w/v Solka floc), for each gram of cellulase enzyme produced, 1.6 g of carbon dioxide (CO_2) are emitted and 4.3 g of cellulose are consumed (Saez et al. 2002). Although the enzyme cost and activity are often important considerations in assessing the feasibility of the overall cellulosic bioethanol production process, there are very few detailed analyses of the energy requirements, CO_2 gas emissions, and diversion of cellulosic carbon associated with the enzyme production process. Although the agricultural energy inputs (e.g., energy associated with the production of fertilizers, herbicides, farm machinery and biomass transportation) have been carefully delineated, in many net energy analyses (EBAMM 2007; Schmer et al. 2008) for cellulosic ethanol production the energy requirements for the biorefinery plant (including the pretreatment, hydrolysis, fermentation, and biofuel recovery steps) are assumed to be satisfied by burning/gasification of the biomass residue. The energy associated with the production of the large quantities of enzymes that will be needed for the process are neglected. Although it is expected that enzyme loading will be dramatically reduced because of improvements in specific activity, long-term stability and reuse, pretreatment methods, and enzyme cocktail formulation, enzymes will still need to be produced on a massive scale. Thus there is an urgent need to develop new enzyme production technologies that will minimize energy and carbon nutrient consumption, reduce GHG emissions, and lower capital costs and total production costs. These types of "green" biomanufacturing technologies will not only benefit the biofuels/biorefinery industries but will also have broader effects on the industrial enzyme industry in general.

One approach to this problem is plant-based production of recombinant cellulase and hemicellulase enzymes or plant cell wall modifying proteins in the plant biomass that is intended for biofuel production itself (e.g., to initiate self-deconstruction and/or modify the plant cell wall to facilitate the accessibility of exogenous enzymes). Plants may also serve as an alternative host for production of cellulase and hemicellulase enzymes to be used as additives to other pretreated biomass, or a combination of the two, because the spent biomass that produces the enzymes can also serve as a source of cellulose. This general approach has been referred to as endogenous, endoplant, or *in planta* enzyme production and has been used in plant-made additive enzymes, cell wall loosening modifications, and self-processing/deconstruction biomass.

Production of the enzymes in plants has several advantages over microbial fermentation in large-scale stainless steel bioreactors. These advantages include use of sunlight/photosynthesis and consumption of CO_2 as the energy and carbon source for host cell growth, lower capital and production costs, easier scale-up, and the ability to glycosylate enzymes that may be important for the activity and/or stability for some cellulases. Use of plant-made enzyme preparations as additives to pretreated biomass also allows flexibility in terms of the biomass feedstock because any cellulosic material (including dried materials such as wood chips, corn stover, rice straw, agricultural wastes, etc.) can be used as the source of biofuels. For this application, the enzymes can be produced in a wide variety of plant hosts, not necessarily a crop with favorable characteristics as a bioenergy crop, in stably transformed (e.g., transgenic) plants or using transient expression. Enzymes can be individually produced in different plants/plant tissues, partially purified, and then mixed to generate appropriate enzyme cocktails. The primary considerations for this application are achieving high expression levels of active enzymes [g enzyme/kg fresh weight (FW) plant tissue or % total soluble protein (TSP)]; ease of recovery; solubilization and formulation of active and stable enzyme preparations; and minimal influence of endogenous plant proteins and metabolites on enzyme activity, stability, and subsequent fermentation. If stable transgenic feedstocks are used it is important that the production of the enzymes does not have a deleterious effect on plant growth. This can be achieved using several approaches:

- Producing enzymes that are not active under plant growth conditions (e.g., thermostable enzymes or lacking activity at cellular pH level) or that require further processing (post-growth) to generate a functional form of the enzyme,
- Sequestering the enzymes in a subcellular space/organelle so that the cell wall matrix is not exposed to the enzymes during plant growth,
- Using inducible expression systems so that enzymes are produced just before or after harvest, or
- Choosing a host that does not have a particular substrate as a major component of its cell wall.

In planta production of enzymes or proteins that lead to cell wall loosening or modifications that enhance the accessibility and/or effectiveness of exogenously added enzymes may reduce the stringency (e.g., temperature, time required, acid/basic conditions, and size reduction) of biomass pretreatment steps, thereby reducing the pretreatment costs and environmental impact. For example, expansins and swollinins are thought to disrupt hydrogen bonding between cell wall polysaccharides without hydrolyzing them. For this application, the plant host should have favorable characteristics as a bioenergy crop and, to avoid detrimental influence on plant growth and viability, the approaches listed above could be used for stable transformants. Alternatively, transient expression of cell-wall-targeted enzymes could be implemented in intact plants after they have reached maturity or in harvested leaves. For the self-deconstructing biomass application, production and use of enzymes in living plant tissue is required. Because the enzymes can be produced within the plant cell, secreted, and localized in the cell wall matrix, higher local enzyme concentrations might be achievable at the substrate site. This could lead to higher efficiency hydrolysis than can be achieved with exogenous application of fermenter-produced enzymes in which nonspecific binding and external and intraparticle mass transfer effects may dominate. However, this application is more challenging because it requires temporal separation of the processes of biomass growth, enzyme production, and enzyme activation to avoid any detrimental effects by the enzymes on biomass yield and/or enzyme production and to properly target the enzymes and minimize nonspecific binding. Additionally, the relative concentrations of each of the enzymes (endoglucanase, exoglucanase, beta-glucosidase, and hemicellulases) must be controlled to achieve optimal hydrolysis. Some form of biomass pretreatment may also be required before enzyme activation, so the robustness of the enzymes used for each given pretreatment method would need to be assessed. For example, Teymouri et al. (2004) have shown that there is a 65% loss in endoglucanse activity after transgenic tobacco leaves expressing thermostable *Acidothermus cellulolyticus* E1 endoglucanse catalytic domain undergo ammonia fiber explosion (AFEX) pretreatment. However, even if the plant-made enzymes are not able to cause complete hydrolysis, independent of pretreatment and/or addition of exogenous enzymes, expression *in planta* may produce some benefits, including

- A reduction in process costs and/or exogenous enzyme requirements because of reduction in the amount of exogenous enzyme needed to achieve the same glucan conversion,
- An increase in biomass digestability,
- Reduced severity requirements for pretreatment, and/or
- Improved yield of the desired biofuel.

This chapter provides an overview of plant biotechnology approaches and considerations for *in planta* production of cell-wall-degrading enzymes for biofuel applications and provides an updated review of recent work on the production of plant-made cellulase and hemicellulase enzymes produced using stable nuclear transformation, chloroplast transformation, and transient expression. Sainz (2009) and Taylor et al. (2008) also provide recent reviews of plant-expressed glycosyl hydrolase enzymes in stably transformed plants. Expression of enzymes in other photosynthetic systems such as moss, algae, cyanobacteria photobioreactors, and plant cell suspension cultures is also possible but is not included here because these systems would likely not be as scalable or economically feasible for biofuel applications.

4.2 PLANT BIOTECHNOLOGY APPROACHES AND CONSIDERATIONS

Heterologous enzymes can be expressed in plants from transgenes that have been stably inserted into the plant nuclear genome or the chloroplast genome, or from expression vectors that are introduced transiently using recombinant plant viruses or recombinant agrobacterial vectors. These alternative approaches and the implications for plant-produced enzymes for biofuels applications are outlined below.

4.2.1 STABLE NUCLEAR TRANSFORMATION

Foreign DNA can be stably inserted in the plant nuclear genome of a plant cell using various methods such as *Agrobacterium tumefaciens*-mediated transformation, microprojectile bombardment, electroporation, free DNA uptake under certain conditions, and other forced penetration methods. However, the transgene is generally inserted at a random position in the plant nuclear genome resulting in "position effects" as well as the potential for multiple inserts. Both of these effects can influence the expression level observed for the heterologous protein, requiring extensive screening to identify the most productive transgenic line. The efficiency of transformation and the ability to regenerate whole transgenic plants depends strongly on the plant species and choice of explant, the transformation method and conditions, selection media/conditions, and regeneration methods. The main advantages of using stably transformed plants for production of cell-wall degrading and/or modifying enzymes is that once a transgenic line is established, the foreign gene is passed to subsequent generations allowing for relatively inexpensive, easily scalable, long-term production of the enzymes within field-grown plants. The main disadvantages are the long time frames (typically 6 months to several years) required to establish the high expressing transgenic line, the potential negative impact of the transgenes and/or transgenic crop in the environment, concerns about the transfer of transgenes from genetically modified (GM) to non-GM relatives through cross-fertilization, and/or the possibility of introduction of the GM material into the food/feed supply chain. Wolt (2009) provides a recent review on environmental risk assessment for transgenic biofuels crops. For transgenic plants expressing cell-wall-degrading or cell-wall-modifying enzymes, it is particularly important to ensure that these enzymes are not active on their host cell walls to avoid detrimental effects on plant growth and viability. Previous work on heterologous expression of cell-wall-deconstructing enzymes in plants has focused on production in stably transformed plants in which the enzyme is produced constitutively but is sequestered in a subcellular compartment away from the cell wall or is not active under growth conditions (see recent review by Sticklen, 2008). Alternatively, the transgene could be under the control of a chemically or environmentally inducible promoter.

4.2.2 CHLOROPLAST TRANSFORMATION

Chloroplast genetic engineering utilizes site-specific homologous recombination to insert the transgene into the chloroplast genome. Because there are approximately 100 chloroplasts per cell and each chloroplast contains approximately 100 identical copies of the chloroplast genome, this approach has the potential to generate stable transformants that have tens of thousands of copies of the transgene. High copy number, combined with the absence of epigenetic effects such as gene silencing, have led to very high expression levels (>46% TSP) of heterologous proteins using this approach although most of the studies have been done in tobacco. Because the chloroplast is maternally inherited, there is reduced potential for gene flow and biosafety is enhanced. Other advantages of this approach, particularly for large-scale production of cell-wall-modifying/degrading enzymes, is the ease of transgene stacking to produce multiple enzymes in the same plant and the sequestration of the enzyme products in the chloroplast where they cannot harm the cell wall during plant growth. One disadvantage of this approach is that the chloroplast is only capable of bacterial-like processing, so it is not possible to produce glycosylated enzymes. There is evidence, at least in some cases, that glycosylation of fungal glycosyl hydrolase enzymes is important for stability and activity of the enzymes (Jeoh et al. 2008). In addition, the high expression levels that have been achieved in transplastomic plants have often resulted in slow growth and/or other detrimental/lethal effects on the plants.

4.2.3 TRANSIENT VIRAL EXPRESSION

Recombinant plant viruses have been recognized as a very useful means for high level, rapid expression of foreign proteins in plants (Scholthof et al. 1996). In this approach, recombinant plant virions (or alternatively infectious RNA) containing the transgene (usually as an insert or as a replacement of a nonessential viral gene) of interest are used to infect a compatible plant host, typically through a mechanical process such as rubbing or spraying the virion solution, which also contains an abrasive agent such as carborundum or diatomaceous earth, onto plant leaves. The virus infects the host, and during replication it generates large amounts of viral nucleic acids including the transgene insert, which may spread cell to cell and systemically throughout the plant within a few days, and expresses the foreign protein during replication. The major advantages of this approach are that it is fast (recombinant protein is produced in a matter of a week), very efficient (high transgene copy number and the ability to spread systemically allow many cells in the plant to act as hosts, not just those initially infected), and does not require the deployment of transgenic crops in the field (nontransgenic plants are used as the hosts). It is also a preferred method for production of proteins, such as cellulases or cell-wall-modifying proteins, that could be detrimental to plant health. However, for most viruses there is a limitation to the gene insert size of approximately 1 kb. The stability of the transgene during serial propagation and systemic spread can be an issue, plant gene silencing mechanisms can limit recombinant protein production, and biosafety aspects need to be assessed for each system and production method. Production of multiple enzymes within the same plant host may require the use of "noncompeting" recombinant viral vectors as described by Giritch et al. (2006).

4.2.4 TRANSIENT *AGROBACTERIUM*-MEDIATED EXPRESSION

Another transient plant-based heterologous expression process known as "agroinfiltration" allows for the rapid, high-level expression of multiple recombinant proteins in harvested plant tissues as well as tissues of intact plants (Sudarshana et al. 2006; Plesha et al. 2007, 2009). This process utilizes *Agrobacterium*-mediated transformation, a highly efficient process for introducing foreign genes into plant cells, significantly more efficient than purely plant viral based systems (Azhakanandam et al. 2007). *Agrobacterium*-mediated transformation of plants is a commonly used approach in which a specific portion of DNA, transfer DNA (T-DNA), on the bacterial Ti (tumor-inducing) plasmid is transferred to the plant nucleus where it can ultimately be integrated into the plant genome to generate stably transformed plants. However, during the initial period after agroinfection, the genes on the T-DNA are transiently expressed within 2–3 days in the plant nucleus. This transient "burst" of expression can lead to higher levels of recombinant protein compared with levels in stable transgenic plants transformed with the same agrobacteria (Wroblewski et al. 2005). Furthermore, co-infiltration using multiple *Agrobacterium* strains containing different recombinant Ti plasmids allows coordinated expression of multiple recombinant proteins (Voinnet et al. 2003). Recombinant *Agrobacterium* containing appropriate binary Ti expression vectors can be grown in shake flasks or bioreactors. Scale-up of *Agrobacterium tumefaciens* cultures up to the 5000-L scale have been reported with final biomass concentrations reaching over 50 g dry weight/L after 96 h of culture (Ha et al. 2007). *A. tumefaciens* can be grown in Luria-Bertani (LB) media containing appropriate selection antibiotics and supplemented with approximately 40 M acetosyringone and MES buffer. Before infiltration, the agrobacteria are centrifuged and resuspended in distilled water containing magnesium chloride (10 mM), acetosyringone (150 mM), and a surfactant (e.g., Silwet® at 0.01%). Several recombinant agrobacteria can be mixed to achieve desired relative concentrations depending upon the application. Harvested leaf biomass (e.g., *Nicotiana benthamiana*) can be immersed in the agrobacterial solution and a weak vacuum (~530–610 mm Hg below atmospheric pressure or an absolute pressure of 150–230 mm Hg) applied for a short period of time (10 s to several minutes) and then rapidly released. The vacuum pulls air trapped in the stomatal cavities out of the plant leaf, and once the vacuum is released, the agrobacteria solution is infused into the stomata. Alternatively, the entire plant can be turned upside down and immersed in the *Agrobacterium* solution before applying a vacuum. The molecular steps involved in the transfer of T-DNA on the Ti plasmid of the *Agrobacterium* to the plant cell are described in detail by Tzfira and Citovsky (2002), but they basically involve attachment of the agrobacteria to the plant cell wall, excision of a single-stranded portion of the T-DNA from the Ti plasmid called the T-strand, formation of a channel called the T-pilus that connects the cytoplasm of the *Agrobacterium* to the plant cytoplasm, and transfer of the T-strand complex into the plant cytoplasm and ultimately to the plant nucleus where expression of transgenes can take place. After the agroinfiltration step, the plant leaves are typically removed from solution, allowed to dry in ambient air for a certain period of time, and incubated in humid air at room temperature to allow enough time for the agroinfection process and T-DNA transfer to take place. Vacuum infiltration processes have been scaled up with reports of over 100 kg of wild-type tobacco infiltrated (Fischer et al. 2004). Vacuum infiltration of harvested biomass in a contained facility is a promising approach for efficient, scalable, and cost-effective production with high biosafety and low environmental impact. It is a rapid, low-cost process that does not require aseptic operation or the generation or deployment of transgenic crops in the field. Although it will be necessary to grow the agrobacteria in a microbial fermenter, the required fermenter volume is estimated to be significantly smaller than would be required for fungal production of enzymes for several reasons. First, only a single bacterium is needed to transfer the T-DNA to each plant cell (representing a mass ratio of ~1:10,000) and it is the plant cell's biosynthetic machinery (generated via photosynthesis) that performs the transcription, translation, folding, post-translational modifications, and secretion processes. Second, the microbial fermentation is used only to propagate the recombinant agrobacteria; resources are not diverted for the high-level production of product. Thus, the approach offers the benefit of enzyme production in a eukaryotic host grown using photosynthesis without the need to generate or deploy stably transgenic plants in the environment. In addition, the required fermenter volume is reduced compared with that required for submerged fungal fermentation, and the agroinfiltration process will be performed in a contained environment but will not require aseptic operation, which further reduces capital and operating costs.

To increase productivity of heterologous proteins, plant viral amplicon expression systems have been developed (see Gleba et al. 2007, for a recent review). In these systems, replication-competent plant viral components, typically from tobacco mosaic virus (TMV) [e.g., the TMV RNA-based overexpression (TRBO) system (Lindbo 2007)] or cucumber mosaic virus (CMV) [e.g., such as the CMViva system (Sudarshana et al. 2006)], are inserted in the T-DNA to enable the cellular machinery to amplify the number of copies of the transgene during replication of the viral RNAs. Expression levels of recombinant proteins using viral amplicon expression systems often greatly exceed (~10- to 100-fold) that which is possible using traditional constitutive promoters such as the cauliflower mosaic virus (CaMV 35S) promoter. Harvested leaves (Plesha et al. 2009), individual leaves on intact plants (Sudarshana et al. 2006), or even entire plants (Gleba et al. 2005; Marillonnet et al. 2005) can be infiltrated and infected this way, making the process useful for rapid evaluation of novel enzymes as well as rapid, scalable production of cellulase and hemicellulase enzymes. Co-infiltration with *Agrobacterium* containing genes capable of suppressing the plant's innate ability to recognize and shut down foreign gene expression (i.e., gene-silencing suppressors) has also proven useful to enhance target protein production in the CMViva and TRBO systems.

In addition to the expression technology (stable nuclear, stable chloroplast, transient viral, or transient agroinfiltration) method used, there are several factors that also influence the enzyme productivity, activity, stability, and ease of recovery. These include the type of plant promoter, subcellular targeting/localization, design of the gene construct, post-translational modifications, and gene silencing.

4.2.5 Plant Promoters

For many biofuel applications, high-level expression of the heterologous enzymes is generally the goal, particularly for applications such as production of plant-made additive enzymes. An efficient plant transgene expression system, consisting of a promoter, targeting signal peptide, target gene that has been optimized for expression in the plant host, and transcription terminator, is essential for high-level production of the target enzyme(s). The choice of promoter system significantly influences the production yield by affecting the transcription rate of the target gene. Plant promoters can be divided into several categories: constitutive, tissue-specific, developmentally regulated, and inducible promoters. Constitutive promoters directly drive the expression of the target gene in all tissues and are largely independent of developmental factors. An example is the well known CaMV 35S promoter. Constitutive expression of a recombinant protein could result in an additional metabolic burden during plant cell growth and hence reduce the plant growth rate. The characteristics of the recombinant protein might also affect the plant cells' physiology because of intrinsic toxic properties of the product on the host cells or interference with host cell metabolism.

Inducible promoters are modulated by the presence of specific external factors or compounds such as light, temperature, wounding, or the concentrations of metal ions, alcohols, steroids, herbicides, etc. Such regulated expression systems are advantageous because they allow the plant growth and protein production phases to be independently optimized. This approach is particularly attractive when product synthesis is deleterious to plant growth and/or viability. Furthermore, because expression of a foreign gene linked to an inducible promoter can be induced at a specific stage during the plant growth, there is less potential for post-transcriptional gene silencing (PTGS) found in gene expression systems that use constitutive promoters in transgenic plants (Vaucheret et al. 2001; Vaucheret and Fagard 2001). Various inducible promoters have been developed for use in plants (for example, see Boetti et al. 1999; Zuo and Chua 2000; Huang et al. 2001; Padidam 2003; Tang et al. 2004). For field production of lignocellulolytic enzymes, chemically inducible promoters are likely to be costly and may have negative environmental impacts depending on the system. However, chemically inducible promoters may have use in applications in which enzyme expression is initiated after harvest in a production facility.

4.2.6   SUBCELLULAR TARGETING/LOCALIZATION

The choice of subcellular localization of the recombinant protein is dictated by the application. For *in planta* production of cell-wall-modifying enzymes and self-deconstructing biomass, the product would need to be localized outside of the cell membrane (e.g., in the apoplast). However, subcellular targeting/localization can also affect production levels and product stability as well as ease of recovery and solubilization; several studies have indicated the effects of subcellular targeting on heterologous protein production. If the recombinant protein is stable in the apoplast, this provides significant advantages in terms of recovery and purification because (1) the protein content of the apoplastic fluid typically corresponds to only a small fraction of the total protein content of the biomass (<5%), so there are fewer contaminating proteins that must be removed; (2) intracellular proteases that would be released during extraction/cell disruption are avoided; and (3) low-molecular-weight intracellular compounds such as phenolics and/or alkaloids that can interfere with protein recovery are also avoided. Demonstrated initially by Klement in 1965, vacuum infiltration has been used in many studies to selectively remove extracellular proteins from the intercellular spaces of plant leaves on a small scale.

Subcellular targeting of proteins to organelles (e.g., chloroplast, vacuole, mitochondria, peroxisome) or to the apoplast is typically accomplished through the addition of a transit peptide or secretion signal peptide at the amino or carboxy terminus of the protein, which is then cleaved during the transport process (Mackenzie 2005). If no secretion signal peptide or transit peptide is included, the soluble protein remains in the cytosol.

4.2.7   OPTIMIZATION OF GENE CONSTRUCTS

Because of the degeneracy in the genetic code (multiple codons encode for the same amino acid), there are many different DNA sequences that can result in the same amino acid sequence. Many of the target cellulase, hemicellulase, and cell-wall-modifying enzymes come from bacterial or filamentous fungal hosts that have different preferences in terms of the codons they use for a particular amino acid and a different GC content than what is found in plants. Because of advances in speed and reduction in costs, gene constructs for expression of heterologous proteins are now often chemically synthesized, which has opened up opportunities for optimization of gene constructs that not only take into consideration codon usage but also other factors such as mRNA structure and stability. Although there has been significant progress in identifying synthetic gene design rules for prokaryotic cells such as *Escherichia coli*, design algorithms for reliable, predictable, and high-level expression of a specific heterologous protein in a particular plant host are not yet available. However, codon optimization algorithms coupled with synthetic gene design are often used when expressing a bacterial or fungal protein in plants.

4.3   PLANT-MADE CELL-WALL DEGRADING ENZYMES

Table 4.1 lists the endoglucanase, exoglucanases, β-glucosidases, xylanases, and hybrid β-(1-3,1-4)-glucanases that have been expressed in plants. As can be seen from the table, most of the work has been done in tobacco, corn, barley, rice, and *Arabidopsis*, generally using stable nuclear transformation or chloroplast transformation methods. In most cases the enzyme expression level is presented in terms of percent TSP or units of activity per TSP; however, it is also important to know the enzyme yield on a gram per kilogram FW or dry weight (DW) tissue basis. For enzymes that contain a cellulose binding domain, it is also useful to assess the "residual" activity associated with the plant material after homogenization/extraction. Maximum yields are presented in the table, but expression levels can range over 2–3 orders of magnitude for different independent transformation events.

It is interesting to note that in most of these cases, a significant level of active enzyme can be produced through judicious choice of the host, subcellular targeting to sequester the enzyme

TABLE 4.1
Cellulase, Hemicellulase, and Cell Wall Modifying Enzymes Expressed in Plants

Enzyme/ Source	Native or Modified	Host/Tissue/ Phenotype	Expression Strategy/ Targeting	Promoter Type/ Source	Maximum Yield	Activity Assay/ Temperature and pH Optima	References
Endoglucanases							
E1 endoglucanase/ *Acidothermus cellulolyticus*	Native	Tobacco/Leaves/ Normal	Stable nuclear/ Chloroplasts	Light responsive/ Tomato RbcS-3C	1.35% TSP	MUCase/81°C and pH 5.25	Dai et al. 2000a
E1 endoglucanase/ *Acidothermus cellulolyticus*	Native	Potato/Leaves/ Normal	Stable nuclear/ Chloroplasts	Light responsive/ Tomato RbcS-3C	2.6% TSP	MUCase/ ND	Dai et al. 2000b
E1 endoglucanase CD/ *Acidothermus cellulolyticus*	Native	*Arabidopsis thaliana*/ Leaves/Normal	Stable nuclear/ Apoplast	Constitutive/ CaMV 35S	26% TSP	MUCase/ ND	Ziegler et al. 2000
E1 endoglucanase/ *Acidothermus cellulolyticus*	Native	Tobacco/Leaves/ Normal	Stable nuclear/ Chloroplasts	Constitutive/ CaMV 35S	ND	MUCase ND	Jin et al. 2003
E1 endoglucanase/ *Acidothermus cellulolyticus*	Native	Tobacco/Leaves Normal Normal Normal	Stable nuclear/ Apoplast Chloroplast Cytosol	Constitutive/ CaMV 35S	0.33% TSP 0.0007% TSP 0.0006% TSP	MUCase/ ND	Ziegelhoffer et al. 2001
E1 endoglucanase CD/ *Acidothermus cellulolyticus*	Native	Tobacco/Leaves Normal Normal Normal	Stable nuclear/ Apoplast Chloroplast Cytosol		0.4% TSP 0.074% TSP 0.0046% TSP		
E1 endoglucanase/ *Acidothermus cellulolyticus*	Native	Tobacco/Leaves	Stable nuclear/ Apoplast	Constitutive/ CaMV35S	0.25% TSP	MUCase	Dai et al. 2005
E1 endoglucanase CD/ *Acidothermus cellulolyticus*	Native	Rice/Leaves/ Normal	Stable nuclear/ Apoplast	Constitutive/ CaMV35S	4.9% TSP	MUCase AFEX-treated corn stover AFEX-treated rice straw CMC Avicel	Oraby et al. 2007

Enzyme/Source	Native/Codon optimized	Host/Tissue/Condition	Localization	Promoter	Expression level	Assay	Reference
E1 endoglucanase CD/ *Acidothermus cellulolyticus*	Native	Corn/Leaves Normal	Stable nuclear/ Apoplast	Constitutive/ CaMV 35S	1.16% TSP	MUCase	Ransom et al. 2007
E1 endoglucanase CD/ *Acidothermus cellulolyticus*	Native	Corn/Leaves Normal	Stable nuclear/ Apoplast	Constitutive/CaMV 35S	2.1% TSP	MUCase	Biswas et al. 2006
E1 endoglucanase/ *Acidothermus cellulolyticus*	Codon optimized first 40 codons	Corn/Seed Normal	Stable nuclear/ ER Apoplast Vacuole	Seed specific/ Globulin 1	17.9% TSP / 0.5% TSP* / 16% TSP	MUCase	Hood et al. 2007
E1 endoglucanase/ *Acidothermus cellulolyticus*	Native	Duckweed/whole plant/ Normal	Stable nuclear/ Cytosol	Constitutive/ CaMV 35S	0.24% TSP / 3.5 mg/g FW	MUCase/ 80°C and pH 5	Sun et al. 2007
E1 endoglucanase/ *Acidothermus cellulolyticus*	Native	Corn/Leaves/ Normal	Stable nuclear/ ER mitochondria	Light responsive/ RbcS1-P	2% TSP / 0.2% TSP	MUCase	Mei et al. 2009
E1 endoglucanase/ *Acidothermus cellulolyticus*	Codon optimized	*Nicotiana benthamiana*/ harvested leaves NA	Transient agroinfiltration/ Apoplast	Constitutive/ CaMV 35S	4.3 mg/kg FW at 6 days post infiltration	MUCase	McDonald et al. 2008
E2 endoglucanase/ *Thermomonospora fusca*	Native	Alfalfa/Leaves Normal Potato/Leaves Normal Tobacco/Leaves Normal	Stable nuclear/ Cytosol	Constitutive/ Hybrid Mac	0.01% TSP / ND / 0.1% TSP	CMC	Ziegelhoffer et al. 1999
EG1 endoglucanase/ *Trichoderma reesei*	Native	Barley/Seed Normal	Stable nuclear/ Cytosol	Developmentally regulated and tissue specific/ Hybrid high-pI α-amylase	0.025% TSP	None	Nuutila et al. 1999
Cel5A endoglucanase/ *Thermotoga maritima*	Native	Tobacco/Leaves/ Normal	Stable nuclear/ Chloroplasts	Light responsive/ Alfalfa RbcSK-1A	5.2% TSP	CMC	Kim et al. 2010

(Continued)

TABLE 4.1 (Continued)
Cellulase, Hemicellulase and Cell Wall Modifying Enzymes Expressed in Plants

Enzyme/Source	Native or Modified	Host/Tissue/Phenotype	Expression Strategy/Targeting	Promoter Type/Source	Maximum Yield	Activity Assay/Temperature and pH Optima	References
Cel6A endoglucanase/ *Thermobifida fusca*	Native	Tobacco/Leaves Normal	Chloroplast transformation/ Chloroplast	Constitutive/ plastid rrn16	2% TSP	CMC	Yu et al. 2007
DB-Cel6A endoglucanase/ *Thermobifida fusca*	Native	Tobacco/Leaves Normal	Chloroplast transformation/ Chloroplast	Constitutive/ Plastid rrn16	10.7% TSP	CMC% T	Gray et al. 2008
CelA/Cel6G endoglucanse/ *Neocallimastix patriciarumPiromyces spp.*	Codon optimized	Barley/Seed endosperm Normal	Stable nuclear/ Cytosol	Developmentally regulated and tissue specific/ Rice GluB-1	1.5% TSP	AZCL-barley glucans AZCL-hydroxyethyl cellulose	Xue et al. 2003
CelD endoglucanase/ *Clostridium thermocellum*	Native	Tobacco/Leaves Normal	Chloroplast transformation/ Chloroplast	Constitutive/ psbA	4900 U/g FW leaf	CMC 60C and pH 6	Verma et al. 2010
Exoglucanases							
Cel6B exoglucanase/ *Thermobifida fusca*	Native	Tobacco/Leaves Normal	Chloroplast transformation/ Chloroplast	Constitutive/ plastid rrn16	3% TSP	BMCC	Yu et al. 2007
CelO exoglucanase/ *Clostridium thermocellum*	Native	Tobacco/Leaves Normal	Chloroplast transformation/ Chloroplast	Constitutive/ psbA	442 U/mg TSP	β-D-glucan (1%)	Verma et al. 2010
CBHI exoglucanase/ *Trichoderma reesei*	Codon optimized first 40 codons	Corn/Seed Normal	Stable nuclear/ ER Apoplast Vacuole	Seed specific/ Globulin 1	17.8% TSP 16.3% TSP 0% TSP	MUCase	Hood et al. 2007
CBHI exoglucanase/ *Trichoderma reesei*	Native	Tobacco/Leaves Normal	Stable nuclear/ Native signal	Constitutive/ CaMV 35S	0.11% TSP	MUCase	Dai et al. 1999

Enzyme/Organism	Type	Host/Tissue	Localization	Promoter	Expression Level	Assay	Reference
E3 exoglucanase/ *Thermomonospora fusa*	Native	Alfalfa/Leaves Normal Potato/Leaves Normal Tobacco/Leaves Normal	Stable nuclear/ Cytosol	Constitutive/ Hybrid Mac	0.002% TSP ND 0.02% TSP	None	Ziegelhoffer et al. 1999
β-Glucosidases							
BglB β-glucosidase *Thermotoga maritima*	Native	Tobacco/Leaves Normal	Stable nuclear/ Cytosol Chloroplast	Light regulated/ Alfalfa RbcSK-1A	4.5% TSP 5.8% TSP	pNPG/ 80°C and pH 4.5	Jung et al. 2010
Bgl1 β-glucosidase *Trichoderma reesei*	Native	Tobacco/Leaves Normal	Chloroplast transformation/ Chloroplast	Constitutive/ psbA	14 U/mg TSP	pNPG	Verma et al. 2010
Bgl1 β-glucosidase *Apergillus niger*	Native	Tobacco/Leaves Normal	Stable nuclear/ Apoplast ER Vacuole	Constitutive/ CaMV 35S	3 U/g FW 4.3 U/g FW 7 U/g FW	pNPG	Wei et al. 2004
Xylanases							
XynA xylanase *Bacillus subtilis*	Native	Tobacco/Leaves Normal	Chloroplast transformation/ Chloroplast	Constitutive Rice psbA	6% TSP	Reducing sugars released from xylan	Leelavathi et al. 2003
XynA xylanase CD *Clostridium thermocellum*	Native	Rice/Leaves and Seeds Normal	Stable nuclear/ Cytosol	Constitutive/ Modified CaMV 35S	ND	Reducing sugars released from xylan	Kimura et al. 2003
XynA xylanase *Neocallimastix patriciarum*	Native	Barley/Seed Normal	Stable nuclear/ Cytosol	Developmentally regulated and tissue specific/ Rice GluB-1 Barley B1 hordein	0.006% TSP 0.037% TSP	AZCL-xylan	Patel et al. 2000

(Continued)

TABLE 4.1 (Continued)
Cellulase, Hemicellulase and Cell Wall Modifying Enzymes Expressed in Plants

Enzyme/Source	Native or Modified	Host/Tissue/Phenotype	Expression Strategy/Targeting	Promoter Type/Source	Maximum Yield	Activity Assay/Temperature and pH Optima	References
XynC xylanase/oleosin fusion *Neocallimastix patriciarum*	Native	Canola/Seed Normal	Stable nuclear/Oil bodies	Tissue specific/Oleosin promoter	2000 U/kg seed	RBB-xylan 40°C and pH 5.5	Liu et al. 1997
XylII xylanase *Trichoderma reesei*	Native	*Arabidopsis*/Leaves Normal	Stable nuclear/Cytosol Chloroplast Peroxisome Chloroplast and Peroxisome	Light responsive/Alfalfa RbcSK-1A	1.2% TSP 3.0% TSP 1.7% TSP 4.8% TSP	Reducing sugars released from xylan	Hyunjong et al. 2006
XynII xylanase *Trichoderma reesei*	Native	*Arabidopsis*/Leaves Normal	Stable nuclear/Cytosol Chloroplast	Constitutive/CaMV 35S	1.4% TSP 3.2% TSP	Reducing sugars released from xylan	Bae et al. 2008
XynB xylanase *Streptomyces olivaceoviridis*	Native	Potato/Leave Normal	Stable nuclear/Cytosol	Constitutive/Double CaMV 35S	5%	Reducing sugars released from xylan	Yang et al. 2007
XynZ xylanase *Clostridium thermocellum*	Native	Tobacco/Leaves Normal	Stable nuclear/Apoplast	Constitutive/CaMV 35S	4.1% TSP	Reducing sugars released from xylan	Herbers et al 1995
XylD xylanase CD *Ruminococcus flavefaciens*	Native	Tobacco/Leaves Normal	Stable nuclear/Apoplast	Constitutive/CaMV 35S	NR	Reducing sugars released from xylan	Herbers et al. 1996
Glucanase							
Hybrid β-(1-3,1-4) glucanase Baccillus sp.	Codon optimized	Barley/Seed Normal	Stable nuclear/Vacuole	Developmentally regulated and tissue specific/Barley D hordein	5.4% TSP	Zymogram	Horvath et al. 2000

| Hybrid β-(1-3,1-4) glucanase Baccillus sp. | Codon optimized | Barley/Seed Normal | Stable nuclear/ Apoplast | Developmentally regulated and tissue specific/ Barley glucanse | NR | Reducing sugars released from xylan | Jensen et al. 1996 |

Source: Adapted from Sainz, MB., *Vitro Cell Dev Biol-Plant*, 45, 314–329, 2009 and Taylor II LE., et al., *Trends Biotechnol*, 26, 413–424, 2008.
ND, Not Determined; NA, Not Applicable; MUC, 4-methylumbelliferyl-β-D-cellobioside; pNPG, p = nitrophenyl β-D-glucopyranoside; CD, catalytic domain; CMC, carboxymethyl cellulose; BMCC, bacterial microcrystalline cellulose; RBB-xylan, remazol brilliant blue xylan; DB, downstream box fusion.
*Mean instead of maximum

away from the cell wall matrix, or choosing a thermostable enzyme that is inactive under plant growth conditions. The E1 endoglucanase, a thermostable enzyme from *A. cellulolyticus*, is one of the most well studied of all of the cellulase enzymes and has been expressed in many different plant species with various subcellular localizations. The fact that high expression levels of E1 are achieved even when localized to the apoplast is likely because of the high temperature optimum (80°C) as well as the fact that generally single enzymes are not sufficient to cause hydrolysis of cellulose within the cell wall matrix. Cleavage of full-length enzymes can sometimes be a problem; for example, cleavage of full-length E1 occurred in several of the expression systems (Dai et al. 2000a; Hood et al. 2007; Sun et al. 2007) and was also cleaved in barley (Nuutila et al. 1999).

Although most of the expression studies performed to date have focused on stably transformed nuclear or chloroplast transformation, our group has been focusing on transient expression in harvested plant tissues using vacuum agroinfiltration (McDonald et al. 2008). This method, which allows for the use of green agricultural wastes, is rapid compared with the time it takes to generate (and to be approved through the regulatory approval process) a GM bioenergy feedstock, and expression of the enzyme is performed in a contained environment. We have produced full-length active E1 in harvested *N. benthamiana* leaves at levels of approximately 4 mg/kg FW 6 days after agroinfiltration.

In many of the studies reported, the authors have demonstrated that the plant-made enzymes, when supplemented with the additional enzymes necessary to complete the hydrolysis, are capable of hydrolyzing substrates to generate glucose. In most cases the enzymes did not need to be purified—a crude extract could be used directly. For example, Oraby et al. (2007) showed that TSP extracts from rice leaves expressing the E1 catalytic domain supplemented with commercial β-glucosidase were able to hydrolyze realistic substrates, AFEX-treated corn stover, and AFEX-treated rice straw into glucose with 30% and 22% conversion, respectively, demonstrating the feasibility of using plant-expressed enzymes as additives for biological deconstruction of biomass feedstocks. Ransom et al. (2007) did a similar study using TSP extracts from corn leaves expressing the E1 catalytic domain supplemented with commercial β-glucosidase on AFEX-pretreated corn stover as a feedstock and demonstrated that sugar production increased with increasing TSP concentration. Jung et al. (2010) also showed that crude extracts from transgenic tobacco producing β-glucosidase (BglB), when mixed with extracts from transgenic tobacco producing Cel5A, could hydrolyze cellulose in pretreated rice straw. Verma et al. (2010) demonstrated that cocktails composed of mixing various plant-produced enzymes and cell-wall-modifying proteins could be used to produce fermentable sugars from pine wood and citrus waste.

Reuse of enzymes is important to the overall economic feasibility of biological methods for lignocellulosic conversion to biofuels. Liu et al. (1997) showed that a xylanase fused to oleosin (proteins that localize to the periphery of oil bodies) successfully targeted the enzyme to oil bodies where it retained its activity. This allowed the use of the enzyme in an "immobilized" oil body format, and they also showed that the enzymes could also be recycled through floatation centrifugation and reused several times.

4.4 CONCLUSIONS

There has been significant progress in the last 10–15 years in the development of plant-based systems for production of cell-wall-degrading enzymes and cell-wall-modifying proteins. Various expression technologies, plant promoters, plant hosts, and subcellular targeting methods have been investigated. It is clear that functional enzymes can be produced in plants in such a way that they do not harm the plant and that crude protein extracts can be used for lignocellulose hydrolysis. In the short term, these approaches are likely to lead to more efficient, cost-effective, and environmentally friendly sources of enzymes for biofuel production and in the longer term they could lead to triggered self-deconstructing bioenergy crops.

REFERENCES

Azhakanandam K, Weissinger KM, Nicholson JS, Qu RD, Weissinger AK (2007) Amplicon-plus Targeting Technology (APTT) for rapid production of a highly unstable vaccine protein in tobacco plants. *Plant Mol Biol* 64:619–619

Bae, H-J, Kim, HJ, Kim, YS (2008) Production of a recombinant xylanase in plants and its potential for pulp biobleaching applications. *Bioresource Technology* 99: 3513–3519

Biswas, GCG, Ransom C, Sticklen M (2006) Expression of biologically active *Acidothermus cellulolyticus* endoglucanase in transgenic maize plants. *Plant Sci* 171:617–623

Boetti H, Chevalier L, Denmat LA, Thomas D, Thomasset B (1999) Efficiency of physical (light) or chemical (ABA, tetracycline, CuSO4 or 2-CBSU)-stimulus-dependent gus gene expression in tobacco cell suspensions. *Biotechnol Bioeng* 64:1–13

Dai ZY, Hooker BS, Anderson DB, Thomas SR (2000a) Expression of *Acidothermus cellulolyticus* endoglucanase E1 in transgenic tobacco: Biochemical characteristics and physiological effects. *Transgen Res* 9:43–54

Dai Z, Hooker BS, Anderson DB, Thomas SR (2000b) Improved plant-based production of E1 endoglucanase using potato: Expression optimization and tissue targeting. *Mol Breed* 6:277–285

Dai Z, Hooker BS, Quesenberry RD, Gao J (1999) Expression of *Trichoderma reesei* exocellobiohydrolase I in transgenic tobacco leaves and calli. *Appl Biochem Biotechnol* 77–79:689–699

Dai ZY, Hooker BS, Quesenberry RD, Thomas S (2005), Optimization of *Acidothermus cellulolyticus* endoglucanase (E1) production in transgenic tobacco plants by transcriptional, post-transcription and post-translational modification. *Transgen Res* 14:627–643

EBAMM (2007) *Energy and Resources Group Biofuel Analysis Meta-Model.* University of California, Berkeley, CA

Fischer R, Stoger E, Schillberg S, Christou P, Twyman RM (2004) Plant-based production of biopharmaceuticals. *Curr Opin Plant Biol* 7:152–158

Giritch, A, Marillonnet S, Engler C, van Eldik G, Botterman J, Klimyuk V, Gleba Y (2006) Rapid high-yield expression of full-size IgG antibodies in plants coinfected with noncompeting viral vectors. *Proc Natl Acad Sci USA* 103:14701–14706

Gleba Y, Klimyuk V, Marillonnet S (2005) Magnifection—A new platform for expressing recombinant vaccines in plants. *Vaccine* 23:2042–2048

Gleba Y, Klimyuk V, Marillonnet S (2007) Viral vectors for the expression of proteins in plants. *Curr Opin Biotechnol* 18:134–141

Gray BN, Ahner BA, Hanson MR (2008) High-level bacterial cellulase accumulation in chloroplast-transformed tobacco mediated by downstream box fusions. *Biotechnol Bioeng* 102:1045–1054

Ha SJ, Kim SY, Seo JH, Oh DK, Lee JK (2007) Optimization of culture conditions and scale-up to pilot and plant scales for coenzyme Q(10) production by *Agrobacterium tumefaciens*. *Appl Microbiol Biotechnol* 74:974–980

Herbers K, Flint HJ, Sonnewald U (1996) Apoplastic expression of the xylanase and beta(1-3,1-4) glucanase domains of the xyn D gene from *Ruminococcus flavefaciens* leads to functional polypeptides in transgenic tobacco plants. *Mol Breed* 2:81–87

Herbers K, Wilke I, Sonnewald U (1995) A thermostable xylanase from *Clostridium thermocellum* expressed at high-levels in the apoplast of transgenic tobacco has no detrimental effects and is easily purified. *Bio/Technology* 13:63–66

Hood EE, Love R, Lane J, Bray J, Clough R, Pappu K, Drees C, Hood KR, Yoon S, Ahmad A, Howard JA (2007) Subcellular targeting is a key condition for high-level accumulation of cellulase protein in transgenic maize seed. *Plant Biotechnol J* 5:709–719

Horvath H, Huang J, Wong O, Kohl E, Okita, Kannangara CG, von Wettstein D (2000) The production of recombinant proteins in transgenic barley grains. *Proc Natl Acad Sci USA* 97:1914–1919

Huang J, Sutliff TD, Wu L, Nandi S, Benge K, Terashima M, Ralston AH, Drohan W, Huang N, Rodriguez RL (2001) Expression and purification of functional human alpha-1-Antitrypsin from cultured plant cells. *Biotechnol Progr* 17:126–133

Hyunjong B, Lee DS, Hwang I (2006) Dual targeting of xylanase to chloroplasts and peroxisomes as a means to increase protein accumulation in plant cells. *J Exp Bot* 57:161–169

Jensen LG, Olsen O, Kops O, Wolf N, Thomsen KK, Von Wettstein D (1996) Transgenic barley expressing a protein-engineered, thermostable (1,3-1,4)-beta-glucanase during germination. *Proc Natl Acad Sci USA* 93:3487–3491

Jeoh T, Michener W, Himmel ME, Decker SR, Adney WS (2008) Implications of cellobiohydrolase glycosylation for use in biomass conversion. *Biotechnol Biofuels* 1:10

Jin R, S Richter, Zhong R, Lamppa GK (2003) Expression and import of an active cellulase from a thermophilic bacterium into the chloroplast both in vitro and in vivo. *Plant Mol Biol* 51:493–507

Jung S, Kim S, Bae H, Lim SB, Bae HJ (2010) Expression of thermostable bacterial beta-glucosidase (BglB) in transgenic tobacco plants. *Bioresour Technol* 101:7144–7150

Kim S, Lee DS, Choi IS, Ahn SJ, Kim YH, Bae HJ (2010) *Arabidopsis thaliana* Rubisco small subunit transit peptide increases the accumulation of *Thermotoga maritima* endoglucanase Cel5A in chloroplasts of transgenic tobacco plants. *Transgen Res* 19:489–497

Kimura T, Mizutani T, Tanaka T, Koyama T, Sakka K, Ohmiya K (2003) Molecular breeding of transgenic rice expressing a xylanase domain of the *xynA* gene from *Clostridium thermocellum*. *Appl Microbiol Biotechnol* 62:374–379

Klement Z (1965) Method of obtaining fluid from the intercellular spaces of foliage and the fluid's merit as substrate for phytobacterial pathogens. *Phytopathol Notes* 55:1033–1034

Lindbo JA (2007) TRBO: A high efficiency tobacco mosaic virus RNA based overexpression vector. *Plant Physiol* 145:1232–1240

Liu JH, Selinger LB, Cheng KJ, Beauchemin KA, Moloney MM (1997) Plant seed oil-bodies as an immobilization matrix for a recombinant xylanase from the rumen fungus *Neocallimastix patriciarum*. *Mol Breed* 3:463–470

Mackenzie SA (2005) Plant organellar protein targeting: A traffic plan still under construction. *Trends Cell Biol* 15:548–554

Marillonnet S, Thoeringer C, Kandzia R, Klimyuk V, Gleba Y (2005) Systemic *Agrobacterium tumefaciens*-mediated transfection of viral replicons for efficient transient expression in plants. *Nat Biotechnol* 23:718–723

McDonald KA, Dandekar AM, Falk BW, Lindenmuth BE (2009) Production of Cellulase Enzymes in Plant Hosts Using Transient Agroinfiltration. U.S. Patent Application PCT/US2009/054359, WO 2010/022186 A1

Mei C, Park SH, Sabzikar R, Qi C, Ransom C, Sticklen M (2009) Green tissue-specific production of a microbial endo-cellulase in maize (*Zea mays* L.) endoplasmic-reticulum and mitochondria converts cellulose into fermentable sugars. *J Chem Technol Biotechnol* 84:689–695

Nuutila AM, Ritala A, Skadsen RW, Mannonen L, Kauppinen V (1999) Expression of fungal thermotolerant endo-1,4-beta-glucanase in transgenic barley seeds during germination. *Plant Mol Biol* 41:777–783

Oraby H, Venkatesh B, Dale B, Ahmad R, Ransom C, Oehmke J, Sticklen M (2007) Enhanced conversion of plant biomass into glucose using transgenic rice-produced endoglucanase for cellulosic ethanol. *Transgen Res* 16:739–749

Padidam M (2003) Chemically regulated gene expression in plants. *Curr Opin Plant Biol* 6:169–177

Patel M, Johnson JS, Brettell RIS, Jacobsen J, Xue GP (2000) Transgenic barley expressing a fungal xylanase gene in the endosperm of the developing grains. *Mol Breed* 6:113–123

Perlack RD, Wright LL, Turhollow AF, Graham RL, Stokes BJ, Erbach DC (2005) *Biomass as Feedstock for a Bioenergy and Bioproducts Industry: The Technical Feasibility of a Billion-Ton Annual Supply.* U.S. Department of Energy and U.S. Department of Agriculture, Washington, DC

Plesha MA, Huang TK, Dandekar AM, Falk BW, McDonald KA (2007) High-level transient production of a heterologous protein in plants by optimizing induction of a chemically inducible viral amplicon expression system. *Biotechnol Progr* 23:1277–1285

Plesha MA, Huang TK, Falk BW, Dandekar AM, McDonald KA (2009) Optimization of the bioprocessing conditions for scale-up of transient production of a heterologous protein in plants using a chemically inducible viral amplicon expression system. *Biotechnol Progr* 25:722–734

Ransom C, Balan V, Biswas G, Dale B, Crockett E, Sticklen M (2007) Heterologous *Acidothermus cellulolyticus* 1,4-beta-endoglucanase E1 produced within the corn biomass converts corn stover into glucose. *Appl Biochem Biotechnol* 137:207–219

Saez JC, Schell DJ, Thorlunder A, Farmer J, Hamilton J, Colucci, JA, McMillan JM (2002) Carbon mass balance evaluation of cellulase production on soluble and insoluble substrates. *Biotechnol Progr* 18:1400–1407

Sainz MB (2009) Commercial cellulosic ethanol: The role of plant-expressed enzymes. *In Vitro Cell Dev Biol-Plant* 45:314–329

Schmer, M R, Vogel KP, Mitchell RB, Perrin RK (2008) Net energy of cellulosic ethanol from switchgrass. *Proc Natl Acad Sci USA* 105:464–469

Scholthof HB, Scholthof KBG, Jackson AO (1996) Plant virus gene vectors for transient expression of foreign proteins in plants. *Annu Rev Phytopathol* 34:299–323

Sticklen MB (2008) Plant genetic engineering for biofuel production: Towards affordable cellulosic ethanol. *Nat Rev Genet* 9:433–443

Sudarshana MR, Plesha MA, Uratsu SL, Falk BW, Dandekar AM, Huang TK, McDonald KA (2006) A chemically inducible cucumber mosaic virus amplicon system for expression of heterologous proteins in plant tissues. *Plant Biotechnol J* 4:551–559

Sun Y, Cheng JJ, Himmel ME, Skory CD, Adney WS, Thomas SR, Tisserat B, Nishimura Y, Yamamoto YT (2007) Expression and characterization of *Acidothermus cellulolyticus* E1 endoglucanase in transgenic duckweed Lemna minor 8627. *Bioresour Technol* 98:2866–2872

Tang W, Luo XY, Samuels V (2004) Regulated gene expression with promoters responding to inducers. *Plant Sci* 166:827–834

Taylor II LE, Dai Z, Decker SR, Brunecky R, Adney WS, Ding SY, Himmel ME (2008) Heterologous expression of glycosyl hydrolases in planta: A new departure for biofuels. *Trends Biotechnol* 26:413–424

Teymouri F, Alizadeh H, Laureano-Perez L, B Dale, Sticklen M (2008) Effects of ammonia fiber explosion treatment on activity of endoglucanase from *Acidothermus cellulolyticus* in transgenic plant. *Appl Biochem Biotechnol* 113–116:1183–1191

Tzfira T, Citovsky V (2002) Partners-in-infection: Host proteins involved in the transformation of plant cells by *Agrobacterium*. *Trends Cell Biol* 12:121–129

U.S. DOE (2009) Bioenergy research centers: An overview of the science. Genome Management Information Systems, Oak Ridge National Laboratory

Vaucheret H, Beclin C, Fagard M (2001) Post-transcriptional gene silencing in plants. *J Cell Sci* 114:3083–3091

Vaucheret H, Fagard M (2001) Transcriptional gene silencing in plants: Targets, inducers and regulators. *Trends Genet* 17:29–35

Verma D, Kanagaraj A, Jin R, Singh ND, Kolattukudy PE, Daniell H (2010) Chloroplast-derived enzyme cocktails hydrolyse lignocellulosic biomass and release fermentable sugars. *Plant Biotechnol J* 8:332–350

Voinnet OS, Mestre RP, Baulcombe D (2003) An enhanced transient expression system in plants based on suppression of gene silencing by the p19 protein of tomato bushy stunt virus. *Plant J* 33:949–956

Wei S, Marton I, Dekel M, Shalitin D, Lewinsohn M, Bravdo BA, Shoseyov O (2004) Manipulating volatile emission in tobacco leaves by expressing Aspergillus niger β-glucosidase in different subcellular compartments. *Plant Biotechnology Journal* 2:341–350

Wolt JD (2009) Advancing environmental risk assessment for transgenic biofeedstock crops. *Biotechnol Biofuels* 2:27

Wroblewski TA, Tomczak A, Michelmore R (2005) Optimization of *Agrobacterium*-mediated transient assays of gene expression in lettuce, tomato and Arabidopsis. *Plant Biotechnol J* 3:259–273

Xue GP, Patel M, Johnson JS, Smyth DJ, Vickers CE (2003) Selectable marker-free transgenic barley producing a high level of cellulase (1,4-beta-glucanase) in developing grains. *Plant Cell Rep* 21:1088–1094

Yang P, Wang Y, Bai Y, Meng K, Luo H, Yuan T, Fan Y, Yao B (2007) Expression of xylanase with high specific activity from *Streptomyces olivaceoviridis* A1 in transgenic potato plants (*Solanum tuberosum* L.). *Biotechnol Lett* 29:659–667

Yu LX, Gray BN, Rutzke CJ, Walker LP, Wilson BD, Hanson MR (2007) Expression of thermostable microbial cellulases in the chloroplasts of nicotine-free tobacco. *J Biotechnol* 131:362–369

Ziegelhoffer T, Raasch JA, Austin-Phillips S (2001) Dramatic effects of truncation and sub-cellular targeting on the accumulation of recombinant microbial cellulase in tobacco. *Mol Breed* 8:147–158

Ziegelhoffer T, Will J, Austin-Phillips S (1999) Expression of bacterial cellulase genes in transgenic alfalfa (*Medicago sativa* L.), potato (*Solanum tuberosum* L.) and tobacco (*Nicotiana tabacum* L.). *Mol Breed* 5:309–318

Ziegler MT, Thomas SR, Danna KJ (2000) Accumulation of a thermostable endo-1,4-beta-D-glucanase in the apoplast of *Arabidopsis thaliana* leaves. *Mol Breed* 6:37–46

Zuo, JR, Chua NH (2000) Chemical-inducible systems for regulated expression of plant genes. *Curr Opin Biotechnol* 11:146–151

5 From Plant Cell Walls to Biofuels—*Arabidopsis thaliana* Model

Sivakumar Pattathil, Utku Avci, and Ajaya Kumar Biswal
University of Georgia

Ajay K. Badhan
University of Alberta

CONTENTS

5.1 BIOFUEL PRODUCTION AND CHALLENGES

5.1.1 NEED FOR BIOFUELS

Recently, interest in biofuels has increased because of socioeconomic and environmental concerns. Because biofuel production from plant material is a more carbon dioxide (CO_2)-neutral source of energy, biofuels are an environmentally friendly fuel alternative. The fundamental source for bioenergy lies in the unique potential of plants to harvest light energy from the sun through photosynthesis and to use that energy to capture atmospheric CO_2 and fix it into the plant biomass. Under optimal growth conditions, *Miscanthus gigantious* could harvest more than 2% of annual incident solar radiation (Beale and Long 1995; Somerville 2007). By cultivating miscanthus on approximately 3.2% of the terrestrial surface area, Somerville (2007) again states that given the average rate of solar radiation [i.e., 120,000 terrawatts (TW)], 2% solar conversion efficiency, and an energy recovery value of 50%, all human energy needs could be met by biofuels (at the usage level of 11.73 TW). Thus, suitable plant crops can be grown on a large scale for capturing and storing solar energy efficiently. Again, this makes plant biomass a carbon neutral energy source for the production of fuels.

Chow et al. (2003) reported that nearly all (~95%) of the global energy market [370 exajoules of energy per year is equivalent to approximately 170 million barrels of oil per day or about 11.73 TW per hour (Chow et al. 2003; Somerville 2007)] comes from fossil fuels. The International Energy Agency (IEA) also suggested that the use of plant biomass could meet approximately one-third of the energy demands in Africa, Asia, and Latin America, and up to 80–90% in the poorest countries of these regions (Chow et al. 2003; Somerville 2007). According to a recent estimate by Somerville (2007), biofuels currently supply approximately one-tenth of all human energy consumption. The net CO_2 assimilation by land plants per year is approximately 56×10^9 t (Field et al. 1998; Pauly and Keegstra 2008), and the biomass production by land plants worldwide is $170–200 \times 10^9$ t (Lieth 1975; Pauly and Keegstra 2008). Approximately three-fourths of this quantity is estimated to be plant cell wall biomass (Duchesne and Larson 1989; Poorter and Villar 1997; Pauly and Keegstra 2008). Currently, humans use only a negligible part (~2%) of plant cell-wall-based biomass in the form of wood for heat production, timber for building materials, pulp in the paper industry (Fenning and Gershenzon 2002; Pauly and Keegstra 2008), and as raw material in the textile industry. Moreover, plant cell-wall-based biomass does not serve as food for animals and humans to the extent that starch does. Therefore, interest in using this plant cell wall biomass resource as a material for biofuels has increased substantially in recent years (Schubert 2006; Pauly and Keegstra 2008).

5.1.2 CURRENT AND POTENTIAL RAW MATERIALS FOR BIOFUEL PRODUCTION

The preferred properties for the ideal bioenergy crop are quite different from those of a food crop; carbon content, rather than nitrogen content, is the trait of interest when harvesting biomass for energy (Porter et al. 2007). Biomass crops such as willow and elephant grass (*Miscanthus* spp.) have approximately one-tenth of the nitrogen concentration of cereal crops (1.5% willow and 2.5% elephant grass of grain dry weight), thus avoiding the requirement of a large supply of nitrogen fertilizer (Porter et al. 2007). Additionally, high levels of low-molecular-weight carbohydrates in unpolymerized states are a requirement for fermenting biomass crops so as to reduce energy inputs in the refining process (Porter et al. 2007; Pauly and Keegstra 2008). Furthermore, to attain better water-use efficiency, leaf size and shape in a biomass crop should be effective in competition against weeds (Porter et al. 2007). The use of perennial biomass crops to attain better weed management has been reported previously (Buhler et al. 1998). Perennials rather than annuals would be advantageous in reducing crop-establishment energy costs. However, because perennial crops cannot be routinely substituted in a plantation, field resistance against diseases and pests should be given utmost priority (Porter et al. 2007). Therefore, cultivating willows and perennial grasses for energy production will be beneficial.

Use of crops like corn and cane in the bioethanol industry in the developed world is well documented. However, the cost-effectiveness of this approach has always been a concern because of the enormous use of water and heat (Somerville 2007). In one interesting report, Lal (2008) summarized an update on plants that can be used for biofuel production that is based on their efficient growth in tropical conditions and cost-effectiveness. These include, according to the list from Lal (2008), biomass from the following groups of plants with examples given in parentheses:

- *Warm season grasses*: Switchgrass (*Panicum virgatum* L.), big bluestem (*Andropogan gerardi* Vitman), big bluestem (*Andropogan gerardi* Vitman), Indian grass (*Sorghastrum nutans* (L.) Nas), giant reed (*Arundo donax*), bluejoint grass (*Calamagrostis canadensis* (Michx.) Beau. L.), bluejoint grass (*Calamagrostis canadensis* (Michx.) Beau. L.), cord grass (*Spartina pectinata* Link), kallar grass (*Leptochloa fusca*), guinea grass (*Panicum maximum*), setaria (*Setaria sphcelate*), molasses grass (*Melinis minutiflora*), and elephant grass (*Pennisetum purpureum* Schm.);
- *Legumes*: Alfalfa (*Medicago sativa*), mucana (*Mucuna utilis*), kudzu (*Pueraria phaseoloides*), and stylo (*Stylosanthes guianensis*);
- *Broad leaf species*: Cup plant (*Silphium perfoliatum* L.);
- *Short rotation woody perennials*: Poplar (*Populus* spp.), willow (*Salix* spp.), black locust (*Robinia pseudoacacia* L.), mesquite (*Prosopis juleflora*), birch (*Onopordum nervosum*), and eucalyptus (*Eucalyptus* spp.); and
- *Herbaceous* spp.: Miscanthus (*Miscanthus* spp.), reed canary grass (*Phalaris arundinacea* L.), and cynara (*Cynara cardunculus*).

5.1.3 Challenges in Biofuel Production from Plant Cell Walls

During the process of evolution, plants have developed efficient mechanisms for resisting attack on their cell walls from the microbial and animal kingdoms. This intrinsic property underlies what has been termed "recalcitrance" (the innate resistance of plant cell walls to deconstruction). This recalcitrance creates technical barriers to the cost-effective transformation of lingo-cellulosic biomass into fermentable sugars. According to Himmel et al. (2007), there are several factors that cause the recalcitrance of lingo-cellulosic feedstocks against their deconstruction. Most of these factors are directly associated with properties of plant cells and cell walls, such as presence of cuticle and epicuticular waxes on the epidermal tissue of the plant body, the vascular bundle arrangement and density, the relative amount of thick-walled tissues (sclerenchymatous), lignification, and the structural heterogeneity and complexity of cell wall components such as microfibrils and matrix polymers. Recalcitrant biomass with these structural and chemical properties often exhibits reduced liquid permeability and/or enzyme accessibility and activity posing higher conversion costs (Himmel et al. 2007). Cell wall microfibrils that contain a core of crystalline cellulose exhibit elevated resistance to chemical and biological hydrolysis (Nishiyama et al. 2002), likely because of biophysical qualities such as the formation of a dense layer of water near the hydrated cellulose surface, which contributes by the hydrophobic face of cellulose sheets (Matthews et al. 2006), and the presence of strong interchain hydrogen-bonding networks in them (Nishiyama et al. 2002). In contrast, hemicellulose and amorphous cellulose lack similar features and are easily digestible. Another contributory factor to plant cell wall recalcitrance is its highly complex structure: the crystalline cellulose core of microfibrils is protected by coatings of hemicelluloses and amorphous cellulose, making the microbial accessibility to the crystalline cellulose difficult (Himmel et al. 2007). Overall, of all the aforementioned factors, the most salient contributing to plant cell wall recalcitrance are the occurrence of lignin, waxes, and phenolic compounds as well as the multilayered complex structure.

5.2 PLANT CELL WALL

5.2.1 Structural Overview of Plant Cell Walls

The cell wall provides a plant cell with protection, structure, and shape. Physiological functions of the cell wall are imperative in facilitating the normal growth and development of a plant cell; for example, they are significant in cell-to-cell interactions, generation of biologically active signaling molecules, regulation of diffusion of materials through apoplast, storage of carbohydrate reserves, and protection from biotic and abiotic stresses. Therefore, research on the understanding of cell wall biosynthesis is an important aspect of basic plant research. The understanding of plant cell walls is also imperative for improving human life because it is a key source for dietary fibers, the textile/paper industries, and the growing interest in the future biofuel industry. Our understanding of plant cell wall structure and biosynthesis is becoming clearer with the onset of technological advancements (Roberts 2001). Generally, plant cell walls are classified as type I and type II cell walls. Type I plant cell walls are characteristic of dicots and noncommelinoid monocots and are mainly composed of significant proportions of pectic polysaccharides and hemicellulosic polysaccharide, xyloglucans (Pauly and Keegstra 2008). In contrast, arabinoxylan is the main hemicellulose in type II walls (i.e., the walls of the Poales, which include grasses) (Carpita 1996). In addition, type II walls have an increased proportion of cellulose and only lesser amounts of pectins and proteins (Carpita 1996).

From a structural and functional point of view, there are two classes of plant cell walls: primary (essentially present in growing and dividing cells) and secondary cell walls (essentially present in woody tissues). As a part of growing and dividing cells, primary cell walls have to be flexible and plastic to contain the cell growth (Cosgrove 2003). The first well-defined model for the structure of primary cell walls came from the work of Peter Albersheim and his colleagues (Keegstra et al. 1973). Progress in research since that first model has allowed us to reach a much more detailed understanding of the macromolecular framework that makes up the plant cell wall, although many details are still missing. The primary plant cell wall contains numerous interconnected matrices composed of diverse polysaccharides and (glyco) proteins. The polysaccharides in these matrices include cellulose microfibrils, hemicelluloses (e.g., xyloglucan, xylan, and galactomannans), and pectins. Cellulose microfibrils are noncovalently cross-linked by hemicelluloses. This intricate network of cellulose and associated hemicelluloses are embedded in the pectic matrix, which is formed of several kinds of pectins. Pectins are yet another important class of plant cell wall molecules that are primarily composed of homogalacturonan, rhamnogalacturonan I (RG-I), and rhamnogalacturonan II (RG-II) (O'Neill and York 2003).

Mature plant cells develop secondary walls. This wall is much stronger and accounts for most of the carbohydrate in biomass. The secondary walls also contain primary wall components such as cellulose, hemicelluloses, and pectins. However, their relative proportion may vary. For instance, along with increased cellulose content, xylan also accounts for up to 30% of the mass in secondary walls of wood and grasses. It is this increased cellulose content that makes secondary walls a good target for the biofuel industry. In secondary walls, lignin largely replaces water, thus making them nearly inaccessible to solutes and degrading enzymes. Lignin is an important component of the cell wall and the main contributor for cell wall recalcitrance against degradation. Once incorporated, lignin further strengthens the secondary walls in most tissues. In brief, secondary walls are composed of cellulose microfibrils embedded in cross-linking hemicelluloses that are strengthened by lignin. The major hurdle for producing bioethanol out of plant cell walls is making the cellulose impregnated in xylan and lignin accessible to degrading enzymes such as cellulases.

5.2.2 *Arabidopsis* Cell Wall Research and Current Understanding of Plant Cell Walls

Much of our increased understanding of plant cell wall structure and biosynthesis has benefited from the complete sequencing of the genomes of several plants, but most notably of *Arabidopsis thaliana*.

This increased knowledge about plant cell walls is of direct benefit to the biofuel industry because detailed understanding of wall structure and biosynthesis will be necessary to optimize procedures for cell wall deconstruction into fermentable sugars. The next section of this review will provide an overview of these advancements by individually covering the major components of plant cell walls.

5.2.3 LIGNIN

Lignin is a plant cell-wall-associated complex polymer composed of aromatic subunits. It is usually derived from phenylalanine (Whetten and Sederoff 1995). Lignin is present in most vascular plants and plays significant roles in mechanical support, transportation of water, and defense against pathogens (Lewis and Yamamoto 1990; Whetten and Sederoff 1995; Douglas 1996). Lignin biosynthesis regulation is dependent on the plant's growth and development. Thus, deposition of lignin is usually limited to specific plant cell types. Some good examples for these cell types are tracheary elements in the xylem and sclerenchyma, where lignin is usually deposited in the secondary thickened walls. Lignin biosynthesis is significantly scaled up in woody plants, where secondary xylem constitutes most of the plant body.

There are three monomeric lignin subunits, namely *p*-hydroxyphenyl (H), guaiacyl (G), and syringyl (S) subunits. The relative amounts these monomeric subunits can vary depending on the plant species and cell type. Lignin composition varies with plant types, with the G and S monomers usually predominating in most lignin. For example, most monomer subunits constituting lignin in dicotyledonous angiosperms are G and S, whereas in conifers, lignin consists almost entirely of G units (Lee et al. 1997). G and S monomer composition also varies according to the cell types and the developmental stage of the tissue (Whetten and Sederoff 1995). Tracheary elements in vascular bundles of *A. thaliana* contain lignin that is mainly composed of G subunits, whereas the bordering heavily lignified sclerenchyma cells contain substantial amounts of S subunits (Chapple et al. 1992). Thus, one of the major challenges in lignin biosynthesis research is unraveling the mechanisms that regulate differential carbon flow during the synthesis of these different lignin precursors and subsequent synthesis of lignin polymers of differing subunit composition.

Cloning and biochemical characterization of most major lignin biosynthetic enzymes have been reported previously by several groups (Lewis and Yamamoto 1990; Whetten and Sederoff 1995; Boudet and Grima-Pettenati 1996; Campbell and Sederoff 1996; Douglas 1996). A detailed scheme for lignin biosynthesis by these enzymes has been previously reported (Whetten and Sederoff 1995). Table 5.1 lists the major lignin biosynthetic enzymes that current research has focused on and their proposed functions.

TABLE 5.1

Major Lignin Biosynthetic Enzymes Focused in *Arabidopsis thaliana* Research

Enzyme	Abbreviation	Function
Cinnamate-4-hydroxylase	C4H	Hydroxylation
Cinnamate-3-hydroxylase	C3H	Hydroxylation
Ferulate-5-hydroxylase	F5H	Hydroxylation
Caffeic acid/5-hydroxyferulic acid *O*-methyl-transferase	COMT	Methylation
4-Coumarate:coenzymeA ligase	4CL	Ferulic acid/Sinapic acid reduction step
Cinnamyl-CoA reductase	CCR	Reduction
Cinnamyl alcohol dehydrogenase	CAD	Reduction
Caffeoyl-CoA-*O*-methyl transferase	CCoOMT	Methylation
Coumaroyl-CoA-3-hydroxylase	CCo3H	Hydroxylation

The pathways, models, and mechanisms in lignin biosynthesis were delineated by earlier studies on phenylpropanoid metabolism by dynamic groups such as Hahlbrock's laboratory in Germany and others as reviewed earlier (Hahlbrock and Scheel 1989). Recent studies with mutants in the *A. thaliana* lignin biosynthesis pathway have revised and refined these models.

Cinnamate-4-hydroxylase (C4H) and ferulate-5-hydroxylase (F5H) are the two cytochrome P450 (CytP450)-dependent monooxygenases that catalyze hydroxylation reactions in the plant phenylpropanoid pathway (Meyer et al. 1996). The first CytP450-dependent monooxygenase of the phenylpropanoid pathway in *Arabidopsis* is C4H. The expression patterns of *C4H* had been thoroughly studied earlier in *A. thaliana* (Bell-Lelong et al. 1997). These studies reported that *C4H* is extensively expressed in different tissues of *Arabidopsis*. *C4H* expression is specifically high in roots and cells undergoing lignification (Bell-Lelong et al. 1997). Furthermore, accumulation of C4H message (mRNA level), according to these studies, was observed to be dependent on light. Another interesting observation made by Bell-Lelong et al. (1997) is the sequence similarity of several putative regulatory motifs in the *C4H* promoter with motifs in promoters of other phenylpropanoid pathway genes. More recently, Chen et al. (2007) successfully cloned two genes encoding C4H from oilseed rape (*Brassica napus*) using sequence identity analysis of the corresponding *Arabidopsis C4H* gene.

Studies show that the expression of the other CytP450-dependent monooxygenase, F5H (that catalyzes an irreversible step of the hydroxylation reaction in the lignin biosynthesis pathway redirecting ferulic acid away from G lignin biosynthesis to sinapic acid and S lignin), does have an influence on the monomeric composition of lignin. Evidence of this was provided by ectopic F5H overexpression studies in *A. thaliana* (Meyer et al. 1998). It was found that tissue specificity of lignin monomer accumulation was negatively influenced by ectopic F5H overexpression (Meyer et al. 1998). Again, when F5H was overexpressed under the control of a lignification-linked promoter (C4H promoter) instead of the more routinely used 35S constitutive promoter, a lignin with almost entirely syringylpropane units was made (Meyer et al. 1998). Structural studies using nuclear magnetic resonance (NMR) imaging on isolated lignins from *Arabidopsis* mutants deficient in F5H (*f5h*) and *f5h* mutants overexpressing the *F5H* gene were done earlier (Marita et al. 1999) and confirmed the compositional and structural differences between these isolated lignins. These studies indicate that altering F5H expression in biofuel crops could result in plants containing lignin with an altered composition that might possess less recalcitrant traits.

Another phenylpropanoid pathway enzyme that has not been fully functionally characterized in plants yet is *p*-coumarate-3-hydroxylase (C3H) or CYP98A3. Franke et al. (2002) reported that lack of C3H function resulted in the reduced epidermal fluorescence phenotype ("ref" phenotype). These mutants were called *ref8* mutants. The *ref8* plants contained lignin with significantly altered composition causing developmental defects in plants and susceptibility to fungal pathogen attack (Franke et al. 2002). These results therefore suggest that the phenylpropanoid pathway products downstream of C3H do hold significance in normal plant development and disease resistance.

4-Coumarate:coenzymeA ligase (4CL) is a critical enzyme in the lignin biosynthesis pathway that is involved in the synthesis of precursor molecules such as CoA thiol esters of 4-coumarate and additional hydroxycinnamates. A *4CL* cDNA was first cloned from Parsley using enzyme purification and a cDNA library screening approach (Ragg et al. 1981). Studies on stress and developmentally regulated expression of the *A. thaliana* 4-coumarate:CoA ligase (*4CL*) gene and the identification of its cDNA sequence opened the way to the discovery of other family members of this gene family (Lee et al. 1995). Subsequently, Ehlting et al. (1999) reported the cloning and characterization of three members of the *4CL* gene family (namely *At4CL1*, *At4CL2*, and *At4CL3*) from *A. thaliana*. Their studies further established that these three members belonged to two divergent evolutionary classes and encode isozymes with different substrate preferences and specificities. More recently, Hamberger and Hahlbrock (2004) identified the fourth and final member of the *At4CL* gene family that encodes the enzyme At4CL4. According to these studies, At4CL4 might be carrying out a distinct metabolic function because this enzyme seems to be efficiently activating sinapate in addition to the regular 4CL substrates such as 4-coumarate, caffeate, and ferulate. Studies are underway to fully elucidate the functions of all four At4CL gene family and other genes with homologous sequences in *Arabidopsis*. A notable study in this regard is the expression and functional characterization of 11 genes (including the four At4CL members) that were identified to be a putative multigene 4-coumarate:CoA ligase network that could be taking part in the formation of syringyl lignin and a sinapate/sinapyl alcohol derivative (Costa et al. 2005). Another enzyme functional characterization study by Ehlting et al. (2001) investigated the domain structures in At4CL1 and two isoforms and the functions of these domains. These studies led to the identification of two adjoining putative functional domains, substrate-binding domains I (sbd I) and II (sbd II), in these isoforms and to the understanding of their roles in substrate recognition and binding. Few studies have been done toward elucidating the expression patterns of the *A. thaliana 4CL* gene family members except for the report by Soltani et al. (2006), which describes the regulatory roles played by multiple cis-regulatory elements in complex patterns of 4CL expressions developmentally or up on induction by wounds.

The first step in lignin monomer synthesis in the lignin branch biosynthetic pathway is catalyzed by cinnamoyl CoA reductase (CCR). Isolation of CCR cDNA was first carried out from *Eucalyptus gunnii* (Boudet and Grima-Pettenati 1996). The same group later carried out its cloning, expression, and in-depth phylogenetic relationship analysis (Lacombe et al. 1997). In *A. thaliana*, two different cDNA clones, namely *AtCCR1* and *AtCCR2*, were reported later (Lauvergeat et al. 2001). AtCCR1 and AtCCR2 proteins shared approximately more than 80% sequence identity. The *A. thaliana* CCR isoform, *AtCCR1*, is engaged in the constitutive lignification process and *AtCCR2* is important for disease resistance in plants by its possible roles in the biosynthesis of phenolics (Lauvergeat et al. 2001). The AtCCR1 *Arabidopsis* mutants [i.e., *irregular xylem4* (*irx4*) mutants] are severely lignin–deficient, with a dramatic wall phenotype that affects the mechanical properties of the stem (Jones et al. 2001). This *Arabidopsis irx4* mutant has also been reported to have a much delayed but normal lignification process (Laskar et al. 2006). More recent studies using *Arabidopsis* transfer DNA (T-DNA) knockout mutants for *CCR1* emphasize phenotype differences in these mutants (Mir Derikvand et al. 2007). Interestingly, the phenylpropanoid pathway in these mutants is redirected to feruloyl malate, resulting in the formation of lignin with altered structure. These results, from a bioenergy research point of view, are significant because they show that the phenylpropanoid metabolism in *Arabidopsis* is flexible, thereby encouraging strategies for developing plants with modified and less recalcitrant lignin structures.

The last step in the biosynthesis of the lignin precursors is the conversion of cinnamaldehydes to the corresponding alcohols. This step is catalyzed by the enzyme cinnamyl alcohol dehydrogenase (CAD). Isolation of the full-length cDNA of CAD was first reported by Walter et al. (1988) from elicitor-treated cell cultures of bean (*Phaseolus vulgaris* L.). Later, using poplar cDNA sequences, Baucher et al. (1995) screened the *Arabidopsis* genome to report the genomic nucleotide sequence of an *AtCAD* gene encoding a CAD. The most popular *Arabidopsis* genome database, The *Arabidopsis* Information Resource (TAIR), annotates approximately 17 genes with putative CAD functions that belong to a CAD multigene family. However, subsequent studies by Kim et al. (2004) functionally reclassified this putative *Arabidopsis* gene family and demonstrated that only 9 of these 17 members could actually be considered as members of the *AtCAD* family. These studies further explored the functions of all nine *AtCADs* and found that the two *Arabidopsis* CAD genes, AtCAD5 and AtCAD4, had the peak activity and shared approximately 83% homology with previously characterized CADs from various plant species. Additionally, functional redundancy of various *AtCADs* in *Arabidopsis* lignin biosynthetic pathways was confirmed using studies with AtCAD5, AtCAD6, and AtCAD9 knockout mutants for various growth and developmental stages (Kim et al. 2004). The reports on such functional redundancy in a set of *CAD* genes thus directs the attempts of making lignin modified plants (that may turn out as less recalcitrant crops suitable to the biofuel industry) by focusing on targeting multiple CAD knockouts to attain a significant change in the lignin composition. More functional studies on *AtCAD4* and *AtCAD5* genes hint as to their roles as main genes in lignin biosynthesis in inflorescence of *A. thaliana* by providing coniferyl and sinapyl alcohols (Sibout et al. 2005). These facts are supported by a limp floral stem phenotype exhibited by the mature *Arabidopsis CAD* double mutant (*cad-4 cad-5*), and these mutants show a modified pattern of lignin staining. These results suggest that *AtCAD4* and *AtCAD5* genes do possess a developmental role in *A. thaliana*. In a recent gene functional analysis study, AtCAD1 was been shown to play a role in the lignification of elongating stems in *Arabidopsis* (Eudes et al. 2006).

5.2.4 CELLULOSE MICROFIBRILS

Secondary xylem or wood accounts for most of the biomass in trees, 42–50% of which is cellulose. Other main components of wood are hemicellulose (25–30%) and lignin (20–25%) (Suzuki et al. 2006). This biomass is one of the major carbon sinks on Earth and provides renewable material that can be used for manufacturing or as an energy source. The latter, in the context of conversion to biofuels, has recently received increased attention, with cellulose being the most abundant and valuable material for conversion processes. Therefore, understanding the biosynthesis of cellulose along with the other cell wall components is very important.

Cellulose is composed of linear chains of β-1,4-linked glucans that are synthesized by cellulose synthase (CesA) complexes at the plasma membrane. Parallel glucan chains can form extensive hydrogen bonds between each other, which gives rise to cellulose microfibrils. Each microfibril is predicted to contain 36 crystalline glucan chains (Mutwil et al. 2008; Taylor 2008). Cellulose microfibril orientation in the cell wall is critical to the strength of the walls (Somerville 2006).

5.2.4.1 Cellulose Synthase Genes

Cellulose synthase genes (*CesAs*) are presumed to encode catalytic subunits of cellulose synthase, which is part of the enzyme complex that is responsible for the synthesis of cellulose. Freeze-fracture electron microscopy of plasma membranes allows visualization of cellulose synthases within rosettes of six intramembrane protein aggregates (Kimura et al. 1999). Each subunit synthesizes β-1,4-glucan chains, which then crystallize to form cellulose microfibrils. These microfibrils provide the main strengthening components of plant cell walls, helping to maintain turgor pressure and controlling the extension direction of a plant cell (Somerville 2006).

The first plant *CesA* gene was identified in cotton (*Gossypium hirsutum*) fiber by subdomain similarity to a bacterial gene with low sequence similarity, and CesA proteins were predicted to be membrane-bound proteins (Pear et al. 1996). Availability of a genomic database and mutant collections in *Arabidopsis* allowed further genetic identification and characterization of *CesA* genes in *Arabidopsis* (Arioli et al. 1998; Somerville et al. 2004). *Arabidopsis* has 10 *CesA* genes (Richmond 2000), some of which are required for primary wall cellulose synthesis (*AtCesA1*, *AtCesA3*, and *AtCesA6*) and some of which are required for secondary wall cellulose synthesis (*AtCesA4*, *AtCesA7*, *AtCesA8*) (Doblin et al. 2002). Although the evidence to date strongly suggests that the CesA proteins are β-1,4-glucosyl transferases, the precise biochemical function and activity of *CesA* genes in what may be a multistep cellulose synthetic process still remain to be unambiguously determined (Doblin et al. 2002).

5.2.4.2 Effects of Mutations on *CesA* Genes

Mutational analysis and recent advancements have given us clues about how cellulose synthesis is regulated. Null mutants of *AtCesA1* and *AtCesA3* are lethal to embryos, indicating they are not redundant with one another (Persson et al. 2007b). Null mutations for *AtCesA6* showed reduced cellulose synthesis and anisotropic cell swelling (MacKinnon et al. 2006). AtCesA2, AtCesA5, and AtCesA9 are partially redundant with AtCesA6 during the different stages of cellular development (Desprez et al. 2007). Together, these data provide evidence that CesA1 and CesA3 are invariable components of the primary wall cellulose synthase complexes where other CesAs (CesA2, 5, 6, 9) may substitute for each other in the complex formation (Desprez et al. 2007).

Irregular xylem mutants *irx1*, *irx3*, and *irx5*, corresponding to genes *AtCesA8*, *AtCesA7*, and *AtCesA4*, respectively, showed collapsed xylem phenotypes and approximately 30% cellulose reduction in secondary walls (Taylor et al. 1999, 2000, 2003). In these mutants, primary walls remained unaffected. All of these single gene mutations resulted in severe phenotypes with defects in the secondary cell wall structures. This indicates that these are nonredundant genes with independent functions for secondary wall synthesis.

Korrigan is a β-1,4-glucanase required for cellulose synthesis in primary and secondary cell walls. It is predicted to be a membrane-bound protein, but the exact localization of the protein is unknown. Several mutants show dwarfism, swelling of root tips, and collapse of xylem vessels (Lane et al. 2001; Sato et al. 2001; Szyjanowicz et al. 2004). In summary, studies to date suggest that Korrigan has a role in the processing of growing microfibrils or the release of cellulose synthase complexes (Szyjanowicz et al. 2004). Null mutants of Cobra, a glycosyl-phosphatidylinositol (GPI)-anchored protein, showed disorganization in the orientation of cellulose microfibrils and reduction in the crystalline cellulose, which makes Cobra a possible regulator of cellulose biosynthesis during cell expansion (Roudier et al. 2005). Kobito mutants (*kob1*) are dwarfed and deficient in cellulose during the elongation phase of the cell. Kobito is a plasma membrane-bound protein, and although overexpression did not result in any phenotype, mutants showed altered microfibril orientation. Pagant et al. (2002) stated that KOB1 may be part of the cellulose synthesis machinery in elongating cells and/or may play a role in the coordination between cellulose synthesis and cell expansion.

In summary, cellulose is considered to be a significant source of biomass raw material for biofuel production because it is rich in carbon. Because cellulose synthesis is highly conserved among plants, using the knowledge obtained from *Arabidopsis* research will likely be applicable to other plant systems. Better understanding of the regulation of cellulose synthesis in *Arabidopsis* will help us recognize one of the key steps in cell wall biosynthesis, which will eventually allow scientists to modify the cell walls for sustainable biofuel production. The knowledge from *Arabidopsis* cellulose biosynthesis research will also help us to focus on nonfood crops instead of food crops (e.g., corn) as target raw materials for bioenergy because many nonfood biomass energy crops can be produced with lower energy input than food crops.

5.2.5 Pectin

5.2.5.1 Pectin Structure

Pectin is a complex group of polysaccharides that contain covalently linked galacturonic acid residues. Homogalacturonan (HG) is the most abundant pectic polysaccharide and is a homopolymer of α-1,4-linked galacturonic acid (GalA). HG comprises approximately 65% of pectin in the primary walls of dicots and nongraminaceous monocots (Mohnen 2008). HG can be acetylated on C_2 or C_3 of the GalA residues; however, the degree of acetylation varies from species to species.

RG-I, another component of pectin, consists of a backbone of alternating GalA and Rha (-α-1,4-GalA-α-1,2-Rha). It represents approximately 20–35% of pectin, and L-rhamnose residues have side chains, which are linear or branched and largely composed of β-D-galactose and α-L-arabinose residues (Mohnen 2008). The galactan branches are mostly linear chains of β-1,4–linked D-galactose residues, whereas arabinan chains include α-1,5 linked L-arabinan residues that are frequently branched with O-2 and O-3 (Nakamura et al. 2002). Nakamura et al. (2002) showed for the first time that pectic polysaccharides RG-I and HG are linked via their backbones. RG-I is usually decorated with arabinogalactan side chains that are thought to be of two basic types. Arabinogalactan I side chains are β-1,4-linked D-galactans chains with β-3-linked L-arabinose or arabinan branches, and highly branched arabinogalactan II side chains are β-1,3-linked D-galactan chains with β-6-linked galactan or arabinogalactan branches (Mohnen 2002). In some species, ferulic acid esters substitute the position of arabinose and galactose residues in RG-I side chains (Fry 1982). Chemical and immunological studies have shown that there is a large variation in RG-I structures found in different groups of plants and even in different tissues of a single plant.

The most complex pectic-polysaccharide is RG-II. Its structure is highly conserved in all vascular plants and it comprises approximately 10% of total pectin (O'Neill et al. 2004). RG-II consists of an HG backbone with side branches consisting of at least 12 different monosaccharides with more than 20 different types of linkages. The monosaccharides found in RG-II include some that are rarely found in other polysaccharides such as 2Me-Fuc, 2Me-Xyl, Dha, and Kdo (2-keto-3-deoxy-D-manno-octulosonic acid), D-apiose, L-aceric acid (3-C-carboxy 5-deoxy-L-xylose) and D-Dha (3-deoxy-D-lyxo-2-heptulosaric acid). RG-II in plant cell walls exists predominantly as dimers that are cross-linked by borate diesters at a 1:2 ratio. Despite its relatively low abundance, RG-II appears to be very important for cell wall structure and function, given that it is highly conserved throughout the plant kingdom (O'Neill et al. 2004) and plants having mutations affecting RG-II structure show growth abnormalities (O'Neill et al. 2001, 2004).

The other substituted galacturonan is xylogalacturonan (XGA). XGA is an HG substituted at O-3 with a β-1,3-linked xylose and this xylose is further substituted at O-4 with an additional β-1,3-linked xylose in *A. thaliana* (Zandleven et al. 2006). XGA has also been detected in cell walls from all *Arabidopsis* tissues with the most predominant expression in flowers (Zandleven et al. 2007). Recently, Henrik Scheller and his group identified a mutant with much lower levels of XGA, called *xgd1* (xylogalacturonan deficient 1), indicating a role of At5g33290 in XGA biosynthesis (Jensen et al. 2008).

5.2.5.2 Localization, Function, and Mechanism of Pectin Biosynthesis

There are several reports suggesting that pectin is biosynthesized in the Golgi vesicles (Staehelin and Moore 1995; Willats et al. 2001) and is then transported to the wall via membrane vesicles. This localization study is now supported by studies showing that RGXT1/2 (Egelund et al. 2006), ARAD1 (arabinan deficient 1; Harholt et al. 2006), XGD1 (xylogalacturonan deficient 1, Jensen et al. 2008) and galacturonosyltransferase 1 (GAUT1; Dunkley et al. 2004) are located in the Golgi vesicles. All published data to date suggest that pectin is synthesized in the Golgi lumen by glycosyltransferases (GTs).

Pectin polysaccharide synthesis occurs in different Golgi cisternae as pectin moves from the cis-, through the medial-, then to the trans-Golgi. Some antibody-based studies suggest that HG and RG-I synthesis begins in the cis-Golgi and moves into the trans-Golgi through the medial Golgi (Staehelin and Moore 1995). It is believed that HG is transported to the plasma membrane and inserted into the wall as a highly methylesterified polymer, where it then goes through various degrees of de-esterification by pectin methylesterases, which convert HG to a more negatively charged form (Pelloux et al. 2007).

5.2.5.3 Genes Involved in HG Biosynthesis

Mohnen and coworkers have successfully proven enzymatically that a HG α-1,4-galacturonosyltransferase (HG:α1,4GalAT) called *GAUT1* (galacturonosyltransferase 1) and found in *Arabidopsis* (Sterling et al. 2006) is involved in HG synthesis. They also showed that *GAUT1* belongs to the GT8 family in the CAZY classification system (www.cazy.org). In *Arabidopsis*, the *GAUT1*-related gene family is made up of 15 *GAUT* genes with 56–100% sequence similarity to *GAUT1* and 10 galacturonosyltransferase-like (*GATL*) genes with 43–53% identity with *GAUT1* (Sterling et al. 2006). The exact role of GAUT1 in pectin synthesis is not clear and is currently under investigation.

Several mutations in genes related to *GAUT1* that affect HG and/or xylan synthesis have been identified in *Arabidopsis*. Although the proteins encoded by these genes are putative pectin biosynthetic glycosyltransferases on the basis of their sequence similarity to *GAUT1*, proof of their enzyme activity is still needed to confirm their role in pectin synthesis. Bouton et al. (2002) isolated two allelic mutants, named *quasimodo1* (*qua1-1* and *qua1-2*). Both of these mutants are dwarfed and show reduced cell adhesion, particularly between epidermal cells in seedlings and young leaves, and show a approximately 25% reduction in GalA content in isolated cell walls, suggesting a possible role for QUA1/GAUT8 in pectin biosynthesis. Independently, Henrik Scheller's group further investigated QUA1 and found for the first time that protein extracts from *qua1* stems have reduced HG:α-1,4-GalAT (~33%) and β-1,4-xylosyltransferase activities (~40%) relative to wild type, suggesting a connection between pectin HG and xylan synthesis (Orfila et al. 2005). The only putative HG-methyltransferase characterized so far is Quasimodo2 (Qua2) (At1g78240) in *Arabidopsis*, which encodes a Golgi-localized protein. The *qua2* mutants showed a reduced cell adhesion, a 50% reduction in HG levels without affecting other cell wall polysaccharides, and are dwarf phenotype-like common pectin mutants. The QUA2 protein contains a putative methyltransferase domain and may function as a HG-methyltransferase (Mouille et al. 2007). However, enzymatic evidence of function is still needed to confirm the role of QUA2 in HG biosynthesis.

5.2.5.4 Genes Involved in RG-I Biosynthesis

Henerik Scheller's group first identified a mutant called *arad1* (At2g35100; *ARABINAN DEFICIENT 1*) through screening of a T-DNA insertion population of *Arabidopsis* (Harholt et al. 2006). The *arad1* mutant showed a 25% reduction in Ara in leaves and 54% less Ara in stems compared with its wild-type counterpart, which suggests that *arad1* affects the incorporation of Ara into plant cell wall glycans. RG-I isolated from the *arad1* mutant has a 68% reduction in Ara content compared with that isolated from wild-type plants, suggesting a role for ARAD1 in RG-1 biosynthesis. However, the enzymatic function of ARAD1 has yet to be proven experimentally.

5.2.5.5 Genes Involved in XGA Biosynthesis

Recently an insertion mutation in the At5g33290 locus of the *Arabidopsis* genome was identified and named *xylogalacturonan deficient 1* (*xgd1*) (Jensen et al. 2008). In the *xgd1* mutant, XGA was almost completely absent except at the root tip and septa of siliques. Transient expression of *XGD1* in *Nicotiana benthamiana* yielded a protein that catalyzes the transfer of xylose from UDP-α-D-xylose onto HG oligosaccharides, confirming that XGD1 is a xylosyltransferase. Expression of a fluorescently tagged fusion protein in *N. benthamiana* confirmed that XGD1 is a Golgi-localized type II membrane protein (Jensen et al. 2008). XGD1 belongs to CAZy family 47.

5.2.5.6 Genes Involved in RG-II Biosynthesis

The Geshi group (Egelund et al. 2006) identified two homologous plant-specific *A. thaliana* genes, *RGXT1* (At4g01770) and *RGXT2* (At4g01750), that belong to a new family of glycosyltransferases (CAZy GT-family-77) and encode cell wall RG-II-α-D-1,3-xylosyltransferases (RG-II-α-1,3XylTs). Heterologously expressed RGXT1 and RGXT2 proteins catalyze the transfer of D-xylose from UDP-D-xylose to L-fucose with an α-glycosidic linkage (Egelund et al. 2006). The product of the reaction was confirmed by biochemical analysis using specific xylosidases and NMR spectroscopy. On the basis of these results, the authors hypothesized that RGXT1 and RGXT2 function in the synthesis of RG-II side chain A, which contains 2-*O*-methyl-D-Xyl attached in a α-1,3 linkage to α-L fuc. RG-II isolated from the *rgxt1* and *rgxt2* mutants serves as an acceptor for the enzyme, thus providing strong evidence that RGXT1 and RGXT2 function in RG-II synthesis. To our knowledge this is the first identification of enzymatically proven RG-II biosynthetic genes.

5.2.6 Cross-Linking Glycans (Hemicelluloses)

Hemicelluloses are branched cell wall polysaccharides with a backbone that is composed of neutral sugars. As an integral part of complex plant cell wall make up, hemicelluloses cross-link cellulose microfibrils by forming hydrogen bonds to their surface. Approximately, one-third of the lignocellulosic biomass is estimated to be composed of various hemicelluloses (Ragauskas et al. 2006). Both hemicelluloses and cellulose microfibrils do share a structural resemblance that in turn creates

TABLE 5.2

General Classification of Hemicelluloses

Polysaccharide Types	Subtypes	Composing Sugars and Linkages
Xylans	Homoxylans (X)	D-Xylopyranosyl (Xylp) linked by β-(1-3) linkages (X3); β-(1-4) linkages (X4) and/or mixed β-(1-3, 1-4) linkages (Xm)
	Glucuronoxylans (GX or MGX)	α-D-Glucoronic acid (GA) and/or its 4-*O*-methyl derivative (MeGA) attached at position 2 of the Xylp monomeric unit
	(Arabino)glucuronoxylans (AGX)	Single α-L-arabinofuranosyl residue attached at position 3 of the β-(1-4)-xylopyranan backbone of GX
	Arabinoxylans (AX)	α-L-arabinofuranosyl residue attached at position 2, 3, or both of the same Xylp monomer unit of β-(1-4)-xylopyranan backbone
	(Glucurono)arabinoxylan (GAX)	α-D-Glucoronic acid and α-L-arabinofuranosyl residues linked to position 2 and 3 of monomeric xylp residues of β-(1-4)-xylopyranan backbone
	Heteroxylans (HX)	β-(1-4)-xylopyranan backbone heavily substituted with single and oligosaccharide side chains
Mannans	Galactomannan (GaM)	Backbone of D-Mannopyranose (Manp) linked by a β-(1-4) linkage that is branched variously at position 6 by single α-galacto pyranose residues
	Glucomannans (GM)	Backbone contain D-Mannopyranose (Manp) and D-Glucopyranose (Glcp) linked by β-(1-4) linkage that is branched variously at position 6 by single α-galacto pyranose residues
Xyloglucan	Type I (XXXG)	Backbone composed of D-Glucopyranose (Glcp) linked by β-(1-4) linkage with D-Xylopyranosyl (Xylp) attached at position 6 of first three Glcp units and fourth Glcp nonsubstituted
	Type II (XXGG)	Backbone composed of D-Glucopyranose (Glcp) linked by β-(1-4) linkage with D-Xylopyranosyl (Xylp) attached at position 6 of first two Glcp units and next two Glcp nonsubstituted
Mixed linked β-glucans		Unbranched backbone composed of D-Glucopyranose (Glcp) linked by β-(1-3, 1-4) linkages

Source: Ebringerová, A., *Macromol Symp.*, 232, 1–12, 2006.

a "conformational homology" between these molecules (O'Neill and York 2003). This causes strong noncovalent association between them (O'Neill and York 2003). This noncovalent interaction also contributes to the cell wall recalcitrance against various agents of degradation. Treatment with strong alkali (i.e., 0.1–4 M KOH/NaOH) typically solubilizes hemicelluloses from pectin-removed cell walls. A broad classification of the hemicellulosic polysaccharides, according to Ebringerová (2006), is shown in the following table (Table 5.2).

5.2.6.1   Xylans

In the plant kingdom, xylans form the most abundant hemicellulosic polysaccharide, given that they are a prominent component of woody biomass (Ebringerová 2006). Like most of the hemicellulosic polysaccharides, xylans render integrity to the cell walls by cross-linking with cellulose microfibrils. One of the major challenges faced by researchers in the biofuel industry is removing these cross-linking xylans, thereby exposing the cellulose fibers to bioconverting enzymes. On the other hand, xylans themselves can be considered a key raw material for bioethanol production through D-xylose fermentation (Lachke 2002). During the past decade, several researchers, using *A. thaliana* as a model, have gained new insights into the biosynthesis and chemistry of xylans.

Recent work on several *Arabidopsis* genes has revealed in more detail the mechanisms of xylan biosynthesis. Two important genes, *IRX7/FRA8* (At2g28110) and *GAUT12/IRX8* (At5g54690), were discovered that are involved in xylan biosynthesis in *A. thaliana* plants (Zhong et al. 2005; Peña et al. 2007; Persson et al. 2007a). The above studies also reported that *fragile fiber8* (*fra8* or *irx7*) and *irx8* mutants of *Arabidopsis* plants show defects in their xylan structures. Xylan polysaccharides isolated from these mutants lack the complex oligosaccharide sequence normally found at the reducing ends, suggesting that IRX7 and IRX8 proteins might have important roles in synthesizing this reducing end oligosaccharide of xylan (York and O'Neill 2008). Also, reduced levels of glucuronoxylan and HG were also noted in these two mutants (Persson et al. 2007a). Specifically, the *fra8* mutation causes a notable reduction in fiber wall thickness and decreased stem strength. However, *irx8* mutants were dwarf mutants with impaired secondary cell wall integrity. The Golgi-complex-localized FRA8 (IRX7) and IRX8 share identical expression patterns in *Arabidopsis* tissues (Peña et al. 2007). In brief, all of the above studies show that *IRX7* and *IRX8* genes can be used for identifying potential candidate genes in biofuel crops to develop xylan-deficient plants with less recalcitrance.

Simultaneous research efforts (Brown et al. 2007; Peña et al. 2007) discovered other important genes, *IRX9* (At2g37090) and *IRX14* (At4g36890), that are required for xylan synthesis in *Arabidopsis*. Research using respective *Arabidopsis* mutants clearly demonstrates that IRX9 and IRX14 proteins, although not functionally redundant, are involved in the xylan backbone elongation process (York and O'Neill 2008). Recent studies reported yet another couple of *Arabidopsis* genes that are critical in xylan biosynthesis, *IRX10* and *IRX10-like* (Brown et al. 2009; Wu et al. 2009). Like IRX9 and IRX14, IRX10 and IRX10-like proteins also contribute to xylan backbone elongation. However, in contrast to IRX9 and IRX14, a functional redundancy exists between IRX10 and IRX10-like proteins (Brown et al. 2009; Wu et al. 2009).

Another important gene that contributes to xylan biosynthesis is the *PARVUS/GLZ1* gene, which encodes a putative family 8 glycosyl transferase (Lao et al. 2003; Shao et al. 2004). Its gene expression is clearly demonstrated to be linked with secondary cell wall synthesis, and *parvus* mutants exhibited thinner cell walls (Lee et al. 2007). Further studies showed that PARVUS/GLZ1, also known as GATL1, has putative galacturonosyltransferase activity and is again speculated to be involved in the formation of the reducing end primer sequence of xylan similar to IRX7 and IRX8 (Lee et al. 2007; York and O'Neill 2008). As mentioned earlier, studies by Brown et al. (2007) described the identification of a xylan-deficient mutant *irx14*. This study (Brown et al 2007) revealed more details on the intricate mechanism of xylan biosynthesis. Mutants of five xylan biosynthesis-associated genes (*irx7*, *irx8*, *irx9*, *parvus*, and *irx14*) were similar in that they showed a significantly reduced amount of xylan and drastically diminished GlcUA to Me-GlcUA side chains ratio (Brown et al. 2007). Only *irx7*, *irx8*, and *parvus* showed the absence of complex xylan oligosaccharide, unlike *irx14* plants, which did possess this structure (Brown et al. 2007). Taken together, these studies give insight into the specific physiological contributions of distinct biosynthetic genes in *Arabidopsis* xylan biosynthesis. Ongoing works such as those mentioned above will be instrumental to altering cell wall xylan networks to create more suitable biofuel crops in the future.

In addition to the above-mentioned genes, *ATCSLD5*, a member of the large family of cellulose synthase-like genes, has been recently characterized (Bernal et al. 2007). The knockout *atcsld5* mutants show reduced xylan content, which causes a negative growth effect (Bernal et al. 2007). Although these results hint that ATCSLD5 might be playing a role in xylan synthesis, the detailed physiological functions of this protein remain unknown.

5.2.6.2 Xyloglucans

The most abundant and most studied hemicellulose in the dicotyledonous primary wall is xyloglucan (XG) (Cosgrove 1997). Xyloglucan contents in primary cell walls vary among dicotyledonous plants (20–25%), grasses (2–5%), and soft woods (10%) (Ebringirova 2006). In most land plants, XGs noncovalently cross-link and coat cellulose microfibrils. Studies on XG-cellulose interactions, because of their potential to allow modification of cellulose microfibers, are immensely important in making ligno-cellulosic bioenergy crops suitable for industrial applications (Zhou et al. 2007). Additionally, understanding the biosynthesis of XGs is equally important in biofuel research because genes underlying the pathways could be targeted to reduce XG cross-linking in cellulose microfibrils.

The branched hemicellulosic polymer XG has a backbone of 1,4-linked β-D-glucopyranose residues. This backbone bears short side chains that contain residues of xylose, galactose, and often a terminal fucose (Cosgrove 1997). The pattern of branching from the XG backbone is peculiar in that two or three of the four backbone glucosyl residues carry an α-D-Xyl residue substitution at O-6 of the glucosyl residues. On the basis of these branching patterns, xyloglucans are classified as "XXXG-type" or "XXGG-type" depending on the number of substituted glucosyl residues on the backbone as described (Table 5.2) (Fry et al. 1993; Vincken et al. 1997). Main structural features of XG are generally conserved through evolution within plant species. Depending on the plant species, many of the xylosyl residues attached to the 1,4-linked β-D-glucopyranosyl backbone are further linked by residues at the O-2 position with either a single β-D-Gal or an α-L-Fuc-(1,2)-β-D-Gal disaccharide. *Arabidopsis* cell walls are characterized by an XXXG-type xyloglucan that contains mainly XXXG, XLXG, XXFG, XLLG, and XLFG subunits (Lerouxel et al. 2006), where "G" represents Glc, "X" represents α-D-Xyl-(1,6)-Glc, "L" represents the β-D-Gal-(1,2)-α-D-Xyl side chain, and "F" stands for α-L-Fuc-(1,2)-β-D-Gal-(1,2)-α-D-Xyl (nomenclature of subunits introduced by Fry et al. 1993; Pauly et al. 1999). Other types of XGs (e.g., those found in the *Poaeceae* and *Solanaceae*) include xylose units that bear α-L-Ara instead of β-D-Gal residues (Jia et al. 2005). Furthermore, variations in the residues linked to the O-2 position of xylosyl residues are also found in different plant tissues within a given plant species (Freshour et al. 1996; O'Neill and York 2003). These properties of plant XGs show that XG biosynthesis might involve the functioning of distinct biosynthetic genes at different developmental stages. *A. thaliana* genome-based research has given many important clues regarding XG biosynthesis in plants. Table 5.3 lists the major *A. thaliana* genes reported to date as playing important roles in XG biosynthesis.

TABLE 5.3

Major Gene Categories Characterized for Xyloglucan Biosynthesis

Classes	Arabidopsis Gene	Functions
Backbone synthases	AT3G28180 (*AtCSLC4*) (Cocuron et al. 2007)	β-1,4 Glucan synthase
Side-chain synthases	AT2G03220 (**MUR2, FUT1**) (Perrin et al. 1999; Faik et al. 2000; Sarria et al. 2001). 3	Fucosyl transferase-1
	AT1G74380 (**XXT5**) (Zabotina et al. 2008)	UDP-xylosyl transferase activity
	AT3G62720 (**XXT1**) (Faik et al. 2002; Cavalier and Keegstra 2006; Cavalier et al. 2008)	
	AT4G02500 (**XXT2**) (Cavalier and Keegstra 2006; Cavalier et al. 2008)	
XG modification/ restructuring	AT1G10550 (**XET, XTH33**) (Divol et al. 2007)	Putative xyloglucan endotransglycosylase/ hydrolases for cell wall modification
	AT1G14720 (**XTR2**) (Kurasawa et al. 2009)	
	AT2G01850 (**XTH27**) (Matsui et al. 2005)	
	AT4G30280 (**XTH18**) (Osato et al. 2006)	

XTR or the *XTH* gene family comprises groups of genes encoding XG endotransglucosylases/ hydrolases, which are responsible for XG modifying/rearranging in plants. Xu et al. (1996) first reported on the *Arabidopsis XET-related* (*XTR*) gene family and their potential roles in cell wall modification. Their studies gave a detailed insight into the regulation of expression of the *XTR* gene family by hormonal and environmental stimuli and demonstrated that distinct but related wall-modifying enzymes (XTRs) might be recruited at various developmental stages and environmental conditions (Xu et al. 1996). Yokoyama and Nishitani (2001), by examining *XTH* gene family (all 33 genes) expression profiles using quantitative real-time reverse transcriptase-polymerase chain reaction, were able further understand the *XTR* gene's tissue specificity and hormonal regulation of expression. According to this work, most XTR genes are distinct in their expression patterns, although with some showing resemblance, as relates to tissue specificity or in response to a hormonal signal. Recent findings show that *AtXTH* members, *AtXTH17*, -18, -19, and -20, are homologous genes that could have resulted from a duplication events (Vissenberg et al. 2005). These members have varied expression patterns in *Arabidopsis* roots emphasizing specific physiological roles of each of these members in cell wall biosynthesis (Vissenberg et al. 2005). Osato et al. (2006) functionally analyzed RNAi plants with downregulated expression of the *AtXTH* gene and subsequently reported the principal role of AtXTH18 in *A. thaliana* root growth. A comprehensive review is available describing the biochemical and functional diversity of XTHs with an overview of the structure and evolutionary organization of the *Arabidopsis XTH* gene family (Rose et al. 2002). Previous reports by Mellerowicz and Sundberg (2008), Mellerowicz et al. (2008), and Bourquin et al. (2002) explain the potential roles of XETs and XTHs in the formation of secondary cell wall layers in trees, such as *Populus*, that are popular current targets for biofuel production.

As summarized in Table 5.3, XG synthesis has three main parts, XG backbone synthesis XG side-chain synthesis, and XG modification. *AtCSLC4* is the only *Arabidopsis* gene characterized so far that plays a role in XG backbone synthesis by encoding a β-1,4 glucan synthase (Cocuron et al. 2007). Most of the other XG biosynthetic enzymes characterized are those that synthesize side chains. Reiter et al. (1993) reported on specific *Arabidopsis* plants, termed *mur1* mutants, with abnormal growth and altered cell walls due to a fucose-deficient mutation. These *mur1* plants lack a gene that encodes an enzyme, GDP-D-mannose-4,6-dehydratase 2, which catalyzes the initial critical step in the de novo synthesis of GDP-L-fucose (Bonin et al. 1997). The MUR1 protein thus may be important for xyloglucan fucosylation. The MUR2 protein is encoded by fucosyltransferase 1 (*FUT1* or *AtFUT1*). Plants deficient in the FUT1 protein are called *mur2* mutants. These plants also show altered cell wall structures and abnormal plant growth (Perrin et al. 1999; Reiter 2002). Recent studies by Cavalier et al. (2008) and Zabotina et al. (2008) conclude that *A. thaliana* XXT1, XXT2, and XXT5 proteins encode xylosyltransferases that are necessary for XG biosynthesis. Mutation of these genes resulted in plants with characteristic abnormal root hair phenotypes (Zabotina et al. 2008; Cavalier et al. 2008). The lack of measurable XG in the *xxt1* X *xxt2* double mutants results in diminished plant growth, emphasizing the critical role played by these genes in XG biosynthesis (Cavalier et al. 2008).

5.2.6.3 Cell-Wall-Associated Proteins

Cell walls must be extremely flexible and elastic to cope with cell growth (Cosgrove 1999). This necessitates many controlled, concerted, and coordinated actions between various cell-wall-associated proteins and wall polysaccharides. Various models depicting these mechanisms have been explained previously (Cosgrove 1999). It is known that a large proportion of cell-wall-associated proteins studied thus far contribute to cell wall loosening and expansion. A broad classification of the cell-wall-associated proteins divides them into families such as hydroxyproline-rich glycoproteins (HRGPs) that include extensins (Showalter 1993), proline-rich proteins (PRPs; Showalter 1993), glycine-rich proteins (GRPs; Keller 1993), arabinogalactan proteins (AGPs; Oxley and Bacic 1999), wall-associated kinases (WAKs; He et al. 1999), lectins (Herve et al. 1999), and expansins (Cosgrove 1999). Because the major challenge in the biofuel industry is identification of the means to degrade and loosen cell walls, the research on these cell-wall-associated proteins could be considered a potential area of interest.

The cell walls of higher plants contain a family of HRGPs called extensins (Showalter 1993). An excellent review is available focusing on structural details and distribution of extensins (Showalter 1993). Extensins contain considerable amounts of hydroxyproline, serine, and various proportions of amino acids like valine, tyrosine, lysine, and hystidine (Showalter 1993). The presence of repeating pentapeptide motifs, like Ser-Hyp$_4$, is a characteristic of HRGPs (Showalter 1993). ATEXT1 proteins from *Arabidopsis* have plant extensin characteristics and are well characterized (Merkouropoulos et al. 1999; Roberts and Shirsat 2006). However, not much work has been conducted in this area. One additional report in this area of research is on the characterization of the second prolyl 4-hdroxylase enzyme (At-P4H-2) from *Arabidopsis* (Tiainen et al. 2005). Experimental results showing enhanced growth rates in plant cells, when exposed to acidic solutions, led to the identification of a class of proteins that were later called expansins (Rayle and Cleland 1992). Expansins were first defined based on their pH dependency and their unique action on the rheological behavior of isolated cell walls (Cosgrove 1999). Current understanding about these proteins is mostly based on their amino acid sequences and not on their biological activities (Cosgrove 1999). The α- and β-expansin families, which share approximately 25% amino acid similarity, are the only ones recognized so far (Cosgrove 1999). Although these classes are identical in their effect on cell wall properties, the exact mechanism of their action is not fully known (Cosgrove 1999). Detailed comparative genome analysis studies of *Arabidopsis* identified 26 putative α-expansins and 5 β-expansin genes (Lee et al. 2001). The gene structural regions shared by expansins contain three exon regions that encode three potential functional domains of the expansin protein, such as a signal peptide and the N-terminus of the mature protein, an endoglucanase-like core region, and a domain with structural resemblance to microbial cellulose binding domains (CBDs) (Cosgrove 1999). Several other studies using *Arabidopsis* as a model system additionally prove that expansins have significant roles in cell wall growth and development (Cho and Cosgrove 2000; Cosgrove 2000; Choi et al. 2006). These outcomes recommend considering expansins as target genes in biomass enhancement research.

5.2.6.4 Arabinogalactan Proteins

Arabinogalactan proteins (AGPs) are extracellular-wall-associated hydroxyproline-rich proteoglycans. Showalter (2001) reviewed the important roles of AGPs in plant growth and development. Some of the important physiological roles of AGPs include their functions in vegetative/reproductive growth, xylem development, programmed cell death and signaling (Showalter 2001). The protein backbone of these molecules is rich in proline/hydroxyproline, serine, alanine and threonine (Sommer-Knudsen et al. 1998). The family members of most AGPs share less than 40% amino acid sequence similarity (Schultz et al. 2002). Taking advantage of the completed *Arabidopsis* genome, Schultz et al. (2002) identified several classes of AGP molecules that included 21 fasciclin-like AGPs, 13 AGPs, 10 AG peptides, and 3 basic AGPs with short lysine-abundant regions. Recent studies have also shown that all classical genes of arabinogalactan backbone proteins, especially *AGP6* and *AGP11*, are expressed in pollen tubes of *Arabidopsis* plants (Pereira et al. 2006; Coimbra et al. 2009). Therefore, considering the above information and the vital function of AGPs in plant growth and development, further research on AGPs in bioenergy crops are equally important to biomass enhancement for biofuel production.

5.3 CONCLUSIONS

For a future sustainable and reliable biofuel industry, it is necessary to understand the ins and outs of plant cell walls, which are the most abundant component of the natural biomass raw material. The most challenging property of plant cell walls is the integrity of its structures, which are mainly composed of cellulose microfibrils cross-liked with hemicelluloses that bind with lignin. This property of cell walls, which inhibits its use in the biofuel industry, is now popularly termed "cell wall recalcitrance." Without cell wall recalcitrance, a sustainable and reliable biofuel industry could have been realized by now. Hence, the major goal of current biofuel research is to develop cell walls with reduced recalcitrance. To achieve rapid progress toward reduced recalcitrance, it is important that we know the key steps and identify all of the contributors involved in cell wall biosynthesis. In other words, knowledge of how plant cell walls are constructed would facilitate the means for deconstructing them. *Arabidopsis* research has been contributing heavily toward increasing our understanding of cell wall synthesis. As described in this review, studies on genes and other key contributors to the biosynthesis of most major cell wall polysaccharides and other components like lignin are rapidly progressing. It is obvious that *A. thaliana* cell wall research and its subsequent application to bioenergy crops of interest will contribute to improving biofuel production and thereby render a sustainable, timely, and cost-effective solution to the global energy crisis.

ACKNOWLEDGMENTS

Supported by grants from U.S. Department of Energy (BioEnergy Science Center and grant #DE PS02-06ER64304) and the National Science Foundation Plant Genome Program (DBI-0421683). The authors thank Michael Hahn for his comments and discussion and Karen Howard, Virginia Brown, and Ashley Grove for formatting the manuscript.

REFERENCES

Arioli T, Peng LC, Betzner AS, Burn J, Wittke W, Hearth W, Camilleri C, Höfte H, Plazinski J, Birch R et al. (1998) Molecular analysis of cellulose biosynthesis in Arabidopsis. *Science* 279:717–720

Baucher M, Van Doorsselaere J, Gielen J, Van Montagu M, Inze D, Boerjan W (1995) Genomic nucleotide sequence of an *Arabidopsis thaliana* gene encoding a cinnamyl alcohol dehydrogenase. *Plant Physiol* 107:285–286

Beale CV, Long SP (1995) Can perennial C4 grasses attain high efficiencies of radiant energy conversion in cool climates? *Plant Cell Environ* 18:641–650

Bell-Lelong DA, Cusumano JC, Meyer K, Chapple C (1997) Cinnamate-4-hydroxylase expression in Arabidopsis. Regulation in response to development and the environment. *Plant Physiol* 113:729–738

Bernal AJ, Jensen JK, Harholt J, Sorensen S, Moller I, Blaukopf C, Johansen B, de Lotto R, Pauly M, Scheller HV, Willats WG (2007) Disruption of ATCSLD5 results in reduced growth, reduced xylan and homogalacturonan synthase activity and altered xylan occurrence in Arabidopsis. *Plant* J 52:791–802

Bonin CP, Potter I, Vanzin GF, Reiter WD (1997) The MUR1 gene of *Arabidopsis thaliana* encodes an isoform of GDP-D-mannose-4,6-dehydratase, catalyzing the first step in the de novo synthesis of GDP-L-fucose. *Proc Natl Acad Sci USA* 94:2085–2090

Boudet A, Grima-Pettanati J (1996) Lignin genetic engineering. *Mol Breed* 2:25–39

Bourquin V, Nishikubo N, Abe H, Brumer H, Denman S, Eklund M, Christiernin M, Teeri TT, Sunberg B, Mellerowicz EJ (2002) Xyloglucan endotransglycosylases have a function during the formation of secondary cell walls of vascular tissues. *Plant Cell* 14:3073–3088

Bouton S, Leboeuf E, Mouille G, Leydecker MT, Talbotec J, Granier F, Lahaye M, Hofte H, Truong HN (2002) QUASIMODO1 encodes a putative membrane-bound glycosyltransferase required for normal pectin synthesis and cell adhesion in Arabidopsis. *Plant Cell* 14:2577–2590

Brown DM, Goubet F, Wong VW, Goodacre R, Stephens E, Dupree P, Turner SR (2007) Comparison of five xylan synthesis mutants reveals new insight into the mechanisms of xylan synthesis. *Plant J* 52:1154–1168

Brown DM, Zhang Z, Stephens E, Dupree P, Turner SR (2009). Characterization of IRX10 and IRX10-like reveals an essential role in glucuronoxylan biosynthesis in Arabidopsis. *Plant J* 57:732–746

Buhler DD, Netzer DA, Riemenschneider DE, Hartzler RG (1998) Weed management in short rotation poplar and herbaceous perennial crops grown for biofuel production. *Biomass Bioenergy* 14:385–394

Campbell MM, Sederoff RR (1996) Variation in lignin content and composition (mechanisms of control and implications for the genetic improvement of plants). *Plant Physiol* 110:3–13

Carpita NC (1996) Structure and biogenesis of the cell walls of grasses. *Annu Rev Plant Physiol Plant Mol Biol* 47:445–476

Cavalier DM, Keegstra K (2006) Two xyloglucan xylosyltransferases catalyze the addition of multiple xylosyl residues to cellohexaose. *J Biol Chem* 281:34197–34207

Cavalier DM, Lerouxel O, Neumetzler L, Yamauchi K, Reineck A, Freshour G, Zabotina OA, Hahn MG, Burgert I, Pauly M, et al. (2008) Disrupting two *Arabidopsis thaliana* xylosyltransferase genes results in plants deficient in xyloglucan, a major primary cell wall component. *Plant Cell* 20:1519–1537

Chapple C, Vogt T, Ellis BE, Somerville CR (1992) An Arabidopsis mutant defective in the general phenylpro-panoid pathway. *Plant Cell* 4:1413–1424

Chen AH, Chai YR, Li JN, Chen L (2007) Molecular cloning of two genes encoding cinnamate 4-hydroxylase (C4H) from oilseed rape (*Brassica napus*). *J Biochem Mol Biol* 40:247–260

Cho HT, Cosgrove DJ (2000) Altered expression of expansin modulates leaf growth and pedicel abscission in *Arabidopsis thaliana. Proc Natl Acad Sci USA* 97:9783–9788

Choi D, Cho HT, Lee Y (2006) Expansins: Expanding importance in plant growth and development. *Physiol Plant* 126:511–518

Chow J, Kopp RJ, Portney PR (2003) Energy resources and global development. *Science* 302:1528–1531

Cocuron JC, Lerouxel O, Drakakaki D, Alonso AP, Liepman AH, Keegstra K, Raikhel N, Wilkerson CG (2007) A gene from the cellulose synthase-like C family encodes a β-1,4 glucan synthase. *Proc Natl Acad Sci USA* 104:8550–8555

Coimbra S, Costa M, Jones B, Mendes MA, Pereira LG (2009) Pollen grain development is compromised in Arabidopsis agp6 agp11 null mutants. *J Exp Bot* 60:3133–3142

Cosgrove DJ (1997) Assembly and enlargement of the primary cell wall in plants. *Annu Rev Cell Dev Biol* 13:171–201

Cosgrove DJ (1999). Enzymes and other agents that enhance cell wall extensibility. *Annu Rev Plant Physiol Plant Mol Biol* 50:391–417

Cosgrove DJ (2000) New genes and new biological roles for expansins. *Curr Opin Plant Biol* 3:73–78

Cosgrove DJ (2003) Expansion of the plant cell. In: Rose JKC (ed), *Plant Cell Wall*. Blackwell Publishers Oxford, United Kingdom, pp 237–263

Costa MA, Bedgar DL, Moinuddin SG, Kim KW, Cardenas CL, Cochrane FC, Shockey JM, Helms GL, Amakura Y, Takahashi H, et al. (2005) Characterization in vitro and in vivo of the putative multigene 4-coumarate:CoA ligase network in Arabidopsis: Syringyl lignin and sinapate/sinapyl alcohol derivative formation. *Phytochemistry* 66:2072–2091

Desprez T, Juraniec M, Crowell EF, Jouy H, Pochylova Z, Parcy F, Hofte H, Gonneau M, Vernhettes S (2007) Organization of cellulose synthase complexes involved in primary cell wall synthesis in *Arabidopsis thaliana. Proc Natl Acad Sci USA* 104:1 5572–5577

Divol F, Vilaine F, Thibivilliers S, Kusiak C, Sauge MH, Dinant S (2007) Involvement of the xyloglucan endo-transglycosylase/hydrolases encoded by celery XTH1 and Arabidopsis XTH33 in the phloem response to aphids. *Plant Cell Environ* 30:187–201

Doblin MS, Kurek I, Jacob-Wilk D, Delmer DP (2002) Cellulose biosynthesis in plants: From genes to rosettes. *Plant Cell Physiol* 43:1407–1420

Douglas CJ (1996) Phenylpropanoid metabolism and lignin biosynthesis: From weeds to trees. *Trends Plant Sci* 1:171–178

Duchesne LC, Larson DW (1989) Cellulose and the evolution of plant life. *Bioscience* 39:238–241

Dunkley TP, Watson R, Griffin JL, Dupree P, Lilley KS (2004) Localization of organelle proteins by isotope tagging (LOPIT). *Mol Cell Proteom* 3:1128–1134

Ebringerová A (2006) Structural diversity and application potential of hemicelluloses. *Macromol Symp* 232:1–12

Egelund J, Petersen BL, Motawia MS, Damager I, Faik A, Olsen CE, Ishii T, Clausen H, Ulvskov P, Geshi N (2006) *Arabidopsis thaliana* RGXT1 and RGXT2 encode Golgi-localized (1,3)-α-D-xylosyltransferases involved in the synthesis of pectic rhamnogalacturonan-II. *Plant Cell* 18:2593–2607

Ehlting J, Büttner D, Wang Q, Douglas CJ, Somssich IE, Kombrink E (1999) Three 4-coumarate:coenzyme A ligases in *Arabidopsis thaliana* represent two evolutionarily divergent classes in angiosperms. *Plant J* 19:9–20

Ehlting J, Shin JJ, Douglas CJ (2001) Identification of 4-coumarate:coenzyme A ligase (4CL) substrate recog-nition domains. *Plant J* 27:455–465

Eudes A, Pollet B, Sibout R, Do CT, Sequin A, Lapierre C, Jouanin L (2006) Evidence for a role of AtCAD 1 in lignification of elongating stems of *Arabidopsis thaliana. Planta* 225:23–39

Faik A, Bar-Peled M, DeRocher AE, Zeng W, Perrin RM, Wilkerson C, Raikhel NV, Keegstra K (2000) Biochemical characterization and molecular cloning of an α-1,2-fucosyltransferase that catalyzes the last step of cell wall xyloglucan biosynthesis in pea. *J Biol Chem* 275:15082–15089

Faik A, Price NJ, Raikhel NV, Keegstra K (2002) An Arabidopsis gene encoding an α-xylosyltransferase involved in xyloglucan biosynthesis. *Proc Natl Acad Sci USA* 99:7797–7802

Fenning TM, Gershenzon J (2002) Where will the wood come from? Plantation forests and the role of biotechnology. *Trends Biotechnol* 20:291–296

Field CB, Behrenfeld MJ, Randerson JT, Falkowski P (1998) Primary production of the biosphere: Integrating terrestrial and oceanic components. *Science* 281:237–240

Franke R, Hemm MR, Denault JW, Ruegger MO, Humphreys JM, Chapple C (2002) Changes in secondary metabolism and deposition of an unusual lignin in the *ref8* mutant of Arabidopsis. *Plant J* 30:49–57

Freshour G, Clay RP, Fuller MS, Albersheim P, Darvill AG, Hahn MG (1996) Developmental and tissue-specific structural alterations of the cell-wall polysaccharides of *Arabidopsis thaliana* roots. *Plant Physiol* 110:1413–1429

Fry SC (1982) Phenolic components of the primary cell wall. Feruloylated disaccharides of D-galactose and L-arabinose from spinach polysaccharide. *Biochem J* 203:493–504

Fry SC, Aldington S, Hetherington PR, Aitken J (1993) Oligosaccharides as signals and substrates in the plant cell wall. *Plant Physiol* 103:1–5

Hahlbrock K, Scheel D (1989) Physiology and molecular biology of phenylpropanoid metabolism. *Annu Rev Plant Physiol Plant Mol Biol* 40:347–369

Hamberger B, Hahlbrock K (2004) The *4-coumarate:CoA ligase* gene family in *Arabidopsis thaliana* comprises one rare, sinapate-activating and three commonly occurring isoenzymes. *Proc Natl Acad Sci USA* 101:2209–2214

Harholt J, Jensen JK, Sorensen SO, Orfila C, Pauly M, Scheller HV (2006) ARABINAN DEFICIENT 1 is a putative arabinosyltransferase involved in biosynthesis of pectic arabinan in Arabidopsis. *Plant Physiol* 140:49–58

He ZH, Cheeseman I, He DZ, Kohorn BD (1999) A cluster of five cell wall-associated receptor kinase genes, Wak1-5, are expressed in specific organs of Arabidopsis. *Plant Mol Biol* 39:1189–1196

Herve C, Serres J, Dabos P, Canut H, Barre A, Rouge P, Lescure B (1999) Characterization of the Arabidopsis lecRK-a genes: Members of a superfamily encoding putative receptors with an extracellular domain homologous to legume lectins. *Plant Mol Biol* 39:671–682

Himmel ME, Ding SY, Johnson DK, Adney WS, Nimlos MR, Brady JW, Foust TD (2007) Biomass recalcitrance: Engineering plants and enzymes for biofuels production. *Science* 315:804–807

Jensen JK, Sørensen SO, Harholt J, Geshi N, Sakuragi Y, Møller I, Zandleven J, Bernal AJ, Jensen NB, Sørensen C et al. (2008) Identification of a xylogalacturonan xylosyltransferase involved in pectin biosynthesis in Arabidopsis. *Plant Cell* 20:1289–1302

Jia ZH, Cash M, Darvill AG, York WS (2005) NMR characterization of endogenously O–acetylated oligosaccharides isolated from tomato (*Lycopersicon esculentum*) xyloglucan. *Carbohydr Res* 340:1818–1825

Jones L, Ennos AR, Turner SR (2001) Cloning and characterization of irregular xylem4 (irx4): A severely lignin-deficient mutant of Arabidopsis. *Plant J* 26:205–216

Keegstra K, Talmadge KW, Bauer WD, Albersheim P (1973) Structure of plant cell walls. III. A model of the walls of suspension-cultured sycamore cells based on the interconnections of the macromolecular components. *Plant Physiol* 51:188–197

Keller B (1993) Structural cell wall proteins. *Plant Physiol* 101:1127–1130

Kim SJ, Kim MR, Bedgar DL, Moinuddin SGA, Cardenas CL, Davin LB, Kang CH, Lewis NG (2004) Functional reclassification of the putative cinnamyl alcohol dehydrogenase multigene family in Arabidopsis. *Proc Natl Acad Sci USA* 101:1455–1460

Kimura S, Laosinchai W, Itoh T, Cui XJ, Linder CR, Brown RM (1999) Immunogold labeling of rosette terminal cellulose-synthesizing complexes in the vascular plant *Vigna angularis*. *Plant Cell* 11:2075–2085

Kurasawa K, Matsui A, Yokoyama R, Kuriyama T, Yoshizumi T, Matsui M, Suwabe K, Watanabe M, Nishitani K (2009) The AtXTH28 gene, a xyloglucan endotransglucosylase/hydrolase, is involved in automatic self-pollination in *Arabidopsis thaliana*. *Plant Cell Physiol* 50:413–422

Lachke A (2002) Biofuel from D-xylose—The second most abundant sugar. *Resonance* 7:50–58

Lacombe E, Hawkins S, Van Dorsselaere J, Piquemal J, Goffner D, Poeydomenge O, Boudet AM, Grima-Pettenati J (1997). Cinnamoyl CoA reductase, the first committed enzyme of the lignin branch biosynthetic pathway: Cloning, expression and phylogenetic relationships. *Plant J* 11:429–441

Lal R (2008) Crop residues as soil amendments and feedstock for bioethanol production. *Waste Manag* 28:747–758

Lane DR, Wiedemeier A, Peng L, Hofte H, Vernhettes S, Desprez T, Hocart CH, Birch RJ, Baskin TI, Burn JE, et al. (2001) Temperature-sensitive alleles of RSW2 link the KORRIGAN endo-1,4-beta-glucanase to cellulose synthesis and cytokinesis in Arabidopsis. *Plant Physiol* 126:278–288

Lao NT, Long D, Kiang S, Coupland G, Shoue DA, Carpita NC, Kavanagh TA (2003) Mutation of a family 8 glycosyltransferase gene alters cell wall carbohydrate composition and causes a humidity-sensitive semi-sterile dwarf phenotype in Arabidopsis. *Plant Mol Biol* 53:687–701

Laskar DD, Jourdes M, Patten AM, Helms GL, Davin LB, Lewis NG (2006) The Arabidopsis cinnamoyl CoA reductase irx4 mutant has a delayed but coherent (normal) program of lignification. *Plant J* 48:674–686

Lauvergeat V, Lacomme C, Lacombe E, Lasserre E, Roby D, Grima-Pettenati J (2001) Two cinnamoyl-CoA reductase (CCR) genes from *Arabidopsis thaliana* are differentially expressed during development and in response to infection with pathogenic bacteria. *Phytochemistry* 57:1187–1195

Lee C, Zhong R, Richardson E, Himmelsbach DS, McPhail BT, Ye ZH (2007) The PARVUS gene is expressed in cells undergoing secondary wall thickening and is essential for glucuronoxylan biosynthesis. *Plant Cell Physiol* 48:1659–1672

Lee D, Ellard M, Wanner LA, Davis KR, Douglas CJ (1995) The *Arabidopsis thaliana* 4-coumarate:CoA ligase (4CL) gene: Stress and developmentally regulated expression and nucleotide sequence of its cDNA. *Plant Mol Biol* 28:871–884

Lee D, Meyer K, Chapple C, Douglas CJ (1997) Antisense suppression of 4-coumarate:coenzyme A ligase activity in Arabidopsis leads to altered lignin subunit composition. *Plant Cell* 9:1985–1998

Lee Y, Choi D, Kende H (2001) Expansins. Ever-expanding numbers and functions. *Curr Opin Plant Biol* 4:527–532

Lerouxel O, Cavalier DM, Liepman AH, Keegstra K (2006) Biosynthesis of plant cell wall polysaccharides—A complex process. *Curr Opin Plant Biol* 9:621–630

Lewis NG, Yamamoto E (1990) Lignin: Occurrence, biogenesis and biodegradation. *Annu Rev Plant Physiol Plant Mol Biol* 41:455–496

Lieth H (1975) Primary production of the major vegetation units of the world. In: Lieth H, Whittaker RH (eds), *Primary Production of the Biosphere*. Springer, Berlin, Germany, pp 203–215

MacKinnon IM, Sturcová A, Sugimoto-Shirasu K, His I, McCann MC, Jarvis MC (2006) Cell-wall structure and anisotropy in procuste, a cellulose synthase mutant of Arabidopsis thaliana. *Planta* 224:438–448

Marita JM, Ralph J, Hatfield RD, Chapple C (1999) NMR characterization of lignins in Arabidopsis altered in the activity of ferulate 5-hydroxylase. *Proc Natl Acad Sci USA* 96:12328–12332

Matsui A, Yokoyama R, Seki M, Ito T, Shinozaki K, Takahashi T, Komeda Y, Nishitani K (2005) AtXTH27 plays an essential role in cell wall modification during the development of tracheary elements. *Plant J* 42: 525–534

Matthews JF, Skopec CE, Mason PE, Zuccato P, Torget RW, Sugiyama J, Himmel ME, Brady JW (2006) Computer simulation studies of microcrystalline cellulose Iβ. *Carbohydr Res* 341:138–152

Mellerowicz EJ, Sundberg B (2008) Wood cell walls: Biosynthesis, developmental dynamics and their implications for wood properties. *Curr Opin Plant Biol* 11:293–300

Mellerowicz EJ, Immerzeel P, Hayashi T (2008) Xyloglucan: The molecular muscle of trees. *Ann Bot* 102:659–665

Merkouropoulos G, Barnett DC, Shirsat AH (1999) The Arabidopsis extensin gene is developmentally regulated, is induced by wounding, methyl jasmonate, abscisic and salicylic acid, and codes for a protein with unusual motifs. *Planta* 208:212–219

Meyer K, Cusumano JC, Somerville C, Chapple C (1996) Ferulate-5-hydroxylase from *Arabidopsis thaliana* defines a new family of cytochrome P450-dependent monooxygenases. *Proc Natl Acad Sci USA* 93:6869–6874

Meyer K, Shirley AM, Cusumano JC, Bell-Lelong DA, Chapple C (1998) Lignin monomer composition is determined by the expression of a cytochrome P450-dependent monooxygenase. *Proc Natl Acad Sci USA* 95:6619–6623

Mir Derikvand M, Sierra JB, Ruel K, Pollet B, Do CT, Thévenin J, Buffard D, Jouanin L, Lapierre C (2007) Redirection of the phenylpropanoid pathway to feruloyl malate in Arabidopsis mutants deficient for cinnamoyl-CoA reductase 1. *Planta* 227:943–956

Mohnen D (2002) Biosynthesis of pectins. In: Seymour GB, Knox JP (eds), *Pectins and Their Manipulation*. Blackwell Publishers, Oxford, United Kingdom, CRC Press, Boca Raton, FL pp 52–98

Mohnen D (2008) Pectin structure and biosynthesis. *Curr Opin Plant Biol* 11:1–12

Mouille G, Ralet MC, Cavelier C, Eland C, Effroy D, Hématy K, McCartney L, Truong HN, Gaudon V, Thibault JF, et al. (2007) Homogalacturonan synthesis in *Arabidopsis thaliana* requires a Golgi-localized protein with a putative methyltransferase domain. *Plant J* 50:605–614

Mutwil M, Debolt S, Persson S (2008) Cellulose synthesis: A complex complex. *Curr Opin Plant Biol* 11:252–257

Nakamura A, Furuta H, Maeda H, Takao T, Nagamatsu Y (2002) Structural studies by stepwise enzymatic degradation of the main backbone of soybean soluble polysaccharides consisting of galacturonan and rhamnogalacturonan. *Biosci Biotechnol Biochem* 66:1301–1313

Nishiyama Y, Langan P, Chanzy H (2002) Crystal structure and hydrogen-bonding system in cellulose Ibeta from synchrotron X-ray and neutron fiber diffraction. *J Am Chem Soc* 124:9074–9082

O'Neill MA, York WS (2003) The composition and structure of primary cell walls. In: Rose JKC (ed), *The Plant Cell*, Vol 8. CRC Press, Boca Raton, FL pp 1–54

O'Neill MA, Eberhard S, Albersheim P, Darvill AG (2001) Requirement of borate cross-linking of cell wall rhamnogalacturonan II for Arabidopsis growth. *Science* 294:846–849

O'Neill MA, Ishii T, Albersheim P, Darvill AG (2004) Rhamnogalacturonan II: Structure and function of a borate cross-linked cell wall pectic polysaccharide. *Ann Rev Plant Biol* 55:109–139

Orfila C, Sørensen SO, Harholt J, Geshi N, Crombie H, Truong HN, Reid JS, Knox JP, Scheller HV (2005) QUASIMODO1 is expressed in vascular tissue of *Arabidopsis thaliana* inflorescence stems, and affects homogalacturonan and xylan biosynthesis. *Planta* 222:613–622

Osato Y, Yokoyama R, Nishitani K (2006) A principal role for AtXTH18 in *Arabidopsis thaliana* root growth: A functional analysis using RNAi plants. *J Plant Res* 119:153–162

Oxley D, Bacic A (1999) Structure of the glycosylphosphatidylinositol anchor of an arabinogalactan protein from Pyrus communis suspension-cultured cells. *Proc Natl Acad Sci USA* 96:14246–14251

Pagant S, Bichet A, Sugimoto K, Lerouxel O, Desprez T, McCann M, Lerouge P, Vernhettes S, Höfte H (2002) KOBITO1 encodes a novel plasma membrane protein necessary for normal synthesis of cellulose during cell expansion in Arabidopsis. *Plant Cell* 14:2001–2013

Pauly M, Albersheim P, Darvill AG, York WS (1999) Molecular domains of the cellulose/xyloglucan network in the cell walls of higher plants. *Plant J* 20:629–639

Pauly M, Keegstra K (2008) Cell-wall carbohydrates and their modification as a resource for biofuels. *Plant J* 54:559–568

Pear JR, Kawagoe Y, Schreckengost WE, Delmer DP, Stalker DM (1996) Higher plants contain homologs of the bacterial celA genes encoding the catalytic subunit of cellulose synthase. *Proc Natl Acad Sci USA* 93:12637–12642

Pelloux J, Rustérucci C, Mellerowicz EJ (2007) New insights into pectin methylesterase structure and function. *Trends Plant Sci* 12:267–277

Peña MJ, Zhong R, Zhou GK, Richardson EA, O'Neill MA, Darvill AG, York WS, Ye ZH (2007) Arabidopsis irregular xylem8 and irregular xylem9: Implications for the complexity of glucuronoxylan biosynthesis. *Plant Cell* 19:549–563

Pereira LG, Coimbra S, Oliveira H, Monteiro L, Sottomayor M (2006) Expression of arabinogalactan protein genes in pollen tubes of *Arabidopsis thaliana*. *Planta* 223:374–380

Perrin RM, DeRocher AE, Bar-Peled M, Zeng W, Norambuena L, Orellana A, Raikhel NV, Keegstra K (1999) Xyloglucan fucosyltransferase, an enzyme involved in plant cell wall biosynthesis. *Science* 284:1976–1979

Persson S, Caffall KH, Freshour G. (2007a). The Arabidopsis irregular xylem8 mutant is deficient in glucuronoxylan and homogalacturonan, which are essential for secondary cell wall integrity. *Plant Cell* 19:237–255

Persson S, Paredez A, Carroll A, Palsdottir H, Doblin M, Poindexter P, Khitrov N, Auer M, Somerville CR (2007b). Genetic evidence for three unique components in primary cell-wall cellulose synthase complexes in Arabidopsis. *Proc Natl Acad Sci USA* 104:15566–15571

Poorter H, Villar R (1997) The fate of acquired carbon in plants: Chemical composition and construction costs. In: Bazzaz FA, Grace J (eds), *Plant Resource Allocation*. Academic Press, San Diego, CA pp 39–72

Porter JR, Kirsch MMN, Streibig J, Felby C (2007) Choosing crops as energy feedstocks. *Nature Biotechnol* 25:716–717

Ragauskas AJ, Williams CK, Davison BH, Britovsek G, Cairney J, Eckert CA, Frederick WJ Jr, Hallett JP, Leak DJ, Liotta CL, et al. (2006) The path forward for biofuels and biomaterials. *Science* 311:484–489

Ragg H, Kuhn DN, Hahlbrock K (1981) Coordinated regulation of 4-coumarate:CoA ligase and phenylalanine ammonia-lyase mRNAs in cultured plant cells. *J Biol Chem* 256:10061–10065

Rayle DL, Cleland RE (1992). The acid growth theory of auxin-induced cell elongation is alive and well. *Plant Physiol* 99:1271–1274

Reiter WD (2002) Biosynthesis and properties of the plant cell wall. *Curr Opin Plant Biol* 5:536–542

Reiter WD, Chapple CCS, Somerville CR (1993) Altered growth and cell walls in a fucose-deficient mutant of Arabidopsis. *Science* 261:1032–1035

Richmond T (2000) Higher plant cellulose synthases. *Genome Biol* 4:30011–30016

Roberts K (2001) How the cell wall acquired a cellular context. *Plant Physiol* 125:127–130

Roberts K, Shirsat AH (2006) Increased extensin levels in Arabidopsis affect inflorescence stem thickening and height. *J Exp Bot* 57:537–545

Rose JKC, Braam J, Fry SC, Nishitani K (2002) The XTH family of enzymes involved in xyloglucan endo-transglucosylation and endohydrolysis: Current perspectives and a new unifying nomenclature. *Plant Cell Physiol* 43:1421–1435

Roudier F, Fernandez AG, Fujita M, Himmelspach R, Borner GHH, Schindelman G, Song S, Baskin TI, Dupree P, Wasteneys GO, et al. (2005). COBRA, an Arabidopsis extracellular glycosyl-phosphatidyl inositol-anchored protein, specifically controls highly anisotropic expansion through its involvement in cellulose microfibril orientation. *Plant Cell* 17:1749–1763

Sarria R, Wagner TA, O'Neill MA, Faik A, Wilkerson CG, Keegstra K, Raikhel NV (2001) Characterization of a family of Arabidopsis genes related to xyloglucan fucosyltransferase1. *Plant Physiol* 127:1595–1606

Sato S, Kato T, Kakegawa K (2001) Role of the putative membrane-bound endo-1,4-beta-glucanase KORRIGAN in cell elongation and cellulose synthesis in *Arabidopsis thaliana*. *Plant Cell Physiol* 42:251–263

Schubert C (2006) Can biofuels finally take center stage? *Nat Biotechnol* 24:777–784

Schultz CJ, Rumsewicz MP, Johnson KL, Jones BJ, Gaspar YM, Bacic A (2002) Using genomic resources to guide research directions: The arabinogalactan protein gene family as a test case. *Plant Physiol* 129:1448–1463

Shao M, Zheng H, Hu Y, Liu Y, Jang JC, Ma H, Huang H (2004) The GAOLAOZHUANGREN1 gene encodes a putative glycosyltransferase that is critical for normal development and carbohydrate metabolism. *Plant Cell Physiol* 45:1453–1460

Showalter AM (1993) Structure and function of plant cell wall proteins. *Plant Cell* 5:9–23

Showalter AM (2001) Arabinogalactan-proteins: Structure, expression and function. *Cell Mol Life Sci* 58:1399–1417

Sibout R, Eudes A, Mouille G, Pollet B, Lapierre C, Jouanin L, Séquin A (2005). CINNAMYL ALCOHOL DEHYDROGENASE-C and –D are the primary genes involved in lignin biosynthesis in the floral stem of Arabidopsis. *Plant Cell* 17:2059–2076

Soltani BM, Ehlting J, Hamberger B, Douglas CJ (2006) Multiple cis-regulatory elements regulate distinct and complex patterns of developmental and wound-induced expression of *Arabidopsis thaliana* 4CL gene family members. *Planta* 224:1226–1238

Somerville CR (2006). Cellulose synthesis in higher plants. *Ann Rev Cell Dev Biol* 22:53–78

Somerville CR (2007) Biofuels. *Curr Biol* 17:115–119

Somerville C, Bauer S, Brininstool G, Facette M, Hamann T, Milne J, Osborne E, Paredez A, Persson S, Raab T, et al. (2004). Towards a systems approach to understanding plant cell walls. *Science* 306:2206–2211

Sommer-Knudsen J, Bacic A, Clarke AE (1998) Hydroxyproline-rich plant glycoproteins. *Phytochemistry* 47:483–497

Staehelin LA, Moore I (1995) The plant Golgi apparatus: Structure, functional organization and trafficking mechanisms. *Ann Rev Plant Physiol Plant Mol Biol* 46:261–288

Sterling JD, Atmodjo MA, Inwood SE, Kumar Kolli VS, Quigley HF, Hahn MG, Mohnen D (2006) Functional identification of an Arabidopsis pectin biosynthetic homogalacturonan galacturonosyltransferase. *Proc Natl Acad Sci USA* 103:5236–5241

Suzuki S, Li L, Sun YH, Chiang VL (2006) The cellulose synthase gene superfamily and biochemical functions of xylem-specific cellulose synthase-like genes in Populus trichocarpa. *Plant Physiol* 142:1233–1245

Szyjanowicz PM, McKinnon I, Taylor NG, Gardiner J, Jarvis MC, Turner SR (2004) The irregular xylem 2 mutant is an allele of korrigan that affects the secondary cell wall of *Arabidopsis thaliana*. *Plant J* 37:730–740

Taylor NG (2008) Cellulose biosynthesis and deposition in higher plants. *New Phytol* 178:239–252

Taylor NG, Howells RM, Huttly AK, Vickers K, Turner SR (2003) Interactions among three distinct CesA proteins essential for cellulose synthesis. *Proc Natl Acad Sci USA* 100:1450–1455

Taylor NG, Laurie S, Turner SR (2000) Multiple cellulose synthase catalytic subunits are required for cellulose synthesis in Arabidopsis. *Plant Cell* 12:2529–2540

Taylor NG, Scheible WR, Cutler S, Somerville CR, Turner SR (1999) The irregular xylem3 locus of Arabidopsis encodes a cellulose synthase required for secondary cell wall synthesis. *Plant Cell* 11:769–780

Tiainen P, Myllyharju J, Koivunen P (2005) Characterization of a second *Arabidopsis thaliana* prolyl 4-hydroxylase with distinct substrate specificity. *J Biol Chem* 280:1142–1148

Vincken JP, York WS, Beldman G, Voragen AGJ (1997) Two general branching patterns of xyloglucan, XXXG and XXGG. *Plant Physiol* 114:9–13

Vissenberg K, Oyama M, Osato V, Yokoyama R, Verbelen JP, Nishitani K (2005) Differential expression of AtXTH17, AtXTH18, AtXTH19 and AtXTH20 genes in Arabidopsis roots. Physiological roles in specification in cell wall construction. *Plant Cell Physiol* 46:192–200

Walter MH, Grima-Pettenati J, Grand C, Boudet AM, Lamb CJ (1988) Cinnamyl alcohol dehydrogenase, a molecular marker specific for lignin synthesis: cDNA cloning and mRNA induction by fungal elicitor. *Proc Natl Acad Sci USA* 85:5546–5550

Whetten R, Sederoff R (1995) Lignin biosynthesis. *Plant Cell* 7:1001–1013

Willats WGT, McCartney L, Mackie W, Knox JP (2001) Pectin: Cell biology and prospects for functional analysis. *Plant Mol Biol* 47:9–27

Wu AM, Rihouey C, Seveno M, Hörnblad E, Singh SK, Matsunaga T, Ishii T, Lerouge P, Marchant A (2009) The Arabidopsis IRX10 and IRX10-LIKE glycosyltransferases are critical for glucuronoxylan biosynthesis during secondary cell wall formation. *Plant J* 57:718–731

Xu W, Campbell P, Vargheese AK, Braam J (1996) The Arabidopsis XET –related gene family: Environmental and hormonal regulation of expression. *Plant J* 9:879–889

Yokoyama R, Nishitani K (2001) A comprehensive expression analysis of all members of a gene family encoding cell-wall enzymes allowed us to predict cis –regulatory regions involved in cell-wall construction in specific organs of Arabidopsis. *Plant Cell Physiol* 42:1025–1033

York WS, O'Neill MA (2008) Biochemical control of xylan biosynthesis—Which end is up? *Curr Opin Plant Biol* 11:1–8

Zabotina OA, van de Ven WT, Freshour G, Drakakaki G, Cavalier D, Mouille G, Hahn MG, Keegstra K, Raikhel NV (2008) Arabidopsis XXT5 gene encodes a putative alpha-1,6-xylosyltransferase that is involved in xyloglucan biosynthesis. *Plant J* 56:101–115

Zandleven J, Beldman G, Bosveld M, Schols HA, Voragen AGJ (2006) Enzymatic degradation studies of xylogalacturonans from apple and potato, using xylogalacturonan hydrolase. *Carbohydr Polymers* 65:495–503

Zandleven J, Sørensen SO, Harholt J, Beldman G, Schols HA, Scheller HV, Voragen AJ (2007) Xylogalacturonan exists in cell walls from various tissues of *Arabidopsis thaliana*. *Phytochemistry* 68:1219–1226

Zhong R, Peña MJ, Zhou GK, Nairn CJ, Wood-Jones A, Richardson EA, Morrison WH 3rd, Darvill AG, York WS, Ye ZH (2005) Arabidopsis *Fragile Fiber8*, which encodes a putative glucuronyltransferase, is essential for normal secondary wall synthesis. *Plant Cell* 17:3390–3408

Zhou GK, Zhong R, Himmelsbach DS, McPhail BT, Ye ZH (2007). Molecular characterization of PoGT8D and PoGT43B, two secondary wall-associated glycosyltransferases in poplar. *Plant Cell Physiol* 48:689–699

6 Ecologically Sustainable Bioenergy Cropping Systems
Species Selection and Habitat Considerations

Christopher R. Webster and David Flaspohler
Michigan Technological University

Steve Pawson and Eckehard Brockerhoff
New Zealand Forest Research Institute

CONTENTS

6.1 INTRODUCTION

Bioenergy at its most fundamental level is about burning the accumulated proceeds of photosynthesis. Fossil bioenergy, in the form of coal, fueled the early industrial revolution and today, along with oil, sustains contemporary human civilization. In the early 20th century, these concentrated and conveniently packaged fossil fuels largely replaced wood, tallow, and dung as sources of heat and animal traction as a source of power; 200 years ago, 20% of U.S. agricultural land was devoted to growing "fuel" to feed livestock (Sexton et al. 2007). Today, many in the world recognize the many environmental, geopolitical, and economic costs of fossil fuel dependence, and the growing immediacy of the exhaustion of our fossil fuel reserves. As a result of this recognition, we find ourselves reconsidering bioenergy, primarily from living plants, as a partial solution to these problems. Despite the recent attention paid to liquid biofuels, all biomass allocated to energy worldwide currently represents only approximately 10% of the total of 11,410 million tons of oil equivalent used per annum (IEA 2007).

The potential advantages of contemporary bioenergy over fossil bioenergy include reduced or neutral greenhouse gas (GHG) emissions, a renewable and sustainable energy source, and an invigorated agricultural and/or forestry sector. Proponents of the recent surge in interest in bioenergy often tout it as a benign alternative to fossil energy. The broader public, at least in the United States and Europe, was eager to embrace this new energy source, which of course could be grown in-country and so had the potential of reducing dependence on imported fuels. Brazil and several nations, had adopted bioenergy much earlier and, in the new era of the bioeconomy, were hailed as models for the rest of the world (Morgan 2005). As the exhaustion of our fossil fuel reserves became clearer and petrol prices rose in 2002 and then soared in 2006, industries and governments around the world quickly poured resources into bioenergy research and production, tripling worldwide biofuel production between 2002 and 2007 (FAO 2008). Today, there is vigorous scientific debate on a host of issues related to biofuels including how to calculate the actual GHG reductions from different bioenergy options and the effects of food-based biofuel feedstocks on world food prices. Although these are important questions, they are beyond the scope of this chapter, which will focus on how various bioenergy production systems affect ecological systems and wildlife.

One lesson that has become increasingly clear from recent investments in bioenergy is that energy policy and production are intertwined in virtually all aspects of the world economy. When European Union policies meant to provide incentives for biodiesel use raised fears of rainforest clearing for oil palm plantations in Indonesia (Figure 6.1), and when the expansion of corn-ethanol production in Iowa contributed to food riots in Mexico, it became obvious that a better understanding of the linkages between bioenergy and the forces driving land-use change was needed. Just as rapid fluctuations in the price of oil have left bioenergy policy-makers and investors humbled and wary, biologists concerned with maintaining ecological services find themselves guardedly hoping that bioenergy can become a step toward a more sustainable way of producing energy. It is our hope that rather than rejecting bioenergy as trading one set of problems for another, we can improve the ecological footprint of bioenergy on the landscape and in natural ecosystems. If we are successful, it will become a smaller and smarter footprint that is more thoughtfully distributed over the landscape and leaves room for healthy soils, clean water, and enduring rich biodiversity; in some cases, bioenergy crops may even represent an improvement in habitat over current intensive land use.

6.1.1 The Mirage of a Bioenergy Panacea

In the early years of renewed interest in bioenergy (in the United States, 2006–2008), the unintended consequences of an expanded bioenergy economy were rarely considered or at least rarely reflected in policy. Well-established agricultural systems and government subsidies ensured that in the United States, *Zea mays* (corn) would be the initial dominant feedstock for liquid biofuel. Today corn accounts for more than 90% of the biofuel produced in the United States, although it represents at best a modest reduction in GHG compared with gasoline and demands large inputs of fertilizer, herbicides, pesticides, and water (NASS/USDA 2007; Fargione et al. 2008;

FIGURE 6.1   An oil palm (*Elaeis* spp.) plantation established on former rainforest land. (Courtesy of Rhett Butler.)

Searchinger et al. 2008). Currently, most corn is grown for food for people or livestock using well-established cropping systems. The challenge for biodiversity is that like many monoculture systems, cornfields provide little or no habitat for other species. However, some "monocultures" such as plantation forests may provide habitat for numerous understory plants and animals (Brockerhoff et al. 2008).

Humans have appropriated approximately 37% of the terrestrial habitats on the planet for food production, and an estimated 10^9 ha of additional land will need to be put into production by 2050 to meet the anticipated 50% increase in food demand as the world population grows and economic development raises the standard of living of many countries (Vitousek et al. 1997). Of all forms of human land use responsible for species listing under the U.S. Endangered Species Act, modern agriculture ranks number one with 38% of species listed as a result of habitat loss to agriculture (Stein et al. 2000). Clearly, any expansion of agriculture at the expense of natural habitat will negatively affect biodiversity. Society expects clean water, fertile soils, food, fiber, nutrient cycling, and aesthetics from the planet's soils and waters. A new service is being added: fuel for electricity and transportation. It is therefore not surprising that ecologists were among the first to recognize that in addition to the potential benefits of bioenergy, its expansion also had great potential to harm ecosystems, habitats, and species (Hill et al. 2006, 2007; Tilman et al. 2006; Fargione et al. 2008, 2009; Searchinger et al. 2008; Flaspohler et al. 2009; Malakoff 2009). In fact, we believe that land-use changes associated with a worldwide expansion of some bioenergy production systems are amongst the greatest current threats to biodiversity. The degree of this threat will be a function of the form that bioenergy production takes on the landscape. In short, any replacement of diverse native habitats with traditional agronomic monocultures will represent a loss of biodiversity and some associated ecosystem services. However, there are several promising alternatives to unsustainable models that could allow bioenergy production and a functioning ecosystem to coexist. In fact, there are opportunities for win-win scenarios in which biofuel production can benefit biodiversity conservation. Another important consideration in gauging bioenergy effects on ecosystems is the type of habitat (e.g., native diverse vs. non-native simplified) being displaced by expanding bioenergy plantings.

We will examine various bioenergy alternatives and assess key tradeoffs associated with each. In this chapter, we address the following questions:

- What are the current dominant agricultural and forestry systems used to grow plants for bioenergy?
- In what ways do these managed systems interact with the physical and biological systems that support life on earth?
- Can managed bioenergy production systems be designed in ways that preserve the ecosystem services furnished by intact ecological systems?

Addressing such questions at this time is of paramount importance because bioenergy production is expanding worldwide, and several nations have adopted policies that will promote the continued growth of the share of energy that originates from plants (Commission of European Communities 2005). The scope of bioenergy systems that we discuss in this chapter include all plant-based energies, including direct electricity generation from burning plant material and conversion of plant material to liquid biofuels such as ethanol and biodiesel. This spans a wide range of target plants from traditional food crops like corn and sugar cane, to grasses such as *Panicum virgatum* (switchgrass) and miscanthus (*Miscanthus* spp.), to woody species including *Pinus* spp. (pines), *Eucalyptus* spp. (eucalypts), hybrid and other *Populus* spp. (poplars), and even numerous species of algae. What all plant-based bioenergy systems have in common is a reliance on photosynthesis as the biochemical means of gathering carbon into a convenient form for use as energy.

Green plants have been sequestering carbon for at least 3 billion years. Our current modern industrial world has grown and become utterly dependent on burning fossil fuels, which are the direct or indirect products of photosynthesis accumulated over many millions of years. A second characteristic of current plant-based bioenergy systems is the intensification of production systems compared with the ecosystems they replace. Like food-based agricultural systems, the focus for these systems has been on maximizing short-term per-hectare plant productivity. Simply put, this means removing all plants but the target species, which is then planted in a systematic array and typically maintained with high inputs of water, fertilizer, herbicides, and insecticides. Fossil fuels are often used at virtually every stage of production and distribution as well as in the manufacture of the diverse agricultural inputs.

It is largely because of the high emissions of GHGs during land conversion or during production and distribution of target plants that some bioenergy systems compare unfavorably to petrol (Fargione et al. 2008; Charles 2009). In fact, some recent analyses suggest that if GHG reduction were the primary goal, conversion of cropland to forest would generate higher carbon sequestration rates than the avoided emissions from the use of ethanol and biodiesel generated from any of the common feedstocks (e.g., sugarcane, corn, sugarbeets, rapeseed, and woody biomass) (Righelato and Spracklen 2007). Clearly, refinements and standardization of life-cycle assessment methods will help clarify the true GHG costs and benefits associated with first- and second-generation biofuels (FAO 2008; Searchinger et al. 2008; Purdon et al. 2009). Until such analyses are available and widely accepted, several well-tested principles from ecological science can be used to guide bioenergy toward more sustainable models.

6.2 DEFINING ECOLOGICAL SUSTAINABILITY

Ecology is the branch of science concerned with understanding how the biological and physical world interacts and function. Ecology is concerned with many spatial scales from microscopic to global: examining processes such as energy flows through ecosystems, biogeochemical cycling; the role of plants, animals, fungi, and microorganisms in supporting these cycles; and the ways that landscape pattern and composition influence biodiversity. During the last century and particularly the last 50 years, ecology has produced a wealth of new understanding, much of which has great relevance for bioenergy and the ecological systems used to produce it. In this section, we briefly describe a few of these insights, focusing on those most germane to bioenergy. It is our belief that for any system to be ecologically sustainable, we need a substantial understanding of how each of the following processes support and are affected by a given bioenergy production system.

6.2.1 KEY TERMS AND PROCESSES

6.2.1.1 Nutrient Cycling

The compounds that form the cells of all living things are primarily composed of carbon, nitrogen, phosphorus, sulfur, and water. Over time, each of these essential nutrients circulates through the local ecosystem and wider biosphere. The rates at which these nutrients move through living and nonliving ecosystem components can be influenced by human activities. Well-documented instances of anthropogenic disruptions in nutrient cycles include increased nitrogen deposition from industrial agriculture (resulting in eutrophication of many aquatic ecosystems) and increased atmospheric carbon dioxide (CO_2) from combustion of fossil fuels. One of the primary attractions of bioenergy is its potential to reduce fossil fuel carbon emissions by closing the carbon cycle with sequestration by green plants instead of the net carbon pollution from coal, oil, and natural gas. However, land-use changes associated with intensified management such as monoculture agriculture and plantation forestry can have enormous effects on other key cycles such as water circulation and the maintenance of soil fertility. The CO_2 and other GHGs released as a direct result of conversion of natural vegetation to bioenergy has been called "the carbon debt" to be "paid back" by reductions in GHG emissions by bioenergy as compared with the fossil fuels being displaced (Fargione et al. 2008). However, some recent analyses suggest that such initial debt can be so high that it can take centuries or longer to make up for carbon losses from ecosystem conversion (Fargione et al. 2008).

6.2.1.2 Biomass and Trophic Structure

For all terrestrial and most marine ecosystems, solar radiation transformed by green plants into sugars produces the living tissues that support life. The total amount of living and dead biomass is greatest at lower trophic levels and declines as one moves up trophic levels from green plant producers, to primary consumers like herbivorous animals, to primary predators like wolves and hawks. This is sometimes called "a pyramid of biomass" because a large amount of primary plant production (i.e., biomass) is needed to support relatively less biomass of herbivores. It is also the reason why bioenergy systems focus on using plants as a carbon source when carbon is also present in rabbits, deer, and bison; growing bison as a biofuel feedstock would be highly inefficient and ethically questionable. On average, there is an approximately 90% reduction in biomass as one moves up one trophic level, so that 10,000 kg of grass could support 1000 kg of grasshoppers, which could support 100 kg of insectivorous songbirds. For bird-eating hawks then, such a system would support only 10 kg, about the weight of ten red-tailed hawks (*Buteo jamaicensis*). For bioenergy systems to meet ecological definitions of sustainability, they must operate with recognition that a dynamic balance must be maintained between trophic levels including soils and plants, and the animals, fungi, and bacteria that depend on continued manufacture and recycling of biomass into the ecosystem. For these reasons, the amount of biomass that can be sustainably removed is often not well known.

6.2.1.3 Vegetation Diversity and Structure

Diversity begets diversity is a simple rule of thumb in ecology. This means that higher diversity at lower trophic levels usually supports higher diversity at the next level up and so on. Thus a temperate forest in North America has relatively few tree species and supports relatively few leaf-chewing caterpillars and a relatively simple bird community that feeds on these caterpillars. In contrast, a similar size patch of rainforest in Brazil may have a hundred times the tree species, a thousand times the diversity of arthropods, and far more arthropod predators including many taxa entirely absent from temperate forests. For this reason, the conversion of native habitats nearer the equator is often more destructive to biodiversity than habitat loss in the temperate zone. Bioenergy cropping systems that include a diversity of species as feedstocks or where the target species are grown in concert with other plants are almost always preferable to monocultures. The cultivation of native species is preferable over the use of exotics, and it is also beneficial for biodiversity when numerous native species are growing along with the feedstock. The greater structural diversity provided by the varying plant architecture (i.e., forest canopy, subcanopy, shrubs, and herbs) provides more niches to allow coexistence of many more species. For woody (e.g., trees and shrubs) and grass-based feedstocks, polycultures will usually support more species of birds, mammals, amphibians, reptiles, and arthropods than monocultures.

6.2.1.4 Species-Area Relationships

One of the most consistent relationships in ecology is the finding that as area increases, the number of species found in this area also increases. The rate of increase as one moves from a square meter to a square kilometer obviously differs for a patch of arctic tundra compared with a tropical rainforest, but the general positive relationship is virtually universal. This simple empirically verified pattern leads to some surprisingly profound insights. Generally speaking, from a species-area perspective, larger patches of undisturbed or less disturbed habitats support more native species. If more intensive systems for growing bioenergy feedstocks allowed larger areas to be set aside for conservation, this would benefit species dependent on native habitats; the less land that is disturbed, the greater the mean and median size of remaining habitat blocks. Less intensive systems sometimes require more area to produce equal yields so that production systems and the amount of area under production are linked. Many species, such as large predators and herbivores, require large areas to persist. Additionally, some biomass production systems, especially those involving re-establishment of native warm season grasses or trees where appropriate, may, if purposefully placed on the landscape, create corridors, increase native habitat patch sizes, and reduce fragmentation. Nevertheless, habitat loss resulting from conversion of diverse native habitats into intensive biomass cropping systems would result in a reduction in patch size and increasing fragmentation and isolation of native habitats. Simply put, along with pressures to convert land for more traditional uses such as agriculture, housing, and commercial development, an expanding bioenergy market will likely be a mixed bag for biodiversity.

6.3   BIOENERGY CROPPING SYSTEMS AND SPECIES SELECTION

6.3.1   Residues

Cellulosic "residues" and "waste" have been touted as bioenergy feedstocks with little to no ecological downside and high compatibility with traditional agricultural and forestry practices. In the context of municipal organic waste (e.g., grass clippings, trimmings, leaves, scrap wood, and paper products), this generalization is quite tenable (Fargione et al. 2008; Koh et al. 2008). However, in the context of agricultural crops and forest residues left after traditional management, this generalization is much more context specific (Harmon 2001; Lal and Pimentel 2007; Robertson et al. 2008).

Organic residues are an important source of soil organic matter; food for soil organisms (Burger 2002); and, in minimum tillage and no-till row crop agricultural systems, provide dormant season cover to the soil surface and are an important driver of soil organic matter dynamics (Six et al. 1999). Soil organic matter content influences soil tilth, erodibility, and water-holding and cation-exchange capacity (Fisher and Binkley 2000; Vance 2000; Blanco-Canqui et al. 2005). Accumulation of carbon from organic residues into the soil also provides an important sink for atmospheric CO_2 (Batjes 1998; Harmon 2001; Post et al. 2004; Omonode and Vyn 2006). Cover of organic material on the soil surface reduces erosive potential by dampening the effect of rain drops and decelerating and decentralizing surface flow (Dabney et al. 2004; Sayer 2006; Montgomery 2007). The quantity of organic residue required to maintain soil health and reduce erosion risk may vary greatly depending on soil physical properties, fertility, topographic position, crop type, and management history. For example, Dabney et al. (2004) found that removal of corn residues doubled erosion rates in conventional and no-till tillage systems on silt loam soils in northern Mississippi. However, Lafond et al. (2009) were unable to find a significant long-term (50-year) correlation between retention of above-ground residues in wheat (*Triticum* spp.) cropping systems and soil organic matter or organic nitrogen under conventional tillage on more level terrain in the thin-black soil zone of Saskatchewan, Canada.

In managed forests, logging slash (i.e., branches and nonmerchantable bole wood; Figure 6.2) and dead standing and down trees are often collectively referred to as forest "residues." Small nonmerchantable trees left after harvest, which are undesirable as future growing stock, may also be included in this category. Estimates suggest that 2.7 million t of forest residues may be available annually in Michigan alone (Froese 2007). Most of the current discussion related to the sustainability of harvesting this material stems from concerns regarding the depletion of soil nutrients and site productivity (Mann and Tolbert 2000), especially on coarse textured soils or under very short rotations. Available research contrasting contemporary stem-only with whole-tree harvesting systems has failed to find lasting effects on site productivity unless harvesting activities result in significant disturbance to the forest floor (Johnson 1992; Johnson and Curtis 2001). However, on some sites, there is a strong potential for calcium depletion (Boyle et al. 1973; Mann et al. 1988; Federer et al. 1989), which may require remediation possibly with ash fertilization (Vance 1996). Less attention has been given to the role of these materials as wildlife habitat and substrates for plant establishment.

FIGURE 6.2 Logging residues (tops, branches, and nonmerchantable bole wood) left after a partial harvest of a mixed stand of aspen (*Populus* spp.) and conifers. If managed sustainably, this material may provide an important source of biomass for bioenergy from traditional forest management activities. (From Janowiak, M.K. and Webster, C.R., *J Forestry*, 108, 16–23, 2010.)

Down dead wood and standing snags provide important habitat for small and large mammals (Zollner and Crane 2003; McCay and Komoroski 2004), birds (Rosenberg et al. 1988), amphibians (Butts and McComb 2000), arthropods (Jonsell and Weslien 2003; Jabin et al. 2004; Ulyshen et al. 2004; Brin et al. 2009), and a host of microorganisms (Allen et al. 2000). Down wood is also an important substrate for the regeneration of many species of forest trees (Noguchi and Yoshida 2004; O'Hanlon-Manners and Kotanen 2004) and other plants (Santiago 2000). Down wood may provide transient refugia for tree seedlings and herbaceous plants in areas with high abundances of native or exotic ungulates (Casabon and Pothier 2007; de Chantal and Granström 2007). Large and highly decayed downed dead wood and snags are significantly less abundant in second-growth and managed forests than in areas of primary or old-growth forest (e.g., Goodburn and Lorimer 1998; Fridman and Walheim 2000). Increasingly, forest managers have recognized the importance of this material, and its conservation and creation are often codified in management plans (Hunter 1990). Consequently, an inherent danger in the rush to increase the utilization of forest residues is the potential further depletion of this foundational habitat element of forested ecosystems.

The use of nonmerchantable living shrubs and trees that are undesirable as future growing stock might incentivize activities that are currently unprofitable, such as timber stand improvement, invasive species control, and fuel reduction treatments. However, provisions should be made for the retention of some low-value cull trees with cavities or other desirable habitat features as well as native species that are poorly represented but add structural or compositional diversity.

Ecologically sustainable residue harvesting will require a thoughtful assessment of the critical thresholds of these materials that are needed to maintain site productivity, reduce erosion hazard, and support biodiversity. Further consideration should also be given to the role of this material as a source and sink for atmospheric CO_2. In the case of forest residues, for the first few years after harvest, logging slash is a net emitter of CO_2 as a result of microbial respiration associated with the decay of fine materials and surface tissues (Ganjegunte et al. 2004; Birdsey et al. 2006). However, depending on the recalcitrance of the species, temperature, and fragment size, the generally slow decay rates associated with coarse woody debris may make them an important carbon sink and long-term source of nutrients (Ganjegunte et al. 2004).

6.3.2 BIOENERGY PLANTATIONS AND DEDICATED CROPS/TREES

Traditional forest management has favored long rotation times that often exceed 75 years for timber (as low as 26 years in New Zealand) but can be as low as 5–10 years for intensive pulp or charcoal production in tropical or subtropical regions. In contrast, arable row crops are frequently harvested on an annual basis, require more intense management than forestry, and consequently place greater demands on ecosystem services. To date, most "first-generation" biofuel feedstocks have been annual crops [e.g., corn and rapeseed (*Brassica napus*) oil], which have replaced existing croplands or displaced native habitat. New "second-generation" feedstocks, such as cellulosic conversion of wood residues and perennial grasses, are likely to increase dramatically in the near future in response to the renewable energy targets of many countries. Cellulosic bioenergy has the potential to dramatically alter current land-use patterns, with flow-on effects that change crop rotation patterns throughout the landscape. Studies that examine different levels of logging residue removals on soils with different potential erodability would provide valuable guidance for future growth of this sector (EEA 2006).

6.3.2.1 Implications of Changing Crop Rotation Lengths

6.3.2.1.1 Soils and Nutrient Cycling

The phenomenon of harvest-induced site nutrient depletion has been of concern to foresters for more than 50 years, and a shift to shorter rotation lengths for bioenergy production will exacerbate this effect. Young trees have a greater proportion of sapwood and crown material, which has higher nutrient concentrations than the heartwood that comprises the bulk of long-rotation stem only harvesting (Kimmins 1997). By reducing rotation length, nutrient losses are magnified by the increased harvest intensity induced by short rotations and the compounding effect of higher nutrient concentrations in young trees. In addition to purpose-grown short-rotation bioenergy forests, the extraction of wood residues from traditional logging is often the first step used by wood-based lignocellulosic refineries because it is comparatively simple to utilize such a "waste" product (Scion 2008). Although residue removal does not change the rotation length per se, it does increase the utilization intensity because whole-tree harvesting removes more nutrients than traditional stem-only harvesting for timber (Kimmins 1997). Furthermore, removing residual woody biomass from forests may create an ecological trap, in which dead-wood dependent species are "mass trapped" by the delayed removal of woody debris (Hedin et al. 2008).

Bioenergy crops may in some cases increase rotation lengths (e.g., the replacement of annual row crops with short-rotation coppice crops of willow and poplar) or perennial grasslands. Annual row crops have been shown to deplete soil organic matter (most within the first few years of cultivation; Bowman et al. 1990) and thus require substantial fertilizer inputs to maintain productivity. However, expansion of new cellulosic bioenergy crops will shift rotations from annual to perennial crops, allowing soils to rejuvenate and facilitating increases in soil organic matter and fertility (Cook and Beyea 2000; Rowe et al. 2009). For example, below-ground biomass is reportedly much higher in *P. virgatum* (switchgrass; 7.74 Mg/ha) than cultivated cropland (4.35 Mg/ha) (Liebig et al. 2005). Similar increases in soil organic matter have been reported after a shift from annual cropping to fast-growing woody biomass crops such as poplars (Hansen 1993; Rowe et al. 2009), although temporary losses as a result of erosion and mineralization are known to occur before canopy closure, and in some cases no increases have been recorded (Grigal and Berguson 1998).

6.3.2.1.2 Wildlife Habitat

Old-growth forests sustain high levels of biodiversity, often with species unique to ancient forests, and the large-scale conversion of virgin forest to short-rotation woody biomass crops would be highly detrimental to forest biodiversity. Wind, fire, disease outbreaks, and other natural disturbance factors initiate natural regeneration in parts of forests, replacing older trees and forest communities with assemblages of disturbance-adapted colonizers and young trees. Over time, the process of succession (Connell and Slatyer 1977) leads to a gradual return of a forest community with shade-tolerant understory plants and canopy trees and the associated fauna, with a considerable exchange of species. This recovery of natural communities takes decades or centuries, depending on a wide range of factors, and is also apparent in modified production ecosystems such as plantation forests (e.g., Brockerhoff et al. 2003). On land where the natural land cover is (or was) forest, the habitat becomes increasingly suitable for forest species and communities as the time span between disturbances (or harvesting) increases. The same applies to disturbance effects in grassland and other nonforest ecosystems. Therefore, short-rotation cropping systems offer generally little opportunity for the establishment of a diverse range of species, except for disturbance-adapted specialists. In forests currently managed for timber production, a shift to shorter rotations or away from management activities that attempt to emulate natural disturbance regimes will further decrease their habitat suitability for forest-specialist species and could conceivably cause localized extinctions of some species, although in some cases shorter forest rotation times may be beneficial because they increase the proportion of early successional habitat that is important for some species. For example, Eycott et al. (2006) found the greatest diversity of understory plants in young stands (<10 years) and concluded that shorter rotations provide greater biodiversity benefits for heath-land landscapes. However, short rotation bioenergy forests are likely to have much higher stocking rates than traditional timber forests, which could suppress understory plant diversity. Increasing crop rotation lengths by planting perennial species rather than annual crops has significant potential to enhance habitat for biodiversity, particularly if ecosourced native crop species are used (Cook and Beyea 2000). Perennial short-rotation woody biomass crops that replace row crops in formerly forested areas have been shown to benefit forest-dependent bird species in some cases (Christian et al. 1997, 1998; Hanowski et al. 1997), but they do not provide a satisfactory substitute for original native forest (Christian et al. 1998). Similarly, perennial native grasses such as *P. virgatum* can provide suitable habitat for some but not all grassland-dependent bird species (Murray and Best 2003; Murray et al. 2003; Roth et al. 2005), which are in severe decline in part because of native habitat loss (With et al. 2008).

6.3.2.1.3 Water Quality

The decreasing quality and quantity of fresh water is a global problem that is, in part, a result of the progressive replacement of native habitat with "thirsty" human-dominated productive ecosystems. For example, plantation forests established in inappropriate areas (e.g., historically nonforested landscapes) can result in decreased regional water availability (Jackson et al. 2005). On the basis of current information, present-day biofuel production in the United States is not thought to be a significant stress on water availability; however, future expansion of bioenergy crops (particularly corn) could place significant demands on scarce water resources with concurrent effects on water quality because of fertilizer and pesticide application (NRC 2007). However, it should be acknowledged that quantification of U.S. water supplies is currently inadequate (OSTP 2007). Although plantations may use more water than some arable crops, if planted in historically forested areas, regional water balances are likely to return to their "natural state". For example, plantation forests that have appropriate riparian buffers to mitigate against unwanted sediment introduction and solar-induced temperature increases after harvesting have better water quality relative to adjacent pastoral areas in New Zealand (Quinn et al. 1997). However, reduced forest rotation lengths for woody biomass crops will increase the frequency of catchment disturbance rates and enhance the risks of detrimental effects such as stream sedimentation and solar-induced temperature increases on aquatic fauna.

6.3.2.2 System Stability and Landscape Considerations

In addition to the unintended consequences of the establishment of biofuel plantations on biodiversity conservation, there are concerns that this could have other unforeseen consequences. Although Lindenmayer's (2009) review addressed concerns about the establishment of plantations for general climate change mitigation, they are equally applicable to plantations for biofuel production. Plantations that are more liable to destruction by wildfire than the vegetation they replaced (Thompson et al. 2007) are potentially counterproductive to the aims of increased carbon sequestration. Likewise, monoculture plantations are potentially more susceptible to insect and disease problems than mixed plantings (Jactel and Brockerhoff 2007; Jactel et al. 2008). This may be the case with native and exotic species. Such unintended potential consequences should be carefully considered when crops are selected for biofuel production.

Apart from habitat loss, the fragmentation of remaining areas of natural habitat is considered a main cause of biodiversity loss (Hunter 1990; Murcia 1995; Henle et al. 2004). Fragmentation can alter species richness and abundance, increase biological invasions, and alter community structure and ecosystem processes (Laurance et al. 2002). Small, isolated patches of habitat may be inadequate for larger species that require large territories to have access to sufficient food sources. Some species avoid forest edges and occur only in core areas of habitat that are not present in smaller fragments. For some migratory songbirds in the northern Great Lakes region, forest edges created by logging increased breeding density and nesting failure, creating an ecological trap (Flaspohler et al. 2001). Even small areas of nonhabitat can become barriers that are impassable. For example, many Amazonian rainforest species avoid clearings of as little as 100 m width, which can lead to greatly reduced dispersal (Laurance et al. 2002). Combined with the fact that fragmentation generally reduces the size of local populations, this affects reproduction and increases the risk of genetic deterioration and extinction (Nason and Hamrick 1997; Fahrig 2001). For example, lizards in habitat fragments have been found to exhibit reduced levels of allelic diversity, which is likely to cause inbreeding depression or reduced recruitment (Stow and Briscoe 2005).

Past land development for agricultural and fiber crops was the main cause of habitat loss and fragmentation, and future land development for the production of biofuels is likely to further aggravate the detrimental effects of fragmentation on biodiversity. However, it is possible to mitigate fragmentation effects. To avoid detrimental effects of development, it is of the utmost importance to protect remnants of natural vegetation and, wherever possible, to restore suitable habitat, preferably in such a configuration that it improves "connectivity" and enables the movement and exchange of species between habitat patches (Lindenmayer and Franklin 2002). However, production land is not necessarily entirely unsuitable as habitat, and some crops may provide certain key habitat features (Fischer and Lindenmayer 2006; Kupfer et al. 2006). For example, plantation forests can alleviate fragmentation of natural forest remnants by providing supplementary habitat for forest species, by increasing connectivity between remnants, and by buffering remnant forests against edge effects (Brockerhoff et al. 2008). These considerations are particularly important in regions that have already been heavily affected by habitat loss and fragmentation. However, increased connectivity can potentially also increase the risk of wildfire and disease and pest spread (Lindenmayer 2009).

6.3.3 Native versus Exotic Species

Intuitively, the selection of native species as a foundation for bioenergy feedstocks is more likely to create conditions and provide resources required by other native species present in the wider landscape. A recent example is the rise of the perennial biofuel plant, *P. virgatum*, which when planted in its native range is known to provide significant habitat for some native grassland birds in the United States (Murray et al. 2003; Roth et al. 2005). This is not to say that exotic crop species cannot provide opportunities for biodiversity; on the contrary, we know that potential exotic bioenergy crop species provide a significant alternative habitat for a proportion of native species. For example, exotic tree plantations can support numerous native birds (Clout and Gaze 1984), invertebrates (Brockerhoff et al. 2005; Mesibov 2005; Barlow et al. 2007; Berndt et al. 2008; Pawson et al. 2008; Rowe et al. 2009), and plants (Brockerhoff et al. 2003; Eycott et al. 2006), whereas the perennial grass crop miscanthus also supports significant numbers of various native species (Semere and Slater 2007a, 2007b). However, we do not in any way advocate the replacement of native habitat with exotic biofuel crops because many species are habitat specialists and are restricted to pristine native habitat (Gardiner et al. 2008).

Given that native plantings should be better for biodiversity, do native species have the desired ecological traits that have been proposed as ideal for economically viable bioenergy crops? Examples of these traits are C_4 photosynthesis, relative freedom from pests and disease, perennial growth, and monocultural cropping (Heaton et al. 2004). Several of these criteria suggest native species will make inferior biofuel feedstocks compared with exotic alternatives. First, bioenergy crops planted in their native bioregion will retain their full suite of pests and diseases, whereas exotic species planted in novel habitats have the potential to escape some, if not all, of their natural enemies (herbivores and plant pathogens) resulting in increased growth rates (Ebeling et al. 2008). However, exotic plants are not always free of natural enemies because similarities among congeners may facilitate host shifts of native herbivores to exotic plants (Frenzel and Brandl 2003). For example, *Pinus radiata* is exotic in New Zealand and Europe. In New Zealand it grows spectacularly fast and has comparatively few pests and diseases, whereas in Europe *P. radiata* can suffer heavily from host shifts from congeneric *Pinus* spp., as shown by extreme attacks of shoot-pruning by the bark beetle, *Tomicus piniperda* (Amezaga 1997). Secondly, some native species are not adapted to a monocultural cropping system, which is the predominant current biofuel paradigm. Native species tend to inhabit complex ecosystems that retain their stability through species diversity [i.e., the diversity = stability hypothesis; see reviews by Hooper at al. (2005) and Jactel and Brockerhoff (2007)]. In the past, polycultures were thought to reduce productivity and intuitively do so if they are grown with the desire to maximize a specific resource (e.g., a fruit, seed, or timber that can only be derived from one species). However, recent studies have shown that the production of a more generic resource (such as "biomass") can be higher in perennial mixtures of native species than monocultures of exotic species (Tilman et al. 2006), although these findings have been controversial (Russelle et al. 2007; Tilman et al. 2007).

The same traits that identify exotic species as the most useful bioenergy crops will also raise additional ecological questions if large-scale planting proceeds, e.g., the invasive potential of exotic species (Keane and Crawley 2002; Colautti et al. 2004; Raghu et al. 2006; Barney and Ditomaso 2008). However, invasiveness is not exclusively the realm of exotic species because political regions are frequently disjunct relative to a species' actual biogeographic range. Thus, species native to one political district may actually represent an exotic species to the ecosystem in question.

6.3.3.1   Case Study: Exotic Conifers as Biofuel Crops and Invasive Weed Risks

Conifers are currently being considered as a biomass feedstock for cellulosic conversion to fuels (Scion 2008). However, previous experience with such species suggests that additional widespread planting of exotic conifers may result in significant negative consequences for biodiversity. Exotic conifers were introduced for forestry and revegetation purposes in the southern hemisphere and have subsequently become a significant invasive weed in South Africa, Australia, and New Zealand (Ledgard 2004; Richardson and Rejmanek 2004), with emerging issues in South America (Pena et al. 2008; Richardson et al. 2008). The extent of these invasions is not trivial, with at least 2000 km^2 affected in New Zealand (Ledgard 2004) (Figure 6.3), whereas in South Africa one species alone, *Pinus pinaster*, has spread over approximately 3256 km^2 and has severely affected the native fynbos vegetation (Richardson et al. 1994). The family *Pinaceae* exhibits strong invasive tendencies (28 of 225 species are known invasives) of conifers, and the *Cupressaceae* (6 of 66 species known invasives) are also of concern (Richardson et al. 1994). The species that have invaded the largest areas are *P. contorta, P. halepensis, P. nigra, P. patula, P. ponderosa, P. pinaster, P. radiata,* and *P. sylvestris* (Richardson et al. 1994; Ledgard 2004). Interestingly, non-native conifers have yet to yield many naturalization events, and they represent very few invasive populations in the northern hemisphere as compared with southern hemisphere countries despite wide-scale planting of exotic species (Mortenson and Mack 2006).

The simplest solution to preventing unwanted invasive species problems from new bioenergy forests is to plant native species. However, in many instances exotic species will be preferred when they produce greater biomass in shorter periods than available native species. Thus, we are left

FIGURE 6.3   Corsican pine (*Pinus nigra*) has spread from a single farm shelter belt to cover Mount Barker in New Zealand's Southern Alps. The tail end of the wilding conifer spread is now 10 km from the original plantings. (Courtesy of Nick Ledgard, Scion.)

with preventative options, in which attempts are made to screen for and restrict the spread of potentially invasive species. However, predicting which species will be invasive is a significant challenge and remains an imprecise science (Hulme 2006). For example, in conifers traits such as seed size (<50 mg), seedling establishment, juvenile survival, length of the juvenile period (<10 years), and the periodicity of large seed crops are proven indicators of invasive potential (Richardson and Rejmanek 2004; Buckley et al. 2005). However, these traits are not infallible given that other external factors such as propagule pressure can be significant (Rouget and Richardson 2003; Richardson and Rejmanek 2004). Indeed, complex genotype-specific screening methods are now being suggested to reduce the risk of future invasive weed problems (Barney and Ditomaso 2008).

6.3.4 Perennial Polycultures

Although considerable emphasis has been placed on the use of monotypic plantings of short rotation woody plants and perennial grasses, a growing body of evidence suggests that diverse plantings may convey greater energetic efficiency, stability, and habitat value (Tilman 1996; Tilman et al. 1996, 2006; Lehman and Tilman 2000; Vilà et al. 2007). For example, Tilman et al. (1996) have shown that ecosystem productivity increases with species diversity in grasslands dominated by native warm-season grasses and tall-grass prairie forbs. An additional advantage of using diverse plantings of native species is that fertilizer and water inputs are lower than those required for many monocultures (Tilman et al. 2006). In fact, low-input/high-diversity grasslands may actually be carbon negative because their potential biofuel yield and ecosystem carbon sequestration well exceeds the fossil CO_2 emitted during their production (Tilman et al. 2006). Although data are lacking from forested systems, recent work in Mediterranean forests suggests wood production may also be positively associated with tree species richness (Vilà et al. 2007). In the eastern United States, forest productivity has been shown to be positively correlated with invertebrate biomass and songbird reproductive rates (Seagle and Sturtevant 2005). Additionally, animal species diversity is strongly correlated with habitat structural diversity (Tews et al. 2004), which can be enhanced in plantings by combining species with contrasting growth forms. For example, arthropod diversity tends to increase with plant diversity in restored grasslands (Siemann et al. 1998).

To some extent, design elements based on the relationship between species and structural diversity and plant and animal productivity could be incorporated into short rotation woody crop plantations in addition to the possibility of using diverse plantings of natives or well-managed native vegetation. For example, planting a patchwork pattern of various species would promote vertical and horizontal heterogeneity across the planting site. Species would need to be paired based on compatible growth and harvesting characteristics. The concept of area control from even-aged silviculture could also be incorporated at a spatial scale relevant to focal wildlife species to provide a diversity of age classes across the planting unit. In area control, harvesting is done in strips or blocks at a rate that allows the entire unit to be harvested over several years so that by the time the last patch is harvested, the first patch is ready to be harvested again.

In forested ecosystems, treatments that enhance structural and compositional diversity and conserve and create biological legacies (e.g., cavity trees and down dead wood) may also be used to achieve many of the same ends as perennial polycultures of native prairie species or short rotation woody crops. Possible techniques include precommercial and commercial thinning in even-aged stands, variable retention and/or legacy tree retention within even-aged treatment blocks, and uneven-aged management.

6.4 OTHER CONSIDERATIONS

Our review has highlighted that it is important to compare the merits of biofuels in terms of net carbon sequestration with other land uses and land-use change and to weigh up all components of carbon sequestration as well as emission. Ideally, this should be done by way of a life-cycle assessment to determine the net benefits of biofuel production in terms of carbon stocks and fluxes. However, there is another important consideration that adds further complexity to the assessment of net effects on climate change mitigation, which is ultimately the key concern. The net effects of biofuel production also need to be assessed in relation to their effects on global warming, which is not solely dependent on atmospheric concentrations of CO_2 and other climate change gases. A recent modeling study by Bala et al. (2007) determined that the difference in albedo between different types of vegetation can have a considerable effect on global warming. Particularly in temperate and boreal regions, forests may reflect less radiation than nonforest land cover. This is because forests usually have a darker hue than grasslands or bare ground and because there is less apparent snow cover in forests. However, this "albedo effect" is much less relevant in subtropical and tropical regions (Bala et al. 2007). Although there is still some uncertainty about the net effects of albedo versus carbon sequestration between different land uses that may be considered for biofuel production or climate change mitigation, it is clear that this is an important factor that needs to be taken into consideration.

6.5 TOWARD A DIVERSE BIOENERGY PORTFOLIO

Biofuels have the potential to provide win–win scenarios for energy production and the maintenance and enhancement of biodiversity. However, as clearly indicated in this chapter, this relationship is not a given and will require the thoughtful and deliberate selection of species and cropping systems. Species and cropping systems will need to be tailored to site and landscape-specific limitations and opportunities. From a practical production efficiency standpoint, there is a strong desire to identify a narrow suite of easy to work generalist species as major feedstocks. However, a diverse portfolio of dedicated bioenergy crops and residue streams would be more advantageous from a systems ecology perspective. The more species used and the more closely bioenergy plantations resemble natural ecosystems, the lower the risk of catastrophic crop losses due to new and emerging pests and diseases and/or drought and environmental change.

As the emerging bioeconomy continues to gather momentum, second-generation biofuels will likely have a distinctive advantage on the ecological balance sheet. These feedstocks, as currently envisioned, may be less likely to lead to habitat loss because they use residues from contemporary forestry and agricultural practices and in many cases have dedicated plantings that can be grown on land that has already been cleared for agriculture but is marginal for that use. When based on native species, conservation and production objectives may be achieved simultaneously (e.g., restoration of native grasslands as a source of cellulosic ethanol). In formerly forested regions, reforestation/plantation establishment may be preferable and could provide additional habitat to declining forest species.

REFERENCES

Allen RB, Buchanan PK, Clinton PW, Cone AJ (2000) Composition and diversity of fungi on decaying logs in a New Zealand temperate beech (Nothofagus) forest. *Can J For Res* 30:1025–1033

Amezaga I (1997) Forest characteristics affecting the rate of shoot pruning by the pine shoot beetle (*Tomicus piniperda* L) in *Pinus radiata* D Don and *P. sylvestris* L plantations. *Forestry* 70:129–137

Bala G, Caldiera K, Wickett M, Phillips TJ, Lobell DB, Delire C, Mirin A (2007) Combined climate and carbon-cycle effects of large-scale deforestation. *Proc Natl Acad Sci USA* 104:6550–6555

Barlow J, Gardner TA, Araujo IS, Ávila-Pires TC, Bonaldo AB, Costa JE, Esposito MC, Ferreira LV, Hawes J, Hernandez MIM, et al. (2007) Quantifying the biodiversity value of tropical primary, secondary, and plantation forests. *Proc Natl Acad Sci USA* 104:18555–18560

Barney JN, Ditomaso JM (2008) Nonnative species and bioenergy: Are we cultivating the next invader? *Bioscience* 58:64–70

Batjes NH (1998) Mitigation of atmospheric CO_2 concentrations by increasing carbon sequestration in the soil. *Biol Fertility Soil* 27:230–235

Berndt LA, Brockerhoff EG, Jactel H (2008) Relevance of exotic pine plantations as a surrogate habitat for ground beetles (Carabidae) where native forest is rare. *Biodiver Conserv* 17:1171–1185

Birdsey RT, Pregitzer K, Lucier A (2006) Forest carbon management in the United States: 1600-2100. *J Environ Qual* 35:1461–1469

Blanco-Canqui H, Lal R, Lemus R (2005) Soil aggregate properties and organic carbon for switchgrass and traditional agricultural systems in the southeastern United States. *Soil Sci* 170:998–1012

Bowman RA, Reeder JD, Lober RW (1990) Changes in soil properties in a central plains rangeland soil after 3, 20, and 60 years of cultivation. *Soil Sci* 150:851–857

Boyle JR, Phillips JJ, Ek AR (1973) Whole-tree harvesting: Nutrient budget evaluation. *J For* 71:760–762

Brin A, Brustel H, Jactel H (2009) Species variables or environmental variables as indicators of forest biodiversity: A case study using saproxylic beetles in Maritime pine plantations. *Ann For Sci* 66; doi: 10.1051/forest/2009009

Brockerhoff EG, Berndt L, Jactel H (2005) Role of exotic pine forests in the conservation of the critically endangered New Zealand ground beetle *Holcaspis brevicula* (Coleoptera: Carabidae). *NZ J Ecol* 29:37–44

Brockerhoff EG, Ecroyd CE, Leckie AC, Kimberley MO (2003) Diversity and succession of adventive and indigenous vascular understory plants in *Pinus radiata* plantation forests in New Zealand. *For Ecol Manag* 185:307–326

Brockerhoff EG, Jactel H, Parrotta JA, Quine CP, Sayer J (2008) Plantation forests and biodiversity: Oxymoron or opportunity? *Biodiver Conserv* 17:925–951

Buckley YM, Brockerhoff E, Langer L, Ledgard N, North H, Rees M (2005) Slowing down a pine invasion despite uncertainty in demography and dispersal. *J Appl Ecol* 42:1020–1030

Burger JA (2002) Soil and long-term site productivity values. In: Richardson J, Björheden R, Hakkila P, Lowe AT, Smith CT (eds), *Bioenergy from Sustainable Forestry: Guiding Principles and Practice*. Kluwer Academic Publishers, Dordrecht, The Netherlands, pp 165–189

Butts SR, McComb WC (2000) Association of forest-floor vertebrates with coarse woody debris in managed forests of western Oregon. *J Wildlife Manag* 64:95–104

Casabon C, Pothier D (2007) Browsing of tree regeneration by white-tailed deer in large clearcuts on Anticosti Island, Quebec. *For Ecol Manag* 253:112–119

Charles D (2009) Corn-based ethanol flunks key test. *Science* 324:587

Christian DP, Collins PT, Hanowski JM, Niemi GJ (1997) Bird and small mammal use of short-rotation hybrid poplar plantations. *J Wildlife Manag* 61:171–182

Christian DP, Hoffman W, Hanowski JM, Niemi GJ, Beyea J (1998) Bird and mammal diversity on woody biomass plantations in North America. *Biomass Bioenergy* 14:395–402

Clout, MN, Gaze PD (1984) Effects of plantation forestry on birds in New Zealand. *J Appl Ecol* 21:795–815

Colautti RI, Ricciardi A, Grigorovich IA, MacIsaac HJ (2004) Is invasion success explained by the enemy release hypothesis? *Ecol Lett* 7:721–733

Commission of the European Communities (2005) Communication from the commission: Biomass action plan. SEC

Connell JH, Slatyer RO (1977) Mechanisms of succession in natural communities and their role in community stability and organization. *Amer Naturalist* 111:1119–1144

Cook J, Beyea J (2000). Bioenergy in the United States: Progress and possibilities. *Biomass Bioenergy* 18:441–455

Dabney SM, Wilson GV, McGregor KC, Foster GR (2004) History, residue, and tillage effects on erosion of loessial soil. *Trans ASAE* 47: 767–775

de Chantal M, Granström A (2007) Aggregations of dead wood after wildfire act as browsing refugia for seedlings of *Populous tremula* and *Salix caprea*. *For Ecol Manag* 250:3–8

Ebeling SK, Hensen I, Auge H (2008) The invasive shrub *Buddleja davidii* performs better in its introduced range. *Diver Distribut* 14:225–233

EEA (2006) How much bioenergy can Europe produce without harming the environment? EEA Report No 7/2006, European Environment Agency, available at www.eea.europa.eu/publications/eea_report_2006_7/at_download/file (accessed June 2009)

Eycott AE, Watkinson AR, Dolman PM (2006) Ecological patterns of plant diversity in a plantation forest managed by clearfelling. *J Appl Ecol* 43:1160–1171

Fahrig L (2001) How much habitat is enough? *Biol Conserv* 100:65–74

FAO (2008) *The State of Food and Agriculture 2008*, U.N. Food and Agriculture Organization, New York

Fargione JE, Cooper TR, Flaspohler DJ, Hill J, Lehman C, McCoy T, McLeod S, Nelson EJ, Oberhauser KS (2009) Bioenergy and wildlife: Threats and opportunities for grassland conservation. *BioScience* 59:767–777

Fargione J, Hill J, Tilman D, Polasky S, Hawthorne P (2008) Land clearing and the biofuel carbon debt. *Science* 319:1235–1238

Federer CA, Hornbeck JW, Tritton LM, Martin CW, Pierce RS (1989) Long-term depletion of calcium and other nutrients in Eastern US forests. *Environ Manag* 13:593–601

Fischer J, Lindenmayer DB (2006) Beyond fragmentation: The continuum model for fauna research and conservation in human-modified landscapes. *Oikos* 112:473–480

Fisher RF, Binkley D (2000) *Ecology and Management of Forest Soils*, John Wiley and Sons, New York

Flaspohler DJ, Temple SA, Rosenfield R (2001) Species-specific edge effects on nest success and breeding bird density in a forested landscape. *Ecol Appl* 11:32–46

Flaspohler DJ, Webster CR, Froese RE (2009) Bioenergy, biomass, and biodiversity: A review of key issues for terrestrial and aquatic ecosystems. In: Solomon BD, Luzadis VA (eds), *Renewable Energy from Forest Resources in the United States*. Routledge, New York, pp 133–162

Frenzel M, Brandl R (2003) Diversity and abundance patterns of phytophagous insect communities on alien and native host plants in the Brassicaceae. *Ecography* 26:723–730

Fridman J, Walheim M (2000) Amount, structure, and dynamics of dead wood on managed forestland in Sweden. *For Ecol Manag* 131:23–36

Froese RE (2007) Biomass, biofuels, and bioenergy: Feedstock opportunities in Michigan. Michigan Department of Agriculture, available at http://www.mda.state.mi.us/renewablefuels/documents/biomass_feedstock_michigan.pdf (accessed June 2009)

Ganjegunte GK, Condron LM, Clinton PW, Davis MR, Mahieu N (2004) Decomposition and nutrient release from radiate pine (*Pinus radiata*) coarse woody debris. *For Ecol Manag* 187:197–211

Gardiner TA, Hernández MIM, Barlow J, Peres CA (2008) Understanding the biodiversity consequences of habitat change: The value of secondary and plantation forests for neotropical dung beetles. *J Appl Ecol* 45:883–893

Goodburn JM, Lorimer CG (1998) Cavity trees and coarse woody debris in old-growth and managed northern hardwood forests in Wisconsin and Michigan. *Can J For Res* 28:427–438

Grigal DF, Berguson WE (1998) Soil carbon changes associated with short-rotation systems. *Biomass Bioenergy* 14:371–377

Hanowski JM, Niemi CJ, Christian DC (1997) Influence of within-plantation heterogeneity and surrounding landscape composition on avian communities in hybrid poplar plantations. *Conserv Biol* 11:936–944

Hansen EA (1993) Soil carbon sequestration beneath hybrid poplar plantations in the north central United States. *Biomass Bioenergy* 5:431–436

Harmon ME (2001) Carbon sequestration in forests: Addressing the scale question. *J For* 99:24–29

Heaton EA, Clifton-Brown J, Voigt TB, Jones MB, Long SP (2004) Miscanthus for renewable energy generation: European union experience and projections for Illinois. *Mitigat Adapt Strat Global Change* 9:433–451

Hedin J, Isacsson G, Jonsell M, Komonen A (2008) Forest fuel piles as ecological traps for saproxylic beetles in oak. *Scan J For Res* 23:348–357

Hill J (2007) Environmental costs and benefits of transportation biofuel production from food- and lignocellulose-based energy crops: A review. *Agron Sustain Dev* 27:1–12

Hill J, Nelson E, Tilman D, Polasky S, Tiffany D (2006) Environmental, economic, and energetic costs and benefits of biodiesel and ethanol biofuels. *Proc Natl Acad Sci USA* 103:11206–11210

Hooper DU, Chapin FS, Ewel JJ, Hector A, Inchausti P, Lavorel S, Lodge DM, Loreau M, Naaem S, Schmid B, et al. (2005) Effects of biodiversity on ecosystem functioning: A consensus of current knowledge. *Ecol Monogr* 75:3–35

Hulme PE (2006) Beyond control: Wider implications for the management of biological invasions. *J Appl Ecol* 43:835–847

Hunter Jr ML (1990) *Wildlife, Forests, and Forestry: Principles of Managing Forests for Biological Diversity*, Prentice-Hall, Upper Saddle River, NJ

IEA (2007) *World Energy Outlook 2007*, International Energy Agency, Paris

Jabin M, Mohr D, Kappes H, Topp W (2004) Influence of deadwood on density of soil macro-arthropods in a managed oak-beech forest. *For Ecol Manag* 194:61–69

Jackson RB, Jobbagy EG, Avissar R, Boy SB, Barrett DJ, Cook CW, Farley KA, le Maitre DC, McCarl BA, Murray BC (2005) Trading water for carbon with biological sequestration. *Science* 310:1944–1947

Jactel H, Brockerhoff EG (2007) Tree diversity reduces herbivory by forest insects. *Ecol Lett* 10:835–848

Jactel H, Brockerhoff E, Piou D (2008) Le risque sanitaire dans les forêts mélangées (Forest health risks in mixed forests). *Rev For Franç* 60:168–180

Janowiak MK, Webster CR (2010) Promoting ecologically sustainability in woody biomass harvesting. *J Forestry* 108:16–23

Johnson DW (1992) Effects of forest management on soil carbon storage. *Water Air Soil Pollut* 64:83–120

Johnson DW, Curtis PS (2001) Effects of forest management on soil C and N storage: Meta analysis. *For Ecol Manag* 140:227–238

Jonsell M, Weslien J (2003) Felled or standing retained wood—It makes a difference for saproxylic beetles. *For Ecol Manag* 175:425–435

Keane RM, Crawley MJ (2002) Exotic plant invasions and the enemy release hypothesis. *Trends Ecol Evol* 17:164–170

Kimmins JP (1997) *Forest Ecology*. Prentice Hall, Upper Saddle River, NJ

Koh LP, Tan HTW, Sodhi NS (2008) Biofuels: Waste not want not. *Science* 320:1419

Kupfer JA, Malanson GP, Franklin SB (2006) Not seeing the ocean for the islands: The mediating influence of matrix-based processes on forest fragmentation processes. *Glob Ecol Biogeogr* 15:8–20

Lafond GP, Stumborg M, Lemke R, May WE, Holzapfel CB, Campbell CA (2009) Quantifying straw removal through baling and measuring the long-term impact on soil quality and wheat production. *Agro J* 101:529–537

Lal R, Pimentel D (2007) Biofuels from crop residues. *Soil Till Res* 93:237–238

Laurance WF, Lovejoy TE, Vasconcelos HL, Bruna EM, Didham RK, Stouffer PC, Gascon C, Bierregaard RO, Laurance SG, Sampaio E (2002) Ecosystem decay of Amazonian forest fragments: A 22-year investigation. *Conserv Biol* 16:605–618

Ledgard N (2004) Wilding conifers—New Zealand history and research background. In: Hill RL, Zydenbos SM, Bezar CM (eds), *Managing Wilding Conifers in New Zealand: Present and Future*, New Zealand Plant Protection Society, Christchurch, New Zealand, pp 1–26

Lehman CL, Tilman D (2000) Biodiversity, stability, and productivity in competitive communities. *Amer Naturalist* 156:534–552

Liebig MA, Johnson HA, Hanson JD, Frank AB (2005) Soil carbon under switchgrass stands and cultivated cropland. *Biomass Bioenergy* 28:347–354

Lindenmayer DB (2009) Forest wildlife management and conservation. The year in ecology and conservation biology, 2009. *Ann NY Acad Sci* 1162:284–310

Lindenmayer DB, Franklin JF (2002) *Conserving Forest Biodiversity: A Comprehensive Multiscaled Approach*, Island Press, Washington, DC

Malakoff D (2009) Biofuels déjà vu. *Conserv Magaz* 10:21–25

Mann LK, Johnson DW, West DC, Cole DW, Hornbeck JW, Martin CW, Riekerk H, Smith CT, Swank WT, Tritton LM, et al. (1988) Effects of whole-tree and stem-only clearcutting on postharvest hydrologic losses, nutrient capital, and regrowth. *For Sci* 34:412–428

Mann L, Tolbert V (2000) Soil sustainability in renewable biomass plantings. *Ambio* 29:492–498

McCay TS, Komoroski MJ (2004) Demographic responses of shrews to removal of coarse woody debris in a managed pine forest. *For Ecol Manag* 189:387–395

Mesibov R (2005) Native species dominate the millipede fauna in a second-rotation *Pinus radiata* plantation in Tasmania, Australia. *Pacif Conserv Biol* 11:17–22

Montgomery DR (2007) Soil erosion and agricultural sustainability. *Proc Natl Acad Sci USA* 104:13268–13272

Morgan D (2005) Brazil biofuel strategy pays off as gas prices soar. *Washington Post*, June 18

Mortenson SG, Mack RN (2006) The fate of alien conifers in long-term plantings in the USA. *Diver Distribut* 12:456–466

Murray LD, Best LB (2003) Short-term bird response to harvesting switchgrass for biomass in Iowa. *J Wildlife Manag* 67:611–621

Murray LD, Best LB, Jacobsen TJ, Braster ML (2003) Potential effects on grassland birds of converting marginal cropland to switchgrass biomass production. *Biomass Bioenergy* 25:167–175

Nason JD, Hamrick JL (1997) Reproductive and genetic consequences of forest fragmentation: Two case studies of Neotropical canopy trees. *J Hered* 88:264–276

NASS/USDA (2007) Crop production, National Agricultural Statistics Service/United States Department of Agriculture, available at http://books.nap.edu/openbook.php?isbn = 0309084946 (accessed May 13, 2009)

Noguchi M, Yoshida T (2004) Tree regeneration in partially cut conifer-hardwood mixed forests in northern Japan: Roles of establishment substrate and dwarf bamboo. *For Ecol Manag* 190: 335–344

NRC (2007) *Water Implications of Biofuels Production in the United States*. National Research Council of the National Academies, The National Academy Press, Washington, DC

O'Hanlon-Manners DL, Kotanen PM (2004) Logs as refuges from fungal pathogens for seeds of eastern hemlock (*Tsuga canadensis*). *Ecology* 85:284–289

Omonode RE, Vyn TJ (2006) Vertical distribution of soil organic carbon and nitrogen under warm-season native grasses relative to croplands in west-central Indiana, USA. *Agric Ecosyst Environ* 117:159–170

OSTP (2007) A strategy for U.S. science and technology policy to support water availability and quality in the United States. President's Office of Science and Technology Policy, Washington DC

Pawson SM, Brockerhoff EG, Meenken ED, Didham RK (2008) Non-native plantation forests as alternative habitat for native forest beetles in a heavily modified landscape. *Biodiver Conserv* 17:1127–1148

Pena E, Hidalgo M, Langdon B, Pauchard A (2008) Patterns of spread of *Pinus contorta* Dougl. ex Loud. invasion in a Natural Reserve in southern South America. *For Ecol Manag* 256:1049–1054

Post WM, Izaurralde RC, Jastrow JD, et al. (2004) Enhancement of carbon sequestration in US soils. *BioScience* 54:895–908

Purdon M, Bailey-Stamler S, Samson R (2009) Better bioenergy: Rather than picking bioenergy "winners", effective policy should let a lifecycle analysis decide. *Altern J* 35:23–29

Quinn JM, Cooper AB, Davies-Colley RJ, Rutherford JC, Williamson RB (1997) Land use effects on habitat, water quality, periphyton, and benthic invertebrates in Waikato, New Zealand, hill-country streams. *NZ J Marine Freshwater Res* 31:579–597

Raghu S, Anderson RC, Daehler CC, Davis AS, Wiedenmann RN, Simberloff D, Mack RN, (2006) Adding biofuels to the invasive species fire? *Science* 313:1742–1742

Richardson DM, Rejmanek M (2004) Conifers as invasive aliens: A global survey and predictive framework. *Diver Distribut* 10:321–331

Richardson DM, van Wilgen BW, Nunez MA (2008) Alien conifer invasions in South America: Short fuse burning? *Biol Invas* 10:573–577

Richardson DM, Williams PA, Hobbs RJ (1994) Pine invasions of the Southern Hemisphere: Determinants of spread and invadability. *J Biogeogr* 21:511–527

Righelato R, Spracklen DV (2007) Carbon mitigation by biofuels or by saving and restoring forests? *Science* 317:902

Robertson GP, Dale VH, Doering OC, Hamburg SP, Melillo JM, Wander MM, Parton WJ, Adler PR, Barney JN, Cruse RM, et al. (2008) Sustainable biofuels redux. *Science* 322:49–50

Rosenberg DK, Fraser JD, Stauffer DF (1988) Use and characteristics of snags in young and old forest stands in southwest Virginia. *For Sci* 34:224–228

Roth AM, Sample DW, Ribic CA, Paine L, Undersander DJ, Bartelt GA (2005) Grassland bird response to harvesting switchgrass as a biomass energy crop. *Biomass Bioenergy* 28:490–498

Rouget M, Richardson DM (2003) Inferring process from pattern in plant invasions: A semimechanistic model incorporating propagule pressure and environmental factors. *Amer Naturalist* 162:713–724

Rowe RL, Street NR, Taylor G (2009) Identifying potential environmental impacts of large-scale deployment of dedicated bioenergy crops in the UK. *Renew Sustain Energy Rev* 13:271–290

Russelle MP, Morey RV, Baker JM, Porter PM, Jung H-JG (2007) Comment on "Carbon-negative biofuels from low-input high-diversity grassland biomass." *Science* 316:1567b

Santiago LS (2000) Use of coarse woody debris by the plant community of a Hawaiian montane cloud forest. *Biotropica* 32:633–641

Sayer EJ (2006) Using experimental manipulation to assess the roles of leaf litter in the functioning of forest ecosystems. *Biol Rev* 81:1–31

Scion (2008) *Bioenergy Options for New Zealand: Pathway Analysis*, Scion, Rotorua, New Zealand

Seagle SW, Sturtevant BR (2005) Forest productivity predicts invertebrate biomass and ovenbird (*Seiurus aurocapillus*) reproduction in Appalachian landscapes. *Ecology* 86:1531–1539

Searchinger T, Heimlich R, Houghton RA, Dong F, Elobeid A, Fabiosa J, Tokgoz S, Hayes D, Yu (2008) Use of US croplands for biofuels increases greenhouse gases through emissions from land-use change. *Science* 319:1238–1240

Semere T, Slater FM (2007a) Ground flora, small mammal and bird species diversity in miscanthus (*Miscanthus × giganteus*) and reed canary-grass (*Phalaris arundinacea*) fields. *Biomass Bioenergy* 31:20–29

Semere, T, Slater FM (2007b) Invertebrate populations in miscanthus (*Miscanthus × giganteus*) and reed canary-grass (*Phalaris arundinacea*) fields. *Biomass Bioenergy* 31:30–39

Sexton S, Rajagopal D, Zilberman D, Roland-Holst D (2007) The intersections of energy and agriculture: Implications of rising demand for biofuels and the search for the next generation. *ARE Update* 10:4–7

Siemann E, Tilman D, Haarstad J, Ritchie M (1998) Experimental tests of the dependence of arthropod diversity on plant diversity. *Amer Naturalist* 152:738–750

Six J, Elliott ET, Paustian K (1999) Aggregate and soil organic matter dynamics under conventional and no-tillage systems. *Soil Sci Soc Amer J* 63:1350–1358

Stein BA, Kutner LS, Adams JS (2000) *Precious Heritage: The Status of Biodiversity in the United States.* Oxford University Press, New York

Stow AJ, Briscoe DA (2005) Impact of habitat fragmentation on allelic diversity at microsatellite loci in Cunningham's skink (*Egernia cunninghami*): A preliminary study. *Conserv Genet* 6:455–459

Tews J, Brose U, Grimm V, Tielbörger K, Wichmann MC, Schwager M, Jeltsch F (2004) Animal species diversity driven by habitat heterogeneity/diversity: The importance of keystone structures. *J Biogeogr* 31:79–92

Thompson JR, Spies TA, Ganio LM (2007) Re-burn severity in managed and unmanaged vegetation in the Biscuit Fire. *Proc Natl Acad Sci USA* 104:10743–10748

Tilman D (1996) Biodiversity: Population versus ecosystem stability. *Ecology* 77:350–363

Tilman D, Hill J, Lehman C (2006) Carbon-negative biofuels from low-input high-diversity grassland biomass. *Science* 314:1598–1600

Tilman D, Hill J, Lehman C (2007) Response to comment on "Carbon-negative biofuels from low-input high-diversity grassland biomass." *Science* 316:1567c

Tilman D, Wedln D, Knops J (1996) Productivity and sustainability influenced by biodiversity in grassland ecosystems. *Nature* 379:718–720

Ulyshen MD, Hanula JL, Horn S, Kilgo JC, Moorman CE (2004) Spatial and temporal patterns of beetles associated with coarse woody debris in managed bottomland hardwood forests. *For Ecol Manag* 199:259–272

Vance ED (1996) Land application of wood-fired and combination boiler ashes: An overview. *J Environ Qual* 25:937–944

Vance ED (2000) Agricultural site productivity: Principles derived from long-term experiments and their implications for intensively managed forests. *For Ecol Manag* 138:369–396

Vilà M, Vayreda J, Comas L, Ibáñez JJ, Mata T, Obón B (2007) Species richness and wood production: A positive association in Mediterranean forests. *Ecol Lett* 10:241–250

Vitousek PM, Mooney, Lubchenco J, Melillo JM (1997) Human domination of Earth's ecosystems. *Science* 277:494–499

With KA, King AW, Jensen WE (2008) Remaining large grasslands may not be sufficient to prevent grassland bird declines. *Biol Conserv* 141:3152–3167

Zollner PA, Crane KJ (2003) Influence of canopy closure and shrub coverage on travel along coarse woody debris by eastern chipmunks (*Tamias striatus*). *Amer Midland Naturalist* 150:151–157

7 Biomass Harvest and Logistics

Tom L. Richard, Doug Brownell, Kusumal Ruamsook, Jude Liu, and Evelyn Thomchick
Pennsylvania State University

CONTENTS

7.1 INTRODUCTION

The harvest, handling, and transport of biomass feedstocks represents a massive materials handling challenge, requiring rapid innovation and investment if bioenergy systems as a whole are to achieve policy goals. The International Energy Agency estimates biomass will need to provide 23% of global primary energy by 2050 to reduce greenhouse gas emissions by 50% (IEA 2008). This target would require sufficient biomass to supply 150 EJ/year (1 EJ = 10^{18} J), which translates to 15 billion metric tons (Mg) annually (Richard 2010). This projected biomass tonnage is more than 7 times greater than the approximately 2 Mg of grains and oilseeds currently traded worldwide (World Agricultural Outlook Board 2011). Without effective densification, this biomass would exceed the current trade volumes of grain, oil, and coal combined by a factor of 4 assuming typical woodchip densities or a factor of 12 assuming typical herbaceous biomass densities respectively (Richard 2010). In the U.S. context, the 1.3 billion Mg of biomass expected to meet domestic biofuel targets by 2030 (Perlack et al. 2005) exceeds the 800 million Mg of total material flow of the U.S. food system (Heller and Keolian 2000) by a factor of 1.6. The material handing equipment and logistics needed to deliver this biomass to energy facilities represents a great opportunity for agricultural equipment and truck manufacturers, but an equally great challenge for rural business development and transportation infrastructure.

The current generation of biomass harvest and logistics systems has been adapted from those developed for agricultural (food, animal feed) and forest (lumber, paper). Although existing harvest technologies and procedures for hay and straw in agricultural production systems can be utilized for herbaceous biomass, they have not been optimized for the distances and scales that bioenergy facilities are expected to require. To supply sufficient feedstock for large facilities, bioenergy crops are likely to be transported over longer distances than traditional on-farm uses such as animal feeding or bedding; these materials may also need to be stored in large centralized facilities for future use. Integrated systems that minimize material handling steps and maximize feedstock flexibility are needed to reduce the costs of harvesting, storing, and transporting agricultural feedstocks (Hess et al. 2007).

Such integrated systems are beginning to be established for woody biomass energy, building on technologies and equipment developed for timber management and the pulp and paper industry. This sector has long dealt with very large volumes as well as long-distance transport, including intercontinental shipments of pulp, sawlogs, and woodchips for manufacturing paper and lumber products. With expanding bioenergy markets, chips are now hauled hundreds of kilometers to biomass power facilities, whereas pellets are shipped from North America to Europe. Typical woody biomass operations include felling, delimbing, debarking, chipping, and transporting. For short-rotation trees such as willow and poplar, agricultural machinery manufacturers have developed oversized forage harvesters and wood choppers to harvest short-rotation trees. Woody biomass materials are naturally denser than herbaceous plants, so woody biomass tends to store more easily and is less expensive to transport. Woody biomass materials also have less need for storage because of a longer harvest season relative to agricultural crop residues and herbaceous energy crops. These advantages simplify woody biomass systems relative to herbaceous biomass management, as illustrated in Figure 7.1.

7.2 HERBACEOUS BIOMASS HARVEST

Herbaceous biomass refers to lignocellulosic plant matter that is nonwoody and includes agricultural residues such as corn stover and wheat straw and dedicated energy crops such as switchgrass and miscanthus. Dual-use crops such as alfalfa have a relatively short harvesting window if used as forage because of changing levels of protein during growth (Muck et al. 2010). When forages are fed to livestock, nutrient levels are of utmost importance. Although the criteria will likely be somewhat different, crop maturity and quality will also be critical for biochemical conversion to energy. But for combustion and thermochemical conversion, biomass crops can be harvested based upon convenience and yield. Depending on the location, weather greatly affects the harvesting window, crop quality, and mass loss of herbaceous biomass feedstocks. Thus, the harvest and logistics systems must take into account the crop, its intended use, and local conditions to maintain an efficient and cost-effective value chain.

7.2.1 MOWING OR SWATHING

Mowing is the initial step in harvesting herbaceous crops. Mowing can be performed with a trac-tor mounted implement or a self-contained unit. In the United States, disc mowers are common

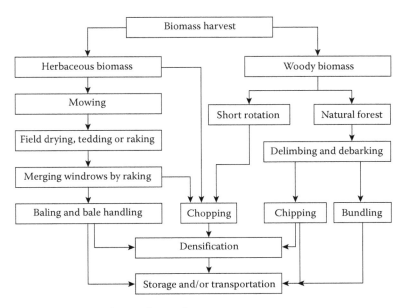

FIGURE 7.1 Biomass harvest options and procedures.

and are powered by the tractor power takeoff. Conventional hay mowing leaves the material in a swath behind the mower, and fresh cut grasses require 3–5 days to dry, depending on crop, soil, and weather conditions (Srivastava et al. 2006). This swath contains the leaves and the stems of the plant; the leaves dry faster because of their larger surface area and unprotected outer layer. When the swath is dried and handled, the leaves will be inherently more brittle and prone to blowing away. To encourage faster stem-drying, many mowers are now equipped with rubber-coated steel rollers that crush the stems to allow greater airflow. This conditioning of the stems can speed drying time to 2–4 days, but it requires an additional energy input of 2–4 kWh/Mg dry matter (DM) (Srivastava et al. 2006). Conditioning is not required for biomass materials that can be mowed dry such as corn stover, soybean straw, cereal straw, spring-harvested switchgrass, and cover crops. Popular in Canadian Prairies, windrowers or swathers, which are equipped with cutter bars, create a crop windrow in a single pass from dry standing biomass in the field.

7.2.2 FIELD DRYING, TEDDING, RAKING, OR MERGING

Raking accumulates the mowed swaths into a large windrow or elongated pile that can be fed into a baler with higher efficiency as compared with a small swath. Tedding is used to achieve faster drying and a uniform moisture level within a swath. Rakes and tedders can also be used to invert a swath to encourage faster drying times. A typical rake will cause mass losses of 3–6% because of breaking of the fragile material (Srivastava et al. 2006). More leaves than stems are lost during the raking or tedding process because of the drier and more brittle nature of the leaves.

7.2.3 BALING AND BALE HANDLING

The use of balers to densify and contain herbaceous biomass is common on most farms that harvest dry material. Currently available equipment includes round balers, small square balers, and large square balers. Round and square balers are equipped with a pick-up header to feed windrowed material into a precompression chamber. A cutting rotor forces material through a series of stationary knifes to reduce the length of material. The cutting mechanism can be engaged or disengaged. Round balers densify biomass by packing and rolling the material. Square balers densify "flakes" of biomass and then compress the flakes into a bale. Large square balers have the highest field efficiency among these three alternatives in terms of mass throughput. Prewitt et al. (2007) baled corn stover and achieved densities up to 176 kg/m^3, whereas Cundiff and Marsh (1996) found the density of switchgrass bales to be 143 and 200 kg/m^3 for round and large square bales, respectively. Round bales generally work well for in-field storage in more humid climates because their shape can shed water. Round balers require less energy to run and also require a lower initial investment cost. Sokhansanj and Turhollow (2004) also found that round balers reduced cost as compared with large square balers when harvesting corn stover. Square bales are inherently easier to transport and store compactly because of the ease in which they can be stacked. However, square bales have the disadvantage of large mass loss because of deterioration if stored in a field without adequate covering and protection from the ground surface. Baling also depends on local weather, conditions of biomass harvesting, and the end requirements of bioenergy production. If the biomass cannot be dried because of weather restrictions, the harvesting and handling systems must be able to process wet materials. For high-volume harvest operations that cannot risk wet weather delays, ensiled storage appears to have significant logistical advantages, with the caveat that drying would be required later for thermochemical processing downstream.

7.2.4 HERBACEOUS BIOMASS CHOPPING

Pull-type and self-propelled forage harvesters are currently used on farms to harvest corn and hay silage. A forage harvester picks up, chops, and transfers the mowed windrow into a truck or wagon after the harvester. These machines are designed to harvest materials at a range of moisture contents. Because the material is loaded and leaves the field immediately as it is harvested, several forage trucks are needed to drive side-by-side along with a forage harvester to collect and transport harvested materials. When these pieces of equipment are used to harvest biomass or dry crops, the forage harvester may experience large mass loss, friction, and combustion problems. A forage harvester was evaluated in Europe for harvesting miscanthus (*Miscanthus x giganteus*) at 11- and 44-mm chop length and resulted in a bulk density of 95 and 70 kg DM/m^3, respectively (Venturi et al. 1998).

Another option for harvesting high moisture biomass materials is a self-loading forage wagon. This system combines a chopping mechanism with a transport wagon, eliminating the separate forage truck required for a typical forage harvester system. Self-loading wagons rely on tractor power to harvest and transport the biomass and thus do not have an independent power unit. Self-loading wagons have been shown to reduce labor and fuel costs compared with forage harvesters (Savoie et al. 1992; Brownell et al. 2009). Despite these benefits of self-loading wagons, the self-propelled forage harvester remains the most commonly used machine for silage harvest at most farms in the United States. This is due to its low cost, high field efficiency, and low risk of forage quality problems and nutrient loss.

Corn and other grain combines can be modified to harvest grain and chopped crop residue in a single-pass operation. This operation can be implemented with single-stream or split-stream collection of the grain and residue. With single-pass, single-stream collection, grain and residue are collected and transported together to a centralized processing location where grain may be separated in a stationary combine, or the combined stream can be made into energy and fuels (Kadam et al. 2000). With single-pass, split-stream collection, the combine separates the grain in the field, and the stover/crop residue is immediately chopped and blown into a receiving truck or trailer. Single-pass, split-stream collection reduces field traffic, thus minimizing compaction, and it also reduces the soil contamination that inevitably occurs when biomass is dried in the field (Shinners et al. 2007a).

7.3 BIOMASS STORAGE

The moisture content of agricultural biomass will greatly influence storage options as well as appropriate strategies for subsequent transportation. Wet materials can be ensiled and stored anaerobically in bunker silos or bags, whereas dry material can be baled and stored like straw and hay. For dry storage, hay tools and hay harvesting technologies are available in most agricultural operations and can be easily adapted for herbaceous biomass harvest. Typical on-farm hay harvest procedures include mowing, raking and/or tedding, merging, baling, and bale handling. Bale moisture content must be maintained below 20% to minimize dry matter loss as well as self-heating and the potential for spontaneous combustion. Since indoor storage is expensive, in humid climates bales can alternatively be stored outdoor in plastic bale wraps to minimize moisture absorption, whereas in more arid climates bales can simply be stacked outside for the most part of the year. For wet materials, (i.e., >35% water on a wet basis) ensilage can be used. Ensilage has been used to store chopped moist biomass such as sorghum, corn stover, and wet chopped grasses as bioenergy crops (Montross and Crofcheck 2004; Ren et al. 2006; Shinners et al. 2007b; Bennett and Anex 2008) and has been used for other forages crops and foods for over a thousand years. At moisture levels above 35% (wet basis), biomass feedstocks need to be stored in a bunker silo, plastic "ag bags," or another anaerobic environment immediately after harvested to ensure quality. Intermediate moisture levels between 20% and 40% can be dried or moistened, with an aim to achieve appropriate dry or wet storage conditions for safe and stable storage. Although dry handling and storage has clear advantages for energy systems based on combustion or thermochemical conversion, wet storage may be more cost-effective for biochemical platforms (Bennett and Anex 2008).

7.4 FOREST BIOMASS HARVEST

As previously indicated, there is considerably more experience today with the harvesting, handling, and managing logistics of forest-based biomass at scales much greater than 500,000 Mg/year than there is with agricultural energy crops and perennials. Forest biomass can come from logging operations, forest thinning for wildfire prevention, or forest improvement projects. Typical feedstocks are limbs, tops, and other byproducts of forest management. Along with being suitable for biofuels, forest biomass can be used for engineered wood products and composite lumber (Dooley et al. 2006). Woody biomass harvest processes may include felling, piling, forwarding, chipping and/or bundling, and trucking. Products could be woodchips, bundles, round wood logs or a combination there of.

7.4.1 FELLING, DELIMBING, AND DEBARKING

In a typical forest harvest operation, round wood products for building materials are cut from a forest as large whole trees. Trees are cut and piled and then forwarded by a feller/buncher to a location where they are delimbed and debarked. Woodchip production for pulp mills has very similar field operations to those described, although pulp-mill procurement specialists are more willing to purchase woodchips and small-diameter "low-use" wood that loggers could collect and sell. These pulp and paper industry employees and former employees represent a legacy of experience and often know of idle equipment in the many communities where the pulp and paper industry is in decline. After typical timber harvest operations, 13% of total softwood volume and 24% of total hardwood volume is left as logging residue, most of which (75%) is in tops and limbs (Hartsough 1992). This residue may be left in the forest or harvested for use as a biomass product. Before removing this material it is important to understand the carbon and nutrient cycle dynamics of the site because up to 30% of the thinnings should be left on the forest floor to replenish soil carbon and nutrients.

Woody biomass is typically processed at the roadside landing area into woodchips, which can then travel considerable distances because of their higher energy density and mass density.

7.4.2 Chipping

Woody biomass materials have a high mass and energy density compared with similarly processed herbaceous material. Although the high density of woodchips improves storage and transport, it also leads to higher costs to grind the woody materials. Size reduction is one of the major uses of energy in wood harvesting operations (Naimi et al. 2006). Wood chipping is achieved by mechanical force applied to the biomass, commonly using a set of knives, hammers, or an auger. Knives are mounted on the surface of a flywheel that is powered by an engine.

Woody biomass chipping, combined with a traditional lumber harvest, can turn treetops and other nonuniform wood into uniform chips, which can be sold for profit. Chips have ready markets and sufficient density and other desirable flow characteristics that make it inexpensive to transport and handle.

7.4.3 Forest Residue Bundling

After delimbing, limbs can be bundled using a mechanical bundler to make trunk-sized bundles for more efficient storing or hauling. Some feller/bunchers are designed for thinning small-diameter wood and have an integral bundler system. The bundler can also be used for loading trucks that line-haul woody biomass. Bundling may be considered as a form of densification of forest residues, much like baling is for agricultural crops. Further size reduction and/or chipping operations can be conducted at a satellite preprocessing facility, a biorefinery, or an industrial bioenergy power plant.

7.5 SHORT-ROTATION WOODY BIOMASS HARVEST

Short-rotation woody biomass crops are trees, commonly willow and poplar, grown specifically for use as biomass. In a process called "coppicing", these fast-growing trees are harvested every 3–5 years, being cut off at the base while leaving the root system intact, which allows the trees to regrow for future cuttings. Because they are grown in even-age plantations, these trees are more uniform in size and shape than forest residues. This uniformity allows the material to be chipped, bundled, or baled, with yields from 10–20 Mg DM/ha per year (Savoie et al. 2006). Bundled woody biomass can be baled wet (40–50% moisture) and will still air-dry, unlike herbaceous biomass that must be baled dry or ensiled if wet. These woody crops can dry slowly in bales with minimal degradation because of their relatively slow moisture transfer through the bark and evenly distributed porosity in the bales, which allows rapid surface drying while the interior of the wood is still moist. Short-rotation woody crops can also be harvested using a forage harvester (called a wood chopper) with a modified head to produce woodchips. As with forage harvesting and silage chopping, trucks are needed to follow the wood chopper to collect chips. Chips are often stored in piles for weeks or months, during which time there may be some biomass degradation and dry matter loss, depending on the tree species, local climate, ventilation, soil contact, and nutrient content.

7.6 DENSIFICATION

Densification and preprocessing refer to the handling steps that occur at the farm or a regional satellite preprocessing facility (Carolan et al. 2007) to minimize storage and transport costs. Preprocessing usually indicates size reduction whereas densification indicates an increase in bulk density. Densification is currently used in almost all farm-level hay harvesting procedures during baling or ensiling. Densification can be used in any herbaceous biomass handling system to minimize handling and storage costs.

TABLE 7.1
Biomass Densification Characteristics

Biomass Feedstock	Mass Density (kg/m³ dry basis)	Energy Density (GJ/m³)
Loose herbaceous biomass	50–95	0.8–1.5
Bales	120–200[a]	2–3
Woodchips	200–250	4–5
Pellets	600–700[a]	8–11
Torrefied pellets	750–850	15–18
Pyrolysis oil	1200–1300	15–16

Source: Cundiff, JS. and Marsh, LS., *Bioresour Technol,* 56, 95–101, 1996; Venturi, P., Huisman, W., and Molenaar, J., *J Agric Eng Res.,* 69, 209–215, 1998; Brown, RC., *Biorenewable Resources: Engineering New Products from Agriculture,* Blackwell Publishers, Boston, MA, 2003; Sokhansanj, S. and Turhollow, AF., *Appl Eng Agric.,* 20, 495–499, 2004; Bergman, PCA., *Combined Torrefaction and Pelletisation—The TOP Process,* ECN Report, ECN-C-05-073, ECN Biomass, Petten, The Netherlands, 2005; Shinners, KJ., et al., *Biomass Bioenergy,* 31, 211–221, 2007b; Uslu, A., Faaij, APC., and Bergman, PCA., *Energy,* 33, 1206–1223, 2008.

[a] Significantly higher values are reported in the text using new equipment that is not widely available.

Herbaceous biomass has a low bulk density relative to wood and very much so relative to coal, oil, and other energy carriers. Densification, including baling, palletizing, and briquetting has been recognized as an effective way to significantly reduce transport and storage costs and associated handling costs (Badger and Fransham 2006; Uslu et al. 2008; Petrolia 2008). Bulk density depends on material composition, particle size, shape, distribution, and moisture content (Lam et al. 2007). Table 7.1 gives the ranges of typical biomass mass and energy densities for fresh biomass and several densification strategies.

7.6.1 PELLETIZING AND BRIQUETTING

Pelletizing and briquetting are traditionally used for producing animal feed and compressing coals and minerals. Biomass is first ground to the size of sawdust, then compressed and forced through a metal die, where friction or supplemental heating seals the surface to keep the individual pellet or briquette intact. The most popular shape of pellets and briquettes is cylindrical. The main difference between pellets and briquettes is size, which affects material handling and downstream conversion processes. The diameter of pellets typically ranges from 5 to 12 mm whereas briquettes have a diameter of 50–100 mm or more. These technologies and equipment can be used for densifying biomass materials with the end goal of bioenergy production.

In a drum pelletizer, many holes are drilled through the drum, and a roller inside squeezes material through these holes. Another piece of metal outside of the drum may be mounted to shear off the pellets at a desired length. Sometimes starch is added as a binder, but most biomass materials can be pelletized with a small amount of electrical heat or steam applied while the material is forced through the die; no binding material is needed. Appropriate moisture content of biomass materials is between 8% to 15% dry basis (d.b.) depending on the type of biomass material. The bulk density of pellets or briquettes normally ranges from 600 to 700 kg/m³, but it can reach as high as 1000 kg/m³ (Mani et al. 2006). One cubic meter of wood pellets is equivalent to approximately 3 m³ of woodchips and 2 m³ of solid wood.

Pellets can also be made with herbaceous biomass feedstocks, although few commercial manufacturers currently exist. Moisture content of the feedstock and temperature control of the die are critical because the lower lignin content in herbaceous material does not bind as well as sawdust from wood. However, because herbaceous biomass is less dense than woodchips even when baled, the densification advantages that pellets offer for these feedstocks warrant a close evaluation.

Biomass pellets are popular in applications such as gasification, flash pyrolysis, biofuels, and syngas production. However, as with other harvest and logistics considerations, the high costs associated with pelletizing and briquetting could be a serious constraint on increasing cellulosic biofuels production (Lam et al. 2007). Increasing energy efficiency and reducing operational cost are current challenges that need to be addressed. To create highly efficient biomass-to-energy chains, biomass torrefaction in combination with densification could be a promising step to overcome logistics economics in large-scale green energy solutions (e.g., Bergman 2005). Torrefaction is a roasting process that drives off volatiles and hemicellulose at temperatures between 200 and 300°C, resulting in a much more energy-dense product that is easily pelletized (see Table 7.1). Although the costs and energy losses associated with combining these two processes are considerable, the savings potential, especially for long-range transport, are also considerable (Uslu et al. 2009).

7.6.2 Bale Compression

Although most hay bales are consumed on the farm where they were produced, some square hay bales are currently densified for transport overseas. This densification is expensive, but it is justified because of the high demand for quality hay in Middle Eastern countries. To increase bale density and, therefore, reduce handling cost, bale compression technology has been tested with several biomass feedstocks and appears to have considerable potential (Brownell et al. 2009). The bulk density of normal hay bales is approximately 160 kg/m^3, but a bale compressor can densify the bale to 2–3 times of this density (Steffen Systems 2009). However, the cost of bale compression is approximately $27 per wet Mg (Miles 2008) and, therefore, is currently only applied to international hay shipping industries with high value markets. Some large square bales of hay are dense enough to be transported at legal truck limits (truck capacity is limited by weight not volume), but loading and unloading of the bales is often time-consuming. These issues need to be addressed and cost efficiencies achieved if biomass energy feedstocks are to compete successfully with low-cost fossil fuels like coal and natural gas.

7.7　SUPPLY CHAIN AND LOGISTICS MANAGEMENT

To achieve system-level efficiencies and economies of scale, the individual unit operations described above need to be knit together into integrated systems. One such approach is the uniform format, solid feedstock supply strategy of Hess et al. (2009). This approach would process diverse biomass feedstocks as close to the harvest locations as possible. Local processing allows the biomass supplier to produce a uniform particle size, consistent flowability, and a consistent moisture content. With these and other related characteristics rationalized and homogenized, downstream transport and logistics operators can manage a diverse range of biomass feedstocks with common equipment and efficient operations. In addition to these coordinated technical systems, there is also a need for coordinated business systems.

Biomass supply chains are generally composed of many interconnected business units that perform a range of different activities. They include landowners and contractors who manage ground preparation and planting of a mix of species in the field and forest. There are also biorefineries and other clients that may range from harvesting, handling, and managing logistics of to producers of chemicals and liquid fuels that, in turn, serve different customers. Various other entrepreneurs are also involved in harvesting, handling, processing, storing, and line-haul transporting biomass for use at biomass energy facilities. The interaction of these businesses at a strategic network and relationship level constitute supply chain management.

A biomass supply chain performs two important physical functions: production and logistics. The production function produces and collects raw biomass and converts raw biomass (such as woody biomass and agricultural wastes) into processed biomass (such as woodchips and pellets) and eventually finished goods (such as heat, electricity, and transportation fuels). The logistics function involves transportation and storage of these items from one point in the supply chain to the next. Thus, logistics function spans key activities such as transportation, industrial packaging, materials handling, warehousing and storage, and inventory control (Fisher 1997).

Frequently, the movement and storage of raw materials in an organization is different from that of finished goods. Accordingly, the logistics function can be further distinguished into inbound logistics (materials management) and outbound logistics (physical distribution). Put in the context of biomass logistics systems from a biorefinery organization perspective, inbound logistics involve movement and storage of biomass feedstock to a biorefinery plant, whereas outbound logistics manage movement and storage of refined bioproducts such as biofuels and electricity. Biomass logistics systems can be characterized as heavy inbound in that a biorefinery has a very heavy inbound flow of biomass feedstock from various sources and a relatively simple outbound flow of finished bio-based products (e.g., biofuels distributed to a few major petroleum storage and distributors or electricity grid operators). Accordingly, much research and managerial focus has been on the inbound side of the logistics system. However, it is important to note that although inbound and outbond logistics activities differ in system designs and requirements, close coordination between the two is critical to achieve the seven R's of logistics performance. The seven R's underscore getting the right product, to the right customer, in the right quantity, in the right condition, at the right place, at the right time, and at the right cost (Coyle et al. 2008).

The biomass supply markets and inbound biomass logistics are evolving systems, ones in which the processes and the underlying technologies are still under early development and are rapidly changing. As a result, the supplier base may be limited in size and/or experience, and uncertainties are not uncommon with respect to yield, process reliability, and lead time. In general, to hedge against such uncertainties, inventory may be increased, lead times cut, process flexibility increased, and/or multiple supply bases or alternative supply resources developed. By using these strategies, the costs of holding more inventory and managing the multiple supply bases may be higher, but the risk of supply outages can be reduced (Fisher 1997; Lee 2002). Logistics activities play a major role in implementing these risk hedging strategies and hence warrant further discussion, particularly in terms of factors that affect the cost of biomass logistics and interrelationships among them.

7.8 FACTORS AFFECTING BIOMASS LOGISTICS COSTS

Factors that influence the cost of biomass logistics can be classified broadly into three categories: competitive factors, product factors, and spatial factors (Coyle et al. 2008).

7.8.1 COMPETITIVE FACTORS

Competitive factors span price and customer service. Although price is frequently a basis of competition, in many markets customer service can be a significant form of competition. A well-accepted principle of logistics management associated with customer service is that of order cycle time. Order cycle time (OCT) can be defined as the time that elapses from when a customer places an order until the order is received. OCT influences product availability and customer inventories. A longer OCT usually requires higher customer inventories, suggesting that if biomass suppliers can improve customer service by shortening its OCT, its customers should be able to operate with less inventory. Variability of OCT also affects customer safety stock inventory levels and stockout costs. A customer can minimize its inventory levels if OCT is constant. That is, a buyer who knows with 100% assurance that the OCT is 10 days could adjust its inventory levels to correspond to the average demand during the 10 days with no need to hold safety stock to guard against stockouts that may result from inconsistent OCT. It then follows that such a cost reduction made possible by shorter and more consistent OCT could be as important as a reduction in prices of biomass itself. The significance of OTC as a competitive basis is particularly the case for biomass markets, given their evolving nature and associated uncertainties, including weather and yield risk previously discussed (Coyle et al. 2008, 2010).

7.8.2 PRODUCT FACTORS

Several characteristics of biomass product have a direct bearing on the cost of logistics, notasly the density and the physical form of biomass.

7.8.2.1 Density

Density refers to the weight-to-space ratio of the product. An item that is lightweight in relation to the space it occupies, as in the case of biomass, has low density. Transportation providers generally consider how much weight they can fit into their vehicles when establishing their prices because they quote their prices in dollars and cents per unit weight. On high-density items, these providers can afford to charge a low price per metric ton, while ensuring acceptable revenue, because they can fit more weight into their vehicle (Coyle et al. 2008, 2010). Hence, the low bulk density characteristic of biomass means less mass of material can be transported in any given trip, resulting in higher costs of transportation. Increasing bulk densities by processing the biomass into the previously described chips, bundles, pellets or bales, and/or increasing load size per trip by, for example, using larger transport vehicles, can help to reduce transportation costs (Allen et al. 1998; Frisk et al. 2010).

7.8.2.2 Physical Form

The physical form of biomass also affects transportation, material handling, and storage requirements and, thus, total logistics cost incurred. Storage requirements vary depending on the type of biomass. Typically, storage facilities can store nonprocessed and processed biomass, albeit separate capacity will be required. For example, although nonprocessed forest biomass such as tree tops, branches, and slashes can be stored on any surface, woodchips have to be protected against rain and must be stored on a hard (e.g., concrete) surface. Similar distinctions apply to transportation and handling requirements. For example, types of trucks used to carry nonprocessed forest biomass are different from those used to carry forest biomass that has been chipped or bundled. Chip vans are typically used for landing-to-market hauling of chipped forest biomass, whereas bundled biomass is transported using log trucks (Sokhansanj and Fenton 2006; Eksioglu et al. 2009). Examples of biomass transportation options are summarized in Table 7.2.

7.8.3 SPATIAL FACTORS

Spatial factors, or the location of fixed points in the logistics system in terms of demand and supply points, are particularly important factors that affect transportation cost because these costs tend to increase with distance (Coyle et al. 2008, 2010). Biomass supply sources are geographically dispersed, and transportation infrastructure networks serving between those source locations and a biorefinery may be limited. Adding these spatial factors to the product factors discussed earlier, one can appreciate the underlying challenges in managing biomass feedstock logistics.

The spatial factors also influence the question of whether an intermediate storage depot or satellite preprocessing facility will be used in the logistics system. Whether biomass products should be taken directly to the biorefinery or to intermediate storage/preprocessing locations before being delivered to the biorefinery depends on a range of factors and involves tradeoffs. If preprocessing generates co-products or byproducts such as animal feed or nutrient streams that should be redistributed to the source farms, a distributed satellite infrastructure may well make sense (Carolan et al. 2007). Because each satellite facility has a specified total capacity, available capacity will affect logistics costs associated with duration of biomass storage, transport planning and arrangements, and biomass processing possibilities. Where storage capacity is limited, small,

TABLE 7.2

Transportation Options for Biomass Feedstocks

Transportation Options	Description
Bags, totes, and bales	Bags and totes are commonly used for wood pellets, pucks, or briquettes. Small bags (15–30 kg) can be stacked on pallets for retail sale, whereas totes (1 m^3; 100–500 kg) are more appropriate for commercial/industrial operations. Bales can be in a small or large format, with small bales typically rectangular and large bales round or cubed. All of these units are typically transported by flat-bed truck or rail, but they can also be containerized for long distance and shipping transport.
Flat-bed trailer	Flat-bed trailers are widely available and work well for a diverse range of materials and applications. They can be used for palleted bags, totes, or bales and can also be used to transport large logs. Forklifts or grapples are needed for loading/unloading.
Dump trucks and tipper trailers	Dump trucks are also commonly used for transport of many bulk commodities. This system can be appropriate for woodchips, pellets, and various other forestry, agricultural, and industrial biomass feedstocks that are sufficiently dense. Biomass is delivered by gravity, dumping the load into receiving hoppers or storage bunkers, thus minimizing the need for additional handling or equipment.
Container and container truck	Containers can be used for transport and storage. Intermodal trailers can be used for truck, rail, and ship and are an efficient way to transport biomass long distances if it is suitably densified. Modified containers can also be used for biomass storage on site, but require sufficient ventilation and a pneumatic delivery system or other mechanism for fuel unloading.
Walking floor trailer	Walking floor trailers have a hydraulic floor that pushes material out for rapid unloading. They are an efficient delivery system for woodchips in bulk and would also be appropriate for other loose but dense biomass. Direct rapid loading is possible from high-throughput woodchip blowers, other pneumatic mechanisms, or overhead through retractable roofs.
Rail	Biomass can be transported via rail in boxcars or flat-bed cars (bags, totes, or bales), hopper cars (bulk sawdust, chips, or pellets), or containers. Rail transport is highly energy-efficient and low-cost where available.
Ship and barge	Palleted bags, totes, and bales can also be transported by barge or ship, but containers and bulk transport in holds will typically be lower cost. Water is the most energy-efficient means of very long-distance transport.
Pipeline	Although usually used for liquids or gases, pipelines can also be a low-cost means of moving woodchip slurries long distances in high volumes (»500,000 Mg/year). Like rail and ship, pipelines require terminals at both ends of transport and thus may be more appropriate for plantation feedstock production and large end-users.

Source: Biomass Energy Centre, available at http://www.biomassenergycentre.org.uk (accessed February 6, 2011), 2011; Kumar, A., Cameron, J., and Flynn, P., *Appl Biochem Biotechnol.,* 113, 27–39, 2004; Uslu, A., Faaij, APC., and Bergman, PCA., *Energy,* 33, 1206–1223, 2008.

short-term on-site stock level (e.g., a few days' supply) can be stored, small processing equipment can be used, and more regular, evenly spread deliveries of biomass will be required. Thus, low levels of stockholding at the facility will increase the importance of reliable and flexible transport services (Allen et al. 1998; Sokhansanj and Fenton 2006; Eksioglu et al. 2009; Frisk et al. 2010).

In contrast, a facility with a large capacity permits prolonged storage of larger on-site stock levels, and cost advantages are made possible by economies of scale and use of large, powerful processing equipment. In this case, transportation of biomass can be arranged less frequently and typically in a less expensive, larger load size. In some cases, trains may be involved to support the transportation to and from storage depots, offering a relatively less expensive transportation alternative to trucks. However, prolonged storage of biomass creates costs in the form of cost of capital invested in the biomass feedstock that remains unused during storage periods. This cost represents the interest rate cost and the lost value cost due to deterioration of biomass during storage. This is where the biorefinery plant can use an intermediate depot to avoid such costs. However, offsetting the advantages of using an intermediate depot is the fact that the biomass will need to be loaded/ unloaded from the vehicles multiple times, and special loaders/unloaders designed for different vehicle types may be required. The types of loading/unloading systems and the time spent loading/ unloading vehicles will ultimately affect transportation costs (Allen et al. 1998; Sokhansanj and Fenton 2006; Eksioglu et al. 2009; Frisk et al. 2010). An integrated, uniform format solid feedstock supply system would address many of these concerns (Hess et al. 2009).

7.9   PATHS FORWARD

This chapter reviews biomass harvest, storage, and densification technology; logistics system characteristics; and the inherent tradeoffs that underscore the importance of a holistic system view of logistics management. Although it is important to have innovative, effective technology and appropriate strategies at the unit operation level, it is also equally important that these individual operations interact as a whole. The objective has to be to operate the whole system effectively, not just the individual parts. In this respect, Coyle et al. (2008) accentuate an important aspect of logistics management: that logistics decisions must be made with an understanding that multiple levels of optimality must be considered and that some levels of suboptimization may occur. An organization may achieve organizational optimality by balancing various elements of logistics systems (e.g., harvest and transportation versus warehousing and inventory and customer service) as well as against other subsystems of the organization (e.g., logistics versus marketing, production, and finance). At the next level, the organization may achieve supply-chain optimality by trading off the effects that its decisions have on the other members within the supply chain, and vice versa, to optimize the operation of the entire supply chain. This will be particularly important in biomass supply chains, which are likely to be highly decentralized with limited vertical or horizontal integration. Finally, operated in a society, a supply chain has imposed upon it, by society, various constraints under social, political, and economic influences. For the biomass industry, this would clearly include environmental sustainability and rural economic development, but also issues of environmental health and social equity. At this level of societal optimality, decisions must be made that optimize the organization and the supply chain subject to the requirements of society. Given the multiple levels of optimality that are critical to logistics decisions, it is clear that each successful organization in a biomass supply chain must understand all of the constituencies affected by its activities and then optimize at the levels that are appropriate.

Sustainability criteria will be an increasingly important factor in this optimization, requiring source identification and chain-of-custody documentation throughout the supply chain to satisfy societal and organizational goals and constraints. Voluntary certification programs are developing multidimensional criteria and standardized metrics, whereas governmental agencies set their own criteria to meet mandates for low-carbon fuels. As specific sustainability targets emerge, they will drive innovation to reduce energy losses, greenhouse gas emissions, nutrient losses, and other negative effects of biomass harvest and logistics. Since transport contributes a much larger fraction of the overall greenhouse gas effects of lignocellulosic biomass than it does for grains and oilseeds, more efficient harvest, densification, and delivery systems are clearly required.

With increased understanding of how different biomass characteristics perform during downstream conversion processes, one can also expect optimization of biomass quality/performance relationships. It will be important to track the intrinsic qualities of biomass that result from special crop varieties or production practices as well as postharvest exposure to moisture, heat, or biodegradation. Tracking these characteristics and exposures will also require source identification and chain-of-custody documentation as well as new biomass quality measurement tools. This optimization will also drive research and innovation to develop integrated supply-chain strategies that improve product quality and performance at an affordable cost.

In pursuing these optimized decisions and paths, it is important to remember that biomass feedstock supply chains are composed of an interconnected network of organizations and individuals, from loggers, farmers, and truckers to bankers, managers, biorefineries, and ultimately biofuel and biopower customers. Because each of these entities optimizes its own operations, it is important to establish professional networks to link people together to inform the best possible decisions. It is also important to address the technical challenges, including harvest fractionation, densification and storage, and efficient transportation, to reduce biomass feedstock prices so that the resulting energy products can compete with natural gas and coal. Recognizing that a billion tons is just a downpayment on the biomass our society needs, it will take significant investments in people, businesses, technology, and infrastructure to successfully address the challenges of ramping up the biomass feedstock supply chain to meet growing demand.

REFERENCES

Allen J, Browne M, Hunter A, Boyd J, Palmer H (1998) Logistics management and costs of biomass fuel supply. *Int J Phys Distribut Logist Manag* 28:463–477

Badger PC, Fransham P (2006) Use of mobile fast pyrolysis plants to densify biomass and reduce biomass handling costs—A preliminary assessment. *Biomass Bioenergy* 30:321–325

Bennett AS, Anex P (2008) Farm-gate production costs of sweet sorghum as a bioethanol feedstock. *Trans ASABE* 51:603–613

Bergman PCA (2005) *Combined Torrefaction and Pelletisation—The TOP Process*, ECN Report, ECN-C-05-073, ECN Biomass, Petten, The Netherlands

Biomass Energy Centre, available at http://www.biomassenergycentre.org.uk (accessed February 6, 2011)

Brown RC (2003) *Biorenewable Resources: Engineering New Products from Agriculture*, Blackwell Publishers, Boston, MA

Brownell, DK, Liu J, Hilton JW, Richard TL, Cauffman GR, Macafee BR (2009) *Evaluation of Two Forage Harvesting Systems for Herbaceous Biomass Harvesting*, ASABE Paper No 097390, American Society of Agricultural and Biological Engineers, St. Joseph, MI

Carolan, JE, Joshi SV, Dale BE (2007) Technical and financial feasibility analysis of distributed bioprocessing using regional biomass pre-processing centers. *J Agric Food Indust Org* 5, Article 10, available at http://www.bepress.com/jafio/vol5/iss2/art10 (accessed February 6, 2011)

Coyle, JJ, Langley Jr CJ, Gibson BJ, Novack RA, Bardi E (2008) *Supply Chain Management: A Logistics Perspective*, 8th ed, South-Western, Cengage Learning, Mason, OH

Coyle JJ, Novack RA, Gibson, BJ, Bardi E (2010) *Transportation*, 7th ed, South-Western, Cengage Learning, Mason, OH

Cundiff JS, Marsh LS (1996) Harvest and storage costs for bales of switchgrass in the southeastern United States. *Bioresour Technol* 56:95–101.

Dooley JH, Fridley JL, DeTray MS, Lanning DN (2006) Large rectangular bales for woody biomass. ASABE Paper No. 068054, American Society of Agricultural and Biological Engineers, St. Joseph, MI

Eksioglu SD, Acharya A, Leightley LE, Arora S (2009) Analyzing the design and management of biomass-to-biorefinery supply chain. *Comput Indust Eng* 57:1342–1352

Fisher ML (1997) What is the right supply chain for your product? *Harvard Business Rev* 75:105–116.

Frisk M, Göthe-Lundgren M, Jörnsten K, Rönnqvist M. (2010) Cost allocation in collaborative forest transportation. *Eur J Oper Res* 205:448–458

Hartsough BR (1992) Production/harvesting options for agroforestry plantations in the San Joaquin Valley, California. *Trans ASAE* 35:1987–1993.

Heller MC, Keoleian GA (2000) *Life Cycle-Based Sustainability Indicators for Assessment of the U.S. Food System*, University of Michigan, Ann Arbor, MI, available at http://css.snre.umich.edu/publication/life-cycle-based-sustainability-indicators-assessment-us-food-system (accessed February 6, 2011)

Hess JR, Wright CT, Kenney KL (2007) Cellulosic biomass feedstocks and logistics for ethanol production. *Biofuels Bioprod Biorefin* 1:181–190

Hess JR, Wright CT, Kenny KL, Searcy EM (2009) *Uniform-Format Solid Feedstock Supply System: A Commodity-Scale Design to Produce an Infrastructure-Compatible Bulk Solid from Lignocellulosic Biomass*, Technical Report INL/EXT-09-15423, U.S. Department of Energy Idaho National Laboratory, http://www.inl.gov/bioenergy/uniform-feedstock (accessed February 6, 2011)

IEA (2008) *Energy Technology Perspectives 2008—Scenarios and Strategies to 2050*, Paris, France, available at http://www.iea.org/techno/etp/index.asp (accessed February 6, 2011)

Kadam, KL, Forrest LH, Jacobson WA (2000) Rice straw as a lignocellulosic resource: Collection, processing, transportation, and environmental aspects. *Biomass Bioenergy* 18:369–389

Kumar A, Cameron J, Flynn P (2004) Pipeline transport of biomass. *Appl Biochem Biotechnol* 113:27–39

Lam PS, Sokhansanj S, Bi X, Mani S, Womac AR, Hoque M, Peng J, JayaShankar T, Naimi LJ, Narayan S (2007) *Physical Characterization of Wet and Dry Wheat Straw and Switchgrass—Bulk and Specific Density*, ASABE Paper No. 076058, American Society of Agricultural and Biological Engineers, St. Joseph, MI

Lee HL (2002) Aligning supply chain strategies with product uncertainties. *Calif Manag Rev* 44:105–119

Mani S, Sokhansanj S, Bi X, Turhollow A (2006) Economics of producing fuel pellets from biomass. *Appl Eng Agric* 22:421–426

Miles T (2008) Personal communications with T.R. Miles, Technical Consultants, http://www.trmiles.com

Montross MD, Crofcheck CL (2004) Effect of stover fraction and storage method on glucose production during enzymatic hydrolysis. *Bioresour Technol* 92:269–274

Muck RE, Shinners KJ, Duncan JA (2010) Ensiling characteristics of alfalfa leaves and stems, ASABE Paper No. 1008613, American Society of Agricultural and Biological Engineers, St. Joseph, MI

Naimi LJ, Sokhansanj S, Mani S, Hoque M, Bi X, Womac AR, Narayan S (2006) Cost and performance of woody biomass size reduction for energy production. Presented at the 2006 CSBE-SCGAB Annual Meeting, Edmonton, Alberta, Canada

Perlack RD, Wright LL, Turhollow AF, Graham RL, Stokes BJ, Erbach DC (2005) *Biomass Feedstock for a Bioenergy and Bioproducts Industry: The Technical Feasability of a Billion-Ton Annual Supply*, DOE/GO-102995-2135; ORNL/TM-2005/66, U.S. Department of Energy Oak Ridge National Laboratory, available at https://bioenergykdf.net/content/billiontonupdate (accessed October 4, 2011)

Petrolia DR (2008) The economics of harvesting and transporting corn stover for conversion to fuel ethanol: A case study for Minnesota. *Biomass Bioenergy* 32:603–612

Prewitt RM, Montross MD, Shearer SA, Stombaugh RS, Higgins SF, McNeill SG, Sokhansanj S (2007) Corn stover availability and collection efficiency using typical hay equipment. *Trans ASABE* 5:705–711.

Ren HY, Richard TL, Chen ZL, Kuo ML, Bian YL, Moore KJ, Patrick P (2006) Ensiling corn stover: Effect of feedstock preservation on particleboard performance. *Biotechnol Progr* 22:78–85

Richard TL (2010) Challenges in scaling up biofuels infrastructure. *Science* 329:793–796

Savoie P, D'Amours L, Lavoie F, Lechasseur G, Joannis H (2006) Development of a cutter-shredder-baler to harvest long-stem willow. ASABE Paper No. 061016, American Society of Agricultural and Biological Engineers, St. Joseph, MI

Savoie P, Tremblay D, Tremblay GF, Wauthy JM, Quevillon M, Theriault R (1992) Silage harvest with a self-loading wagon on dairy farms. *Trans ASABE* 35:1385–1392

Shinners KJ, Adsit GS, Binversie BN, Digman MF, Muck RE, Weimer PJ (2007a) Single-pass, split-stream harvest of corn grain and stover. *Trans ASABE* 50:355–363

Shinners KJ, Binversie BN, Muck RE, Weimer PJ (2007b) Comparison of wet and dry corn stover harvest and storage. *Biomass Bioenergy* 31:211–221

Sokhansanj S, Fenton J (2006) *Cost Benefit of Biomass Supply and Pre-Processing, Synthesis Paper, A BIOCAP Research Integration Program, March 2006*, available at http://www.biocap.ca/rif/report/Sokhansanj_S.pdf (accessed April 3, 2011)

Sokhansanj S, Turhollow AF (2004) Biomass densification—Cubing operations and costs for corn stover. *Appl Eng Agric* 20:495–499

Srivastava AK, Goering CE, Rohrbach RP, Buckmaster DR (2006) *Engineering Principles of Agricultural Machines. Engineering Principles of Agricultural Machines*, 2nd ed., American Society of Agricultural and Biological Engineers, St. Joseph, MI, pp 1–14

Steffen Systems (2009) High density bale compression systems, available at http://www.steffensystems.com/Products/Bale_Press/index.htm (accessed February 6, 2011)

Uslu A, Faaij APC, Bergman PCA (2008) Pre-treatment technologies, and their effect on international bioenergy supply chain logistics: Techno-economic evaluation of torrefaction, fast pyrolysis and pelletisation. *Energy* 33:1206–1223

Venturi P, Huisman W, Molenaar J (1998) Mechanization and costs of primary production chains for *Miscanthus* x *giganteus* in the Netherlands. *J Agric Eng Res* 69:209–215

World Agricultural Outlook Board (2011) January 12, 2011 report. Office of the Chief Economist, U.S. Department of Agriculture, Washington, DC, available at www.usda.gov/oce/commodity/wasde/latest.pdf (Accessed February 6, 2011)

8 Chemical Engineering for Bioenergy Plants
Concepts and Strategies

David R. Shonnard and Michael J. Brodeur-Campbell
Michigan Technological University

Abraham R. Martin-Garcia
University of Sonora

Tom N. Kalnes
UOP LLC

CONTENTS

8.1 INTRODUCTION

Although global interest in liquid biofuels for transportation has recently increased, these fuels are not something new. Nearly a century ago automobiles were designed to run on gasoline, ethanol, and blends of these fuels; these were the forerunners of modern flexible-fueled vehicles. The motivation for biofuel use a century ago was different than today. Previously, biofuels were one of a diversity of fuel sources whereas in the modern era of petroleum, biofuels are meant to address a list of issues including global warming, rural economic development, domestic jobs, energy security, and balance of trade. The list of potential transportation biofuels today is much larger than in previous eras, mostly because of advances in biochemical and thermochemical processing technologies, which have also increased the list of possible biomass feedstocks that can be converted.

This chapter will first give a brief overview of the technologies for converting biomass feedstocks into usable liquid transportation fuels. After that will be an introduction into the main processing steps used in commercial practice and a review of current research and development in this field. Afterward, a short summary will give the current status for global biofuel production.

8.1.1 An Overview of Biomass-to-Biofuels Processing Technologies

Processing routes for conversion of biomass to biofuels has traditionally been organized into two categories depending on the agents for transformation and reaction conditions: biochemical and thermochemical. Biochemical conversion processes use biological catalysts (e.g., enzymes) at mild temperatures to produce sugars from the original biomass and then microorganisms to ferment sugars into oxygenated biofuels. The choice of microorganism should take into account the types of sugars to be fermented and the desired fermentation products. For example, hydrolysis of woody biomass will yield a mixture of five- and six-carbon sugars, but not many microorganisms are able to readily ferment five-carbon sugars. Recent advances in metabolic engineering of microorganisms have created unique metabolic pathways within microorganisms so that mixtures of sugars obtained from lignocellulosic biomass can be fermented into oxygenated biofuels (e.g., ethanol and butanol) but more recently into true hydrocarbon fuels. Thermochemical conversion processes use chemical catalysts and are, for the most part, carried out at higher temperatures and pressures. These reactions exhibit much shorter reaction times than biochemical conversion processes, but selectivity to a particular biofuel product is not as specific as for biochemical conversions.

Figure 8.1 is an overview diagram showing the main conversion steps for conventional biofuels (such as ethanol from starch crops and cane, biodiesel from triglycerides in soybeans) and for

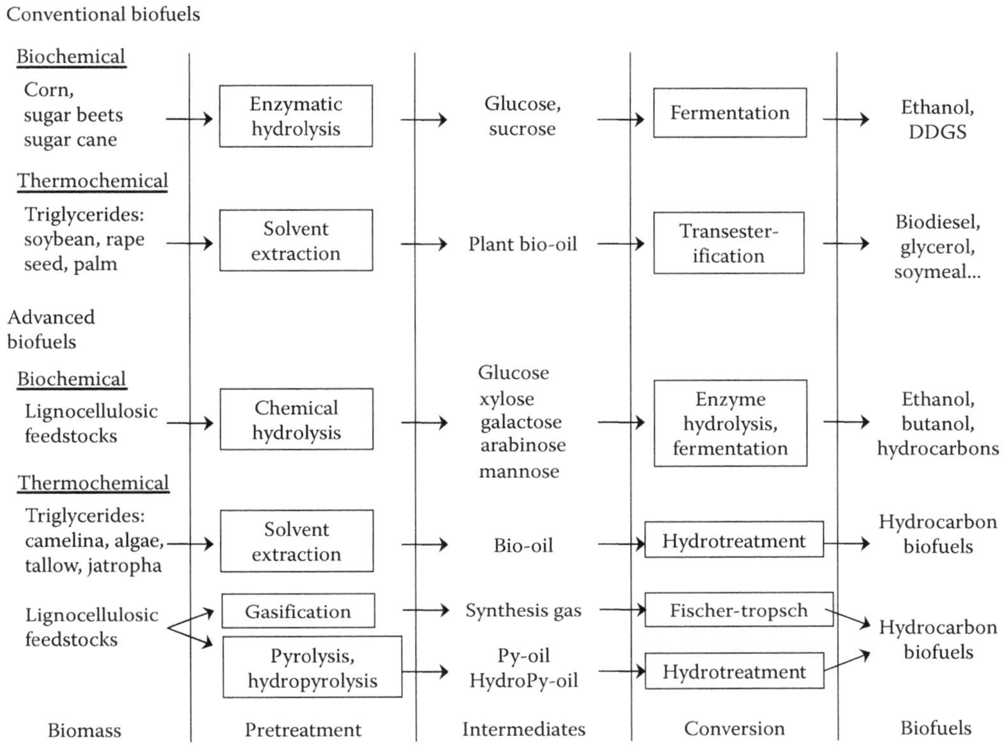

Conventional biofuels

FIGURE 8.1 Processing routes for conventional and advanced biofuels.

advanced biofuels. In both cases, biomass feedstock is first converted into an intermediate product through initial reaction or separation steps. The intermediates are then transformed to final biofuel product and co-products through additional reaction and separation steps. The vast majority of global production is conventional biofuels [2008 data, 18.7 billion gallons ethanol/year, 4.4 billion gallons biodiesel/year (EIA 2010)]. Dry and wet mill corn ethanol facilities produce an intermediate glucose sugar product and a final product of ethanol plus dry distiller grains solids (DDGS), which is marketed as an animal feed. The overall fermentation reaction is given by $C_6H_{12}O_6 \rightarrow 2\ C_2H_5OH + 2\ CO_2$. Two carbons from the sugar molecule are emitted as carbon dioxide, but the energy content of the two ethanol molecules is substantially higher than the sugar feedstock. Process energy for corn ethanol production is typically from natural gas for steam production and electricity is from the local grid. Because of the importing of these energy resources, corn ethanol has a relatively large fossil energy demand (ratio of fossil energy required for all processing steps per unit of energy in ethanol produced) of approximately 0.5–0.8 (Wu et al. 2006; Shapouri et al. 2010). Current (as of 2010) production rates of corn grain ethanol in the United States are approximately 50% that of global production at 12 billion gallons/year (RFA 2010), and from sugar cane in Brazil it is approximately 8 billion gallons ethanol/year (EIA 2010). Biodiesel is a methyl ester of fatty acids derived from plant oils. The biodiesel reaction can be simply described as triglyceride + methanol → 3 fatty acid methyl esters + glycerol ($CH_2OH–CHOH–CH_2OH$). Methanol is almost always produced from natural gas (fossil origin), and co-products of biodiesel production include glycerol and a residue from the oil extraction step (soymeal for example), which is often marketed as animal feed. The key intermediate is a plant oil obtained from the oil extraction step.

Advanced biofuels take advantage of the most plentiful biomass resource on Earth, lignocellulosic (woody) feedstocks, and of unconventional plant oils and algae. Processing routes for advanced biofuels are more complicated than for starch ethanol and biodiesel, and less is known about the costs and environmental impacts of these biofuels. This is because very few demonstration- or commercial-scale facilities exist for these biofuels at this time, yet this situation may soon change because of recent industry and government initiatives in research and commercial development. For example, as of 2010 there are 26 cellulosic ethanol facilities in various stages of construction in the United States with capacities ranging from a low of 20,000 gallons/year to 100 million gallons/ year (RFA 2010). Additionally, in the United States, demonstration-scale facilities of similar scale are in the planning and construction phase for production of some other types of advanced biofuels (Figure 8.1).

The remaining sections of this chapter describe some of the main features of conventional and advanced biofuel technologies. Technological barriers are discussed and recent approaches to overcome them are presented. The chapter ends with a discussion of the relative advantages and disadvantages of each conversion technology from cost and environmental impact perspectives.

8.2   PLANT OILS: CONVERSION TO BIODIESEL AND HYDROCARBONS

8.2.1   INTRODUCTION TO THE PROCESSING CONCEPTS

This section focuses on the conversion of plant-derived oils to distillate transportation fuels. Two distinctly different processing pathways are commercially available: (1) base-catalyzed transesterification of triacylglycerol (TAG)-rich feedstocks with methanol to produce fatty acid methyl esters (FAME), referred to as biodiesel; and (2) catalytic reaction of TAG and free fatty acid (FFA)-rich feedstocks with hydrogen to produce hydrocarbon-based fuels such as hydrotreated renewable diesel (HRD) and hydrotreated renewable jet (HRJ) fuel. Biodiesel can also be produced from FFA-rich feedstocks using a more severe acid-catalyzed esterification process.

To date, most of the existing capacity for producing diesel fuel from plant oils has centered on biodiesel. Although its use in blends with petroleum diesel is already widespread, there are some limitations to biodiesel that may ultimately limit its future use. These limitations include fuel storage stability, cold flow properties in winter climates, and high solvency that can lead to engine compatibility issues. The implementation of an ASTM specification for biodiesel (ASTM D6751-09) has helped to overcome issues related to fuel contamination with methanol, glycerol, and partially converted glycerides and fatty acids. To minimize the chance of engine problems, some car manufacturers have recommended capping the amount of biodiesel in petroleum blends at approximately 5% (McCormick 2009).

In the hydroprocessing pathway to biofuels, plant oils can be processed in pure form or co-processed with petroleum feedstock. HRD and HRJ are hydrocarbon liquids produced from plant oil feedstock using advanced hydroprocessing technology. Their physical properties (e.g., viscosity, flash point, distillation, freeze point) are similar to petroleum-derived fuels and their chemical composition is similar to paraffin-rich Fischer–Tropsch (FT)-derived fuels (Koers et al. 2009). Because these fuels are fully deoxygenated, they have excellent storage stability and can be used in any proportion with existing petroleum fuels.

Although commercial production of hydrocarbon-based biofuels is not yet widespread, sufficient production capacity has been brought on stream to demonstrate fuel quality and compatibility with existing engine technology. A significant increase in the HRD and HRJ production rate is expected over the next 3–5 years.

A more detailed discussion of the chemistry and processing steps for each processing pathway is provided in the paragraphs that follow. Process yields and conversion efficiencies are compared as well as fossil fuel and total energy requirements. A discussion of the current state of commercialization is also provided, including a brief discussion of sustainable feedstock selection.

8.2.2 Process Descriptions

A simplified diagram of a modern biodiesel production plant is provided in Figure 8.2. Dry, refined plant oil is subjected to a two-stage transesterification reaction. Fresh and recycled methanol in excess of stoichiometric requirements are fed to the mix tank, where an alkali catalyst such as sodium hydroxide is added. The mixture is fed along with the oil to a first-stage esterification reactor, which is typically heated by a steam coil. Effluent from the first stage is separated in a settling tank, with the upper phase going to the second-stage esterification reactor and the lower phase routed to methanol recovery. Additional methanol and catalyst are fed to the heated second-stage reactor and the effluent is sent to a second separator. The lower phase from the second separator can be recycled to the first reaction stage or sent to the glycerine recovery whereas the ester phase is sent to a water wash column for removal of water-soluble impurities. The water phase is routed to a methanol recovery column where excess methanol is recovered, dried, and recycled to the first reaction stage. After drying, a biodiesel product can be recovered from crude feedstock at a yield of approximately 96%. Co-product crude glycerol and a wastewater stream containing soaps and neutralized catalyst are recovered.

An example of an advanced hydroprocessing technology is the UOP/ENI Ecofining™ process. A simplified block flow diagram of the process is provided in Figure 8.3 (Kalnes et al. 2008).

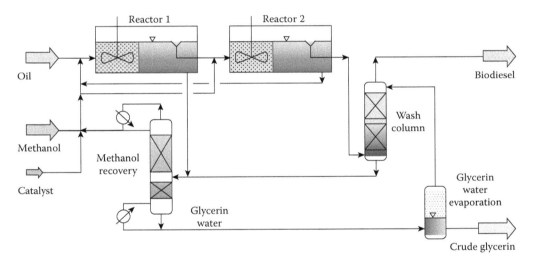

FIGURE 8.2 A modern biodiesel production facility.

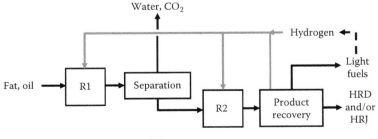

Eni/UOP ecofining™ process

FIGURE 8.3 Advanced process for HRD and HRJ production. (Kalnes, T., Marker, T., and Shonnard, DR., *Biofuels Technol Quart,* 2008, available at www.biofuels-tech.com.)

Similar to biodiesel plants, a pretreatment step is typically required to remove insoluble materials and trace metals, and two reaction stages are used. Unlike base-catalyzed transesterification, hydroprocessing is robust to high concentrations of FFAs, thus allowing other lower-cost materials such as tallow oil and waste greases to also be used as feedstock. Advanced hydroprocessing also produces a low-cloud-point product that can be used neat or blended with petroleum diesel without limitation.

In the Ecofining process, pressurized feedstock is mixed with recycled hydrogen and then sent to a multistage adiabatic catalytic hydrodeoxygenation reactor (R1) where the feedstock is saturated and completely deoxygenated. A vapor–liquid separator is used to recover excess hydrogen, and gas recycle to R1 is set to achieve a minimum hydrogen partial pressure. Conversion of feed is complete, and the volumetric yield of deoxygenated hydrocarbon products is greater than 100%. Selectivity to diesel boiling-range paraffin is very high. The primary deoxygenation reaction byproducts are propane, water, and carbon dioxide.

The effluent from R1 is separated to remove carbon dioxide, water, and low-molecular-weight hydrocarbons. The resultant diesel is mixed with additional recycle gas and then routed to a catalytic hydroisomerization reactor (R2) wherein a branched paraffin-rich diesel fuel is produced. In this manner, the cold flow properties of the diesel are adjusted to meet required specifications. The isomerization reaction is selective and consumes very little hydrogen.

Isomerized product is separated from excess hydrogen in a conventional gas/liquid separator. After purification, the excess hydrogen is recycled back to both reactors to maintain the minimum required hydrogen partial pressure. Make-up hydrogen is added to the process to balance chemical consumption and solution losses. The liquid product is sent to the product recovery section of the process where conventional distillation steps are used to separate co-products such as propane and naphtha.[5]

Table 8.1 contrasts Ecofining™ inputs and outputs to those of conventional biodiesel production (Kalnes et al. 2007). The Ecofining process for producing HRD (green diesel) operates at conditions similar to a petroleum hydrodesulfurization unit and integrates well within existing petroleum refineries. If required, a portion of the light fuel co-product can be steam-reformed to generate all of the hydrogen consumed in the process (Kalnes et al. 2007). Estimates of fossil energy demand (FED), cumulative energy demand (CED), and life-cycle greenhouse gas (GHG) emissions for biodiesel and HRD produced in the UOP/ENI Ecofining process, green diesel (GD), are compared to petroleum diesel in Table 8.2.

TABLE 8.1
Comparison of Biodiesel and HRD Process Inputs and Outputs

	Ecofining Green Diesel			Biodiesel	
	wt%	vol%		wt%	vol%
Feeds			Feeds		
Vegetable oil	100	100	Vegetable oil	100	100
Hydrogen	1.5–3.8		Methanol	10	11
			Chemicals[a]	4	
Products			Products		
Propane	5	9	FAME	96	100
Butane	0–2	0–3	Glycerol	10	7
Naphtha	<1–7	1–10			
Green diesel	75–85	88–99			

[a] Chemicals include sodium hydroxide (NaOH) and an acid to neutralize the products.

TABLE 8.2

Comparison of Biodiesel and HRD Energy Requirements and GHG Emissions

	Petroleum Diesel	SBO Biodiesel	SBO GD	RSO Biodiesel	RSO GD	Tallow Biodiesel	Tallow GD
FED, MJ/MJ	1.25	0.41	0.34	0.41	0.37	0.24	0.11
CED, MJ/MJ	1.27	1.53	1.42	1.88	1.82	1.26	1.14
GHG gCO$_2$-eq/MJ	85	48	40	46	41	17	6

SBO, soybean oil; RSO, rapeseed oil; GD, HRD from UOP/ENI ecofining process.

Three renewable feedstocks are considered in Table 8.2: soybean oil, rapeseed oil, and tallow. It can be seen that plant-derived fuels exhibit a higher CED and lower FED when compared with petroleum diesel. The higher CED is due to the additional energy required to cultivate, harvest, and transport the feedstock. Because carbon dioxide (CO_2) generated in the combustion of biomass-derived fuel is considered carbon neutral, their FED and GHG emissions are significantly lower than petroleum diesel for all of the biofuels. For tallow, a waste material derived from meat production, the CED is lower than that for petroleum diesel. As a waste, this feedstock is free of the upstream energy burdens assigned to the primary meat product.

It is important to note that land-use change (LUC) effects are not included in the GHG emissions shown in Table 8.2. Although LUC GHG impacts can be negative or positive, depending on the prior condition of the land, controversial indirect land-use change (ILUC) impacts often consider a worst-case scenario in which food crop conversion to energy crop production leads to rainforest destruction. Long-term concerns associated with diverting food crops to fuel production or carbon-rich forests to cultivated land has shifted the commercial focus away from first-generation feedstocks like soybean and rapeseed oil and toward next-generation (inedible) oils such as jatropha and camelina, which can be intercropped, grown on marginal land, or grown as a rotation crop (Shonnard et al. 2009). Ultimately, algae could become the major source of TAG-rich feedstock for biofuel production.

8.2.3 CURRENT STATE OF COMMERCIALIZATION

The production of biodiesel is widespread, with most of the existing production capacity located in Europe, mainly Germany and France. In 2007, total world biodiesel production was approximately 5–6 million tons, with 4.9 million tons processed in Europe (of which 2.7 million tons was from Germany) and most of the rest from the United States (NBB 2008). For comparison, total world production of vegetable oil for all purposes in 2005–2006 was approximately 110 million tons, with approximately 34 million tons each of palm oil and soybean oil (FEDIOL 2008). A significant increase in the production of biodiesel is expected over the next few years in South America. Table 8.3 provides an estimate of existing and future biodiesel capacity in countries with large production rates (Thurmond 2008).

There are numerous biodiesel technology providers, and the required process scheme is strongly dependent upon feedstock properties. Special pretreatment or acid-catalyzed esterification is required for feeds (such as tallow) containing significant amounts of FFAs.

Although not as widespread, hydroprocessing for the production of renewable transportation fuels is also a commercial technology practiced in existing petroleum refineries. Several commercial facilities for HRD production are in operation or in latter stages of design and construction. These include the NExBTL process (Neste Oy), Synfining process (Syntroleum), and Co-processing (Petrobras and ConocoPhillips) as well as the UOP/ENI Ecofining and UOP Renewable Jet Fuel process. Table 8.4 provides a partial summary of feedstocks, plant locations, production capacity, and approximate startup date. Over 800 million gallons/year (~3 million tons) of production capacity is expected to be on stream before 2012.

TABLE 8.3

Estimate of Biodiesel Production Capacity 2005–2010

					South America	
Country	EU	Germany	United States	Canada	Brazil	Argentina
Primary feed	85% rapeseed	Rapeseed	Soybean	Canola	Soybean	Soybeans
Percent crop to fuels	N20%		N13%		Tallow	
2005 operating units		137	110	4		10
New units planned			179		Many	Many
2005 capacity, MGPY[a]	960		250	30		18
2006 capacity, MGPY[a]		900	400			
"2010" capacity, MGPY[a]					528	164

[a] million gallons per year

TABLE 8.4

Partial List of Announced Capacity for Commercial Production of HRD

Feedstock	Location	Producer	Technology	Mgal/Year	Start-Up[d]
Jatropha	Portugal	GALP	Ecofining[c]	95	2011
Mixed	Italy	Eni Spa	Ecofining[c]	95	2012
Palm oil	Finland	Neste	NExBTL[a]	56	2008
Palm oil	Singapore	Neste	NExBTL[a]	264	2011
Animal fats	Louisiana, United States	Dynamic Fuels	Bio-SynFining[b]	75	2010
Mixed/petroleum	Brazil	PetroBras	H-Bio[e]	NA	2006
Mixed/petroleum	Ireland	ConocoPhillips	ConocoPhillips[e]	14	2006
5% tallow	Australia	BP	Co-processing	34	2008

NA, not applicable.
[a] Neste Oil Company
[b] Syntroleum
[c] Eni/UOP
[d] estimated from press releases
[e] PetroBras and ConocoPhillips are co-processing vegetable oil (soy, sunflower, palm, animal fat, etc.)

Most of the early units to come on stream (ConocoPhillips, PetroBras, and BP Australia) co-processed small percentages of fats and oils in existing petroleum HDS units. The remaining units are advanced hydroprocessing technologies designed to process feedstock that is 100% renewable fat and oil.

Although the social, economic, and regulatory issues associated with expanded production of HRD and HRJ are outside of the scope of this chapter, it is crucial that future commercialization efforts focus on sustainable methods of producing feedstock.

To support this effort, industry participants including commercial airlines have formed a consortium that have set criteria that are complementary to emerging internationally recognized standards such as those being developed by the Roundtable on Sustainable Biofuels. These criteria are

- Feedstock sources should be developed in a manner that is noncompetitive with food and in which biodiversity impacts are minimized. In addition, the cultivation of those plant sources should not jeopardize drinking water supplies.
- Total life-cycle GHG emissions from plant growth, harvesting, processing, and end use should be significantly reduced compared with those associated with hydrocarbon fuels from fossil sources.

- In developing economies, development projects should include provisions or outcomes that improve socioeconomic conditions for small-scale farmers who rely on agriculture to feed themselves and their families and that do not require the involuntary displacement of local populations.
- High conservation value areas and native ecosystems should not be cleared and converted for fuel plant source development.

Current and future producers are targeting sustainable production scenarios that, in addition to minimizing impact on LUC and food and water resources, provide an energy alternative that is economically competitive with current petroleum-based fuels. Market growth will require a coordinated effort between feedstock producers, refiners, and industry regulators to ensure that environmental impacts are minimized.

If done responsibly, increasing HRD and HRJ usage in the transportation sector can significantly reduce GHG emissions as well as diversify energy sources, enhance energy security, and stimulate the rural agricultural economy.

8.3 THERMOCHEMICAL CONVERSION OF PLANT WOODY BIOMASS

Archeological findings show the first evidence of thermochemical processing of woody biomass by humans to have occurred approximately 1.9 million years ago, the first controlled use of fire by humans approximately 400,000 years ago, and charcoal production and controlled burns some tens of thousands of years ago (Bowman 2009). Although the above examples are thermochemical reactions, this review will focus on more advanced thermochemical processing routes to upgrade woody biomass into liquid transportation and gaseous energy products.

The second major platform covered in this review is for conversion of woody biomass into biofuels and high-value chemicals through the use of thermochemical processing steps conducted at high temperature and pressure, often in the presence of chemical catalysts. Gasification and pyrolysis of wood are the two most important thermochemical processing routes. In both cases, wood is thermally decomposed into intermediate compounds of small molecular weight relative to the starting polymeric carbohydrate and lignin wood fractions. The predominant products of gasification constitute a synthesis gas, whereas in pyrolysis, depending on reaction conditions, the major products could be a crude bio-oil, a synthesis gas, or a solid carbonaceous char. The following sections will introduce these major thermochemical processing platforms, present the main features of reaction chemistries, and estimate energy efficiencies from conversion.

8.3.1 GASIFICATION-BASED CONVERSION OF PLANT BIOMASS

Gasification of woody biomass provides a means for production of renewable energy products, notably synthesis gas for liquid fuels production, fuel gas for heat or power applications, and hydrogen for fuel cells. Figure 8.4 shows a process flow diagram for gasification of woody feedstocks for production of these three main energy products. Wood chips are transferred to a storage operation, and upon entering the process pass through a size reduction step before gasification.

Gasification is a partial oxidation/thermal decomposition reaction carried out at high temperature (600–900°C) in the presence of a gasification medium (air, oxygen, steam) to yield a synthesis gas containing major products carbon monoxide (CO), H_2, CO_2, and H_2O but with significant amounts of minor products, mainly ash, ammonia, hydrogen sulfide (H_2S), tars, and particulate char (Torres et al. 2007) as well as trace contaminants such as hydrogen cyanide (HCN), halogens [e.g., hydrogen chloride HCl], alkali metals, and other metals (Pb, As, Hg) (NSF 2008). These minor products, or impurities, must be removed (gas cleanup) before using the synthesis gas for electricity generation, heat production, or catalytic conversion to the high-value fuels and chemicals, as shown in Figure 8.4. For example, (1) tars can coat surfaces within downstream processes and

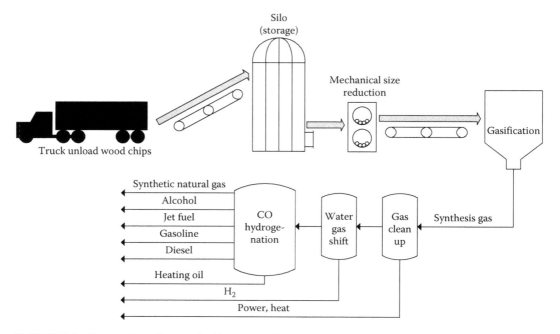

FIGURE 8.4 Process flow diagram for biomass gasification-based processing to energy products.

interfere with rotating power generation equipment, (2) ammonia and other nitrogen-containing gases will be converted to oxides of nitrogen (NO_x) in engines or turbines, and (3) H_2S poisons catalysts and ammonia blocks catalyst active sites and therefore both must be removed.

Conversion to fuels and chemicals (CO hydrogenation in Figure 8.4) occurs in the presence of specific metal catalysts and requires unique ratios of CO to H_2 to form the different fuel and chemical products. The CO/H_2 ratio can be adjusted through a catalytic water gas shift reaction in which CO is combined with H_2O to form additional H_2 and CO_2. Other approaches are possible to adjust the final CO/H_2 ratio, including the use of methane in a steam-reforming reaction ($CH_4 + H_2O \rightarrow CO + 3 H_2$) or a H_2-selective membrane separation step. Depending on application, separation of final products and co-products is accomplished through distillation, membranes, and other devices.

8.3.1.1 Gasification of Biomass

Gasification of biomass can be carried out in several reactor configurations. Biomass chips can be introduced at the top of the reactor to be contacted with a co-current or countercurrent stream of the gasification medium (air, oxygen, and water vapor). Another configuration utilizes a bubbling or circulating fluidized bed reactor in which heat is transferred by a circulating stream of solid media such as sand. The gasification product composition is dependent on process conditions, gasification medium, and feedstock type, but with CO/H_2 ratio usually less than 1 and with a heating value between 5 and 15 MJ/Nm³ (Nm³ is normal cubic meter). Thus, biomass-derived synthesis gas is considered a low to medium fuel gas compared with natural gas (35 MJ/Nm³).

The gasification reaction itself is endothermic and requires a source of external heat. Volatile matter from heating of biomass is initially oxidized forming CO_2 and H_2O and releasing heat, which leads to further volatilization of biomass and H_2O. Partial oxidation and thermal decomposition reactions also take place leading to the product mixture discussed in Section 8.3.1.

Ash from the biomass is mostly bound up with particulate char and can be separated easily from the gas product using filtration or other means. The char can be oxidized yielding additional process heat or partially oxidized with O_2 or reformed with H_2O to increase carbon conversion in synthesis gas. Tars are high-molecular-weight aromatic hydrocarbons and must be removed to protect downstream equipment and catalysts. Tar formation in the gasifier can be minimized by increasing the O_2-to-fuel ratio, raising gasification temperature, or using low-cost dolomite ($CaMg(CO_3)_2$) which acts as a base catalyst in steam reforming ($C_nH_m + H_2O \rightarrow nCO + ((m/n)/2 + 1) H_2$) of tars. For example, Gil et al. (1999) showed that addition of dolomite to a biomass gasifier decreased tars from approximately 20 to 2 g tar/Nm³, but was accompanied by higher production of dolomite attrition dust. Higher gasifier temperatures have disadvantages that lead to decreased carbon yield in syngas or with higher slag production (NSF 2008). Even with prevention steps in the gasifier, aftertreatment is required to reduce tars to tolerable levels, as discussed in the next section. Other minor products such as H_2S and ammonia are derived from biomass-bound S and N and their concentrations in gasifier effluent are feedstock dependent for the most part.

8.3.1.2 Synthesis Gas Clean-Up from Biomass Gasifier

Tars are produced at levels between 10,000 and 20,000 mg/Nm³ under typical biomass gasifier operating conditions (Torres et al. 2007; NSF 2008). Tars are best removed using steam-reforming (Ni catalyst) or cracking (dolomite, carbonate, or Ni catalysts) reactions, but cracking requires higher temperatures (900°C) than normally achieved at gasifier exit (<800°C) and Ni catalysts can be poisoned by low amounts of H_2S. Tar concentrations as low as 2 mg/Nm³ were observed over a 50-h trial when dolomite was used in the gasifier followed by two reactors containing Ni catalyst to remove tar (Caballero et al. 2000). However, a recent review concludes that no tar removal process has demonstrated long-term performance at a gasifier's normal exit temperature of less than 800°C (NSF 2008).

Ammonia is produced at levels between 2000 and 4000 ppmv in a biomass gasifier (Torres et al. 2007) and is removed through decomposition to N_2 and H_2 using supported Ni, Ru, and Fe catalysts, but carbides and nitrides of W, V, and Mo can also be used. Much attention has been given to NH_3 decomposition in synthesis gas in the presence and absence of H_2S (Jothimurugesan and Gangwal 1998). NH3 can be effectively removed with commercial reforming catalysts at 650°C, but removal of H_2S below 10 ppmv is required.

Because sulfur content of biomass is relatively low compared with other solid gasification feedstocks, the H_2S (and COS-carbonyl sulfide) concentration is normally between 20 and 600 ppmv in biomass gasifier outlets (Torres et al. 2007). But this H_2S concentration is too high to avoid poisoning of Ni catalysts for tar and ammonia removal, and although higher temperature operation can mitigate this S poisoning effect in Ni catalysts, sintering/deactivation of catalyst and materials-of-construction issues become important (NSF 2008). The most feasible option for sulfur removal is to use a high temperature adsorbent/regeneration process using ZnO, but Zn ferrites and Zn titanates are also effective.

$$ZnO + H_2S \rightarrow ZnS + H_2O \text{ (adsorption step)} \qquad (8.1)$$

$$ZnS + 3/2\ O_2 \rightarrow ZnO + SO_2 \text{ (regeneration step)} \qquad (8.2)$$

Although gasification hot gas clean-up processes have been extensively studied in the past, significant research and development is needed to overcome barriers to commercialization of biomass gasification (Torres et al. 2007).

8.3.1.3 Water–Gas-Shift Reaction

The water–gas-shift (WGS) reaction is used to adjust the CO/H_2 ratio in synthesis gas by the mechanism:

$$CO + H_2O \leftrightarrows CO_2 + H_2 \qquad (8.3)$$

Commercial catalysts are available for WGS reactions in the temperature ranges of low (225–250°C, Cu–Zn), medium (350–375°C, Co–Mo sulfide form), and high (450–475°C, Fe–Cr). Because the

Cu–Zn catalysts require very low H_2S concentrations (<60 ppbv) and the Co-Mo catalyst tolerates very high H_2S concentrations (several thousands of parts per million by volume), perhaps the best catalyst for WGS reactions for biomass synthesis gas is the Fe–Cr catalyst, which can tolerate moderate H_2S levels of 50–100 ppmv. At this high temperature, remaining tars might continue in the vapor rather than deposit on the catalyst surface.

8.3.1.4 CO Hydrogenation Reaction

The final step in production of fuels and chemicals from biomass is the CO hydrogenation reaction. Conversion to hydrocarbons, alcohols, and other organic compounds is dependent on catalyst type, CO/H_2 ratio, and other reaction conditions. Fischer–Tropsch synthesis (FTS) converts CO and H_2 to gaseous and liquid hydrocarbons through a highly exothermic polymerization mechanism. FTS reaction begins with adsorption and dissociation of CO and H_2 on active sites on the catalyst surface (NSF 2008). A single C atom and multiple H atoms combine on a single active site to initiate chain growth of the hydrocarbon. Further chain growth occurs when adjacent intermediates combine in a C–C bond. Chain termination occurs when an adjacent adsorbed H combines with the growing chain to form a terminal C–H bond. The likelihood of chain growth is governed by the Anderson-Schultz-Flory chain growth probability, α, which is dependent on catalyst type, promoters present in the catalyst, and reaction conditions (Bartholomew and Farrauto 2006). Products ranging from C_1 to C_{60} + can be achieved, and the actual distribution of carbon number in the hydrocarbon products is dependent on α, with high values of $\alpha > 0.9$ being preferred for liquid hydrocarbon products.

Common catalysts for FTS include those based on Fe and Co. Iron-based FTS catalysts are favored for converting low H_2/CO (0.6–1), non-WGS synthesis gas from biomass because these catalysts also exhibit significant WGS activity. Cobalt-based catalysts are used on higher H_2/CO (2.0–2.2) synthesis gas from WGS biomass synthesis gas. Co catalysts achieve much higher activities compared to Fe (5–10 times higher), and Co catalysts are more selective to higher-molecular-weight hydrocarbons (NSF 2008).

Gasoline products are favored in the high-temperature (HT) FTS range of 300–350°C, whereas diesel and jet fuel are prominent in the low-temperature (LT) FTS at 200–250°C. In a large biorefinery based on FTS, LT and HT reactors would be in operation along with refinery-type FTS liquid and wax upgrading steps (oligomerization, catalytic reforming, hydrotreating, and hydrocracking/hydroisomerization) and separations. Furthermore, aromatic content of the FTS jet fraction must be increased to provide desired properties, such as low freeze point (–47°C).

Because of the exothermic FTS reaction, the reactor design must remove heat effectively to avoid catalyst thermal degradation and maintain product selectivity. A thorough review of FTS reactor configurations is available in Bartholomew and Farrouto (2006), but the choices are between tubular fixed-bed reactors (TFBRs), fluidized bed reactors, and slurry bubble column reactors (SBCRs). SBCRs are reported to have advantages over TFBRs because of the simplicity, lower cost, higher volumetric productivity, heat removal efficiency, and more favorable catalyst productivities and handling. However, recent improvements of TFBR designs achieve performance similar to large SBCRs, as reported by Hoek and Kersten (2004).

Interest in alcohols as transportation fuels stems from their favorable engine performance and lower emissions compared with hydrocarbon fuels (Verbeek and Van der Weide 1997; Phillips and Reader 1998; NSF 2008). A blend of 10% ethanol in petroleum gasoline decreases GHG emissions, lowers CO and particulate emissions, and increases octane rating (NSF 2008). Dimethyl ether (DME) produced from methanol is reported to decrease particulate and NO_x emissions (Sorenson and Mikkelsen 1995), and DME can easily form blends with petroleum diesel.

Methanol production from synthesis gas is practiced commercially and occurs at high pressure (50–80 atm) and at 225–250°C in the presence of Cu–Zn catalyst with nearly 100% selectivity, but with low conversions, and with a requirement of syngas recycling (Bartholomew and Farrauto 2006). Other catalysts are also used (silicoaluminophosphate) at similar temperatures with similarly high selectivity (Williams et al. 1995; Shonnard et al. 2006). Presence of a small amount of CO_2 (4% vol.) helps productivity because the reaction mechanism is thought to involve hydrogenation of CO_2. Cu–Zn catalysts are very susceptible to poisoning by sulfur and arsine in syngas, thus requiring gas clean-up. Methanol synthesis is carried out in various reactor configurations, as discussed in Bartholomew and Farrouto (2006).

Higher alcohols are also produced by catalytic conversion of syngas; for example, producing the octane enhancer methyl tertbutylether (MTBE) from methanol and isobutanol. One alcohol of interest is ethanol, the direct synthesis of which from synthesis gas can be accomplished using a rhodium- or copper-based catalyst. Hu et al. (2007) report ethanol selectivity of more than 50% using an Rh catalyst at high pressure but with low conversion. Side reaction byproducts include methane, C_2–C_5 alkanes and alkenes, and low-molecular-weight oxygenated organics. A few pilot-scale processes for higher alcohol synthesis are in place, but no commercial facilities currently exist (NSF 2008).

Although catalytic CO hydrogenation reactions have been investigated in the past, and there is commercial production of select products, there is a need for further catalyst and reactor innovation. A comprehensive list of research recommendations leading to improved commercial production from catalytic CO hydrogenation is included in a recent benchmark study (NSF 2008).

8.3.2 Pyrolysis-Based Conversion of Plant Biomass

Pyrolysis is a thermal depolymerization and molecular fragmentation process carried out in the absence of oxygen (or air) and at moderate temperature (~450–700°C) (Mohan et al. 2006). These thermal reactions occur in stages, each corresponding to higher reaction severity, in which increasing severity refers to longer reactor residence times and higher reaction temperatures. Primary reactions at low severity yield gases (CO_2, CO, H_2O), organic vapors, and liquid products. Secondary reactions act on primary products decomposing larger molecules into low-molecular-weight gaseous and liquid species as well as char (a carbonaceous solid). Tertiary reactions further the degradation process to produce synthesis gas (CO_2, CO, H_2O, H_2) and soot (NSF 2008). Three primary co-products are present at the reactor exit: synthesis gas (CO, CO_2, H_2O), bio-oil, and char. The proportion of each of these is dependent on reaction severity.

A process flow diagram showing pyrolysis-based conversion of woody biomass into liquid transportation fuel is shown in Figure 8.5. In a sand fluidized bed pyrolysis reactor (a common configuration), char and gaseous co-product generated in the pyrolysis reaction are combusted in an integrated recycle vessel to maintain the sand at the required pyrolysis reactor temperature. Liquid bio-oil from the pyrolysis unit is then subjected to hydrotreating and hydrocracking catalytic reactions. Gaseous products from the hydrotreater can be converted to hydrogen and CO_2 using a steam reformer and gas shift reactor. Heavy oil product from the hydrotreater will feed a hydrocracker reactor to generate additional light oil components. The light oil products are then separated into different product blends such as gasoline, diesel, and, potentially, aviation fuel.

The following sections describe the processing steps in more detail and provide information on the properties and stability of pyrolysis bio-oil.

8.3.2.1 Pyrolysis Reactions

Pyrolysis reactions on biomass feedstocks have been carried out under various temperature regimes and in the presence of different solvents (see Mohan et al. 2006, Table 8.5). Carbonization occurs when biomass is heated slowly over several days at approximately 400°C, yielding charcoal as the main product, but also significant amounts of gas (CO_2, H_2O, CO). Conventional pyrolysis features biomass residence times of between 5 and 30 min, a slow heating rate, and 600°C temperature in which the products include synthesis gas, bio-oil, and char. Fast pyrolysis produces mostly bio-oil by using high heating rates, short residence times of 0.5–5 s, and temperatures of 425–650°C and it produces mostly bio-oil, but small amounts of synthesis gas and char. Ultrapyrolysis produces mostly gaseous products and chemicals through very high heating rates, high temperatures of

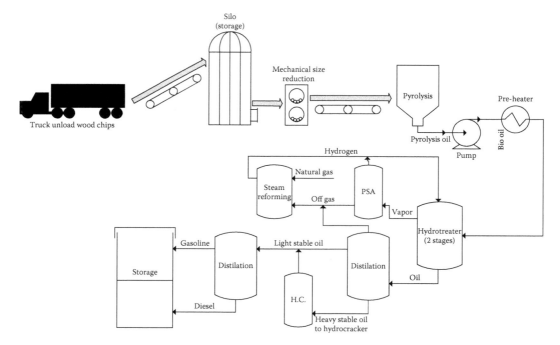

FIGURE 8.5 Pyrolysis-based process for production of liquid transportation fuel from woody biomass.

approximately 1000°C, and ultrashort residence times (< 0.5 s). Vacuum pyrolysis, similar to fast pyrolysis, produces predominantly bio-oil at 400°C, with a 2- to 30-s residence time, and medium heating rate. Hydropyrolysis uses water as a reducing reaction medium to convert woody biomass to bio-oil (mostly at < 500°C), using a high heating rate and a 10-s residence time. Because water is the reaction medium, hydropyrolysis has the advantage that it can accept biomass with field moisture without a drying step. Methanol is used as the reaction medium in methano pyrolysis, but the main products are chemicals rather than fuels, and reactor conditions include a high heating rate, a residence time of less than 10 s, and temperatures of more than 700°C.

8.3.2.2 Fast Pyrolysis

Because the main focus of this volume is energy products, fast pyrolysis and utilization of bio-oil in energy and fuels applications will be covered in more detail, whereas additional information on other aspects of pyrolysis can be found in several excellent reviews (Bridgewater and Peacocke 2000; Mohan et al. 2006) and the references listed therein. Fast pyrolysis achieves high yields of bio-oil from biomass and lower recoveries of synthesis gas and char by carefully controlling the reaction environment. Typical yields for fast pyrolysis are bio-oil (60–75%), synthesis gas (10–20%), and char (15–25%). A further analysis of bio-oil reveals that water comprises approximately 20–30% by weight, the remaining being oxygenated organic compounds of varying molecular weight. Organic compounds in bio-oil from the carbohydrate fraction of wood are more water soluble, whereas those from lignin are considered more hydrophobic and are higher in molecular weight (NSF 2008). Requirements for fast pyrolysis are rapid heat transfer rates; a finely ground dry (5%) moisture biomass feedstock to achieve high heat transfer and reaction rates; and rapid cooling of pyrolysis vapors and aerosols to avoid degradation, dehydration, and fragmentation.

8.3.2.3 Properties of Pyrolysis Bio-Oil

Before discussing conversion of pyrolysis bio-oil into liquid transportation fuels or the use of bio-oil in heat or power applications, it is necessary to describe the chemical nature and properties of bio-oil

TABLE 8.5

Properties of Fast Pyrolysis Bio-Oil Compared with Petroleum Heavy Fuel Oil

Properties	Fast Pyrolysis Bio-Oil	Petroleum Heavy Fuel Oil
Bio-oil yield, % by weight of biomass	17–75	NA
Water content, % by weight of bio-oil	0–40	0.1
pH	2.3–5.5	NA
Specific gravity	0.91–1.29	0.94
Elemental composition, % by weight		
Carbon	46–78	85
Hydrogen	5–12	11
Oxygen	11–47	1
Nitrogen	0–10	0.3
Sulfur	Trace	0.5–3.0
Ash	Trace–1.5	0–0.1
Viscosity, centistokes at 50°C	1–50	50
Higher heating value, MJ/kg	14–41	40
Pour point, °C	–36 to 25	–18
Solids, % by weight	0.2–3	1
Stability	Poor	Good

Source: Czernik, S. and Bridgwater, AV., *Energy Fuel,* 18, 590–598, 2004 and Mohan, D., Pittman, CU., and Steele, PH., *Energy Fuel,* 20, 848–889, 2006.

NA, not applicable.

because these will determine the processing steps needed for upgrading. In addition to H_2O content, which was described earlier, fast pyrolysis bio-oil is composed of the following classes of oxygenated organic compounds in descending order of occurrence (Bridgewater et al. 2001): pyrolytic lignin, 15–25%; aldehydes (formaldehyde, acetaldehyde, hydroxyacetaldehyde, and glyoxal), 10–20%; carboxylic acids (formic, acetic, propionic, and butyric, etc.), 10–15%; carbohydrates (cellobiosan, levoglucosan, and oligosaccharides); 5–10%; phenols (phenol, cresols, guaiacols, and syringols), 2–5%; alcohols (methanol and ethanol), 2–5%; ketones (acetol and cyclopentanone), 1–5%; and furfurals, 1–4%. The key challenge in upgrading these compounds to hydrocarbon transportation fuels is efficient de-oxygenation through hydrotreating, but to limit consumption of hydrogen, saturation of aromatic rings should be avoided (NSF 2008).

Properties of pyrolysis bio-oil and comparisons to petroleum heavy fuel oil are shown in Table 8.5. The ranges of property values in this table are a compilation from a wide diversity of biomass feedstocks, such as hardwoods, softwoods, bark, bagasse, straw, oil seed feedstocks, microalgae, and other sources (Mohan et al. 2006). Additional factors that affect the bio-oil properties listed in Table 8.5 are reaction temperature (400–650°C) and reaction residence time (0.3 s to 30 min).

8.3.2.4 Stability of Bio-Oil

Diebold (2000) reviewed reactions occurring in pyrolysis bio-oil during long-term storage and provided recommendations on improving storage stability. Reactions during aging of bio-oil appear to be catalyzed by organic acids (low pH) and elements found in char. These aging reactions result in changes to the molecular weight distribution of the bio-oil components, increase in bio-oil viscosity, and phase separation of higher-molecular-weight oligomers and polymers. Oxygen from air reacts with organics in bio-oil to form peroxides that catalyze polymerization of olefins and addition of mercaptans to olefins. Organic acids react with alcohols to form esters and water, and aldehydes react with components in bio-oil such as water, alcohols, other aldehydes, phenolics, and proteins to form hydrates, ethers, resins, oligomers, and dimers. In addition, olefins polymerize to form oligomers and polymers. The addition of methanol or ethanol to fast-pyrolysis bio-oil at the 10 wt% level was found to be effective in largely retarding these reactions (Diebold and Czernik 1997). Furthermore, elimination of contact with air, addition of antioxidants, and mild hydrogenation are effective measures to increase storage stability of pyrolysis bio-oils.

8.3.2.5 Catalytic Upgrading of Pyrolysis Bio-Oil

Pyrolysis bio-oil has limited applications as a fuel other than direct combustion in furnaces to provide heat or power. As a result, bio-oil must be upgraded to serve as fuel in vehicular transportation. The goal of upgrading is to increase volatility through molecular-weight reduction, enhance storage stability, and eliminate oxygen to raise product fuel heating value. Catalytic upgrading using hydrogen is accomplished using two main methods: hydroprocessing and catalytic cracking. These upgrading reactions are similar in nature to hydrotreating reactions that occur in conventional petroleum refinery processes. The advantage of producing hydrocarbon fuels from pyrolysis bio-oils, similar to FT reactions of biomass gasification synthesis gas, is the compatibility of the final fuel product with distribution infrastructure in pipelines and with conventional and high-efficiency engines.

Recent reviews have appeared on upgrading pyrolysis bio-oils to liquid hydrocarbon fuels through hydrotreatment and hydrocracking (Furimsky 2000; Huber et al. 2006; Elliott 2007). In hydroprocessing, oxygen is removed from compounds in pyrolysis bio-oils through reactions with hydrogen to produce water plus hydrocarbons, which can be isolated as immiscible separate phases. Various heterogeneous catalyst materials to carry out hydroprocessing reactions on bio-oils include sulfide catalysts found in the petroleum refining industry and precious metal catalysts.

The use of acidic cracking catalysts is well known in petroleum refining to reduce molecular weight and convert fuel components to more aromatic structures. When these catalysts are applied to pyrolysis bio-oils, similar results are observed, but with high levels of coke formation. Zeolitic acid catalysts, such as HZSM-5, yield high ratios of aromatic to aliphatic compounds and have been used to crack the product of pyrolysis bio-oil hydrotreatment. This yields a 5:1 ratio of aromatic to aliphatic product with 30–50% conversion to coke. Coke is burned to recover catalyst and generate process heat. However, a continuing challenge is achieving a good heat balance (NSF 2008).

Long-term catalyst stability and life has not been demonstrated in pyrolysis bio-oils that have been catalytically upgraded. Currently the longest test lasted 8 days before significant hydrotreating catalyst deactivation was observed. Deactivation of cracking catalysts has limited in-process life to only a few cycles in recent experiments. Catalyst deactivation has been attributed to the presence of water on oxide structures of catalysts and to the presence of trace contaminants such as minerals in biomass feedstocks, but the actual mechanisms are not currently well understood (NSF 2008).

8.3.2.6 Power Uses of Pyrolysis Bio-Oil

Pyrolysis bio-oil can be combusted similarly to fossil fuels in boilers, gas turbines, and diesel engine power applications. Co-firing with coal, natural gas, and fuel oil are also options for utilization of bio-oil as a renewable feedstock. Bio-oil has an advantage compared with wood chips and pellets because of its wider application in co-firing applications, which allows easier adaptation to higher-efficiency power generation (conventional natural gas) and advanced power (combined cycle power). A review of pyrolysis bio-oil power generation can be found in a recent publication by Czernik and Bridgwater (2004), showing that bio-oil can be successfully used as boiler fuel and in diesel engine and gas turbine applications.

8.4 BIOCHEMICAL CONVERSIONS PLANT WOODY BIOMASS

Biochemical conversion of lignocellulosic biomass to fuels and other chemicals generally focuses primarily on the isolation of sugars in their monomeric form from hemicellulose and cellulose. The four main processing steps are mechanical size reduction, chemical pretreatment, enzymatic hydrolysis, and fermentation. Although mechanical size reduction has obvious effects on downstream equipment and processes and may come before or after chemical pretreatment, the focus of this chapter will be on the remaining three steps: pretreatment, enzymatic hydrolysis, and fermentation.

8.4.1 Overview of Biochemical Conversion Processes

The purpose of pretreatment and enzymatic hydrolysis is to fractionate and recover monomeric sugars from lignocellulosic biomass, whereas fermentation takes those sugars and converts them into valuable products. Cellulose is a linear homopolymer of the six-carbon sugar glucose. Hemicellulose is a branched heteropolymer of five- and six-carbon sugars, primarily xylose, arabinose, galactose, glucose, and mannose. Six-carbon (hexose) sugars are readily fermented by many microorganisms, whereas the five-carbon (pentose sugars) are fermented by only a few native strains (Mosier et al. 2005). Figure 8.6 shows the major unit operations found in a biochemical conversion process for production of ethanol or other fermentation products from woody biomass.

Pretreatment serves two basic purposes: it begins the process of breaking down the more-easily hydrolyzable (mostly hemicellulose) sugars, and it opens up the structure of the lignocellulosic biomass to make the cellulose more accessible for enzymatic hydrolysis. Pretreatment must also preserve the hemicellulose sugars, limit the formation of degradation products, and make it easier to reduce biomass particle size (Mosier et al. 2005).

Separate hydrolysis and fermentation (SHF) is when enzymatic hydrolysis takes place separately from fermentation. However, certain processes can be consolidated in different configurations with some advantages. Simultaneous saccharification and fermentation (SSF) is the term for hydrolysis taking place in the presence of fermenting microorganisms. This reduces the effect of product inhibition during hydrolysis as the fermenting organisms consume the products (sugars) as they are produced If pentose and hexose sugars are being fermented in the same vessel as enzymatic hydrolysis, it is known as simultaneous saccharification and co-fermentation (SSCF). This can be done with a mixed culture of organisms that can ferment five- and six-carbon sugars, or with a single organism that can utilize all sugars.

Ultimately, genetic engineering maybe able to produce custom organisms capable of converting raw biomass into value-added product without pretreatment or enzymatic hydrolysis. This goal is known as consolidated bioprocessing (CBP). SSF and SSCF are preferred to SHF, because minimizing unit operations leads to lower costs. They are generally viewed as realistic near-term possibilities, whereas CBP is a longer-term goal (Mosier et al. 2005; Lynd et al. 2002; Wright 1988).

8.4.2 Pretreatment Hydrolysis Conversions

Pretreatment is among the most costly steps in processing cellulosic biomass, and it has effects on upstream (size reduction) and downstream (enzymatic hydrolysis) processes. For example, more efficient pretreatment can lead to lower enzyme loading requirements for enzymatic hydrolysis. Therefore, the interactions between pretreatment and enzymatic hydrolysis are of critical importance (Mosier et al. 2005; Wyman et al. 2005b). Several pretreatment technologies are available; however, their relative attributes differ. Table 8.6 shows several of the most promising biomass pretreatment methods along with their advantages and disadvantages.

In general, low pH treatments give liquid fractions containing most of the hemicellulose sugars and a solid residue containing most of the cellulose and lignin. High pH conditions remove lignin while leaving a solid residue that contains most of the cellulose and hemicellulose. Although AFEX (ammonia fiber explosion) is a high pH treatment, it does not generate a liquid stream, and therefore essentially 100% of the feedstock is recovered as dry matter with a disrupted structure suitable for

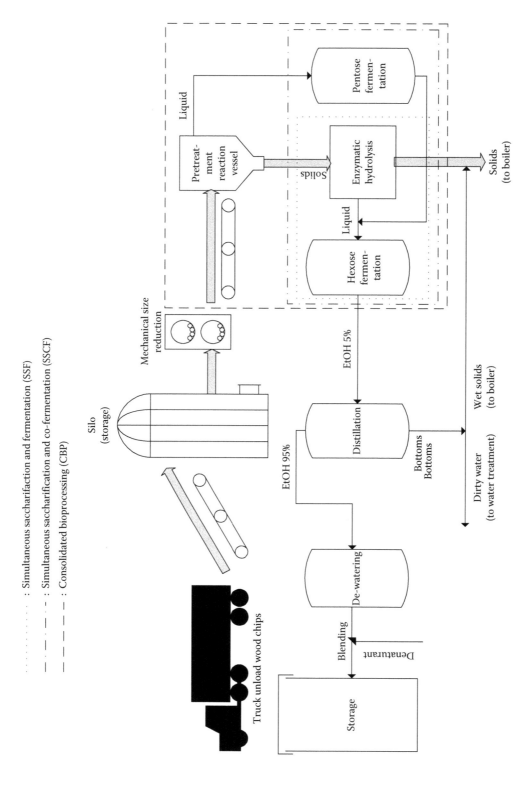

FIGURE 8.6   Process diagram for biochemical conversions of plant woody biomass.

TABLE 8.6
Promising Biomass Pretreatment Technologies

Method	Description	Advantages	Disadvantages
		Low pH Treatments	
Dilute acid, co-current	0.5–3% sulfuric acid, 130–200°C, 3–15 atm, 2–30 min., 10–40 wt% solids	Monomeric hemicellulose sugars, increases cellulose hydrolysis to near 100%	Costly materials of construction, high operating pressures, hydrolysate conditioning (neutralization required)
Dilute acid, flow-through	0–0.1% sulfuric acid, 190–200°C, 20–24 atm, 12–24 min., 2–4 wt% solids	Enhances hemicellulose and lignin removal	Challenging to implement commercially, high water use results in high energy requirements
Hot water/steam autolysis	Hot water, 160–190°C, 6–14 atm, 10–30 min, 5–30 wt% solids	No chemical additions, low capital and operation cost	Incomplete hydrolysis of hemicellulose
		High pH Treatments	
AFEX/FIBEX	Anhydrous ammonia, 70–90°C, 15–20 atm, 5 min, 60–90 wt% solids	Lower-cost vessels, hydrolysate conditioning unnecessary	Incomplete hydrolysis of hemicellulose
ARP	10–15 wt% ammonia, 150–170°C, 9–17 atm, 10–20 min, 15–30 wt% solids	Enhances cellulose digestion, lignin removal	Less effective at softwood pulps
Lime	0.05–0.15 g Ca(OH)$_2$/g biomass, 70–130°C, 1–6 atm, 1–6 hours, 5–20 wt% solids	Low temperature, atmospheric pressure; low-cost and inherently safe materials	Longer reaction times required
		Solvent Pretreatments	
Organosolv	50–80% methanol or ethanol, 180–210°C, approximately 1 h, 2 wt% solids	Easy solvent recovery, lignin isolated as a solid, carbohydrates isolated as a syrup	Complicated washing arrangements, higher energy costs, explosion hazards
Ionic Liquids	5 wt% solids 130–150°C	Nonflammable, low vapor pressure, high recoverability	Expensive, large antisolvent requirement, little known about long-term prospects

Source: Wyman, CE., et al. *Bioresour Technol*, 96, 1959–1966, 2005b; Larsen, et al. 2008; Liu, ZL., *Appl Microbiol Biotechnol*, 73, 27–36, 2006 and Dadi, AP., Varanasi, S., and Schall, CA., *Biotechnol Bioeng*, 95, 904–910, 2006.

enzymatic hydrolysis. Furthermore, high pH conditions can reduce subsequent cellulase loading requirements for enzymatic hydrolysis, provided that hemicellulase activity is present in sufficient quantity during the enzymatic hydrolysis. Flow-through pretreatments can remove as much as 75% of the lignin in addition to the hemicellulose, leaving behind mostly cellulose in the solids. (Wyman et al. 2005a) A detailed discussion of pretreatment technologies follows.

8.4.2.1 Low pH Pretreatments

8.4.2.1.1 Dilute Acid

Strong acids at 0.5–3 wt% and 130–200°C will hydrolyze most hemicellulose into soluble sugars as well as disrupting and removing some lignin. Dilute acid pretreatment removes and recovers hemicellulose sugars as monomers and increases glucose yields from cellulose to near 100%, but requires costly construction materials, high operating pressures, and neutralization and conditioning of the hydrolyzate before fermentation.

8.4.2.1.2 Hot Water/Steam Autolysis

Hot water under pressure without the addition of acid will hydrolyze hemicellulose in a process similar to a dilute acid co-current pretreatment. Water under pressure has a lower pH, penetrates cell walls, dissolves hemicellulose and some lignin, directly breaks hemiacetal linkages, and catalyzes the breakage of ether linkages through released organic acids. This process has lower capital and operating costs, reduces corrosion, and requires no chemical additions compared with dilute acid, but it suffers from the disadvantage of incomplete hydrolysis of hemicellulose and requires hemicellulases later on.

8.4.2.2 High pH Pretreatments

8.4.2.2.1 AFEX/Fiber Extrusion

AFEX is a batch process. The continuous process is called FIBEX (fiber extrusion). This is a dry-to-dry process that produces no liquid stream but a solid and a vapor phase, and it produces no hydrolysis products but only physical disruption of the biomass for increased enzymatic hydrolysis. Biomass is incubated with anhydrous ammonia at elevated temperatures. Rapid depressurization volatilizes ammonia for recovery and increases biomass surface area for enzymatic hydrolysis. Little lignin is removed, but lignin structure is modified, thus diminishing interference with enzymatic hydrolysis. AFEX can be performed in lower-cost vessels than those required for dilute acid pretreatment. The hydrolyzate is compatible with fermentation without conditioning, and it decrystallizes the cellulose, thus increasing downstream enzymatic hydrolysis. Most hemicellulose is recovered as oligomers, which require important hemicellulases in the enzymatic hydrolysis for efficient sugar recovery.

8.4.2.2.2 Ammonia Recycle Percolation/Soaking in Aqueous Ammonia

Ammonia recycle percolation (ARP) is a process in which aqueous ammonia is passed over biomass at elevated temperatures and pressures. Ammonia swells biomass, depolymerizes lignin, and breaks lignin–hemicellulose bonds. A lower-cost alternative is soaking in aqueous ammonia (SAA), which treats biomass in a batch reactor at 25–60°C under atmospheric pressure. ARP enhances cellulose digestion, reduces nonproductive binding of cellulase enzymes, and improves microbial activity, largely because of the removal of lignin, but it is less effective at treating softwood pulps.

8.4.2.2.3 Lime

Typical conditions are 0.1 g lime plus 5 g water per gram biomass. Additional water has little effect on the process. Temperatures can vary from 25 to 130°C, corresponding to treatment times ranging from weeks to hours. For high-lignin materials, additional lignin removal can be accomplished through the addition of oxygen or air. Because this is a low temperature pretreatment, biomass can be treated without a pressure vessel. Approximately 33% of lignin is removed and 100% of acetyl groups are removed. It is low cost and uses inherently safe materials, but it requires extremely long reaction times.

8.4.2.3 Solvent Pretreatments

8.4.2.3.1 Organic Solvents

Biomass is treated with an ethanol:water or methanol:water solvent to remove 70–90% of the lignin in the aqueous phase. The solvent is then recovered by distillation, and the condensed black liquor is diluted with water to precipitate lignin. The solvent can also contain a catalyst composed of acids, bases, or mineral salts. Solvents are easy to recover, and lignin is isolated as a solid, whereas carbohydrates are produced as syrup. However, pretreated solids must be washed with solvent before water washing, requiring complicated washing arrangements and higher energy costs. Organic solvents also have to be tightly controlled because of fire and explosion hazards.

8.4.2.3.2 Ionic Liquids

Ionic liquids are salts in which the ions are only loosely coordinated, resulting in a liquid at or near room temperature. Some ionic liquids have been shown to be a good solvent for cellulose, disrupting the crystalline structure, then precipitating out amorphous cellulose after the addition of an antisolvent. Although expensive, ionic liquids have many advantages such as recoverability, low vapor pressure, and nonflammability. Research into ionic liquid pretreatments is still at an introductory stage.

An economic analysis performed by Eggeman et al. (2005) found that major economic effects on plant operations from pretreatment include yield of five- and six-carbon sugars, solids concentration (and subsequent ethanol concentration), enzyme loading, and hemicellulase activity. There is little overall economic differentiation among the pretreatment technologies; low-cost pretreatment reactors were often counterbalanced by higher costs associated with catalyst or product recovery. The process was modeled using ASPEN Plus 10 and implemented in four parts: (1) capital cost estimate, (2) operating cost estimate (in which feed pricing was assumed to be $35/dry t, enzyme pricing is assumed to be $0.15/gallons of ethanol), (3) revenue (from ethanol sales and electricity sales at a price of $0.04/kW-h), (4) and discounted cash flow (2.5 years construction, 0.5 years start-up, and 20 years operation with cash flows discounted at 10%/year). Ethanol pricing is done on a rational pricing basis in which the minimum ethanol selling price (MESP) to achieve a zero net present value is determined. The MESP for the listed pretreatment technologies ranges from $1.34 (dilute acid) to $1.67 (hot water).

8.4.3 Enzymatic Hydrolysis of Cellulose and Hemicellulose

After pretreatment, a solid residue remains that contains cellulose, plus varying amounts hemicellulose and lignin depending on the pretreatment strategy used. Cellulose, and any hemicellulose that may be present, is then hydrolyzed to monomeric sugars for fermentation. The enzymatic hydrolysis step provides the greatest opportunity for biomass ethanol production to be cost-competitive with that of other liquid biofuels (Mosier et al. 2005). SHF is more costly than SSF for several reasons, an important one being that using SSF eliminates the cost of a reaction vessel and reduces the enzyme loading necessary because fermentation consumes the glucose product, which is a strong inhibitor of further enzymatic hydrolysis.

8.4.3.1 Cellulases

Cellulose degrading enzymes are classified by sequence homology into families 1, 3, 5–9, 12, 44, 45, 48, 61, and 74 of the glycoside hydrolases. On the basis of modes of action, they can be classified into three groups: exo-1,4-β-D-glucanases (cellobiohydrolases, EC 3.2.1.91), endo-1,4-β-D-glucanases (endoglucanases, EC 3.2.1.4), and β-glucosidases (EC 3.2.1.21) (McFarland et al. 2007).

In natural systems, these enzymes work synergistically to rapidly break down cellulose into glucose for metabolism. Endoglucanases cleave cellulosic bonds at random, mostly amorphous regions in the cellulose chain, thereby producing more chain ends for cellobiohydrolase attachment. Cellobiohydrolases produce cellobiose and some cellotriose from mostly crystalline regions of cellulose, attaching to chain ends and working progressively along the cellulose chain. The product of this action, cellobiose, is a strong inhibitor of cellobiohydrolase activity. β-glucosidases work on the soluble cellobiose and cellotriose products, breaking them into glucose monomers. All of these enzymes can be produced as complexed enzyme (attached to the surface of the cell) or noncomplexed (released into the environment) according to species (Lynd et al. 2002). To function effectively, cellulolytic enzymes must be able to associate with the insoluble substrate, disrupt the structure, and guide a single polymer chain through the catalytic domain (Eijsink et al. 2008).

Cellulase activity can be determined on several different substrates, natural and artificial, each with its own advantages and disadvantages. Table 8.7 provides a summary of many of the potential substrates. Filter paper is readily available and has a well-established assay method. Avicel can be used to effectively measure exoglucanase activity because it is largely crystalline with a low degree of polymerization. Carboxymethylcellulose (CMC) can be used to measure endoglucanase activity, but methylation blocks progressive action of exoglucanases. Lignocellulosic biomass is a heterogeneous matrix of diverse linkages of polysaccharides and aromatic compounds, and hydrolysis of such biomass is much more complex than for pure cellulose. Activities determined for enzymes on pure cellulose do not necessarily correlate with activities on biomass (King et al. 2009).

TABLE 8.7

Substrates Containing β-1,4-Glucosidic Bonds Hydrolyzed by Cellulases and Their Detection

Substrate	Detection	Enzymes
Soluble		
Short chain (low DP cellodextrins	RS, HPLC, TLC	Endo, Exo, BG
radiolabeled cellodextrins	TLC plus liquid scintillation	Endo, Exo, BG
b-methylumbelliferyl-oligosaccharides	Fluorophore liberation, TLC	Endo, Exo, BG
β-nitrophenol-oligosaccharides)	Chromophore liberation, TLC	Endo, Exo, BG
CMC, HEC	RS, viscosity	Endo
Dyed CMC	Dye liberation	Endo
Insoluble		
Crystalline		
Cotton, microcrystalline cellulose (Avicel,		
Valonia cellulose, bacterial cellulose	RS, TSS, HPLC	Total, Endo, Exo
Amorphous cellulose		
PASC, alkali-swollen cellulose, RAC	RS, TSS, HPLC, TLC	
Dyed cellulose	Dye liberation	Total, Endo, Exo
Fluorescent cellulose	Fluorophore liberation	Total, Endo
Chromogenic and fluorephoric derivatives		Total
TNP-CMC	Chromophore liberation	Endo
Fluram-cellulose	Fluorophore liberation	Endo, Total
Pretreated biomass	HPLC, RS	Total

Source: Adapted from Zhang, Y-HP., Himmel, ME., and Mielenz, JR., *Biotechnol Adv*, 24, 452–481, 2006.

RS, reducing sugars; TSS, total soluble sugars; HPLC, high-performace liquid chromatography; TLC, thin layer chromatography; HEC, hydroxyethylcellulose; PASC, phosphoric acid swollen cellulose; RAC, regenerated amorphous cellulose.

Reaction conditions for cellulose hydrolysis are frequently in the ranges of 40–50°C, acetate or citrate buffer pH 4.8–5, 15–60 FPU/g-glucan plus excess β glucosidase, 1–20 wt% solids, 24- to 72-h reaction time.

8.4.3.2   Hemicellulases

Hemicellulose degrading enzymes act specifically on the hemicellulose. Because hemicellulose is a branched heteropolymer, many different hemicellulases are required for efficient enzymatic hydrolysis. Hemicellulases include the xylanases (E.C. 3.2.1.8), β-mannanases (EC 3.2.1.78), α-L-arabinofuranosidases (EC 3.2.1.55) and α-L-arabinanases (EC 3.2.1.99), α-D-glucuronidases (E.C. 3.2.1.139), β-xylosidases (EC 3.2.1.37), xylan esterases (EC 3.1.1.72), and eruloyl esterases (EC 3.1.1.73) (Shallom and Shoham 2003).

8.4.4   FERMENTATION OF BIOMASS-DERIVED SUGARS TO BIOFUELS

8.4.4.1   Overview of Fermentation-Based Conversions

Fermentation refers to the anaerobic conversion by microorganisms of substrates into products such as acetone, 2-propanol, butanol, hydrogen, methanol, and hydrocarbons chains. Amongst the potential products that could be obtained from fermentations, this chapter focuses mainly on ethanol obtained from five- and six-carbon sugars. Traditional sources of sugars for ethanol production such as corn, sugarcane, and sugarbeets are mostly composed of disaccharides or polysaccharides of six-carbon sugars, but the most abundant feedstocks for ethanol production, lingocellulosic materials, contain six-carbon sugars as well as five-carbon sugars in considerable amounts. The nature of having five- and six-carbon sugars in these feedstocks raises the need of acquiring or developing one or more organisms, or a mixture thereof, to release and ferment these sugars sequentially or simultaneously. The main source of six-carbon sugars, specifically glucose, is cellulose (Zaldivar et al. 2001). Hemicellulose is the source of five-carbon sugars such as arabinose and xylose as well some six-carbon sugars, such as glucose, mannose, and galactose (Badal 2003). Glucose, the main component of cellulose, and xylose, the main component of hemicellulose, are the most abundant carbohydrates on Earth (Zaldivar et al. 2001; Badawl 2003). As stated previously, these carbohydrates can be fermented into several products. The biochemical reactions that summarize their biosynthesis are shown in Table 8.8. Biochemical fermentation of five- and six-carbon sugars is presented in greater detail in the next sections.

8.4.4.2   Fermentation of Six-Carbon Sugars

The preferred biochemical substrate for obtaining energy in the cell is glucose. When glucose is used as a source of energy under anaerobic conditions, it undergoes a process called glycolysis. Glycolysis consists of a series of reactions that mainly produce adenosine triphosphate (ATP), which is the carrier and energy source of all biochemical processes inside of cells. Glucose fermentation or glycolysis consists in a series of ten reactions that break down this carbohydrate into two molecules of pyruvate, which could undergo subsequent reactions to finally obtain acetone, 2-propanol, ethanol (Hahn-Hägerdal et al. 1994; Ostergaard et al. 2000; Dien et al. 2003), and others compounds. Once that pyruvate is produced, it could undergo a series of transformations to acetone, 2-propanol, and ethanol. To produce ethanol, pyruvate is first decarboxylated by the enzyme pyruvate decarboxylase in the presence of magnesium(II) and thiamine pyrophosphate (TPP); as a product of this process, acetaldehyde (2C) is produced. Acetaldehyde is further reduced to ethanol by means of the enzyme alcohol dehydrogenase in the presence of NADH and H^+. The overall process of ethanol production is given by the following reaction with a theoretical yield of 0.51 g of ethanol per gram of glucose (Nelson and Cox 2000):

$$Glucose \rightarrow 2[Ethanol] + 2CO_2 + 2H_2O \qquad (8.4)$$

During the treatment of biomass for biofuel production, the six-carbon sugars mannose and glucose are also produced. Also the feedstock for biofuels production could contain fructose. All of

TABLE 8.8
Products Obtained by Fermentation

Product	Biochemical Reaction	Theoretical Yield (g product/ g substrate)	Reference
Butanol	$C_6H_{12}O_6 \rightarrow C_4H_{10}O + 2CO_2 + H_2O$	0.41	Nelson and Cox 2000; Liu et al. 2009
Acetone, hydrogen	$C_6H_{12}O_6 + H_2O \rightarrow C_3H_6O + 3CO_2 + 4H_2$	0.32 for acetone 0.04 for hydrogen	Nelson and Cox 2000; Liu et al. 2009
Ethanol	$C_6H_{12}O_6 \rightarrow 2C_2H_6O + 2CO_2$	0.51	Nelson and Cox 2000; Liu et al. 2009
Butyric acid	$C_6H_{12}O_6 \rightarrow C_4H_8O_2 + 2CO_2 + 2H_2$	0.49	Antonopoulou et al. 2008; Liu et al. 2009
Acetic acid	$C_6H_{12}O_6 + 2H_2O \rightarrow 2CH_3COOH + 2CO_2 + 4H_2$	0.66	Nelson and Cox 2000; Liu et al. 2009; Antonopoulou et al. 2008
Gasoline-like hydrocarbons	Renewable sugars are metabolized into fatty acids, which are subsequently converted into hydrocarbons	Not available	LS9, Inc.

the carbohydrates mentioned before are metabolized as depicted in the previous section after some modifications, entering in the glycolysis cycle in different stages. Mannose, galactose, and fructose are phosphorylated and isomerized to fructose-6-phosphate, which is then fed to glycolysis.

8.4.4.3 Fermentation of Five-Carbon Sugars

Five-carbon carbohydrates such as xylose and arabinose are transferred into the cells of microorganisms and converted into xylulose, a five carbon sugar. Xylulose is then fed into the pentose phosphate pathway, where six units of xylulose are converted into five units of glucose, or its isomer fructose (McMillan 1993; Ostergaard et al. 2000; Jeffries and Jin 2004; Dmutruk et al. 2008), which are then fed to the glycolysis pathway. The intermediaries of this process are sedoheptulose, a seven-carbon carbohydrate, erythrose, a four-carbon carbohydrate, and glyceraldehyde. The six-carbon sugars are then fed to the glycolysis process. Overall the process of fermenting six units of xylose to ethanol, by means of the pentose phosphate pathway followed by glycolysis, is summarized by the following reaction with a theoretical yield of xylose to ethanol of 0.51 g of ethanol per g of xylose (McMillan 1993):

$$6[\text{Xylose}] \rightarrow 10[\text{Ethanol}] + 10CO_2 + 10H_2O \qquad (8.5)$$

8.4.4.4 Fermentation of Five- and Six-Carbon Sugars to Ethanol and Other Biofuels

Ethanol production from lignocellulosic materials deals with the fermentation of xylose (derived from hemicelluloses) and glucose (mostly derived from cellulose). The fermentation of these sugars could be achieved separately or simultaneously. When the fermentation of xylose and glucose occurs in different vessels, it is possible to select a specific microorganism to ferment xylose and a specific microorganism to ferment glucose. When the fermentation is simultaneous, it is possible to achieve the fermentation of both carbohydrates by the use of a mixture of microorganisms or a single specially engineered microorganism (DOE 2006). Also, during the selection of the microorganisms, it is important to consider the technology used to make available the carbohydrates because of the possibility of generation of some inhibitors of fermentation. When lignocellulosic material is subjected to dilute acid hydrolysis at high temperature, furfural and 5-hydroxymethylfurfural are formed. *Saccharomyces cerevisiae* is inhibited by these compounds (Liu 2006). Table 8.9 shows information related to ethanol production from lignocellulosic materials.

TABLE 8.9
Ethanol Production from Lignocellulosic Materials

Microorganism	Substrate	Fermentation Type	Yield (g ethanol/ g substrate)	Process Configuration	Reference
Saccharomyces cerevisiae Y-1528	Softwood-derived water-soluble fraction	Batch	0.47	Fermentation only	Keating et al. 2004
Saccharomyces cerevisiae Tembec T1	Softwood-derived water-soluble fraction	Batch	0.44	Fermentation only	Keating et al. 2004
Candida shehatae NCL-3501	Xylose in the range of concentrations from 10 to 80 g/L	Batch	0.40–0.43	Fermentation only	Abbi et al. 1996
Escherichia coli FBR5	Sugars from wheat straw by lime pretreatment and enzyme hydrolysis	Batch	0.26	Simultaneous enzymatic hydrolysis and fermentation	Badal and Cotta 2007a
Escherichia coli	Alkaline peroxide pretreated rice hulls	Batch	0.49 0.48	SHF SSF	Badal and Cotta 2007b
Saccharomyces cerevisiae and *Candida shehatae*	Mixture of glucose at a concentration of 35 g/L and xylose at a concentration of 15 g/L	Continuous with immobilized microorganisms	0.48	Fermentation only	Lebeau et al. 1998
Recombinant xylose and cellooligosaccharides-assimilating yeast strain[a]	Sulfuric acid hydrolysate of lignocellulosic biomass with a concentration of total sugars of 73 g/L	Batch	0.41	Fermentation only	Katahira et al. 2006
Zymomonas mobilis	Corn stover	Batch	0.30	Co-current dilute acid prehydrolysis and enzymatic hydrolysis	Aden et al. 2002

[a] *Saccharomyces* strain with expression of xylose reductase and xylitol dehydrogenase from *Pichia stipitis*, xylulokinase from *Saccharomyces cerevisiae*, and β-glucosidase from *Aspergillus acleatus*.

Utilizing mixtures of microorganisms or recombinant microorganisms (DOE 2006), together with enzyme technology for hydrolysis, can contribute to making cellulosic ethanol cost-effective (Solomon et al. 2007) and environmentally beneficial, because its energy output is greater that its energy input (Kemppainen and Shonnard 2005). For unit energy outputted from lignocellulosic material as ethanol, 14% of the energy must be added from fossil fuels, showing a global thermal efficiency of 86% (Kemppainen and Shonnard 2005). Because of the cost- effectiveness and thermal efficiency of cellulosic ethanol technology, there are presently four demonstration plants in operation, three in North America, and the other in Europe. Iogen, located in Ottawa, Canada, has operated since 2004 and has a production capacity of 3000 m³/year of ethanol using wheat, oat, and barley straw. Operating since 2007, ClearFuels Technology of Kauai, Hawaii in the United States has a production capacity of 11,400 m³/year of ethanol using bagasse and wood residues. Celunol, located in Jennings, LA, in the United States, processes bagasse and rice hulls and has produced 5000 m³/year of ethanol since 2007. Scheduled to open in 2009, Etek EtanolTeknik will have a capacity of producing 30,000 m³/year of ethanol from softwood residues of spruce and pine (Solomon et al. 2007).

As stated previously, several products, including biofuels, may be obtained through anaerobic degradation of glucose. The biofuels that could be produced in addition to ethanol are hydrogen and butanol. Antonopoulou and collaborators (2008) produced hydrogen from whey cheese, obtaining a yield of 0.9 mol per mol of glucose. Also, butanol has been produced achieving a yield of 0.4 g of butanol per gram of glucose (Lee et al. 2008). Other biofuels produced by fermentation of sugars are shown in Table 8.10.

8.4.4.5 Fermentation of Five- and Six-Carbon Sugars to Hydrocarbons and High-Value Products

The fermentation of five- and six-carbon sugars can produce hydrocarbon chains that could be used instead of gasoline and diesel and allow for the production of other high-value products such as solvents (acetone in Table 8.10), intermediaries to produce plastics, and sweeteners. Park and collaborators (2005) reported that *Vibrio furnissii* M1, when grown in a 50-mL scale with a carbon source of 3 mmol provided by glucose, xylose, starch, or sucrose, yielded between 10 and 27 mg of a mixture of alkanes and alkenes. The chain length of the alkanes and alkenes produced was in the range of C_{14}–C_{27}. LS9, Inc. of San Francisco, CA, produces gasoline-like hydrocarbons from renewable sugars. These sugars are first converted to fatty acids and subsequently to hydrocarbons by certain microorganisms. Roa-Engel and collaborators (2008) reported the

TABLE 8.10
Biofuels from Fermentation of Five- and Six-Carbon Sugars

Microorganism	Product and Substrate	Fermentation Type	Yield (g product/ g substrate)	Reference
Clostridium saccharoperbutylacetonicum	Hydrogen from cheese whey	Batch	0.01	Antonopoulou et al. 2008
Clostridium beijerinckii NCIMB 8052	Butanol from glucose	Batch and continuous	0.40	Lee et al. 2008
Clostridium sp.	Hydrogen from wheat bran	Batch	0.01	Pan et al. 2008
Citrobacter sp.	Hydrogen from glucose	Batch	0.03	Oh et al. 2003
Clostridium sp.	Acetone and butanol from hydrolyzed agricultural waste	Continuous	For acetone 0.05 For butanol 0.09	Zverlov et al. 2006

production of fumaric acid by fermentation. Fumaric acid could be used as raw material for the polymer industry. This compound is produced from glucose, with a yield of 0.85 g per gram of glucose, by *Rhizopus* species. Also, through fermentation it is possible to produce succinic acid, which is a raw material involved in the making of surfactants, detergents, pharmaceuticals, and foods. Succinic acid is produced by *Actinobacillus succinogenes*. When this microorganism is grown on wheat flour, it produces 0.19 g of succinic acid per gram of substrate (Du et al. 2007). Polyhydroxyalkanoic acids and polyhydroxyalkanaotes are biopolymers stored intracellularly as a energy source. These biopolymers present attractive properties in the fields of biodegradation and thermoplasticity. These biopolymers are produced by microorganisms such as *Pseudomonas guezenni* (Simon-Colin et al. 2008) and *Pseudomonas aeruginosa* ATCC 9027 (Rojas-Rosas et al. 2007) using glucose as carbon source. Finally xylitol, a sweetener, is produced from xylose fermentation by *Candida tropicalis*. This microorganism yields 0.75 g of xylitol per gram of xylose (Kim et al. 2002). Xylitol possesses a higher sweetening power than sucrose and promotes oral health and prevents cavities.

8.5 COMPARISON OF ENERGY EFFICIENCIES AND COSTS OF BIOMASS PROCESSING TECHNOLOGIES

Technology energy efficiency and GHG emissions for biomass-derived energy products are of great interest to industry, policy-makers, and government regulators. Energy efficiency here is defined as the ratio of the energy content of the product electricity or biofuel to the energy content of biomass feedstock to the conversion process. GHGs include CO_2 released from the combustion of fossil carbon, nitrous oxide (N_2O), methane, solvents, and refrigerants.

For biofuels especially, GHG emissions over the product life-cycle are of interest to confirm the magnitude of reduction compared with conventional and next-generation fossil-derived fuels. For example, the U.S. Congress set Renewable Fuels Standards (RFS) for each year up to 2022 in the Energy Independence and Security Act (EISA) of 2007 (Congress 2007). The RFS calls for 36 billion gallons of renewable fuels to be produced by 2022, or approximately 25% of current gasoline consumption in the United States (Brodeur-Campbell et al. 2008). Advanced biofuels, according to EISA, are renewable fuels other than corn ethanol, for which the life-cycle emissions of GHG remain less than 50% of fossil gasoline baseline GHG emissions. Of the 36.0 billion gallons of renewable fuels in 2022, it will be required that 21.0 billion gallons be advanced biofuels. Cellulosic ethanol is the nearest to commercialization of the advanced biofuels technologies and is expected to provide a significant fraction of the 21 billion gallons of advanced biofuel mandated by the RFS. Alternatives to cellulosic ethanol exist, including biomass-to-liquid (BTL) diesel and importing ethanol from Brazil. The most likely fulfillment scenario includes a mixture of all of these options (DOE/EIA 2008).

Table 8.11 shows technology energy efficiency of several biomass-to-energy carrier technologies taken from a recent report (Shonnard et al. 2006; Shonnard and Koers 2009). For liquid and gaseous energy products, efficiencies range from 43 to 81% and electricity generation efficiencies are from 27.7 to 38%. Biomass-derived fuels are less efficient than fossil fuels, such as petroleum gasoline or diesel, the conversion efficiencies of which are approximately 80–90%. However, the life-cycle fossil energy savings of biomass-derived fuels are still very large because process energy is largely provided by the renewable biomass feedstock itself rather than from fossil resources (Kalnes et al. 2007; Koers et al. 2009). Thus, the lower technology energy efficiency of biofuels is compensated by substantial savings of fossil energy over the life-cycle.

8.6 CONCLUSIONS

The chemical engineering processing routes for converting bioenergy crop plants into liquid transportation fuels are quite diverse, as demonstrated by the broad range of processing technologies

TABLE 8.11

Overall Investment and Process Energy Efficiency of Technologies Used to Produce Biomass Derived Energy Carriers

Energy Carrier	Production Route	Biomass Resource	Total Investment Cost[a] 2005 $/kW	Technology Energy Efficiency (% LHV[b])
Electricity	Combustion (direct, co-firing, gasification)	Any type of biomass, lignocellulosic is preferred	397–926	27.7[d] 32.5[d] 37[d]
Hydrogen	Thermal conversion (gasification)	Any type of biomass, lignocellulosic is preferred	439–586	60[c]
Methanol	Thermal conversion (gasification)	Any type of biomass, lignocellulosic is preferred	647–732	55[c]
FT liquids	Thermal conversion (gasification)	Any type of biomass, lignocellulosic is preferred	659–879	45[c]
Ethanol from wood	Biochemical conversion (fermentation)	Lignocellulosic biomass	219–428	46[c]
Ethanol from sugar	Biochemical conversion (fermentation)	Sugarcane Sugarbeet	208–354	43[c]
Biodiesel RME	Mechanical/chemical (extraction)	Oily seeds	135–184	45[c]
Green diesel	Mechanical/chemical (extraction)/catalytic hydrotreatment	Algae, SBO, SFO, RSO, Jatropha, Camelina	170–230 (2008 $)	55–81[e]

Source: Shonnard, DR., et al., *Evaluation of Low Greenhouse Gas Bio-Based Energy Technologies,* Michigan Technological University, Houghton, MI, 2006.

[a] Original data given in €/kW. The 2003 average exchange rate of 1.15 was used to convert into 2003 U.S. dollars. Exchange rates were taken from http://epp.eurostat.cec.eu.int (accessed March 15, 2006). The quantities in 2003 U.S. dollars were updated to 2005 U.S. dollars using the harmonized Consumer Price Index.

[b] LHV, low heating value is defined as the amount of energy released when a fuel is burned completely in a steady-flow process and the products are returned to the state of reactants, except water, which remains in the vapor form. This efficiency is only for the manufacturing part of the energy product life-cycle.

[c] Faaij, APC., *Energy Policy,* 34:322–342, 2006.

[d] DeMeo E, et al. (1997) *Renewable Energy Technology Characterizations,* Energy Efficiency and Renewable Energy and Electric Power Research Institute.

[e] Range of values represents differences in data sources for cultivation, fuel production, and transport. (Shonnard, DR., Williams, L., and Kalnes TN., *Environ Prog Sustain Energy,* 29:382–392, 2009)

presented in this chapter. Some of these processing technologies are being currently applied in the commercial production of biofuels, but most described here are still in the research and development (R&D) pipeline. However, most of the described chemical engineering technologies are being aggressively developed using R&D investments by government and industry sources. It is the opinion of the authors of this chapter that the future growth of a renewable liquid transportation biofuel industry will depend on the success of these R&D efforts, the goals of which are to increase conversion efficiencies and lower production costs. Commercial success will also depend on advancements in efficient and cost-effective production of bioenergy plant crops as described in other chapters in this handbook.

REFERENCES

Abbi M, Kuhad RC, Singh A (1996) Fermentation of xylose and rice straw hydrolysate to ethanol by Candida shehatae NCL-3501. *J Indust Microbiol Biotechnol* 17:20–23

Aden A, Ruth M, Ibsen K, Jechura J, Neeves K, Sheehan J, Wallace B (2002) *Lignocellulosic Biomass to Ethanol Process Design and Economics Utilizing Co-Current Dilute Acid Prehydrolysis and Enzymatic Hydrolysis for Corn Stover*, National Renewable Energy Laboratory, Golden, CO

Antonopoulou G, Stamatelatou K, Venetsaneas N, Kornaros M, Lyberatos G (2008) Biohydrogen and methane production from cheese whey in a two-stage anaerobic process. *Indust Eng Chem Res* 47:5227–5233

Badal CS (2003) Hemicellulose bioconversion. *J Indust Microbiol Biotechnol* 30:279–291

Badal CS, Cotta MA (2007a) Enzymatic hydrolysis and fermentation of lime pretreated wheat straw to ethanol. *J Chem Tech Biotechnol* 82:913–919

Badal CS, Cotta MA (2007b) Enzymatic saccharification and fermentation of alkaline peroxide pretreated rice hulls to ethanol. *Enz Microb Technol* 41:528–532

Bartholomew CH, Farrauto RJ (2006) *Fundamentals of Industrial Catalytic Processes*, 2nd ed. Wiley-Interscience, Hoboken, NJ

Bowman DMJS (2009) Fire in the earth system. *Science* 324:481

Bridgwater AV, Peacocke GVC. (2000) Fast pyrolysis processes for biomass. *Renewable and Sustainable Energy Reviews* 4:1–73

Bridgwater AV, Czernik, S, Piskorz, J (2001) Progress in thermochemical biomass conversion. *An Overview of Fast Pyrolysis*, A.V. Bridgwater, ed., Blackwell Science Ltd, Oxford, Vol. 1, pp 977–997.

Brodeur-Campbell MJ, Jensen JR, Eatmon TJ, Sutherland JW, Shonnard DR (2008) Political feasibility, environmental sustainability, and economic efficiency analysis of conventional and advanced biofuels technologies in the US. In: *Global Conference on Sustainable Product Development and Life Cycle Engineering*, Busan, South Korea, pp 46–51

Congress (2007) Energy Independence and Security Act. U.S. Congress, Washington DC

Czernik S, Bridgwater AV (2004) Overview of applications of biomass fast pyrolysis oil. *Energy Fuel* 18:590–598

Dadi AP, Varanasi S, Schall CA (2006) Enhancement of cellulose saccharification kinetics using an ionic liquid pretreatment step. *Biotechnol Bioeng* 95:904–910

DeMeo E, Schweizer T, Bain R, Craig K, Comer K, et al. (1997) *Renewable Energy Technology Characterizations,* Energy Efficiency and Renewable Energy and Electric Power Research Institute

Diebold JP (2000) *A Review of the Chemical and Physical Mechanisms of the Storage Stability of Fast Pyrolysis Bio-Oils*, National Renewable Energy Laboratory, U.S. Department of Energy, Golden, CO

Diebold JP, Czernik S (1997) Additives to lower and stabilize the viscosity of pyrolysis oils during storage. *Energy Fuel* 11:1081–1091

Dien BS, Cotta MA, Jeffries TW (2003) Bacteria engineered for fuel ethanol production: Current status. *Appl Microbiol Biotechnol* 63:258–266

DOE (2006) *Breaking the Biological Barriers to Cellulosic Ethanol: A Joint Research Agenda*, U.S. Department of Energy Office of Science and Office of Energy Efficiency and Renewable Energy, Washington, DC

DOE/EIA (2008) annual energy outlook 2008: With projections to 2030, available at http://www.trb.org/Main/Blurbs/157084.aspx

Du C, Lin SKC, Koutinas A, Wang R, Webb C (2007) Succinic acid production from wheat using a biorefining strategy. *Appl Microbiol Biotechnol* 76:1263–70

Eggeman T, Elander RT (2005) Process and economic analysis of pretreatment technologies. *Bioresour Technol* 96:19–25

EIA (2010) US Energy Information Administration, International energy statistics, available at http://tonto.eia.doe.gov/cfapps/ipdbproject/iedindex3.cfm (accessed on April 20, 2010)

Eijsink VGH, Vaage-Kolstad G, Varum KM, Horn SJ (2008) Towards new enzymes for biofuels: Lessons from chitinase research. *Trends Biotechnol* 26:228–235

Elliott DC (2007) Historical developments in hydroprocessing bio-oils. *Energy Fuel* 21:1792–815

Faaij APC (2006) Bio-energy in Europe: Changing technology choices. *Energy Policy* 34:322–342

FEDIOL (2008) Major commodities (EU oil and protein meal industry), available at http://www.fediol.be/2/index.php (accessed on April 21, 2010)

Furimsky E (2000) Catalytic hydrodeoxygenation. *Appl Catal A*: Gen 199:147–90

Gil J, Caballero MA, Martin JA, Aznar M-P, Corella J (1999) Biomass gasification with air in a fluidized bed: Effect of the in-bed use of dolomite under different operation conditions. *I&EC Res* 38:4226–4235

Hahn-Hägerdal B, Jeppsson H, Skoog K, Prior BA (1994) Biochemistry and physiology of xylose fermentation by yeasts. *Enz Microb Technol* 16:933–943

Hoek A, Kersten LBJM (2004) The Shell Middle Distillate Synthesis process: Technology, products and perspective. *Stud Surf Sci Catal* 147 (Natural Gas Conversion VII):25–30

Huber GW, Iborra S, Corma A (2006) Synthesis of transportation fuels from biomass: Chemistry, catalysts, and engineering. *Chem Rev* 106:4044–4098

Jeffries TW, Jin YS (2004) Metabolic engineering for improved fermentation of pentoses by yeasts. *Appl Microbiol Biotechnol* 63:495–509

Jothimurugesan K, Gangwal S (1998) Simultaneous H_2S and NH_3 removal from hot coal gas. *Adv Environ Res* 2:116

Kalnes T, Marker T, Shonnard DR (2007) Green diesel: A second generation biofuel. *Int J Chem React Eng* 5(article A48), available at http://www.bepress.com/ijcre/vol5/A48 (accessed on April 19, 2010)

Kalnes T, Marker T, Shonnard DR (2008) Green diesel production by hydrorefining renewable feedstock. *Biofuels Technol Quart*, available at www.biofuels-tech.com (accessed on April 19, 2010)

Katahira S, Mizuike A, Fukuda H, Kondo A (2006) Ethanol fermentation from lignocellulosic hydrolysate by a recombinant xylose- and cellooligosaccharide-assimilating yeast strain. *Appl Microbiol Biotechnol* 72:1136–1143

Keating JD, Robin J, Cotta MA, Saddler JN, Mansfield SD (2004) An ethanologenic yeast exhibiting unusual metabolism in the fermentation of lignocellulosic hexose sugars. *J Indust Microbiol Biotechnol* 31:235–244

Kemppainen AJ, Shonnard DR (2005) Comparative life-cycle assessments for biomass-to-ethanol production from different regional feedstocks. *Biotechnol Progr* 21:1075–1084

Kim JH, Han KC, Koh YH, Ryu YW, Seo JH (2002) Optimization of fed-batch fermentation for xylitol production by Candida tropicalis. *J Indust Microbiol Biotechnol* 29:16–19

King BC, Donnelly MK, Bergstrom GC, Walker LP, Gibson DM (2009) An optimized microplate assay system for quantitative evaluation of plant cell wall-degrading enzyme activity of fungal culture extracts. *Biotechnol Bioeng* 102:1033–1044

Koers KP, Kalnes TN, Marker T, Shonnard DR (2009) Green diesel: A technoeconomic and environmental life cycle comparison to biodiesel and syndiesel. *Environ Prog Sustain Energy* 28:111–120

Larsen J, Petersen MO, Thirup L, Li HW, Iversen FK (2008) The IBUS process—Lignocellulosic bioethanol close to a commercial reality. *Chemical Engineering & Technology* 31:765–72

Lebeau T, Jouenne T, Junter GA (1998) Continuous alcoholic fermentation of glucose/xylose mixtures by co-immobilized Saccharomyces cerevisiae and Candida shehatae. *Appl Microbiol Biotechnol* 50:309–313

Lee SM, Cho MO, Park CH, Chung YC, Kim JH, Sang BI, Um Y (2008) Continuous butanol production using suspended and immobilized clostridium beijerinckii NCIMB 8052 with supplementary butyrate. *Energy Fuel* 22:3459–3464

Liu J, Wu M, Wang M (2009) Simulation of the process for producing butanol from corn fermentation. *Indust Eng Chem Res* 48:5551–5557

Liu ZL (2006) Genomic adaptation of ethanologenic yeast to biomass conversion inhibitors. *Appl Microbiol Biotechnol* 73:27–36

Lynd LR, Weimer PJ, Zyl WHv, Pretorius IS (2002) Microbial cellulose utilization: fundamentals and biotechnology. *Microbiol Molecul Biol Rev* 66:506–577

McCormick B (2009) *Biodiesel Market and Market Barriers*. National Renewable Energy Laboratory, Golden, CO

McFarland KC, Ding H, Teter S, Vlasenko E, Xu F, Cherry J (2007) Development of improved cellulase mixtures in a single production organism. In: *American Chemical Society Symposium Series*. American Chemical Society, Washington, DC, pp 19–31

McMillan JD (1993) *Xylose Fermentation to Ethanol: A Review*, National Renewable Energy Laboratory, Golden, CO

Mohan D, Pittman CU, Steele PH (2006) Pyrolysis of wood/biomass for bio-oil: A critical review. *Energy Fuel* 20:848–889

Mosier N, Wyman C, Dale B, Elander R, Lee YY, Holtzapple M, Ladisch M (2005) Features of promising technologies for pretreatment of lignocellulosic biomass. *Bioresource Technol* 95:673–686

NBB (2008) US Biodiesel Demand, available at http://www.biodiesel.org/pdf_files/fuelfactsheets/Production_Graph_Slide.pdf (accessed on April 22, 2010).

Nelson DL, Cox MM (2000) *Lehninger Principles of Biochemistry*, 3rd ed, Worth, New York

NSF (2008) *Breaking the Chemical and Engineering Barriers to Lignocellulosic Biofuels: Next Generation Hydrocarbon Biorefineries*, National Science Foundation, Chemical-Bioengineering-Environmental and Transport Systems Division, Washington, DC

Oh YK, Seol EH, Kim JR, Park S (2003) Fermentative biohydrogen production by a new chemoheterotrophic bacterium Citrobacter sp. Y19. *Int J Hydrogen Energy* 28:1353–1359

Ostergaard S, Olsson L, Nielsen J (2000) Metabolic engineering of *Saccharomyces cerevisiae*. *Microbiol Mol Biol Rev* 64:34–50

Pan C, Fan Y, Hou H (2008) Fermentative production of hydrogen from wheat bran by mixed anaerobic cultures. *Indust Eng Chem Res* 47:5812–5818

Park MO, Heguri K, Hirata K, Miyamoto K (2005) Production of alternatives to fuel oil from organic waste by the alkane-producing bacterium, Vibrio furnissii M1. *J Appl Microbiol* 98:324–331

Phillips JG, Reader GT (1998) The use of DME as a transportation fuel—A Canadian perspective. Paper presented at the ASME Fall Technical Conference, Clymer, NY, September 26–28

RFA (2010) Ethanol industry statistics, available at http://www.ethanolrfa.org/pages/statistics (accessed on April 24, 2010)

Roa-Engel CA, Straathof AJJ, Zijlmans TW, Gulik WMv, Wielen LAMvd (2008) Fumaric acid production by fermentation. *Appl Microbiol Biotechnol* 78:379–389

Rojas-Rosas O, Villafaña-rojas J, López-Dellamary FA, Nungaray-arellano J, González-Reynoso O (2007) Production and characterization of polyhydroxyalkanoates in *Pseudomonas aeruginosa* ATCC 9027 from glucose, an unrelated carbon source. *Can J Microbiol* 53:840–851

Shallom D, Shoham Y (2003) Microbial hemicellulases. *Curr Opin Microbiol* 6:219–228

Shapouri H, Gallagher PW, Nefstead W, Schwartz R, Noe S, Conway R (2010) 2008 Energy balance for the corn-ethanol industry. In: *Agricultural Economic Report 846*: U.S. Department of Agriculture, Washington, DC

Shonnard DR, Koers KP (2009) *Expanded Life Cycle Assessments of Biofuels, Petroleum Diesel, and Synthetic Diesel: Effects of Plant Oil Feedstocks, N₂O Emissions, and Land Use Change*. Michigan Technological University, Houghton, MI

Shonnard DR, Williams L, Kalnes TN (2009) Camelina derived jet fuel and diesel: Sustainable advanced biofuels. *Environ Prog Sustain Energy* 29:382–392

Shonnard DR, Zhang Q, Johnson DM, Froese RE, Sutherland JW, Mártin-García AR, Miller CA, Jenkins TL, Wright, GJ (2006) *Evaluation of Low Greenhouse Gas Bio-based Energy Technologies*. Michigan Technological University, Houghton, MI

Simon-Colin C, Alain K, Colin S, Cozien J, Costa B, Guezennec JG, Raquénès GH (2008) A novel mcl PHA-producing bacterium, Pseudomonas guezennei sp. nov., isolated from a 'kopara' mat located in Rangiroa, an atoll of French Polynesia. *J Appl Microbiol* 104:581–586

Solomon BD, Barnes JR, Halvorsen KE (2007) Grain and cellulosic ethanol: History, economics, and energy policy. *Biomass Bioenergy* 31:416–425

Sorenson SC, Mikkelsen S (1995) *Performance and emissions of a DI diesel engine fueled with neat dimethyl ether*, Paper 950064, Society of Automotive Engineers, Warrendale, PA

Thurmond W (2008) *Global Biodiesel Market Trends, Outlook and Opportunities*: Canola Council of Canada, Winnipeg, Manitoba, Canada

Torres W, Pansare SS, Goodwin JG (2007) Hot gas removal of tars, ammonia, and hydrogen sulfide from biomass gasification gas. In: Bell AT, Klier K (eds), *Catalysis Reviews: Science and Engineering*. 49:407–456

Verbeek R, Van der Weide J (1997) *Global Assessment of Dimethyl Ether: Comparison with Other Fuels*. Society of Automotive Engineers, Warrendale, PA

Williams RH, Larson ED, Katofsky RE, Chen J (1995) Methanol and hydrogen from biomass for transportation. *Energy Sustain Dev* 1:18–34

Wright JD (1988) Evaluation of enzymatic hydrolysis processes. In *Energy from Biomass and Wastes XII*. New Orleans, LA: Institute of Gas Technology, Chicago IL, pp 1247–77

Wu M, Wang M, Huo H (2006) *Fuel-Cycle Assessment of Selected Bioethanol Production Pathways in the United States*, Reference No. ANL/ESD/06-7, Argonne National Laboratory, Argonne, IL

Wyman CE, Dale BE, Elander RT, Holtzapple M, Ladisch MR, Lee YY (2005a) Comparative sugar recovery data from laboratory scale application of leading pretreatment technologies to corn stover. *Bioresour Technol* 96:2026–2032

Wyman CE, Dale BE, Elander RT, Holtzapple M, Ladisch MR, Lee YY (2005b) Coordinated development of leading biomass pretreatment technologies. *Bioresour Technol* 96:1959–1966

Zaldivar J, Nielsen J, Olsson L (2001) Fuel ethanol production from lignocellulose: A challenge for metabolic engineering and process integration. *Appl Microbiol Biotechnol* 56:17–34

Zhang Y-HP, Himmel ME, Mielenz JR (2006) Outlook for cellulase improvement: Screening and selection strategies. *Biotechnol Adv* 24:452–481

Zverlov VV, Berezina O, Velikodvorskaya GA, Schwarz WH (2006) Bacterial acetone and butanol production by industrial fermentation in the Soviet Union: Use of hydrolyzed agricultural waste for biorefinery. *Appl Microbiol Biotechnol* 71:587–597

9 International Fuel Quality

Georgios Sarantakos
Swiss Federal Institute of Technology

CONTENTS

9.1 INTRODUCTION

The quality of biofuels varies because of differences in feedstock, production process, or regional factors such as climate, fuel and emissions regulations, car fleet, and so on. The existing biofuel quality specifications reflect these variations. However, the continuing growth of the biofuel market from the one side and the pressures from vehicle and engine manufacturers from the other side requires the establishment of international biofuel standards. These standards should guarantee that the use of biofuels shall not harm human health or the environment, should not influence proper engine operation and performance, and should keep the safety of use and the fuel economy at an acceptable level.

This chapter addresses only standards pertaining to the biofuels being currently traded: bioethanol and biodiesel. Table 9.1 lists the main fuel properties for bioethanol and biodiesel, which will be explained later in Sections 9.2 and 9.3 of this chapter.

TABLE 9.1

Properties of Biofuels (Biodiesel and Bioethanol)

Biodiesel	Bioethanol
Flash point	Color and appearance
Viscosity	Density
Sulfated ash	Sulfate content
Sulfur	Sulfur content
Copper strip corrosion	Copper content
Cetane number	Electrical (electrolytic) conductivity
Cloud point, and cold filter plugging point	Ethanol content
Carbon residue	pHe
Acid number	Acidity
Total and free glycerin and mono-, di-, and triglycerides	Phosphorus content
Phosphorus content	(Unwashed) gum/evaporation residue/nonvolatile material
Distillation temperature, T90	Chloride content
Oxidation stability	Water content
Alkali and alkaline-earth metals	Iron and sodium content
Water and sediment (water content and total contamination)	
Ester content	
Iodine number, linolenic acid methyl ester, and polyunsaturated biodiesel	
Density	
Ethanol and methanol content	

The analysis presented here was based on the specifications, methods, and limits on biofuel quality set by ASTM International (United States) and the European Committee for Standardization (CEN; European Union). The existing standards from ASTM International and CEN are mainly used as examples because of their worldwide use. In the case of bioethanol, the Brazilian specifications will also be mentioned because of the lion's share that Brazil owns in the bioethanol market. At the same time, the current and future quality of biofuels is based on the requirements of these specifications, and at the same time these specifications are adjusted to the fuel quality that producers manage to achieve as a result of several influences, such as the demand of the market, the climate conditions of each region, the existing technology and infrastructure, and so on.

A general observation is that for the control of biodiesel quality, more sophisticated and challenging standards are required than for that of bioethanol. This could be explained from the following factors:

- Bioethanol consists of a single chemical compound (i.e., ethanol), whereas biodiesel can be derived from several feedstocks, such as animal fat, vegetable oil, and used fried oil, with various chemical compositions (e.g., from saturated to polyunsaturated fatty acids with different length chains). These variations are reflected in the final product.
- Bioethanol feedstock is derived almost exclusively from one country, Brazil. The feedstocks of biodiesel are more widely distributed, so biodiesel has many different regulations that have to be respected and met, as well as climate conditions that influence some of its characteristics.
- Fatty acid methyl esters (FAME) and fatty acid ethyl esters (FAEE) can be called biodiesels. However, these two chemicals do not have the same properties (see Section 9.2.19) and, as a result, nor do the final products.

Given that some fuel specifications (e.g., sulfur content and cloud or cold filter plugging point) depend on given national regulations, the examination of the fuel quality in several countries also has to be presented. This overview provides a better view of the current fuel quality situation and future trends as well as demonstrates the differences of specifications related to the climate and the feedstock preferred as well as the biofuel trade pathways in-between.

9.2 BIODIESEL

9.2.1 Flash Point

Flash point is a measure of the flammability of fuels; that is, the tendency of fuel to form a flammable mixture with air (Prankl et al. 2004; White Paper 2007). At the same time, the flash point for biodiesel is an indirect control of the level of the nonreacted alcohol remaining in the finished fuel (i.e., the free monoalcohol; Mittelbach 1996; Costenoble 2006; Bacha et al. 2007; White Paper 2007; De Klerk 2008; Rilett and Gagnon 2008) given that a very small amount of residual alcohol will reduce the flash point greatly (Prankl et al. 2004; Foon et al. 2005; White Paper 2007). Thus, the flash point for biodiesel is an important safety criterion of its handling and storage (Prankl et al. 2004; Costenoble 2006; White Paper 2007) that is not directly related to engine performance (Bacha et al. 2007). Moreover, alcohol residues of biodiesel can also affect fuel pumps, seals, and elastomers (Foon et al. 2005). It should be noted that the flash point of biodiesel is approximately double the value of that for diesel (Prankl et al. 2004; White Paper 2007; see Table 9.2).

The main reason for this difference is that the composition of the naturally occurring oils and fats of biodiesel (i.e., long fatty acid chains) means that it has a high flash point. The composition of biodiesel varies within a small range regardless of the feedstock (Table 9.2; Prankl et al. 2004), so the flash point is process dependent rather than feedstock dependent. Hence, the flash point serves

TABLE 9.2
Typical Properties of Biodiesel and ULSD

Property	Biodiesel	ULSD
Flash point, °C	130	60
Cetane number	55	44
Sulfur, ppm	<15	15
Relative density, 15°C	0.88	0.85
Kinematic viscosity at 40°C, mm²/s	6.0	2.6
Heating value, net, Btu/gal (kJ/kg)	128,000 (40,600)	130,000 (42,700)

Source: Bacha, J., Freel, J., and Gibbs, A., *Diesel Fuels Technical Review,* Chevron Corporation, San Ramon, CA, 2007.

as an indicator of the efficiency of that process for residual methanol removal (Foon et al. 2005; Rilett and Gagnon 2008).

To ensure that the manufacturers remove the excess methanol used in production, the European Union (EU) sets a low limit for the flash point of biodiesel—120°C (EN 14214), whereas the United States (ASTM D6571) sets the same limit at 130°C* (Prankl et al. 2004; Foon et al. 2005; White Paper 2007). At the same time, the Brazilian specification (NBR 14598[†]) defines a flash point minimum of 100°C, which correlates with 0.2% v/v alcohol, because this limit value is compatible with the NFPA nonhazardous category code (White Paper 2007). Although the limits of the EU and United States were set at these values because of the high uncertainty of their test methods (EN 14110 and ASTM D93, respectively), the development of improved or alternative test methods will allow them soon to reduce their flash point limit to 100°C (Bacha et al. 2007; White Paper 2007).

9.2.2 VISCOSITY

The injection of diesel or biodiesel fuel into a compression-ignition engine is controlled by positive displacement (volumetrically) or by a solenoid valve. The viscosity is one of the fuel parameters that determine these injection systems. A high or a low viscosity can influence the engine performance and durability, as well as increase exhaust emissions (Prankl et al. 2004; WWFC 2006; McGill et al. 2008). On one side, low viscosity can lead to loss of power and fuel economy (Bacha et al. 2007) because of injection pump and injector leakage (Rilett and Gagnon 2008); on the other side, high viscosity is limited by the characteristics of the injection system (Rilett and Gagnon 2008) and reduces fuel flow rates, resulting in inadequate fueling (WWFC 2006; White Paper 2007).

The viscosity of biodiesel is higher than that of fossil diesel (Prankl et al. 2004; see Table 9.2). At the same time, the viscosity of biodiesel is much lower than that of the pure oil it is derived from (Zhang et al. 2003; WWFC 2006). The viscosity is related more to molecular weight than to hydrocarbon class [i.e., for a given carbon number, naphthenes generally have slightly higher viscosities than paraffins or aromatics (Bacha et al. 2007)]. The high content of high molecular compounds such as nonreacted glycerides and polymers can increase the viscosity, which is the case when used frying oil is used as feedstock (Mittelbach 1996; Wörgetter et al. 1998; Prankl et al. 2004), or palm oil for biodiesel (Bacha et al. 2007). In contrast, free methylesters (in polyolesters) can reduce the viscosity of the fuel (Singh 2005), e.g., a high amount of mid-chain fatty acids in coconut resulted in lower viscosity (Bacha et al. 2007). Additionally, the viscosity of biodiesel can be used

---

* Only if methanol is not measured directly; otherwise, 93°C.
[†] Based on ASTM D93, but considers ASTM D93 and EN ISO 3679.

TABLE 9.3
Viscosity Limits at 40°C (mm²/s) for Several Countries

Countries	EU	United States	Japan	India	Thailand		South Africa	Brazil
Fuel quality standards	EN 14214	ASTM D 6751	JIS K 2390:2008	IS 15607	Agricultural engines	FAME	SANS 342	ANP No 42 Act 05/2005
Viscosity at 40°C (mm²/s)	3.5–5.0	1.9–6.0	3.5–5.0	2.5–6.0	1.9	3.5	2.2–5.3	To report

Source: Rehnlund, B., *Outlook on Standardization of Alternative Vehicle Fuels: Global, Regional and National Level.* Prepared by: Atrax Energi AB, Sweden for IEA Advanced Motor Fuels Implementing Agreement, 2008.

as an indicator for the control of the residual heavy impurity removal during the production process (Rilett and Gagnon 2008).

Another parameter that influences the viscosity of biodiesel is the ambient temperature (Prankl et al. 2004; WWFC 2006). At low temperatures (less than –20°C) some additives are required to reduce viscosity (Prankl et al. 2004); otherwise, the high viscosity may compromise the mechanical integrity of the injection pump drive system (White Paper 2007).

Because viscosity is a very critical factor in engine durability (Bacha et al. 2007), a range of viscosity is a fuel quality requirement. The EU sets this range between 3.5 and 5.0 mm²/s (EN 1421: 2003)* and the United States between 1.9 and 6.0 mm²/s (ASTM D6751). Table 9.3 lists the viscosity limits for biodiesel by country. Although Brazil currently has no limits, a range of 3.5–6.0 mm²/s will be established soon (White Paper 2007).

9.2.3 SULFATED ASH

The inorganic matter content of biodiesel, mainly abrasive solids, soluble metallic soaps, and unremoved catalysts (Bacha et al. 2007; White Paper 2007; Rilett and Gagnon 2008), is estimated by the value of the (sulfated) ash content (Mittelbach 1996). Abrasive solids and unremoved catalysts are oxidized during the combustion of the fuel, which forms an ash that wears the injector, fuel pump, pistons, and rings of the engine and may contribute to engine deposits. Soluble metallic soaps, with their oxidation to ash, contribute to filter plugging and engine deposits (Mittelbach 1996; Prankl et al. 2004; Foon et al. 2005; WWFC 2006). According to the Worldwide Fuel Charter (WWFC 2006), these effects can worsen from the use of ash-forming additives and therefore should be avoided.

The sulfated ash content varies according to process rather than to feedstock, that is, control of the efficiency of the manufacturing process in removing soaps formed during production (Rilett and Gagnon 2008) as well as the excess catalyst used in the process (Prankl et al. 2004). In the case of transesterification of fully refined oils using alkaline conditions, the sulfate ash content is mainly determined by the remaining soaps (Mittelbach 1996). In the case of unrefined oils, the value of sulfate ash correlates with the content of phosphorus (Mittelbach 1996). In this case, the sulfate ash will exceed the minimum allowed level of 0.04% m/m, and additional purification steps are required for the biodiesel to reach the fuel quality limits (Mittelbach 1996).

The ash limit of biodiesel cannot be compared with that of diesel because the latter is expressed in oxide and not sulfate ash (Prankl et al. 2004). This difference results from the fact that for the determination of ash in biodiesel, sulfuric acid is added to the sample before the combustion

---

* If CFPP is –20°C or lower, the viscosity measured at –20°C shall not exceed 48 mm²/s.

(and oxidation) to convert metallic impurities into the corresponding sulfates (Prankl et al. 2004), e.g., the remaining alkaline catalysts into alkali sulfates (Mittelbach 1996). With this process, the sodium and potassium sulfate produced from catalyst residues are less volatile than the corresponding oxides, and as a result the material loss at higher temperatures is eliminated. At the same time, the ash determination method that is used for fossil diesel samples is considered sufficient and does not need to change (Prankl et al. 2004). Most international fuel quality standards require the sulfured ash content in the biodiesel to be equal to or below 0.02% m/m and the oxide ash content of diesel to be equal to or below 0.01% m/m (Prankl et al. 2004).

9.2.4 SULFUR

Sulfur is one of the "heteroatoms" of (bio-)diesel that, although present only in small amounts, plays a catalytic role in the environmental performance of the fuel (Bacha et al. 2007) as well as in the engine durability. The combustion of high-sulfur fuel, which emits more sulfur dioxide (Bacha et al. 2007; White Paper 2007) and particulate matter (PM), is harmful to the environment by contributing to atmospheric pollution and acid rain (Foon et al. 2005) and to human health, given that these emissions have high mutagenic potential (Prankl et al. 2004; White Paper 2007). At the same time, the use of high-sulfur fuel can wear the engine [i.e., the sulfuric acid produced corrodes the cylinder liner and piston (Foon et al. 2005)) and reduce the efficiency and durability of the emissions control system (White Paper 2007), although this effect is highly correlated with the operating conditions (WWFC 2006; Rilett and Gagnon 2008).

By nature, biodiesel is an ultra-low-sulfur fuel (ULSF) (Körbitza et al. 2003; Costenoble 2006; White Paper 2007), that is, <0.001% for biodiesel delivered from palm oil (Foon et al. 2005) because it contains just traces of sulfur.* Thus, it can be used in a diesel engine with a catalyst without the risk of reducing the efficiency of the catalyst, unlike sulfur-containing diesel (Prankl and Wörgetter et al. 2000). The "poisoning" of the catalyst from the sulfur results in increased emissions of carbon monoxide (CO), hydrocarbons (HCs), and particles but is avoided by the use of a ULSF such as biodiesel (Prankl and Wörgetter 2000; WWFC 2006).

Moreover, biodiesel could be used as an additive for ultra-low-sulfur diesel (Prankl and Wörgetter 2000). Using hydrotreating process (an upgrading process of desulfurization), the sulfur can be removed from the diesel to meet the required levels of a ULSF[†] (Bacha et al. 2007). However, hydrotreating also tends to destroy minor constituents, mainly nitrogen and oxygen compounds, that provide lubricity to the fuel (Prankl et al. 2004) as well as naturally occurring antioxidants by increasing the risk of peroxide formation (Bacha et al. 2007). Hence, an additive will need to be added to ULSF diesel (Prankl et al. 2004). A blending of 2% v/v biodiesel can play the role of this additive by substituting fossil additives and significantly improving the lubricating properties of diesel fuels with low sulfur content (Prankl and Wörgetter et al. 2000).

These multiple and serious impacts from the combustion of sulfur increase the need for strict limits on the sulfur contamination of (bio)diesel. Hence, the EU defined the maximum sulfur content as 10.0 mg/kg[‡] and the United States as 15/500 mg/kg[§] (on road/off road fuel). Moreover, although many countries that have rapidly increased their share of the biodiesel market have not yet set limits for sulfur content, such as Brazil, they are planning to set them in the near future (White

---

* Although B100 from most feedstocks is essentially sulfur free, this parameter is feedstock and process dependent, i.e., used restaurant grease and tallow-derived biodiesel can contribute to higher than typical sulfur levels in biodiesel (Rilett 2008).

[†] The term "ultralow sulfur diesel" may refer to different levels of sulfur in different parts of the world; that is, in the United States it means less than 15 ppm sulfur and in Europe and the Asia-Pacific region it means less than 10 ppm sulfur (Bacha et al. 2007).

[‡] The standardized analytical methods either involve ultraviolet fluorescence spectrometry (EN ISO 20846)—applicable to sulfur concentrations of 3–500 ppm—or wavelength-dispersive X-ray fluorescence spectrometry (EN ISO 20884).

[§] The standardized analytical methods either involve ultraviolet fluorescence spectrometry (ASTM D 5453)—applicable to sulfur concentrations of 3–500 ppm—or wavelength-dispersive X-ray fluorescence spectrometry (ASTM D 4294).

Paper 2007). It should be noted that the national standards for biodiesel often reflect the national regulatory requirements for maximum sulfur content in fossil diesel (White Paper 2007).

9.2.5 COPPER STRIP CORROSION

The copper strip corrosion parameter of a biodiesel sample indicates its tendency to cause corrosion to copper, zinc, and bronze parts of the engine, the storage tank (Prankl et al. 2004; White Paper 2007), and the fuel system (Rilett and Gagnon 2008). The presence of acids or sulfur-containing compounds can accelerate the aging process (McGill et al. 2008) and tarnish the copper strip (Foon et al. 2005; Rilett and Gagnon 2008). From this oxidation, sediments will be created (WWFC 2006; DEWHA 2008; McGill et al. 2008) that may plug fuel filters (WWFC 2006). Additionally, even small quantities of copper can lead to significant injector fouling and as a result cause power loss of the engine and increase exhaust gas PM (WWFC 2006). Thus, at the transitioning from diesel fuel to biodiesel blends, fuel system parts must be specially chosen for their compatibility with biodiesel properties (WWFC 2006; DEWHA 2008).

More specifically, copper, as a dissolved metal, contributes as an oxidation catalyst to the conversion of precursors to species of higher molecular weight, more often nitrogen- and sulfur-containing compounds, organic acids, and reactive olefins (Bacha et al. 2007). The sulfur and acid compounds of biodiesel accelerate this oxidation. As a result, the copper strip corrosion parameter can be used as an indicator of the acid number of the biodiesel (White Paper 2007). The use of metal deactivators can tie up (chelate) the copper and neutralize its catalytic effect (Bacha et al. 2007).

The effects noted during copper strip corrosion demonstrate the need for an adequate limit with regard to corrosion (Foon et al. 2005; Rilett and Gagnon 2008). However, the current limits of this parameter [i.e., Class 1 for EU (EN ISO 2160) and Class 3 for the United States (ASTM D130)] are under discussion because the results are unlikely to give a rating higher than Class 1. Hence, a steel strip corrosion test would be more realistic for the present fuel systems (White Paper 2007).

9.2.6 CETANE NUMBER

The cetane number measures the ignition quality of the fuel (WWFC 2006; Bacha et al. 2007; Rilett and Gagnon 2008; Crown and Warfield 2009). It is defined as the volume percent of n-hexadecane in a blend of n-hexadecane and isocetane that gives the same ignition delay period as the test sample; a high value represents fuels that readily ignite (WWFC 2006). From this definition, it is demonstrated that the cetane number depends on engine design, size, nature of speed and load variations, and on starting and atmospheric conditions (Rilett and Gagnon 2008).

Moreover, the cetane number of a biodiesel sample depends on the molecular properties of the fatty acid esters of the fuel (Mittelbach 1996; Wörgetter et al. 1998). The cetane number increases with the length of both the fatty acid chain and the ester group and decreases with the number of double bonds (Wörgetter et al. 1998; White Paper 2007); that is, saturated fatty acids such as those delivered from animal fats and used vegetable oils have a higher cetane number than unsaturated acids (Singh 2005; McGill et al. 2008; Table 9.4).

In general, blends of biodiesel score a better cetane number than their base diesels* (Prankl et al. 2004; White Paper 2007; Rilett and Gagnon 2008). At the same time, the ethyl esters have a cetane number slightly higher than methyl esters (Knothe et al. 2003). A high cetane number indicates a short ignition delay period, and as a result good cold-start behavior (reduces white smoke on startup) and smooth combustion [reduces oxides of nitrogen (NO_x) and PM emissions as well as fuel consumption] (WWFC 2006; Bacha et al. 2007; Rilett and Gagnon 2008). In contrast, a low cetane number in a fuel indicates an incomplete combustion, which results in engine knock and increased

---

* Most of the biodiesel in the United States has cetane numbers higher than 47, compared with a low of 40 for highway diesel fuel (average for United States is 42–44; in Europe this is 51 min) (McGill et al. 2008).

TABLE 9.4
Correlation between Biodiesel Properties and Feedstock Saturation Level

	Saturated	Monounsaturated	Polyunsaturated
Fatty acid	12:0, 14:0, 16:0, 18:0, 20:0, 22:0	16:1, 18:1, 20:1, 22:1	18:2, 18:3
Cetane number	High	Medium	Low
Cloud point	High	Medium	Low
Stability	High	Medium	Low

Source: McGill, R., Aakko-Saksa, P., and Nylund, N.O., *Final Report, Annex XXXIV: Biomass-Derived Diesel Fuels, Task 1: Analysis of Biodiesel Options.* IEA Advanced Motor Fuels Implementing Agreement, Wieselburg, Austria, 2008 and Adapted from National Renewable Energy Laboratory (NREL), *Biodiesel Handling and Use Guidelines*, 3rd ed. U.S. Department of Energy, Golden, CO, 2006. Available at http://www.nrel.gov/vehiclesandfuels/npbf/pdfs/40555.pdf.

exhaust emissions (Prankl et al. 2004; Bacha et al. 2007; White Paper 2007). However, a reduction of these emissions can be achieved with the use of cetane improver additives, such as 2-ethylhexyl nitrate (McCormick 2005; Bacha et al. 2007).

Concerning the fuel quality requirements, the requested cetane number changes from one regulation to the other. This is the result of the strong dependence of this parameter on vehicle technology, climate conditions, and the regional diesel fuel regulations (White Paper 2007). However, because of the high cetane number of biodiesel, even the strict requirements of the EU [minimum 51 cetane (ISO 5165)] do not eliminate any known feedstock.

9.2.7 CLOUD POINT AND COLD FILTER PLUGGING POINT

Cloud point (CP) and cold filter plugging point (CFPP) are two properties used to control a fuel's cold-temperature behavior. CP expresses the temperature at which a cloud or haze of crystals is formed within the fuel sample when it is cooled (Prankl et al. 2004; McGill et al. 2008; Rilett and Gagnon 2008). CFPP describes the fuel filterability at low ambient temperatures (Prankl et al. 2004) and expresses the lowest temperature at which the fuel can pass through the filter in a standardized filtration test (WWFC 2006). These parameters are mainly an issue with middle distillate fuels that contain straight- and branched-chain hydrocarbons (e.g., paraffin waxes). When the ambient temperature approaches the CFPP, the viscosity of the fuel increases and at the point that falls below the cloud point, these waxes become solids (WWFC 2006; Bacha et al. 2007). This partial solidification in cold weather may cause blockages of fuel lines and filters, leading to fuel starvation and problems during engine startup and driving and engine damage due to inadequate lubrication (White Paper 2007; McGill et al. 2008). Even when the fuel reaches the CP it may still be used as long as the temperature remains at this point and the filters do not clog (McGill et al. 2008).

Cold-temperature properties of biodiesel vary according to the feedstock and the process (Rilett and Gagnon 2008). The biodiesel derived from long-chain saturated fatty acids, such as animal fats and frying oils, have higher cold-temperature values than the unsaturated fatty acids, such as most vegetable-oil derived biodiesel fuels (Foon et al. 2005; White Paper 2007; McGill et al. 2008; Table 9.4).

In general, biodiesel has a higher CP and CFPP than petroleum-based diesel fuel (Foon et al. 2005; Rilett and Gagnon 2008). However, improvement of cold-flow properties can be achieved with the use of special additives or by blending it with winter-grade diesel fuel (WWFC 2006; McGill et al. 2008). Most of these additives are polymers that interact with the wax crystals that form in diesel fuel when it is cooled below the cloud point (Bacha et al. 2007). Moreover, engine design changes to address this problem include locating the fuel pump and filter at the spot where they will receive the most heat from the engine or the pumping of more fuel to the injectors than the engine requires (Bacha et al. 2007).

Because of the dependence of the cold-temperature properties on a given climate, it is difficult to define a specific value of cold-flow performance even on a national level. The specification for this parameter is under discussion to avoid the elimination of any feedstocks and to prevent an overly large difference in cold-flow requirements between biodiesel and fossil diesel (White Paper 2007).

9.2.8 Carbon Residue

The carbon residue value of a fuel defines its residue-depositing tendencies (Prankl et al. 2004; Bacha et al. 2007; White Paper 2007; Rilett and Gagnon 2008). The limit of this parameter refers to the amount of carbonaceous matter left after evaporation and pyrolysis of a fuel sample under specified conditions (Prankl et al. 2004; White Paper 2007; Soriano 2008). Although this residue is composed of more than carbon, the term "carbon residue" is a commonly accepted phrase (Prankl et al. 2004; White Paper 2007).

Carbon residue has proven to be one of the most important indicators of the quality of biodiesel because this parameter has a direct correlation with the content of glycerides (Mittelbach et al. 1992), free fatty acids, soaps, and remaining catalysts or contaminants (Mittelbach 1996). Additionally, the carbon residue is influenced by high concentrations of polyunsaturated FAME and polymers (Prankl et al. 2004; White Paper 2007). Hence, the carbon residue can be an indicator for the coking tendency of fuel. Moreover, the carbon residue is indirectly correlated with engine deposits (Bacha et al. 2007; Rilett and Gagnon 2008).

Because of the significance of this parameter, its limitation is included in biodiesel specifications. However, the method that has to be used is under discussion. Some countries, such as the United States [0.050% m/m maximum (ASTM D4530)], India [0.05% m/m maximum (ASTM D4530)], and Brazil [0.10% m/m maximum (EN ISO 10370/ASTM D4530)], have decided to use a 100% sample in the place of 10% because of the similar boiling point of most biodiesel fuels that makes it difficult to leave a 10% residual on distillation. Alternatively, countries such as the EU [0.30% m/m maximum (EN ISO 10370)], Japan [0.3% m/m maximum (JIS K 2270)], and Australia [0.30% m/m maximum (EN ISO 10370)] acknowledge this difficulty and are still skeptical because the precision of the 100% method is still uncertain (White Paper 2007).

9.2.9 Acid Number

The acid number (or neutralization number) of a biodiesel determines the level of mineral and free fatty acids contained (Prankl et al. 2004; White Paper 2007; Rilett and Gagnon 2008). For the case of diesel, in which there are neither free fatty acids nor degradation of byproducts, the acid number determines the amount of mineral acids (Rilett and Gagnon 2008). The acid number is defined as the milligrams of potassium hydroxide (KOH) required to neutralize 1 g of FAME (Prankl et al. 2004; White Paper 2007). The acid number may be correlated with the copper strip corrosion number (see 9.2.5) because some sulfur and acid compounds can induce the latter (White Paper 2007).

The acid number can ensure the absence of free fatty acids (Prankl et al. 2004; Singh 2005) and is directly related to oxidation stability (Rilett and Gagnon 2008). A high acid number in a biodiesel sample is associated with fueling system deposits (mainly on pumps and filters) and leads to filter clogging and other fuel system malfunctions (Foon et al. 2005; Rilett and Gagnon 2008).

Acid number varies according to process and feedstock (Rilett and Gagnon 2008). The type of feedstock used to deliver the biodiesel and the degree of refinement are two of the factors that influence the acid number. Moreover, the production process (mineral acids or degradation by-products) and aging during storage influence the acid number of biodiesel (Cvengros 1998).

To avoid low oxidation stability because of the presence of acids in biofuel, most specifications set the maximum limit at 0.50 mg KOH/g of FAME. However, the methods on which most standards are based, namely EN 14104* and the ASTM D664[†], present several differences. As a result, the comparison between the different limits is not feasible (White Paper 2007).

9.2.10 TOTAL AND FREE GLYCERIN AND MONO-, DI-, AND TRIGLYCERIDES

Glycerin, or else glycerol, is formed as a by-product in transesterification (van Walwijk 2005; Nylund et al. 2008) during the production of biodiesel. The level of glycerin in the fuel, including free and bound glycerin, is measured by the total glycerin method (Bacha et al. 2007; White Paper 2007; Rilett and Gagnon 2008). This measurement can ensure the conversion of the fat or oil to monoalkyl esters (Prankl et al. 2004; Singh 2005); that is, a low level of total glycerol ensures that high conversion has taken place (Bacha et al. 2007; Rilett and Gagnon 2008). The level of free glycerin alone in the fuel is estimated by the free glycerin method (Rilett and Gagnon 2008) and can ensure the removal of glycerin from the fuel (Prankl et al. 2004; Singh 2005). Additionally, the level of glycerin bound in the form of mono-, di-, and triglycerides is measured by the relative method. The contents of mono-, di-, and triglycerides indicate the degree of conversion from oil to methyl esters or completion of the esterification/transesterification process (Foon et al. 2005).

The content of glycerin in biodiesel is process dependent (White Paper 2007; Rilett and Gagnon 2008). High values of free glycerin may result from the insufficient separation or washing of the ester product (Prankl et al. 2004; White Paper 2007), which causes the glycerin to separate during storage once the solvent (methanol) has evaporated (White Paper 2007). Alternatively, glycerin may also be formed because of hydrolysis of the remaining mono-, di-, and triglycerides in stored fuel (Mittelbach 1996).

Glycerin is a clear, viscous, nontoxic, sweet-tasting liquid used in cosmetics, soaps, food production, and pharmaceuticals. However, when it is mixed with fuel, it can harm the engine. Fuels out of specifications for total glycerin with respect to these parameters are prone to coking and may thus cause the formation of deposits on injector nozzles, pistons, and valves (Mittelbach et al. 1983; Bacha et al. 2007; White Paper 2007).

High levels of free glycerin can cause problems during storage by settling to the bottom of storage tanks or damage the fuel injection system by clogging fuel systems and injector deposits (Mittelbach 1996; Prankl et al. 2004; Bacha et al. 2007; Rilett and Gagnon 2008). Moreover, high levels of mono-, di-, and triglycerides can cause injector deposits and may adversely affect cold weather operation and filter plugging (Bacha et al. 2007). This is particularly true for high-saturated monoglycerides (Rilett and Gagnon 2008). For these reasons free glycerin is limited in the specifications.

A maximum content of 0.02% m/m of free glycerin is consistently defined in all of the standards discussed. The European standard limits the amounts of mono-, di-, and triglycerides to 0.80, 0.20, and 0.20% m/m, respectively, and defines a maximum amount of 0.25% m/m for total glycerin (EN ISO 14105). The United States, Brazil, Australia, and India are the only countries with national standards that do not have explicit limits for the contents of partial glycerides. However, they have a limit for total glycerin, set at approximately 0.25% m/m[‡], with the exception of Brazil, which is at 0.38% m/m (ABNT NBR 15344/EN 14105/ASTM D6584). Brazil is planning to reduce this limit, but first a new method for the measurement of total glycerin has to be developed that will also be suitable for castor oil (White Paper 2007).

---

* EN 14104.
[†] ASTM D664.
[‡] Only the United States has this limit to 0.24% (m/m) for total glycerol. However, this difference is not important.

9.2.11 Phosphorus Content

Phosphorus in FAME stems from phospholipids (animal and vegetable material) and inorganic salts (used frying oil) contained in the feedstock (Prankl et al. 2004; White Paper 2007). The phosphorus content of the vegetable oil mainly varies according to the process (Mittelbach 1996; Prankl et al. 2004). Cold-pressed plant oils usually contain less phosphorus than hot-pressed plant oils (Prankl et al. 2004). Moreover, this parameter increases with the decrease of the grade of refined oil; that is, the phosphorus content of fully refined oils is only several parts per million, whereas that of unrefined or water-degummed oil is more than 100 ppm (Mittelbach 1996). The phosphorus content is highly correlated with the sulfate ash content (see Section 9.2.3) (Mittelbach 1996).

A high level of phosphorus in the fuel can increase particulate emissions and clog the filter (Nylund et al. 2008) of the emissions control system and as a result damage the operation of the catalytic converter (Mittelbach 1996; Gray 2005; Bacha et al. 2007; White Paper 2007; Rilett and Gagnon 2008). However, its value can be reduced by various forms of degumming before transesterification (Prankl et al. 2004) as well as by using alkaline catalysts during transesterification. Distillation of the final product can eliminate the residual phosphorus of the fuel (Prankl et al. 2004).

A phosphorus content of 10 ppm is suspected of being the critical point from where the efficiency of oxidation catalytic converters starts to decrease (Prankl et al. 2004). The wide use of catalytic converters on diesel-powered equipment as well as the tight emission limits justifies the importance of setting limits for the phosphorus content of biodiesel (Rilett and Gagnon 2008). Currently, the CEN specification (EN 14107) is aligned with the ASTM specification (ASTM D4951) to this critical limit of 10 ppm phosphorus content. However, the CEN specification will shortly be changed to limit phosphorus to 4 mg/kg to ensure that all biodiesel, regardless of the source, has a low enough phosphorus content (Rilett and Gagnon 2008) to avoid long-term effects on the emission treatment systems (White Paper 2007).

9.2.12 Distillation Temperature, T90

Distillation temperature, T90, is an indication of the purity of the diesel. In the case of biodiesel, this parameter is used to make sure unscrupulous blenders did not adulterate B100 with heavy petroleum components (White Paper 2007). The distillation temperature is mainly part of the ASTM specification (ASTM D1160) and is similar to the ester content limit used by the EU (White Paper 2007). T90 is the temperature at which 90% of a particular biodiesel fuel distills in a standardized distillation test (Bacha et al. 2007). A high T90 may dilute the engine oil and cause difficulties during cold start (Nylund et al. 2008). Reducing T90 slightly decreases NO_x emissions but increases HC and CO emissions. PM emissions are unaffected (Bacha et al. 2007).

Contrary to diesel, biodiesel exhibits a narrow boiling range rather than a distillation curve (Bacha et al. 2007; Rilett and Gagnon 2008). This fact can be explained by the difference between the composition of diesel (many compounds boiling at differing temperatures) and biodiesel (contains only a few compounds) (Prankl et al. 2004). The fatty acid chains in the raw oils and fats from which biodiesel is produced are mainly composed of straight-chain HCs with 16–18 carbons (Table 9.A1) that boil at approximately the same temperature (Prankl et al. 2004; Bacha et al. 2007).

The atmospheric boiling point of biodiesel generally ranges from 330 to 357°C, thus the specification value of 360°C maximum for the T90 (ASTM D1160) is reasonable. This specification was incorporated as an added precaution to ensure that the fuel has not been adulterated with high boiling contaminants (Bacha et al. 2007; Rilett and Gagnon 2008). However, the method used is difficult, costly, and has dubious precision. The T90 limit plays the same role as with ester content limit (see Section 9.2.16), and currently the choice of only one method, the upgrade of the existing methods or the replacement of both by another method, such as content of unsaponifiable material, is discussed (White Paper 2007).

9.2.13 OXIDATION STABILITY

The oxidation stability value of biodiesel measures the formation degree of undesirable breakdown products (DEWHA 2008). Low oxidation stability (i.e., high concentration of acids and polymers in biodiesel) affects the performance and durability of fuel system components (DEWHA 2008) because it can cause fuel system deposits and lead to filter clogging and fuel system malfunctions (Rilett and Gagnon 2008). However, these effects can be avoided with the use of additives (White Paper 2007; Rilett and Gagnon 2008), either antioxidants or stabilizers (Bacha et al. 2007).

In general, biodiesels are more sensitive to oxidative degradation than diesel because of their chemical composition (FAMEs) (Prankl et al. 2004). This oxidation sensitivity increases with the length and the saturation of the fatty acids that constitute biodiesel; that is, a fuel composed primarily of C18:3 is 100 times more unstable than a fuel made of C18:1 (McGill et al. 2008). Fuels highly saturated are less chemically reactive with oxygen, so they have high oxidation stability (Rilett and Gagnon 2008). Alternatively, the unsaturated esters of biodiesel are easily oxidized and form insoluble sediments and gums (Prankl et al. 2004; White Paper 2007). These residues are associated with fuel filter plugging and deposit within the injection system and the combustion chamber (Mittelbach and Gangl 2001).

The content of natural antioxidants, such as tocopherols and carotenes (Simkovsky 1997), in biodiesel can improve the oxidation stability of the fuel (Prankl et al. 2004; McGill et al. 2008). Although these antioxidants are removed during distillation (Prankl et al. 2004; Rilett and Gagnon 2008) or destroyed in the case of used frying oil, the use of synthetic antioxidants is equally efficient (Mittelbach and Schober 2003). However, they have to be added in high concentrations to compensate for those destroyed during fuel storage (Mittelbach and Schober 2003).

For the protection of biodiesel blends and engine and fuel injection equipment from the deleterious effects of poor oxidation stability, biodiesel standardizations have set a specific limit on oxidation stability for B100 at 6 h (Rilett and Gagnon 2008) measured by the EN 14112 method. Only the United States is using a 3-h limit; however, this is for the B20 blend (White Paper 2007), whereas the EU is for B100 fuel. Another concern is where the standards must be met: At the point of production or at the point of delivery? Consideration must be given that the oxidation stability degrades with time (White Paper 2007). However, the induction period of oxidation stability is well correlated with other biodiesel quality parameters, such as kinematic viscosity, ester content, acid value, and polymer content (Lacoste and Lagardere 2003; Soriano 2008). Hence, if a biodiesel has low oxidation stability, then it will score out of the specification range and also these parameters.

9.2.14 ALKALI AND ALKALINE-EARTH METALS

Sodium and potassium (alkali metals) are common catalysts for the biodiesel reaction (Prankl et al. 2004; White Paper 2007; Rilett and Gagnon 2008) and are removed during processing; otherwise sodium/potassium can form soap and cause adverse effects on new particulate trap technologies being considered for future diesel engines (Rilett and Gagnon 2008).

Calcium and magnesium (alkaline-earth metals) may originate from the use of soft water in the washing step of biodiesel production (Prankl et al. 2004; White Paper 2007; Rilett and Gagnon 2008). These can be detected in biodiesel as abrasive solids or soluble metallic soaps. If these compounds are not removed, abrasive solids can wear the engine by damaging the injector, fuel pump, piston, and ring and thereby contribute to engine deposits. From their side, soluble metallic soaps can contribute to filter plugging and engine deposits. Moreover, a high calcium and magnesium value may damage the vehicle exhaust system because these compounds are collected in exhaust particulate removal devices and they are not removed during regeneration. In this way, the backpressure increases, thus reducing the time between service maintenance (Mittelbach 1996; Rilett and Gagnon 2008).

The parameters are interrelated with several other fuel quality criteria, such as sulfated ash content and carbon residue (Mittelbach 1996), therefore they are partially limited by these parameters. However, some engine and fuel injection equipment manufacturers have requested a specific limit on these metal ions, especially for vehicles with particulate traps (Prankl et al. 2004; Bacha et al. 2007; White Paper 2007). Hence, for alkali and alkaline metals, their concentration should be below 5 ppm as measured by the EN 14108 and EN 14109 methods and the EN 14538 method, respectively.

9.2.15 Water and Sediment (Water Content and Total Contamination)

The water content and sediment parameters are treated together or as separate parameters, with the sediment being treated by the total contamination property.

Water is introduced into biodiesel at the final washing step of the production process. Then the water content value has to be reduced by drying (Wörgetter et al. 1998), otherwise even a very small quantity of water on the production point could inhibit the biodiesel and cause it to fail to meet the water content specification at the combustion stage (White Paper 2007; ARDD 2009). Biodiesel can fix relatively higher quantities of water than mineral diesel (Mittelbach 1996; Prankl et al. 2004; Rilett and Gagnon 2008) and can absorb water up to a concentration of approximately 1000 ppm during storage (Mittelbach 1996).

Free water in the biodiesel is the ideal environment for the growth of microorganisms because they can take the nutrients required for their growth from the fuel. The higher ambient temperatures also favor their growth. As a result, during fuel storage microorganisms can produce enough acidic by-products to accelerate tank corrosion and enough biomass (microbial slime) to plug filters and fuel lines (Prankl et al. 2004; Bacha et al. 2007; Rilett and Gagnon 2008). Additionally, high water content can cause filter blockages by hydrolyzing FAME to free fatty acids. Moreover, the water may damage the engine and injection system by causing corrosion to their chromium and zinc parts (Kobmehl and Heinrich 1997). To avoid these effects, seals and valves of storage tanks should be checked to prevent humidity from entering the fuel. If the water content is already high, water can be drained (Prankl et al. 2004).

Total contamination is defined as the concentration of insoluble material retained after filtration of a fuel sample under standardized conditions (Prankl et al. 2004; White Paper 2007). The total contamination has turned out to be an important quality criterion because biodiesel with a high concentration of insoluble impurities tends to cause filter blocking at filling pumps and on vehicles, resulting in premature wear of the injection system components (Prankl et al. 2004). These effects result from high soap and sediment concentrations (Mittelbach 2000).

To prevent biodiesel fuel from acquiring a high water content and the resulting risk of corrosion and microbial growth, special limits were set at 0.050% v/v maximum (ASTM D2709) for water content and sediment when they are treated together, and 500 mg/kg maximum (EN ISO 12937) for water content and 24 mg/kg maximum (EN ISO 12662) for total contamination when they are treated separately. Currently, ASTM is working toward the separation of water standards from sediment standards (White Paper 2007). Although only the upper limit of water content is defined, a very low water content in biodiesel also has to be avoided because it may result in a separation phase in blends with fossil diesel (Prankl et al. 2004; White Paper 2007).

9.2.16 Ester Content

Ester content is an indication of biodiesel's purity and sometimes a confirmation that the fuel meets the legal definition of biodiesel (i.e., monoalkyl esters) (White Paper 2007). Biodiesel that contains various minor components within the original fat or oil source or originates from inappropriate reaction conditions scores a low ester content value (Prankl et al. 2004; White Paper 2007). The ester content value of biodiesel increases with distillation, as impurities such as sterols, residual alcohols, partial glycerides, and unseparated glycerin are removed (Prankl et al. 2004; Rilett and Gagnon 2008).

It is demonstrable that ester content, as a parameter, is similar to the distillation temperature, T90 parameter. However, this is mainly used by the EU and the biodiesel should have an ester content value equal or higher than 96.5% m/m (EN 14103) to meet the fuel specifications. The United States limits biodiesel purity with the use of the T90 parameter (see Section 9.2.12). The main concern for this parameter is the precision of the methodology used, which may be considered unacceptable given that this method is not suitable for lauric oils such as coconut or palm kernel oil. Currently, the upgrade or the removal of the test method is being discussed. In the second case, the ester content values would be replaced by another parameter such as the content of unsaponifiable material (White Paper 2007).

9.2.17 IODINE NUMBER, LINOLENIC ACID METHYL ESTER, AND POLYUNSATURATED BIODIESEL

Iodine value is indicative of total unsaturation within a mixture of fatty material, regardless of the relative shares of mono-, di-, tri-, and polyunsaturated compounds (White Paper 2007; Rilett and Gagnon 2008) and only dependent on the origin of the vegetable oil (Mittelbach 1996). Biodiesel derived from oil with high unsaturation will also have high iodine values (Rilett and Gagnon 2008). Unsaturated esters blended with engine oil may form high-molecular-weight compounds that can negatively reduce the lubricating quality and as a consequence damage the engine (Schäfer et al. 1997).

Fuels with high iodine number (i.e., consisting of many unsaturated esters) tend to polymerize and form deposits on injector nozzles, piston rings, and piston ring grooves when they are heated (Kobmehl et al. 1997; Heinrich 1997). Actually, polymerization reactions are directly related to the number of double bonds and become significant in fatty acid esters containing at least three double bonds (Prankl et al. 1999, 2004). However, three- or more-fold unsaturated esters are not a very common component for all seed oils, and as a result various promising seed oils are incorrectly excluded from the biodiesel market according to some national standards because of their high iodine value (e.g., soybean oil with an iodine value of 125–140 g I$_2$/100 g). Therefore, it is better to limit the content of linolenic acid methyl esters and polyunsaturated FAMEs rather than the total degree of unsaturation (iodine value) (Knothe and Dunn 1998; Prankl et al. 2004; White Paper 2007).

European specification limits the iodine number to 120 g of iodine/100 g of biodiesel maximum (EN ISO 14111), the linolenic acid to 12.0 mg/kg maximum (EN 14103), and polyunsaturated (four or more double bonds) methyl ester to 1 mg/kg maximum (method under development). Some countries, such as Brazil and the United States, disagree with this limit because it is very similar to the oxidation stability limit. Moreover, the iodine number eliminates soybean, sunflower, and other unsaturated oils from meeting specifications regardless of the use of stability additives or use as a blend stock rather than as a standalone fuel. EU experts are currently discussing increasing iodine number combined with other precautions such as reinforced limits on linolenic acid, polyunsaturates, and oxidation stability (White Paper 2007).

9.2.18 DENSITY

Density is the mass of a unit volume of material at a selected temperature. The density of the fuel determines the heating value per volume of this fuel given that the other fuel properties are unchanged (Bacha et al. 2007). Moreover, the density of the fuel directly influences the injection timing of mechanically controlled injection equipment as well as the volume of injected fuel, both of which are related to the engine power and, consequently, engine emissions and fuel economy (WWFC 2006).

TABLE 9.5
The Density (15°C) of Biodiesel in Europe

	Value	Unit
EN 14214	860–900	kg/m^3
Minimum	879.3	mm^2/s
Maximum	885.7	mm^2/s
Average	883.4	mm^2/s
Standard deviation	1.2	mm^2/s
Range 95% maximum	885.8	mm^2/s
Range 95% minimum	881.0	mm^2/s
Out of specification	0	–

Source: European Biodiesel Board (EBB), *EBB European Biodiesel Quality Report (EBBQR)— Results of the Third Round of Tests (Winter 2007/2008),* European Biodiesel Board, Brussels, Belgium, 2008.
Samples from 39 biodiesel plants in Europe. Sampling period January to February 2008.

Variations in fuel density therefore result in nonoptimal exhaust gas recirculation rates for a given load and speed point in the engine map and, as a consequence, influence the exhaust emission characteristics (WWFC 2006). Low fuel density reduces PM and slightly reduces carbon dioxide (CO_2) emissions from all diesel vehicles and NO_x emission from heavy-duty vehicles but increases fuel consumption and reduces power output (WWFC 2006). However, the emission influence of fuel density concerns mainly older technology engines and not modern engines with electronic injection and computer control (Lee et al. 1998).

The density of biodiesel depends on the feedstock (Mittelbach 1996), and in general the density of a biodiesel blend is higher than that of the base diesel (White Paper 2007; Nylund et al. 2008). Density increases with decreasing chain length and increasing number of double bonds of biodiesel fatty acid composition (Prankl et al. 2004; White Paper 2007); that is, high-density values for fuels are derived from feedstocks with many unsaturated compounds, such as sunflower oil and linseed oil (Prankl et al. 2004). Also, the density of biodiesel depends on the fuel purity and is decreased by the presence of low-density contaminants, such as methanol (White Paper 2007).

In general, density limits are in the range of 860–900 kg/m^3 at 15°C (ISO 3675/12185). However, fuels that meet the other specifications fall between these limits (Table 9.5), and some countries (e.g., the United States and Brazil) doubt the need of this specification (White Paper 2007).

9.2.19 ETHANOL AND METHANOL CONTENT

During the transesterification process for the production of biodiesel, vegetable oil or animal fat reacts with a catalyst (usually sodium hydroxide) and methanol or ethanol to form one of two fatty acid esters, FAME or FAEE, respectively, and glycerin (also called glycerol) (Combs 2008). The remaining alcohol (methanol or ethanol) is removed by distillation or by repeated aqueous washing steps (Prankl et al. 2004). Methanol is most commonly used and the least expensive (McGill et al. 2008), has a short chain, and is polar (Khan 2002). However, ethanol is less toxic (Singh 2005; Nylund et al. 2008), less corrosive, and more soluble than methanol (Nylund et al. 2008); and biodiesel produced has improved cold-flow properties (McGill et al. 2008).

Even a low methanol or ethanol content can greatly reduce the flash point (Foon et al. 2005), posing safety risks in biodiesel handling and storing (Prankl et al. 2004) (see impacts from the flash point reduction in Section 9.2.1). Methanol or ethanol has high volatility that can cause fuel system corrosion, low lubricity, and adverse effects on injectors as well as harm some materials in fuel distribution and vehicle fuel systems (White Paper 2007), such as fuel pumps, seals, and elastomers (Foon et al. 2005).

The ethanol and methanol content is strongly related to the flash point (Foon et al. 2005), which is already limited in fuel specialization. Namely, biodiesel with a flash point value higher than 100°C will have a methanol concentration less than 0.2% m/m (Mittelbach 1996). For low-methanol residue biodiesel, typical flash point values exceed 160°C (Rilett and Gagnon 2008). However, a more precise estimation of these parameters is required to guarantee the safe use of the fuel.

The CEN specification currently limits the ethanol and methanol content of the biodiesel to 0.2% m/m (EN 14110). The ASTM specification has the same limit but just for methanol. For the case of ethanol, the flash point is measured and is limited to 130°C minimum. ASTM is working to fill this gap (White Paper 2007).

9.3 BIOETHANOL

9.3.1 COLOR AND APPEARANCE

Color is an easy and fast way to identify bioethanol (IFQC 2004) because proteins of ethanol give a natural yellow color to the fuel (White Paper 2007). In some countries, such as Brazil, the use of orange dyes was proposed. However, their use will not be mandated and in general they will not be allowed for bioethanol intended for export (White Paper 2007).

Appearance specification aims to ensure that the ethanol is free of suspended and precipitated contaminants (Chevron 2008; RFA 2009). If these contaminants are not eliminated, they can damage the engine by increasing wear and shortening filter life by causing blockage of the fuel system (IFQC 2004).

Color specification may vary according to national regulation (WWFC 2008). Appearance specifications are examined with visual inspection at ambient temperature (WWFC 2008). If a fuel fails to meet these specifications, it will likely fail to meet the rest because this would indicate poor fuel handling practices (WWFC 2008).

9.3.2 DENSITY

Density is a natural characteristic of ethanol (IFQC 2004). It is not strongly linked to the engine operation or protection and influences only slightly the energy value of the fuel and consequently vehicle performance (De Klerk 2008). With the use of high-purity ethanol with a low water content, this parameter is even less important (IFQC 2004). However, density measurement is an easy and quick test (ebio 2006) that can be used for quality monitoring purposes of the blendstock (WWFC 2008). Density can be used to estimate the purity of the fuel with high precision in cases when the fuel ethanol contains small amounts of methanol (that has lower density than ethanol) and/or high alcohols (i.e., C_3–C_5 alcohols), which have higher density than ethanol (White Paper 2007).

Currently, CEN and ASTM do not have density specifications for bioethanol. Given that bioethanol contains negligible quantities of water, methanol, and higher alcohols, its density should be approximately the same as that of pure ethanol. However, a density measurement is needed to correct volume to a reference temperature and for calculating % v/v results from reported % m/m results. A recommended value of 793 kg/m³ maximum at 20°C (ASTM D4052/ABNT NBR 5992) is provided by the Worldwide Fuel Charter (WWFC 2008).

9.3.3 SULFATE CONTENT

Sulfate content and the electrical conductivity of a fuel are interrelated. By regulating electrical conductivity, detrimental effects of sulfates [present as sulfite (SO_3) and sulfate (SO_4)], such as injector fouling due to deposits, could be reduced. Moreover, sulfate, even in a small concentration, accelerates corrosion of vehicle fuel system parts (IFQC 2004; WWFC 2008).

ASTM and the Agency of Petroleum, Natural Gas, and Biofuels (ANP) specifications limit the sulfate content to 4 mg/kg maximum (ASTM D7319 and ASTM D7328 and ANP NBR 10894 methods, respectively). CEN has recently identified the problem of injector clogging due to sulfate deposits and is currently working to determine an acceptable upper limit of sulfate in bioethanol (White Paper 2007).

9.3.4 Sulfur Content

Sulfur reduces the efficiency of catalysts and therefore it has a significant impact on vehicle emissions. Moreover, sulfur adversely affects advanced on-board diagnostic system requirements, such as heated exhaust gas oxygen sensors. Reductions in sulfur will provide immediate reductions of emissions from all catalyst-equipped engines (WWFC 2006).

CEN and ASTM have specifications for sulfur content. The EU specification limits the sulfur content to 10 mg/kg for undenatured ethanol (EN 15485 or EN 15486 methods), whereas U.S. specifications set the limit at 30 mg/kg for denatured ethanol (ASTM D2622 and ASTM D5453 methods), which could be 5 mg/kg for undenatured ethanol if the hydrocarbon denaturant is neglected. Brazil is expected to establish a sulfur specification for ethanol in the future (IFQC 2004; Gray 2005; White Paper 2007).

However, engine manufacturers are pushing for reduction of the sulfur content limit so that they can introduce new technologies into the market, such as lean-burn fuel-efficient technologies or fuel-level sender units (WWFC 2006). Because of the tightening of the petrol sulfur content specification, that of bioethanol will also likely soon decrease (Gray 2005; Costenoble 2006). Either way, the natural concentration of sulfur in the ethanol is 1 or 2 mg/kg, much below the current limit (White Paper 2007).

9.3.5 Copper Content

Copper is an active catalyst that even in very low concentration [0.012 ppm (ASTM D4806-03)] in bioethanol can decrease fuel stability by accelerating the low-temperature oxidation of HCs (IFQC 2004; Gray 2005; RFA 2009). The gum formed by the oxidation of hydrocarbons can result in scale in engine pipes and injector deposits (RFA 2009). The copper and the electrical conductivity are interrelated, and by controlling the former, the presence of the latter can be minimized (RFA 2009). The use of metal deactivators in bioethanol can inhibit copper's catalytic activity (Chevron 2008).

CEN and ASTM specifications currently limit the copper to 0.1 mg/kg maximum [methods EN 15488 and ASTM D1688 modified (http://goo.gl/Hkldt), respectively]. At the same time, Brazilian specifications are slightly stricter (0.07 mg/kg for anhydrous ethanol, method ANP NBR 10893). Although copper contamination during bioethanol production is prevented by prohibiting copper tubes and stills, this test should remain in place to test ethanol derived from the alcoholic beverage industry, where copper stills are commonly used (White Paper 2007). To reduce measurement costs, inductively coupled plasma (ICP) spectrometry could be used to measure copper (Cu), sodium (Na), iron (Fe), and phosphorus (P) in one test (White Paper 2007).

9.3.6 Electrical (Electrolytic) Conductivity

Electrical conductivity of bioethanol represents the concentration of metallic ions, such as chloride, sulfate, sodium, and iron in the fuel (WWFC 2008) and is an indicator of the corrosive properties of the fuel. Water content is another parameter that influences electrical conductivity (IFQC 2004). A high electrical conductivity signifies a high risk of corrosion and thus clogging of the fuel systems and injector deposits (WWFC 2008).

In contrast to bioethanol and bioethanol blends, gasoline is an electrical insulator. As a result, this specification is not important for low percentage blends but is significant for higher content ethanol, such as E75 and E85 (IFQC 2004). In practice, flexible-fuel vehicles (FFVs) use this difference of conductivity between ethanol and gasoline and are equipped with a sensor for the measurement of electrical conductivity in the fuel. In this way, they determine the ratio of the bioethanol blend and optimize its combustion parameters, such as injection, ignition time, and quantity of air (IFQC 2004).

The test of bioethanol electrical conductivity is a relatively simple and cheap way to measure its corrosiveness and ionic contamination (WWFC 2008). Brazilian specifications limit this parameter to the level of 500 μS/m maximum for anhydrous and hydrous ethanol (method NBR 10547:2006). Although, CEN and ASTM specifications do not yet include any limit for the electrical conductivity of bioethanol, the EU and United States are considering adding the same limit as Brazil (White Paper 2007).

9.3.7 ETHANOL CONTENT

The ethanol content is an indicator of bioethanol quality (WWFC 2008) and identifies the presence of contaminants such as water, methanol, and higher alcohols in bioethanol (White Paper 2007). Although bioethanol is originally a pure product, during the production process several contaminants may decrease its ethanol content (White Paper 2007). Moreover, the denaturing process reduces the ethanol content (e.g., undenatured bioethanol with 99.0% v/v will fall to 94.0% v/v after denaturing) (Gray 2005).

Bioethanol with slightly lower ethanol content than normal may influence the engine performance rather than harm any of its parts (White Paper 2007). This effect is less important for FFVs than optimizing the fuel combustion by adjusting the air-to-fuel ratio according to the ethanol content of the fuel (see Section 9.3.6). Sometimes, lower ethanol content is even suggested to provide better cold-start and warm-up performance by increasing fuel volatility (RFA 2009). From the other side, very low ethanol content may affect the lubricating properties of the fuel and its water tolerance (IFQC 2004). Moreover, low ethanol content usually indicates high methanol content. Methanol is toxic and when it is present in bioethanol in concentrations higher than 2.5% v/v it may cause corrosive problems (RFA 2009), lower the water tolerance, and increase the vapor pressure of the fuel (IFQC 2004).

Currently, the United States has set its limit at 93.9% v/v minimum (ASTM D5501 method), whereas the EU limits the ethanol content to 96.8% v/v (indirectly from the ethanol + C_3–C_5 alcohols limit, i.e., ethanol + C_3–C_5 at 98.8% v/v and C_3–C_5 alcohols at 2% v/v) and Brazil at 99.6% v/v (ASTM D5501 method). However, all three countries have agreed to move toward a common 98% v/v (ASTM D5501 method) limit (White Paper 2007).

Apart from ethanol content, there are some more specifications that determine the ethanol or other categories of alcohol in bioethanol. These specifications are in use for the moment, but they will be eliminated in the near future because their importance is questionable because of the use of ethanol and water content specifications (White Paper 2007). Briefly, as already mentioned, the EU limits ethanol plus heavier (C_3–C_5) alcohol content (EC/2870/2000 Method I, Appendix II, Method B) and C_3–C_5 alcohol content (method EC/2870/2000—Method IIIb) separately (Costenoble 2006). The United States also has a limit for C_3–C_5 alcohol content (4.5% v/v maximum on the basis of the ethanol, water, and methanol content limits). The EU and United States have a specification for methanol content (i.e., 1.0% v/v, maximum; ASTM D5501) and 0.5% v/v maximum (EC/2870/2000 Method III), respectively. Brazil specifies HCs at 3.0% v/v maximum (ABNT NBR 13993 method) and total alcohol (ethanol plus methanol plus heavier alcohols) content of ethanol produced by fermentation of sugarcane at 99.6% v/v minimum (density method similar to ASTM D4052). By calculating from the rest of the specifications, the total alcohol content allowed by the EU and United States is 99.8% v/v minimum and 99.0% v/v minimum, respectively (White Paper 2007).

9.3.8 pHe

pHe is used to ensure that some strong acids, such as sulfuric acid (H_2SO_4), hydrochloric acid (HCl), and phosphoric acid (H_3PO_4), are within acceptable limits. These acids enter the bioethanol production process in the manufacturing step and carryover in the final product (IFQC 2004). Although even a low level of these strong acids in bioethanol can cause corrosion problems (RFA 2009), the acidity of the fuel is not always influenced so much as to be detectable by some other test methods (e.g., the acidity test; Gray 2005) that better measure weak acids (IFQC 2004).

The pHe of bioethanol is tested after the addition of denaturant and corrosion inhibitor (Gray 2005) and should be limited to 6.5–9. The use of bioethanol with pHe below this range may damage the engine because of corrosion of fuel pump and injector equipment. When pHe values exceed this range, the fuel contains more alkaline components and the plastic parts of the fuel system may be damaged. These effects of bioethanol's pHe increase with the level of bioethanol blend; that is, are more important for E75 and E85 blends than for E5 or E10 (IFQC 2004). pHe can be adjusted by the use of sodium hydroxide (NaOH). However, this can have some side effects and a sodium content test is required (IFQC 2004) (see Section 9.3.14).

Currently, the United States and Brazil limit the pHe range to 6.5–9 for anhydrous denatured bioethanol (ASTM D6423) and 6–8 for hydrated bioethanol (NBR 10891). However, given that the ASTM method (ASTM D6423) is not for nonaqueous solutions, such as denatured ethanol, its use for the estimation of the pHe of this fuel has to be based on empirical estimation. Consequently, the repeatability and reproducibility of the results are under investigation (IFQC 2004; WWFC 2008). For this reason, CEN has excluded this test from its bioethanol quality specification (White Paper 2007). Another workable test method for pHe is under development (White Paper 2007).

9.3.9 Acidity

Acidity specification is used to ensure that some weak acids (e.g., acetic acid) are within acceptable limits (IFQC 2004). This method also measures acidity, as does pHe (see Section 9.3.8), but it is effective mainly for the measurement of the very dilute aqueous solutions of low-molecular-weight organic acids (Gray 2005). These solutions are corrosive to a wide range of metals and alloys (RFA 2009) and may result in some long-term wear problems in vehicles. Nevertheless, this correlation between acidity by acetic acids and corrosion is doubted by some experts (White Paper 2007). Contrary to the strong acids, there is no negative effect from the low concentration of these acids in bioethanol (IFQC 2004) and consequently bioethanol blenders should keep their acidity value at a very low level (WWFC 2008).

The United States currently limits the acidity by acetic acid at 0.0074% m/m (ASTM D1613). This limit is approximately the same as that of the EU [0.007% m/m (EN 15491)] and almost double that of the Brazilian limit [0.0038% m/m (ABNT NBR 9866)]. The lower limit of Brazil can be explained by the higher concentration of ethanol in its blends (White Paper 2007). These limits are not expected to change in the near future, and this specification can be eliminated after the development of a more accurate pHe test method (White Paper 2007).

9.3.10 Phosphorus Content

Phosphorus enters the bioethanol production chain from certain performance additives or during the fermentation process, especially when synthetic ethanol is produced from ethylene and the H_3PO_4

is used as a catalyst (IFQC 2004). The feedstock is also another source of phosphorus in bioethanol (White Paper 2007). Phosphorus is a powerful catalyst poison and can increase a vehicle's emissions by deactivating the exhaust catalyst system (IFQC 2004; White Paper 2007). Additionally, phosphorus can serve as a nutrient source for microbes (IFQC 2004) and thus it increases bioethanol impurity, accelerates tank corrosion, and plugs filters and fuel lines.

Because of these effects, international fuel quality standards are becoming more strict concerning the phosphorus content of fuel. However, it is not yet clear whether these specifications should concern the fuel ethanol or the final fuel product to be more effective (IFQC 2004).

CEN specification includes a 0.50-mg/L maximum (EN 15487) limit for the phosphorus content of bioethanol. The United States and Brazil doubt the need for this requirement for current bioethanol but acknowledge that this could be an issue for bioethanol derived from nontraditional feedstock and processes. To achieve this, the United States may consider adding a phosphorus specification whereas Brazil, which aims to continue deriving bioethanol from its current feedstock, sugarcane, by fermentation, will not follow (White Paper 2007).

9.3.11 (Unwashed) Gum/Evaporation Residue/Nonvolatile Material

Gum, evaporation residue, and nonvolatile (involatile) material are three different procedures that measure the residue of ethanol evaporation (Costenoble 2006; WWFC 2008). These residues come principally from additives, carrier oils used with additives, and diesel fuels that contain some heavy components, such as iron (Fe), copper (Cu), and sulfates (present as SO_3 and SO_4) (IFQC 2004). These components form gum and can damage engines by contributing to deposits on the surface of carburetors, fuel injectors, and intake manifolds as well as ports, valves, and valve guides (Gray 2005; Chevron 2008; WWFC 2008).

The United States currently limits heptane washed gum at 5.3 mg/100 mL (ASTM D381* method), whereas Brazil limits unwashed gum at 5.0 mg/100 mL (NBR 8644[†] method) and the EU limits the unwashed residue at 10 mg/100 mL (procedure from Annex II of ECD/2870/2000). Although the comparison between these processes is not feasible, the difference between the U.S. and Brazilian methods could determine the presence of nonvolatile materials (RFA 2009). However, the United States is planning to change its method and to measure the gum concentration without washing it with heptane (White Paper 2007).

9.3.12 Chloride Content

The chloride content of bioethanol varies according to the chloride concentration of the feedstock from which it is derived as well as according to the quantity of HCl that may be used in the production process (IFQC 2004). Chloride ions in water can form HCl that, even in a low concentration, causes corrosion problems (RFA 2009) in the stainless steel equipment involved in production processing or storage or in the vehicle, such as stainless steel exhaust systems and fuel injection equipment (IFQC 2004; WWFC 2008).

Because of this effect on the whole fuel chain, biofuel producers and vehicle manufacturers press for a low chloride content limit in the fuel quality specification (IFQC 2004; White Paper 2007). Brazil has currently set its chloride content limit at 1 mg/kg maximum for hydrous ethanol (NBR 10894[‡] method) (White Paper 2007). In any case, Brazilian distilleries have no problem meeting the chloride specification because they use H_2SO_4 in place of HCl to avoid corrosion of their equipment (IFQC 2004).

---

* ASTM D381 Standard Test Method for Gum Content in Fuels by Jet Evaporation.
[†] ABNT NBR 8644 Fuel Ethylic Alcohol—Determination of Residues by Evaporation.
[‡] ABNT NBR 10894 Ethyl Alcohol—Determination of Chloride and Sulphate—Ion Chromatography Method.

CEN and ASTM specifications are still keeping the chloride content limit for bioethanol too high at 42 mg/kg maximum (ASTM D7319* and ASTM D7328[†] methods) and 25 mg/kg maximum (EN 15492[‡] method), respectively, for undenatured ethanol (White Paper 2007). However, the U.S. limit for E85 is already at 1 mg/kg maximum, and the EU is also planning to reduce its limit, but first the new, more precise Ion Chromatography (IC) method must be adopted (White Paper 2007).

9.3.13 WATER CONTENT

Water can enter ethanol during production, fuel distribution, and storage through condensation (WWFC 2008). Bioethanol with high water content should not be blended with gasoline (Gray 2005), otherwise a phase separation may occur, and the free water can influence the performance or even damage the engine (WWFC 2008). Additionally, water in the fuel can cause corrosion and microbial growth. The water content limit is used to protect the vehicle from these effects when they use bioethanol even at low-ethanol blends.

Ethanol is hydroscopic and can collect water from ambient air and the distribution system (IFQC 2004; White Paper 2007). By blending bioethanol (that contains mainly ethanol and water) with gasoline, ethanol dissolves in gasoline and water, but water dissolves very little in gasoline. As a result, the blend has a higher water content than its base fossil fuel (WWFC 2008); that is, gasoline can dissolve up to 150 ppm water at 21°C and E10 up to 7000 ppm at the same temperature (Chevron 2008).

The solubility of the fuel increases proportionately with ethanol content, the aromaticity of the base gasoline, and temperature (IFQC 2004; Costenoble 2006). At low temperatures, a phase separation will occur (separation of the ethanol from water) and form an aqueous lower phase in the storage tank and vehicle fuel tank (White Paper 2007). The upper gasoline layer will have a lower octane number and will be less volatile and consequently may cause serious operating problems for spark-ignition engines (Chevron 2008). The lower layer is incapable of running the engine (Chevron 2008). The risk is higher for low-ethanol blends (i.e., E5 or E10).

The handling and distribution practices of bioethanol can influence the water content of the fuel (RFA 2009). There are local parameters that also influence the water content, such as temperature and humidity. All of these factors must be considered for determination of regional water content limits. A compilation of the influence of all of these effects could be indicative, reflected by the fact that the maximum water content of bioethanol in Brazil is in the range of 0.4% v/v[§], half that of the United States (between 0.6 and 0.7% v/v) (White Paper 2007).

Currently, the EU limits the water content to 0.24% v/v, whereas the United States sets the limit at 1.0% v/v. Brazil has no water specification. The United States and Brazil found the EU limit to be conservative (White Paper 2007). The difference between these specifications can be explained by the varying ethanol concentration permitted in gasoline and the difference in gasoline handling and distribution. For example, the EU, unlike the other two countries, uses no higher than E5 blends and has a wet logistics infrastructure that enhances the risk of such problems occurring (White Paper 2007). At the same time, Brazilian blends are between E20 and E25 and therefore can hold more water without phase separation (White Paper 2007). For this specification, the additional cost must be taken into consideration because the additional drying required increases the cost of production and can reduce productivity at the mill by up to 7% (White Paper 2007).

---

* ASTM D7319 Standard Test Method for Determination of Total and Potential Sulfate and Inorganic Chloride in Fuel Ethanol by Direct Injection Suppressed Ion Chromatography.
[†] ASTM D7328-07e1 Standard Test Method for Determination of Total and Potential Inorganic Sulfate and Total Inorganic Chloride in Fuel Ethanol by Ion Chromatography Using Aqueous Sample Injection.
[‡] EN 15492:2009 Ethanol As a Blending Component for Petrol—Determination of Inorganic Chloride and Sulfate Content - Ion Chromatographic.
[§] Based on a minimum alcohol content of 99.6 vol% (White Paper 2007).

9.3.14 IRON AND SODIUM CONTENT

Iron (Fe) as well as copper (Cu) and sulfates (present as SO_3 and SO_4) form a gum with petrol and result in scale in engine pipes.

Sodium (Na) accumulates in the vehicle combustion chamber and causes corrosion. Na is used in the form of NaOH to regulate the pHe of the fuel when it is too low to meet pHe specifications (IFQC 2004).

The Fe and Na contents of hydrated ethanol are limited only by Brazilian specifications. Although Fe and Na contents are important parameters, they will probably be eliminated because they are covered by electrical conductivity and chloride content specifications (White Paper 2007) (see Section 9.3.6). ICP spectrometry could be used to measure Cu, Na, Fe, and phosphorus (P) in one test (WWFC 2008).

9.4 FUEL QUALITY VARIATION BY FEEDSTOCK AND BY COUNTRY

9.4.1 FUEL QUALITY BY FEEDSTOCK

In general, the fuel quality required for bioethanol depends on the process, which is in contrast to biodiesel, for which some quality specifications are influenced only by the feedstock or by the process or by both. The main reason for this difference is that bioethanol is one chemical product; that is, the ethanol content is more than 93.9% v/v (see Section 9.3.7), whereas biodiesel is a blend of many FAMEs or FAEEs.

The chain length and the number and place of chemical bonds of esters that compose biodiesel vary, in addition to the percentage of each of them present in the raw material (the feedstock) that derives the final product (Table 9.A1). The fuel quality of the fuel derived varies according to these variations of the feedstock. The biodiesel quality is improved with lower levels of polyunsaturated fatty acids [e.g., linolenic acid (18:3)] or saturated fatty acids [e.g., palmitic (16:0) and stearic acid (18:0)] and by higher levels of monounsaturated fatty acids [e.g., oleic acid (18:1)]. The more that these criteria are met, the higher the oxidation stability will be and the more improved winter operability the fuel will have (Körbitza et al. 2003).

Currently, there are several feedstocks commonly used for the production of biodiesel. All of them could be clustered into two categories mostly on the basis of the saturation of biodiesel:

1. Vegetable oil such as rapeseed oil, sunflower oil, soy oil, and palm oil.
2. Used oil and animal fat such as tallow, grease, poultry fats, and fish oils.

Table 9.6 presents the main properties of biodiesel by the feedstock used.

There are other raw materials used as a feedstock for the production of biodiesel. However, their production is, at least for the moment, very low; they have some promising results, but they are not currently traded or they are still on the research level. These include oils from algae, artichoke, coconut, cottonseed, flaxseed, hemp, jojoba, karanj, kukui nut, milk bush, pencil bush, mustard, neem, olive, peanut, radish, rice bran, safflower, sesame, and tung (http://www.bdpedia.com/biodiesel/plant_oils/plant_oils.html). The blend of more than one of these oils could improve the quality of the biodiesel produced.

9.4.2 FUEL QUALITY/STANDARDS BY COUNTRY

The fuel quality of biofuels varies from country to county to meet the corresponding variations in fuel regulation. The fuel regulation of each country is based on the specifications of standardization organizations. These can have an international (i.e., ISO, CEN, and ASTM) or national range of application.

TABLE 9.6
Biodiesel Properties by Feedstock

Feedstock	Properties
Rapeseed oil	Relatively high oxidation stability, IV < 120, acceptable cold weather operability
Sunflower oil	For countries with warm and dry climatic conditions, IV > 120[a] (should be blended with low IV-oils to be used in Europe)
Soy oil	Relatively low oxidation stability, relatively low cetane number, IV > 120 (should be blended with low IV-oils to be used in Europe), acceptable cold-weather operability
Palm oil	High cetane number, unfavorable cold-weather operability (CFPP at + 11°C) (not suitable for colder climatic conditions unless it is blended with low CFPP oils)
Tallow	High cetane number, unfavorable cold-weather operability (CP > 10°C), carbon residues and ether content risk, high viscosity
Grease	High cetane number, unfavorable cold-weather operability, high viscosity
Used (recycling) oil	Some may have high polymer content (careful and clean recycling practices are needed), high viscosity

Source: Adapted from Körbitza, W., et al., Worldwide review on biodiesel production. Prepared by Austrian Biofuels Institute for IEA Bioenergy Task 39, Subtask Biodiesel, 2003; Rilet, J. and Gagnon, A., *Renewable Diesel Characterization Study.* Climate Change Central. Calgary, Alberta, Canada, 2008; Kinast, J.A., *Production of Biodiesels from Multiple Feedstocks and Properties of Biodiesels and Biodiesel/Diesel Blends—Final Report (Report 1 of 6).* National Renewable Energy Laboratory, Oak Ridge, TN, 2003.

IV, Iodine value. Table contains only the biodiesel properties that vary according to the feedstock and not the common parameters of biodiesel (e.g., lubricity, high flash point, and low vapor pressure) that have already been mentioned.

[a] European Standard EN 14214, American standard ASTM D-6751-02 does not contain any IV limit.

9.4.2.1   International

9.4.2.1.1   International Organization for Standardization

International Organization for Standardization (ISO) is the world's largest standard-developing and publishing organization. ISO is a nongovernmental organization that is based in Geneva, Switzerland, and consists of 161 members, all national standards institutes of different countries (http://www.iso.org/iso/home.htm).

The aim of ISO is to support the "facilitation of global trade, (the) improvement of quality, safety, security, environmental and consumer protection, as well at the rational use of natural resources, and (the) global dissemination of technologies and good practices all of which contribute to economic and social progress" (ISO 2004).

ISO was established in 1947 and has developed more than 17,500 international standards dealing with various subjects. Every year approximately 1100 new ISO standards are published (http://www.iso.org/iso/home.htm).

ISO has covered many fuels and energy-related activities, such as coal, gas, petrol, nuclear, hydrogen, and solar energy on one side and fuel-consuming products such as road vehicles and gas turbines on the other. However, concerning biofuels, ISO is mainly participating as a technical collaborator of the CEN standardization organization (see Section 9.4.2.1.3) by providing standards for methods applied to the estimation of some of the characteristics of biofuel quality.

9.4.2.1.2   ASTM International

ASTM International was founded as the American Society for Testing and Materials (ASTM) in 1898 in Philadelphia, PA. ASTM started as a voluntary standards development organization addressing the U.S. need for technical standards for materials, products, systems, and services. However, in 1981 ASTM opened a standards distribution center in Europe, and in 2001 changed its name to ASTM International to reflect the range of ASTM activities (http://www.astm.org/HISTORY/index. html). Currently, ASTM has more than 30,000 ASTM members from different sectors, such as producers, users, consumers, government, and academia from more than 120 countries (http://www. astm.org/ABOUT/aboutASTM.html).

The aim of ASTM International is to develop and provide voluntary consensus standards and the related technical information and services that promote public health and safety, support the protection and sustainability of the environment, and the overall quality of life; contribute to the reliability of materials, products, systems, and services; and facilitate international, regional, and national commerce (http://www.astm.org/NEWS/Mission2.html).

Biofuels is one of the domains in which ASTM International is active. So far, the ASTM D6751 standard specification for biodiesel fuel blend stock (B100) is only for biodiesel used in up to 20% v/v blends (B20) and is not in neat form. If the neat biodiesel meets ASTM D6751, and the base diesel, used for the B20 blend, meets the relevant standards (ASTM D975), then the final product is acceptable for use (Table 9.A2).

ASTM D4806 is currently the standard specification for denatured fuel ethanol used as a blend with gasoline to fuel an automotive spark-ignition engine. For a high percentage (75–85% v/v) of denatured bioethanol into the fuel blend (Ed75–Ed85), the Ed75–Ed85 should meet the property limits set in ASTM D5798. These blends are for the fuel of FFVs and limit change according to the season; that is, standards of class 1 for summer grade, class 2 for interseasonal, and class 3 for winter grade (RFA 2009) (Table 9.A3).

9.4.2.1.3 European Committee for Standardization

CEN aims to "facilitate the exchange of goods and services by eliminating technical barriers" (CEN 2008) in Europe. Despite this, CEN is mainly addressed to the European market, and its close collaboration with ISO [i.e., more than 30% of CEN standards are identical to international standards (CEN 2008)] promotes the adoption of CEN standards from countries beyond Europe.

Currently, CEN has 30 national members with more than 60,000 technical experts as well as business federations and consumer and other social interest organizations (http://www.cen.eu/ cenorm/aboutus/index.asp). Through May 2009, CEN had produced 13,501 documents, 1144 of these in 2008 (http://goo.gl/Hkldt).

Regarding biodiesel in Europe, EN 14214 describes the specifications required for automotive fuels—FAMEs for diesel engines. These standards concerns B100 and biodiesel in diesel blends. When biodiesel is blended with diesel, the final fuel should meet the diesel standards (i.e., EN 590) that allow no more than 5% v/v of FAME into the blend. CEN is planning to increase this limit to 10% v/v; however, this has to be first incorporated in the EU fuel legislation (Table 9.A2).

Concerning bioethanol, EN 15376* sets the specifications for automotive blends of ethanol in gasoline. However, this blend should contain a maximum of 5% v/v bioethanol to meet the EN 228 gasoline specification. Because of the Renewable Energy Directive [COM (2008) 19] on the promotion of the use of energy from renewable sources that set a target of 10% v/v biofuels in transport in each member state (COM 2008), this limit is planned to be increased (Saunders 2009). Before that, the impact assessment of this increase of the percentage has to precede and to satisfy any concern (http://goo.gl/SRKoI) (Table 9.A3).

So far, CEN has a workshop agreement[†] on E85 specifications for use in FFV. This agreement has already been applied in Germany and the Netherlands (Maniatis et al. 2009).

At the same time, some EU member countries have set additional standards. Their scope is to cover categories of bioethanol used as automotive fuel that are out of the range of EU regulation. Some examples are

---

* EN 15376:2007 Automotive Fuels—Ethanol As a Blending Component for Petrol—Requirements and Test Methods.
[†] This is one step before the standards setting (http://goo.gl/RipqI).

- *ÖNORM C 1114 (Austria):* Automotive fuels—Petrol superethanol E85—Requirements and test methods
- *CSN 656511 (Czech Republic):* Fermentation denatured ethanol determined for application in automotive petrol—Requirements and test methods
- *SS 155437 (Sweden):* Motor fuels—Fuel alcohols for high-speed diesel engines
- *SS 155480 (Sweden):* Automotive fuels—Ethanol E85—Requirement and test methods.

9.4.2.1.4 Asia-Pacific Economic Cooperation

Asia-Pacific Economic Cooperation (APEC), established in 1989, aims to "further enhance economic growth and prosperity for the region and to strengthen the Asia-Pacific community." APEC has 21 members including Australia, Brunei Darussalam, Canada, Chile, People's Republic of China, Hong Kong China, Indonesia, Japan, Republic of Korea, Malaysia, Mexico, New Zealand, Papua New Guinea, Peru, the Republic of the Philippines, the Russian Federation, Singapore, Chinese Taipei, Thailand, United States, and Vietnam (http://www.apec.org/apec/about_apec.html; Novianto 2008).

APEC has 11 working groups, one of them working on the energy sector. One of the projects of this working group is the establishment of the Guidelines for the Development of Biodiesel Standards in the APEC Region (APEC 2009). This project aims to establish the guidelines for development of biodiesel standards in the APEC region for enhancing the trade of biodiesel among APEC member economies (http://goo.gl/lGiMf).

9.4.2.1.5 Worldwide Fuel Charter

The Worldwide Fuel Charter (WWFC) was created in 1998 to represent automobile and engine manufacturers worldwide. WWFC aims to "promote greater understanding of the impact of fuel quality on engine and vehicle emissions and performance and to promote harmonization of fuel quality worldwide in accordance with engine and vehicle needs in different markets."

In March 2009, the WWFC presented the first edition of two guidelines, one for biodiesel (WWFC 2009a) (Table 9.A3) and another one for bioethanol (WWFC 2009b) (Table 9.A3) quality. Both of them are focused on the quality of the biofuel blended with fossil fuels to produce a 5 or 10% v/v biodiesel or bioethanol blend, respectively. The finished product should meet the WWFC specifications for diesel and gasoline, respectively. It should be noted that the last edition of WWFC on fuels (4th edition, released in 2006) establishes four different categories of specifications to reflect market conditions and engine and vehicle requirements (WWFC 2006).

9.4.2.2 National

Worldwide there are several national standardization organizations charged with the definition of specifications that biofuels are required to meet when they are used in the local market. Brazil, Canada, India, Japan, South Africa, and Thailand are some of the countries that have at least one of these standardization organizations.

9.4.2.2.1 Brazil

The Brazilian Association for Technical Standards (known as ABNT*) was founded in 1940 and consists of the national standardization body of Brazil, responsible for the identification of the technical national standards (http://goo.gl/STWp6). ABNT has a close collaboration with the National Institute of Metrology, Standardization, and Industrial Quality (known as INMETRO[†]). INMETRO was founded in 1973 to support Brazilian enterprises and to increase their productivity and the quality of goods and services (http://www.inmetro.gov.br/english/institucional/index.asp).

---

* Associação Brasileira de Normas Técnicas.
[†] Instituto Nacional de Metrologia, Normalização e Qualidade Industrial.

However, the regulatory measures, the contracting, and the monitoring economic activities related to energy industries, including biofuels, are controlled by a third organization, the Agency of Petroleum, Natural Gas, and Biofuels (ANP*) (http://www.anp.gov.br/index.asp).

The Brazilian specification on the biodiesel quality required is defined by the ANP n° 07/2008 regulation (Aranjo 2009). This specification is addressed to the B100 that is blended with diesel. Since July 2008, diesel sold must be a blend of at least 3% v/v of biodiesel, and currently Brazil is the third largest biodiesel producer and consumer in the world (http://www.anp.gov.br/biocombustiveis/biodiesel.asp) (Table 9.A2).

Regarding bioethanol, Brazil is one of the first (since 1974) and most successful examples of the implementation of a national bioethanol program. Currently, 20–25% v/v of anhydrous ethanol blended with gasoline is obligatory, but the use of hydrous ethanol in E100 cars and FFVs is still voluntary. Brazil regulations have two different specifications that account for the difference between the hydrous and anhydrous ethanol. These specifications are described by the ANP Act 36/2005 regulation (Table 9.A3).

9.4.2.2.2 Canada

In Canada, the Canadian General Standards Board (CGSB) is the standardization organization charged with setting national biofuels standards. CGSB was founded in 1934 and aims to offer "client-centred, comprehensive standards development and conformity assessment services in support of the economic, regulatory, procurement, health, safety and environmental interests of our stakeholders—government, industry and consumers" (http://www.tpsgc-pwgsc.gc.ca/ongc/home/index-e.html).

Because of the cold climate of Canada, special care is given to the adaptability of the biodiesel from different feedstocks to low temperature. On February 2009, the Alberta Renewable Diesel Demonstration (ARDD) released a report concerning this (ARDD 2009) as a follow-up of the research of Climate Change Central (August 2008) (Rilett and Gagnon 2008). For this reason, the automotive low-sulfur diesel fuel can currently contain no more than 5% v/v of biodiesel and should meet the CAN/CGSB 3.520-2005 specification. At the same time, the automotive low-sulfur diesel fuel used for the blend should meet the CAN/CGSB 3.517-2007 specification.

Respectively, the oxygenated unleaded automotive gasoline that contains ethanol should meet the CAN/CGSB 3.511-2005 specifications[†] at the time when the unleaded automotive gasoline should meet the CAN/CGSB 3.5-2004 specification[‡] (Table 9.A3).

9.4.2.2.3 India

In India, the Bureau of Indian Standards (BIS) is the body that sets the quality standards of biofuels traded in the country. BIS was founded in 1987, but it is considered as the successor of the Indian Standards Institution (ISI), set up in 1947. The aims of the BIS are the "harmonious development of standardization, marking and quality certification, to provide new thrust to standardization and quality control, as well as, to evolve a national strategy for according recognition to standards and integrating them with growth and development of production and exports" (http://www.bis.org.in/org/obj.htm).

The biodiesel specification of India, namely IS 15607:2005, is addressed to the neat biodiesel blended with diesel to result in an up to B20 final fuel. Although a specified blend may contain a higher percentage of biodiesel than is allowed in other countries, according to the current Indian fuel legislation the fuel blends cannot contain more than 5% v/v biodiesel. However, this 5% v/v biodiesel is also allowed by the specification standards of diesel, namely IS 1460:2005[§] (Table 9.A2).

---

* Agência Nacional do Petróleo, Gás Natural e Biocombustíveis.
† CAN/CGSB 3.511-2005 AMEND. 2 Oxygenated Unleaded Automotive Gasoline Containing Ethanol (Incorporates Amendment 1).
‡ CAN/CGSB 3.5-2004 AMEND. 2: Unleaded Automotive Gasoline (Incorporates Amendment 1).
§ IS 1460:2005—Automotive Diesel Fuels—Specification (Fifth Revision).

The Indian legislation concerning bioethanol is similar to that of biodiesel by allowing up to 5% v/v anhydrous ethanol in gasoline. The ethanol should meet the IS 15464:2004 quality specification,and the final product should meet the IS 2796:2000 specification.

In 2007, India set a 10% v/v target for bioethanol and biodiesel in a national fuel market for October 2008. This target was not achieved (BiofuelsDigest 2008, Tyagi 2008). As a result, the modification of national bioethanol quality specifications was also postponed.

9.4.2.2.4 Japan

The Japanese Standards Association (JSA), formed on December 1945 through the merger of the Dai Nihon Aerial Technology Association and the Japan Management Association, is charged with the standardization of conventional and alternative fuels in Japan. In general, JSA sets the national Japanese standards and aims "to educate the public regarding the standardization and unification of industrial standards, and thereby to contribute to the improvement of technology and the enhancement of production efficiency" (http://www.jsa.or.jp/eng/about/about.asp).

At the same time, the Japanese Industrial Standards Committee (JISC) sets the company- and industrial-level standards. The main tasks of the JISC are the "establishment and maintenance of JIS (abbreviation of Japanese Standards), administration of accreditation and certification, participation and contribution in international standardization activities, and development of measurement standards and technical infrastructure for standardization" (http://www.jisc.go.jp/eng/). As a result, there is a close collaboration between these two bodies; namely, the JSA and the JISC.

Currently, according to the Law of Quality Control of Gasoline, the diesel used in Japan has to meet the diesel fuel quality specification JIS 2204:2007. According to this specification, up to 5% m/m content of FAME is allowed in diesel oil as long as the methanol content is less than 0.01% m/m. The standards of the neat biodiesel used in these blends are optional and are described by JIS K 2390:2008* (Table 9.A2).

Concerning bioethanol automotive use in Japan, the fuel quality specification of gasoline, JIS K 2202:2007, sets the limit of the ethanol content in gasoline to 3% v/v, considering that the oxygen content is less than 1.3% v/v (Table 9.A3) (Numata 2009).

9.4.2.2.5 South Africa

In South Africa, the South African Bureau of Standards (SABS) establishes and monitors the standards for automotive fuels. SABS, founded in 1945, is the national institution of South Africa for "the promotion and maintenance of standardization and quality in connection with commodities and the rendering of services" (http://goo.gl/dwgtK).

SABS's fuel quality specifications are adapted to the local conditions of the environment and market in South Africa. More specifically, South Africa produces synthetic fuels from coal and gas and as a result the fuel qualities have to take into consideration the synthetic vehicle fuels (e.g., the sulfur content is set at 500 ppm maximum). Additionally, the use of standardized fuels ensures the vehicles' operation at high altitudes (>1600 m) on which a large part of the road network in South Africa runs (Rehnlund 2008, Prins 2009).

Currently, the neat biodiesel should meet the SANS 1935:2004 standards. Moreover, the SANS 342:2006† quality specification of diesel allows 5% v/v maximum of biodiesel content (Table 9.A2). The neat bioethanol should meet the SANS 465:2005 standards, and at the same time the gasoline quality traded in South Africa should be aligned with the SANS 1598:2006 quality specification. The standard on bioethanol quality concerns the anhydrous denatured ethanol intended to be blended at up to 10 % v/v gasoline (Table 9.A3).

---

* JIS K 2390:2008—Automotive Fuels—Fatty Acid Methyl Ester (FAME) As Blend Stock.
† SANS 342:2006—Automotive Diesel Fuel.

9.4.2.2.6 Thailand

The Thai Industrial Standards Institute (TISI) founded in January 1969 is charged with the standardization of biofuels. The TISI aims mainly "to develop national standards and monitor quality of products and services to be in line with the requirements and international practices" (http://www.tisi.go.th/eng/tisi.html).

In 2005, the Department of Energy Business (DEB), Ministry of Energy, classified diesel into high- and low-speed diesel categories. The high-speed diesel was further classified into the normal high-speed diesel and B5 high-speed diesel types. Additionally, DEB specified the diesel characteristics for the different types of diesel.* For the case of B5 high-speed diesel, the characteristics of biodiesel used for the blend were also separately specified[†] (Table 9.A2).

In the same year, DEB classified bioethanol blends (known in Thailand as "gasohol") into gasohol E10 octane 91 and gasohol E10 octane 95 types and specified their characteristics. The characteristic of the gasoline used was also defined by first distinguishing the gasoline in type I and type II fuel.

---

* Notification of the Department of Energy Business (DOEB) on Characteristic and Quality of Diesel (No.3) B.E. 2548(2005). 2005. Department of Energy Business (DOEB), Ministry of Energy.
[†] Notification of the Department of Energy Business on Characteristic and Quality of Biodiesel— Fatty Acid Methyl Ester B.E. 2548(2005). 2005. Department of Energy Business (DOEB), Ministry of Energy.

APPENDIX 1: FATTY ACIDS OF BIODIESEL FEEDSTOCK

TABLE 9.A1
Component Fatty Acids of Selected Biodiesel Feedstock

Names	Lauric 12:0 and Smaller	Myristic 14:0	Palmitic 16:0	Palmitoleic 16:1	Stearic 18:0	Oleic 18:1	Linoleic 18:2	Linolenic 18:3	Arachidic 20:0	Gadoleic 20:1	Erucic 22:1
Canola	—	—	4	—	2	56	26	10	—	—	—
Cottonseed	—	—	27	—	2	18	51	Trace	—	—	—
Peanut	—	—	13	—	3	38	41	Trace	—	3	3
Olive	—	—	10	—	2	78	7	—	—	—	—
Rice bran	—	—	16	—	2	42	37	1	—	—	—
Soybean	—	—	11	—	4	22	53	8	—	—	—
Sunflower	—	—	5	—	5	20	69	—	—	—	—
Palm	—	—	44	—	4	39	11	—	—	—	—
Cocoa butter	—	—	26	—	34	35	3	—	—	—	—
Rapeseed	—	—	4	—	2	33	18	9	—	12	22
Mustard	—	—	4	—	—	22	24	14	—	12	20
Coconut	**Caproic (6:0) : 0.5** **Caprylic (8:0): 9** **Capric (10:0): 6.8** **Lauric (12:0): 46.4**	18	9	—	1	7.6	1.6	—	—	—	—
Palm kernel	**Caprylic (8:0): 2.7** **Capric (10:0): 7.0** **Lauric (12:0): 46.9**	14.1	8.8	—	1.3	18.5	0.7	—	—	—	—
Jatropha curcas	—	—	12.8	—	7.8	44.8	34	—	—	—	Other:1.1
Pig	—	1	24	3	13	41	10	1	—	—	—
Beef	—	4	25	5	19	36	4	Trace	—	—	—
Sheep	—	3	21	2	25	34	5	3	—	—	—
Chicken	—	1	24	6	6	40	17	1	—	—	—

(Continued)

TABLE 9.A1 (Continued)
Component Fatty Acids of Selected Biodiesel Feedstock

Names	Lauric 12:0 and smaller	Myristic 14:0	Palmitic 16:0	Palmitoleic 16:1	Stearic 18:0	Oleic 18:1	Linoleic 18:2	Linolenic 18:3	Arachidic 20:0	Gadoleic 20:1	Erucic 22:1
Turkey		–	20	6	6	38	24	2	–	–	–
Lard	Capric (10:0): Trace Lauric (12:0): < 0.5	1.5	24–30	2.3	12.18	36–52	10–12	1	0.5	0.5–1	<0.5
Beef tallow	Trace	2–4	23–29	2–4	20–35	26–45	2–6	1	<0.5	<0.5	Trace
Yellow grease	–	2.4	23.2	–	13.0	44.3	7.0	0.7	–	–	–

Source: Bacha, J., Freel, J., and Gibbs, A., *Diesel Fuels Technical Review,* Chevron Corporation, San Ramon, CA, 2007.

APPENDIX 2: SPECIFICATIONS

TABLE 9.A2
Biodiesel Specifications

	EU (EN 14214)[a]	United States (ASTM D6751-07a)[b]	WWFC (B100 Guidelines)[c]	Brazil (ANP 42)[d]	Japan (JIS K 2390)[e]	South Africa (SANS 1935)[f]	India (IS 15607: 2005)[g]	Thailand (B100-FAME)[h]	Indonesia (SNI 04-7182-2006)[i]	Malaysia (Spec of Palm Methyl Esters)[j]
Ester content	≥96.5 %m/m (EN 14103)	—	≥96.5 %m/m (EN 14103 mod other: ABNT NBR 15342)	Report (ABNT NBR 15342/ EN 14103)	≥96.5 %m/m (EN 14103)	≥96.5 %m/m (EN 14103)	≥96.5 %m/m	≥96.5 %m/m	≥96.5 %m/m	≥96.5 %m/m
Density	0.86–0.90 g/mL at 15°C (EN ISO 3675, EN ISO 12185)	—	Report g/mL (EN ISO 3675, ASTM D4052, JIS K2249 Other: EN ISO 12185, ABNT NBR 7148/14065)	Report at 20°C (ABNT NBR 7148/ ABNT NBR 14065/ ASTM D1298/ ASTM D4052)	0.86–0.90 g/mL (JIS K 2249)	0.86—0.90 g/mL (ISO 3675, ISO 12185)	0.86–0.90 g/mL (ASTM D4052)	0.86—0.90 g/mL	0.85–0.90 g/mL (40°C)	0.86–0.90 g/mL
Kinematic viscosity	3.5–5.0 mm²/s (EN ISO 3104)	1.9–6.0 mm²/s (ASTM D445)	2.0–5.0 mm²/s (EN ISO 3104, ASTM D445, JIS K2283 Other: ABNT NBR 10441)	Report (ABNT NBR 10441/ EN ISO 3104/ ASTM D445)	3.5–5.0 mm²/s (JIS K 2283)	3.5–5.0 mm²/s (ISO 3104)	2.5–6.0 mm²/s (ASTM D445)	3.5–5.0 mm²/s	2.3–6.0 mm²/s	3.5–5.0 mm²/s

(Continued)

TABLE 9.A2 (Continued)
Biodiesel Specifications

	EU (EN 14214)[a]	United States (ASTM D6751-07a)[b]	WWFC (B100 Guidelines)[c]	Brazil (ANP 42)[d]	Japan (JIS K 2390)[e]	South Africa (SANS 1935)[f]	India (IS 15607: 2005)[g]	Thailand (B100-FAME)[h]	Indonesia (SNI 04-7182-2006)[i]	Malaysia (Spec of Palm Methyl Esters)[j]
Flash point	≥120°C (EN 3679)	≥130°C (ASTM D93) (if methanol not measured directly, otherwise 93 min.)	≥100°C (ISO 3679, ASTM D93)	≥100°C (ABNT NBR 14598/EN 3679/ASTM D93)	≥120°C (JIS K 2265)	≥120°C (ISO 3679)	≥120°C (IP-170)	≥120°C	≥100°C	≥120°C
Sulfur	≤10 mg/kg (EN 20846/ EN 20884)	≤15 mg/kg (ASTM D5453/ ASTM D4294)	≤10 mg/kg (EN 20846/ EN 20884, ASTM D5453/D2622 JIS K3541-1, −2, −6 or −7)	Report (EN 20846/EN 20884/ ASTM D5453)	≤10 mg/kg (JIS K 2541-1/2/6/7)	≤10 mg/kg (ISO 20846, ISO 20884)	≤50 mg/kg (ASTM D5453)	≤10 mg/ kg	≤100 mg/ kg	≤10 mg/kg
Carbon residue (on 10% sample)	≤0.3 %m/m (EN ISO 10370)	≤0.05 %m/m (on 100% sample) (ASTM D4530)	≤0.05 %m/m (ASTM D4530) (on 100% distillation residue)	≤0.1 %m/m (on 100% sample) (EN 10370/ ASTM D4530)	≤0.3 %m/m (JIS K 2270)	≤0.3 %m/m (ISO 10370)	≤0.05 %m/m (ASTM D524) and ≤0.05 %m/m (on 100% sample) (ASTM D4530)		≤0.05 %m/m (on 100% sample)	≤0.05 %m/m (on 100% sample)

Cetane number	≥51 (EN ISO 5165)	≥47 (ASTM D 613)	≥51 (ISO 5165, ASTM D613, JIS K2280)	Report (EN 5165 / D613)	≥51 (JIS K 2280)	≥51 (ISO 5165)	≥48 (ASTM D 613)	≥51	≥51	≥51
Sulfated ash	≤0.02 %m/m (ISO 3987)	≤0.02 %m/m (ASTM D874)	≤0.005 %m/m (ISO 3987, ASTM D874 Other: ABNT NBR 984) and ≤0.001 %m/m (ash content) (ISO 6245, ASTM D482, JIS K2272)	≤0.02 %m/m (ABNT NBR 6294/ISO 3987/ASTM D874)	≤0.02 %m/m (JIS K 2272)	≤0.02 %m/m (ISO 3987)	≤0.02 %m/m (ASTM D874)	≤0.02 %m/m	≤0.02 %m/m	≤0.02 %m/m
Water	≤500 mg/kg (EN 12937)	≤0.050% v/v (water and sediment) (ASTM D2709)	≤500 mg/kg (EN 12937) and ≤0.05 vol% (water and sediment) (ASTM D2709)	≤0.050% v/v (Water and Sediment) (ASTM D2709)	≤500 mg/kg (JIS K 2275)	≤500 mg/kg (ISO 12937)	≤0.050% v/v (IP 386)		≤0.05% v/v (water and sediment)	≤0.05% v/v
Total contamination	≤24 mg/kg (EN 12662)	—	≤24 mg/kg (EN 12662, ASTM D2276, ASTM D5452, ASTM D6217)	Report (EN 12662)	≤24 mg/kg (EN 12662)	≤24 mg/kg (EN 12662)	—	≤24 mg/ kg	—	≤24 mg/kg
Copper corrosion	Class 1 (EN 2160)	Class 3 (ASTM D130)	Light rusting (ASTM D665 Procedure A)	Class 1 (ABNT NBR 14359/EN 2160/ASTM D130)	≤1 (JIS K 2513)	Class 1 (ISO 2169)	Class 1 (IP 154)	Class 1	Class 3	Class 1

(Continued)

TABLE 9.A2 (Continued)
Biodiesel Specifications

	EU (EN 14214)[a]	United States (ASTM D6751-07a)[b]	WWFC (B100 Guidelines)[c]	Brazil (ANP 42)[d]	Japan (JIS K 2390)[e]	South Africa (SANS 1935)[f]	India (IS 15607: 2005)[g]	Thailand (B100-FAME)[h]	Indonesia (SNI 04-7182-2006)[i]	Malaysia (Spec of Palm Methyl Esters)[j]
Total acid number	≤0.5 mg KOH/g (EN 14104)	≤0.5 mg KOH/g (ASTM D664)	≤0.5 mg KOH/g (ISO 6618 ASTM D664, D974 JIS K2501 Other: ABNT NBR 14448)	≤0.8 mg KOH/g (ABNT NBR 14448/ EN 14104/ ASTM D664)	≤0.5 mg KOH/g (JIS K 2501, K 0070)	≤0.5 mg KOH/g (EN 14104)	≤0.5 mg KOH/g (ASTM D974)	≤0.5 mg KOH/g	≤0.8 mg KOH/g	≤0.5 mg KOH/g
Oxidation stability	≥6 h (EN 14112)	≥3 h (EN 14112)	≥10 h (prEN 15751 or EN 14112 as alternative)	≥6 h (EN 14112)	(After blending) (—)	≥6 h (EN 14112)	≥1.5 h (ASTM D 5304)	≥6 h	—	≥6 h
Iodine number	≤120 gI/100 g (EN 14111)	—	≤130 gI/100g (EN 14111)	Report (EN ISO14111)	≤120 gI/100g (JIS K 0070)	≤140 gI/100g (EN 14111)	≤115 gI/100g proposed	Report	≤115 gI/100g proposed	≤110 gI/100g proposed
Methyl linolenate	≤12.0 %m/m (EN 14103)	—	≤12.0 %m/m (EN 14103)	—	≤12.0 %m/m (EN 14103)	≤12.0 %m/m (EN 14103)	—	—	—	≤12.0 %m/m
Methanol	≤0.20 %m/m (EN 14110)	≤0.20 %m/m (EN 14110)	≤0.20 %m/m (EN 14110 JIS K2536 Other: ABNT NBR 15343)	≤0.50 %m/m (ABNT NBR 15343/ EN 14110)	≤0.20 %m/m (JIS K 2536, EN14110)	≤0.20 %m/m (EN 14110)	≤0.20 %m/m by GC	≤0.20 %m/m	—	≤0.20 %m/m

Monoglyceride	≤0.80 %m/m (EN 14105)	—	≤0.80 %m/m (EN 14105 ASTM D6584 Other: ABNT NBR 15342)	Report (ABNT NBR 15342/ EN 14105)	≤0.80 %m/m (EN 14105)	≤0.80 %m/m proposed	≤0.80 %m/m (EN 14105)	—	≤0.80 %m/m
Diglyceride	≤0.20 %m/m (EN 14105)	—	≤0.20 %m/m (EN 14105 ASTM D6584 Other: ABNT NBR 15342)	Report (ABNT NBR 15342/ EN 14105)	≤0.20 %m/m (EN 14105)	—	≤0.20 %m/m (EN 14105)	—	≤0.20 %m/m
Triglyceride	≤0.20 %m/m (EN 14105)	—	≤0.20 %m/m (EN 14105 ASTM D6584 Other: ABNT NBR 15342)	Report (ABNT NBR 15342/ EN 14105)	≤0.20 %m/m (EN 14105)	—	≤0.20 %m/m (EN 14105)	—	≤0.20 %m/m
Free glycerol	≤0.02 %m/m (EN 14105/ EN 14106)	≤0.02 %m/m (ASTM D6584)	≤0.02 %m/m (EN 14105 ASTM D6584 Other: ABNT NBR 15341)	≤0.02 %m/m (ABNT NBR 15341/ EN 14105/ EN 14106)	≤0.02 %m/m (EN 14105,14106)	≤0.02 %m/m (ASTM D6584)	≤0.02 %m/m (EN 14105, EN 14106)	≤0.02 %m/m	≤0.02 %m/m
Total glycerol	≤0.25 %m/m (EN 14105)	≤0.24 %m/m (ASTM D6584)	≤0.25 %m/m (EN 14105 ASTM D6584 Other: ABNT NBR 15344)	≤0.38 %m/m (ABNT NBR 15344/EN 14105/ASTM D6584)	≤0.25 %m/m (EN 14105)	≤0.25 %m/m (ASTM D6584)	≤0.25 %m/m (EN 14105)	≤0.24 %m/m	≤0.25 %m/m
Metals (Na + K)	≤5 mg/kg (EN 14108/ EN 14109)	≤5 mg/kg (EN 14538)	≤5 mg/kg (EN 14108/ EN 14109, EN 14538)	10 mg/kg (EN 14108/ EN 14109)	≤5 mg/kg (EN 14108,14109)	Report by ICP-AES	≤5 mg/kg (EN 14108,14109)	—	≤5 mg/kg
Metals (Ca + Mg)	≤5 mg/kg (EN 14538)	≤5 mg/kg (EN 14538)	≤5 mg/kg (EN 14538)	Report (EN 14538)	≤5 mg/kg (EN 14538)	Report by ICP-AES	≤5 mg/kg (EN 14538)	—	≤5 mg/kg

(Continued)

TABLE 9.A2 (Continued)
Biodiesel Specifications

	EU (EN 14214)[a]	United States (ASTM D6751-07a)[b]	WWFC (B100 Guidelines)[c]	Brazil (ANP 42)[d]	Japan (JIS K 2390)[e]	South Africa (SANS 1935)[f]	India (IS 15607: 2005)[g]	Thailand (B100-FAME)[h]	Indonesia (SNI 04-7182-2006)[i]	Malaysia (Spec of Palm Methyl Esters)[j]
Phosphorous	≤10 mg/kg (EN 14107)	≤10 mg/kg (ASTM D4951)	≤4 mg/kg (EN 14107, ASTM D4951, ASTM D3231)	Report (EN 14107/ ASTM D4951)	≤10 mg/kg (EN 14107)	≤10 mg/kg (EN 14107)	≤10 mg/kg by ICP-AES	≤10 mg/kg	≤10 mg/kg	≤10 mg/kg
Polyunsaturated methyl esters	1 max (in development)	—	1 max (prEN 15779)	—		≤1 %m/m (—)	—	—	—	—
Distillation Temperature, 90% Recovered		≤360°C (ASTM D1160)	—	≤360°C (ASTM D1160)						
Cloud Point	Based on National Specifications (EN 23015)	Report (ASTM D2500)	—	—	—					
Pour point	—			—	(After blending) (—)	—(—)				

| CFPP | — | (5 max (Grade A) 0 max (Grade B) –5 max (Grade C) –10 max (Grade D) –15 max (Grade E) –20 max Grade F) (EN 116) | — | Based on National Specifications (ABNT NBR14747/ ASTM D6371) | (After blending) (—) | —(—) | — | — | — | — |

a   White Paper 2007.
b   White Paper 2007.
c   WWFC 2009a.
d   White Paper 2007.
e   Numata 2009.
f   Prins 2009.
g   Tyagi 2008.
h   APEC 2009.
i   APEC 2009, Novianto 2008.
j   APEC 2009.

TABLE 9.A3
Ethanol Specifications

Parameter	CEN Anhydrous (EN 15376)[a]	ASTM Anhydrous denatured (D4806)[b]	WWFC (E100 Guidelines)[c]	Brazil Anhydrous (RESOLUÇÃO N°36)[d]	Brazil Hydrated (RESOLUÇÃO N°36)[e]	Canada Anhydrous (CGSB- 3.511-93)[f]	Japan (JASO M 361)[g]	South Africa (SANS 465)[h]
Ethanol	—	—	—	≥99.6 %v/v	≥95.1 %v/v (ASTM D5501)	—	—(—)	≥92.1 %v/v
Ethanol and higher alcohols	≥98.7 %m/m (EC/2870/2000)	≥92.1 %v/v (ASTM D5501)	≥99.2 %m/m (EN 15721, ASTM D5501 Other: JAAS001–6.2)	≥99,3 %m/m (NBR 5992)	≥92.6/ 93.8* %m/m (ASTM D5501)	≥98.75 %m/m	≥99.5 %v/v (JAAS001, 6.2)	—
Higher alcohols C_3–C_5	≤2,0 %m/m (EC/2870/2000, EN 13132, EN 1601)	—	≤2.0 %m/m (EN 15721)	—	—	—	—	—
Hydrocarbons	—	—	—	≤3.0 %v/v (NBR 13993)	≤3.0 %v/v (NBR 13993)	—	—	—
Methanol	≤1.0 %m/m (EC/2870/2000, EN 13132, EN 1601)	≤0.5 %m/m (ASTM D1152)	≤0.5 %m/m (EN 15721, ASTM D5501)	—	—	—	≤4.0 g/L (JAAS001, 6.4)	≤0.5 %v/v
Existent gum content (solvent washed)	—	≤5.0 mg/100mL (ASTM D381)	—	—	≤5 mg/100mL (unwashed) (NBR 8664)	—	—	≤5.0 mg/100 mL (ASTM D381)
Water content	≤0.24 %m/m (EN 15489)	≤1 %v/v = 1.6 %m/m (ASTM D1193)	≤0.3 %m/m (EN 15489, ASTM E203, JIS K8101)	—	—	≤0.1 %m/m	≤0.70 %m/m (JIS K 8101)	≤1 %v/v
Denaturant content	Set by country	1.96–5.0 %v/v (ASTM D5580)	—	—	—	≥5 %v/v	—	≥1.96 %v/v

Property								
Inorganic chloride content	≤25 mg/L (EN 15484, EN 15492)	≤40 ppm (32 mg/l) (ASTM D512)	—	≤10 mg/L (EN 15484 or EN 15492, ASTM D7319, D7328 Other: ABNT NBR 10894/10895)	≤1 ppm (chloride ion) (NBR 10894, NBR 10895, ASTM D 512)	≤40 mg/kg Chlorine	—	≤40 mass ppm
Halogen	—	—	—	—	—	≤10 (chlorine)	—	
Copper content	≤0.1 mg/kg (EN 15488)	≤0.1 mg/kg (ASTM D1688)	—	≤0.100 mg/kg (EN 15488, ASTM D1688 modified, Method A, JIS K 0101 Other: ABNT NBR 10893)	≤0.07 mg/kg (NBR 10893)	≤0.1 mg/L	≤0.10 mg/kg (JIS K 0101, 51.2/3)	≤0.1 mg/kg
Iron content	—	—	—	Heavy metals: Nondetectable; no intentional addition (Other: ICP-AES)	≤5 mg/kg (NBR 11331)	—	—	—
Phosphorus	≤0.5 mg/L (EN 15487)	—	—	≤0.50 mg/L (EN 15487, ASTM D3231)	—	—	—	—
Sulfur	≤10.0 mg/kg (EN 15485, EN 15486)	≤30 ppm (ASTM D5453)	—	≤10.0 mg/kg (EN 15486, ASTM D5453 (<20 ppm), JIS K2541)	—	—	≤10 mg/kg (JIS K 2541-6/7)	≤30 mg/kg

(Continued)

TABLE 9.A3 (Continued)
Ethanol Specifications

Parameter	CEN Anhydrous (EN 15376)[a]	ASTM Anhydrous denatured (D4806)[b]	WWFC (E100 Guidelines)[c]	Brazil Anhydrous (RESOLUÇÃO N°36)[d]	Brazil Hydrated (RESOLUÇÃO N°36)[e]	Canada Anhydrous (CGSB- 3.511-93)[f]	Japan (JASO M 361)[g]	South Africa (SANS 465)[h]
Sulfate	—	≤4 ppm	≤4 mg/kg (EN 15492, ASTM D7318, D7319, D7328 Other: ABNT NBR 10894/12120)	—	≤4 mg/kg (NBR 10894/ NBR 12120)	—	—	—
Sodium content, maximum	—	—	—	—	≤2 mg/kg (NBR 10422)	—	—	—
Acidity (as acetic acid CH_3COOH)	≤0.007 %m/m (EN 15491)	≤0.007 %m/m (ASTM D1613)	≤0.007 %m/m (EN 15491, ASTM D1613 Other: ISO 1388/2; ABNT NBR 9866)	≤30 mg/L (NBR 9866/ ASTM D1613)	≤30 mg/L (NBR 9866/ASTM D1613)	≤30 mg/L (NBR 9866/ ASTM D1613)	≤0.0070 %m/m (ISO 1388/2)	≤0.007 %m/m
pHe	(dropped)	6.5–9.0 (ASTM D6423)	6.5–9 (ASTM D6423) and 6–8 (pHe-like) (EN 15490, JIS JASO M361-6.10 Other: ABNT NBR 10891)	—	6.0–8.0 (NBR 10891)	—	7.0 ± 1.0 (pHe) (ASTM D 6423)	6.5–9.0 (pHe)
Appearance	Clear and bright	Clear and bright	Clear and bright, no visible impurities (visual inspection)	Clear and impurity free	Clear and impurity free	—	Transparent, no turbidity (naked eye)	Visible free of suspended or precipitated contaminants (clear and bright)

Property								
Density at 20°C	—	—	≥789 kg/m^3 (ASTM D4052)	(807.6–811.0)* kg/m^3 (ASTM D4052)	≥791.5 kg/m^3 (ASTM D4052)	Report kg/m^3 (ASTM D 4052 Other: ABNT NBR 5992)	—	—
Electrical conductivity	—	≤500 µS/m (JIS K 0130)	—	≤500 µS/m (NBR 10547 ASTM D 1125)	≤500 µS/m (NBR 10547, ASTM D 1125)	≤500 µS/m (ASTM D 1125, JIS K0130 Other: ABNT NBR 10547)	—	—
Involatile material	—	≤5.0 mg/100 mL (JAAS 001, 6.3)	≤30 mg/L (at 100°C)	—	—	≤5 mg/100 mL (prEN 15691, ASTM D381, JIS JAAS001– 6.3 Other: ABNT NBR 8644)	—	≤190 mg/100mL (EC/2870/2000)
Organic impurity (except methanol)	—	≤10 g/L (JAAS001, 6.4)	—	—	—	≤10 mg/L (JAAS001, 6.4)	—	—

a   White Paper 2007.
b   White Paper 2007.
c   WWFC 2009b.
d   White Paper 2007.
e   White Paper 2007.
f   Costenoble 2006.
g   Numata 2009.
h   Prins 2009.

APPENDIX 3: MEASUREMENT METHODS AND STANDARDS

ASTM D1152, Standard Specification for Methanol (Methyl Alcohol)

ASTM D1160, Standard Test Method for Distillation of Petroleum Products at Reduced Pressure

ASTM D1193, Standard Specification for Reagent Water

ASTM D1266, Standard Test Method for Sulfur in Petroleum Products (Lamp Method)

ASTM D1298, Standard Test Method for Density, Relative Density (Specific Gravity), or API Gravity of Crude Petroleum and Liquid Petroleum Products by Hydrometer Method

ASTM D130, Standard Test Method for Detection of Copper Corrosion from Petroleum Products by the Copper Strip Tarnish Test

ASTM D1552, Standard Test Method for Sulfur in Petroleum Products (High-Temperature Method)

ASTM D1613, Standard Test Method for Acidity in Volatile Solvents and Chemical Intermediates Used in Paint, Varnish, Lacquer, and Related Products

ASTM D1688, Standard Test Methods for Copper in Water

ASTM D1796, Standard Test Method for Water and Sediment in Fuel Oils by the Centrifuge Method (Laboratory Procedure)

ASTM D189, Standard Test Method for Conradson Carbon Residue of Petroleum Products

ASTM D2274, Standard Test Method for Oxidation Stability of Distillate Fuel Oil (Accelerated Method)

ASTM D240, Standard Test Method for Heat of Combustion of Liquid Hydrocarbon Fuels by Bomb Calorimeter

ASTM D2500, Standard Test Method for Cloud Point of Petroleum Products

ASTM D2622, Standard Test Method for Sulfur in Petroleum Products by Wavelength Dispersive X-Ray Fluorescence Spectrometry

ASTM D2709, Standard Test Method for Water and Sediment in Middle Distillate Fuels by Centrifuge

ASTM D287, Standard Test Method for API Gravity of Crude Petroleum and Petroleum Products (Hydrometer Method)

ASTM D3117, Standard Test Method for Wax Appearance Point of Distillate Fuels

ASTM D3242, Standard Test Method for Acidity in Aviation Turbine Fuel

ASTM D381, Standard Test Method for Gum Content in Fuels by Jet Evaporation

ASTM D4052, Standard Test Method for Density and Relative Density of Liquids by Digital Density Meter

ASTM D4294, Standard Test Method for Sulfur in Petroleum and Petroleum Products by Energy Dispersive X-Ray Fluorescence Spectrometry

ASTM D445, Standard Test Method for Kinematic Viscosity of Transparent and Opaque Liquids (the Calculation of Dynamic Viscosity)

ASTM D4530, Standard Test Method for Determination of Carbon Residue (Micro Method)

ASTM D4737, Standard Test Method for Calculated Cetane Index by Four Variable Equation

ASTM D4806-03, Standard Specification for Denatured Fuel Ethanol for Blending with Gasolines for Use As Automotive Spark-Ignition Engine Fuel

ASTM D4806-06, Specification for Denatured Fuel Ethanol for Blending with Gasolines for Use As Automotive Spark-Ignition Engine Fuel

ASTM D4814-06, Specification for Automotive Spark-Ignition Engine Fuel

ASTM D4815, Standard Test Method for Determination of MTBE, ETBE, TAME, DIPE, Tertiary-Amyl Alcohol and C_1 and C_4 Alcohols in Gasoline by Gas Chromatography

ASTM D482, Standard Test Method for Ash from Petroleum Products

ASTM D4928, Standard Test Methods for Water in Crude Oils by Coulometric Karl Fischer Titration

ASTM D4951, Determination of Additive Elements in Lubricating Oils by Inductively Coupled Plasma Atomic Emission Spectrometry

ASTM D4951, Standard Test Method for Determination of Additive Elements in Lubricating Oils by Inductively Coupled Plasma Atomic Emission Spectrometry

ASTM D512, Standard Test Method for Chloride Ion in Water

ASTM D524, Standard Test Method for Rams Bottom Carbon Residue of Petroleum Products

ASTM D5452, Standard Test Method for Particulate Contamination in Aviation Fuels by Laboratory Filtration

ASTM D5453, Standard Test Method for Determination of Total Sulfur in Light Hydrocarbons, Motor Fuels, and Oils by Ultraviolet Fluorescence

ASTM D5501, Standard Test Method for Determination of Ethanol Content of Denatured Fuel Ethanol by Gas Chromatography

ASTM D5580, Standard Test Method for Determination of Benzene, Toluene, Ethylbenzene, p/m-Xylene, o-Xylene, C_9 and Heavier Aromatics, and Total Aromatics in Finished Gasoline by Gas Chromatography

ASTM D5798-99, Standard Specification for Fuel Ethanol (Ed75–Ed85) for Automotive Spark-Ignition Engines

ASTM D5863, Standard Test Methods for Determination of Nickel, Vanadium, Iron, and Sodium in Crude Oils and Residual Fuels by Flame Atomic Absorption Spectrometry

ASTM D5949, Standard Test Method for Pour Point of Petroleum Products (Automatic Pulsing Method)

ASTM D613, Standard Test Method for Cetane Number of Diesel Fuel Oil

ASTM D6217, Standard Test Method for Particulate Contamination in Middle Distillate Fuels by Laboratory Filtration

ASTM D6371, Standard Test Method for Cold Filter Plugging Point of Diesel and Heating Fuels

ASTM D6423, Standard Test Method for Determination of pHe of Ethanol, Denatured Fuel Ethanol, and Fuel Ethanol (Ed75–Ed85)

ASTM D6468, Standard Test Method for High Temperature Stability of Distillate Fuels

ASTM D6584, Standard Test Method for Determination of Free and Total Glycerine in B100 Biodiesel Methyl Esters by Gas Chromatography

ASTM D6585, Standard Specification for Unsintered Polytetrafluoroethylene (PTFE) Extruded Film or Tape

ASTM D664, Standard Test Method for Acid Number of Petroleum Products by Potentiometric Titration

ASTM D6731, Standard Test Method for Determining the Aerobic, Aquatic Biodegradability of Lubricants or Lubricant Components in a Closed Respirometer

ASTM D6751, Specification for Biodiesel Fuel Blend Stock (B100) for Middle Distillate Fuels

ASTM D7042, Standard Test Method for Dynamic Viscosity and Density of Liquids by Stabinger Viscometer (and the Calculation of Kinematic Viscosity)

ASTM D86, Standard Test Method for Distillation of Petroleum Products at Atmospheric Pressure

ASTM D874, Standard Test Method for Sulfated Ash from Lubricating Oils and Additives

ASTM D93, Standard Test Methods for Flash Point by Pensky–Martens Closed Cup Tester

ASTM D95, Standard Test Methods for Water in Petroleum Products and Bituminous Materials by Distillation

ASTM D97, Standard Test Method for Pour Point of Petroleum Products

ASTM D974, Standard Test Method for Acid and Base Number by Color Indicator Titration

ASTM D975-06, Specification for Diesel Fuel Oils

ASTM D976, Standard Test Methods for Calculated Cetane Index of Distillate Fuels

ASTM D1125, Standard Test Methods for Electrical Conductivity and Resistivity of Water

ASTM D1160, Standard Test Method for Distillation of Petroleum Products at Reduced Pressure

ASTM D1613, Standard Test Method for Acidity in Volatile Solvents and Chemical Intermediates Used in Paint, Varnish, Lacquer, and Related Products

ASTM D2622, Standard Test Method for Sulfur in Petroleum Products by Wavelength Dispersive X-Ray Fluorescence Spectrometry

ASTM D2709, Standard Test Method for Water and Sediment in Middle Distillate Fuels by Centrifuge

ASTM D381, Standard Test Method for Gum Content in Fuels by Jet Evaporation

ASTM D4530, Standard Test Method for Determination of Carbon Residue (Micro Method)

ASTM D5453, Standard Test Method for Determination of Total Sulfur in Light Hydrocarbons, Spark Ignition Engine Fuel, Diesel Engine Fuel, and Engine Oil by Ultraviolet Fluorescence

ASTM D5501 Standard Test Method for Determination of Ethanol Content of Denatured Fuel Ethanol by Gas Chromatography

ASTM D5797-96, Specification for Fuel Methanol (M70–M85) for Automotive Spark-Ignition Engines

ASTM D6423, Standard Test Method for Determination of pHe of Ethanol, Denatured Fuel Ethanol, and Fuel Ethanol (Ed75–Ed85)

ASTM D6571, Standard Test Method for Determination of Compression Resistance and Recovery Properties of Highloft Nonwoven Fabric Using Static Force Loading

ASTM D6584, Standard Test Method for Determination of Free and Total Glycerin in B100 Biodiesel Methyl Esters by Gas Chromatography

ASTM D7319, Standard Test Method for Determination of Total and Potential Sulfate and Inorganic Chloride in Fuel Ethanol by Direct Injection Suppressed Ion Chromatography

ASTM D7328, 07e1 Standard Test Method for Determination of Total and Potential Inorganic Sulfate and Total Inorganic Chloride in Fuel Ethanol by Ion Chromatography Using Aqueous Sample Injection

ASTM E203, Standard Test Method for Water Using Volumetric Karl Fisher Titration

CAN/CGSB 3.511-2005 AMEND. 2, Oxygenated Unleaded Automotive Gasoline Containing Ethanol (Incorporates Amendment 1)

CAN/CGSB 3.517-2007, Automotive Low Sulfur Diesel Fuel

CAN/CGSB 3.5-2004 AMEND. 2, Unleaded Automotive Gasoline (Incorporates Amendment 1)

CAN/CGSB 3.520-2005 AMEND. 2, Automotive Low-Sulphur Diesel Fuel Containing Low Levels of Biodiesel Esters (B1-B5) (Incorporates Amendment 1)

EC/2870/2000—Method I, Appendix II, Method B, Determination of Real Alcoholic Strength by Volume of Spirit Drinks—Measurement by Electronic Densimetry (Based on the Resonant Frequency Oscillation of a Sample in an Oscillation Cell, Commission Regulation (EC) No. 2870/2000 of 19 December 2000, Laying down Community Reference Methods for the Analysis of Spirit Drinks

EC/2870/2000—Method III, Determination of Volatile Substances and Methanol of Spirit Drinks, Commission Regulation (EC) No. 2870/2000 of 19 December 2000, Laying down Community Reference Methods for the Analysis of Spirit Drinks

EN 116, Diesel and Domestic Heating Fuels—Determination of Cold Filter Plugging Point

EN 12662, Liquid Petroleum Products—Determination of Contamination in Middle Distillates

EN 13132, Liquid Petroleum Products—Unleaded Petrol—Determination of Organic Oxygenate Compounds and Total Organically Bound Oxygen Content by Gas Chromatography Using Column Switching

EN 14103, Fat and Oil Derivatives—Fatty Acid Methyl Esters (FAME)—Determination of Ester and Linolenic Acid Methyl Ester Contents

EN 14104, Fat and Oil Derivatives—Fatty Acid Methyl Esters (FAME)—Determination of Acid Value

EN 14105, Fat and Oil Derivatives—Fatty Acid Methyl Esters (FAME)—Determination of Free and Total Glycerol and Mono-, Di-, Triglyceride Contents

EN 14106, Fat and Oil Derivatives—Fatty Acid Methyl Esters (FAME)—Determination of Free Glycerol Content

EN 14107, Fat and Oil Derivatives—Fatty Acid Methyl Esters (FAME)—Determination of Phosphorus Content by Inductively Coupled Plasma (ICP) Emission Spectrometry

EN 14108, Fat and Oil Derivatives—Fatty Acid Methyl Esters (FAME)—Determination of Sodium Content by Atomic Absorption Spectrometry

EN 14109, Fat and Oil Derivatives—Fatty Acid Methyl Esters (FAME)—Determination of Potassium Content by Atomic Absorption Spectrometry

EN 14110, Fat and Oil Derivatives—Fatty Acid Methyl Esters (FAME)—Determination of Phosphorus Content by Inductively Coupled Plasma (ICP) Emission Spectrometry

EN 14111, Fat and Oil Derivatives—Fatty Acid Methyl Esters (FAME)—Determination of Iodine Value

EN 14112, Fat and Oil Derivatives—Fatty Acid Methyl Esters (FAME)—Determination of Oxidation Stability (Accelerated Oxidation Test)

EN 14214, Automotive Fuels—Fatty Acid Methyl Esters (FAME) for Diesel Engines—Requirements and Test Methods

EN 14538, Fat and Oil Derivatives—Fatty Acid Methyl Esters (FAME)—Determination of Ca, K, Na and Mg Content by Optical Emission Spectral Analysis with Inductively Coupled Plasma (ICP OES)

EN 15376, Automotive Fuels—Ethanol As a Blending Component for Petrol—Requirements and Test Methods

EN 15484, Ethanol As a Blending Component for Petrol—Determination of Inorganic Chloride—Potentiometric Method

EN 15485, Ethanol As a Blending Component for Petrol—Determination of Sulphur Content —Wavelength Dispersive X-Ray Fluorescence Spectrometric Method

EN 15486, Ethanol As a Blending Component for Petrol—Determination of Sulphur Content—Ultraviolet Fluorescence Method

EN 15487, Ethanol As a Blending Component for Petrol—Determination of Phosphorus Content—Ammonium Molybdate Spectrometric Method

EN 15488, Ethanol As a Blending Component for Petrol. Determination of Copper Content. Graphite Furnace Atomic Absorption Spectrometric Method

EN 15489, Ethanol As a Blending Component for Petrol—Determination of Water Content—Karl-Fischer Coulometric Titration Method

EN 15490, Ethanol As a Blending Component for Petrol—Determination of pHe

EN 15491, Ethanol As a Blending Component for Petrol. Determination of Total Acidity. Color Indicator Titration Method

EN 15492, Ethanol As a Blending Component for Petrol—Determination of Inorganic Chloride and Sulfate Content—Ion Chromatographic

EN 1601, Liquid Petroleum Products—Unleaded Petrol—Determination of Organic Oxygenate Compounds and Total Oxygen Content by Gas Chromatography (O-FID)

EN 228, Automotive Fuels—Unleaded Petrol—Requirements and Test Methods

EN 589, Automotive Fuels—LPG—Requirements and Test Methods

EN 590, Automotive Fuels—Diesel—Requirements and Test Methods

EN ISO 2160, Petroleum Products. Corrosiveness to Copper. Copper Strip Test

IS 1460, Automotive Diesel Fuel—Specification

IS 15464, Anhydrous Ethanol for Use in Automotive Fuel—Specification

IS 15607, Bio-Diesel (B 100) Blend Stock for Diesel Fuel—Specification

IS 2796, Motor Gasolines—Specification (Third Revision)

ISO 10370, Petroleum Products. Determination of Carbon Residue. Micro Method

ISO 12156-1, Diesel Fuel—Assessment of Lubricity Using the High-Frequency Reciprocating Rig (HFRR)—Part 1: Test Method

ISO 12185, Crude Petroleum and Petroleum Products. Determination of Density. Oscillating U-tube Method

ISO 12205, Petroleum Products. Determination of the Oxidation Stability of Middle-Distillate Fuels

ISO 12662, Liquid Petroleum Products—Determination of Contamination in Middle Distillates

ISO 12937, Petroleum Products. Determination of Water. Coulometric Karl Fisher Titration Method

ISO 14596, Petroleum Products—Determination of Sulphur Content—Wavelength Dispersive X-Ray Fluorescence Spectrometry

ISO 20846, Petroleum Products. Determination of Sulphur Content of Automotive Fuels. Ultraviolet Fluorescence Method

ISO 20884, Petroleum Products. Determination of Sulphur Content of Automotive Fuels. Wavelength—Dispersive X-Ray Fluorescence Spectrometry

ISO 2160, Petroleum Products. Corrosiveness to Copper. Copper Strip Test

ISO 2719, Determination of Flash Point—Pensky-Martens Closed Cup Method

ISO 3015, Petroleum Products—Determination of Cloud Point

ISO 3104, Petroleum Products. Transparent and Opaque Liquids. Determination of Kinematic Viscosity and Calculation of Dynamic Viscosity

ISO 3405, Petroleum Products. Determination of Distillation Characteristics at Atmospheric Pressure

ISO 3675, Crude Petroleum and Liquid Petroleum Products. Laboratory Determination of Density. Hydrometer Method

ISO 3679, Petroleum Products—Determination of Flash Point—Rapid Equilibrium Closed Cup

ISO 3733, Petroleum Products and Bituminous Materials—Determination of Water—Distillation Method

ISO 3987, Petroleum Products. Lubricating Oils and Additives. Determination of Sulfated Ash

ISO 4260, Petroleum Products and Hydrocarbons. Determination of Sulphur Content. Wickbold Combustion Method

ISO 4264, Petroleum Products. Calculation of Cetane Index of Middle. Distillate Fuels by the Four-Variable Equation

ISO 5165, Petroleum Products. Determination of the Ignition Quality of Diesel Fuels. Cetane Engine Method

ISO 6245, Petroleum Products—Determination of Ash

ISO 6296, Petroleum Products—Determination of Water—Potentiometric Karl Fischer Titration Method

ISO 6618, Petroleum Products and Lubricants. Determination of Acid or Base Number. Colour-Indicator Titration Method

ISO 8754, Petroleum Products. Determination of Sulphur Content. Energy-Dispersive X-Ray Fluorescence Spectrometry

JIS K 0070, Test Methods for Acid Value, Saponification Value, Ester Value, Iodine Value, Hydroxyl Value, and Unsaponifiable Matter of Chemical Products

JIS K 2202, Motor Gasoline

JIS K 2204, Diesel Fuel

JIS K 2249, Crude Petroleum and Petroleum Products—Determination of Density and Petroleum Measurement Tables Based on a Reference Temperature (15 Centigrade Degrees)

JIS K 2265, Crude Oil and Petroleum Products—Determination of Flash Point

JIS K 2270, Crude Petroleum and Petroleum Products—Determination of Carbon Residue

JIS K 2272, Crude Oil and Petroleum Products—Determination of Ash and Sulfated Ash

JIS K 2275, Crude Oil and Petroleum Products—Determination of Water Content

JIS K 2280, Petroleum Products—Fuels—Determination of Octane Number, Cetane Number and Calculation of Cetane Index

JIS K 2283, Crude Petroleum and Petroleum Products—Determination of Kinematic Viscosity and Calculation of Viscosity Index from Kinematic Viscosity

JIS K 2390, Automotive Fuels—Fatty Acid Methyl Ester (FAME) As Blend Stock

JIS K 2501, Petroleum Products and Lubricants—Determination of Neutralization Number

JIS K 2513, Petroleum Products—Corrosiveness to Copper—Copper Strip Test

JIS K 2536, Liquid Petroleum Products—Testing Method of Components

JIS K 2541-1, Crude Oil and Petroleum Products—Determination of Sulfur Content Part 1: Wickbold Combustion Method

JIS K 2541-2, Crude Oil and Petroleum Products—Determination of Sulfur Content Part 2: Oxidative Microcoulometry

JIS K 2541-6, Crude Oil and Petroleum Products—Determination of Sulfur Content Part 6: Ultraviolet Fluorescence Method

NBR 10422, Álcool Etílico—Determinação do Teor de Sódio por Fotometria de Chama

NBR 10441, Produtos de Petróleo—Líquidos Transparentes e Opacos—Determinação da Viscosidade Cinemática e Cálculo da Viscosidade Dinâmica

NBR 10547, Álcool Etílico—Determinação da Condutividade Elétrica

NBR 10891, Álcool Etílico Hidratado—Determinação do pH

NBR 10893, Álcool Etílico - Determinação do Teor Do Cobre Por Espectrofotometria De Absorção Atômica

NBR 10894, Álcool Etílico—Determinação dos Ions Cloreto e Sulfato por Cromatografia Iônica

NBR 10895, Álcool Etílico—Determinação do Teor de Ion Cloreto por Técnica Potenciométrica

NBR 11331, Álcool Etílico—Determinação do Teor de Ferro por Espectrofotometria de Absorção Atômica

NBR 12120, Álcool Etílico—Determinação do Teor de Sulfato por Volumetria

NBR 13993, Álcool Etílico—Determinação do Teor de Hidrocarbonetos

NBR 14065, Destilados de Petróleo e Óleos Viscosos—Determinação da Massa Específica e da Densidade Relativa Pelo Densímetro Digital

NBR 14359, Produtos de Petróleo – Determinação da Corrosividade—Método da Lâmina de Cobre

NBR 14448, Produtos de Petróleo—Determinação do Indice de Acidez Pelo Método de Titulação Potenciométrica

NBR 14598, Produtos de Petróleo—Determinação do Ponto de Fulgor Pelo Aparelho de Vaso Fechado Pensky-Martens

NBR 14747, Óleo Diesel—Determinação do Ponto de Entupimento de Filtro a Frio

NBR 15344, Determinação de Glicerina Total e do Teor de Triglicerídeos em Biodiesel de Mamona

NBR 5992, Determinação da Massa Específica e do Teor Alcoólico do Álcool Etílico e Suas Misturas com Água

NBR 7148, Petróleo e Produtos de Petróleo—Determinação da Massa Específica, Densidade Relativa e API—Método do Densímetro

NBR 8644, Álcool Etílico Combustível—Determinação do Resíduo por Evaporação

NBR 9842, Produtos de Petróleo—Determinação do Teor de Cinzas

NBR 9866, Álcool Etílico—Verificação Da Alcalinidade e Determinação da Acidez Total

Notification of the Department of Energy Business (DOEB) on Characteristic and Quality of Diesel (No.3) B.E. 2548(2005). 2005. Department of Energy Business (DOEB), Ministry of Energy

Notification of the Department of Energy Business on Characteristic and Quality of Basic Gasoline B.E. 2548(2005). 2005. Department of Energy Business (DOEB), Ministry of Energy

Notification of the Department of Energy Business on Characteristic and Quality of Biodiesel—Fatty Acid Methyl Ester B.E. 2548(2005). 2005. Department of Energy Business (DOEB), Ministry of Energy

Notification of the Department of Energy Business on Characteristic and Quality of Gasohol (No. 4) B.E. 2548(2005). 2005. Department of Energy Business (DOEB), Ministry of Energy

Resolução Nº 36:2005—Regulamento Técnico Anp Nº 7/2005, Especificações Do Álcool Etílico Anidro Combustível (Aeac) E Do Álcool Etílico Hidratado Combustível (AEHC)

Resolução Nº 42:2004—Regulamento Técnico Anp Nº 4/2004, Especificaçõe De Biodiesel

SANS 1598, Unleaded Petrol

SANS 1935, Automotive Biodiesel Fuel

SANS 342, Automotive Diesel Fuel

SANS 465, Standard Specification for Denatured Fuel Ethanol for Blending with Gasolines for Use As Automotive Spark-Ignition Engine Fuel

REFERENCES

Alberta Renewable Diesel Demonstration (ARDD) (2009) *Final Report*. Climate Change Central, Alberta, Canada

Araujo M (2009) The regulation of biofuels quality in Brazil and the developments regarding global fuel specifications. In: *Proceedings of the 2nd International Conference on Biofuels Standards and Measurements for Biofuels: Facilitating Global Trade*, Brussels, Belgium, March 20, 2009

Asia-Pacific Economic Cooperation (APEC) (2009) *Establishment of the Guidelines for the Development of Biodiesel Standards in the APEC Region*. APEC Secretariat, Singapore

Bacha J, Freel J, Gibbs A (2007) *Diesel Fuels Technical Review*. Chevron Corporation, San Ramon, CA

BiofuelsDigest (2008) The top 10 biofuels stories of 2008: #9, India sets a 5 percent biofuels mandate, raises to 10 Percent, misses the First Target Badly, available at http://goo.gl/IWhOC

Chevron (2008) *Motor Gasolines Technical Review*. Chevron Corporation, San Ramon, CA, available at http://www.chevron.com/products/ourfuels/prodserv/fuels/documents/Motor_Fuels_Tch_Rvw_complete.pdf

Combs S (2008) *The Energy Report*. Texas Comptroller of Public Accounts, Austin, TX, available at http://www.window.state.tx.us/specialrpt/energy/pdf/14-Biodiesel.pdf

Commission of the European Communities (COM) (2008) Proposal for a Directive of the European Parliament and of the Council on the Promotion of the Use of Energy from Renewable Sources. Commission of the European Communities, Brussels, Belgium, available at http://ec.europa.eu/energy/climate_actions/doc/2008_res_directive_en.pdf

Costenoble O (2006) *Worldwide Fuels Standards: Overview of Specifications and Regulations on (Bio)fuels*. NEN—The Netherlands Standardization Institute. Delft, The Netherlands

Crown L, Warfield L (2009) *Uniform Laws and Regulations in the Areas of Legal Metrology and Engine Fuel Quality*. National Institute of Standards and Technology Weights and Measures Division, Washington, DC

Cvengros J (1998) Acidity and corrosiveness of methyl esters of vegetable oils. *Fett/Lipid* 100:41–44

De Klerk A (2008) Fischer-Tropsch refining. PhD Thesis, University of Pretoria, South Africa

Department of the Environment, Water, Heritage, and the Arts (DEWHA) (2008) *Setting National Fuel Quality Standards—Proposed Management of Diesel/Biodiesel Blends—Position Paper*. Australian Government, Department of the Environment, Water, Heritage and the Arts, Canberra, Australia

ebio (2006) ANNEX IX specifications of reference fuel. Paper presented to the Motor Vehicle Emissions Group—Subgroup on Euro 5 and Euro 6 6th Meeting

European Biodiesel Board (EBB) (2008) *EBB European Biodiesel Quality Report (EBBQR)—Results of the Third Round of Tests (Winter 2007/2008)*. European Biodiesel Board, Brussels, Belgium, available at http://goo.gl/JeRDP

European Committee for Standardization (CEN) (2008) *CEN Compass*. The CEN Management Centre, Brussels, Belgium, available at ftp://ftp.cen.eu/cen/AboutUs/Publications/Compass.pdf

Foon CS, May CY, Liand YC, Ngan MA, Basiron Y (2005) Palm biodiesel: Gearing towards Malaysian biodiesel standards. *Palm Oil Develop* 42:28–34

Gray N (2005) *Setting National Fuel Quality Standards—Proposed Fuel Quality Standard for Fuel Grade Ethanol—Australian Government Position*. Australian Government, Department of the Environment, Water, Heritage, and the Arts, Canberra, Australia

International Fuel Quality Center (IFQC) (2004) Setting a quality standard for fuel ethanol report. Paper presented to the Department of the Environment and Heritage, International Fuel Quality Center, Houston, TX

Khan KA (2002) Research into biodiesel kinetic and catalytic development. PhD Thesis, University of Queensland, Australia

Kinast JA (2003) *Production of Biodiesels from Multiple Feedstocks and Properties of Biodiesels and Biodiesel/ Diesel Blends—Final Report (Report 1 of 6)*. National Renewable Energy Laboratory, Oak Ridge, TN

Knothe G, Dunn RO (1998) Recent results from biodiesel research at the national center for agricultural utilization research. Landbauforschung Völkenrode. Sonderheft 190 (Biodiesel—Optimierungspotentiale und Umwelteffekte): pp 69–78

Knothe G, Matheaus AC, Ryan III TW (2003) Cetane numbers of branched and straight-chain fatty esters determined in an ignition quality tester. *Fuel* 82:971–975

Kobmehl SO, Heinrich H (1997) The automotive industry's view on the standards for plant oil-based fuels. In: Martini N, Schell J (eds), *Plant Oils as Fuels. Present State of Science and Future Developments*. Springer. Berlin, Germany, pp 18–28

Körbitza W, Friedricha S, Wagingerb E, Wörgetter M (2003) Worldwide review on biodiesel production. Prepared by Austrian Biofuels Institute for IEA Bioenergy Task 39, Subtask Biodiesel

Lacoste F, Lagardere L (2003) Quality parameters evolution during biodiesel oxidation using Rancimat test. *Eur J Lipid Sci Technol* 105:149–155

Lee R, Hobbs CH, Pedley JF (1998) *Fuel Quality Impact on Heavy Duty Diesel Emissions: A Literature Review*. SAE International, Warrendale, PA

Maniatis K, May W, Brandi H (2009) Results of the White Paper and future prospects. Paper presented to the 2nd International Conference on Biofuels Standards and Measurements for Biofuels: Facilitating Global Trade, Brussels, Belgium, March 20, 2009

McCormick RL (2005) Effects of biodiesel on NO$_x$ emissions. Paper presented to the ARB Biodiesel Workgroup, Sacramento, CA, June 8, 2005

McGill R, Aakko-Saksa P, Nylund NO (2008) *Final Report, Annex XXXIV: Biomass-Derived Diesel Fuels, Task 1: Analysis of Biodiesel Options*. IEA Advanced Motor Fuels Implementing Agreement, Wieselburg, Austria

Mittelbach M (1996) Diesel fuel derived from vegetable oils, VI: Specifications and quality control of biodiesel. *Bioresour Technol* 56:7–11

Mittelbach M (2000) *Chemische und Motortechnische Untersuchungen der Ursachen der Einspritzpumpenverklebung bei Biodieselbetrieb*. Bund-Bundesländer-kooperationsprojekt, Graz, Austria

Mittelbach M, Gangl S (2001) Long storage stability of biodiesel made from rapeseed and used frying oil. *J Amer Oil Chem Soc* 78:573–577

Mittelbach M, Pokits B, Silberholz A (1992) Diesel fuel derived from vegetable oils, IV: production and fuel properties of fatty acid methyl esters from used frying oil. In: *Liquid Fuels from Renewable Resources, Proceedings Alternative Energy Conference*, American Society of Agricultural Engineers, St. Joseph, MI, p 74

Mittelbach M, Schober S (2003) The influence of antioxidants on the oxidation stability of biodiesel. *J Amer Oil Chem Soc* 80:817–823

Mittelbach M, Wörgetter M, Pernkopf J, Junek H (1983) Diesel fuel derived from vegetable oils: Preparation and use of rape oil methyl ester. *Energy Agric* 2:369–384

National Renewable Energy Laboratory (NREL) (2006) *Biodiesel Handling and Use Guidelines*, 3rd ed. U.S. Department of Energy, Golden, CO, available at http://www.nrel.gov/vehiclesandfuels/npbf/pdfs/40555. pdf

Novianto A (2008) Biofuel development: Indonesia. Paper presented at the APEC Workshop on Establishment of the Guidelines for the Development of Biodiesel Standards in the APEC Region, Chinese Taipei, July 16–18, 2008

Numata M (2009) Prospects of biofuels in Japan, and preliminary studies on biofuel CRMs by the National Metrology Institute of Japan (NMIJ). Paper presented to the 2nd International Conference on Biofuels Standards and Measurements for Biofuels: Facilitating Global Trade. Brussels, Belgium, March 19, 2009

Nylund NO, Aakko-Saksa P, Sipilä K (2008) *Status and Outlook for Biofuels, Other Alternative Fuels and New Vehicles*. VTT Technical Research Centre of Finland, Espoo, Finland

Prankl H, Körbitz W, Mittelbach M, Wörgetter M (2004) *Review on Biodiesel Standardization World-wide*. Prepared by Bundesanstalt für Landtechnik for IEA Bioenergy Task 39, Subtask "Biodiesel"

Prankl H, Wörgetter M (2000) *The Introduction of Biodiesel as a Blending Component to Diesel Fuel in Austria*. BLT Wieselburg, Wieselburg, Austria

Prankl H, Wörgetter M, Rathbauer J (1999) Technical performance of vegetable oil methyl esters with a high iodine number. Paper presented at the 4th Biomass Conference of the Americas, Oakland, CA, August 29, 1999

Prins S (2009) Biofuels in South Africa and the tripartite White Paper. Paper presented at the 2nd International Conference on Biofuels Standards and Measurements for Biofuels: Facilitating Global Trade, Brussels, Belgium, March 19, 2009

Rehnlund B (2008) *Outlook on Standardization of Alternative Vehicle Fuels: Global, Regional and National Level.* Prepared by: Atrax Energi AB, Sweden for IEA Advanced Motor Fuels Implementing Agreement

Renewable Fuels Association (RFA) Technical Committee (2009) *E 85 Fuel Ethanol Industry Guidelines, Specifications and Procedures.* Renewable Fuels Association, Washington DC

Rilett J, Gagnon A (2008) *Renewable Diesel Characterization Study.* Climate Change Central. Calgary, Alberta, Canada

Saunders B (2009) CEN biofuels specification. Paper presented to the 2nd International Conference on Biofuels Standards and Measurements for Biofuels: Facilitating Global Trade, Brussels, Belgium, March 19, 2009

Schäfer A, Naber D, Gairing M (1997) Biodiesel als alternativer Kraftstoff für Mercedes-Benz-Dieselmotoren. *Mineralöltechnik* 43:1–32

Simkovsky N (1997) Oxidationsstabilität von fettsäuremethylestern. PhD Thesis, Vienna University for Technology

Singh R (2005) Literature review on biodiesel. The South Pacific Geoscience Commission, Suva, Fiji Islands

Soriano NU (2008) Evaluation of biodiesel derived from Camelina sativa oil. Paper presented at the 99th American Oil Chemists Society Meeting, Seattle WA, May 18–21, 2008

Tyagi OS (2008) Biodiesel situation in India. Paper presented to the APEC Workshop on Establishment of the Guidelines for the Development of Biodiesel Standards in the APEC Region. Chinese Taipei, July 16–18, 2008

van Walwijk M (2005) *Biofuels in France 1990–2005.* PREMIA report, Paris, France

White Paper (2007) *White Paper on Internationally Compatible Biofuel Standards.* Tripartite Task Force Brazil, European Union and United States of America

Wörgetter M, Prankl H, Rathbauer J (1998) Eigenschaften von Biodiesel. Landbauforschung Völkenrode. Sonderheft 190 (Biodiesel- Optimierungspotentiale und Umwelteffekte): pp 31–43

World-Wide Fuel Charter (WWFC) (2006) World-Wide Fuel Charter. Published by ACEA, Alliance, EMA and JAMA. European Automobile Manufacturers Association, available at http://www.autoalliance.org/index.cfm?objectid = 96BB28DD-1D09-317F-BB80B643C3B8E837

World-Wide Fuel Charter (WWFC) (2008) World-Wide Fuel Charter guidelines for ethanol. Published by ACEA, Alliance, EMA and JAMA. European Automobile Manufacturers Association, available at http://www.acea.be

World-Wide Fuel Charter (WWFC) (2009a) Biodiesel guidelines. Published by ACEA, Alliance, EMA and JAMA. European Automobile Manufacturers Association, available at http://www.acea.be

World-Wide Fuel Charter (WWFC) (2009b) Ethanol guidelines. Published by ACEA, Alliance, EMA and JAMA. European Automobile Manufacturers Association, available at http://www.acea.be

Zhang Y, Dube MA, McLean DD, Kates M (2003) Biodiesel production from waste cooking oil: Process design and technological assessment. *Bioresour Technol* 89:1–16

WEBSITES

Agency of Petroleum, Natural Gas, and Biofuels (ANP). Biodiesel webpage, available at http://www.anp.gov.br/biocombustiveis/biodiesel.asp

Agency of Petroleum, Natural Gas, and Biofuels (ANP). Homepage, available at http://www.anp.gov.br/index.asp

American Society for Testing and Materials (ASTM) International. About ASTM International webpage., available at http://www.astm.org/ABOUT/aboutASTM.html

American Society for Testing and Materials (ASTM) International. History of ASTM International webpage., available at http://www.astm.org/HISTORY/index.html

American Society for Testing and Materials (ASTM) International. Mission of ASTM International webpage., available at http://www.astm.org/NEWS/Mission2.html

Asia-Pacific Economic Cooperation (APEC). Homepage, available at http://www.apec.org/apec/about_apec.html

Asia-Pacific Economic Cooperation (APEC). Energy Working Group webpage, available at http://www.ewg.apec.org/

BDpedia Biodiesel WWW Encyclopedia. Plant oils used for bio-diesel, available at http://www.bdpedia.com/biodiesel/plant_oils/plant_oils.html (accessed February 2, 2009)

Brazilian Association for Technical Standards (ABNT). Homepage, available at http://goo.gl/STWp6

Bureau of Indian Standards (BIS). Homepage, available at http://www.bis.org.in/org/obj.htm

Canadian General Standards Board (CGSB). Homepage, available at http://www.tpsgc-pwgsc.gc.ca/ongc/home/index-e.html

EurActiv. The review of the EU's Fuel Quality Directive, available at http://goo.gl/SRKoI

European Committee for Standardization (CEN). About us webpage, available at http://www.cen.eu/cenorm/aboutus/index.asp

European Committee for Standardization (CEN). Statistics webpage, available at http://goo.gl/Hkldt

European Committee for Standardization (CEN). Guidance—Characteristics of the CEN Workshop Agreement and CEN Workshop Guidelines, available at http://goo.gl/RIpqI.

International Organization for Standardization (ISO) (2004) ISO Strategic Plan 2005-2010: Standards for a sustainable world. ISO Central Secretariat, Geneva, Switzerland, available at http://www.iso.org/iso/isostrategies_2004-en.pdf

International Organization for Standardization (ISO) Homepage, available at http://www.iso.org/iso/home.htm

Japanese Industrial Standards Committee (JISC) Homepage, available at http://www.jisc.go.jp/eng/

Japanese Standards Association (JSA). History and organization webpage, available at http://www.jsa.or.jp/eng/about/about.asp

National Institute of Metrology, Standardization and industrial quality (Inmetro). About Inmetro webpage, available at http://www.inmetro.gov.br/english/institucional/index.asp

South African Bureau of Standards (SABS). About us webpage, available at http://goo.gl/dwgtK

Thai Industrial Standards Institute (TISI). Homepage, available at http://www.tisi.go.th/eng/tisi.html

10 Biofuel Use from Bioenergy Crops
Internal Combustion Engines in Transportation

Jaclyn E. Johnson, Jeremy Worm,
Scott Miers, and Jeffrey Naber
Michigan Technological University

CONTENTS

10.1 POWERTRAIN—PROPULSION SYSTEMS

Internal combustion (IC) engines are inherently linked to the specific physical, thermodynamic, and chemical properties of the fuel and the effects of these properties on fuel–air preparation and mixing, combustion initiation, combustion rates, combustion anomalies, and emissions formation. Throughout history, different engine technologies have been developed and used to operate with different fuels (Cummins 1989), including the 1908 Ford Model T, which operated on gasoline, ethanol, and their blends.

In the United States, spark-ignited engines fueled with gasoline (gasoline engines) prevail as the primary mover in light-duty vehicles (LDVs), representing 95% of the total LDVs in the United States in 2008 (EIA 2009a). Spark-ignition (SI) engines are able to meet the most stringent emissions standards, including U.S. Tier II Bin 2 and California partial zero emissions vehicle (PZEV) standards, but they have a lower thermal efficiency with a peak at approximately 35% (U.S. DOE 2010a) in comparison to their counterpart, compression-ignition (CI) engines. CI engines fueled by diesel fuel (diesel engines) are found in a high percentage of medium-duty vehicles (MDVs) and dominate heavy-duty vehicle (HDV) and heavy equipment applications. See

Table 10.1 for definitions of vehicle classifications for light-duty applications (trucks and passenger vehicles) and heavy-duty engines (U.S. DOT RITA 2010). Diesel engines are characterized as having a higher efficiency, with a peak of 45% (U.S. DOE 2010a), but are hampered by high engine-out particulate matter (PM) and oxides of nitrogen ($NO + NO_2 = NO_x$) emissions, which require complex exhaust aftertreatment systems to meet emissions standards. There are many places that energy is "lost" from the vehicle as an automobile travels down the road. A summary of these losses, as a percentage of the fuel's total energy, is shown in Figure 10.1 (U.S. DOE and EPA 2008), where it is seen that the most losses originate from the engine. Current LDV hybrid electric vehicles (HEVs) (e.g., Toyota Prius, and hybrid versions of the Ford Escape and Chevy Tahoe) include on-board storage of electrical energy and electric drives, yet still derive nearly all required energy for vehicle operation from the combustion of fuel in the on-board IC engine. However, HEVs are designed to reduce the losses in several of the areas shown in Figure 10.1. HEVs achieve reductions in energy consumption by permitting the vehicle system to manage power requirements such that the IC engine is operated near its peak efficiency more frequently than a nonhybrid powertrain (targeting the engine losses), including allowing the engine to be shut off when not needed on decelerations and when the vehicle is stopped (targeting the idle losses). Additionally, hybrid electric systems are able to recapture kinetic and potential energy when a vehicle is stopping or descending a hill (targeting the braking losses). This energy, which would otherwise be lost to heat in the friction brakes,

TABLE 10.1
Vehicle Categories Used in Standards

	Gross Vehicle Weight Rating (GVWR)		
Category	Passenger Vehicle Classifications	Light-Duty Truck Classifications	Heavy-Duty Engine Classifications
Light	<8500 lb	<6000 lb	8500 to 19,500 lb
Medium	8500–10,000 lbs		19,500 to 33,000 lb
Heavy		6000–8500 lb	>33,000 lb

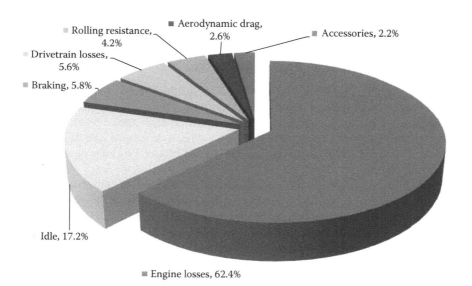

FIGURE 10.1 The losses in a typical LDV shown as a percentage of the fuel's total energy, assuming an SI engine. (From U.S. DOE and EPA, Advanced technologies & energy efficiency, 2008.)

can be used to assist in propelling the vehicle at a later time. Together, these effects result in fuel consumption reductions on the order of 30% (Jones 2008).

Electrified vehicles, including "plug-in" hybrids, "extended range" hybrids (e.g., GM's Chevrolet Volt), and full electric vehicles (e.g., Nissan's Leaf), can use electrical energy from the municipal power grid for vehicle propulsion, directly displacing energy produced from the liquid-fueled IC engine. However, despite high mile per gallon (MPG) ratings based on fuel from the tank, the overall energy conversion process from source to wheels must still be considered. When analyzing the electrical generation processes, it is found that nearly 70% of the source energy is wasted (EIA 2006a), and furthermore, 71% of the U.S. electrical power production is from fossil fuels (EIA 2006b). Considering this, it becomes clear that even when electricity is used as the primary source for power, these vehicles are still contributing to the depletion of fossil fuels and the increase of atmospheric carbon dioxide (CO_2). However, as the penetration of renewable sources of electricity production increases, including raw biomass, biomass-based syngas, hydroelectric, solar, and wind, these vehicles will certainly show added benefit. Nevertheless, debate will likely remain as to whether electrified vehicles running on electricity generated from renewable sources or conventional vehicles running on renewable liquid fuels is the optimal long-term solution.

10.1.1 U.S. AND WORLD TRANSPORTATION FUEL USAGE

Transportation fuel usage continues to increase worldwide. Transportation accounts for almost 30% of the total global energy delivered in 2007, making up more than 50% of global liquid fuel consumption (EIA 2010). Of the 97.9 quadrillion (10^{15}) Btu (103 quadrillion kilojoules) of energy consumed for transportation in the world, the United States uses nearly 30% of this (EIA 2010). To put this into perspective, if the petroleum used for transportation in the United States in 1 day was placed in 55-gallon oil drums, and these drums were placed next to each other, they would form a line from New York to Los Angeles, passing through Detroit and Houston (Figure 10.2). Petroleum makes up the largest portion of transportation energy consumption, providing nearly

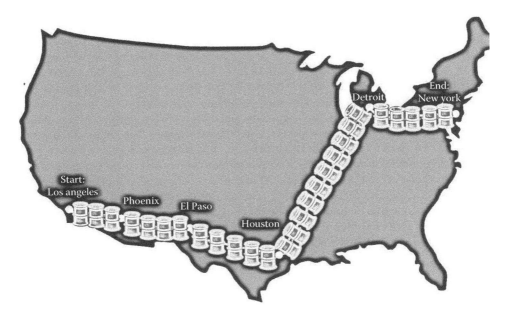

FIGURE 10.2 Schematic representation of fuel used for transportation in the United States in one day as barrels of fuel lined up end to end across the nation.

94% of transportation energy in 2009, with only 3.4% of transportation energy consumption being provided by renewables (Davis et al. 2010). This current dependence on petroleum is detrimental to the U.S. economy and national security, especially as shortages and high oil prices become increasingly prevalent (EIA 2009b). In 2009, renewable energy represented only 8% of the total U.S. energy consumption, with biofuels composing 20% of this renewable energy use (EIA 2009b). Efforts are in place, including the U.S. Renewable Fuel Standard (RFS) 2, which requires 36 billion gallons of renewable fuels to be used for transportation by 2022. (U.S. EPA 2007a, 2010a) This will promote the development and incorporation of renewable fuels into the transportation sector to overcome the current low percentage of consumption, assist with greenhouse gas reduction, and decrease reliance on imported petroleum. However, biofuel and engine operational challenges must be understood and overcome to ensure the success of renewable transportation fuels.

Vehicle and engine original equipment manufacturers (OEMs) have developed engines to use ethanol–gasoline and methyl–ester biodiesel–petroleum diesel blends. These two biofuels have received significant attention and development in the United States. As of 2010, nearly 8 million E85 (85% volume ethanal, 15% volume gasoline) ethanol flex-fuel vehicles (FFVs) were estimated to be on the road in the United States (U.S. DOE 2010b), and automotive manufacturers have committed to making one half of their vehicles ethanol flex-fuel capable by 2012 (GM 2007). These vehicles are able to utilize gasoline–ethanol blends in the range of 0 to 85% ethanol by volume. Additionally, 10% ethanol is approved for blending with gasoline for use in vehicles with conventional SI gasoline engines, with higher blend ratios likely being approved soon. Similarly, the use of biodiesel is approved at levels of 5–20% concentrations by many of the diesel engine and vehicle OEMs (Cummins 2007). Standards are placed upon these fuels to ensure they do not significantly deteriorate the IC engine performance or emissions (ASTM D7467-09a and ASTM D6751-10). However, other fuels, including Fisher–Tropsch (green or synthetic) diesel (Goodger 1975), dimethyl ether (DME) (Silva-Petrobras 2006), methanol, butanol, biogas [a mixture of methane (CH_4) and CO_2 produced via anaerobic digestion of biodegradable matter (IEA 2005)], and hydrogen (Naber and Siebers 1998) are also notable alternatives for transportation fuels for use in IC engines (SAE 2007) and other power generation systems including fuel cells.

10.1.2 BIOFUEL CHARACTERISTICS

Combustion fuels are primarily composed of the atomic elements hydrogen, carbon, and oxygen, although trace amounts of other elements, including sulfur and nitrogen, can also be present (Goodger 1975). These fuels are suitable for SI or CI engines, depending on the fuel characteristics, with varying requirements for modification of the fuel, engine, control, and aftertreatment systems. SI and CI fuels are subject to detailed specifications (ASTM D4814-10a–SI Fuels, ASTM D975-10a–CI fuels) to ensure they meet minimum requirements for operation of modern IC engines. Petroleum hydrocarbon fuels (gasoline and diesel) are a complex mixture of hundreds of compounds of different molecular structures and atomic weights. The composition is made up of straight-chained paraffins, cycloparaffins, alkenes, and aromatics. Ranges of atomic weights result in a nearly continuous distillation curve. Gasoline is composed of C_4–C_{14} molecules with a 50% distillation point of 99°C (Totten et al. 2003). Diesel is composed of higher-molecular-weight compounds (C_7–C_{24}) with a 50% distillation point of 256°C (Totten et al. 2003). Ratios of the molecular components affect autoignition (Taylor et al. 2004), combustion, and emissions of soot (Svensson 2005). Limits are placed on aromatics in fuels because they decrease the hydrogen-to-carbon ratio and result in high soot emissions (ASTM D975-10a). Regulations as documented in the standards also require reduced sulfur in the fuel to enable and improve performance of advanced exhaust aftertreatment systems.

Table 10.2 lists characteristics of petroleum-based fuels and biofuels along with their applications. Included are the hydrogen- (column A) and oxygen (B)- to carbon ratios, mass density (C), and specific energy and specific CO_2 emissions. As can be seen in Table 10.2, oxygenated fuels have a lower energy density. For example, the energy density of gasoline is 44 MJ/kg, whereas that of

TABLE 10.2

Transportation Fuel Properties Including Specific Energies and CO_2 Production

		A	B	C	D	E	F	G	H	I
Application	Fuel	H/C (–)	O/C (–)	Density (kg/m³)	Energy Unit Mass (MJ/kg)	Energy Unit Volume (–)	CO_2/Fuel (g_{CO_2}/g_{fuel})	CO_2/Energy (g_{CO_2}/MJ)	Octane Number[h] (–)	Cetane Number (–)
Spark ignition (liquid fuels)	Gasoline[a]	1.87	0.00	760	44.0	1.00	3.17	72	87–93	14–20
	Methanol	4.00	1.00	792	20.0	0.47	1.37	69	99	12
	Ethanol	3.00	0.50	785	26.9	0.63	1.91	71	98	14
	Butanol[b]	2.50	0.25	810	32.0	0.78	2.37	74	96	15
Compression ignition	Diesel	1.86	0.00	827	43.2	1.07	3.17	73	33–54	40–50
	Biodiesel[c]	1.83	0.11	885	37.3	0.99	2.83	76	15–31	52–58
	FT diesel	1.74	0.01	761	44.6	1.01	3.14	70	–	70–90
	DME[d]	3.00	0.50	668	28.4	0.57	1.91	67	–	55–60
Spark ignition (gaseous fuels)	Methane[e]	4.00	0.00	165	50.0	0.25	2.74	55	120	–
	Natural gas[f]	3.80	0.00	185	45.0	0.25	2.78	62	130	–
	Hydrogen[g]	Inf	–	29	120.0	0.10	0.00	0	106	–

A H/C ratio is the molar ratio of hydrogen to carbon atoms.

B O/C ratio is the molar ratio of oxygen to carbon atoms.

C Density is the mass of fuel per unit volume.

D Energy content on a mass basis. Known as the heating value [e.g., lower heating value (LHV)] and is the quantity of energy contained per unit mass of fuel.

(Continued)

TABLE 10.2 (Continued)
Transportation Fuel Properties Including Specific Energies and CO_2 Production

E   Energy content on a volume basis. The product of the energy per unit mass times the density normalized to the value for gasoline.

F CO_2 production in terms of grams of CO_2 produced from each gram of fuel.

G CO_2 production in terms of grams of CO_2 produced per unit energy of the fuel.

[a] Petroleum-based gasoline has typically only trace amounts of oxygen; however, current regulations allow for blending with 10% ethanol. The table lists properties typical of petroleum-derived gasoline.

[b] Butanol properties are those of n-butanol (or 1-butanol). Various forms of butanol exist that have the same chemical formula ($C_4H_{10}O$) but different chemical structure and hence thermodynamic properties and combustion characteristics. These include sec-butanol, $tert$-butanol, and iso-butanol (Szwaja and Naber 2009).

[c] Soy-based methyl-ester biodiesel.

[d] DME is a gas at ambient temperature (20°C) and pressure (1 bar), but it is a liquid at relatively low pressures (5 bar) (Arcoumanis et al. 2008).; Specific energies in the table are for liquid DME. DME can be mixed with diesel fuel, but it requires a specialized fuel system and vapor handling not needed for diesel or biodiesel (Chapman et al. 2003; Arcoumanis et al. 2008).

[e] Pressurized to 25.0 MPa.

[f] Pressurized to 25.0 MPa.

[g] Pressurized to 34.5 MPa.

[h] (R + M)/2

methanol, an oxygenated fuel also used in SI engines, has an energy density of only 20 MJ/kg, a 55% reduction. Despite this energy density reduction, a benefit of oxygenated fuels is that they require less air (oxygen) during combustion and provide carbon monoxide (CO) and PM emission reductions. The use of oxygenated fuels in SI engines was mandated as part of the 1990 Clean Air Act (CAA) (U.S. EPA 2004) in regions of CO emission nonattainment because oxygen in the fuel assists with more complete oxidation of the carbon to CO_2, thereby reducing CO emissions (CFDC 2008).

Specific energy on a unit volume basis relative to gasoline is given in column E of Table 10.2. Examining the alcohol liquid fuels, it is seen that all have a lower energy density, which indicates that under the same fuel conversion efficiency, their specific fuel consumption (mass or volume of fuel per unit output) will be higher. Neat ethanol has 63% of the energy density of gasoline, and E85 has 70% of the energy density of gasoline. This is the cause of the reduced fuel economy (MPG) observed in FFVs. The alternative CI fuels have energy densities that are more similar to petroleum diesel, except for DME. The gaseous fuels have low energy densities on a volumetric basis because of their low mass densities, even when compressed. As Table 10.2 shows, hydrogen, even when compressed to 34.5 bar (5000 psi), has a lower energy density, only one tenth that of gasoline. This brings to light that we should not be paying for fuels based upon a volumetric basis (dollars per gallon), but on an energy basis (dollars per MJ), which can be computed based on the ratio in column E.

Columns F and G characterize the specific emission of CO_2 of these fuels. Column F is the mass-specific CO_2 produced per unit mass of fuel consumed, and column G is the energy-specific CO_2 in grams of CO_2 per MJ of fuel energy. This energy-specific CO_2 provides a useful metric for comparing the energy/CO_2 tradeoff of the fuels. Examining the values for the liquid fuels, their energy-specific CO_2 values cover a relatively small range and are within 7% of gasoline. On first approximation, engine efficiency will not change significantly for these different fuels when used in the same IC engine technology. Thus, a good estimate of the relative CO_2 impact for alternative fuels on a tank-to-wheels analysis would be the ratio of the biofuel to the reference fuel (JRC/IES 2007). For example, comparing ethanol to gasoline would be 71/72 = 0.99, or a 1% estimated reduction in tank-to-wheels CO_2. Therefore, to produce less CO_2 at this stage of the life-cycle, aside from the solution of decreasing the miles driven, the efficiency at which the IC engine converts the chemical energy in the fuel to useful mechanical work at the wheels needs to be improved and/or the amount of mechanical work required to propel the vehicle must be reduced. One may note that although some gaseous fuels, and in particular hydrogen, have a high energy content per unit mass (120 MJ/kg for H_2 compared with just 44 MJ/kg for gasoline), they still cannot deliver the same vehicle range as liquid fuels such as gasoline or diesel because of the low energy per unit volume in the gas state. Technologies do exist to store hydrogen at low temperatures (–253°C or –423°F) in a liquid state (BNL 2008) and at high pressures. However, in addition to adding considerable cost, these solutions require a significant amount of additional energy to compress and cool the hydrogen, and this extra energy must be taken into account when evaluating the overall energy requirements throughout the life-cycle of hydrogen (see Figure 10.3 for comparisons of vehicle driving ranges for several types of fuel). From Figure 10.3, it is apparent that the typical CI liquid fuels of diesel, FT diesel, and biodiesel yield the largest distance range, followed by liquid SI fuels of gasoline, butanol, and ethanol. The gaseous SI fuels (CH_4, hydrogen, and biogas) yield the lowest driving range. These trends all relate back to the energy content of the fuel on a volume basis, determined by the energy contained in the fuel on a mass basis and the fuel density. The gaseous fuels, although having a reasonably high energy content, have low densities that yield reductions in driving ranges. CI and SI fuels have comparable energy contents and densities that provide similarities in driving ranges, with exceptions to this including methanol, ethanol, and DME, each of which has a lower energy content relative to the other fuels considered. It should also be noted that electrical energy storage can be used, but considering that gasoline has a volumetric energy density that is over 60 times greater than that of a nickel metal hydride battery (Komatsu et al. 2008), the issue becomes the ability to physically store enough energy onboard the vehicle. Furthermore, electric vehicles currently need batteries that are 12 times the size of conventional plug-in hybrid batteries to provide

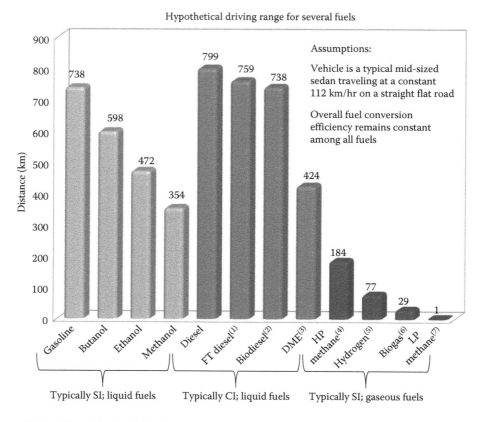

(1) Fisher–Tropsch (synthetic) diesel.
(2) Soy-based methyl-ester biodiesel.
(3) Dimethylether.
(4) High-pressure methane. Pressurized to 25 MPa absolute @ 20°C and completely consumed during drive.
(5) Pressurized to 34.5 MPa absolute @ 20°C and completely consumed during drive.
(6) Wood-based (producer) syngas with typical constituent compositions from Borman and Ragland (1998).
(7) Low-pressure methane. Barometric pressure (101.3 kPa) @ 20°C and completely consumed during drive.

FIGURE 10.3 A hypothetical comparison of maximum driving range between fueling stops for different types of fuels, assuming no changes to the vehicle, including constant fuel tank volume and constant overall fuel conversion efficiency.

the same vehicle range (Komatsu et al. 2008). It is factors such as these that led to 93.6% of LDVs sold in the United States in 2005 being designed to run on only gasoline or diesel. When adding to this number hybrids and FFVs that use gasoline or diesel as one of the on-board energy sources, the percentage increases to 99.96 (EIA 2007). However, with current increases in alternative energy mandates, this percentage is likely to decrease in coming years.

10.1.3 Fuel Economy and CO_2 Regulations

It is through the reaction of fuel with air to produce CO_2 and water that chemical energy is converted to sensible (thermal) energy that elevates the product gas temperature and enables a heat engine such as the IC engine to extract energy from the working fluid. Therefore, CO_2 is an unavoidable byproduct of combustion when fuel contains carbon, and for a given fuel, CO_2 production is directly proportional to fuel consumption (every carbon atom in the fuel produces one CO_2 molecule). On a mass basis, every kilogram of carbon (molecular weight of 12.01) in the fuel produces 3.67 kg of

CO_2 (molecular weight of 44.01). For modern IC engines that are regulated by toxic emission standards and operate under normal conditions, the conversion of the chemical energy to sensible energy is typically greater than 97% (Heywood 1988). For on-road vehicles, over 99.9% of the carbon in the fuel is converted to CO_2 through combustion and aftertreatment devices such as catalytic converters (Heck and Farrauto 2002). However, because of several losses and limitations, not all of that released chemical energy results in usable work transmitted to the output shaft of the engine. Therefore fuel consumption and CO_2 emissions are directly coupled and need to be considered jointly.

There are two common measures of fuel usage in vehicles. The United States uses a rating of MPG whereas Europe and many other countries use a rating of liters per 100 kilometers (L/100 km). Per above, CO_2 specific emissions in grams of CO_2 per kilometer (g(CO_2)/km) is proportional to L/100km for a given fuel. The relationships between these metrics are shown in Figure 10.4 for gasoline on the basis of the properties in Table 10.2. MPG is a measure of efficiency in that it is the useful output in miles divided by the input in gallons of fuel, whereas L/100 km is a specific consumption metric (input over output). It is then clear that they are inversely proportional (L/100 km = 234.2/MPG). Following through, the g(CO_2)/km-specific emissions metric is proportional to L/100 km and is dependent on the fuel properties [density (column C) and CO_2/fuel (column F) in Table 10.2]. Here we note that because of the inverse relationship between MPG and L/100 km or g(CO_2)/km, when changes are discussed, a specific percentage change in one does not correspond to the same percentage change in the other. For example a 42% increase in MPG from 35 to 50 corresponds only to a 30% decrease in L/100 km or g(CO_2)/km of CO_2 emissions.

In addition to the specific emission scale in Figure 10.4, also shown are the related standards and targets for CO_2 and CO_2-equivalent greenhouse gases (GHGs). With this figure, comparisons can be made between U.S. Corporate Average Fuel Economy (CAFE) standards, including the newly adopted U.S. CAFE standard (Sissine 2007), European CO_2-specific emissions targets (Brink et al. 2005), and California CO_2 equivalent GHG specific emissions standards (California Legislation 2002; ARB 2005). From this figure we can see that these standards, including the CAFE fuel economy standard, are in application all regulating CO_2-specific emissions. Additionally, although the measured emissions are determined on different vehicle test cycles, it can be seen that the U.S. 2016 MPG CAFE regulation is set at a higher CO_2-specific emission level than the European targets and California limits.

From the previous discussion, it can be seen that regulating fuel economy not only provides benefits regarding reduction in oil consumption, but also directly reduces CO_2 emissions. With the implementation of the latest LDV CAFE standards, U.S. GHG emissions could decrease by 960 million metric tones, saving 1.8 billion barrels of oil over the lifetime of the vehicles sold between 2012 and 2016 (U.S. EPA 2010c). This is a measureable savings for the economy and the environment, as well as providing increased energy surety by reducing reliance on imported oil. However these estimates do not account for increases in miles driven that have been observed with increased vehicle fuel economy. Although the original CAFE standards in 1975 increased vehicle fuel economy by 37.5% from 1980 to 2007 (U.S. DOT 2004), over this time the total driven miles in the United States increased by 199% (U.S. DOT RITA 2010), resulting in an overall increase in fuel consumption. Hence, fuel consumption is not decreasing as rapidly as desired from these CAFE standards. One method proposed to compensate for this trend is to increase fuel costs via gasoline taxes as an example, to motivate the demand for fuel-efficient vehicles meeting stringent CAFE standards while also reducing miles driven on the basis of the increased cost of fuel (NRC 2002, 2010a).

Various methods have been proposed to improve fuel economy for SI and CI engines, as will be discussed further for both engine technologies in subsequent sections. Options to reduce fuel consumption applying to SI and CI engines include vehicle improvements such as mass reduction, decreased rolling resistance and friction, improved aerodynamics, advanced materials and body designs, transmission modifications (Jones 2008; NRC 2010b), and hybridization. In the case of MDVs and HDVs, proposed methods are similar to those mentioned, but other options include automated manual transmissions, wide-base low-rolling resistance tires, and intelligent transportation systems that include appropriately training drivers, modifying truck size and weight restrictions,

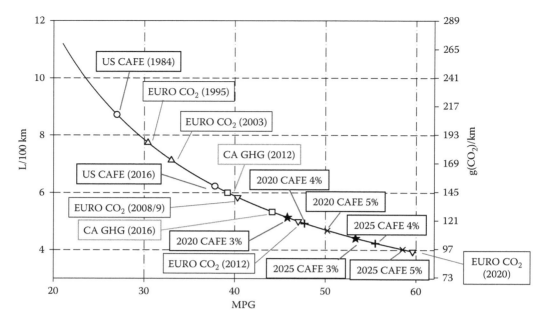

FIGURE 10.4 Relationships between fuel efficiency metric (MPG) and fuel consumption metric (L/100 km) with U.S. CAFE standards (O), European g(CO$_2$)/km specific emissions levels achieved (Δ) and targets (∇), and California (CA)-proposed CO$_2$ equivalent GHG emissions standards ($\square$). Also included are U.S. CAFE standards scenarios, which call for 3, 4, and 5% reductions in GHG emissions per year from 2016 levels (225 g(CO$_2$)/mi for passenger vehicles) to 2025. (From ARB, *Regulations to Control Greenhouse Gas Emissions from Motor Vehicles: Final Statement of Reasons.* California Environmental Protection Agency, Air Resources Board, available at http://www.arb.ca.gov/regact/grnhsgas/fsor.pdf (accessed November 16, 2010), 2005; Brink, P., et al., *Service Contract to Carry Out Economic Analysis and Business Impact Assessment of CO$_2$ Emissions Reduction Measures in the Automotive Sector,* Institute for European Environmental Policy, Brussels, Belgium, IEEP/TNO/CAIR, 2005; European Union, *Setting Emission Performance Standards for New Passenger Cars as Part of the Community's Integrated Approach to Reduce CO$_2$ Emissions from Light-Duty Vehicles,* Regulation (EC) No 443/2009 of the European Parliament and of the Council, 2009. available at http://eur-lex.europa.eu/LexUriServ/LexUriServ.do?uri = OJ:L:2009:140:0001:0015:EN:PDF (accessed November 16, 2010); U.S. DOT, *Automotive Fuel Economy Program, Annual Update, Calendar Year 2003,* National Highway TrafficSafety Administration, DOT HS 809 512, U.S. Department of Transportation, 2004. available at http://www.nhtsa.gov/cars/rules/cafe/FuelEconUpdates/2003/index.htm (accessed October 31, 2010); U.S. EPA DOT, *Light-Duty Vehicle Greenhouse Gas Emission Standards and Corporate Average Fuel Economy Standards; Final Rule,* National Highway Traffic SafetyAdministration, 40 CFR Parts 85, 86 and 600; 49 CFR Parts 531, 533, 536, et al., U.S. Environmental Protection Agency and U.S. Department of Transportation, available at http://www.nhtsa.gov/staticfiles/rulemaking/pdf/cafe/CAFEGHG_MY_2012-2016_Final_Rule_FR.pdf (accessed November 16, 2010), 2010, and U.S. EPA, *Interim Joint Technical Assessment Report: Light-Duty Vehicle Greenhouse Gas Emission Standards and Corporate Average Fuel Economy Standards for Model Years* 2017–2025. Office of Transportation and Air Quality, available at http://www.epa.gov/otaq/climate/regulations/ldv-ghg-tar.pdf (accessed October 21, 2010), 2010).

and idle reduction (NRC 2010a). In addition, manufacturers have other options to assist in meeting fuel economy standards. This includes credits that can be received and transferred between car and truck MPG ratings or applied to prior or future standards. Credits can be earned for improvements in air conditioning systems by reducing hydrofluorocarbon refrigerant losses. Credits can also be earned for reducing indirect CO$_2$ emissions and for advanced technology and alternative fuel vehicles. Credits have also been proposed for overcompliance to standards (U.S. EPA 2010c).

Current CAFE standards apply to LDVs, including passenger cars and light trucks because they are responsible for almost 60% of GHG emissions from transportation (U.S. EPA 2010d). However, heavy-duty engines in commercial trucks do not currently have regulations on fuel economy or GHG emissions, despite being the second-largest transportation sector oil consumer and GHG emission producer, accounting for approximately 20% of transportation GHG emissions (U.S. EPA 2010e). This is changing; the EPA and U.S. Department of Transportation (DOT) recently proposed fuel efficiency and GHG standards to be phased in beginning with model year 2014 for heavy-duty engines (U.S. EPA 2010f). In addition to CO_2 GHG regulation, two other GHGs, nitrous oxide (N_2O) and CH_4, will also be regulated under this proposed program (U.S. EPA 2010e). Similar to CO_2, N_2O and CH_4 are suspected of contributing to global warming, but they have a higher impact than CO_2 for the same level of emissions concentration (U.S. EPA 2006). It is estimated that these proposed new standards for heavy-duty engines could reduce GHG emissions by 250 million metric tones, saving approximately 500 million barrels of oil over the vehicle life, for those sold in 2014–2018 (U.S. EPA 2010e).

10.1.4 EMISSIONS REGULATIONS

The combustion process inside of an IC engine leads to the production of harmful compounds, including CO, NO_x, unburned hydrocarbons (UBHCs or HCs), and PM. These chemicals, commonly referred to simply as "emissions", are harmful to humans and the environment, and as a result are regulated in the United States by the EPA. The development of advanced engines and aftertreatment systems has reduced emissions to comparatively extremely low levels. For automotive LDVs, the trends in emissions reduction are shown in Figure 10.5 for U.S. and European standards for NO_x and HC (CO has been similarly reduced, but not shown). As can been seen, the emissions have decreased by more than a factor of 100 over this period. For heavy-duty diesel engine applications, similar reductions have been mandated. The latest regulations in 2007 and 2010 require a reduction

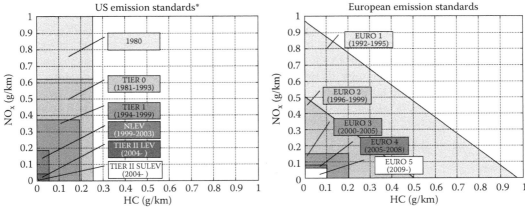

NLEV = National low emission vehicle
PZEV = SULEV levels with no evaporative emissions
* Standards in g/mi converted to g/km

FIGURE 10.5   U.S. and European NO_x and HC emission standards for LDVs (From NRC, *Evaluating Vehicle Emissions Inspection and Maintenance Programs,* Board on Energy and Environmental Systems Transportation Research Board, National Research Council, National Academy Press, Washington, DC, 2001. available at http://www.nap.edu/openbook.php?record_id=10133&page=R1 (accessed November 11, 2010); U.S. EPA, *Summary of Current and Historical Light-Duty Vehicle Emission Standards,* Office of Transportation and Air Quality, U.S. Environmental Protection Agency, available at http://www.epa.gov/greenvehicles/detailedchart. pdf (accessed November 11, 2010), 2007c; U.S. EPA, *Light-Duty Vehicle and Light-Duty Truck – Tier 0, Tier 1, National Low Emission Vehicle (NLEV), and Clean Fuel Vehicle (CFV) Exhaust Emission Standards*, U.S. Environmental Protection Agency, available at http://www.epa.gov/oms/standards/light-duty/tiers0-1-ldstds.htm (accessed November 11, 2010), 2010h; Dieselnet, Emission Standards–European Union–Cars and Light Trucks, 2010. Available at http://www.diesel-net.com/standards/eu/ld.php (accessed November 11, 2010).

by a factor of 10 in PM (from 0.1 to 0.01 g/bhp-h) and NO_x (from 2 to 0.2 g/bhp-h), respectively (U.S. EPA 2010g). In the case of diesel CI engine combustion, there is an inherent NO_x-soot tradeoff that is based on the temperature-equivalence ratio path that combustion follows, making emission control difficult. To meet the 2007–2010 standards, diesel engines now incorporate complex after-treatment systems that typically include an oxidation catalyst, a continuously regenerating PM trap, and a lean NO_x reduction system such as a selective-catalyst-reduction/urea system.

Evaporative emissions are those that can come from fuel in storage tanks during various drive and park cycles and as a result of refueling the vehicle. These, in addition to the tailpipe emissions outlined above, have been regulated for gasoline-fueled vehicles because of the fuel's high volatility. Evaporative emissions are composed of light volatile organic compounds (VOCs) in the fuel and cause ground-level ozone problems and human health issues (U.S. EPA 2007b; de Nevers 2000). Tier 2 evaporative emission standards for LDVs include a three-diurnal test combined with a hot soak (after engine operates above ambient temperatures and is shut down) along with running losses, being 0.50 g/test and 0.05 g/mi for LDVs in model year 2009 (U.S. EPA 2009).

To maintain these standards, vehicles must include on-board-diagnostic (OBD) systems to continuously monitor all components that would result in the failure of the vehicle to meet these standards. If a failure occurs, a diagnostic check engine light informs the driver of a problem and diagnostic codes on the faulty component(s) are stored and used for servicing purposes.

10.2 IC ENGINES

10.2.1 BASIC ENGINE OPERATION

The IC engine has existed since the 19th century, when pioneers including Jean Lenoir, Nicholas Otto, and Rudolf Diesel proved that the concept of a heat engine relying upon IC was a viable mechanism (Cummins 1989). An IC engine is a mechanical device that converts energy contained within chemical bonds in the fuel into kinetic energy in the form of a rotating shaft that can be used to do useful work. The four-stroke cycle engine, which is used in nearly all transportation vehicles, consists of a piston moving up and down within a cylinder inside of the engine. The piston is connected to a crankshaft; thus, the reciprocating motion of the piston is translated via the connecting rod and crankshaft into rotational kinetic energy at the shaft. The engine operates on a cycle, wherein the piston makes four distinct strokes of the cylinder over two revolutions of the crankshaft as shown in Figure 10.6. These four strokes are described as follows:

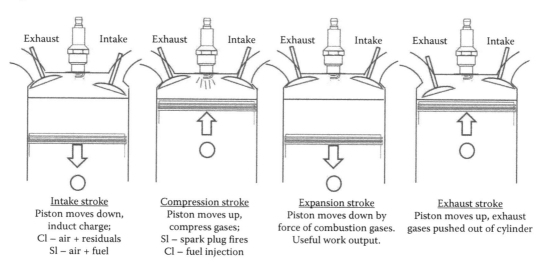

Intake stroke	Compression stroke	Expansion stroke	Exhaust stroke
Piston moves down, induct charge;	Piston moves up, compress gases;	Piston moves down by force of combustion gases.	Piston moves up, exhaust gases pushed out of cylinder
CI – air + residuals	SI – spark plug fires	Useful work output.	
SI – air + fuel	CI – fuel injection		

FIGURE 10.6 Four-stroke engine operating cycle. SI engine shown with spark-plug for ignition.

- *Intake stroke*: Fuel and air are brought into the cylinder through intake valve(s) as the piston moves down the cylinder [from top dead center (TDC) to bottom dead center (BDC)].
- *Compression stroke*: As the piston moves up the cylinder (from BDC to TDC), the intake valves are closed and gases are compressed, thus increasing the temperature and pressure.
- *Expansion stroke*: The combustion of fuel and air creates CO_2, H_2O, and other products at high temperature and pressure. The high pressure pushes the piston down the cylinder (TDC to BDC). This is the only stroke in which work is extracted from the engine.
- *Exhaust stroke*: The exhaust valves open and the piston moves up the cylinder (from BDC to TDC), pushing the combustion products to the exhaust system. After the exhaust stroke is complete, the intake stroke begins again, signifying the start of another cycle.

Although all four-stroke engines operate on this same basic cycle, there are two distinctly different variations. These two variations, the SI and CI engine, differ in how the fuel is introduced and how the fuel and air mixture is ignited and burned. These are shown visually in Figure 10.7. This includes SI engines that use a premixed fuel–air mixture and exhibit a homogeneous propagating flame. CI combustion is based on a diffusion flame with nonpremixed combustion. The SI and CI engine are discussed in the following two sections, along with effects of fuels on their performance and emissions.

10.2.2 SI ENGINES

Inside of the combustion chamber of the engine, the correct fuel–air mixture must be present to initiate and sustain combustion in a controlled manner over a portion of the cycle. In SI engines, as

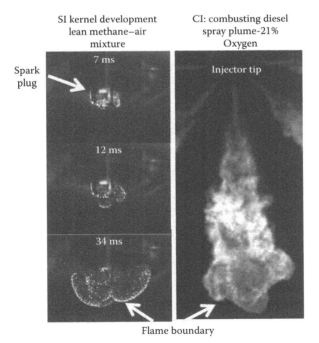

FIGURE 10.7 SI and CI combustion images. SI combustion exhibits a homogeneous propagating flame of the premixed air–fuel mixture, with CI combustion utilizing diffusion flame combustion based on fuel injection into a hot environment of ambient gases. Images were acquired in the Michigan Technological Univeristy optically accessible CV combustion vessel. (From Johnson, S., et al., *Proceedings of the ASME 2010 International Combustion Engine Division Fall Technical Conference* ICEF 2010, September 12–15, 2010; San Antonio, TX and Nesbitt, J., et al., *Proceedings of the ASME 2010 International Combustion Engine Division Fall Technical Conference ICEF 2010,* September 12–15, 2010, San Antonio, TX.)

the name indicates, combustion is initiated by a high-energy electrical discharge, a spark, which causes the fuel–air mixture locally to ignite. Once the fuel–air mixture starts to burn, the flame propagates through the combustion chamber, consuming the premixed fuel and air in a controlled manner. To avoid uncontrolled combustion, the autoignition of the fuel–air mixture at any stage of this must be prevented. Gasoline and other SI fuels must have a high resistance to autoignition, which is measured by the octane rating of the fuel. The higher the octane, the more resistant the fuel is to autoignition (see values in Table 10.2, column H). If the octane rating is not high enough, combustion knock (unwanted autoignition) will occur, reducing efficiency and potentially damaging the engine. The SI biofuels in Table 10.2 all exhibit octane numbers higher than gasoline, indicating that they are better than the standard petroleum gasoline regarding resisting autoignition. However, additional fuel properties need to be considered as they influence performance, startability, and emissions.

10.2.2.1 Influence of Fuel on Operating Characteristics of SI Engines

Two very important pathways for improving the efficiency of an engine are (1) to increase the compression ratio (CR), which is a measure of how much the fuel and air mixture is compressed before ignition, and (2) to reduce the amount of work that is wasted pumping air and fuel into the cylinder and out the exhaust port.

In SI engines, the maximum compression ratio is limited by the fuels octane number. A higher compression ratio leads to higher temperatures during the combustion process, and this can cause a portion of the fuel to autoignite and burn rapidly at the wrong time in the engine cycle, leading to combustion knock. Fuels with a higher octane rating, such as alcohol-based fuels like ethanol, methanol, or some isomers of butanol, will resist autoignition, which allows engine developers to increase an engine's compression ratio and thus improve efficiency. However, because these alternative fuels, even ethanol, remain a low portion of the overall fuel supply, vehicle OEMs must produce engines that are capable of running on gasoline in addition to alternative fuels. Unfortunately, the requirement that these vehicles, called FFVs, operate on gasoline limits the engines compression ratio, and thus limits the full benefit of the higher octane of the alternative fuel. Engines with variable compression ratio (VCR) systems are a developing technology that shows promise in alleviating this limitation because these systems would allow the compression ratio to be automatically optimized as the fuel in the tank changes (Drangle 2002; Moteki 2003; Rosso 2006).

The load in SI engines is controlled by throttling the incoming air and controlling the prior cycle residuals in the cylinder by continuously adjusting the phase of the intake and exhaust valves. By restricting the air, less air enters the cylinder during the intake stroke. To maintain the proper air-to-fuel ratio, less fuel is injected, resulting in less heat being released during combustion and resultant power being reduced. Similarly, by increasing the amount of prior cycle residuals, which displace air, less fuel is injected. The disadvantage of the throttling process is that although the engine ingests less air, the added restriction in the intake air path causes the engine to do negative work during the intake stroke to bring the air into the cylinder, thus reducing the efficiency. The alternative method to control residuals via valve phasing reduces this pumping work within engines. The burned exhaust gases (primarily CO_2, H_2O, and N_2) do not participate in the combustion reaction, but instead fill space in the cylinder that would have otherwise been occupied by the air and fuel mixture. To compensate and maintain engine load, the throttling effect is reduced, which in turn reduces the pumping work. A limiting factor in this is the stability of the combustion process. As more residuals are added to the cylinder, the combustion process becomes prolonged and less stable, eventually leading to misfire. The type of fuel and its properties affect this in that some fuels can tolerate greater amounts of dilution than others, and some fuels such as gaseous fuels require less throttling because of the low density of the fuel.

Another approach to decreasing pumping losses is to reduce the size (displacement) of the engine. A smaller engine requires less throttling to achieve the same load as a large engine and thus runs more efficiently. However, to get the same maximum power out of the engine, intake air boosting with a turbocharger or supercharger is required. Boosting under high loads leads to high temperatures in the cylinder and thus fuels with high octane ratings are more desirable for boosted applications. A related technique is to operate a larger engine with variable displacement.

SI engines are able to meet the lowest levels of regulated pollutant emissions standards (U.S. Tier II, bin 2 and CA PZEV) with relatively inexpensive three-way catalysts coupled with advanced sensing and fully electronic control. However, there is a tradeoff in efficiency to meet these stringent emissions standards: three-way catalysts require stoichiometric operation of the fuel and air to simultaneously reduce HC, CO, and NO_x emissions. A limited number of SI gasoline-fueled engines have been produced, primarily in Europe. They have lean combustion systems in which the NO_x emissions are not as stringent and fuel sulfur levels are lower, enabling lean NO_x aftertreatment devices to be more effective.

SI engines are continuing to evolve, improving engine efficiency, performance, and power density. New technologies including direct injection, turbocharging with downsized engines, optimized valve timing, lift, and duration, homogeneous-charge CI technology, variable displacement, and VCRs (Jones 2008) will continue to enable significant improvements over the next decade.

10.2.3 CI ENGINES (DIESEL ENGINES)

To initiate combustion in CI engines, the compression process heats the gases (including air, prior cycle residuals, and exhaust gas), and when the fuel, which is injected near the end of compression, mixes with the hot gases, it undergoes rapid exothermic reactions (autoignition and combustion). Conventional CI engines (often called diesel engines in recognition of their inventor, Rudolf Diesel) operate using diesel fuel, which has a high cetane number (see Table 10.2, column I). The cetane number is a measure of the fuel's ignition delay. The cetane number can be thought of as the opposite of an octane number, and fuels with a high cetane number have a low octane rating. The higher the cetane number, the easier (lower temperature) and faster autoignition will occur. In Table 10.2 it is seen that diesel has the lowest cetane rating, indicating that at least from an autoignition standpoint, the alternative diesel fuels are better fuels than the base petroleum diesel fuels.

Several factors typically give CI engines a higher efficiency than SI engines. CI engine compression ratios are higher than SI engines because the compression ratio in a SI engine is limited by combustion knock due to limited octane ratings. Compression ratios for CI engines range from 16 to 24, as opposed to 9 to 13 for a typical SI engine, as shown in Figure 10.8. Higher compression ratios lead to higher efficiency on the basis of the thermodynamics of the cycle. The two lines shown in Figure 10.8 correspond to heat addition at constant volume (CV) and constant pressure (CP) for the

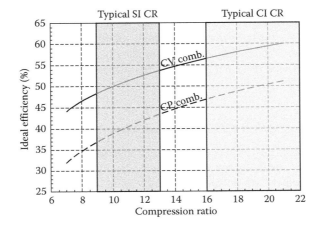

FIGURE 10.8 Compression ratio effect on ideal efficiency and ranges of current gasoline and diesel engine operation.

respective ideal thermodynamic cycles. In practice, engines use a limited pressure cycle in which combustion occurs at a rate somewhere between these, yielding ideal efficiencies that fall between these lines. Increasing the compression ratio (from SI to CI levels) yields improvements in efficiency, with actual engine efficiencies falling below these lines because of the nonideal cycles with various losses including heat transfer, rate-limited induction and exhaust of the working gases, and friction. Another unique aspect of CI engines is the load control mechanism. Unlike an SI engine, in which engine load is controlled by restricting the flow of air into the engine, a CI engine controls load by injecting more or less fuel into the cylinder. By controlling load in this manner, the engine does not waste energy across a restriction as it does in the SI engine during throttling. Furthermore, at light and even moderate loads, when a relatively small amount of fuel is being injected, the overall air-to-fuel mixture is lean of stoichiometric (excess air), which also leads to the higher efficiency of CI engines relative to SI engines. Further options to improve CI engine efficiency and fuel economy include advanced technologies such as turbocharging and injection pressure increases (Jones 2008). It should be noted that the addition of exhaust gas recirculation (EGR) typical in 2004 and newer engine systems may require the addition of an intake air restriction to draw relatively low pressure exhaust gas into the intake manifold. This restriction acts like a throttle on an SI engine and reduces the pumping efficiency. New technology using high-pressure EGR loops and variable geometry turbochargers alleviate the addition of an intake throttle for EGR implementation.

10.2.3.1   Diesel Engine Emissions: NO_x and PM

Despite higher efficiencies, CI engines fueled with diesel exhibit a disadvantage in terms of regulated toxic emissions and their reduction in aftertreatment systems. Of particular concern are NO_x and PM composed of dry soot and soluble organic compounds (SOCs). There is an inherent tradeoff in these two emissions in a diesel-fueled CI engine: when one is reduced the other will increase. NO_x is formed in the higher-temperature, near-stoichiometric combustion regions with soot forming in the lower temperature, fuel-rich regions of the combustion zone. These emissions can be reduced using various means internal and external to the engine. Internal reduction methods include changes in fuel injection pressure and timing as well as fuel–air mixture dilution. External reduction methods include aftertreatment systems consisting of an oxidation catalyst for HC and CO reduction, a particulate filter to reduce PM emissions, and a selective catalytic reduction (SCR) catalyst to reduce NO_x emissions.

The use of oxygenated fuels, such as biodiesel, in a CI engine reduces PM emissions with resultant smaller increases in NO_x. As a means for decreasing NO_x emissions with biodiesel, EGR can be added or combustion phasing can be delayed, resulting in similar NO_x emissions but still lower PM emissions relative to diesel fuel, as seen in Figure 10.9 (Polonowski et al. 2010).

In addition to biodiesel, DME is a promising fuel for CI engines, with low exhaust emissions including very low PM. However, because of it is high volatility (gas at ambient temperature–pressure) and low lubricity, the fuel and injection system must be modified from a conventional design. DME can be blended with conventional diesel, but this still results in a high-volatility fuel requiring specialty systems.

10.2.4   Impact of Biofuels on Current Engine/Vehicle Operation

It should be apparent that biofuels affect engines and transportation vehicles in many ways. Biofuels affect emissions (as previously discussed) and can have a significant impact on full load power. The impact areas that vehicle owners and operators tend to be most sensitive to are fuel economy, maximum driving range between fueling, and initial purchase cost. Additionally, vehicle owners are affected by the significant variation in implementation costs for alternative fuels. From the manufacturers' point of view, enabling a vehicle originally designed to run on petroleum-based fuel to operate on biofuel can be a significant challenge. For example, the fuel system materials may

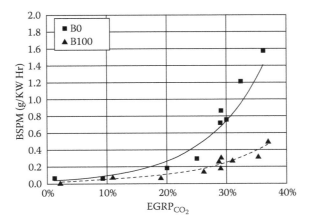

FIGURE 10.9 Impact of EGR on PM emissions with petroleum diesel fuel (B0) and 100% biodiesel (B100).

need to be changed to be compatible with the alternative fuel as is the case in enabling a gasoline vehicle to run on ethanol or methanol blends or a diesel engine to run on biodiesel. Piston rings and valve seats are often upgraded to be compliant with the alternative fuel, which can also add cost. Electronic determination of the type of fuel currently in use can be required. In some cases this can be done with relatively low cost through the use of "virtual sensors," which are software algorithms operating within the vehicles controller that utilize inputs from existing sensors to determine fuel type. There is an additional cost for increased engineering effort because of the additional calibration and federal certification required. However, if production volumes are sufficiently high, this increased cost is low on a per vehicle basis.

One issue with some fuels is their poor cold-start characteristics, on the basis of their fuel properties. In the case of diesel engines, DME has a higher cetane number than petroleum diesel (Table 10.2), which promotes ignition by reducing the required compression energy for autoignition. In conjunction with its low boiling point, this helps reduce ignition delay, improve fuel–air mixing, and therefore minimize cold-start issues. On the other hand, biodiesel exhibits fuel gelling in colder environments, a characteristic common in petroleum ultralow sulfur diesel (ULSD). In the case of SI engines, ethanol fuel results in different cold-starting issues. Ethanol does not vaporize significantly in low-temperature conditions because of lower vapor pressure and higher heat of vaporization (17 kPa vapor pressure vs. 62 kPa at 100°F for gasoline, and 900 kJ/kg heat of vaporization vs. 400 kJ/kg for gasoline) and hence the fuel–air mixture is not rich enough (attributed to limited vaporization) to combust (Davis et al. 2000). Hence, cold-start issues are common to alcohol fuels. Adding gasoline to the mixture (e.g., yielding E85) helps to reduce, but not overcome, this issue because the vaporized gasoline can assist with initial combustion; however, cold-start emissions can be higher (Davis et al. 2000).

10.3 SUMMARY

In this chapter, we discussed the effect and interdependencies of fuels, engine type, aftertreatment, and vehicle technologies with a focus on biofuels and where they fit into current and future technologies. The energy densities on a volumetric basis for the oxygenated biofuels are lower than their comparable petroleum fuels. Furthermore, combustion of a carbon-based fuel, be it gasoline, ethanol, diesel, or biodiesel, in the presence of oxygen, produces CO_2. When normalized, the CO_2 produced from combustion on an energy basis for the petroleum and bio-based liquid fuels are within ±7% of gasoline.

Biodiesel has the closest energy density to petroleum fuels, with butanol the next closest. For the gaseous fuels (hydrogen and CH_4), although high in energy on a gravimetric basis, the volumetric energy density is low even at high pressures. Significant advancements in storage will be required to match the energy density of the liquid fuels. Furthermore, fuels such as biodiesel and DME are better suited to CI engines, whereas other fuels such as ethanol and gaseous fuels are better suited to SI engines. Even so, it is difficult to take full advantage of biofuels such as ethanol with a high octane rating when designing a FFV because many design parameters have to be specified for the "worse-case" fuel mixture in the vehicle tank.

REFERENCES

ARB (2005) *Regulations to Control Greenhouse Gas Emissions from Motor Vehicles: Final Statement of Reasons.* California Environmental Protection Agency, Air Resources Board, available at http://www.arb.ca.gov/regact/grnhsgas/fsor.pdf (accessed November 16, 2010).

Arcoumanis C, Bae C, Crookes R, Kinoshita E (2008) Review article: The potential of di-methyl ether (DME) as an alternative fuel for compression-ignition engines: A Review. *Fuel* 87:1014–1030.

ASTM (2009) *Standard Specification for Diesel Fuel Oil, Biodiesel Blend (B6 to B20)*, Designation D7467-09a, American Society for Testing and Materials, West Conshohocken, PA

ASTM (2010) *Standard Specification for Automotive Spark-Ignition Engine Fuel*, Designation D4814-10a, American Society for Testing and Materials, West Conshohocken, PA

ASTM (2010) *Standard Specification for Biodiesel Fuel Blend Stock (B100) for Middle Distillate Fuels*, Designation D6751-10, American Society for Testing and Materials, West Conshohocken, PA

ASTM (2010) *Standard Specifications for Diesel Fuel Oils*, Designation D975-10a, American Society for Testing and Materials, West Conshohocken, PA

BNL (2008) *Hydrogen Storage*, U.S. Department of Energy, Energy Sciences and Technology Department, Brookhaven National Laboratory, available at http://www.bnl.gov/est/erd/hydrogenStorage (accessed November 1, 2010)

Borman LG, Ragland KW (1998) *Combustion Engineering*, 1st ed. McGraw- Hill

Brink P, Skinner I, Fergusson M, Haines D, Smokers R, van der Burgwal E, Gense R, WeLLS P, Nieuwenhuis P, (2005) *Service Contract to Carry Out Economic Analysis and Business Impact Assessment of CO_2 Emissions Reduction Measures in the Automotive Sector*, Institute for European Environmental Policy, Brussels, Belgium, IEEP/TNO/CAIR

California Legislation (2002) Vehicular emissions: Greenhouse gases, Assembly Bill No. 1493, available at http://www.calcleancars.org/ab1493.pdf (accessed March 20, 2008)

CFDC (2008) *Oxygenates Fact Book: A Compilation of Information on the Benefits of Oxygenates in Gasoline*, Clean Fuels Development Coalition, available at http://www.cleanfuelsdc.org/pubs/documents/oxyfact-book.pdf (accessed November 10, 2010)

Chapman EM, Boehman A, Wain K, Lloyd W, Perez JM, Stiver D, Conway J (2003) *Impact of DME-Diesel Fuel Blend Properties on Diesel Fuel Injection Systems*, DOE Contract Technical Report TRN: US200409%%31, available at http://www.osti.gov/bridge/servlets/purl/821275-ho3SG7/native/ (accessed October 19, 2010)

Cummins CL (1989) *Internal Fire: The Internal-Combustion Engine 1673-1900*, 2nd ed. Carnot Press, Wilsonville, OR

Cummins Inc. (2007) Cummins announces approval of B20 biodiesel blends, Cummins Press Release March 21, 2007, available at http://www.biodiesel.org/resources/PR_supporting_docs/20070321_cummins_b20.pdf (accessed March 20, 2008)

Davis GW, Heil ET, Rust R (2000) *Ethanol Vehicle Cold Start Improvement When Using a Hydrogen Supplemented E85 Fuel*, AIAA-2000-2849, American Institute of Aeronautics and Astronautics, available at http://ieeexplore.ieee.org/stamp/stamp.jsp?tp=&arnumber=870702&userType=inst (accessed November 1, 2010)

Davis, SC, Diegel SW, Boundy RG (2010) *Transportation Energy Data Book: Edition 29*, ORNL-6985, U.S. Department of Energy, available at http://cta.ornl.gov/data/tedb29/Edition29_Full_Doc.pdf (accessed October 15, 2010)

de Nevers, N (2000) *Air Pollution Control Engineering,* 2nd ed. McGraw-Hill, Boston

Dieselnet (2010) Emission standards—European Union—Cars and light trucks, available at http://www.diesel-net.com/standards/eu/ld.php (accessed November 11, 2010)

Drangel H, Reinmann R, Olofsson E (2002) *The Variable Compression (SVC) and the Combustion Control (SCC)—Two Ways to Improve Fuel Economy and Still Comply with World-Wide Emission Requirements.* SAE Technical Paper Series 2002-01-0996, Society of Automotive Engineers, Warrenton, PA

EIA (2006a) *Annual Energy Review 2006*, DOE/EIA-0384, U.S. Department of Energy, Energy Information Administration, Washington, DC, available at http://www.eia.doe.gov/emeu/aer/contents.html (accessed March 20, 2008)

EIA (2006b) *Electric Power Annual*, DOE/EIA-0348, U.S. Department of Energy, Energy Information Administration, Washington, DC, available at http://www.eia.doe.gov/cneaf/ electricity/epa/epa_sum. html (accessed March 20, 2008)

EIA (2007) *Regional and Other Detailed Tables: A Supplement to Annual Energy Outlook 2007 with Projections to 2030*, DOE/EIA-0383, U.S. Department of Energy, Energy Information Administration, Washington, DC, available at http://www.eia.doe.gov/oiaf/aeo/ supplement/index.html (accessed March 20, 2008)

EIA (2009a) *Supplemental Tables to the Annual Energy Outlook 2010*, U.S. Department of Energy, Energy Information Administration, Washington DC, available at http://www.eia.doe.gov/oiaf/aeo/supplement/ suptab_58.xls (accessed October 17, 2010)

EIA (2009b) *Annual Energy Review*, DOE/EIA-0384, U.S. Department of Energy, Energy Information Administration, Washington, DC, available at http://www.eia.doe.gov/emeu/aer/pdf/aer.pdf (accessed October 17, 2010)

EIA (2010) *International Energy Outlook 2010*, DOE/EIA-0484, U.S. Department of Energy, Energy Information Administration, Washington, DC, available at http://www.eia.doe.gov/oiaf/ieo/pdf/0484%282010%29. pdf (accessed 15 Oct 2010).

European Union (2009) *Setting Emission Performance Standards for New Passenger Cars as Part of the Community's Integrated Approach to Reduce CO_2 Emissions from Light-Duty Vehicles*, Regulation (EC) No 443/2009 of the European Parliament and of the Council, available at http://eur-lex.europa. eu/LexUriServ/LexUriServ.do?uri=OJ:L:2009:140:0001:0015:EN:PDF (accessed November 16, 2010)

GM (2007) *2006 Annual Report*, Report 002CX-13758, General Motors Corporation, Detroit, MI

Goodger EM (1975) *Hydrocarbon Fuels—Production, Properties and Performance of Liquids and Gases*, Wiley, New York

Heck RM, Farrauto, RJ (2002) *Catalytic Air Pollution Control.* Wiley, New York

Heywood JB (1988) *Internal Combustion Engine Fundamentals.* McGraw Hill, New York

IEA (2005) *Biogas Production and Utilisation*, IEA Bioenergy: T37: 2005:01, International Energy Association, available at http://www.iea-biogas.net/Dokumente/Brochure%20final.pdf (accessed October 17, 2010).

Johnson S, Nesbitt J, Naber JD (2010) Mass and momentum flux measurements with a high pressure common rail diesel fuel injector. In: *Proceedings of the ASME 2010 International Combustion Engine Division Fall Technical Conference* ICEF 2010, September 12-15, 2010, San Antonio, TX

Jones TO (2008) *Assessment of Technologies for Improving Light Duty Vehicle Fuel Economy: Letter Report'* The National Academies, available at http://www.nap.edu/catalog/12163.html (accessed October 17, 2010)

JRC/IES (2007) *Well-to-Wheels Analysis of Future Automotive Fuels and Powertrains in the European Context*, Tank-to-Wheels Report Version 2C, Joint Research Centre/Institute for Environment and Sustainability, available at http://ies.jrc.ec.europa.eu/uploads/media/TTW_Report_010307.pdf (accessed November 15, 2010)

Komatsu M, Takaoka T, Ishikawa T, Gotouda Y, Suzuki N, Ozawa T (2008) *Study on the Potential Benefits of Plug-In Hybrid Systems.* SAE Technical Paper Series 2008-01-0456, Society of Automotive Engineers, Warrenton, PA

Moteki K, Aoyama S, Ushijima K, Hiyoshi R, Takemura S, Fujimoto H, Arai T (2003) A study of a variable compression ratio system with a multi-link mechanism. SAE Technical Paper Series 2003-01-0921, Society of Automotive Engineers, Warrenton, PA

Naber JD, Siebers DL (1998) Hydrogen combustion under diesel engine conditions. *Int J Hydrogen Energy* 23:363–371

Nesbitt J, Lee SY, Naber JD, Arora R (2010) An optical study of spark ignition and flame kernel development near the lean limit at elevated pressure. In: *Proceedings of the ASME 2010 International Combustion Engine Division Fall Technical Conference ICEF 2010*, September 12-15, 2010, San Antonio, TX

NRC (2001) *Evaluating Vehicle Emissions Inspection and Maintenance Programs*, Board on Energy and Environmental Systems Transportation Research Board, National Research Council, National Academy Press, Washington, DC, available at http://www.nap.edu/openbook.php?record_id=10133&page=R1 (accessed November 11, 2010)

NRC (2002) *Effectiveness and Impact of Corporate Average Fuel Economy (CAFE) Standards*, Board on Energy and Environmental Systems Transportation Research Board, National Research Council, National Academy Press, Washington, DC, available at http://books.nap.edu/openbook.php?record_id=10172&page=R1 (accessed November 1, 2010)

NRC (2010a) *Summary: Technologies and Approaches to Reducing the Fuel Consumption of Medium- and Heavy-Duty Vehicles*, Board on Energy and Environmental Systems Transportation Research Board, National Research Council, National Academy Press, Washington, DC, available at http://www.nap.edu/nap-cgi/report.cgi?record_id=12845&type=pdfxsum (accessed November 16, 2010)

NRC (2010b) *Summary: Assessment of Fuel Economy Technologies for Light-Duty Vehicles*, Board on Energy and Environmental Systems Transportation Research Board, National Research Council, National Academy Press, Washington, DC, available at http://www.nap.edu/nap-cgi/report.cgi?record_id=12924&type=pdfxsum (accessed November 16, 2010)

Polonowski C, Lecureux M, Miers S, Naber J, Worm J, Shah J (2010) *The Effects of Oxygenated Biofuel on Intake Oxygen Concentration, EGR, and Performance of a 1.9L Diesel Engine*, SAE Technical Paper Series 2010-01-0868, Society of Automotive Engineers, Warrenton, PA

Rosso P, Beard J, Blough JR (2006) *A Variable Displacement Engine with Independently Controllable Stroke Length and Compression Ratio*, SAE Technical Paper Series 2006-01-0741, Society of Automotive Engineers, Warrenton, PA

SAE (2007) *Alternative Automotive Fuels*, SAE Technical Report, Society of Automotive Engineers, Warrenton, PA

Silva-Petrobras DF (2006) *DME As Alternative Diesel Fuel: Overview*, SAE Technical Paper Series 2006-01-2916, Society of Automotive Engineers, Warrenton, PA

Sissine F (2007) *Energy Independence and Security Act of 2007: A Summary of Major Provisions*, Congressional Research Service Report for Congress, Order Code RL34294, Washington, DC

Svensson KI (2005) Effects of fuel molecular structure and composition on soot formation in direct-injection spray flames. PhD Thesis, Brigham Young University, Provo, UT

Szwaja S, Naber JD (2009) Combustion of n-butanol in a spark-ignition IC engine. *Fuel* 89:1573–1582

Taylor J, McCormick R, Clark W (2004) Report on the relationship between molecular structure and compression ignition fuels, both conventional and HCCI, NREL Report NREL/MP-540-36726, National Renewable Energy Laboratory, Golden, CO

Totten GE, Westbrook SR, Shah RJ (2003) *Fuel and Lubricants Handbook: Technology, Properties, Performance and Testing*, Volume 1, ASTM International, West Conshohocken, PA

U.S. DOE (2010a) Diesel power: Clean vehicles for tomorrow, U.S. Department of Energy, Energy Efficiency and Renewable Energy Vehicle Technologies Program, available at http://www1.eere.energy.gov/vehiclesandfuels/pdfs/diesel_technical_primer.pdf (accessed November 1, 2010)

U.S. DOE (2010b) Flexible fuel vehicles: Providing a renewable fuel choice, DOE Fact Sheet, DOE/GO-102010-3002, U.S. Department of Energy, available at http://www.afdc.energy.gov/afdc/pdfs/47505.pdf (accessed October 16, 2010)

U.S. DOE and EPA (2008) Advanced technologies & energy efficiency, U.S. Department of Energy and U.S. Environmental Protection Agency, available at http://www.fueleconomy.gov/feg/atv.shtml (accessed November 15, 2010)

U.S. DOT (2004) *Automotive Fuel Economy Program, Annual Update, Calendar Year 2003*, National Highway Traffic Safety Administration, DOT HS 809 512, U.S. Department of Transportation, available at http://www.nhtsa.gov/cars/rules/cafe/FuelEconUpdates/2003/index.htm (accessed October 31, 2010).

U.S. DOT RITA (2010) *Motor Vehicle Fuel Consumption and Travel*, Bureau of Transportation Statistics (BTS) TP, U.S. Department of Transportation, Research and Innovative Technology Administration, available at http://www.bts.gov/publications/national_transportation_statistics/html/table_04_09.html (accessed October 31, 2010)

U.S. EPA (2004) *The Clean Air Act*, U.S. Environmental Protection Agency, available at http://epw.senate.gov/envlaws/cleanair.pdf (accessed November 2, 2010)

U.S. EPA (2006) The U.S. inventory of greenhouse gas emissions and sinks: Fast facts, Office of Atmospheric Programs (6207 J), U.S. Environmental Protection Agency, available at http://www.epa.gov/climatechange/emissions/downloads06/06FastFacts.pdf (accessed November 15, 2010)

U.S. EPA (2007a) Regulation of fuels and fuel additives: Renewable Fuel Standard program, *Federal Register* 72(83), available at http://www.epa.gov/otaq/renewablefuels/rfs-finalrule.pdf (accessed October 16, 2010)

U.S. EPA (2007b) *Evaporative Emissions*, U.S. Environmental Protection Agency, available at http://www.epa.gov/otaq/evap/index.htm (accessed October 15, 2010)

U.S. EPA (2007c) *Summary of Current and Historical Light-Duty Vehicle Emission Standards*, Office of Transportation and Air Quality, U.S. Environmental Protection Agency, available at http://www.epa.gov/greenvehicles/detailedchart.pdf (accessed November 11, 2010)

U.S. EPA (2009) *Light-Duty Vehicle, Light-Duty Truck and Medium-Duty Passenger Vehicle—Tier 2 Evaporative Emission Standards*, available at http://epa.gov/otaq/standards/light-duty/tier2evap.htm (accessed October 18, 2010)

U.S. EPA (2010a) Regulation of fuels and fuel additives: Changes to Renewable Fuel Standard program, *Federal Register* 75(58), available at http://www.regulations.gov/search/Regs/contentStreamer?objectId=0900006480ac93f2&disposition=attachment&contentType=pdf (accessed October 16, 2010)

U.S. EPA (2010c) *Regulatory Announcement: EPA and NHTSA Finalize Historic National Program to Reduce Greenhouse Gases and Improve Fuel Economy for Cars and Trucks*, Office of Transportation and Air Quality, EPA-420-F-10-014, U.S. Environmental Protection Agency, available at http://www.epa.gov/otaq/climate/regulations/420f10014.pdf (accessed October 19, 2010)

U.S. EPA (2010d) *Interim Joint Technical Assessment Report: Light-Duty Vehicle Greenhouse Gas Emission Standards and Corporate Average Fuel Economy Standards for Model Years 2017– 2025*, Office of Transportation and Air Quality, available at http://www.epa.gov/otaq/climate/regulations/ldv-ghg-tar.pdf (accessed October 21, 2010)

U.S. EPA (2010e) *Regulatory Announcement: EPA and NHSTA Propose First-Ever Program to Reduce Greenhouse Gas Emissions and Improve Fuel Efficiency of Medium- and Heavy-Duty Vehicles: Regulatory Announcement*, EPA-420-F-10-901, U.S. Environmental Protection Agency, available at http://www.epa.gov/otaq/climate/regulations/420f10901.pdf (accessed October 29, 2010)

U.S. EPA (2010f) *Regulatory Announcement: EPA and NHTSA to Propose Greenhouse Gas and Fuel Efficiency Standards for Heavy-Duty Trucks; Begin Process for Further Light-Duty Standards*, Office of Transportation and Air Quality, EPA-420-F-10-038, U.S. Environmental Protection Agency, available at http://www.epa.gov/otaq/climate/regulations/420f10038.pdf (accessed October 19, 2010)

U.S. EPA (2010g) *Heavy-Duty Highway Compression-Ignition Engines and Urban Busses—Exhaust Emission Standards*, U.S. Environmental Protection Agency, available at http://www.epa.gov/oms/standards/heavy-duty/hdci-exhaust.htm (accessed November 15, 2010)

U.S. EPA (2010h) *Light-Duty Vehicle and Light-Duty Truck – Tier 0, Tier 1, National Low Emission Vehicle (NLEV), and Clean Fuel Vehicle (CFV) Exhaust Emission Standards*, U.S. Environmental Protection Agency, available at http://www.epa.gov/oms/standards/light-duty/tiers0-1-ldstds.htm (accessed November 11, 2010)

U.S. EPA DOT (2010) *Light-Duty Vehicle Greenhouse Gas Emission Standards and Corporate Average Fuel Economy Standards; Final Rule*, National Highway Traffic Safety Administration, 40 CFR Parts 85, 86 and 600; 49 CFR Parts 531, 533, 536, et al., U.S. Environmental Protection Agency and U.S. Department of Transportation, available at http://www.nhtsa.gov/staticfiles/rulemaking/pdf/cafe/CAFE-GHG_MY_2012-2016_Final_Rule_FR.pdf (accessed November 16, 2010)

11 Life-Cycle Energy and Greenhouse Gas Impacts of Biofuels and Biomass Electricity

Alphonse Anderson and Gregory Keoleian
University of Michigan

CONTENTS

11.1   THE CRITICAL ROLE OF LIFE-CYCLE ASSESSMENT

11.1.1   INTRODUCTION

Biofuels have caused a great deal of controversy in recent years. Opponents point to studies indicating that biofuels have increased global food prices and can cause adverse land-use changes, whereas proponents point out domestic economic benefits and increased energy security. Policymakers are responsible for passing laws that will guide society either toward or away from biofuels. Ideally, the direction would be chosen with complete understanding of the economic, social, and environmental impacts associated with biofuel production. Similarly, biomass electricity generation has the potential to scale significantly and its impacts must be understood. Life-cycle assessment (LCA) provides a comprehensive framework for evaluating the environmental impacts of biofuels and biomass electricity.

LCA is a systematic method for evaluating the environmental performance of products and processes by tracking the energy and material flows related to their production and consumption activities. These energy and material flows generate environmental, economic, and social impacts, both positive and negative. This chapter focuses on the environmental impacts, particularly energy and greenhouse gas emissions, from the life-cycles of liquid biofuels, biomass electricity, and comparable conventional fuels such as petroleum, coal, and natural gas. Through examples and a review of the literature, the most important concepts, methods, and challenges in LCA will be discussed.

LCA has four basic components (Keoleian and Spitzley 2006): goal and scope definition, a life-cycle inventory (LCI), life-cycle impact assessment, and interpretation. Goal definition and scoping is essentially a roadmap for how the life-cycle model will be constructed and executed. The system boundaries, allocation method, metrics, data characteristics, and functional unit are all established in this phase. Once these parameters are defined and a model of the life-cycle system is constructed, an LCI can be generated. The inventory is a quantification of system inputs and outputs, including energy, material, air and water pollutant emissions, and wastes. Life-cycle impact assessment can then be used to translate the energy, emissions, and material flows catalogued in the inventory into environmental impacts. Results from LCI analysis and impact assessment studies are interpreted to inform policy, engineering design decisions, and consumer choice.

11.1.2   FUNCTIONAL UNIT

The functional unit is the fundamental basis upon which life-cycle model results are calculated. Critically, if different life-cycle studies are to be compared, the functional unit must allow for comparison on an equivalent basis. For example, directly comparing the life-cycle impacts of 1 gallon of ethanol fuel to 1 gallon of gasoline is not appropriate because the fuels have different energy contents by volume. A functional unit of 1 mile driven on each of the fuels (in a single, well-defined vehicle) would provide a better perspective. The functional unit should be measurable, meaningful to the intended audience, and relevant to data collection (Keoleian and Spitzley 2006).

11.1.3 METRICS

LCA provides metrics that can be used to measure progress toward environmental sustainability (Keoleian and Spitzley 2006). Metrics are established in the goal and scope definition phase of an LCA. In this chapter, the net energy ratio (NER) is often used in biomass energy life-cycle studies and is defined as the ratio of output energy (e.g., delivered energy such as transport fuel energy or electricity) to input energy (total nonrenewable energy invested). The NER is a unitless intensity factor that gives an idea of overall system efficiency. However, the NER gives no indication of cropland productivity (Liska and Cassman 2008).

Net energy value (NEV) is another commonly used energy metric in biofuel life-cycle studies. An example formula used to calculate this metric, taken from a study presented later in this chapter (Farrell et al. 2006), is shown in Equations 11.1–11.3.

$$NEV = \text{Output Energy} - \text{Input Energy} \qquad (11.1)$$

$$\text{Input Energy} = (\text{Agricultural Energy/Net Yield}) + \text{Biorefinery Energy} \qquad (11.2)$$

$$\text{Output Energy} = \text{Fuel Energy} + \text{Co-product Energy} \qquad (11.3)$$

Agricultural Energy is the farm input, machinery, and labor energy required per hectare of cultivated cropland. It is calculated in units of megajoules (MJ) of input energy per hectare and includes the following upstream and on-site energy types: embedded energy in seed, fertilizer, and farm machinery; transport energy of farm inputs; farm input packaging energy; and on-site farm energy consumption. The *Net Yield* is in units of liters of ethanol fuel per hectare. *Agricultural Energy* divided by *Net Yield* is in units of MJ input energy per liter of ethanol. This is added to *Biorefinery Energy*, which considers transport of corn to the refinery, refinery energy consumption, energy embedded in refinery water, and equipment manufacturing (capital) energy—all per liter of ethanol delivered.

Critics of net energy metrics point out that they do not account for the quality of different energy inputs (Dale 2007). For example, 1 MJ of coal is not equivalent to 1 MJ of petroleum in terms of energy density or delivery requirements. Energy metrics that are more relevant to policy could be in terms of petroleum displacement or fuel production per area of land used. However, no policy decision should be based solely on the performance of a system in terms of a single metric. For example, Farrell et al. 2006 determined that the life-cycle petroleum requirements of average corn grain ethanol are less than those of cellulosic ethanol (note that technology for cellulosic ethanol is in an earlier stage than corn ethanol). However, according to results from the same model, cellulosic ethanol has a much higher life-cycle net energy value.

11.1.4 ALLOCATION METHODS

As biofuels are converted from biomass to refined liquid fuels through industrial processes, usable or otherwise marketable co-products are generated. These co-products required energy inputs and created impacts. Allocation is the assignment of responsibility for positive and negative impacts to processes, products, and co-products. Allocation methods vary; different studies have used mass, market value, energy content, equivalent product displacement, and service time as bases.

One problem with using mass or market value is the assumption that process inputs, such as energy or materials, are concentrated in particular outputs based on one variable (e.g., retail value). A heavier product does not necessarily require more energy, and a more expensive product will not necessarily create more impacts, but use of these methods can help relate impacts to products in socially meaningful ways.

Biofuel life-cycle studies have been moving away from market value and adopting the displacement approach as an allocation method (Kendall and Chang 2009). The displacement method allows the modeled product to receive credit for burdens that would have been generated had it not replaced an equivalent conventional product. For example, dried distiller's grain with solubles (DDGS) is a co-product of corn-grain ethanol production. It is high in protein and can be used as feed for livestock. When DDGS is used in place of conventional feed, such as corn and soybean meal, less conventional feed production is required [A 2007 U.S. Department of Agriculture (USDA) survey found that 13.6% of the dairy and hog farmers who responded use DDGS as livestock feed, and the single biggest reason it was not used was lack of availability (USDA/NASS 2007)]. An LCA that uses the displacement method for allocation would consider the impacts of DDGS and the avoided impacts of not using conventional feed. This allocation method was used in a biofuel life-cycle study to examine the effects of ethanol fuel displacing gasoline (Kim and Dale 2005). Kim and Dale (2005) used the International Organization for Standardization (ISO) 14041 standard system expansion approach (ISO 1998) and included alternative products by expanding the life-cycle system boundaries to allow functionally equivalent product comparisons. The life-cycle greenhouse gas emissions related to four cropping systems are presented in Table 11.1. The bold avoided system values in Table 11.1 are credits allocated to the biofuel systems. Ideally, the displacement method would use a general equilibrium model to simulate complex market interactions and accurately determine the displacement ratio of conventional products by introduced products (e.g., Does 1 lb of DDGS displace 1 lb of conventional feed? How does broad introduction of DDGS affect other market prices, and thus consumer choices?). Currently, general

---

TABLE 11.1
Example Life-Cycle GHG Emissions Results with Displacement Method for Modeling Co-Products

	GHG Emissions (Mg CO_2-eq/ha)			
	CS	CC	CC50	CwC70
Agricultural Process	23	28.1	55.4	12
Wet milling	79.5	160	158	167
Avoided co-product systems	**1.3**	**–137**	**–136**	**–143**
Soybean milling	13.7	–	–	–
Biodiesel production	6.2	–	–	–
B20 driving	127	–	–	–
Avoided B20 driving system	**–157**	–	–	–
Corn stover conversion	–	–	4.2	6.2
Avoided electricity	–	–	**–38.1**	**–56.1**
E10 driving	1794	3618	5067	5967
Avoided E10 driving system	**–1923**	**–3877**	**–5431**	**–6395**
Total	–35.4	–209	–320	–442

Source: Kim, S. and Dale, B.E., *Biomass Bioenergy,* 29, 426–439, 2005.
Note: CS, corn–soybean crop rotation (also produced biodiesel in the form of B20 blended fuel); CC, corn only; CC50, corn only with 50% residue removal; and CwC70, corn only, except wheat in the winter, with 70% residue removal.

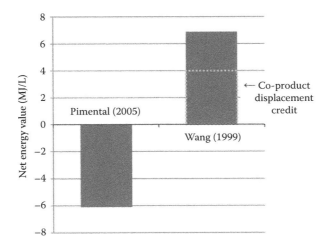

FIGURE 11.1   NEV of two corn ethanol life-cycle studies. (From Farrell, A.E. et al., *Science*, 311, 506–508, 2006.)

equilibrium modeling is not integrated into life-cycle analysis, but it remains an active area of research (Delucchi 2008; Kammen et al. 2008).

To get an idea of the relative effect of allocation, consider Figure 11.1, which shows the life-cycle energy of two corn ethanol LCAs (Farrell et al. 2006). Pimentel's study calculated an NEV of −6.1 whereas results from Wang's GREET model show an NEV of +6.9. Wang's study used the displacement method to calculate the benefits of co-product use in the market, whereas Pimentel's study did not use any allocation method. The importance of selecting an appropriate allocation method is likely to grow as further optimization of biorefinery processes will potentially lead to a more diverse and efficiently produced portfolio of high-value co-products (Lyko et al. 2009).

11.1.5 DATA

Data collection is one of the greatest challenges in LCA because data may be proprietary, aggregated, or inconsistently available. Because biofuel production technology changes relatively rapidly, age of data is very important. For example, most current ethanol production facilities have been built since 2003, and fuel production has more than tripled (RFA 2009b). The energy efficiency of newer corn grain-to-ethanol conversion facilities has also increased (Perrin et al. 2009). In general, energy data tend to be more accurate than air emissions data because differences in regulatory limits, technology, and measurement practices cause greater variance in the values (Keoleian and Spitzley 2006).

Aside from age, data can also be characterized in terms of source and quality. Primary data are collected directly (e.g., by taking measurements of process inputs or obtaining information from a facility manager), and ISO 14041 documentation standards should be observed. Frequently, primary data are unavailable and secondary data, available from LCI databases and literature sources, must be used. Secondary data may represent industry-wide average values for process inputs or outputs, which can account for the mix of older and newer technologies in use.

Data quality is another important consideration, addressable using the following indicators:

* *Precision:* Measure of data variability.
* *Completeness:* Fraction of total primary sources reporting data.
* *Representativeness:* Temporal, geographic, and technology characteristics.
* *Consistency:* A qualitative assessment of uniformity of analysis methodology (i.e., a standardized data collection method).
* *Reproducibility:* The study can be independently repeated and achieve the same result.

11.1.6 System Boundaries

Defining system boundaries is part of an iterative process, in which a balance must be struck between achieving study goals, analysis feasibility, and data availability (Keoleian and Spitzley 2006). In practice, system boundaries are defined with a cutoff rule—flows below a specified mass, energy amount, or environmental relevance criteria are deemed negligible and ignored (Keoleian and Spitzley 2006).

One contributing factor to the differing NEVs in Figure 11.1 is the Pimentel and Patzek (2005) study's inclusion of farm worker food and transportation energy. These parameters are generally considered to be outside system boundaries in other biofuel LCAs. Recently, biofuel studies have controversially (Sylvester-Bradley 2008) examined the effects of indirect land-use change on greenhouse gas (GHG) emissions. This issue is discussed in greater detail in Section 11.4. It should be noted that the federal Renewable Fuel Standard, which mandates biofuel production levels for the United States using life-cycle GHG reduction criteria, does include indirect land-use change.

11.1.7 Modeling

Life-cycle models are constructed according to system boundaries, allocation rules, data characteristics, and assumptions. Several well-developed models were utilized in the life-cycle studies reviewed in this chapter. Argonne National Energy Laboratory's peer-reviewed Greenhouse Gases, Regulated Emissions, and Energy Use in Transportation (GREET) series 1 model is capable of simulating the well-to-wheel fuel production pathways of more than 100 fuels (ANL 2009). GREET calculates an LCI of energy consumption, GHG emissions, and criteria pollutant emissions based on user-specified scenarios.

DAYCENT, another well-developed model used in the following studies, is a daily time-step version of the CENTURY dynamic soil model (Del Grosso 2009). It allows nitrous oxide emissions to be estimated from cultivated farm fields on the basis of inputs that include climate and soil parameters. Models can be one of the greatest sources of uncertainty because they may represent systems that are complex and not well understood.

11.1.8 Life-Cycle Inventory and Impact Assessment

The LCI is a quantification of the material, energy, and emissions inputs and outputs that pass through the system boundary. The LCI is generated by simulating a scenario with the life-cycle model. The ISO (ISO 1998, 2000) and the U.S. Environmental Protection Agency (EPA) (EPA 2008) offer guidance on preparing LCIs. To characterize the environmental and social effects of LCI flows, life-cycle impact assessment is required (Keoleian and Spitzley 2006).

The purpose of impact assessment is to better understand how the modeled product or system affects the surrounding environment. The impacts can be measured by indicators related to resources, human health, and ecological health. This assessment depends on classification and characterization of LCI results. For example, carbon dioxide (CO_2) and methane (CH_4) emissions can be classified as causing global warming impacts and characterized in terms of kg CO_2-equivalent (CO_2e) emissions by using their global warming potential values, available from Intergovernmental Panel on Climate Change assessment reports (IPCC 2009).

11.2 BIOFUELS LIFE-CYCLE ENERGY AND GHG EMISSIONS

11.2.1 Introduction

Over half of the world's fuel ethanol production is based in the United States, as shown in Figure 11.2, and greater than 93% of this ethanol is derived from corn grain (EERE 2006). In 2007, nearly 23% of the U.S. corn crop was used to supply the ethanol market (USDA 2009). The second-largest producer of ethanol is Brazil, with most of their ethanol derived from sugarcane. Although biodiesel production is an order of magnitude less than ethanol production in the United States (EIA 2009),

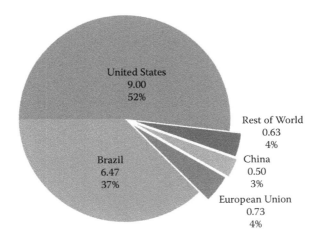

FIGURE 11.2 World fuel ethanol production in billion gallons, 2008. (Data from RFA, Industry statistics: 2008 world fuel ethanol production table, Renewable Fuel Association, available at http://www.ethanolrfa.org/industry/statistics/#E (accessed July 27, 2009), 2009a.)

its life-cycle impacts are significant. Soybeans are the primary U.S. biodiesel feedstock, whereas Europe uses rapeseed and sunflower oil, and Malaysia uses palm oil (EIA 2007). Other biofuel feedstock sources include sugarbeets, wheat, and potatoes for ethanol, and jatropha, palm oil, coconuts, and canola for biodiesel (Quirin et al. 2004). The following section primarily focuses on corn ethanol, sugarcane ethanol, and soybean biodiesel fuel production pathways.

11.2.2 Life-Cycle Energy

11.2.2.1 Ethanol

One of the most discussed biofuel topics is life-cycle energy. Does it cost more net energy to produce liquid fuel from corn than from crude petroleum, and which feedstock crops have the lowest life-cycle energy requirements? Direct comparison of results across studies is often difficult because of variance in system boundaries, allocation methods, and data sources. Here we will discuss the results and key assumptions from some significant biofuel life-cycle studies.

The general consensus on corn ethanol is that it is energetically beneficial compared with gasoline, although the degree of this benefit varies and is sensitive to modeling assumptions. Farrell et al. (2006) performed a meta-analysis using the ERG Biofuel Analysis Meta-Model (EBAMM) to analyze six corn ethanol biofuel studies that best represented the perspectives of the original authors. Some of the studies included or excluded different parameters and allocation methods, making comparison difficult, or used low-quality and unverifiable data. System boundaries were redrawn to correct for these inequalities and where necessary add the following parameters: farm machinery embodied energy, inputs packaging, capital equipment embodied energy, process water, effluent restoration, and co-product credits. Although Farrell et al. (2006) pointed out the inadequacy of net energy as a metric, they used it to compare previous study results because it was used by the original authors.

Figure 11.3 shows the EBAMM analysis life-cycle energy results. In the figure, the hollow circles represent the results originally published by the respective authors and the filled circles represent the EBAMM-adjusted values. The vertical axis metric was purposely designed by Farrell et al. (2006) to address the U.S. policy goal of reducing fossil fuel dependence. Only the Pimentel (Pimentel and Patzek 2005) and Patzek (Patzek 2004) biofuel studies calculated net energy values less than zero (also less than gasoline), although both studies still show a significant reduction in net petroleum consumption in comparison to gasoline. The cellulosic scenario shows a high net energy balance, but no significant reduction in petroleum input requirements per megajoule of ethanol fuel produced

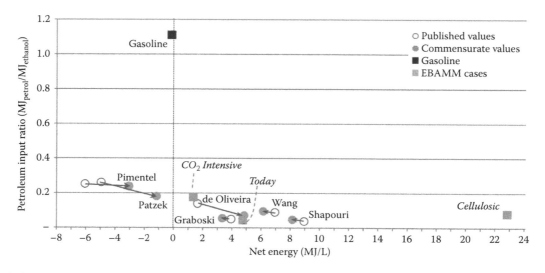

FIGURE 11.3 EBAMM net energy and petroleum requirements of corn and cellulosic ethanol. (From Farrell, A.E. et al., *Science*, 311, 506–508, 2006.)

TABLE 11.2
Cropping System Scenarios

	Cropping System Scenarios		
	Crop Rotation	Winter Cover Crop	Residue Removal (%)
CS	Corn–soybean	No	0
CC	Continuous corn	No	0
CC50	Continuous corn	No	50
CwC70	Continuous corn	Wheat	70

Source: Kim, S. and Dale, B.E., *Biomass Bioenergy,* 29, 426–439, 2005.

(cellulosic technology is still in development). According to the Farrell et al. (2006) analysis, the most sensitive life-cycle parameters were biofuel refinery energy, farm yield, biofuel refinery yield, co-product credits, nitrogen fertilizer energy, and nitrogen fertilizer application rate. Agriculture was found to demand 45–80% of the petroleum inputs across the six analyzed studies. Policies targeted at the agriculture sector could effectively reduce life-cycle petroleum consumption.

Kim and Dale (2005) examined four cropping scenarios, including two that utilized corn stover (residue), as summarized in Table 11.2. Stover refers to the nongrain portion of the plant above the ground. When stover was collected, it was used to produce ethanol and burned for electricity generation, which was assumed to offset grid electricity. It was also assumed that on a mass per acre basis, equal amounts of corn grain and stover were available for collection (i.e., 50% residue collection in the CC50 scenario means that for each unit of harvested grain, a half-unit of stover was also harvested). The corn–soybean rotation scenario is important to consider because it represents the most common cropping system used in the United States. The study's functional unit was 1 hectare (ha) of arable land considered over a 40-year period.

The nonrenewable energy impacts of each cropping system over the 40-year study period are presented in Table 11.3. The CS system consumed the least nonrenewable energy in the agriculture production phase. This is due to reduced field operations and nitrogen fertilizer requirements.

TABLE 11.3
Nonrenewable Energy Requirements

	Nonrenewable Energy (TJ/ha)			
	CS	CC	CC50	CwC70
Agricultural process	0.46	0.72	0.8	0.91
Wet milling	0.81	1.63	1.61	1.7
Avoided co-product systems[a]	−0.07	−1.89	−1.87	−1.98
Soybean milling	0.22	–	–	–
Biodiesel production	0.12	–	–	–
B20 driving[b]	1.83	–	–	–
Avoided B20 driving system	−2.27	–	–	–
Corn stover conversion	–	–	0.13	0.2
Avoided electricity	–	–	−0.41	−0.6
E10 driving[b]	26.6	53.7	75.1	88.4
Avoided E10 driving system	−28.5	−57.5	−80.5	−94.8
Total	**−0.77**	**−3.38**	**−5.14**	**−6.17**

Source: Kim, S. and Dale, B.E., *Biomass Bioenergy,* 29, 426–439, 2005.

[a] Co-products: corn oil, corn gluten meal, corn gluten feed, and soybean meal and glycerine (CS system).

[b] Transportation and distribution included.

The energy values in the avoided rows are negative because they were credited to the system using displacement allocation by system expansion. Clearly, the volume of biofuel production (E10 driving) and consequent offset of nonbiofuel driving (avoided E10 driving system) had the greatest effect on life-cycle nonrenewable energy consumption. The CC50 and CwC70 systems outperformed the others primarily because the collected corn stover provided more feedstock for conversion to ethanol at a relatively low increase in agricultural input energy (see Table 11.3).

Stover electricity conversion efficiency and co-product allocation were the most sensitive parameters affecting nonrenewable energy results. When the efficiency of converting stover to electricity was increased from 15% to 32%, as suggested by one study (Stahl 1998), nonrenewable energy consumption was reduced by 16% in the CC50 system and 20% in the CwC70 cropping system. Kim and Dale (2005) also considered allocation by mass and found that it improved (decreased) the nonrenewable energy consumption of the CS system but reduced the benefits of the CC, CC50, and CwC70 cropping systems by 55%, 36%, and 31%, respectively, compared with the baseline (displacement allocation).

Switchgrass is a promising alternative ethanol feedstock because it is not a food crop, can be grown on agriculturally marginal land, and has relatively low fertilizer and pesticide requirements (Hill et al. 2006). A large body of data exists on annual corn yields, but not for switchgrass yields. A 5-year USDA study (Schmer et al. 2008) was carried out to evaluate switchgrass yield potential in several midwestern U.S. locations through larger-scale field trials on marginal cropland, rather than on small research plots. Reported agricultural data from the study were entered into the EBAMM model to calculate well-to-wheel energy. Only 1 MJ of petroleum was required to produce 13.2 MJ of ethanol over the fuel's life-cycle. This result, although presented differently in Schmer et al. (2008), approximately agrees with Farrell et al.'s (2006) EBAMM determined result for cellulosic ethanol in Figure 11.3; recalculating Schmer et al.'s (2008) result yields 1 MJ petroleum/13.2 MJ ethanol ≈ 0.08 MJ petroleum/MJ ethanol.

The United States produces more than half of the world's fuel ethanol, followed by Brazil. In Brazil, most ethanol is derived from sugarcane, and laws mandate that internally sold gasoline contain at least 25% ethanol. Wang et al. (2007) examined the well-to-wheel life-cycle of sugarcane ethanol and compared the results to those of corn ethanol using the GREET model. The baseline

TABLE 11.4
GREET Analysis of Sugarcane Ethanol Scenarios

Scenario	Key Assumptions
Sugarcane 1 (SC1)	Base case. Sugarcane ethanol produced in Brazil, used in the United States. Farm equipment manufacture and sugarcane mill construction energy excluded.
Sugarcane 2 (SC2)	Same as 1, except farm equipment manufacture and sugarcane mill construction energy are included.
Sugarcane 3 (SC3)	Same as 1, except farm equipment manufacture energy is included.
Sugarcane 4 (SC4)	Same as 3, except sugarcane ethanol is consumed in Brazil.
Petroleum	Petroleum produced and consumed in the United States. All infrastructure activities excluded.
Corn	Corn ethanol produced and consumed in the United States. Farm machinery energy included.
Cellulosic	Cellulosic ethanol from switchgrass, produced and consumed in the United States. Farm equipment manufacture included.

Source: Wang, M., et al., *Well-to-Wheels Energy Use and Greenhouse Gas Emissions of Brazilian Sugarcane Ethanol Production Simulated by Using the GREET Model,* Argonne National Laboratory, Argonne, IL, 2007

scenario was defined as U.S. consumption of sugarcane that was produced and refined into ethanol in Brazil (SC1). Table 11.4 summarizes the scenarios that were simulated in this study.

Sugarcane is bulky and heavy—70% water by weight—so sugarcane mills are typically close to the fields. Thermal energy constitutes most energy demand at the processing plant. Wang et al. (2007) assumed no external thermal energy or electricity input was required to power the conversion process. Electricity and heat co-products can be derived by burning the residue (bagasse) that remains after the sugarcane is crushed and the juice is extracted.

In this study, over 40% of the electricity generated at the fuel processing plant was assumed to be surplus, which is marketable for export to the electric grid. In reality, sugarcane ethanol processing facilities may not be connected to the grid, so electricity export was modeled as an option.

In the first three sugarcane scenarios, ethanol was produced in Brazil and shipped to the United States for consumption. Wang et al. (2007) modeled transport in Brazil to include pipeline, rail, and small amounts by truck. The ethanol fuel was then shipped from Brazil to New York or Los Angeles and consumed near the East and West Coasts (the interior was assumed to consume midwestern U.S. corn ethanol). A general overview of the life-cycle boundaries is presented in Figure 11.4.

This study also examined the effects of using five different energy sources to power a corn ethanol refinery. Figure 11.5 shows the well-to-wheel net energy balance for the base case sugarcane ethanol scenario (SC1), cellulosic ethanol, and five corn ethanol scenarios: coal, electricity grid average (Average), natural gas (NG), distiller's grains and solubles (DGS), and biomass.

We can interpret the results as follows: A corn ethanol system that uses a grid average-powered refinery produces fuel with a 20% net renewable energy gain [Btu content of ethanol − Btu fossil fuel inputs ≈ (200,000 Btu)/(1 million Btu) = 20%]. On a life-cycle energy basis, the scenarios that used DGS and biomass to power the fuel production facility outperformed the other modeled systems because DGS and biomass displaced fossil fuel inputs to the production facility. Similarly, combustion of bagasse can completely power the sugarcane ethanol production facility and result in export of surplus electricity to the grid. The surplus electricity was assumed to displace natural gas-generated electricity, which was credited to the sugarcane fuel life-cycle system.

Wang et al. (2007) found little difference amongst the four sugarcane scenarios considered in this study (see Table 11.4) in terms of fossil energy input requirements. Even when comparing scenarios SC1, where the ethanol was shipped to the United States for consumption, and SC4, where the ethanol was consumed in Brazil, the life-cycle fossil energy requirements were less than 2% different. This may seem surprising considering the shipping distance from Brazil to America, but the energy required for shipping, per functional unit, was a relatively small fraction of total life-cycle energy.

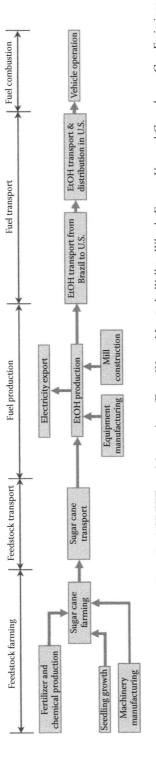

FIGURE 11.4 Brazilian sugarcane well-to-wheel GREET model overview. (From Wang, M., et al., *Well-to-Wheels Energy Use and Greenhouse Gas Emissions of Brazilian Sugarcane Ethanol Production Simulated by Using the GREET Model*, Argonne National Laboratory, Argonne, IL, 2007.)

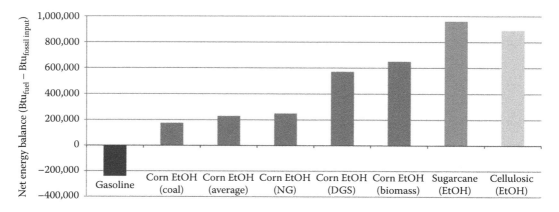

FIGURE 11.5 Net energy balance of gasoline and various bioethanol feedstocks (in Btu). (From Wang, M., et al., *Well-to-Wheels Energy Use and Greenhouse Gas Emissions of Brazilian Sugarcane Ethanol Production Simulated by Using the GREET Model,* Argonne National Laboratory, Argonne, IL, 2007.)

Wang et al. (2007) found that sugarcane ethanol can reduce life-cycle fossil energy consumption by 97% relative to gasoline. The savings are largely due to the combustion of bagasse for heat and electricity in the sugarcane processing facility, which displaces purchased natural gas-generated electricity. Wang et al. (2007) also found that sugarcane ethanol offers even greater energy benefits than cellulosic ethanol (see Figure 11.5), although the latter technology is less mature.

11.2.2.2 Biodiesel

In this section we will discuss the results of a comparative LCA of biodiesel and petroleum diesel (Sheehan et al. 1998). Biodiesel results were calculated for 100% biodiesel (B100), and a 20% bio-diesel/80% petroleum blend (B20). The functional unit in this study was 1 brake horsepower-hour (bhp-h) of work delivered by an urban bus engine. In the past, concerns have been raised regarding biodiesel's cold weather performance. Since the Sheehan study, additives and ASTM blending standards have been developed to mitigate cold weather performance issues (NBB 2008). In Minnesota, B2 and B5 (2% and 5% biodiesel, respectively) are currently used, and the mandated blend fraction is planned to increase to 20% by 2015 (Minnesota 2008). However, legislation dictates that the 20% blend is only for summer months—a 5% biodiesel blend is allowed for winter months until the cold weather issues are sufficiently addressed (MDA 2009). Sheehan et al. (1998) used #2 low-sulfur diesel as a baseline, which also exhibits reduced cold weather performance, although it has a better tolerance for low temperatures than biodiesel (MDA 2009). The life-cycle model of #2 low-sulfur petroleum diesel is outlined in Figure 11.6. Sheehan et al.'s (1998) study is slightly dated, but it is described in this section because of its comprehensiveness, transparency, and role as a foundation for the numerous biofuel LCAs that followed.

11.2.2.2.1 Diesel Model

Sheehan et al. (1998) focused their comparative biodiesel study on petroleum diesel, but it should also be noted that 44% of the crude oil refinery's output by mass was gasoline (co-product in Figure 11.6). Sheehan et al. (1998) assumed that foreign and domestic crude petroleum was extracted by conventional onshore (73%), conventional offshore (20%), and advanced methods (7%). The energy requirements for producing crude oil using steam injection (an advanced method) are an order of magnitude higher than those of conventional methods.

Crude oil pumped to the surface is typically mixed with natural gas and water, which must be separated before the crude can be refined. The natural gas recovered during this production stage is treated as a co-product; Sheehan et al. (1998) used allocation by mass to assign it 30% of the extraction process burdens. Crude oil production was modeled as evenly split between foreign and

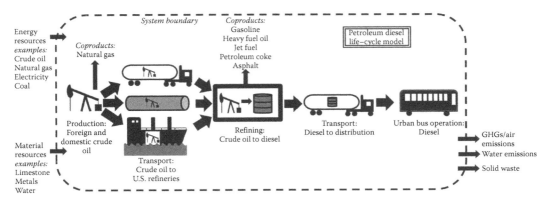

FIGURE 11.6 Life-cycle of petroleum diesel for use in an urban bus. (From Sheehan, J., et al., Life cycle inventory of biodiesel and petroleum diesel for use in an urban bus. NREL/SR-580-24089, National Renewable Energy Laboratory, Golden, CO, 1998.)

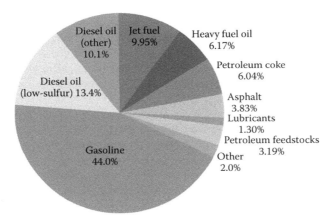

FIGURE 11.7 Petroleum diesel refinery products, by mass. (From Sheehan, J., et al., Life cycle inventory of biodiesel and petroleum diesel for use in an urban bus. NREL/SR-580-24089, National Renewable Energy Laboratory, Golden, CO, 1998.)

domestic sources. All crude petroleum was transported to diesel refineries in the United States. This assumption is fair, given that only 4% of petrodiesel entered the United States in a refined state at the time of this study. On average, foreign crude traveled 7 times farther than domestic crude. Transportation of foreign crude was 4 times more primary energy-intensive (1.09 MJ/kg for foreign vs. 0.27 MJ/kg for domestic), mainly attributed to tanker ship transportation.

The refinery model was based on U.S. average data. Over 90% of the refinery's energy was supplied by the crude petroleum entering for conversion to diesel. Refinery products and co-products are shown in Figure 11.7. This study assumed low-sulfur diesel will be used by the bus. One problem with the mass allocation method is that it does not identify refinery inputs uniquely required for diesel production (e.g., a chemical input exclusively needed for diesel production) and specifically assign their burdens to the diesel fuel. This allocation method may be selected when detailed sub-process data are not available.

11.2.2.2.2 Biodiesel Model

The Kim and Dale (2005) ethanol study discussed earlier considered the dominant corn–soybean crop rotation system, but Sheehan et al. (1998) focused solely on soybean production over one growing season (per year) with no crop rotations. Energy and emissions related to production of capital

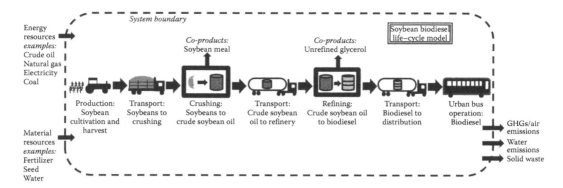

FIGURE 11.8   Soybean biodiesel life-cycle model. (From Sheehan, J., et al., Life cycle inventory of biodiesel and petroleum diesel for use in an urban bus. NREL/SR-580-24089, National Renewable Energy Laboratory, Golden, CO, 1998.)

equipment for farming were not included within the system boundaries. Crop yield data came from the 14 major soybean-producing states in the United States, which accounted for 86% of national soybean production at the time of the study. Phosphate and potash (potassium) were the most heavily used fertilizers, with lesser amounts of nitrogen also applied. Soybeans, unlike corn, are able to capture and fix nitrogen from the atmosphere. Herbicide and insecticide (agrochemicals) application were also factored into the model. The Sheehan et al. (1998) biodiesel life-cycle model is shown in Figure 11.8.

For the agriculture phase, diesel tractor use was attributed with 37% of the primary energy, followed by agrochemical production (19%), nitrogen fertilizer production (10%), gasoline tractor and gasoline truck use (10% each), phosphate fertilizer production (7%), and potash fertilizer production (4%). On average, soybeans were transported 75 miles from the fields to crushing mills, and the primary energy associated with this phase was only 5% of what was required for agricultural production.

Modeling of soybean crushing was based on past detailed studies, updated with recent data for key components such as the thermal efficiency of soybean dryers. The crushing process consists of five steps: (1) whole soybeans were stripped of their hull and ground into smaller flakes; (2) solvent was applied to chemically extract the oils; (3) the remaining bean, which no longer contained oil, was ground into a marketable meal product; (4) the mixture of solvent (hexane) and soybean oil (triglycerides) was separated for solvent reuse and further purification of the soybean oil; and (5) the soy oil was washed with water to remove unwanted gums (phosphatides).

The relative magnitudes of the electricity, natural gas, and steam requirements of the soybean crushing subprocesses are shown in Figure 11.9. Soybean drying, during receiving and storage, used natural gas-powered equipment and was one of the most energy-intensive steps at the crushing facility. The final crude soy oil product contained 92.5% of the oil originally contained in the delivered beans. The soybean crushing facility produces 18% soy oil and 82% meal, by mass. The crude soy oil is then sent to a refinery, where glycerine and methanol are recaptured for sale and reuse, respectively. The refinery produces 18% unrefined glycerol and 82% biodiesel, by mass. Detailed analysis of this production phase can help identify key opportunities for improvement.

11.2.2.2.3 Urban Bus Operation

The ethanol life-cycle studies discussed earlier were generally normalized on a fuel energy content basis to facilitate comparison with gasoline. Although a gallon of biodiesel contains approximately 10% less energy than a gallon of diesel, supporting data examined by Sheehan et al. (1998) found that biodiesel fuel economy varies only slightly in comparison to petroleum diesel. In this study, it was assumed that the biodiesel and conventional diesel buses operated at the same fuel economy.

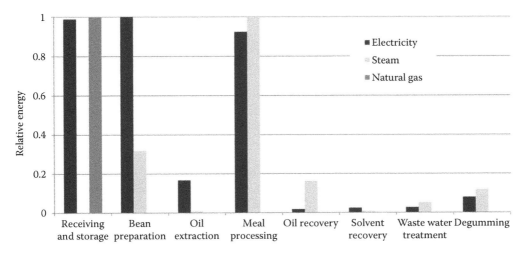

FIGURE 11.9 Relative energy requirements of soybean crushing processes. (From Sheehan, J., et al., Life cycle inventory of biodiesel and petroleum diesel for use in an urban bus. NREL/SR-580-24089, National Renewable Energy Laboratory, Golden, CO, 1998.)

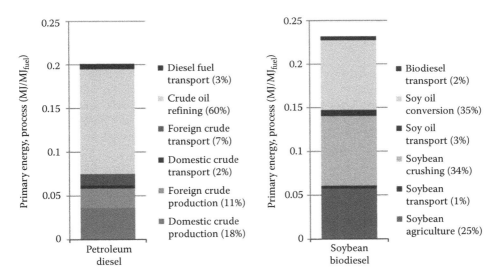

FIGURE 11.10 Primary energy inputs of petroleum diesel and soybean biodiesel production. (From Sheehan, J., et al., Life cycle inventory of biodiesel and petroleum diesel for use in an urban bus. NREL/SR-580-24089, National Renewable Energy Laboratory, Golden, CO, 1998.)

11.2.2.2.4   Results

More primary energy (60%) is associated with crude oil refining than any other phase in the life-cycle of petroleum diesel, per megajoule of fuel delivered. In the soybean biodiesel system, conversion of the oil to biodiesel required more than one-third of the total primary energy (excluding the energy content of the fuel itself). The primary energy requirements of each system are shown in Figure 11.10.

Although crude oil refining required most of the primary energy in the petroleum diesel model, it was not examined with sensitivity analysis. A recent assessment (Wang 2008) reported average U.S. petroleum refinery energy efficiency as 88% and refinery diesel output by mass fraction as 25%; both values are close to the assumed values in Sheehan et al. (1998; see Figure 11.7). However, petroleum refinery energy efficiency could improve in the future. Since the above Sheehan analysis, a Lawrence Berkeley National Laboratory report (Worrel and Galitsky 2005) concluded that the typical petroleum refinery could economically increase their energy efficiency by 10–20%.

Sheehan et al. (1998) found that using biodiesel in place of petroleum diesel reduces life-cycle energy consumption and can also significantly reduce life-cycle petroleum consumption; B20 and B100 were found to reduce life-cycle petroleum consumption by 19 and 95%, respectively, when compared with petroleum diesel.

11.2.2.3 Soybean Biodiesel and Renewable Gasoline

A more recent well-to-wheel analysis by Huo et al. (2008) of soybean biodiesel examined six different fuel production pathways: conventional petroleum gasoline, conventional petroleum low-sulfur diesel, soybean biodiesel produced by three different methods, and soybean-derived gasoline. The authors applied the allocation methods listed in Table 11.5 to each of the four soybean fuel systems.

Other studies have discussed the benefits of the displacement allocation method, but there are reasons for avoiding this approach. For one, it can be difficult to accurately determine the life-cycle impacts of the displaced conventional products, which must be known to calculate credits for avoided burdens. Second, if a large amount of co-products are generated per unit of primary product, they will be assigned a large fraction of the burdens. When this occurs and only the primary product is considered, the cumulative impacts of the production system may be masked.

Huo et al. (2008) used a hybrid allocation method that considered the energy value of fuel products and market displacement of nonfuel products (i.e., soybean meal) in three of the four soybean fuel systems (excluding the soybean transesterification system). Life-cycle calculations used the GREET model to simulate performance in year 2010. Soybean agriculture modeling was updated with 2007 USDA data, and nitrous oxide (N_2O) emissions values were updated with 2006 IPCC data. Potential land-use change was not considered in this study.

After the soybeans were cultivated and crushed to extract the crude soy oil using a hexane solvent, the soy oil was processed using one of four pathways: (1) transesterification with methanol produced biodiesel along with glycerin; (2) hydrogenation using hydrogen produced supercetane along with fuel gas and heavy oils; (3) a modified hydrogenation process using hydrogen and different inputs than in pathway 2 produced green diesel along with a propane fuel mix; and (4) catalytic cracking of soy oil produced renewable gasoline along with product gas, light-cycle oil, and clarified slurry oil. The novelty of the Huo et al. (2008) GREET analysis was modeling of catalytic cracking and two hydrogenation pathways. The difference between the two hydrogenation processes is that

---

TABLE 11.5
Allocation Methods for Soybean-Based Biofuels

Method	Description
Displacement	New products with equivalent function replace conventional products. Burdens of conventional products are avoided and credited to the system.
Energy allocation	Burdens assigned according to energy content of products. Values are constant. Combustion energy of fuel for primary product, animal digestion energy for co-product.
Market allocation	Burdens assigned according to market value of products. Dynamic values, projected for the future based on past trends.
Hybrid	Burdens assigned by either displacement or energy allocation depending on whether the subsystem output is a product or fuel (i.e., glycerin vs. biodiesel)

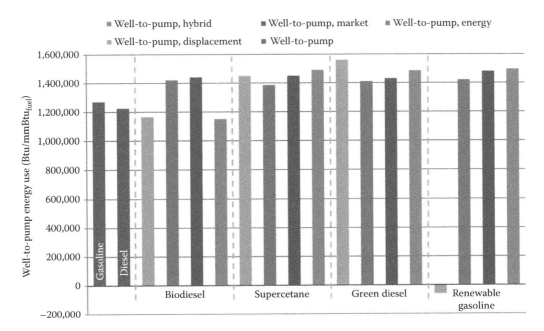

FIGURE 11.11 Well-to-pump energy requirements of six fuel production systems for various allocation methods. (From Huo, H., et al., *Life-Cycle Assessment of Energy and Greenhouse Gas Effects of Soybean-Derived Biodiesel and Renewable Fuels.* ANL/ESD 08-2, Argonne National Laboratory, Argonne, IL, 2008.)

in pathway 2 the co-product fuel gas and heavy oils were used to generate steam for supercetane production, whereas green diesel required natural gas to supply steam. Internally used supercetane co-products were subtracted from net co-product production (net co-products are those which exit the system boundary). Several commercial hydrogenation biodiesel production facilities are already in operation around the world.

The energy content of the four soybean-based fuels, petroleum gasoline, and low-sulfur diesel varies by 12%, which was accounted for by normalizing to 1 million Btu of fuel produced and consumed. Figure 11.11 shows the well-to-pump energy requirements of all six fuel pathways. Pump-to-wheel energy is not shown in the figure because it is the same for each fuel.

Production of green diesel resulted in the least energy co-products, which is why its well-to-pump energy calculated by displacement allocation was relatively high; compared with the other biofuel systems, fewer conventional products were displaced. Renewable gasoline production resulted in vast amounts of energy co-products, leading to the lowest well-to-pump energy use per million Btu of fuel produced and most petroleum displaced, as seen in Figure 11.12. The pump-to-wheel petroleum use of gasoline and diesel includes the fuel itself, which of course is zero for the bio-based fuels.

11.2.3 Life-Cycle GHG Emissions

11.2.3.1 Modeling Uncertainty

The following are examples of modeling uncertainties in biofuel LCA that can affect net GHG emissions calculations.

N_2O is a potent GHG, produced naturally from mineral nitrogen in the soil by microbes. Soil nitrogen (N) content depends on agricultural practices, such as fertilizer application, production of nitrogen-fixing crops, and retention of crop residues. For the states analyzed by the USDA in a recent survey, nitrogen fertilizer was applied to 96% of U.S. corn acres (USDA/ERS 2008b). In total,

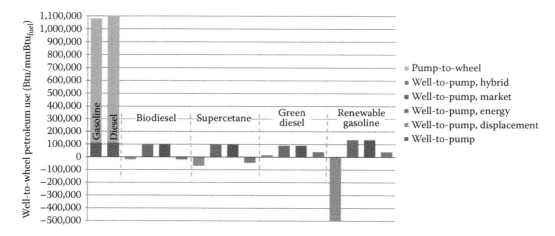

FIGURE 11.12 Full well-to-wheel petroleum requirements of six fuel production systems. (From Huo, H., et al., *Life-Cycle Assessment of Energy and Greenhouse Gas Effects of Soybean-Derived Biodiesel and Renewable Fuels.* ANL/ESD 08-2, Argonne National Laboratory, Argonne, IL, 2008.)

approximately 69% of U.S. N_2O emissions came from croplands in 2007 (EPA 2009c). Nitrogen fertilizer use creates environmental impacts, but it can also increase crop yields; it is a necessary component of agriculture system life-cycle models. However, nitrogen fertilizer application rates vary by more than 200% from state to state (USDA/ERS 2008a), and aside from the variables affecting soil N content, the conversion rate of soil N to N_2O is highly uncertain and depends on many factors (Snyder et al. 2009). It has been suggested that the N_2O conversion rate should be 3–5 times higher than the value currently recommended by the IPCC and used in many biofuel LCAs (Crutzen et al. 2007). Modeling of N_2O field emissions requires further work to reduce the uncertainty (Kendall and Chang 2009).

On the basis of current models, when indirect land-use change is included within system boundaries, life-cycle GHG emissions from corn-grain ethanol can be 93% higher than gasoline GHG emissions over a 30-year period (Searchinger et al. 2008). Conversion of previously uncultivated land is likely to increase soil organic carbon losses (Anderson-Teixeira et al. 2009), but indirect land-use change studies such as Searchinger et al. (2008) have been questioned regarding assumptions of free market behavior and geographic selection of land-use change areas (Sylvester-Bradley 2008). A lengthier discussion follows in Section 11.4.

Conservation tillage, also known as "no-till farming," is an agricultural management practice in which 30% or more of the soil surface is left covered with crop residue, rather than being plowed under after harvest. This practice can reduce soil erosion, agricultural production costs, and fossil fuel consumption. No-till farming is also thought to increase the carbon sequestration potential of soil, although this claim may be insufficiently supported by evidence (Baker et al. 2007).

11.2.3.2 Ethanol

The net greenhouse gas emissions found in the studies assessed by Farrell et al. (2006) with the EBAMM model are presented in Figure 11.13. Pimentel and Patzek calculated net energy losses and some of the highest GHG emissions—even higher than those of the EBAMM CO_2-intensive scenario. These studies have been criticized for using outdated agricultural data as well as inappropriate system boundaries and allocation methods. The ethanol today scenario is an EBAMM specific case created using the best available data from the six studies.

The largest causes of variation in GHG emissions were limestone application rates and farm machinery embodied energy assumptions. Also, several parameters affecting N_2O emissions and nitrogen displacement were not considered in the six studies: manure application, crop residue left

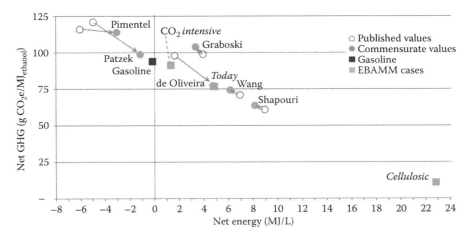

FIGURE 11.13 EBAMM net energy vs. net GHG emissions for corn and cellulosic ethanol. (From Farrell, A.E., et al., *Science,* 311, 506–508, 2006.)

on the field, growing legumes, wetland conversion to farmland, and use of a corn–soybean rotation system (soybeans capture nitrogen from the atmosphere, thus reducing fertilizer demand).

Farrell et al.'s (2006) EBAMM analysis of the six studies indicates that agriculture generates 34–44% of life-cycle GHG emissions. On the basis of sensitivity analysis results, Farrell et al. (2006) concluded that major reductions in net GHG emissions are only likely to be achieved with a cellulosic ethanol fuel system, as shown in Figure 11.13.

Kim and Dale (2005) evaluated four cropping scenarios, including a corn stover utilization option. One important feature of their study was its detailed consideration of soil dynamics, particularly nitrogen flux. The peer-reviewed DAYCENT model was used to incorporate organic carbon and nitrogen soil dynamics. The authors pointed out that 90% of corn stover is left on fields in the United States, which was assumed to increase soil organic carbon content (sequestration) but also increases N_2O emissions.

Table 11.6 shows Kim and Dale's (2005) GHG emissions results. Just as in the nonrenewable energy results in Table 11.3, life-cycle GHG emissions are also dominated by driving impacts for each cropping system scenario. Although removing corn stover from the field requires more energy and results in less soil carbon sequestration, the benefits of increased ethanol production and surplus electricity generation outweigh these effects. In the CwC70 system, winter wheat crops sequester more carbon and elevate soil fertility, which increases ethanol production.

The Schmer et al. (2008) analysis of ethanol derived from switchgrass found that this cellulosic feedstock would reduce life-cycle GHG emissions by 94% compared with gasoline. Combustion of lignocellulosic plant residue for energy, which displaces fossil fuel electricity, was cited as the main reason for the GHG savings. This study was significant because agriculture data were obtained from field-scale plots of marginal cropland rather than small research plots. Modeling was performed using EBAMM, and the switchgrass was assumed to sequester soil carbon over a 100-year period on converted croplands.

Sugarcane ethanol, as modeled by Wang et al. (2007), has a different agriculture process than corn ethanol, including burning of fields before harvest to clear out pests and sharp leaves. The authors assumed that open-field burning of sugarcane plant remnants was practiced pre- and post-harvest at a rate of 80%. Field burning emissions include: carbon monoxide (CO), CH_4, oxides of nitrogen (NO_x), N_2O, particulate matter less than 2.5 µm in aerodynamic diameter ($PM_{2.5}$), particulate matter less than 10 µm in aerodynamic diameter (PM_{10}), volatile organic carbons (VOCs), and oxides of sulfur (SO_x). Additionally, N_2O emissions result from nitrogen fertilizer application. These emissions are uncertain (see Section 11.2.3.1), but the authors assumed that 1.5% of the

TABLE 11.6
GHG Emissions for 100-Year Global Warming Potential in CO_2-eq

	GHG Emissions (Mg CO_2-eq/ha)			
	CS	CC	CC50	CwC70
Agricultural process	23	28.1	55.4	12
Wet milling	79.5	160	158	167
Avoided co-product systems	1.3	−137	−136	−143
Soybean milling	13.7	−	−	−
Biodiesel production	6.2	−	−	−
B20 driving	127	−	−	−
Avoided B20 driving system	−157	−	−	−
Corn stover conversion	−	−	4.2	6.2
Avoided electricity	−	−	−38.1	−56.1
E10 driving	1794	3618	5067	5967
Avoided E10 driving system	−1923	−3877	−5431	−6395
Total	**−35.4**	**−209**	**−320**	**−442**

Source: Kim, S. and Dale, B.E., *Biomass Bioenergy,* 29, 426–439, 2005.

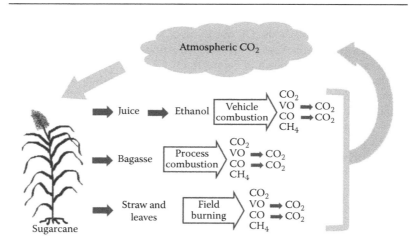

FIGURE 11.14 Brazilian sugarcane carbon cycle, as modeled by GREET. (From Wang, M., et al., *Well-to-Wheels Energy Use and Greenhouse Gas Emissions of Brazilian Sugarcane Ethanol Production Simulated by Using the GREET Model*, Argonne National Laboratory, Argonne, IL, 2007.)

nitrogen applied to the field, by weight, would be released as N_2O. The organic carbon content of the soil was assumed to remain constant. However, the resulting net carbon release could increase if sugarcane ethanol production causes direct or indirect land-use change—an issue discussed in greater detail in Section 11.4. The sugarcane carbon cycle, as modeled by Wang et al. (2007), is shown in Figure 11.14. Note that VOCs and CO were assumed to oxidize and become CO_2 within days of release to the atmosphere. Only CH_4 emissions (see Figure 11.14) were included in the calculation of GHG emissions from field burning.

According to the analysis, nearly 70% of life-cycle greenhouse gas emissions were due to growing and harvesting sugarcane. Figure 11.15 shows the base case well-to-wheel GHG emissions of Brazilian sugarcane ethanol. Burning of sugarcane fields contributed 24% of total life-cycle GHG emissions, in the form of CH_4. It should be noted, again, that this practice is being phased out, and emissions from this source are assumed to be zero by 2020 in the GREET model.

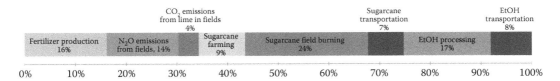

FIGURE 11.15 Brazilian sugarcane ethanol life-cycle GHG emissions. (From Wang, M., et al., *Well-to-Wheels Energy Use and Greenhouse Gas Emissions of Brazilian Sugarcane Ethanol Production Simulated by Using the GREET Model,* Argonne National Laboratory, Argonne, IL, 2007.)

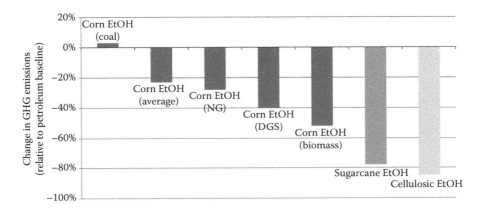

FIGURE 11.16 Change in life-cycle GHG emissions, relative to petroleum. (From Wang, M., et al., *Well-to-Wheels Energy Use and Greenhouse Gas Emissions of Brazilian Sugarcane Ethanol Production Simulated by Using the GREET Model,* Argonne National Laboratory, Argonne, IL, 2007.)

Wang et al.'s (2007) life-cycle GHG results shown in Figure 11.16 are relative to petroleum. According to the study, if the ethanol refinery used coal as its process fuel, GHG emissions would be higher than those of the petroleum baseline. In the two corn ethanol scenarios with the greatest GHG reductions, the DGS co-product or field-collected biomass residue were burned in the processing plant to supplement energy requirements, much like cane residue use in the sugarcane scenario. Surplus electricity from bagasse combustion in the sugarcane mill was assumed to displace natural gas electricity from the grid. Wang et al.'s (2007) study found these avoided emissions to be the primary reason sugarcane ethanol can substantially reduce life-cycle GHGs.

11.2.3.3   Biodiesel

Because of the high uncertainty involved in estimating N_2O emissions, Sheehan et al. (1998) did not include this parameter in their model. Petroleum diesel life-cycle CO_2 was dominated by tailpipe emissions, which contributed 87% of the total. The next-largest source of GHG emissions was crude oil refining. Most crude petroleum extraction was by conventional onshore and offshore methods. If economic or policy drivers caused wider adoption of either the energy-intensive steam injection or CO_2 injection methods, crude oil production GHGs could significantly change. However, even if CO_2 emissions from crude production doubled or tripled, fuel combustion would still dominate life-cycle CO_2 emissions (see Figure 11.17).

In the biodiesel life-cycle system, vehicle tailpipe CO_2 emissions are offset by CO_2 uptake during soybean plant growth, as shown in Figure 11.18. Sheehan et al. (1998) found that B20 and B100 reduce life-cycle CO_2 emissions by 16 and 78%, respectively, when compared with petroleum diesel. Note that this study predates the more recent considerations of incorporating indirect land-use change GHG emissions, as in the EPA Renewable Fuel Standard modeling (EPA 2010) discussed

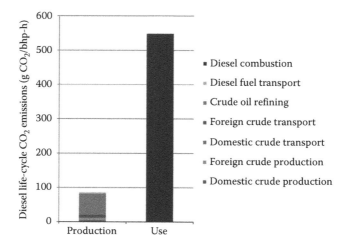

FIGURE 11.17 Petroleum diesel life-cycle CO_2 emissions. (From Sheehan, J., et al., Life cycle inventory of biodiesel and petroleum diesel for use in an urban bus. NREL/SR-580-24089, National Renewable Energy Laboratory, Golden, CO, 1998.)

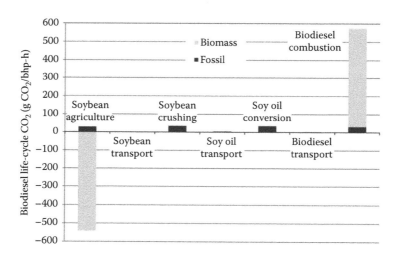

FIGURE 11.18 Soybean biodiesel life-cycle CO_2 emissions. (From Sheehan, J., et al., Life cycle inventory of biodiesel and petroleum diesel for use in an urban bus. NREL/SR-580-24089, National Renewable Energy Laboratory, Golden, CO, 1998.)

later in this chapter. However, Sheehan et al. (1998) is still one of the more transparent and well-documented biodiesel LCAs in publication.

11.2.3.4 Soybean Biodiesel and Renewable Gasoline

Huo et al. (2008) used an updated version of the GREET model to calculate the life-cycle GHG emissions of four soybean biofuel production pathways, petroleum gasoline, and low-sulfur petroleum diesel. In this study, GHG emissions are the sum of CO_2, CH_4, and N_2O emissions, converted to CO_2e using IPCC 100-year global warming potential factors (IPCC 2007).

Figure 11.19 shows the well-to-wheel GHG emissions determined for the six fuels by each of the allocation methods considered in the study and described in Table 11.5. The greatest GHG savings were calculated with displacement allocation, except in the green diesel system. With the

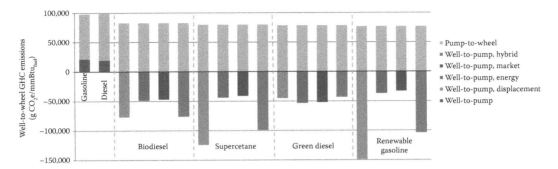

FIGURE 11.19 GHG emissions for six fuel production pathways. (From Huo, H., et al., *Life-Cycle Assessment of Energy and Greenhouse Gas Effects of Soybean-Derived Biodiesel and Renewable Fuels.* ANL/ESD 08-2, Argonne National Laboratory, Argonne, IL, 2008.)

displacement method, supercetane reduces GHG emissions by 130% compared with petroleum diesel, and renewable gasoline reduces GHG emissions by 174% compared with gasoline. The reason for these savings is that not only do the primary products (i.e., supercetane and renewable gasoline) displace liquid petroleum fuels, but their co-products also displace substantial amounts of fossil fuels. Using the energy, market, and hybrid allocation methods, life-cycle GHG emissions were calculated as declining by 57–94% across the four modeled biofuel pathways.

11.2.4 OTHER BIOFUEL SOURCES

U.S. corn and Brazilian sugarcane dominate the global ethanol production market, but other biofuel crops have received attention around the world on the basis of their suitability to regional climates and resources. A wide range of biofuel feedstocks, although not all, have been evaluated with LCA—a few alternative feedstock sources are discussed in Sections 11.2.4.1 and 11.2.4.2.

11.2.4.1 Ethanol

One proposed source (Kalogo et al. 2007) of ethanol is municipal solid waste (MSW). It has been estimated that approximately 2–3.6 billion gallons of ethanol could be produced from MSW annually (compare this with the 9 billion gallons of ethanol produced in the United States in 2008; see Figure 11.2), and that conversion of MSW to ethanol for vehicle use would reduce fossil fuel use by displacing gasoline and corn ethanol. It could also reduce GHG emissions by 65 and 58%, compared with gasoline and corn ethanol, respectively. As of July 2009, the EPA intends to disallow MSW as an ethanol fuel source under the Energy Policy Act of 2007, but it is seeking further comment on this issue because exclusion would narrow available options for meeting legislative transportation fuel GHG reduction targets (EPA 2009a).

11.2.4.2 Biodiesel

One study (Reijnders and Huijbregts 2008) of the production of palm oil, which can be converted to biodiesel, assessed agriculture and palm oil production processes in South Asia. The authors point out that large amounts of organic residue are generated during palm oil production. These residues could be combusted to produce steam and electricity, although this is rarely done in practice; they assumed 75% of palm oil production energy was provided by fossil fuels, in accordance with actual practice. Wang et al. (2007) found that sugarcane bagasse displacement of grid electricity was the largest contributor to life-cycle GHG emissions reductions. However, the dominant GHG emissions source in Reijnders and Huijbregts (2008) was the direct conversion of natural lands to palm oil plantations. Total life-cycle GHG emissions were estimated to be in the broad range of 2.6–18.2 t CO_2e/t of palm oil. Conversion of forests and peat lands contributed between 1.5 and 17 t CO_2e/t palm oil, or 58–93% of life-cycle GHG emissions, respectively. Processing palm oil into biodiesel was estimated to produce an additional 0.7 t CO_2/t biodiesel.

Reijnders and Huijbregts (2008) assumed 75% fossil fuel use during palm production, but if the processing facility were 98% powered by palm residue, as de Vries (2008) suggests to be more accurate, life-cycle GHG emissions would be 3–21% lower than those reported above. However, other recent findings (Science Daily 2007) suggest that CO_2 emissions from peat land conversion may be more than 4 times higher than the values used in Reijnders and Huijbregts (2008), significantly increasing life-cycle GHGs.

Jatropha is a tropical plant that can be converted to biodiesel. Except for promising select varieties that are not currently used, most *Jatropha curcas* varieties are toxic. This makes it more difficult to process the seeds and not possible to generate co-product animal feeds from the meal (King et al. 2009). However, several Asian and African countries, most notably China and India, have shown interest in developing jatropha as a biodiesel feedstock (Fairless 2007).

11.3 BIOMASS ELECTRICITY

11.3.1 INTRODUCTION

In 1978, the Bioenergy Feedstock Development Program began at Oak Ridge National Laboratory and soon recognized the potential value of herbaceous bioenergy crops such as switchgrass. Research in soil science, management techniques, biotechnology, and other areas led to yield increases of 50%, projected production cost reductions of 25%, and nitrogen fertilizer reductions of 40% (McLaughlin and Kszos 2005). Different species of switchgrass were identified and optimized regionally. Switchgrass is prized for its stress tolerance and ability to grow on marginal lands. As will be discussed in Section 11.4, bioenergy feedstocks that are noncompetitors for cropland can avert significant life-cycle GHG emissions.

Short rotation woody crops (SRWC) are another biomass source with large potential. In northern temperate areas, SRWC development has focused on willow shrubs and hybrid poplar, whereas eucalyptus has been studied for warmer climate applications. Willow has several advantageous characteristics: high biomass production in short time periods, a broad genetic base and ease of breeding, and the ability to re-sprout after multiple harvests (Keoleian and Volk 2005). Like switchgrass, willow is able to reach its peak height and mass after only a few years.

The biomass electricity systems considered in this section utilize perennial feedstocks—crops are not replanted each year, only harvested. This is beneficial from life-cycle energy and GHG perspectives because it reduces farm operations and retains the soil carbon stored by plant root systems. Next-generation biofuel feedstocks (e.g., switchgrass and giant miscanthus) are also perennial, but they result in an end product that is not necessarily comparable because electricity and liquid fuel are not functionally equivalent (pending the coming electric vehicle market). However, as will be shown below, biomass electricity systems are better at leveraging fossil fuel inputs than biofuel systems.

11.3.2 LIFE-CYCLE ENERGY

11.3.2.1 Willow

Willow biomass can be burned to directly generate heat or produce electricity and could possibly serve as an ethanol feedstock in the future. It is typically harvested on a 3- or 4-year cycle, with seven to ten harvests per system before replanting. As an example, see Table 11.7 (Heller et al. 2003). The biomass is not available for harvest until the third year and is subsequently cut and collected every 3 years to allow the plants to reach sufficient density. Biomass density factors include genetic variety, length of growing season, soil conditions, and climate conditions. Root systems remain in the ground until the 23rd year, with only the upper portion of the plant harvested during this period.

TABLE 11.7

Willow Biomass Cropping System Management and Harvest Schedule

Year	Season	Activity
0	Fall	Mow, contact herbicide, plow, disk, seed covercrop, cultipack
1	Spring	Disk, cultipack, plant, pre-emergent herbicide, mechanical and/or herbicide weed control
1	Winter	First year coppice
2	Spring	Fertilize
3	Spring	
4	Winter	First harvest
5	Spring	Fertilize
6	Spring	
7	Winter	Second harvest
(8–22)		(Repeat 3-year cycle for 3rd–7th harvest)
23	Spring/summer	Elimination of willow stools

Source: Keoleian, G.A. and Volk, T.A., *Crit Plant Sci Rev*, 24, 385–406, 2005.

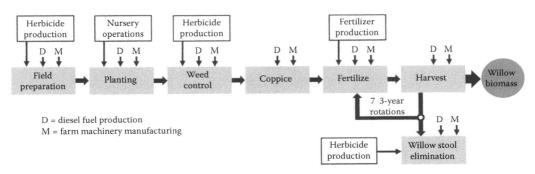

FIGURE 11.20 Willow biomass production agriculture model. (From Keoleian, G.A. and Volk, T.A., *Crit Plant Sci Rev*, 24, 385–406, 2005.)

Keoleian and Volk (2005), building on previous work (Heller et al. 2003, 2004), provided a comprehensive assessment of the energy, GHG, air pollutant, land requirements, and economic impacts of a willow biomass-to-electricity conversion system in New York state. Herein, the focus is placed on the system's life-cycle energy balance.

The base case agriculture scenario in Keoleian and Volk (2005) followed the schedule described in Table 11.7. Willow biomass yields were expected to increase 30–40% by later harvests, but willow plants devote significant amounts of energy to establishing their roots in their first few years. The willow biomass was assumed to contain 50% moisture by dry weight at harvest. Combustion of this biomass would subsequently cause the power plant's efficiency to slightly decrease. The agricultural model is outlined in Figure 11.20.

Combined biomass and coal combustion at a power plant (i.e., co-firing) was examined for two 90% coal/10% biomass scenarios, in addition to a baseline 100% coal input scenario. Co-firing above 2% requires power plant modification; the impacts of these material requirements were included in this study (retrofit material production in Figure 11.21). Manufacture of the original power plant equipment was excluded because prior studies have shown it affects system life-cycle indicators by less than 1% (Spath et al. 1999). The biomass-to-electricity system model is outlined Figure 11.21. The functional unit in this analysis was 1 MWh of electricity delivered to the grid.

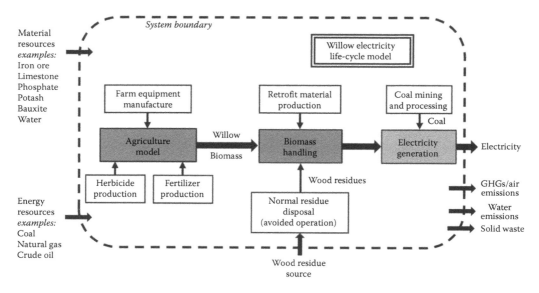

FIGURE 11.21 Life-cycle model of willow biomass for electricity generation. (From Keoleian, G.A. and Volk, T.A., *Crit Plant Sci Rev*, 24, 385–406, 2005.)

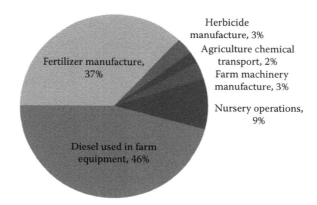

FIGURE 11.22 Primary energy distribution for the agriculture phase of willow biomass production. (From Keoleian, G.A. and Volk, T.A., *Crit Plant Sci Rev*, 24, 385–406, 2005.)

Fertilization and harvesting account for the majority of the primary energy required during the agriculture phase of willow biomass production. Primary energy requirements by activity are shown in Figure 11.22.

Heller et al. (2003) reported the agriculture NER (energy in the biomass at farm exit divided by input fossil energy) to be 16.6 after the first crop rotation, but it cumulatively increased to 55.3 after seven rotations. This value was higher than what was found in previous biomass studies because of yield assumptions, fertilizer application rates, and slightly different boundaries. When prior study results were recalculated to account for these differences and compared to the Heller et al. (2003) results, the agriculture net energy ratios were proximate.

Heller et al. (2004) calculated the NER of the entire willow-to-electricity system as the power plant electricity output divided by total life-cycle fossil fuel input. This metric offers the means to compare the renewable energy benefit of different energy systems. The NERs for the co-fire scenarios are approximately 9% better than the coal only case. However, the systems that relied on direct-fired and gasified willow biomass produced approximately 40–50 times more electricity per unit of fossil fuel input energy than the U.S. grid average, as shown in Table 11.8. For comparison purposes, building-integrated photovoltaic (BIPV) and wind energy systems are included in the table (Keoleian and Lewis 2003; Schleisner 2000).

11.3.2.2 Mixed Wood Waste

A similar LCA of biomass co-fired electricity compared a pulverized coal power plant to 5 and 15% biomass co-fire facilities (Mann and Spath 2001; see Figure 11.23). The LCA boundaries included surface coal mining operations, power plant capital equipment production, transport of biomass and coal, upstream grid electricity production, and displacement credits for avoiding the assumed typical fate of the biomass feedstock. The model considered biomass from various sources, including urban wood waste, mill waste, construction residue, and industrial wood residue. The physical properties of this mixed feedstock were defined similarly to those of the willow studied by Keoleian and Volk (2005), except that the energy content was approximately 8% lower, by mass, and the ash content was approximately 50% higher. Higher ash content can increase solid waste generation, although solid waste is not discussed in this chapter.

TABLE 11.8
NER for Various Electricity Production Systems

Electricity Source	NER (electricity out/fossil energy in)
Coal only (pulverized coal)	0.313
10% Blend co-fire (9.5% wood waste, 0.5% willow, 90% coal)	0.341
10% Willow co-fire (10% willow, 90% coal)	0.342
U.S. grid average	0.257
Gasified willow (100% willow gasified)	12.9
Direct-fired willow (100% willow combusted)	9.9
BIPV	4.3
Wind	30.3

Source: Keoleian, G.A. and Volk, T.A., *Crit Plant Sci Rev,* 24, 385–406, 2005.

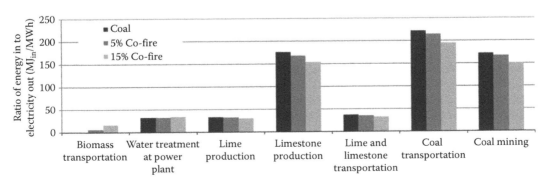

FIGURE 11.23 Energy inputs per MWh of electricity output for various life-cycle activities. (From Mann, M.K. and Spath, P.L., *Clean Prod Processes,* 3, 81–91, 2001.)

TABLE 11.9
NER of Coal and Co-Fired Wood
Waste Electricity Generation

Electricity Source	NER (electricity out/fossil energy in)
Coal	0.31
5% Co-fire	0.32
15% Co-fire	0.35

Source: Mann, M.K. and Spath, P.L., *Clean Prod Processes*, 3, 81–91, 2001.

Mann and Spath's (2001) energy results were calculated in terms of a NER and ratio of energy inputs to electricity produced, as shown in Figure 11.23. Coal consumption at the power plant is not included in Figure 11.23. In this study, the NER was defined as the electricity delivered to the grid divided by the system fossil energy inputs. Mann and Spath's (2001) NER values agree very well with the willow NER reported by Heller et al. (2004). The NER increases with the biomass co-fire fraction, as shown in Table 11.9, directly reflecting an increase in overall system efficiency.

Mann and Spath (2001) found coal transport to be highly energy-intensive compared with moving harvested biomass. Production of limestone and lime, key components of sulfur dioxide (SO_2) power plant emissions control, were also significant contributors to life-cycle energy requirements. In this study, biomass co-firing at rates of 5% and 15% reduced total energy consumption by 3.5% and 12.4%, respectively.

11.3.2.3 Hybrid Poplar

Hybrid poplar was studied as a feedstock source for a hypothetical biomass gasification combined-cycle (IGCC) power plant located in the midwestern United States (Mann and Spath 1997). The biomass was assumed to be supplied to the power plant in the form of wood chips that were transported by truck and rail. Mann and Spath (1997) found that transportation had a relatively minor effect on life-cycle results compared with the agriculture and power generation stages. The study included three major components: agricultural production (including farm capital equipment production and use, fertilizer and herbicide production and use, and biomass preparation), transportation (including production and use of equipment and fuel), and electricity production (including capital equipment construction and use). Raw material extraction and waste disposal options were also included within the life-cycle boundaries.

The poplar was assumed to be grown on land adjacent to the power plant, thus minimizing transportation distance. The biomass yield was nearly identical to the value found by Heller et al. (2003). Nitrogen, phosphorus, and potassium were all applied as fertilizers, and herbicide was also deemed necessary. Poplar growth and harvest occurred on a 7-year rotation, with no fertilizer applied until year 4. Transport of the harvested biomass was 70% by truck and 30% by train. A simulated gasification combined-cycle power plant was assumed to receive the poplar biomass as wood chips and operate for 30 years. The first batch of poplar was planted and harvested seven years before the power plant came online. Additional acreage was assumed to be brought into production according to a carefully sequenced schedule such that harvest cycles would occur annually and keep the power plant sufficiently supplied.

When ignoring parasitic losses at the power plant, production of the poplar feedstock required 77% of life-cycle energy. The distribution of energy requirements is shown in Figure 11.24. Farm

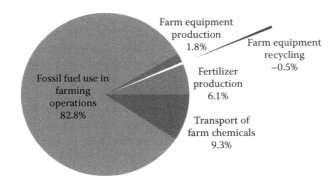

FIGURE 11.24   Poplar agricultural production energy distribution. (From Mann, M.K. and Spath, P.L., *Life Cycle Assessment of a Biomass Gasification Combined-Cycle System*, National Renewable Energy Laboratory, 1997. Available at http://www.nrel.gov/docs/legosti/fy98/23076.pdf)

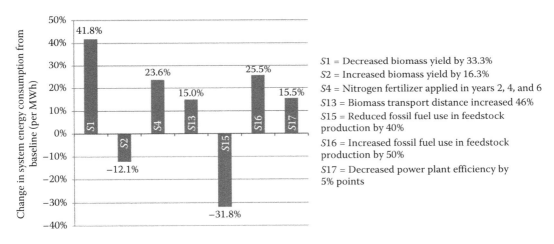

FIGURE 11.25   Change in system energy for alternative poplar biomass electricity scenarios. (From Mann, M.K. and Spath, P.L., *Life Cycle Assessment of a Biomass Gasification Combined-Cycle System*, National Renewable Energy Laboratory, 1997. Available at http://www.nrel.gov/docs/legosti/fy98/23076.pdf)

equipment recycling energy is negative because of a credit received for avoiding disposal in a landfill, which would have consumed more energy than recycling. Fossil fuel use not only dominates farming operations, but it also makes up 64% of total life-cycle energy.

Mann and Spath (1997) calculated NER (electricity delivered to the grid divided by fossil energy consumed in the system) of 15.6 for the life-cycle of poplar biomass electricity generation. Sensitivity analysis of select parameters determined that biomass yield, nitrogen fertilizer application, transportation distance, fossil fuel use on the farm, and power plant efficiency all significantly affected system energy consumption, as seen in Figure 11.25. The percent changes in energy consumption per megawatt-hour of electricity generation shown in the figure are with respect to the base case, as described above. The seven scenarios shown in the figure had the strongest influence on system energy consumption in Mann and Spath's (1997) study.

11.3.2.4 Carbon Sequestration

Power plant carbon capture and sequestration is an important emerging technology. A $1.5 billion (U.S.) commercial-scale IGCC coal power plant with carbon capture and sequestration technology is in development (as of August 2009) in the United States (DOE 2009; FutureGen Alliance 2009), and a similar project is in development in China (GreenGen 2009; WRI 2009a). Although carbon sequestration can reduce power plant stack emissions, it does not reduce upstream life-cycle impacts. For example, consider the upstream energy requirements of coal mining and transport identified by Mann and Spath (2001) in Figure 11.23.

Another study by Spath and Mann (2004) compared coal (baseline), natural gas combined-cycle, co-fired biomass, direct-fired biomass, and biomass gasification combined-cycle electricity generation with and without carbon capture and sequestration (CO_2-seq). The biomass co-fired and biomass gasification combined-cycle systems are the same as those described in Sections 11.3.3.2 and 11.3.3.3 of this chapter, respectively, and the direct-fired system is modeled as using the same biomass source as the co-fire system. Each of the systems were required to produce 600 MW of electricity in this study; for the biomass systems, achieving this output rate required more than one facility (to reduce the biomass transportation distance, smaller dispersed plants were assumed—refer to Figure 11.25, case S13). Chemical absorption using monoethanolamine was selected as the carbon capture technology and was assumed to capture 90% of the power plant's CO_2 emissions. Using CO_2-seq technology, which includes capturing and compressing the CO_2 gas, with a coal power plant required additional energy. The plant's electricity production capacity was reduced from 600 to 457 MW. To compensate for this reduction, supplemental power from a natural gas combined-cycle power plant was assumed to fill the gap. A small amount of additional energy—less than 4% of what was required for carbon capture and compression—was taken from the grid to transport the compressed CO_2 (up to 1800 km) by pipeline to an underground sequestration site.

Figure 11.26 shows the life-cycle fossil energy requirements, per kilowatt-hour of electricity produced, for the five power plant technologies, with and without CO_2-seq. CO_2-seq increased the fossil energy requirements of every system in Spath and Mann (2004). Although capturing CO_2 in the biomass systems led to a relatively large increase in energy consumption, net fossil fuel consumption was more than 80% lower than that of coal, even without CO_2-seq.

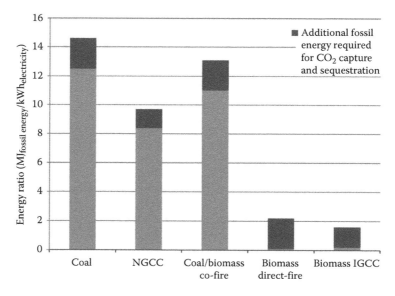

FIGURE 11.26 Life-cycle fossil energy requirements of electricity generation with and without carbon capture and sequestration. (From Spath, P.L. and Mann, M.K., *Biomass Power and Conventional Fossil Systems with and without CO_2 Sequestration—Comparing the Energy Balance, Greenhouse Gas Emissions and Economics.* NREL/TP-510-32575, National Renewable Energy Laboratory, Golden, CO, 2004.)

11.3.3 Life-Cycle GHG Emissions

11.3.3.1 Willow

As in the biofuel studies, the CO_2 taken in from the atmosphere by willow as it grows is significant. However, because electricity generation is a central component of Keoleian and Volk's (2005) analysis, and biomass provides only 10% of the power plant's fuel in the co-fire case, stack emissions (mostly in the form of CO_2) dominate life-cycle GHG emissions. Agriculture-related GHG emissions are shown in Figure 11.27 but do not include the carbon that was stored in the above-ground portion of the willow plants and subsequently released from the power plant stack. Carbon sequestration in Figure 11.27 refers to the carbon captured by plants and stored in their roots, which remain in the ground until the end of the 23-year willow farm management schedule (see Table 11.7).

The results in Figure 11.27 represent Keoleian and Volk's (2005) willow base case, which used ammonium sulfate as fertilizer. N_2O emissions are highly uncertain, as shown by the representative dashed lines in the figure, and are influenced by fertilizer type and application rate. If N_2O emissions were as high as the upper range of uncertainty, the net global warming potential would increase by 240%. If the lower-end N_2O emissions are more representative of the fertilizer type and amount used, the agriculture phase of the willow electricity life-cycle system would result in a net negative global warming potential. Also note that 75% of the global warming potential from agriculture inputs was due to fertilizer manufacture.

Table 11.10 provides the global warming potential of several biomass electricity systems along with that of the U.S. electricity grid average, wind systems, and solar photovoltaic electricity. Biomass electricity systems reduce global warming impacts, but not as well as other renewable technologies. Wind energy systems can decrease the life-cycle global warming potentials by 90–99% compared with biomass co-fire, coal, and average U.S. grid GHG emissions. As in the system net energy value comparison, given in Table 11.8, the gasified and direct-fired willow biomass systems offer much higher benefits than the co-fire systems. Note that the global warming potential factors calculated for wind and BIPV systems only include CO_2 emissions.

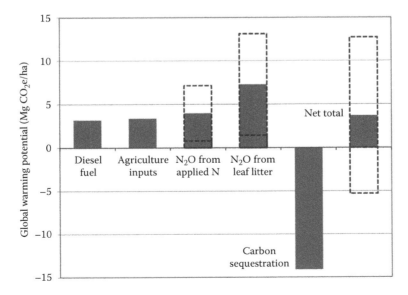

FIGURE 11.27 Agriculture life-cycle GHG emissions of willow excluding primary biomass carbon capture. (Form Keoleian, G.A. and Volk, T.A., *Crit Plant Sci Rev*, 24, 385–406, 2005.)

TABLE 11.10
Life-Cycle Global Warming Potential of
Electricity Generated from Various Sources

Electricity Source	Global Warming Potential (kg CO_2 eq./MWh$_{elec}$)
10% Blend co-fire	905.7
10% Willow co-fire	882.7
Average U.S. grid	989.1
Gasified willow	40.2
Direct-fired willow	52.3
BIPV	59.4
Wind	9.7

Source: Keoleian, G.A. and Volk, T.A., *Crit Plant Sci Rev*, 24, 385–406, 2005.

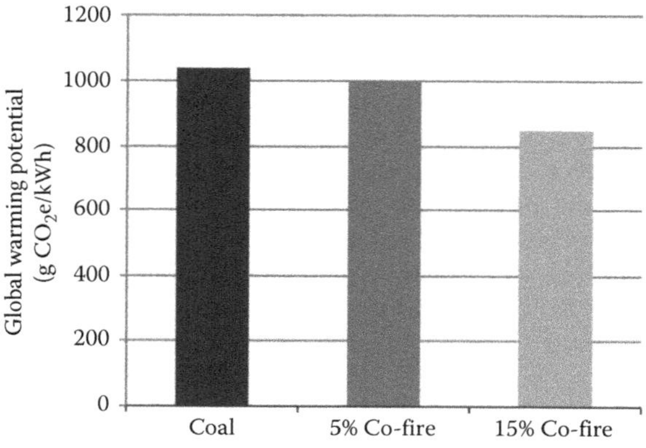

FIGURE 11.28 Life-cycle GHG emissions of coal and two coal/biomass co-fire electricity systems. (From Mann, M.K. and Spath, P.L., *Clean Prod Processes*, 3, 81–91, 2001.)

11.3.3.2 Mixed Wood Waste

One notable assumption in Mann and Spath (2001) was the exclusion of a CO_2 credit for carbon sequestered in biomass during growth. This credit was commonly used in the biofuel studies discussed in this chapter. No credit was assigned in this study because the biomass was not originally intended for energy production. Instead, carbon reduction credits were calculated for what was assumed to be the biomass' avoided fate—disposal in a landfill or conversion to mulch. The fraction that would have become mulch would eventually decompose and release carbon. Ten percent of the carbon would have become CH_4, and 90% would have become CO_2. As for the biomass sent to the landfill, Mann and Spath (2001) used the literature and assumed that 26% of the carbon would be released to the atmosphere as CO_2 and 9% as CH_4. The wood was not assumed to be usable in durable goods, so it could not displace these products and receive a GHG emissions avoidance credit. Critically, actual system operations must be understood to make appropriate modeling assumptions and yield accurate and defensible results.

The Mann and Spath (2001) life-cycle GHG emissions are presented in Figure 11.28. Compared with coal, 5% and 15% co-firing of wood waste biomass reduces life-cycle global warming potential

(GWP) by 5.4% and 18.2%, respectively. Nearly 90% of life-cycle CO_2 emissions in this study are released during combustion of coal in the power plant. Again, the biomass CO_2 emissions from the power plant are not offset by CO_2 sequestration during biomass growth, as typically assumed in other studies.

Sensitivity analysis of the co-firing rate showed that as the biomass percentage increased, GWP decreased faster than the decreasing rate of CO_2 emissions because CH_4, a potent GHG, was avoided by not landfilling or mulching the biomass. Additional sensitivity analysis results show that varying the landfill CH_4 emission rate could change the system GWP by as much as a 28% decrease to a 3% increase, although neither extreme is probable.

11.3.3.3 Hybrid Poplar

In Mann and Spath's (1997) assessment of poplar biomass production for electricity, the carbon from poplar wood chip combustion was assumed to be offset by carbon uptake during poplar growth, just as in Keoleian and Volk (2005) but differing from Mann and Spath (2001). It was assumed that 2% of nitrogen applied to the poplar fields would be released to the atmosphere as N_2O emissions. Also, the base case assumed no net gain or loss of soil carbon.

Only 5% of the total life-cycle CO_2 emissions were released to the atmosphere in Mann and Spath's (1997) model, whereas the remaining 95% of CO_2 was assumed to remain within system boundaries via biomass carbon uptake. Of the 5% fraction released, 62% was from feedstock production, 26% from power plant construction, and 12% from transportation. The total global warming potential, in terms of CO_2e, was calculated as the sum of CO_2, CH_4, and N_2O GHG emissions. Mann and Spath (1997) found a life-cycle GWP of 49 g CO_2e/kWh for poplar biomass electricity, which is close to the gasified and direct-fired values for willow found in Keoleian and Volk (2005) (Table 11.10). The eight alternative scenarios that most strongly influenced net CO_2 emissions are shown in Figure 11.29. These scenario numbers precisely correspond to those in Figure 11.25.

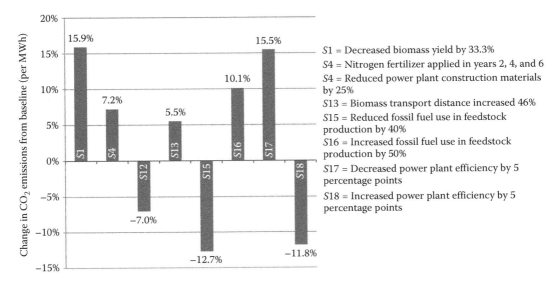

FIGURE 11.29 Change in CO_2 emissions for alternative poplar biomass electricity scenarios. (From Mann, M.K. and Spath, P.L., *Life Cycle Assessment of a Biomass Gasification Combined-Cycle System*, National Renewable Energy Laboratory, 1997. Available at http://www.nrel.gov/docs/legosti/fy98/23076.pdf)

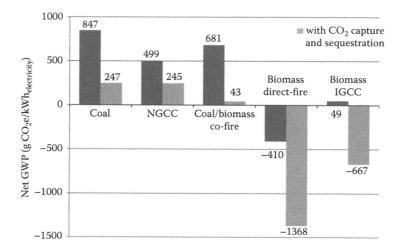

FIGURE 11.30 Life-cycle GHG emissions of five electricity sources with and without CO_2-seq. (From Spath, P.L. and Mann, M.K., *Biomass Power and Conventional Fossil Systems with and without CO_2 Sequestration—Comparing the Energy Balance, Greenhouse Gas Emissions and Economics*. NREL/TP-510-32575, National Renewable Energy Laboratory, Golden, CO, 2004.)

11.3.3.4 Carbon Sequestration

Spath and Mann (2004) assessed the life-cycle GHG emissions of five electricity generation technologies, with and without CO_2-seq: coal, natural gas combined-cycle, biomass co-fire, biomass direct-fire, and biomass gasification combined-cycle. CO_2, CH_4, and N_2O all contributed to the GWP of each system, summed in terms of CO_2e. Although the nitrogen content of coal is high (72%), most was assumed to become NO_x, which can contribute to acid rain and photochemical smog but is not a GHG, rather than N_2O. The co- and direct-fire systems used waste wood, as described in Section 11.3.3.2. Each of the biomass electricity systems considered thus far produced lower life-cycle GHG emissions than the comparative fossil fuel baseline systems. Figure 11.30 shows the life-cycle GHG emissions of the five fuel pathways assessed in Spath and Mann (2004). CO_2-seq reduced GHG emissions in every system by at least 50% and by up to 262% when comparing across systems to the coal baseline (no CO_2-seq).

11.4 LAND-USE IMPACTS

11.4.1 INTRODUCTION

In 2007, biofuels were estimated to use 1.87% of global arable land, or 26.6 million ha (Ravindranath et al. 2008). In the United States, 3.7% of arable land, or 10 million ha, is devoted to corn production for ethanol (Goettemoeller and Goettemoeller 2007). Crop yield per land area, market demand, policy, and natural resource constraints are some of the determining factors of land use for biomass feedstock crops. In this section, we will discuss production potentials, land requirements for modeled biomass systems, and GHG impacts based on modeling results.

Different biofuel feedstock crops are grown around the world based on geographic suitability. Yield is an important part of determining total scalability. Some crops yield more biofuel per hectare, and the same type of crop can produce different yields regionally depending on many factors, including rainfall, soil characteristics, plant genetics, and nutrient availability. The liquid fuel yields of several biofuel feedstocks, in gallons per acre, are presented in Figure 11.31. The figure gives an indication of the land-use benefits that can be achieved with future biofuel feedstocks (i.e., algae and switchgrass). In Europe, Giant Miscanthus grass has been a focal point of biofuel research, whereas switchgrass has received more attention in the United States. A recent study suggests that Giant Miscanthus can

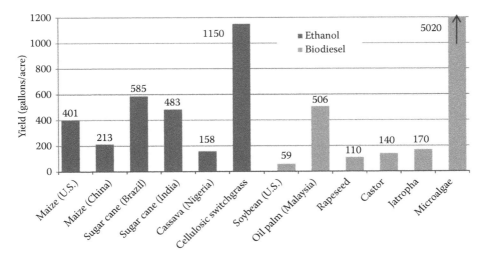

FIGURE 11.31 Potential biofuel yields of first-, second-, and third-generation feedstocks. (From U.N. Food and Agriculture Organization, *The State of Food and Agriculture*, 2008. Available at http://www.fao. org/docrep/011/i0100e/i0100e00.htm; Fulton, L., *Biodiesel: Technology Perspectives.* Geneva UNCTAD Conference, 2006 and Chisti, Y., *Biotechnol Adv*, 25, 294–306, 2007; ORNL, *Biofuels from Switchgrass: Greener Energy Pastures*, Oak Ridge National Laboratory, available at http://bioenergy.ornl.gov/papers/misc/ switgrs.html, 2005.)

be used to produce 2.6 times more ethanol per acre than corn (Heaton et al. 2008). High-yield micro-algae strains have been estimated to only require 1–2.5% of arable land in the United States to meet half of the nation's transportation fuel demand, whereas corn and soybeans would require several times more than the current arable land area to meet the same target (Chisti 2007).

A 2005 joint analysis by the U.S. Department of Energy (DOE) and U.S. Department of Agriculture (USDA) known as the Billion Ton Study projected that 1.3 billion tons of biomass could be produced annually in the United States by 2030 (USDOE/USDA 2005). The sources and amounts of biomass projected to be available in 2030 for a high-yield scenario (where agriculture yields were 70% greater than the moderate yield increase scenario) are shown in Figure 11.32. According to the study, three-quarters of the biomass would come from agricultural sources. The analysis of forestry stocks excluded areas without roads and environmentally sensitive habitats. In the agriculture sector, it was assumed that corn, wheat, and small grain crop yields would increase by 50%; that 75% of crop residues were recoverable; and that 55 million acres of land could be devoted to perennial bioenergy crop production.

In Brazil, sugarcane was cultivated on 7.8 million hectares in 2009, and an estimated 34 million hectares of pastureland could be targeted for sugarcane expansion (USDA/FAS 2010). In Asia and Africa, jatropha is a biodiesel crop of interest because it can grow on marginal land and does not require large amounts of water or fertilizer. In order to satisfy biofuel/petroleum blending targets in India, jatropha may occupy upwards of 11–13 million hectares (USDA/FAS 2011). Land use considerations will likely play an important role in national agriculture plans.

11.4.2 Life-Cycle GHG Emissions

Converting forested land to other uses releases CO_2 that was removed from the atmosphere. Tropical deforestation throughout the 1990s accounted for approximately 20% of anthropogenic GHG emissions (Gullison et al. 2007). Globally, many countries have established short-term national targets or mandates for biofuel consumption (Peterson 2008). As biofuel production increases, where will the additional crops be grown? Land-use change is an important component of the biofuels discussion,

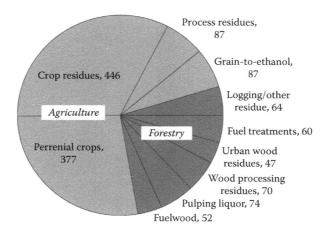

FIGURE 11.32 U.S. biomass production potential in million dry tons per year by 2030 from agricultural and forestry sources. (From USDOE/USDA, *Biomass as a Feedstock for a Bioenergy and Bioproducts Industry: The Technical Feasibility of a Billion-Ton Annual Supply*, 2005. U.S. Department of Energy/ U.S. Department of Agriculture, available at http://www1.eere.energy.gov/biomass/pdfs/final_billionton_vision_report2.pdf)

with the impacts typically categorized as direct or indirect. Direct land-use impacts are those that result from the actions of the biofuel supply chain. An example would be expansion of corn acreage for ethanol production. Indirect land-use change is more difficult to measure because it is caused by market forces and can act over vast distances. An example would be farmland expansion in Brazil that reduces available rangeland for cattle and causes ranchers to clear rainforest, driven by increased commodity crop prices in the United States. Some have suggested that indirect land-use change should not be included in life-cycle calculations (Kim et al. 2009). When indirect land-use change is modeled within LCA system boundaries, the increase in GHG emissions can be substantial. Integration of land-use impacts into LCA is an active area of research (Kløverpris et al. 2008b). More generally, indirect land-use change studies are examples of the recent shift from attributional to consequential LCAs and require global economic models to understand complex ripple effects (Sheehan 2009). Since this chapter was written many new studies have been conducted on the controversial and complex topic of land use change. The following studies represent a few of the first that brought major attention to this issue.

Searchinger et al. (2008) assessed the GHG emissions of corn ethanol, in comparison to petroleum, with a focus on the effects of indirect land-use change. Key assumptions were that increased ethanol consumption in the United States would increase planting of replacement crops because total crop demand is inelastic, and subsequently cropland would expand globally because doing so is cost-effective, fast, and convertible lands are relatively abundant. This market-driven land clearing for agriculture would result in a large initial release of the carbon stored in soil and plants—a carbon debt—as well as reduce the annual carbon sequestering capacity of the land, with the magnitude of both dependent on the type of natural habit lost (e.g., forest, grassland, bog). Searchinger's results agree with Wang et al. (2007) for corn ethanol when land-use change is not considered; GHG emissions would be reduced by approximately 20% for the average corn ethanol case, as shown in Figure 11.16. However, when land-use change is considered, life-cycle GHG emissions for corn ethanol are 93% higher those of petroleum over the 30-year study period, according to Searchinger et al. (2008). Because the corn ethanol production that is indirectly responsible for this cropland expansion would displace GHG-intensive petroleum, the large initial carbon debt allocated to the corn ethanol will slowly be paid back over the life of the project. Searchinger et al. (2008) calculated a payback period of 167 years—the time required to negate the land-use change GHG emissions and return to a carbon-neutral state.

The Searchinger et al. (2008) study has been criticized regarding assumptions on crop prices and co-product displacements, land-use change locations, and oversimplified modeling of free market interactions. Indirect land-use change caused by biofuel production is an active area of research that will continue to evolve through research on modeling of global market interactions and allocation methods. For example, Kendall et al. (2009) developed a method for reporting the life-cycle emissions intensity of a large initial CO_2 emission pulse, such as from clearing new cropland, which reflects the relative effect of those early emissions. As will be discussed later in this section, the U.S. Renewable Fuel Standard policy—a central driver of U.S. biofuel production—does include indirect land-use in life-cycle GHG calculations.

Fargione et al. (2008) examined land-use change to determine upper bounds on potential GHG emissions due to conversion of various types of habitats. The authors estimated that conversion of rainforest could incur a carbon debt that would take from dozens to hundreds of years to repay, whereas using abandoned or marginal cropland for agriculture would incur little to no carbon debt. The fraction of the carbon debt allocated to biofuels produced from these lands (allocation to co-products was included) was as low as 39% for soybeans and as high as 100% for sugarcane on the basis of a relative market value allocation approach. However, Kim et al. (2009) contend that better modeling of land management practices, which are more representative of the current state of tillage practices, significantly reduces the carbon debt payback period.

The Gallagher review concluded that Europe has sufficient land available for biofuel production, but the current production growth rate should be reduced until the impacts of increased production are better understood (RF Agency 2008). The review also recommended that marginal and idle lands be targeted for biofuel crop production and expressed concern over the findings of the Searchinger et al. (2008) study, although they found the results to be unreliable because of numerous assumptions that compounded uncertainty. Evaluation of marginal lands for feedstock crop production is an active area of research (Gutierrez and Ponti 2009).

The U.S. Energy Independence and Security Act of 2007 (EISA 2007) is federal legislation that includes mandates for biofuel production through year 2022, known as the Renewable Fuel Standard. In the EISA (2007), biofuels are categorized according to their ability to reduce life-cycle GHG emissions relative to a 2005 petroleum baseline, as shown in Table 11.11. For example, the advanced biofuels category is defined as renewable fuels other than ethanol derived from corn starch with 50% lower life-cycle GHG emissions. The EPA was charged with implementing a life-cycle system model to calculate the life-cycle GHG emissions of various biofuel production pathways.

TABLE 11.11
Life-Cycle GHG Emission Criteria and Example Biofuels

Fuel Category	GHG Reduction (%)[a]	Example Qualifying Fuels
Renewable fuel	20	Ethanol produced from corn starch at a new natural gas-fired facility using advanced efficient technologies
Advanced biofuel	50	Ethanol produced from sugarcane
Biomass-based diesel	50	Biodiesel from soy oil and renewable diesel from waste oils, fats, and greases; diesel produced from algal oils
Cellulosic biofuel	60	Cellulosic ethanol and cellulosic diesel (based on currently modeled pathways)

Source: EPA, *EPA Fuels and Additives—Renewable Fuel Standard Program*, 2010. U.S. Environmental Protection Agency, available at http://www.epa.gov/OMS/renewablefuels/index.htm#regulations

[a] Life-cycle GHG reduction requirement compared with 2005 petroleum baseline.

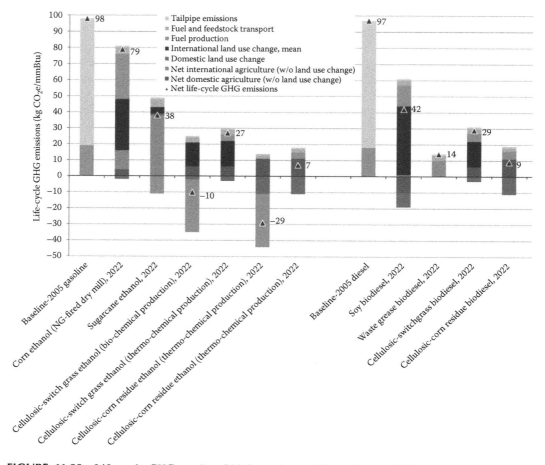

FIGURE 11.33 Life-cycle GHG results of biofuels that qualify under the EISA 2007 Renewable Fuel Standard as modeled by the EPA.

One of the most-debated features of the Renewable Fuel Standard final rule, as published by the EPA, is the inclusion of indirect land-use change GHG emissions. The EPA's model of indirect land-use change used improved techniques since Searchinger et al. (2008) to include a more comprehensive inventory of the lands that would be affected. Figure 11.33 shows the life-cycle GHG emissions in the year 2022 of Renewable Fuel Standard qualifying biofuel production pathways, as determined by the EPA's peer-reviewed model (EPA 2010). Note the magnitude of GHGs attributed to indirect land-use change relative to total emissions in Figure 11.33.

11.4.3 LIFE-CYCLE LAND-USE METRICS

The life-cycle land area requirements of several electricity sources, including biomass, are shown in Figure 11.34 (Spitzley and Keoleian 2005). The results presented in the figure considered acquisition of the input fuel, material acquisition and distribution, use, and end of life for different energy systems per kilowatt-hour of electricity generated. The time period of land use was also considered. For example, the biomass systems use a relatively large amount of land the entire time they are in production, whereas the natural gas mining operation may have a relatively small surface footprint, and the land may be restored once the natural gas has been extracted. However, this figure does not account for the quality of the utilized land. Biomass may be grown on marginal farmlands, fallow

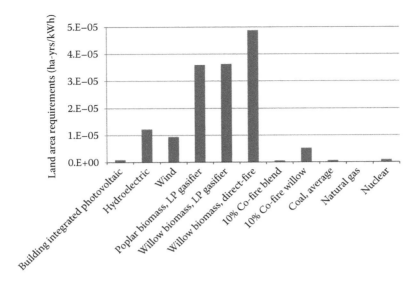

FIGURE 11.34 Land requirements of various electricity sources. (From Spitzley, D.V. and Keoleian, G.A., *Life Cycle Environmental and Economic Assessment of Willow Biomass Electricity: A Comparison with Other Renewable and Non-Renewable Sources*, 2005. Available at http://css.snre.umich.edu/main.php?control=detail_proj&pr_project_id = 19)

land, and Conservation Reserve Program land, whereas natural gas extraction points may be more constrained.

11.5 OVERALL CONCLUSIONS

LCA provides a framework for quantifying the sustainability of products and systems, and the results can be used to improve system performance. The biofuel and biomass electricity studies discussed in this chapter used different metrics, allocation methods, system boundaries, and data. Results are contingent upon these differences. Despite this, the following conclusions can be drawn from the biofuel studies considered in this chapter:

(1) using corn ethanol rather than gasoline can significantly reduce life-cycle petroleum consumption but only modestly reduce GHG emissions when indirect land-use change is not considered; (2) sugarcane ethanol production in Brazil requires less fossil fuel inputs and results in less GHG emissions than corn ethanol, with a primary reason being the use of renewable bagasse (sugarcane plant residue) to power the conversion facility and displace grid electricity; (3) switchgrass-derived ethanol can be produced at an estimated ratio of 13.1 MJ ethanol/MJ petroleum and reduce life-cycle GHG emissions by 94%; and (4) producing and consuming soybean biodiesel in the United States could reduce life-cycle petroleum use by 95% and CO_2 emissions by 78% compared with petroleum diesel when land-use change is not considered (Sheehan et al. 1998; Farrell et al. 2006; Wang et al. 2007; Schmer et al. 2008; Searchinger et al. 2008, respectively). The recently revised U.S. Renewable Fuel Standard, a component of the EISA 2007, regulates biofuel production levels in the United States through the year 2022. Life-cycle GHG emissions reduction criteria are central to the legislation, and indirect land-use change effects were incorporated into the final peer-reviewed life-cycle model, which was used to determine qualifying fuels. U.S. corn ethanol and soybean biodiesel produced in modern, efficient plants qualify under this program and are projected to be significant components of U.S. biofuel consumption through 2022. Modeling results also show that sugarcane ethanol offers greater GHG reductions than

corn ethanol. In the future, cellulosic biofuels will play a greater role as their production costs are reduced; they should offer significantly greater GHG reduction benefits compared with corn ethanol and petroleum.

Biomass electricity was found to greatly reduce GHG emissions. Using 100% hybrid poplar in an IGCC power plant can reduce life-cycle GHG emissions by 94%, and adding CO_2-seq technology can reduce life-cycle GHG emissions by 179%, both with respect to coal. The benefits of willow biomass energy become more apparent when examining direct-fired or gasified willow electricity compared with co-firing with coal. The NER was found to be 9.9 and 12.9 for direct-firing and gasification, respectively. Hybrid poplar was found to have a NER of 15.6. The high NERs demonstrate the tremendous fossil energy leveraging of this renewable energy resource. For example, 13 units of electricity are generated for every 2 units of fossil energy consumed across the full life-cycle of willow gasification (Keoleian and Volk 2005). The NER is a useful indicator of overall life-cycle energy performance. When life-cycle boundaries and assumptions are equivalent, the NER also allows comparison between biofuels and biomass as an energy source.

This chapter focused on liquid fuel production for transportation using internal combustion (IC) vehicles and biomass production for electricity generation. The NERs for biomass electricity are generally much higher than those of liquid biofuels. This demonstrates that biomass electricity is a significantly more efficient method for leveraging fossil energy resources and displacing GHG emissions. In the future, electric vehicles (EVs) and plug-in hybrid electric vehicles (PHEVs) are likely to play a role in transportation. Ohlrogge et al. (2009) compared the efficiencies of converting biomass to electricity with conversion to biofuel for the purpose of powering EVs and IC vehicles, respectively. Overall, the biomass electricity system was found to be nearly twice as efficient at delivering the energy originally contained in the biomass to the end user.

Life-cycle energy and GHG emissions were the focal point of this chapter, but other air emissions, such as criteria pollutants, are often included in life-cycle inventories. As an example, consider biodiesel tailpipe emissions, which are important because of their prevalence in highly populated areas (e.g., biodiesel city buses). Smog is a recurring problem in major urban areas, formed by CO, total hydrocarbon (THC), and NO_x interactions with sunlight. Sheehan et al. (1998) found that biodiesel releases fewer CO and THC emissions but increases NO_x emissions, each relative to petroleum diesel. Tailpipe emissions of CO and PM_{10} are reduced by 46 and 68%, respectively, when B100 is used in place of diesel fuel. However, the increase in NO_x is troubling because it is known to adversely affect the human respiratory system (EPA 2009b). Research on reducing biodiesel NO_x tailpipe emissions has achieved results through re-engineering of engine timing, use of fuel additives, and varying fuel properties (McCormick/NREL 2005). Other recent research has concluded that because of high variance in NO_x emissions for different engine types and test conditions, no definitive conclusions on biodiesel NO_x tailpipe emissions can be drawn (McCormick et al./NREL 2006). There is also concern over gasoline emissions. Preliminary results from a gasoline combustion focused life-cycle study indicate that reducing particulate emissions can significantly reduce health impacts (LBNL 2009). Hill et al. (2009) found that only cellulosic ethanol derived from corn stover, switchgrass, mixed prairie vegetation, and miscanthus can reduce the combined costs of GHG impacts and criteria pollutant $PM_{2.5}$-induced health impacts. Corn ethanol, at best, was found to cause negative economic impacts equivalent to those of gasoline, and only when produced using efficient advanced technologies fueled by natural gas (Hill et al. 2009).

The recent trend in LCA of biomass systems has been a move away from attributional and toward consequential studies (Sheehan 2009). Attributional LCAs evaluate the collective impacts of a system and assign a fraction of these impacts to the system's products. Consequential LCAs attempt to project the net effects of marginal changes within a system, e.g., considering the dynamic effects increased production would have on a market. Consequential LCA, such as Searchinger et al.'s (2008) indirect land-use impacts study, seem to offer a more desirable perspective to policy-makers, but studies of this nature are inherently more complex. Two assumptions often made in biofuel LCAs are perfect market elasticity (i.e., consumption of one product will lead to an equivalent increase in its production without affecting other products) and linear substitution (i.e., functionally equivalent products are completely interchangeable) (Kloverpris et al. 2008a). Both of these mechanisms can affect land-use change results. Kloverpris et al. (2008a) propose that calculations of global agricultural land utilization, driven by biofuel consumption, should carefully consider marginal land data on a local level, policy constraints on fertilizer use, and natural resource constraints. Categorizing regions according to type of vegetation could also increase the accuracy of land-use change modeling (Kloverpris 2009). Modeling of indirect land-use change, soil carbon content, and soil nitrogen flux are active areas of research. As life-cycle models continue to improve in accuracy and comprehensiveness, additional impacts, such as the risk of biofuel feedstocks spreading and becoming invasive (Raghu et al. 2006), will likely be included to enable better decision-making. The inclusion of life-cycle GHG reduction requirements in the Energy Policy Act of 2007 was a significant step forward in policy-making. LCA offers a framework for the holistic evaluation of products and systems and can help guide society toward more sustainable biomass energy systems.

REFERENCES

Anderson-Teixeira KJ, Davis SC, Masters MD, DeLucia EH (2009) Changes in soil organic carbon under biofuel crops. *Glob Change Biol Bioenergy* 1:75–96

ANL (2009) *The Greenhouse Gases, Regulated Emissions, and Energy Use in Transportation (GREET) Model*, Argonne National Laboratory, available at http://www.transportation.anl.gov/modeling_simulation/ GREET/ (accessed June 18, 2009)

Baker JM, Ochsner TE, Venterea RT, Griffis TJ (2007) Tillage and soil carbon sequestration—What do we really know? *Agric Ecosyst Environ* 118:1–5

Chisti Y (2007) Biodiesel from microalgae. *Biotechnol Adv* 25:294–306

Crutzen PJ, Mosier AR, Smith KA, Winiwarter W (2007) N_2O release from agro-biofuel production negates global warming reduction by replacement of fossil fuels. *Atmos Chem Phys Discuss* 8:389–395

Dale BE (2007) Thinking clearly about biofuels: Ending the irrelevant 'net energy' debate and developing better performance metrics for alternative fuels. *Biofuels Bioprod Biorefin* 1:14–17

Del Grosso SJ, Ojima DS, Parton WJ, Stehfest E, Heistemann M, DeAngelo B, Rose S (2009) Global scale DAYCENT model analysis of greenhouse gas emissions and mitigation strategies for cropped soils. *Glob Planet Change* 67:44–50

Delucchi M (2008) Working paper: Incorporating price Effects into lifecycle analysis, available at http:// yosemite.epa.gov/ee/epa/wpi.nsf/7e8e12c2a0e34c0585256c2c00577d69/344b66c4d3d6761685257408 006f6ba3!OpenDocument (accessed July 24, 2009)

de Vries SC (2008) The bio-fuel debate and fossil energy use in palm oil production: a critique of Reijnders and Huijbregts 2007. *J Cleaner Prod* 16:1926–1927

DOE (2009) Fossil energy: DOE's FutureGen initiative, U.S. Department of Energy, available at http://fossil. energy.gov/programs/powersystems/futuregen/index.html (accessed July 1, 2009)

EERE (2006) *Biomass Energy Data Book: Edition 1*, U.S. Department of Energy, Energy Efficiency and Renewable Energy, available at http://cta.ornl.gov/bedb/index.shtml (accessed July 1, 2009)

EIA (2007) *Biofuels in the U.S. Transportation Sector*, U.S. Department of Energy, Energy Information Administration, available at http://www.eia.doe.gov/oiaf/analysispaper/biomass.html#1

EIA (2009) *Monthly Energy Review. Tables 10.3 and 10.4*, U.S. Department of Energy, Energy Information Administration, available at http://www.eia.doe.gov/emeu/aer/renew.html (accessed June 21, 2009)

EISA (2007) H.R. 6 – 110th Congress: Energy Independence and Security Act of 2007 In GovTrack.us (database of federal legislation), available at http://www.govtrack.us/congress/bill.xpd?bill = h110-6 (accessed March 2, 2010)

EPA (2008) *Life Cycle Assessment Research*, U.S. Environmental Protection Agency, http://www.epa.gov/ nrmrl/lcaccess (accessed June 15, 2009)

EPA (2009a) Regulation of fuels and fuel additives: Changes to Renewable Fuel Standard program; Proposed rule. *Federal Register* 74:24904–24952

EPA (2009b) Air & radiation: Nitrogen dioxide, U.S. Environmental Protection Agency, available at http:// www.epa.gov/air/nitrogenoxides/index.html (accessed July 11, 2009)

EPA (2009c) *Inventory of U.S. Greenhouse Gas Emissions and Sinks: 1990-2007*, U.S. Environmental Protection Agency, Washington, DC

EPA (2010) *EPA Fuels and Additives—Renewable Fuel Standard Program*, U.S. Environmental Protection Agency, available at http://www.epa.gov/OMS/renewablefuels/index.htm#regulations (accessed May 22, 2010)

Fairless D (2007) The little shrub that could—Maybe. *Nature* 449:652–655

Fargione, J, Hill J, Tilman D, Polasky S, Hawthorne P (2008) Land clearing and the biofuel carbon debt. *Science* 319:1235–1238

Farrell AE, Plevin RJ, Turner BT, Jones AD. O'Hare M, Kammen DM (2006) Ethanol can contribute to energy and environmental goals. *Science* 311:506–508

Fulton L (2006) Biodiesel: Technology perspectives. Geneva UNCTAD Conference

FutureGen Alliance (2009) available at http://www.futuregenalliance.org

Goettemoeller J, Goettemoeller A (2007) *Sustainable Ethanol: Biofuels, Biorefineries, Cellulosic Biomass, Flex-Fuel Vehicles, and Sustainable Farming for Energy Independence*. Prairie Oak Publishing, Maryville, MO

GreenGen (2009), available at http://www.greengen.com.cn/en/aboutgreengenproject.htm (accessed June 15, 2009)

Gullison RE, Frumhoff PC, Canadell JG, Field CB, Nepstad DC, Hayhoe K, Avissar R, Curran LM, Friedlingstein P, Jones CD, Nobre C (2007) Tropical forests and climate policy. *Science* 316:985–986

Gutierrez, AP, Ponti L (2009) Bioeconomic sustainability of cellulosic biofuel production on marginal lands. *Bull Sci Technol Soc* 29:213–225

Heaton, EA, Dohleman FG, Long SP (2008) Meeting U.S. biofuel goals with less land: The potential of Miscanthus. *Glob Change Biol* 14:2000–2014

Heller MC, Keoleian GA, Mann MK, Volk TA (2004) Life cycle energy and environmental benefits of generating electricity from willow biomass. *Renew Energy* 29:1023–1042

Heller MC, Keoleian GA, Volk TA (2003) Life cycle assessment of a willow bioenergy cropping system. *Biomass Bioenergy* 25:147–165

Hill J, Nelson E, Tilman D, Polasky S, Tiffany D (2006) Environmental, economic, and energetic costs and benefits of biodiesel and ethanol biofuels. *Proc Nat Acad Sci USA* 103:11206–11210

Hill J, Polasky S, Nelson E, Tilman D, Huo H, Ludwig L, Neumann J, Zheng H, Bonta D (2009) Climate change and health costs of air emissions from biofuels and gasoline. *Proc Nat Acad Sci USA* 106:2077–2082

Huo H, Wang M, Bloyd C, Putsche V/Argonne National Laboratory (2008) *Life-Cycle Assessment of Energy and Greenhouse Gas Effects of Soybean-Derived Biodiesel and Renewable Fuels*. ANL/ESD 08-2, Argonne National Laboratory, Argonne, IL

IPCC (2007) *Fourth Assessment Report: The Physical Science Basis*, Intergovernmental Panel on Climate Change, available at http://www.ipcc.ch/ipccreports/ar4-wg1.htm (accessed April 5, 2009)

IPCC (2009) *The IPCC Assessment Reports*, available at http://www.ipcc.ch (accessed April 5, 2009)

ISO (1998) *ISO 14041 Environmental Management: Life Cycle Assessment—Goal and Scope Definition and Inventory Analysis*, International Organization for Standardization, Geneva, Switzerland

ISO (2000) *ISO TR 14049 Environmental Management: Life Cycle Assessment—Examples of Application of ISO 14041 to Goal and Scope Definition and Inventory Analysis*, International Organization for Standardization, Geneva, Switzerland

Kalogo Y, Habibi S, MacLean HL, Joshi SV (2007) Environmental implications of municipal solid waste-derived ethanol. *Environ Sci Technol* 41:35–41

Kammen DM, Farrell AE, Plevin RJ, Jones AD, Nemet GF, Delucchi MA (2008) *Energy and Greenhouse Gas Impacts of Biofuels: A Framework for Analysis*, UC Berkeley Transportation Sustainability Research Center, available at http://repositories.cdlib.org/its/tsrc/UCB-ITS-TSRC-RR-2008-1/

Kendall A, Chang B (2009) Estimating life cycle greenhouse gas emissions from corn-ethanol: A critical review of current U.S. practices. *J Cleaner Prod* 17:1175–1182

Kendall A, Chang B, Sharpe B (2009) Accounting for time-dependent effects in biofuel life cycle greenhouse gas emissions calculations. *Environ Sci Technol* 43:7142–7147

Keoleian GA, Lewis GM (2003) Modeling the life cycle and environmental performance of amorphous silicon BIPV roofing in the US. *Renew Energy* 28:271–293

Keoleian GA, Spitzley DV (2006) Life cycle based sustainability metrics. In: Abraham MA (ed), *Sustainability Science and Engineering*, Elsevier, San Diego, pp 127–159

Keoleian GA, Volk TA (2005) Renewable energy from willow biomass crops: Life cycle energy, environmental, and economic performance. *Crit Plant Sci Rev* 24:385–406

Kim H, Kim S, Dale BE (2009) Biofuels, land use change, and greenhouse gas emissions: Some unexplored variables. *Environ Sci Technol* 43:961–967

Kim S, Dale BE (2005) Life cycle assessment of various cropping systems utilized for producing biofuels: Bioethanol and biodiesel. *Biomass Bioenergy* 29:426–439

King AJ, He W, Cuevas JA, Freudenberger M, Ramiaramanana, Graham IA (2009) Potential of *Jatropha curcas* as a source of renewable oil and animal feed. *J Exp Bot* 60:2897–2905

Kløverpris J, Wenzel H, Banse MAH, Mila I, Canals L, Reenberg A (2008a) Global land use implications of biofuels: State-of-the-art. *Int J Life Cycle Assess* 13:178–183

Kløverpris J, Wenzel H, Nielsen PH (2008b) Life cycle inventory modelling of land use induced by crop consumption. *Int J Life Cycle Assess* 13:13–21

Kløverpris J (2009) Identification of biomes affected by marginal expansion of agriculture land use induced by increased crop consumption. *J Cleaner Prod* 17:463–470

LBNL (2009) The coming of biofuels: Study shows reducing gasoline emissions will benefit human health, Lawrence Berkeley National Laboratory, available at http://newscenter.lbl.gov/feature-stories/2009/05/27/biofuels-and-human-health (accessed June 1, 2009)

Liska A, Cassman K (2008) Towards standardization of life-cycle metrics for biofuels: greenhouse gas emissions mitigation and net energy yield. *J Biobased Materials Bioenergy* 2:187–203

Lyko H, Deerberg G, Weidner E (2009) Coupled production in biorefineries—Combined use of biomass as a source of energy, fuels and materials. *J Biotechnol* 142:78–86

Mann MK, Spath PL (1997) *Life Cycle Assessment of a Biomass Gasification Combined-Cycle System*, National Renewable Energy Laboratory, available at http://www.nrel.gov/docs/legosti/fy98/23076.pdf (accessed May 1, 2009)

Mann MK, Spath PL (2001) A life-cycle assessment of biomass co-firing in a coal-fired power plant. *Clean Prod Processes* 3:81–91

McCormick B/NREL (2005) *Effects of Biodiesel on NO$_x$ Emissions*. ARB Biodiesel Workgroup, National Renewable Energy Laboratory, available at http://www.nrel.gov/vehiclesandfuels/npbf/pdfs/38296.pdf (accessed July 8, 2009)

McCormick B, Alleman T, Barnitt R, Clark W, Hayes B, Ireland J, Proc K, Ratcliff M, Thornton M, Whitacre S, Williams A/NREL (2006) Biodiesel R&D at NREL. In: *Proceedings of the National Biodiesel Conference & Expo*, San Diego, CA, February 6, 2006, available at http://www.nrel.gov/vehiclesandfuels/npbf/pdfs/39538.pdf (accessed June 12, 2009)

McLaughlin SB, Kszos LA (2005) Development of switchgrass (*Panicum virgatum*) as a bioenergy feedstock in the United States. *Biomass Bioenergy* 28:515–535

MDA (2009) *Report to the Legislature: Petroleum Diesel Fuel and Biodiesel Technical Cold Weather Issues*, Minnesota Department of Agriculture, available at http://www.state.mn.us/mn/externalDocs/Commerce/Biodiesel_Cold_Weather_Issues_022009043331_BiodieselColdWeather.pdf

Minnesota (2008) Laws of Minnesota for 2008, Chapter 297, § 239.77, available at https://www.revisor.leg.state.mn.us/data/revisor/law/2008/0/2008-297.pdf (accessed June 22, 2009)

NBB (2008) Biodiesel blends and cold weather, National Biodiesel Board, available at http://www.biodiesel.org/pdf_files/fuelfactsheets/COLD_BIOFuelDistFactShtNOSOY.pdf (accessed June 16, 2009)

ORNL (2005) *Biofuels from Switchgrass: Greener Energy Pastures*, Oak Ridge National Laboratory, available at http://bioenergy.ornl.gov/papers/misc/switgrs.html (accessed May 19, 2009)

Ohlrogge J, Allen D, Berguson B, Dellapenna D, Shachar-Hill Y, Stymne S (2009) Driving on biomass. *Science* 324:1019–1020

Patzek TW (2004) Thermodynamics of the corn-ethanol biofuel cycle. *Crit Rev Plant Sci* 23: 519–567

Perrin RK, Fretes NF, Sesmero JP (2009) Efficiency in midwest U.S. corn ethanol plants: A plant survey. *Energy Policy* 37:1309–1316

Peterson J-E (2008) Energy production with agricultural biomass: Environmental implications and analytical challenges. *Eur Rev Agric Econ* 35:385–408

Pimentel D, Patzek TW (2005) Ethanol production using corn, switchgrass, and wood; Biodiesel production using soybean and sunflower. *Nat Resour Res* 14:65–76

Quirin M, Gärtner SO, Pehnt M, Reinhardt GA/IFEU Institute for Energy and Environmental Research GmbH (2004) *CO$_2$ Mitigation through Biofuels in the Transport Sector*, available at http://www.biodiesel.org/resources/reportsdatabase/reports/gen/20040801_gen-351.pdf (accessed June 9, 2009)

Raghu S, Anderson RC, Daehler CC, Davis AS, Wiedenmann RN, Simberloff D, Mack RN (2006) Adding biofuels to the invasive species fire? *Science* 313:1742.

Ravindranath NH, Manuvie R, Fargione J, Canadell P, Berndes G, Woods J, Watson H, Sathaye J (2008) GHG implications of land use change and land conversion to biofuel crops. In: Howarth RW, Bringezu S (ed), *Biofuels: Environmental Consequences and Interactions with Changing Land Use*, Gummersbach, Germany, pp 111-125, available at http://cip.cornell.edu/biofuels

Reijnders L, Huijbregts MAJ (2008) Palm oil and the emission of carbon based greenhouse gases. *J Cleaner Prod* 16:477–482

RF Agency (2008) *The Gallagher Review of the Indirect Effects of Biofuels Production*, Renewable Fuels Agency, available at http://www.renewablefuelsagency.org/_db/_documents/Report_of_the_Gallagher_review.pdf (accessed July 5, 2009)

RFA (2009a) Industry statistics: 2008 world fuel ethanol production table, Renewable Fuel Association, available at http://www.ethanolrfa.org/industry/statistics/#E (accessed July 27, 2009)

RFA (2009b) Ethanol industry overview table, Renewable Fuel Association, available at http://www.ethanolrfa.org/industry/statistics/#C (accessed June 22, 2009)

Schleisner L (2000) Life cycle assessment of a wind farm and related externalities. *Renew Energy* 20:279–288

Schmer MR, Vogel KP, Mitchell RB, Perrin RK (2008) Net energy of cellulosic ethanol from switchgrass. *Proc Natl Acad Sci USA* 105:464–469

Science Daily (2007) Carbon dioxide expelled from peatland when natural swamp forest is converted to oil palm, available at http://www.sciencedaily.com/releases/2007/12/071206235448.htm (accessed May 5, 2009)

Searchinger T, Heimlich R, Houghton RA, Dong F, Elobeid A, Fabiosa J, Tokgoz S, Hayes D, Yu T-H (2008) Use of U.S. croplands for biofuels increases greenhouse gases through emissions from land-use change. *Science* 319: 1238–1240

Sheehan J, Camobreco V, Duffield J, Graboski M, Shapouri H (1998) Life cycle inventory of biodiesel and petroleum diesel for use in an urban bus. NREL/SR-580-24089, National Renewable Energy Laboratory, Golden, CO

Sheehan JJ (2009) Biofuels and the conundrum of sustainability. *Curr Opin Biotechnol* 20:318–324

Snyder CS, Bruulsema TW, Jensen TL, Fixen PE (2009) Review of greenhouse gas emissions from crop production systems and fertilizer management effects. *Agric Ecosyst Environ* 133:247–266

Spath PL, Mann MK (2004) *Biomass Power and Conventional Fossil Systems with and without CO_2 Sequestration—Comparing the Energy Balance, Greenhouse Gas Emissions and Economics*. NREL/TP-510-32575, National Renewable Energy Laboratory, Golden, CO

Spath PL, Mann MK, Kerr DR (1999) *Life Cycle Assessment of Coal-fired Power Production*, NREL/TP-570-25119, National Renewable Energy Laboratory, Golden, CO

Spitzley DV, Keoleian GA (2005) *Life Cycle Environmental and Economic Assessment of Willow Biomass Electricity: A Comparison with Other Renewable and Non-Renewable Sources*, available at http://css.snre.umich.edu/main.php?control = detail_proj&pr_project_id = 19 (accessed April 1, 2009)

Stahl K, Neergaard M (1998) IGCC power plant for biomass utilization. *Biomass Bioenergy* 15:205–211

Sylvester-Bradley R (2008) Critique of Searchinger (2008) & related papers assessing indirect effects of biofuels on land-use change. *The Gallagher Biofuels Review*, ADAS UK Ltd., Wolverhampton, UK

U.N. Food and Agriculture Organization (2008) *The State of Food and Agriculture*, available at http://www.fao.org/docrep/011/i0100e/i0100e00.htm

USDA (2009) Growing crops for biofuels has spillover effects. In: *Amber Waves*, U.S. Department of Agriculture/National Agricultural Statistics Service, available at http://www.ers.usda.gov/AmberWaves/March09/Features/Biofuels.htm (accessed May 25, 2009)

USDA/ERS (2008a) U.S. fertilizer use and price. Table 10: Nitrogen used on corn, rate per fertilized acre receiving nitrogen, selected states, 1964-2005, available at http://www.ers.usda.gov/Data/FertilizerUse (accessed June 22, 2009)

USDA/ERS (2008b) U.S. fertilizer use and price. Table 9: Percentage of corn acreage receiving nitrogen fertilizer, selected states, 1964-2005, available at http://www.ers.usda.gov/Data/FertilizerUse (accessed June 22, 2009)

USDA/FAS (2010) *Brazil Biofuels Annual 2010*. U.S. Department of Agriculture/Foreign Agriculture Service

USDA/FAS (2011) *Peoples Republic of China, Malaysia, Indonesia Biofuels Annual 2011*. U.S. Department of Agriculture/Foreign Agriculture Service

USDA/NASS (2007) *Ethanol Co-Products Used for Livestock Feed Survey*, U.S. Department of Agriculture/National Agricultural Statistics Service, available at https://www.msu.edu/~gtonsor/Documents/Ethanol%20Co-Products%20Used%20for%20Livestock%20Feed.pdf (accessed May 1, 2009)

USDOE/USDA (2005) *Biomass As a Feedstock for a Bioenergy and Bioproducts Industry: the Technical Feasibility of a Billion-Ton Annual Supply*, U.S. Department of Energy/U.S. Department of Agriculture, available at http://www1.eere.energy.gov/biomass/pdfs/final_billionton_vision_report2.pdf (accessed April 16, 2009)

Wang M (2008) *Estimation of Energy Efficiencies of U.S. Petroleum Refineries*, Argonne National Laboratory, http://greet.es.anl.gov/publication-hl9mw9i7 (accessed June 24, 2009)

Wang M, Wu M, Huo H, Liu J (2007) *Well-to-Wheels Energy Use and Greenhouse Gas Emissions of Brazilian Sugarcane Ethanol Production Simulated by Using the GREET Model*, Argonne National Laboratory, Argonne, IL

Worrel E, Galitsky C (2005) *Energy Efficiency Improvement and Cost Saving Opportunities for Petroleum Refineries*, LBNL-56183, Lawrence Berkeley National Laboratory, Berkeley, CA

WRI (2009a) A first-hand view of China's carbon capture and storage actions, World Resources Institute, available at http://www.wri.org/stories/2009/07/first-hand-view-chinas-carbon-capture-and-storage-actions. (accessed July 22, 2009)

WRI (2009b) Biofuels and the time value of carbon: Recommendations for GHG accounting protocols. working paper, World Resources Institute, available at http://www.wri.org/publication/biofuels-and-time-value-of-carbon (accessed July 23, 2009)

12 Public Policies, Economics, Public Perceptions, and the Future of Bioenergy Crops

Barry D. Solomon
Michigan Technological University

Nicholas H. Johnson
Pennsylvania State University

CONTENTS

12.1 INTRODUCTION

Over the past decade, a renewed interest in biofuels has arisen as concerns about energy security, fuel prices, and the adverse impacts of global climate change have grown. Government support has been especially strong for biofuel programs in the US, European Union (EU), and Brazil, although not without controversy. Environmental groups are concerned about possible deforestation from an increase in cropland, and concerns have arisen over the use of food-crops for fuel instead of food.

Crops have been used to make synthetic fuels for many years. For instance, engines were developed in the 1800s by designers such as Henry Ford and Nicholas Otto that used grain ethanol as fuel, and the 1908 Model T had an adjustable carburetor that could use alcohol, gasoline, or a blend of the two (Solomon et al. 2009). Rudolph Diesel, the inventor of the diesel engine, used peanut oil in an engine (Shay 1993). In 1919, the British government experimented with using Jerusalem artichokes to produce fuel alcohol because of an oil shortage, and near the end of World War II Japan was using potatoes, sugar, and rice wine to produce alcohol for fuel (Yergin 1992). After the oil crises of the 1970s, Brazil and the United States enacted policies that successfully encouraged ethanol production as a gasoline substitute. More recently, legislation in the United States set an ambitious goal of 1.36×10^{11} L/year of ethanol production by 2022, with most of that to be made from cellulose. In the past decade, the EU, especially Germany, has also greatly increased its production of biofuels, particularly biodiesel. In 2008, 67×10^9 L of ethanol and 12×10^9 L of biodiesel were produced worldwide, compared with 39×10^9 and 6×10^9 L, respectively, produced in 2006 (REN21 2009). Table 12.1 shows the top 15 producing countries of biofuels as of 2008.

There are many types of biofuels, the two most common of which are the liquid fuels ethanol and biodiesel (Solomon and Johnson 2009a). Ethanol is produced from grains (especially corn) and crops with high sugar content, such as sugarcane and sugarbeets. To process the starch in grains to alcohol, seven steps are required: milling, liquefaction, saccharification, fermentation, distillation, dehydration, and denaturing. Processing crops with high sugar content is cheaper than processing grains and it requires five steps: milling, pressing, fermentation, distillation, and dehydration (Solomon et al. 2009). Other alcohols, especially methanol, propanol, and butanol, can also be manufactured. Alcohol can also be processed from cellulosic-based feedstocks, including wood, agricultural residues, straws, and grasses. Although little cellulosic ethanol is currently produced commercially because of the immature technology and high costs involved, capacity to produce cellulosic ethanol will grow rapidly in the next decade.

Biodiesel is produced primarily from vegetable oils, but it can also be created from animal fats. Common crops include rapeseed, palm, soy, sunflower, and peanut. After oil is extracted from

TABLE 12.1

Global Biofuels Production in 2008 for the Top 15 Producing Countries

Country	Billion Liters		
	Fuel Ethanol	Biodiesel	Total
United States	34	2.0	36
Brazil	27	1.2	28
France	1.2	1.6	2.8
Germany	0.5	2.2	2.7
China	1.9	0.1	2.0
Argentina	–	1.2	1.2
Canada	0.9	0.1	1.0
Spain	0.4	0.3	0.7
Thailand	0.3	0.4	0.7
Colombia	0.3	0.2	0.5
Italy	0.13	0.3	0.4
India	0.3	0.02	0.3
Sweden	0.14	0.1	0.2
Poland	0.12	0.1	0.2
United Kingdom	–	0.2	0.2
EU total	**2.8**	**8**	**10.8**
World total	**67**	**12**	**79**

Source: REN21, Renewables Global Status Report: 2009 update. Renewable Energy Policy Network for the 21st Century Secretariat, Paris, 2009. Available at www.martinot.info/RE_GSR_2009_Update.pdf (accessed July 17, 2009). With permission.

crops, three reversible steps are used to chemically modify the oil so that its properties are similar to diesel fuel: transesterification, pyrolysis, and emulsification (Meher et al. 2006). In addition to ethanol and biodiesel, bioenergy crops, especially from forestlands, are used to produce many other types of energy, including heat, electricity, and syngas (Solomon and Johnson 2009a).

This chapter begins by presenting the policies that are helping drive the rapid increase in the worldwide production of biofuels, especially in the United States, EU, and Brazil, with an examination of the current economic situations in each of these places. Results of a case study of public perceptions in the upper midwestern United States are presented, along with some ideas about what the near future might hold for biofuels.

12.2 GOVERNMENT POLICIES

12.2.1 UNITED STATES

The United States is the world's leading ethanol producer. Government support for ethanol development in the United States has existed at the federal and state levels. Three basic federal subsidies fueled the early years of the modern fuel ethanol industry, which was focused on corn until 2007. The first and most important of these was a partial exemption from the federal gasoline excise tax for "gasohol" (defined as a fuel containing at least a 10% component of biomass-derived ethanol). This exemption was approved by the Energy Tax Act of 1978 and implemented in 1979 (Solomon 1980). A fuel blender's tax credit and a pure alcohol fuel credit were added in 1980. These new initiatives were basically the same subsidy as the fuel excise tax exemption but recouped through a different system and available to a small number of companies who were unable to claim the fuel tax exemption. Through later years, all of these tax provisions were periodically renewed and altered in terms of the benefit magnitude, with changes in one being mirrored by changes in the others. The excise tax exemption has been by far the most widely used incentive (double crediting with the fuel blender's tax credit is not allowed) with total government revenue impacts estimated at between 16 and 56 times those of the other two tax credits combined (GAO 2000). Thus, this exemption was the most important of the early ethanol support mechanisms and it remains of critical importance to this industry in the United States today.

Further federal support came in 1990 with passage of the Small Ethanol Producer Tax Credit, which provided small refineries (less than 1.1×10^8 L/year production capacity) with an additional $0.026/L income tax credit for volumes up to 5.7×10^7 L/year (California Energy Commission 2004). The Energy Policy Act of 2005, which is discussed below, redefined small producers as those producing up to 2.3×10^8 L/year. In recent years the total federal and state support for ethanol and biodiesel has equaled a taxpayer subsidy of $10 + billion/year (Koplow 2007), including a tariff on ethanol imports, although these subsidies arguably offset an even larger subsidy to U.S. farmers and the cost of foreign oil imports (Goldemberg 2007).

From 1978 to 2004, there was little change to the main component of federal support, the excise tax exemption. Benefit levels changed several times, culminating in a progressive reduction from $0.14/L ($0.54/gal) to $0.13/L ($0.51/gal) of ethanol during 1998–2005 as a result of the Transportation Equity Act of 1998. However, in 2004 the basic mechanics of the subsidy were changed by the introduction of the Volumetric Ethanol Excise Tax Credit (VEETC). The VEETC (or "blender's credit") streamlined the system by (1) making it volume-based rather than limited to specific blends; (2) eliminating negative impacts on the Highway Trust Fund by taking the credit from general government revenues; and (3) renewing the subsidy at $0.13/L ($0.51/gal) of ethanol, which was lowered to $0.12/L ($0.45/gal) for 2009 by the 2008 Farm Bill (RFA 2011). Similarly, biodiesel was given a Volumetric Biodiesel Excise Tax Credit (VBETC) of $0.13/L ($0.50/gal) for biodiesel from waste cooking oils and $0.26/L ($1.00/gal) for biodiesel made from virgin agricultural feedstock (Koplow 2007). Most of the biodiesel produced in the United States is soybean based.

The Energy Policy Act (EPAct) of 2005 approved several major incentives to usher in a new era of renewable transportation fuels (Public Law 109-58 2005). The most widely publicized provision of EPAct, the Renewable Fuel Standard (RFS), applies to corn and cellulosic ethanol and will operate in the place of the now eliminated oxygenate requirement for reformulated gasoline. Implementation of the RFS by the U.S. Environmental Protection Agency (EPA) began in 2006 at 1.5×10^{10} L/year (which was nearly met in 2005) and was scheduled to increase to 2.8×10^{10} L/year in 2012. In light of the current market for ethanol, EPAct has provided only a modest production boost. Additional provisions of EPAct were designed to improve commercialization prospects for cellulosic ethanol through increased research and development funding in all aspects of the industry, including feedstock development, processing technology, co-product production, and systems optimization (Wyman 2003). Project financing and funding, a major bottleneck (Hamelinck et al. 2005), received attention as well through a series of large grants and loan guarantees for biorefinery development and commercialization (Public Law 109-58 2005). Overall, EPAct provides a short-term boost to accelerate commercialization and technological development while also attempting to cement a place for the new technology in the longer-term ethanol market. Implicit in this is the assumption that cellulosic ethanol is capable of providing larger societal benefits than corn ethanol, although in the near future, corn ethanol will still dominate the market.

The prospects for cellulosic ethanol received a much larger boost through passage of the Energy Independence and Security Act in December 2007. This law revises and extends the RFS beginning in 2008 at 3.4×10^{10} L/year, up to a rather ambitious 1.36×10^{11} L/year by 2022. Of this total, no more than 5.7×10^{10} L/year will come from cornstarch, with the remaining 7.9×10^{10} L/year to come from advanced biofuels with greatly reduced greenhouse gas emissions (including biodiesel). Over 75% of all advanced biofuels will eventually come from cellulose, and this part of the mandate could be met by any combination of ethanol and other alcohols (Sissine 2007).

During the revival of the domestic ethanol industry in the late 1970s, more than a dozen state governments quickly approved partial or total gasohol exemptions from state road use taxes (Solomon 1980; Solomon et al. 2009). These state programs are generally similar to the federal programs, and as of 2010, 32 states were supporting ethanol development (Voegele 2010). The most important state policies included mandates and various levels of excise tax exemptions.

The policy environments in Minnesota and Iowa have been especially strong, combining measures that support production and consumption of ethanol (Solomon et al. 2009). Instrumental to this effort in Minnesota has been a 1997 state requirement that all gasoline sold in the state must have a 10% ethanol content. An increase to a 20% mandate was approved in Minnesota in 2005 (which would take effect in 2013) pending the granting of a waiver by EPA under the Clean Air Act. Similar laws that mandate a 10% ethanol-blended fuel were passed in Hawaii and Montana in 2005, and Missouri, Washington and Oregon in 2006–2007, and Florida in 2008 (RFA 2011). The requirements have varying phase-in dates. Kansas has a 10% ethanol blend mandate that only applies to state fleet vehicles. Iowa joined Minnesota in 2006 by approving a comprehensive state RFS—a 10% mandate for 2009 that increases to 25% in 2020. In addition, a retail tax credit on E85 (a blend of 85% ethanol and 15% gasoline) of $0.07/L ($0.25/gal) was approved, although it is being lowered over time. This compares with Minnesota, which has a state fuel tax exemption on E85 of $0.015/L ($0.058/gal) and an ethanol production payment of $0.05/L ($0.20/gal). Minnesota also has the most extensive E85 infrastructure in the country, with 360 retail fueling outlets (DOE 2011).

12.2.2 European Union

The European Union (EU) also provides support for a domestic biofuels industry, most notably through its Biofuels Directive of 2003 and the Renewable Energy Directive of 2008 (Rosch and Skarka 2008). Only a small amount of ethanol is produced, mostly from sugarbeets, wine, wheat, and corn in France, Germany, and Spain. More significant biofuel production is in the form of biodiesel from rapeseed, especially from Germany (the world leader) and to a lesser extent from

France, Spain, Italy, and Austria. This pattern is not surprising given the greater use of diesel vehicles in Europe than in the United States.

The EU Biofuels Directive, officially 2003/30/EC—the directive on the promotion of the use of biofuels and other renewable fuels for transport—entered into force in May 2003. The Biofuels Directive initially set a target of 2% biofuels for 2005, which was not met. This measure mandated that national measures among member countries must be taken that will result in the replacing of 5.75% of all transport fuels (i.e., gasoline and diesel) with biofuels by December 31, 2010 and 10% by 2020. Differentiation of national target levels was allowed for various reasons, such as limited national production potential and allocation of resources to biomass production in sectors in addition to transport. For example, France had a 7% target for 2010 whereas Italy's was 2.5%. This became moot, however, when controversy over its effects led to repeal of this Directive in 2009. The replacement EU Directive 2009/28/EC promotes biofuels that can significantly reduce greenhouse gas emissions through a shift to second-generation biofuels, similar to that required in the United States, and that sustainability criteria be met. The latter criteria are being developed internationally, although the Swedish firm SEKAB already has a sustainable ethanol standard (Solomon 2010).

The greenhouse gas reduction and sustainability criteria for biofuels in the EU will be instituted in response to the Fuel Quality Directive 98/70/EC approved in Brussels in December 2008. This directive focuses on the incorporation of sustainability criteria for biofuels to reduce life-cycle greenhouse gas emissions, among other considerations, and will be coordinated with the broader EU Renewable Energy Directive (Rosch and Skarna 2008).

12.2.3 ELSEWHERE

Brazil's modern sugarcane ethanol industry is older than that of the United States because it began as the Pro-Alcool (National Alcohol) Program in 1975 to wean the nation off of oil imports. Today Brazil is the world's second-largest ethanol producer, after the United States, and has a modest but growing biodiesel program based on soybeans and castor oils. Brazil is also the world's leading exporter of ethanol, mostly to the United States and the Netherlands.

Brazil has supported its ethanol industry with various policies (Reel 2006). First, since 1976 the federal government has mandated the blending of anhydrous ethanol with gasoline at levels that have varied from 10 to 25%. In addition, the Brazilian government has guaranteed ethanol purchases by Petrobras (the state-owned oil company), subsidized sugarcane growers, provided low-interest loans and credit guarantees to ethanol refiners, mandated that gasoline service station owners in towns of more than 1500 people install ethanol pumps, and instituted price controls that made ethanol available for 59% of gasoline prices. Ethanol-only automobiles (technically ethanol with a E95 blend with gasoline) were even promoted in the 1980s by tax breaks, and over half of the nation's cars ran on the fuel in the late 1980s. Most of the ethanol price supports and subsidies were eliminated by the mid 1990s because greater efficiency and lower production costs have allowed Brazilian ethanol to compete without government support (Gallagher 2007).

There are small but growing biofuels programs elsewhere in South America, most notably in Colombia and most recently in Venezuela, Argentina, and Uruguay. Another new player to the biofuels scene is Africa. For example, in South Africa, seven corn ethanol refineries are planned as well as one each based on sorghum and sugarbeets, plus a biodiesel plant based on waste vegetable oil. Other ethanol plants are at various stages of development in Nigeria, Tanzania, Kenya, and Mozambique that are based on cassava, sugarcane, and sorghum along with a biodiesel plant in Zimbabwe that is based on cotton and sunflower seeds, soybeans, and possibly jatropha oil (Solomon 2009).

There are also several important biofuels markets in Asia. The leading ethanol producer in the region is China, although its output in 2008 was only 5.6% that of the United States and 7.8% that of Brazil (RFA 2011). Like the United States, China subsidizes its ethanol industry. China has had plans to expand its ethanol production by 650% from 2007 to 2020, although in the summer of 2007 (and again in summer 2008) the government announced restrictions on ethanol made from food grains, corn in particular (Koizumi and Ohga 2007). Even so, approximately 80% of ethanol production in China is made from cornstarch, with the rest derived from wheat and rice. Other feedstocks are now getting greater attention, such as cassava, sweet potato, sweet sorghum, and sugarcane, and a demonstration plant that converts corn stover to ethanol (Leng et al. 2008; Solomon 2009). However, short-term ethanol production goals are not being met because of a shortage of raw materials during this transition period. India has also been a major producer of sugarcane-based ethanol, but its industry has fallen on hard times and output has dropped dramatically since 2006.

Several countries in Southeast Asia have growing biofuels industries. For example, Indonesia, Malaysia, and even Singapore are all planning major expansion of palm oil production to make biodiesel. Serious concerns have been raised in Europe and elsewhere about these plans because of their contribution to tropical deforestation (Koh and Wilcove 2008; Danielsen et al. 2009). A fourth country in the region, Thailand, is also expanding it biodiesel capacity, as well as ethanol production, from sugarcane and cassava feedstocks (Solomon 2009).

12.3   ECONOMICS

Recent years have seen a spike of oil, crop, and seed prices. From the beginning of 2007 to the middle of 2008 global seed prices and oil prices doubled, and meal prices also experienced a sharp increase (FAO 2009). At the same time, the global recession caused numerous biofuel producers to file for bankruptcy and others to not produce at full capacity (HGCA 2009). For biofuels to be competitive on the market, crude oil prices need to be between $50 and $100 per barrel (Lange 2007), a level they have been at since March 2009, following 3 months below that range.

12.3.1   United States

The primary bioenergy crop in the United States is corn, accounting for over 95% of total ethanol production (RFA 2011). As discussed in the aforementioned policies, the United States has seen widespread implementation of a corn-based ethanol market, an increasingly common additive to gasoline. In 2010, ethanol production totaled approximately 50×10^9-/L, or 9% of all gasoline-type fuel sold in the United States (EIA 2011). This accounted for just over half of the global output, and there is capacity to produce over 54×10^9L/year (RFA 2011). In September 2011 there were 2442 E85 fuel stations, mostly in the Midwest, an increase of over 1000 in the past three years (DOE 2011; Solomon et al. 2009).

Since the late 1970s, gasoline has consistently been less expensive than corn ethanol. This has been the case although the price of grain ethanol has been partially offset by the production and sale of co-products, such as distillers dried grains with solubles. Over 80% of U.S. production, including the most recently built plants, has come from anhydrous mills with the rest made in the more costly hydrous mills (Urbanchuk 2006). However, the actual price differential between ethanol and gasoline has been relatively small, mainly because of the large government subsidies noted earlier, especially the federal excise tax exemption and the fact that most ethanol sold in the United States is through a 10% ethanol blend with 90% gasoline (E10). Even if the retail price of the ethanol blend is less expensive than gasoline, as is typically the case for E85, the real economic efficiency of corn ethanol is much lower because the energy density of ethanol is only two-thirds that of gasoline (Lide 1992).

Perhaps the largest controversy involved in the large quantity of corn-to-ethanol production is the impact this has had on food prices. The increase of ethanol production has come with an increase of corn production—from 9.5×10^9 bushels produced in 2001 to 12.1×10^9 bushels produced in 2009 (DOA 2009). Approximately 11 L (2.8 gal) of ethanol is produced from a bushel of corn (de Gorter and Just 2009). Corn converted into fuel alcohol from 2001 to 2009 expanded from 0.7×10^9 bushels to 4.1×10^9 bushels. Thus, the additional ethanol production came largely from an increase in crop

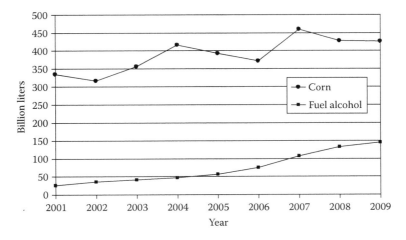

FIGURE 12.1 Total U.S. production of corn and fuel ethanol between 2001 and 2009. (Adapted from USDA, Feed grains database: Yearbook tables. U.S. Department of Agriculture, Economic Research Service, 2009. Available at www.ers.usda.gov/Data/FeedGrains/FeedYearbook.aspx (accessed July 17, 2009)) feed grains database—yearbook tables.

yield, as show in Figure 12.1. A study by the U.S. Department of Agriculture (USDA) concludes that food prices have risen very modestly because of the growth of corn-based ethanol, and the price rises have been largely in the beef sector (because most corn grown in the United States is used as feed), not in corn-based products such as cereals. It was calculated that even if corn prices were to increase 50%, overall food prices would raise less than 1% (Leibtag 2008). However, the debate is far from over because a diversity of opinions remains on this subject (cf. Pimentel et al. 2009). Ultimately, the constraining factor on increasing ethanol production will be the land available for feedstock supply growth as the ethanol market expands.

Although almost all ethanol in the United States is currently made from corn, there is substantial interest in producing cellulosic ethanol, i.e., ethanol that is produced from wood, farming, and forestry residues or grasses rather than the starch in grains. According to the optimistic 2005 "Billion Ton Study," cellulosic ethanol has the potential to offset at least 30% of the current petroleum consumption in the United States (Perlack et al. 2005). Thus, the potential for cellulosic ethanol is much higher than that of corn ethanol.

However, cellulosic ethanol is not yet market-efficient, and a widespread cellulosic ethanol market has yet to be created. This could change in the near future; worldwide, approximately 15 or so commercial cellulosic ethanol plants are expected to be commissioned in the next few years, with capacities ranging from 3.8×10^6 to 3.8×10^8 L/year. Most of these plants will be located in the United States. The cellulosic ethanol plants will use various feedstocks, including corn stover, bagasse, wheat straw, wood waste and residues, wood chips, municipal solid wastes, and switchgrass. About a dozen demonstration plants and 20 pilot plants have also been built (Solomon et al. 2009). The wood chips would probably be produced from fast-growing trees such as willow shrubs or poplar (Volk et al. 2006).

Cellulosic ethanol is expensive to produce for three main reasons. First, achieving economies of scale is difficult because processing plants must be located near ethanol feedstocks since transportation of the feedstocks is expensive. Thus, processing plant necessarily have to be limited in size because the maximum shipping range for the feedstock is approximately 120 km (Froese et al. 2008). Interestingly, this could provide an opportunity for numerous small-scale markets to be created, rather than one or several large markets, as is the case with grain. Second, the initial capital costs for constructing processing plants are much larger than for grain ethanol plants. This is because of the need for pretreatment of the biomass and the high cost of utilities. For example,

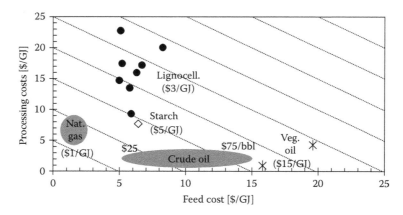

FIGURE 12.2 Feed and processing cost of transportation fuels derived from lignocellulose and fossil resources. (Reprinted from Lange, J.P., *Biofuels Bioprod Biorefi*, 1, 39–48, 2007. With permission.)

Wyman (1999) calculated that a 2.2×10^8-L/year processing plant would cost 250×10^6 (1997 dollars). Third, the technology and processes (both biochemical and thermochemical) involved in creating ethanol from various feedstocks are still immature and are being actively researched (Ruth 2008; Solomon et al. 2009). This can be seen in Figure 12.2, which shows the relative costs of feedstocks and processing for various transportation fuels. Access to capital investment and the need to create transportation and market infrastructures are other challenges that will need to be overcome for cellulosic ethanol to be commercially viable in large quantities.

The United States is the second-largest biodiesel producer in the world, behind Germany (Canakci and Sanli 2008). In the United States, biodiesel is made primarily from soybeans. The U.S. industry experienced rapid growth, expanding from 1.9×10^6 L produced in 1999 to 2.65×10^7 L in 2008 (NBB 2009). Production plummeted in 2009-2010 due to the recession, but has since rebounded because of the RFS. The major cost of biodiesel is the cost of the feedstock, which accounts for nearly 90% of the total fuel cost (Haas et al. 2005). Because the feedstock costs alone can be 1.5–3.0 higher than the price of diesel fuel, only lower feedstock costs or higher diesel prices will make biofuels a more attractive option. The prices of plant-based feedstocks such as soybeans, canola, sunflower, and rapeseed are typically more expensive than various animal fats (Canakci and Sanli 2008).

12.3.2 EUROPEAN UNION

The EU is the world's leading biodiesel producer, with approximately two-thirds of the world's market (REN21 2009). Unlike the United States, biodiesel is the primary biofuel produced in the EU; biodiesel makes up approximately 80% of the total biofuels produced in the region (Bozbas 2008). Germany is by far the largest producer, producing half of all production. France, Italy, Spain, and Austria are the next three largest producers. In total, the EU has 241 plants with a total capacity of 16×10^9 L/year, suggesting strong growth in the next few years (EBB 2008; REN21 2009). On the other hand, in 2009 Germany and the United Kingdom lowered their mandatory blend rate by 1 and 0.5%, respectively, for a year in response to market conditions (FAO 2009).

12.3.3 BRAZIL

Brazil is the world's second-largest ethanol producer and has recently started producing biodiesel as well. Between 2006 and 2008 Brazil increased ethanol production from 18×10^9 to 27×10^9 L, and there are now 400 ethanol mills and 60 biodiesel mills in operation (REN21 2009). A total of 4.2×10^6 ha of sugarcane is harvested yearly. One reason for this massive increase is that sugar prices have been dropping, whereas ethanol prices have been increasing (Moreira 2009).

The main advantage that Brazil has over the United States is that it is much cheaper to produce ethanol from sugarcane than corn because of the simplified processing, the availability of free bagasse, and the favorable agricultural environment in Brazil (Moreira 2000). Furthermore, sugarcane has an energy density approximately three times greater than corn (Ruth 2008).

12.4 PUBLIC PERCEPTIONS: RESULTS OF AN OPINION SURVEY IN THE UPPER MIDWESTERN UNITED STATES

Concerns about climate change, energy security, and environmental impacts have revived interest in using biofuels to produce energy. Although certain bioenergy crops, such as corn, and their associated biofuels have established markets, many do not. Forest residues and agricultural waste are plentiful in many areas of the world and could be used to provide liquid biofuels or electricity through combustion. However, markets that use these products are not widespread and are limited in scale.

A public opinion survey was sent to 1500 households in the states of Michigan, Minnesota, and Wisconsin to gauge attitudes about global climate change and to determine people's willingness to pay (WTP) for cellulosic ethanol as a substitute for gasoline. As part of the survey, responses were elicited as to whether or not biofuel production would be good for the respondents' local economies. This section of the chapter will present descriptive results from some of the survey questions relevant to biofuels. It was found that residents generally believe that local biofuel production would be good for the local economies. Although owners of farm and forestland were more supportive than non-owners, they thought that they themselves would only be marginally affected. There was strong support for the idea that the United States should produce all of its own energy requirements. Overall, farmers and forestland owners did not have significantly different attitudes toward a potential biofuel market and climate change than did other survey respondents.

12.4.1 PREVIOUS RESEARCH

Little research has been conducted on the public perceptions of biofuel markets in the United States. However, there have been several recent studies completed that provide a good indication of current public attitudes. For example, a telephone survey of over 1000 people in the United States done by Wegener and Kelly (2008) found that although 78.6% of people agreed with the statement "using biofuels such as ethanol is a good idea," only 70.6% of people agreed that "using corn to produce ethanol is a good idea" and 55.5% of people agreed that "using wood or wood chips to produce ethanol is a good idea." In comparison, 58.6 and 56.6% of people agreed with the statements "using coal-based energy is a good idea" and "using energy from oil is a good idea," respectively. Nearly 24% of the respondents felt "not at all informed" about biofuels, and about double this rate felt "not at all informed" about cellulosic ethanol (depending on the feedstock).

In a nationwide web and phone-based WTP study, which focused on support for energy research and development of nonfossil fuel technologies, Li et al. (2009) found very high support for spending on research and development of renewable energy technologies including biofuel crops. A dummy variable that accounted for corn-producing states was shown to be insignificant, thus suggesting attitudes about biofuel markets in the Midwest are not significantly different than those in the rest of the country. In another study, Borchers et al. (2007) found that solar power and wind power are favored more than bioenergy as renewable energy sources.

In a mail survey sent to over 15,000 Tennessee farmers, Jensen et al. (2007) examined their willingness to grow switchgrass as a bioenergy feedstock. Approximately 3500 surveys were returned completed. Only approximately 21% of the respondents had heard of switchgrass being used as a biofuel. Although there was a large amount of uncertainty for many respondents as to whether or not they would be willing to grow it (~47%), approximately 30% said they would be willing to try it if it were profitable.

What is more typical than the measurement of the public perceptions of biofuel markets is the measurement of WTP for green electricity, of which biomass is one possible fuel. Several studies have shown a large gap between the number of people who say they are willing to pay more for green electricity and the number who actually do so when given the option by their electric power provider (e.g., Borchers et al. 2007; Wiser 2007; Hite et al. 2008).

Another typical way that public perceptions of biofuel markets in the United States are measured is through studies gauging public perceptions of climate change. For instance, Dietz et al. (2007) examined people's attitudes toward eight possible climate change mitigation policies, including several that would make biomass combustion and liquid biofuel consumption in vehicles more economically viable. Policies that had the least direct economic impact on consumers, such as the reduction of government subsidies for fossil fuels, had the most support, whereas direct taxes, such as a 60-cent per gallon gasoline tax, had the least. Our three-state public opinion survey included sections on public perception and attitudes toward global climate change similar to Dietz et al. (2007), but also explicitly addressed cellulosic ethanol as a possible solution plus a section that addressed the WTP for this fuel (Solomon and Johnson 2009b).

12.4.2 Methodology

The survey was mailed to 1500 households throughout Michigan, Minnesota, and Wisconsin. Rural addresses received 60% of the surveys so as to increase the number of farm and forest owners in the sample. An expanded version of Dillman's tailored design method was used to increase the response rate (Clendenning et al. 2004). This entailed sending out a presurvey notification letter, multiple survey mailings, and reminder postcards between mailings. A $2 bill was sent with the first survey as a thank you. The surveys were sent in three rounds between November 2007 and January 2008. After accounting for bad addresses, the final survey size was 1432. A total of 745 households returned a survey for an overall response rate of 52%. For this analysis, a further 75 surveys were discarded for not providing complete answers to all of the relevant questions. The survey responses were then weighted to correct for the aforementioned oversampling and for slightly varying response rates from different states, adjusting the results based on the actual populations on the basis of data from the U.S. Census Bureau for the rural and urban subpopulations in the three states.

As noted earlier, the survey had sections. The first section included 44 Likert scale questions that asked respondents about their opinions of climate change, energy cost and consumption, and the environment. The second section was a spilt sample contingent valuation method and "fair share" survey that measured WTP for cellulosic ethanol fuel in the states surveyed (Solomon and Johnson 2009b). Owners of farms or forests were asked several additional non-Likert scale questions to determine if the owners had forest residues or waste biomass they would want to sell. The average time to complete the survey was approximately 30 min.

12.4.3 Results

Table 12.2 shows the raw average responses and the weighted responses to the questions regarding public attitudes and beliefs toward cellulosic ethanol. Respondents who chose "don't know" for answers were excluded from the average. There was some agreement that making biofuels would be helpful to local economies. Farmers and forestland owners exhibited greater support for a local biofuel economy than did the general population. It is interesting to note that not a single farm or forest owner ($n = 93$) was indecisive ("neither agree nor disagree") to the statement "America should produce its own energy." Similarly, only one farm owner and one forest owner responded with "neither agree nor disagree" to the statement "Making more biofuels, like corn ethanol, would be good for my area's economy." For non-farm and forest owners, this option was selected 16% and 14%, respectively.

TABLE 12.2
Public Perceptions Survey Responses Comparing Farm and Forest Owners to the Total Population

Question	Total Population (n = 670)			Farm Owners (n = 42)			Forest Owners (n = 51)		
	Weighted Mean	Raw Mean	n	Weighted Mean	Raw Mean	n	Weighted Mean	Raw Mean	n
Making more biofuels, like corn ethanol, would be good for my area's economy.	3.70	3.72	636	3.80	3.83	42	3.87	3.78	51
America should produce all of its own energy.	4.02	4.02	659	4.11	4.02	42	4.02	4.10	51
Climate change is *not* going to happen.	1.82	1.87	638	1.91	1.92	40	1.75	1.86	50
Climate change is *not* likely to be a serious problem.	2.12	2.16	636	2.30	2.52	42	2.14	2.08	51
Burning fossil fuels is one of the primary causes of climate change.	3.59	3.60	573	3.55	3.47	38	3.42	3.48	46

1, strongly disagree; 2, somewhat disagree; 3, neither agree nor disagree; 4, somewhat agree; 5, strongly agree; 0, don't know

Question	Total Population (n = 670)			Farm Owners (n = 42)			Forest Owners (n = 51)		
How well would removing leftover materials or residues, such as leaves and branches or corn and wheat stalks, fit with current management of your land?	—	—	—	2.46	2.48	42	2.28	2.20	42

1, very well; 2, somewhat well; 3, poorly; 4, very poorly; 0, don't know

Question	Total Population (n = 670)			Farm Owners (n = 42)			Forest Owners (n = 51)		
How much do you think having a regional cellulosic or "cellulosic" ethanol-based market for these leftover materials or residues benefit you economically?	—	—	—	2.59	2.62	42	2.60	2.69	42

1, not at all; 2, very little; 3, somewhat; 4, significantly; 0, don't know

Although there was a weak preference for the idea that making biofuels would be good for the local economy, neither farm nor forest owners thought that the removal of waste material on their property would fit with their current land management strategies. Similarly, there was only marginal support for the idea that a nearby cellulosic ethanol plant to which they could sell waste material would be economically beneficial.

To ensure that respondents were familiar with the subject of climate change, the first two questions of the survey were "I have heard the term climate change" and "I have heard the term global warming." Of the 670 respondents, 635 (93%) answered either "strongly agree" or "somewhat agree" to one of these two questions. The subset of farm owners and forestland owners were similarly aware of the terms. It is interesting to note that although there was a high degree of disagreement with the statement "climate change is *not* going to happen," 97 respondents expressed uncertainty ("don't know") about the statement "Burning fossils is one of the primary causes of climate change." This indicates that although there is agreement that climate change is happening, the public is unsure of the causes.

12.5   THE FUTURE

The future proliferation of bioenergy crops will be highly dependent upon government policies supporting (or hindering) biofuels and the economics of both biofuels and the oil products they are market substitutes for, namely gasoline and diesel. In the near term, corn-based ethanol will reach a saturation point in the United States because of the limited quantity of additional available corn and the cap imposed by the RFS, although it is not known exactly what this saturation point is. As of September 2011, an additional 1×10^9 L/year of ethanol capacity was under construction in the United States (RFA 2011). However, the recent global economic downturn has had a toll. VeraSun Energy, formerly the largest commercial ethanol producer in the United States, entered bankruptcy in 2008 because of the credit crisis and falling oil prices (HGCA 2009). On the other hand, Brazil is in a flexible position because of its ability to produce relatively cheap ethanol and the availability of a significant amount of land that can be dedicated to additional sugarcane production.

As for cellulosic ethanol, the 2007 Energy Security and Independence Act, combined with decreasing industry costs as indicated by the number of commercial plants being built, indicate that it will become a major competitor to gasoline. World cellulosic ethanol production capacity is expected to be at least 1×10^9 L/year by 2012, primarily in the United States (Solomon et al. 2009). However, this is by no means assured because cellulosic ethanol is still an emerging market. Additionally, high fragmentation of land ownership, especially in the Midwest, will pose a barrier to the availability of feedstock (e.g., Potter-Witter 2005). Ethanol has to be shipped by road, rail, or barge because ethanol attracts water and transportation in oil pipelines can contaminate the pipeline. A major technological breakthrough in transportation infrastructure came in 2008 when ethanol was successfully shipped through a gasoline pipeline in Florida (UPI 2008). This should allow for a reduction in transportation costs and could allow for (cellulosic) ethanol production in regions with large amounts of feedstock but low population densities, such as federal lands in the North American West.

The EU is expected to maintain its strong growth in biodiesel production, with 3×10^9 L/year additional capacity under construction, although there has been some recent retrenchment, especially in Germany. Argentina is poised to be one of the world's largest biodiesel producers. Currently, the country is the world's third-largest producer of soy oil and the largest exporter; half of all cultivated land in Argentina is soy. There is currently approximately 3×10^9 L/year of biodiesel production capacity, with an additional 2×10^9 L/year of biodiesel production under construction there. Almost all of the biodiesel is expected to be exported (Mathews and Goldsztein 2009; REN21 2009). Biodiesel production is expanding in Brazil as well, and Asian markets continue to grow. Thus, although the near-term future of bioenergy crops is mixed, the long-term prospects remain bright.

REFERENCES

Borchers AM, Duke JM, Parsons GR (2007) Does willingness to pay for green energy differ by source? *Energy Policy* 25:3327–3334

Bozbas K (2008) Biodiesel as an alternative motor fuel: Production and policies in the European Union. *Renew Sustain Energy Rev* 12:542–552

California Energy Commission (2004) Ethanol fuel incentives applied in the U.S., reviewed from California's perspective. CEC Staff Report P600-04-001, Sacramento, CA

Canakci M, Sanli H (2008) Biodiesel production from various feedstocks and their effects on the fuel properties. *J Ind Microbiol Biotechnol* 35:431–441

Clendenning G, Field DR, Jensen D (2004) A survey of seasonal and permanent landowners in Wisconsin's northwoods: Following Dillman and then some. *Soc Natur Resour* 17:431–442

Danielsen F, Beukema H, Burgess ND, Parish F, Brüehl CA, Donald PF, Murdiyarso D, Phalan B, Reijnders L, Struebig M, Fitzherbert EB (2009) Biofuel plantations on forested lands: Double jeopardy for biodiversity and climate. *Conserv Biol* 23:348-358

de Gorter H, Just DR (2009) The welfare economics of a biofuels tax credit and the interaction effects with price contingent farm subsidies. *Amer J Agric Econ* 91:477–488

Dietz T, Dan A, Shwom R (2007) Support for climate change policy: Social psychological and social structural influences. *Rural Sociol* 72:185–214

DOE (2011) Alternative fuels data center, U.S. Department of Energy, available at http://www.eere.energy.gov/afdc/fuels/stations_counts.html (accessed September 27, 2011)

EIA (2011) Supply and disposition of crude oil and petroleum products, U.S. Energy Information Administration, available at http://tonto.eia.doe.gov/dnav/pet/pet_sum_snd_d_nus_mbbl_m_cur-1.htm (accessed September 27, 2011)

EBB (2008) European Biodiesel Board, available at www.ebb-eu.org/ (accessed July 16, 2009)

FAO (2009) Monthly price and policy update (March 2009), Food and Agriculture Organization of the United Nations, available at www.fao.org/es/esc/en/15/120/highlight_573.html (accessed July 17, 2009)

Froese RE, Waterstraut JR, Johnson DM, Shonnard DR, Whitmarsh JH, Miller CA (2008) Lignocellulosic ethanol: Is it economically and financially viable as a fuel source? *Environ Qual Manag* 18:23–45

Gallagher PW (2007) A look at US-Brazil ethanol trade and policy. *Biofuels Bioprod Biorefin* 1:9–13

GAO (2000) *Petroleum and Ethanol Fuels: Tax Incentives and Related GAO Work*. RCED-00-301R, U.S. General Accounting Office, Division of Resources, Community, and Economic Development, Washington, DC

Goldemberg J (2007) Ethanol for a sustainable energy future. *Science* 315:808–810

Haas MJ, McAloon AJ, Yee WC, Foglia TA (2005) A process model to estimate biodiesel production costs. *Bioresour Technol* 97:671–678

Hamelinck CN, van Hooijdonk G, Faaij APC (2005) Ethanol from lignocellulosic biomass: Techno-economic performance in short-, middle- and long-term. *Biomass Bioenergy* 28:384–410

HGCA (2009) Biofuel and industrial news, Home-Grown Cereals Authority, available at http://www.hgca.com/content.output/2948/2948/Markets/Market%20News/Biofuel%20and%20Industrial%20News%20Update.mspx (accessed May 29, 2009)

Hite D, Duffy P, Bransby D, Slaton D (2008) Consumer willingness-to-pay for biopower: Results from focus groups. *Biomass Bioenergy* 32:11–17

Jensen K, Clark CD, Ellis P, English B, Menard J, Walsh M, de la Torre Ugarte D. (2007) Farmer willingness to grow switchgrass for energy production. *Biomass Bioenergy* 31:773–781

Koh LP, Wilcove DS (2008) Is oil palm agriculture really destroying tropical biodiversity? *Conserv Lett* 1:60–64

Koizumi T, Ohga K (2007) Biofuels policies in Asian countries: Impacts of the expanded biofuels programs on world agricultural markets. *J Agric Food Indust Organ* 5(no 2): Article 8, available at http://www.bepress.com/jafio/vol5/iss2/art8 (accessed July 17, 2009)

Koplow D (2007) *Biofuels—At What Cost? Government Support for Ethanol and Biodiesel in the United States: 2007 Update*. Prepared by Earth Track, Inc. for The Global Subsidies Initiative of the International Institute for Sustainable Development, Geneva, Switzerland

Lange JP (2007) Lignocellulose conversion: An introduction to chemistry, process and economics. *Biofuels Bioprod Biorefin* 1:39–48

Leibtag E (2008) Corn prices near record high but what about food costs? *Amber Waves*. February. Economic Research Service. Department of Agriculture, available at www.ers.usda.gov/AmberWaves/February08/Features/CornPrices.htm (accessed July 17, 2009)

Leng R, Wang C, Zhang C, Dai D, Pu G (2008) Life cycle inventory and energy analysis of cassava-based fuel ethanol in China. *J Clean Prod* 16:374–384

Li H, Jenkins-Smith HC, Silva CL, Berrens RP, Herron KG (2009) Public support for reducing US reliance on fossil fuels: Investing household willingness-to-pay for energy research and development. *Ecol Econ* 68:731–742

Lide DR (ed) (1992) *CRC Handbook of Chemistry and Physics*, 73rd ed. CRC Press, Boca Raton, FL

Mathews JA, Goldsztein H (2009) Capturing latecomer advantages in the adoption of biofuels: The case of Argentina. *Energy Policy* 37:326–337

Meher LC, Vidya Sagar D, Naik SN (2006) Technical aspect of biodiesel production by transeterification—A review. *Renew Sustain Energy Rev* 10:248–268

Moreira JR (2000) Sugarcane for energy—Recent results and progress in Brazil. *Energy Sustain Dev* 4:43–54

NBB (2009) National Biodiesel Board, available at www.biodiesel.org (accessed May 25, 2009)

Perlack RD, Wright LL, Turhollow AF, Graham RL, Stokes BJ, Erbach DC (2005) Biomass as Feedstock for a Bioenergy and Bioproducts Industry. Oak Ridge National Laboratory Report TM-2005/66. Oak Ridge, TN

Pimentel D, Marklein A, Toth MA, Karpoff MN, Paul GS, McCormack R, Kyriazis J, Krueger T (2009) Food verses biofuels: Environmental and economic costs. *Hum Ecol* 37:1-12

Potter-Witter K (2005) A cross-sectional analysis of Michigan nonindustrial private forest landowners. *N J Appl For* 22:132–138

Public Law 109-58 (2005) The Energy Policy Act of 2005. U.S. Congress, Washington, DC

Reel M (2006) Brazil's road to energy independence. *Washington Post*, August 20, 2006, A01

REN21 (2009) Renewables Global Status Report: 2009 update. Renewable Energy Policy Network for the 21st Century Secretariat, Paris, available at www.martinot.info/RE_GSR_2009_Update.pdf (accessed July 17, 2009)

RFA (2011) Renewable Fuels Association, available at www.ethanolrfa.org (accessed September 27, 2011)

Rosch C, Skarka J (2008) European biofuel policy in a global context: The trade-offs and strategies. *Gaia* 17:378–836

Ruth L (2008) Bio or bust? The economic and ecological cost of biofuels. *EMBO Rep* 2:130–133

Shay EG (1993) Diesel fuel from vegetable oil: Status and opportunities. *Biomass Bioenergy* 4:227–242

Sissine F (2007) *Energy Independence and Security Act of 2007: A Summary of Major Provisions*. Congressional Research Service Report for Congress, Order Code RL34294, Washington, DC

Solomon BD (1980) Gasohol, economics, and passenger transportation policy. *Transp J* 20:57–64

Solomon BD (2009) Alternative energy & land use change scenarios. Paper presented at the Workshop on Soil Diversity & Ecosystem Services, Wageningen, the Netherlands

Solomon BD (2010) Biofuels and sustainability. *Ecol Econ Rev* 1:119–134

Solomon BD, Barnes JR, Halvorsen KE (2009) From grain to cellulosic ethanol: History, economics and policy. In: Solomon, BD, Luzadis, VA (eds), *Renewable Energy from Forest Resources in the United States*. Routledge, London, UK, pp 49–66

Solomon BD, Johnson NH (2009a) Introduction. In: Solomon BD, Luzadis VA (eds), *Renewable Energy from Forest Resources in the United States*. Routledge, London, UK, pp 3–27

Solomon BD, Johnson NH (2009b) Valuing climate protection through willingness to pay for biomass ethanol. *Ecol Econ* 68:2137–2244

UPI (2008) Ethanol test-shipped in gasoline pipeline. United Press International Incorporated, October 16, 2008, available at http://www.upi.com/Top_News/2008/10/16/Ethanol-test-shipped-in-gasoline-pipeline/UPI-24441224197853/ (accessed July 17, 2009)

Urbanchuk JM (2006) Contribution of the ethanol industry to the economy of the United States. Prepared by LECG LLC for the Renewable Fuels Association, Washington DC

USDA (2009) Feed grains database: Yearbook tables. U.S. Department of Agriculture, Economic Research Service, available at www.ers.usda.gov/Data/FeedGrains/FeedYearbook.aspx (accessed July 17, 2009)

Voegele E (2010) Incentives: It's all about location, location, location. *Ethanol Producer Magazine*, available at www.ethanolproducer.com/articles/6248/incentives-it's-all-about-location-location-location

Volk TA, Abrahamson LP, Nowak CA, Smart LB, Tharakan PJ, White EH (2006) The development of short-rotation willow in the northeastern United States for bioenergy and bioproducts, agroforestry, and phytoremediation. *Biomass Bioenergy* 30:715–727

Wegener DT, Kelly JR (2008) Social psychological dimensions of bioenergy development and public acceptance. *BioEnergy Res* 1:107–117

Wiser RH (2007) Using contingent valuation to explore willingness to pay for renewable energy: A comparison of collective and voluntary payment vehicles. *Ecol Econ* 62:419–432

Wyman CE (1999) Biomass ethanol: Technical progress, opportunities and commercial challenges. *Annu Rev Energy Env* 24:189–226

Wyman CE (2003) Potential synergies and challenges in refining cellulosic biomass to fuels, chemicals, and power. *Biotechnol Progr* 19:254–262

Yergin D (1992) *The Prize: The Epic Quest for Oil, Money, & Power*. Free Press, New York

Section II

13 Cassava

Satya S. Narina
Virginia State University

Damaris Odeny
Agricultural Research Council

CONTENTS

13.1 INTRODUCTION

Cassava (*Manihot esculenta* Crantz), also known as *yuca, manihoc,* and *mandioca* in Spanish, French, and Portuguese, respectively, belongs to the family Euphorbiaceae. The Euphorbiaceae family contains many flowering plants with 300 genera and approximately 7500 species. The other important members of this family include castor oil plant (*Ricinus communis*), barbados nut (*Jatropha curcas*), para rubber tree (*Hevea brasiliensis*), and many ornamental plants.

Cassava ranks sixth among crops in global production, with Africa being the largest center of production (Table 13.1). The enlarged starch-filled roots contain nearly the maximum theoretical concentration of starch on a dry-weight basis among food crops (Cock 1985). Cassava is considered a staple root crop for approximately 900 million people living in the developing tropical countries (Nassar 2006). It is the third largest source of carbohydrates for human food in the world (Phillips 1983) and the primary source of carbohydrates in sub-Saharan Africa (Okogbenin et al. 2007). The leaves and tender shoots are eaten in many parts of Africa as a source of vitamins, minerals, and proteins (Cock 1982; Balagopalan 2002; Nweke et al. 2002) and also as animal feed.

Apart from its conventional role as a food crop, cassava has gained high importance as a fuel commodity. It is one of the few non cereal sources of commercial starch (Ceballos et al. 2007) important for ethanol production. Although fossil fuels are considered as the major sources of energy available for world consumption, bioethanol has been recognized as environmental friendly because of less greenhouse gas emissions (Nguyen and Gheewala 2008). Recent assessments of the potential of alternative crops as sources of bioethanol production reported that sweet potato [*Ipomoea batatas* (L.) Lam] and cassava (*Manihot esculentum*) have greater potential as ethanol sources than the existing sources, including maize (*Zea mays* L.) (Ziska et al. 2009; Sriroth et al. 2010).

In general, the plants of the Euphorbiaceae family have been reported to have great potential as renewable sources of energy. The *J. curcas* L. is valued for its rich oil obtained from seeds known in the trade as curcas oil. This oil can be used in place of kerosene and diesel and as a substitute for fuel wood (Augustus et al. 2002). Castor oil plant (*R. communis*), another Euphorbiaceae member, is the source

TABLE 13.1
World Cassava Production (2008)

	Area Harvested (ha)	Production Quantity (t)
Africa	11,988,993	118,049,214
Asia	3,967,563	78,754,445
Americas	2,718,461	35,903,872
Oceania	20,145	242,649
World	18,695,162	232,950,180

Source: Calculated from FAOSTAT, 2010. Available at http://faostat.fao.org/
site/339/default.aspx (accessed June 19, 2010).

of castor oil, which has also been proven to be a biodegradable and environmentally friendly fuel. The large-scale production of these economically important Euphorbiaceae plants, including cassava, presents an opportunity for agricultural development in arid and impoverished areas (Gressel 2008).

13.2 BOTANICAL DESCRIPTION AND CULTURAL PRACTICES

Cassava is a perennial woody shrub with a plant height that ranges from 2 to 4 m (Fregene et al. 2001). Being monoecious, male and female flowers of cassava are located on the same plant. Male flowers develop near the tip and the female flowers develop closer to the base of the inflorescence (Ekanayake et al. 1997). Flowering is controlled by genotype and environmental factors, including temperature and photoperiod, which in turn influence the breeding potential of cassava. The fruits are capsular with three locules whereas the seeds have a caruncle, which varies in size (Nassar 2000). Root yield in cassava is dependent upon the balance between leaf area and root production (Cock et al. 1979). Root yield also determines starch yield, which is important for ethanol production.

Propagation in cassava is carried out by stem cuttings called "stakes" and by seed, but the use of stakes is most common. Spacing depends on the variety and the cropping system. Under sole cropping systems, a 1 × 1 m and 1 × 0.8 m is recommended for the branching types and nonbranching types, respectively, whereas for intercropping, 1 × 1.5 m and 1 × 1 m is recommended for the branching and nonbranching types, respectively. Although cassava can be grown in all types of soils, including infertile and arid lands, the growing demand for cassava in the bioenergy production increases the need for farmers to adopt fertilizer use to improve cassava productivity. Under intensive cultivation, cassava depletes the soil off the nutrients especially potassium (Carsky and Toukourou 2005). Fertilizer input has been shown to substantially increase yields (Howeler 2008; Fermont et al. 2010). Exchangeable potassium is known to be especially critical for cassava production (Howeler 1985) due to its cultivation by small resource -farmers on low fertility soils. Out of fourteen cultivars of cassava tested for response to potassium fertilization, CM507-37 found to be the best genotype that tolerates low potassium soils (EL-Sharkawy and cadavid 2000).

`Loosen-structure soil such as light sandy loams and loamy sands are optimal for cassava root formation (Sriroth et al. 2010). The plant can cause severe erosion when grown on sloppy land because of the wide spacing and the slow initial development (Suyamto and Howeler 2001). Although it is grown in a wide range of climates (between 30° north and 30° south latitude), from drought-prone to well-watered regions, it is commonly cultivated in areas receiving less than 800 mm rainfall per year with a dry season of 4–6 months and where tolerance to water deficit is an important attribute (El-Sharkawy 1993). Cassava is naturally tolerant to acidic soils (Jaramillo et al. 2005) but will not tolerate waterlogging and high pH associated with sodium salts. In general, cassava does not require irrigation and pesticide application (Nguyen et al. 2008). Weeding is required during the first few months until cassava plants develop shade large enough to compete for sunlight.

13.3 STARCH YIELD AND ETHANOL PRODUCTION FROM CASSAVA

Under favorable environments, cassava yields more than 15 tons/ha of oven-dried roots (El-Sharkawy 2006). Cassava starch is obtained from the roots, and approximately 73.7–84.9% of the dry root weight of cassava is starch (Rickard et al. 1991). Ethanol is conventionally produced from starch in a sequential two-step process, which involves two main stages. The first stage is the enzymatic hydrolysis of starch to glucose, and the second stage involves the fermentation of glucose to ethanol (Lyubenova et al. 2007). Basically, starch is first hydrolyzed by adding a liquefying enzyme, α-amylase, to avoid gelatinization and is then cooked at high temperatures (140–180°C) (Chen et al. 2008). The liquefied starch is then converted to glucose using a saccharifying enzyme, glucoamylase (Khaw et al. 2006). The glucose is finally converted to ethanol by yeast cells.

However, the traditional two-step bioethanol production process from starch is an expensive one because it requires addition of large amounts of amylolytic enzymes. The yeast *Saccharomyces cerevisiae* cannot utilize starchy material. The starchy material also needs to be cooked at high temperatures to obtain high ethanol yield. Development of more efficient one-step processes has been the interest of several major studies (Khaw et al. 2006; Chen et al. 2008) in the recent past. Through the use of recombinant strains, it is now possible to conduct the process in a one-step mode doing the simultaneous saccharification and fermentation of starch to ethanol (Lyubenova et al. 2007). Mixed cultures of fungal enzymes such as Koji enzymes (Ueda et al. 2004) or *Endomycopsis fibuligera* NRRL 76 and *Zymomonas mobilis* ZM4 (Reddy and Bassappa 1996) have also been shown to directly and more efficiently ferment cassava starch in a one-step process than the monocultures.

Cell immobilization process (Chen et al. 2008) coupled with continuous removal of ethanol from the fermentation broth is another effective method of improving the efficiency of substrate utilization and productivities of various fermentation processes. Immobilizing yeast cells on several support types can provide high cell densities in the bioreactor, which in combination with high flow rates leads to short residence times (Verbelen et al. 2006). On the other hand, continuous production is achieved through simultaneous saccharification and fermentation. Recently, Choi et al. (2010) evaluated a continuous process by using flocculating yeast, *S. cerevisiae* CHFY0321, which showed excellent fermentation results under continuous ethanol production.

Cassava remains one of the richest fermentable sources for the production of alcohol. It is also considered an attractive raw material for bioethanol production because it is inexpensive and not affected by feed or food shortage concerns. In general, the cost of bioethanol production using cassava for 1 ton of ethanol is approximately $40–60 less than that using maize or wheat (Rubo et al. 2008). Cassava starch is easily extractable from the roots because it contains low levels of protein and fat (Ceballos et al. 2007). These special characteristics of cassava starch make it less complicated to set up ethanol factory because of lower investment and simpler processing technology. Moreover, energy inputs of cassava represent 5–6% of the final energy content of the total biomass, showing an energy profit of 95%, assuming complete utilization of the energy content of the total biomass (Jain 2006).

The actual amount of cassava needed is dependent upon the starch content, but as a guide, cassava at 30% starch content will produce approximately 280 L of alcohol per ton (http://www.efairtrade.net/mp.php3?cat_id=428). For example, at Alvan Blanch, United Kingdom, fresh cassava is brought in, washed and peeled, grated, cooked in a jet cooker, fermented, distilled, and then bottled. The plant produces approximately 5000 L/day of alcohol at 96%, which is equivalent to 7500 L/day at 40%. The plant needs good water supply and continuous electrical supply (~50 kVA). The steam requirements are approximately 1500 kg/h. The cost of shipping and installing the equipment, including commissioning and operator training, is £290,000 (http://www.bioenergywm.org/documents/Biofuels%20study.pdf).

13.4 CASSAVA BREEDING FOR IMPROVED BIOENERGY PRODUCTION

The importance of cassava as a global food and fuel source demands the production of improved germplasm. The International Center for Tropical Agriculture (CIAT), headquartered in Columbia, established a cassava-breeding program in the early 1970s with the aim of improving yield potential and tolerance to pests and diseases. Cassava is naturally outbreeding, which when coupled with possible introgression of wild species germplasm has resulted in highly diverse genotypes. A lot of genetic diversity has been shown to exist for most traits examined, but extensive inbreeding depression and the long life cycle make it difficult to develop appropriate stocks for classical genetic studies (Fregene et al. 2001). As a consequence, studies on the genetics of cassava have been very limited and genetics of most traits are not well understood. This is especially the case for traits with relevance to bioenergy production, which are mainly quantitative.

Cassava improvement has continued to tap genetic variation for a long time, mainly through conventional breeding and to a less extent, using advanced molecular techniques. Clonal selection has been the predominant method in conventional breeding for cassava improvement at national centers in Africa and Brazil. The only exception to this is the production of cassava mosaic disease (CMD)-resistant clones, which has been done by hybridizing *M. glaziovii* with cassava (Storey and Nichols 1938). The search for varieties that are resistant to various pests and diseases has formed a major part of cassava research over the last three decades (Ceballos et al. 2004). CMD is the most important disease of cassava and is the most widespread cassava disease in Africa (Akano et al. 2002). Breeding has tended toward the development of varieties with CMD resistance (Thresh and Cooter 2005). Several quantitative trait loci (QTL) associated with polygenic and recessive sources of resistance to CMD have been identified (Lokko et al. 2007).

A few attempts that have been made to understand the genetics of starch, and starch-related traits have been reported to be polygenically controlled. Kawano et al. (1987) reported polygenic additive control of root dry matter content in cassava, whereas a nonadditive gene control was demonstrated for dry matter and starch content (Easwari Amma et al. 1995; Easwari Amma and Sheela 1998) using classical studies. Several attributes of cassava carbohydrate metabolism suggest that there is unrealized potential for enhanced starch production (Ihemere et al. 2006). The recent development of linkage maps and use of molecular markers has made it possible to study and localize quantitative traits of agronomic importance using QTL mapping. Several QTL for dry matter content have also been identified in cassava (Kizito et al. 2007). Dry matter content has been found to have a positive correlation with starch content.

There are currently known expressed sequence tags (ESTs) for specific genes involved in the starch biosynthesis pathway (Sakurai et al. 2007) that could be used to develop single nucleotide polymorphisms (SNPs) to improve our understanding of the genetics of starch content and yield in cassava. The application of the complementary DNA (cDNA)-AFLP (amplified fragment length polymorphism) technique to generate polymorphic transcript-derived fragments (TDFs) (Suarez et al. 2000) between the parents of a mapping population has been shown to be a potentially powerful way of identifying candidate loci controlling agronomic traits in cassava and could be used to identify more candidate genes for starch content and yield. There are major efforts toward the development of SNPs from ESTs and bacterial artificial chromosome (BAC)-end clones (Lopez et al. 2005; Sakurai et al. 2007) that could be further used for fine-mapping and cloning of the important traits. Amplification of simple sequence repeats (SSRs) in wild relatives of cassava has been tested with success (Roa et al. 2000), creating an opportunity for possible exploitation of important starch traits in the wild species of cassava.

Mutation genetics using irradiation looks promising for creating genetic variation for starch content in cassava, although it has not yet been exploited. The probable recessive nature of such mutations will require the need for ways of overcoming inbreeding depression. CIAT has further implemented different approaches to develop and identify clones with novel starch properties

(Ceballos et al. 2006) and these have resulted in the recent discovery of amylose-free starch mutant (Ceballos et al. 2007). Efforts are currently underway to introduce this gene into germplasm adapted to the most important cassava growing environments. Compared with several mutations reported for starches from other crops like maize and potato (*Solanum tuberosum*), cassava offers very little variation because it is seldom self-pollinated yet most mutations are recessive in nature.

The development of transgenic technologies in cassava could circumvent many of the problems inherent in traditional improvement programs. By manipulating the embryonic culture systems of cassava, four different techniques have been reported for recovering transgenic cassava plants with integration of transgenes for desirable agronomic traits (Schöpke et al. 1996; Taylor et al. 2004). These four transformation systems include production of somatic embryos from friable embryogenic callus, cotyledon fragments, and immature leaf explants (Taylor et al. 2004). Recently, Ihemere et al. (2006) succeeded in improving starch production in cassava by genetically modifying AGPase activity using the bacterial AGPase gene, *glgC*. AGPase in plants plays a critical role in the regulation of starch synthesis and is also involved in the rate-limiting step in starch synthesis.

Smith (2008) has recently reviewed prospects for increasing starch and sucrose yields for bioethanol production in second-generation biofuel crops, under which cassava is classified. She recommends the manipulation of starch degradation in organs in which starch turnover is occurring and introduction of starch synthesis into the cytosol as possible options for increasing starch production in the future.

13.5 COMMERCIALIZATION OF CASSAVA AS A BIOENERGY CROP

Cassava has been promoted for introduction in most parts of Africa and Asia as a cash crop and an alternative crop for food and industrial use. In Latin America, where the crop originates, there is an increasing entrepreneurial entry into the cassava sector by small to medium growers as well as investors (Hershey et al. 2004). In fact, cassava has been proposed as an ideal focus for development-oriented research in Latin America (Hershey et al. 2004). Although cassava is not yet competitive with sugarcane for bioethanol production in Brazil, there are indications that bioethanol produced from cassava may be less expensive than that produced from sugarcane (Energy and Energy Conservation News 2009). This is likely to shift producers' interest into more investment in cassava. Production of cassava in Venezuela is also likely to increase after the recent development of a biofuel production process by Venezuelan scientists (Global Bio-Energy Industry News 2009). Currently, most of the cassava production in Latin America remains largely for food production.

Several African countries are also promoting commercialization of cassava for food and alternative uses. Although the use of cassava for biofuel production in Africa is currently only exploited in Nigeria, some African countries are reported to have biofuel research underway (Chege 2010). There are signs of growing interest in using locally made cassava starch as an import substitute in Uganda, Tanzania, Madagascar, and Malawi (Spotlight 2006). Evidence suggests that volumes of traded cassava have been increasing about twice as fast as production in Zambia (Haggblade and Nyembe 2008), probably as a result of erratic rainfall that has increasingly led to low maize yields (Barratt et al. 2006). A task force has been initiated by the Zambia's Agricultural Consultative Forum (ACF) to help accelerate commercial development of cassava and cassava-based products (Haggblade and Nyembe 2008) such as bioenergy production. In West Africa, a renewed interest in cassava as an alternative crop has been developed in the last decade (Camara et al. 2001). Much more potential for use of cassava in Nigeria exists and is yet to be exploited (Kehinde 2006). Although central Africa remains the biggest consumer of cassava in the continent, there has been production decline in the region as a result of lack of resources to control biotic and abiotic stresses (Aerni 2006). Availability of improved cassava cultivars adapted to the central African region is likely to increase cassava production and commercialization.

In Thailand, cassava became an important industrial crop after World War II. Cassava production in Thailand has since progressed from basic starch production to the world's largest producer of bioethanol. Such progress was made possible by the announcement in 2003 by the Thai cabinet to include bioethanol as a renewable energy (Suksri et al. 2007). Although molasses is the major source of ethanol in Thailand, the country already has a cassava-derived ethanol pilot plant and the government was aiming to build 12 full-scale plants by the end of 2008. Thailand's leading petroleum company also announced a feasibility study that would use cassava to produce 1 million L of bioethanol per day (Spotlight 2006).

In southern China, cassava was initially used as a food crop but has now become an important crop for on-farm feeding of animals and for processing into various industrial products such as native starch, modified starch, sweeteners, and alcohol (Hershey et al. 2001). Cassava production in China remained unattractive to farmers for a long time and it was only until the late 1980s when many new cassava-based products were developed successfully that farmers started to increase their production. The government of China supports cassava breeding and production and this has encouraged a lot of farmers to venture into cassava farming. In India, cassava was initially introduced as a food crop but has now changed its status to a commercial crop in the states of Tamil Nadu and Andhra Pradesh (Srinivas 2009) as a result of intensive research and development efforts of different research institutes. In Vietnam, cassava has been reported to have great potential as food and feed crop in the north and as a cash crop in the south (Van Bien et al. 2009).

13.6 PROBLEMS AND CONCERNS OF CURRENT BIOFUEL PRODUCTION

Record oil prices have boosted demand for biofuels as consumers and companies look for cheaper and cleaner energy sources. Biofuels like ethanol and biodiesel have been widely viewed as the answer to reducing greenhouse gas emissions (up to 40%), but critics warn that a reliance on them could lead to higher food prices, deforestation, and ultimately do more damage to the environment than the fossil fuels they are supposed to replace (Lumpar 2007).

A major hurdle for widespread adoption of biofuels is the challenge of growing enough crops to meet demand. "Replacing only five percent of the nation's diesel consumption with biodiesel would require diverting approximately 60% percent of today's soy crops to biodiesel production," says Matthew Brown, an energy consultant and former energy program director at the National Conference of State Legislatures (Dand 2010).

Another dark cloud looming over biofuels is whether producing them actually requires more energy than they can generate. After factoring in the energy needed to grow crops and then convert them into biofuels, turning plants such as cassava, corn, soybeans, and sunflowers into fuel uses much more energy than the resulting ethanol or biodiesel generates (Lang 2005).

Resilience and sustainability of cassava in marginal lands, drought-prone locations, or acid soils explains its importance as a stable crop in areas with poor resources for soil fertility and crop management. However, weeds compete for the nutrient resources in these areas and can significantly affect the crop yields. Although the larger leaf canopy of cassava plants in later stages of growth helps in suppressing the weeds, the slow initial growth rate of cassava renders the crop particularly vulnerable to weed interference soon after planting. Weed control during tuberization stage is particularly very critical for higher crop yields (Melinfonwu 1994). These weeds not only increase crop management costs but also serve as reservoirs for new and invasive pests. The cassava root scale (*Protortonia navesi*) is a recent pest in the Brazilian "Cerrado" that causes qualitative and quantitative damage by sucking plant sap and is hosted by 13 weed species during the growing season and five weed species after harvest (Oliviera and Fontes 2008).

New migratory plant pests increase risks over global food security, and cassava mealy bug (*Phenacoccus manihoti*) in Africa is one such example. Cassava mealybug (CMB) was introduced into many African nations along with new germplasm resources imported from other countries and turned into invasive species from 1970 to 1980 because of optimum growing conditions, wider host range, and lack of natural predators in the new area. The cassava production area was significantly reduced because of this mealy bug infestation. A research team led by Hans R. Herren and Peter Neuenschwander at International Institute of Tropical Agriculture (IITA) successfully identified the presence of the same mealy bug species and its natural predator species, *Anagyrus lopezi* (a predator wasp), in South America, the center of diversity for cassava. Importing of these natural enemies, rearing them at laboratory conditions, and their dispersal in infested areas in Africa by IITA researchers in collaboration with CIAT significantly controlled mealy bug infestation.

To increase the effect of biological control of CMB, a complex of natural enemies that include *Acerophagus coccois* and *Aenasius vexans*, were later introduced to contain cassava mealy bug populations by laying eggs on the mealy bugs and parasitizing them. The spectacular biological control of the cassava mealy bug was the first of many successes in the history of the Biological Control Center for Africa set up by IITA (Neuenschwander 2001).

Cassava green mite (*Mononychellus tanajoa*), another sucking pest of cassava, was introduced into the cassava belt causing an estimated 30–50% reduction in yield. This pest was also brought under biological control by identifying, introducing, and establishing predatory mites (*Typhlodromalus aripo* and *T. manihoti*) and later an acaropathogenic fungus (*Neozygites tanajoae*) from climatically similar areas of Brazil. These efforts have resulted in a 30–65% reduction in green mite density and a 15–35% increase in the yield of cassava (Mégevand et al. 1987).

Cassava mosaic virus is another such invasive pathogen that threatens this key crop in East Africa, and the Food and Agricultural Organization (FAO) is currently assisting countries' response to such transboundary plant pest and disease emergencies and also in developing disease-resistant cultivars. Stringent regulatory restrictions for germplasm transfer, transgenic crops, and improved crop management practices alleviate the risks imposed by transboundary and invasive pests and super weeds.

13.7 FUTURE PROSPECTS

Global food demand has been projected to double by the year 2050, and research for increased agricultural production has been suggested as key to feeding the increasing world population. Cereal crops have been the major and preferred food sources across many nations, but their abilities to grow in difficult climatic and soil conditions are relatively poor. Cassava currently fits as the best alternative to cereals because it has the ability to tolerate longer periods of drought, disease, and insect pressure and yet produce economical yields for the poor farming communities. Bringing marginal lands and poor soils into cassava cultivation and exploring the additional uses of cassava can help meet global food demand. Exploitation of cassava production not only for human consumption and industrial use but also as feed in the poultry and livestock industry will likely ease the global pressure on cereal crops.

Diminishing fossil fuel resources has no doubt led to the increased global interest in biofuel production efforts in many nations. The use of cereal crops for ethanol production in recent years has been blamed as the major cause of the global rise in food commodity prices. High starch content in cassava makes it one of the best alternative candidates for ethanol production. The technology for converting cassava into biofuel ethanol is currently being perfected and very soon will be applicable anywhere in the world. Brazil, the world's leading producer of these substitute fuels or biofuels, makes more than 120 million hL/year from sugarcane (*Saccharum* spp.) and cassava. Establishing ethanol/biofuel production plants in cassava-growing areas not only aids in meeting fuel demands but also economically helps the local farming communities. For example, cassava production in Vietnam (9.39 million tons in 2008) has rapidly changed its role from food crop to an industrial one (FAO 2008).

To boost the role of cassava as food security and biofuel crop, there is need for increased research to improve and stabilize yields by developing genetic resistance to major pests and diseases. More efforts toward collection and conservation of diverse genetic stocks will help combating insects and diseases by providing sources of genetic resistance and tolerance. Pest- and disease-resistance traits can be easily introgressed into cultivated genotypes by traditional and molecular breeding approaches. Global organizations such as FAO, CIAT, and IITA are actively advocating for cassava crop improvement, and effective utilization of biological control agents for mealy bug and green mite controls in Africa are two examples of such global efforts. Global collaborative and conservation efforts will help preserve the natural diversity of cassava, enhance the development of high-yielding disease- and pest-resistant varieties, and improve crop management and postharvest practices while avoiding duplicated efforts. To bring the fruits of these global collaborations to farmers, multiplication and distribution of the improved cassava vegetative stocks should be taken up by local, national, and international government agencies.

Cultivation of transgenic crops is rapidly spreading across the world. Accessibility of novel transgenic technology at a nominal licensing fee from commercial and academic institutions around the world to researchers would enhance the current crop improvement efforts and crop returns from cassava. Increased investment and international collaboration in global cassava research will definitely help meet the global food and biofuel demand at a rapid pace.

REFERENCES

Aerni P (2006) Mobilizing science and technology for development: The case of the Cassava Biotechnology Network (CBN). *AgBioForum* 9:1–14

Akano AO, Dixon AGO, Mba C, Barrera E, Fregene M (2002) Genetic mapping of dominant gene conferring resistance to cassava mosaic disease. *Theor Appl Genet* 105:521–525

Augustus GDPS, Jayabalan M, Seiler GJ (2002) Evaluation and bioinduction of energy components of *Jatropha curcas*. *Biomass Bioenergy* 23:161–164

Balagopalan C (2002) Cassava utilization in food, feed and industry. In: Hillocks RJ, Thresh JM, Bellotti AC (eds), *Cassava: Biology, Production and Utilization*, CABI Publishers, Wellingford, Oxon, United Kingdom, pp 301–318

Barratt N, Chitundu D, Dover O, Elsinga J, Eriksson S, Guma L, Haggblade M, Haggblade S, Henn TO, Locke FR, O'Donnell C, Smith C, Stevens T (2006) Cassava as drought insurance: Food security implications of cassava trials in central Zambia. *Agrekon* 45:106–123

Camara Y, Staatz JM, Eric C (2001) Comparing the profitability of cassava-based production systems. In: *Three West African Countries: Cote D'Ivoire, Ghana and Nigeria*, Staff Papers 11593, Department of Agricultural, Food, and Resource Economics, Michigan State University, Lansing, MI

Carsky RJ, Touskourou MA (2005) Identification of nutrients limiting cassava yield maintenance on a sedimentary soil in southern Benin, West Africa. *Nutr Cycl Agroecosyst* 71:151–162

Ceballos H, Iglesias CA, Pérez JC, Dixon AGO (2004) Cassava breeding: Opportunities and challenges. *Plant Mol Biol* 56:506–516

Ceballos H, Sánchez T, Chávez AL, Iglesias C, Debouck D, Mafla G, Tohme J (2006) Variation in crude protein content in cassava (*Manihot esculenta* Crantz) roots. *J Food Comp Anal* 19:589–593

Ceballos H, Sánchez T, Morante N, Fregene M, Dufuor D, Smith AM, Denyer K, Perez JC, Calle F, Mestres C (2007) Discovery of an amylase-free starch mutant in cassava (*Manihot esculenta* Crantz). *J Agric Food Chem* 55:7469–7476

Chege K (2010) Biofuel: Africa's new oil? SciDevNet, available at http://www.Scidev.net/en/ features/ biofuel-africas-new-oil.html (accessed June 19, 2010)

Chen JP, Wu KW, Fukuda H (2008) Bioethanol production from uncooked raw starch by immobilized surface-engineered yeast cells. *Appl Biochem Biotechnol* 145:59–67

Choi GW, Kang HW, Moon SK, Chung BW (2010) Continuous ethanol production from cassava through simultaneous saccharification and fermentation by self-flocculating yeast *Saccharomyces cerevisiae* CHFY0321. *Appl Biochem Biotechnol* 160:1517–1527

Cock JH (1982) Cassava: A basic energy source in the tropics. *Science* 218:755–762

Cock JH (1985) *Cassava: New Potential for a Neglected Crop*, Westview Press, Boulder, CO

Cock JH, Franklin D, Sandoval G, Juri P (1979) The ideal cassava plant for maximum yield. *Crop Sci* 19:271–279

Dand J (2010) The pros and cons of biofuels: Dear EarthTalk: What are the environmental pros and cons of switching to plant-based "biofuels" to reduce our reliance on oil?, available at http://www.foodreference. com/html/a-pc-biof-21707.html (accessed June 19, 2010)

Easwari Amma CS, Sheela MN (1998) Genetic analysis in a diallel cross of inbred lines of cassava. *Madras Agri J* 85:264–268

Easwari Amma CS, Sheela MN, Thankamma PPK (1995) Combining ability analysis in cassava. *J Root Crops* 21:65–71

Ekanayake IJ, Osiru DSO, Porto MCM (1997) Physiology of cassava. IITA Research Guide 60. Training Program, IITA, Ibadan, Nigeria, 22

El-Sharkawy MA (1993) Drought-tolerant cassava for Africa, Asia and Latin America. *BioScience* 43:441–451

El-Sharkawy MA (2006) International research on cassava photosynthesis, productivity, eco-physiology and responses to environmental stresses in the tropics. *Photosynthetica* (Prague) 44:481–512

El-Sharkawy MA, Cadavid LF (2000) Genetic variation within cassava germplasm in response to potassium. *Expl Agric* 36:323–334

Energy and Energy Conservation News (2009) Perspective Brazilian ethanol from cassava. Science Niche, available at http://scienceniche.com/type/news/perspective-brazilian-ethanol-from-cassava.html (accessed June 19, 2010)

FAO (2008), Food and Agricultural Organization of the United Nations, available at www.FAO.org (accessed April 10, 2010)

FAOSTAT (2010), available at http://faostat.fao.org/site/339/default.aspx (accessed June 19, 2010)

Fermont AM, Tittonell PA, Baguma Y, Ntawuruhunga P, Giller KE (2010) Towards understanding factors that govern fertilizer response in cassava: lessons from East Africa. *Nutr Cycl Agroecosyst* 86:133–151

Fregene M, Okogbenin E, Mba C, Angel F, Suare, MC, Gutierrez J, Chavarriaga P, Roca W, Bonierbale M, Tohme J (2001) Genome mapping in cassava improvement: Challenges, achievements and opportunities. *Euphytica* 120:159–165

Global Bio-Energy Industry News (2009) Biofuel from cassava extract, available at www.thebio-energysite. com/news/4244/biofuel-from-cassava-extract (accessed June 19, 2010)

Gressel J (2008) Transgenics are imperative for biofuel crops. *Plant Sci* 174:246–263

Haggblade S, Nyembe M (2008) Commercial dynamics. In: *Zambia's Cassava Value Chain*. International Development Collaborative Working Papers ZM-FSRP-WP-32, Department of Agricultural Economics, Michigan State University, Lansing, MI

Hershey C, Henry G, Best R, Iglesias C (2004) Cassava in Latin America and the Caribbean: resources for global development. In: *A Review of Cassava in Latin America and the Caribbean with Country Case Studies on Brazil and Colombia*, Vol 4. Proceedings of the Validation Forum on the Global Cassava Development Strategy. FAO and IFAD, Rome, Italy

Hershey CH, Henry G, Best R, Kawano K, Howeler R, Iglesias C (2001) Cassava in Asia—Expanding the competitive edge in diversified markets. In: *A Review of Cassava in Asia with Country Case Studies on Thailand and Vietnam*, FAO and IFAD, Rome, Italy, pp 1–62

Howeler RH (1985) Potassium nutrition of Cassava. In: Munson RD (ed), *Potassium in Agriculture*. Proceedings of the International Symposium, Atlanta, GA, July 7–10, 1985, American Society of Agronomy, Madison, WI, pp 819–841

Howeler RH (2008) Results, achievements and impact of the Nippon Foundation Cassava Project. In: Howeler RH (ed), *Integrated Cassava-Based Cropping Systems in Asia*. Proceedings of the Workshop of the Nippon Foundation Cassava Project in Thailand, Vietnam and China. Thai Nguyen, Thailand, October 27–31, 2003

Ihemere U, Arias-Garzon D, Lawrence S, Sayre R (2006) Genetic modification of cassava for enhanced starch production. *Plant Biotechnol J* 4:453–465

Jain SM (2006) Biotechnology and mutagenesis in genetic improvement of cassava. In: Ortiz R, Nassar NMA (eds), *Cassava Improvement to Enhance Livelihoods in Sub-Saharan Africa and Northeastern Brazil*. 1st International Meeting on Cassava Breeding, Biotechnology and Ecology, Brasilia, Brazil, November 11–15, 2006. Universidade de Brasilia, Brasilia, Brazil

Jaramillo G, Morante N, Perez JC, Calle F, Ceballos H, Arias B, Bellotti AC (2005) Diallel analysis in cassava adapted to the midaltitude valleys environment. *Crop Sci* 45:1058–1063

Kawano K, Goncalves WMF, Cempukdee U (1987) Genetic and environmental effects on dry matter content of cassava root. *Crop Sci* 27:69–74

Kehinde T (2006) Utilization potentials of cassava in Nigeria: The domestic and industrial products. *Food Rev Int* 22:29–42

Khaw TS, Katakura Y, Koh J, Kondo A, Ueda M, Shioya S (2006) Evaluation of performance of different surface-engineered yeast strains for direct ethanol production from raw starch. *Appl Genet Mol Biotechnol* 70:573–579

Kizito EB, Ro AC, Stjung NW, Egwang T, Gullberg U, Fregene M, Westerbergh A (2007) Quantitative trait loci controlling cyanogenic glucoside and dry matter content in cassava (*Manihot esculenta* Crantz) roots. *Hereditas* 144:129–136

Lang SS (2005) Cornell ecologist's study finds that producing ethanol and biodiesel from corn and other crops is not worth the energy. Cornell University News Service, July 5, 2005

Lokko Y, Okogbenin E, Mba C, Dixon A, Raji A, Fregene M (2007) Cassava. In: Kole C (ed), *Genome Mapping and Molecular Breeding in Plants. Vol 3: Pulses, Sugar and Tuber Crops*. Springer, Berlin, Heidelberg, Germany, pp 249–269

Lopez C, Piegu B, Cooke R, Delseny M, Tohme J, Verdier V (2005) Using cDNA and genomic sequences as tools to develop SNP strategies in cassava (*Manihot esculenta* Crantz). *Theor Appl Genet* 110:425–431

Lumpar K (2007) Record oil price boosts demand for biofuels but critics question the cost. *AFX News Ltd* October 19, 2007

Lyubenova V, Ochoa S, Repke J, Ignatova M, Wozny G (2007) Control of on stage bioethanol production by recombinant strain. *Biotechnol Biotechnol Equip* 21:372–376

Mégevand B, Yaninek JS, Freese DD (1987) Classical biological control of cassava green mite. *Insect Sci Appl* 8:871–874

Melinfonwu AA (1994) Weeds and their control in Africa. *Afr Crop Sci J* 2:519–530

Nassar NMA (2000) Cytogenetics and evolution of cassava (*Manihot esculenta* Crantz). *Genet Mol Biol* 23:1003–1014

Nassar NMA (2006) Cassava genetic resources: Extinct everywhere in Brazil. *Genet Resour Crop Evol* 53:975–983

Neuenschwander P (2001) Biological control of the cassava mealybug in Africa: A review. *Biol Control* 21:214–229

Nguyen TLT, Gheewala SH (2008) Life cycle assessment of fuel ethanol from cassava in Thailand. *Int J Life Cycle Assess* 13:147–154

Nguyen TLT, Gheewala SH, Bonnet S (2008) Life cycle cost analysis of fuel ethanol produced from cassava in Thailand. *Int J Life Cycle Assess* 13:564–557

Nweke F/I, Spencer DSC, Lynam JK (2002) *The Cassava Transformation: Africa's Best-Kept Secret*, Michigan State University Press, East Lansing, MI

Okogbenin E, Porto MCM, Egesi C, Mba C, Espinosa E, Santos LG, Ospina C, Marín J, Barrera E, Gutiérrez J, Ekanayake I, Iglesias C, Fregene MA (2007) Marker-assisted introgression of resistance to cassava mosaic disease into Latin American germplasm for the genetic improvement of cassava in Africa. *Crop Sci* 47:1895–1904

Oliviera CM, Fontes JRA (2008) Weeds as hosts for new crop pests: The case of *Protortonia navesi* Fonseca (Hemiptera: Monophlebidae) on cassava in Brazil. *Weeds Res* 48:197–200

Phillips TP (1983) An overview of cassava consumption and production. In: Delange F, Ahluwalia R (eds), *Cassava Toxicity and Thyroid*. Proceedings of a Workshop on International Development Research Centre Monograph, Ottawa (1982) 207e, pp 83–88, International Development Research Centre, Canada

Reddy OVS, Basappa SC (2004) Direct fermentation of cassava starch to ethanol by mixed cultures of *Endomycopsis fibuligera* and *Zymomonas mobilis*: Synergism and limitations. *Biotechnol Lett* 18:1315–1318

Roa A, Chavarriaga-Aguirre P, Durque MC, Maya MM, Bonierbale MW, Iglesias C, Tohme J (2000) Cross-species amplification of cassava (*Manihot esculenta*) (Euphorbiaceae) microsatellites: Allelic polymorphism and degree of relationship. *Am J Bot* 87:1647–1655

Rubo L, Wang C, Zhang C, Dai D, Pu G (2008) Life cycle inventory and energy analysis of cassava-based fuel ethanol in China. *J Cleaner Prod* 16:374–384

Sakurai T, Plata G, Rodriguez F, Seki M, Salcedo A, Toyoda A, Ishiwata A, Tohme J, Sakaki Y, Shinozaki K, Ishitani M (2007) Sequencing analysis of 20.000 full-length cDNA clones from cassava reveals lineage specific expansions in gene families related to stress response. *BMC Plant Biol* 7:66

Schöpke C, Taylor N, Cárcamo R, Konan NK, Marmey P, Henshaw GG, Beachy R, Fauquet C (1996) Regeneration of transgenic cassava plants (*Manihot esculenta* Crantz) from microbombarded embryogenic suspension cultures. *Nat Biotechnol* 14:731–735

Smith A (2008) Prospects for increasing starch and sucrose yields for bioethanol production. *Plant J* 54:546–558

Spotlight (2006) Starch market adds value to cassava, available at http://www.fao.org/ag/magazine/0610sp1.htm (accessed June 19, 2010)

Srinivas T (2009) Impact of research investment on cassava production technologies in India. *Aust J Agri Res Econ* 53:367–383

Sriroth K, Piyachomkwana K, Wanlapatita S, Nivitchanyongc S (2010) The promise of a technology revolution in cassava bioethanol: From Thai practice to the world practice. *Fuel* 89:1333–1338

Storey HH, Nichols RFW (1938) Studies on the mosaic diseases of cassava. *Ann Appl Biol* 25: 790–806

Suarez MC, Bernal A, Gutierrez J, Tohme J, Fregene M (2000) Developing expression sequence tags (ESTs) from polymorphic transcript-derived fragments in cassava (*Manihot esculenta* Crantz). *Genome* 43:62–67

Suksri P, Moriizumu Y, Hondo H, Wake Y (2007) An introduction of bio-ethanol to Thai economy (I)—A survey on sugarcane and cassava fields. Digital Asia Discussion Paper series, DP07-03

Suyamto, Howeler RH (2001) Cultural practices for soil erosion control in cassava-based cropping systems in Indonesia. In: Barker DH, Watson A, Sombatpanit S, Northcut B, Maglinao AR (eds), *Ground and Water Bioengineering for the Asia-Pacific Region*, International Erosion Control Association, Science Publishers, Enfield, NH

Taylor NJ, Kent L, Fauquet CM (2004) Progress and challenges for the deployment of transgenic technologies in cassava. *AgBioForum* 7:article 10

Thresh JM, Cooter RJ (2005) Strategies for controlling cassava mosaic disease in Africa. *Plant Pathol* 54:587–614

Ueda S, Zenin CT, Monteiro DA, Park YK (2004) Production of ethanol from raw cassava starch by a nonconventional fermentation method. *Biotechnol Bioeng* 23:291–299

Van Bien P, Kim H, Wang JJ, Howeler RH (2009) Present Situation of Cassava Production and the Research and Development Strategy in Vietnam. CBN-V Video Archives-S1-24, available at http://www.danforthcenter.org/media/video/cbnv/session1/S1-24.htm (accessed March 1, 2010)

Verbelen PJ, De Schutter DP, Delvaux F, Verstrepen KJ, Delvaux FR (2006) Immobilized yeast cell systems for continuous fermentation applications. *Biotechnol Lett* 28:1515–1525, doi:10.1007/s10529-006-9132-5

Ziska LH, Runion GB, Tomecek M, Prior SA, Torbet HA, Sicher RC (2009) An evaluation of cassava, sweet potato and field corn as potential carbohydrate sources for bioethanol production in Alabama and Maryland. *Biomass Bioenergy* 33:1503–1508

14 Jatropha

Hifjur Raheman
Indian Institute of Technology

CONTENTS

14.1   INTRODUCTION

Jatropha (*Jatropha curcas* L.) is a drought-resistant shrub or tree with many attributes, multiple uses, and considerable potential to meet future energy requirements. The plant can be grown to prevent and/ or control soil erosion, reclaim land, act as a live fence especially to prevent farm animals, and as a commercial crop. The plant originated from tropical America, but it is now available in the arid, semi-arid, tropical, and subtropical regions of the world (Hikwa 1995; Henning 1996; Makkar et al. 1997; Haas and Mittelbach 2000; Openshaw 2000; Sirisomboon et al. 2007). This plant is able to thrive in a number of climatic zones with wide range of rainfall ranging from 200 to over 1200 mm/year.

All parts of the jatropha plant can be used for various purposes. Its bark can be used for making dyes, and its leaves can be used for rearing of silkworm and in the preparation of dyes and medicines. Its leaves can also used as an anti-inflammatory substance. Its latex has a high medicinal value and can be used as a pesticide and for control of molluscs. Its flowers attract bees and thus it has a potential use in honey production. Its fruit hulls contain tannin and can be used as a green manure. Fruit hulls are also combustible and can be used as feedstock for production of producer gas and active carbon. Its seeds can be used for controlling insects and pests. These seeds are rich in oil and the seed oil can be used for the preparation of soap, biodiesel, lubricant, insecticide, and medicine. When seed oil is mixed with iron oxide, it can be used in varnish. The oil cake left after expelling of oil from the seed is useful as fertilizer or it can be used as feedstock in the biogas- and producer-gas-generating plants. Briquettes can be prepared from the seed shells for using as fuel. Lastly, the roots contain yellow oil with strong antihelminthic properties (Heller 1996; Gübitz et al. 1999; Augustus et al. 2002; Sirisomboon 2007). A schematic view of multiple uses of jatropha is shown in Figure 14.1.

Among its various uses, biodiesel from jatropha oil has been the center of attraction of many countries to substitute diesel fuel because of diminishing petroleum sources, rising energy consumption, and environmental issues. Hence, detailed information on its cultivation, management, processing for biodiesel production, characterization, and engine performance with jatropha biodiesel are of utmost importance to scientists, researchers, plant-growers, and planners.

14.2   BOTANICAL DESCRIPTION OF THE PLANT

The term jatropha is derived from two Greek words—*jatros*, meaning "doctor," and *trophe*, meaning "nutrition." Botanically this is known as *Jatropha curcas* L. and it belongs to the family

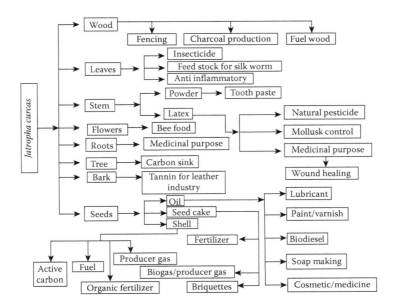

FIGURE 14.1 Multiple uses of jatropha.

Euphorbiaceae. Jatropha has as many as 476 species and is distributed throughout the world. Among these species, 12 species are recorded in India. Jatropha is known by nearly 200 different names, indicating its occurrence in various countries (Anonymous 2009a). It is a deciduous small tree or a large shrub that can grow up to a height of 3–5 m with a productive life of 30–50 years (Gandhi et al. 1995). The plant has a smooth gray bark that exudes watery and sticky latex when cut. When the plant is trimmed and pruned, it sprouts readily and grows rapidly, making it suitable for fencing (Anonymous 2009a). The plant develops a deep taproot system and has four shallow lateral roots. The taproots stabilize the soil against landslides, whereas the shallow roots prevent and control soil erosion caused by wind and water. The leaves are smooth with four to six lobes, 10–15 cm long and wide, and are usually pale green in color. The plant produces flowers once in a year during the rainy season. In permanently humid regions or under irrigated conditions, the flowers come throughout the year. Fruits are produced in winter or throughout the year depending on temperature and soil moisture. Seeds are encased within green shells of the fruit, and the color of the shells changes to yellow once the fruit attains maturity. The dry *curcas* fruit contains 37.5% shell and 62.5% seed. Seeds of jatropha resemble castor seeds in shape and are black in color. The seeds contain 42% husk/hull and 58% kernel (Singh et al. 2008).

14.3 CULTURAL PRACTICES

14.3.1 ECOLOGICAL REQUIREMENTS

Jatropha is well adapted to arid and semiarid conditions. It grows almost anywhere except on waterlogged lands. It can even grow on gravelly, sandy, and saline soils (Munch and Kiefer 1989; Anonymous 2009b). It can thrive on the poorest stony soil and can grow even in the crevices of rocks. It also grows well in shallow soils and adapts to low-fertility sites. Better yields on poor quality sites are obtained if fertilizer or manure containing small amounts of calcium, magnesium, and sulfur is applied. Tigere et al. (2007) reported that jatropha can be established on marginal land, paddocks, and contour ridges, hilly slopes, and gullies. For economic returns, a soil with moderate fertility is preferred. The leaves shed during the winter months and form mulch around the base of the plant. The organic matter from shed leaves enhances earthworm activity in the soil around the root-zone of the plants, which improves the fertility of the soil. It grows under a wide range of rainfall regimes from 250 to over 1200 mm/year (Katwal and Soni 2003; Agarwal and Agarwal 2007). In low-rainfall areas and in prolonged dry periods, the plant sheds its leaves to counter drought. Its water requirement is extremely low, and it can withstand long periods of drought by shedding most of its leaves to reduce transpiration losses. On heavy soils, root formation is reduced.

14.3.2 PROPAGATION METHODS

Jatropha is a fast-growing plant and can be propagated either by generative (direct seedlings or precultivated seedlings) or vegetative (direct planting of cuttings) methods. For quick establishment of living fences and plantations for erosion control, direct planting of cuttings is preferred (Heller 1996). Jatropha plants propagated from cuttings yield early (Heller 1996) but do not develop a taproot system. The plants only develop thin roots, which fail to penetrate deep into the soil, making them susceptible to uprooting by wind (Severino et al. 2007). Further, propagation by cuttings lowers the longevity of the plant and reduces its resistance to drought and disease as compared with those propagated by seeds. In agroforestry and intercropping systems, direct seeding is preferred over precultivated jatropha plants because the taproot of directly seeded plants has the ability to penetrate into deeper soil layers, where it can access extra nutrient resources and competes less with the roots of the other crops (Heller 1996; Jongschaap et al. 2007). If early seed yields are to be achieved, direct planting of stakes can be used as well (Heller 1996).

Recommendations on vegetative propagation vary. Cuttings of 25–30 cm length from 1-year-old branches (Gour 2006) or longer cuttings up to 120 cm (Henning 2000) are among the options. Kaushik and Kumar (2006) reported that the survival percentage depends on the origin of the source material (top, middle, or base of the branch) and the length and diameter combination of the cutting. Their study showed a survival percentage of 42% when the tops of the branches were used as cuttings, whereas cuttings from the middle and base showed significantly better survival results (i.e., 72 and 88%, respectively). The product of the length and diameter of the cuttings used was also positively correlated with the survival percentage. The longer and larger a cutting, the higher is its survival rate. Survival percentages higher than 80% were obtained for the length-diameter combinations from 105 to 2.5, 45 to 3.5, 45 to 4.5 cm and onward. Thicker cuttings form more roots than the thinner ones (Heller 1996). Cuttings can be planted directly in the field, in nursery beds, or in polyethylene bags for first root development (Heller 1996; Henning 2000; Gour 2006; Kaushik and Kumar 2006). They have to be planted at 10- to 20-cm depth in the soil depending on their length and diameter. The rainy season is the best period for the planting of cuttings (Ghosh et al. 2008).

In the generative propagation method, seeds are directly sown at the beginning of the rainy season, after the first shower when soil is wet, because it helps in the development of a healthy taproot system (Gour 2006). Seedlings can be precultivated in polythene bags or tubes or on seed beds under nursery conditions. The use of plastic bags or tubes induces root node formation and spindly growth (Severino et al. 2007). In the nursery, seeds should be sown 3 months before the rainy season on the soil with a high concentration of organic material [sandy loam soil/compost ratio of 1:1 (Kaushik and Kumar 2006); in case of more heavy soils, sand is added at a sand/soil/compost ratio of 1:1:2 (Gour 2006); sand/soil/farmyard manure ratio of 1:1:1 (Singh et al. 2006)] and should be well watered (Henning 2000). Presoaked seeds (24 h in cold water) germinate in 7–8 days in hot humid environments, whereas the process generally continues for 10–15 days (Gour 2006). A study on the germination enhancement of jatropha seeds (96% germination) showed the best results with presoaking in a cow-dung slurry for 12 h. The traditional 24-h cold-water treatment showed 72% germination. Nicking yields similar germination rates, whereas pretreatments using hot water or sulfuric acid (H_2SO_4) (0.5 M) did not enhance germination (Brahmam 2007).

14.3.3 PLANTATION

Field preparation for plantations mainly consists of land clearing and preparation of the planting pits for the seedlings. Although planting can be done without any clearing, for oil production purposes it is advisable to clear the land at least partially (Gour 2006). Tall trees can be left, but shrubs and bushes that cover the soil should be cut. After clearing, planting pits of 30–45 cm × 30–45 cm × 30–45 cm should be dug before the rainy season (Gour 2006; Singh et al. 2006). For good establishment, the pits are refilled with a mixture of the local soil/sand/organic matter (compost and/or artificial fertilizer) in the ratio of 1:1:1, respectively. Seeds or cuttings can be directly planted in the pits. Transplanting of prerooted cuttings (grown in poly bags) in the pits gives better results. The best time for planting is the warm season—if watering can be provided, otherwise it has to be carried out during the onset of the monsoon or monsoon season (Openshaw 2000). Gour (2006) reported that seedlings require irrigation, especially during the first 2–3 months after planting. Of course, the water demand depends on local soil and climatic conditions.

Actual plant spacing depends on the end use of the plant, seed quality, humidity, rainfall, intercropping, etc. Narrow spacing is preferred if the plant is grown as a hedgerow, live fence, or for soil conservation purposes. Plant spacing of 2 × 2 m (2500 plants/ha), 2.5 × 2.5 m (1600 plants/ha) or 3 × 3 m (1111 plants/ha) are common practice (Heller 1996). Kaushik and Kumar (2006) proposed wider spacing (4 × 2 and 4 × 3 m) and spacing of 5 × 2 and 6 × 6 m as the optimum for an agroforestry system to obtain high yield from individual jatropha plants. In 2.5-year-old plantations, it was observed that with the increase in spacing, seed yield per tree increased significantly, whereas the seed yield per hectare decreased (Chikara et al. 2007). Openshaw (2000) recommended spacing in hedgerows for soil conservation as 15–25 cm within and between rows (in case of double fence), resulting in 4000–6700 plants/km. Wider spacing is reported to give larger yields of fruit, at least in early years.

14.3.4 CARE AND MAINTENANCE

14.3.4.1 Trimming and Pruning

Pruning and canopy management are important because they help in the growth of more branches and the stimulation of abundant and healthy inflorescence, eventually enhancing good fruit setting and seed yield (Gour 2006). Trimming of branches during the period from February to March for the first 5 years is necessary to give bushy shape to the plant. Pruning at the age of 6 months is useful to pinch off the terminal shoots to induce lateral branching (Gour 2006; Kaushik and Kumar 2006). Pruning the main branch at 30- to 45-cm height depending on the growth rate is ideal (Gour 2006). At the end of the first year, the secondary and tertiary branches should be pruned to induce more branches. During the second year, each side branch should be pruned up to two-thirds of the top portion, retaining one-third of the branches on the plant. Pruning should be done in the dry or winter period after the trees have shed their leaves. This results in a lower and wider tree shape, helps in earlier seed production, and facilitates manual harvesting. Once in every 10 years, the entire plant has to be cut low, leaving a stump of 45 cm. The regrowth is quick, and the trees start yielding again within approximately 1 year. This intervention induces new growth and helps to stabilize the yield (Gour 2006; Kaushik and Kumar 2006). In addition to trimming hedgerows and pruning plantations annually, periodic thinning of plantations is also recommended. Starting from 1600 seedlings/ha, stand density should be thinned to 400–500 trees/ha in the final mature stand (Openshaw 2000).

14.3.4.2 Fertilizer Application

Optimal application of fertilizer and irrigation water can increase the seed and oil yield. Permanent humid situations and/or situations with high irrigation and fertilizer application can induce high biomass but low seed production. The optimal levels of input to achieve high harvest index in a given situation are yet to be quantified. At present, quantitative data on water need, water productivity, and water-use efficiency of jatropha plants are not available. In general, application of super phosphate or NPK fertilizer is reported to increase the yield. The optimal levels of application of inorganic nitrogen and phosphorus fertilizers are observed to vary according to the age of the plantation (Patolia et al. 2007). On degraded sites, jatropha plants are found to respond better to organic manure than to mineral fertilizers (Francis et al. 2005). On the basis of the nutrient composition of jatropha fruit, it is estimated that harvesting the equivalent amount of fruits for a yield of 1 t of seeds per hectare results in a net removal of 14.3–34.3 kg of nitrogen, 0.7–7.0 kg of phosphorus, and 14.3–31.6 kg of potassium per hectare (Jongschaap et al. 2007). Hence, this needs to be compensated for by applying an appropriate quantity of fertilizer (artificial or organic).

Application of manures at 3–5 kg/plant along with NPK should be done near the crown during the monsoon season. Literature shows that the application of super phosphate at 150 kg/ha and alternating with one dose of 20:120:60 kg NPK/year from the second year onward improves the yield. From the fourth year onward, 150 kg of super phosphate should be added to the above dose (Anonymous 2009a). Jatropha requires fertilizers containing NPK at 46:48:24 per hectare apart from the organic manure mixture. For direct planting, application of 20 g urea, 120 g single super phosphate, and 16 g murate of potash near the planting hole has been recommended. In the case of a transplanted crop, the above-mentioned fertilizer should be applied at the time of transplanting or immediately after the plant establishment in the pits. The remaining dose of urea should be applied in two splits at 10 g per plant. The first split should be applied one month after basal dressing and the second split at one month thereafter (Anonymous 2009c).

14.3.4.3   Intercultivation

The field should be kept free from weeds at all times by carrying out regular weeding operations. Approximately three to four weedings in the initial period are enough to keep the field free from weeds until the crop crosses the grand growth period. Light harrowing is beneficial for this purpose (Anonymous 2009d). Uprooted weeds can be left on the field as mulch. Jatropha being a perennial crop, intercrops can be raised in between the rows. The shade-loving plants, pulses such as black gram and a few vegetables such as tomato, bitter gourd, pumpkin, and cucumber can be profitably grown under jatropha plants for the first two years (Anonymous 2009a).

14.3.4.4   Irrigation

In the case that the monsoon season is proper and well distributed, additional irrigation during the rainy season is not required. During the dry period, the crop should be irrigated as required. Usually from the second year onward, irrigation is not required unless the soils are shallow and sandy (Anonymous 2009c).

14.3.4.5   Pests and Diseases

Some species of jatropha are known to be infected with leaf spots, powdery mildew, leaf curl, and leaf-distorting diseases. However, none of these diseases cause considerable yield losses. Likewise, the common pests such as leaf miner, leaf webber, and beetles found on certain species of jatropha are not serious. Collar rot disease may be the problem in the beginning that can be controlled by applying 0.2% copper oxychloride (COC) or by drenching of the plant with 1% Bordeaux mixture (Anon 2009a).

14.3.5   Seed Yield

The plants produce flowers nine months after sowing. However, plants established through cuttings produce flowers from six months onward. Wherever jatropha is cultivated under irrigated conditions, flowering occurs throughout the year. For best oil yields, the seeds should be harvested at maturity. Seeds are assumed to be matured when the color of the fruits changes from green to yellow. Maturity is attained 90 days after flowering (Heller 1996). Because all of the fruits do not mature at the same time, they have to be harvested manually at regular intervals, making this operation very labor-intensive (Heller 1996; Singh et al. 2006). The time and length of the harvest period is likely to vary according to the seasonal conditions of the locality. In semiarid regions, the harvesting spans over a period of two months, which implies daily or weekly harvests. In permanently humid situations, weekly harvest is necessary throughout the year. Separation of the seeds from the husks can be done manually or mechanically (Gour 2006).

Apparently, the yield depends on a range of factors such as water availability, soil conditions, altitude, sunshine hours, and temperature. No systematic research seems to have been conducted yet to determine the influence of these factors and their interactions on seed yield. The seed yields reported for different countries and regions range from 0.4 to over 12 t/ha per year after 5 years of growth (Jones and Miller 1992). In relatively poor soils, the yields have been reported to be 1 kg per plant, whereas in lateritic soils the seed yields have been reported as 0.75–1.00 kg/plant (Openshaw 2000). Thus, the average yield can be considered as 4.00–6.00 t/ha per year depending on the agroclimactic zone and agricultural practices (Biswas et al. 2007). Harvesting of the seeds takes place during the dry season, which is normally a quiet period for agricultural laborers. The ripened fruits are plucked from the trees manually, and the seeds are sun-dried (three weeks) or oven-dried (105°C) to reduce the moisture content up to a certain level. The dried pods are collected, and the seeds are separated manually or mechanically. Seeds are once again dried under sunlight for four days until the moisture reaches approximately 6–10% before oil extraction. Drying of seeds is necessary because moist seeds can develop mold and can jam the pressing equipment.

14.3.4 Oil Extraction Methods

The seeds contain on an average approximately 34% oil. The oil contains a toxic substance, curcasin, which is a strong purgative (Chachage 2003). To prepare the kernels for oil extraction, they should be solar heated for several hours or roasted for 10 min. This process breaks down the cells containing the oil and eases the oil flow. The heat also liquefies the oil, which improves the extraction process. The seeds should not be overheated. The oil from jatropha seeds can be extracted by mechanical extraction using a screw press or chemical extraction (solvent extraction). Figure 14.2

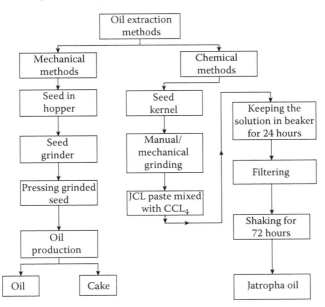

FIGURE 14.2 Flowchart for mechanical and chemical methods of oil extraction.

shows the procedure of mechanical and chemical methods followed for extraction of oil from jatropha seeds.

14.3.4.1 Mechanical Extraction Using a Screw Press

In this method, seeds are pressed in a manual ram press or in a powered screw press to expel the oil, leaving seed cake. For mechanical extraction of oil from the seeds, the extraction rate of the ram press is low (60–65% of the available oil) as compared with the powered screw press (75–80% of the available oil). The leftover seed cake still contains some oil. One liter of oil can be expelled from approximately 3 kg of seeds (Henning 2004). The seeds can be subjected to a number of extractions by passing through the expeller. Up to three passes is common practice. Pretreatment of the seeds (e.g., cooking) increases the oil yield of screw pressing up to 89% after a single pass and 91% after two passes (Beerens 2007). The theoretical maximal amount of oil in a jatropha seed has been reported to be approximately 48.6% (Kandpal and Madan 1995).

14.3.4.2 Chemical Extraction

The reaction temperature, reaction pH, time requirement, and oil yield of different chemical extraction methods tested on jatropha seeds are summarized in Table 14.1. The n-hexane method is the most common chemical extraction method. It results in the highest oil yield, but it also takes the most time. In aqueous enzymatic oil extraction, the use of alkaline protease gave the best results (Winkler et al. 1997; Bryant et al. 2008). Furthermore, it is shown that the ultrasonication pretreatment is a useful step in aqueous oil extraction. Adriaans (2006) opined that solvent extraction is economical only for the large-scale production of more than 50 t biodiesel per day. It was reported that the conventional n-hexane solvent extraction is hazardous to the environment (generation of wastewater, higher specific energy consumption, and higher emissions of volatile organic compounds) and human health (working with hazardous and inflammable chemicals). Using aqueous enzymatic oil extraction greatly reduces these problems (Adriaans 2006), as does the use of supercritical

TABLE 14.1
Effect of Reaction Parameters on Different Chemical Extraction Methods on Oil Yields from Jatropha Seed

Extraction Method	Reaction Temperature (°C)	Reaction pH	Time Consumption (h)	Oil Yield (%)
n-Hexane oil extraction (soxhelt apparatus) first acetone, second n-hexane	– –	– –	24 48	95–99 –
AOE	– 50	– 9	2 6	38 38
AOE with 10 min of ultrasonication as pretreatment	50	9	6	67
AEOE (hemicellulase or cellulase) AEOE (alkaline protease)	60 60 50	4.5 7 9	2 2 6	73 86 64
AEOE (alkaline protease) with 5 min of ultrasonication as pretreatment	50	9	6	74
Three-phase partitioning	25	9	2	97

AOE, aqueous oil extraction; AEOE, aqueous enzymatic oil extraction.
Source: Achten, W.M.J., et al. *Biomass Bioenergy,* 32, 1063–1084, 2008.

TABLE 14.2

Oil Extraction Efficiency of Different Methods Used for Obtaining Oil from Jatropha Seed

Method		Oil Yield (%)
Theoretical maximum		44.00
Pressing	Hand press	22.55
	Motor press	22.98
	Industrial press	27.00
Aqueous oil extraction (AOE)	Basic AOE	16.72
	AOE with sonication	29.48
	Aqueous enzymatic oil extraction (AEOE)	28.16
	AEOE with sonication	32.56
Three-phase partitioning (TPP)	Basic TPP	36.08
	Enzyme-assisted three-phase partitioning (EATPP)	40.48
	EATPP with sonication	42.68

Source: Bryant C, et al., *Jatropha Curcus L.: Biodiesel Solution or All Hype? A Scientific, Economic and Political Analysis of the Future Energy Crop*, 2008. Energy and Energy Policy, available at http://humanities.uchicago.edu/orgs/institute/bigproblems/Energy/BP-Energy-Jatropha.doc (accessed June 9, 2009)

solvents (mainly supercritical carbon dioxide) or biorenewable solvents such as bioethanol and isopropyl alcohol. Although new-generation *n*-hexane extraction units are far more efficient and cause far less environmental burdens than the older units, further research on these alternative solvents needs to be conducted on their commercial viability. Foidl and Mayorga (2007) reported the use of supercritical isopropanol or carbon dioxide (CO_2) in a continuous mechanical oil extraction system, which left only 0.3% oil (weight basis) in the cake.

Bryant et al. (2008) extracted oil by soxhlet extraction with hexane solvent. There are several processes for oil extraction that range greatly in cost and efficiency. The oil extraction methods can be divided into three primary categories: crushing the seeds with a press, aqueous enzymatic oil extraction, and three-phase partitioning. The oil yield efficiency of different methods used for oil extraction from jatropha seeds are summarized in Table 14.2.

14.3.4.3 Jatropha Oil

The fatty acid compositions and characteristics of the crude jatropha oil are given in Tables 14.3 and 14.4, respectively. It is important to note that the values of the free fatty acids, unsaponifiables, acid number, and carbon residue show wide variations, although it is a small data set. This indicates that the oil quality is dependent on the interaction effect of environment and genetics, with the former having a higher impact than the latter.

The average saturated and unsaturated fatty acids constitute 20.1% and 79.9% of the oil, respectively, which is reflected in the pour and cloud points of the oil. Among the fatty acids present, oleic and linoleic are the major constituents. The maturity stage of the fruits at the time of collection is reported to influence the fatty acid composition of the oil (Raina and Gaikwad 1987).

14.3.4.4 Jatropha Seed Cake

The average crude protein content of the seed cake is 58.1% by weight and has an average gross energy content of 18.25 MJ/kg (Figure 14.3). On the basis of the extraction efficiency and the average oil content of the whole seed (34.4% on a mass basis), the seed cake contains 9–12% oil by weight. This oil content will of course influence the gross energy value of the seed cake. In addition to high-quality proteins (Figure 14.3), this seed cake contains various toxins and therefore it cannot be used as fodder (Francis et al. 2005).

TABLE 14.3
Fatty Acid Composition of Jatropha Oil from Different Regions

Fatty Acid Type	Caboverde Variety	Nicaragua Variety	Indian Variety
Capric acid (C10:0)	0.1	0.1	–
Myristic acid (C14:0)	0.1	0.1	0.1
Palmitic acid (C16:0)	15.1	13.6	14.1–15.3
Palmitoleic acid (C16:1)	0.9	0.8	1.3
Stearic acid (C18:0)	7.1	7.4	3.7–9.8
Oleic acid (C18:1)	44.7	38.3	34.3–45.8
Linoleic acid (C18:2)	31.4	43.2	29–44.2
Linolenic acid (C18:3)	0.2	0.2	0.3
Arachidic acid (C20:0)	0.2	0.3	–
Behenic acid (C22:0)	0.2	–	–
Investigator	Foidl et al. 1996		Kumar et al. 2003a

TABLE 14.4
Characterization of Jatropha Oil

Properties	Range	Mean	SD
Density (g/cm^3)	0.860–0.933	0.914	00.018
Kinematic viscosity at 30–40°C (cSt)	37.00–54.80	46.82	7.24
Calorific value (MJ/kg)	37.83–42.05	38.63	1.52
Cetane number	38–51	49.63	6.2
Flash point (°C)	210–240	235	11
Acid value (mg KOH/g)	0.92–6.16	3.71	2.17
Pour point (°C)	–3, 15[a]	–	–
Water content (%)	1.4[b]		
Carbon residue % (kg/kg × 100)	0.07–0.64	0.38	0.29
Saponification number (mg/g)	102.9–209.0	182.8	34.3
Unsaponifiable % (kg/kg × 100)	0.79–3.80	2.03	1.57
Iodine number (mg iodine/g)	0.92–112	101	7
Monoglycerides % (kg/kg × 100)	1.7	–	–
Diglycerides % (kg/kg × 100)	2.50–2.70	–	–
Triglycerides % (kg/kg × 100)	88.20–97.30	–	–
Sulfur content % (kg/kg × 100)	0–0.13		

Source: Modified from Achten, W.M.J., et al., *Biomass Bioenergy,* 32, 1063–1084, 2008.

[a] Vyas, A.P., Subrahmanyam, N., and Patel, P.A., *Fuel,* 88, 625–628, 2009.

[b] Tiwari, A.K., Kumar, A., and Raheman H., *Biomass Bioenergy,* 31, 569–575, 2007.

However, the raw kernel or seed cake can be valuable as a organic nutrient source because it contains more nutrients than chicken and cattle manure (Francis et al. 2005). An overview of the experiments shows that jatropha seed cake is useful as fertilizer and it is summarized in Table 14.5.

When the seed cake is applied as fertilizer, the presence of biodegradable toxins (mainly phorbol esters) in it makes the seed cake simultaneously serve as a biopesticide/insecticide and molluscicide (Rug et al. 1997; Francis et al. 2005). Although the phorbol esters decompose completely within six days, it is advisable to check that no phorbol esters are present in the crops grown for human consumption using jatropha seed cake as fertilizer (Rug et al. 1997). Heller (1996) warned about

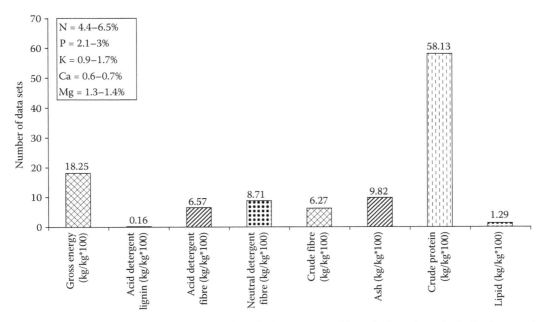

FIGURE 14.3 Average kernel cake composition of solvent-extracted jatropha kernels on the basis of reported data sets. (From Achten, W.M.J., et al., *Biomass Bioenergy,* 32, 1063–1084, 2008.)

TABLE 14.5
Summary of Case Studies Using Jatropha Seed Cake as Fertilizer

Country	Crop	Dosage (t/ha)	Comments
Mali	Pearl millet	5	46% yield increase in comparison to zero input
Zimbabwe	Cabbage	2.5–10	40%–113% yield increase in comparison to zero input. Free from pest and disease, whereas cutworm infestation occurred with cow manure application
Nepal	Rice	10	11% yield increase in comparison to zero input
India	*Jatropha curcas* L.	0.75–3	13%–120% yield increase in comparison to zero input

Source: Achten, W.M.J., et al., *Biomass Bioenergy,* 32, 1063–1084, 2008.

phytotoxicity of overapplication of jatropha seed cake. One study showed phytotoxicity to tomatoes, expressed in reduced germination, when seed cake was applied at high rates of up to 5 t /ha (Achten et al. 2008).

The cake is also used as feedstock for biogas production through anaerobic digestion before using it as a soil amendment. Staubman et al. (1997) obtained 0.446 m^3 of biogas, containing 70% methane (CH_4), per kilogram of dry seed press cake using pig manure as inoculum. Radhakrishna (2007) reported a total production of 0.5 m^3 biogas/kg of solvent-extracted kernel cake and 0.6 m^3 biogas/kg of mechanically de-oiled cake using specifically developed microbial consortia as inoculum. The fact that jatropha seed cake can be used for different purposes makes it an important byproduct. Recycling of wastes as a fertilizer can help to reduce inputs needed for jatropha cultivation and other crop cultivation or it can produce extra energy in the form of biogas. From an environmental point of view, anaerobic digestion of the cake in the biogas plant and application of the effluent to the agricultural field is thought to be the best practice at present. A number of questions concerning the long-term and cumulative impacts of jatropha seed cake on soils have not been addressed. In the event that detoxification becomes viable, the use of seed cake as animal feed will be more beneficial.

14.4 PRODUCTION OF BIODIESEL FROM JATROPHA OIL

The use of straight jatropha seed oil in compression ignition (CI) engines is restricted by some unfavorable physical properties, particularly its viscosity. Because of higher viscosity, the straight jatropha oil causes poor fuel atomization, incomplete combustion, and carbon deposition on the injector and valve seats resulting in serious engine fouling. It has been reported that when direct injection engines are run with neat vegetable oil as fuel, injectors get choked up after few hours and lead to poor fuel atomization, less efficient combustion, and dilution of lubricating oil by partially burnt oil (Ma and Hanna 1999). One possible method to overcome the problem of higher viscosity is blending of jatropha oil with diesel in the proper proportion, and the other method is transesterification of oil to produce biodiesel. It is reported that the transesterification process has been proven worldwide as an effective means of biodiesel production and viscosity reduction of vegetable oils (Peterson et al. 1992). Biodiesel is now gaining more attention as an alternative fuel for substituting diesel because it is renewable in nature, produces less emissions, and it can be produced from locally available oils/fats.

14.4.1 Preparation of Oil for Biodiesel Production

Raw jatropha oil contains wax and gums, such as phospholipids, due to which its viscosity becomes high. The phospholipids can possibly deactivate the alkaline catalyst in biodiesel production (Freedman et al. 1984). Also it was reported that 50 ppm phosphorus in oil reduced the yield of methyl esters by 3–5% (Gerpen 2005). Higher phosphorus content also causes higher sulfate ash of biodiesel and thus leads to higher particulate emissions, which in turn may influence the performance of the catalytic converters (Mittelbach 1996). Therefore, conventional biodiesel processes often have a degumming operation to remove phospholipids. Jatropha oil has less than 1% free fatty acids (FFAs), and, taking into consideration the other fuel properties such as pour point and flash point, it should be degummed and made free from impurity by using toluene or by means of filtration with 5-μm filter paper before transesterification. This will help in easy completion of the process and production of good-quality biodiesel. But, if a two-step process (esterification/pretreatment followed by transesterification) is followed for producing biodiesel when the FFA level is more than 1%, there is no need for degumming because the esterification or pretreatment is more or less similar to the degumming operation. Phospholipids are hydrophilic and tend to aggregate when moisture is present, especially in the presence of acidic water. The sedimentation and washing may remove most of the phospholipids. Lu et al. (2009) studied the effect of phospholipids on the esterification/pretreatment of jatropha oil and investigated the process by adding phospholipids into the refined oil. From this study it was found that phospholipids did not show much influence on the final FFA conversion during esterification/pretreatment, as shown in Table 14.6. The phospholipid content decreased from the original 0.33% to the final 0.04% (on an average 92% reduction) after the pretreatment process.

14.4.2 Biodiesel Production Technology

A number of technologies have been developed to obtain biodiesel from jatropha oil either having low or high level of FFAs. A brief review of different methods followed for producing biodiesel from jatropha oil is presented here. Production of biodiesel from any feedstock is largely dependent on the origin of plant, type of oil extraction process involved, quality of feedstock, FFA contents of oil, production technology adopted, postproduction parameters, and the desired biodiesel quality

TABLE 14.6

Effect of Phospholipids on Pre-Esterification

Particulars	Level		
	I	II	III
Phospholipid content before pretreatment, %	0.33	1.03	1.62
Phospholipid content after pretreatment, %	0.04	0.05	0.12
Removal of phospholipids, %	87.9	95.1	92.6
Initial acid value of oil, mg KOH/g	20.5	20.9	20.6
Acid value after pretreatment, mg KOH/g	1.18	1.34	2.04
Pretreatment conversion, %	94.2	93.6	90.1

Source: Data from Lu, H., et al., *Comput Chem Eng*, 33, 1091–1096, 2009.

(Lu et al. 2009). However, the FFA contents of the feedstock is the important criterion for selecting a suitable processing technology for production of biodiesel because production process parameters are dependent on it and the yield of biodiesel is influenced by the methanol-to-oil ratio, catalyst-to-oil ratio, reaction temperature, reaction time, and agitation speed.

It is reported that the FFA level of jatropha oil varies between 0.29% and 0.4% (Caboverde variety), 0.60% and 1.27% (Nicaragua variety), 7 and 15% (China and Indonesia variety), and 1.76 and 14% (India variety) (Foidl et al. 1996; Akintayo 2004; Chitra et al. 2005; Tiwari et al. 2007; Berchmans and Hirata 2008; Lu et al. 2009; Vyas et al. 2009). A process such as acid- or base-catalyzed transesterification may be followed for production of biodiesel from oil having an FFA level less than 1% (Ma and Hanna 1999; Canakci and Gerpen 2001; Ghadge and Raheman 2005); for oils with higher FFAs (>1%), a two-step process (i.e., an acid pretreatment process followed by base transesterification process, supercritical fluid process, and a solid catalysis process) may be followed for producing biodiesel (Canakci and Gerpen 2001; Ghadge and Raheman 2006; Wang et al. 2006; Sahoo et al. 2007; Tiwari et al. 2007; Berchmans and Hirata 2008; Shi and Bao 2008; Hawash et al. 2009).

The yield of biodiesel following the two-step process is between 90 and 99%. Heterogeneous catalysis processes such as enzymatic catalysis and solid catalysis are reported to be time-consuming and give an average yield of biodiesel between 70% and 92.8% (Shah et al. 2004; Modi et al. 2007a, 2007b; Tamalampudi et al. 2008; Vyas et al. 2009). The supercritical methanol process was found to be better among the processes because of its homogeneity and better yield of biodiesel up to 100% (Hawash et al. 2009), but the process requires higher temperature, pressure, and a large amount of methanol, hence it involves a higher production cost (Saka and Kusdiana 2001). The detailed technologies followed and developed for producing biodiesel from jatropha oil are summarized in Table 14.7.

14.4.2.1 Reaction Chemistry for Transesterification

Vegetable oils and animal fats are principally composed of triacylglycerols (TAGs) consisting of long-chain fatty acids chemically bound to a glycerol (1,2,3- propanetriol) backbone (Moser 2009). Transesterification or alcoholysis is the process in which triglycerides or TAGs of oil react with an alcohol to produce esters (known as biodiesel) and glycerin (glycerol) in the presence of a catalyst, as shown in Figure 14.4, where R_1, R_2, and R_3 are long hydrocarbon chains, sometimes called fatty acid chains (Gerpen 2005). The catalyst could be alkali catalyst, acid catalyst, or enzymatic catalyst. The whole process is normally a sequence of three consecutive steps that are reversible reactions (Figure 14.5). In the first step, diacylglycerol is obtained from TAG (step 1); from diacylglycerol, monoacylglycerol is produced in the second step (step 2); and in the last step, glycerol is obtained from monoacylglycerol (step 3). In all of these steps, esters are produced, which are called biodiesel (Marchetti et al. 2007).

TABLE 14.7
Different Technologies for Producing Biodiesel from Jatropha Oil

Investigator(s)	Process	Alcohol-to-Oil Ratio		Catalyst-to-Oil Ratio		Reaction Time, min	Reaction Temperature, °C	Yield, %
		Type	Ratio	Type	Concentration			
Foidl et al. 1996	Base catalyzed	Methanol	4.45:1 molar	KOH	0.23 molar	60	30	92
		Ethanol	5.79:1 molar	KOH	0.44 molar	90	75	88.4
Kumar et al. 2003a	Base catalyzed	Methanol	0.45 v/v	KOH	0.01 v/v	60	70	NA
Chitra et al. 2005	Base catalyzed	Methanol	20%	NaOH	1%	90	60	98
Sivaprakasam and Saravanan 2007	Base catalyzed	Methanol	0.18 v/v	NaOH	8 g/500 mL	60	70	91
Kalbande et al. 2008	Base catalyzed	Methanol	20% v/w	KOH	1% w/w	60	90	91.5
Tiwari et al. 2007	Two-step acid/base process	Methanol	0.28 v/v	H_2SO_4	1.43% v/v	88	60	
		Methanol	0.16 v/v	KOH	5 g/1000 mL	24	60	99
Berchmans and Hirata 2008	Two-step pretreatment process	Methanol	0.60 w/w	H_2SO_4	1% w/w	60	50	90
		Methanol	0.24 w/w	NaOH	1.4% w/w	120	65	
Sahoo and Das 2009a	Two-stage transesterification	Methanol, toluene, *ortho*-phosphoric acid	350 ml/L, 4 ml/L, 4 ml/L	No catalyst	—	120	66	93
		Methanol	110 ml/L	KOH	8 g/L	180	66	
Shah et al. 2004	Lipase catalyzed	Ethanol	4:1 molar	Free tuned Chromobacterium with addition of 1% w/v water				73
				Immobilized Chromobacterium with addition of 0.5% w/v water	40 mg/0.5 g	480	40	92
Modi et al. 2007a	Lipase catalyzed	Ethyl acetate	11:1 molar			720	50	91.3
Modi et al. 2007b		Propan-2-ol	4:1 molar	Novozym-435	10% w/w	480	50	92.8

Tamalampudi et al. 2008	Lipase catalyzed	Methanol	3:1 molar	Whole cell	0.2 g	3600	30	80
				Novozym 435		5400		70
Wang et al. 2008	Solid catalyst	Methanol	10:1	X/Y/MgO/γ-Al$_2$O$_3$	1% w/w	180	500	96.5
Vyas et al. 2009	Solid catalyst	Methanol	12:1	KNO$_3$/Al$_2$O$_3$ solid catalyst	6% w/w	360	70	84
Hawash et al. 2009	Supercritical methanol	Methanol	43:1	No catalyst	–	4 min	320 at 8.4 MPa	100

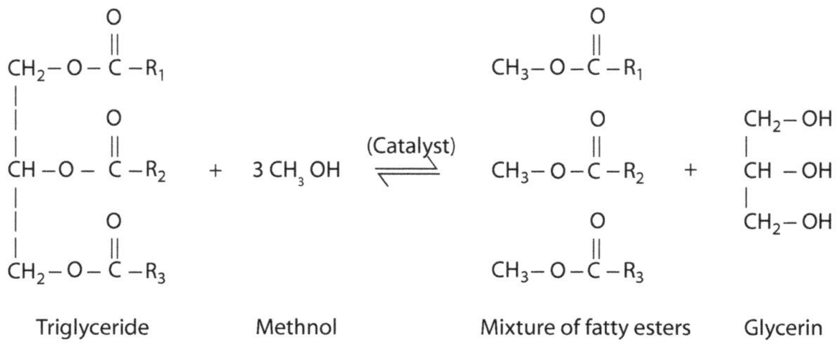

FIGURE 14.4 Transesterification of vegetable oil.

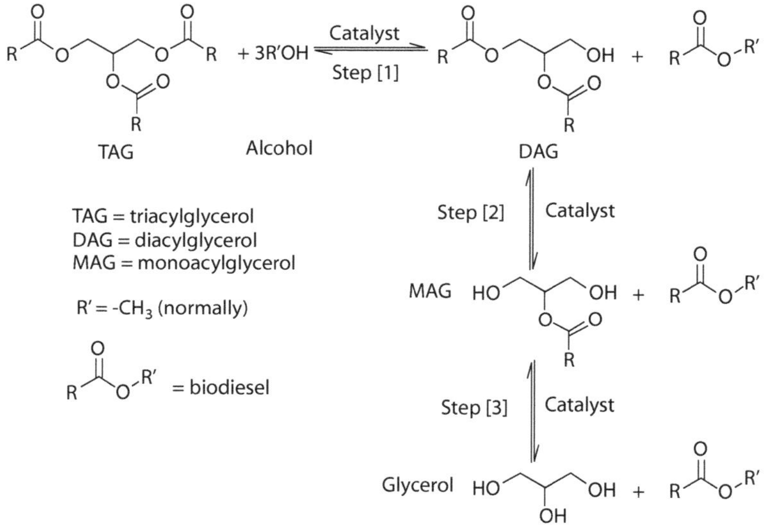

FIGURE 14.5 Common steps in transesterification process. (From Moser, B.R., *In Vitro Cell Dev Biol—Plant*, 2009. doi: 10.1007/s11627-009-9204-z.)

The transesterification process is given various names such as alcoholysis, methanolysis, ethanolysis, etc. depending upon the type of alcohol used for the reaction (Romano 1982; Clark et al. 1984; Freedman et al. 1984; Schwab et al. 1987; Peterson et al. 1992; Ma and Hanna 1999; Srivastava and Prasad 2000). Among the alcohols that can be used in the transesterification process are methanol, ethanol, propanol, butanol, or amyl alcohol. Methanol and ethanol are used most frequently, especially methanol because of its low cost and its physical and chemical advantages. Being polar and the shortest chain alcohol, methanol can quickly react with triglycerides. Among the catalysts, alkali-catalyzed transesterification using potassium hydroxide (KOH)/sodium hydroxide (NaOH) or their corresponding alkoxides was found to be used most often because they are much faster, more efficient, and less corrosive (Freedman et al. 1984; Ma and Hanna 1999; Azam et al. 2005; Leung and Guo 2006; Marchetti et al. 2007).

14.4.3 BIODIESEL CHARACTERIZATION

In addition to mono-alkyl esters, glycerol, alcohol, catalyst, FFAs, and tri-, di-, and monoglycerides comprise the final output of the biodiesel production process. These contaminants can lead to severe operational and environmental problems. Therefore, the quality control of biodiesel is highly essential for the success of its commercialization and market acceptance. Some important issues on the biodiesel quality control involve monitoring of the transesterification reaction, quantification of mono-alkyl esters, free and bonded glycerol, and determination of residual catalysts and alcohol. Chromatography and spectroscopy are the most commonly used analytical methods for biodiesel analyses.

Biodiesel can be blended with diesel fuels and used in diesel engines with few or no modifications. Depending upon the volume of biodiesel in a blend, it is designated (i.e., for 20% volume of biodiesel, the blend is designated as B20, for 40% biodiesel in the blend, it is designated as B40, and so on). But, the fuel properties of biodiesel-diesel fuel blends change with the amount of biodiesel in the fuel mixture because biodiesel has different fuel properties compared with conventional diesel fuel. There are several key properties of biodiesel that need to be characterized before using biodiesel-diesel fuel blends in a diesel engine. These properties include kinematic viscosity, density, pour point, flash point, acid value, sulfated ash, total and free-glycerin, water content, carbon residue, etc. and are summarized in Table 14.8. Jatropha biodiesel has comparable fuel properties with those of diesel and it conforms to the latest U.S. and European standards for biodiesel.

14.4.3.1 Density

The densities of jatropha oil (JO) and jatropha biodiesel (JB) are approximately 7.52 and 2.94% higher than that of diesel, respectively (Tables 14.4 and 14.8). Thus, the density of JO is reduced by approximately 4.26% in its conversion to JB. The higher density of JO and JB as compared with diesel may be attributed to the higher molecular weights of the triglyceride molecules present in them.

TABLE 14.8
Fuel Characteristics of Jatropha Biodiesel

Fuel Properties	Diesel	JB Range	JB Mean	Jatropha Ethyl Ester	Biodiesel Standards ASTM D6752	Biodiesel Standards DIN V 51606
Density (g/cm³)	0.850	0.864–0.880	0.875	0.89	0.86–0.90	0.87–0.90
Kinematic viscosity at 30–40°C (cSt)	2.6	4.84–5.65	5.11	5.54	3.5–5.0	3.5–5.0
Calorific value (MJ/kg)	42	38.45–41.00	39.65	–	–	–
Cetane number	47.68	50–56.1	52.3	59	Min 51	Min 49
Flash point (°C)	68	170–192	186	190	Min 120	Min 110
Acid value (mg KOH/g)	–	0.06–0.5	0.27	0.08	Max 0.8	Max 0.5
Pour point (°C)	−20 to −10	4.2–6[a]	–	–	–	–
Water content (%)	0.02	0.07–0.10	–	0.16	Max 0.5	Max 0.3
Ash content (%)	0.01	0.005–0.01	0.013	–	Max 0.02	Max 0.03
Carbon residue (%)	0.17	0.02–0.50	0.18	–	Max 0.3	Max 0.3
Saponification number (mg/g)	–	202.6	–	–	–	–
Iodine number (mg/g)	–	93–106	99.5	–	Max 120	Max 115
Sulfur content (%)	–	0.0036	–	–	Max 0.01	Max 0.01
Monoglycerides (%)	–	0.24	–	0.55	Max 0.8	Max 0.8
Diglycerides (%)	–	0.07	–	0.19	Max 0.2	Max 0.4
Methyl ester content (%)	–	99.6	–	99.3	Min 96.5	–
Methanol (%)	–	0.06–0.09	–	0.05	Max 0.2	Max 0.3
Free glycerol (%)	–	0.015–0.030	–	–	Max 0.02	Max 0.02
Total glycerol (%)	–	0.088–0.100	–	0.17	Max 0.25	Max 0.25

Source: Modified from Achten, W.M.J., et al., *Biomass Bioenergy*, 32, 1063–1084, 2008.

[a] Sahoo, P.K. and Das, L.M., *Fuel*, 88, 1588–1594, 2009a and Vyas, A.P., Subrahmanyam, N., and Patel, P.A., *Fuel*, 88, 625–628, 2009.

14.4.3.2 Kinematic Viscosity

The main problem of using JO as a fuel is its high viscosity, which is approximately 46.82 cSt, which is 18 times more than that of diesel at 30°C (Table 14.4). The high viscosity of JO in contrast with diesel may be due to the greater intermolecular attraction of the long chains of its glyceride molecules. This higher viscosity of JO hinders its smooth flow through pipelines, injector nozzles, and orifices. A viscosity that is too high can cause excessive heat generation in the injection equipment because of viscous shear in the clearance between the pump plunger and cylinder of a fuel injector. Higher viscosity reduces the injector spray cone angle, fuel distribution, and penetration while increasing the droplet size, thus affecting injection timing. On conversion to JB, the kinematic viscosity is reduced (i.e., it becomes 1.97 times higher than that of diesel). The viscosity of biodiesel and oil blends with diesel increases with the increase in concentration of biodiesel/oil in the blends.

14.4.3.3 Calorific Value

The amount of heat generated by combustion of a unit weight of fuel is expressed in terms of calorific value. The gross calorific values of JO and JB are approximately 38.63 and 39.65 MJ/kg, respectively, as compared with 42 MJ/kg for diesel (Pramanik 2003; Tiwari et al. 2007; Sahoo and Das 2009a). This could be due to the difference in their chemical composition from that of diesel or the difference in the percentage of carbon and hydrogen content, or the presence of oxygen molecules in the molecular structure of JO and JB. The oxygen molecule present unites with the hydrogen of the oil to form water vapor even before secondary air or oxygen supplied for combustion reaches the hydrogen. This results in a decrease in the available hydrogen, thus decreasing the calorific value. The calorific value of the JO and JB blends with diesel proportionately decreases with the increase in biodiesel or oil percentage in the blends.

14.4.3.4 Cetane Number

The cetane number is a measure of the ignition quality of fuels. The higher the cetane number, the more efficient the fuel will be , and it will ignite easily when injected into the engine (Demirbas 2008). The cetane number of JO and JB is 4.08 and 9.69% higher than that of diesel, respectively (Achten et al. 2008; Vyas et al. 2009). This higher cetane number is due to the higher oxygen content in the fuel.

14.4.3.5 Flash Point

The flash point of JO and JB is 235 and 186°C, respectively, as compared with 68°C for diesel (Tables 14.4 and 14.8). This might be due to the presence of components of longer chains in their molecules and a higher degree of unsaturation (79.9%). Generally, a material with a flash point of approximately 90°C or higher is considered as nonhazardous from a storage and fire-hazard point of view.

14.4.3.6 Acid Value

The presence of FFAs in the fuel leads to corrosion as well as gum and sludge formation. The acid value of JO has been reported to be vary between 0.92 and 29.8 mg KOH/g (Tiwari et al. 2007; Achten et al. 2008; Berchmans and Hirata 2008; Vyas et al. 2009). On conversion to JB, its acid value is reduced to 0.27 mg KOH/g.

14.4.3.7 Pour Point

The pour point is used to characterize the cold flow operability of a fuel because this property of fuel affects the utility of the fuel, especially in the cold climate condition. The pour point of JO, JB, and their blends with diesel are higher than that of diesel. This might be due to the presence of wax and the higher amount of saturated fatty acids (Alptekin and Canakci 2009), which begins to crystallize with a decrease in temperature.

14.4.3.8 Water Content

Water present in fuel interferes with its smooth flow through lines and combustion inside of the cylinder. The water content of JO has been reported as 1.4% (Tiwari et al. 2007), which is much higher than the ASTM prescribed limits of less than 0.03%. However, the water content of JB is approximately 0.16%, which is well within the limits prescribed by the ASTM and DIN standards.

14.4.3.9 Ash Content

Ash is the incombustible material that remains after burning and correlates with the amount of deposits formed in the combustion chamber. Therefore, fuels having less ash content are preferred for better engine operation and maintenance. JO has a very high level of ash content, which is one of the main reasons for its unsuitability as a fuel in the long run (Tiwari 2007). However, upon conversion of the JO to JB, the ash content is reduced to 0.013%, which is very similar to that of diesel.

14.4.3.10 Carbon Residue

The property that correlates with the amount of carbonaceous deposits, the fuel that will form in the combustion chamber of the engine is known as carbon residue—the non-volatile residue left when fuel is heated to a high temperature in the absence of air. The higher this value is, the greater the expected carbon deposits in the combustion chamber. JO has a high carbon residue level of 0.38%. However, the carbon residue of JB is very similar to that of diesel and is within the specified limit of 0.3%.

All of these fuel properties except flash point and cetane number are improved by increasing the percentage of diesel in the blend. However, a reverse trend has been reported for the flash point and cetane number with increasing diesel in the blend.

The influence of the quality of biodiesel on major fuel properties and on engine performance is summarized in Table 14.9.

14.5 ENGINE PERFORMANCE USING JATROPHA BIODIESEL AND OIL

To get more market acceptance of JO/JB, engine performance evaluation is required along with quality checking of biodiesel. The specific need will decide whether to go for JB (B100, i.e., pure biodiesel) or a particular blend to substitute diesel fuel with little sacrifice on engine output and economy. In this section, efforts have been made to review different aspects such as combustion, performance, and emission characteristics of diesel engine while running with JO, JB, and jatropha biodiesel blend (JBB).

14.5.1 COMBUSTION OF BIODIESEL

Combustion in CI engines is a three-dimensional, unsteady, and heterogeneous process and is affected by type of fuel, fuel characteristics, design of combustion chamber, fuel injection system, and the engine's operating conditions (Heywood 1988). The main aspects of combustion are a rise in cylinder pressure, heat release rate, and ignition delay. Ignition delay of the fuel is extremely important in CI engines primarily because of its effect on combustion rate and detonation. It may influence the starting ability of the engine. The combustion characteristics for JB, JBB, and JO in CI engines are briefly summarized in Table 14.10.

14.5.1.1 Cylinder Pressure

The peak cylinder pressure in a CI engine depends upon the rate of combustion at the initial stage, which is influenced by the fuel taking part in the uncontrolled heat release phase. The high viscosity and low volatility of JO leads to poor atomization and mixture preparation with air during

TABLE 14.9

Influence of Major Biodiesel Properties on Quality of Biodiesel and Engine Performance

Major Properties	Influence on Fuel Quality	Influence on Engine Performance	Investigator(s)
Density	Influence on cetane number and heating value of fuel	Efficiency of fuel atomization and engine output power	Ryan et al. 1984; Bahadur et al. 1995; Tat and Gerpen 2000; Tyson 2001; Demirbas 2008
Kinematic viscosity	Flow characteristics.	Atomization quality, combustion quality, leakage in the fuel system. Increases the engine deposits. Needs more energy to pump the fuel and wears fuel pump elements and injectors.	Schwab et al. 1987; Heywood 1988; Tat and Gerpen 1999; Kinast 2001; Encinar et al. 2005; Tate et al. 2006; Knothe and Steidley 2007
Cetane number	Fuel ignition quality and increase level of unsaturation.	Affects the ignition delay time and knocking	De Oliveira et al. 2006; Knothe et al. 2006; Knothe et al. 2003
Flash point	Safety in storage, fuel handling, and transportation	No direct effect on combustion in diesel engine	Owen and Coley 1995; Caro et al. 2001
Acid value	Presence of FFAs. Oxidative stability, kinematic viscosity and lubricity	Damage to injector. Deposits in fuel lines. Affects the life of pumps and filters. Not suitable at low climatic temperature.	Miyashita and Takagi 1986; Frankel 2005; Knothe and Steidley 2005a
Water and sediment	Deterioration of oil due to growth of microbes	Fuel fouling, deposits clog fuel filters	Knothe 2005
Ash content	Residual catalyst	Wear of engine components. Filter plugging and injector deposits.	Knothe and Steidley 2005b
Total glycerin and free glycerin	Incomplete conversion, residual mono-, di-, and triglycerides. Affects the kinematic viscosity	Filter clogging and injector deposits. Low-temperature operatibility.	Yu et al. 1998; Knothe and Steidley 2005a

the ignition delay period, resulting in lower peak cylinder pressure and rate of pressure rise. The peak cylinder pressure in CI engines is higher for JB and JBB as compared with JO because of improvement in the preparation of the air-fuel mixture as a result of lower viscosity (Kumar et al. 2003a). Sivprakasam and Saravanan (2007) while studying the combustion characteristics of a 7.46-kW diesel engine when fueled with JB, JBB, and diesel reported that the peak pressure was maximum at full load for a B20 blend (70 bar) followed by diesel (68 bar). Sahoo and Das (2009b) studied combustion characteristics of JB in a 6-kW diesel engine, and this is presented in Figure 14.6. As compared with diesel, a 7.6% higher peak pressure has been reported for JB at full load of the engine. A similar trend has also been observed during the entire range of engine operation at no-load and half-load conditions. It is clear from Figure 14.6 that while running with biodiesel and their blends, the peak pressure occurred definitely after top dead center (TDC) for safe and efficient operation. Otherwise, a peak pressure occurring very close to TDC or before that causes severe engine knock and thus affects engine durability.

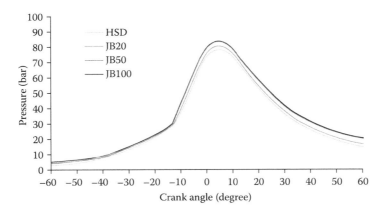

FIGURE 14.6 Pressure versus crank angle for diesel, JB, and JBB at full engine load. (From Sahoo, P.K. and Das, L.M., *Fuel,* 88, 994–999, 2009b.)

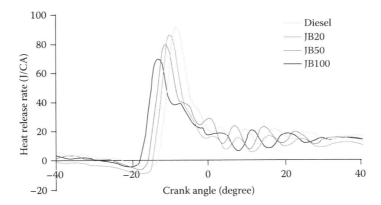

FIGURE 14.7 Heat release rate for diesel, JB, and JBB at full engine load. (From Sahoo, P.K. and Das, L.M., *Fuel,* 88, 994–999, 2009b.)

Pradeep and Sharma (2007) analyzed the combustion parameters of a 3.7-kW diesel engine using JB with and without exhaust gas recirculation (EGR). No significant deterioration in cylinder pressure was observed for JB with 15% EGR and without EGR. The peak cylinder pressure was found to be 52.5, 53.9, and 53 bar for diesel, JB, and JB with 15% EGR, respectively. The rate of pressure rise at full load was found to be 5.8 and 6.2 bar/deg for diesel and JB, respectively, as compared with 5.7 bar/deg for JB with 15% EGR.

14.5.1.2 Heat Release Rate

While studying the combustion characteristics of a 7.46-kW diesel engine when operated with JB, JBB, and diesel, Sivprakasam and Saravanan (2007) reported that heat release rate was maximum for the B20 blend [80 kJ/m³ °crank angle (CA)], followed by diesel (75 kJ/m³ °CA). The heat release rate for JB, JBB, and diesel is shown in Figure 14.7 (Sahoo and Das 2009b). The maximal heat release rate of biodiesel and their blends is lower (69.97 J/°CA for JB) than that for diesel (90.96 J/° CA). This is because the premix combustion phase for biodiesel and their blends is less intense. On the other hand, while running with diesel, increased accumulation of fuel during the relatively longer delay period resulted in a higher rate of heat release. Because of the shorter delay, the maximal heat release rate occurs earlier for biodiesel and their blends as compared with neat diesel. The heat release during the later combustion phase for biodiesel and their blends is marginally lower than that of diesel. This is because the constituents with higher oxygen content are adequate to ensure complete combustion of the fuel that is left over during the main combustion phase, and they continue to burn in the later combustion phase.

The maximal rate of rise in pressure and heat release rate were found to be lower for JB with 15% EGR in comparison to JB without EGR (Pradeep and Sharma 2007).

14.5.1.3 Ignition Delay

The ignition delay has been compared among diesel, JB, and JBB at various loads in Table 14.10. A shorter ignition delay has been found in the case of JB and JBB. The delay is consistently shorter for JB, i.e., 4.2° CA lower than diesel at peak load, and the difference increases with an increase in engine load for a 6-kW diesel engine (Sahoo and Das 2009b). Biodiesel usually includes a small percentage of diglycerides having higher boiling points than diesel. However, the chemical reactions during the injection of biodiesel at high temperature results in the breakdown of the high-molecular-weight esters. These complex chemical reactions lead to the formation of gases of low molecular weight. Rapid gasification of this lighter oil in the fringe of the spray spreads out the jet, and thus volatile combustion compounds ignite earlier and reduce the delay period.

However, while studying the combustion characteristics of JOME, SOME, and HOME in a four-stroke, 5.7-kW, single-cylinder diesel engine, Banapurmath et al. (2008) observed that the injection delay of biodiesels was longer than that of diesel, and it decreased with an increase in the brake power. The values of the ignition delay for JOME, SOME, and HOME were found to be 11.5, 10.5 and 11° CA, respectively, as compared with 9.91° CA with diesel operating at 80% load. Among the methyl esters tested, SOME had a shorter ignition delay when compared with HOME and JOME. The combustion duration for all esters as compared with diesel was found to increase because of a longer diffusion combustion phase. SOME and HOME resulted in an improved heat release rate as compared with JOME, which resulted in better brake thermal efficiency. On the whole, it was observed that operation of the engine was smooth on SOME, HOME, and JOME. While analyzing the combustion parameters of a 3.7-kW diesel engine using JB with and without EGR, the combustion duration for JBB with an optimized value of 15% EGR was found to be increased by 1° than with the no-EGR condition. An overall analysis of combustion parameters indicates comparable heat release rates, cylinder pressures, cumulative heat release, combustion duration, and noise-free operation with and without EGR (Pradeep and Sharma 2007).

Kumar et al. (2003a) used JB, JO, and blends of JO with methanol to study the combustion characteristics of a 3.7-kW single-cylinder diesel engine. The ignition delay was found to be 11° CA with JO. The same trend was observed with the 30% blend of methanol with JO and in dual-fuel operation, i.e., ignition delays of 12° and 13° CA were observed for the JO methanol blend and in dual-fuel operation, respectively. In the dual-fuel mode, ignition delay was increased because of a cooling effect produced by the methanol as it vaporized. Combustion duration increases with a rise in power output due to an increase in the quantity of fuel injected. Higher combustion duration was observed with JO than diesel. The increase in combustion duration is mainly due to the slow combustion of the injected fuel. However, it is reduced with JB and the blend as compared with JO. In the dual-fuel operation due to burning of the inducted methanol by flame propagation, the combustion duration is increased. The premixed burning is more with diesel. Once the autoignition of the fuel commences, the pressure rises, entering into the rapid or premixed combustion phase. The diffusion-burning phase indicated under the second stage is greater for JO. At the time of ignition, less fuel-air mixture is prepared for combustion with JO; therefore, more burning occurs in the diffusion phase rather than the premixed phase.

14.5.2 Performance Characteristics

The performance parameters such as brake specific fuel consumption (BSFC), brake thermal efficiency (BTE), and exhaust gas temperature (EGT) of CI engines when operated with JB, JBB, and JO are summarized in Table 14.11.

TABLE 14.10

Combustion Characteristics of JO, JB, and Their Blends with Diesel in a CI Engine

Investigators	Engine Specifications	Fuel	Injection Timing (°CA) and Speed (rpm)	Effect on Combustion Parameters				
				Ignition Delay (°CA)	Cylinder Peak Pressure (bar)	Combustion Duration (°CA)	Maximal Rate of Pressure Rise (bar/°CA)	Heat Release Rate (J/°CA)
Kumar et al. 2003a	3.7-kW, four-stroke, water-cooled, single-cylinder	Diesel	27 bTDC 1500	8.6	83.2	41	3.9	44
		JB	29 bTDC 1500	10	62.4	44	3.8	38
		JO	29 bTDC 1500	11	61.6	45	3.6	34
Kumar et al. 2003b	3.7-kW, four-stroke kirloskar AVL	Diesel	27 bTDC 1500	9	64	–	–	–
		JO	29 bTDC 1500	10	62	–	–	–
		JO with 7% hydrogen mass	29 bTDC 1500	13	64	–	–	–
Reddy et al. 2006	4.48-kW (6 hp), single-cylinder, air-cooled, open-chamber CI	Diesel	32 bTDC plunger diameter 9 mm, IOP – 220 bar,	–	85	–	9	–
		JO		8.2	70	–	5	62.9
Pradeep and Sharma 2007	3.7-kW, single-cylinder, DI, water-cooled CI	Diesel	27 bTDC 1500	–	52.5	80	5.8	48.4
		JB	28 bTDC, 1500	–	53.9	78	6.2	51.7
		JB + 15% EGR	1500	–	53	79	5.7	47.7
Sivprakasam and Saravanan 2007	7.46-kW (10 hp), two-cylinder, water-cooled, vertical CI	Diesel	–	–	68	–	–	79
		B20	24 bTDC 1500	–	70	–	–	80.2
Banapurmath et al. 2008	5.2-kW, single-cylinder, four-stroke, HS DE	Diesel	23 bTDC 1500	9.9	–	36		90
		JB	27 bTDC 1500	11.5	–	49	–	61

(Continued)

TABLE 14.10 (Continued)
Combustion Characteristics of JO, JB, and Their Blends with Diesel in a CI Engine

Investigators	Engine Specifications	Fuel	Injection Timing (°CA) and Speed (rpm)	Ignition Delay (°CA)	Cylinder Peak Pressure (bar)	Effect on Combustion Parameters			
						Combustion Duration (°CA)	Maximal Rate of Pressure Rise (bar/°CA)	Heat Release Rate (J/°CA)
Sahoo and Das	6-kW, single-cylinder,	Diesel	26 bTDC	8.5	78.7	–	–	90.96
2009b	four-stroke	B20		7.5	80.7	–	–	86.79
		B50	23 bTDC	6.2	83.71	–	–	80.52
		JB	1500	4.3	84.7	–	–	69.97

TABLE 14.11

Performance of Diesel Engine When Fueled with JO, JB, and Their Blends

Investigators	Engine Specifications	Fuel	Findings			
			Power (kW)	BTE (%)	BSFC (kg/kWh)/BSEC	EGT (°C)
Kumar et al. 2003a	3.7-kW, four-stroke, water-cooled, single-cylinder	Diesel	–	30.3	–	402
		JO	3.6	27.4	–	428
		JB		29	–	415
Kumar et al. 2003b	3.7-kW, four-stroke kirloskar AVI	Diesel	–	30.3	–	404
		JO	–	27.3	–	428
		JO with 7% hydrogen mass	–	29.4	–	435
Pramanik 2003	3.7-kW, single-cylinder, open-chamber CI	Diesel	3.078	27.11	0.316	425
		JO and blends	3.078	26.09 for 30:70 J/D and 24.36 for 40:60 J/D	0.338 for 30:70 J/D and 0.365 for 40:60 J/D	435 with 20:80 J/D (Close to diesel)
Reddy et al. 2006	4.48-kW (6 hp), single-cylinder, air-cooled, open-chamber CI	Diesel	–	32.1	–	
		JO	–	28.9	–	400
Pradeep and Sharma 2007	3.7-kW, single-cylinder, DI, water-cooled CI	Diesel + 15% EGR	–	32.4	12	
		JB	–	31	11.6	–
		JB + 15% EGR	–	30.1	11.9	–
Agarwal and Agarwal 2007	7.4-kW, vertical four-stroke, DI, water-cooled Injection pressure 200 bar	Diesel				
		Preheated jatropha oil to 100°C	–	30.71	0.300	
Sivprakasam and Saravanan 2007	7.46-kW (10 hp), two-cylinder, water-cooled, vertical CI	Diesel		28.56	0.300	–
		JB	7.5	25.63	0.320	–
		B20	–	28.20	0.280	–
Tiwari 2007	10.3-kW, four-stroke, single-cylinder, water-cooled	Diesel	–	29.80	0.320	380
		B20	–	28.98	0.351	375
Banapurmath et al. 2008	5.2-kW, single-cylinder, four-stroke	Diesel	–	31.25	–	–
		JB	–	29	–	–

14.5.2.1 BSFC

The BSFC of a CI engine when operated with JB, JBB, and JO decreases with increases in engine load. However, it increases with an increase in JB/JO in the blend. This is mainly due to the combined effects of the relative fuel density, viscosity, and heating value of the blends. Further, this reduction could be attributed to a higher percentage of increase in brake power with load, as compared with fuel consumption, and a lesser percentage of heat loss (Pramanik 2003; Agarwal and Agarwal 2007; Sivprakasam and Saravanam 2007; Tiwari 2007). The difference between the BSFC of JBB (B20) and diesel is almost negligible at higher engine loads. The BSFC for B20 was found to be even less than diesel (Sivprakasam and Saravanan 2007). While testing a 7.47-kW, two-cylinder, four-stroke diesel engine, Sivprakasam and Saravanan (2007) reported the fuel consumption to be 7% lower for a B20 blend and 4% higher for JB (B100) as compared with the diesel fuel. The fuel consumption of other blends of biodiesel had a small variation between these two fuels.

Pramanik (2003) evaluated the performance of a single-cylinder 3.7-kW diesel engine using JO and its blends with diesel under the break load ranging from 0 to 3.078 kW. The specific fuel consumption was found to be close to diesel at 0.338 and 0.365 kg/kWh at a load of 3.078 kW for blends containing 30:70 JO/diesel and 40:60 JO/diesel, respectively. The corresponding value for diesel was found to be 0.316 kg/kWh. For blends of JO and diesel, BSFC was found to be higher compared with diesel for a 7.4-kW diesel engine (Agarwal and Agarwal 2007).

14.5.2.2 BTE

BTE is the ratio of brake power to the fuel energy. It is considered important because it is ultimately the brake power that is of the most concern. The efficiency of any internal combustion engine depends on the physical process involved, such as atomization, evaporation, combustion temperature, etc., which in turn are governed by the type of fuel and operating conditions of the engine. The BTE of a CI engine increases with an increase in engine load for JB, JBB, and JO, and maximal BTE is obtained at maximal brake power. Tiwari (2007) studied the performance of a 10.3-kW, single-cylinder, four-stroke diesel engine when fueled with B20 and diesel. The maximal BTE was found to be 28.98 and 29.80% for B20 and diesel, respectively. The difference in BTE between B20 and diesel is almost negligible. However, with a further increase in biodiesel percentage in the JBB, the BTE decreased. This drop in thermal efficiency is due to poor combustion characteristics because of the higher viscosity and lower volatility of blend fuels.

Kumar et al. (2003a) evaluated the performance of a 3.7-kW diesel engine using single-fuel operation (JO and JO/methanol blends) and dual-fuel operation (induction of methanol while running the engine with JO). The thermal efficiency for JO and JO/methanol blends was found to be slightly reduced as compared with diesel. Maximal BTE was found to be 27.4 and 30.2% with JO and diesel, respectively. Use of a JO methanol blend as compared with neat JO resulted in an increase in BTE from 27.4 to 28.1%. In the dual-fuel operation with methanol induction and JO as the pilot fuel, an increase of peak BTE from 27.4 to 28.7% was observed. While testing a 3.7-kW, single-cylinder, four-stroke diesel engine, Pradeep and Sharma (2007) reported that the BTE with JB was comparable with diesel at all loads with and without EGR.

The BTE of a single-cylinder 3.7-kW diesel engine when operated with JO and its blends with diesel was found to be lower than that of diesel for the entire range of loads (Pramanik 2003). Among the blends tested, thermal efficiency and maximal power output of 30:70 JO/diesel and 40:60 JO/diesel blends were found to be closer to those of diesel.

The thermal efficiency of a 7.4-kW diesel engine was found to be lower for unheated JO compared with heated JO and diesel (Agarwal and Agarwal 2007). However, the BTE was found to be closer to that of diesel for JO and diesel blends (J20).

14.5.2.3 EGT

The EGT gives an indication of the thermal efficiency of the engine. Heat loss is important for engines because it leads to lower working temperature and ultimately to reduced efficiency. The combustion is supposed to be complete at the end of the constant pressure burning in the case of CI engines. However, in actual practice, it continues up to approximately half of the expansion stroke. This "after-burning" may continue up to an exhaust stroke in the case of improper ignition characteristics because of faulty engine settings. In general, approximately 30–33% of the energy supplied to a diesel engine is lost as exhaust heat and another 28–30% is lost in cooling (Thipse 2008). The EGT for a given engine depends upon the fuel-air ratio and the combustion characteristics of the fuel, which in turn depend upon the type of fuel and loading conditions. The EGT of CI engines when operated with JB, JBB, and JO is summarized in Table 14.11 and it increases with an increase in engine load. The minimal and maximal EGT for a 10.3-kW diesel engine when fueled with B20 were found to be 182 and 375°C, respectively, as compared with 166 and 380°C for diesel for no load to full load. For both of the fuels tested, EGT was increased with an increase in engine load; however, at higher loads, it was comparable for both of the fuels (Tiwari 2007). For higher blends, it was higher than that of diesel.

Agarwal and Agarwal (2007) studied the performance of a 7.4-kW diesel engine when fueled with unheated and heated JO and found that EGT for unheated JO was found to be higher as compared with diesel and heated JO. A higher EGT was also reported for the blends of JO and diesel having a high percentage of JO.

14.5.3 Emission Characteristics

Exhaust emissions due to combustion of fuels are becoming increasingly important in view of the present-day awareness of environmental pollution. If the combustion were complete, the exhaust would consist only of CO_2 and water vapors plus air that did not enter into the combustion process. However, for several reasons, oxidation of fuel during combustion remains incomplete and produces carbon monoxide (CO), a deadly poisonous gas, unburned hydrocarbons, and oxides of nitrogen (NO_x) other than particulate matter in the form of smoke. These emissions formed during burning of the air/fuel mixture depend on the conditions during combustion, expansion stroke, and especially before the exhaust valve opening. Emission characteristics of diesel engines operated with diesel, JO, JB, and their blends are summarized in Table 14.12.

14.5.3.1 CO and CO_2

CO is a product of the incomplete combustion due to either inadequate oxygen or insufficient time for completion of the reaction. Generally, the CO values are found to vary between 0 and 2% in the case of diesel engines. It is also said that complete elimination of CO is not possible and 0.5% CO should be considered as a reasonable goal. Although diesel engines always operate with considerable excess air, CO emissions are still considered a significant problem regarding environmental pollution. CO emission levels were found to be higher with JB and JO as compared with diesel at all engine loads of a 3.7-kW diesel engine (Kumar et al. 2003a). However, a reverse trend was reported by Tiwari (2007) while studying exhaust emissions from a single-cylinder 10.3-kW diesel engine when operated with B20, i.e., the minimal and maximal CO produced was 0.38% at no load and 0.55% at full load as compared with 0.57% at no load and 0.74% at full load for diesel. Banapurmath et al. (2008) reported slightly more CO for JB (0.155%) than the diesel (0.1125%) for a four-stroke 5.2-kW CI engine. Tiwari (2007) reported that the amount of CO_2 produced for B20 was comparable with diesel at higher loads, i.e., the maximal CO_2 produced was 7.79 and 7.8% at 100% load for B20 and diesel, respectively.

Kumar et al. (2003b) reported that induction of 7% hydrogen in a jatropha-fueled diesel engine was found to be the optimum for reducing CO from 0.26 to 0.17%. The corresponding reduction in CO with diesel as a pilot fuel was found to be from 0.2 to 0.1%. Agarwal and Agarwal (2007) studied the emissions of a 7.4-kW diesel engine fueled with unheated and heated JO and found that CO and CO_2 were higher for JO compared with those produced with diesel. These emissions were found to be closer to diesel for preheated JO and were increased with an increasing proportion of JO in the blends as compared with diesel.

14.5.3.2 NO_x

Higher combustion temperature and availability of oxygen favors the formation of NO_x emissions. Unlike diesel, biodiesel and its blend with diesel contain more oxygen. This, factored together with higher combustion temperature, favors the production of higher NO_x. The other reason could be the presence of higher quantities of nitrogen in the fuel. The NO_x emission for B20 is 8% higher,

TABLE 14.12

Emission Characteristics of Diesel Engine Fueled with JO, JB, and Their Blends

			Findings				
Investigators	Engine Specifications	Fuel	CO (%)	CO_2 (%)	NO_x (ppm)	HC (ppm)	Smoke Level (BSU or HSU)/ Smoke Opacity (%)
Kumar et al. 2003a	3.7-kW, four-stroke, water-cooled, single-cylinder	Diesel	0.20	–	780	100	3.8 BSU
		JO	0.26	–	740	130	4.4 BSU
		JB	0.22	–	760	110	4 BSU
Kumar et al. 2003b	3.7-kW, four-stroke kirloskar AVI	Diesel	0.2	–	785	100	3.9 BSU
		JO	0.26	–	735	130	4.4 BSU
		JO with 7% hydrogen mass	0.17	–	875	100	3.7 BSU
Reddy et al. 2006	4.48 kW (6 hp), single-cylinder, air-cooled, open-chamber CI	Diesel	–	–	1760	798	2.7 BSU
		JO	–	–	1162	532	2 BSU
Pradeep and Sharma 2007	3.7-kW, single-cylinder, DI, water-cooled CI	Diesel	0.03	–	1255	20	58.8
		JB	0.01	–	1350	10	36.8%
		JB with 15% EGR	0.03	–	780	20	58%
Sivprakasam and Saravanam 2007	7.46-kW (10 hp), two-cylinder, water-cooled, vertical CI	JB	–	–	525	–	14.75 HSU
		B20	–	–	475	–	12.5 HSU
Tiwari 2007	10.3-kW, single-cylinder, four-stroke	Diesel	0.74	7.80	714	–	–
		B20	0.55	7.79	724	–	–
Banapurmath et al. 2008	5.2-kW, single-cylinder, four-stroke	Diesel	0.1125		1080	40.5	53 HSU
		JB	0.115	–	970	67	70 HSU

whereas for JB (B100) it is 7% lower than diesel in a 7.46-kW diesel engine (Sivprakasam and Saravanan 2007). However, 23.66% higher NO_x emissions have been reported for a 10.3-kW, single-cylinder, four-stroke diesel engine when operated with B20 as compared with diesel at no load (Tiwari 2007). The minimal and maximal NO_x produced for B20 were 115 ppm (at no load) and 724 ppm (at 100% load), respectively, as compared with 93 and 714 ppm for diesel, respectively.

Pradeep and Sharma (2007) reported that the use of 15% EGR in a 3.7-kW diesel engine fueled with JB was effective in reducing nitric oxide (NO) emissions as compared with diesel at all loads without many adverse effects on the performance, smoke, and other emissions.

The NO level was lower with JO as compared with diesel for a 3.7-kW, single-cylinder, four-stroke engine (Kumar et al. 2003a). It was further reduced when the engine was operated with a JO and methanol blend. Similar reduced NO emissions were reported by Reddy and Ramesh (2006) for a 4.5-kW, single-cylinder, four-stroke diesel engine when operated with JO under optimal operating conditions.

14.5.3.3 Smoke and Hydrocarbon

The smoke density for JB and JBB is lower than that of diesel, it increases with an increase in engine load, and it is nearly same at the maximal load (Sivprakasam and Saravanan 2007). However,

Banapurmath et al. (2008) reported the hydrocarbon (HC) emissions with SOME, HOME, and JOME in a 5.2-kW, single-cylinder, four-stroke diesel engine to be slightly more than that of the diesel. This is attributed to the incomplete combustion because of their lower volatility and higher viscosity. Pradeep and Sharma (2007) reported that smoke and HC emissions from JB were lower than diesel at peak loads with and without EGR.

Kumar et al. (2003a) found that the smoke emissions for JO increased particularly at higher loads because of poor atomization of the fuel. The smoke level at the maximal power output of 3.7 kW was found to be 4.4 BSU (Bosch's smoke unit) with pure JO as compared with 4 and 3.8 BSU with JB and diesel, respectively. Also, it was observed that an increase in the methanol admission beyond 30% along with JO leads to a further reduction in smoke emissions up to the value of 2.6 BSU at 67% of methanol substitution, which was at the best efficiency point. The HC emissions were found to be 100 ppm with diesel and 130 ppm with neat JO at maximal power output and they increased to 150 ppm in the dual-fuel operation. Pradeep and Sharma (2007) reported that the smoke emissions from JB were found to be lower than diesel at peak loads with and without EGR. Kumar et al. (2003b) reported that induction of 7% hydrogen in JO was found to be the optimum for reducing HCs from 130 to 100 ppm in a diesel engine. The corresponding reduction in HCs with diesel as the pilot fuel was found to be from 100 to 70 ppm. The smoke level also dropped from 4.4 to 3.7 BSU with JO and from 3.9 to 2.7 BSU with diesel. However, a significant rise in NO level from 735 to 875 ppm with JO and from 785 to 894 ppm with diesel was reported because of the higher combustion temperature.

Reddy and Ramesh (2006) reported that the emissions from a 4.5-kW, single-cylinder, four-stroke diesel engine when operated with JO under optimal conditions were 2 BSU whereas with diesel they were 2.7 BSU. However, while evaluating the emissions of a 7.4-kW diesel engine when fueled with unheated and heated JO, Agarwal and Agarwal (2007) found that HCs and smoke opacity were higher for JO compared with diesel and they increased with an increasing proportion of JO in the blends. These emissions were found to be closer to diesel for preheated JO.

14.6 OTHER USES OF JATROPHA BYPRODUCTS

14.6.1 USE OF JATROPHA SHELL/HULL

Crude enzymes can be prepared by decomposing jatropha hull (Sharma et al. 2009). It was reported that inoculation of lignocellulolytic fungi resulted in better compost of jatropha hulls within 1 month. However, phytotoxic compounds present in the compost resulted in the low germination of cress seed (*Lepidium sativum*). When this compost was matured after 4 months, the phytotoxicity was reduced in terms of the germination index ($\approx 80\%$). Therefore, it would be advisable to continue the composting of jatropha hulls for 4 months to reduce the phytotoxicity of compost by the action of enzymes secreted by lignocellulolytic fungi. Because compost has alkaline pH, it can be applied to acidic soil as manure to neutralize soil pH. The potential of lignocellulolytic fungi used in this study to produce higher quantities of cellulolytic enzymes can be tapped in an effective manner by using them for the solid state fermentation or submerged fermentation of jatropha hulls. It was also reported that the resultant crude enzyme preparations can be exploited for fermentation of hulls to produce ethanol that can be used as an additional source of biofuel.

Jatropha seed husk can be used to obtain producer gas in a down-draft gasifier. The laboratory study conducted by Singh et al. (2008) in an open-core down-draft gasifier with jatropha seed husk as feedstock found that the maximal gasification efficiency was 68.31% at a gas flow rate of 5.5 m³/h and a specific gasification rate of 270 kg/h m². The calorific value of producer gas, the concentration of CO in the producer gas, and the gasification efficiency in general increased with the increase in gas flow rate.

14.6.2 USE OF SHELL BRIQUETTES

The very high ash content (14.88%) of the jatropha shell fuses the ash at temperature above 750°C. As the temperature in the oxidation zone of the gasifier reaches 900–1000°C, the jatropha shell becomes unfit for use in a gasifier (Singh et al. 2008). They prepared the briquettes from jatropha shell powder in the improved metal cooking stove (Chullah) having a thermal efficiency of approximately 24%. The time required for the complete combustion of 1-kg briquettes was found to be 35 min, and the temperature in the range of 525–780°C was achieved during the combustion period. Briquettes did not crumble, and their original shape was maintained during combustion.

14.6.3   USE OF OIL CAKE FOR BIOGAS AND MANURE

Jatropha cake can be used as a source of feedstock for biogas production. In a study conducted by Singh et al. (2008), biogas produced from jatropha cake was measured and compared with that generated from cattle dung. The quantity of biogas produced from seed cake was observed to be 60% higher than that produced from cattle dung. It was reported that the total biogas production after the incubation period of 40 days was 348 L/kg total solids (TS) from the reactors having jatropha cake at 10% TS, whereas it was 241 L/kg TS from the reactors having jatropha cake at 15% TS. Methane in biogas was found to be 66 ± 2% in both of the cases.

14.6.4   PRODUCTION OF SILVER NANOPARTICLES

Present green synthesis shows that the environmentally benign and renewable latex of jatropha can be used as an effective capping and reducing agent for the synthesis of silver nanoparticles. Silver nanoparticles synthesized by the above method are quite stable and no visible changes were observed, even after a month or so, when the nanoparticle solutions were kept in light-proof conditions (Bar et al. 2009). Synthesis of metallic nanoparticles using green resources such as jatropha latex is a challenging alternative to chemical synthesis because this novel green synthesis is a pollutant-free and ecofriendly synthetic route for silver nanoparticles.

14.6.5   PRODUCTION OF ACTIVATED CARBON

Activated carbon, also known as porous carbon, has been widely used as an adsorbent in the separation and purification of gas or liquid. In addition, high-porosity carbons have recently been applied in the manufacture of high-performance layer capacitors. Sricharoenchaikul et al. (2008) prepared activated carbon from pyrolyzed jatropha waste char in a laboratory-scale facility. The carbon content of the activated carbon was found to be in the range of 80.4–90.3% depending on the activation method. The activated carbon prepared by chemical activation of the pyrolyzed jatropha residue at 800°C with KOH attained a maximal Brunauer-Emmett-Teller (BET) surface area of 532.3 m^2/g, as compared with the surface area of those activated with phosphoric acid (H_3PO_4) and CO_2. Pores of activated carbon from the jatropha residue were found to be mainly mesopores. The adsorption capacity of the prepared activated carbon confirmed the feasibility of the production of quality activated carbon from plant oil (JO) waste.

14.6.6   WASTEWATER TREATMENT

The jatropha seed coat powder can be used as an adsorbent for removal of Cu(II) from wastewater. Jain et al. (2008) reported that the time required to reach adsorption equilibrium was 80 min, and 82–89% of Cu(II) was removed by the jatropha seed coat at initial Cu(II) concentrations of 20–50 mg/L. The most plausible mechanism of adsorption seemed to be the electrostatic attraction of Cu(II) toward lignocellulosic polar groups of the jatropha seed coat.

14.7   CONCLUSIONS

The interest in using jatropha oil as feedstock for the production of biodiesel is rapidly growing throughout the world. The properties of the jatropha plant and its oil have persuaded investors, policy-makers, and clean development mechanism (CDM) project developers to consider jatropha oil as a substitute for fossil fuels to reduce greenhouse gas emissions and overcome the shortage in fossil fuel supply to meet the ever-increasing demand. However, jatropha is still a wild plant, the basic agronomic properties of which are not thoroughly understood, and the environmental effects have not been thoroughly investigated. In this chapter, an overview of the currently available information on jatropha plantation, its propagation, and the processes to extract the maximal benefit in terms of production of alternative fuel and byproduct utilization, performance, and emissions of CI engines when fueled with JO, JB, and JBB have been given.

However, this state-of-the-art literature review on the production and use of jatropha seeds, oil, and biodiesel detected several knowledge gaps that need to be bridged before larger-scale cultivation can be undertaken. Currently, growers are unable to achieve the optimal economic benefits from the plant, especially for its various uses. The markets of different products from this plant have not been properly explored or quantified. Consequently, the actual or potential growers, including those in the subsistence sector, do not have an adequate information base about the potential and economics of this plant to exploit it commercially. The information provided in this chapter will help to explore the true potential risks and benefits of jatropha plants as well as the production and utilization of JO/JB in CI engines.

REFERENCES

Achten WMJ, Verchot L, Franken YJ, Mathijs E, Singh VP, Aerts R, Muys B (2008) Review: Jatropha biodiesel production and use. *Biomass Bioenergy* 32:1063–1084

Adriaans T (2006) Suitability of solvent extraction for Jatropha curcas, available at www.fact-foundation.com/media_en/FACT_(2006)_Suitability_of_solvent_extraction_for_*Jatropha_curcas* (accessed June 12, 2009)

Agarwal D, Agarwal AK (2007) Performance and emissions characteristics of jatropha oil (preheated and blends) in a direct injection compression ignition engine. *Appl Therm Eng* 27:2314–2323

Akintayo ET (2004) Characteristics and composition of *Parkia biglobbossa* and *Jatropha curcas* oils and cakes. *Bioresour Technol* 92:307–310

Alptekin E, Canakci M (2009) Characterization of the key fuel properties of methyl ester–diesel fuel blends. *Fuel* 88:75–80

Anonymous (2009a) *Cultivation of Jatropha curcas L.* Centre for Jatropha Promotion & Biodiesel, B-132, Sainik Basti, Churu, Rajasthan, India, available at www.biodiesel.org (accessed June 12, 2010)

Anonymous (2009b) *Cultivation of Jatropha Curcas L.* NEDFi Biodiesel Portal. Basundhara Enclave, Ulubari, Guwahati, Assam, India, available at www.biodiesel.nedfi.com (accessed June 12, 2010)

Anonymous (2009c) *Jatropha Production Technology.* Tamilnadu Agricultural Institute, Coimbatoor, India, available at http://www.tnau.ac.in/tech/swc/evjatropha.pdf (accessed June 10, 2010)

Anonymous (2009d) *Jatropha Carcus L.* Hosting and Development Company in Southern California, available at http://www.geocities.com/biodieselindia/jatropha.doc (accessed May 04, 2010)

Augustus GDPS, Jayabalan M, Seiler GJ (2002) Evaluation and bioinduction of energy components of *Jatropha curcas*. *Biomass Bioenergy* 23:161–164

Azam MM, Waris A, Nahar NM (2005) Prospects and potential of fatty acid methyl esters of some non-traditional seed oils for use as biodiesel in India. *Biomass Bioenergy* 29:293–302

Bahadur NP, Boocock DGB, Konar SK (1995) Liquid hydrocarbons from catalytic pyrolysis of sewage sludge lipid and canola oil: evaluation of fuel properties. *Energy Fuels* 9:248–256

Banapurmath NR, Tewari PG, Hosmath RS (2008) Performance and emission characteristics of a decompression ignition engine operated on honge, jatropha and sesame oil methyl esters. *Renew Energy* 33:1982–1988

Bar H, Bhui DK, Sahoo PG, Sarkar P, De SP, Misra A (2009) Green synthesis of silver nanoparticles using latex of *Jatropha Curcas*. *Colloids and Surfaces A* 339:134–139

Beerens P (2007) Screw-pressing of jatropha seeds for fueling purposes in less developed countries. MSc Dissertation, Eindhoven University of Technology, AZ Eindhoven, The Netherlands

Berchmans HJ, Hirata S (2008) Biodiesel production from crude *Jatropha curcas* L. seed oil with a high content of free fatty acids. *Bioresour Technol* 99:1716–1121

Biswas S, Kaushik N, Srikanth G (2007) *A Report on Biodiesel: Technology & Business Opportunities—An Insight.* Technology Information, Forecasting and Assessment Council (TIFAC), available at http://knowledge.cta.int/en/Dossiers/S-T-Issues-in-Perspective/Biofuels/Documents-online/Worldwide/Biodiesel-Technology-Business-Opportunities-An-Insight (accessed June 9, 2009)

Brahmam M (2007) Enhancing germination percentage of *Jatropha curcas* L. by pre-sowing. In *Proceedings of the Fourth International Biofuels Conference*, New Delhi, India, Feb 1–2, pp 175–180

Bryant C, Landsman-Roos N, Naughton R, Olenyik K (2008) *Jatropha Curcus L.: Biodiesel Solution or All Hype? A Scientific, Economic and Political Analysis of the Future Energy Crop.* Energy and Energy Policy, available at http://humanities.uchicago.edu/orgs/institute/bigproblems/Energy/BP-Energy-Jatropha.doc (accessed June 9, 2009)

Canakci M, Gerpen JV (2001) Biodiesel production from oils and fats with high free fatty acids. *Trans ASAE* 44:1429–1436

Caro PS, Mouloungui Z, Vaitilingom G, Berge JC (2001) Interest of combining an additive with diesel-ethanol blends for use in diesel engines. *Fuel* 80:565–574

Chachage B (2003) Jatropha oil as a renewable fuel for road transport. Policy implications for technology transfer in Tanzania. MSc Thesis, International Institute for Industrial Environmental Economics, Sweden

Chikara J, Ghosh A, Patolia JS, Chaudhary DR, Zala A (2007) Productivity of *Jatropha curcus* under different spacing. In *FACT Seminar on Jatropha curcus L. Agronomy and Genetics*, Wageningen, Netherlands, March 26–28, Article no. 9

Chitra P, Venkatachalam P, Sampathrajan A (2005) Optimization of experimental conditions for biodiesel production from alkali-catalysed transesterification of Jatropha curcas oil. *Energy Sustain Dev* 9:13–18

Clark SJ, Wagner L, Schrock MD, Piennaar PG (1984) Methyl and ethyl soybean esters as renewable fuels for diesel engines. *J Amer Oil Chem Soc* 61:1632–1638

Demirbas A (2008) Mathematical relationship derived from biodiesel fuels. *Energy Sources: A* 30:56–69

De Oliveira E, Quirino RL, Suarez PAZ, Prado AGS (2006) Heats of combustion of biofuels obtained by pyrolysis and by transesterification and of biofuel/diesel blends. *Thermochim Acta* 450:87–90

Encinar JM, Gonzalez JF, Rodriquez RA (2005) Biodiesel from used frying oil variables affecting the yields and characteristics of the biodiesel. *Indust Eng Chem Res* 44:5491–5499

Foidl N, Foidl G, Sanchez M, Mittelbach M, Hackel S (1996) *Jatropha curcas* L. as a source for the production of biofuel in Nicaragua. *Bioresour Technol* 58:77–82

Foidl N, Mayorga L (2007) Producción de Ester Metilico del tTempate (EMAT) como Sustituto del Combustible Diesel. Presented at the FACT Seminar on *Jatropha curcas* L. Agronomy and Genetics, Wageningen, The Netherlands, March 26–28, 2007

Francis G, Edinger R, Becker K (2005) A concept for simultaneous wasteland reclamation, fuel production, and socioeconomic development in degraded areas in India: Need, potential and perspectives of Jatropha plantations. *Nat Resour Forum* 29:12–24

Frankel EN (2005) *Lipid Oxidation*, 2nd ed., The Oily Press, Bridgewater, Somerset, United Kingdom

Freedman B, Pryde EH, Mounts TL (1984) Variables affecting the yields of fatty esters from transesterified vegetable oils. *J Amer Oil Chem Soc* 61:1638–1643

Gandhi VM, Cherian KM, Mulky MJ (1995) Toxicological studies on ratanjyot oil. *Food Chem Toxic* 33:39–42

Gerpen JV (2005) Biodiesel processing and production. *Fuel Process Technol* 86: 1097–1107

Ghadge SV, Raheman H (2005) Biodiesel production from mahua (*Madhuca indica*) oil having high free fatty acids. *Biomass Bioenergy* 28:601–605

Ghadge SV, Raheman H (2006) Process optimization for biodiesel production from mahua (*Madhuca indica*) oil using response surface methodology. *Bioresour Technol* 97:379–384

Ghosh L, Singh L, Singh AK (2008) Variation in rooting response in cuttings of *Jatropha curcas* L. *Indian J Agrofor* 10:49–53

Gour VK (2006) Production practices including post-harvest management of *Jatropha curcas*. In: Singh B, Swaminathan R, Ponraj V (eds). *Proceedings of the Biodiesel Conference toward Energy Independence—Focus of Jatropha*. Hyderabad, India, New Delhi, June 9–10, 2006, pp 223–251

Gübitz GM, Mittelbach M, Trabi M (1999) Exploitation of the tropical oil seed plant *Jatropha curcas* L. *Bioresour Technol* 67:73–82

Haas W, Mittelbach M (2000) Detoxification experiments with the seed oil from *Jatropha curcas* L. *Indust Crops Prod* 12:111–118

Hawash S, Kamal N, Zaher F, Kenawi O, Diwani GE (2009) Biodiesel fuel from jatropha oil via non-catalytic supercritical methanol transesterification. *Fuel* 88:579–582

Heller J (1996) *Physic Nut, Jatropha curcas L. Promoting the Conservation and Use of Underutilized and Neglected Crops 1*. International Plant Genetic Resources Institute (IPGRI), Rome, Italy

Henning R (1996) Combating desertification: The jatropha project of Mali, West Africa, available at http://ag.arizona.edu/oals/ALN/aln40/jatropha.html (accessed May 10, 2009)

Henning R (2004) The Jatropha system. Integrated rural development by utilization of *Jatropha curcas* L. as raw material and as renewable energy, Paper presented at Studientag Möglichkeiten und Grenzen erneuerbarer Energien in Tansania-Erfahrungen in der Partnerschaftsarbeit', Hamburg, Germany, available at http://www.tanzanianetwork.de/downloads/referate_hh_energien/reinhard_henning_Jatropha_presentation.pdf (accessed June 9, 2009)

Henning RK (2000) The Jatropha booklet—A guide to the Jatropha system and its dissemination, available at http://www.jatropha.de/documents/jcl-booklet-o-b.pdf (accessed June 5, 2009)

Heywood JB (1988) *Internal Combustion Engines Fundamentals.* McGraw Hill, New York

Hikwa D (1995) *Jatropha curcas L.* Agronomy Research Institute, Department of Research and Specialist Services (DR and SS), Harare, Zimbabwe

Jain N, Joshi HC, Dutta SC, Kumar S, Pathak H (2008) Biosorption of copper from wastewater using jatropha seed coat. *J Sci Indust Res* 67:154–160

Jones N, Miller JH (1992) *Jatropha curcas: A Multipurpose Species for Problematic Sites.* Land Resources Series No 1. Asia Technical Department, World Bank, Washington, DC, Annex 1–6:12

Jongschaap REE, Corré WJ, Bindraban PS, Brandenburg WA (2007) Claims and facts on *Jatropha curcas* L.: Global *Jatropha curcas* evaluation, breeding and propagation programme. Plant Research International BV, Wageningen Report no.158, pp 1–42

Kalbande SR, More GR, Nadre RG (2008) Biodiesel production from non-edible oils of jatropha and karanj for utilization in electrical generator. *Bioenergy Resour* 1:170–178, doi 10.1007/s12155-008-9016-8

Kandpal JB, Madan M (1995) A technical note on *Jatropha curcas*: A renewable source of energy for meeting future energy needs. *Renew Energy* 6:159–160

Katwal RPS, Soni PL (2003) Biofuels: An opportunity for socioeconomic development and cleaner environment. *Indian For* 129:939–949

Kaushik N, Kumar S (2006) *Jatropha curcas L. Silviculture and Uses.* 2nd revised ed., Agrobios, Jodhpur, India

Kinast AJ (2001) *Production of Biodiesel from Multiple Feedstocks and Properties of Biodiesels and Biodiesel-Diesel Blends.* National Renewable Energy Laboratory Report, Des Plaines, IL

Knothe G (2005) Dependence of biodiesel fuel properties on the structure of fatty acid alkyl esters. *Fuel Process Technol* 86:1059–1070

Knothe G, Steidley KR (2005a) Kinematic viscosity of biodiesel fuel components and related compounds. Influence of compound structure and comparison to petrodiesel fuel components. *Fuel* 84:1059–1065

Knothe G, Steidley KR (2005b) Lubricity of components of biodiesel and petrodiesel: The origin of biodiesel lubricity. *Energy Fuel* 19:1192–2000

Knothe G, Steidley KR (2007) Kinematic viscosity of biodiesel components (fatty acid alkyl esters) and related compounds at low temperatures. *Fuel* 86:2560–2567

Knothe G, Matheaus AC, Ryan III TW (2003) Cetane numbers of branched and straight-chain fatty esters determined in an ignition quality tester. *Fuel* 82:971–975

Knothe G, Sharp CA, Ryan III TW (2006) Exhaust emissions of biodiesel, petrodiesel, neat methyl esters, and alkanes in a new technology engine. *Energy Fuels* 20:403–408

Kumar MS, Ramesh A, Nagalingam B (2003a) An experimental comparison of methods to use methanol and Jatropha oil in a compression ignition engine. *Biomass Bioenergy* 25:309–318

Kumar MS, Ramesh A, Nagalingam B (2003b) Use of hydrogen to enhance the performance of a vegetable oil fuelled compression ignition engine. *Int J Hydrogen Energy* 28:1143–1154

Leung DYC, Guo Y (2006) Transesterification of neat and used frying oil: Optimization for biodiesel production. *Fuel Process Technol* 87:883–890

Lu H, Liu Y, Zhou H, Yang Y, Chen M, Liang B (2009) Production of biodiesel from *Jatropha curcas* L. oil. *Comput Chem Eng* 33:1091–1096

Ma F, Hanna MA (1999) Biodiesel production: A review. *Bioresour Technol* 70:1–15

Makkar HPS, Beckerand K, Schmook B (1997) *Edible Provenances of Jatropha Curcas from Quintna Roo State of Mexico and Effect of Roasting on Antinutrient and Toxic Factors in Seeds.* Institute for Animal Production in the Tropics and Subtropics, University of Hohenheim, Stuttgart, Germany, available at http://ec.europa.eu/research/agriculture/pdf/events/edible_provenances_of_jatropha_curcas.pdf (accessed June 9, 2009)

Marchetti JM, Miguel VU, Errazu AF (2007) Possible methods for biodiesel production. *Renew Sustain Energy Rev* 11:1300–1311

Mittelbach M (1996) Diesel fuel derived from vegetable oils. VI: Specifications and quality control of biodiesel. *Bioresour Technol* 56:7–11

Miyashita K, Takagi T (1986) Study on the oxidative rate and pro-oxidant activity of free fatty acids. *J Amer Oil Chem Soc* 63:1380–1384

Modi MK, Reddy JRC, Rao BVSK, Prasad RBN (2007a) Lipase-mediated conversion of vegetable oils into biodiesel using ethyl acetate as acyl acceptor. *Bioresour Technol* 98:1260–1264

Modi MK, Reddy JRC, Rao BVSK, Prasad RBN (2007b) Lipase-mediated transformation of vegetable oils into biodiesel using propan-2-ol as acyl acceptor. *Biotechnol Lett* 28:637–640, doi 10.1007/s10529-006-0027-2

Moser BR (2009) Invited review: Biodiesel production, properties, and feedstocks. *In Vitro Cell Dev Biol—Plant*, doi: 10.1007/s11627-009-9204-z

Munch E, Kiefer J (1989) Purging nut (*Jatropha curcas* L.) Multipurpose use plant as a source of fuel in future. *Schriftenreihe der GTZ* 209:1–32

Openshaw K (2000) A review of *Jatropha curcas*: An oil plant of unfulfilled promise. *Biomass Bioenergy* 19:1–15

Owen K, Coley T (1995) *Automotive Fuels Reference Book*, 2nd ed. Society of Automotive Engineers, Warrendale, PA

Patolia JS, Ghosh A, Chikara J, Chaudhary DR, Parmar DR, Bhuva HM (2007) Response of *Jatropha curcas* grown on wasteland to N and P fertilization. Presented at the *FACT Seminar on Jatropha curcas L. Agronomy and Genetics*, Wageningen, The Netherlands, March 26–28, 2007, article no. 34

Peterson CL, Reece DL, Cruz R, Thompson J (1992) A comparison of ethyl and methyl esters of vegetable oils as diesel fuel substitute. In: *Proceedings of Alternate Energy Conference of the ASAE*, Nashville, TN, pp 99–110

Pradeep V, Sharma RP (2007) Use of HOT EGR for NO_x control in a compression ignition engine fuelled with bio-diesel from Jatropha oil. *Renew Energy* 32:113–154

Pramanik K (2003) Properties and use of *Jatropha curcas* oil and diesel fuel blends in compression ignition engine. *Renew Energy* 28:239–248

Radhakrishna P (2007) Contribution of de-oiled cakes in carbon sequestration and as a source of energy, in Indian agriculture-need for a policy initiative. In: *Proceedings of the 4th International Biofuels Conference*, New Delhi, India, February 1–2, 2007, pp 65–70

Raina AK, Gaikwad BR (1987) Chemobotany of *Jatropha* species in India and further characterisation of *curcas* oil. *J Oil Technol India* 19:81–85

Reddy JN, Ramesh A (2006) Parametric studies for improving the performance of a jatropha oil-fuelled compression ignition engine. *Renew Energy* 31:1994–2016

Romano S (1982) Vegetable oils—A new alternative. Vegetable oil fuels. *Trans ASAE* 4:106–115

Rug M, Sporer F, Wink M, Liu SY, Henning R, Ruppel A (1997) Molluscicidal properties of *J. curcas* against vector snails of the human parasites *Schistosoma mansoni* and *S. japonicum*. In: Gurbitz GM, Mittelbach M, Trabi M (eds), *Biofuels and Industrial Products from Jatropha curcas—Proceedings from the Symposium "Jatropha 97"* Managua, Nicaragua, February 23–27, 1997, Dbv-Verlag, Graz, Austria, pp 227–232

Ryan TW, Dodge LG, Callahan TJ (1984) The effects of vegetable oil properties on injection and combustion in two different diesel engines. *J Amer Oil Chem Soc* 61:1610–1619

Sahoo PK, Das LM (2009a) Process optimization for biodiesel production from jatropha, karanja and polanga oils. *Fuel* 88:1588–1594

Sahoo P K, Das LM (2009b) Combustion analysis of jatropha, karanja and polanga based biodiesel as fuel in a diesel engine. *Fuel* 88:994–999

Sahoo PK, Das LM, Babu MKG, Naik SN (2007) Biodiesel development from high acid value polanga seed oil and performance evaluation in a CI engine. *Fuel* 86:448–454

Saka S, Kusdiana D (2001) Biodiesel fuel from rapeseed oil as prepared in supercritical methanol. *Fuel* 80:225–231

Schwab AW, Bagby MO, Freedman B (1987) Preparation and properties of diesel fuels from vegetable oils. *Fuel* 66:1372–1378

Severino LS, de Lourdes Silva de Lima R, Leao AB, de Macedo Deltrao NE (2007) Root system characteristics of *Jatropha curcas* plants propagated through five methods. Presented at the *FACT Seminar on Jatropha curcas L. Agronomy and Genetics*, Wageningen, The Netherlands, March 26–28, 2007, article no. 31

Shah S, Sharma S, Gupta MN (2004) Biodiesel preparation by lipase-catalyzed transesterification of jatropha oil. *Energy Fuels* 18:154–159

Sharma DK, Pandey AK, Lata (2009) Use of *Jatropha curcas* hull biomass for bioactive compost production. *Biomass Bioenergy* 33:159–162

Shi H, Bao Z (2008) Direct preparation of biodiesel from rapeseed oil leached by two-phase solvent extraction. *Bioresour Technol* 99:9025–9028

Singh L, Bargali SS, Swamy SL (2006) Production practices and post-harvest management. In: Singh B, Swaminathan R, Ponraj V (eds), *Proceedings of the Biodiesel Conference toward Energy Independence—Focus of Jatropha*. Hyderabad, India, June 9–10, 2006, pp 252–267

Singh RN, Vyas DK, Srivastava NSL, Narra M (2008) SPRERI experience on holistic approach to utilize all parts of *Jatropha curcas* fruit for energy. *Renew Energy* 33:1868–1873

Sirisomboon P, Kitchaiya P, Pholpho T, Mahuttanyavanitch W (2007) Physical and mechanical properties of *Jatropha curcas* L. fruits, nuts and kernels. *Biosyst Eng* 97:201–207

Sivaprakasam S, Saravanan CG (2007) Optimization of the trans-esterification process for biodiesel production and use of biodiesel in a compression ignition engine. *Energy Fuels* 21:2998–3003

Sricharoenchaikul V, Chiravoot P, Duangdao A, Atong D (2008) Preparation and characterization of activated carbon from the pyrolysis of physic nut (*Jatropha curcas* L.) waste. *Energy Fuels* 22:31–37

Srivastava A, Prasad R (2000) Triglycerides-based diesel fuels. *Renew Sustain Energy Rev* 4:111–133

Staubmann R, Foidl G, Foidl N, Gurbitz GM, Lafferty RM, Valencia AVM (1997) Production of biogas from *J. curcas* seeds press cake. Gurbitz GM, Mittelbach M, Trabi M (eds), *Biofuels and Industrial Products from Jatropha curcas—Proceedings of the Symposium "Jatropha 97"*. Managua, Nicaragua, February 23–27, 1997, Dbv-Verlag, Graz, Austria, pp 123–131

Tamalampudi S, Talukder MR, Hamad S, Numatab T, Akihiko K, Fukuda H (2008) Enzymatic production of biodiesel from Jatropha oil: A comparative study of immobilized-whole cell and commercial lipases as a biocatalyst. *Biochem Eng J* 39:185–189

Tat ME, Gerpen JHV (1999) The kinematic viscosity of biodiesel and its blends with diesel fuels. *J Amer Oil Chem Soc* 76:1511–1513

Tat ME, Gerpen JHV (2000) The specific gravity of biodiesel and its blends with diesel fuels. *J Amer Oil Chem Soc* 77:115–119

Tate RE, Watts KC, Allen CAW, Wilkie KI (2006) The viscosities of three biodiesel fuels at temperatures up to 300°C. *Fuel* 85:1010–1015

Thipse SS (2008) *A Textbook on Internal Combustion Engines*, 1st ed. Jaico Publishing House, Mumbai, India

Tigere TA, Gatsi TC, Mudita II, Chikuvire TJ, Thamangani S, Mavunganidze Z (2007) Potential of *Jatropha curcas* in improving smallholder farmer's livelihoods in Zimbabwe. An exploratory study of Makosa Ward, Mutoko District, available at http://www.jsd-africa.com/Jsda/Fall2006/PDF/Arc_Potential%20of%20JC%20in%20Improving%20Smallholder%20Farmers's%20livlihoods%20in%20Zimbabwe.pdf (accessed May 10, 2010)

Tiwari AK (2007) Jatropha biodiesel production and its effect on engine wear and lubricant contamination. MTech Thesis, Indian Institute of Technology, Kharagpur, India

Tiwari AK, Kumar A, Raheman H (2007) Biodiesel production from jatropha (*Jatropha curcas*) with high free fatty acids: An optimized process. Biomass Bioenergy 31:569–575

Tyson SK (2001) *Biodiesel Handling and Use Guidelines*. National Renewable Energy Laboratory Report, Golden, CO

Vyas AP, Subrahmanyam N, Patel PA (2009) Production of biodiesel through transesterification of Jatropha oil using KNO_3/Al_2O_3 solid catalyst. *Fuel* 88:625–628

Wang R, Yang S, Yin S, Song B, Bhadury PS, Xue W, Tao S, Jia Z, Liu D, Gao L (2008) Development of solid base catalyst X/Y/MgO/γ-Al$_2$O$_3$ for optimization of preparation of biodiesel from *Jatropha curcas* L. seed oil. *Front Chem Eng China* 2:468–472, doi:10.1007/s11705-008-0074-4

Wang Y, Ou S, Liu P, Xue F, Tang S (2006) Comparison of two different processes to synthesize biodiesel by waste cooking oil. *J Molec Catal A* 252:107–112

Winkler E, Gurbitz GM, Foidl N, Staubmann R, Steiner W (1997) Use of enzymes for oil extraction from *J. curcas* seeds. In: Gurbitz GM, Mittelbach M, Trabi M (eds), *Biofuels and Industrial Products from Jatropha curcas—Proceedings of the Symposium "Jatropha 97,"* Managua, Nicaragua, February 23–27, 1997, Dbv-Verlag, Graz, Austria, pp 184–189

Yu L, Lee L, Hammond EG, Johnson LA, Gerpen JHV (1998) The influence of trace components on the melting point of methyl soyate. *J Amer Oil Chem Soc* 75:1821–1824

15 Forest Trees

Donald L. Rockwood and Matias Kirst
University of Florida

Judson G. Isebrands
Environmental Forestry Consultants

J.Y. Zhu
U.S. Department of Agriculture

CONTENTS

15.1   INTRODUCTION

Interest in renewable biomass for fuel, chemicals, and materials is high (e.g., Rocha et al. 2002), as many products currently derived from petrochemicals can be produced from biomass (Sims et al. 2006). Biomass can be converted into many energy products and chemicals: e.g., alcohol by fermenting cellulose, charcoal, bio-oil, and gases by biomass pyrolysis (Khesghi et al. 2000). Biomass and biofuels technologies with the most potential in the United States include co-firing in coal-fired power plants, integrated gasification combined-cycle units in forestry, and ethanol from hydrolysis of lignocellulosics (Sims 2003). A wide range of products from woody biomass has been demonstrated in New Zealand: "value-added" chemicals, hardboards, activated carbon, animal feed, and bioenergy feedstock (Sims 2003). Using harvested biomass to replace fossil fuels has long-term significance in using forest lands to prevent carbon emissions, and bioenergy projects can contribute to slowing global climate change (Swisher 1997). The potential importance and cost-effectiveness of bioenergy measures in climate change mitigation require evaluation of cost and performance in increasing terrestrial carbon storage.

Biomass-based heat, electricity, and liquid fuels are about 14% of the world's primary energy supply (IEA 1998), with about 25% of that in developed countries and 75% in developing countries (Parikka 2004). From an estimated 3.87 billion ha of forest worldwide, global production and use of wood fuel was about 1.753 billion m^3 in 1999 (Table 15.1), 90% of which was produced and consumed in developing countries.

The total sustainable worldwide annual bioenergy potential is about 100 EJ (Table 15.1), about 30% of total current global energy consumption (Parikka 2004). Annual woody biomass potential is 41.6 EJ, or 12.5% of total global energy consumption. Worldwide, less than 40% of the existing bioenergy potential is used. In all regions except Asia, current biomass use is less than the available

TABLE 15.1

Plantation and Total Forest Areas (10^6 ha), Total Woody Biomass (10^9 t), Wood Fuel Produced Annually (10^6 m^3), and Bioenergy Potential Per Year (10^{18} EJ) from Wood, All Biomass, and Currently Used (%) by World Region

Region	Forest Area		Woody Biomass	Wood Fuel	Bioenergy Potential		
	Plantation	Total			Wood	All	Used
North America	2	549	61	133	12.8	19.9	16
South America	10	885	179	168	5.9	21.5	12
Asia	115	547	44	883	7.7	21.4	108
Africa	8	649	70	463	5.4	21.4	39
Europe	32	1039	61	95	4.0	8.9	22
World	171	3869	421	1753	41.6	103.8	38

Source: Parikka, M., *Biomass and Bioenergy*, 27, 613–620, 2004.

potential, especially in North America and South America. Therefore, increased woody biomass use is possible in most countries.

Although most wood currently used for bioenergy comes from natural forests, the greatest potential for increasing bioenergy production is in establishing plantations, which currently are only 5% of the world's forest area (Table 15.1). Forest plantations have dramatically increased for the last 25 years, mainly in Asia. The area of productive forest plantations increased by 2 million ha/year from 1990 to 2000 and by 2.5 million ha/year from 2000 to 2005. *Eucalyptus*, *Pinus*, and *Populus* species are particularly suited for short rotation woody crop (SRWC) plantings, although many other forest species may also be used. Accordingly, we review current activity with, and the potential of, the main species in these three genera, with additional focus on the role and potential of biotechnology to increase *Eucalyptus* biomass production and enhance conversion to bioenergy products.

15.2   *EUCALYPTUS*

Many of over 700 *Eucalyptus* species, native primarily to Australia, have potential as biomass crops. Eucalypts are successful exotics because of their fast growth and environmental tolerance because of attributes such as indeterminate growth, coppicing, lignotubers, drought/fire/insect resistance, and/or tolerance of soil acidity and low fertility. Many eucalypts have desirable wood properties, such as high density, for bioenergy production.

With 19.6 million ha of plantations in some 90 countries (Trabedo and Wilshermann 2008; Figure 15.1), *Eucalyptus* is the most valuable and widely planted hardwood worldwide. Eucalypts

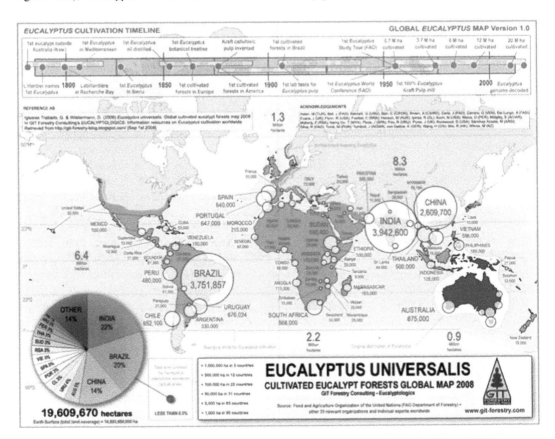

FIGURE 15.1 (See color insert)   Worldwide *Eucalyptus* planting. (From Trabedo, G.I. and Wilshermann, D., *Eucalyptus universalis*. Global Cultivated Eucalypt Forests Map, 2008, available at www.git.forestry.com)

are grown in SRWCs in tropical and subtropical Africa, South America, Asia, and Australia, and in temperate Europe, South America, North America, and Australia (Figure 15.1), but 12 countries account for nearly all of the almost 12 million ha classified as productive (FAO 2006; Table 15.1, Figure 15.1). Eucalypt planting is still intensifying, especially in tropical countries.

Four species and their hybrids from the subgenus *Symphyomyrtus* (*Eucalyptus grandis* (*EG*), *E. urophylla* (*EU*), *E. camaldulensis,* and *E. globulus*) constitute about 80% of eucalypt SRWCs worldwide. *EG, EU,* and *EG × EU* hybrids are favored in tropical and subtropical regions. *EG* has been planted extensively in countries such as India and South Africa, and it is grown in California, Hawaii, and Florida in the United States. *E. globulus* is best for temperate countries such as Portugal, Spain, Chile, and Australia. Eucalypt products worldwide include pulp for high quality paper (Tournier et al. 2003), poles, lumber, plywood, veneer, flooring, landscape mulch (Aaction Mulch 2007), fiberboard (Krysik et al. 2001), composites (Coutts 2005), essential oils (Barton 2007), firewood, and charcoal.

15.2.1 *Eucalyptus* Bioenergy Use and Potential Worldwide

Although eucalypts are major bioenergy contributors worldwide, their utilization by individual countries reflects a number of factors. Renewable energy incentives, greenhouse gas (GHG) emission targets, synergism with industrial waste management projects, and high oil prices are major drivers of SRWCs for bioenergy.

Bioenergy consumption is greatest in countries with heavy subsidies or tax incentives, such as Brazil and China (Wright 2006). Total bioenergy consumption in China, EU, the United States, Brazil, Canada, and Australia is 17.1 EJ. SRWCs in Brazil, New Zealand, and Australia total about 5 million ha, and SRWCs in China may be as large as 10 million ha, whereas SRWCs and other energy crops in the United States and EU are less than 100,000 ha. SRWCs have mainly been established for other than bioenergy production. Australia, Brazil, Canada, New Zealand, Sweden, the United Kingdom, and the United States are among the countries conducting SRWC research and development.

15.2.1.1 Brazil

Brazil is the world's largest bioenergy producer and consumer, with 29.6% of its energy coming from biomass in 2003, including 13.0% from firewood and charcoal (Pelaez-Samaneigo et al. 2008). With 3 million ha of *Eucalyptus* SRWCs (Rosillo-Calle 2004), it may have the most SRWCs grown specifically for energy. *Eucalyptus* planting occurred primarily from 1966 to 1989 under government incentives (Couto et al. 2004). The pulp and paper industry has more than 1.4 million ha of mainly *Eucalyptus* and pine plantations.

Wood production, especially SRWC *Eucalyptus*, is well established in Brazil (Couto et al. 2002). Brazil has about 5 million ha of sustainable forests (afforestation and reforestation), with 56% of these forests in the Southeast, where *Eucalyptus* is 64% of the area (BMME 2003). SRWC establishment targets from 2004 to 2007 included 0.8 and 1.2 million ha through small/medium farmers and through medium/large companies, respectively.

SRWC utilization is 33% for charcoal production, 13% for industrial energy, 31% for pulp and paper mills, and 23% for timber (BMME 2003). Brazil produced 7.3 million tons (225 PJ) of charcoal in 2003, of which the steel industry consumed 85% (Walter et al. 2006). SRWC *Eucalyptus* accounted for 87%, 97%, and 74% in 1997, 1999, and 2003, respectively, of the charcoal consumed by steel mills (ABRACAVE 2003).

Brazil's forests produce about 55 million tons of firewood (Walter et al. 2006). Of this, over 40% is for charcoal (almost 580 PJ), 29% is for residential use, and 23% is for industrial and commercial uses such as steam generation. Charcoal production was 22% of industrial roundwood consumption from SRWCs in Brazil in 2006 (Pelaez-Samaniego et al. 2008). Wood chip production for electricity and heating in Europe in 2004 was expected to be 250,000 green mt, or 3 PJ, using acacia and bark of *Eucalyptus* and *Pinus*.

SRWCs are benefitting from decades of research and development of forest management practices including genetic improvement, spacing, fertilization, planting techniques, control of pests and diseases, and coppicing. Mean yields of *Eucalyptus* SRWCs went from 14 m^3 ha per year during the 1970s to the current approximately 40 m^3/ha per year through breeding and silvicultural practices. Up to three harvests are possible from one planting of *Eucalyptus*. In comparison to its own average productivity of 450 GJ/ha per year, one Brazilian company's highest *Eucalyptus* productivity is 1000 GJ/ha per year, whereas in the United States, commercial forests average less than 100 GJ/ha per year and switchgrass may achieve 430 GJ/ha per year. Stumpage prices for *Eucalyptus* in Brazil were just 0.5–0.6 U.S.$/GJ (ABRACAVE 2003).

Vegetative propagation of superior clones has enhanced *Eucalyptus* productivity. Starting in the 1960s, rooted cuttings provided local site adaptation with cost and wood property advantages. Optimized and efficient transformation and recovery procedures exist for some *Eucalyptus* genotypes. Various transgenic *Eucalyptus* plantlets have been regenerated from stem or leaf segments. Micropropagation and transformation have combined with efforts to engineer novel or alter existing traits. Elite hybrid clones with superior wood quality, rapid growth, and disease resistance are also used extensively in tropical and subtropical regions of South Africa, Congo, and China.

Under Brazil's Code of Best Practices for Planted Forests, several desirable SRWC practices are encouraged (Piketty et al. 2008). No-tillage or reduced tillage is the usual method of planting. Reduced input of chemicals is recommended. For SRWC licensing, general BMPs must be followed. Small farmers have incentives for reforesting marginal lands. In an "integrated system," industry provides clones (seedlings) and all other agricultural materials; farmers provide land and labor.

A transition is taking place from "conventional" biomass (e.g., firewood used for cooking) to "modern" biomass (industrial heat and electricity and biofuels). In Brazil's advanced programs for bioenergy (Lora and Andrade 2008), biomass gasification is widely applied and encouraged. At 11.3%, "modern" biomass in the Brazilian energy matrix is practically the same as traditional biomass (12.5%). On a global scale, modern biomass is only 1.73% of the whole energy consumption.

In Brazil, bioenergy is implemented at three levels: (1) low, 1–25 kWs (small communities); (2) up to 5 MWs (small communities, sawmills, furniture factories, and rice treatment plants), and (3) over 5–10 MW (sugar and alcohol plants, pulp and paper mills, and biomass thermal power plants).

Brazil ranked first in a global assessment of land potentially available to supply charcoal to the steel industry (Piketty et al. 2008). Increasing charcoal making efficiency from 330 to 450 kg/tons of wood (+36%), using 100% SRWCs by 2010, cleaner carbonization techniques, and by-products such as liquid fuels and chemicals through gases and liquids recovery will insure the continued use of charcoal (Rosillo-Calle and Bezzon 2000). A simple upgrade of traditional charcoal production can significantly increase liquid fuel output. Slow pyrolysis bio-oil can be an excellent, cost-effective and renewable liquid fuel (Stamatov and Rocha 2007). A biochar-refinery for production of charcoal, activated carbon, liquid fuel and variety of chemicals presents a possible approach for the development of biomass-based industry. Under the current levels of best practices in *Eucalyptus* SRWCs, carbonization and charcoal use, 50 million tons of steel would require 6.5 million ha of SRWCs.

When new plantations and reforestation resulting mainly from the National Programs are ready to be harvested, Brazil will have forest biomass for export. A least three big energy and pulp companies have their own ports.

Estimates of the total installed potential for electricity generation from biomass range from 2680 to 4740 MW. The total installed power is 96.63 GW of which 4.74 GW (4.9%) is from biomass. Several cogeneration plants have been built in the sugar/alcohol sector. The pulp and paper industry had potential in 2003 for 1740 MW and had excellent prospects for electric self-sustainability through cogeneration using renewable energy (wood residue, bark and black liquor). Six thermal power plants using biomass totaling 80.35 MW are being built, and 42 other units with a total power of 673.6 MW have been authorized. About 5% of SRWC *Eucalyptus* could be used to generate electricity, increasing potential output from 4000 to 8000 MW.

Some 85 million ha may be suitable for natural regeneration, farm forestry and SRWCs, but only 20 million ha could be planted by 2030 (Piketty et al. 2008). Other estimates of potential SRWC extent included 30 and 41.2 million ha. Brazilian states with available grasslands and savannas for SRWC planting include Mato Grosso, Maranhao, Tocantins, Piaui, Goias, Para, Minas Gerais, Mato Grosso do Sul, and Bahia, with Piaui preeminent for sustainable *Eucalyptus* SRWCs involving small and medium landowners.

Brazil has very favorable land, climate, policies, and experience in highly productive SWRCs for the pulp and paper industry and increasingly for bioenergy. Given the prominence of current and future *Eucalyptus* SRWCs (Table 15.1) in areas with large energy needs and production potential, *Eucalyptus* can be a significant contributor of a range of energy products. *Eucalyptus* wood is even used for making activated carbon adsorbents for liquid-phase applications such as water and waste-water treatment. In the tropics and subtropics where *Eucalyptus* species are already widely planted, their further deployment for energy products is especially likely.

15.2.1.2 China

In 2002, China's 7.5 EJ of bioenergy consumption (16.5% of total energy), including approximately 200 million tons of firewood, more than doubled that of any other country and is increasing (Wright 2006). In 2000, China led the world in afforestation. Plantations totaled 4.67 million ha in 2002, though only a portion was likely SRWCs for energy. In 2006, SRWCs were up to 10 million ha, mostly in southern China. Goals for 2010 and 2015 were some 13.5 million ha of "fast growing plantations." China's 24 million ha of new plantations and natural regrowth transformed a century of net carbon emissions by forestry to net gains of 0.19 Pg C per year, offsetting 21% of its fossil fuel emissions in 2000 (Canadell and Raupach 2008).

Commercial biomass energy is only about 14% of the total energy consumed in China (Wright 2006). China has very abundant but inefficiently used bioenergy resources. Bioenergy is mainly used in rural areas, accounting for 70% of rural energy consumption. Started in 1981, firewood plantations totaled 4.95 million ha by 1995 and provided 20–25 million tons of firewood each year that largely alleviated the rural energy crisis. Now, 210 million m^3 of firewood are produced annually, equivalent to 120 million tons of standard coal.

China's forest area of 175 million ha in 2003 contained a forest volume of approximately 12 billion m^3. Plantations were approximately 53 million ha with a volume of 1.5 billion m^3. Still, China faced a deficiency of industrial timber, as industrial roundwood consumption in 2004 was 310 million m^3 and was expected to reach 472 million m^3 in 2020. Fast-growing, high-yielding trees for industrial plantations are needed to ease the deficiency.

Eucalyptus species are very important for industrial plantations in China (Xiong 2007). China now has the second largest *Eucalyptus* plantation in the world, just behind Brazil. First introduced into China in 1890 as ornamental and landscape trees, 70 species have been successful, and over 1.8 million ha of *Eucalyptus* are established in the Southeast. More than 20 years of selection and breeding in collaboration with Australian researchers resulted in 135 seed sources and more than 600 clones for Guangxi province. Through tests, approximately 2000 elite trees in 500 families were selected for the first breeding populations of 10 species including *EU, E. tereticornis, EG, EU × EG, E. pellita, E. camaldulensis,* and *E. dunnii.* Genetic gains of 15–25% were realized after the first generation. By crossing 12 species, good hybrids were selected, with the best families and clones growing substantially better. Cold hardy crosses such as *EU × E. camaldulensis* and *E. tereticornis* have also been created. Concurrent with a broad research program, seed orchards and clonal gardens were created for selected species. The biggest propagation center is in Guangxi, which provides superior seed or vegetative propagules for southern *Eucalyptus* plantations.

In China, eucalypt plantations are harvested for pulpwood, fiberboard, sawlogs, roundwood, veneer, fuelwood and oil. Residues and leaf litter are used for fuelwood. Eucalypt oil production is primarily confined to cooler, temperate regions, where up to 150,000 ha of plantations are used for oil.

Although China has considerable bioenergy potential, many factors may constrain its effective use. To overcome these problems, China has formed a bioenergy development strategy. In China's strategic framework for energy development, bioenergy would become the major component in a sustainable energy system that may account for 40% or more of total energy consumption by 2050 (Wright 2006).

15.2.1.3   India

Bioenergy accounts for nearly 25% of total primary energy in India and over 70% of rural energy (Ravindranath et al. 2007). Consumption of fuelwood, the dominant biomass fuel, is 162–298 million tons. Only about 8% of the fuelwood is considered sustainable, and private plantations and trees account for almost 50%.

Although India's current bioenergy use is relatively inefficient, modern bioenergy technologies are opportunities for meeting energy needs, improving the quality of life, and protecting the environment. These technologies include biomass combustion and gasification for power generation and liquid (biodiesel and ethanol) and gaseous (biogas) fuels. The dominant bioenergy option is power generation using woody biomass, largely from SRWCs.

India has a large renewable energy promotion program with several financial and policy incentives. Total installed biomass combustion and gasification capacity is 738 MW, with a potential of over 20,000 MW. However, technical, financial and policy barriers limit the large-scale production of biomass for power generation.

Critical to realizing the technical and economic potential of bioenergy is a sustainable biomass supply, including woody biomass. Forests occupy approximately 64 million ha of India, and forest cover, including SRWCs (Table 15.2), home gardens, and agroforestry, is 77 million ha (23.5% of the country). India has implemented one of the largest afforestation programs in the world, using largely *Eucalyptus*, *Acacia*, *Casuarina*, *Populus*, and teak to meet fuelwood, industrial wood, and timber needs. From 1980 to 2005, approximately 34 million ha were afforested at an annual rate of 1.32–1.55 million ha, whereas deforestation was 0.272 million ha per year (Ravindranath et al. 2001). Consequently, the 10 GtC of carbon in Indian forests has nearly stabilized or is increasing.

TABLE 15.2
Area of Productive Eucalypt Plantations and Semi-Natural Forests (*) in 2005 for Selected Countries, Species, and Age Classes

Country	Species	Area (1000 ha) by Age Class (years)					
		0–5	5–10	10–20	20–30	30–40	>40
RSA	nitens	109.7	99.3	19.4	0.7	1.8	
	grandis	144.1	140.7	44.9	3.7	1.7	
Sudan	spp.	118.2	189.1	165.5	8.0		
China	spp.	683.0	576.4	982.7	154.4		
India	spp.	43.0	64.4	103.2			
	spp.*	656.1	984.2	1576.0			
Vietnam	spp.	222.4	286.5	67.1	7.0	3.0	
Australia	globulus	131.2	260.1	48.7	1.1	0.4	
Brazil	spp.	2118.1	756.5	121.0	30.3		
Chile	spp.	353.4	204.1	85.4	7.2	2.0	

Source: FAO, Global planted forests thematic study: results and analysis. Planted forests and trees. FAO Working Paper FP38E, 2006. Available at http://www.fao.org/forestry/webview/-media?mediaId=12139&langId=1

A sustainable forestry scenario aimed at meeting the projected biomass demands, halting deforestation, and regenerating degraded forests was developed (Ravindranath et al. 2001). Excluding the land required for traditional fuelwood, industrial wood, and timber production, and considering only potential land categories suitable for plantation forestry, 41–55 million ha is available for SRWCs. About 12 million ha is adequate to meet the incremental fuelwood, industrial wood, and timber requirements projected to 2015. An additional 24 million ha is available for SRWCs with sustainable biomass potential of 158–288 million t annually. Other estimates vary from 41–130 million ha. Marginal cropland and long-term fallow lands are also available for SRWCs. Assuming a conservative 35 million ha of SRWCs producing 6.6–12 tons/ha per year, the additional woody biomass production potential is estimated to be 230–410 million t annually, and the corresponding total annual power generation potential, at 1 MWh/tons of woody biomass, would be 228–415 TWh (Sudha et al. 2003), equivalent to 36.5–66.5% of total power generated during 2006 (623.3 TWh) in India. Because the photosynthetic efficiency of forest trees, rarely above 0.5%, could be genetically and silviculturally improved to 6.6% (Hooda and Rawat 2006), there is an ample opportunity to improve efficiency up to 1%, thus doubling forest productivity.

Thus, biomass for power generation has a large potential to meet India's power needs sustainably. The life-cycle cost of bioenergy is economically attractive compared to large coal-based power generation (Ravindranath et al. 2006). Clean biomass, coupled with other changes, could reduce GHG emissions by approximately 640 MT CO_2 per year in 2025 (~18 % of India's emissions). Sustainable forestry could increase carbon stock by 237 million Mg C by 2012, whereas commercial forestry could sequester another 78 million Mg Carbon. The sustainable biomass potential of 62–310 million tons per year could provide about 114% of the total electricity generation in 2000 (Bhattacharya et al. 2003).

Sustainable bioenergy potential has also been estimated for Malaysia, Philippines, Sri Lanka, and Thailand (Bhattacharya et al. 2003). Sustainable SRWC production could be 0.4–1.7, 3.7–20.4, 2.0–9.9, and 11.6–106.6 million tons per year, respectively. Using advanced technologies, maximum annual electricity generation may be about 4.5, 79, 254, and 195% of total generation in 2000, respectively. Biomass production cost varies from U.S.$381 to 1842/ha and from U.S.$5.1 to 23 tons.

15.2.1.4 Australia

Australia has abundant energy resources, a dispersed population highly dependent on fossil fuel based transport, and relatively fast population growth (Baker et al. 1999). Its total energy consumption has recently increased by 2.6% per year and was estimated to be 4810 PJ in 1997/98. Australia's per capita GHG emissions ranked third among industrialized countries and were projected to be approximately 552 million Mg CO_2-e in 2010, a 43% increase from 1990. In 1995/96, energy production from renewable sources (~263 PJ) was mainly from bagasse (90.3 PJ), residential wood (82.1 PJ), hydroelectric (54.8 PJ), and industrial wood (27.6 PJ).

Its biomass sources including bagasse, paper pulp liquor, forestry and wood processing residues, energy crops, crop residues, and agricultural and food processing wastes may contribute significantly to a 2% renewables target (Baker et al. 1999). Sugarcane, one of the least costly forms of biomass, could alone meet the entire 2% target. Gasification of cotton-gin residues could generate 50 MW of electricity. Rice hulls could support approximately 5 MW of electricity.

The use of SRWCs for bioenergy in Australia is being explored. Interest in the establishment of forest plantations on irrigated and nonirrigated sites in the southern Australia has increased (e.g., Bren et al. 1993, Baker et al. 1994, Stackpole et al. 1995, Baker et al. 1999). Municipal wastewater systems are opportunities for bioenergy production, as land, irrigation infrastructure, and water costs would be part of the total water treatment costs, not part of the costs of biofuel production.

The potential for wastewater-irrigated plantations for bioenergy production is evident in Victoria (Baker et al. 1999). Assuming growth rates of 30 Mg/ha per year, approximately 400,000 Mg per year could generate 50 MW. *EG* at Wodonga (irrigated with municipal effluent) and Kyabram (first irrigated with freshwater and then groundwater) had growth ranging from 15 to 45 m^3/ha per year at 10 years (Baker 1998) when irrigation water salinity and initial soil salinity were not problematic.

At Wodonga, species growth to 4 years was *E. saligna* (134 m³/ha stem volume, 84 Mg/ha aboveground biomass), *EG* (126, 80), *Populus deltoides × nigra* (85, 47), *Pinus radiata* (61, 42), *Casuarina cunninghamiana* (43, 49), and *E. camaldulensis* (40, 52) (Stewart et al. 1988). At Kyabram, species ranking at 4 years was *EG, E. saligna, E. globulus*, and *E. camaldulensis* (Baker 1998).

At Werribee with *E. globulus, EG, E. saligna*, and *E. camaldulensis* and densities of 1333, 2500, and 4444 trees per ha, stem volume growth to 4 years across planting densities varied: *E. globulus* (57–108 m³/ha), *EG* (46–89), and *E. saligna* (37–77) (Delbridge et al. 1998). Tripling planting density approximately doubled potential yield.

At Bolivar (Boardman et al. 1996, Shaw et al. 1996), initial tree growth rates were relatively high, with species ranking to 4 years of *E. globulus* (139–182 m³/ha overbark stem volume), *C. glauca* (85–111), *E. occidentalis* (90–92), *EG* (75–91), and *E. camaldulensis* (59–71).

A Wagga Wagga project developed national guidelines for sustainable management of water, nutrients, and salt in effluent-irrigated plantations. The project studied biophysical processes under three effluent irrigation rates: medium (M, rate of water use less rainfall), high (H, about 1.5 times M), and low (L, about 0.5 times M). The main species studied were *EG* and *P. radiata*. Tree spacing was 3×2 m. Among several conclusions, the project identified best eucalypt species and radiata pine clones for effluent-irrigated plantations for sustainable biomass production, salt sensitivity in tree species, and water use efficiency (Myers et al. 1996). After 6 years, halving irrigation (M vs. L) led to only a 10% decrease in volume production—20 to 18 m³/ha per year.

At Shepparton (Baker et al. 1994), both *E. globulus* and *EG* grew well through 5 years, with *E. globulus* larger than *EG* (172 vs. 128 m³/ha) (Duncan et al. 1999). Stem volume MAIs were 30 and 38 m³/ha for *E. globulus* and 23 and 31 m³/ha for *EG* for 1333 and 2667 trees/ha, respectively. Coppice growth at 2 years varied between 31 and 42 Mg/ha and was greater than that of the planted seedlings. Biomass production for a 12-year cycle of 3-, 6-, and 12-year rotations was projected to be 330, 390, and 350 Mg/ha, respectively, for *E. globulus* growing at an estimated peak of 45 m³/ha per year.

In 1998, Australia had approximately 1.25 million ha of plantations, 23% hardwoods (*E. globulus, E. nitens, E. regnans*, and *EG*) and 77% softwoods (*Pinus radiata, P. elliottii, P. caribaea, P. pinaster*, and *Araucaria cunninghammii*) (Baker et al. 1999). Although softwood plantations managed in 20–40 year rotations for veneer logs, sawlogs, posts, poles, and pulpwood have bioenergy potential from silvicultural and product residues, the rapidly developing 10–15 year rotation hardwood pulpwood industry also has potential. Australia has a plantation goal of approximately 3 million ha by 2020.

New hardwood plantations in Australia totaled 49,000 ha per year in 1998 and were expected to increase (Baker et al. 1999). These plantations have been mostly for pulpwood on agricultural land in southwestern Western Australia, southeastern South Australia, Victoria, Tasmania, and north-coastal New South Wales. Assuming a growth rate of 20 m³/ha per year and ultimately 500,000 ha, potentially 500,000 Mg per year of wood residue will be available for bioenergy production.

In southern Australia, SRWCs with 3–5 year rotations may be used for bioenergy after alleviating salinization of land and water in dryland (300–600mm annual rainfall) farming systems (Sochacki et al. 2007). Planting density and slope position had strong influences on biomass yield. Mean 3-year yields of *E. globulus, E. occidentalis*, and *P. radiata* planted at 500, 1000, 2000, and 4000 trees/ha at upper-, mid-, and lower-slope positions ranged from 12 to 14 tons/ha. Biomass yield consistently increased with planting density, generally greatest at 4000 trees/ha. The best *E. globulus* and *E. occidentalis* yields, 16.6 and 22.2 tons/ha, respectively, were at lower slope positions at 4000 trees/ha, whereas the best yield of *P. radiata* was 15.4 tons/ha in an upper slope position. *E. globulus* did not perform well on the upper-slope site. *E. occidentalis* has some salt tolerance. *P. radiata* yields were relatively low in the lower landscape being relatively small. Using high planting densities and different species for different hydrological settings can optimize biomass productivity. Higher yields are expected under more normal rainfall conditions.

Root:shoot (R:S) ratios that did not vary significantly between planting density and slope position but varied between species (*E. occidentalis*—0.51, *E. globulus*—0.31, *P. radiata*—0.33) have implications for harvesting systems, in terms of biomass yield and stump removal (Sochacki et al.

2007). R:S ratios of 2.5-year-old *EG* planted at 100–2000 trees/ha decreased with increasing planting density (Eastham and Rose 1990). Roots may be unsuitable for bioenergy because of soil contamination. If roots are not utilized and need to be removed, species with high R:S ratios, e.g., *E. occidentalis*, may not be desirable.

The whole tree could be harvested and roots removed (Harper et al. 2000). Alternative harvesting scenarios include separation and retention of leaves to maintain site fertility (Sochacki et al. 2007). In the case of root retention in the ground for pasture rather than cereal cropping, *P. radiata* would be preferred as root decay would allow cereal cropping sooner. Methods to harvest tree crops with roots may be affected by tree size and their respective root systems. Trees grown at 500 trees/ ha may be difficult to harvest with typical harvesters (Mitchell et al. 1999).

With high stocking densities and optimal slope position, 3-year biomass yields of 15–22 tons/ha were possible, dependent on species. When averaged across the landscape, yields were more modest and ranged from 12 to 14 tons/ha. These were achieved in lower than normal rainfall conditions; with normal conditions and higher planting densities, higher yields may be possible. To maximize biomass production and water use, planting density, water availability, and species must be matched to site. There are significant opportunities to expand forest bioenergy in Australia through distributed electricity generation and production of ethanol and bio-oil. Utilizing the large amounts of readily available forest residues would generate greenhouse benefits, assist forest regeneration, and improve forest management. New forests in low rainfall environments would also provide residues for energy production, thus enhancing their overall viability. A recent mandate that electricity retailers increase renewable energy production by 9.5 TWh annually has created a small, relatively high value ($10–12/delivered green t) biomass market (Wright 2006).

Producing biochar from farm or forestry waste could have many benefits: generation of renewable electricity, liquid and gas biofuels, activated carbon, *eucalyptus* oil, heat or low-pressure steam, and a net sequestration of CO_2 (McHenry 2009). With new policies and initiatives, the profitability of these various products is likely to improve, especially if integrated into existing agricultural production and energy systems.

Higher rates of soil sequestration and lower uncertainties in carbon asset verification, coupled with lower risks of storing carbon in soils, make integrating biochar applications and agricultural SOC into carbon markets appealing. Carbon markets that include agricultural SOC will enable farmers to trade sequestered biochar soil applications and facilitate expanding new technologies that improve farm productivity and energy security.

Growing SRWCs on surplus and degraded agricultural land could be environmentally beneficial (Bartle et al. 2007). Dryland salinity in southern Australia could be ameliorated using SRWCs. At A$35/green ton and a water use efficiency of 1.8 dry g/kg of water, SRWCs could produce 39 million per year of dry biomass from 1.5% and 8% of farmland in the 300–400 and 400–600 mm rainfall zones, respectively, of the southern Australian wheatbelt.

The relatively low cost of fossil fuels in Australia generally limits the development of bioenergy (Baker et al. 1999). The greatest prospects are therefore crops that yield a commercial product (e.g., eucalypt oil) or provide environmental benefits (beneficial re-use of wastewaters, salinity control in catchments) as well as bioenergy. The economics of bioenergy projects are highly site-specific.

The "Integrated Oil Mallee" project in Western Australia involves more than heat and power generation (Baker et al. 1999; Sims et al. 2006; Bartle et al. 2007). Biomass will come from growing SRWC *eucalyptus* mallee to help solve the dryland salinity problem on croplands. The mallees are *Eucalyptus* species (e.g., *E. horistes*, *E. kochii*, *E. angustissma*, *E. loxophleba*, and *E. polybractea*) with a multi-stemmed habit, lignotubers, and high oil concentration in their leaves, usually 90% cineole; elite lines have total oil content of 3.2% of leaf fresh weight. Mallee oil has short-term fragrance and pharmaceutical markets and a potential long-term market as a solvent degreaser. By 1999, 12 million oil mallee had been planted on approximately 9000 ha. Linear hedges (twin rows) will be harvested on a 2-year cycle yielding approximately 20 Mg/km FW. Harvesting trees on a 3 to 4 year cycle will provide pharmaceutical oils, activated carbon, heat and power, renewable energy credits, and even a carbon credit. An Integrated Mallee Processing plant in each mallee growing area could convert 100,000 Mg per year of mallee (i.e., 50,000 Mg each of leaves and wood) into approximately (1) *Eucalyptus* oil 1600 Mg, charcoal 8300 Mg, and activated carbon 5000 Mg; (2) electrical energy 2.3 MWh and energy 8.6 MWh; or (3) electrical energy 5.1 MWh, depending on the conversion process.

A harvester/chipper was developed to produce material suitable for large-volume materials handling systems and low-cost feedstock for oil, charcoal, and thermal energy (Baker et al. 1999). The machine can harvest and chip mallee stems up to 6 m height and 150 mm basal diameter. The leaf material is separated from the other material before oil distillation. Although the system is for the oil mallee industry, it may be applied to any SRWC. Processes and conversion techniques for utilizing biomass for energy included cogeneration, cofiring, gasification, charcoal, gas, liquids, digestion, and fermentation.

Larger scale plants may follow because the dryland salinity problem extends over millions of hectares. Direct liquefaction of *eucalyptus* mallee costs less and has higher transportation efficiency to a central user or processing facility (Bridgwater et al. 2007).

SRWCs and their bioenergy products in the South Australian River Murray Corridor could have both local and global environmental benefits (Bryan et al. 2008). Some 360,000 ha could produce over 3 million tons of green biomass annually and reduce annual carbon emissions by over 1.7 million tons through bioenergy production and reduced coal-based electricity generation. River salinity could be reduced by 2.65 EC (mS/cm) over 100 years, and over 96,000 highly erosive ha could be stabilized.

Despite these significant opportunities, forest bioenergy has developed little in Australia, except for firewood for domestic heating. Public acceptance and support are lacking, especially for the use of natural forest residues, the main biomass source.

15.2.1.5 New Zealand

New Zealand has 18 million ha of short rotation hardwoods for multiple use, but only 6.3% of its energy is obtained from biomass (Wright 2006). Of several *Eucalyptus* species evaluated as SRWCs, after five 3-year rotations, *E. brookerana* and *E. ovata* were the most productive, achieving yields as high as 50 dry ton/ha per year (Sims et al. 1999a, 1999b). At an even higher initial planting density, *E. pseudoglobulus* yielded more after the second 3-year rotation than the top performing *E. viminalis* clone did in the first (Sims et al. 2001). SRWC genotype selection appears to require evaluation over several coppice rotations.

15.2.1.6 United States

In the United States, which at 103 EJ is the most energy-consumptive country, bioenergy contributes approximately 2.8%, with approximately 60% of this produced and consumed by the forest products industry (Wright 2006). Annually, approximately 40% of 250 million dry ton of wood is used for energy. By 2030, forest bioenergy could double (1.7 EJ) with improvements in forest productivity and biomass conversion. At present, approximately 50,000 ha of SRWCs are planted in the United States, with *Eucalyptus* deployed in California, Hawaii, and Florida.

Hawaii. An extensive, broad research and development program examined how to grow SRWC *Eucalyptus* for bioenergy on up to 100,000 ha of former sugarcane land on the island of Hawaii (Whitesell et al. 1992). Techniques were developed for seedling production and plantation site preparation, weed control, planting, fertilization, and yield estimation. Tree biomass equations were derived for *EG, E. saligna, Albizia falcataria, Acacia melanoxylon, E. globulus, E. robusta,* and *EU* (Schubert et al. 1988, Whitesell et al. 1988). Mean annual increment of SRWC *Eucalyptus* did not peak before significant competition-related mortality, slightly beyond likely harvest age. Although stand densities of 3364–6727 trees/ha had the highest production, their trees did not reach the minimum tree diameter, suggesting that yields at lower densities will equal or surpass higher density yields in longer rotations. To achieve a minimum tree DBH of 15 cm in 5 years, stand density must be less than 1500 trees/ha.

Certain *Eucalyptus/Albizia* mixes appeared promising for maintaining *Eucalyptus* productivity without fertilization. Because *Eucalyptus* yields in 50% and 66% *Albizia* mixes approached that of fertilized pure *Eucalyptus*, and the *Eucalyptus* in mixes were so large, the mixes may be more economical than pure *Eucalyptus* because of lower harvesting costs. The suggested mixed species configuration was alternate rows spaced 3.0 m apart, with *Albizia* planted 2.1 m and *Eucalyptus* 3.0 m apart within their respective rows.

Because efficient harvesting was key to SRWC *Eucalyptus* feasibility, three slightly modified conventional pulpwood harvesting systems were examined. Cable yarding did little damage to the site or stumps but was inefficient because of inexperience and undersized and underpowered equipment. Mechanized equipment, including wheeled and tracked feller-bunchers, a skidder, and a whole tree chipper, handled the trees without difficulty and with relatively inexpensive but damaged stumps. Overall, logistics as well as tree size were major determinants of productivity and cost of harvest. Smaller, less expensive equipment might be more appropriate.

Soil erosion and nutrient depletion with SRWC *Eucalyptus* were minor issues. In spite of high rainfall, pre- and postharvest soil erosion was minimal when a litter layer was developed and retained or when a post-harvest cover crop was used (Schultz 1988). Initial total N levels were inadequate for good growth, presumably because of intensive sugarcane cropping, mineralization and leaching, and depletion of organic matter. N applications were substantially lower than for sugarcane, however. N deficiency could be less when the trees are older because soil N levels after 4 years met or exceeded initial levels, *Eucalyptus* is very efficient at internal recycling of nutrients (Florence 1986), or *Eucalyptus/Albizia* (or other N-fixer) mixes improve N levels and other soil properties. In general, SRWC *Eucalyptus* impacts seem substantially smaller and less frequent than those of agricultural crops, e.g., sugarcane.

Although superior *EG* and *E. saligna* Australian provenances were used, tree improvement would probably increase *Eucalyptus* yields (Skolmen 1986). Short- and long-term improvement programs were proposed but could not be implemented before program termination in 1988. Subsequently, efforts by various public and private agencies have identified promising genotypes in several species.

Costs of production of the three most promising SRWC *Eucalyptus* alternatives were compared. A 5-year rotation on former sugarcane land with periodic fertilization produced the minimum acceptable 15-cm tree at a rate of 20.2 dry ton/ha per year. A 6-year rotation resulted in a 20 cm tree with product quality advantages and a total biomass yield of 18.6 ton/ha per year. An 8-year rotation of *Eucalyptus/Albizia* mix gave a larger tree size at a reduced fertilizer cost, an *Eucalyptus* yield of 22.4 ton/ha per year, and *Eucalyptus/Albizia* yield of 26.9 or more ton/ha per year. Harvesting costs varied with tree size, decreasing by one-third as tree size doubled. Consequently, total costs per dry ton of chipped *Eucalyptus* biomass were highest for the 5-year rotation and lowest for the 6-year rotation. Overall, the information developed by the program provided valuable guidelines for future SRWC *Eucalyptus* ventures in Hawaii.

About 9000 ha of *Eucalyptus* plantations have been established in Hawaii since 1996 (Forest Solutions 2009). Using management procedures from around the world, these plantations are producing over 40 m³/ha per year in the most productive areas.

Florida. Because of Florida's challenging climatic and edaphic conditions, much SRWC emphasis has been placed on *Eucalyptus* tree improvement for adaptability to infertile soils and damaging freezes. U.S. Forest Service research from 1965 to 1984 focused on the best of 67 species for southern Florida (Geary et al. 1983), resulting in >1500 selected *EG* (Meskimen et al. 1987) that were subsequently widely tested to develop four recently released freeze resilient, fast-growing clones. Since 1979, the University of Florida also assessed nine species, producing desirable genotypes of *E. amplifolia* (*EA*) for areas of frequent freezes (Rockwood et al. 1987, 1993). *EG* is now grown commercially in southern Florida for mulchwood (Aaction Mulch 2007) and can be used in central Florida (Rockwood et al. 2008), whereas *EA* is suitable from central Florida into the lower Southeast. *EG* is more productive, largely because of five generations of genetic improvement (Meskimen et al. 1987).

These *EG* and *EA* genotypes are desirable for SRWC systems in Florida and similar regions for many bioenergy applications. On suitable sites and/or with intensive culture, *EG* and *EA* may reach harvestable size in as few as three years (Rockwood 1997, Langholtz et al. 2007). Whole-tree chips of 9-year-old *EG* produced 70% char and oil and 21% noncondensed volatile oil and low-energy gas (Purdy et al. 1978). *EA* and *EG* SRWCs are promising for cofiring in coal-based power plants in central Florida (Segrest et al. 2004), but little is known about their suitability for a wider range of value-added products. Even when used for dendroremediation (Rockwood et al. 1995, 2004, 2005) and windbreaks (Rockwood et al. 2005), *EG* and *EA* may be bioenergy resources.

SRWC opportunities for renewable bioenergy have recently gained momentum in the State's public policy and in research and media coverage. By combining superior clones (Meskimen et al. 1987, Rockwood 1991), suitable culture (Rockwood et al. 2006, 2008), innovative harvesting (Rockwood et al. 2008), and efficient conversion, *EG* and *EA* are poised to meet bioenergy needs in Florida. As in other SRWC development situations, research and development on genetic material, spacing, fertilization, planting, control of pests and diseases, forest management, etc., will be essential for achieving high SRWC productivity.

Biomass-derived electricity and liquid fuels may compete with fossil fuels in the short-term, most likely by using integrated gasifier/gas turbines to convert biomass to electricity (Sims et al. 2003). Biomass production and conversion into modern energy carriers must be more fully developed, and favorable policy options such as subsidies and carbon taxes are also needed to support bioenergy expansion.

15.2.1.7 Summary

Overall, bioenergy could be the highest contributor to global renewable energy in the short to medium term with SRWC *Eucalyptus* providing a large portion of the biomass (Sims et al. 2006). *Eucalyptus* species can be widely planted to produce abundant biomass (Table 15.3). Several conversion technologies are operational, and more are being developed. Biomass characteristics, difficulty in securing adequate and cost effective supplies early in project development, and planning constraints currently constrain *Eucalyptus* bioenergy development. However, increased biomass productivity and quality, carbon trading, distributed energy systems, multiple high-value products from biorefining, and government incentives should foster *Eucalyptus* use for bioenergy. Opportunities for energy crops include development of biorefineries, carbon sequestration, and small, distributed energy systems.

Many other *Eucalyptus* species may be grown for bioenergy (NAS 1980, 1982). By broad climatic region, these include *E. brassiana*, *E. deglupta*, and *E. pellita* for humid tropics, *E. globulus*, *E. robusta*, and *E. tereticornis* for tropical highlands, and *E. citriodora*, *E. gomphocephala*, *E. microtheca*, and *E. occidentalis* for arid and semiarid regions.

Brazilian experience suggests that *Eucalyptus* bioenergy can be produced sustainably at low cost. With reduced production costs, bioenergy could be commercialized widely and reduce carbon

TABLE 15.3

Biomass Energy Consumption and Share and SRWC Base for Some Large Countries That Grow *Eucalyptus*

Country	Biomass (EJ)	Biomass (%)	SRWC base (ha)
China	7.5	16.4	7–10 million
United States	2.9	2.8	50,000
Brazil	2.0	27.2	3 million
Australia	0.2	3.8	~6000

Source: Wright, L., *Biomass Bioenergy*, 30, 706–714, 2006.

emissions while enhancing local economies. The Brazilian experience can also be transferred to other developing countries, thus enabling locally produced bioenergy worldwide.

15.2.2 *Eucalyptus* Bioenergy Enhancement through Biotechnology and Genomics

As the main SRWCs in the southern hemisphere, several *Eucalyptus* species have been the subject of studies in genomics and biotechnology, targeted at improving wood and growth properties for bioenergy, or related applications. The most significant developments for enhancement of wood and productivity properties began almost two decades ago, with the production of the first dense genetic maps. Before that, the available tools were severely restricted to the analysis of few genetic loci. Since then, most research efforts in *Eucalyptus* genomics have been dedicated to the discovery of polymorphisms that confer superior phenotype to elite trees. The expectation is that these polymorphisms identify genetic markers for indirect selection in breeding programs, but may also define a gene for modification using transgenic approaches.

As a first step in the establishment of the tools necessary to identify genes of economic value for selection, the first high-density *Eucalyptus* genetic maps were established in the mid-1990's (Grattapaglia and Sederoff 1994). Development of genetic maps was soon followed by the mapping of genomic regions, of quantitative trait loci (QTL) that regulate a portion of the phenotypic variation for wood quality and growth (Grattapaglia et al. 1995, 1996). These early developments were highly restricted because of the lack of sequence information for *Eucalyptus*, and most other woody species. This deficiency started to diminish in the past decade, initially with the high-throughput sequencing of fragments of expressed genes (expressed sequence tags, or ESTs) (Paux et al. 2004; Vicentini et al. 2005; Novaes et al. 2008), and more recently with the perspective of the availability of the first *Eucalyptus* genome. Here we review the developments in the use of biotechnology and genomics, aimed at improving bioenergy traits in *Eucalyptus*. For general reviews about *Eucalyptus* genomics and its applications to breeding, the readers are directed to other recent publications (Poke et al. 2005; Myburg et al. 2007; Grattapaglia and Kirst 2008).

15.2.2.1 Genetic Mapping and Quantitative Traits Loci Analysis

Most mapping studies in agricultural crops or model plants have relied on the analysis of inbred lines, near-isogenic lines or backcross progenies. In outbred tree species like *Eucalyptus*, the long-generation time and high genetic load have limited the development of these types of segregating populations (Kirst et al. 2004a). To address these limitations, new mapping strategies that use half-sib, full-sib and pseudo-backcross populations were developed and successfully implemented (Grattapaglia and Sederoff 1994; Myburg et al. 2003). Isozyme markers were first applied to genetic mapping of *Eucalyptus* (Moran and Bell 1983) but were of limited use for genetic studies that require high-density, saturated maps. Restriction fragment polymorphism markers were also used for the development of second-generation genetic maps (Byrne et al. 1995; Thamarus et al. 2002). However, the most significant advances came with the development of PCR-based markers such as random amplified polymorphic DNA (RAPD) markers (Grattapaglia and Sederoff 1994; Verhaegen and Plomion 1996; Bundock et al. 2000; Gan et al. 2003), amplified fragments length polymorphism (AFLP) and microsatellite markers (Brondani et al. 1998, 2002; Marques et al. 1998, 2002; Bundock et al. 2000; Gion et al. 2000; Thamarus et al. 2002; Gibbs et al. 2003). A few genes have also been added to existing maps (Bundock et al. 2000; Gion et al. 2000; Thamarus et al. 2002).

Development of high-coverage genetic maps established the foundation for the quantitative genetic analysis of bioenergy traits in *Eucalyptus* and identification of genes that regulate them. QTL analyses for traits associated with biomass growth have been numerous in *Eucalyptus*, including in pure *EG* crosses (Grattapaglia et al. 1996), *E. nitens* (Byrne et al. 1997), *EG* and *EU* hybrids (Verhaegen et al. 1997) and *EG* and *E. globulus* crosses (Kirst et al. 2004b). The first report (Grattapaglia et al. 1996) identified three QTL, explaining 13.7% of the phenotypic variation for circumference at breast height in a large half-sib population of *EG*. Similarly, other studies have also typically reported few QTL with moderate to high effect in growth and biomass traits, suggesting an immediate value of this information for marker-assisted selection. QTL analysis of chemical and physical property traits (e.g., lignin and cellulose content, wood specific gravity), which are critical for efficient biomass conversion to biofuels, was initially severely hampered by the cost, labor and time required for sample analysis. However, the development of novel methods for high-throughput phenotyping, such as near-infrared spectrometry, SilviScan (x-ray densitometry combined with automated scanning x-ray diffraction and image analysis) and mass spectrometry, computer tomography x-ray densitometry (CT scan) and pyrolysis molecular beam mass spectrometry (pyMBMS) has modified that scenario drastically in the past decade, and analysis of these traits is now commonplace. For example, the application of indirect, high throughput phenotyping of wood quality traits by NIR was demonstrated in *Eucalyptus*, and the information used for QTL mapping in a pseudo-backcross of *EG* and *E. globulus* (Myburg 2001). Approximately 300 individuals that had been previously genotyped with AFLP markers were analyzed by NIR, and predictions were made for pulp yield, alkali consumption, basic density, fiber length and coarseness, and several wood chemical properties (lignin, cellulose and extractives).

In summary, efforts to identify regions of the genome that regulate biomass growth and wood quality in *Eucalyptus* have been largely successful. However, the use of this information in breeding programs was rapidly shown to be unreliable because of two main factors: (1) rapid linkage disequilibrium (LD) decay among unrelated individuals and (2) the extensive level of genetic heterogeneity in diverse populations, as previously predicted (Strauss et al. 1992). The low extent of LD meant that significant marker-trait associations detected in specific segregating populations were not detectable when unrelated genotypes were considered. The genetic heterogeneity of existing populations, where multiple alleles at a large number of loci may contribute to trait variation, signified that marker-trait associations detected in one pedigree were not relevant in all backgrounds. Therefore, it became clear that the identification of makers associated with traits in one or few segregating populations was not sufficient, but that the causative polymorphisms, or at least knowledge of the specific gene that regulated trait variation, was necessary. However, success in positionally cloning QTL in forest tree species (Stirling et al. 2001), as it had been done in some agricultural crops like tomato (Paterson et al. 1988; Martin et al. 1993), was not immediately achieved.

15.2.2.2 Genomics and the Identification of Genes Regulating Bioenergy Traits

Identification of genes that regulate quantitative variation in plants, animals and humans, went through significant advances in large part because of the development of genomic technologies. The high-throughput sequencing of expressed genes initiated in the beginning of the decade, with the release of the first ESTs of *Eucalyptus* species (Kirst et al. 2004b). This initial effort was rapidly followed by more extensive EST sequencing projects, which surveyed the pool of genes expressed in several *Eucalyptus* tissues, and identified putative orthologs for the suite of genes involved in metabolic and regulatory pathways associated with biomass growth and wood quality (Kirst et al. 2004b). The sequencing of expressed genes lead to the development of the first studies that characterized the expression of large number of genes (i.e., transcriptomics) in *Eucalyptus* and hybrid populations (Voiblet et al. 2001; Kirst et al. 2004b). A genetic genomics study, which combined the information from QTL of biomass growth and sequence and expression of genes in differentiating xylem suggested that the genetic elements that regulated traits related to bioenergy and other properties could be rapidly unraveled (Kirst et al. 2004b, 2005). Similarly, novel approaches to identify polymorphisms that regulate complex traits were also developing rapidly in *Eucalyptus*. Thumma and colleagues were the first to demonstrate the power of association genetics in a tree species (Thumma et al. 2005), which relies on the detection of polymorphisms associated with quantitative variation in populations of unknown ancestry, in a woody species. Specifically, the study identified cinnamoyl CoA-reductase (CCR), a known gene in the phenylpropanoid pathway, as being a significant determinant of fiber properties in *Eucalyptus*. Although this first study was focused on a specific gene, several studies are underway that explore the identification of genes of value for bioenergy for several hundred genes, targeting particularly those that code for enzymes of the lignin and carbohydrate/cellulose pathways.

15.2.2.3 Future Developments

Biotechnology and genomics research have allowed for achievements in the past few years that were inconceivable at the beginning of the millennium. Procedures for genome sequencing, as well as transcriptome, proteome, and metabolome characterization have all gained in efficiency by two to three orders of magnitude within less than a decade. At the same time, the costs per data point have been reduced by the same proportion. As a consequence of the decrease in cost, and the growing interest of the U.S. Department of Energy (DOE) in bioenergy crops, an *Eucalyptus* genotype was selected recently for sequencing by the Joint Genome Institute (JGI)/DOE, for completion in early 2010. The sequence will provide the foundation on which QTL cloning should become achievable. The genome sequencing and the complete catalogue of genes will also allow for the development of genomic tools such as whole-transcriptome microarrays, for characterization of gene expression variation. The current sequence has coverage of 4 × (meaning that, on average, every nucleotide has been sequenced 4 times) and is expected to reach 8 × by the summer of 2009. An assembly based on the existing sequence data has already captured almost 80% of the genome sequence. This suggests that—with the added sequencing to complete 8 ×—the final genome information will be close to completion. The individual being sequenced is an elite *EG* genotype (Brasuz1). To support the assembly and annotation of the genome sequence, two bacterial artificial chromosomes (BACs) have been developed and over one hundred thousand BAC-ends have been sequenced. Furthermore, a number of pedigrees are being genotyped with diversity array technology (DArT), microsatellite, gene expression (GEM) and single-feature polymorphisms (SFP) markers, for the development of hyper-saturated genetic maps that will be invaluable for the sequence assembly. To support the annotation of the genome sequence (i.e., identification and definition of function for genes in the genome), JGI is also sequencing a large number of random gene sequence fragments (expressed sequence tags).

Perhaps even more exciting are the anticipated developments in the years ahead, particularly in genome sequencing. Platforms in development have recently demonstrated the capacity to generate sequencing data—although not yet fully interpretable—over 10–20 kilobases within a few hours (Korlach et al. 2008). At the same time, miniaturization of devices and single molecule detection methods now permit sequencing of several million molecules in parallel (Eid et al. 2009). In that scenario, sequencing a moderate size genome such as *Eucalyptus* (~500–600 Mbp) could be achieved in less than one day. Ultra low sequencing reaction volumes also suggest that the costs of such a task will be a few hundred U.S. dollars, rather than the current several hundred thousand.

In summary, in the next decade, genetic and genomics studies will likely discover the majority of genes that regulate a significant part of the heritable variation of biomass productivity and wood property traits in *Eucalyptus* and the most commercially important tree crops, such as *Populus* and *Pinus*. Consequently, it will be possible to identify superior genotypes based in large part on their genotype across multiple critical loci. The challenge will be to develop genotyping assay methods that will be sufficiently cost effective to permit rapid screening of large progenies in breeding programs. That will allow the development of genotypes that combine the optimal alleles for each specific end-use purpose, including plants optimized for bioenergy purposes.

15.3 PINE

Worldwide, *Pinus* species are currently widely used for bioenergy and have considerable potential for future use. The *Pinus* genus has over 100 species (Syring et al. 2005), and pines have high genetic and phenotypic diversity that has bioenergy ramifications. Many pine species growing in natural stands and/or established in plantations provide bioenergy opportunities.

In the United States, the forest products industry is currently the largest producer of bioenergy, much of which is in the southeastern United States. The approximately 82 million ha of forestland in the Southeast produce 18% and 25% of the world's roundwood and pulp, respectively (FAO 2004). Loblolly, *P. taeda*, and slash, *P. elliottii*, are the most important southern pines because of their broad natural ranges that constitute the majority of the approximately 15 million ha of plantations in the Southeast (Peter 2008). The main products for loblolly and slash pine plantations have been pulpwood, wood composites, sawtimber, and poles/pilings. Loblolly pine and slash pine are in closely related clades (Dvorak et al. 2000). Loblolly pine is a model pine because of its economic importance and well characterized reproduction and genetics (Lev-Yadun and Sederoff 2000).

Silvicultural intensity for loblolly and slash pines varies depending on plantation objectives and initial investments. Silvicultural treatments and genetic improvement greatly enhance tree growth and stand productivity (Fox et al. 2004, 2007) and shorten rotations. Seed orchards provide >95% of the seed for commercial nurseries, and aggressive breeding and genetic testing are underway.

Tree improvement has focused on growth, stem form, and disease resistance. Improving volume growth and yield has been emphasized (White et al. 1993, McKeand and Bridgwater 1998). In the first two breeding cycles of loblolly pine, gains of 30–40% in stem volume per cycle were achieved (Li et al. 1999). Resistance to fusiform rust and pitch canker may also be improved in loblolly and slash pines (Kayihan et al. 2005). Loblolly and slash pines each have only one breeding zone each because genetic by environmental interactions are not significant (McKeand et al. 2006). Southern pines can be clonally propagated by rooted cuttings and somatic embryogenesis (Nehra et al. 2005). Varietal lines of elite germplasm selected and propagated by somatic embryogenesis have been developed, and loblolly pine varieties are now being commercially deployed in the Southeast.

Although wood properties are important in the traditional utilization of southern pines (Peter 2007; Peter et al. 2007) and presumably also for bioenergy, improvement programs are not actively breeding for these traits. Wood density is under moderate to high heritability in loblolly and slash pines. Strong correlations between juvenile and mature wood suggest that early selection can be used. In loblolly pine, it may be more difficult to improve both density and growth simultaneously. Significantly less is known about genetic control of wood chemical composition in loblolly and slash pines, even though these traits are significant for both chemical pulp production and bioconversion to ethanol. Loblolly pine wood chemical composition is under weak genetic control in juvenile and mature wood, and was not correlated, or only weakly so, with growth (Sykes et al. 2006).

Loblolly pine's economic importance, genetic material, and easily studied wood characteristics have stimulated significant biotechnology research, e.g., Sewell et al. (1999, 2000, 2002), Brown et al. (2004), Kirst et al. (2003), and Lorenz et al. (2006). A large loblolly pine resequencing project discovered single nucleotide polymorphisms (SNP) in 8000 unigenes (Neale 2007), which will identify gene candidates that control disease resistance and wood properties, potentially leading to a genome sequence. Genetic engineering methods can genetically transform loblolly pine. Transgenic plants have been derived from an organogenic method starting with mature zygotic embryos (Tang et al. 2001) and from somatic embryos (Connett-Porceduu and Gulledge 2005).

Advanced generation southern pines in the Southeast have high bioenergy potential (Peter 2007). Tree improvement programs and management systems coupled with clonal propagation, genetic engineering, and genomic research customizing trees for bioenergy and chemicals make southern pine even more promising for bioenergy production through (1) integrated forest biorefineries that produce bioenergy and biofuels in addition to pulp and paper (Van Heiningen 2006; Chambost et al. 2007a, 2007b; Towers et al. 2007) or (2) co-firing, wood pellets, biofuels, or gasification. Overall, harvesting and transportation account for approximately two-thirds of the total delivered wood costs (Peter et al. 2007).

Although harvesting and transporting small diameter trees is a significant cost barrier to growing and using southern pines for bioenergy, slash and sand (*P. clausa*) pines have been evaluated as SRWCs with more than 4000 trees/ha and harvests within 8 years (Campbell 1983; Campbell et al. 1983; Frampton and Rockwood 1983; Rockwood et al. 1983, 1985; Rockwood and Dippon 1989).

Yields from high-density slash pine stands peaked at 9800 trees/ha at 80 mt/ha at age 6 years and 98 mt/ha at 10 years at 6200 trees/ha (Campbell 1983; Campbell et al. 1983). Through three years on a site with a higher P level, slash pine yields nearly tripled when stand densities tripled even up to 43,300 trees/ha, but on a less fertile site, a similar tripling of yield only extended up to 14,600 trees/ha (Rockwood et al. 1983). High density sand pine appeared less productive and required longer rotations, with maximum yields up to 8 mt/ha per year at almost 20 years. In spite of favorable energy output/input ratios of as high as 28 and 26 for slash and sand pine SRWCs with 5–20 year rotations, respectively, only slash pine SRWC systems generated suitable break-even prices (Rockwood et al. 1985). Although biomass yields from these early SRWC tests were low, genetic variation within slash and sand pines for traits including survival and tree biomass quantity and quality may be utilized to increase their SRWC productivity (Frampton and Rockwood 1983). Still, because southern pines do not coppice, they are not well suited to SRWC systems.

Combustion, pyrolysis, gasification, and bioconversion convert wood into heat, electricity and liquid fuel (Peter 2007). The well-established infrastructure and extensive plantations in the forest products industry are huge advantages for using southern pine for bioenergy. European demand for renewable sources of electricity is driving wood pellet production using southern pine roundwood (Kotra 2007), and the U.S. forest products industry is actively researching integrated forest biorefineries (Amidon 2006; Larson et al. 2006; Van Heiningen 2006; Chambost et al. 2007a, 2007b; Towers et al. 2007).

Wood gasification facilities to convert wood into energy and power are planned in the Southeast. For example, a northern Florida facility to produce electricity and gas began construction in 2008. Also in 2008, a facility to produce ethanol from syngas was begun in south Georgia. Oglethorpe Power Corporation (OPC), the United States' largest power supply cooperative, is planning to build as many as three 100 MW biomass-fired electric generating facilities in Georgia. The facilities will provide power to OPC's members, which supply electricity to nearly 50% of Georgia's population. The steam-electric power plants will use fluidized bed boiler/steam turbine technology for a woody biomass mixture.

Other pine species elsewhere in the world that exemplify additional bioenergy opportunities from natural stands and plantations include radiata pine (*P. radiata*), jack pine (*P. banksiana*), and *P. halepensis*. Radiata pine is widely planted as an exotic in New Zealand, Chile, and other temperate regions. Jack pine is a wide ranging species in northern North America. *P. halepensis* is common to semiarid Mediterranean areas.

Renewable energy, particularly bioenergy, can be important for reducing New Zealand's greenhouse gas emissions back to 1990 levels by 2012 (Hall et al. 2001). Currently, biomass provides less than 5% (28 PJ) of New Zealand's primary energy supply. However, large quantities of current and future forest residues have potential to fuel biomass power generation. New Zealand has approximately 1.7 million ha of forest plantations of which 91% is *P. radiata*. By 2010, the annual log harvest is expected to be over 30 million m^3. Residue delivery costs largely depended on the delivery system chosen, the site characteristics and the transport distance. The cheapest system ranged from 22 to 37 NZ$/ton (1.2–2.0 NZ$/GJ) for residues at the landing and from 29 to 42 NZ$/ton (1.6–2.2 NZ$/GJ) for residues collected from the cutover. The cheapest option was the simplest system because extra handling added cost. Use of landing residues for the generation of heat and/or electricity could be feasible, particularly for sites with short transport distances on private roads that have no legal restrictions for payloads. Biomass delivery systems have also been assessed elsewhere for forest residues (Bjorheden 2000), willow short rotation coppice (Gigler et al. 1999), and biomass fuel mixes (Allen et al. 1997; Sims and Culshaw 1998).

Potential forest bioenergy costs in central Chile were estimated (Faundez 2003), as approximately 70% of Chile's energy comes from imported fossil fuels, and approximately 70% of its electricity comes from hydroelectricity. Biomass does contribute 19% of the total energy (mainly as firewood from native forests), but increasing its share would have economic and environmental benefits. Sustainable production of firewood would reduce the use of native forests.

The potential production costs of four silvicultural regimes for *Populus*, *Salix*, *Pinus*, and *Eucalyptus* considered the costs of cultivation, harvest, and soil use, yields expected at each site, and the energy value of biomass (Faundez 2003). *Populus* and *Salix* grow quickly and have been widely used for bioenergy, whereas *Pinus* and *Eucalyptus* plantations occupy >2 million ha in central Chile. Because of a nonlinear relationship between soil use cost and productivity, the minimum costs for both nonintensive *Pinus* and *Eucalyptus* regimes were for high productivity sites (more expensive land), whereas for both intensive *Populus* and *Salix* regimes, the minimum costs were for low productivity sites (less expensive land). Minimum costs of production of the nonintensive regimes of 0.0355–0.1662UF/GJ for *Eucalyptus* and 0.0626–0.3822 UF/GJ for *Pinus* were at most one-half those of the intensive regimes of 0.1201–0.1325 UF/GJ for *Populus* and 0.1387–0.1503 UF/GJ for *Salix*, which were fairly insensitive to site productivity.

Given the large area of land available in central Chile for nonintensive compared to intensive forestry and the broad experience with nonintensive *Pinus* and *Eucalyptus*, these regimes appear best (Faundez 2003). Because their energy production costs were comparable to those for oil and natural gas, forest bioenergy could be a feasible alternative to fossil fuels and a way to avoid CO_2 emissions. If forests are to be used for bioenergy in Chile, selection of appropriate clones and species will be essential for increasing productivity.

Jack pine SRWC production in Wisconsin peaked at age 5 in the densest planting and progressively later in more open spacings (Zavitkovski and Dawson 1978). Biomass production was two or more times higher than in plantations grown under traditional silvicultural systems. For 10-year-old irrigated, intensively cultured plantations in northern Wisconsin, energy inputs (site preparation, fertilization, weed control, irrigation, harvesting, chipping, and drying) were approximately 20% of the total energy (Zavitkovski 1979). The net energy of 1863 MBtu/ha, equivalent to 340 barrels of oil, was 13% more energy than reported for highly productive, non-irrigated, intensively cultured stands in eastern United States. Net energy returns were linearly and positively correlated with energy invested in both irrigated and nonirrigated, intensively cultured plantations and a naturally regenerated forest, indicating that intensive culture brings commensurate returns.

In association with fire hazard reduction policies, forest biomass from extensive natural pine forests, such as in northern California, southwestern Oregon, and Oregon east of the Cascade Mountains, can fuel power plants (Barbour et al. 2008). For three hazard reduction scenarios, the mix of species and sizes removed was similar, and average yields were quite high. Sawlogs were 67 to 79% of the weight removed. Tops and limbs of commercial species >25.4 cm DBH and noncommercial species provided most of the biomass chips. Low value conifers (17.8 to 40.6 cm DBH) were also an important biomass source, whereas trees <17.8 cm DBH were a relatively minor component. To pay for fire hazard reduction treatments, 9.1 to 18.2 mt/ha of sawlogs plus perhaps a quarter of that in biomass chips need to be removed. Considerable emphasis has been placed on finding uses for small trees (USDA FS 2005; USDE and USDA 2005). Fire hazard reduction treatments could provide enough raw materials to fuel one or more 20 MW wood-fired electrical power generation plants.

In arid and semiarid Mediterranean regions, *P. halepensis* and other pines are potential bioenergy sources. For example, Catalonia in northeast Spain has 12,146 km² of forests (38% of Catalonia, 10% of Spain's forests) composed primarily of *Quercus ilex*, *P. sylvestris*, *P. halepensis* and *P. nigra* (Puy et al. 2008). The Catalan Energy Plan 2006–2015 estimates 197% more forest and agricultural biomass consumption, mainly as heating for household and industrial uses, such as sawmills. Catalonia's biomass potential of forest, agricultural, and sawmill residues, and other bulky wastes is approximately 2.6 million tons, approximately 1 Mtoe. If forest biomass is to be an important bioenergy source in Mediterranean countries, key factors are property regimes, low productivity, weak institutional capacity, logistics and supply difficulties, and forest product profitability. Technological solutions alone do not guarantee a prominent role for forest biomass in southern Europe.

15.4 POPLAR BIOENERGY USE AND POTENTIAL WORLDWIDE

15.4.1 Introduction and Background

The use of poplars for bioenergy is not new (Anderson et al. 1983; Dickmann and Stuart 1983; Dickmann 2006). In fact, there is archaeological and historical evidence that poplars have been used for cooking and fuel throughout civilization. For example, archaeological evidence suggests that indigenous people in North America including the Paleo-Indian, Hohokam and Ojibwe used poplar for cooking and heating as well as many other uses as early as 3000 BC (Logan 2002). Over a thousand years ago Euphrates poplars growing along the Tigris and Euphrates rivers, now in Iraq, were used for charcoal and many other practical uses by the third Dynasty of Ur in Mesopotamia (Gordon 2001). Moreover, the ancient Chinese before the Han Dynasty also used poplars for cooking and heating at Youmulakekum in Western China from 700 to 200 BC; and during the Xian period in China there is archaeological evidence that poplars were used for fuel and other practical uses around 600 AD (Zhang J, personal communication).

Poplars have been used for fuel since antiquity throughout Europe, the Middle East, near East, and Mediterranean in close association with agriculture (Zsuffa 1993). And, according to the diaries of Lewis and Clark and David Thompson in the early 19th century, the native cottonwood growing along the rivers was used for cooking and fuel during the exploration and settlement of the western United States and Canada (DeVoto 1953; Richardson et al. 2007).

Thus, it is well documented that poplar has been a source of bioenergy throughout early civilization. However, as the world became industrialized, both industry and home owners became dependent upon inexpensive fossil fuels for energy, and the use of wood for fuel declined. But, in the early 1970s when fossil fuels became more expensive and scarce, modern societies began to seek alternative sources of energy such as biomass (Rockwood et al. 2004). Over the years, the forest products and pulp and paper industries have been the largest user of poplar biomass for energy in developed industrialized countries; in fact, now they have largely become energy self-sufficient through their use of biomass as a fuel (Konig and Skog 1987).

The aforementioned oil crisis and shortages in the 1970s sparked a worldwide emphasis on bioenergy research and development from trees, especially poplar (Fege et al. 1979; Ranney et al. 1987). But, after only a few years when fossil fuels became more available and inexpensive again, bioenergy research and development on poplars declined. Recent oil shortages and increased costs of fossil fuels in the 21st century have prompted yet another new wave of emphasis on alternative fuels and bioenergy from poplar biomass.

Poplars have received worldwide attention for bioenergy use because of their high biomass production rates and genetic improvement potential (Stettler et al. 1996; Davis 2008). We review worldwide poplar biomass production rates and discuss the advantages and limitations of using poplars for bioenergy, with emphasis on the current and potential future use of poplars for bioenergy in developed countries. We fully realize that poplar is still an important source for fuel for cooking and heating in many developing countries, and will likely remain so for years to come (FAO 2009).

15.4.2 Poplar Biomass Production

15.4.2.1 North America

The origin of the use of poplar plantations for biomass production in North America dates back to the late 1800s and early 1900s when governmental agencies became concerned that the forests of the northeastern United States and Canada were not sustainable because of overharvesting (McKnight and Biesterfeldt 1968). This concern led to the establishment of the U.S. Forest Service in 1905 and the Forest Survey in 1928 (LaBau et al. 2007). At the same time, Stout et al. (1927) working at the New York Botanical Garden outlined opportunities for meeting the growing needs for wood through poplar tree breeding. Stout and Schreiner (1933) reported on their pioneering breeding effort in hybridizing poplars that was patterned after the classical work of A. Henry in the United Kingdom (Henry 1914). Also, in the 1920s and 1930s Canadian poplar breeder pioneers Frank Skinner and Carl Heimburger began their pioneering poplar breeding programs in Canada (Richardson et al. 2007). These four pioneering tree breeders laid the foundation for the next 100 years of poplar breeding for biomass production in North America.

In the United States from 1900 to the 1940s, poplar plantations were mostly made up of native poplar species, including eastern cottonwood (*Populus deltoides*) and black cottonwood (*P. trichocarpa*) (Bearce 1918; McKnight and Biesterfeldt 1968; Smith and DeBell 1973). In Canada during that same period, there were thousands of hectares of abandoned farm land planted to native poplars and imported hybrid poplars from Europe (Smith 1968). These plantations had varying degrees of success depending upon conditions (Smith and Blom 1966). There were also thousands of hectares of both native poplars and hybrids planted in shelterbelts and windbreaks during early settlement of the Canadian and American west (Cram 1960).

Initially, these poplars were grown for shelter, but they were also used for fuelwood and other wood products (Richardson et al. 2007). The multitude of hybrid poplars developed by Stout and Schreiner (1933) were tested throughout North America during the 1930s and 1940s to determine their adaptability to varying conditions including survival, growth, as well as pest and disease resistance. These hybrids, known as Oxford Paper (OP) or Northeast (NE), showed varying degrees of growth and survival largely because of climatic factors, cultural practices, and disease susceptibility (Blow 1948; Smith and Blom 1966; Maisenhelder 1970).

In the 1950s, Ernst Schreiner, then with the U.S. Forest Service, distributed the better performing NE hybrid poplars to cooperating landowners throughout North America to gain information on their performance. He also shared the NE hybrid poplar clones with many other countries so they might be tested under worldwide conditions (Garrett 1976). Many of them did not live up to their expectations in Canada, Europe, and Eurasia (Pryor and Willing 1965; FAO 1980). Canadian poplar breeders in the meantime were developing their own poplar hybrids for use in Canada (Smith 1968; Richardson et al. 2007). Moreover, an active program of selection and breeding of eastern cottonwood was initiated in the eastern and southern United States in the late 1950s and 1960s (Wilcox and Farmer 1967; Schreiner 1971). These breeding programs focused on improving adaptability, growth and disease resistance with primary emphasis on producing wood for sawtimber and pulp and paper (Thielges and Land 1976).

In the 1960s, there was a change in the focus on sawtimber and pulp and paper to biomass of the forest. This change was largely due to the efforts of production ecologists interested in determining the primary production of the vegetation of forests including dry matter production and energy accumulation. An initiative was known as the National Academy of Science, International Biological Program (IBP) had one of their focal points on production of the eastern deciduous forest biome (Newbould 1967). This program resulted in a number of studies of the standing crop biomass and caloric values of native poplar stands and plantations (Peterson et al. 1970; Switzer et al. 1976; Zavitkovski 1976; Crow 1978). The change in emphasis to weight of forest products rather than volume prompted the U.S. Forest Service to start tallying biomass beginning in 1964 (LaBau et al. 2007). Also, there was recognition by many that intensive forest utilization required an understanding of functional forest ecosystem processes, including dry matter, energy and chemicals (Young 1973). Moreover, biomass studies were essential to realizing the full potential of the forest for energy (Young 1971, 1973, 1976, 1977).

The concept of short rotation forestry or short rotation woody corps developed in parallel with the biomass and bioenergy studies of forests (Larson and Gordon 1969; Gordon 1975). Schreiner (1970) called it "mini-rotation forestry." Short rotation forestry involves growing trees on short rotations (i.e., <15 years) using agronomic principles including planting genetically improved stock, fertilization, irrigation, weed and pest control (Drew et al. 1987; Dickmann 2006). Early efforts to identify suitable species for short rotation forestry included many species (Dawson 1976), but after many field trials *Populus* emerged as one of the top candidates for further research. Poplars emerged because they are genetically diverse, amenable to genetic improvement, fast growing, easy to propagate and well-studied (Dickmann et al. 2001). A key factor in the acceleration of research on poplars for bioenergy was the OPEC oil embargo of 1973. Soon after the huge increase in oil prices, the oil importing countries began seeking alternative sources of energy including biomass (Dickmann 2006). Research and development efforts were soon initiated on bioenergy by many governmental agencies. For example, the International Energy Agency (IEA) was formed to foster international research and development on energy in 1974. IEA includes a Bioenergy Task 30— Short Rotation Crops for Bioenergy Systems. In 1978, the U.S. Department of Energy (DOE) initiated a Short Rotation Woody Crops Program, which was later was renamed the Biofuels Feedstock Development Program. This program provided funds for poplar cultivation, genetics, physiology, pest management, growth and yield for 25 years. This continuous source of funding provided the research continuity needed to advance poplar bioenergy research significantly (Stettler et al. 1992). Similar bioenergy funding for poplars became available in the Canadian Forest Service and Energy Agency. In 1995 the USDA Forest Service, the U.S. DOE, and the Electric Power Research Institute (EPRI) also established the Short Rotation Woody Crop Operations Working Group to pursue efficient development of practices of culture, harvest and handling of woody biomass plantations. Further details of the history of poplar production and bioenergy in the United States and Canada are provided by Thielges and Land (1976), Dickmann (2006) and Richardson et al. (2007).

Published data on biomass production rates in North America range from 2 to 35 mt/ha per year (Table 15.4). Realistic yields are probably in the range of 5–20 mt/ha per year (Stanturf et al. 2001; Dickmann 2006; Davis 2008). The reasons for the large variation are many. Some reports were from small plot experiments and unreplicated demonstrations whereas others are from larger replicated field studies, or genetic trials with adequate border rows, or plots within commercial operational

TABLE 15.4

Chronological Synthesis of Poplar Biomass Production Field Studies and Plantations by Location in North America

Location	Clone/ Species*	Age (years)	Spacing (m)	Productivity (mt/ha per year)	Reference
W US	T	2	0.3	13.4–20.9	Heilman et al. (1972)
W US/Canada	T	Multiple	0.3–1.2	9.0–11	Smith and DeBell (1973)
S US	D	5–20	3.0	10–11	Switzer et al. (1976)
NE US	DN	4	0.12–0.76	7.7	Bowersox and Ward (1976)
NC US	DN	4	0.23–0.61	11.3–13.8	Ek and Dawson (1976)
E Canada	DN	Multiple	Multiple	5–19	Anderson (1979)
W US	T	8	0.3–1.2	5.8–9.7	Heilman and Peabody (1981)
NC US	B	5	1.2	4.2	Isebrands et al. (1982)
W US	T,TD	4	1.2	5.2–27.8	Heilman and Stettler (1985)
W US	T,TD	4	1.2	11.3–12.6	Heilman and Stettler (1990)
NC US	DN,B	Multiple	Multiple	4.9–12.8	Strong and Hansen (1993)
W US	TD	5	0.18–0.3	6.4–30	DeBell et al. (1993)
W US	TD	7	0.5–2.0	10.1–18.2	DeBell et al. (1996)
W US	T,TD	7	1.0	11–18	DeBell et al. (1997)
W US	T,TD	4	1.0	14–35	Scarascia et al. (1999)
NC US	DN	7–12	2.4	4.7–10	Netzer et al. (2002)
W Canada	TD	4–13	3.0	9.2–13.6	Zabek and Prescott (2006)

*B, *P. balsamifera*; D, *P. deltoides*; N, *P. nigra*, T, *P. trichocarpa*; DN, D × N; TD, T × D; N, north; E, east; S, south; W, west; C, central; US, United States.

plantings. Yield data from small plots are higher than larger ones because of an edge effect bias (Zavitkovski 1981). The yields across the continent also vary with factors such as species, clone, site conditions, region, cultural methods, spacing, harvest strategy, pests and diseases (Hansen et al. 1992; Dickmann 2006). Poplars are particularly sensitive to competition for available light, nutrients and water. Therefore, weed control and nutrition were determined to be essential for maintaining survival, growth and yield of planted poplars (Aird 1962; McKnight and Biesterfeldt 1968; van Oosten 2006; Isebrands 2007). For example, Kennedy (1975) found weed control essential for eastern cottonwood plantings. Czapowskyj and Stafford (1993) found that mowed and fertilized poplar plantations in Maine grew 4 times as fast as the untreated control areas. And Coleman et al. (2006) found that post-establishment fertilized operational poplar plantations grew 3.5 times as fast as unfertilized stands and that fertilizer increased biomass by 40% in the third year.

The early plantations of poplars in North America were planted largely with improved native species (Smith and Blom 1966; McKnight and Biesterfeldt 1968; Heilman et al. 1972). Black cottonwood plantations in the Pacific Northwest grew from 2 to 7 mt/ha per year at close spacing and from 9 to 11 mt/ha per year after coppicing (i.e., resprouting) (Smith and DeBell 1973). Mean annual biomass accumulation of cottonwood plantations in the lower Mississippi Valley averaged approximately 10–13 mt/ha per year at 3×3 m spacing with the maximum rate occurring at age 5 without thinning (Krinard and Johnson 1975; Switzer et al. 1976). Francis and Baker (1981) reported annual biomass production of a commercial plantation of a selected cottonwood clone at 3×3 m spacing to be 7.6 mt/ha per year in 4 years.

With the advent of the short rotation forestry concept, researchers were anxious to test all of the aforementioned yield factors; many started with small plots. Moreover, forest geneticists quickly designed genetic trials throughout North America to test their new and sometimes old poplar clones (Schreiner 1959; Garrett 1976; Randall 1976; Stettler et al. 1992). In some cases, the native species outgrew the ill-adapted new clones (Blow 1948; Smith 1968; Maisenhleder 1970). There was an early trend toward very close spacing trials with repeated coppicing (Blake 1983; Ferm and Kauppi 1990). Spacing was determined as an important factor for biomass yields, and it had to be coupled properly with rotation age to maximize yield. Early results from Wisconsin in the midwestern United States with a Canadian hybrid poplar by Ek and Dawson (1976) planted at 0.22, 0.3 and 0.6 m² spacing were 11.3, 12.6 and 13.8 mt/ha per year, respectively after 4 years. After 10 years the highest biomass production from a clone and spacing study in Wisconsin was 10.4 mt/ha per year at a 2.4 m² spacing (Zavitkovski 1983). These Wisconsin studies were with poplar clones from the NE poplars and from Canadian cooperators Cram (1960) and Zsuffa (1975). Strong and Hansen (1993), in an overview of 16 years of the Wisconsin studies, stated the maximum mean annual biomass production was 12.8 mt/ha per year by an NE clone planted at 1×1 m spacing. They concluded that biomass yield was independent of spacing except that time to maximum biomass (i.e., rotation) and tree size will vary according to spacing.

Block demonstration plantings throughout the midwestern United States of the NE clones planted at 2.4×2.4 m spacing yielded from 4.7 to 9.5 mt/ha per year in 8 yrs. The mean annual biomass production peaked at year 10 at 7 mt/ha per year and the "best" sites averaged 9.2 mt/ha per year (Netzer et al. 2002). Berguson (2008) recently presented results of yield tests for commercially managed poplar plantations in the Midwest; yields of current clones produce from 7.8 to 12.3 mt/ha per year.

L. Zsuffa worked in Ontario on poplar breeding and short-rotation forestry throughout his career (Richardson et al. 2007). The "best" poplar clones from his program yielded between 5 and 19 mt/ha per year (Zsuffa 1975; Zsuffa et al. 1977; Anderson 1979). The Ontario Ministry of Forestry growers guide for hybrid poplar presents clonal descriptions and biomass tables for Canadian clones (Boysen and Strobl 1991). At the same time as the Ontario work, G. Vallee and co-workers established a poplar breeding program and conducted biomass trials in Quebec. Results from that breeding program are presented in Dickmann et al. (2001). Labreque and Teodoreseu (2005) reported hybrid poplar yields of 17–18 mt/ha per year at 0.3×1.7 m spacing in a coppice study in Quebec.

In the U.S. Northwest, R.F. Stettler of the University of Washington and P.E. Heilman of Washington State University collaborated on a hybridization and cloning program of *P. tricho-carpa* that has had a major influence on poplar biomass production in the United States and Canada (Heilman and Stettler 1985; Stettler et al. 1988). The new *P. trichocarpa* × *P. deltoides* hybrids from that research and development program have been studied widely throughout the region and world and with industrial collaborators has led to extensive poplar commercialization in the Pacific Northwest of North America (Stanton et al. 2002).

Early studies conducted at 1.2 × 1.2 m spacing compared the new hybrids with select native black cottonwood clones. Average biomass production of the black cottonwood varied by clone and was from 5.2 to 23.1 mt/ha per year. However, the new hybrid *P. trichocarpa* × *P. deltoides* clones were impressive and yielded from 15.6 to 27.8 mt/ha per year. These high yields prompted a series of further research studies on the productivity of the new clones (Heilman and Stettler 1985; Stettler et al. 1988). Heilman and Stettler (1990) studied coppicing of the new hybrids compared to native black cottonwood. Mean yields were 11.3 mt/ha per year for the coppice and 12.6 mt/ha per year for the initial harvest. However, many clones had higher yields after coppice, and there was significant clonal variation in coppice yields that suggested that superior coppicing clones can be selected. D. DeBell and coworkers at the U.S. Forest Service conducted a series of field studies with the Stettler/Heilman clones in the 1990s in western Washington State. They found that poplar clones had uneven yields in monoclonal versus polyclonal blocks at close spacing (i.e., 0.5, 1.0, and 1.5 m spacings). Biomass yields for individual clones ranged from 11.7 to 18 mt/ha per year for these spacings. Monoclonal blocks yielded more biomass than polyclonal blocks. The highest biomass was produced in the closest spacing (i.e., 0.5 m) (DeBell and Harrington 1993; DeBell and Harrington 1997). DeBell et al. (1993) also studied the so-called "wood grass" concept versus more conventional spacings. Two hybrid poplar clones were planted at 0.2 and 0.3m spacing and at wider spacings (i.e., 0.5, 1.0 and 2.0 m). The fast growing Stettler clone, H 11-11, harvested annually produced 7.0 mt/ha per year over 5 years. The yield of the same clone at wider spacing was 18.8 mt/ha per year. During the 5 years, H 11-11 produced over 30 mt/ha per year for a single rotation. They concluded wider spacings and longer cutting cycles outperformed "wood grass." DeBell et al. (1996) also found biomass differences in the two aforementioned clones planted in monoclonal blocks for 7 years. Clone H 11-11 averaged 18.2 mt/ha per year at 1m spacing for 7 years. DeBell et al. (1997) also studied four clones including black cottonwood and H 11-11 monoclonal blocks at 1 × 1 m spacing. Biomass productivity ranged from 11 to 18 mt/ha per year after 5 years. Hybrid poplar H 11-11 out-produced the native black cottonwood, but production peaked at 4 years at close spacing. Ceulemans et al. (1992) and Scarascia-Mugnozza et al. (1997) also studied the productivity of two Stettler hybrids compared to parental *P. trichocarpa* and *P. deltoides* clones at 1 × 1 m spacing. Clonal ranking for biomass production of 4 years showed the hybrid clones outperformed the parental clones. H 11-11 produced over 35 mt/ha per year whereas the parental clones yielded 16 and 14 mt/ha per year, respectively, for the *P. trichocarpa* and *P. deltoides* clones.

Operational plantings are not as productive as field research studies because of soil heterogeneity, weather, and pests and diseases. DeBell et al. (1997) have reported biomass production rates for operational poplar plantings of 12–17 mt/ha per year in western Washington. Industry researchers Stanton et al. (2002) reported commercial yields of improved poplars averaging 13.8 mt/ha per year with selected hybrid poplars yielding over 20 mt/ha per year in 6 years. These results highlighted the importance of genetic improvement in short rotation poplar culture. When clones from the early stages of hybridization and cloning program produce over 20 mt/ha per year in commercial plantations, even higher yields can be expected when future improved clones become available. These results suggest a "green revolution" of short rotation biomass production is possible with poplar. Just 30 years ago Cannell and Smith (1980) had calculated in their review and appraisal of close spaced hardwoods in temperate regions that the maximum theoretical working production was between just 10–12 mt/ha per year.

15.4.2.2 Europe

Europe obviously has a long rich history of poplar plantations. Notably much of the European interest in poplar came when European explorers returned from North America with handsome poplar specimens that they subsequently planted in their gardens. Spontaneous hybrids that resulted gave rise to poplars known as "intercontinental" or "Canadian" poplars. Another popular European introduction was the "Carolina" poplar—a cottonwood from southeast United States (FAO 1980; Dickmann 2006). The new poplar hybrids were first planted along waterways and roadways. Remarkably Claude Monet, the French impressionist painter, loved poplars and became a poplar plantation owner in 1891 when his favorite poplar models depicted in his famous "Poplars" series were about to be cut by the village (Tucker 1989). The "Euramerican" poplars, a natural hybrid of *P. deltoides,* and *P. nigra* (now known as *P. canadensis)*, were planted widely throughout Europe between 1900 and World War II (FAO 1980).

An important milestone in the advancement of European poplar culture was the establishment of the Instituto Sperimentazione Pioppicoltura at Casale Monferrato, Italy, during World War II (Dickmann 2006). This institute, still active today, was involved in creating many new poplar clones (such as I-214) and development of poplar culture. The clones produced in Casale were planted throughout Europe and are still widely planted throughout the world. Another important European milestone was the founding of the International Poplar Commission (IPC) in 1947 in Rome under the auspices of the United Nations Food and Agriculture Organization (FAO) (Pourtet 1976; Viart 1976). This organization also exists today with a mission to promote all aspects of poplar culture for improving rural livelihoods worldwide. In recent years, the IPC is becoming more active in promoting the uses of poplars for bioenergy.

After WW II, there was a revival of breeding and growing poplars in Europe albeit for conventional wood products. Poplar breeding began in the United Kingdom in the early 20th century (Henry 1914), but breeding programs took some time to evolve (Schreiner 1959). Pauley (1949) and Muhle-Larsen (1970) reviewed advances in poplar breeding in Europe. There were many active poplar breeding programs/breeders in Italy (Piccarolo), Netherlands (Koster), Belgium (Steenackers), Germany (Weisberger), and Spain (FAO 1980). Most of these breeders actively cooperated with North American breeders in the exchange of information and materials.

Thousands of hectares were planted to poplars in the ensuing years in Europe. According to the IPC (FAO 2008), there are currently 236,000 ha of poplar plantations in France, 118,500 ha in Italy, 100,000 ha in Germany, 98,500 ha in Spain, and thousands of hectares in other countries such as Croatia and Romania. Many of these plantations were planted in the 1950s and 1960s.

Short rotation plantations of poplar for pulp and paper and for bioenergy did not become popular in Europe until after the OPEC oil embargo of 1973. Avanzo (1974) promoted the genetic improvement of poplars for biomass harvests in Italy. The bibliographic review of international research on "short rotation forests" detailed widespread interest in growing poplars on short rotations in Europe and worldwide (Louden 1976).

In Europe, poplars have been largely grown at wide spacings for timber and traditional wood products in modern times. Notably, FAO (1980) made little mention of the use of poplars for energy except in some eastern countries where poplars were grown for fuelwood. In the 1960s, this tradition began to change toward growing poplars at close spacings and on shorter rotations. The goal was to produce more wood per unit of land area in a shorter time frame for wood fiber and energy (Mitchell et al. 1992). Eastern European countries embraced short rotations very early. Markovic et al. (2000) reported short rotation poplar test plantations as early as 1960 with many poplar clones and varieties and many spacings and cutting cycles. Marosvolgyi et al. (1999) reported long term studies of poplars for bioenergy in Hungary. European biomass studies differed from North American in that more poplar species and species hybrids were tested, and coppicing was early on accepted as a means of increasing poplar biomass production per hectare. Avanzo (1974) reported tests of 600 *P. deltoides* clones, 148 clones of *P. nigra*, and 73 D × N hybrids for biomass production on 3–5 year rotations. Most of the European studies of short rotation poplars were still at traditional wide spacings (Louden 1976).

Ceulemans and Deraedt (1999) reviewed the potential of poplars grown under short rotation culture for bioenergy. They outline that the way to optimize biomass productivity is to optimize plant genotype or cultural management regime, or the interaction between both. Their review compares many poplar clones, spacings and climates under coppicing (Ferm and Kauppi 1990) and noncoppicing regimes. They found that there are clonal differences in biomass yields under different coppicing regimes and reported yields of 20–30 mt/ha per year. In some cases the noncoppiced clones yielded more than the coppiced (Proe et al. 2002). Cannell and Smith (1980) concluded that the "working maximum" for short rotation forests was theoretically 10–12 mt/ha per year with a possible increase of 10–20% with coppicing. At the time, this upper limit seemed reasonable with existing clones. We know now that yields can be increased further with genetically improved material and coppicing. For example, Pontailler et al. (1999) reported biomass yields of 4 poplar clones at 0.8 m spacings over 5 coppices and found the highest biomass yield was over 30 mt/ha per year! Improved genetic material does make a difference (Ceulemans 2004). Benetka et al. (2007) found that native *P. nigra* clones produced much less than an NE U.S. hybrid. However, in some regions the use of native plant material is required. They cited disease resistance as an important factor in biomass yield.

R. Ceulemans and co-workers conducted a series of studies in the late 1990s to identify coppicing differences in biomass production in short rotation poplar (Ceulemans 2004). Laureysens et al. (2003, 2004) reported results of the first rotation coppice study in Belgium. After 6 years the biomass production of 17 poplar clones varied with clone; mortality of stools was also clone related. Biomass production was 1.6–10.8 mt/ha per year (Table 15.5). The "best" clone was a T × D hybrid. Laureysens et al. (2005) reported results of the second rotation of short rotation coppice culture of poplar of the above study. Notably the results of the second rotation were significantly different than the first. The *P. nigra* had biomass yields at 9.7 mt/ha per year, but the first rotation "best" T × D clones performed poorly. *Melampsora* rust played an important part in determining

TABLE 15.5
Chronological Synthesis of Poplar Biomass Production Field Studies and Plantations by Location in Europe

Location	Clone/ Speciesa	Age (years)	Spacing (m)	Productivity (mt/ha per year)	Reference
United Kingdom	T	5	0.5	9–10	Cannell and Smith (1980)
Finland	B	6	0.7–1.4	4.2	Ferm et al. (1989)
France	TD	2–3	Multiple	0.6–3.5	Auclair and Bouvarel (1992)
United Kingdom	T, TD	4	1.0–2.0	13.6	Armstrong et al. (1999)
Belgium	TD	Multiple	0.8	30	Pontailler et al. (1999)
United Kingdom	B	5	1.5	10.2–16.2	Proe et al. (2002)
Belgium	Multiple	6	0.75–1.5	10.8	Laureysens et al. (2003)
Belgium	Multiple	4	1.0	2.8–11.4	Laureysens et al. (2004)
Belgium	Multiple	Multiple	1.0	9.7	Laureysens et al. (2005)
Sweden	T,TD	13	1.2	8.0	Christersson (2006)
Czech Republic	N,MT	4–7	2.1	7.6–9.4	Benetka et al. (2007)
Belgium	Multiple	3	0.75–1.5	2.8–9.7	Al Afas et al. (2007)
Belgium	Multiple	11	0.75–1.5	13.3–14.6	Al Afas et al. (2008)

[a] B, *P. balsamifera*; D, *P. deltoides*; N, *P. nigra*, T, *P. trichocarpa*, DN, D × N; TD, T × D; M, *P. maximowiczii*; MT, M × T.

biomass production in the second rotation. Al Afas et al. (2008) summarized the biomass yields over 3 rotations (11 years) from the above studies at Boom, Belgium. Biomass production varied with clone and rotation. Biomass increased from year to year within a rotation, but declined with number of rotations. The N × T hybrids were the "best" overall performing clones and were "best" adapted to multiple coppice rotations.

Recent EU policy changes have had a major effect on efforts to find high yielding poplar biomass systems. The goal is to double EU renewable energy use from 6 to 12% by 2010 (Klasnja et al. 2006). Therefore, there has been an increase in the poplar biomass production studies in Europe. Sixto et al. (2006) reported 18 new experimental studies of poplars with very high density plantings (i.e., 15,000–33,000 trees/ha) of many poplar clones. Their goal is to identify high biomass producers while providing environmental benefits and rural employment. Scarascia-Mugnozza et al. (2006) reported 5000 new ha of poplar plantations in northern Italy. New clones improve yields by 100%, yielding 18–24 mt/ha per year. The maximum yield was 20 mt/ha per year. The studies also investigated the use of animal and urban wastewater for irrigating short rotation poplar plantations. More recent results are even more favorable. New clones with N fertilization are producing an average of 23 mt/ha per year and have achieved up to 50 mt/ha per year (Paris et al. 2008). The Italian Ministry of Agriculture has also recently financed new research on poplar biomass supply chains. The goal is to use poplar biomass for feeding district heating plants. Early results confirm biomass yields of 20–25 mt/ha per year planted at different spacings and rotations (Facciotto et al. 2008; Nervo et al. 2008).

15.4.2.3   Other Countries

The largest poplar plantations in the world are in China. China has over 4.3 million ha of poplar plantations and an additional 2.5 million ha of poplars used in agroforestry (FAO 2008). India also has vast areas of poplars used for agroforestry (Puri and Nair 2004), and Turkey has over 138,000 ha of planted poplars; again, large areas are planted under agroforestry with poplars interplanted with beans, corn and melons (Toplu 2008). There are large areas of poplars used for agroforestry and silvo-pasture in New Zealand (FAO 2008), and in the 1960s there were plantations of poplar in Australia and South Africa (Pryor and Willing 1965). There are also an increasing number of poplar plantations in South America (Eaton 2008b).

Small farmers in China, India and Turkey have been using poplars for fuelwood since antiquity, but until recently poplar plantations were used mostly for traditional wood products and not bioenergy. Fang et al. (1999) reported 6 year results of short rotation poplars in China with annual biomass increments of 17 mt/ha per year. Biomass yields differed considerably with planting density, clone and rotation length. In later reports, Fang et al. (2006, 2007) reported biomass yields of 14.6 mt/ha per year for three poplar clones at age 10. Das and Chatuverdi (2009) reviewed biomass production of poplars in agroforestry systems in India and reported biomass yields of 5–10 mt/ha per year. Dhanda and Kaur (2000) also reported on poplar biomass production in an agri-silviculture system in Punjab, India. There is great potential in South America for growing poplars for biomass production. Eaton (2008b) reported test yields of genetically improved poplars of 30 mt/ha per year in Chile.

15.4.3   Bioenergy from Poplars

Interest in biomass for energy accelerated after WW II (Young 1973). Wood energy plantations were seen as a way of achieving energy independence. Szego and Kemp (1973) were early proponents of the "energy forests" or fuel plantations. Up until that time, short rotation plantations were viewed as a solution for maximizing fiber yield and taking the pressure off of native forests to help provide the nation's pulpwood supply (Dawson 1976). Then, came the OPEC oil embargo in 1973. The embargo changed the public and government attitude toward alternative energy sources, such as biomass. Notably, Stephens (1976) gave the keynote address at an International Cottonwood Symposium promoting poplar biomass as an energy source. Tillman (1976) reviewed the potential of wood as an alternative fuel at that time. Clearly, high biomass yields of plantations were needed to make fuel plantations feasible.

Zavitkovski et al. (1976) outlined the potential of using poplar hybrids as a source of wood and energy. Soon after, Isebrands et al. (1979) examined the potential of an integrated utilization strategy for biomass from short rotation poplars. The strategy provided alternatives, including 1) pulp only, 2) energy only, 3) pulp and fuel and 4) pulp, fuel and animal feed. The case study used was a 5-year-old hybrid poplar plantation grown at 1.2m spacing. The biomass yield was 8.4 mt/ha per year for pulp or 200 MKcal/ha for energy. Energy values determined for NE poplar clones in Wisconsin (Strong 1980) were used to calculate the high heating value of the biomass. The caloric values of poplar clones were relatively similar, ranging from 4636 to 4755 cal/g or $1.9–2.0 \times 10^4$ J/g (i.e., 19 MJ/kg). These results are similar to those of Bowersox et al. (1979) with other clones. Their values also agreed closely with those provided by Sastry and Anderson (1980) for juvenile poplar clones grown in Canada and more recent findings (Klasnja et al. 2006). There are 4.2 joules in a calorie in biomass. So, the poplar biomass plantation produced over 200 Mkcal/ha (8.4×10^4 J/ha). A high heating value of 200 Mkcal/ha in 5 years is equivalent to 40 Mkcal/ha per year or 27 barrels of oil equivalent per ha. Their analysis assumed each barrel of oil equals 1.48 Mkcal or 6.1×10^9 J. With each ton of poplar biomass equating to more than 3 barrels of oil equivalent, 30 mt/ha equals 90 barrels of oil!

For the energy plantation concept to work successfully, the energy output/input ratio must be positive. Therefore, biomass energy produced must exceed the energy spent on production and harvesting (Anderson et al. 1983). Zavitkovski and Isebrands (1985) analyzed the energy output/input ration in short rotation hybrid poplar and found them favorable. Anderson et al. (1983) in a thorough review of energy plantations concluded that one must maximize bioecological factors, genetic materials, and cultural factors to maximize biomass yields of an energy plantation. Therefore, there is a premium placed upon achieving as much biomass per ha as possible to be successful. Kauter et al. (2003) conducted an analysis of quantity and quality of *Populus* short rotation coppice systems for energy in European systems. Notably, their energy content was very similar to those found in the North American studies. And, the energy content of the short rotation poplar biomass was similar to caloric values of native aspen stands in Canada (Peterson et al. 1970). Christersson (2008) made an energy comparison of short rotation poplar with agricultural crops in Sweden. According to his analysis, the poplar biomass crops compare more favorably in energy balance than sugar beet and wheat in Sweden.

15.4.4 Limitations and Challenges

Despite the promise of using poplar biomass for bioenergy, there are many limitations and challenges that must be overcome. These limitations can be classified into several categories including biological, technical, economic, political and psychological (Anderson et al. 1983; Hoffman and Wieh 2005; van Rees 2008). These categories are closely interrelated and often have multiple interactions. Despite the huge genetic gains that have been made in biomass yields through genetic hybridization and cloning programs, the major factor affecting the production costs of bioenergy is biomass yield. Thus, the emphasis placed on biomass production in this chapter. The consensus is that biomass yields will have to be at least 15–20 mt/ha per year to compete with fossil fuels (Anderson et al. 1983), and some analysts believe 20 mt/ha per year may not be enough given recent economic conditions (Gallagher et al. 2006). They concluded that yields will need to be increased by 40% through improvements in genetics and silvicultural practices to make short rotation biomass plantations cost effective. Genetic improvements take considerable time, but "best management practices" for growing poplars have recently been developed (van Oosten 2006; Isebrands 2007). However, climate effects and diseases are also very important (Anderson et al. 1983; Kauter et al. 2003). Other important variables in economic analysis of growing poplars have been known for many years. They include: land prices, planting stock, fuel, labor, site preparation, maintenance, fertilizer, irrigation, harvesting, drying, handling, transportation, taxes, and insurance as well as market values at harvest (Lothner 1983; Yemshanov and McKenney 2008). These variables have not changed over the years, but the absolute costs and revenues have changed markedly (Isebrands 2007). Economic variables also vary with the technology used to produce fuels and chemicals from biomass. There are many conversion technologies possible for use with woody biomass- new and old. They include pyrolysis, gasificaton, liquefaction, ethanol production by hydrolysis, and other biochemical processes (Phelps 1983). Unfortunately, although all are technically feasible, none are presently economical on a commercial basis without subsidy (Christersson 2008). Recent developments suggest cellulosic ethanol will become economic (Decker 2009).

A major limitation can be the availability of markets and proximity to markets (Hoffman and Wieh 2005). For example, without a viable market, nurserymen are reluctant to risk investment in scaling up genetically improved clones, and without large numbers of improved plant materials, large biomass plantations are not possible (i.e., "the chicken and egg" effect). With such uncertainty, financial credit is often difficult to secure. These factors affect the mindset or psychology of farmers who might change from traditional crops to biomass crops (van Rees 2008).

When it comes to cellulosic ethanol from biomass, there are a number of hurdles to commercializing the process (Alexander and Gordon 2009). One is difficulty in raising capital for a project without a cash flow. Another hurdle is the need to convince farmers that a market for their product will emerge. Moreover, there is uncertainty about the production and transportation costs of a new crop and product. In addition, the land availability for cellulosic ethanol biomass may be distant from the access to populated areas where the ethanol is needed, thereby increasing costs. Finally, there is a political hurdle that is termed the: "blend wall", i.e., the percent of ethanol blend allowed in gasoline. Unless the blend percentages are increased, this factor could be a major hurdle. All of these factors are inherent in the development of a new crop with new conversion technology (Sims et al. 2009). Dallenmand et al. (2008) summarized the challenges faced with implementing biomass energy in the EU. These challenges include (1) supply industry with raw material year-round, (2) ensure a harvest window, (3) ensure harvest and collection efficiency, (4) improve energy processing technologies, (5) develop reliable storage and transport systems, (6) optimize feedstock quality, (7) optimize biomass pellet technology, and (8) develop efficient logistical structure such as fuel depots. All of the limitations and challenges are a part of the development of any new crop and conversion technology.

15.4.5 Future Considerations and Conclusions

Despite the numerous limitations and challenges to commercializing poplar bioenergy, there are a number of promising developments and opportunities for poplar biomass ahead. For example, there are some important policy developments that will affect biomass for energy. At the 17th European Biomass Conference in Germany in July, 2009, high officials from the German Ministry of the Environment and the U.S. Department of Energy spoke about their country's commitment to biomass as an alternative energy source. Moreover, the United States is now participating in International Climate Change talks. This development should help short rotation woody crops reach their potential for part of a carbon management framework (Tuskan and Walsh 2001). Zabek and Prescott (2005) have shown that the carbon content of poplar biomass plantations ranges from 74 to 89 Mg/ha in 14 years. Moreover, in China, which has the largest number of poplar plantations in the world, Fang et al. (2006) has shown that the carbon sequestration potential in China alone is 3.8×10^7 mt/ha per year.

As the CO_2 concentration in the world increases, there are some potential opportunities for poplar culture. Liberloo et al. (2006) working at the EUROFACE site has shown that poplars are growing much faster in a CO_2 rich environment. So, it is hoped that landowners worldwide someday will be able to receive carbon incentives to help make poplar biomass plantations more economically feasible (Yemshanov and McKenney 2008).

Another encouraging development is the recent sequencing of the poplar genome by molecular biologists (Tuskan et al. 2006). This new development should allow geneticists to improve growth, disease resistance and chemical composition of poplars in the future (Taylor 2002, 2008; Wullschleger et al. 2005). There now may be opportunities for genetic engineering of herbicide resistance into poplars (i.e., Roundup Ready poplars) and improved cellulosic biosynthesis needed to make cellulosic ethanol more economical (Joshi et al. 2004; Sims et al. 2009).

There are some possible policy decisions that could make cellulosic ethanol more competitive, but they are not yet available (Carling 2009). But there are still other limitations that include production and transportation costs as well as access to land in populated areas (Alexander and Gordon 2009). Notably, Greenwood Resources, Inc., and ZeaChem are currently building the first poplar demonstration biorefinery to produce cellulosic ethanol in Boardman, OR. It is scheduled to open in 2010 (Eaton 2008a).

Europe is ahead of the rest of the world in development of poplar biomass crops for power production, probably because of incentives. In Italy, four power plants of the energy company Power Crop that will use poplar biomass as a feedstock are under construction. These facilities will require 30,000 ha of land dedicated to short rotation poplar coppice plantations to feed them. The commercial plantations were scheduled to become operational in 2009. Another power company near Milano, Italy, is constructing a small power plant that will utilize poplar biomass from 3000 ha of poplar plantations nearby. Another European power company, RWE, is planting 10,000 ha of short rotation poplar coppice to feed a co-fired coal power plant in Germany. RWE has plans to develop other facilities in Romania and Spain that will utilize poplar biomass. A French power company also has plans to use 1000 ha of short rotation poplar coppice for a co-fired power plant in Hungary. And, in the United Kingdom, the Drax power company is planning on building three power plants that will utilize local poplar biomass. They presently have a power plant that utilizes imported wood pellets for fuel. These power plants are able to economically use poplar biomass because of the EU tax policies aimed at reducing $CO2$ emissions to the atmosphere. There have already been several successful biomass fired power plants in Northern Ireland and Sweden using willow biomass. The short rotation willow coppice system was developed by a research group at the Swedish University of Agricultural Science in Uppsala, Sweden (Perttu 1989).

Poplar biomass will likely be used for fuel in developing countries for years to come. However, there is a worldwide increase in the use of poplar wood residues for energy production in small forest products industries and businesses in developing countries, especially China and India. Moreover, there is a trend toward co-firing poplar biomass with coal for electricity generation in some countries to decrease costs and improve air quality (De and Assadi 2009). This trend will likely continue to increase in the future because of the shrinking world economy.

The cellulose ethanol industry in Europe is still undergoing development as in North America (Slade et al. 2009). Poplars are one of the several preferred biomass feedstocks under study, and North America companies may build or acquire poplar based cellulosic ethanol plants in Europe once this technology has been commercialized. Thus, the successes in the next few years will likely determine the future success of commercialization of poplar bioenergy throughout the industrialized world.

There are also encouraging developments for the use of poplars for their environmental benefits. Short rotation woody crops can provide ways of cleaning up (i.e., phytoremediation) polluted water and soils (Perttu 1989; Licht and Isebrands 2004). There are many intangible benefits that short rotation poplars for biomass production provide including soil erosion control, protection, wildlife benefits, air quality improvement, carbon sequestration, aesthetics and psychological benefits (Licht and Isebrands 2004; Isebrands 2007). All of the aforementioned developments point toward a promising future for the use of poplars as a bioenergy source.

15.5 OTHER FOREST SPECIES BIOENERGY USE AND POTENTIAL WORLDWIDE

In addition to *Eucalyptus*, *Pinus*, and *Populus* species, many other trees have promise for bioenergy production in various parts of the world. *Salix* species covered in Section 3, Chapter 4 of this book are well suited to wet temperate climates. Numerous others have been proposed for use in different climatic zones (Little 1981; NAS 1977, 1979, 1980, 1982; NRC 1983a, 1983b, 1983c, 1984).

Humid Tropics—*Acacia auriculiformis, A. mangium, Albizia falcataria, Bursera simaruba, Calliandra calothyrsus, Casuarina equisetifolia, Coccoloba uvifera, Derris indica, Gliricidia sepium, Gmelina arborea, Guazuma ulmifolia, Hibiscus tiliaceus, Leucaena leucocephala, Maesopsis eminii, mangrove genera, Mimosa scabrella, Muntingia calabura, Psidium guajava, Sesbania bispinosa, S. grandiflora, Syzygium cumini, Terminalia catappa, Trema* spp.

Tropical Highlands—*Acacia decurrens, A. mearnsii, Ailanthus altissima, Alnus acuminata, A. nepalensis, A. rubra, Gleditsia triacanthos, Grevillea robusta, Inga vera, Melaleuca quinquenervia, Melia azedarach, Robinia pseudoacacia, Sapium sebiferum*

Arid and Semiarid Regions—*Acacia brachystachya, A. cambagei, A. cyclops, A. nilotica, A. saligna, A. senegal, A. seyal, A. tortilis, Adhatoda vasica, Ailanthus excelsa, Albizia lebbek, Anogeissus latifolia, Azadirachta indica, Balanites aegyptiaca, Cajanus cajan, Cassia siamea, Colophospermum mopane, Combretum micranthum, Conocarpus lancifolius, Dalbergia sissoo, Emblica officinalis, Haloxylon aphyllum, H. persicum, Parkinsonia aculeata, Pithecellobium dulce, Propsopis alba, P. chilensis, P. cineraria, P. juliflora, P. pallida, P. tamarugo, Sesbania sesban, Tamarix aphylla, Tarchonanthus camphoratus, Ziziphus mauritiana*, and *Z. spina-christi*

Many of these are multipurpose trees that have more than bioenergy uses. They tend to have many favorable characteristics such as wide adaptability, easy establishment, low maintenance requirements, tolerance of difficult environments, nitrogen-fixing ability, rapid growth, coppicing, and high energy content. Those that are multistemmed, poorly formed, and short-lived may be best for family and small-scale use. Most are suitable for plantations to meet larger-scale bioenergy needs. Because these species are aggressive, grow rapidly, and often seed early and prolifically, they are potentially invasive and should be used carefully. Native species should always be strongly considered for bioenergy plantations.

15.6 PRETREATMENT OF FOREST BIOMASS

Biochemical conversion of lignocellulosic biomass through enzymatic saccharification and fermentation is a major pathway for liquid fuel production from biomass (USDE 2005; NSF 2007). In this approach, biomass cellulose is converted to glucose through microbial or enzymatic actions. The glucose is then fermented into alcohols, such as ethanol. Residues resulting from microbial and enzymatic actions contain mainly lignin and can be converted to energy thermochemically. Whereas starch functions as an energy storage material in plants, wood functions as a structural material. As a result, woody biomass has natural resistance—often called "recalcitrance"—to microbial and enzymatic deconstruction (Himmel et al. 2007). Forest (woody) biomass differs from agriculture biomass physically, structurally, and chemically. Specifically, forest biomass is physically large and structurally very strong. Its density is higher than agricultural biomass. It has higher lignin, higher cellulose, and lower hemicellulose contents than most agricultural biomass. As a result, forest biomass has much greater recalcitrance than does agricultural biomass. On the other hand, the high density and high lignin and cellulose content increase energy content and reduce transportation cost for forest biomass, which is favorable for advanced energy production.

Recalcitrance of lignocellulosic biomass is a major barrier to economical development of bio-based fuels and products through the biochemical pathway. This is especially true for forest biomass because it has greater recalcitrance than does agriculture biomass. The technical approach to overcome this recalcitrance has been pretreatment to make cellulose more accessible to hydrolytic enzymes for conversion to glucose. Both physical and chemical pretreatments have been used to achieve satisfactory conversion of lignocellulose. Physical pretreatment refers to the reduction of physical size of forest biomass feedstock to increase enzyme-accessible surface areas (Lynd 1996; Zhu et al. 2009) and decrease the crystallinity of cellulose. Chemical pretreatment refers to the process of using chemicals to remove or modify key chemical components that protect cellulose in biomass, mainly hemicellulose and lignin. Chemical pretreatment often provides a good separation of hemicellulose in the form of sugars from cellulose for high-value utilization, including liquid fuel through fermentation.

15.6.1 Physical Pretreatment of Forest Biomass—Physical Size Reduction

The issue of physical size reduction has been largely overlooked in the lignocellulosic ethanol research community. The reason is likely in part that most lignocellulosic ethanol research has focused on using agricultural biomass that does not need a significant amount of mechanical energy to achieve satisfactory size reduction. However, size reduction is very energy intensive for forest biomass. In wood-fiber production, size reduction is in two steps. The first step is coarse size reduction, reducing wood logs to chips of 10–50 mm in two dimensions and 2–10 mm in the third dimension. The second step is to further reduce the wood chips to fibers of millimeters in length. Energy consumption in the first step is much lower than that in the second step.

A simple energy balance calculation using forest biomass demonstrates the importance of size reduction for biomass refining. Assume that ethanol yield from wood is about 300 L/tonne of ovendried wood with current technology. Higher heating value of ethanol is about 24 MJ/L, which gives total wood ethanol energy of 7.2 MJ/kg wood. Typical energy consumption to produce wood chips is about 50 Wh/kg; energy consumption in the second step through disk milling can be anywhere from 150 to 700 Wh/kg (Schell and Harwood 1994), depending the fiberization process and the degree of milling. With these assumptions, total size-reduction cost is 200–600 Wh/kg, which is equivalent to 0.72–2.16 MJ/kg, or 10% to 30% of the wood ethanol energy available.

Three factors affect energy consumption during size reduction: the degree of size reduction, the fiberization mechanism, and chemical or biological pretreatment before size reduction. All of these factors also affect enzymatic cellulose saccharification. Most of the existing literature on size reduction relates to pellet, fiber, and wood flour production. Few studies on biomass size reduction have taken an integrated approach to examining energy consumption, enzyme-accessible substrate surface, and chemical pretreatment efficiency in terms of enzymatic cellulose conversion. Most reported work on size reduction has not involved cellulose conversion (Schell and Harwood 1994; Cadoche and Lopez 1989; Mani et al. 2004) and has addressed only energy consumption and substrate size. On the other hand, reports on enzymatic hydrolysis using size-reduced substrates did not provide information about energy consumed to produce the substrate and/or a careful and complete characterization of substrate size (Allen et al. 2001; Zhu et al. 2005; Nguyen et al. 2000). At most, substrates were characterized by sieving or screen methods (Sangseethong et al. 1998; Chundawat et al. 2007; Dasari and Berson 2007; Hoque et al. 2007). Consequently, there is a knowledge gap linking energy consumption, substrate surface, and pretreatment efficiency.

15.6.1.1 Degree of Size Reduction and Substrate-Specific Surface

To address the degree of size reduction, proper characterization of forest biomass substrate is necessary. The geometric mean diameter of the substrate particles measured by traditional sieving and screen methods has been almost exclusively used for biomass substrate size characterization (Mani et al. 2004). This size measure is significantly affected by biomass substrate morphology, such as particle aspect ratio (Zhu et al. 2009). Most size-reduction processes produce fibrous substrate with a wide range of particle (fiber) aspect ratio of 5:100. As a result, existing data on substrate size characterization have limited value. Enzyme-accessible surface area is of most interest for saccharification. Holtzapple et al. (1989) calculated specific surface area based on a spherical particle assumption to correlate energy consumption for comparing the efficiencies of several size-reduction processes. The spherical

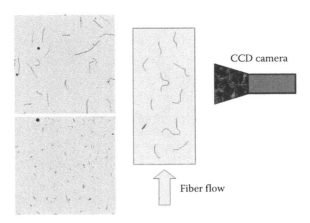

FIGURE 15.2 (See color insert) Schematic diagram showing the wet-imaging technique and typical images acquired.

model for specific surface calculation is justifiable for particles with close to unity-aspect ratio, such as wood sawdust, but is questionable for most forest biomass substrates consisting of fibers. Recently, we developed a wet-imaging technique for the characterization of forest biomass substrate (Zhu et al. 2009). In this technique, the two dimensions of each substrate fiber are measured in a flowing water channel by an optical microscopy using a charge-coupled device (CCD) camera (Figure 15.2).

The total surface and volume of the substrate can be calculated using a cylinder model for each fiber. The volumetric specific surface can be determined by dividing the total surface of a sample by its volume [eq. (15.1)] (Zhu et al. 2009). For most mechanically derived substrates, the cylinder assumption is reasonable, as confirmed by scanning electron microscope pictures of the substrates (Zhu et al. 2009). If the substrate particles are spherical like sawdust, then a spherical model can be used to determine specific surface [eq. (15.2)].

$$S_f^V = \frac{A_f}{V_f} = \frac{2\sum_i n_i(d_i^2 + 2 \cdot d_i \cdot L_i)}{\sum_i n_i d_i^2 \cdot L_i} = \frac{4\sum_i n_i d_i (L_i + d_i/2)}{\sum_i n_i L_i \cdot d_i^2} \approx 4 \cdot \frac{\sum_i n_i L_i \cdot d_i}{\sum_i n_i L_i \cdot d_i^2} = \frac{4}{D_{L21}} \quad (15.1)$$

$$S_p^V = \frac{A_p}{V_p} = 6 \cdot \frac{\sum_i n_i d_i^2}{\sum_i n_i d_i^3} = \frac{6}{D_{32}} \quad (15.2)$$

S and A are the specific surface and total surface area of the substrate, respectively. Subscript f and p are fiber (cylinder model) and particle (sphere model), respectively. D_{L21} is fiber-length weighted-surface-length mean fiber diameter or "width." D_{32} is often called Sauter mean diameter (SMD). With this wet-imaging technique, we were able to objectively compare the efficiencies of different size-reduction and chemical pretreatment processes in terms of size-reduction energy consumption and cellulose to glucose conversion.

The specific surface can be used to measure the degree of size-reduction. Zhu et al. (2009) used specific surface to successfully measure the efficiency of various size-reduction processes.

15.6.1.2 Fiberization Mechanism

Energy consumption in mechanical pulping of wood depends significantly on how wood chips are fiberized. Refiner mechanical pulps (RMP) are produced under atmospheric refining conditions,

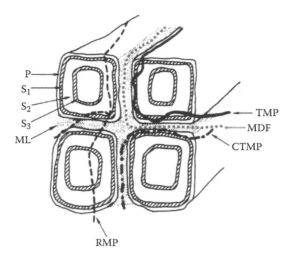

FIGURE 15.3 A schematic diagram showing various fiberization mechanisms of wood. (Adapted from Franzen, R., *Nordic Pulp Paper Res J,* 1, 4, 1986; Salmen, L., *Fundamentals of Mechanical Pulping,* in Book 5: Mechanical Pulping, Papermaking Science and Technology, Fapet Oy, Finland, 1999. Used with permission of Nordic Pulp and Paper Research Journal and Finland Paper Engineers' Association.)

with wood chips fractured through the lumen of wood tracheid. Thermomechanical pulps (TMP) are produced using low-pressure steam ($\sim 2.4 \times 10^5$ Pa, $\sim 134°C$) to soften wood chips before disk refining. The wood chips are fractured in the S1 and S2 layer of cell wall.

Medium-density fiberboard pulps (MDF) are produced under increased steam pressure of $>5 \times 10^5$ Pa. In the MDF production process, wood chips are fractured in the lignin-rich middle lamella (ML). This is because the steam temperature reaches the glass transition temperature of lignin (Irvine 1985). Figure 15.3 is a schematic of various fracture mechanisms of wood chips during fiberization. Energy consumption of different pulping processes varies significantly. Typical energy consumptions for producing RMP, TMP, and MDF are about 600, 450, 150 Wh/kg oven-dried wood, respectively. Energy consumption for chemical-thermomechanical pulp (CTMP) is just lower than that for TMP. The surface chemical compositions of these pulps are very different. RMP exposes mostly cellulose on fiber surfaces. MDF fibers are lignin-coated on their surfaces. This can be clearly seen from the color of these pulps, with RMP being the lightest and MDF being brown. The difference in surface chemical composition certainly affects cellulose enzymatic conversion to glucose, as revealed in our previous study (Zhu et al. 2009). The significant variations in mechanical energy consumption of these different pulping processes may provide avenues for potential energy savings in biomass size reduction. However, attempts have not yet been taken to explore this potential.

15.6.1.3 Effect of Chemical Pretreatment

The third factor affecting size-reduction energy consumption and enzymatic cellulose saccharification is chemical pretreatment. Most of the enzymatic cellulose saccharification research has used size-reduced substrate for the purpose of reducing substrate recalcitrance to achieve high cellulose conversion (Nguyen et al. 2000; Allen et al. 2001; Zhu et al. 2005). In fact, to achieve good chemical penetration and therefore effective pretreatment, size reduction before chemical pretreatment is necessary for the dilute acid process (Lynd 1996), one of the most investigated chemical pretreatment processes for lignocellulosic ethanol production. Chemical pretreatment alters the chemical composition and physical structure of biomass by partly removing some cell-wall components, such as hemicellulose and lignin. As a result, size reduction after chemical pretreatment can reduce energy consumption. This energy savings may be insignificant for some agricultural biomass, such as corn stover or switchgrass, but can be very significant for forest biomass (Zhu et al. 2010). This suggests

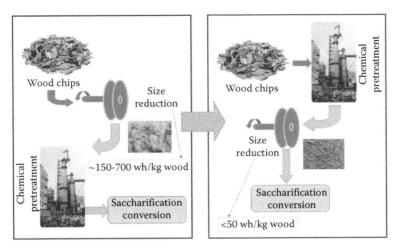

FIGURE 15.4 (See color insert) Required order for size reduction operation from pre- to postchemical pretreatment to reduce energy consumption.

that post-chemical pretreatment size reduction is not only preferred (Figure 15.4) to take advantage of chemical pretreatments for economical size reduction, but also necessary for the economical biochemical conversion of forest biomass as discussed previously. Furthermore, different chemical pretreatments alter biomass structure to various degrees and therefore affect post-pretreatment size-reduction energy savings (Zhu et al. 2010). For example, the SPORL pretreatment (Zhu et al. 2009) of wood chips is much more effective to reduce size-reduction energy consumption than the dilute acid pretreatment conducted at the same pretreatment time, temperature, and acid charge conditions (Zhu et al. 2010).

This post-chemical pretreatment size-reduction process flow design has several benefits: (1) it takes advantage of chemical pretreatment to alter wood structure to reduce energy consumption in the subsequent size-reduction process; (2) it avoids the difficulties and high-energy consumption for mixing high-consistency pulp with chemicals in pretreatment when size-reduced substrate is used; (3) it can reduce thermal energy consumption in chemical pretreatment; and (4) it can potentially produce a concentrated hemicellulose sugar stream to reduce concentration cost. The rationale for benefits 3 and 4 is that a lower liquid-to-wood ratio can be used in the chemical pretreatment of wood chips than that in pretreatment of fiberized wood pulps. Liquid uptake of fiberized wood pulps is much higher than that of wood chips because of the porous and hydrophilic nature of wood.

15.6.2 CHEMICAL PRETREATMENT OF FOREST BIOMASS

Existing enzymes cannot effectively convert lignocellulose to fermentable sugars without chemical pretreatment. Few pretreatment technologies are capable of effectively removing recalcitrance of forest biomass, especially softwood, to achieve satisfactory cellulose conversion to glucose. Because of the large amount of energy required in size reduction for forest biomass, a viable pretreatment process for forest biomass needs to be not only effective in removing recalcitrance but also capable of doing so before wood size reduction (i.e., on wood chips, not on fiberized materials). This makes pretreatment of forest biomass much more difficult than and different from agriculture biomass, which has not been well recognized by the research community. The most studied and currently widely adopted process in commercial demonstration (i.e., dilute acid process) developed a century ago has been shown to be ineffective for almost all wood species. Acid-catalyzed steam explosion, organosolv, and sulfite pretreatment to overcome recalcitrance of lignocellulose (SPORL) (Zhu et al. 2009) are the only three processes that have demonstrated effectiveness for forest biomass bioconversion.

15.6.2.1 Acid-Catalyzed Steam Explosion

Steam pretreatment was derived from a failed steam explosion pulping process (Kokta and Ahmed 1998). In acid-catalyzed steam explosion, SO_2 (De Bari et al. 2007) or sulfuric acid (Ballesteros et al. 2006; Sassner et al. 2008) has been used as a catalyst. Wood chips are often first impregnated with acid catalyst, either in gas phase with SO_2 or in the aqueous phase with sulfuric acid, before steam pretreatment. The further size reduction to fiber or fiber bundle level is accomplished through steam explosion. Acid-catalyzed steam pretreatment is actually another form of dilute acid pretreatment in which the pretreatment is carried out in the vapor phase rather in the aqueous phase. The explosion feature has now been used by many dilute acid operations for further size reduction. Therefore, the difference between dilute acid and acid-catalyzed steam pretreatment is becoming less clear. Catalyzed steam pretreatment works well with hardwood when pretreatment is conducted at an elevated temperature of around 210°C (De Bari et al. 2007; Sassner et al. 2008). The effectiveness on hardwood is achieved at the expense of the large amount of energy consumption in steam explosion. Typical energy consumption for the pretreatment at 210°C is about 1.8 MJ/kg oven-dried wood, even after accounting for low-quality steam recovery. The conversion of softwood cellulose is less satisfactory than that of hardwood (Duff and Murray 1996). Furthermore, total hemicellulose and glucose yield from pretreatment and enzymatic hydrolysis is about 70% (Gable and Zacchi 2002). Typical pretreatment conditions for wood are temperatures around 210°C and SO_2 or sulfuric acid charge of 1–2% on oven-dried wood. Pretreatment time varies from 3 to 10 min. Steam explosion can produce a relatively concentrated hemicellulose sugar stream from the pretreatment hydrolysate when the washing water is limited to the minimum (e.g., less than two times the biomass solids). Just like dilute acid pretreatment, steam explosion has a relatively low hemicellulose recovery of about 65% (Lynd 1996). The scalability of the process has not yet been addressed for commercialization.

15.6.2.2 Organosolv Pretreatment

The development of organosolv pretreatment technology is directly related to organosolv pulping (Kleinert 1974; Aziz and Sarkanen 1989; Paszner and Cho 1989). The chemistry of organosolv pulping is fairly well understood (McDonough 1993). Early work using organosolv pretreatment for fermentable sugar production was mostly conducted in the 1980s, with some success (Holtzapple and Humphrey 1984; Chum et al. 1988). The ethanol organosolv process was originally designed to produce clean biofuel for gas turbine combustors and was further developed into the Alcell[R] process for pulp production from hardwood (Williamson 1988; Pye and Lora 1991; Stockburger 1993). Ethanol is now the preferred solvent in organosolv process for biomass fractionation and pretreatment (Pan et al. 2005). The main advantages of the ethanol organosolv process are that 1) it can be directly applied to wood chips to produce a readily digestible cellulose substrate from almost all kinds of feedstock, including softwood and hardwood (Pan et al. 2005, 2006), therefore it eliminated the need for wood size reduction, and 2) it also produces very high purity and quality lignin with the potential of high-value applications (Kadla et al. 2002). Typical pretreatment conditions for woody biomass are temperatures of 175–195°C, pretreatment time around 60 min, ethanol concentration in pretreatment liquor of 50%, pretreatment liquor pH 2–3, and liquid to biomass solid ratio of 4–7. In pretreating poplar wood (Pan et al. 2006), about 70% of the lignin was removed from the substrate and recovered as high-purity lignin. Approximately 80% of the xylan was separated from the substrate, with 50% recovered as monomeric xylose in the soluble stream. About 88% of the glucan was retained in the substrate, and almost all of it was converted to glucose. Despite the excellent cellulose conversion, the xylose recovery rate was low. Furthermore, the relatively high liquor-to-solid ratio used in pretreatment produces a lower hemicellulose sugar concentration in the pretreatment hydrolysate. It also increases thermal energy consumption in pretreatment. Because the recovery of solvent ethanol is also expensive, the organosolv process is expensive.

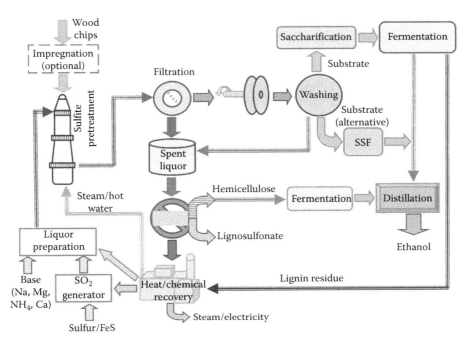

FIGURE 15.5 A schematic process flow diagram of the SPORL.

15.6.2.3 Sulfite Pretreatment—SPORL

The recently developed SPORL process (Wang et al. 2009; Zhu et al. 2009) is based on the fundamental understanding of sulfite wood pulping. The degrees of hemicellulose dissolution, cellulose depolymerization, and lignin sulfonation and condensation are controllable by varying pulping conditions, such as temperature and pH (Yorston 1942; Hall and Stockman 1958; Bryce 1980). By properly controlling reaction temperature, pH, and time, lignocellulose recalcitrance can be removed through a mild sulfite pretreatment process.

A schematic process flow diagram of the SPORL is shown in Figure 15.5. Wood chips first react with a solution of sodium bisulfite (or calcium or magnesium or other bisulfite) at 160–190°C and pH 2–5 for about 30 min and then are fiberized using a disk refiner to generate fibrous substrate for subsequent saccharification and fermentation. The removal of the recalcitrance of lignocellulose by SPORL is achieved by the combined effect of dissolution of hemicelluloses, depolymerization of cellulose, partial delignification (less than 30%), sulfonation of lignin, and increasing surface area by defiberization through disk refining. Lignin sulfonation increased the hydrophilicity of SPORL-pretreated substrates and may have promoted the enzyme processes. The pretreatment liquor to biomass ratio is typically in a range of 2–3, significantly lower than that used in dilute acid and organosolv processes. Therefore, SPORL can produce a relatively concentrated hemicellulose sugar stream. The dissolved hemicellulose stream (a mixture of hexoses and pentoses) can be further fermented to ethanol. The fermentation of spent sulfite pulping liquor (SSL) has been in industrial practice for commercial cellulosic ethanol production for decades (Helle et al. 2008). The dissolved lignin sulfonate or five-carbon hemicellulose sugars can be used to produce value-added co-products that can be directly marketed and has been practiced in the industry. Therefore, SPORL has the advantage of valuable co-product commercial pathways from dissolved hemicellulsoe sugars and lignin, important to the economics of the process.

Typical pretreatment conditions of SPORL are temperatures of 170–190°C and pH 2–5. For laboratory batch operation, retention time is about 30 min. The bisulfite charge on biomass depends on the feedstock species. For example, bisulfite of about 8% is required for softwood, whereas only 2–4% is required for hardwood. Cellulose-to-glucose conversion over 90% can be easily achieved even for softwood with low enzyme loading. The pretreatment is directly applied to wood chips without further size reduction. Furthermore, size-reduction energy consumption after pretreatment was only 20–50 Wh/kg oven-dried wood (Zhu et al. 2009; Zhu et al. 2010). With excellent cellulose conversion and very low size-reduction energy consumption, SPORL fits the requirement for forest biomass conversion. SPORL

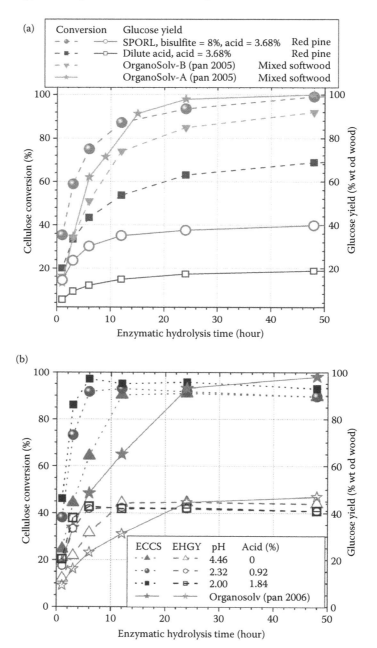

FIGURE 15.6 Comparison of enzymatic hydrolysis cellulose conversion and glucose yield between sulfite pretreatment to overcome recalcitrance of lignocellulose (SPORL) and the organosolv processes: (a) softwood and (b) aspen.

also has low formation of hydroxymethylfurfural (HMF) and furfural, two of the main inhibitors to fermentation. The combined severity factor under optimal SPORL pretreatment conditions ranges from 1.3 to 1.7. When SPORL was applied to spruce, a softwood, under the conditions for optimal cellulose conversion (Zhu et al. 2009), HMF and furfural formation levels were about 7 and 3 mg/g oven-dried wood, respectively. This is an order of magnitude lower than the 50 and 25 mg/g produced using steam-catalyzed pretreatment when glucose yield was maximized at a combined severity factor between 3 and 3.4 (Larsson et al. 1999). SPORL has an overall glucose recovery of 93% from spruce. SPORL also has excellent hemicellulose recovery. Major saccharide—arabinose, galactose, xylose, and mannose—yields are about 56, 86, 76, and 88%, respectively, for spruce (Zhu et al. 2009).

Because ethanol organosolv pretreatment is the most robust process in terms of removing lignocellulose recalcitrance, Figure 3.6 a and b compare cellulose conversion between SPORL and organosolv pretreated softwood (Zhu et al. 2009) and hardwood (Wang et al. 2009). Very similar enzymatic hydrolysis conditions were used (i.e., 2% solid consistency, enzyme loadings of about 20 FPU cellulase and 30 CBU ~β-glucosidase/g cellulose). SPORL effectively removed lignocellulose recalcitrance and achieved cellulose conversion rates that match those of ethanol organosolv pretreatment with equivalent glucose yield.

Because SPORL is developed based on a wood pulping process, it has excellent process scalability, which is one of the key challenges in commercialization of lignocellulosic ethanol technologies. Most processes have not been demonstrated at commercial scales. Capital equipment required for commercial demonstrations of steam explosion and organosolv processes does not exist. On the other hand, the pulp and paper industry has the capability of handling biomass on the scale of 1000 tons/day, equivalent to the scale of future cellulosic ethanol production of 10^8 L per year. The SPORL process can make full use of the capital equipment, process technologies, and human capital in the pulp and paper industry, which can significantly reduce technological and environmental barriers for commercialization. Specifically, a pulping digester can be used for the sulfite pretreatment, and a disk refiner can be used for size reduction after the pretreatment. Furthermore, fermentation of SSL is a mature technology and has been practiced in the pulp and paper industry for many decades. With the good performance of a newly developed yeast strain on SSL (Helle et al. 2004, 2008), the prospect of achieving good ethanol yield through fermenting SPORL pretreatment hydrolysate is excellent. Therefore, SPORL is one of the most promising pretreatment processes for lignocellulosic ethanol production from forest biomass.

15.7   CONCLUSIONS

Eucalyptus, *Pinus*, *Populus*, and other promising tree species have considerable current and potential use for bioenergy worldwide. *Eucalyptus*, the most widely planted hardwood with 19.6 million ha, has many bioenergy applications in Brazil, China, India, Australia, and other tropical to freeze-infrequent countries. In the next decade, genetic and genomics studies will likely identify most genes regulating heritable variation of biomass productivity and wood property traits in *Eucalyptus*. *Pinus* species have similar bioenergy potential worldwide, especially loblolly and slash pines in the southeastern United States and *P. radiata* in New Zealand, Chile, and other temperate countries. *Populus* species are very promising for temperate regions of North America, Europe, and Asia. Biotechnology may also enhance the bioenergy production and qualities of *Pinus* and *Populus* species. SPORL pretreatment of forest biomass from eucalypts, pines, poplars, and other species promises to increase bioethanol production efficiency.

ACKNOWLEDGMENTS

J. G. Isebrands gratefully acknowledges the clerical and moral support of Sharon K. O'Leary in preparing this manuscript. D. L. Rockwood appreciatively recognizes the patience and support of Joanne Rockwood in this chapter's development.

REFERENCES

Aaction Mulch, Inc (2007), available at http://www.aactionmulch.com

ABRACAVE–Associação Brasileira de Florestas Renováveis (2003), available at http://www.abracave.com.br

Aird PL (1962) Fertilization, weed control, and the growth of poplar. *For Sci* 8:413–428

Al Afas N, Marron N, Ceulemans R (2007) Variability in *Populus* leaf anatomy and morphology in relation to canopy position biomass production and varietal taxon. *Ann For Sci* 64:521–532

Al Afas N, Marron N, Van Dongen S, Laureysens L, Ceulemans R (2008) Dynamics of biomass production in a poplar coppice culture over three rotations (11 years). *For Ecol Manag* 255:1883–1891

Alexander T, Gordon L (2009) What's stopping us? The hurdles to commercializing cellulose ethanol. Renewable Energy World, March 6, 2009, available at www.renewablenergyworld.com

Allen J, Browne M, Hunter A, Boyd J, Palmer H (1997) Supply systems for biomass fuels and their delivered costs. *Asp Appl Biol* 49:369–378

Allen SG, Schulman D, Lichwa J, Antal Jr MJ (2001) A comparison of aqueous and dilute–acid single-temperature pretreatment of yellow poplar sawdust. *Indust Eng Chem Res* 40:2352–2361

Amidon TE (2006) The biorefinery in New York: Woody biomass into commercial ethanol. *Pulp Pap-Canada* 107:47–50

Anderson HW (1979) Biomass production of hybrid poplar in mini-rotation. In: *Poplar Research Management and Utilization in Canada*. Ontario Ministry of Natural Resources, Forest Research Information Paper No 102

Anderson HW, Papadopol CS, Zsuffa L (1983) Wood energy plantations in temperate climates. *For Ecol Manag* 6:281–306

Armstrong A, Johns C, Tubby I (1999) Effects of spacing and cutting cycle on the yield of poplar grown as an energy crop. *Biomass Bioenergy* 17:305–314

Auclair D,. Bouvarel L (1992) Intensive or extensive coppice cultivation of short rotation hybrid poplar coppice on forest land. *Bioresour Technol* 42:53–59

Avanzo A (1974) Possibility of genetic improvement of production of dry substance through short rotation harvesting of poplars of the section Aigeiros. *Cellul Carta* 25:22–27

Aziz, S, Sarkanen K (1989) Organosolv pulping—A review. *TAPPI J* 1989:169–175

Baker T, Bail I, Borschmann R, et al. (1994) *Commercial Tree-Growing for Land and Water Care: 2. Trial Objectives and Design for Pilot Sites.* Trees for Profit Research Centre, University of Melbourne, Victoria, Australia.

Baker T, Bartle J, Dickson R, Polglase P, Schuck S (1999) Prospects for bioenergy from short-rotation crops in Australia. Proc of the Third Meeting of IEA Bioenergy Task, available at http://www.p2pays.org/ref/17/16274/baker.pdf

Baker TG (1998) *Tree Growth in Irrigated Plantations at Wodonga and Kyabram, Victoria (1976–1994).* Centre for Forest Tree Technology, Victoria, Australia, Department of Natural Resources and Environment (unpublished).

Ballesteros I, Negro MJ, Oliva JM, Cabanas A, Manzanares P, Ballesteros M (2006) Ethanol production from steam-explosion pretreated wheat straw. *Appl Biochem Biotechnol* 129–132:69–85

Barbour RJ, Fried JS, Daugherty PJ, Christensen G, Fight R (2008) Potential biomass and logs from fire-hazard-reduction treatments in southwest Oregon and northern California. *For Policy Econ* 10:400–407.

Bartle J, Olsen G, Cooper D, Hobbs T (2007) Scale of biomass production from new woody crops for salinity control in dryland agriculture in Australia. *Int J Glob Energy Iss* 27:115–137

Barton A (2007) Industrial uses of eucalyptus oil. White Paper, available at http://www.oilmallee.com.au/docs/BARTON.doc

Bearce GD (1918) The commercial culture of cottonwood. *Ohio For* 10:3–8

Benetka V, Vratny F, Salkova I (2007) Comparison of productivity of *Populus nigra* with an interspecific hybrid in a short rotation coppice on marginal areas. *Biomass Bioenergy* 31:367–374

Berguson WE (2008) Woody biomass crops in the midwestern U.S.: Past, present and future. USDA Forest Service, GTR NRS-P-31, p.4.

Bhattacharya SC, Salam PA, Pham HL, Ravindranath NH (2003) Sustainable biomass production for energy in selected Asian countries. *Biomass Bioenergy* 25:471–482

Bjorheden R (2000) Integrating production of timber and energy—A comprehensive view. *NZ J For Sci* 30:67–78

Blake TJ (1983) Coppice systems for short rotation intensive forestry: The influence of cultural, seasonal and plant factors. *Aust For Res* 13:279–291

Blow FE (1948) Hybrid poplar performance in tests in the Tennessee Valley. *J For* 46:493–497

Boardman R, Shaw S, McGuire DO, Ferguson T (1996) Comparison of crop foliage biomass, nutrient contents and nutrient-use efficiency of crop biomass for four species of Eucalyptus and *Casuarina glauca* irrigated with secondary sewage effluent from Bolivar, South Australia. In: Polglase PJ, Tunningley WM (eds) *Land Application of Wastes in Australia and New Zealand: Research and Practice*. Proc 14th Land Treatment Collective Meeting. CSIRO Forestry and Forest Products, Canberra, Australia, pp 47–158

Bowersox TW, Blankenhorn PR, Murphey WK (1979) Heat of combustion, ash content, nutrient content and chemical content of Populus hybrids. *Wood Sci* 11:257–261

Bowersox TW, Ward WW (1976) Growth and yield of close spaced young hybrid poplars. *For Sci* 22:449–454

Boysen B, Strobl S (1991) *A Growers Guide to Hybrid Poplar*. Ontario Ministry of Natural Resources, Toronto, Ontario, Canada. 148 p.

Brazil Ministry of Mines and Energy. *Brazilian Energy Balance* (2003), available at www.mme.gov.br/site/menu/select_main_menu_item.do?channelId=1432&pageId=1501.

Bren L, Hopmans P, Gill B, Baker T, Stackpole D (1993) *Commercial Tree-Growing for Land and Water Care: 1. Soil and Groundwater Characteristics of the Pilot Sites*. Trees for Profit Research Centre, University of Melbourne. Melbourne, Australia.

Bridgwater AV, Carson P, Coulson M (2007) A comparison of fast and slow pyrolysis liquids from mallee. *Int J Glob Energy* Iss 27:204–216

Brondani RPV, Brondani C, Grattapaglia D (2002) Towards a genus-wide reference linkage map for Eucalyptus based exclusively on highly informative microsatellite markers. *Mol Genet Genom* 267:338–347

Brondani RPV, Brondani C, Tarchini R, Grattapaglia D (1998) Development, characterization and mapping of microsatellite markers in *Eucalyptus grandis* and *E. urophylla*. *Theor Appl Genet* 97:816–827

Brown GR, Gill GP, Kuntz RJ, Langley CH, Neale DB (2004) Nucleotide diversity and linkage disequilibrium in loblolly pine. *Proc Natl Acad Sci USA* 101:15255–15260

Bryan BA, Ward J, Hobbs H (2008) An assessment of the economic and environmental potential of biomass production in an agricultural region. *Land Use Policy* 25:533–549

Bryce JRG (1980) Sulfite pulping. In: Casey J (ed) *Pulp and Paper: Chemistry and Chemical Technology*, 3rd ed. John Wiley, New York, pp 291–376

Bundock PC, Hayden M, Vaillancourt RE (2000) Linkage maps of *Eucalyptus globulus* using RAPD and microsatellite markers. *Silvae Genet* 49:223–232

Byrne M, Murrell JC, Allen B, Moran GF (1995) An integrated genetic linkage map for Eucalyptus using RFLP, RAPD and isozyme markers. *Theor Appl Genet* 91:869–875

Byrne M, Murrell JC, Owen JV, Kriedemann P, Williams ER, Moran GF (1997) Identification and mode of action of quantitative trait loci affecting seedling height and leaf area in *Eucalyptus nitens*. *Theor Appl Genet* 94:674–681

Cadoche L, Lopez GD (1989) Assessment of size-reduction as a preliminary step in the production of ethanol from lignocellulosic wastes. *Biol Wastes* 30:153–157

Campbell MSF (1983) Biomass yields of young, heavily-stocked slash pine in north Florida. MS Thesis, University of Florida, 56 p

Campbell MSF, Comer CW, Rockwood DL, Henry C (1983) Biomass productivity of slash pine in young, heavily-stocked stands. Proc 1983 S Forest Biomass Workshop, pp 77–82

Canadell JG, Raupach MR (2008) Managing forests for climate change mitigation. *Science* 320:1456–1457

Cannell M, Smith RL (1980) Yields of mini rotation in close-spaced hardwoods in temperate regions: Review and appraisal. *For Sci* 26:415–428

Carling B (2009) Study: viable, sustainable biofuel production can reduce petroleum dependence. *Renewable Energy World*, February 13, 2009, available at www.renewableenergyworld.com

Ceulemans R (2004) The short rotation coppice in Boom (Antwerp): Productivity, phytoremediation, and population dynamics. University Antwerpen, Antwerp, Belgium

Ceulemans R, Deraedt W (1999) Production physiology and growth potential of poplars under short rotation forestry culture. *For Ecol Manag* 121:9–23

Ceulemans R, Scarascia-Mugnozza G, Wiard, B et al. (1992) Production physiology and morphology of *Populus* species and their hybrids under short rotation I. Clonal comparisons of 4-year growth and phenology. *Can J For Res* 22:1937–1948

Chambost V, Eamer B, Stuart P (2007a) Forest biorefinery: Getting on with the job. *Pulp Pap-Canada* 108:19–22

Chambost V, Eamer R, Stuart PR (2007b) Systematic methodology for identifying promising forest biorefinery products. *Pulp Pap-Canada* 108:30–35

Christersson L (2006) Biomass production of intensively grown poplars in the southern most part of Sweden: observations of characters, traits, and growth potential. *Biomass Bioenergy* 30:497–508

Christersson L (2008) Poplar plantations for paper and energy in the south of Sweden. *Biomass Bioenergy* 32:997–1000

Chum HL, Johnson DK, Black S, et al. (1988) Organosolv pretreatment for enzymatic-hydrolysis of poplars 1. Enzyme hydrolysis of cellulosic residues. *Biotechnol Bioeng* 31:643–649

Chundawat SPS, Venkatesh B, Dale BE (2007) Effect of particle size based separation of milled corn stover on AFEX pretreatment and enzymatic digestibility. *Biotechnol Bioeng* 96:219–231

Coleman MD, Tolsted D, Nichols T, Johnson WD, Wene EG, Houghtaling T (2006) Post-establishment fertilization of Minnesota hybrid poplar plantations. *Biomass Bioenergy* 30:740–747

Connett-Porceduu M, Gulledge J (2005) Enhanced selection of genetically modified pine embryogenic tissue. MeadWestvaco Corporation, U.S. Patent 6,964,870

Couto L, Muller MC, Barcellos DC, Couto MMF (2004) Eucalypt based agroforestry systems as an alternative to produce biomass for energy in Brazil (extended abstract). In: Proc of Conference on Biomass and Bioenergy Production for Economic and Environmental Benefits. Charleston, SC, November 2004, available at www.woodycrops.org/publications.html

Coutts RSP (2005) A review of Australian research into natural fibre cement composites. *Cement Concrete Composites* 27:518–526

Cram WH (1960) Performance of seventeen poplar clones in south central Saskatchewan. *For Chron* 36:204–208

Crow TR (1978) Biomass and production in three contiguous forests in northern Wisconsin. *Ecology* 59:265–273

Czapowskyj MM, Safford LO (1993) Site preparation, fertilization, and 10 year yields of hybrid poplar on a clear cut site in eastern Maine. *New For* 7:331–344

Dallenmand JF, Petersen J, Karp A (2008) *Short Rotation Forestry, Short Rotation Coppice, and Perennial Grasses in the European Union*. European Institute Energy, Ispra, Italy, 166 p

Das DK, Chaturvedi OP (2009) Energy dynamics and bioenergy production of *Populus deltoides* plantation in eastern India. *Biomass Bioenergy* 33:144–148

Dasari RK, Berson RE (2007) The effect of particle size on hydrolysis reaction rates and rheological properties in cellulosic slurries. *Appl Biochem Biotechnol* 137:289–299

Davis JM (2008) Genetic improvement of poplar (*Populus* spp.) as a bioenergy crop. In: Vermerris W (ed) *Genetic Improvement of Bioenergy Crops*. Springer Science, New York, pp 397–419

Dawson DH (1976) *Intensive Plantation Culture—5 Years Research*. USDA Forest Service, GTR NC–21, 117 p

De S, Assadi M (2009) Impact of cofiring biomass with coal in power plants–a techno-economic assessment. *Biomass Bioenergy* 33:283–293

De Bari I, Nanna F, Braccio G (2007) SO2-catalyzed steam fractionation of aspen chips for bioethanol production: optimization of the catalyst impregnation. *Industrial & Engineering Chemistry Research* 46:7711–7720.

DeBell DS, Clendenen GW, Zasada JC (1993) Populus biomass: comparison of wood grass versus wider spaced short rotation systems. *Biomass Bioenergy* 4:305–313

DeBell DS, Harrington CA (1993) Deploying genotypes in short rotation plantations: mixtures and pure clones and species. *For Chron* 69:705–713

DeBell DS, Harrington CA (1997) Productivity of Populus in monoclonal and polyclonal blocks at three spacings. *Can J For Res* 27:978–985

DeBell DS, Harrington CA, Clendenen GW, Zasada JC (1996) Tree growth and stand development in short rotation Populus stands: 7-year results of two clones at three spacings. *Biomass Bioenergy* 11:253–269

DeBell DS, Harrington CA, Clendenen GW, Zasada JC (1997) Tree growth and stand development of four Populus clones in large monoclonal blocks. *New For* 14:1–18

Decker J (2009) Going against the grain: Ethanol from lignocellulosics. *Renew Energy World* 11:116

Delbridge JB, Baker TG, Stokes RC (1998) Growth in sewage irrigated eucalypt Plantations at Werribee, Victoria: Progress report to 8 years of age. Centre for Forest Tree Technology, Victoria, Australia, Department of Natural Resources and Environment (unpublished)

DeVoto B (1953) *The Journals of Lewis and Clark*. Mariner Books, Boston, MA

Dhanda RS, Kaur I (2000) Production of poplar (*Populus deltoides*) in agri-silviculture system in Punjab. *Indian J Agrofor* 2:93–97

Dickmann DI (2006) Silviculture and biology of short rotation woody crops in temperate regions: then and now. *Biomass Bioenergy* 30:606–705

Dickmann DI, Isebrands JG, Eckenwalder JE, Richardson J (2001) *Poplar Culture in North America*. NRC Research Press. Ottawa, Canada

Dickmann DI, Stuart KW (1983) *The Culture of Poplars in Eastern North America*. Michigan State University, Eeast Lansing, MI

Drew AP, Zsuffa L, Mitchell CP (1987) Terminology relating to woody plant biomass and its production. *Biomass* 12:79–82

Duff SJB, Murray WD (1996) Bioconversion of forest products industry waste cellulosics to fuel ethanol: a review. *Bioresour Technol* 55:1–33

Duncan M J, Baker TG, Stackpole DJ, Stokes RC (1999) Demonstration and development of fast growing irrigated eucalypt plantations in Northern Victoria: Progress report 1998/99. Centre for Forest Tree Technology, Victoria, Australia, Department of Natural Resources and Environment (unpublished)

Dvorak WS, Jordon A, Hodge GP, Romero JL (2000) Assessing evolutionary relationships of pines in the Oocarpae and Australes subsections using RAPD markers. *New For* 20:163–192

Eastham J, Rose CW (1990) Tree/pasture interactions at a range of tree densities in an agroforestry experiment. I Rooting patterns. *Aust J Agric Res* 41:683–695

Eaton JA (2008a) Agroforestry and cellulosic ethanol from sustainable poplar tree farms. USDA Forest Service Northern Research Station, GTR NRS-P-31, 12 p

Eaton JA (2008b) Renewable energy from sustainable poplar tree farms. In: FAO Working Paper IPC/5. FAO, Rome, Italy, 57 p

Eid J, Fehr A, Gray J, et al. (2009) Real-Time DNA sequencing from single polymerase molecules. *Science* 323:133–138.

Ek AR, Dawson DH (1976) Actual and projected yields of *Populus tristis* #1 under intensive culture. *Can J For Res* 6:132–144

Facciotto G, Nervo G, Bergante S, Lioia C (2008) New poplar and willows clones selected for short rotation coppice in Italy. In: FAO Working Paper IPC/5. Rome, Italy, 27 p

Fang, S, Xu X, Lin S, Tang L. 1999. Growth dynamics and biomass production in short rotation poplar plantations: 6 year results for three clones at four spacings. *Biomass Bioenergy* 17:417–425

Fang S, Xu X, Tang L (2006) Above ground biomass production and nutrient removal in short rotation plantations of poplar clones. *Proc Int Poplar Symposium IV*. Nanjing, China, 5–9 June 2006, p 99

Fang S, Xue J, Tang L (2007) Biomass production and carbon sequestration potential in poplar plantations with different management patterns. *J Environ Manag* 85:672–679

FAO (1980) *Poplars and Willows in Wood Production and Land Use*. FAO Forestry Series #10. Rome, Italy

FAO (2004) *Forest Products Yearbook*, 2002. Forestry Series No 37/FAO Statistics Series, No 179. Food and Agricultural Organization of the United Nations (FAO), Rome Italy

FAO (2006) Global planted forests thematic study: results and analysis. Planted forests and trees. FAO Working Paper FP38E, available at http://www.fao.org/forestry/webview/-media?mediaId=12139&langId=1

FAO (2008) Poplars, willows, and the peoples' wellbeing. Abstracts International Poplar Commission 23rd Session. Working Paper IPC/5. Beijing, China, 27–30 Oct 2008

FAO (2009) Small-scale bioenergy helps rural development in poor countries. April 25, 2009, available at www.iowafarmertoday.com

Faundez P (2003) Potential costs of four short-rotation silvicultural regimes used for the production of energy. *Biomass Bioenergy* 24:373–380

Fege AS, Inman RE, Salo DJ (1979) Energy farms for the future. *J For* 77:358–361

Ferm A, Hytonen J, Vuori J (1989) Effect of spacing and nitrogen fertilization on the establishment and biomass production of short rotation poplar in Finland. *Biomass* 18:95–108

Ferm A, Kauppi A (1990) Coppicing as a means for increasing hardwood biomass production. *Biomass* 22:107–121

Florence RG (1986) Cultural problems of Eucalyptus as exotics. *Commonwealth For Rev* 65:141–165

Forest Solutions (2009) Commercial forestry, available at www.hawaiiforest.com/commercial_forestry.htm

Fox TR, Jokela EJ, Allen H (2004) The evolution of pine plantation silviculture in the southern United States. In: General Technical Report SRS-75, U.S. Department of Agriculture, Forest Service, Southern Research Station, Asheville, NC, pp 63–82

Fox TR, Jokela EJ, Allen HL (2007) The development of pine plantation silviculture in the southern United States. *J For* 15:337–347

Frampton LJ Jr, Rockwood DL (1983) Genetic variation in traits important for energy utilization of sand and slash pines. *Silvae Genet* 32:18–23

Francis JK, Baker JB (1981) Biomass and nutrient accumulation in a cottonwood plantation–the first four years. USDA Forest Service Southern Forest Experiment Station Res Note SO-278, pp 250–278

Franzen R (1986) General and selective upgrading of mechanical pulps. *Nord. Pulp Paper Res J* 1(3):7

Gable M, Zacchi G (2002) A review of the production of ethanol from softwood. *Appl Microbiol Biotechnol* 59:618–628

Gallagher T, Shaffer B, Rummer B (2006) An economic analysis of hardwood fiber production on dryland irrigated sides in the U.S. Southeast. *Biomass Bioenergy* 30:794–802

Gan SM, Shi JS, Li M, Wu KM, Wu JY, Bai JY (2003) Moderate-density molecular maps of *Eucalyptus urophylla* S. T. Blake and E-tereticornis Smith genomes based on RAPD markers. *Genetica* 118:59–67

Garrett PW (1976) Interspecific hybridization: the American experience. In: *Proc Symposium Eastern Cottonwood and Related Species*, Greenville, MS, 28 Sept–02 Oct, 1976, pp 156–164

Geary TF, Meskimen GF, Franklin EC (1983) Growing eucalypts in Florida for industrial wood production. USDA For Serv Gen Tech Rpt SE-23

Gibbs RA, Belmont JW, Hardenbol P, et al. (2003) The international HapMap Project. *Nature* 426:789–796.

Gigler JK, Meerdink G, Hendrix EMT (1999) Willow supply strategies to energy plants. *Biomass Bioenergy* 17:185–98

Gion JM, Rech P, Grima-Pettenati J, Verhaegen D, Plomion C (2000) Mapping candidate genes in Eucalyptus with emphasis on lignification genes. *Mol Breed* 6:441–449

Gordon JC (1975) The productive potential of woody plants. *Iowa State J Res* 49:267–274

Gordon JC (2001) Poplars: Trees of the people, trees of the future. *For Chron* 77:217–219

Grattapaglia D, Bertolucci FLG, Penchel R, Sederoff RR (1996) Genetic mapping of quantitative trait loci controlling growth and wood quality traits in *Eucalyptus grandis* using a maternal half-sib family and RAPD markers. *Genetics* 144:1205–1214

Grattapaglia D, Bertolucci FL, Sederoff RR (1995) Genetic-mapping of QTLs controlling vegetative propagation in *Eucalyptus grandis* and *E. urophylla* using a pseudo-testcross strategy and RAPD markers. *Theor Appl Genet* 90:933–947

Grattapaglia D, Kirst M (2008) Eucalyptus applied genomics: from gene sequences to breeding tools. *New Phytol* 179:911–929

Grattapaglia D, Sederoff R (1994) Genetic linkage maps of *Eucalyptus grandis* and *Eucalyptus urophylla* using a pseudo-testcross mapping strategy and RAPD markers. *Genetics* 137:1121–1137

Hall L, Stockman L (1958) Sulfitkokning vid Olika Aciditet. *Svensk Papperstidning* 61:871–880

Hall P, Gigler JK, Sims REH (2001) Delivery systems of forest arisings for energy production in New Zealand. *Biomass Bioenergy* 21:391–399

Hansen EA, Heilman P, Strobl S (1992) Clonal testing and selection for field plantations. In: Mitchell CP, Hinckley TM, Sennerby-Forsse L, Ford-Robert JB (eds) *Ecophysiology of Short Rotation Forest Crops*. Elsevier Applied Science, Essex, United Kingdom, pp 124–145

Harper RJ, Hatton TJ, Crombie DS, et al. (2000) Phase farming with trees: a scoping study of its potential for salinity control, soil quality enhancement and farm income improvement in dryland areas of southern Australia. RIRDC publication no 00/48. Canberra: Rural Industries Research and Development Corporation, 53 p

Heilman P, Peabody DV Jr (1981) Effect of harvest cycle and spacing on productivity of black cottonwood in intensive culture. *Can J For Res* 11:118–123

Heilman PE, Peabody DV Jr, DeBell DS, Strand RF (1972) A test of close-spaced, short rotation culture of black cottonwood. *Can J For Res* 2:456–459

Heilman PE, Stettler RF (1985) Genetic variation and productivity of *Populus trichocarpa* and its hybrids II. Biomass production of a 4 year plantation. *Can J For Res* 15:384

Heilman PE, Stettler RF (1990) Genetic variation and productivity of *Populus trichocarpa* and its hybrids. IV Performance of short rotation coppice. *Can J For Res* 20:1257–1264

Helle SS, Lin T, Duff SJB (2008) Optimization of spent sulfite liquor fermentation. *Enz Microb Technol* 42:259–264

Helle SS, Murry A, Lam J, Cameron DR, Duff SJB (2004) Xylose fermentation by genetically modified *Saccharomyces cerevisiae* 259ST in spent sulfite liquor. *Bioresour Technol* 92:163–171

Henry A (1914) Note on *Populus generosa*. Card Chronicle, 56 p

Hillring B (2006) World trade in forest products and wood fuel. *Biomass Bioenergy* 30:815–825

Himmel ME, Ding SY, Johnson DK, Adney WS, Nimlos MR, Brady JW, et al. (2007) Biomass recalcitrance: engineering plants and enzymes for biofuels production. *Science* 315:804–807

Hoffman D, Wieh M (2005) Limitation and improvement of the potential utilization of woody biomass derived from woody biomass crops in Sweden and Germany. *Biomass Bioenergy* 28:267–279

Holtzapple MT, Humphrey AE (1984) The effect of organosolv pretreatment on the enzymatic hydrolysis of poplar. *Biotechnol Bioeng* 26:670–676

Holtzapple MT, Humphrey AE, Taylor JD (1989) Energy-requirements for the size-reduction of poplar and aspen wood. *Biotechnol Bioeng* 33:207–210

Hooda N, Rawat V (2006) Role of bio-energy plantations for carbon-dioxide mitigation with special reference to India. *Mitigat Adaptat Strat Glob Change* 11:437–459

Hoque M, Sokhansanj S, Naimi LJ, Bi T, Lim J, Womac AR (2007) Review and analysis of performance and productivity of size equipment for fibrous materials. Paper No. 076164. In: 2007 ASABE Annual International Meeting. *Amer Soc of Agric and Biol Engineers*, St. Joseph, MI

International Energy Agency (IEA) (1998) World Energy Outlook, 1998 Edition: ww.iea.org.

Irvine GM (1985) The significance of the glass transition of lignin in the thermomechancial pulping. *Wood Sci Technol* 19:139–149

Isebrands JG (2007) Best management practices poplar manual for agroforestry supplications in Minnesota, available at http://www.regionalpartnershipsiumn.edu/public

Isebrands JG, Ek AR, Meldahl RS (1982) Comparison of growth model and harvest yields of short rotation intensively cultured Populus: a case study. *Can J For Res* 12:58–63

Isebrands JG, Sturos JA, Crist JB (1979) Integrated utilization of biomass. A case study of short rotation intensively cultured Populus raw material. *TAPPI J* 62:67–70

Joshi CP, Bhandar S, Ranjan P, Kalluri UC, Liang X, Fujino T, et al. (2004) Genomics of cellulose biosynthesis I. Poplars. *New Phytol* 164:53–61

Kadla JF, Kubo S, Venditti RA, Gilbert RD, Compere AL, Griffith W (2002) Lignin-based carbon fibers for composite fiber applications. *Carbon* 40:2913–2920

Kauter D, Lewandowski I, Claupein W (2003) Quantity and quality of harvestable biomass from Populus short rotation coppice for solid fuel use–a review of the physiological basis and management influences. *Biomass Bioenergy* 24:411–427

Kayihan GC, Huber DA, Morse AM, White TL, Davis JM (2005) Genetic dissection of fusiform rust and pitch canker disease traits in loblolly pine. *Theor Appl Genet* 110:948–958

Kennedy Jr HE (1975) Proper cultivation need for good survival and growth of planted cottonwood. USDA Forest Service Res Paper SO-60, 17 p

Kheshgi HS, Prince RC, Marland G (2000) The potential of biomass fuels in the context of global climate change: Focus on transportation fuels. *Annu Rev Energy Environ* 25:199–244

Kirst M, Basten CJ, Myburg AA, Zeng ZB, Sederoff RR (2005) Genetic architecture of transcript-level variation in differentiating xylem of a eucalyptus hybrid. *Genetics* 169:2295–2303

Kirst M, Johnson AF, Baucom C, et al. (2003) Apparent homology of expressed genes from wood-forming tissues of loblolly pine (*Pinus taeda* L). with *Arabidopsis thaliana*. *Proc Natl Acad Sci USA* 100:7383–7388

Kirst M, Myburg AA, De Leon JPG, Kirst ME, Scott J, Sederoff R (2004a) Coordinated genetic regulation of growth and lignin revealed by quantitative trait locus analysis of cDNA microarray data in an interspecific backcross of Eucalyptus. *Plant Physiol* 135:2368–2378

Kirst M, Myburg A, Sederoff R (2004b) Genetic mapping in forest trees: markers, linkage analysis and genomics. In: Setlow JK (ed) *Genetic Engineering, Principles and Methods*. Kluwer Academic/Plenum Publishers, New York, pp 105–142

Klasnja B, Orlovic S, Galic Z (2006) Poplar biomass of short rotation plantations as renewable energy raw material. In: Brenes M (ed) *Biomass Bioenergy: New Research*. Nova Science Publ, New York, pp 35–66

Kleinert TN (1974) Organoslov pulping with aqueous alcohol. *TAPPI J* 57:99–102

Kokta BV, Ahmed A (1998) Steam explosion pulping. In: Young RA, Ahmed A (eds) *Environmentally Friendly Technologies for the Pulp and Paper Industry*. John Wiley, New York, pp 191–214

Konig JW, Skog KE (1987) Use of wood energy in the United States–An opportunity. *Biomass* 12:27–36

Korlach J, Marks PJ, Cicero RL, et al. (2008) Selective aluminum passivation for targeted immobilization of single DNA polymerase molecules in zero-mode waveguide nanostructures. *Proc Natl Acad Sci USA* 105:1176–1181

Kotra R (2007) Closing the energy circle. In November, http://biomassmagazine.com/article

Krinard RM, Johnson RL (1975) Ten year results in cottonwood plantations spacing study. USDA Forest Service Research Paper SO-106

Krzysik AM, Muehl JH, Youngquist JA, Franca FS (2001) Medium density fiberboard made from *Eucalyptus saligna*. *For Prod J* 51:47–50

LaBau VJ, Bones JT, Kingsley NP, Lund HG, Smith WB (2007) History of the forest survey in the United States: 1830–2004. USDA Forest Service FS-877, 82 p

Labreque M, Teodoreseu TI (2005) Field performance and biomass production of 12 willow and poplar clones in short rotation coppice in southern Quebec (Canada). *Biomass Bioenergy* 29:1–9

Langholtz M, Carter D, Alavalapati J, Rockwood D (2007) The economic feasibility of reclaiming phosphate mined lands with short-rotation woody crops in Florida. *J For Econ* 12: 237–249

Larson ED, Consonni S, Katofsky S, Iisa K, Frederick WJ (2006) Benefit assessment of gasification-based biorefining in the kraft pulp and paper industry, 4 volumes, Department of Energy, Washington DC

Larson PR, Gordon JC (1969) Photosynthesis and wood yield. *Agric Sci Rev* 7:7–14

Larsson S, Palmqvist E, Hahn-Hagerdal B, et al. (1999) The generation of fermentation inhibitors during dilute acid hydrolysis of softwood. *Enz Microb Technol* 24:151–159

Laurenysens I, Bogaert J, Blust R, Ceulemans R (2004) Biomass production of 17 poplar clones in a short rotation coppice culture on a waste disposal site and its relation to soil characteristics. *For Ecol Manag* 187:295–309

Laureysens I, Deraedt W, Indeherberge T, Ceulemans R (2003) Population dynamics in a 6-year-old coppice culture of poplar. I. Clonal differences in stool mortality, shoot dynamics, and shoot diameter distribution in relation to biomass production. *Biomass Bioenergy* 24:81–95

Laureysens I, Pellis A, Willems J, Ceulemans. 2005. Growth and production of short rotation coppice culture of poplar. III. Second rotation results. *Biomass Bioenergy* 29:10–21

Lev-Yadun, S. and R. Sederoff R (2000) Pines as model gymnosperms to study evolution, wood formation, and perennial growth. *J Plant Growth Regul* 19:290–305

Li BL, McKeand S, Weir R (1999) Tree improvement and sustainable forestry impacts of two cycles of loblolly pine breeding in the USA. *For Genet* 8:213–224

Liberloo M, Calfapietra C, Lukac M, et al. (2006) Woody biomass production during the second rotation of a bioenergy Populus plantation increases in a future high CO_2 world. *Glob Change Biol* 12:1094–1106

Licht LA, Isebrands JG (2004) Linking phytoremediated pollutant removal to biomass economic opportunities. *Biomass Bioenergy* 28:203–218

Little EL Jr (1983) *Common Fuelwood Crops: A Handbook for Their Identification*. Communi-Tech Associates, Morgantown, WV

Logan MF (2002) *The Lessening Stream*. University of Arizona Press, Tucson, AZ

Lora ES, Andrade RV (2008) Biomass as energy source in Brazil. *Renew Sustain Energy Rev* 13:777–788

Lorenz WW, Sun F, Liang C, et al. (2006) Water stress-responsive genes in loblolly pine (*Pinus taeda*) roots identified by analyses of expressed sequence tag libraries. *Tree Physiol* 26:1–16

Lothner DC (1983) Economic investigations of short rotation intensively cultured poplars In: USDA Forest Service GTR NC-91, pp 139–148

Louden L (1976) *Short Rotation Trees*. Institute Paper Chemistry Bibiographic Series #273. Appleton, WI

Lynd LR (1996) Overview and evaluation of fuel ethanol from cellulosic biomass: technology, economics, the environment, and policy. *Annu Rev Energy Environ* 21:403–465

Maisenhelder LC (1970) Eastern cottonwood selections outgrow hybrids on southern sites. *J For* 68:300–301

Mani S, Tabil LG, Sokhansanj S (2004) Grinding performance and physical properties of wheat and barley straws, corn stover and switchgrass. *Biomass Bioenergy* 27:339–352

Markovic J, Roncevic S, Andrasen S (2000) Poplar biomass production in short rotations. In: USDA Forest Service GTR NC-215, pp 114

Marosvolgyi B, Halupa L, Wesztergom I (1999) Poplars as biological energy sources in Hungary. *Biomass Bioenergy* 16:245–247

Marques CM, Araujo JA, Ferreira JG, et al. (1998) AFLP genetic maps of *Eucalyptus globulus* and *E. tereticornis*. *Theor Appl Genet* 96:727–737

Marques CM, Brondani RPV, Grattapaglia D, Sederoff R (2002) Conservation and synteny of SSR loci and QTLs for vegetative propagation in four Eucalyptus species. *Theor Appl Genet* 105:474–478

Martin GB, Brommonschenkel SH, Chunwongse J, Frary A, Ganal MW, Spivey R, et al. (1993) Map-based cloning of a protein-kinase gene conferring disease resistance in tomato. *Science* 262:1432–1436

McDonough TJ (1993) The chemistry of organosolv delignification. *TAPPI J* 1993:186–193

McHenry MP (2009) Agricultural bio-char production, renewable energy generation and farm carbon sequestration in Western Australia: Certainty, uncertainty and risk. *Agric Ecosyst Environ* 129:1–7

McKeand SE, Bridgwater FE (1998) A strategy for the third breeding cycle of loblolly pine in the Southeastern U.S. *Silvae Genet* 47:223–234

McKeand SE, Jokela EJ, Huber DA, et al. (2006) Performance of improved genotypes of loblolly pine across different soils, climates, and silvicultural inputs. *For Ecol Manag* 227:178–184

McKnight JS, Biesterfeldt RC (1968) Commercial cottonwood planting in Southern United States. *J For* 66:670–675

Meskimen GF, Rockwood DL, Reddy KV (1987) Development of eucalyptus clones for a summer rainfall environment with periodic severe frosts. *New For* 3:197–205

Mitchell CP, Ford-Robertson JB, Hinckley T, Sennerby-Forsse L (1992) *Ecophysiology of Short Rotation Forest Crops*. Elsevier Applied Science, Essex, United Kingdom, 308 p

Mitchell CP, Stevens EA, Watters WP (1999) Short-rotation forestry - operations, productivity and costs based on experience gained in the UK. *For Ecol Manag* 121:123–36

Moran GF, Bell JC (1983) Eucalyptus. In: Tanksley SD, Orton TJ (eds) *Isozymes in Plant Genetics and Breeding*. Elsevier, Amsterdam, Netherlands, pp 423–441

Muhle-Larsen C (1970) Recent advances in poplar breeding. *Int Rev For Res* 3:1–67

Myburg AA (2001) Genetic architecture of hybrid fitness and wood quality traits in a wide interspecific cross of eucalyptus tree species. Forestry Department. North Carolina State University, Raleigh, NC

Myburg AA, Griffin AR, Sederoff RR, Whetten RW (2003) Comparative genetic linkage maps of *Eucalyptus grandis*, *Eucalyptus globulus* and their F_1 hybrid based on a double pseudo-backcross mapping approach. *Theor Appl Genet* 107:1028–1042

Myburg AA, Potts B, Marques CM, et al. (2007) Eucalyptus. In: Kole C (ed) *Genome Mapping & Molecular Breeding in Plants. Vol 7: Forest Trees*. Springer, Heidelberg, Berlin, New York, Tokyo, pp 115–160

Myers BJ, Theiveyanathan S, O'Brien ND, Bond WJ (1996) Growth and water use of effluent-irrigated *Eucalyptus grandis* and *Pinus radiata* plantations. *Tree Physiol* 16:211–219

National Academy of Sciences (1977) *Leucaena: Promising Forage and Tree Crop for the Tropics*. Washington DC for U.S. Agency for International Development, 115 p

National Academy of Sciences (1979) *Tropical Legumes: Resources for the Future*. Washington DC for U.S. Agency for International Development, 331 p

National Academy of Sciences (1980) *Firewood Crops: Shrub and Tree Species for Energy Production*. Washington DC for U.S. Agency for International Development, 1980 PB81–150716 (NTIS): http://sleekfreak.ath.cx:81/3wdev/CD3WD/APPRTECH/B28FIE/INDEX.HTM

National Academy of Sciences (1982) *Firewood Crops: Shrub and Tree Species for Energy Production*, Vol 2. Washington DC for U.S. Agency for International Development, 92 p

National Research Council (1983a) *Sowing Forests from the Air*. Washington DC for U.S. Agency for International Development

National Research Council (1983b) *Mangium and Other Fast-Growing Acacias of the Humid Tropics*. Washington DC for U.S. Agency for International Development, 41 p

National Research Council (1983c) *Calliandra: A Versatile Small Tree for the Humid Tropics*. Washington DC for U.S. Agency for International Development, 52 p

National Research Council (1984) *Casuarinas: Nitrogen-Fixing Trees for Adverse Sites*. Washington DC for U.S. Agency for International Development, 118 p

National Science Foundation (NSF) (2007) Breaking the chemical and engineering barriers to lignocellulosic biofuels: next generation biorefineries, a research road map for making lignocellulosic biofuels a practical reality, based on a workshop sponsored by Nat Sci Found, Amer Chem Soc, and U.S. Dept of Energy, June 25–26, 2007. Nat Sci Foundation, Washington DC

Neale DB (2007) Genomics to tree breeding and forest health. *Curr Opin Genet Dev* 17:1–6

Nehra NS, Becwar MR, Rottmann WH, et al. (2005) Forest biotechnology: Innovative methods, emerging opportunities. *In Vitro Cell Dev Biol-Plant* 41:701–717

Nervo G, Facciotto G, Bisoffi S (2008) Poplar activities in the Italian project on biomass for energy use. Proc IPC 23rd Session. FAO Working paper IPC/5. Rome, Italy, 139 p

Netzer DA, Tolsted DN, Ostry ME, Isebrands JG, Riemenschneider DE, Ward KT (2002) Growth, yield, and disease resistance of 7 to 12 year old poplar clones in the north central United States. USDA Forest Service GTR-NC-229

Newbould PJ (1967) *Methods of Estimating Primary Production in Forests*. International Biological Program. Blackwell Science Publishers, London

Nguyen QA, Tucker MP, Keller FA, Eddy FP (2000) Two-stage dilute acid pretreatment of softwoods. *Appl Biochem Biotechnol* 70–72:77–87

Novaes E, Drost DR, Farmerie WG, Pappas GJ, Grattapaglia D, Sederoff RR, et al. (2008) High-throughput gene and SNP discovery in *Eucalyptus grandis*, an uncharacterized genome. *BMC Genomics* 9:312

Pan X, Arato C, Gilkes N, et al. (2005) Biorefining of softwoods using ethanol organosolv pulping: preliminary evaluation of process streams for manufacture of fuel-grade ethanol and co-products. *Biotechnol Bioeng* 90:473–481

Pan XJ, Gilkes N, Kadla J, et al. (2006) Bioconversion of hybrid poplar to ethanol and co-products using an organosolv fractionation process: optimization of process yields. *Biotechnol Bioeng* 94:851–861

Parikka M (2004) Global biomass fuel resources. *Biomass Bioenergy* 27:613–620

Paris P, Mareschi L, Sabatti M, Ecosse A, Nardin F, Scarascia-Mugnozza G (2008) Comparing Populus clones for short rotation forestry in Italy after two-year rotations: Survival, growth and yield. Proc IPC 23rd Session. FAO Working Paper IPC/5. Rome, Italy

Paszner L, and Cho HJ (1989) Organosolv pulping: Acidic catalysis options and their effect on fiber quality and delignification. *TAPPI J* 1989:135–142

Paterson AH, Lander ES, Hewitt JD, Peterson S, Lincoln SE, Tanksley SD (1988) Resolution of quantitative traits into Mendelian factors by using a complete linkage map of restriction fragment length polymorphisms. *Nature* 335:721–726

Pauley SS (1949) Forest genetics research: *Populus* L. *Econ Bot* 3:299–330

Paux E, Tamasloukht M, Ladouce N, Sivadon P, Grima-Pettenati J (2004) Identification of genes preferentially expressed during wood formation in Eucalyptus. *Plant Mol Biol* 55:263–280

Pelaez-Samaniego MR, Garcia-Perez M, Cortez LB, Rosillo-Calle F, Mesa J (2008) Improvements of Brazilian carbonization industry as part of the creation of a global biomass economy. *Renew Sustain Energy Rev* 12:1063–1086

Perttu KL (1989) Short rotation forestry: an alternative energy resource? In: Perttu KL, Kowalik PJ (eds) *Modelling of Energy Forestry: Growth, Water Relations, and Economics Simulation,* Monographs 30 Pudoc. Wageningen, Netherlands, pp 181–186

Peter G (2007) *Developments in Biological Fibre Treatment.* Pira International, Surrey, United Kingdom

Peter G, White D, de la Torre R, Singh R, Newman R (2007) The value of forest biotechnology: A cost modeling study with loblolly pine and kraft linerboard in the Southeastern USA. *Int J Biotechnol* 9:415–435

Peter GF (2008) Southern Pines: A Resource for Bioenergy. In: Vermerris W (ed) *Genetic Improvement of Bioenergy Crops.* Springer, New York, pp 421–449

Peterson EB, Chan YH, Cragg JB (1970) Aboveground standing crop leaf area, and caloric value in an aspen clone near Calgary, Alberta. *Can J Bot* 48:1459–1469

Phelps JE (1983) Biomass from intensively cultured plantations as an energy, chemical and nutritional feedstock. In: Hansen EA (ed) *Intensive Plantation Culture: 12 Years Research.* USDA Forest Service NC-91, pp 131–138

Piketty M-G, Wichert M, Fallot A, Aimola L (2008) Assessing land availability to produce biomass for energy: The case of Brazilian charcoal for steel making. *Biomass Bioenergy* 33:180–190

Poke FS, Vaillancourt RE, Potts BM, Reid JB (2005) Genomic research in Eucalyptus. *Genetica* 125:79–101

Pontailler JY, Ceulemans R, Guittet J (1999) Biomass yield of poplar after five 2-year coppice rotations. *Forestry* 72:157–163

Pourtet J (1976) The International Poplar Commission–its role in poplar culture. In: Thielges B, Land S (eds) *Proc Symposium on Eastern Cottonwood and Related Species.* Greenville, MS, 28, Sept - 02 Oct 1976, pp 31–37

Proe M, Griffiths JH, Craig J (2002) Effect of spacing, species, and coppicing in leaf area, light interception, and photosynthesis in short rotation forestry. *Biomass Bioenergy* 23:315–326

Pryor LD, Willing RR (1965) The development of poplar clones suited to the low latitudes. *Silvae Genet* 14:123–127

Purdy KR, Elston LW, Hurst DR, Knight JA (1978) Pyrolysis of *Eucalyptus grandis* and *melaleuca* whole-tree chips. Final Report, Project A-2148. Georgia Institute of Technology, Engineering Experiment Station, Atlanta, GA

Puri S, Nair PK (2004) Agroforestry research for development in India: 25 years of experiences of a national program. *Agrofor Syst* 61:437–452

Puy N, Tabara D, Molins JB, Almera JB, Rieradevall R (2008) Integrated Assessment of forest bioenergy systems in Mediterranean basin areas: The case of Catalonia and the use of participatory IA-focus groups. *Renew Sustain Energy Rev* 12:1451–1464

Pye EK, Lora JH (1991) The alcell process - a proven alternative to kraft pulping. *TAPPI J* 74:113–118

Randall WK (1976) Progress in breeding the Aigeiros poplars In: Thielges B, Land S (eds) *Proc Symposium Eastern Cottonwood and Related Species*, Greenville, MS, 28 Sept–02 Oct, 1976, pp 140–150

Ranney JW, Wright LL, Layton PA (1987) Hardwood energy crops: The technology of intensive culture. J For 85:17–28

Ravindranath NH, Balachandra P, Dasappa S, Rao KU (2006) Bioenergy technologies for carbon abatement. *Biomass Bioenergy* 30:10

Ravindranath NH, Deepak P, Najeem S (2007) Biomass for Energy; Resource Assessment in India. In: *Phase I of a Project on Linking Climate Policy with Development Strategy in Brazil, China, and India*, pp 224–247: http://regserver.unfccc.int/seors/attachments/file_storage/xn6304o8mxf1el4.pdf

Ravindranath NH, Sudha P, Rao S (2001) Forestry for sustainable biomass production and carbon sequestration in India. *Mitigat Adapt Strat Glob Change* 6:233–256

Richardson J, Cooke JE, Isebrands JG, Thomas BR, van Rees KCJ (2007) Poplar research in Canada—A historical perspective with a view to the future. *Can J Bot* 85:1136–1146

Rocha JD, Coutinho AR, Luengo CA (2002) Biopitch produced from eucalyptus wood pyrolysis liquids as a renewable binder for carbon electrode manufacture. *Braz J Chem Eng* 19:127–132

Rockwood DL (1991) Freeze resilient *E. grandis* clones for Florida, USA. In: Proc. *IUFRO Symposium on Intensive Forestry: The Role of Eucalypts*, 02–06 Sept 1991, Durban, South Africa, 1:455–466

Rockwood DL (1997) Eucalyptus—Pulpwood, mulch, or energywood? Florida Cooperative Extension Service Circular 1194, available at http://edis.ifas.ufl.edu/FR013

Rockwood DL, Carter DR, Langholtz MH, Stricker JA (2006) Eucalyptus and Populus short rotation woody crops for phosphate mined lands in Florida USA. *Biomass Bioenergy* 30:728–734

Rockwood DL, Carter DR, Stricker JA (2008) Commercial tree crops on phosphate mined lands. Florida Institute of Phosphate Research. FIPR Publication #03–141–225

Rockwood DL, Comer CW, Conde LF, et al. (1983) Final report: Energy and chemicals from woody species in Florida. Oak Ridge National Laboratory. ORNL/Sub/81-9050/1

Rockwood DL, Dippon DR (1989) Biological and economic potential of *Eucalyptus grandis* and slash pine as biomass energy crops. *Biomass* 20:155–166

Rockwood DL, Naidu CV, Carter DR, et al. (2004) Short-rotation woody crops and phytoremediation: Opportunities for agroforestry? In: Nair PKR, Rao MR, Buck LE (eds) *New Vistas in Agroforestry, A Compendium for the 1st World Congress of Agroforestry 2004*. Kluwer Academic Publishers, Dordrecht, Netherlands, pp 51–63

Rockwood DL, Pathak NN, Satapathy PC (1993) Woody biomass production systems for Florida. *Biomass Bioenergy* 5:23–34

Rockwood DL, Peter GF, Langholtz MH, Becker B, Clark A III, Bryan J (2005) Genetically improved eucalypts for novel applications and sites in Florida. In: *Proc 28th South For Tree Imp Conf*, 21–23 June 2005, Raleigh, NC, pp 64–75

Rockwood DL, Reddy KV, Warrag EI, Comer CW (1987) Development of *Eucalyptus amplifolia* for woody biomass production. *Aust For Res* 17:173–178

Rockwood DL, Snyder GH, Sprinkle RR (1995) Woody biomass production in wastewater recycling systems. Southeastern Reg Biomass Energy Prog Pub No 91327, TVA, Muscle Shoals, AL

Rosillo-Calle F (2004) A brief account of Brazil's biomass energy potential. *Biomassa Energia* 1:225–236

Rosillo-Calle F, Bezzon G (2000) Production and use of industrial charcoal. In: Rosillo-Calle F, Bajay S, Rothman H (eds) *Industrial Uses of Biomass Energy—The Example of Brazil*. Taylor & Francis, London, pp 183–199

Salmen L, Lucander M, Harkonen E, Sundholm J (1999) *Fundamentals of Mechanical Pulping*, Book 5: *Mechanical Pulping* (ed: Sundholm, J), p 36. Papermaking Science and Technology (Eds: Gullichsen J, Paulapuro H), Fapet Oy, Finland

Sangseethong K, Meunier-Goddik L, Tantasucharit U, Liaw ET, Penner MH (1998) Rationale for particle size effect on rates of enzymatic saccharification of microcrystalline cellulose. *J Food Biochem* 22:321–330

Sassner P, Martensson CG, Galbe M, Zacchi G (2008) Steam pretreatment of H_2SO_4-impregnated Salix for the production of bioethanol. *Bioresour Technol* 99:137–145

Sastry CBR, Anderson HW (1980) Clonal variation in gross heat of combustion of juvenile Populus hybrids. *Can J For Res* 10:245–249

Scarascia-Mugozza GE, Ceulemans R, Heilman R, Isebrands JG, Stettler RF, Hinckley TM (1997) Production physiology and morphology of Populus species and their hybrids grown under short rotation II. Biomass components and harvest index of hybrid and parental species clones. *Can J For Res* 27:285–294

Scarascia-Mugnozza GE, Maresch L, Sabatti M, et al. (2006) Poplars to produce bioenergy in Italy: Early results on new high yielding poplar clones for short rotation forestry and phytoremediation. *Proc Int Poplar Symposium IV*, Nanjing, China 05–09 June 2006

Schell DJ, Harwood C (1994) Milling of lignocellulosic biomass—Results of pilot-scale testing. *Appl Biochem Biotechnol* 45–66:159–168

Schreiner EJ (1959) Production of poplar timber in Europe and its significance and application to the United States. USDA Ag Handbook # 150

Schreiner EJ (1970) Mini-rotation forestry. USDA Forest Service Res Paper NE–174

Schreiner EJ (1971) Genetics of eastern cottonwood. USDA Forest Service Research Paper WO-11

Schubert TH, Strand RF, Cole TG, McDuffie KE (1988) Equations for predicting biomass of six introduced tree species, island of Hawaii. Res Note PSW-401. Berkeley, CA: PSW Forest and Range Experiment Station, Forest Service, U.S. Department of Agriculture

Schultz JM (1988) Erosional relations in a short-rotation eucalyptus plantation on a typic hydrandept. Thesis, Department of Agronomy and Soil Science, College of Tropical Agriculture and Human Resources, University of Hawaii at Manoa

Segrest SA, Rockwood DL, Carter DR, Smith WH, Green AES, Stricker JA (2004) Short rotation woody crops for cofiring in central Florida. In: Proc. 29th International Technical Conference on Coal Utilization & Fuel Systems, 18–22 April 2004, Clearwater, FL. Coal Technology Association CD, ISBN No. 0–932066–29–54, Paper 12

Sewell MM, Bassoni DL, Megraw RA, Wheeler NC, Neale DB (2000) Identification of QTLs influencing wood property traits in loblolly pine (*Pinus taeda* L.). I. Physical wood properties. *Theor Appl Genet* 101:1273–1281

Sewell MM, Davis MF, Tuskan GA, Wheeler NC, Elam CC, Bassoni DL, et al. (2002) Identification of QTLs influencing wood property traits in loblolly pine (*Pinus taeda* L.). II. Chemical wood properties. *Theor Appl Genet* 104:214–222

Sewell MM, Sherman BK, Neale DB (1999) A consensus map for loblolly pine (*Pinus taeda* L.). I. Construction and integration of individual linkage maps from two outbred three-generation pedigrees. *Genetics* 151:321–330

Shaw S, Boardman R, McGuire D (1996) Early growth of native hardwood species under effluent irrigation at Bolivar, South Australia. In: *Farm Forestry and Plantations: Investing in Future Wood Supply*. Proceedings. Australian Forest Growers Conference. Mount Gambier, South Australia, 9–12 Sept 1996, pp 286

Sims REH (2003) Bioenergy to mitigate for climate change and meet the needs of society, the economy and the environment. *Mitigat Adapt Strat Glob Change* 8:349–370

Sims, R. E. H. and D. Culshaw. 1998. Fuel mix supply reliability for biomass-fired heat and power plants. In: Kopetz H, Weber T, Palz W, Chartier P, Ferrero GL (eds) *Biomass for Energy and Industry—Proceedings of the 10th European Bioenergy Conference*, 8–11 June 1998, CARMEN, Wurzburg, Germany, pp 188–191

Sims REH, Hastings A, Schlamadinger B, Taylor G, Smith P (2006) Energy crops: current status and future prospects. *Glob Change Biol* 12:2054–2076

Sims REH, Rogner H-H, Gregory K (2003) Carbon emission and mitigation cost comparisons between fossil fuel, nuclear and renewable energy resources for electricity generation. *Energy Policy* 31:1315–1326

Sims REH, Senelwa K, Maiava T, Bullock BT (1999a) Eucalyptus species for biomass energy in New Zealand– Part I: Growth screening trials at first harvest. *Biomass Bioenergy* 16:199–205

Sims REH, Senelwa K, Maiava T, Bullock BT (1999b) Eucalyptus species for biomass energy in New Zealand– Part II: Coppice performance. *Biomass Bioenergy* 17:333–343

Sims REH, Senelwa K, Maiava T, Bullock BT (2001) Short rotation coppice tree species selection for biomass energy in New Zealand. *Biomass Bioenergy* 20:329–335

Sims R, Taylor M, Saddler J, Mabee W (2009) IEA's report on 1st and 2nd generation biofuel technologies. Renewable Energy World (11 March 2009), available at www.renewableenergyworld.com

Sixto H, Barias M, Alba N, Hernadez MJ, Montot JL, Roig S, et al. (2006) Poplar trials in Spain for biomass as a renewable energy source. *Proc Int Poplar Symposium IV*, Nanjing, China, 05–09 June 2006, p 113

Skolmen RG (1986) Performance of Australian provenances of *Eucalyptus grandis* and *Eucalyptus saligna* in Hawaii. Res. Paper PSW-181. Forest Service, U.S. Department Agriculture

Slade R, Bauen A, Shah N (2009) The commercial performance of cellulosic ethanol supply-chains in Europe. *Biotechnol Biofuels* 2:DOI:10.1186/1754–6834–2-3

Smith JHG (1968) Silviculture and management of poplar plantations. In: Maini JS, Cayford JH (eds) *Growth and Utilization of Poplars in Canada*. Ministry of Forestry and Rural Development, Ottawa, Canada, pp 101–112

Smith JHG, Blom G (1966) Decade of intensive cultivation of poplars in British Columbia shows need for long term research to reduce risks. *For Chron* 42:359–376

Smith JHG, DeBell DS (1973) Opportunities for short rotation culture and complete utilization of seven Northwestern tree species. *For Chron* 49:31–34

Sochacki SJ, Harper RJ, Smettem KRJ (2007) Estimation of woody biomass production from a short-rotation bio-energy system in semi-arid Australia. *Biomass Bioenergy* 31:608–616

Stackpole D, Borschmann R, Baker T (1995) *Commercial Tree-Growing for Land and Water Care: 3. Establishment of Pilot Sites*. Trees for Profit Research Centre, University of Melbourne, Melbourne, Australia

Stamatov V, Rocha JD (2007) Bio-char refineries: an accessible approach for the development of biomass-based industry. *Int J Global Energy Issues* 27:217–230

Stanton B, Eaton J, Johnson J, Rice D, Schuette B, Moser B (2002) Hybrid poplar in the Pacific Northwest: the effects of market-driven management. *J For* 100:28–33

Stanton BJ, Shuren RA (2008) Populus hybridization for the renewable transportation industry: Integration of the genomic tools into a varietal development program. In: FAO Working Paper IPC/5 Rome, Italy, p 169

Stanturf JA, van Oosten C, Netzer DA, Coleman MD, Portwood CJ (2001) Ecology and silviculture of poplar plantations. In: Dickmann DI, Isebrands JG, Eckenwalder JE, Richardson J (eds) *Poplar Culture in North America*. NRC Research Press, Ottawa, Canada, pp 153–206

Stephens EP (1976) Populus in perspective. In: Thielges B, Land S (eds) *Proceedings of the Symposium on Eastern Cottonwood and Related Species*. Greenville, MS, 28 Sept–02 Oct 1976, pp 1–5

Stettler RF, Bradshaw HD, Heilman PE, Hinckley TM (1996) *Biology of Populus and Its Implications for Management and Conservation*. NRC Research Press, Ottawa, Canada

Stettler RF, Bradshaw HD, Zsuffa L (1992) The role of genetic improvement in short rotation forestry. In: Mitchell CP, Hinckley TM, Sennerby-Forsse L, Ford-Robert JB (eds) *Ecophysiology of Short Rotation Forest Crops*. Elsevier Applied Science, Essex, United Kingdom, pp 285–308

Stettler RF, Fenn R, Heillman PE, Stanton BJ (1988) *Populus trichocarpa* x *Populus deltoides* hybrids for short rotation culture: Variation patterns and 4 year field performance. *Can J For Res* 18:745–753

Stewart HTL, Hopmans P, Flinn DW, Hillman TJ, Collopy J (1988) Evaluation of irrigated tree crops for land disposal of municipal Effluent at Wodonga. Technical Report No. 7. Albury, NSW, Australia: Albury-Wodonga Development Corporation

Stirling B, Newcombe G, Vrebalov J, Bosdet I, Bradshaw HD (2001) Suppressed recombination around the MXC3 locus, a major gene for resistance to poplar leaf rust. *Theor Appl Genet* 103:1129–1137

Stockburger P (1993) An overview of near-commercial and commercial solvent-based pulping processes. *TAPPI J* 76:71–74

Stout AB, McKee RH, Schreiner EJ (1927) The breeding of forest trees for pulpwood. *J NY Bot Gard* 28:49–63

Stout AB, Schreiner EJ (1933) Results of project in hybridizing poplar. *J Heredity* 24:216–229

Strauss SH, Lande R, Namkoong G (1992) Limitations of molecular marker-aided selection in forest tree breeding. *Can J For Res* 22:1050–1061

Strong TF (1980) Energy values of nine Populus clones. USDA North Central Forest Experiment Station Res Note NC–257. St. Paul, MN

Strong TF, Hansen EA (1993) Hybrid spacing /productivity relations in short rotation intensive culture plantations. *Biomass Bioenergy* 4:255–261

Sudha P, Somashekhar HI, Rao S, Ravindranath NH (2003) Sustainable biomass production for energy in India. *Biomass Bioenergy* 25:501–515

Swisher JN (1997) Incremental costs of carbon storage in forestry, bioenergy and land-use. *Crit Rev Environ Sci Technol* 27:S335-S350

Switzer GL, Nelson LE, Baker JB (1976) Accumulation and distribution of dry matter and nutrients in Aegeiros poplar plantations In: Thielges B, Land S (eds) *Proc Symposium Eastern Cottonwood and Related Species*. Greenville, MS, 28 Sept–2 Oct 1976, pp 359–369

Sykes R, Li BL, Isik F, Kadla J, Chang HM (2006) Genetic variation and genotype by environment interactions of juvenile wood chemical properties in *Pinus taeda* L. *Ann For Sci* 63:897–904

Syring J, Willyard A, Cronn R, Liston A (2005) Evolutionary relationships among Pinus (*Pinaceae*) subsections inferred from multiple low-copy nuclear loci. *Amer J Bot* 92:2086–2100

Szego GD, Kemp CC (1973) Energy forests and fuel plantations. *Chem Technol* 3:257–64

Tang W, Sederoff R, Whetten R (2001) Regeneration of transgenic loblolly pine (*Pinus taeda* L.) from zygotic embryos transformed with *Agrobacterium tumefaciens*. *Planta* 213:981–989

Taylor G (2002) Populus: Arabidopsis for forestry: Do we need a model tree? *Ann Bot* 90:681–689

Taylor G (2008) Poplar and willow–sustainable second generation biofuel crops? *Comp Biochem Physiol–Part A. Mol Integr Physiol* 150(Suppl 1): S180

Thamarus KA, Groom K, Murrell J, Byrne M, Moran GF (2002) A genetic linkage map for *Eucalyptus globulas* with candidate loci for wood, fibre, and floral traits. *Theor Appl Genet* 104:379–387

Thielges BA, Land SB Jr (eds) (1976) *Proceedings: Symposium on Eastern Cottonwood and Related Species*, Greenville, MS, 28 Sept–02 Oct 1976

Thumma BR, Nolan MR, Evans R, Moran GF (2005) Polymorphisms in cinnamoyl CoA reductase (CCR) are associated with variation in microfibril angle in Eucalyptus spp. *Genetics* 171:1257–1265

Tillman DA (1976) *Wood as an Energy Resource*. Academic Press, New York

Toplu F (2008) Poplar development in Turkey. In: FAO Working Paper IPC/5. Rome, Italy, p 178

Tournier V, Grat S, Marque C, El Kayal W, Penchel R, de Andrade G, et al. (2003) An efficient procedure to stably introduce genes into an economically important pulp tree (*Eucalyptus grandis* × *Eucalyptus urophylla*). *Transgen Res* 12:403–411

Towers M, Browne T, Kerekes M, Paris J, Tran H (2007) Biorefinery opportunities for the Canadian pulp and paper industry. *Pulp Pap-Canada* 108:26–29

Trabedo GI, Wilshermann D (2008) *Eucalyptus universalis*. Global Cultivated Eucalypt Forests Map 2008, available at www.git.forestry.com

Tucker PH (1989) *Monet in the 90s*. Yale University Press, New Haven, CT, pp 127–143

Tuskan GA, DiFazio S, Jannson S, et al. (2006) The genome of black cottonwood, *Populus trichocarpa*. *Science* 313:1596–1604

Tuskan GA, Walsh M (2001) Short rotation woody crop systems, atmospheric carbon dioxide, and carbon management: A case study. *For Chron* 77:259–264

USDA Forest Service (2005) A strategic assessment of forest biomass and fuel reduction treatments in Western States. Gen Tech Rep RMRS-GTR-149. Rocky Mountain Research Station, Fort Collins, CO

USDE (2005) Breaking the biological barriers to cellulosic ethanol: a joint research agenda, a research road map resulting from the Biomass to Biofuel Workshop sponsored by the Department of Energy. 7–9 Dec 2005. U.S. Department of Energy, Washington DC

USDE and USDA (2005) Biomass as a feedstock for a bioenergy and bioproducts industry: The technical feasibility of a billion ton annual supply. Washington DC

Van Heiningen A (2006) Converting a kraft pulp mill into an integrated forest biorefinery. *Pulp Pap-Canada* 107:38–43

Van Oosten C (2006) *Hybrid Poplar Crop Manual for the Prairie Provinces*, Saskatchewan Forest Centre. Prince Albert, SK, Canada

Van Rees KCJ (2008) Wood bioenergy systems in Canada. In: USDA Forest Service GTR NRS-P-31, p 62.

Verhaegen D, Plomion C (1996) Genetic mapping in *Eucalyptus urophylla* and *Eucalyptus grandis* using RAPD markers. *Genome* 39:1051–1061

Verhaegen D, Plomion C, Gion JM, Poitel M, Costa P, Kremer A (1997) Quantitative trait dissection analysis in Eucalyptus using RAPD markers. 1. Detection of QTL in interspecific hybrid progeny, stability of QTL expression across different ages. *Theor Appl Genet* 95:597–608

Viart M (1976) Importance of *Populus deltoides* to poplar silviculture in France. In: Thielges B, Land S (eds) *Proc Symposium Eastern Cottonwood and Related Species*. Greenville, MS, 28 Sept 28–02 Oct 1996, pp 38–43

Vicentini R, Sassaki FT, Gimenes MA, Maia IG, Menossi M (2005) In silico evaluation of the Eucalyptus transcriptome. *Genet Mol Biol* 28:487–495

Voiblet C, Duplessis S, Encelot N, Martin F (2001) Identification of symbiosis-regulated genes in *Eucalyptus globulus-Pisolithus tinctorius* ectomycorrhiza by differential hybridization of arrayed cDNAs. *Plant J* 25:181–191

Walter A, Dolzan P, Piacente E (2006) Biomass energy and bio-energy trade: Historic developments in Brazil and current opportunities, available at http://www.bioenergytrade.org/downloads/brazilcountryreport.pdf

Wang GS, Pan XJ, Zhu JY, Gleisner R, Rockwood D (2009) Sulfite pretreatment to overcome recalcitrance of lignocellulose (SPORL) for robust enzymatic saccharification of hardwoods. *Biotechnol Progr* 25:1086–1093

White TL, Hodge GR, Powell GL (1993) An advanced-generation tree improvement plan for slash pine in the Southeastern United States. *Silvae Genet* 42:359–371

Whitesell CD, DeBell DS, Schubert TH, Strand RF, Crabb TB (1992) Short-rotation management of Eucalyptus: guidelines for plantations in Hawaii. Gen Tech Rep PSW-GTR-137. Albany, CA: PSW Research Station, Forest Service, U.S. Department of Agriculture

Whitesell CD, Miyasaka SC, Strand RF, Schubert TH, McDuffie KE (1988) Equations for predicting biomass in 2-to 6-year-old *Eucalyptus saligna* in Hawaii. Res Note PSW-402, Berkeley, CA: PSW Forest and Range Experiment Station, Forest Service, U.S. Department of Agriculture

Wilcox JR, Farmer RE Jr (1967) Variation and inheritance of juvenile characters of eastern cottonwood. *Silvae Genet* 16:162–165

Williamson PN (1988) Repaps-alcell-process—New demonstration facility shows how pulpmills can be cheaper. *Svensk Papperstidning-Nordisk Cellulosa* 91:21–23

Wright L (2006) Worldwide commercial development of bioenergy with a focus on energy crop-based projects. *Biomass Bioenergy* 30:706–714

Wullschleger SD, Yin TM, DiFazio SP, et al. (2005) Phenotypic variation in growth and biomass distribution for two advanced generation pedigrees of hybrid poplar. *Can J For Res* 35:1779–1789

Xiong JS (2007) *Progress of Tree Breeding in China*: cnr.ncsu.edu/fer/intl/documents/report_xiong.pdf

Yemshanov D, McKenney D (2008) Fast growing poplar plantations as a bioenergy supply source in Canada. *Biomass Bioenergy* 32:185–197

Yorston FH (1942) Studies in sulphite pulping. *Dominion For Serv Bull* No 97. Dominion Forest Service, Ottawa, Canada

Young HE (1971) *Forest Biomass Studies*. University of Maine. Orono, MN

Young HE (1973) *IUFRO Biomass Studies*. University of Maine. Orono, MN

Young HE (1976) *Oslo Biomass Studies*. University of Maine, Orono, MN

Young HE (1977) Forest biomass survey: the basis for complete utilization. In: *Proc TAPPI Biol Conf*, Atlanta, GA, pp 119–124

Zabek LM, Prescott CE (2006) Biomass equations and carbon content of aboveground leafless biomass of hybrid poplar in coastal British Columbia. *For Ecol Manag* 223:291–302

Zavitkovski J (1976) Biomass studies in intensively managed forest stands. In: *Intensive Plantation Culture.* USDA Forest Service GTR NC-21, pp 32–38

Zavitkovski J (1979) Energy production in irrigated, intensively cultured plantations of Populus 'Tristis #1' and jack pine. *For Sci* 25:383–392

Zavitkovski J (1981) Small plots with unplanted border can distort data in biomass production studies. *Can J For Res* 11:9–12

Zavitkovski J (1983) Projected and actual biomass production of 2 to 10 year old intensively cultured *Populus Tristis* #1. In: Hansen EA (ed) *Intensive Culture Research: 12 Years Research.* USDA Forest Service GTR NC-91, pp 72–76

Zavitkovski J, Dawson DH (1978) Structure and biomass production of one- to seven-year-old intensively cultured jack pine plantation in Wisconsin. Research Paper NC-157. St. Paul, MN: U.S. Dept of Agriculture, Forest Service, North Central Forest Experiment Station

Zavitkovski J, Isebrands JG (1985) Biomass production and energy accumulation in the world's forests. In: *Proceedings FPRS*, Wood Energy Forum, Nashville, TN

Zavitkovski J, Isebrands JG, Dawson DH (1976) Productivity and utilization potential of short rotation Populus in the Lake States. In: Thielges B, Land S (eds) *Proc Symposium Cottonwood and Related Species*, Greenville, MS, 28 Sept 28–02 Oct 1976, pp 392–401

Zhu JY, Pan XJ, Wang GS, Gleisner R (2009) Sulfite pretreatment for robust enzymatic saccharification of spruce and red pine. *Bioresour Technol* 100:2411–2418

Zhu JY, Wang GS, Pan XJ, Gleisner R (2009) Specific surface for evaluating wood size-reduction and pretreatment efficiencies. *Chem Eng Sci* 64:474–485

Zhu Y, Lee YY, Elander RT (2005) Optimization of dilute-acid pretreatment of corn stover using a high-solids percolation reactor. *Appl Biochem Biotechnol* 121:1045–1054

Zsuffa L (1975) Some problems with hybrid poplar selection and management in Ontario. *For Chron* 51:240–242

Zsuffa L (1993) Strategies for clonal forestry with poplars, aspens, and willows. In: Ahuja MR, Libby WJ (eds) *Clonal Forestry II.* Springer, Berlin, Germany, pp 91–119

Zsuffa L, Anderson HW, Jaciw P (1977) Trends and prospects in Ontario's poplar plantation management. *For Chron* 53:195–200

Zhu W, Zhu JY, Glesiner R, Pan XJ (2010) On energy consumption for size-reduction and yield from subsequent enzymatic hydrolysis of lodgepole pine. *Bioresource Technology* 101:2782–2792

16 Maize

Shoba Sivasankar, Sarah Collinson,
Rajeev Gupta, and Kanwarpal S. Dhugga
Pioneer Hi-Bred International, Inc.

CONTENTS

16.1 INTRODUCTION

Fossil fuels have finite reserves. For example, oil has been projected to have already reached its peak capacity for production and may run out over the next several decades even at the current pace of use, i.e., not accounting for increased demand from China and India as their respective economies expand (Dhugga 2007). These concerns have intensified the debate on the potential of renewable fuels as supplements to fossil fuels. Biofuels, a collective term used to describe fuels derived from biological sources, offer a partial solution to the energy demand along with solar, wind, geothermal, and hydroelectric power.

Commercial production of ethanol from the maize grain is a growing industry with upward of 50 GL of ethanol already produced in the United States annually. However, a competing demand for the grain from the food and feed industries makes it less likely a source for renewable energy for the long term. This concern is somewhat mitigated if stover, the proportion of which in the plant is approximately equal to grain, can be utilized to produce energy.

In this chapter, we focus on the potential of maize as an energy crop. Some of the key topics discussed are global production of maize in comparison to other cereals, its origin and evolution, history of breeding and yield improvement, residue composition, and potential for its alteration as an improved feedstock for bioenergy. We also discuss the complexity of the plant systems in the context of genotype × environment interactions as an impediment in improving plant performance as well as detecting the effect of a transgene on agronomic performance under field conditions.

16.2 MAIZE BREEDING AND PRODUCTION

16.2.1 GLOBAL PRODUCTION OF MAIZE AND OTHER CEREALS

Global annual production of cereal grains is approximately 2 gigatons (Gt) (Figure 16.1). From the harvest index (ratio of grain to total biomass) of different cereal crops, the estimated crop residue produced is approximately 2.5 Gt (Figure 16.2). Maize, rice, and wheat account for 80% of the annually planted area and 90% of production of all cereals grains (Figure 16.2). Barley, sorghum, oats, and rye, in that order, account for the remainder.

Maize is grown on approximately 160 million hectares (Mha) annually, which is comparable with the area planted to rice at 155 Mha and second to wheat at 220 Mha. Maize accounts for approximately 40% of all of the cereal grains produced annually from 25% of the total area under cereal crops (Figure 16.2). In comparison, wheat constitutes 30% of the total cereal grains produced from 33% of the area. Although maize is a C_4 plant, which perhaps contributes to its high productivity in comparison to other crops, intensive cultural practices account for most of the difference. For example, more than 40% of the world's corn is produced in the United States, where high inputs are used to grow this crop. Sorghum, another C_4 crop, has the lowest productivity of all of the cereals because it is grown mostly on marginal soil and in drought-prone areas.

In developed countries, the crop residue is ploughed into the field, where it contributes to soil organic matter content. In contrast, it is used to supplement cattle fodder in developing countries. In some cases, because of high cropping intensity, it is burnt in the field to make way for the next crop.

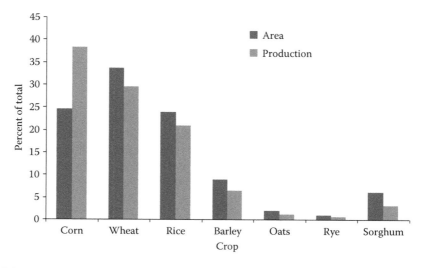

FIGURE 16.1 Land area planted to different cereals and their production. Total global area planted in 2007/2008 was 650 M ha and total production was 2.07 Gt. (From U.S. Department of Agriculture/National Agricultural Statistics Service, *Agricultural Statistics Database*, 2009. Available at http://www.fas.usda.gov/wap/current/toc.asp)

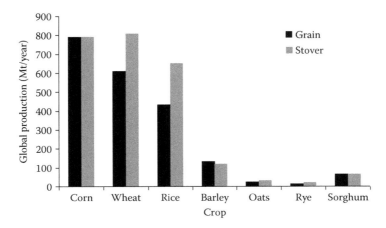

FIGURE 16.2 Global annual production of grain and crop residue from cereal crops. (From U.S. Department of Agriculture/National Agricultural Statistics Service, *Agricultural Statistics Database*, 2009. Available at http://www.fas.usda.gov/wap/current/toc.asp)

This not only contributes to environmental pollution but also adversely affects soil microflora. The availability of a market for the crop residue would help to mitigate these problems. As agriculture becomes more modernized along with industrialization of developing nations and crop productivity increases, correspondingly larger amounts of crop residue will become available.

The advantages of maize as a candidate for a bioenergy crop are numerous, the most important being the depth of genetic knowledge and commercial success of its transgenic products. Another major advantage corn stover holds over alternative crops in that it is already produced with grain as the target and does not require dedicated land (Dhugga 2007).

16.2.2 Origin and Evolution

The center of origin of maize, a member the genus *Zea* in the grass family Poaceae, previously referred to as Gramineae, is agreed upon to be southern and southwestern Mexico. The genus *Zea* has five species: *diploperennis, perennis, luxurians, nicaraguensis,* and *mays* L.; the first two are perennial and the last three annual. All the species are diploid with the exception of *Zea perennis*, which is tetraploid. The species *Zea mays* L. is highly polymorphic and consists of four subspecies: *huehuetenangensis, mexicana, parviglumis,* and *mays*. All *Zea* species except cultivated maize, *Zea mays* L. ssp. *mays*, are collectively referred to as teosinte. Based on multiple pieces of evidence, which include molecular tools and fossil records, maize domestication occurred approximately 10,000 years ago, which was preceded by human migration to the Americas around 15,000 years ago.

Teosinte remarkably differs from maize in morphological features, which led some scientists to propose that the progenitor of maize was extinct. However, the geneticists led by Beadle, arrived at the conclusion that teosinte was the immediate progenitor of maize, mainly based on the ability to obtain fertile crosses of maize with teosinte, but also several other observations, such as chromosome number and positioning of knobs in the chromosomes. The latter view is now generally accepted as additional evidence from evolving molecular technology has provided abundant supporting evidence in its favor.

From screening a large F_2 population derived from a cross between teosinte and maize, Beadle arrived at a conclusion that as few as five loci could explain the major differences between these two subspecies. In keeping with his foresight, single genes, for example, *teosinte branched-1* (*tb1*) and *teosinte glume architecture-1* (*tga1*), have indeed been found to account for two of the remarkable changes from teosinte to maize: suppression of lateral branches that ended in tassels and their transformation into ears instead (*tb1*), and exposure of the kernels on the surface of the ear which were otherwise enclosed in a hardened casing or glumes (*tga1*). Except for *tga1*, which manifests its phenotype more or less independent of the genetic background, the effects of other genes with major influence on morphological features vary quantitatively depending upon the genetic background, indicating the involvement of the epistatic effects.

On average, nucleotide diversity in maize at a given locus is approximately 70% of that of teosinte, which is expected as a consequence of selective pressure. In the *tb1* locus, however, nucleotide diversity was only 30% in the coding region and 2% in the 5'-upstream region, clearly indicating strong selection, particularly for the regulatory region of the gene.

Based on polymorphisms in approximately 800 genes, an estimate of 2–4% was obtained for the genes that were subjected to artificial selection, which translates into approximately 1000–2000 genes across the maize genome assuming ~55,000 total genes.

16.2.3 Maize Breeding over the Past Century

Maize was introduced to North and South America by human migration and, naturally, the farmers selected their own favorite lines that showed improved performance at the place of migration, resulting in the origin of myriad landraces. Prior to initiation of organized breeding by the Old World immigrants in the twentieth century in the USA, the local farmers had been selecting for desirable traits, albeit in a primitive manner, for thousands of years. In the mid-nineteenth century, farmer breeders selected high yielding, open-pollinated varieties (OPV) from the unadapted lines. However, the yield gain stagnated in the beginning of the twentieth century, at a time when the population of the USA was beginning to pick up and demand for food was on the rise (Figure 16.3).

George Shull, a scientist at the Cold Spring Harbor Laboratory, realized in 1908 that the reduced performance of two inbred lines that had undergone repeated selfing was recovered upon crossing them into a hybrid, so much so that in some cases hybrid outyielded the OPV from which the inbred lines were derived. At about the same time, Edward East at Harvard University who conducted

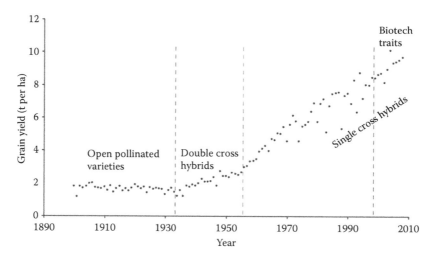

FIGURE 16.3 Grain yield of maize over time in the United States. (From U.S. Department of Agriculture/ National Agricultural Statistics Service, *Agricultural Statistics Database*, 2009. Available at http://www.fas. usda.gov/wap/current/toc.asp)

his experiments at the Connecticut Agricultural Experiment Station also observed that repeated selfing caused a reduction in performance but was of the opinion that, given the poor performance of inbreds, it would be impractical to produce enough seed for commercial use. A solution was provided by Donald Jones, a graduate student of East, in 1922, in the form of double cross hybrids, that is, a cross between two hybrids, to overcome the seed production bottleneck. Soon thereafter, Henry A. Wallace introduced the first commercial hybrid, Copper Cross, a single cross hybrid, in 1924 in the Corn Belt. He went onto found a corn seed company, Hi-Bred Corn Company, which sold 650 bu of corn seed in its first year but eventually went on to become the largest and premier seed company in the world and is currently known as Pioneer Hi-Bred International, Inc., a DuPont Business. It was in the 1930s, however, that farmers adopted the practice of growing hybrids instead of OPV. By 1943, all of the corn acreage in Iowa was planted to hybrids, which contrasted with that of 90% of the Corn Belt and 60% of the area in the entire U.S.

Initial inbred lines developed from landraces were not very productive so the breeders started developing inbred lines from the crosses between two elite lines. Pedigree selection for development of inbred lines from biparental crosses constituted less than 40% of the effort up until the 1970s, with the rest of the inbred lines coming from populations of various types, that is, genetically broad or narrow, or those generated from a mixture of elite inbred lines. The shift to the use of biparental crosses for inbred line development over the last forty years is exemplified by the fact that they account for nearly 80% of the current inbred parents of hybrids across the seed industry.

Coincident with the introduction and acceptance of hybrids in the 1930s, grain yield started to increase at a rate of ~5% year[1] (Figures 16.3 and 16.4). Increased corn production resulted both from an increase in grain yield per unit area as well as increased land area planted to hybrids.

Application of synthetic N as a fertilizer picked up after World War II, which further boosted grain yield (Figure 16.5). Whereas fertilizer application peaked in the 1980s, grain yield kept an upward trend (Figure 16.5). N use efficiency is defined as grain yield as a function of soil N. It has been increasing linearly since the time N fertilizer application peaked, having improved by more than 50% over twenty five years (from ~40 kg kg[-1] N in 1980 to 60 kg kg[-1] N in 2005) (Figure 16.6). A combination of improved N acquisition from the soil and its utilization by modern hybrids as well as improved cultural practices apparently contribute to improved N use efficiency.

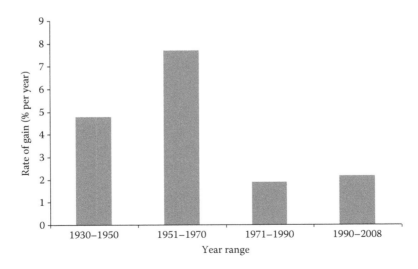

FIGURE 16.4 Rate of gain for grain yield during different eras as measured over 2-decade windows starting in 1930. (From U.S. Department of Agriculture/National Agricultural Statistics Service, *Agricultural Statistics Database*, 2009. Available at http://www.fas.usda.gov/wap/current/toc.asp)

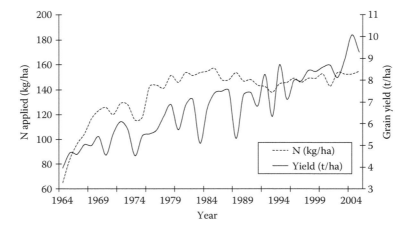

FIGURE 16.5 N application rates and grain yield of maize in the United States since 1964. (From U.S. Department of Agriculture/National Agricultural Statistics Service, *Agricultural Statistics Database*, 2009. Available at http://www.fas.usda.gov/wap/current/toc.asp; U.S. Department of Agriculture/Economic Research Service, *Economic Research Service Database*, 2011. Available at http://www.ers.usda.gov/Data/FertilizerUse/).

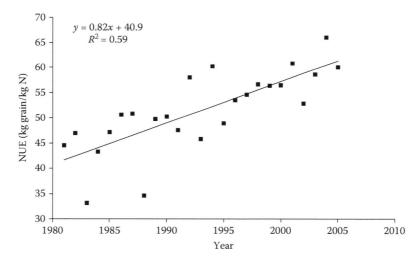

FIGURE 16.6 N use efficiency (NUE) in the United States since 1981 when N fertilizer application peaked. Figure derived from data in Figure 16.5.

Planting density (number of plants per unit land area) also increased linearly, matching more or less an increase in N application up until the mid-1980s. The average density, which was less than 12,000 plants acre^{-1} in the 1930s, reached 16,000 plants acre^{-1} in the 1960s, 24,000 plants acre^{-1} in the 1980s, and currently stands at approximately 34,000 plants acre^{-1}. Planting density has kept an upward trend, however, even after the N application peaked, and still keeps increasing at an average density of 400 plants acre^{-1} year^{-1}, a rate that has been maintained over the last 55 years.

Selection of modern hybrids for adaptation to higher density planting has been accompanied by several favorable agronomic traits, for example, resistance to stresses such as drought, disease and insects, and low soil fertility, improving their relative stability to perform consistently in variable and unpredictable environments. Reductions in barrenness and late-season stalk lodging further contributed to yield improvement. A number of other adaptive changes accompanied selection of modern hybrids at higher planting densities: a more acute leaf angle to reduce shading, thus allowing increased photosynthesis per unit land area, and reduction in tassel size to free up carbon for investment in other productive plant parts.

A change in grain composition has also contributed to grain yield over time. An increased grain starch/protein ratio meant channeling of more photosynthate toward starch formation and thus grain yield. Starch requires half the energy required for its formation than protein (also see section 16.6: Bioenergetic Considerations for Biomass Interconversion). When compared to total biomass productivity per unit land area, however, the contributions of reduced tassel size and starch/protein ratio toward increased grain yield is minor.

The highest recorded grain yield has stayed around 22 Mg ha^{-1} (~350 bu acre^{-1}) for several decades, suggesting that this is likely the potential yield of maize. Environmental stresses are apparently responsible for the gap between the potential and the actual grain yield. No wonder then that the modern hybrids perform better under environmental stresses in comparison to the older ones. With the current average US corn grain yield of 10 Mg ha^{-1}, significant room exists for yield improvement assuming that genetic variation for tolerance to various stresses is not exhausted.

Beginning in 1996, seed companies started marketing maize hybrids containing a transgene that imparted resistance to European Corn Borer (ECB), a major pest in the Corn Belt to which a corn plant is susceptible from early vegetative stage all the way to physiological maturity. These hybrids are referred to as *Bt* hybrids after a gene from the bacterium, *Bacillus thuringiensis* (*Bt*). The first generation of *Bt* hybrids contained the gene that produced the insecticidal protein Cry1Ab (Cryptochrome 1Ab), followed by stacking of additional insecticidal genes for corn rootworm resistance later on. Aside from a substantial reduction in the application of insecticides, the yield advantage of the transgenic hybrids containing the *Bt* genes over the non-transgenic controls can exceed 10% depending upon the insect pressure.

16.2.4 BIOMASS PRODUCTION AND HARVEST INDEX

From the time hybrids were first introduced in the 1930s, grain yield in maize has maintained an upward trend (Figure 16.3) (Tollenaar and Lee 2002; Duvick 2005a, 2005b; Anonymous 2007). Linear regression coefficients of grain yield for 2-decade windows starting in 1931 are 58, 158, 99, and 152 kg/ha per year, respectively. The regression coefficients as a percentage of yield in the beginning of each period are 4.6, 7.7, 1.9, and 2.3, respectively (Figure 16.4). The rate of yield gain appears to be high in the beginning but then waned over time (Figure 16.4). However, the actual rate of gain, which is approximately 150 kg/ha per year, has not changed since the 1950s. The average yield in the beginning of each of the respective 2-decade windows was 1.2, 2.0, 5.2, and 7 Mg/ha. The current average yield of corn is approximately 10 Mg/ha. Because of lower starting yields in the 1930s and 1950s, the apparent rates of gain were relatively higher. The dip in the third decade could be accounted for by a high frequency of low-yielding years because of environmental stresses.

Introduction of dwarfing genes in small grain cereals, such as rice and wheat, increased the harvest index (HI), or the ratio of grain yield to total aboveground biomass (Hay 1995; Sinclair 1998). The resulting dwarf varieties could be grown under extensive inputs but with significantly reduced lodging, which was a recurrent problem with the older, taller varieties that limited their yield potential (Hay 1995; Sinclair 1998). Partitioning of a greater amount of biomass to grain and increased total biomass production under intensive agricultural practices were the key factors that led to the green revolution.

In contrast to small grain cereals, HI in maize has remained unchanged at approximately 50% over the last century (Hay 1995; Sinclair 1998; Tollenaar and Wu 1999). Grain yield improvement has thus resulted from increased total biomass production per unit land area, that has been achieved through selecting modern hybrids to be productive at increasing planting densities (Tollenaar and Lee 2002; Duvick 2005a, 2005b). An unchanged HI as the grain yield increased severalfold implies that the sink/source ratio, unlike in small-grain cereals, was already optimized in maize before the era of modern breeding (Dhugga 2007).

Increased demand for grain from the ethanol industry will most likely result in a consistent increase in the land area planted to corn, which will both displace some other crops as well as bring additional area from the 14 Mha currently held in the Conservation Reserve Program (CRP) into cultivation (Wright et al. 2006).

16.2.5 STOVER REMOVAL AND SOIL ORGANIC MATTER

Corn stover left in the field after grain harvest provides soil cover and contributes to the organic matter content. To facilitate its breakdown, the soil usually undergoes multiple tillings, which erode the organic matter content of the soil through volatilization and loss of top soil. Up to two-thirds of the stover may be removed on a sustainable basis under no-till conditions from some corn-growing regions without an adverse effect on soil organic matter content (Wilhelm et al. 2004; Perlack et al. 2005; Johnson et al. 2006; Graham et al. 2007). In the United States, more than 100 million tons of maize stover could be collected annually within the tolerant limits of soil erosion at the current production levels (Graham et al. 2007).

Roots constitute approximately 20% of total biomass at maturity and contribute to soil organic matter content (Amos and Walters 2006). However, root mass alone is not sufficient to maintain soil organic matter content over time (Wilts et al. 2004). Detailed discussion on this topic can be found in several recent publications (Allmaras et al. 2004; Wilhelm et al. 2004; Wilts et al. 2004; Perlack et al. 2005; Johnson et al. 2006; Graham et al. 2007).

Stover removal will also entail net removal of nutrients that otherwise contribute to soil nutrition. For example, the N harvest index, which is defined as the ratio of grain N to total aboveground plant N, is approximately 65% in maize (Banziger et al. 1999). Thus, an appropriate portion of the one-third of the total plant N sequestered in stover would have to be replenished by an additional fertilizer application depending upon the amount of stover removed (Dhugga 2007).

16.3 BIOMASS STRUCTURE AND COMPOSITION

Maize stalks account for more than half of the stover biomass, followed respectively by leaves, cobs, and husks (Figure 16.7; Atchison and Hettenhaus 2003; Masoero et al. 2006). Most of the stalk biomass is concentrated in the rind tissue, which consists of a mixture of densely packed vascular bundles embedded in a matrix of sclerenchymatous cells on the outer periphery of the internodes. Rind accounts for less than 20% of the cross-sectional area but more than 80% of stalk dry mass (K.S. Dhugga, unpublished data). Most of the remaining 20% of the biomass is likely accounted for by the vascular bundles that are embedded in the ground tissue consisting of parenchymatous cells. Parenchymatous cells have thin primary walls and are nearly devoid of free sugars at maturity so their contribution to biomass is minor.

The chemical composition of corn stover, rice and wheat straw, and switchgrass is relatively similar (Figure 16.8). Corn stover is approximately 38% cellulose, 26% hemicellulose, and 19% lignin. Rice straw stands out in ash content, which is mostly accounted for by silica deposition on the leaf and sheath surfaces (Figure 16.8).

16.4 CELL WALL BIOSYNTHESIS

Plant cells are surrounded by viscoelastic primary cell walls, that are amenable to expansion under turgor pressure. Upon completion of the process of cell expansion, a thick secondary cell wall is deposited in some cell types. Cellulose microfibrils are embedded in a hemicellulosic matrix along with pectins and structural proteins in primary walls whereas secondary walls contain little protein or pectin, but normally contain lignin (Carpita 1996). Grass cell walls are unique in that they contain little pectin. Apparently, the role of pectin is fulfilled by the hemicellulose matrix consisting mainly of glucuronoarabinoxylan (GAX).

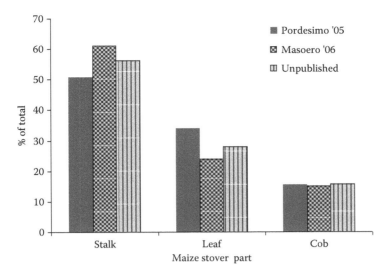

FIGURE 16.7 Components of maize stover. Unpublished data were obtained from a number of hybrids of Pioneer Hi-Bred Intl., Inc. (A DuPont Company) grown in Woodland, CA in 2007 and 2008.

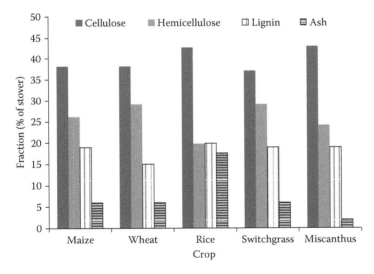

FIGURE 16.8 Chemical composition of vegetative dry matter from various grasses. (Data from Wiselogel et al., *Bioresour Technol*, 56, 103–109, 1996; Christian et al., *Bioresour Technol*, 83, 115–124, 2002; Lewandowski et al., *Agron J*, 95, 1274–1280, 2003a; Lewandowski et al., *Biomass Bioenergy*, 25, 335–361, 2003b; Velasquez et al., *Wood Sci Technol*, 37, 269–278, 2003; Sun et al., *J Agri Food Chem*, 53, 860–870, 2005; DOE (U.S. Department of Energy), Biomass feedstock composition and property database, Department of Energy, Biomass Program, http://www.eere.energy.gov/biomass/progs/search1.cgi, 2006; Mani et al., *Biomass Bioenergy*, 30, 648–654, 2006; Hongzhang, C., and Liying, L., *Bioresour Technol*, 98, 666–676, 2007; Garcia-Barneto et al., *Bioresour Technol,* 100, 3963–3973, 2009.)

16.4.1 Cellulose Synthesis

Cellulose, a major constituent of cell walls, is a paracrystalline form of H-bonded β-1,4-glucan chains and is deposited directly into the apoplast by the plasma membrane-localized cellulose synthase complex. A major breakthrough in our understanding of the biosynthesis of cell wall polysaccharides was the identification of cellulose synthase catalytic subunit (*CesA*). The gene for

CesA was first isolated from the bacterium *Acetobacter xylinum* by enzyme purification and peptide sequencing (Saxena et al. 1990). Similar attempts at identifying plant polysaccharide synthases achieved little success, although a solubilized callose synthase was purified (Meikle et al. 1991; Dhugga and Ray 1994). The first plant *CesA* gene was isolated from developing cotton fibers by screening a few hundred expressed sequence tags (Pear et al. 1996). The *CesA* gene family, which is represented by at least 10 members in various plant species, has at least 13 members in maize (Appenzeller et al. 2004; Djerbi et al. 2005; Somerville 2006). *Arabidopsis*, maize, and rice all have a group of three different *CesA* genes, that are co-expressed in secondary cell wall-forming cells (Tanaka et al. 2003; Taylor et al. 2003; Appenzeller et al. 2004).

The *CesA* genes are believed to encode plasma- membrane-localized catalytic subunits of the rosette complex. These transmembrane complexes are assembled as hexamers, presumably consisting of 36 individual cellulose synthase proteins. After assembly in the Golgi apparatus, the cellulose synthase complexes are exported for integration into the plasma membrane (Somerville 2006). Cellulose is synthesized on the cytosolic side and is subsequently extruded out to the plasma membrane through the pore made by the predicted transmembrane helices of the CesA protein. CesA proteins consist of two transmembrane domains (TMDs) near the amino-terminal (N-terminal) end and six TMDs toward the carboxy-terminus. The zinc-binding N-terminal domain is putatively involved in CesA–CesA interactions (Kurek et al. 2002). The cytoplasmic catalytic domain between TMD2 and TMD3 contains the signature of β-glycosyltransferases, D,D,D,QXXRW (Saxena et al. 1995; Vergara and Carpita 2001).

Several other proteins are known to affect cellulose synthesis: a membrane-associated protein, Kobito, of unknown function; Korrigan, a membrane-anchored β-glucanase; and Cobra, a protein attached to the membrane through a glycophosphatidylinositol (GPI) anchor (Somerville 2006). A member of the Cobra gene family, *Brittle culm-1* (*Bc-1*), was found to specifically affect secondary wall formation in rice (Li et al. 2003). Its orthologs from *Arabidopsis* [*Cobra-like-4* (*CobL4*)] and maize [*Brittle stalk-2* (*Bk2*)] performed orthologous function in respective species (Brown et al. 2005; Ching et al. 2006; Sindhu 2007). Two other enzymes or classes of enzymes that are also known to affect cellulose formation are GDP-mannose pyrophosphorylase, that forms GDP-mannose, and glycosydases. GDP-mannose contributes mannose for the formation of the GPI anchor of the Cobra-like proteins, glycosylation of other proteins, and glucomannan formation. Glycosidases are involved in the glycosylation processing of proteins. Both of these enzymes can thus potentially affect cellulose formation through multiple mechanisms.

Secondary wall-forming genes are in general expressed at significantly higher levels than the primary wall-forming genes, which could explain the rapid deposition of secondary wall on cessation of cell expansion, assuming that there is a correspondence between the levels of CesA proteins and their respective transcripts (Appenzeller et al. 2004).

16.4.2 HEMICELLULOSE SYNTHESIS

The hemicellulosic fraction, unlike pectin, is tightly associated with cellulose and requires harsh chemical treatment to be separated from the cellulosic fraction. Hemicellulosic polysaccharides are divided into four major categories: xyloglucans (XG), (gluco)mannans, glucuronoarabinoxylans (GAX), and mixed-linkage glucans (MLG) (Carpita 2000; Somerville et al. 2004). Corn stover biomass is made of cellulose microfibrils embedded in a hemicellulosic matrix consisting mainly of GAX, a β-1,4-linked xylan backbone that is substituted with glucuronosyl and arabinosyl residues with α-1,2 and α-1,3 linkages, respectively (Figure 16.9) (Carpita 1996). GAX also contains acetate that is esterified to the xylosyl residues at the second and third positions. In addition to GAX, grass cell walls contain MLG, which consists of β-1,4-linked oligosaccharides of varying lengths coupled through β-1,3-linkages. MLG is expressed in expanding cells and is recycled as expansion ceases, whereas it serves as a structural polysaccharide in the walls of the seed endosperm of certain grasses, such as barley, oats, and wheat (Genc et al. 2001). Because the β-1,4 linkage of cellulose is similar

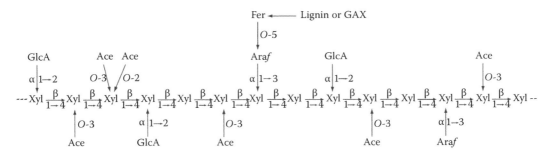

FIGURE 16.9 Chemical structure of glucuronoarabinoxylan, a major wall constituent of grass cell wall. Ace, acetate; Ara*f*, arabinofuranose; Fer, ferulate; GlcA, glucuronate; Xyl, xylose. Arabinose/xylose ratio is ~0.1, glucuronate/xylose is ~0.2, and ferulate ester/arabinose is ~0.4 in a corn stalk (Jung and Casler 2006). All of the arabinose and most of the glucuronate in stover are assumed to occur as GAX, with the remainder of glucuronate present as potentially in other forms (e.g., trace amounts of pectin). Acetate concentration in maize stover has been reported to be 30–50 g/kg of dry matter (Wooley et al. 1999; McAloon et al. 2000). Assuming that all of it occurs in GAX and adjusting for molarity, approximately one-third to one-half of the xylosyl residues on GAX are expected to be acetylated. (Adapted from Dhugga, K.S., *Crop Sci*, 47, 2211–2227, 2007. With kind permission from the Crop Science Society of America.)

to the linkages found in the backbones of various hemicelluloses, it was postulated that cellulose synthase-like (*Csl*) genes might be responsible for the biosynthesis of glycan backbones in the Golgi (Richmond and Somerville 2000). Bioinformatics tools helped to identify *Csl* genes in various plant species, which were then grouped into nine families, *CslA* through *CslH* and *CslJ* (Fincher 2009; Van Erp and Walton 2009). *CslF*, *CslH*, and *CslJ* families are unique to the grasses, whereas *CslB* and *CslG* occur only in dicots. The remaining families are represented in both grasses and dicots.

Genes for the backbone formation of three of the hemicellulosic polysaccharides—XG, MLG, and β-(gluco)mannan—have been identified (Dhugga et al. 2004; Liepman et al. 2005; Burton et al. 2006; Cocuron et al. 2007; Doblin et al. 2009). The first successful identification of a Golgi polysaccharide synthase was achieved through transcriptional profiling, whereby a *CslA* gene was found to make β-1,4-mannan (Dhugga et al. 2004). Guar endosperm consists nearly entirely of galactomannan, a hemicellulosic polysaccharide that is deposited in the cell wall after synthesis in the Golgi apparatus and which serves the function of seed storage carbohydrates. In the expressed sequence tag (EST) database of developing guar endosperm, a particular *CslA* gene was most abundant at a developmental stage when the mannan synthase activity was at its peak. This *CslA* gene was identified and named mannan synthase (*ManS*) (Dhugga et al. 2004). Soybean somatic embryos were used for functional characterization of *ManS* because soybean somatic embryos do not incorporate significant amounts of mannose into polymeric form. Membrane particles derived from the somatic embryos transformed with the *ManS* gene exhibited substantial mannan synthase activity that was coincident with the level of expression of the gene, demonstrating that the candidate gene indeed coded for mannan synthase (Dhugga et al. 2004). Some members of the *CslA* group from *Arabidopsis* were later found to possess (gluco)mannan synthase activity (Liepman et al. 2005).

A similar approach was used to identify a synthase involved in the formation of glucan, which forms the backbone of XG, from the nasturtium (*Tropaeolum majus* L.) seed EST database. Nasturtium seeds accumulate XG as a storage polysaccharide (Cocuron et al. 2007). Expression of the nasturtium *CslC* gene, which was the most abundant *Csl* gene in the EST database, in yeast led the formation of β-1,4-glucan; however, co-expression of a xylosyltransferase did not produce any XG (Cocuron et al. 2007). XG synthesis has been reported to require simultaneous actions of glucan synthase and xylosyltransferase enzymes (Hayashi 1989; Faik et al. 2002). What then could the role be of β-glucan produced by CslC? Purified Golgi fraction has been shown to make β-1,4-glucan independent of xylose addition (Ray 1979). The enzyme catalyzing this reaction is often referred to as glucan synthase-I (GS-I) (Ray 1979). It is possible that CslC is actually GS-I (Sandhu 2009). Adjacent glycosyl residues of the surface chains in cellulose microfibrils are twisted with respect to each other in comparison to the generally linear arrangement of β-1,4-linked residues (Vietor et al. 2002). The CslC product upon deposition after microfibril assembly, which happens at the plasma membrane, could potentially acquire the altered conformation.

A different route was taken to determine the function of two groups of genes specific to grasses—*CslF* and *H*. Both were found to make MLG (Burton et al. 2006; Doblin et al. 2009). MLG was mapped in barley, and a syntenic relationship between rice and barley helped to identify a cluster of *CslF* genes (Burton et al. 2006). Heterologous expression of a *CslF* gene in *Arabidopsis*, a species with no *CslF* and no MLG, led to the accumulation of MLG in the cell wall. Following a similar approach, expression of a *CslH* gene from barley in *Arabidopsis* also resulted in the accumulation of MLG (Doblin et al. 2009).

Molecular components for xylan synthase, as well as arabinosyl and glucuronosyl transferases that add respective residues onto the xylan backbone of GAX, remain unknown. Maize stover contains 3–5% acetate and approximately 20% xylose on a mass basis (Wooley et al. 1999; McAloon et al. 2000). Molecular components for acetylation of GAX also remain to be identified. Apparently, acetylation occurs in the Golgi compartment and the transferase that acetylates GAX may use acetyl-CoA as a substrate. An acetyl transferase that acetylates rhamnogalacturonan has been assayed in cultured potato (*Solanum tuberosum L.*) cells, but the corresponding protein(s) and gene(s) remain to be isolated (Pauly and Scheller 2000).

16.4.3 Lignin Synthesis

Lignin, a major component of the secondary cell wall of vascular plants, is the second-most abundant macromolecule on earth after cellulose (Boerjan et al. 2003). Being hydrophobic, it most likely contributes to the mechanical strength of the tissues by excluding water from the vicinity of cellulose, the strength of which varies with moisture content (drier the stronger). Lignin has high resilience to degradation, a characteristic that is desirable for defense against insects and pests but poses a problem in the production of cellulosic biofuels, forage digestibility, and chemical pulping. Its distinctive polymeric structure is the reason for its non-degradable nature. In addition, because of its covalent linkages in the wall and its role as a matrix for the adsorption of enzymes and proteins, lignin hinders accessibility of the hydrolytic enzymes to cellulose and hemicellulose in the cell wall matrix (Iiyama et al. 1994). Pretreatment of lignocellulosic biomass to remove lignin from cellulose and hemicellulose for biofuel production and paper pulping is a costly step that requires noxious chemicals and results in the formation of compounds that inhibit downstream processes in ethanol production.

Lignin is also of significant interest for silage and forage digestibility in dairy cattle. The high content of cell wall in forages, together with the restriction imposed by lignin on cell wall digestibility, creates a disparity between the available energy of grain and forage although the gross amount of energy per unit dry matter is comparable between the two (Ralph et al. 2004). Although 70% of the energy value of silage maize is derived from the grain, there exists significant genetic variation in the extent of digestibility in maize and other forage grasses, and even among different maize genotypes. These genetic differences are exemplified in the brown midrib (*bmr*) mutants of maize and sorghum, in which natural or chemically induced mutations affect lignin content and composition and ultimately cell wall digestibility.

Lignin is a heteropolymer derived from monomeric units of *p*-coumaryl, coniferyl, and sinapyl alcohols that give rise, respectively, to *p*-hydroxyphenyl (H), guaiacyl (G), and syringyl (S) phenylpropanoid units upon incorporation into the lignin polmer (Boerjan et al. 2003). The monolignols are transported to the cell walls, where dehydrogenative polymerization results in lignification. The three monolignols are generally incorporated during specific stages of cell–wall formation, with H units being deposited first, followed by G and finally S units. In maize, young tissues preferentially accumulate G lignin, whereas the content of S lignin increases with maturity.

On the basis of thioacidolysis, the most lignified organs in maize at silking are the basal internode and roots and the least lignified tissues are leaves (Guillaumie et al. 2007). In the basal internode, with the exception of a few parenchyma layers inside peripheral sclerenchyma, all cell types are lignified, and the predominant lignin is of the S type. Detailed histochemical analyses of different maize tissues in this study have shown that roots of plants at the four- to five-leaf stage contain a wide variety of lignified cells, except for the single layer of hypodermal cells just below the epidermal layer, and that these lignified cells are rich in S units. At the same growth stage, leaf tissue was found to be characterized by a large proportion of sclerenchyma cells that likely synthesize H and/or G units, whereas the parechyma cells associated with the xylem vessels are rich in S units. In maize leaf parenchyma cells, deposition of G lignin has been shown to precede that of S lignin, whereas in vascular bundles S lignin is deposited at the earliest stages of lignification (Wen et al. 2008). In general, grasses have been shown to deposit large amounts of ferulate-polysaccharide esters during the early stages of lignification, and the ferulate esters likely serve as nucleation sites for lignin polymerization (Zhong et al. 2008; Zhou et al. 2009).

In maize, as in dicots, most of the lignin biosynthetic enzymes are encoded by multigene families (Guillaumie et al. 2007). So far, five sequences have been identified for the phenyl ammonia lyase (PAL) enzyme in maize, which has activity toward tyrosine and phenylalanine. These can be grouped into classes I, II, and III, with two sequences each within classes I and III and one in class II. All are expressed at higher levels in stems and roots, with relatively low expression in leaves, and the highest expression being in the sixth internode at silking. The expression pattern agrees with the role of PAL in secondary wall formation. The class II *PAL* gene has the highest expression of all three classes in all tissues, followed sequentially by class I and III genes.

Polymerization of monolignols occurs through the action of peroxidases and laccases. Thirteen maize peroxidase sequences have been reported in the Maize Genetics and Genomics database (Guillet-Claude et al. 2004), but only three have been characterized. Of the three, *ZmPox1* occurs in the epidermal cells of root tips where lignification does not occur, whereas *ZmPox2* and *ZmPox3* are observed in vascular tissues of the elongation zone of young roots, with the expression of *ZmPox2* being much more abundant than that of *ZmPox3*. Analysis of a retrotransposon insertion in exon 2 of the *ZmPox3* gene provided further evidence for the role of this gene in monolignol polymerization (Guillet-Claude et al. 2004). A deficiency in *ZmPox3* activity has a negative effect on cell wall digestibility. Genetic diversity analysis of this peroxidase indicates that it could be a relevant target for improving digestibility through the use of specific allele introgressions (Guillet-Claude et al. 2004).

16.5   CELL WALL CROSS-LINKING AND MECHANICAL STRENGTH

Lignin has long been a target for reduction because of its known adverse effect on rumen digestibility as well as on ethanol production from stover biomass (Grabber 2005). The cross-links that lignin forms with other wall polymers are believed to increase the recalcitrance of vegetative tissues to hydrolytic enzymes (Jung 2003; Jung and Casler 2006).

Activated forms of the hydroxycinnamic acids ferulate and *p*-coumarate are, respectively, feruloyl-CoA and *p*-coumaroyl-CoA, which, aside from being intermediates in the monolignol biosynthetic pathway, act as substrates for feruloylation or coumaroylation of GAX (Campbell and Sederoff 1996; Anterola and Lewis 2002). Ferulate and coumarate are linked to the arabinosyl residues of GAX through an ester linkage in the Golgi compartment by a transferase that remains to be identified (Figure 16.9; Iiyama et al. 1994). These feruloyl moieties can form linkages with other reactive groups in the cell wall through a peroxidase-mediated reaction that results in GAX-GAX and GAX-lignin linkages (Iiyama et al. 1994). Some cross-linking may also occur in the Golgi itself (Fry et al. 2000). Another type of cross-linking is hypothesized to occur among wall polysaccharides through a class of enzymes that have the ability to cut and paste glycans (Fry 2004). A transglycosylase/hydrolase has recently been reported to mediate the formation of linkages among diverse cell wall polysaccharides, potentially augmenting the cell wall network (Hrmova et al. 2007). However, occurrence of many of these interpolymer linkages in the cell wall remains to be demonstrated. Feruloyl transferase and transglycosylase/ hydrolase are attractive targets for potentially altering the degree of cell wall cross-linking for improved digestibility. A candidate gene for feruloyl transferase from rice was proposed using a bioinformatics approach (Mitchell et al. 2007). However, the protein encoded by this gene is devoid of a transmembrane domain. Golgi transferases identified thus far are type II membrane proteins with a transmembrane domain near the N-terminus (Edwards et al. 1999; Perrin et al. 1999; Faik et al. 2002).

The effect of lignin on increasing mechanical strength of cellulose is indirect and is achieved by keeping the latter relatively dry. Because of the plasticizing effect of water, moist cellulose is mechanically weaker than dry cellulose (Tolstogurov 2000). In agreement, the amount of cellulose in a unit length of maize stalk nearly completely explained the mechanical strength of the internodes below the ear (Appenzeller et al. 2004). In an independent study involving a mutation in the maize *bk2* gene that specifically and dramatically reduced cellulose accumulation in the secondary walls, lignin and hemicellulose amounts were unaltered but mechanical strength was dramatically reduced, further indicating that cellulose is the primary determinant of mechanical strength and that any contribution of lignin is through alteration of the aqueous environment around cellulose in the cell wall (Figure 16.10) (Ching et al. 2006). Selection for reduced lignin and increased cellulose in stover can increase mechanical strength and ethanol yield (Appenzeller et al. 2004; Ching et al. 2006).

16.6 BIOENERGETIC CONSIDERATIONS FOR BIOMASS INTERCONVERSION

Ethanol production from renewable resources is a direct function of the available carbohydrate content. The grain in maize is the richest source of carbohydrates in the plant (Earle et al. 1946). Protein and oil content in the grain can be reduced to further increase carbohydrate content, but because of the reduced calorific content in these plants they will be suitable only for ethanol production and not animal feed.

Stover-only corn, after elimination of the ear, is another option to increase carbohydrate content (Dhugga 2007). Although technically feasible, this avenue is not very attractive because (1) grain,

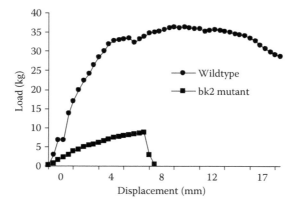

FIGURE 16.10 Difference in internodal mechanical strength between the brittle stalk (*bk2*) mutant and its wild-type sibling. Internodal flexural strength was measured below the ear 1 week after flowering. The seeds from the same selfed ear segregating for *bk2* were grown in the greenhouse. These data are for the fifth internode below the ear node. *Brittle stalk 2* encodes a putative glycosylphosphatidylinositol-anchored protein that affects the mechanical strength of maize tissues by altering the composition and structure of secondary cell walls. (Figure from Ching A., et al. *Planta*, 224, 1174–1184, 2006. With kind permission from Springer Science+Business Media B.V.)

because of its high density, is easier to store and transport than stover; (2) ethanol from grain starch is produced close to the theoretically predicted efficiency (Patzek 2006); (3) grain ethanol is already a mature industry whereas cellulosic ethanol is still several to many years away from being commercially profitable; and (4) conversion of grain to stover will also mean more total lignin, assuming its concentration in the stover does not change, which will more than offset the expected biomass gain from converting protein and oil into carbohydrates (see below).

Comparative bioenergetic costs associated with constructing tissues of different compositions can be used in estimating the amount of biomass that can be produced from a given amount of photosynthate. It was estimated from the underlying biochemical pathways that 1 g of photosynthate could be used to make 826 mg of storage or structural carbohydrates, 465 mg of lignin, 404 mg of protein, or 330 mg of lipid (Penning de Vries et al. 1974; Sinclair and de Wit 1975; Bhatia and Rabson 1976).

Maize varieties adapted to tropical regions when grown in temperate zones may not flower and keep producing vegetative biomass, resulting in increased biomass production. However, energy-rich lignin in stover would negate most of the expected gain in biomass from the elimination of oil and protein in the form of grain. For example, bioproductivity of a typical maize grain is approximately 720 mg/g photosynthate (Figure 16.11) (Dhugga and Waines 1989). Similarly, for maize stover of a typical composition, bioproductivity would be 704 mg/g of photosynthate (Figure 16.11). Additional biomass gain can be realized from a slightly prolonged growing season because of an expected delay in leaf senescence and hence continued photosynthesis from the lack of a strong N sink in the form of grain. Reducing lignin, which has been difficult to accomplish because of undesirable pleiotropic effects, can further aid in increasing biomass in stover-only maize. For example, a reduction in lignin content by 50% (from 18% of dry mass to 9%) may produce an additional 6–7% biomass (Figure 16.11).

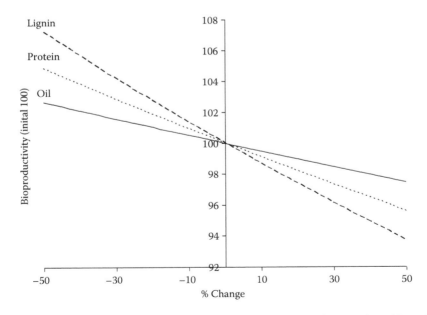

FIGURE 16.11 Predicted effects of altering oil or protein (grain) or lignin (stover) on bioproductivity in maize. Starting composition for grain was taken as 90 g/kg protein, 40 g/kg oil, 20 g/kg ash, and (by subtraction) 850 g/kg carbohydrates (starch and fiber). The initial composition for stover was 730 g/kg carbohydrates, 180 g/kg lignin, 40 g/kg protein, and 50 g/kg ash. Replacement of lignin, oil, or protein only by carbohydrates is assumed. Although oil requires the most energy for its synthesis, relatively steeper slopes for lignin and protein are due to their higher initial amounts in the respective tissues. (Adapted from Dhugga, K.S., *Crop Sci.*, 47, 2211–2227, 2007.)

Theoretical conversions, which help define upper limits for productivity of biomass of a particular composition, do not always lead to expected results and may require substantial breeding effort in selecting desirable recombinants (Dhugga and Waines 1989). For example, a reduction in lignin is expected to be accompanied by an increased bioproductivity but an opposite outcome is observed because of the associated pleiotropic effects (Figure 16.11) (Pedersen et al. 2005).

Increasing starch in the grain is yet another option, but the starch content of the grain is already high at more than 70% (Earle et al. 1946; Belyea et al. 2004). Additional starch can be accumulated at the expense of embryo size, grain protein in the endosperm, and/or grain fiber. Reductions in endosperm protein as well as embryo size below a certain threshold may cause viability problems and a reduction in fiber content may affect grain strength, causing increased breakage. Most of the fiber in the grain is in the pericarp, which is rich in GAX (Hazen et al. 2003). Incrementally replacing GAX with cellulose, which is mechanically stronger but also more crystalline, may allow for maintenance of grain strength yet allocate more carbon to additional starch formation (Dhugga 2007).

16.6.1 MUTANTS RELEVANT TO BIOENERGY

Improved cell wall degradability through the alteration of lignin content and/or composition has been observed in the naturally occurring brown midrib mutants of maize, the understanding of which has led to the exploration of transgenic manipulation to improve digestibility. Several QTL have been identified for cell wall lignin content, the amount of *p*-coumarate and ferulate esters, and for digestibility. Some of these overlap with QTL for insect tolerance and with the localization of lignin biosynthetic genes. Brown midrib mutants have been reported in maize, sorghum, and millet, in which tissues associated with lignification, such as leaf midrib, exhibit a reddish brown coloration (Barrieire et al. 2004). This coloration is intrinsic to the lignin polymer itself because it remains in the tissue after soluble metabolites and cell wall polysaccharides are removed. The actual reason for the coloration and the interesting fact that, to date, *bmr* mutants have only been reported in diploid Panicoideae C4 grass species, are aspects that remain unanswered. The *bmr* mutants reported in maize are naturally occurring, whereas those in sorghum and millet were obtained through chemical mutagenesis. Four *bmr* mutants are known in maize and 19 in sorghum.

The most well characterized of the *bmr* mutants of maize is the *bm3* mutant, where the mutation is known to be caused by a partial reduction in the activity of the enzyme, cinnamoyl *o*-methyltransferase (COMT) (Morrow et al. 1997). The *bm3* mutant exhibits a reduction in lignin content on the order of 25–40%, a reduction in the S unit composition, and also the occurrence of unusual lignin units, specifically, 5-hydroxyguaiacyl units (Barrieire et al. 2004). The S/G ratio is reduced in the *bm3* mutants by more than 3-fold. These mutants have fewer *p*-coumarate esters, which is consistent with the preferential acylation of S units by *p*-coumaric acid, and noticeably elevated levels of ferulate in younger tissues (Marita et al. 2003). Disruption of the COMT gene in the *bm3* mutant has been shown to result in an overexpression of other monolignol biosynthetic genes in the basal and ear internodes, with the exception of PAL (Guillaumie et al. 2008). These mutants also showed overexpression of several transcription factors, cell signaling genes, and transport and detoxification genes in these tissues. Of the four *bm* mutants reported in maize, *bm3* causes most improvement in plant stover digestibility.

The maize *bm1* mutation reduces the expression of the cinnamoyl alcohol dehydrogenase (CAD) gene, the activity of which is reduced 60–70% in aerial tissues and 90–97% in roots of these mutants (Halpin et al. 1998). The amounts of lignin, *p*-coumarate esters, and ferulate ethers are significantly reduced in the stem of *bm1* mutants, whereas the amount of ferulate esters is only slightly reduced. There is no alteration in the lignin composition of *bm1* mutants, and the amount of S, G, and H units remain similar to that of the wild type. The lignin of *bm1* mutant incorporates coniferaldehyde and sinapaldehyde units as a result of CAD deficiency (Barrieire et al. 2004).

The genes responsible for the *bm2* and *bm4* mutations have not been identified yet. In both the cases, the amount of lignin and ferulate ether is reduced, whereas the amounts of *p*-coumarate and ferulate esters are slightly increased.

Several studies on QTL analyses for cell wall digestibility and lignification in maize have been reported (Mechin et al. 2001; Roussel et al. 2002; Cardinal et al. 2003; Barrieire et al. 2004). Five major QTL clusters can be found in bins 6.06, 3.05/06, 1.03, 8.05, and 9.02, whereas additional locations and clusters involved in these traits have been located in nine other bins. The genes underlying these QTL have not been identified yet, although some of the lignin biosynthesis genes colocalize with these QTL. The QTL for cell wall digestibility colocalize with lignification and cross-linking traits, and in several cases they also colocalize with QTL for tolerance to European corn borer. Thus, breeding for improved digestibility through modification of lignin biosynthesis should consider the implications to the plant's defense against insect pests. Yet, two important QTL for cell wall digestibility (i.e., bins 1.03 and 6.06) that do not colocalize with any published lignin-biosynthetic pathway sequence do not show co-localization with European corn borer tolerance either.

16.6.2 BIOENERGY AND AGRONOMIC TRAITS

The goal of higher ethanol productivity, if only looked at from the viewpoint of reducing lignin, is at odds with that of stalk standability, an important agronomic trait. Vascular bundles and sclerenchymatous cells have lignified, cellulose-rich walls. Lignin is generally believed to contribute to mechanical strength; however, it plays a role in compression rather than tensile strength (Ching et al. 2006). To maximize mechanical resistance to breakage, for example, cellulose microfibrils in the stem tissue subjected to mechanical bending are oriented along the axis of tension (Green 1962). Similarly, cellulose microfibrils are oriented perpendicular to the axis of growth in an expanding cell, again to maximize radial strength so as to allow longitudinal expansion (Green 1962). Lignin synthesis is downregulated in the cells subjected to tension but is upregulated in those subjected to compression, suggesting that it is not lignin but cellulose that is primarily responsible for tensile strength (Donaldson et al. 1999; Joseleau et al. 2004; Andersson-Gunneras et al. 2006; Schmitt et al. 2006).

The relationship between planting density and biomass increase is nonlinear, particularly at higher densities, i.e., biomass increase is less than would be expected from the number of plants per unit land area. Eventually, the reduction in individual plant biomass that accompanies increasing planting density will make the plants mechanically unstable to the extent that any gain in grain yield will be negated by an increase in crop lodging (Appenzeller et al. 2004). At that point, about the only option left to the breeders would be to sacrifice HI to maintain standability, which again will offset the gain in grain yield from increased planting density (Dhugga 2007).

16.7 SYSTEMS APPROACH IN IMPROVING MAIZE BIOPRODUCTIVITY

Plant breeders, by selecting for grain yield, essentially use a systems approach because they indirectly select simultaneously for many gene combinations that underlie the desired phenotype. However, it is challenging to model plant performance using theoretical knowledge of the metabolic pathways because of the enormous plasticity of central metabolism (Carrari et al. 2003).

Grain yield and biomass are complex traits that result from an integration of many developmental steps during the ontogeny of the plant. Throughout its development, a plant is exposed to variable environmental factors, which vary across geographic locations. Additional unpredictability results from the year-to-year variation of these environmental factors. For example, moisture content of the soil and soil temperature can affect germination, which eventually manifests in the final crop stand in the field. Available nutrients in the soil can vary depending upon the soil type and unpredictable moisture content, which affects plant growth at different stages of development. Mineralization, which means release of organically bound nutrients for utilization by the plant, also varies as a function of soil nutrient content, moisture level, and soil temperature, which affects nutrient availability in an unpredictable manner. Unpredictable abiotic stresses around flowering are known to most severely affect grain yield.

The picture that emerges from the aforementioned discussion is that in the light of the unpredictable components of the environmental variation, which make it extremely difficult to discern the effects of a single or a few genetic or transgenic factors on increasing grain yield, testing varieties over multiple years and geographic locations is required. Of course, it is easier to determine the effect of a single gene or a few genes on drastically reducing grain yield (for example, through an increase in barrenness, lodging, or a severe reduction in plant height). What is challenging is to measure the effect of a single or a few genes on increasing grain yield. A diagrammatic representation of genetic and environmental factors influencing plant growth and development is shown in Figures 16.12 and 16.13.

16.8   BIOTECHNOLOGICAL OPPORTUNITIES, OUTCOMES, AND FUTURE PROSPECTS

Rapid advances in genomic sciences have provided an abundance of molecular tools that can potentially be used to alter plant structure and function. Despite considerable progress in the area of cell wall synthesis over the last decade, some of the problems remain unsolved. For example, molecular components of GAX formation as well as its acetylation remain to be identified. A discussion on prospects and attempts to alter plant biomass and composition follows. Recent reviews cover other complementary aspects of transgenic manipulation to improve maize for ethanol production (Torney et al. 2007; Gressel 2008).

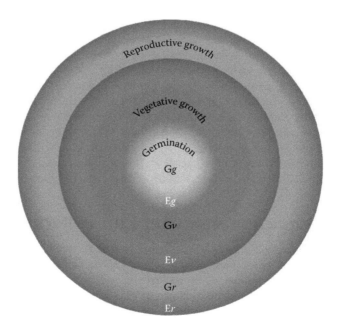

FIGURE 16.12 (See color insert)   Illustration of complex genotype × environment effects on plant growth. G, genotype; E, environment; subscripts *g*, germination; *v*, vegetative growth; and *r*, reproductive growth. Dark-to-light gradient is indicative of genotypic effects (dark), genotype × environment effects (transition from dark to light), and environmental effects (light, at the edge of each developmental stage). Environmental factors that affect germination are different from the ones that affect vegetative growth, which are different from the ones that the plant experiences during reproductive growth. Environment can vary independently during each of these phases, increasing the unpredictability of plant performance. For further details, see Figure 16.13.

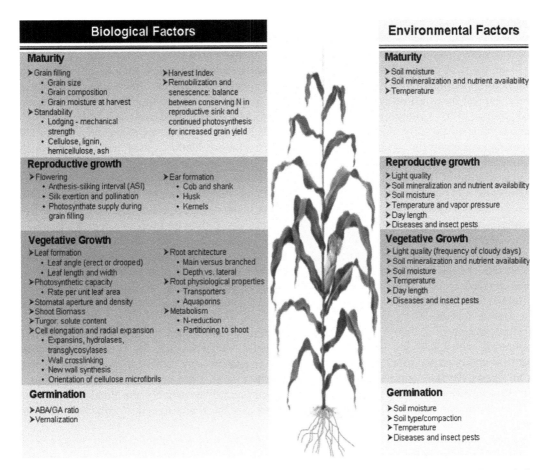

FIGURE 16.13 (See color insert) Diagrammatic representation of genetic and environmental factors influencing plant growth and development.

Introduction of genes for herbicide, insect, and rootworm resistance into maize has indirectly led to increased biomass by controlling weeds that would otherwise compete for nutrients and water, and maintaining healthy roots and leaves for continued nutrient absorption, and improved photosynthesis, respectively.

Attempts at improving biomass by altering the expression of metabolic genes have given mixed results. Transgenic expression of single genes has been reported to increase biomass in several plant species (Smidansky et al. 2002, 2007; Biemelt et al. 2004; Lefebvre et al. 2005). These types of results are often obtained with experimental lines and in highly controlled environments. Any of these types of discoveries has yet to lead to a commercial product, in part perhaps because of the inherent plasticity of the complex plant metabolism to maintain homeostasis despite transgenic perturbations (Carrari et al. 2003). It is possible that the experimental line being used in a study claiming the positive effect of a transgene on biomass is indeed deficient in the particular function that the transgene enhances under the specific experimental conditions used. The positive effect of a transgene on biomass observed in an experimental line disappeared when it was backcrossed into agronomically elite genetic backgrounds (Meyer et al. 2007).

Grain sink in maize could potentially be made limiting by space-planting the hybrids adapted to high planting density. Reduced shading allows plants to increase photosynthate production that may exceed the amount needed to fill the available grains. Progressive farmers deliberately increase planting density to the extent that the kernels at the ear tip do not fill, a phenotype referred to as "nose-backing," which is indicative of sink not being limiting. This apparently ensures capturing of the maximal amount of available photosynthate for grain production. Biomass of space-planted individual plants or trees is not thus an indicator of the potential biomass production per unit land area.

Complexity of the cellulose synthase system poses a challenge to increasing its activity through transgenic manipulation. For an overexpressed component to have an effect, a parallel increase in the expression of all other components of the complex must be assumed (Sere 1987; Barreiro and Dhugga 2007). Alternatively, the component being overexpressed must indeed be limiting in the assembly of the cellulose synthase complex. Regardless of these apprehensions, preliminary experiments suggest that it may be possible to affect cellulose synthesis with the overexpression of individual *CesA* genes (Dhugga et al. 2005).

Association of the rate of cellulose synthesis or tissue cellulose concentration with available haplotypes and specifically with allelic variants of different genes known to be involved in cellulose formation offers perhaps more promise in accomplishing the objective of increasing cellulose in plant tissues. The associated haplotypes could then be used to select for increased cellulose production in breeding populations. Crystalline cellulose can be determined by a simple chemical method that is amenable to scaling-up for high throughput analysis (Updegraff 1969). Direct measurement of cellulose would allow for the mapping of this trait using recombinant lines derived from divergent parents, which might also reveal as yet unknown effectors of cellulose formation. Mutational genetics has revealed a number of non-*CesA* genes to be involved in cellulose formation.

Upregulation of cellulose synthesis may eventually lead to altered composition of stover by competing with GAX and lignin in secondary walls, which account for most of the biomass. For this approach to affect biomass production, vegetative sink, which may be approximated as an aggregate of nonphotosynthetic, nongrain plant parts, must be assumed to be limiting in the current varieties. HI in maize is held nearly constant even when grain yield varies widely under varying environmental conditions, suggesting that the source-to-grain sink ratio is already optimized (Sinclair et al. 1990).

Overexpression of a cellulose-binding domain (CBD) from microbial cellulose hydrolases in poplar has been claimed to enhance growth (Levy et al. 2002). Microbial and fungal cellulases possess a CBD and a hydrolytic domain. CBD enhances hydrolysis of cellulose by acting as an anchor for the hydrolytic domain and also perhaps by depolymerizing the cellulose crystal into glucan chains. Separated from the hydrolytic domain, CBD still retains the ability to bind crystalline cellulose. The explanation for enhanced growth rate in poplar overexpressing CBD was that by binding to the nascent glucan chains coming out of the cellulose synthase complex, CBD slowed the rate of crystallization and thus enhanced the rate of polymerization (Levy et al. 2002). The same group claimed that overexpression of an endo-β-1,4-glucanase accelerated growth in tobacco and *Arabidopsis*. Implicit in these studies is the assumption that carbon supply (source) is nonlimiting and only cellulose polymerization or wall properties limit growth. Follow-up long-term studies are not available yet, so it is not possible to determine if the reported increase in cellulose synthesis or accelerated growth indeed resulted in increased biomass per unit land area per unit time.

Disruption of the enzymes that potentially add side chains onto the xylan backbone of GAX causes a severe disruption in secondary wall synthesis (Zhong et al. 2005). A number of *fragile fiber* (*fra*) and *irregular xylem* (*irx*) mutants are caused by defects in genes encoding the enzymes potentially involved in the addition of a side group onto the xylan backbone (Burk and Ye 2002; Zhong et al. 2004, 2005; Persson et al. 2007). However, complete knockouts caused severe phenotypes, and it remains to be seen whether partial downregulation of these genes can quantitatively reduce GAX content without adversely affecting plant growth (Burk and Ye 2002; Zhong et al. 2004, 2005; Persson et al. 2007).

A candidate gene for feruloylation of GAX has been proposed based on comparative genomics (Mitchell et al. 2007). It can be downregulated by transgenic means to determine whether it reduces the degree of cross-linking in the wall and, if so, whether and to what extent it is biologically tolerable. Once the molecular components for acetylation of GAX are identified, they could similarly be manipulated to potentially reduce the acetyl content of the dry matter for improved fermentation.

Lignin has been down regulated by the introgression of spontaneous *bmr* mutations into elite germplasm and by transgenically silencing a number of genes from the monolignol biosynthetic pathway (Anterola and Lewis 2002; Burk and Ye 2002; Zhong et al. 2004, 2005). Reduction in lignin is generally accompanied by a concomitant reduction in biomass (Pedersen et al. 2005). In addition, plants with lowered lignin are generally more susceptible to insects and diseases that likely result from a weakening of the physical barrier that lignin poses in accessing the wall polysaccharides that the pathogens need for growth (Pedersen et al. 2005). Variable expression of the *bmr* mutations in different genetic backgrounds suggests that it may be possible to reduce lignin to some extent without a remarkable effect on total biomass (Grabber et al. 2004; Pedersen et al. 2005; Pichon et al. 2006). For further details, see Section 16.7.

Lignin-rich residue remaining after biomass hydrolysis in the current cellulosic ethanol protocols is projected to be burnt to generate electricity and heat (Wright et al. 2006). Various technologies to generate high-value products from this residue are being explored, which, if commercially successful, might help boost the value of stover (Hahn-Hagerdal et al. 2006).

An increase in cellulose concentration in lignin-downregulated aspen was reported to be accompanied by increased growth rate (Hu et al. 1999). This conclusion has been questioned for various reasons, the main ones being (1) a lack of relationship between the degree of lignin reduction and growth rates of various transgenic events and (2) non-normalization of the proportions of different wall constituents to compensate for reduced lignin, a major constituent itself (Anterola and Lewis 2002).

The residue from the grain remaining after ethanol distillation is referred to as distiller's dried grains with solubles (DDGS), which is rich in oil, protein, and fiber (Belyea et al. 2004). Currently, DDGS is sold mainly as cattle feed. Because of the high amount of the pentose-rich polysaccharide fraction, which adversely affects digestibility in monogastric animals, it is blended only in small amounts with poultry and swine feed (Fastinger et al. 2006; Amezcua and Parsons 2007). Lowering the amount of GAX in the grain and replacing it with cellulose will expand the uses of DDGS and increase its suitability as a cellulosic feedstock for ethanol production.

Hydrolytic enzymes have been successfully expressed in plants, apparently with the objective of reducing the cost of their production (Dai et al. 2005). The idea of these enzymes eliminating or reducing the addition of exogenous enzyme cocktails is unlikely to be valid under the currently used protocols to process biomass for ethanol production because these enzymes are unlikely to withstand the harshness of the pretreatment steps. Whether production of these enzymes in plants proves to be more economical than the microbially produced enzymes remains to be seen.

Exogenous addition of an expansin protein at a level of 10 mg/g (or, in practical terms, 10 kg/Mg) of isolated maize cell walls caused significantly greater swelling than the control walls (Yennawar et al. 2006). Apparently, this occurs because of facilitated access of the wall polysaccharides to water. Whether increased swelling improves digestibility of the cell walls by the hydrolytic enzymes is not yet known. Also, the cell walls were prepared from the silk tissue that hardly has any lignin. It would be interesting to see if the expansin proteins can be accumulated in plant to such high levels as used in this study without affecting plant biomass. Secondary walls may be the preferred sites for the in planta expression of these proteins because their deposition begins as the cell expansion begins to cease.

As is obvious from the aforementioned discussion, genetic engineering of polysaccharide biosynthetic pathways for cell wall formation is still in its infancy. Complete understanding of these pathways will greatly expand the tool kit needed for biotechnological manipulation of the cell wall. Enough critical information is already available for exploring the use of a combination of genetic and transgenic methods to alter stover and grain composition for improved stover digestibility and increased ethanol yield.

16.9 CONCLUSIONS

With increasing demand for fossil fuels from developing countries and limited scope for a further increase in oil production, it is inevitable that alternative sources of energy be explored. Ethanol or electricity produced from cellulosic biomass is such an alternative. Whereas ethanol production from corn grain is a mature industry, cellulosic ethanol is still in development, the main bottlenecks being the costs associated with stover pretreatment and hydrolytic enzymes needed to convert stover into soluble sugars. Considerable progress has been made in lowering the cost inputs associated with these bottlenecks, but the technology to convert cellulosic biomass into ethanol on a commercial scale is still years away.

Direct selection will continue to contribute to increased grain yield and biomass production and perhaps to increased cellulose content of the stover biomass. There is no dearth of reports claiming a positive effect of reduced lignin on stover digestibility. To be commercially viable, any such improvement must be reflected in the increased efficiency of ethanol or electricity production per unit land area, not simply per unit dry mass. Biotechnology offers considerable scope to alter the chemical composition of stover through alteration of GAX amounts and composition and perhaps lignin cross-linking. The degree to which these alterations can be tolerated by the plant without adverse pleiotropic effects will determine the extent of success of the biotechnological approach.

REFERENCES

Allmaras RR, Linden DR, Clapp CE (2004) Corn-residue transformations into root and soil carbon as related to nitrogen, tillage, and stover management. *Soil Sci Soc Amer J* 68:1366–1375

Amezcua CM, and Parsons CM (2007) Effect of increased heat processing and particle size on phosphorus bioavailability in corn distillers dried grains with solubles. *Poultry Sci* 86:331–337

Amos B, Walters DT (2006) Maize root biomass and net rhizodeposited carbon: An analysis of the literature. *Soil Sci Soc Amer J* 70:1489–1503

Andersson-Gunneras S, Mellerowicz EJ, Love J, Segerman B, Ohmiya Y, Coutinho PM, Nilsson P, Henrissat B, Moritz T, Sundberg B (2006) Biosynthesis of cellulose-enriched tension wood in *Populus*: Global analysis of transcripts and metabolites identifies biochemical and developmental regulators in secondary wall biosynthesis. *Plant J* 45:144–165

Anterola AM, Lewis NG (2002) Trends in lignin modification: A comprehensive analysis of the effects of genetic manipulations/mutations on lignification and vascular integrity. *Phytochemistry* 61:221–294

Appenzeller L, Dhugga KS, Doblin M, Barreiro R, Wang H, Niu X, Kollipara K, Carrigan L, Tomes D, Chapman M (2004) Cellulose synthesis in maize: Isolation and expression analysis of the cellulose synthase (CesA) gene family. *Cellulose* 11:287–299

Atchison JE, Hettenhaus JR (2003) *Innovative Methods for Corn Stover Collecting, Handling, Storing and Transportation*. National Renewable Energy Laboratory Subcontractor Report NREL/SR-510-33893, National Renewable Energy Laboratory, Golden, CO

Banziger M, Edmeades GO, Lafitte HR (1999) Selection for drought tolerance increases maize yields across a range of nitrogen levels. *Crop Sci* 39:1035–1040

Barreiro R, Dhugga KS (2007) From cellulose to mechanical strength: Relationship of the cellulose synthase genes to dry matter accumulation in maize. In: Brown RM Jr, Saxena IM (eds), *Cellulose: Molecular and Structural Biology*. Springer, Dordrecht, The Netherlands, pp 63–83

Barrieire Y, Ralph J, Meichin V, Guillaumie S, Grabber JH, Argillier O, Chabbert B, Lapierre C (2004) Genetic and molecular basis of grass cell wall biosynthesis and degradability. II. Lessons from brown-midrib mutants. *Comptes Rendus Biol* 327:847–860

Belyea RL, Rausch KD, Tumbleson ME (2004) Composition of corn and distillers dried grains with solubles from dry grind ethanol processing. *Bioresour Technol* 94:293–298.

Bhatia CR, Rabson R (1976) Bioenergetic considerations in cereal breeding for protein improvement. *Science* 194:1418–1421

Biemelt S, Tschiersch H, Sonnewald U (2004) Impact of altered gibberellin metabolism on biomass accumulation, lignin biosynthesis, and photosynthesis in transgenic tobacco plants. *Plant Physiol* 135:254–265

Boerjan W, Ralph J, Baucher M (2003) Lignin biosynthesis. *Annu Rev Plant Biol* 54:519–546

Brown DM, Zeef LAH, Ellis J, Goodacre R, Turner SR (2005) Identification of novel genes in Arabidopsis involved in secondary cell wall formation using expression profiling and reverse genetics. *Plant Cell* 17:2281–2295

Buckler ES, Stevens NM (2006) Maize origins, domestication, and selection. In: Motley TJ, Zerega N, Cross H (eds) *Darwin's Harvest: New Approaches to the Origin, Evolution, and Conservation of Crops.* Columbia Univ. Press, New York, pp 67–90

Burk DH, Ye ZH (2002) Alteration of oriented deposition of cellulose microfibrils by mutation of a katanin-like microtubule-severing protein. *Plant Cell* 14:2145–2160

Burton RA, Wilson SM, Hrmova M, Harvey AJ, Shirley NJ, Medhurst A, Stone BA, Newbigin EJ, Bacic A, Fincher GB (2006) Cellulose synthase-like CslF genes mediate the synthesis of cell wall (1,3;1,4)-β-ᴅ-glucans. *Science* 311:1940–1942

Campbell MM, Sederoff RR (1996) Variation in lignin content and composition: Mechanisms of control and implications for the genetic improvement of plants. *Plant Physiol* 110:3–13

Cardinal AJ, Lee M, Moore KJ (2003) Genetic mapping and analysis of quantitative trait loci affecting fiber and lignin content in maize. *Theor Appl Genet* 106:866–874

Carpita N, McCann M (2000) The cell wall. In: Buchanan BB, Gruissem W, Jones RL (eds), *Biochemistry and Molecular Biology of Plants.* American Society of Plant Physiology, Rockville, MD, pp 52–110

Carpita NC (1996) Structure and biogenesis of the cell walls of grasses. *Annu Rev Plant Physiol Plant Mol Biol* 47:445–476

Carrari F, Urbanczyk-Wochniak E, Willmitzer L, Fernie AR (2003) Engineering central metabolism in crop species: Learning the system. *Metabol Eng* 5:191–200

Ching A, Dhugga KS, Appenzeller L, Meeley R, Bourett TM, Howard RJ, Rafalski A (2006) Brittle stalk 2 encodes a putative glycosylphosphatidylinositol-anchored protein that affects mechanical strength of maize tissues by altering the composition and structure of secondary cell walls. *Planta* 224:1174–1184

Christian, DG, Riche AB, and Yates NE (2002) The yield and composition of switchgrass and coastal panic grass grown as a biofuel in Southern England. *Bioresour Technol* 83:115–124

Cocuron JC, Lerouxel O, Drakakaki G, Alonso AP, Liepman AH, Keegstra K, Raikhel N, Wilkerson CG (2007) A gene from the cellulose synthase-like C family encodes a beta-1,4 glucan synthase. *Proc Natl Acad Sci USA* 104:8550–8555

Crow JF (1998) 90 years ago: The beginning of hybrid maize. *Genetics* 148:923–928

Dai Z, Hooker BS, Quesenberry RD, Thomas SR (2005) Optimization of *Acidothermus cellulolyticus* endo-glucanase (E1) production in transgenic tobacco plants by transcriptional, post-transcription and post-translational modification. *Transgen Res* 14:627–643

Dhugga KS (2007) Maize biomass yield and composition for biofuels. *Crop Sci* 47:2211–2227

Dhugga KS, Barreiro R, Whitten B, Stecca K, Hazebroek J, Randhawa GS, Dolan M, Kinney AJ, Tomes D, Nichols S, Anderson P (2004) Guar seed beta-mannan synthase is a member of the cellulose synthase super gene family. *Science* 303:363–366

Dhugga KS, Ray PM (1994). Purification of 1,3-beta-D-glucan synthase activity from pea tissue. Two polypeptides of 55 kDa and 70 kDa copurify with enzyme activity. *Eur J Biochem* 220:943–953

Dhugga KS, Waines JG (1989). Analysis of nitrogen accumulation and use in bread and durum wheat. *Crop Sci* 29:1232–1239

Dhugga K, Wang H, Tomes D, Helentjaris T (2005) Maize cellulose synthases and uses thereof. U.S. Patent 9,630,225 B2

Djerbi S, Lindskog M, Arvestad L, Sterky F, Teeri TT (2005) The genome sequence of black cottonwood (Populus trichocarpa) reveals 18 conserved cellulose synthase (CesA) genes. *Planta* 221:739–746

Doblin MS, Pettolino FA, Wilson SM, Campbell R, Burton RA, Fincher GB, Newbigin E, Bacic A(2009) A barley cellulose synthase-like CSLH gene mediates (1,3;1,4)-beta-D-glucan synthesis in transgenic Arabidopsis. *Proc Natl Acad Sci USA* 106:5996–6001

DOE (U.S. Department of Energy) (2006) Biomass feedstock composition and property database. Department of Energy, Biomass Program. http://www.eere.energy.gov/biomass/progs/search1.cgi

Doebley J (2004) The genetics of maize evolution. *Annu Rev Genet* 38:37–59

Doebley J, Stec A, Hubbard L (1997) The evolution of apical dominance in maize. *Nature* 386:485–488

Donaldson LA, Singh AP, Yoshinaga A, Takabe K(1999). Lignin distribution in mild compression wood of Pinus radiata. *Can J Bot* 77:41–50

Duvick DN (2005a) The contribution of breeding to yield advances in maize (*Zea mays* L.). *Adv Agron* 86:83

Duvick DN (2005b) Genetic progress in yield of United States maize (*Zea mays* L.). *Maydica* 50:193–202

Duvick DN, Cassman, KG (1999) Post-green revolution trends in yield potential of temperate maize in the north-central United States. *Crop Sci* 39:1622–1630

Duvick DN, Smith JSC, Cooper M (2004) Long-term selection in a commercial hybrid maize breeding program. *Plant Breed Rev* 24 (2 Part):109–151

Earle FR, Curtis JJ, Hubbard JE (1946) Composition of the component parts of the corn kernel. *Cereal Chem* 23:504–511

Edwards ME, Dickson CA, Chengappa S, Sidebottom C, Gidley MJ, Reid JSG (1999) Molecular characterisation of a membrane-bound galactosyltransferase of plant cell wall matrix polysaccharide biosynthesis. *Plant J* 19:691–697

Faik A, Price NJ, Raikhel NV, Keegstra K (2002) An Arabidopsis gene encoding an alpha-xylosyltransferase involved in xyloglucan biosynthesis. *Proc Natl Acad Sci USA*:7797–7802

Fastinger ND, Latshaw JD, Mahan DC (2006) Amino acid availability and true metabolizable energy content of corn distillers dried grains with solubles in adult cecectomized roosters. *Poult Sci* 85:1212–1216

Fincher, GB (2009) Revolutionary times in our understanding of cell Wall biosynthesis and remodeling in the grasses. *Plant Physiol* 149:27–37

Fry SC (2004). Primary cell wall metabolism: Tracking the careers of wall polymers in living plant cells. *New Phytol* 161:641–675

Fry SC, Willis SC, Paterson AEJ (2000) Intraprotoplasmic and wall-localised formation of arabinoxylan-bound diferulates and larger ferulate coupling-products in maize cell-suspension cultures. *Planta* 211:679–692

Garcia-Barneto A, Ariza J, Martín JE, Jiménez L (2009) Use of autocatalytic kinetics to obtain composition of lignocellulosic materials. *Bioresour Technol* 100:3963–3973

Genc H, Ozdemir M, Demirbas A (2001) Analysis of mixed-linked $(1\rightarrow3)$, $(1\rightarrow4)$-β-D-glucans in cereal grains from Turkey. *Food Chem* 73:221–224

Grabber JH (2005). How do lignin composition, structure, and cross-linking affect degradability? A review of cell wall model studies. *Crop Sci* 45:820–831

Grabber JH, Ralph J, Lapierre C, Barrieiere Y (2004) Genetic and molecular basis of grass cell-wall degradability. I. Lignin-cell wall matrix interactions. *Comptes Rendus Biologies* 327:455–465

Graham RL, Nelson R, Sheehan J, Perlack RD, Wright LL (2007). Current and potential U.S. corn stover supplies. *Agron J* 99:1–11

Green PB (1962) Mechanism for plant cellular morphogenesis. *Science* 138:1404–1405

Gressel J (2008) Transgenics are imperative for biofuel crops. *Plant Sci* 174:246–263

Guillaumie S, Goffner D, Barbier O, Martinant JP, Pichon M, Barriere Y (2008) Expression of cell wall related genes in basal and ear internodes of silking brown-midrib-3, caffeic acid O-methyltransferase (COMT) down-regulated, and normal maize plants. *BMC Plant Biol* 8:71

Guillaumie S, Pichon M, Martinant JP, Bosio M, Goffner D, Barrière Y (2007) Differential expression of phenylpropanoid and related genes in brown-midrib bm1, bm2, bm3, and bm4 young near-isogenic maize plants. *Planta* 226:235–250

Guillet-Claude C, Birolleau-Touchard C, Manicacci D, Rogowsky PM, Rigau J, Murigneux A, Martinant JP, Barriere Y (2004) Nucleotide diversity of the ZmPox3 maize peroxidase gene: Relationship between a MITE insertion in exon 2 and variation in forage maize digestibility. *BMC Genet* 5:19

Hahn-Hagerdal B, Galbe M, Gorwa-Grauslund MF, Liden G, Zacchi G (2006) Bio-ethanol—The fuel of tomorrow from the residues of today. *Trend Biotechnol* 24:549–556

Hallauer AR, Russell WA, Lamkey KR (1988) Corn breeding. In: Sprague GF, Dudley JW (eds), *Corn and Corn Improvement*. American Society of Agronomy, Madison, WI, pp 463–564

Halpin C, Holt K, Chojecki J, Oliver D, Chabbert B, Monties B, Edwards K, Barakate A, Foxon GA (1998) Brown-midrib maize (bm1)—A mutation affecting the cinnamyl alcohol dehydrogenase gene. *Plant J* 14:545–553

Hay RKM (1995) Harvest index: A review of its use in plant breeding and crop physiology. *Ann Appl Biol* 126:197–216

Hayashi T (1989) Xyloglucans in the primary cell wall. *Annu Rev Plant Physiol Plant Mol Biol* 40:139–168

Hazen SP, Hawley RM, Davis GL, Henrissat B, Walton JD (2003) Quantitative trait loci and comparative genomics of cereal cell wall composition. *Plant Physiol* 132:263–271

Hongzhang C, Liying L (2007) Unpolluted fractionation of wheat straw by steam explosion and ethanol extraction. *Bioresour Technol* 98:666–676

Hrmova M, Farkas V, Lahnstein J, Fincher GB (2007) A barley xyloglucan xyloglucosyl transferase covalently links xyloglucan, cellulosic substrates, and (1,3;1,4)-beta-D-glucans. *J Biol Chem* 282:12951–12962

Hu WJ, Harding SA, Lung J, Popko JL, Ralph J, Stokke DD, Tsai CJ, Chiang VL (1999) Repression of lignin biosynthesis promotes cellulose accumulation and growth in transgenic trees. *Nat Biotechnol* 17:808–812

Iiyama K, Lam TBT, Stone BA (1994) Covalent cross-links in the cell wall. *Plant Physiol* 104:315–320

Johnson JMF, Reicosky D, Allmaras R, Archer D, Wilhelm W (2006) A matter of balance: Conservation and renewable energy. *J Soil Water Conserv* 61:120A–125A

Joseleau JP, Imai T, Kuroda K, Ruel K (2004) Detection in situ and characterization of lignin in the G-layer of tension wood fibres of *Populus deltoides*. *Planta* 219:338–345

Jung HG, Casler MD (2006) Maize stem tissues: Cell wall concentration and composition during development. *Crop Sci* 46:1793–1800

Jung HJG (2003) Maize stem tissues: Ferulate deposition in developing internode cell walls. *Phytochemistry* 63:543–549

Kurek I, Kawagoe Y, Jacob WD, Doblin M, Delmer D (2002) Dimerization of cotton fiber cellulose synthase catalytic subunits occurs via oxidation of the zinc-binding domains. *Proc Natl Acad Sci USA* 99:11109–11114

Lefebvre S, Lawson T, Zakhleniuk OV, Lloyd JC, Raines CA (2005) Increased sedoheptulose-1,7-bisphosphatase activity in transgenic tobacco plants stimulates photosynthesis and growth from an early stage in development. *Plant Physiol* 138:451–460

Levy I, Shani Z, Shoseyov O (2002) Modification of polysaccharides and plant cell wall by endo-1,4-beta -glucanase and cellulose-binding domains. *Biomol Eng* 19:17–30

Lewandowski I, Clifton-Brown JC, Andersson B, Basch G, Christian DG, Jorgensen U, Jones MB, Riche AB, Schwarz KU, Tayebi K, Teixeira F (2003a) Environment and harvest time affects the combustion qualities of Miscanthus genotypes. *Agron J* 95:1274–1280

Lewandowski I, Scurlock JMO, LindvallE, Christou M (2003b) Development and current status of perennial rhizomatous grasses as energy crops in the US and Europe. *Biomass Bioenergy* 25:335–361

Li YH, Qian O, Zhou YH, Yan MX, Sun L, Zhang M, Fu ZM, Wang YH, Han B, Pang XM, Chen MS, Li JY (2003) BRITTLE CULM1, which encodes a COBRA-like protein, affects the mechanical properties of rice plants. *Plant Cell* 15:2020–2031

Liepman AH, Wilkerson C, Keegstra K (2005) Expression of cellulose synthase-like (Csl) genes in insect cells reveals that CslA family members encode mannan synthases. *Proc Natl Acad Sci USA* 102:2221–2226

Mani S, Tabil LG, Sokhansanj S (2006) Effects of compressive force, particle size and moisture content on mechanical properties of biomass pellets from grasses. *Biomass Bioenergy* 30:648–654

Marita JM, Ralph J, Hatfield RD, Guo D, Chen F, Dixon RA (2003) Structural and compositional modifications in lignin of transgenic alfalfa down-regulated in caffeic acid 3-O-methyltransferase and caffeoyl coenzyme A 3-O-methyltransferase. *Phytochemistry* 62:53–65

Masoero F, Rossi F, Pulimeno AM (2006) Chemical composition and in vitro digestibility of stalks, leaves and cobs of four corn hybrids at different phenological stages. *Ital J Anim Sci* 5:215–227

McAloon A, Taylor F, Yee W, Ibsen K, Wooley R (2000) *Determining the Cost of Producing Ethanol from Corn Starch and Lignocellulosic Feedstocks*. NREL/TP-580-28893, National Renewable Energy Laboratory, Golden, CO

Mechin V, Argillier O, Hebert Y, Guingo E, Moreau L, Charcosset A, Barriere Y (2001) Genetic analysis and QTL mapping of cell wall digestibility and lignification in silage maize. *Crop Sci* 41:690–697

Meikle PJ, Ng KF, Johnson E, Hoogenraad NJ, Stone BA (1991) The beta-glucan synthase from Lolium multiflorum: Detergent solubilization, purification using monoclonal antibodies, and photoaffinity labeling with a novel photoreactive pyrimidine analogue of uridine 5'-diphosphoglucose. *J Biol Chem* 266:22569–22581

Meyer FD, Talbert LE, Martin JM, Lanning SP, Greene TW, Giroux MJ (2007) Field evaluation of transgenic wheat expressing a modified ADP-glucose pyrophosphorylase large subunit. *Crop Sci* 47:336–342

Mikel MA (2011) Genetic composition of contemporary U.S. commercial dent corn germplasm. *Crop Sci* 51:592–599

Moll RH, Kamprath EJ, Jackson WA (1982) Analysis and interpretation of factors which contribute to efficiency of nitrogen utilization. *Agron J* 74:562–564

Mitchell RA, Dupree P, Shewry PR (2007) A novel bioinformatic approach identifies candidate genes for the synthesis and feruloylation of arabinoxylan. *Plant Physiol* 144:43–53

Morrow SL, Mascia P, Self KA, Altschuler M (1997) Molecular characterization of a brown midrib3 deletion mutation in maize. *Mol Breed* 3:351–357

Nelson OE (1993) A notable triumvirate of maize geneticists. *Genet* 135:937–941

Park JR, McFarlane I, Hartley Phipps R, Ceddia G (2011) The role of transgenic crops in sustainable development. *Plant Biotechnol J* 9:2–21

Patzek TW (2006) A statistical analysis of the theoretical yield of ethanol from corn starch. *Nat Resour Res* 15:205–212

Pauly M, Scheller HV (2000) O-acetylation of plant cell wall polysaccharides: Identification and partial characterization of a rhamnogalacturonan O-acetyl-transferase from potato suspension-cultured cells. *Planta* 210:659–667

Pear JR, Kawagoe Y, Schreckengost WE, Delmer DP, Stalker DM (1996) Higher plants contain homologs of the bacterial celA genes encoding the catalytic subunit of cellulose synthase. *Proc Natl Acad Sci USA* 93:12637–12642

Pedersen JF, Vogel KP, Funnell DL (2005) Impact of reduced lignin on plant fitness. *Crop Sci* 45:812–819

Peferoen M (1997) Progress and prospects for field use of Bt genes in crops. *Trend Biotechnol* 15:173–177

Penning de Vries FWT, Brunsting AHM, van Laar HH (1974) Products, requirements and efficiency of biosynthesis: A quantitative approach. *J Theor Biol* 45:339–377

Perlack RD, Wright LL, Turhollow AF, Graham RL, Stokes BJ, Erbach DC (2005) *Biomass as Feedstock for a Bioenergy and Bioproducts Industry: The Technical Feasibility of a Billion-Ton Annual Supply*. U.S. Department of Energy, Oak Ridge, TN

Perrin RM, DeRocher AE, Bar-Peled M, Zeng W, Norambuena L, Orellana A, Raikhel NV, Keegstra K (1999) Xyloglucan fucosyltransferase, an enzyme involved in plant cell wall biosynthesis. *Science* 284:1976–1979

Persson S, Caffall KH, Freshour G, Hilley MT, Bauer S, Poindexter P, Hahn MG, Mohnen DCS (2007) The Arabidopsis irregular xylem8 mutant is deficient in glucuronoxylan and homogalacturonan, which are essential for secondary cell wall integrity. *Plant Cell* 19:237–255

Pichon M, Deswartes C, Gerentes D, Guillaumie S, Lapierre C, Toppan A, Barrieire Y, Goffner D (2006) Variation in lignin and cell wall digestibility in caffeic acid O-methyltransferase down-regulated maize half-sib progenies in field experiments. *Mol Breed* 18:253–261

Ralph J, Bunzel M, Marita JM, Hatfield RD, Lu F, Kim H, Schatz PF, Grabber JH, Steinhart H (2004) Peroxidase-dependent cross-linking reactions of p-hydroxycinnamates in plant cell walls. *Phytochem Rev* 3:79–96

Ray PM (1979) Maize coleoptile cellular membranes bearing different types of glucan synthetase activity. In: Reid E (ed), *Plant Organelles*. Halsted Press/John Wiley, Chichestor, United Kingdom, pp 135–146

Richmond TA, Somerville CR (2000) The cellulose synthase superfamily. *Plant Physiol* 124:495–498

Roussel V, Gibelin C, Fontaine AS, Barriere Y (2002) Genetic analysis in recombinant inbred lines of early dent forage maize. II—QTL mapping for cell wall constituents and cell wall digestibility from per se value and top cross experiments. *Maydica* 47:9–20

Sandhu APS, Randhawa GS, Dhugga KS (2009) Plant cell wall matrix polysaccharide biosynthesis. *Mol Plant* 5:840–850

Saxena IM, Lin FC, Brown RM Jr (1990) Cloning and sequencing of the cellulose synthase catalytic subunit gene of *Acetobacter xylinum*. *Plant Mol Biol* 15:673–684.

Saxena IM, Brown RM, Fevre M, Geremia RA, Henrissat B (1995) Multidomain architecture of beta-glycosyl transferases: Implications for mechanism of action. *J Bacteriol* 177:1419–1424

Schmitt U, Singh A, Frankenstein C, Moller R (2006) Cell wall modifications in woody stems induced by mechanical stress. *NZ J Forest Sci* 36:72–86

Sere PA (1987) Complexes of sequential metabolic enzymes. *Annu Rev Biochem* 56:89–124

Sinclair TR (1998) Historical changes in harvest index and crop nitrogen accumulation. *Crop Sci* 38:638–643

Sinclair TR, Bennett JM, Muchow RC (1990) Relative sensitivity of grain yield and biomass accumulation to drought in field-grown maize. *Crop Sci* 30:690–693

Sinclair TR, de Wit CT (1975) Photosynthate and nitrogen requirement for seed production by various crops. *Science* 189:565–567

Smidansky ED, Clancy M, Meyer FD, Lanning SP, Blake NK, Talbert LE, Giroux MJ (2002) Enhanced ADP-glucose pyrophosphorylase activity in wheat endosperm increases seed yield. *Proc Natl Acad Sci USA* 99:1724–1729

Smidansky ED, Meyer FD, Blakeslee B, Weglarz TE, Greene TW, Giroux MJ (2007) Expression of a modified ADP-glucose pyrophosphorylase large subunit in wheat seeds stimulates photosynthesis and carbon metabolism. *Planta* 225:965–976

Somerville C (2006) Cellulose synthesis in higher plants. *Annu Rev Cell Dev Biol* 22:53–78

Somerville C, Bauer S, Brininstool G, Facette M, Hamann T, Milne J, Osborne E, Paredez A, Persson S, Raab T, Vorwerk S, Youngs H (2004) Toward a systems approach to understanding plant cell walls. *Science* 306:2206–2211

Sun XF, Sun RC, Fowler P, Baird MS (2005) Extraction and characterization of original lignin and hemicelluloses from wheat straw. *J Agri Food Chem* 53:860–870

Tanaka K, Murata K, Yamazaki M, Onosato K, Miyao A, Hirochika H (2003) Three distinct rice cellulose synthase catalytic subunit genes required for cellulose synthesis in the secondary wall. *Plant Physiol* 133:73–83

Taylor NG, Howells RM, Huttly AK, Vickers K, Turner SR (2003) Interactions among three distinct CesA proteins essential for cellulose synthesis. *Proc Natl Acad Sci USA* 100:1450–1455

Tollenaar M, Lee EA (2002) Yield potential, yield stability and stress tolerance in maize. *Field Crop Res* 75:161–169

Tollenaar M, Wu J (1999) Yield improvement in temperate maize is attributable to greater stress tolerance. *Crop Sci* 39:1597–1604

Tolstogurov VB (2000) The importance of glassy biopolymer components in food. *Nahrung* 2:76–84

Torney F, Moeller L, Scarpa A, Wang K (2007) Genetic engineering approaches to improve bioethanol production from maize. *Curr Opin Biotechnol* 18:1–7

Updegraff DM (1969) Semimicro determination of cellulose in biological materials. *Anal Biochem* 32:120–124

USDA/NASS (2009) *Agricultural Statistics Database*, U.S. Department of Agriculture/National Agricultural Statistics Service, available at http://www.fas.usda.gov/wap/current/toc.asp

USDA/ERS (2011) *Economic Research Service Database*, U.S. Department of Agriculture/Economic Research Service, available at http://www.ers.usda.gov/Data/FertilizerUse/

Velasquez JA, Ferrando, F, Farriol C, Salvado J (2003) Binderless fiberboard from steam exploded *Miscanthus sinensis*. *Wood Sci Technol* 37:269–278

Van Erp H, Walton JD (2009) Regulation of the cellulose synthase-like gene family by light in the maize mesocotyl. *Planta* 229:885–897

Vanholme R, Morreel K, Ralph J, Boerjan W (2008) Lignin engineering. *Curr Opin Plant Biol* 11:278–285

Vergara CE, Carpita NC (2001) β-D-Glycan synthases and the CesA gene family: Lessons to be learned from the mixed-linkage (1,3),(1,4) β-D-glucan synthase. *Plant Mol Biol* 47:145–160

Vietor RJ, Newman RH, Ha MA, Apperley DC, Jarvis M-C (2002) Conformational features of crystal-surface cellulose from higher plants. *Plant J* 30:721–731

Wang H, Nussbaum-Wagler T, Li B, Zhao Q, Vigouroux Y, Faller M, Bomblies K, Lukens L, Doebley JF (2005) The origin of the naked grains of maize. *Nature* 436:714–719

Wen F, Celoy RM, Nguyen T, Zeng W, Keegstra K, Immerzeel P, Pauly M, Hawes MC (2008) Inducible expression of *Pisum sativum* xyloglucan fucosyltransferase in the pea root cap meristem, and effects of antisense mRNA expression on root cap cell wall structural integrity. *Plant Cell Rep* 27:1125–1135

Wilhelm WW, Johnson JMF, Hatfield JL, Voorhees WB, Linden DR (2004) Crop and soil productivity response to corn residue removal: A literature review. *Agron J* 96:1–17

Wilts AR, Reicosky DC, Allmaras RR, Clapp CE (2004) Long-term corn residue effects: Harvest alternatives, soil carbon turnover, and root-derived carbon. *Soil Sci Soc Amer J* 68:1342–1351

Wiselogel AE, Agblevor FA, Johnson DK, Deutch S, Fennell JA, Sanderson MA (1996) Compositional changes during storage of large round switchgrass bales. *Bioresour Technol* 56:103–109

Wooley R, Ruth M, Sheehan J, Ibsen K (1999) *Lignocellulosic Biomass to Ethanol Process Design and Economics Utilizing Co-Current Dilute Acid Prehydrolysis and Enzymatic Hydrolysis Current and Futuristic Scenarios*. DOE-NREL Technical Report 580-26157, U.S. Department of Energy, National Renewable Energy Laboratory, Golden, CO

Wright SI, Bi IV, Schroeder SG, Yamasaki M, Doebley JF, McMullen MD, Gaut BS (2005) The effects of artificial selection of the maize genome. *Science* 308:1310–1314

Wright LL, Boundy B, Perlack B, Davis S, Saulsbury B (2006) *Biomass Energy Databook*, 1st ed. U.S. Department of Energy, Oak Ridge, TN

Yennawar NH, Li AC, Dudzinski DM, Tabuchi A, Cosgrove DJ (2006) Crystal structure and activities of EXPB1 (Zea m 1), a β-expansin and group-1 pollen allergen from maize. *Proc Natl Acad Sci USA* 103:14664–14671

Zhong R, Burk DH, Morrison Iii WH, Ye ZH (2004) FRAGILE FIBER3, an Arabidopsis gene encoding a type ii inositol polyphosphate 5-phosphatase, is required for secondary wall synthesis and actin organization in fiber cells. *Plant Cell* 16:3242–3259

Zhong R, Lee C, Zhou J, McCarthy RL, Ye ZH (2008) A battery of transcription factors involved in the regulation of secondary cell wall biosynthesis in Arabidopsis. *Plant Cell* 20:2763–2782

Zhong R, Pena MJ, Zhou GK, Nairn CJ, Wood-Jones A, Richardson EA, Morrison III WH, Darvill AG, York WS, Ye ZH (2005) Arabidopsis fragile fiber8, which encodes a putative glucuronyltransferase, is essential for normal secondary wall synthesis. *Plant Cell* 17:3390–3406

Zhou Y, Li S, Qian Q, Zeng D, Zhang M, Guo L, Liu X, Zhang B, Deng L, Liu X, Luo G, Wang X, Li J (2009) BC10, a DUF266-containing and Golgi-located type II membrane protein, is required for cell-wall biosynthesis in rice (*Oryza sativa* L.). *Plant J* 57:446–462

17 Oil Palm

Yuen May Choo, Chee-Liang Yung, and Ah-Ngan Ma
Malaysian Palm Oil Board

CONTENTS

17.1 INTRODUCTION

Quickly diminishing energy reserves, greater environmental awareness, and increasing energy consumption as a result of rapid industrialization have led to the search for alternative energy resources. Alternative energy resources from silviculture or solar and wind power appear to be promising from the environmental and renewable resource perspectives. However, from the Malaysian perspective, there are several technical and socioeconomic uncertainties involved in their utilization. Nevertheless, their long-term feasibility and use are important assumptions of the nation's research and development efforts, but their immediate applications on a commercial scale are limited. What is needed is a renewable energy resource for which the utilization system is already proven and operational as well as easily amenable to expansion.

In 2001, the Malaysian government included renewable energy resources as fuel in its energy policy, and oil palm biomass has been identified as having great potential as a renewable energy resource. For example, the vast areas of planted oil palm ensure an abundant and readily available supply of material.

There are many energy-generating technologies available for the purpose of converting palm oil into biodiesel (chemically known as palm oil methyl esters) via esterification and transesterification processes. The biodiesel produced meets the international specifications for biodiesel EN 14214 and ASTM D 6751. Being the cheapest among the vegetable oils, palm oil is the most favored feedstock for biodiesel production. Palm oil can also be used directly as boiler fuel. This paper will discuss and highlight various bioenergy fuels derived from palm.

17.2 MALAYSIAN ENERGY SCENARIO

In 2007, total final energy demand in Malaysia was 44,268 kilotons of energy equivalent (ktoe). Energy demand was highest for the industrial sector (43%), followed by the transport sector (36%) and the residential and commercial sector (14%), with the remaining in the nonenergy sector (7%) and the agriculture sector (<1%) (Figure 17.1). By fuel source, petroleum products contributed the largest share with 56.1 % of total energy demand, followed by natural gas at 23.1%, 17.3% for electricity, and 3.1% for coal and coke (Figure 17.2). The total demand for all types has shown an upward trend since 2006.

In 2007, the total installed electricity capacity in Malaysia was 21,815 MW (PTM 2009). Electricity gross generation registered 101,325 GWh, an increase of 8.7% as compared with 2006 (Table 17.1). At the same time, the electricity consumption was 89,298 GWh, 88% of the total electricity generated with the rest as nonutilized power. The peak demand for Peninsular Malaysia was recorded at 13,620 MW and Sarawak at 833 MW (Figure 17.3). The calculated reserve margin for Peninsular Malaysia and Sarawak was 47% and 16%, respectively.

Approximately 90% of the electricity is currently generated by natural gas and coal-fired power plants (Figure 17.4). The percentage of coal in the fuel-mix ratio has increased because of its competitive market price. The renewable components of the fuel-supply chain have yet to gain prominence. The supply of renewable energy to industrial clusters areas can potentially reduce the dependence of petroleum in the country.

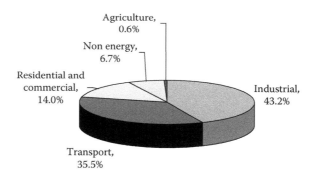

FIGURE 17.1 (See color insert)   Final energy demand in Malaysia in 2007 in various sectors. (From PTM, National Energy Balance, 2007, 2009.)

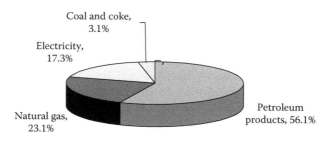

FIGURE 17.2 (See color insert)   Final energy demand in Malaysia in 2007 by fuel source. (From PTM, National Energy Balance, 2007, 2009.)

TABLE 17.1

Installed Capacity, Peak Demand, and Reserve Margin (2007)

Region	Electricity Gross Generation		Electricity consumption		Installed Capacityc	Peak Demand	Reserve Margin
	GWh	%	GWh	%	MW	MW	%
Peninsular Malaysia	**92,055**	**90.9**	**81,710**	**91.5**	**20,047**	**13,620**	**47**
Sarawak	**5,274**	**5.2**	**4,271**	**4.8**	**968**	**833**	**16**
Grid	5,043	5.0	4,087	4.6	900	791	14
Non-grid	231	0.2	184	0.2	68	42	62
Sabah	**3,996**	**3.9**	**3,317**	**3.7**	**–**	**–**	**–**
WC grid	1,251	1.2	1,075	1.2	520	406	28
EC grid	686	0.7	542	0.6	315	206	53
Sabah grid	2,016	2.0	1,669	1.9	838	625	34
Non-grid	43	0.0	31	0.0	3	10	–
Total	**101,325**	**100.0**	**89,298**	**100.0**	**–**	**–**	**–**

Source: PTM, National Energy Balance, 2007, 2009.
WC grid, West Coast grid; EC grid, East Coast grid.

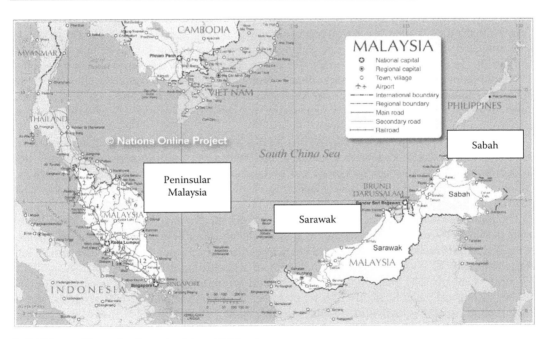

FIGURE 17.3 (See color insert)   Map of Malaysia.

17.3   MALAYSIA'S OIL PALM INDUSTRY

The palm tree was first introduced into Malaysia from West Africa in the late 1870s (Darus 2002). The palm species, *Elaeis guineensis*, was originally planted as an ornamental plant, and the first commercial planting was undertaken on a small scale in 1917. The oil palm was not commercially cultivated in large scale until the 1960s when the government's crops diversification program promoted it to avoid overdependence on another commodity, natural rubber. Since then, the palm

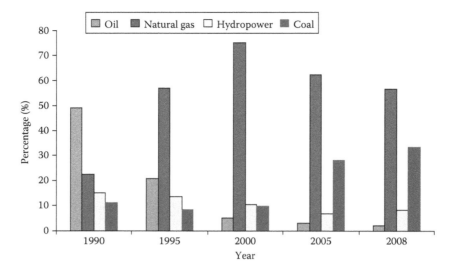

FIGURE 17.4 (See color insert) Fuel mix in electricity generation (1990–2010). (From PTM, National Energy Balance, 2007, 2009.)

oil industry has grown by leaps and bounds and has emerged as the most remunerative agricultural commodity, overtaking the position of natural rubber.

In 1960, the total planted area for oil palm was only 55,000 ha. Ten years later, 261,199 ha of land were cultivated with oil palm. This figure continued to surge—in 1980 to 1,023,306 ha and in 1990 to 1,984,167 ha. In 2008, there were more than 4.48 million ha of oil palm cultivation area in the country.

The oil palm grows well in tropical climates within 5° latitude of the equator (Basiron 1996). Adequate rainfall of over 2000 mm per year spread evenly throughout the year and adequate sunshine of over 2000 h per annum with moderately high temperatures of 25–33°C are ideal growing conditions for the oil palm tree.

The current oil palm planted in Malaysia is known as *Tenera* produced by crossbreeding *Dura* and *Pisifera*. The world production of palm oil in 2008 was 43.1 million tons, 27% of the world production of 17 oils and fats. Indonesia and Malaysia are the two largest producers, producing 19.3 and 17.7 million tons, respectively, and 86% of the total world palm oil production in 2008 (MPOB 2009a).

The palm tree starts bearing fruit 2.5–3 years after field planting. Harvesting of the oil palm fruits in bunches is carried out at approximately 30 months after field planting, when the ripeness of the bunch has reached such that at least some loose fruits drop from the bunch (Pantzaris 2000; Tang 2009; MPOB 2009b). The usual frequency of a harvesting round is 10–15 days, which is 2–3 times a month. The plant continues to bear fruit for up to 25 years of its economic life and reaches its optimal yield during its 10th year (Ong et al. 1995). The palm fruit is about the size of a small plum, weighing, individually, approximately 8–20 g (Wood 1987). It grows large bunches weighing 10–50 kg, and each bunch consists of up to 2000 fruits (Pantzaris 2000; Tang 2009). Each fruit contains an exocarp (skin), mesocarp (which contains palm oil and water in a fibrous matrix), endocarp (shell), and kernel (the seed) (Figure 17.5). The mesocarp contains approximately 49 wt% palm oil and the kernel contains approximately 50 wt% palm kernel oil.

17.4 PALM BIOMASS

Palm biomass is obtained under two situations in the field and in the palm oil mill. During the 25 productive years of the palm, the common products available daily are fresh fruit bunches (FFB) (Figure 17.6) and pruned fronds, whereas trunk together with fronds were obtained during

FIGURE 17.5 (See color insert)   Typical Malaysian palm fruits.

FIGURE 17.6 (See color insert)   Fresh fruit bunches.

replanting. Most of the pruned fronds will be used back in the field as mulch to maintain the fertility of soil. During the processing of FFB in the palm oil mill, the core products are palm oil and palm kernel. Useful co-products include empty fruit bunches (EFB; stalks no longer containing fruit) (Figure 17.7), fibers, shells and palm oil mill effluents (POME).

17.5   AVAILABILITY OF TRUNKS AND FRONDS

Approximately 115 kg of dry fronds per palm tree are obtained during replanting (Chan et al. 1980). Chan reported that 75.4 tons of trunks and 14.5 tons of fronds (each dry weight) are obtained while replanting each hectare of oil palm. However, these figures are based on planting densities common in the 1980s (126 oil palm trees/ha), which is only 85–93% of the current recommended densities

FIGURE 17.7 (See color insert)   Empty fruit bunches.

of 136–148 palm trees/ha of land in Malaysia. On the basis of the total matured planted area of oil palm in 2008, 3.91 million ha, there are currently approximately 531–578 million matured palm trees in the country. Assuming 5% of these matured oil palm planted areas are replanted annually and assuming 148 palm trees/ha of land, 28.9 million palm trees will be available each year with a contribution of 17.3 million tons of trunk biomass and 3.3 million tons of frond biomass.

On the basis of the above estimates and 10.4 tons of biomass per hectare produced during the pruning (Chan 2009) and normal harvesting of fronds, a total of 40.66 million tons of frond biomass was obtained in 2008.

17.6   AVAILABILITY OF EFB, SHELL, FIBER, AND POME

In general, FFB contain approximately 20–22% palm oil, 6–7% palm kernel, 14% fiber, 7% shell, and 23% EFB (Ma 2002). During the processing of FFB in the palm oil mill, the main products are palm oil and palm kernel with other co-products including EFB, fiber, shells, and POME. These co-products are obtained at different stages of the milling and nut crushing processes.

The FFB received are subjected to sterilization in a vessel with steam at 140°C at 3 bar. After that, the sterilized FFB are threshed to remove the palm fruits from their spikelets. Sterilizer condensate is produced as a byproduct.

The crude palm oil is mechanically pressed out from the sterilized palm fruits by a screw press. The oil is then clarified and an effluent is produced from the sludge and centrifugal waste. A press cake containing fibers and nuts is another product from the screw press. The nuts are separated from fibers in a vertical column having an upward airflow.

The nuts are cracked in either centrifugal nut crackers or a ripple mill, and the shells are separated from the cracked mixture by a pneumatic dry air separator, a wet hydrocyclone, or clay bath. Here, hydrocyclone washings are generated as waste. The sludge and centrifugal waste, the sterilizer condensate and hydrocyclone washings are mixed together to form POME, which normally receives further treatment through a pond system. The fibers and shells generated are used as boiler fuel.

Malaysia had more than 4.48 million ha of land under oil palm cultivation. In 2008, the country produced 17.73 million tons of crude palm oil from the processing of 87.74 million tons of FFB (MPOB 2009a). On the basis of these figures, the type and amount of biomass generated and its heat value was calculated (Table 17.2).

TABLE 17.2

Biomass Generated by Palm Oil Mills in 2008

Biomass	Quantity (million tons)	Moisture Content (%)	Oil Content (%)	Heat Value (dry) (kJ/kg)
EFB	20.18	67	5	18,883
Fiber	12.28	37	5	19,114
Shell	6.14	12	1	20,156
POME	59.23	93	1	17,044

17.7 ENERGY FROM FIBER AND SHELL

Palm oil mills in Malaysia use fiber and shell as the boiler fuel to produce steam for electricity generation, which in turn powers the palm oil and kernel production processes, and for other subsectors within the mill complex. The fiber and shell alone can supply more than enough electricity to meet the energy demand of a palm oil mill. It is estimated that 20 kWh (lower kWh for higher capacity mill) of electrical energy is required to process 1 ton of FFB. Thus, in 2008, approximately 1755 million kWh of electricity was generated and consumed by the palm oil mills. Assuming that each mill operates on average 400 h/month, the palm oil mills together will have a generating capacity of 365.6 MW. This constituted approximately 0.34% of the energy demand of the country in 2007. It must be mentioned here that the palm oil mills generally have excess fiber and shell that are not used and have to be disposed off separately.

Assuming that a diesel power generator consumes 0.34 L of diesel for every kilowatt-hour of electricity output, the oil palm industry in 2008 is estimated to have saved the country approximately 596.7 million L of diesel which amounted to approximately Malaysian Ringgit (MYR) 1014 million (price of diesel at MYR1.70/L).

The oil palm industry is fortunate that the fiber and shell can be used directly as boiler fuel without further treatment. With proper control of combustion, emissions usually associated with the burning of solid fuel can be reduced. An additional advantage of using both of these residues as fuel is that it helps dispose of these bulky materials that otherwise would create a solid waste burden. Unless these materials can be more beneficially utilized, it is envisaged that they will continue to be used as boiler fuel for the foreseeable future. It has generally been considered that energy is free in the palm oil mills. This has undoubtedly contributed greatly to the success of the palm oil industry.

17.8 ENERGY FROM EMPTY FRUIT BUNCHES

Apart from fiber and shell, EFB are another valuable biomass that can be readily converted into energy. However, this material has only been used on a limited basis because there is already enough energy available from fiber and shell. Also, during pretreatment, its bulkiness must be reduced and its high moisture content (67%) brought to below 50% to render it more easily combustible (Jorgensen 1985; Chua 1991).

The EFB has a calorific value of 18,883 kJ/kg (dry weight). Thus, the total heat energy obtainable from EFB in 2008 would be 125.75×10^{12} kJ. This is sufficient to generate approximately 31.4 million tons of steam (at 65% boiler efficiency and 2604 kJ/kg of steam) and 1147 million kWh of electricity, which saves the country 390 million L of diesel or MYR 663 million. The above calculations were based on a standard noncondensing turbo-alternator working against a back pressure of 3 bars gauge. More than double this energy could be obtained if condensing turbines working at a vacuum of 0.25 bar (absolute) were used for power generation (Chua 1991).

The above estimation represents the total obtainable energy from all of the 410 palm oil mills distributed all over the country. Thus, it can be said that the energy generated from a single palm oil mill will not be significant in volume, and it may not be viable for commercial consideration or for supplying electricity to the national grid. However, the EFB, unlike fiber, can be easily collected and transported. The possibility of producing electricity at a central power generating plant could be a viable proposition. The central power plants can be sited at locations where there is a high concentration of palm oil mills so that the EFB and the surplus fiber and shell from the mills can be transported at a reasonable distance and cost to the respective central power plants. Also, because the power plants would be independent entities, they could be operated year-round. The energy data are analyzed and presented for the various palm biomasses in Table 17.3.

17.9 PALM BIOMASS BRIQUETTE

Palm biomass briquette is a green fuel that can be used either as household fuel or for large thermo-chemical energy conversion such as in biomass-fired boilers. Briquetting is a process of converting low bulk density biomass into uniform and higher density solid fuels. Treated EFB, either in fibrous or pulverized form, is compressed and exposed to high pressures and temperatures to form compressed and high density solid biomass fuel (Nasrin et al. 2008). Currently, local briquette plants use sawdust as the raw material for briquette production. The technology used for briquette production involves a commercial-scale screw extruder. Future development includes promotion of briquette production in the integrated plants in palm oil mills as household fuel or potential boiler fuel for electricity generation for rural electrification and for the overseas market.

The surface of the briquette is partially carbonized for easy ignition and to minimize the absorption of water. The product is designed with a hole through the center for better air circulation during combustion. The briquette can be made of 100% EFB or a mixture of EFB and sawdust. The properties of these briquettes are given in Table 17.4.

Palm briquettes offered two times more combustion stability and heat output compared with the raw material. An average combustion rate of 0.43 g/m^3, capable of generating 0.13 kW of thermal output, was recorded at normal environmental conditions.

TABLE 17.3
Energy Database for Palm Biomass

Sample	Calorific Value (kJ/kg)	Ash (%)	Volatile Matter (%)	Moisture (%)	Hexane Extractable (%)
EFB	18,795	4.60	87.04	67.00	11.25
Fibers	19,055	6.10	84.91	37.00	7.60
Shell	20,093	3.00	83.45	12.00	3.26
Palm kernel Cake	18,884	3.94	88.54	0.28	9.35
Nut	24,545	4.05	84.03	15.46	4.43
CPO	39,360	0.91	1.07	1.07	95.84
Kernel oil	38,025	0.79	0.02	0.02	95.06
Liquor from EFB	20,748	11.63	78.50	88.75	3.85
Palm oil mill effluent	16,992	15.20	77.09	93.00	12.55
Trunk	17,471	3.39	86.73	76.00	0.80
Petiole	15,719	3.37	85.10	71.00	0.62
Root	15,548	5.92	86.30	36.00	0.20

Source: Chow, M.C., Subramaniam, V., and Ma, A.N., *Proceedings of 2003 MPOB International Palm Oil Congress*, Hotel Marriott, Putrajaya, Malaysia, August 24–28, 2003.

17.10 BIOGAS FROM POME

In addition to the solid residues, palm oil mills also generate large quantities of liquid waste in the form of POME, which, because of its high biochemical oxygen demand (BOD), is required by law to be treated to acceptable levels before it can be discharged into watercourses or onto land. In a conventional palm oil mill, approximately 0.675 m^3 of POME is generated for every ton of FFB processed. Hence, in 2008, approximately 59.23 million m^3 of POME was generated in Malaysia. Anaerobic processes have been adopted by palm oil mills to treat their POME. The biogas produced during the decomposition is a valuable energy source. It contains approximately 60–70% methane, 30–40% CO_2, and a trace amount of hydrogen sulfide. Its fuel properties are shown in Table 17.5 together with other gaseous fuels.

Approximately 28 m^3 of biogas are generated for every cubic meter of POME treated. However, most of the biogas is not recovered. So far, only a few palm oil mills use biogas for heat and electricity generation (Quah et al. 1982; Gillies and Quah 1985; Chua 1991; Ma et al. 2009). In a gas engine, it has been reported that approximately 1.8 kWh of electricity could be generated from 1 m^3 of biogas (Quah et al. 1982). The potential energy from biogas generated by POME is shown in Table 17.6. Because palm oil mills already meet their energy needs by combusting fiber and shell, there is no outlet for this surplus energy. Considering the costs of storing and transporting the biogas, perhaps the most viable proposition is to encourage industries to locate in the vicinity of the palm oil mills where the biogas energy can be directly utilized. This could result in substantial savings in energy costs (Chua 1991).

TABLE 17.4
Properties of Palm Briquettes

Product	Calorific Value (kJ/kg)	Moisture Content (%)	Ash Content (%)
Pulverized EFB	17,823	7.39	2.85
Pulverized EFB + sawdust (50:50)	18,273	7.22	2.22
Pulverized EFB + sawdust (40:60)	18,775	7.32	2.99
Sawdust	18,936	6.81	1.63
DIN 51731	17,500 (min)	<10.0 (min)	0.7 (min)

DIN 51731: Test of Solid Fuels—Compressed Untreated Wood Requirements and Testing.

TABLE 17.5
Some Properties of Gaseous Fuels

	Biogas	Natural Gas	LPG
Gross calorific value (kJ/Nm^3)	19,908–25,830	3797	100,500
Specific gravity	0.847–1.002	0.584	1.5
Ignition temperature (°C)	650–750	650–750	450–500
Inflammable limits (%)	7.5–21	5–15	2–10
Combustion air required (m^3/m^3)	9.6	9.6	13.8

Source: Quah, S.K. and Gillies, D., *Proceedings of National Workshop on Oil Palm By-Product Utilization.* Palm Oil Research Institute of Malaysia, Kuala Lumpur, pp. 119–125, 1981.

All gases evaluated at 15.5°C, atmosphere pressure and saturated with water vapor. LPG, liquefied petroleum gas.

TABLE 17.6
Potential Energy from Biogas

Year	Palm Oil Production (million t)	POME (million m³)	Biogas (million m³)	Electricity (million kWh)
2002	13.35	45.63	1,278	2,300
2006	15.9	53.55	1,499	2,698
2008	17.73	59.23	1,658	2,984

It is estimated that 1 m³ of biogas is equivalent to 0.65 L of diesel in terms of electricity generation. Hence, the total potential biogas energy could substitute for 1077 million L of diesel per year (based on 2008). This amounted to MYR1.83 billion. Again, the amount of biogas generated by an individual palm oil mill is not significant for commercial exploitation. However, it may be economically viable and indeed profitable if palm oil mills could collectively use their fiber, shell, EFB, and biogas for steam and electricity generation beyond their own needs.

17.11 PRODUCER GAS (SYNGAS)

EFB, which are produced in abundance at palm oil mills, can be a good feedstock for gasification to yield gaseous biofuel. EFB was fed into a biomass gasification system at temperatures ranging from 700 to 1000°C to yield bioproducer gases (i.e., CO, hydrogen, and CH_4). The producer gases can be compressed for household cooking, power generation, and as intermediate material for the production of Fischer–Tropsch liquid, methanol, and ammonia. MPOB has set up a semi-pilot-scale downdraft biomass gasification system with a capacity of 50 kg/h at its Palm Oil Milling Technology Centre (POMTEC) (Zulkifli et al. 2006; Zulkifli and Halim 2008). This system consists of a gasifier and a gas purification system. The concentration of the component gases was as follows: CO, 18.5%; hydrogen, 10.9%; and CH_4, 3.4% at a maximum gasification efficiency of 76% at 99 m³/h gas flow.

17.12 BIO-OIL

Bio-oil can be derived from oil palm biomass, particularly by pyrolyzing powdered EFB via a quartz fluidized fixed-bed reactor at temperatures in the range of 300 to 700°C (Chow and Li 2003; Mohamad et al. 2009). The char, bio-oils, and gases were collected via the pyrolysis of EFB. The highest bio-oil yield (42%) was obtained at an optimal temperature of 500°C with a particle size of 91–106 µm and a heating rate of 100°C/min. The optimum char production was 42% at 300°C and gas production reached an optimum at 46% at 700°C. The calorific values of bio-oil ranged from 16 to 23 MJ/kg. The moisture content of bio-oil varied between 18 and 22%. The ash content in the bio-oil varied from 0.2 to 0.65%. The pH values of the bio-oil varied between 2.6 and 3.9. A great range of chemical functional groups of phenol, alcohols, ketones, aldehydes, and carboxylic acids were indicated in a Fourier transform infrared (FTIR) spectrum.

17.13 PALM OIL AS A DIESEL SUBSTITUTE

Many researchers have investigated the possibility of using vegetable oils (straight or blended) as a diesel substitute. A good account of their attempts was reported in the 1983 JAOCS Symposium on Vegetable Oils as Diesel Fuels (Klopfenstein and Walker 1983; Pryde 1983; Strayer et al. 1983). The symposium revealed that vegetable oils have good potential as alternative fuels if the following problems could be overcome satisfactorily. These include high viscosity, low volatility, and the reactivity (polymerization) of the unsaturated hydrocarbon chains present in highly unsaturated oil.

TABLE 17.7

Fuel Characteristics of CPO, MFO, and Blends of CPO/MFO

Property	Method ASTM	Unit	MFO	CPO	CPO/MFO (50:50)
Gross heat of combustion	D 240	Btu/lb	18,350 min	17,064	17,692
		kJ/kg	42,680 min	39,690	41,150
Sulfur	D 4294	Wt %	3.5 max	0.03	1.55
Viscosity at 50°C	D 445	mm²/s	180 max	25.6	67.3
Flash point	D 93	°C	66 min	268	99
Ash	D 482	wt %	0.1 max	NA	0.012
Pour point	D 97	°C	21 max	21.0	−6
Carbon residue	D 4530	wt %	13.0 max	8.5	7.0
Density at 15°C	D 1298	kg/L	0.98 max	0.9140	0.9408
Sediment by extraction	D 473	wt %	0.10 max	NA	0.02
Water by distillation	D 95	vol %	0.5 max	NA	0.25

NA, not available.

These will give rise to coking on the injectors, carbon deposits, oil ring sticking, and thickening and gelling of the lubricating oil as a result of contamination with vegetable oil.

It is possible to reduce the viscosity of the vegetable oil by incorporating a heating device to the diesel engine as has been successfully demonstrated by the engine manufacturer, Elsbett (Basiron and Hitam 1992; Hitam and Jahis 1998). Other factors that may have long-term effects on the engine are free fatty acids and gummy substances which are found in the crude vegetable oils. Incomplete combustion residues may cause undesirable deposits on the engine components whereas the gummy substances may cause filter plugging problems. This will call for more regular and frequent servicing and maintenance of the engine.

Various blends of crude palm oil (CPO) and palm oil products such as refined, bleached, and deodorized palm oil (RBDPO); refined, bleached, and deodorized palm olein (RBDPOo) with medium fuel oil (MFO); and petroleum diesel have been evaluated as boiler fuel and a diesel substitute (Hitam et al. 2004). CPO was blended with MFO whereas RBDPOo and RBDPO were blended with petroleum diesel, each at various ratios by volume. The resultant fuel blends, CPO/MFO, RBDPO/petroleum diesel, and RBDPO/petroleum diesel exhibit superior fuel characteristics compared with those of the individual blend components (Tables 17.7–17.9) (Basiron and Choo 2004).

Evaluation of CPO/MPO blended fuel for power generation was conducted at the Tenaga Nasional Berhad (TNB) power station at Prai, Penang. A blend of 20% CPO and 80% MFO was evaluated. During testing, the boiler hopper area, the flame characteristics, the air heater elements, and flue gas emissions appeared normal. However, the emission of CO, SO_x, and NO_x were reduced. On the basis of trials and commercial runs using blended CPO and MFO fuels for power generation and steam production, there are no technical difficulties in boiler operation using these blended fuels. Additionally, blended fuels were found to be less viscous and thus easier for fuel flow (Basiron and Choo 2004).

Field trials using 19 Malaysian Palm Oil Board vehicles (of various makes) were conducted to evaluate blends of RBDPOo/petroleum diesel (up to 10% RBDPOo) and RBDPO/petroleum diesel (up to 5% RBDPO) as a diesel substitute. The highest mileage per vehicle of 400,000 km was recorded, and no technical problems were reported.

17.14 PALM OIL METHYL ESTERS AS DIESEL SUBSTITUTE

Biodiesel has gained much attention in recent years because of increasing environmental awareness. Biodiesel is produced from renewable resources, and, more importantly, it is a clean-burning fuel

TABLE 17.8

Fuel Characteristics of RBD Palm Oil Olein (RBDPO), Petroleum Diesel, and Blends of RBDPOo/Petroleum Diesel (RBDPO/Diesel)

Property	RBDPOo	RBDPOo/ Diesel (90:10)	RBDPOo/ Diesel (70:30)	RBDPOo/ Diesel (50:50)	RBDPOo/ Diesel (30:70)	RBDPOo/ Diesel (10:90)	Diesel
				Blends			
Density at 40°C (kg/L) ASTM D 1298	0.9150	0.8940	0.8770	0.8600	0.8435	0.8275	0.8190
Sulfur content (wt %) IP 242	0.035	0.035	0.055	0.060	0.080	0.090	0.100
Viscosity at 40°C (mm²/s) ASTM D 445	39.2	29.5	14.8	8.6	7.0	3.8	3.7
Pour point (°C) ASTM D 97	9	9	12	12	12	15	15
Gross heat of combustion (kJ/kg) ASTM D 240	38,975	39,800	40,625	41,450	42,275	43,100	45,000
Flash point (°C) ASTM D 93	326	142	110	99	93	90	89

TABLE 17.9

Fuel Characteristics of RBD Palm Oil (RBDPO), Petroleum Diesel, and Blends of RBD Palm Oil/Diesel (RBDPO/Diesel)

Property	Diesel	RBDPO/ Diesel (2:98)	RBDPO / Diesel (3:97)	RBDPO/ Diesel (5:95)	RBDPO/ Diesel (6:94)	RBDPO/ Diesel (7:93)	RBD Palm Oil (RBDPO)
				Blends			
Density at 15°C (kg/L) ASTM D 1298	0.8479	0.8492	0.8499	0.8502	0.8521	0.8525	0.9151
Sulfur content (wt %) IP 242	0.16	0.13	0.11	0.11	0.11	0.11	0.12
Viscosity at 40°C (mm²/s) ASTM D 445	0.4248	4.895	4.576	4.656	5.010	5.021	40.68
Pour point (°C) ASTM D 97	9	9	9	9	9	12	24
Gross heat of combustion (kJ/kg) ASTM D 240	45,050	45,340	45,160	45,095	45,085	45,015	39,260
Flash point (°C) ASTM D 93	84.0	84.0	84.0	84.0	85.0	86.0	–
ASTM D 92	–	–	–	–	–	–	322.0

and thus does not contribute to the net increase in atmospheric CO_2. From 1996 to 2008, the biodiesel production capacity in the European Union increased by a factor of 13 from 591,000 tons to a total of 7.755 million tons (Bockey 2002, 2004; EBB 2009).

Further utilization of biodiesel is anticipated because of initiatives around the world aimed at promoting its use. The European Union Directive 2009/28 is aiming at replacing 10% of all transport fossil fuels with biofuels from renewable sources by 2020 (European Union 2009). The current trends and legislation will create momentum for greater biodiesel production and consumption. Thus, there will be an upward course and new market opportunities for biodiesel.

Methyl esters of vegetable oils have been successfully evaluated as a diesel substitute worldwide (Choo et al. 1997; Choo and Ma 2000). For example, the following have been studied: rapeseed methyl esters in Europe, soybean oil methyl esters in the United States, sunflower oil methyl esters in Europe and the United States, and palm oil methyl esters in Malaysia. Because the choice of vegetable oil depends on the cost of production and reliability of supply, palm oil would be the preferred choice because it is the highest oil-yielding crop (4–5 tons/ha per year) among all of the vegetable oils.

Malaysia began its extensive biodiesel program in 1982. This includes development of production technology to convert palm oil to palm oil methyl esters (palm biodiesel), pilot plant studies of palm biodiesel production, and exhaustive evaluation of palm biodiesel as a diesel substitute in conventional diesel engines, both stationary and in vehicles.

CPO can be readily converted to its methyl esters. The production by MPOB (then Palm Oil Research Institute of Malaysia, PORIM)/PETRONAS patented technology (Ong et al. 1992) has been successfully demonstrated in a 3000 tons/year pilot plant (Choo et al. 1995, 1997; Choo and Cheah 2000). The novel aspect of this patented process is the use of solid acid catalysts for the esterification. The product of the reaction mixture, which is neutral, is then transesterified in the presence of an alkaline catalyst. The conventional washing stage or neutralization step after the esterification process is obviated, creating an economic advantage. This patented process can also be adopted for other palm oil products such as crude palm stearin and crude palm kernel oil as well as other raw materials such as used frying oil.

CPO methyl esters (palm biodiesel) were systematically and exhaustively evaluated as a diesel fuel substitute from 1983 to 1994 (Choo et al. 1995, 2002b). This included laboratory evaluations, stationary engine testing, and field trials on many vehicles including taxis, trucks, passenger cars, and buses. All of these tests were successfully completed. It is worth mentioning that the tests also covered stationary engine testing and field trials with 36 Mercedes Benz engines mounted onto passenger buses running on three types of fuels: 100% petroleum diesel, blends of palm biodiesel and petroleum diesel (50:50), and 100% palm biodiesel. Each bus covered 300,000 km, which is the expected life of the engines (total mileage covered by the 10 buses on 100% palm diesel is 3.7 million km). The results from this exhaustive field trial are promising. Fuel consumption of palm biodiesel by volume was comparable to the petroleum diesel. Differences in engine performance are so small that an operator would not be able to detect. The exhaust gas was found to be much cleaner; it contained comparable NO_x, but fewer hydrocarbons, CO, and CO_2. The very obvious advantage is the absence of black smoke and SO_2 from the exhaust.

Methyl esters from CPO produced by MPOB/PETRONAS technology have very similar fuel properties to the petroleum diesel (Table 17.10). They also have a higher cetane number (63) than diesel (<40) (Table 17.11). A higher cetane number indicates shorter ignition time delay characteristics and generally a better fuel. These methyl esters can be used directly as fuel in unmodified diesel engines or, of course, for blending traditional diesel. Compared with CPO, these methyl esters have very much improved viscosity and volatility properties and do not contain gummy substances. However, it has a pour point of 15°C, and this has confined its utilization to tropical countries.

In recent years, research efforts have produced palm biodiesel with a low pour point (without additives) that can be produced to meet seasonal pour point requirements, e.g., spring (–10°C), summer (0°C), autumn (–10°C), and winter (–20°C). The MPOB patented technology (Choo et al.

TABLE 17.10
Fuel Properties of Normal and Low-Pour-Point Palm Biodiesel

Property	Unit	Normal Palm Biodiesel	Low Pour Point Palm Biodiesel	EN 14214:2003	ASTM D6751:07b	MS 2008:2008
Ester content	% (m/m)	98.5	98.0–99.5	96.5 (min)	–	96.5 (min)
Density at 15°C	kg/m³	878.3	870–890	860–900	–	860–900
Viscosity at 40°C	mm²/s	4.415	4.423	3.50–5.00	1.9–6.0	3.50–5.00
Flash point	°C	182	180	120 (min)	130 (min)	120 (min)
Cloud point	°C	15.2	−18 to 0	–	Report	–
Pour point	°C	15	−21 to 0	–	–	–
Cold filter Plugging point	°C	15	−18 to 3	–	–	15
Sulfur content	mg/kg	<10	<10	10 (max)	15 (max) Grade S15 500 (max) Grade S500	10 (max)
Carbon residue (on 10% distillation residue)	% (m/m)	0.02	0.03	0.30 (max)	0.050 (max) (100% sample)	0.30 (max) (10% distillation residue) 0.050 (max) (100% sample)
Acid value	mg KOH/g	<0.5	<0.5	0.50 (max)	0.50 (max)	0.50 (max)
Sulfated ash content	% (m/m)	<0.01	<0.01	0.02 (max)	0.02 (max)	0.02 (max)
Water content	mg/kg	<500	<500	500 (max)	500 (max)	500 (max)
Total contamination	mg/kg	12	14	24 (max)	–	24 (max)
Cetane number	–	58.3	53.0–59.0	51 (min)	47 (min)	51 (min)
Copper strip corrosion (3 h at 50°C)	Rating	1a	1a	Class 1	No. 3 (max)	Class 1
Oxidation stability, 110°C	Hours	16	10.2	6 (min)	3 (min)	6 (min)
Iodine value	g iodine/ 100 g	52	<100	120 (max)	–	110 (max)
Linolenic acid methyl ester	% (m/m)	<0.5	<0.5	12 (max)	–	12 (max)
Polyunsaturated methyl esters (≥4 double bonds)	% (m/m)	<0.1	<0.1	1 (max)	–	1 (max)
Methanol content	% (m/m)	<0.2	<0.2	0.2 (max)	0.2 (max)	0.2 (max)
Monoglyceride content	% (m/m)	<0.8	<0.8	0.8 (max)	–	0.8 (max)
Diglyceride content	% (m/m)	<0.2	<0.2	0.2 (max)	–	0.2 (max)
Triglyceride content	% (m/m)	<0.1	<0.1	0.2 (max)	–	0.2 (max)
Free glycerol	% (m/m)	<0.01	<0.01	0.02 (max)	0.02 (max)	0.02 (max)
Total glycerol	% (m/m)	<0.20	<0.20	0.25 (max)	0.24 (max)	0.25 (max)

TABLE 17.10 (Continued)
Fuel Properties of Normal and Low-Pour-Point Palm Biodiesel

Property	Unit	Normal Palm Biodiesel	Low Pour Point Palm Biodiesel	EN 14214:2003	ASTM D6751:07b	MS 2008:2008
Phosphorus content	mg/kg	<1	<1	10.0 (max)	10.0 (max)	10.0 (max)
Group I metals (Na + K)	mg/kg	3.1	3.1	5.0 (max)	5.0 (max)	5.0 (max)
Group II metals (Ca + Mg)	mg/kg	<1	<1	5.0 (max)	5.0 (max)	5.0 (max)
Distillation temperature, 90% recovered (T90)	°C	<360	<360	–	360 (max)	–

EN14214: 2003—European Biodiesel Standard on Automotive Fuels. Fatty Acid Methyl Esters (FAME) for Diesel Engines. Requirements and Test Methods.

ASTM D6751: 07b—American Standard Specification for Biodiesel Fuel Blend Stock (B100) for Middle Distillate Fuels

MS 2008: 2008—Malaysian Biodiesel Standard on Automotive Fuels—Palm Methyl Esters (PME) for Diesel Engines—Requirements and Test Methods.

TABLE 17.11
Cetane Numbers of CPO Methyl Esters, Petroleum Diesel, and Their Blends

Blends		Cetane Number
CPO Methyl Esters (%)	Petroleum Diesel (%)	ASTM D613
100	0	62.4
0	100	37.7
5	95	39.2
10	90	40.3
15	85	42.3
20	80	44.3
30	70	47.4
40	60	50.0
50	50	52.0
70	30	57.1

2002a) has overcome the pour point problem of palm biodiesel and thus turned it into a more versatile product. With the improved pour point, palm biodiesel can be utilized in temperate countries. In addition to having good low-temperature flow characteristics, the palm biodiesel with the low pour point also exhibits fuel properties comparable to petroleum diesel (Table 17.10).

The storage properties of the CPO methyl esters are very good. After 6 months in a 50-m^3 storage tank, there was little deterioration in the fuel quality parameters except for the color which had changed from red to light yellow. This was due to the breakdown of the high-value colored carotene compounds, which could be recovered.

The uncertainty in feedstock price has prompted the biodiesel producers to search for other cheaper raw materials. These include RBD palm stearin and palm fatty acid distillate (PFAD).

Compared with RBD palm oil and palm stearin, PFAD contains free fatty acids (FFA) in the range of 70–90%. Additional pretreatment facilities must be incorporated into existing biodiesel plants to process such oils. With the knowledge and experience gained by handling oil with a high FFA content in the laboratory and pilot plants (Choo and Goh 1987; Choo and Ong 1989, Choo et al. 1990; Ong et al. 1992), MPOB has further developed a process to produce biodiesel from high-acid oils, thus improving the production viability of the palm biodiesel plant. (Harrison et al. 2009)

The filter blocking problems in vehicles running with biodiesel blends first surfaced in 2007. Researchers discovered that this problem was due to the presence of steryl glucosides (Hoed et al. 2008; Moreau et al. 2008; Lacoste et al. 2009). Since then, ASTM has incorporated a new parameter, the cold soak filtration test (CSFT), into the revised specification for Biodiesel Blend Stock (B100) for Middle Distillate ASTM D6751-08. A CFST time of 360and 200 s was set as the upper limit for normal and low-temperature conditions (at or below –12°C), respectively. MPOB has also developed methods to improve the CSFT of palm biodiesel to fulfil the requirement set by ASTM through various processes (Harrison and Choo 2009; Harrison et al. 2009; MPOB unpublished data).

Considering that transportation is one of the highest energy consumption sectors in the country, and that diesel constitutes approximately 40% of the fuel consumption, the use of renewable and environment-friendly palm oil methyl esters as a diesel substitute merits serious consideration.

The main benefit derived from such renewable sources of energy is the reduced emission of greenhouse gases (GHG) such as CO_2. The production and consumption of palm biodiesel has a closed carbon cycle, thus there is no net accumulation of CO_2 in the atmosphere. Subsequently, palm biodiesel production, because of its lower emissions, is in line with the Clean Development Mechanism (CDM) of 1997 Kyoto Protocol.

Under the terms of 1997 Kyoto Protocol, there is potential financial gain in transacting these GHG benefits to the palm oil industry under the CDM. This mechanism allows emission reduction projects to be implemented, and credits are awarded to the investing parties. Financial incentives such as an attractive carbon credit scheme could further enhance the economic viability of these renewable fuels.

17.15   CONCLUSIONS

The progressive escalation of energy shortages and fuel prices in recent times has led to an intensified search for viable alternative sources of energy globally. As conventional energy resources become more difficult to obtain, efforts must be directed toward developing alternative energy sources.

The palm oil industry is bestowed with a plentiful supply of co-products that can be readily and easily be used as energy resources. When EFB and biogas are properly processed using proven and innovative techniques, a considerable amount of energy can be economically recovered. The use of these co-products from palm oil mills, if implemented by the respective authorities, could help ease these escalating energy shortages. To this end, the production and application technologies have been fully demonstrated.

The energy needs of palm oil mills are met for free. Fiber and shell together can supply more than enough energy to meet their energy demands, with the electricity generated representing approximately 2% of the national electricity demand. Energy from biogas and EFB has so far been ignored, although they represent a hefty 4% of the national energy demand in terms of electricity. Efforts are being made to encourage palm oil mills to sell this excess energy in the form of electricity to the national grid.

Palm oil methyl esters have been fully evaluated as a potential diesel substitute and a diesel/cetane improver. Low pour point palm biodiesel that can meet stringent winter diesel specifications has also been produced. The palm biodiesel is an environmentally benign fuel substitute in terms of exhaust gas emission. Blends of CPO/MFO and RBDPOo/diesel have also been evaluated as potential fuel for boilers.

All of the above-mentioned energy sources are renewable and their supply is readily available and assured. Currently, burning biomass residues is often considered to be a disposal method rather than an energy source. These resources should be commercially exploited. This will have the added benefit of making the palm oil industry more environmentally sustainable.

REFERENCES

Basiron Y (1996) Palm Oil. In: Hui YH (ed) *Bailey's Industrial Oil and Fat Products: Edible Oil & Fat Products: Oils and Oilseeds*, Vol 2, 5th ed. John Wiley, New York, pp 271–375

Basiron Y, Hitam A (1992) Cost effectiveness of the CPO fuel in the Mercedes Elsbett engine car. PORIM Information Series, No 4, July.

Basiron Y, Choo YM (2004) Crude palm oil as a source of biofuel: its impact on price stabilization and environment. 1½-day course on refineries' crude palm oil purchasing, hedging, conceptual & its operational aspects. Subang Jaya, Selangor, Malaysia

Bockey D (2002) Situation and development potential for the production of biodiesel—An international study. Union for Promoting Oilseeds and Protein Plants, available at http://www.bio2power.org/dmdocuments/Situation_Potential_Bockey.pdf (accessed March 4, 2010)

Bockey D (2004) Policy initiative schemes and benefits of biofuel promotion in Germany—Current status of legislation and production. Paper presented at the Conference on Biofuels—Challenges for Asian Future. Queen Sirikit National Convention Center, Bangkok, Thailand, August 30–31, 2004

Chan KW (2009) Biomass production and uses in oil palm industry. In: Singh G, Lim KH, Teo L, Chan KW (eds), *Sustainable Production of Palm Oil A Malaysian Perspective*, Malaysian Palm Oil Association, pp 133–161

Chan KW, Watson I, Lim KC (1980) Use of oil palm waste material for increased production. In: *Proceedings of the Conference of Soil Science & Agricultural Development in Malaysia*, Malaysian Society of Soil Science, Kuala Lumpur, Malaysia, pp 214–241

Choo YM, Cheah KY (2000) Biofuel. In: Yusof B, Jalani BS, Chan KW (eds), *Advances of Oil Palm Research*, Vol II. Malaysian Palm Oil Board, Malaysia, pp 1293–1345

Choo YM, Cheng SF, Yung CL, Lau HLN, Ma AN, Yusof B (2002a) Low pour point palm diesel. Malaysian Patent No PI 20021157

Choo YM, Goh SH (1987) Esterification of carboxylic acids/glycerides mixtures. UK Patent No. 2148897

Choo YM, Ma AN (2000) Plant power. *Chemistry & Industry*, August 2000, pp 530–534, available at http://www.soci.org/Chemistry-and-Industry/CnI-Data/2011/6

Choo YM, Ma AN, Basiron Yusof B (1995) Production and evaluation of palm oil methyl esters as diesel substitute. *Elaeis* (Special Issue): 5–25

Choo YM, Ma AN, Ong ASH (1997) Biofuel. In: Gunstone FD, Padley FB (eds), *Lipids: Industrial Applications and Technology*, Marcel Dekker, New York, 771–785

Choo YM, Ma AN, Yusof B (2002b) Palm diesel. Paper presented at 2002 Oils and Fats International Congress (OFIC), Putra World Trade Centre, Kuala Lumpur, Malaysia, October 7–10, 2002

Choo YM, Ong ASH (1989) Carboxylic acid esterification. UK Patent No 2161809.

Choo YM, Ong ASH, Goh SH, Khor HT (1990) Transesterification of fats and oils. UK Patent No 2188057

Chow MC, Li CZ (2003) Preliminary work on the pyrolysis of empty fruit bunches and reactivity of chars. In: *Proceedings of 2003 MPOB International Palm Oil Congress*, Hotel Marriott, Putrajaya, Malaysia, August 24–28, 2003

Chow MC, Subramaniam V, Ma AN (2003) Energy database of the oil palm. In: *Proceedings of 2003 MPOB International Palm Oil Congress*, Hotel Marriott, Putrajaya, Malaysia, August 24–28, 2003

Chua NS (1991) Optimal utilization of energy sources in a palm oil processing complex. Paper presented at Seminar on Developments in Palm Oil Milling Technology and Environment Management, Genting Highlands, Pahang, Malaysia, May 16–17, 1991

Darus A (2002) Overview of Malaysian palm oil industry. ICS-UNIDO Workshop on Catalytic Technologies for Sustainable Industrial Process Utilising Crop Derived Renewable Raw Materials, Selangor, Malaysia, December 17–19, 2002

EBB (2009) European Biodiesel Board, available at http://www.ebb-eu.org/stats.php (accessed March 4, 2010)

European Union (2009) Directive 2009/28/EC of the European Parliament and of the Council on the Promotion of the Use of Energy from Renewable Sources and Amending and Subsequently Repealing Directives 2001/77/EC and 2003/30/EC. *Official Journal of the European Union*, pp 16–62

Gillies D, Quah SK (1985) Tennmaran biogas project. Paper presented at the 2nd Asian Workshop on Biogas Technology, Kuala Trengganu, Terengganu, Malaysia, October 8–13, 1984

Harrison LLN, Choo YM (2009) Effects of contaminants on cold soak filtration and cold filter plugging point of palm oil methyl esters. In: *Proceedings of Chemistry, Processing Technology & Bio-Energy Conference*, PIPOC 2009, Kuala Lumpur Convention Center, Malaysia, November 9–12, 2009, p 374

Harrison LLN, Nur Sulihartimarsyilla AW, Choo YM (2009) *Production Technology of Biodiesel from Palm Fatty Acid Distillate*. MPOB TOT No 430.

Hitam A, Jahis S (1998) Palm oil as diesel fuel: field trial on cars with Elsbett engine. In: *Proceedings of 1998 PORIM International Biofuel and Lubricant Conference*, PORIM, Bangi, Malaysia, pp 165–174

Hitam A, Hasamuddin WWH, Solah MD, Basiron Y (2004) Blended palm oil and its derivatives as fuel. In: *Proceedings of 2004 MPOB National Seminar on Green and Renewable Biofuel: Future Outlook of Biofuel in Malaysia*, KLIA, Kuala Lumpur, Malaysia, December 6–7, 2004

Hoed VV, Zyaykina N, Greyt WD, Maes J, Verhe R, Demeestere K (2008) Identification and occurrence of steryl glucosides in palm and soy biodiesel. *J Amer Oil Chem Soc* 85:701–709

Jorgensen H K (1985) Treatment of empty bunches for recovery of residues oil and additional steam production. *J Amer Oil Chem Soc* 62:282–284

Klopfenstein WE, Walker HS (1983) Efficiencies of various esters of fatty acids as diesel fuels. *J Amer Oil Chem Soc* 60:1596–1598

Lacoste F, Dejean F, Griffon H, Rouquette C (2009) Quantification of free and esterified steryl glucosides in vegetable oils and biodiesel. *Eur J Lipid Sci Technol* 111:822–829

Ma AN (2002) Carbon credit from palm: Biomass, biogas and biodiesel. *Palm Oil Eng Bull* Issue 65:24–26

Ma AN, Choo YM, Toh TS, Chua NS (2009) Renewable energy from oil palm industry. In: Singh G, Lim KH, Teo L, Chan KW (eds), *Sustainable Production of Palm Oil: A Malaysian Perspective*. Malaysian Palm Oil Association, Kuala Lumpur, pp 403–417

MPOB (2009a) *Malaysian Oil Palm Statistics 2008*, Malaysian Palm Oil Board, Ministry of Plantation Industries and Commodities, Selangor, Malaysia

MPOB (2009b) *Selected Readings on Palm Oil and Its Uses for 29th Palm Oil Familiarization Programme*, Malaysian Palm Oil Board, Ministry of Plantation Industries and Commodities, Selangor, Malaysia

Mohamad AS, Nor KAB, Chow MC (2009) Optimization of pyrolysis of oil palm empty fruit bunches. *J Oil Palm Res* 21:653–658

Moreau RA, Scott KM, Hass MJ (2008) The identification and quantification of steryl glucosides in precipitates from commercial biodiesel. *J Amer Oil Chem Soc* 85:761–770

Nasrin AB, Ma AN, Choo YM, Mohamad S, Rohaya MH, Azali A, Zainal Z (2008) Oil palm biomass as potential substitution raw materials for commercial biomass briquettes production. *Amer J Appl Sci* 5:179–183

Ong ASH, Choo YM, Cheah KY, Bakar A (1992) Production of alkyl esters from oils and fats. Australian Patent No. 43519/89.

Ong ASH, Choo YM, Ooi CK (1995) Developments in palm oil. In: Hamilton RJ (ed), *Developments in Oils and Fats*. Blackie Academic & Professional, Chapman & Hall, London, pp 151–191

Pantzaris TP (2000) *Pocketbook of Palm Oil Uses*, 5th ed. Malaysian Palm Oil Board, Ministry of Primary Industries, Kuala Lumpur, Malaysia

Pryde EH (1983) Vegetable oils as diesel fuels: Overview. *J Amer Oil Chem Soc* 60:1557–1558

PTM (2009) *National Energy Balance 2007*. Pusat Tenaga Malaysia, Malaysia Energy Center, Selangor, Malaysia

Quah SK, Gillies D (1981) Practical experience in production use of biogas. In: *Proceedings of National Workshop on Oil Palm By-Product Utilization*. Palm Oil Research Institute of Malaysia, Kuala Lumpur, pp 119–125

Quah SK, Lim KH, Gillies D, Wood BJ, Kanagaratnam K (1982) Sime Darby POME treatment and land application system. In: *Proceeding of Regional Workshop on Palm Oil Mill Technology and Effluent Treatment*, Palm Oil Research Institute of Malaysia, Kuala Lumpur, pp 193–200

Schöpe M, Britschkat G (2002) Macroeconomic evaluation of rape cultivation for biodiesel production in Germany. Munich, March 2002, available at http://www.ufop.de/downloads/ifo_makroeconomic_Evaluation.pdf (accessed March 4, 2010)

Strayer RC, Blake JA, Craig WK (1983) Canola and high erucic rapeseed oil as substitutes for diesel fuel: Preliminary tests. *J Amer Oil Chem Soc* 60:1587–1592

Tang TS (2009) *Pocketbook of Palm Oil Uses*, 6th ed. Malaysian Palm Oil Board, Ministry of Plantation Industries and Commodities, Kuala Lumpur, Malaysia

Wood BJ (1987) Growth and production of oil palm fruits. In: Gunstone FD (ed), *Palm Oil: Critical Reports on Applied Chemistry*, Vol 15. John Wiley, New York, pp 11–28

Zulkifli AR, Halim KH (2008) Assessment of a pilot palm oil mill biomass gasification unit for bioproducer gas production. In: Paridah MT, Luqman CA, Wan AI, Ainun ZM, Anis M, Wan HWH, Jalaluddin H (eds), *Proceedings of the Utilization of Oil Palm Tree: Strategizing for Commercial Exploitation*, Perpustakaan Negara, Malaysia, pp 289–304

Zulkifli AR, Ropandi M, Ravi M, K Halim KH (2006) Gasification technology using palm oil biomass for producer gas production. In: *Proceedings of the 2006 National Seminar on Palm Oil Milling, Refining Technology, Quality and Environment*, Crowne Plaza Riverside, Kuching, Sarawak, Malaysia, August 14–15, 2006

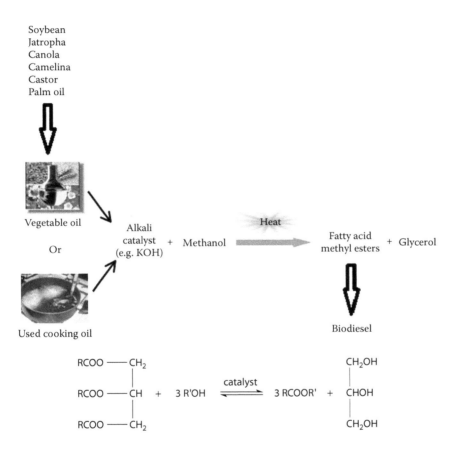

FIGURE 3.2 The biodiesel production process. (Modified from Sustainable Green Technologies, sgth2. com/bio-diesel_faq)

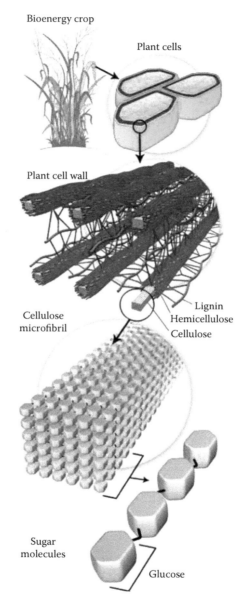

FIGURE 4.1 Plant wall recalcitrance: A key scientific challenge. (From U.S. DOE, Bioenergy research centers: An overview of the science, Genome Management Information Systems, Oak Ridge National Laboratory, 2009.)

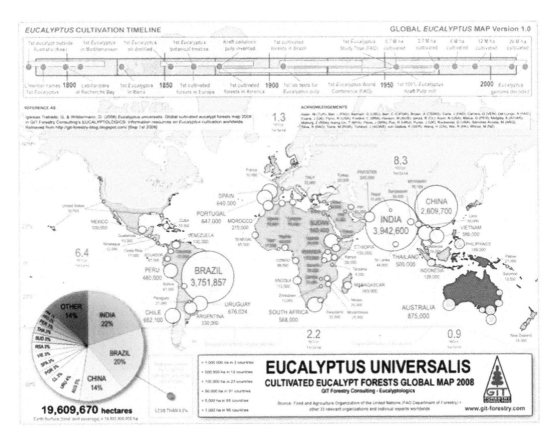

FIGURE 15.1 Worldwide *Eucalyptus* planting (From Trabedo, G.I. and Wilshermann, D., *Eucalyptus universalis*. Global Cultivated Eucalypt Forests Map, 2008, available at www.git.forestry.com).

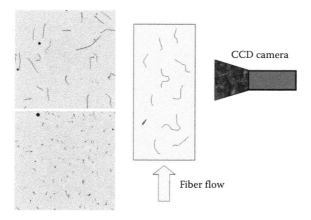

FIGURE 15.2 Schematic diagram showing the wet-imaging technique and typical images acquired.

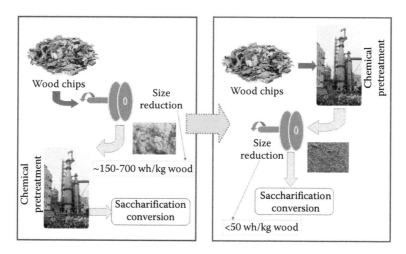

FIGURE 15.4 Required order for size reduction operation from pre- to postchemical pretreatment to reduce energy consumption.

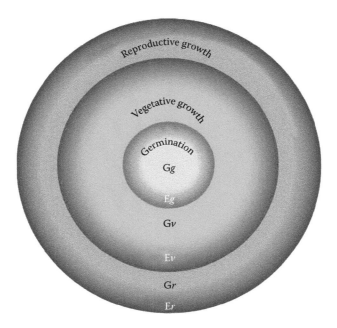

FIGURE 16.12 Illustration of complex genotype × environment effects on plant growth. G, genotype; E, environment; subscripts *g*, germination; *v*, vegetative growth; and *r*, reproductive growth. Dark-to-light gradient is indicative of genotypic effects (dark), genotype × environment effects (transition from dark to light), and environmental effects (light, at the edge of each developmental stage). Environmental factors that affect germination are different from the ones that affect vegetative growth, which are different from the ones that the plant experiences during reproductive growth. Environment can vary independently during each of these phases, increasing the unpredictability of plant performance. For further details, see Figure 16.13.

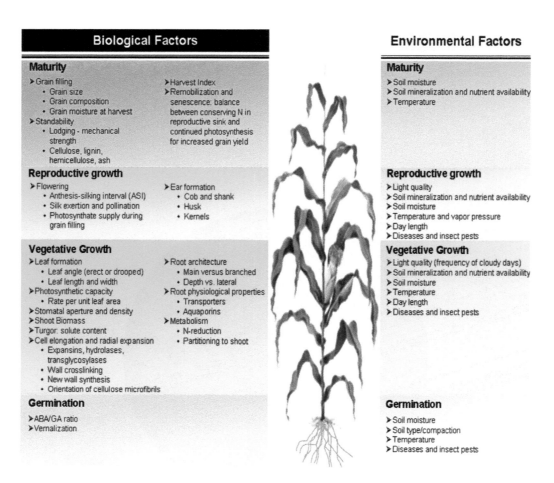

FIGURE 16.13 Diagrammatic representation of genetic and environmental factors influencing plant growth and development.

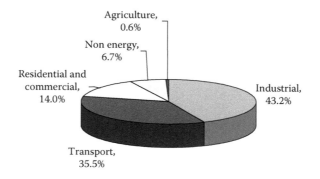

FIGURE 17.1 Final energy demand in Malaysia in 2007 in various sectors. (From PTM, *Proceedings of National Workshop on Oil Palm By-Product Utilization*. Palm Oil Research Institute of Malaysia, Kuala Lumpur, pp 119–125, 2009.)

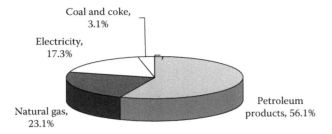

FIGURE 17.2 Final energy demand in Malaysia in 2007 by fuel source. (From PTM, *Proceedings of National Workshop on Oil Palm By-Product Utilization*. Palm Oil Research Institute of Malaysia, Kuala Lumpur, pp 119–125, 2009.)

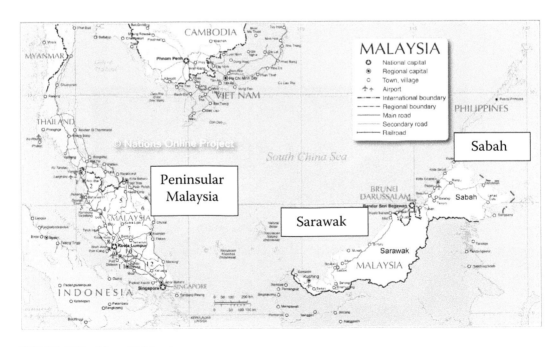

FIGURE 17.3 Map of Malaysia.

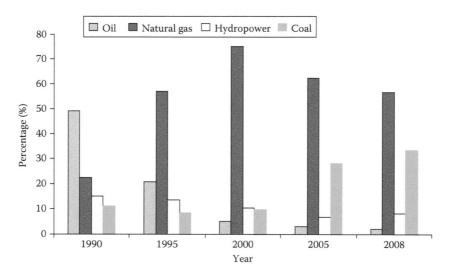

FIGURE 17.4 Fuel mix in electricity generation (1990–2010). (From PTM, National Energy Balance, 2007, 2009.)

FIGURE 17.5 Typical Malaysian palm fruits.

FIGURE 17.6 Fresh fruit bunches.

FIGURE 17.7 Empty fruit bunches.

FIGURE 19.1 Sweet sorghum plants reach 450 cm in September 2007 at Lincoln, NE. The plants were rain-fed and were grown in rotation with soybean.

FIGURE 19.2 Sweet sorghum stalks with different thickness ranges from 15 to 45 mm.

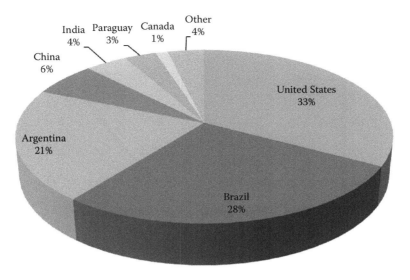

FIGURE 20.1 World oil seed production for major oilseed crops.

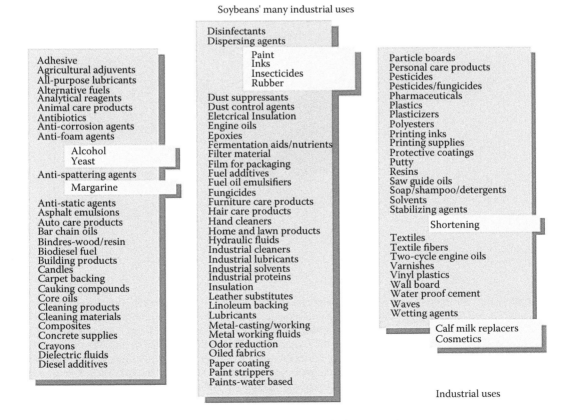

Soybeans' many industrial uses

Adhesive	Disinfectants	Particle boards
Agricultural adjuvents	Dispersing agents	Personal care products
All-purpose lubricants	Paint	Pesticides
Alternative fuels	Inks	Pesticides/fungicides
Analytical reagents	Insecticides	Pharmaceuticals
Animal care products	Rubber	Plastics
Antibiotics	Dust suppressants	Plasticizers
Anti-corrosion agents	Dust control agents	Polyesters
Anti-foam agents	Eletcrical Insulation	Printing inks
Alcohol	Engine oils	Printing supplies
Yeast	Epoxies	Protective coatings
Anti-spattering agents	Fermentation aids/nutrients	Putty
Margarine	Filter material	Resins
Anti-static agents	Film for packaging	Saw guide oils
Asphalt emulsions	Fuel additives	Soap/shampoo/detergents
Auto care products	Fuel oil emulsifiers	Solvents
Bar chain oils	Fungicides	Stabilizing agents
Bindres-wood/resin	Furniture care products	Shortening
Biodiesel fuel	Hair care products	Textiles
Building products	Hand cleaners	Textile fibers
Candles	Home and lawn products	Two-cycle engine oils
Carpet backing	Hydraulic fluids	Varnishes
Cauking compounds	Industrial cleaners	Vinyl plastics
Core oils	Industrial lubricants	Wall board
Cleaning products	Industrial solvents	Water proof cement
Cleaning materials	Industrial proteins	Waves
Composites	Insulation	Wetting agents
Concrete supplies	Leather substitutes	Calf milk replacers
Crayons	Linoleum backing	Cosmetics
Dielectric fluids	Lubricants	
Diesel additives	Metal-casting/working	
	Metal working fluids	
	Odor reduction	
	Oiled fabrics	
	Paper coating	
	Paint strippers	
	Paints-water based	

Industrial uses

FIGURE 20.2 Production of soybeans for major soybean producing countries.

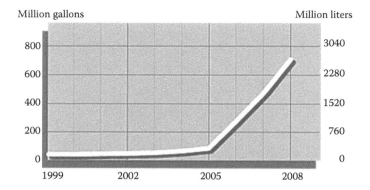

FIGURE 20.4 Biodiesel consumption in the United States.

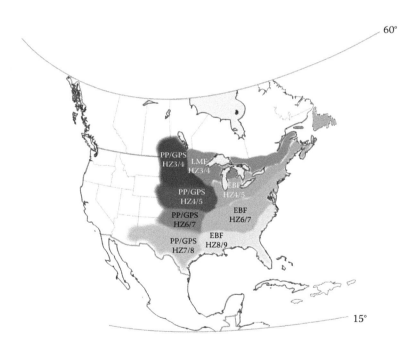

FIGURE 22.2 Proposed gene pools for deployment of regionally adapted switchgrass germplasm and cultivars for use in breeding programs or in conservation and restoration projects. PP, prairie parkland; GPS, Great Plains steppe; LMF, Laurentian mixed forest; EBF, eastern broadleaf forest. (From Bailey, R.G., *Ecoregions: The Ecosystem Geography of the Oceans and Continents,* Springer-Verlag Inc., New York, 1998; Bailey, R. G., *Ecosystem Geography,* 2nd ed, Springer-Verlag Inc., New York, 2009.) HZ, USDA hardiness zone. (From Cathey, H.M., USDA plant hardiness zone map, USDA Misc Pub No 1475. U.S. Department of Agriculture, Washington, DC, 1998; U.S. National Arboretum, 1990. Available at www.usna.usda.gov/Hardzone/ushzmap.html)

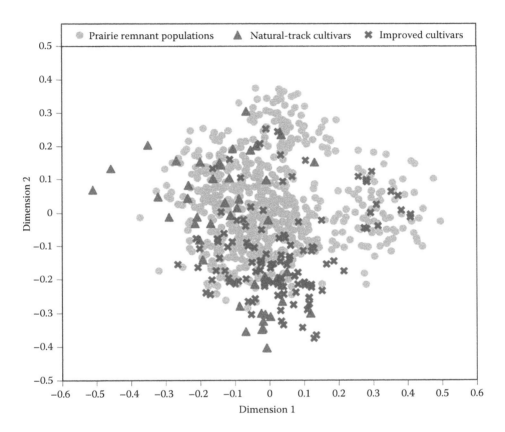

FIGURE 22.3 Multidimensional scale plot of 818 switchgrass plants representing collections made in remnant prairies, public cultivars derived as random seed increases from remnant prairies (natural track), and cultivars improved by selection and breeding, based on random amplified polymorphic DNA (RAPD) markers (Casler et al. 2007b).

FIGURE 23.1 Adult plant comparisons and transformation. Flowering plants of maize (a), switchgrass (b), and Brachypodium (c) compared with the same 32-cm ruler. The small size of Brachypodium is an advantage for a model system. (d–f) Comparison of *B. distachyon* and *B. sylvaticum*. (d) The annual species *B. distachyon* (left) next to its perennial relative *B. sylvaticum* (right). Bar is 15 cm. Inflorescences of *B. distachyon* (e) and *B. sylvaticum* (f). Note the exerted anthers of the outcrossing *B. sylvaticum* and the enclosed anthers of the inbreeding *B. distachyon*. (g) Embryogenic Brachypodium callus. The yellow structured regions are competent to regenerate plants and are used for transformation. Bar is 0.5 cm. (h) Plants regenerating from transgenic callus. After transformation and selection, a mixture of dying (brown) and healthy (yellow) callus can be seen. When placed in the light, healthy callus will turn green and regenerate plantlets. Plate is 10 cm in diameter (i). Close-up of region in regeneration plate designated by the arrow in panel h.

FIGURE 24.1 Identification and tapping of *C. langsdorfii* trees near Nova Odessa, Sao Paulo State, Brazil. (a) A *C. langsdorfii* tree growing near a farm in Nova Odessa, Brazil. The trees grow as single individuals rather than in stands, making it difficult to locate and tap multiple trees. (b) Tapping a *C. langsdorfii* tree with a manual drill. The oils collect in the heartwood and so the hole must be drilled to the very center of the tree, making collection difficult. (c) Botanical characterization of *C. langsdorfii*. Pictures of leaves, seeds with fleshy aril, and seed pods were taken to correctly identify the genus and species of the trees.

FIGURE 25.2 Photographic view of test engine.

FIGURE 25.3 Photographic view of pressure measuring setup.

FIGURE 27.1 Various aspects of *P. elongata* biology. (a) Area falling under green line in the U.S. map is suitable for commercial cultivation of *P. elongate*. (b) A 4-year-old, well-managed plantation of *P. elongata* near Lenox, GA. (c) A 20-week-old *P. elongata* tree. (d) Many new shoots sprout from a coppiced tree. (e) Vigorously growing shoots photographed 1 year after coppicing. (f) *P. elongata* in bloom during early spring. Note that there are no leaves at this time of the year. (g) Immature green fruits. (h) Mature and dehisced fruits (capsules). (i) Winged seed. A cluster of seeds attached to the placenta is shown in the inset. (j) Seed germination on tissue culture medium. (k) Six-week-old planting of *P. elongata* at Fort Valley State University farm in Georgia.

FIGURE 27.2 Steps in the micropropagation of *P. elongata*. (a) Use of nodal explant for multiple shoot induction. (b) Commercial in vitro production at WPI. Each container has 30 rooted plantlets. (c) Elongated shoots can be easily rooted in the presence of 1 mg/L indole butyric acid supplemented in the medium. (d) Leaf with petiole can also be used for developing multiple shoots through an intervening callus phase. (e) Shoot tips also produced multiple shoots. (f) The "Liquid Lab Rocker" system has been found to be useful in our laboratory for rapid multiplication and reduction in the cost of production. (g) Acclimatized uniform planting stock in the greenhouse.

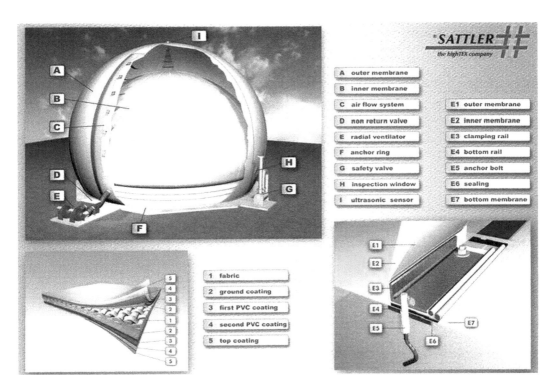

FIGURE 32.9 An example of a commercially available biogas holder. (From Sattler, Biogas storage systems, 2009. http://www.sattler-ag.com/sattler-web/en/products/190.htm)

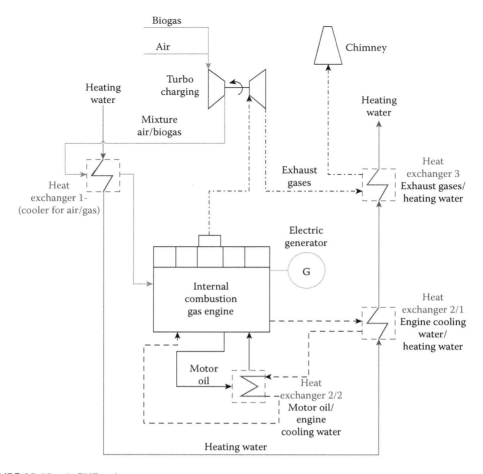

FIGURE 32.10 A CHP unit.

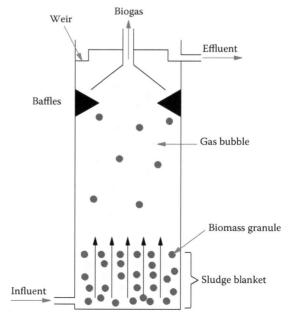

FIGURE 32.12 Scheme of UASB.

18 Oilseed Brassicas

Muhammad Tahir, Carla D. Zelmer, and Peter B.E. McVetty
University of Manitoba

CONTENTS

18.1 UTILIZATION OF VEGETABLE OILS AS ENERGY SOURCES

18.1.1 TRADITIONAL/HISTORICAL ENERGY USES

Vegetable oils provide energy. This is the reason plants tend to pack lipids into their seeds so that the stored energy can be used to support germination. In contrast to animal fats, the lipids stored in the seeds are liquids and are classified as oils or vegetable oils. Several plant species have a remarkable capacity to store large amounts of oils in their seeds, which can be extracted and used in numerous ways. Traditional and historical uses of vegetable oils are wide and varied, including food, lubricants, fuel for paraffin lamps, medicinal and therapeutic uses, spiritual uses, chemical feedstocks, and wood preservatives. With the development of science and technology, uses of vegetable oils expanded, based on their qualities and physical and chemical properties (Walker 2004).

The use of vegetable oils as source of energy (non-food) goes back to ancient times when oils were used in various kinds of lamps to produce light and in some cases burned to produce heat. In modern times, scientists are endeavoring to enhance production and use of vegetable oils which will be able to substitute for traditional fossil fuels. The Hubert Peak Theory (http://www.hubbertpeak.com/summary.htm) for oil reserves suggests that fossil oil production is dropping and it will cease to exist sometime in the future, therefore it is necessary that alternate energy sources are found to supply the world's growing energy demands. In light of these speculations, new concepts, theories, and methods are being developed to utilize vegetable oil as an alternate energy to fossil oil.

Vegetable oil can be used effectively to replace conventional fuel in diesel engines and heating oil burners. The first person to introduce the idea of using vegetable oil in the diesel engines was Rudolf Diesel, who in the early 1900s investigated the idea and presented his theory in front of British Institute of Mechanical Engineers. He remarked, "The fact that fat oils from vegetable sources can be used may seem insignificant today, but such oils may perhaps become in course of time of the same importance as some natural mineral oils and the coal tar products are now" (http://www.biodiesel.org/resources/reportsdatabase/reports/gen/20011101_gen-346.pdf). Whenever there has been a fossil oil crisis in the world due to reduced production and variations in demand/supply as witnessed in 1930s, 1940s, and 1970s, interest in vegetable oil as a potential source of energy resurfaced. Presently, the concept is taken very seriously by numerous countries as well as by the private industrial sector, chiefly driven by energy security issues and environmental concerns. Commercialization of fuels derived from vegetable oils ("biofuels") is being adopted in numerous countries around the world by developing fuels that meet quality standards and putting the appropriate legislation in place. A good example of such adoption is in the transport sector and includes such names as Elsbett AG (http://en.wikipedia.org/wiki/Elsbett), a well-known German-based company, which produces a variant of a diesel engine that can run on pure vegetable oil. An award winning documentary released in 2002, *The Coconut Revolution*, provided evidence of the successful survival of Bougainville people in Papua New Guinea by effectively utilizing coconut oil to replace traditional fuel in their Jeeps (Gerhard 2001). Publicly available information on biofuels is increasing and the list of publications on vegetable oil based fuels is rapidly growing. This article is the result of an effort to synthesize the current knowledge of vegetable oils as sources of energy, with a focus on the oilseed Brassicas.

18.1.2 BIOFUELS AND BIODIESEL

18.1.2.1 Straight Vegetable Oil and Waste Vegetable Oil

Although research on vegetable oil as an alternative form of energy has been an ongoing process for at least a century, it is only now that scientists have been able to understand the necessary engine parameters and related fuel properties for its reliable long-term usage in vehicles. The most basic form of vegetable oil is straight vegetable oil (SVO), which is used primarily for cooking purposes. Although the term generally refers to new, not previously used oils, it has also been used for the cleaned waste vegetable oil discussed later. It is readily available, but it is not recommended as an effective replacement for vehicle fuel due to its viscosity. Published research (U.S. Department of Energy 2006) on the issue indicates that the high viscosity of SVO promotes the deposition of carbon and results in the accumulation of SVO in the engine crankcase. Over time, this can cause diesel engine malfunction. From recent research, it can safely be concluded that SVO is acceptable for short term use but its long term usage will harm diesel engines (Babu and Devaradjane 2003).

To solve this problem, engineers have devised a kit (http://www.journeytoforever.org/biodiesel_svo.html) that preheats the SVO to reduce its viscosity before using it as a fuel in diesel engines. The kit is usually sold at approximately U.S.$1,200 but the prices may vary. This after-market kit normally contains three sections: a heat exchanger, an original diesel fuel tank, and an SVO tank. Initially, the driver has to start the car and wait for the vegetable oil to be heated before switching the engine to the SVO tank. The process is reversed and the power is switched back to diesel before parking the vehicle. These two-tank, SVO systems were the first to be developed. Optimization of this system is not complete. Preheating of the vegetable oil in these systems takes place at approximately 70–80°C, but according to the European Advance Combustion Research for Energy from Vegetable Oil, the viscosity of vegetable oil equals that of diesel oil at 150°C (http://www.biomatnet.org/secure/Other/S1033.htm).

German manufacturers are also offering single-tank conversion kits that do not require the driver to switch between the engine tanks (http://www.noendpress.com/caleb/biodiesel/biodiesel_conversion_mercedes_booklet.pdf). These kits mostly use German produced rapeseed oil which meets the German rapeseed oil fuel standards DIN 51605 and are suitably modified to operate at subfreezing temperatures greater than –10°C. Experts recommend that users of these modified engines should regularly change the oil and keep the engine well maintained. The most successful prototypes of this experiment have been the pre-1990 Mercedes Benz models which are widely used for experimentation in biofuel research.

It is evident that the replacement of petrodiesel fuel with SVO is a work in progress. Enthusiasts and engineers continue to strive for perfection by testing a variety of apparatuses and vegetable oil blends. The Vegetable Oil Fuel Database was produced by a diverse community of people comparing actual road tests with laboratory research. According to the online database, almost 93% of the vehicles have performed adequately in the short term whereas 16 of 341 vehicles observed have logged over 50,000 mi using vegetable oil. Waste vegetable oil (WVO) can also be used in diesel engines, but there are many theories of the best way of preparing it for fuel use. Most agree that the oil should settle for at least two weeks to allow suspended sediments (food particles) and water to settle to the bottom. After settling, a good filtration system in addition to the heating system is all that is required to run most diesel engines on vegetable oil whether it is virgin, unused oil, or waste vegetable oil (WVO). Both WVO and SVO are still experimental fuels (see review by Jones and Peterson (2002); http://journeytoforever.org/biofuel_library/uidaho_rawoils.html) and could be risky for the engine. Therefore they are exciting novel sources of bioenergy but may not be commercially viable.

18.1.2.2 Biodiesel

Vegetable oil is the basis of biodiesel. Although similar to petrodiesel, some modifications are required to utilize this fuel in standard diesel engines. The goal of much research has been the production of an engine that runs on biodiesel without the need for modifications. The very first account of the production of biodiesel traces back to 1937, when Belgian Patent 422,877 was granted to G. Chavanne. He described the formation of biodiesel by separating the fatty acids from glycerol and replacing glycerol with alcohols (Knothe 2001). Biodiesel is usually a blend of petro diesel and vegeoil diesel with a B description. The "B" designation is used to describe the amount of biodiesel present in a fuel mixture compound. A "B20" fuel is one that contains 20% biodiesel whereas "B100" denotes 100% biodiesel. Currently a blend of 20% biodiesel and 80% petroleum diesel is commonly used in large vehicles to reduce greenhouse gas emissions. Such a blend does not require engine modifications. In contrast, use of a 100% biodiesel requires specifically modified engines to provide the needed performance.

Research on biodiesel reveals that a blend of biodiesel and petrodiesel performs much better than petrodiesel alone. The widespread use of biodiesel is limited by production costs and limited availability of raw materials (Radich 2004). In his study, "BioDiesel Performance Costs and Use," Anthony Radich (Radich 2004) defines biodiesel as, "the monoalkyl esters of long chain fatty acids derived from plant or animal matter which meet (A) the registration requirements for fuels and fuel additives established by the Environmental Protection Agency under section 211 of the Clean Air Act (42 U.S.C. 7545), and (B) the requirements of the American Society of Testing and Materials D6751."

The major drawback of using pure biodiesel is its tendency to gel at cold temperatures (–10°C for canola-derived biodiesel), which can clog the fuel lines and filters in the vehicle's fuel system. It is therefore suggested that the long-term efficacy of biodiesel blends remains controversial in low-temperature zones such as in North America and Europe. In addition to low temperature concerns, some believe that using B20 lowers a vehicle's fuel economy. It is generally hypothesized that vehicles running on B20 achieve 2.2% fewer miles per gallon of fuel (Radich 2004).

In contrast, research now indicates that the difference in fuel economy is insignificant during actual road tests. If used under the guidelines of such authoritative bodies as American Society of Testing and Materials (ASTM), B20 might actually enhance the performance of the vehicle. Since 2005, The National Renewable Energy Laboratory in Colorado has been working with Regional Transport District of Denver to test the performance of B20 as compared to conventional diesel. Initial research on buses indicates that after one year of observation and more than 150,000 mi, there seemed to be neither significant difference in the average fuel economy nor any disparity in the maintenance costs (Proc et al. 2005). Furthermore, the laboratory test, conducted at a 99% confidence level, indicated that greenhouse gas emissions from B20 were lower than the conventional diesel bus. Similarly, the St. Louis Metro Biodiesel Transit Bus Evaluation Project helped shatter the myths surrounding the performance of B20. The study involved a comparison between eight transit buses running on B20 and seven using conventional diesel. At the end of a 12-month period, the fuel economy of the diesel buses with a total mileage of 325,407 was calculated to be 3.58 mpg. The numbers were just slightly higher than the buses operating on B20, which after approximately 634,268 cumulative kilometers reported 5.66 km/gal. In fact, the MBRC (miles between road calls) were better for B20 than their counterparts predicting a better performance of engines in buses using B20 (Barnitt et al. 2008). Likewise, the difference between total maintenance costs for both groups of buses was insignificant. Assuming labor costs at U.S.$50/h, the cost for B20 group of vehicles were only $0.0012/mi higher than that of the pure petrodiesel-powered fleet.

The Denver study also revealed minor drawbacks to the use of B20. It provided evidence that using B20 or any kind of vegetable oil blend will necessitate more frequent fuel injector and fuel filter replacement in the vehicles. For the study, filter replacements were done regularly to avoid plugging and to keep the engine running smoothly under subzero temperatures. Without this precaution, the failure of fuel injectors in B20 group occurred within the expected mileage range. The frequent filter replacement can be costly, particularly at a point when no long-term data exist.

Biodiesel is commonly prepared by mixing vegetable oil and methanol or ethanol in the presence of sodium hydroxide, a process called transesterification (www.biodiesel.org/pdf_files/fuelfactsheets/prod_quality.pdf). One of the primary reasons for the limited use of biodiesel is its production cost. Using soybean oil to produce biodiesel is currently the easiest of all methods but according to U.S. Department of Agriculture statistics, the cost of producing a gallon of soybean oil biodiesel will increase from U.S.$2.4 in 2004 to almost $2.8 in 2013, which is approximately four times more expensive than producing a gallon of petroleum. Instead of soybean oil, the use of yellow grease to produce biodiesel is more economical. Yellow grease (waste oil from restaurants) is currently half the cost of soybean oil, but the U.S. Energy Information Administration predicts that yellow grease will only be produced in limited amounts (Pearl 2001).

The total demand for diesel fuel in the United States and Europe amounts to 490 million t whereas the capacity to produce biodiesel totaled 16 million t, with an average annual growth rate of 40% from 2002 to 2006 (http://www.martinot.info/RE2007_Global_Status_Report.pdf). Even with high production costs associated with producing biodiesel, the demand continues to rise, accompanied by growing interest in producing biodiesel around the world. Such interest can be attributed to federal tax credits, public awareness of greenhouse gas emissions, pollution control requirements, government subsidies, and an increase in the production of diesel engines. In Europe, Germany is the largest consumer of biodiesel with a total consumption reaching 34,395 GWh (gigawatt hours) followed by France at 13,506 GWh (http://www.biofuelpowerandlight.com). Because of active Government and Public contributions, the price of biodiesel at fuel pumps around the world has already decreased in comparison to conventional diesel, but its utilization as a replacement to diesel is still hampered by lack of availability, required engine modifications, frequent maintenance of selected engine parts, and a scarcity of conclusive literature on its long-term effects.

Technically, biodiesel can be obtained from any vegetable oil. Therefore, the choice of feedstock mainly depends upon the country and its climate. This is evident from the fact that soybean oil (Kaieda et al. 1999; Samukawa et al. 2000; Watanabe et al. 2000; Wei et al. 2004; Noureddini 2005) is the most commonly used vegetable oil for biodiesel production in the United States, whereas rapeseed/canola (Korbitz 1999; Lang et al. 2001; Massimo et al. 2003; Chang et al. 2005) is the major feedstock in Europe and Canada. Regions with tropical climates utilize tropical oils such as coconut and palm oil (Crabbe et al. 2001; Fukuda 2001; Sulaiman 2007). Cotton seed, peanut, and sunflower oils are also commonly used as source for biodiesel (Soumanou and Bornscheuer 2003; Knothe and Dunn 2005; Orcaire et al. 2006; Sulaiman 2007; Demirbas 2008; Joshi et al. 2008; Keskin and Guru 2008; Sagiroglu 2008). A comparison between the physical and chemical properties of biodiesel fuels (Fukuda et al. 2001; Knothe and Dunn 2005) indicated that palm oil has the highest yield (4,000 kg/ha) as compared with other vegetable oils leading to substantial gains in worldwide production of palm oil. Other vegetable oils that have been studied for the production of biodiesel include sesame oil (Sydut et al. 2008), olive oil (Sanchez and Vasudevan 2006), tobacco seed oil (Giannelos et al. 2002), rubber seed oil (Ikwuawgu et al. 2000), coffee oil (Oliveira et al. 2008), jatropha (Foidl et al. 1996; Shah and Gupta 2006; Berchmans and Hirata 2008), and other nonconventional oils or oils from novel sources (Munavi and Obhiambo 1984; Park et al. 2005, 2007; Holser and Harry-Okuru 2006; Kesari et al. 2008; Naik et al. 2008; Rashid et al. 2008; Razon 2008; Ruan et al. 2008; Santos et al. 2008; Sinha et al. 2008).

18.1.3 SECOND-GENERATION BIOFUELS

The first-generation biofuels are limited in their wide scale production and utility in many ways. The issues of producing enough biofuel without compromising the food supply, disturbing the ecosystem, creating a noncompetitive product in terms of cost, and increasing fossil fuel related emissions generate the need for second-generation fuels. First-generation biodiesel is made mainly from edible vegetable oil crops such as soybeans, rapeseed, sunflowers, and palm. The increase in the demand of these vegetable oils has put pressure on valuable ecosystems such as rainforests and has decreased the area available for food crops. There are examples where old forest land has been burned to cultivate oil palms. Burning of old growth forests not only leads to deforestation and reduces biodiversity but also releases large amounts of carbon dioxide in the air. Because very high quantities of vegetable oil will need to be produced to replace diesel and gasoline, the required methods of cultivation may lead to greater harm of natural resources.

To counter such threats, some experts recommend using jatropha seeds for a vegetable oil supply. Compared with the four most common plant oils suitable for fuel production (soybean, sunflower, rapeseed, oil palm), jatropha is drought resistant and tolerant of high humidity and temperature. It is a non-food crop, bears seed for almost fifty years, and performs very well as a low-input crop on marginal land. Considering its current yield of 5 tons of seed/ha per year with a potential for a much larger production, it can easily provide sustainable environmental solutions for biofuel production. The only major limitation to its large scale production is the current labor-intensive extraction methods used for oil extraction of jatropha seed. With research and improvement of traditional extraction techniques such as the Bielenberg hand press which is used in Tanzania, it is likely that oil extraction time and cost can be reduced to make such seeds more valuable (Benge 2006).

Another improvement for first-generation biofuels lies in the cost-effective and efficient production of vegetable oil on a mass scale. This can be attained by harvesting particular types of algae which contain almost 50% vegetable oil. The process is known as algaculture and can produce much more oil per unit area than current farming methods (Sheehan et al. 1998). Experts are certain that algaculture can produce quantities of vegetable oil sufficient to replace petrodiesel (Christi 2007). It can, therefore, be concluded that production of vegetable oil from algae might just be the solution to a wide range of issues including deforestation and high cost of production. Because scientists and engineers are mostly concerned about replacing petroleum and diesel with vegetable oil, it may be practical to produce large quantities of vegetable oil that could fulfill the transportation needs of a society. Algaculture professionals believe that it can be done.

In the United States, the Office of Fuel Development funded research from 1978 to 1996 to determine the feasibility of using algae for oil production. The results of the Aquatic Species Program concluded that algae farms in the United States have the potential to supply biodiesel that can satisfy not only the transportation needs of the United States but also replace existing home heating oil (Briggs 2004). Optimism in algaculture remains high as various studies have provided evidence that one day vegetable oil from algae may provide the needed feedstock to replace diesel and home heating oil. According to one such report, the yield per acre of oil from algae is over 200 times the yield from the best performing vegetable oil seeds (Sheehan et al. 1998). Another recent study (Wagner 2007) concluded that per unit area yield of oil from algae is estimated to be 5,000–20,000 gal/acre, which is 7–31 times greater than the next best crop, palm oil. The most important aspect of algaculture is that it does not entail a decrease in food production because production of algae is not dependent on any farmland or fresh water. A large number of companies are emerging (http://peswiki.com/index.php/Directory:Biodiesel_from_Algae_Oil) which have plans to produce algae in open ponds or by other methods.

Hydrogenated vegetable oils (HVO) are also being looked at as the second generation of biodiesel (Arvidsson 2008) and have been termed "hydrotreated biodiesel." The HVOs are straight-chain paraffinic hydrocarbons that are free of aromatics, oxygen, and sulfur. They have high cetane numbers and are devoid of detrimental effects (i.e., increased NO_x emission, deposit formation, storage stability problems, more rapid aging of engine oil or poor cold properties) of traditional ester-type biodiesel fuels (Aatola et al. 2009). Proprietary commercial hydrotreated biodiesels are appearing on the market and are claimed to perform better than rapeseed methyl esters or biodiesel (http://www.greencarcongress.com/2006/03/neste_oil_and_o.html).

18.1.4 OTHER BIOENERGY USES OF VEGETABLE OILS

Besides transportation, there are various other energy uses of vegetable oils, which include home heating and electricity generation. At home, either vegetable oil or biodiesel can replace the No. 2 heating oil used in furnaces and boilers. The primary advantage of utilizing vegetable oil in home heating appliances is the lower price which can be significant for a small business or home owner. In a farm energy case study in Old Athens Farm at Westminster in Vermont, the owner replaced the No. 2 heating oil with vegetable oil and was able to save approximately U.S.$3,600 in the winter of 2005/2006 and almost U.S.$9,000 in 2008 by burning 4,000 gals of vegetable oil (Grubinger 2008).

Vegetable oil is also used for electricity generation in small power plants. The use of vegetable oil in this application is significantly limited because of the large deposits of coal in the earth's surface. Although oil reserves in the Earth's crust are limited, scientists are more concerned with replacing liquid fuel such as diesel and petroleum with vegetable oil instead of replacing coal in electricity generating plants. Nevertheless, it is estimated that refined vegetable oil, although viscous, is well suited to large, low-speed engines that are used in power stations (vegetable oil power stations). These diesel reciprocating engines are widely used in power plants ranging in size from 1 to 17 MW and are specifically designed to run on heavy oil similar to refined vegetable oil.

18.1.5 IMPORTANCE OF *BRASSICA* OILSEEDS AS BIOENERGY SOURCE

Like other vegetable oils, Brassica oils have a history spanning over thousands of years as sources of edible oil for humans and protein feed for animals. However, rapeseed oil has also been an important source of energy for many non-food purposes. For example, the oil from rapeseed has been used for light (oil lamps) and heating for centuries in India and China. Today, canola/rapeseed oil is considered an important component of the bioeconomy and/or bioenergy. Some examples are the production of bioenergy heaters, stoves, and furnaces (http://www.thermobile.co.uk/bio.asp;

http://www.freetherm.de/rapseng.html) which have been designed to be operated exclusively on canola/rapeseed oil. The most exciting prospect for canola/rapeseed oil lies in its role as a renewable energy source in temperate regions. With the decline in the Earth's petroleum resources, individuals, the public sector, and industries are under pressure to reduce their consumption of fossil fuels. This pressure is the driving force for advanced technological solutions. These will make use of renewable energy resources and thus maximize the efficient use of remaining nonrenewable energy resources. On the basis of this concept, canola/rapeseed oil has been studied or used to replace mineral oils for manufacturing products such as plastics (http://news.mongabay.com/bioenergy/2007/07/scientists-develop-polyurethane.html), engine oils (Glamser and Widmann 2001), lubricants (Wightman et al. 1999), and surfactants (BioMatNet).

The most important use of Brassica oil as a renewable energy source is in the manufacturing of biodiesel for powering motor vehicles. In general, oil from canola/rapeseed has 90% of the calorific value of mineral oil. However, being 10% heavier than mineral oil, its heat content per liter is similar to mineral oil. Biodiesel from canola/rapeseed oil is obtained by the process called methanolysis (Vicente et al. 2004) which yields a renewable fuel and a glycerol byproduct. Canola/rapeseed oil-derived biodiesel is a renewable and biodegradable fuel which is compatible with commercial diesel engines and has environment friendly properties such as reduced toxicity and lower emissions (Mittelbach et al. 1983). Biodiesel derived from Brassica oil (including rapeseed and canola oil) has been successfully used in pure form in newer diesel engines without engine damage and is also frequently mixed with standard diesel in varying ratios (2–20%). The diesel fuel derived from canola/rapeseed is also used to generate heat, power, and electricity.

The simple extraction process includes mechanical pressing of seeds with mild heating to recover 90% of the oil; titration and washing with hydrocarbon solvents (hexane or heptane); heating to 150°C to evaporate the solvent; filtration to remove gummy fats and carbohydrates; mixing with caustic soda to lower acidity and remove free fatty acids molecules and refining. The refined oil obtained in this way is similar to common diesel fuel except for its 10% higher viscosity. It is suitable for use in power stations to generate electricity. In addition, the seed press cake obtained as a byproduct in the above process can be used as a source of energy in an incineration plant to produce electricity. A privately held company in the United States (http://www.biofuelpowerandlight.com) designs and builds biofuel power plants based on vegetable oil feed stocks including canola/rapeseed oil. The company owns one biodiesel plant that produces 70 million gal/year of B100 biodiesel, some of which is sent to biofuel power plants in Texas that generate green electricity. Combined heat and power units (CHP) fuelled with canola/rapeseed oil are also of interest to industry due to efficient utilization of renewable agricultural energy sources, and feasibility/research studies point to their advantages with possible improvements including a list of these advantages (Glamser and Widmann 2001; Klaus et al. 2005; Pantaleo et al. 2008). The concept of energy cabins heated by biofuels or biodiesel is also emerging and some people are experimenting in this area (http://www.motherearthnews.com/Renewable-Energy/2003-12-01/Heat-Your-Home-with-Biodiesel.aspx; http://www.juwi.com/fileadmin/user_upload/en/pdf/JUWI_Bereichsbrosch%C3%BCre_Bio_EN_1Aufl.pdf). Feasibility studies are also being conducted concerning the use of biodiesel in low-speed diesel engines, gas turbines, and steam generation boilers for generation of electricity (Bolszoa and Mcdonell 2008; http://mydocs.epri.com/docs/public/000000000001014867.pdf).

18.1.6 DOMESTIC AND WORLDWIDE ENERGY USAGE

Oil obtained from Brassica seeds (rapeseed, mustard or canola) is 40% of the total seed weight on average, generating approximately 80% of the cash value of the crop, and is now considered an important feedstock for renewable energy products, mainly biodiesel. A recent study (Johnston and Hollowway 2007) revealed that there are 109 countries in the world which can produce biodiesel on a profitable basis. However, it is now well recognized that profitability in the bioenergy sector depends on the availability of a particular feedstock, its yield potential, and cost of production.

Canola/rapeseed is successfully grown in the cooler agricultural regions and also as a winter crop in temperate climates of Europe. Based on this climatic adaptability, substantial production of canola/rapeseed occurs in China, India, parts of the United States, Canada, northern Europe, and Australia. In recent years, China has been the leading producer of canola/rapeseed oil, accounting for 22% of the total world production of 20,330 thousand metric tons during 2008/2009 (USDA Foreign Agriculture Service 2009). India, Canada, Japan, and the countries of European Union are among the other leading canola/rapeseed oil producers, representing 11, 9, 4, and 41% of the total production, respectively. Accordingly, a common choice of feedstock for the production of biodiesel in these countries is rapeseed and/or canola, especially in the European Union and Canada. A report/ data of the year 2005 (World Biodiesel Production) indicated that 13 countries (Germany, France, United States, Italy, Czech Republic, Austria, Spain, Denmark, Poland, the United Kingdom, Brazil, Australia, and Sweden) produced 97% of the 3,762 million L of biodiesel production in the world, with major contributions coming from Germany (51%), France (15%), United States (8%), and Italy (6%). Except for the United States, where the main feedstock for production is soybean, most countries listed above used canola/rapeseed as a major feedstock for the production of biodiesel (http://www. bioenergywiki.net/index.php/Top_biodiesel_producing_countries). On the basis of 2008/2009 data (USDA Foreign Agriculture Service 2009), Canada and the European Union (EU-27) are the major exporters of canola/rapeseed oil with annual exports of 1,350 and 150 thousand metric tons, respectively. China was the leading importer of canola/rapeseed oil during 2008/2009. The choice of suitable feedstock for the production of biodiesel will influence the global biodiesel potential (USDA Foreign Agriculture Service 2009) and countries like Malaysia, Indonesia, Argentina, and the United States are expected to have the highest volume potential for biodiesel due to large productions/yields of palm and soybean oil in these countries.

Canada is one of the main regions of the world where canola/rapeseed oil is likely to be renewable energy source for the production of biodiesel. The major options for feedstocks to produce biodiesel in Canada presently include canola/rapeseed, flaxseed, tallow or soybean. However, based on current production levels, availability and market stability, canola/rapeseed appears to be the most favorable feedstock to meet the demands for biodiesel. The demand for biodiesel in Canada is expected to increase based on the Canadian government's target to use 2% biodiesel as a renewable fuel content in all diesel fuel sold in the country by the year 2010. Therefore, current biodiesel production in the country (240 million L) is expected to rise to approximately 500 million L by 2012 if the requirement for B2 is reached. These targets can be met considering canola as one of the major oil seeds grown in Canada, covering approximately 6,000 million ha of land in 2007/2008 and produced approximately 10 million t of seed (http://www.canola-council.org). Roughly 0.370 tons of oil can be produced from 1 ton of canola/rapeseed seed. If 10% of the total canola/rapeseed production in Canada is dedicated to producing biodiesel, there is a potential to produce over 4 billion L of biodiesel from the currently available canola/rapeseed feedstock. Of the eight biodiesel production plants in Canada, three, including Topia Energy Inc., Milligan BioTech, and Rothsay, use canola as a feedstock (www.countyofnewell.ab.ca/pdf/economicdev/biodiesel_production_in_ the_county_of_newell.pdf). Dominion Energy Services, LLC is listed as no. 3 among the 12 largest biofuel plants in the world (Ooko 2008) and has a $400 million integrated biodiesel and ethanol refinery in Innisfail, Alberta, Canada. The biofuel plant has a 100 million gal biodiesel production capacity, and will use approximately 900,000 t of canola per year as raw material for feedstock. In contrast to Canada, canola/rapeseed is a minor feedstock for bioenergy or biodiesel production in the United States because 80% of the biodiesel is produced from soybean and sunflower oil. North Dakota is the major producer of canola/rapeseed in the United States, which contributes 90% of the total 1.5 billion pounds of rapeseed used for biodiesel (http://www.biodiesel.org). However, there is an interest in several states (Idaho, Montana, and Colorado) to grow rapeseed as a high yielding feedstock for the production of biodiesel (http://www.canola-council.org).

The countries of the European Union (EU) represent a model for the use of canola/rapeseed as a source of energy as reflected by the large scale application and production of biodiesel from canola/

rapeseed oil. Ever since the opening in 1988 of first biodiesel plant in Austria, canola/rapeseed oil has become the feedstock foundation of biodiesel in Europe. Many EU countries such as France, Germany, Italy, and the Czech Republic have developed biodiesel plants over the past decade and approximately 4 million tons of rapeseed went into biodiesel in 2006 making the EU a global leader in oilseed biodiesel. Presently, Germany is the largest producer of biodiesel in Europe (http://www.ebb-eu.org/stats.php) with a production capacity of 2.89 million tons (2007 data) which is almost 50% of the total biodiesel production in Europe. Other EU countries with significant biodiesel production include France, Italy, and Austria producing 872,000, 363,000, and 267,000 thousand metric tons based on 2007 data (European Biodiesel Board). The EU is heavily subsidizing canola/rapeseed cultivation to meet its carbon dioxide reduction targets and production will need to be doubled by the year 2020 to fulfill its mandate of using 10% of all vehicle fuel from biofuels. According to Klaus Thuneke (Thuneke 2007), 13 industrial and over 300 oil mills in Germany produce rapeseed oil for mainly energetic uses. Germany also has 1,800 CHP plants running on agro-fuels. The plants were initially intended to be run by home-grown canola/rapeseed oil but the majority of these plants gradually shifted to use of cheaper palm oil whereas canola/rapeseed is now mostly used for producing biodiesel.

18.2 CHARACTERIZATION OF *BRASSICA* SPECIES

18.2.1 Major *Brassica* Oilseed Species and Their Habit

The family Brassicaceae (Cruciferae), commonly known as the mustard family, is a large group of about 338 genera and 3,709 species (Warwick and Hall 2009). *Brassica* is a genus within this family comprising about 100 species and includes not only ornamental, vegetable, and weed species but also crops which are an important source of edible and industrial oil. The major species which are widely cultivated as oil seed crops in various parts of the world are *Brassica napus*, *Brassica rapa*, *Brassica juncea*, and *Brassica carinata*, although *Brassica nigra* and the closely related *Sinapis alba* are also used in India for oil extraction. Several varieties of these species are also used as mustard or vegetable crops, but when grown for oilseed the species are generally termed as "rapeseed" or "oil rape."

Because of the economic importance of rapeseed, *Brassica* species have been extensively studied and characterized in terms of their history, botanical features, geographical distribution, and growth habits. A summary extracted from various sources (Hedge 1976; Parkash 1980; Parkash and Hinta 1980; Downny 1983; Kazalowaska 1990; Yan 1990; Downey and Rimmer 1993; Callihan et al. 2000; Martin et al. 2006) is presented here. In general, rapeseed production has an ancient history. Artifacts dating to more than 2,500 YBP and bearing the Chinese word for "rapeseed" are known, and Sanskrit literature from India dating back to 3,500 YBP mentions rapeseed (*B. rapa*). Seeds of *B. juncea* also have been found in Indian archaeological sites from 2,300 YBP.

Cultivation of *B. carinata* A. Braun (Ethiopian mustard) is mainly confined to North East Africa or the Ethiopian Plateau which may be its region of origin. It is a slow growing annual species which is mainly used as a leafy vegetable but also harvested for seed oil.

Brassica juncea L. appears to have originated in the Near East and in southern Iran. Two forms of this annual species are known and can be distinguished by seed size, color, and country of cultivation. Indian mustard, brown mustard, or oriental mustard is grown in India, where the larger brown seeds are used for oil extraction. Yellow mustard, grown mainly in China, has smaller, yellow seeds. The Indian or yellow mustard in China is used as a leafy vegetable, but it is used as an oilseed crop in the Ukraine. The use of *B. juncea* (brown mustard) in western countries is mainly for the production of table mustard, especially in western Canada.

The species *B. rapa* L. (syn. *B. campestris*), also known as field mustard, rapeseed, turnip rape and polish rape, seems to have the widest distribution of the Brassica oilseeds. The primary center of diversity for this species is the Himalayan region but it spread to Europe, Asia, India, and northern Africa about 2000 years ago as turnip, turnip rape, and mustard. It thrives in temperate climates and is now grown as a vegetable or an oil seed crop in China, India, Sweden, Finland, and Canada (Rakow 2004), and it is also found as a weed in the United States. Spring/summer and winter annual cultivars have been developed.

The *Brassica* species most widely grown for its seed oil is *B. napus* L. (also known as rapeseed, oilseed rape, swede rape, and Argentine rape) which is considered to have its origin either in the Mediterranean or western and northern Europe. There are both winter and spring/summer annual forms of *B. napus*. Winter type *B. napus* is the main rapeseed crop in Europe, parts of China, and the eastern United States. Both summer and winter forms are cultivated in Europe and Canada and only the summer form is grown in Australia.

The current rapeseed varieties grown in the United States and Canada are cultivars of either *B. napus* or *B. rapa* species, the former being cultivated on the majority of hectares. Both the species have been used to breed spring or winter varieties and for industrial or edible oil markets. Industrial rapeseed, along with its wild ancestors, contains long chain fatty acids that limit its acceptability as an edible oil. High levels of glucosinolates (500 µM/g) give rapeseed oil an unacceptable sharp taste and are responsible for its thyroid disrupting properties. Erucic acid, a long chained fatty acid (C22:1) is present in the oil, often in concentrations reaching 50–60% of total seed oil.

To address the antinutritional aspects of rapeseed oil, and to increase the value of the protein-rich meal remaining after oil extraction, edible rapeseed varieties were developed by Canadian plant breeders in 1974. They contained very low levels of glucosinolates (30 µ mol/g) and of erucic acid (about 1%). These new varieties were termed "Canola," which is an acronym for Canadian Oil Low Acid (Steffansson and Downey 1995). As in the case of rapeseed varieties, spring and winter canola-quality cultivars have been bred from both *B. rapa* and *B. napus*. Canola therefore refers to *B napus* and (less commonly) *B. rapa* varieties that are low in both glucosinolates and erucic acid. Within each species, the edible oil, canola-quality cultivars are visually indistinguishable from the industrial oil, rapeseed varieties.

18.2.2 FATTY ACID PROFILES

The energy in seed oils is stored in the form of triacylglycerols (TAG), which are a source for carbon storage and synthesis of fatty acids (FA). The FA present in seed oils are diverse in their chemical composition and vary based on the length of carbon chain (C12 to C24) and their degree of saturation/desaturation (presence of 1 to 3 double bonds). They are generally denoted by their carbon chain length followed by the number of double bonds. On the basis of this system of notation, FA are classified as saturated FA (SFA, e.g., C16:0 and C18:0), monounsaturated FA (MUFA, e.g., C18:1 and C20:1), polyunsaturated FA (PUFA, e.g., C18:2, C18:3), and very long-chain FA (VLCFA, e.g., C22:1 and higher). It is the profile of FA in a particular oil which determines its nutritional quality or its industrial value. For example, the FA composition of a superior vegetable oil for human consumption is expected to contain a high ratio of MUFA/SFA, a good proportion (at least 5:1) of PUFA and no VLCFA (FAO/WHO). On the other hand, the oils with a high percentage (20–50%) of VLCFA such as C20:0, C22:0, and C24:0 are desirable for industrial use to manufacture lubricants (McVetty et al. 2008).

Brassica/rapeseed oil is composed of FA with varying carbon chain lengths and levels of desaturation. Typical FA profiles for cultivars of *B. napus*, *B. juncea*, and *B. rapa* (Shahidi 1990; Scarth and Tang 2005) are approximately 5% C16:0 (palmitic acid), 1% C18:0 (stearic acid), 15% C18:1 (oleic acid), 14% C18:2 (linoleic acid), 9% C18:3 (linolenic acid), and 45% C22:1 (erucic acid). A comparison (McVetty and Scarth 2002) of the FA profiles of rapeseed and mustard oils with an FA composition of selected naturally occurring vegetable oils (soybean, sunflower, corn, peanut, palm, and olive) reveals that the composition of Brassica oil is genetically more variable than any of the other vegetable oils included in the study. This variability in the composition of Brassica oil has been successfully manipulated to produce cultivars with specific FA profiles in the seed oil tailored for specific end uses. A classic example of this was the development of canola from rapeseed where the nutritionally undesirable fatty acid (C22:1, erucic acid) has been reduced to a very low level by breeding, thus making the oil suitable for edible uses.

Canola oil typically has a nutritionally desirable FA profile with only traces of erucic acid, low saturated FA (5–8%), and a good balance of MUFA (60–65%) and PUFA (30–35%). However, there are several other profiles of FA in Brassica oil, which are of interest for food, non-food, industrial, and novel uses. From a nutritional point of view the desirable Brassica oils are considered to have reduced or very low levels (<7%) of saturated FA (C16:0, C18:0, C20:0, C22:0), low or less than 3.5% of C18:3 (linolenic acid), mid to high levels (67 to over 75%) oleic acid (C18:1), zero or negligible levels of erucic acid (C22:1), and a C18:3/C18:2 ratio of 1:2 (Shahidi 1990). The detailed nutritional and functional benefits of these FA profiles in Brassica oils have been discussed and reviewed elsewhere (Fitzpatrick and Scarth 1998; McVetty and Scarth 2002).

Brassica oil with high (>50%) or super high (>66%) erucic acid levels are termed as HEAR and SHEAR (McVetty et al. 2008), respectively, and are suitable for industrial oil applications such as the manufacturing of lubricants, slip agents, polymers, paints, inks, cosmetics, and pharmaceuticals. The opportunity to develop a wide range of novel oil profiles in Brassica tailored to end use is increasing. This may be particularly appropriate for Brassica oil used for the production of biodiesel. Although the FA profile for an oil most suitable for making biodiesel is still being debated, it is generally known that FA profiles of oils will have to be modified toward higher carbon chain lengths and reduced branching/degree of unsaturation to improve the ignition quality and cold flow properties of diesel engines run on biodiesel produced from Brassica oil.

Other FA profiles which are being created in Brassica oils using conventional and transgenic breeding approaches (Voelker et al. 1996; Knothe et al. 1997.; Murphy 1999; Scarth and Tang 2006) include oils with high levels of short and medium chain FA such as lauric acid (C12:0), caprylic acid (C8:0), capric acid (C10:0), palmitic acid (C16:0) and stearic acid (C18:0).

In addition, several unusual FA of interest to industry are found in the seed oils of nonagronomic species (Jaworski and Cahoon 2003). The unique functional properties of these oils are due to varied carbon chain length, number and position of double bonds (0–5), and presence of uncommon functional groups, such as hydroxyl, epoxy, or acetylinic groups (Thelen and Ohlrogge 2002). Incorporation of these unusual FA into Brassica oils have been examined as a way to create new industrial oils with novel properties. Profiles of such Brassica oils might include enrichment or additions of monoenoic acid, gamma-linolenic acid, very long-chain (20–22 carbons) polyunsaturated (4–6 double bonds) FA, conjugated FA, and FA with epoxy and hydroxy functional groups (Safford et al. 1993; Liu et al. 2001; Huang 2004; Scarth and Tang 2006).

18.2.3 REGIONAL ADAPTATIONS

Brassica species exhibit adaptation to a wide range of agroclimatic conditions. Due to the plastic nature of their phenotypes they can adapt themselves equally well to low or high input cultivation techniques. *B. juncea* is more drought tolerant compared to other *Brassica* species and is predominantly grown in the Indian subcontinent but is also widely grown throughout Asia. This species, along with *B. napus,* is being developed for Australian climatic conditions and is becoming increasingly popular for cultivation there. Another *Brassica* species, which is quite tolerant to drought and also has resistance to diseases, pests and pathogens, is *B. carinata* that is a major oilseed crop in Ethiopia. Because of its ability to adapt to difficult climatic conditions, this species is of interest for cultivation in arid regions such as southern Europe. *Brassica rapa* is more cold tolerant and early maturing varieties of this species are well adapted to the climatic conditions of western Canada. However, various other ecotypes of this species (brown and yellow sarson and toria) are well adapted for cultivation in the Indian subcontinent. Within Canada, the canola-quality varieties of *B. napus* and *B. rapa* are mainly adapted to the Prairie Provinces (Manitoba, Saskatchewan, and Alberta) with limited cultivation in Ontario and Quebec. Generally, *B. napus* varieties grown in western Canada require more frost-free days and mature in approximately

105 days as compared to *B. rapa* varieties, which are earlier-maturing (approximately 88 days). Therefore, *B. napus* varieties are better adapted to southern Manitoba and central Saskatchewan/ Alberta.

The primary and secondary gene pools of *Brassica* species are very diverse and offer allelic variants for FA profiles as well as for tolerance to various biotic and abiotic stresses. This diversity plays an important role in their adaptation to specialized regional requirements.

18.2.4 GENETICS, GENOMICS, AND BIOCHEMISTRY

The genomic relationships among the cultivated *Brassica* species are well known (Downey and Rimmer 2003). Three diploid species, *B. rapa*, *B. oleracea*, and *B. nigra*, are the primary species with their genomes designated as A ($n = 10$), B ($n = 8$), and C ($n = 9$), respectively. Early cytogenetic studies (188) conducted on the interspecific hybrids of these diploid species established that *B. juncea* ($n = 18$; AB), *B. napus* ($n = 19$; AC), and *B. carinata* ($n = 17$; BC) are amphidiploids resulting from natural crosses between pairs of corresponding species. Extensive studies (Parkash and Hinta 1980; Coulthart and Denford 1982; Palmer 1988; Warwick and Black 1991; Truco and Quiros 1994; Simonsen and Heneen 1995) using cytogenetics such as meiotic chromosome pairing, isozyme patterns, restriction fragment length polymorphism (RFLP) and restriction patterns of chloroplast/mitochondrial DNA have not only verified these genomic relationships among the *Brassica* species but also have elucidated their genomic composition and evolution. These studies reveal that *B. juncea* and *B. carinata* are the recipients of cytoplasm from *B. rapa* and *B. nigra*, respectively, whereas the origin of the *B. napus* cytoplasm is still not known. It is also apparent from these studies that the evolution of three diploid species of *Brassica* involved duplication of chromosomes followed by chromosomal aberrations such as deletion, translocation, and inversion. Phylogenetic relationships as illuminated by molecular analysis of Brassica genomes also pointed toward two evolutionary pathways—one leading to the origin of *B. rapa* and *B. oleracea* and the other leading to the origin of *B. nigra*. Various gene mapping studies (Song et al. 1990, 1991; Truco et al. 1996; Quiros 1999; Inaba and Nishio 2002; Ananga et al. 2008) using molecular markers have thrown some light on the intragenomic and intergenomic homoeology of the chromosomes of the A, B, and C genomes of *Brassica* species. It appears that all three genomes share homologous regions; however this colinearity is confined to some chromosomal regions only and broken for most other chromosomal segments resulting in complex relationships within and between chromosomes of the three diploid species. Based on gene marker arrangements on the conserved chromosomal regions it can be concluded that the C genome is the possible progenitor of the A genome and the genes have undergone extensive reordering during the evolution of *Brassica* species (Quiros 1999).

Oil content of rapeseed is a complex quantitative trait involving several genes. These direct a battery of physiological processes leading to the accumulation of oil in the seeds. This trait is mainly controlled by genetic makeup of the species/variety but is also significantly affected by both environmental conditions and interaction between the genotype and environment (McVetty and Scarth 2002). Various studies based on classical genetic analyses have been conducted to partition the variation in oil content into genetic and environmental components and determine the type of gene action involved in the inheritance of traits. Some findings indicate that (Han 1990; Grami et al. 1997) oil content had higher heritability (both broad and narrow sense) in summer rape than in winter rape. Others (Chen and Beversdorf 1990; McVetty and Scarth 2002) indicated that both additive and dominance effects govern the inheritance of oil content in *Brassica*. The unpredictable or inconclusive gene action for oil content in Brassica has lead to nonconventional approaches for selecting high oil genotypes in breeding programs. These approaches include the detection of major quantitative trait loci (QTL), exploitation of hybrid vigor, and use of high oil mutants (Uzunova et al. 1995; Burns et al. 2003; Zhao et al. 2005; Delourme et al. 2006; Qiu et al. 2006; Gao et al. 2007).

The quality and utility of rapeseed oil is mostly determined by its fatty acid composition, and considerable research efforts have been directed to understand the genetic control of major FA (erucic acid, oleic acid, linolenic acid, palmitic acid, etc.) and the inheritance of genes involved in their synthesis and accumulation. The major biosynthetic pathways (Murphy 1999) and the genes/ enzymes involved in the FA synthesis (Broun et al. 1998), which have been studied or manipulated for FA accumulation in rapeseed oil include plastid specific genes/proteins responsible for increasing the carbon-chain length using acetyl-coA as base; a group of genes/enzymes (desaturases) located in endoplasmic reticulum responsible for the production of double bonds; thioesterases enzymes in the cytosol for carbon chain termination and genes controlling the esterification of glycerol and FA molecules.

Erucic acid (C22:1) is a long-chain FA and requires C18:1-CoA as a substrate for its synthesis, which is catalyzed by the beta ketoacyl-CoA synthase (KCS) enzyme. It was reported (Barret et al. 1998; Han et al. 1998) that alleles at two genetic loci encode for the KCS, and two KCS genes also cosegregate with two major loci controlling erucic acid levels found on independent linkage groups of *B. napus* (Gupta et al. 2004). These results substantiate the genetic analyses (Chen and Beversdorf 1990; Luhs et al. 1999) of erucic acid content in three amphidiploid *Brassica* oilseed species, which indicated that C22:1 level is controlled by two genes with additive effects. The mapping studies searching for QTL for erucic acid also confirmed the presence of two major loci controlling the erucic acid levels in *Brassica* species (Mahmood et al. 2003; Gupta et al. 2004). However, two additional genetic loci were identified, which influence the erucic acid level in *B. carinata* (del Rio et al. 2003) and offer a novel source (other than the KCS gene) for manipulating the erucic acid content in *Brassica* species (Scarth and Tang 2006).

Because of their importance in determining the oil quality of *Brassica* seed oil, the genetic control of monounsaturated fatty acid (MUFA, 18:1) and polyunsaturated fatty acids (PUFA, 18:2 and 18:3) has also been a subject of interest. Gene mapping and gene cloning studies in *B. napus, B, rapa* and *B. juncea* (Sharma et al. 2002; Tanhuanpaa and Schulman 2002; Laga et al. 2004) have shown that allelic variation at two loci controlling the expression of the FA desaturase (fad2) gene are associated with varying levels (up to 32 %) of C18:1. Two genetic loci with additive effects control the level of C18:3 in *B. napus* (Scarth 1995), and were mapped close to two FA desaturase (fad3) genes (Tanhuanpaa and Schulman 2002). Other studies have shown the presence of more than one genetic loci (Mahmood et al. 2003) or minor genes (Pleines and Friedt 1989) influencing the level of C18:3 in *B. rapa* and *B. juncea*.

The exploitation of DNA polymorphism to develop linkage maps for detecting or tagging the genetic loci influencing economically important traits is routinely done in agricultural crops. *Brassica* species offer a high degree of DNA polymorphism, making them well suited to this modern approach for the development of molecular linkage maps. Several linkage maps have been constructed in *Brassica* oilseed species based on a variety of molecular markers (RFLP, AFLP, RAPD, SRAP, SNP, SSR, STS, SINE, ACGMs, microsatellites) of mapping populations such as doubled haploid, F_2, substitution lines, recombinant inbred lines, and backcross inbred lines (for review see Quiros 2003; Quiros and Paterson 2004; Snowdon and Friedt 2004; Mikolajczyk 2007). At least 15 linkage maps of *Brassica* species (Lakshmikumaran and Srivastava 2003) with a conservative estimate of 1000 loci (Quiros and Paterson 2004) have been described and most of them were constructed using *B. napus* populations.

Researchers have also attempted to use markers common to several *Brassica* species (Sun et al. 2007) and markers targeting homologs of defined genes of *A. thaliana* (Qiu et al. 2006) to align oilseed rape maps and maps between *Brassica napus* and *A. thaliana*. Originally, a common nomenclature to represent the linkage groups across species included *B. napus* linkage groups N1-N10 (A genome) corresponding to *B. rapa* linkage groups R1-R10, and linkage groups N-11-N19 (C genome) corresponded to *B. oleracea* O1-O9 (Kim et al. 2006). However, a new nomenclature was recently adapted based on the recommendation of The Steering Committee of the Multinational *Brassica* Genome Project (Suwabe et al. 2006). Based on this, N1-N10 and R1-R10 were replaced by A1-A10, and C1-C9 designation was used in place for O1-O9 and N11-N19 whereas B1-B8 was used to represent *B. nigra* (B genome). In spite of the extensive straight or comparative genetic mapping in *Brassica* species and *Arabidopsis* (for resources see http://*Brassica*.bbsrc.ac.uk; http://www. tigr.org/tdb/e2k1/bog1/) at the genome and microstructure levels, including the bacterial artificial chromosome (BAC) based physical mapping, a comprehensive integration of maps across species and genera is still in its infancy. There is a need to expand the marker analysis at the microstructure level (Suwabe et al. 2008).

The power of the high density genetic maps lies in their use to detect, tag or even clone the genes or quantitative trait loci (QTL) for economically important traits by their association with easily scored markers. There are several examples in oilseed Brassica (Snowdon and Friedt 2004) where genetic maps/markers have been used to discover genes/QTL and to clone genes for disease resistance, oil content/quality, abiotic stresses, male sterility and morphological traits. QTL associated with oil content in oilseed rape have been identified using different populations and different mapping methods (Burns et al. 2003; Zhao et al. 2005; Delourme et al. 2006; Mahmood et al. 2006). The number of QTL involved in oil content has ranged from 1 to 18 in these reports. Moreover, these studies revealed that a single QTL could explain from 1.2% to 15.7% of the phenotypic variance, and collectively the detected QTL could explain up to 51% of the total phenotypic variance for oil content (Ecke et al. 1995), or explain up to 80% of oil content variation based on additive effects of QTL and additive x additive epistasis (Zhao et al. 2005).

QTL for FA composition have also been identified on genetic maps of the *B. napus* genome and reported in several articles (Ecke et al. 1995; Burns et al. 2003; Zhao et al. 2005; Zhang et al. 2007). Although several loci (2–8 QTL) for almost all important FA in Brassica oil (C16:0, C18:0, C18:1, C18:2, C18:3, C20:1, C22:1) were reported in these studies, the results vary widely, probably due to the different genetic materials and marker systems used. This is reflected in the broad range of phenotypic variability for FA which could be ascribed to their linkage with molecular markers. For example, 8% of linolenic acid (C18:3) was significantly associated with random amplified polymorphic DNA (RAPD) markers in *B. napus* as compared to 73.5% of the linolenic acid in *B. rapa* (Tanhuanpaa and Schulman 2002). As with other crop species, *B. napus* has benefited from advances in plant genomics, plant biotechnology and genetic engineering in the past 20 years.

The technologies in these disciplines have been efficiently used in *Brassica* due to their relative ease of tissue culture and transformation as compared to other dicots. Microspore-derived doubled haploids are routinely produced in *B. napus* (Zhao et al. 1996; Murphy and Scarth 1998) to produce homozygous lines for quick transfer of traits across the genotypes or species. Somatic hybridization or protoplast fusion has been successfully used to conduct wide crosses for transfer of disease resistance (Hu et al. 2002), to produce male sterile lines (Liu et al. 1999) and to produce asymmetric hybrids (Yamigashi et al. 2002). Creation of useful genetic variability for breeding in Brassica has been achieved by induced mutation (Kott et al. 1996) and somaclonal variation (Hoffmann et al. 1982). However, most progress for trait improvement/transfer in Brassica has been achieved by transformation technologies. *B. napus* is predominantly used for genetic transformation and herbicide resistance (HR) is the most prominent trait for improvement. Herbicide (glufosinate, glyphosphate, etc.) resistance genes have been successfully transferred into commercial canola varieties through transformation.

Transformation of *Brassica* for modification of FA composition is another area where interesting progress has been made. The targeted genes mostly encode for enzymes (ACCase, KAS, acyl-ACP thioesterases, destaurases, elongases, acyltransferases) involved in FA biosynthesis pathways, and are introduced into *Brassica* not only from related and unrelated plant species but also from yeast, bacteria and mammals (Scarth and Tang 2006). Some notable achievements include the elevation of oleic acid (C18:1) levels up to 89% in *B. napus* and up to 73% in *B. juncea* for plants transformed with desaturase sense or antisense genes (Sivaraman et al. 2004); enhancement in the ratio of C18:2/ C18:3 and C18:1/C22:1 in *B. juncea* transgenic lines engineered with a novel thioesterase from *Diploknema butyracea* (Sinha et al. 2007); production of high (<40%) gamma-linolenic acid (GLA) canola by the introduction of delta-12-destaurase genes from the fungus *Mortierella alpine* (Liu et al.

2001), and up to 40% GLA increase in *B. juncea* transformed with *Pythium irregulare* Buis. delta-6-desaturase gene (Hong et al. 2002); reduction (>3.4%) or elevation (up to 68%) of saturated FA (C16:0, C18:0) in *Brassica* by altering the expression of acyl-acyl carrier protein (KAS), desaturase and thioesterases (Facciotti et al. 1999; Dehesh 2004); production of super-high erucic acid (C22:1) rapeseed with an approximately 10% increase in erucic acid levels in rapeseed plants transformed with the yeast *FAE1* gene (Katavic et al. 2000); and a transcriptome/metabolome analysis of *B. napus* prototypes revealing an increase in TAG (triacylglycerol) accumulation in transgenic DGAT1 (diacylglycerol transferase 1) plants (Sharma et al. 2008).

18.2.5 CULTIVAR DEVELOPMENT

The widespread cultivation of Brassica oilseeds throughout the world as one of the three most important sources of vegetable oils is a result of plant breeders' successes in developing adapted cultivars with reduced levels of erucic acid in the oil and glucosinolates in the meal. The *B. napus* rapeseed cultivars developed before 1970 (Stefansson and Downey 1995) had high concentrations of both erucic acid (>40%, C22:1) in the oil and glucosinolate (>100 μM/g air-dried meal), and were considered potentially harmful for humans and animals (Eskin et al. 1996). However, the status of *B. napus* cultivars changed when plant breeders successfully introduced low levels of C22:1 from the *B. napus* forage cultivar Liho from Europe into adapted cultivars by backcrossing (Stefansson et al. 1961; Stefansson and Downey 1995). This was followed by the release of the first low erucic acid (<5%), high glucosinolate *B. napus* and *B. rapa* cultivars such as Oro, Zephyr and Span (Stefansson and Downey 1995). Although these cultivars produced high quality edible oil, the nutritional quality/value of the meal was limited by the high concentration of glucosinolates. The discovery of a *B. napus* cultivar Broonowski in Poland with a low glucosinolate concentration presented the opportunity to overcome this barrier. The breeders at the University of Manitoba (Canada) released world's first *B. napus* cultivar (Tower) with low erucic acid (<5%) and low glucosinolate (<30 μM/g) concentration in 1974 (Stefansson and Downey 1995). Traditional breeding was then used to transfer the double-low traits (low erucic acid and low glucosinolate) to both *B. napus* and *B. rapa* genetic backgrounds and release hundreds of double-low (canola) cultivars (McVetty et al. 2008).

Canola cultivars derived from *B. napus* and *B. rapa* are now defined as having <2% erucic acid and <30 μM total glucosinolates per gram of oil-free meal at 8.5% moisture (http://www.canola-council.org). According to The Canola Council of Canada new standards are being set for Canadian canola cultivars which will require the cultivars to have less than 1% erucic acid and less than 18 μM total glucosinolates, and triple low (<2% C:22:1, <30 micromoles glucosinolates, <2% acid detergent lignin content) canola cultivars are also a future goal for canola breeders. The canola standard was extended to *B. juncea* in the early 2000 to develop canola quality mustard and the first Canadian *B. juncea* canola commercial cultivars (Arid and Amulet) were released in 2002 in Canada although development of canola quality *B. juncea* started in Australia as early as 1981 (http://www.canola-council.org). Polyunsaturated FA (C18:2, C18:3) are nutritionally beneficial for human health, but studies have shown that reducing the linolenic acid (C18:3) concentration in canola from 10% to 3% can improve the canola oil stability, leading to a longer shelf life (Tanhuanpaa and Schulman 2002). Similarly, it is known that oils with a high oleic acid (C18:1) and low linoleic acid (C18:3) combination show higher oxidative stability at high temperatures and thus produce less undesirable products during deep frying (Fitzpatrick and Scarth 1998). Efforts to develop high-oleic-low-linolenic acid canola cultivars led to the release of world's first low linolenic acid (>3%) cultivar "Stellar" by the University of Manitoba in 1987 (Scarth et al. 1987). The commercial cultivation of Stellar and other low-linolenic canola cultivars (Apollo and Allons) remained limited due to their less than satisfactory agronomic performance (Stefansson and Downey 1995; Scarth et al. 1997). The development of agronomically desirable double-low canola cultivars with very high oleic acid (70–90%) and reduced linolenic (<3%) concentrations have been achieved and several high oleic, low linoleic cultivars are now are commercially grown in Canada (http://www.canola-council.org).

Although canola cultivars were developed to produce high quality edible oil with low to zero erucic acid (C22:1), development of high erucic acid rapeseed cultivars (known as HEAR) have progressed in parallel to produce oil for a large number of erucic-acid-based industrial applications (McVetty et al. 2008). *B. napus* cultivars with seed oil containing up to 55% erucic acid have been developed using traditional breeding techniques (McVetty et al. 1999) with 'Reston' being Canada's first high erucic acid (40–45%), low glucosinolate rapeseed cultivar. It was released in 1982 and was grown for several years with commercial success (Stefansson and Downey 1995). The current HEAR cultivars with erucic acid levels of 50–55% or more are being developed for commercial cultivation by a few breeding organizations around the world [LimaGrain Australia, Danisco Seeds Denmark, the University of Idaho, the United States, the University of Manitoba Canada), although germplasm resources having up to 60% erucic acid are available (McVetty and Scarth 2002)].

18.3 ENERGY BALANCE OF *BRASSICA* OILS

18.3.1 ENERGY BALANCE

The major use of rapeseed oil for energy, as discussed earlier, is in the production of biodiesel. However, the feasibility of using vegetable oils, including rapeseed/canola oil for biodiesel production depends on a positive energy return for the energy used to produce the biodiesel. This is known as the energy balance. For a biodiesel production system, the energy balance can be determined by studying the relationship between the output per unit of biodiesel (energy produced) versus the input per unit of biodiesel (energy consumed) calculated for each unit of product or byproducts (Mootabadi et al. 2008). For this purpose, a life-cycle analysis according to the standards of the International Standard Organization (ISO 14040-14049) is used to measure the impact of potential factors on the product life-cycle. The factors considered are the energies required to produce the raw materials, those used in the production, direct consumption, utilization of waste/byproducts etc. Because rapeseed methyl ester (RME) is the commonly used biodiesel obtained from rapeseed, life-cycle analysis is generally done for RME although REE (rapeseed ethyl ester) has also been subjected to life-cycle analysis (Janulis 2004). In a 2004 study by Janulis, the energy consumption for the production of RME was classified into three main categories including agriculture, oil pressing and transesterification. The total energy consumption to produce 1 t of RME was then calculated by adding the energy consumption in all of the subcategories for agriculture (agromachinery/equipment, fuel and oils, electricity, seeds/chemicals), oil pressing (electricity, equipment) and transesterification (electricity, equipment, chemicals). Similarly, the energy content in products (ester, glycerol, straw, oil cake) for 1 t of ester production was added to estimate the total energy produced. Based on these calculations, the energy balance for rapeseed with a seed yield productivity of 3 t/ha under Lithuanian conditions came out to be 1.43 for RME and 1.62 for REE.

The energy balance or energy ratios for RME are not static values and could greatly vary depending not only on the diverse geographic, social, cultural and economic factors but also on the prevailing conditions within a region. For example, a comparative analysis of two previous studies (Walker 2004) revealed large differences in energy ratios for RME under good and poor production/cultural conditions, and also on the basis of outputs included (RME only, RME + rape meal, RME + rape meal + glycerol, RME + rape meal + glycerol + straw). It was also noted that glycerol and straw did not have significant impact on energy balance. Other studies have placed more importance to the value of byproducts and rapeseed straw was found to have a large impact on energy ratios in some cases (Batchelor 1995; Janulis 2004). Other studies indicated that energy balance is improved when biofertilizers are used, seed drying (dehydration) is substituted with chemical conservation and an energy-efficient biotechnological method of oil extraction is used (Mootabadi et al. 2008). A study conducted in Germany (Rathke and Diepenbrock 2006) revealed that the energy balance of winter oilseed rape was influenced by different nitrogen management strategies and the most favorable N rate for maximizing energy gain was 240 kg/ha.

Energy balance values help to clarify the degree to which a vegetable oil-based biodiesel is renewable. However, the life-cycle analysis for energy balance is usually reduced to a single number or a ratio, and the "renewability" of biodiesel is dependent on whether that number or ratio is less than or greater than one. A value greater than one indicates that more energy is obtained from the product than was used to create it, making it a more sustainable option than those with energy values less than one. A heated debate has been generated regarding the validity or power of the energy balance ratios to reflect the economic significance and renewability of crop oil-based biofuels/biodiesel. The conflicting arguments centre on the methods of energy balance calculation (Hill et al. 2006) and the calculation of input costs (Pimental and Patzck 2005). Some of the considerations (Wesseler 2007) include whether the cost of production of crops should be included in the life-cycle analysis considering it is a fixed cost regardless of the end use, and how to set the boundaries for input cost and energy so that input energy requirements are not inflated. Some believe that the renewability of a biofuel should be adjusted based on the amount of nonrenewable fuel used to produce biofuel for such energy costs as field cultivation, fuel processing, transportation and even the energy associated with the construction of processing facilities. One researcher (Dale 2007) went to the extent of challenging the very concept of net energy balance considering it wrong and misleading. He proposed to develop and use energy matrices based on the ability of a biofuel to displace gasoline/diesel, the contribution of biofuel in producing greenhouse gases per km driven, and the net gain in land use efficiency associated with biofuel production.

A general mass balance for RME production is presented by Walker 2004 from data obtained in Western Australia. According to this, 0.41 t of crude oil and 0.58 t of meal can be produced from 1 t of rapeseed with an oil content of 43%. After refinement of the crude oil, with negligible losses during refinement, 0.38 t of RME can be produced, which can generate 432 L of biodiesel using a refined oil density of 0.88 g/ml. Under this scenario, the biodiesel produced per hectare will amount to 1,300 L assuming a seed yield of 3 t/ha. However, reported results for the production RME under different agronomic and ecological conditions vary considerably. A study conducted in Poland (Jankowski and Budzynski 2003) revealed that the energy obtained from 1 ha of winter oilseed rape cultivation was equivalent to the energy yielded by 3.5 t of diesel oil. It was also found that the highest energy efficiency for this region could be obtained through the cultivation of white mustard instead of winter oilseed rape.

Biodiesel production from canola did not seem to be sustainable in Western Australia (WA) based on a negative net energy return (http://www.chemeca2009.com/abstract/351.asp). A study estimated that more than 0.8 million ha of land was required to replace 15% of the diesel consumption in WA with biodiesel. This would require twice the total land under cultivation for oilseeds in that state. An environmental impact study prepared for the Canola Council of Canada (http://www.canola-council.org) determined that a total of 1 L of petroleum-based diesel is consumed during all of the stages of crop production, biodiesel production and transportation to generate a net 2.5 L of biodiesel. This is not cost-effective because it takes more energy to produce biodiesel than to make diesel. However, it was expected that the life-cycle assessment of canola's energy balance for biodiesel would improve with an increase in canola oil yield per acre. The mechanisms for this increase would likely be due to development of new higher yielding canola varieties. Improved farm practices such as reduction the use of fertilizers to produce the canola would also help to improve the energy balance for this crop.

18.3.2 COMPARISON WITH OTHER VEGETABLE OILS

Vegetable oils can be divided into three main categories (Gunstone et al. 2004) based on the nature of their source plants. The first category includes the oils obtained as byproducts of crop plants such as cotton seed oil (main use fiber), corn (main use as grain) and soybean (main use as high protein meal). Tree crops such as palm, coconut and olive would comprise the second category where oil is produced regularly from the seeds of slow-growing but long-lived plants. Rapeseed/canola, along with other oilseed crops (sunflower, peanut, sesame, castor, linseed, tobacco, etc.), falls into the third and the largest category of annual crop plants producing oil.

In terms of the annual production of vegetable oils throughout the world, rapeseed oil ranks among the four major oils with annual oil productions of approximately 30, 25, 13, and 8 million MT per year for soybean, palm, rapeseed and sunflower, respectively (www.oilworld.biz; annual 2002). The minor vegetable oils, with annual production ranging from 0.4 to 5.3 MT per year, include (in descending order) groundnut, cotton seed, coconut, olive, corn, sesame, linseed and castor. There are many other vegetable oils which may not rank high in terms of global oil production but may have production potential and/or possess unique chemical properties. These include oils from avocado, brazil nut, *Camelina*, coffee, hemp, *Lesquerella*, safflower, rice bran, opium poppy, jojoba, jatropha, and macadamia nut.

Rapeseed also ranks high (among the top eleven) when compared to other vegetable oils in terms of oil yield per unit of area. Table 18.1 (derived from www.journeytoforever.org) presents a comparison of the vegetable oils with highest oil yield capacities per hectare, and is presented in descending order.

The amount of extractable oil also varies widely among the major oil crops and rapeseed is placed near the top (sixth) with a capacity to produce 37 kg of oil from 100 kg of rapeseed (www.journeytoforever.org). This capacity is higher than soybean, sunflower, palm and cotton which can produce 13–36 kg of oil from 100kg of seeds, and lower than the 42–60 kg oil/100 kg of seed for castor, coconut, sesame, and peanut.

Rapeseed and other vegetable oil crops are given the title of "bioenergy crops" mainly due to the use of the seed oil from these crops for the production of biodiesel. Although capacity to produce seed oil is an important consideration when comparing the merits of crops for biodiesel production,

TABLE 18.1
Conservative Estimates of the Crop Oil Yields, Which Can Vary Widely

Rank	Oil Crop	Oil Yield (L/ha)	Rank	Oil Crop	Oil Yield (L/ha)
1	Oil palm	5950	19	Camelina	583
2	Coconut	2689	20	Mustard seed	572
3	Avocado	2638	21	Coriander	536
4	Macadamia nut	2246	22	Pumpkin seed	534
5	Jatropha	1892	23	Euphorbia	524
6	Jojoba	1818	24	Hazelnut	482
7	Pecan nut	1791	25	Flax	478
8	Castor bean	1413	26	Coffee	459
9	Olive	1212	27	Soybean	446
10	Rapeseed	1190	28	Hemp	363
11	Opium poppy	1163	29	Cotton	325
12	Peanut	1059	30	Calendula	305
13	Cocoa	1026	31	Kenaf	273
14	Sunflower	952	32	Lupine	232
15	Tung oil	940	33	Oats	217
16	Rice	828	34	Cashew	176
17	Safflower	779	35	Corn	172
18	Sesame	696			

Source: Data from Addison, K., Straight vegetable oil as diesel fuel. 2001. http://www.journeytoforever.org/biodiesel_svo.html.

other factors also influence the suitability of a crop for bioenergy use. The quality and functionality of the diesel fuel is heavily influenced by the physical/chemical properties and fatty acid composition of the vegetable oils. The cetane number, viscosity, density, flashpoint, pour and cloud points and iodine values are all properties of diesel fuel that are affected by the source vegetable oil (Lang et al. 2001; Ramdhas et al. 2004; Al-Zuhair 2007; Canacki and Sanli 2008). Cetane number, a measure of fuel ignition quality, increases with increasing FA chain length and decreasing unsaturation/branching. A cetane value of between 40 and 60 is considered satisfactory for ignition quality. Viscosity describes a fuel's heaviness. High viscosity interferes with the fuel injection process, and it increases with increasing carbon chain length of the fatty acids and decreases with the increasing unsaturation (double bonds). Density indicates the fuel's density or compressibility. An increase in density improves fuel injection, and density is raised by the chain length/saturation of the fatty acids. A fuel's flashpoint is the temperature at which fuel starts to burn when in contact with fire. This is a critical temperature from an engine safety point of view. It is usually affected by the degree of FA saturation. Pour and cloud points are indicators of a fuel's reaction to low temperatures. They predict the cold flow quality of the fuel and need to be below the freezing points of the vegetable oils used. Feed stocks with large amounts of saturated FA have higher cloud and pour points. Finally, the iodine value is a measure of the number of double bonds in oils and determines fuel stability. Higher iodine values cause the formation of corrosive acids and deposits (oxidation) in the engines. Table 18.2 provides the comparison of the above properties in selected vegetable oils including the rapeseed, whereas the FA compositions of some vegetable oils are compared in Table 18.3.

According to The Canola Council of Canada (http://www.canola-council.org), canola's lowest level of saturated FA and low iodine values make it one of the best feedstocks for biodiesel, imparting greater cold flow properties and oxidative stability than those of many other vegetable oils.

18.3.3 Economic Implications

The economic implications of rapeseed/canola as a bioenergy crop depends on the economic viability of producing methyl or ethyl esters (biodiesel) using rapeseed as a feedstock. Therefore, the main thrust of an economic analysis is always to compare the cost of biodiesel production with the cost of diesel fuel, and such an analysis is highly affected by the fluctuations in market trends for both feedstock and diesel fuel. For example, the cost of diesel fuel in September 2007 was 733 U.S.$/t whereas the cost of RME (rapeseed derived biodiesel) was up to 3 times higher (1,060 U.S.$/t), but with the increase in fuel prices (1,017 U.S.$/t) this price gap narrowed

TABLE 18.2
Properties of Major Vegetable Oils as Related to Biodiesel

Oil Crop	IV	VS	CN	FP	CP	PP
Rapeseed	94–120	37.3	37.5	246	–3.9	–31.7
Soybean	117–143	33.1	38.1	254	–3.9	–12.2
Sunflower	110–143	34.4	36.7	274	7.2	–15
Palm	35–61		42			
Cotton	90–119	33.7	33.7	234	1.7	–15
peanut	80–106	40	34.6	271	12.8	–6.7
Safflower	126–152	31.6	36.7	246	–3.9	–3.17

Source: Derived from Canakci, M. and Sanli, H., *J Indust Microbiol Biotechnol Special Issue*, pp 1–23, 2008; Han, J., *Advances in Plant Lipid Research*. Secretariado de Publicaciones de la Universidad de Sevilla, Sevilla, Spain, pp 665–668, 1998.

IV, iodine value; CN, cetane number; VS, viscosity (mm^2/s); FP, flash point (°C); CP, cloud point (°C); PP, pour point (°C).

TABLE 18.3
FA Composition (wt %) of Major Vegetable Oils

Oil Crop	12:0	14:0	16:0	18:0	18:1	18:2	18:3	22:1
Rapeseed	–	1.5	1–4.7	1–3.5	13–38	9.5–22	1–10	40–64
Soybean	–	–	2.3–11	2.4–6	22–31	49–53	2–10.5	–
Sunflower	–	–	3.5–6.5	1.3–5.6	14–43	44–68.7	–	–
Oil Palm	–	0.6–2.4	32–46.3	4–6.3	37–53	6–11	–	–
Cotton	–	0.8–1.5	22–24	2.6–5	19	50–52.5	–	–
peanut	–	0.5	6–12.5	2.5–6	37–61	13–41	–	1
Safflower	–	–	6.4–7	2.4–29	9.7–13.8	75.3–80.5	–	–

Source: Derived from Canakci, M. and Sanli, H., *J Indust Microbiol Biotechnol Special Issue*, pp 1–23, 2008; Han, J., *Advances in Plant Lipid Research*. Secretariado de Publicaciones de la Universidad de Sevilla, Sevilla, Spain, pp 665–668, 1998.

considerably in September 2008 and RME prices (1,415 U.S.$/t) came close to diesel fuel prices (Pinzi et al. 2009).

The cost of biodiesel is determined by taking the difference between the total cost to produce RME and the total income from the byproducts (Walker 2004) where total cost refers to capital costs, operating costs, labor, power, annual maintenance, capital interests, purchase of feedstock, miscellaneous costs relating to administration/management, and total income including those from the sales of the meal and the glycerol byproduct. However, studies (Zhang et al. 2003; Haas et al. 2005) have shown that a major portion (<85%) of the cost for biodiesel production is attributed to the cost of feedstock. Walker 2004 estimated that the per liter cost of RME in a 22-million-L facility under U.K. conditions was approximately £0.31 or approximately 0.50 U.S. $. It was concluded that the production of biodiesel was not economical when compared to the cost of mineral diesel production (~£0.10 or 0.16 U.S.$/L) at that time.

The EU and Canada are the main regions where food-grade canola is used as feedstock for biodiesel production, however, to produce economically competitive biodiesel in these regions it has been necessary to take measures such as application of tax credits, optimization of processes to maximize the RME yield and use of alternate lower cost feed stock such as nonedible *Brassica* (*B. carinata*) oil. According to The Canola council of Canada (http://www.canola-council.org), in spite of the economic noncompetitiveness of biodiesel with petrodiesel fuel, a canola-based-biodiesel industry would benefit Canadians by creating more than 700 direct jobs in crushing, processing and food production industries. An investment in biodiesel infrastructure will also enhance economic activity in construction and supporting industries. Additional market development for the byproducts would help the economics.

If the oil supply dwindles or regulations insist on the use of biodiesel, it will no longer matter if biodiesel is competitively priced or not. The required infrastructure will already be available if steps are taken now. Some industries may also voluntarily choose to use biodiesel in spite of the economics because of public opinion, concerns about emissions, company green policies or desire to use local resources. Additional market development for the byproducts would help the economics of biodiesel.

18.4 FOOD-VERSUS-FUEL CONSIDERATIONS

A major concern associated with the production of biodiesel is its reliance on vegetable or canola oil. There is a perceived competition between food and fuel uses of the edible oils, posing a putative moral dilemma. The issue is being debated in scientific journals as well as in popular press but there are very few actual scientific studies that help to dissect out the socioeconomic implications of using food crops for both food and fuel.

The main argument in the food-versus-fuel debate is that there are millions of underfed people in the world, mostly in third-world (developing) countries; therefore it is morally wrong to divert a portion of food crop production (mostly in the developed world) to biofuel production while people elsewhere in the world go hungry. But, it is counterargued that, in realistic terms the world produces enough food to feed everyone adequately. Approximately one half of all food produced in the world is wasted—either due to spoilage before use or it is not eaten after it is served. A small reduction in food spoilage or post serving waste would see everyone in the world potentially provided with adequate food. Total food crop production is not the issue—fair and equitable distribution is the real issue.

From the perspective of a developed country such as Canada, reasons could be cited (http://www.canola-council.org) to support that increased demand for canola oil to produce biodiesel will not reduce the availability of canola for food use. For example, 1.3 million t of additional canola seed is required to fulfill the Canadian government's requirement of 2% renewable fuel (biodiesel) added to the normal diesel fuel. This requirement will not affect the supply of edible canola oil because the carryover (unsold) volume of canola was 1.59, 2.02, and 1.58 MT in the years 2004/2005, 2005/2006, and 2006/2007, respectively. It is also pointed out that farmers in Canada produce 75% more canola than is required to meet the needs of Canadian consumers. Therefore, an increase in local consumption of canola is desirable which can be achieved by developing a new market for the crop, such as biodiesel.

Another concern in the food-versus-fuel issue is that bioenergy oil crops not only encourage the destruction of habitat (e.g., clearing of South East Asia forests for palm oil) but also causes the displacement of agriculture production toward grasslands or other "marginal" uncultivated areas causing a loss of biodiversity, GHG savings and local land rights (http://www.dft.gov.uk/rfa/_db/_documents/Report_of_the_Gallagher_review.pdf). Indonesia and Malaysia together meet 88% of the global palm oil demand, and as more palm oil production is diverted to the production of biodiesel, it is likely that the area cropped to oil palm will expand (Cassman and Lska 2007). This expansion may come at the expense of diverse grasslands and forests, disturbing the ecosystems. The land-use change will also result in loss of most of the above ground (vegetation and litter) and below ground (soil, roots) carbon in forests, savannas and grasslands, and the carbon payback in Malaysia (palm to biodiesel) and the United States (soybean to biodiesel) could be as high as 38 years (The Gallagher Review). However, there are contrary views and arguments in this regard emphasizing that land displacement can be avoided by increasing the crop yields per unit area with advanced technologies (better management for inputs, efficient water and land use, high yielding varieties, etc.) in agriculture production systems. According to one study (Wolk et al. 2003), 55% of the total land area under agriculture throughout the world is enough to meet the global food supply to 2050 using the modern agricultural production practices, which leaves the remainder of the agriculture land (45%) to produce crops for non-food uses. Canada is an example where the challenge of increased demand (12–15 MT) for canola (due to its use as fuel crop and resulting attractive prices) has been met not only by a modest increase in the number of acres planted to canola but also by an elevation of canola yields from an average of 26–30 bushels/acre as a result of improved high yielding varieties (http://www.canola-council.org).

Increases in food prices such as for seed oil are also considered to be driven by a rise in the value of commodities as feedstock for fuel. For example, the high demand for biodiesel in EU has led to a large increase in the price of rapeseed oil. However, the consumers in developed countries (EU, the United States, Canada, Japan) generally can afford the small increases in food (canola oil) costs. Even in the poorest regions of the world, the production of biodiesel can in fact improve the economic welfare of the people. One example is Peru where farmers are producing biodiesel from canola for domestic use, and the other is cultivation of jatropha for biodiesel production in parts of Africa, creating new sources of income for farmers (http://www.canola-council.org). Alternate outlets/markets for food crops such as provided by the use of biofuel positively affects average crop prices and encourages more production to meet market needs. These are positive changes for producers. Food crop use is primarily a local area business so crops grown locally are processed into biofuels and then used locally as well. This can be very beneficial for the environment. The fact that biofuels will be more expensive (without distorting government subsidies) than petrofuels is also favorable, because it will encourage the conservation of fuel and the development of more fuel-efficient vehicles.

The logical place to start with crop production for biofuel is the major crops already grown (such as canola/rapeseed) because all of the required knowledge and infrastructure needed to produce these crops is already in place. In the longer term, a switch to non-food crops, especially low input and perennial crops will eliminate the current controversy regarding growing our major traditional food crops for fuel. Modern techniques such as use of GIS-based yield and suitability mapping (Lovett et al. 2009) to study the bioenergy generation potentials of perennial biomass crops and land use implications at regional scales could also aid in countering the issues in food versus food issue.

18.5 FUTURE PROSPECTS

The end of the petro oil era (not more than 30 years from now) will see a major shift in energy production and use in the world. There will be a much greater reliance on biofuels for transportation. Biodiesel is the most logical biofuel based on bioenergetic considerations. It has a positive energy balance, and diesel engines are the most efficient of the internal combustion engines. Small diesel-electric cars, which run primarily on stored electric energy from overnight recharging from the electric grid with the diesel motor as a supplementary power source and portable battery recharging source will likely predominate personal urban transportation (combined with biodiesel fuelled buses).

Current vegetable oil crops, new vegetable oil crops, algal oil production and biomass conversion to biodiesel via pyrolysis will all have a role to play in the future of biofuels. Nitrogen fixing crops—current and new ones, and the conversion of existing crops to Nitrogen fixing crops will improve the energy balance of crops. Production of low input marginal land adapted annual and perennial crops for biomass to be converted via pyrolysis to biofuels, especially biodiesel, will predominate. Perhaps perennial oilseed crops (perennial sunflowers or perennial flax for example) can provide a portion of the vegetable oil needed for biodiesel in the future.

Life-cycle analysis, complete energy balance considerations, and full economic analyses of potential bioenergy crops will be done to choose those oilseed crops most likely to provide the environmental benefits we are hoping to achieve.

18.6 CONCLUSIONS

Brassica oilseeds are a major source of vegetable oil in the world. They are widely adapted and readily bred to produce a wide range of edible and industrial oils. They display high seed productivity, high oil productivity and high meal productivity. Canola oil is an excellent feedstock for biodiesel production especially in temperate climates. New higher yielding, higher oil content rapeseed cultivars will provide the oil required to satisfy both edible and nonedible (biodiesel and industrial oil) markets. Oilseed Brassicas will be the predominant oilseeds grown in temperate climates providing food, feed and fuel for the world. The economics of using Brassica oilseeds as feedstock for energy (biodiesel) can be improved by increasing the oil content (<50%), increasing the seed yield, and by further development of oilseed *Brassica* species including *B. juncea* and *B. carinata*, and thereby increasing oil productivity per unit area. The development of abiotic and biotic stress tolerant Brassica oilseed cultivars especially when combined with enhanced fertilizer and water use efficiency traits will greatly improve the economics of Brassica oilseed production for all uses including energy.

REFERENCES

Aatola H, Larmi M, Sarjovaara T (2009) Hydrotreated vegetable oil (HVO) as a renewable diesel fuel: Trade-off between NOx, particulate emission, and fuel consumption of a heavy duty engine. *SAE Int J Engines* 1:1251–1262

Al-Zuhair S (2007) Production of biodiesel: possibilities and challenges. *Biofuel Bioprod Biorefin* 1:57–66

Ananga AO, Cebert E, Soliman K, Kantety R, Konan K, Ochieng JW (2008) Phylogenetic relationships within and among *Brassica* species from RAPD loci associated with blackleg resistance. *Afr J Biotechnol* 7:1287–1293

Arvidsson R (2008) Life cycle assessment of biodiesel-hydrotreated oil from rape, oil palm or jatropha, available at http://publications.lib.chalmers.se/cpl/record/index.xsql?pubid = 75983 (accessed July 28, 2009)

Babu AK, Devaradjane G (2003) *Vegetable Oils and Their Derivatives as Fuels for CI Engines: An Overview.* SAE Technical Paper No 2003-01-0767

Barnitt R, McCormick RL, Lammert M (2008) *St. Louis Metro Biodiesel (B20). Transit Bus Evaluation. 12 Month Final Report.* National Renewable Energy Laboratory, available at http://www.biodiesel.org/resources/reportsdatabase/reports/tra/20080701_tra-57.pdf (accessed April 8, 2011)

Barret P, Delourne R, Renard M, Domergue F, Lessire R, Delseny M, Roscoe TJ (1998) A rapeseed *FAE1* gene is linked to the *E1* locus associated with variation in the content of erucic acid. *Theor Appl Genet* 96:177–186

Batchelor, SE, Booth EJ, Walker KC (1995) Energy analysis of rape methyl ester (RME) production from winter oilseed rape. *Indust Crops Prod* 4:193–202

Benge M (2006) *Assessment of the Potential of Jatropha Curcas for Energy Production and Other Uses in Developing Countries* (DOC), available at http://www.echotech.org/mambo/index.php?option = com_docman&task = doc_view&gid = 178 (accessed July 28, 2009)

Berchmans HJ, Hirata HJ (2008) Biodiesel production from crude *Jatropha curcus* seed oil with a high content of free fatty acids. *Bioresour Technol* 99:1716–1721

Bolszoa CD, McDonell VG (2008) Emissions optimization of a biodiesel fired gas turbine. *Proc Combust Inst* 32:2949–2956

Briggs M (2004) *Widescale Biodiesel Production from Algae.* University of New Hampshire, Physics Department, available at http://www.unh.edu/p2/biodiesel/article_alge.html (accessed July 28, 2009)

Broun P, Shanklin J, Whittle E, Somerville C (1998) Catalytic plasticity of fatty acid modifications enzymes underlying chemical diversity of plant lipids. *Science* 282:1315–1317

Burns MJ, Barnes SR, Bowman JG, Clarke MHE, Werner CP, Kearsey MJ (2003) QTL analysis of an intervarietal set of substitution lines in *Brassica napus*: I. seed oil content and fatty acid composition. *Heredity* 90:39–48

Callihan B, Brennan J, Miller T, Brown J, Moore M (2000) *Mustards in Mustards. Guide to Identification of Canola, Mustard, Rapeseed and Related Weeds.* Agricultural Publications, University of Idaho, Moscow, ID

Canakci M, Sanli H (2008) Biodiesel production from various feedstocks and their effects on the fuel properties. *J Indust Microbiol Biotechnol* Special Issue, pp 1–23.

Cassman KG, Lska AJ (2007) Food and fuel for all: realistic or foolish? *Biofuels Bioprod Biorefin* 1:18–23

Chang HM, Liao HF, Lee CC, Shieh CJ (2005) Optimised synthesis of lipase-catalyzed biodiesel by Novozym 435. *J Chem Technol Biotechnol* 80:307–312

Chen JL, Beversdorf WR (1990) Fatty acid inheritance in microspore-derived populations of spring rapeseed (*Brassica napus* L.). *Theor Appl Genet* 80:465–469

Christi Y (2007) Biodiesel from microalgae. *Biotechnol Adv* 25: 294–306, available at http://dels.nas.edu/banr/gates1/docs/mtg5docs/bgdocs/biodiesel_microalgae.pdf. (accessed July 28, 2009)

Coulthart M, Denford KE (1982) Isozyme studies in *Brassica*. Electrophoretic techniques for leaf enzymes and comparison of *B. napus*, *B. compastris* and *B. oleracea* using phosphoglucomutase. *Can J Plant Sci* 62:621–630

Crabbe E, Hipolito CN, Kobayashi G, Sonomoto K, Ishizaki A (2001) Biodiesel production from crude palm oil and evaluation of butanol extraction and fuel properties. *Proc Biochem* 37:65–71

Dale B (2007) Thinking clearly about biofuels: ending the irrelevant "net energy" debate and developing better performance metrics for alternate fuels. *Biofuels Bioprod Biorefin* 1:14–17

Dehesh K (2004) Nucleic acid sequences encoding eta-ketoacyl-ACP synthesis and uses thereof, available at http://www.patentstorm.us/patents/6706950/claims.html (accessed July 28, 2009)

Delourme R, Falentin C, Clouet V, Horvais R, Gandon B, Specel S, Hanneton L, Dheu JE, Deschamps M, Margale E, Vincourt P, Renard M (2006) Genetic control of oil content in oilseed rape (*Brassica napus* L.). *Theor Appl Genet* 113:1331–1345 del Río M, de Haro A, Fernandez-Martinez JM (2003) Transgressive segregation of erucic acid content in *Brassica carinata*. A. Braun. *Theor Appl Genet* 107:643–651

Demirbas A (2008) Studies on cottonseed oil biodiesel prepared in non-catalytic SCF conditions. *Bioresour Technol* 99:1125–1130

Dheu M. Deschamps E, Margale P, Vincourt M, (2006) Genetic control of oil content in oilseed rape (*Brassica napus* L.). *Theor Appl Genet* 113:1331–1345

Downey R, Rimmer K (1993) Agronomic improvements in oilseed *Brassicas. Adv Agron* 50: 1–66

Downny RK (1983) The original description of the Brassica oilseed crops. In: Kramer JKG, Saver FD, Pigden WJ (eds), *High and Low Erucic Acid Rapeseed Oils*. Academic Press, Toronto, Canada, pp 61–83

Ecke W, Uzunova M, Weibleder K (1995) Mapping the genome of rapeseed (*Brassica napus* L.). II. Localization of genes controlling erucic acid synthesis and seed oil content. *Theor Appl Genet* 91:972–977

Eskin NAM, McDonald BE, Przybylski R, Malcolmson LJ, Scarth R, Mag T, Ward K, Adolph D (1996) Canola oil. In: Hui YH (ed), *Edible Oil and Fat Products: Oil and Oil Seeds*. John Wiley, New York, pp 1–96

Facciotti MT, Bertain PB, Yuan L (1999) Improved state phenotype in transgenic canola expressing a modified acyl-acyl-carrier protein thioesterase. *Nat Biotechnol* 17:593–597

FAO/WHO (1993) Reports on "Fats and oils in human nutrition", available at http://www.fao.org/docrep/V4700E/V4700E00.htm (accessed July 28, 2009)

Fitzpatrick K, Scarth R (1998) Improving the health and nutritional value of seed oils. *PBI Bulletin*, 15–19 Jan 1998. Saskatoon NRC-CRC, Canada

Foidl N, Foidl G, Sanchez M, Mittelbach M, Hackel S (1996) *Jatropha curcas* L. as a source for the production of biofuel in Nicaragua. *Bioresour Technol* 58: 77–82

Fukuda H, Kondo A, Noda H (2001) Biodiesel fuel production by transesterification of oils. *J Biosci Bioeng* 92:405–416

Gao M, Li G, Yang B, Qiu D (2007) High density *Brassica oleracea* linkage map: identification of useful new linkages. *Theor Appl Genet* 115:277–287

Giannelos PN, Zannikos F, Stournas S, Loias E, Anastopoulos G (2002) Tobacco seed oil as alternative diesel fuel: Physical and chemical properties. *Indust Crops Prod* 16: 1–9

Glamser S, Widmann B (2001) A system for the use of rapeseed oil based engine oils. In: *Proceedings of the 1st World Congress on Biomass for Energy and Industry*, June 5–9, 2001, Sevilla, Spain. James & James Science Publishers Ltd, United Kingdom, pp 1057–1059

Grami B, Baker RJ, Stefansson BR (1997) Genetics of protein and oil content in summer rape: heritability, number of effective factors, and correlations. *Can J Plant Sci* 57:937–943

Grubinger V (2008) On Farm Energy Case Study (PDF). The University of Vermont Extension, available at http://www.uvm.edu/vtvegandberry/Pubs/Waste%20Vegetable%20Oil%20for%20Greenhouse%20Heat.pdf (accessed July 28, 2009)

Gunstone FD (2004) *The Chemistry of Oils and Fats: Sources, Properties and Uses*. Blackwell Publishing, Oxford, United Kingdom, pp 23–32

Gupta, V, Mukhopadhyay A, Arumugam N, Sodhi YS, Pental D, Pradhan AK (2004) Molecular tagging of erucic acid trait in oilseed mustard (*Brassica juncea*) by QTL mapping and single nucleotide polymorphism in *FAE1* gene. *Theor Appl Genet* 108:743–749

Haas MJ, Foglia TA (2005) Alternate feedstocks and technologies for biodiesel production. In: Knothe G, Krahl J, Van Gerpen (eds), *The Biodiesel Handbook*. AOCS Press, Champaign, IL, pp 42–61

Han J, Lühs W, Sonntag K, Borchardt DS, Frentzen M, Wolter FP (1998) A *Brassica napus* cDNA restores the deficiency of canola fatty acid elongation at a high level. In: Sanchez J et al. (eds), *Advances in Plant Lipid Research*. Secretariado de Publicaciones de la Universidad de Sevilla, Sevilla, Spain, pp 665–668

Han JX (1990) Genetic analysis on oil content in rapeseed (*Brassica napus* L.). *Oil Crop China* 2:1–6

Hatje, G (1989) World importance of oil crops and their products. In: Robbelen G, Downey RK, Ashri A (eds), *Oil Crops of the World*. McGraw-Hill Publishing, New York, pp 1–22

Hedge, IC (1976) A systematic and geographical survey of the world cruciferae. In: Vaughn JG, Macleod AJ, Jones BMG (eds), *The Biology and Chemistry of the Cruciferae*. Academic Press, London, pp 331–341

Hill J, Nelson E, Tilman D, Polasky S,Tiffany D (2006) Environmental, economic and energetic costs and benefits of biodiesel and ethanol fuels. *Proc Natl Acad Sci USA* 103:11206–11210

Hoffmann F, Thomas E, Wenzel G (1982) Anther culture as a breeding tool in rape. II. Progeny analysis of androgenetic lines and induced mutants from haploid cultures. *Theor Appl Genet* 61:225–232

Hong H, Datla N, Reed DW, Covello PS, MacKenzie SL, Qiu X (2002) High level production of gamma-linolenic acid in Brassica juncea using a delta-6- desaturase from *Pythium irregulare*. *Plant Physiol* 129:354–362

Holser RA, Harry-Okuru R (2006) Transesterified milkweed (Asclepias) seed oil as a biodiesel fuel. *Fuel* 85:2106–2110

Huang YS, Pereira SL, Leonar AE (2004) Enzymes for transgenic biosynthesis of long chain polyunsaturated fatty acids. *Biochimie* 86:793–798

Hu Q, Anderson SB, Dixelius C, Hansen LN (2002) Production of fertile intergeric somatic hybrids between *Brassica napus* and *Sinapus arvensis* for the enrichment of the rapeseed gene pool. *Plant Cell Rep* 21:147–152

Ikwuagwu OE, Ononogbu IC, Njoku, OU (2000) Production of biodiesel using rubber (*Hevea brasiliensis*) seed oil. *Indust Crops Prod* 12: 57–62

Inaba R, Nishio T (2002) Phylogenetic analysis of Brassicaceae based on the nucleotide sequences of the S-locus related gene, SLRI. *Theor Appl Genet* 95:400–407

Janulis P (2004) Reduction of energy consumption in biodiesel fuel life cycle. *Renew Energy* 29:861–871

Jaworski J, Cahoon EB (2003) Industrial oils from transgenic plants. *Curr Opin Plant Biol* 6:178–184

Johnston M, Hollowway T (2007) A global comparison of national biodiesel production potentials. *Environ Sci Technol* 41:7967–7973

Jones S, Peterso CL (2002) *Using Unmodified Vegetable Oils as a Diesel Fuel Extender—A Literature Review*, available at http://journeytoforever.org/biofuel_library/uidaho_rawoils.html (accessed July 28, 2009)

Joshi HC, Toler J, Walker T (2008) Optimization of cottonseed oil ethanolysis to produce biodiesel high in Gossypol content. *J Amer Oil Chem Soc* 85:357–363

Kaieda M, Samukawa T, Matsumoto T, Ban K, Kondo A, Shimada Y, Noda H, Nomotos F, Ohtsuka K, Izumoto E, Fukuda H (1999) Biodiesel fuel production from plant oil catalyzed by *Rhizopus oryzae* lipase in a water-containing system without an organic solvent. *J Biosci Bioeng* 88:627–631

Katavic V, Friesen W, Barton DL, Gossen KK, Giblin EM, Luciw T, Zou JAJ, MacKenzie SL, Keller WA, Males D, Taylor DC (2000) Utility of the Arabidopsis *FAE1* and yeast *SLC1-1* genes for improvements in erucic acid and oil content in rapeseed. *Biochem Soc Trans* 28:935–937

Kesari V, Krishnamachari A, Rangan L (2008) Systematic characterization and seed oil analysis in candidate plus trees of biodiesel plan, *Pongamia piñata*. *Ann Appl Biol* 152:397–404

Keskin A, Guru M, Altiparmak D, Aydin K (2008) Using of cotton oil soapstock biodiesel-diesel fuel; blends as an alternative diesel fuel. *Renew Energy* 33:553–557

Kim JS, Chung TW, King GJ, Jin M, Yang TJ, Jin YM, Kim H, Park BS (2006) A sequence-tagged linkage map of *Brassica rapa*. *Genetics* 174:29–39

Klaus T, Kathrin S, Heiner L (2005) Exhaust gas particulate filter systems for rapeseed oil fuelled combined heat and power units. *Landtechnik* 60:268–269

Knothe G (2001) Historical perspectives on vegetable oil-based diesel fuels. *Inform* 12:1103–1107

Knothe G, Dunn RO (2005) Biodiesel: An alternative diesel fuel from vegetable oils or animal fats. In: Erhan SZ (ed), *Industrial Uses of Vegetable Oils*. American Oil Chemists Society Press, Champaign, IL

Knothe G, Dunn RO, Bagby MO (1997) Biodiesel: The use of vegetable oils and their derivatives as alternative diesel fuels. In: Badal CS, Jonathan W (eds), *Fuels and Chemicals from Biomass*. ACS Publications, Washington, DC, pp 172–208

Knutzon DS, Thompson SE, Radke SE, Johnson WB, Knauf VC, Kridl JC (1992) Modification of Brassica seed oil by antisense expression of a stearoyl-acyl carrier protein desaturase gene. *Proc Natl Acad Sci USA* 89:2624–2628

Kott L (1996) Double haploid technology accelerates canola breeding. *Agric Food Res Ontario* 19:16–18

Korbitz W (1999) Biodiesel production in Europe and North America, an encouraging prospect. *Renew Energy* 16:1078–1083

Kozlowska H, Naczk M, Shahidi F, Zadernowski R. (1990) Phenolic acids and tannins in rapeseed and canola. In: Shahidi F (ed), *Canola and Rapeseed: Production, Chemistry, Nutrition and Processing Technology*. Van Nostrand Reinhold, New York, pp 193–210

Laga B, Seurinck J, Verhoya T, Lambert B (2004) Molecular breeding for high oleic acid and low linoleic fatty acid composition in *Brassica napus*. *Pflanzenschultz-Nachrichten Bayer* 57:87–92

Lakshmikumaran, MDS, Srivastava PS (2003) Application of molecular markers in *Brassica* coenospecies: comparative mapping and tagging. In: Nagata T, Tabata S (eds), *Biotechnology in Agriculture and Forestry*. Vol. 52: Brassicas and Legumes. Springer, New York, pp 37–68

Lan TH, Delmonte TA, Reischamann KP, Hyman J, Kowalski S (2000) EST-enriched comparative map of *Brassica oleracea* and *Arabidopsis thaliana*. *Genome Res* 10:776–788

Lang X, Dalai AK, Bakhashi NN, Reaney MJ, Hertz PB (2001) Preparation and characterization of biodiesel from various bio-oils. *Bioresour Technol* 80:53–62

Liu CJH, Chevre AM, Landgren M, Glimelius K (1999) Characterization of sexual progenies of male sterile somatic cybrids between *Brassica napus* and *Brassica tourneforti*. *Theor Appl Genet* 99:605–610

Liu JW, DeMichele S, Bergana S, Bobik M, Hastilow C, Chuang LT, Mukherji P, Huang YS (2001) Characterization of oil exhibiting high gamma-linolenic acid from a genetically transformed canola strain. *J Amer Oil Chem Soc* 78:489–493

Liu JW, Huang YS, DeMichele S, Bergana M, Bobik E, Hastilow C, Chuang LT, Mukerji P, Knutzon D (2001) Evaluation of the seed oils from a canola plant genetically transformed to produce high levels of gamma linolenic acid. In: Huang YS, Ziboh VA (eds), *Gamma Linolenic Acid: Recent Advances in Biotechnology and Clinical Applications*. American Oil Chemists Society Press, Champaign, IL, pp 61–71

Liu XP, Tu JX, Liu ZW, Chen BY, Fu TD (2005) Construction of a molecular marker linkage map and its use for QTL analysis of erucic acid content in *Brassica napus* L. *Acta Agron Sin* 31:275–282

Lovett AA, Sunnenberg GM, Richter GM, Dailey AG, Riche AB, Karp A (2009) Land use implications of increased biomass production identified by GIS-based suitability and yield mapping for Miscanthus in England. *Bioenergy Res* 2:17–28

Lühs WW, Voss A, Seyis F, Friedt W (1999) Molecular genetics of erucic acid content in the genus *Brassica*. In: *Proceedings of the 10th International Rapeseed Congress: New Horizons for an Old Crop*. Canberra, Australia, available at http://www.regional.org.au/au/gcirc/4/442.htm (accessed July 30, 2009)

Mahmood, T, Rahman MH, Stringam GR (2006) Identification of quantitative trait loci (QTL) for oil and protein contents and their relationship with other seed quality traits in *Brassica juncea*. *Theor Appl Genet* 113:1211–1220

Mahmood T, Ekuere U, Yeh F, Good AG, Stringam GR (2003) RFLP linkage analysis and mapping genes controlling the fatty acid profile of *Brassica juncea* using reciprocal DH populations. *Theor Appl Genet* 107:283–290

Martin JH, Leonard WH, Stamp DL, Waldren R (2006) *Principals of Field Crop Production*. Prentice Hall, NJ

Massimo C, Marco M, Stefano M, Vittorio R, Adolfo S, Maurizia S, Sandra V (2003) *Brassica carinata* as an alternative oil crop for the production of biodiesel in Italy: agronomic evaluation, fuel production by transesterification and characterization. *Biomass Bioenergy* 25:623–636

McVetty PBE, Fernando WGD, Li G, Tahir M, Zelmer C (2008) High erucic acid, low-glucosinolate rapeseed (HEAR) cultivar development in Canada. In: Hou CT, Shaw JF (eds), *Biocatalysis and Agricultural Biotechnology*. CRC Press, Boca Raton, FL, pp 43–61

McVetty PBE, Scarth R (2002) Breeding for improved oil quality in *Brassica* oilseed species. *J Crop Prod* 5:345–369

McVetty PBE, Scarth R, Rimmer SR (1999) Millennium 01 high erucic acid low glucosinolate summer rape. *Can J Plant Sci* 79:251–252

Mikolajczyk K (2007) Development and practical use of DNA markers. In: Gupta SK (ed), *Advances in Botanical Research. Vol 45: Rapeseed Breeding*. Academic Press, London, pp 100–125

Mittelbach MM, Woergetter MM, Pernkopf J, Junek H (1983) Fuel derived from vegetable oils: preparation and use of rape oil methyl ester. *Energy Agric* 2:369–384

Mootabadi H, Salamatinia B, Razali N, Bhatia S, Abdullah AZ (2008) Energy balance in production of biodiesel. Presented at the International Conference on Environmental Research and Technology (ICERT 2008), Parkroyal Penang, Malaysia, pp 901–905

Munavu RM, Odhiambo D (1984) Physiochemical characterization of nonconventional vegetable oils for fuel in Kenya. *Ken J Sci Technol* 5:45–52

Murphy DJ (1999) Manipulation of plant oil composition for the production of valuable chemicals. Progress, problems and prospects. *Adv Exp Med Biol* 464:21–35

Murphy LA, Scarth R (1998) Inheritance of the vernalization response determined by double haploids in spring oilseed rape (*Brassica napus* L.). *Crop Sci* 38:1463–1467

Naik M, Meher LC, Naik SN, Das LM (2008) Production of biodiesel from high free fatty acid karanja (*Pongamia pinnata*). *Biomass Bioenergy* 32:354–357

Noureddini H, Gao X, Philkana RS (2005) Immobilized *Pseudomonas cepacia* lipase for biodiesel fuel production from soybean oil. Bioresour Technol 96:769–777

Oliveira LS, Franca AS, Camargos, RRS, Ferraz VP (2008) Coffee oil as a potential feedstock for biodiesel production. *Bioresour Technol* 99:3244–3250

Ooko SA (2008) 12 largest biofuel plants in the world, available at http://ecoworldly.com/2008/04/23/12-worlds-largest-biofuel-plants/ (accessed July 28, 2009)

Orcaire O, Buisson P and Pierre AC (2006) Application of silica aerogel encapsulated lipases in the synthesis of biodiesel by transesterification reactions. *J Mol Catal B: Enzy* 42:106–113

Palmer JD (1988) Interspecific variation and multicircularity in Brassica mitochondrial DNAs. *Genetics* 118:341–351

Pantaleo A, Pellerano A, Carone MT (2008) Potential and feasibility assessment of small scale CHP plants fired by energy crops in Puglia region (Italy). *Biosyst Eng* 102:345–349

Park J, Kim D, Wang Z, Lu P, Park S, Lee J (2007) Production and characterization of biodiesel from tung oil. *Appl Biochem Biotechnol* 148:109–117

Park JY, Koo DH, Hong CP, Lee SJ, Jeon JW (2005). Physical mapping and microsynteny of *Brassica rapa* spp. *pekinensis* genome corresponding to a 222 kb gene-rich region of Arabidopsis chromosome 4 and partially duplicated on chromosome 5. *Mol Gen Genet* 274:579–588

Parkash S (1980) Cruciferous oilseeds in India. In: Tsunoda S, Hinata K, Gomez-Campo C (eds), *Brassica Crops and Wild Allies. Biology and Breeding*. Japan Scientific Society Press, Tokyo, Japan, pp 151–163

Parkash S, Hinta K (1980) Taxonomy, cytogenetics and origin of crop Brassicas, a review. *Opera Bot* 55:1–57

Pearl GG (2001) Biodiesel Production in the U.S. Render Magazine, August 2001, available at www.rendermagazine.com/August2001/ TechTopics.html

Pimentel D, Patzek TW (2005) Ethanol production using corn, switchgrass, and wood; biodiesel production using soybean and sunflower. *Nat Resour Res* 14:65–76

Pinzi S, Garcia IL, Lopez-Gimenez FJ, Luque de Castro MD, Dorado G, Dorado MP (2009) The ideal vegetable oil-based biodiesel composition: A review of social, economical and technical implications. *Energy Fuels* 23:2325–2341

Pleines S, Friedt W (1989) Genetic control of linoleic acid concentration in seed oil of rapeseed (*Brassica napus* L.). *Theor Appl Genet* 78:793–797

Proc K, Barnitt R, McCormick RL (2005) *RTD Biodiesel (B20) Transit Bus Evaluation: Interim Review Summary*. National Renewable Energy Laboratory, available at http://www.nrel.gov/vehiclesandfuels/npbf/pdfs/38364.pdf (accessed July 28, 2009)

Qiu D, Morgan C, Shi J, Long Y, Liu J, Li R (2006) A comparative linkage map of oil seed rape and its use for QTL analysis of seed oil and erucic acid content. *Theor Appl Genet* 114:67–80

Quiros CF (1999) Genome structure and mapping. In: Gomez-Campo C (ed), *Biology of Brassica Coenospecies: Developments in Plant Genetics and Breeding*. Elsevier, Amsterdam, Netherlands, pp 217–245

Quiros CF (2000) DNA-based marker maps of Brassica. In: Phillips RL, Vasil IK (eds), *DNA-Based Markers in Plants*. Kluwer Academic Publishers, Dordrecht, Netherlands, pp 240–245

Quiros CF, Paterson AH (2004) Genome mapping and analysis in Brassica. In: Pua EC, Douglas CJ (eds), *Biotechnology in Agriculture and Forestry*, Vol. 54. Springer, New York, pp 31–42

Radich A (2004) *Biodiesel Performance Cost & Use*. U.S. Energy Information Administration, available at http://www.eia.doe.gov/oiaf/analysispaper/biodiesel/ (accessed July 28, 2009)

Rakow G (2004) Species origin and economic importance of Brassica. In: Pua EC, Douglas CJ (eds), *Biotechnology in Agriculture and Forestry*, Vol. 54. Springer, Berlin, Germany, pp 13–28

Ramdhas AS, Jayaraj S, Muraleedharan C (2004) Use of vegetable oils as I.C. engine fuels—A review. *Renew Energy* 29:727–742

Rashid U, Anwar F, Moser BR, Knothe G (2008) *Moringa oleifera* oil: A possible source of biodiesel. *Bioresour Technol* 99:8175–8179

Rathke GW, Diepenbrock W (2006) Energy balance of winter oilseed rape (*Brassica napus* L.) cropping as related to nitrogen supply and preceding crop. *Eur J Agron* 24:35–44

Razon LF (2008) Selection of Philippine plant oils as possible feedstocks for biodiesel. *Phil Agric Sci* 91:278–286

Ruan CJ, Li H, Guo Y, Qin P, Gallagher JL, Seliskar DM, Lutts S, Mahy G (2008) *Kosteletzkya virginica*, an agroecoengineering halophytic species for alternative agricultural production in China's east coast: ecological adaptation and benefits, seed yield, oil content, fatty acid and biodiesel properties. *Ecol Eng* 32:320–328

Safford R, Moran MT, De Silva J, Robinson SJ, Moscow S, Jarman CD, Slabas AR (1993) Regulated expression of the rat medium chain hydrolase genein transgenic rapeseed. *Transgen Res* 2:191–198

Sagiroglu A (2008) Conversion of sunflower oil to biodiesel by alcoholysis using immobilized lipase. *Artif Cells Blood Substit Biotechnol* 36:138–149

Samukawa T, Kaieda M, Matsumoto T, Ban K, Kondo A, Shimada Y, Noda H and Fukuda H (2000) Pretreatment of immobilized *Candida antarctica* lipase for biodiesel fuel production from plant oil. *J Biosci Bioeng* 90:180–183

Sanchez F, Vasudevan PT (2006) Enzyme catalyzed production of biodiesel from olive oil. *Appl Biochem Biotechnol* 135:1–14

Santos ICF, de Carvalho,, Solleti SHV, La Salles JI, La Salles WF, Meneghetti KT (2008) Studies of *Terminalia catappa* L. oil: Characterization and biodiesel production. *Bioresour Technol* 99:6545–6549

Scarth R (1995) Developments in the breeding of edible oil in *Brassica napus* and *B. rapa*. In: *Proceedings of the 9th International Rapeseed Congress: Rapeseed Today and Tomorrow*, Cambridge, United Kingdom, pp 337–382

Scarth R, McVetty PBE, Rimmer SR, Stefansson BR (1987) Steller low linolenic-high linoleic summer rape. *Can J Plant Sci* 68:509–511

Scarth R, McVetty PBE, Rimmer SR, Stefansson BR (1997) Allons low linolenic summer rape. *Can J Plant Sci* 77:125–126

Scarth R, Tang J (2006) Modification of Brassica oil using conventional and transgenic approaches. *Crop Sci* 46:1225–1236

Sebastian RL, Howell EC, King CJ, Marshall DF, Kearsey MJ (2000) An integrated AFLP and RFLP *Brassica oleracea* linkage map from two morphologically distinct double haploid mapping populations. *Theor Appl Genet* 100:75–81

Shah S Gupta MN (2006) Lipase catalyzed preparation of biodiesel from Jatropha oil in a solvent free system. *Proc Biochem* 42:409–413

Sharma N, Anderson M, Kumar A, Zhang Y, Giblin EM, Arams SR, Zaharia LI, Taylor DC, Fobert PR (2008) Transgenic increase in seed oil content are associated with the differential expression of novel *Brassica*-specific transcripts. *BMC Genom* 9:619–647

Sharma R, Aggarwal RAK, Kumar R, Mohapatra T, Sharma RP (2002) Construction of a RAPD linkage map and localization of QTL's for oleic acid level using recombinant inbreds in mustard (*B. juncea*). *Genome* 45:467–472

Sheehan J, Dunahay T, Benemann J, Roessler P (1998) *A Look Back at the U. S. Department of Energy's Aquatic Species Program: Biodiesel from Algae; Close-Out Report*, available at http://www.nrel.gov/docs/legosti/fy98/24190.pdf (accessed July 28, 2009)

Simonsen V, Heneen WK (1995) Inheritance of Isozymes in *Brassica campestris* L. and genetic divergence among different species of *Brassiceae*. *Theor Appl Genet* 91:353–360

Sinha S, Agarwal AK, Garg S (2008) Biodiesel development from rice bran oil: Transesterification process optimization and fuel characterization. *Energy Conver Manag* 49:1248–1257

Sinha S, Jha JK, Maiti MK (2007) Metabolic engineering of fatty acid biosynthesis in Indian mustard (*Brassica juncea*) improves nutritional quality of seed oil. *Plant Biotechnol Rep* 1:185–197

Sivaraman I, Arumugam N, Sodhi YS, Gupta V, Mukhopadhyay A, Pradhan AK, Burma PK, Pental D (2004) Development of high oleic and low linoleic acid transgenics in a zero erucic acid *Brassica juncea L.* (Indian mustard) line by anti-sense suppression of the *fad2* gene. *Mol Breed* 13:365–375

Snowdon RJ, Friedt W (2004) Molecular markers in Brassica oilseed breeding: Current status and future possibilities. *Plant Breed* 123:1–8

Song K, Osborn TC, Williams PH (1990) Brassica taxonomy based on nuclear restriction fragment length polymorphism (RFLPs). 3. Genome relationships in Brassica and related genera and the origin of *B. oleracea*. *Theor Appl Genet* 79:497–506

Song KM, Susuki JY, Slocum MK (1991) A linkage map of *Brassica rapa* based on restriction fragment length polymorphism loci. *Theor Appl Genet* 82:296–304

Soumanou MM, Bornscheuer UT (2003) Improvement in lipase-catalyzed synthesis of fatty acid methyl esters from sunflower oil. *Enz Microbiol Technol* 33:97–103

Stefansson BR, Downey RK (1995) Rapeseed. In: Slinkard AE, Knott DR (eds), *Harvest of Gold*. University Extension Press, University of Saskatchewan, Saskatoon, Canada, p 367

Stefansson BR, Hougen FW, Downey RK (1961) Note on the isolation of rape plants with seed oil free from erucic acid. *Can J Plant Sci* 41:218–219

Sulaiman Al-Zuhair (2007) Production of biodiesel: Possibilities and challenges. *Biofpr* J 1:57–66

Sun Z, Wang Z, Tu J, Zhang J, Yu F, Mcvetty PBE, Li G (2007) An ultra dense genetic recombination map for *Brassica napus*, consisting of 13551 SRAP markers. *Theor Appl Genet* 114:1305–1317

Suwabe K, Morgan C, Bancroft I (2008) Integration of *Brassica* A genome genetic linkage map between *Brassica napus* and *B. rapa*. *Genome* 51:169–176

Suwabe K, Tsukazaki H, Iketani H, Hatakeyama K, Kondo K, Fujimura M (2006) Simple sequence-repeat based comparative genomics between *Brassica rapa* and *Arabidopsis thaliana*: the genetic origin of club root resistance. *Genetics* 173:309–319

Sydut S, Duz AZ, Kaya C, Kafadar AB, Hamamci C (2008) Transesterified sesame (*Sesamum indicum* L.) seed oil as a biodiesel fuel. *Bioresour Technol* 99:6656–6660

Tanhuanpaa PK, Schulman A (2002) Mapping of genes affecting linolenic acid content in *Brassica rapa*. *Mol Breed* 10:51–62

Tanhuanpaa PK, Vilkki JP, Vihinen M (1998) Mapping and cloning of *FAD2* gene to develop allele specific PCR for oleic acid in spring turnip rape (*Brassica rapa*). *Mol Breed* 4:543–550

Thelen JJ, Ohlrogge JB (2002) Metabolic engineering of fatty acid biosynthesis in plants. *Metabol Eng* 4:12–21

Thuneke K (2007) http://www.chemmystery.be/chem3/programma/klaus%20thuneke.pdf (accessed July 28, 2009)

Truco MJ, Hu J, Sadowsky J, Quiros CF (1996) Inter- and intra-genomic homology of the Brassica genomes: implications for their origin and evolution. *Theor Appl Genet* 93:1225–1233

Truco MJ, Quiros CF (1994) Structure and organization of the B genome based on a linkage map in *Brassica nigra*. *Theor Appl Genet* 89:590–598

UN (1935) Genome analysis in Brassica with special reference to the experimental formation of *B. napus* and peculiar mode of fertilization. *Jpn J Bot* 7: 389–452

USDA, Foreign Agriculture Service (2009), http://www.fas.usda.gov (accessed July 28, 2009)

Uzunova M, Ecke W, Weissleder K, Robbelen G (1995) Mapping the genome of rapeseed (*Brassica napus* L.) I. Construction of an RFLP linkage map and localization of QTL's for seed glucosinate content. *Theor Appl Genet* 90:194–204

Vicente G, Martinez GM, Aracil J (2004) Integrated biodiesel production: A comparison of different homogeneous catalysts systems. *Bioresour Technol* 92:297–305

Voelker TA, Worrell AC, Anderson L, Bleibaum J, Fan C, Hawkins DJ, Radke SE, Davies HM (1992) Fatty acid biosynthesis redirected to medium chains in transgenic oilseed plants. *Science* 257:72–74

Voelker TA, Hayes TR, Cramer AM, Turner JC, Davies HM (1996) Genetic engineering of a quantitative trait: Metabolic and genetic parameters influencing the accumulation of laurate in rapeseed. *Plant J* 9:229–241

Walker K (2004) Non-food uses In: Gunstone FD (ed), *Rapeseed and Canola Oil*. CRC Press, Boca Raton, FL, pp 115–185

Warwick SI, Black LD (1991) Molecular systematics of Brassica and allied genera (Subtribe Brassicinae, Brassiceae)—Chloroplast genome and cytodeme congruence. *Theor Appl Genet* 82:81–92

Warwick SI, Hall CJ (2009) Phylogeny of Brassica and wild relatives. In: Gupta SK (ed), *Biology and Breeding of Crucifers*. CRC Press, Boca Raton, FL, pp 19–36

Watanabe Y, Shimada Y, Sugihara A, Tominaga Y (2000) Conversion of degummed soybean oil to biodiesel fuel with immobilized *Candida antarctica* lipase. *J Mol Catal B: Enzy* 17:151–155

Wei D, Yuanyuan X, Dehua L, Jing Z (2004) Comparative study on lipasecatalyzed transformation of soybean oil for biodiesel production with different acyl acceptors. *J Mol Catal B: Enzy* 30:125–129

Wesseler J (2007) Opportunities ('costs) matter: a comment on Pimentel and Patzek "Ethanol production using corn, switchgrass, and wood; biodiesel production using soybean and sunflower". *Energy Policy* 35:1414–1416

Wightman PS, Eavis RM, Walker RM, Batchelor SE, Carruthers SP, Booth EJ (1999) Life-cycle assessment of chainsaw lubricants made from rapeseed oil or mineral oil. 10th International Rapeseed Congress, Camberra, Australia

Wolf J, Bindraban PS, Luitjten JC, Vleeshouwers LM (2003) Exploratory study on the land area required for global food supply and the potential global production of bioenergy. *Agricultural Systems* 76:841–861

World Biodiesel Production by country, 2005, available at http://www.earthpolicy.org/Updates/2006/Update55_data.htm#table4 (accessed July 28, 2009)

Yamagishi H, Landgren M, Forsberg J, Glimelius K (2002) Production of asymmetric hybrids between *Arabidopsis thaliana* and *Brassica napus* utilizing an efficient protoplast culture system. *Theor Appl Genet* 104:959–964

Yan Z (1990) Overview of rapeseed production and research in China. In: *Proceedings of the International Canola Conference on Potash and Phosphate Institute*, Atlanta, GA, pp 29–35

Zhang JF, Cun-Kou QI, Li GY, Hui-Ming PU, Song C, Chen S, Gao JQ, Chen XJ, Gu H, Fu SZ (2007) Genetic map construction and apetalous QTL's identification in rapeseed (*B. napus* L.). *Acta Agron Sin* 33:1246–1254

Zhang JF, Cun-Kou QI, Hui-Ming PU, Song C, Chen F, Gao JQ, Chen XJ, Gu H, Fu SZ (2008) QTL identification for fatty acid content in rapeseed (*B. napus.L.*). *Acta Agron Sin* 34:54–60

Zhang Y, Dube MA, McLean DD, Kates M (2003) Biodiesel production from waste cooking oil: 2. Economic assessment and sensitivity analysis. *Bioresour Technol* 90:229–240

Zhao J, Becker HC, Zhang D, Zhang Y, Ecke W (2005) Oil content in a European–Chinese rapeseed population: QTL with additive and epistatic effects and their genotype–environment interactions. *Crop Sci* 45:51–59

Zhao J, Simmonds DH, Newcomb W (1996) High frequency production of doubled haploid plants of *Brassica napus* cv. Topas derived from colchicine-induced microspore embryogenesis without heat shock. *Plant Cell Rep* 15:668–671

19 Sorghum

Ismail Dweikat
University of Nebraska–Lincoln

CONTENTS

19.1 INTRODUCTION

Energy security and greenhouse gas emissions reductions are perhaps now more than ever one of the most important priorities of many countries. Energy security is a growing concern because of uncertainties in supply coupled with sharp increases in prices due to geopolitical tensions and weather disturbances in oil-producing countries. In addition, maintaining a clean and healthy environment has also gained worldwide attention, even as the Intergovernmental Panel on Climate Change recently confirmed that human activities are to blame for global warming. To address these issues, many oil-importing countries have embarked on programs to develop alternative cost-effective but locally available, unconventional renewable energy sources that would reduce their dependence on oil, especially for transport, as well as minimize adverse impacts on the environment. Advances in technology have opened new opportunities for achieving some of these objectives.

Sorghum [*Sorghum bicolor* (L.) Moench] is the fifth most important cereal crop worldwide (http://apps.fao.org/default.jsp) as well as an important source of feed, fiber, and biofuel (Doggett

1988). Sorghum, like maize and sugarcane, carries out C_4 photosynthesis, a specialization that makes these grasses well adapted to environments subject to high temperature and water limitation (Edwards et al. 2004). Sorghum is an important target of genome analysis among the C_4 grasses because the sorghum genome is relatively small (730 Mbp) (Paterson et al. 2009), the cultivated species is diploid ($2n = 20$), and the sorghum germplasm is diverse (Dje et al. 2000; Menz et al. 2004; Casa et al. 2005). As a consequence, numerous sorghum genetic, physical, and comparative maps have been constructed (Tao et al. 1998; Boivin et al. 1999; Peng et al. 1999; Klein et al. 2000, 2003; Haussmann et al. 2002; Menz et al. 2002; Bowers et al. 2003, 2005), a sorghum expressed sequence tag (EST) project (Pratt et al. 2005) and associated microarray analyses of sorghum gene expression have been carried out (Buchanan et al. 2005; Salzman et al. 2005), a comprehensive analysis of sorghum chromosome architecture has been completed (Kim et al. 2005), and an $8 \times$ draft sequence of the sorghum genome (about twice the size of rice) has been completed by the U.S. Department of Energy Joint Genome Sequencing program (Paterson et al. 2009; http://www.phytozome.net/sorghum). In addition, genetic maps have been assembled at Texas A&M University and the University of Georgia (Menz et al. 2002; Bowers et al. 2003). The U.S. Department of Agriculture (USDA) germplasm system maintains 42,614 accessions, of which more than 800 exotic landraces have been converted to day length-insensitive lines to facilitate their use in breeding programs. In total, there are approximately 168,000 sorghum accessions held at repositories around the world. A set of mutation stocks developed by the USDA Plant Stress and Germplasm Development Unit in Lubbock, TX (Xin et al. 2008) is extensive enough to provide mutations in all of the genes in the sorghum genome. Such genomic tools, already in place, will greatly facilitate the introduction of traits required to optimize sweet sorghum for bioenergy production schemes.

19.2 BIOLOGY OF SORGHUM

19.2.1 ORIGIN

The sorghum plant has undergone selection, domestication, and hybridization by humans to become a crop that can produce grain and forage in low-rainfall, high-temperature environments, thereby meeting the nutritional needs of people living in marginal areas. Vavilov (1951; cited by Mann et al. 1983), Snowden (1936; cited by Mann et al. 1983), Harlan and de Wet 1972, and Mann et al. (1983) have all theorized that Africa is the center of origin for sorghum. Although the exact location is debatable, domestication can be attributed to the selection for sorghums without grain shattering to improve harvest ability (Mann et al. 1983). Early records of sorghum show its existence in India in the first century AD (Bennett et al. 1990), China in the 13th century (Undersander et al. 1990), and in the United States in the Seventeenth century (Agyeman et al. 2002). Currently sorghum is the fifth most important crop produced worldwide (Rooney and Awika 2004) and has widespread production in sub-humid and semi-arid regions in both tropical and temperate climates.

19.2.2 CLASSIFICATION AND DOMESTICATION

Sorghum is a self-pollinating species consists of cultivated and wild species. *S. bicolor* subsp. *bicolor* ($2n = 20$) is the taxon that includes the agronomically important grain races bicolor, caudatum, durra, guinea, and kafir and ten intermediate races. Additionally, hybrid races can be identified from crosses among the basic races. There are more than 35,000 accessions of sorghum that are been maintained at the germplasm collection centers in the United States with similar numbers being maintained at the International Crops Research Institute for the Semi Arid Tropics (ICRISAT). The U.S. and ICRISAT collections have been evaluated phenotypically, and the USDA has developed a description list of more than 75 descriptors (http://www.ars-grin.gov/cgi-bin/npgs/html/desclist.pl?69). Very high levels of genetic diversity exist among and within races. Most U.S. sorghums are derived from Kafirn × milo crosses.

Sorghum bicolor (L.) *Moench*

Kingdom	*Plantae*—Plants
Subkingdom	*Tracheobionta*—Vascular plants
Superdivision	*Spermatophyta*—Seed plants
Division	*Magnoliophyta*—Flowering plants
Class	*Liliopsida*—Monocotyledons
Subclass	*Commelinidae*
Order	*Cyperales*
Family	*Poaceae*—Grass family
Genus	*Sorghum* Moench—sorghum
Species	*Sorghum bicolor* (L.) Moench—sorghum

19.2.3 PLANT DESCRIPTION

Sorghum is a summer annual that is coarse and erect with much variability in growth characteristics. The culms are solid or sometimes with spaces in the pith, 0.6–5 m tall (Figure 19.1), depending on the variety and growing conditions. It is 5 to over 45 mm in diameter (Figure 19.2) and is either dry at maturity or has a sweet, insipid juice.

Vanderlip (1993) has defined the growth of the sorghum plant from emergence (stage 0) to physiological maturity (stage 9). The timing of various stages of development and the condition of the plant at each stage can be affected by soil fertility, insect or disease damage, water stress, plant population, and weed competition. As the total number of leaves increases and gives rise to a greater leaf area index, the growth rate of the sorghum plant increases. Leaves arise alternately at nodes along the stem and can be as much as 1 m long and 10- to 15-cm wide (House et al. 1995). Most leaf growth occurs between stages 3 and 5, when the upper eight to ten leaves develop (Vanderlip 1993). Leaves can number from 7 to 18 or more, and the total number is determined by the length of the vegetative period (Bennett et al. 1990). The flag leaf is the first leaf below the panicle and it holds the boot before extension. Flowering begins at the tip of the panicle and continues to the base, taking 4–5 days. Although primarily known as a self-pollinating crop, 2–20% cross-pollination can

FIGURE 19.1 (See color insert)   Sweet sorghum plants reach 450 cm in September 2007 at Lincoln, NE. The plants were rainfed and were grown in rotation with soybean.

FIGURE 19.2 (See color insert)   Sweet sorghum stalks with different thickness ranges from 15 to 45 mm.

occur, with higher percentages taking place from more open-headed panicles (House et al. 1995). Compared with the other two major spring planted crops, maize (*Zea mays*) and soybean (*Glycine max*), sorghum has the smallest seed (Vanderlip 1993). Sorghum kernels are generally spherical and range in weight from 20 to 30 mg (Hoseney 1994) and can be colored white, cream, pink, yellow, red, buff, brown, and reddish brown (Bennett et al. 1990). The kernels are covered by glumes, which can be black, red, brown, or straw. Most are colored. The sorghum kernel is made up of three major anatomical parts—the pericarp, germ, and endosperm—and each generally accounts for roughly 6, 10, and 84% of the kernel, respectively (Rooney and Serna-Saldivar 2000). The pericarp protects the kernel and is composed mostly of fiber; the endosperm is a major contributor to the kernel's protein (80%), starch (94%), and 50–75% of B-complex vitamins; and the germ contains over 68% of the total mineral matter and 75% of the oil in the whole kernel (FAO 1995).

Sorghum leaves are broad and coarse, similar in shape to those of corn but shorter and wider. The blades of the leaves are glabrous and waxy. Sheaths encircle the culm and have overlapping margins. The panicle is erect, sometimes recurved, and is usually compact in most grain sorghums and more open in forage types. Prop roots may grow from culm nodes, and there is a bud at each node from which a tiller may grow. The seeds are white, yellow, red, or brown and are covered by glumes that may or may not be removed by threshing. It has a spikelet that contains two flowers, only one of which is usually fertile and sets a seed. When threshed, the seed separates from the floral bracts as in wheat, and the panicle has up to 6000 spikelets. The kernels are small with a round to conical shape. Seeds number at 25,000–61,740 per kilogram. Sorghum is most commonly red and hard when ripe, and it is usually dried after harvesting.

19.2.4   ADAPTATION

Temperature, day length, and water needs are three elements that affect sorghum adaptation and growth. Sorghum has the ability to produce a crop in areas with marginal rainfall and high temperatures, where other cereal grains often fail (Cothren et al. 2000), and is found from 40°S to 45° north latitude (Maunder 2002). Sorghum is more tolerant to high temperature (>38°C) and drought than most major agronomic crops. Sorghum is a warm-region crop that requires warm temperature for germination and growth. The optimum germination temperature is between 20°C and 25°C. Temperatures below freezing may kill the plants depending on stage of development. At early seedling stage (1–3 weeks), plants could recover after a short exposure to temperature of 5°C below freezing point; however, at temperatures lower than 5°C, plants are killed. Plants older than 3 weeks are more susceptible and may die at 0°C. The crop can grow where available water is between 360 and 500 mm and can respond positively to higher precipitation (Fribourg 1995). Sorghum can reduce its water losses by its heavy wax cuticle, curling of its leaves, and its relatively small number of leaf stomata (Gardner et al. 1981). When water supply is limited, sorghum has more efficient water transport system than either corn or cotton (Ackerson and Krieg 1977).

Sorghum has a fibrous root system that grows rapidly in deep soils, and it is efficient as a water forager. The adventitious root starts several weeks after emergence and extends rapidly up to 2 m depending on the depth of soil wetting (Sullivan and Blum 1970). Graser (1985) compiled seasonal water use of sorghum at several locations from 1976 to 1981 and reported a range of 179–540 mm under dry-land and 321–645 mm under irrigated conditions depending on the length of the growing season. Erie et al. (1981) reported that consumptive water use increases with plant growth, reaches a peak, and then decreases by harvest time. Water use of sorghum was found to be greatest during the boot and soft dough stage and lower during the seedling, tillering, and ripening stages (Porter et al. 1960). Several factors including temperature, precipitation, solar radiation, humidity, wind movement, and hybrid affect sorghum water-use efficiency. The water-use curve in any one year or at any site will vary from the long-term average because of changes in some of the factors listed above. Water-use efficiency ranging from 13 to 29 kg/ha per mm has been reported in the literature under dry-land and irrigated conditions (Hedge et al. 1976; Sivakumar et al. 1979). The crop has the ability to delay development under water stress during the vegetative growth stages and resume growth when water conditions improve. This drought avoidance mechanism works well under tropical and subtropical conditions with a long growing period. However, this mechanism of drought resistance may result in poor yield because of prolonged drought, insufficient season length, or when it occurs at critical growth stage. Field water capacities of 25–50% and temperatures above 28°C are most favorable for optimal sorghum germination (Fawusi and Agboola 1980). The favorable mean temperature for sorghum growth is approximately 37°C, and the minimal temperature for growth is 15°C (Cothren et al. 2000).

19.2.5 Crop Uses

Sorghum is one of the top five cereal crops in the world, along with wheat, oats, rice, corn, and barley. Worldwide more than 40 million ha are planted to sorghum. It is a genus with many species and subspecies, and there are several types of sorghum, including grain, forage, and sweet sorghums. Therefore, sorghum is extremely versatile in offering multiple pathways to ethanol.

- *Starch-to-ethanol*: Grain sorghum conversion to ethanol is equal to corn. Today, approximately 28% of the U.S. grain sorghum crop currently goes into ethanol production according to industry estimates.
- *Sugar-to-ethanol from sweet sorghum*: The conversion efficiency is similar to that of sugarcane.
- *Cellulosic ethanol*: No other crop or other cellulose source equals sorghum in conversion, production efficiency, or ethanol gallons per acre.

19.3 GRAIN SORGHUMS

Grain sorghums include what are commonly known as kafir, kafir corn, durra, and milo. Grain sorghum has been bred to produce approximately 2- to 5-ft height to facilitate harvesting with standard grain harvesters. Grain sorghum has a wide variety of uses that includes food for humans, feed for livestock, alcohol production, and industrial uses. Sorghum can be sorted into grain sorghums, forage sorghums, and sweet sorghums (Rooney and Serna-Saldivar 2000). Grain sorghums often are short and can be mechanically harvested whereas forage sorghums are tall and provide fodder for livestock. The juice from sweet sorghums can be used to produce syrups and sugar or be fermented into alcohol. An estimated 40% of grain sorghum production worldwide is used for human consumption (Rooney and Waniska 2000). High-yielding white grain, tan-plant grain sorghum hybrids have developed an export market to Japan and other Asian countries for production of snack foods and beer because of their bland tastes, lack of gluten proteins, and genetically modified organism (GMO) traits. Gluten, the protein found in wheat (*Triticum aestivum* L.), barley (*Hordeum vulgare* L.), and rye (*Secale cereale* subsp. Cereale), is indigestible by people with celiac disease (Fasano and Catassi 2001). Sorghum grain can be milled into fractions of bran, germ, meal, flour, and grits of different sizes. The general methods of dry milling have been summarized by Hahn (Rooney et al. 1980) and Munck (1995). Sorghum flour can be used as a blend with wheat flour in baked products and has gained consumer acceptability (Munck 1995; Ragaee and Abdel-Aal 2006). In India and Africa, sorghum is used to make thin or stiff porridges and fermented beverages (Lochte-Watson et al. 2000) whereas in Central America and southern Mexico it is a total or partial replacement for maize in tortilla production (Almeida-Dominguez et al. 1991). A brewer using white food-grade sorghums and waxy sorghum grits can achieve reduced color, shorter conversion and runoff times, and improved yields for brewing (Figueroa et al. 1995).

Sorghum is being evaluated for use in health food supplements because of the presence of antioxidants, phytosterols, and policosanols in the germ and pericarp. Concentrated in the byproducts of milling and alcohol fermentation, these products might be in high enough quantities for commercial use. Sorghum oils have high levels of phytosterols (Singh et al. 2003) which reduce cholesterol absorption (Weller 2006), whereas sorghum wax contains policosanols, which reduce cholesterol production (Rooney and Awika 2004; Weller 2006). In the United States, sorghum is the second most important feed grain behind maize. The feeding value of grain sorghum for feed-lot cattle is 85–100% of maize (Kriegshauser et al. 2006) and 90–95% of maize for swine and poultry (Hulan and Proudfoot 1982). White-grain, tan-plant cultivars are ideal for feeding broilers because the absence of colored glumes reduces the amount of dark specks found on carcasses (Rooney and Waniska 2000). Hicks et al. (2002) reported that hybrids with heavy kernel weights had increased crude protein and fat content with reduced starch, which improved broiler chicken performance to be equal to or better than maize (Kriegshauser et al. 2006). Sorghum provides a source of starch for ethanol production in Nebraska and Kansas, and the byproducts have become important dairy and beef cattle feeds. In all countries except the United States, sorghum is used extensively as a cereal food. The grain is an excellent food source when ground into flour and used to make pancakes, porridge, and flatbreads. Sorghum grain produces edible oil, starch, dextrose, paste, and alcoholic beverages. Sorghum can be puffed, popped, shredded, and flaked to produce ready-to-eat breakfast cereals. Economically, the use of sorghum grits and commercial enzymes is also practical. In the United States, sorghum is used primarily as a corn substitute for livestock feed because their nutritional values are very similar. Some hybrids commonly grown for feed have been developed to deter birds therefore contain a high concentration of tannins and phenolic compounds, which causes the need for additional processing to allow the grain to be digested by cattle. In arid regions in less-developed regions of the world, sorghum is an important food crop especially for subsistence farmers.

19.3.1 GRAIN SORGHUM AS A SOURCE OF ETHANOL

Researchers and ethanol producers have shown that grain sorghum is a good feedstock for ethanol that is comparable to that from corn grain and could make a larger contribution to the nation's fuel ethanol requirements. However, in the past, factors affecting ethanol yield were less well studied for sorghum than for corn. Little research has been conducted on performance of sorghum varieties in ethanol fermentation. Several researchers have investigated the digestibility of sorghum starch and sorghum protein (Duodu et al. 2003; Selle et al. 2010; Wong et al. 2009) as related to its use in feed or food. Others have investigated the isolation of sorghum starch and its properties (Park et al. 2006; Sang et al. 2008). The economic viability of an ethanol production facility depends on several factors, including ethanol yield, efficiency of conversion, and quality of the "distiller's grain" (grain residue and yeast mass remaining after the fermentation process).

Sorghum has the potential for being used in the production of bioindustrial products, including bioethanol. Sorghum is a starch-rich grain with similar composition to maize, and, as with all cereals, its composition varies significantly because of genetics and environment (Rooney and Serna-Saldivar 2000). Starch ranges of 60–77% and 64–78% have been reported for sorghum and maize, respectively (Shelton and Lee 2000). As such, sorghum grain would be appropriate for use in fermentation similar to the use of maize for the production of bioethanol. Its use may be of particular benefit in countries where rainfall is limiting and maize does not grow well. With regard to the United States, approximately 95% of the bioethanol is currently produced from maize starch, primarily in the maize-growing regions. Sorghum production in the United States in 2004 was 11.6 million t (http://faostat.fao.org), equivalent to approximately 457 million bushels, and 10–20% of those were used for ethanol production (http://www.sorghumgrowers.com). In the same year in the United States, 3.4 billion gal of ethanol were produced from 1.22 billion bushels of grain (http://www.ksgrains.com/ethanol/useth.html). From this, it may be calculated that 1.2–2.3 million metric t sorghum was used for ethanol production, 3.7–7.5% of the grain for ethanol production was sorghum, and 0.13–0.25 billion gal of ethanol originated from sorghum. Although significant research into the production of ethanol from maize grain has been conducted, comparatively little research has been done on the conversion of sorghum grain into bioethanol.

19.4 FORAGE SORGHUMS

Forage sorghum differs from grain sorghum primarily in utilization. Forage sorghum is an important annual forage source in the Midwestern and Plains regions of the United States and can be planted later than maize (*Z. mays* L.) to provide fodder for stock between late spring and autumn. The crop may be cut only once (single cut) or several times (multicut) during the growing season, whereas maize can be cut only once. It uses water more efficiently, yields greater biomass, and provides an acceptable yield when exposed to drought. These include sudan grass and tunis grass and are used for pasture and forage. Forage sorghums ranged from 2- to 5-m tall, and whole-plant yields ranged between 3.1 and 10.1 t of dry matter per acre. They are annuals and grow quickly. They are generally used for summer pasture. Johnson grass, perennial grass sorghum, is considered a pest when out of control, but it makes an excellent hay and cattle feed. It is important to remember that forage-sorghum varieties vary widely with respect to agronomic characteristics.

Sweet sorghum is one of the many types of cultivated sorghum, characterized by the high sugar content in its stem juice. Some lines attain juice yields of 78% of total plant biomass and contain 15–23% soluble fermentable sugar (comparable to sugarcane). The sugar is composed mainly of sucrose (70–80%), fructose, and glucose. Most of the sugars are uniformly distributed in the stalk, whereas approximately 2% are in the leaves and inflorescences (Vietor and Miller 1990). Even in dry climates, sweet sorghum can yield high levels of fermentable sugars, together with grain and lignocellulosics (Gnansounou et al. 2005). Stalks of sweet sorghum contain fermentable sugars capable of producing 400–800 gallons of ethanol per acre (Reddy et al. 2008), which is comparable to ethanol yield from corn grain (assuming an average irrigated corn yield of 170 bushels/acre results in 470 gal of ethanol); however, the conversion of sweet sorghum into ethanol does not require the energy-intensive steam cooking step necessary to produce ethanol from corn grain.

Sweet sorghum reproduces by seed and produces tillers, but it has no rhizomes. It is a perennial grass under tropical conditions, but it is winter-killed in areas where frost occurs. Some sweet sorghum cultivars are grown for syrup production, whereas others are grown for forage (silage). Sweet sorghum is adapted to widely differing climatic and soil conditions, rendering it ideally suited as a biofuel crop for marginal land production. Utilization for bioenergy conversion processes is higher for sweet sorghum compared with other crops because it produces high biomass, fermentable carbohydrates yields, and a small amount of grain.

19.5   SWEET SORGHUM AS A BIOENERGY CROP

Sweet sorghum yields high levels of fermentable sugars, together with grain and lignocellulose. Hunter and Anderson (1997) indicated that the sugar produced in sweet sorghum has a potential ethanol yield up to 8000 L/ha, or about twice the ethanol yield potential of maize grain. In addition to producing large amounts of sugar-rich biomass, hybrids can be developed from crosses between grain-type seed parents and sweet-type pollen parents (Hunter and Anderson 1997). The product of these crosses typically increase biomass yields and sugar content when compared with the original grain-type seed parents. Such hybrids can co-produce grain at levels approaching the yields of the grain-type seed parent (Miller and McBee 1993).

Sweet sorghum has already been identified as a preferred biomass crop for fermentation into methanol and ethanol fuel. Its processing takes ethanol and its derivatives from the direct fermentation of sugars present in the stem juices, followed by processing of the bagasse (the remaining part of the stems after juice extraction), to pyrolytic oils, quality fuels, pellets of carbon, synthesis gas, and lignocellulosic materials. An alternative energetic application of the bagasse may be electricity production through combustion of total biomass. In addition, the stillage from sweet sorghum, after extraction of juice, has a higher biological value than the bagasse from sugarcane when used as fodder for animals because of higher levels of micronutrients and minerals. It is also processed as a feed for ruminant animals. Apart from these, the stillage contains levels of cellulose similar to those in sugarcane bagasse, suggesting that it has good potential as a raw material for pulp products.

Sweet sorghum offers several important advantages for bioenergy; remarkable drought tolerance, productivity as a hybrid crop, amenability to genetic transformation, and extensive genomic resources available for its study. Because of its high yield in biomass and fermentable sugars, sweet sorghum can be converted into energy carriers through one of two pathways: biochemical and thermochemical. Through biochemical processes, the crop sugars can be converted to biofuels (ethanol, butanol, and hydrogen). Thermochemical processes such as combustion and gasification can be used for the conversion of the sweet sorghum bagasse to heat and electricity (Table 19.1). Pulp for paper, compost, and composites materials are some other products that can also be derived from sweet sorghum bagasse.

Sorghum's adaptation to drought stress allows it to grow in some of the world's less favorable climates. Morphological and physiological responses under drought stress make this plant unique among the cereals. Sorghum's capacity to fold (rather than roll) its leaves and the deposition of a heavy layer of wax over the leaves reduce evapotranspiration. The root system of sorghum is bigger

TABLE 19.1
Chemical Analysis, Burning Profile, Biomass Production, Potential Ethanol, and Total Land Needed for Different Bioenergy Crops to Reach the 35 Billion Gal U.S. Reachable Goal

Crop	Carbon	Nitrogen	Sulfur	Ash	HHV (Btu/lb)	Total Biomass (Mg/ha)	Ethanol (gal/ha)	Total Land to Produce 35 Billion Gal
Tropical maize	49.00	0.97	0.09	4.78	8058	35.0	3500	10.0[a]
Silage corn	47.42	0.58	0.05	6.84	7718	11.0	1100	31.4
Bioenergy millet	49.18	0.56	0.08	4.02	8161	36.0	3600	9.7
SS M 81 E	46.97	0.35	0.08	4.50	8031	35.0	3500	10.0
Switch grass	42.00	0.59	0.10	7.09	7590	10.4	1040	33.7
Corn stover	43.60	0.83	0.09	6.90	7782	7.4	740	47.2
Miscanthus	41.00	0.39	0.09	5.65	7750	28.0	2800	12.5

SS, sweet sorghum; HHV, high heating value.

[a] Million hectares.

and deeper than that of maize. Sorghum plants have a capacity to remain relatively inactive during drought and renew growth when conditions are favorable (Doggett 1988).

Because sorghum grown for biomass can be harvested before it is fully mature, it is possible to grow it in a double crop sequence with a winter annual legume. Winter annuals are planted in the fall, grow rapidly in the spring, and reach harvest anytime during late spring to early summer. Sorghum, which is well adapted to germination under limited moisture, can be no-till planted into the stubble of a winter annual crop. The primary advantages of a double crop sequence are to maximize use of solar radiation, provide winter cover against wind and water erosion, and increase biomass per hectare yields (Karpenstein-Machan 2001). Because sorghum and winter annuals have differing cardinal temperatures for growth, the double crop sequence can take advantage of a longer growing season than either crop alone.

Given that water availability is poised to become a major constraint to agricultural production in coming years, cultivation of corn becomes difficult. Sweet sorghum would be a logical crop option in lieu of corn in such situations. Sweet sorghum can be grown with less irrigation, rainfall, and inputs compared with corn. In addition to sweet stalk, grain yields of 2–6 t/ha (which can be used as food or feed) could be harvested from sweet sorghum.

A wide range of maturity classes is required to extend the harvest period to meet the requirements of the processing factories. Sweet sorghum's energy-savings and value emerge in several ways.

- The crop only needs 12–15 in. of rain during the growing season to make a crop; therefore, it is suitable for dry-land production or under limited irrigation. If the crop receives more moisture, it will respond positively.
- It requires only 40–60 lb of nitrogen per acre. The crop is long-rooted and can extract residual nitrogen left by previous crops or from nitrogen-fixing soybeans proceeding in rotation.
- Sweet sorghum juice does not require the long fermentation and cooking time needed to process corn ethanol.
- Some of the crop residue left after juice extraction (called bagasse) can be dried and burned for fuel ethanol distillation. These residues can also be used for animal feed, paper, or fuel pellets.
- The crop need not be grown on a farmer's best land which allows farmers to make use of poorer ground.
- The simplicity of ethanol production from sweet sorghum could lend itself to on-farm or small-cooperative efforts at fuel-making.
- Ethanol plants in the state could choose, with some additional equipment, to make seasonal runs of sweet sorghum juice.

19.6 COMPARISON TO OTHER BIOENERGY CROPS

Sorghum is relatively inexpensive to grow to obtain high yields (Chiaramonti et al. 2004). It can produce approximately 30–50 dry Mg/ha of biomass per year on low-quality soils with minimal inputs—fertilizer and water per dry ton of crop. When compared with the input requirements of other crops, sorghum requires half of those needed by sugarbeets, and one-third of the requirements of sugarcane or corn (Soltani and Almodares 1994; Renewable Energy World 2000). Sweet sorghum juice is ideally suited for ethanol production given its higher content of reducing sugars compared with the content of other sources, including sugarcane juice. These important characteristics, along with suitability for seed propagation, mechanized crop production, and ethanol production capacity comparable to sugarcane molasses and sugarcane, make sweet sorghum a viable alternative raw material source for ethanol production. Furthermore, after the extraction of juice, sorghum stover can be used as a fuel for ethanol distillation. The remaining stover can be used as fodder for animals or for additional ethanol production through lignocellulose digestion. It can out-produce most other cereals under marginal environmental conditions, especially under hot and dry conditions. Furthermore, sweet sorghum is frequently grown in environments that are normally too harsh for other C_4 plants.

Also important is the amount of energy used to produce ethanol. Historically, for each unit of energy it took to plant and harvest a crop and process it into ethanol, the fuel returned 0.92 units of energy. Ethanol had a negative "energy balance" of 1 unit in for 0.92 units out (1:0.92). However, steady improvements have been made in corn yield, harvesting, and ethanol processing efficiency. The latest studies show corn ethanol boasts a positive energy balance of 1:1.25—a 25% net increase in net energy (Farrell et al. 2006). Today, corn ethanol is made by converting the starch in corn to sugars and then into alcohol by a fermentation process. Sugarbeets (*Beta vulgaris* L.) are a better ethanol source, producing nearly 2 units of energy for every unit used in production. However, sugarcane is by far the most efficient of the current feedstocks, yielding more than 3 units as much energy as is needed to produce the ethanol derived from it (Hopkinson and Day 1980). Sweet sorghum's positive energy balance, with a ratio of 1:3, is comparable to that of sugarcane (Worley et al. 1992). Given their positive energy balances and higher yields, it makes more sense to produce ethanol from sugar crops than from starchy grains.

Switchgrass (*Panicum virgatum*) has gained a great deal of attention as a biomass crop in North America. However, establishment issues and relatively low annual dry matter yields suggest other crops may be better suited for cellulosic ethanol production in Iowa, Nebraska, and the surrounding states. There have been multiple reports of switchgrass biomass yields from the north-central U.S. states. Second-year switchgrass dry matter at four Nebraska locations was 1.6–7.3 Mg/ha (Schmer et al. 2006). Annual yields of fully established switchgrass swards were 13–21 Mg/ha at Mead, NE and 6.5–11.0 Mg/ha at Arlington, WI (Casler et al. 2004). Switchgrass dry matter yields at Ames, IA were 4.5–14.3 Mg/ha depending on the genotype (Hopkins et al. 1995). Biomass production on Conservation Reserve Program land in South Dakota totaled 3–4 Mg/ha with 56 kg N/ha (Mulkey et al. 2006). Biomass yields of switchgrass fertilized with 120 kg N/ha and harvested at maturity stages R3–R5 averaged 10.5–11.2 Mg/ha at Mead, NE and 11.6–12.6 Mg/ha at Ames, IA (Vogel et al. 2002). The average dry matter yield of 20 switchgrass populations grown in southern Iowa was 9.0 Mg/ha (Lemus et al. 2002). In comparison, final dry matter yields of grain-type winter triticales in Iowa were 8–16 Mg/ha with no more than 33 kg N/ha (Schwarte et al. 2005).

19.7 VARIETIES OF SWEET SORGHUM

Variety selection is an important decision in sweet sorghum production. Improved varieties have been developed in recent years at the U.S. Sugar Crops Field Station near Meridian, MS. Seed of older varieties originating at other places may still be available in some areas. Important varieties are described (Table 19.2).

19.8 BIOENERGY-RELATED TRAITS

19.8.1 FLOWERING AND MATURITY

Sorghum is a short-day species and requires short days (10–11 h) and long nights to stimulate the reproduction phase. Quinby (1967) reported that four genes influenced inheritance of duration of growth in sorghum. Manipulation of delayed flowering is an important trait to obtain a higher total biomass. Photoperiods longer than 11 h promote the vegetative growth. Most U.S. collection is sensitive to the photoperiod. Efforts were made in the early 1960s in Texas to convert more than 500 unique sorghum lines to photoperiod-insensitive lines to enhance the genetic variability Grain sorghum requires less water than corn, under low to modest yield conditions, and it is an alternative to corn in production environments with frequent severe water deficits (Carter et al. 1989; Bennett et al. 1990; Maman et al. 2004; AFRIS-FAO 2006; Wikipedia 2006).

TABLE 19.2
Brix (Sugar Content) and Other Biomass-Related Trait Values of Sweet Sorghum Cultivars/Lines

Cultivar/Line	Days to Anthesis	Plant Height (cm)	Brix Reading	Wet Stalk (g/stalk)	Dry Stalk (g/stalk)	Stalk Moisture (%)
Dale	100	288	16.07	831.67	241.33	71.04
Tracy	105	266	15.20	715.00	209.33	70.65
Della	78	270	16.00	878.33	292.33	66.70
White African	83	266	14.87	720.00	204.33	71.62
Wray	79	216	18.00	383.33	132.66	64.97
M 81E	110	374	12.67	1173.3	306.00	73.87
Theis	111	348	10.93	1026.6	263.00	74.46
Brawley	78	240	17.33	518.33	163.33	68.51
Kansas-Collier	75	194	18.20	393.00	134.00	65.92
White-Collier	71	180	17.80	328.33	107.33	67.32
African-Millet	79	180	12.93	493.33	145.66	70.55
Colman	79	242	12.87	528.33	158.00	70.07
N. Sugarcane	69	160	15.86	545.00	164.66	70.19
Brandes	113	276	12.97	695.00	197.00	71.55
Bailey	105	290	14.60	544.67	146.33	73.17
Smith	96	248	18.60	990.00	309.33	68.60
Ramada	105	286	17.93	815.00	260.66	67.97
Roma	98	280	16.66	813.33	248.66	69.41
Topper 76-6	110	302	12.67	840.00	217.66	74.10
Grassl	110	288	14.13	935.67	276.66	70.18
Keller	99	304	16.13	1041.6	322.66	68.99
Norkan	68	172	16.30	348.33	108.00	68.53
Sugar Drip	82	236	16.73	810.00	241.00	70.23
Early Sumac	71	188	15.33	446.67	129.00	71.11
N. Sugarcane	85	218	19.26	561.67	207.66	60.28
N98	68	104	18.33	316.67	99.00	68.67
N99	62	166	14.73	240.00	76.33	67.38
N100	79	230	17.20	623.33	202.66	67.27
N108	68	122	12.60	440.00	128.00	70.93
N109	68	122	17.00	216.67	73.33	66.03
N110	67	188	16.13	443.33	110.00	73.17
N111	67	184	15.00	458.33	147.00	67.98
Waconia-L	67	180	17.77	408.33	131.00	67.95
Kansas Orange	68	192	17.73	498.33	160.33	67.63
Red-X	67	190	15.53	431.67	129.66	70.00
Mennonite	68	184	17.66	335.00	112.33	65.83
Rox Orange	67	194	15.00	390.00	121.33	68.84
G. B. Ribbon	71	206	13.30	486.67	148.66	69.36
Fremont Cane	59	174	12.07	325.00	92.66	71.53
Simon	63	198	17.33	415.00	134.66	67.66
Sumac	72	200	17.60	498.00	146.00	70.93
Rex	85	210	17.27	390.00	131.66	66.09
Honey	78	222	12.60	481.67	152.67	68.45
Greenleaf	70	188	12.27	95.00	44.00	52.07
Lahoma	79	218	12.33	386.00	125.00	67.67

(Continued)

TABLE 19.2 (Continued)

Brix (Sugar Content) and Other Biomass-Related Trait Values of Sweet Sorghum Cultivars/Lines

Cultivar/Line	Days to Anthesis	Plant Height (cm)	Brix Reading	Wet Stalk (g/stalk)	Dry Stalk (g/stalk)	Stalk Moisture (%)
Rancher	57	166	13.40	138.67	49.33	64.59
Leoti	63	170	13.67	353.33	119.33	66.19
Red Amber	62	190	11.87	305.00	100.00	67.06
Chinese Amber	68	220	16.93	406.67	133.00	67.09
Dakota Amber	67	192	14.27	255.00	87.33	65.64
Minnesota Amber	68.	174	14.00	305.00	108.66	63.70
Black Ambercane	59	190	13.20	268.33	93.00	64.97
Black-Amber	62	200	12.13	275.00	95.00	65.42
Early Folger	70	216	16.47	363.33	125.33	65.50
Cowley	75	202	16.67	508.33	162.33	68.05
Williams	78	224	15.93	670.00	202.00	69.85
Umbrella	71	226	15.47	596.67	172.66	70.92
Blacktop	57	168	7.13	258.33	75.66	70.62
Snowflake	85	216	17.06	583.33	187.66	67.61
Hastings	77	252	14.40	503.33	155.33	69.14
Iceberg Amber	81	204	15.73	606.67	179.667	70.24
White Orn	85	230	10.60	673.00	190.00	70.89

19.8.2 PLANT HEIGHT

Sweet sorghum may grow to 3–5 m tall. Height is desirable because it influence total biomass. There is a strong correlation between photoperiodism and plant height. The longer the plant remains in the vegetative stage, the greater the number of nodes and leaves. There are four unlinked genes that affect the total number of internodes. These genes exhibit partial dominance and were designated *dw1*, *dw2*, *dw3*, and *dw4*. Several studies have identified and mapped quantitative trait loci (QTL) that affect plant height using recombinant inbred populations (Lin et al. 1995; Pereira and Lee 1995). The map locations of these QTL seem to be similar in maize and sorghum and indicated that possible homologies exist with maize QTL and mutations known to affect plant height.

19.8.3 LIGNIN CONCENTRATIONS

Lignin concentration is important component in any bioenergy crop. In sweet sorghum, both high and low lignin level lines could be utilized. Low lignin lines are suitable feedstock for cellulosic-based ethanol, whereas high lignin lines are desirable for co-firing to produce electricity. Brown midrib mutation in cereals affects lignin level. The brown midrib (*bmr*) mutations were first discovered in corn 1926. Early studies revealed the trait resulted in lower fiber and lignin within the plant and could increase the conversion efficiency of sorghum biomass for lignocellulosic bioenergy. In sorghum, more than 19 *bmr* mutants were discovered by Porter et al. (1978). The *bmr* mutants are characterized by the reddish-brown coloration of the vascular tissue of the leaf blade, leaf sheath, and stem that is associated with alteration of secondary cell wall composition, especially lignin. Because of the development of biocatalysts (e.g., genetically engineered enzymes, yeasts, and bacteria), it is possible to produce ethanol from any plant or plant part containing lignocellulose biomass, including cereal crop residues (stovers). Sorghum stover also serves as an excellent feedstock for ethanol production. Stover contains lignin, hemicellulose, and cellulose. The hemicellulose and cellulose are enclosed by lignin (which contains no sugars), making them difficult to convert into ethanol, thereby increasing the energy requirement for processing. The *bmr* mutant sorghum, pearl millet (*Pennisetum glaucum*), and corn lines have significantly lower levels of lignin content (51% less in their stems and 25% less in their leaves). Purdue University research showed 50% higher yield of the fermentable sugars from the stover of certain sorghum *bmr* lines after enzymatic hydrolysis. Therefore, the use of *bmr* cultivars would reduce the cost of biomass-based ethanol production. The *bmr* crop residues have higher rumen digestibility and palatability, making them also good for fodder.

19.8.4 Seedling Cold Tolerance

As a C_4 tropical crop species, most sorghum lines are susceptible to chilling stress during early-season planting, and tolerance to early-season cold temperatures is a major trait that is needed in many of the sorghum production areas of the United States. Stand establishment of sorghum is adversely affected by air and soil temperature below 15°C at germination, emergence, and early seedling growth. Partial credit for a broader range of adaptation for maize can be given to breeding efforts for improved seedling cold tolerance. Similar improvement in sorghum would allow for expansion of this crop into cooler climatic regions. The University of Nebraska holds an extensive, highly variable collection of over 500 lines of sweet sorghum. Cold-tolerant germplasm could serve to expand the geographical range of sorghum cultivation and minimize the inherent risks involved in early-season planting of sorghum within production areas. Additionally, an earlier sowing date offers growers the option of capitalizing on higher levels of available soil moisture and lower evapotranspirative demands in the early spring, potentially serving as a drought-avoidance strategy. Early planting of the crop within the Midwest would allow for multiple harvests. For example, in Nebraska, cold-tolerant sweet sorghum has been planted on April 15, harvested in early August (22 Mt/ha of dry biomass) and the ratoon crop harvested (12 MT ha^{-1}) in mid October. The ratoon crop alone equals a full season harvest of 1 ha of switchgrass. Improved seedling cold tolerance has been attributed to seedling vigor and greater biomass (Cisse 1995).

19.8.5 Sugar Yield

Sweet sorghum is one of the many types of cultivated sorghum and is characterized by high sugar content in the juice of the stem. Some lines attain juice yields of 78% of the total plant biomass comprised of 15–23% soluble fermentable sugar. The sugar is composed mainly of sucrose (70–80%), fructose, and glucose. Most of the sugars are uniformly distributed in the stalk, whereas approximately 2% are in the leaves and inflorescences (Vietor and Miller 1990). The wide range of variability in soluble solids (brix; 15.5–24.9) and sucrose percentage (from 7.2 to 15.5%) indicates the high potential for genetic improvement to produce high sweet-stalk yield coupled with high sucrose percentage sweet sorghum lines. In an early study, Ayyangar et al. (1936) in crosses between gain and sweet sorghum suggested that a single dominant gene controlled the nonsweet phenotypes. A more recent study (Li et al. 2004) suggested that more than one gene with a dominant effect control the level of sugar in the stalks. Genotypic differences for extractable juice, total sugar content, fermentation efficiency, and alcohol production have also been reported. The predominant role of nonadditive gene action total soluble solids, millable sweet-stalk yield, and extractable juice yield indicates the importance of breeding for heterosis for improving these traits.

19.9 SORGHUM JUICE HARVEST SCENARIO

The options for harvesting sorghum include removing the whole or chopped stalk or pressing the sugar-rich juice in the field and removing only the juice or the juice and pressed stalk. Chopped stalks can be collected with traditional forage harvesters that are readily available and can be easily adapted to harvest sweet sorghum. The primary advantage of this approach includes the rapid loss of moisture in the first 24 h after chopping. In-field juice harvesters expel the sugar-rich juice during harvest and can thus eliminate the cost of transporting stalk material. This approach may also permit the use of low-cost, on-farm fermentation as an alternative to large-scale processing/fermentation facilities. A prototype juice harvester has been successfully demonstrated with sweet sorghum varieties. The juice needs to be extracted to reduce the cost of transportation and speed-up the drying of the stalks. The stalks contain from 40 to 50% moisture after extraction, depending on the efficiency of the extraction process. The disadvantage of extracting the juice before burning is that it reduces the Btu per pound by approximately 5–8%.

19.9.1 FACTORS AFFECTING JUICE PRODUCTION

Sugar content was highest in the middle of the plant stalk for sweet sorghum (Janssen et al. 1930). The top 300–450 mm of the stalk could be removed without significant loss of juice and sugar. Plants expressed 3 days after harvest and stripped of their leaves had higher sugar contents in the juice than those expressed immediately after harvest (Janssen et al. 1930). Other experiments have shown that juice yields decreased and sugar contents increased for sweet sorghum stalks stored 48 h between harvest and expression (Broadhead 1972). The change in sugar content and juice yield after storage was attributed to evaporation of water from the plant in both cases. Juice yields from plants are also affected by the amount of moisture in the soil (Janssen et al. 1930). The amount of solar radiation received by sweet sorghum is responsible for 75% of the variation in plant crop yield (Hipp et al. 1970). The top 300 juice yields from plants are also affected by stalk, and juice yield increased linearly as the amount of solar radiation received between the boot and early seed formation stages increased. Row spacing had a highly significant effect on the fresh mass of the stalk and total plant of Rio variety sweet sorghum. Fresh plant mass yields were higher for narrow row spacing. Sweet sorghum planted in a narrow-row spacing had a greater leaf area compared with wide row spacing (Wortmann et al. 2010).

19.9.2 POTENTIAL USES OF SWEET SORGHUM JUICE

There are a number of uses of sweet sorghum juice:

1. Sweet sorghum syrup has been produced in the United States since colonial days. Some sweet sorghum syrup has at one time or another been produced in every one of the contiguous 48 states. Sweet sorghum is grown extensively for syrup production in the southeastern states. Kentucky is one of eight states in the Southeast and Midwest producing approximately 90% of the total U.S. output. Excellent quality syrup could be made when the brix of raw juice was at least 14°C, which was more or less throughout the year. The prepared syrup generally has a final brix of 70–75°C (corresponding to syrup temperatures of approximately 106°C) and a minimum shelf-life of 6–9 months.
2. Ethanol can be produced from sweet sorghum juice using yeast. Immediately after harvesting, the fermentation process must begin. Fermentation can take place in large storage containers in the environment without temperature control.
3. Sweet sorghum juice could be returned back to the soil to provide nutrients. Each 1000 gal of juice contain 10 lb of nitrogen, 10 lb of potassium, and 10 lb of phosphorus, in addition to micronutrients.

19.10 CONCLUSIONS

S. bicolor, a diploid, has a relatively small genome (735 Mbp), which although larger than rice (389 Mbp) is smaller than the other important cereals (wheat 16,900 Mbp and maize 2600 Mbp). The last genome duplication event for the *S. bicolor* genome seems to have occurred much earlier than the divergence of the major cereal crops from a common ancestor. Completion of the whole genome sequencing project in 2009 will exponentially increase the sequence data available for *Sorghum* spp. and will provide valuable information on cereal domestication in the African continent, an event that appears to have occurred independently of other continents although by similar reinforced selective pressures. In a way, sorghum genome sequencing will close a biographic triangle into the knowledge of the polymorphism shared before the divergence of these important grasses and ultimately in the understanding of the evolution in cereals crops among Africa, the Americas, and Asia. The tenets of colinearity and microlinearity of grass genomes mean that our knowledge of other cereals and their evolutionary ties will also greatly improve. Because of their economic and scientific value, cereal genomes have been studied over the last 15 years using highly advanced technologies. The similarity at the DNA level makes it possible to use comparative genetics to look for particular genes of unknown sequence among the genomes with the aim of using that information to develop new varieties or discovering new genes that could have a potential impact on traits that are of global importance (e.g., food quality and drought resistance).

Sweet sorghum has long been known to be an excellent source of sugar which can easily be fermented and distilled into fuel-grade ethanol. The main factor keeping sweet sorghum from competing with corn as a fuel crop is the lack of an established production method. Mechanically harvesting sweet sorghum requires either a specialized harvester capable of extracting the sugary juice from the stalks in the field or a modified sugarcane harvester and a large nearby pressing facility. The juice must be quickly fermented to prevent degradation of the sugars in the juice. Also, the costs associated with transportation of the crop to the mill will be the major limiting factor for where sweet sorghum can be profitably grown. Varieties that have higher sugar contents per ton of biomass will be more efficient to process and haul to the mill.

REFERENCES

Agyeman GA, Loiland J, Karow R, Payne WA, Trostle C, Bean B (2002) *Grain Sorghum. 2002.* Report 8794, Oregon State University Extension Service, Corvallis, OR.

Almeida-Dominguez HD, Sema-Saldivar SO, Rooney LW (1991) Properties of new and commercial sorghum hybrids for use in alkaline-cooked foods. *Cereal Chem* 68:25–30

Ayyangar G, Ayyar M, Rao V, Nambiar A (1936) Mendelian segregation for juiciness and sweetness in sorghum stalk. *Madras Agric. J.* 24, 247

Boivin K, Deu M, Rami JF, Trouche G, Hamon P (1999) Towards a saturated sorghum map using RFLP and AFLP markers. *Theor Appl Genet* 98:320–328

Bowers JE, Abbey C, Anderson S, Chang C, Draye X, Hoppe AH, Jessup R, Lemke C, Lennington J, Li Z, et al. (2003) A high-density genetic recombination map of sequence-tagged sites for Sorghum, as a framework for comparative structural and evolutionary genomics of tropical grains and grasses. *Genetics* 165:367–386

Bowers JE, Arias MA, Asher R, Avise JA, Ball RT, Brewer GA, Buss RW, Chen AH, Edwards TM, Estill JC, Exum HE et al (2005) Comparative physical mapping links conservation of microsyteny to chromosome structure and recombination in grasses. *Proceedings of the National Academy of Sciences of the United States of America* 102:13206–13211

Broadhead DM (1972) Effects of stalk chopping on leaf removal and juice quality of Rio sweet sorghum. *Agron J* 64:306-308.

Buchanan C, Lim S, Salzman RA, Kagiampakis I, Klein RR, Pratt LH, Cordonnier-Pratt M-M, Klein PE, Mullet J (2005) *Sorghum bicolor*'s transcriptome response to dehydration, high salinity and ABA. *Plant Mol Biol* 58:699–720

Carter PR, Hicks DR, Oplinger ES, Doll JD, Bundy LG, Schuler RT, Holmes BJ (1989) *Grain Sorghum (Milo).* Alternative Field Crops Manual

Casa AM, Mitchell SE, Hamblin MT, Sun H, Bowers JE, Paterson AH, Aquadro CF, Kresovich S (2005) Diversity and selection in sorghum: Concurrent studies with simple sequence repeats (SSRs). *Theor Appl Genet* 111:23

Casler MD, Vogel KP, Taliaferro CM, Wynia RL (2004) Latitudinal adaptation of switchgrass populations. *Crop Sci* 44:293–303

Chiaramonti D, Grassi G, Nardi A, Grimm HP (2004) *ECHI-T:Large Bio-Ethanol Project from Sweet Sorghum in China and Italy*. Energia Trasporti Agricoltura, Florence, Italy

Cisse N (1995) Heritability estimates, genetic correlation, and identification of RAPD markers linked to seedling vigor and associated agronomic traits in sorghum. Ph.D. thesis. Purdue University, West Lafayette

Cothren JT, Matocha JE, Clark LE (2000) Integrated crop management for sorghum. In: Smith CW, Frederiksen RA (eds), *Sorghum: Origin, History, Technology, and Production*. John Wiley, New York, pp 409–441

Dje Y, Heuertz M, Lefebvre C, Vekemans X (2000) Assessment of genetic diversity within and among germplasm accessions in cultivated sorghum using microsatellite markers. *Theor Appl Genet* 100:918–925

Doggett H (ed) (1988) *Sorghum*, 2nd ed. John Wiley, New York

Duodu K, Taylor J, Belton P, Hamaker B (2003) Factors affecting sorghum protein digestibility. *J Cereal Sci* 38:117–131

Edwards GE, Franceschi VR, Voznesenskaya EV (2004) Single-cell C_4 photosynthesis versus the dual-cell (Kranz) paradigm. *Annu Rev Plant Biol* 55:173–196

Erie, LJ, French OF, Bucks DA, and Harris K (1981) Consumptive use of water by major crops in the southwestern United States. USDA. Conservation Research Report No. 29. Phoenix, Ariz. 42 pp

FAO (2005), available at http://apps.fao.org/page/collections?subset=agriculture

FAO (1995) Sorghum and millet in human nutritions. *Food and Nutrition Series* 27, U.N. Food and Agricultural Organization, Rome, Italy

Farrell AE, Plevin RJ, Turner BT, Jones AD, O'Hare M, Kammen DA (2006) Ethanol can contribute to energy and environmental goals. *Science* 311:506–508

Fasano A, Catassi C (2001) CUlTent approaches to diagnosis and treatment of celiac disease: An evolving spectrum. *Gastroenterology* 120:636–651

Fawusi M, Agboola A (1980) Soil moisture requirement for germination of sorghum, millet, tomato and Celosia. *Agronomy J.* 72:353–357

Figueroa JDC, Martinez BF, Rios E (1995) Effect of sorghum endosperm type on the quality of adjuncts for the brewing industry. *J Amer Soc Brewing Chem* 53:5–9

Gardner DP, Blad BL, Garrity DP, Watts DG (1981) Relationships between crop temperature and the physiological and phenological development of differentially irrigated corn. *Agron. J.* 73:743–747

Graser EA (1985) Micrometeorology of sorghum at two row spacing. PhD. Thesis. University of Nebraska, Lincoln. 389 pp

Gnansounou E, Dauriat A, Wyman C (2005) Refining sweet sorghum to ethanol and sugar: Economic trade-offs in the context of North China. *Bioresour Technol* 96:985–1002

Harlan JR, de Wet JM (1972) A simplified classification of cultivated sorghum. *Crop Sci* 12:172–176

Haussmann G, Hess E, Seetharama N, Welz G, Geiger H (2002) Construction of a combined sorghum linkage map from two recombinant inbred populations using AFLP, SSR, RFLP, and RAPD markers, and comparison with other sorghum maps. *Theor. Appl. Genet.* 105:629–637

Hedge BR, Major DJ, Wilson OB, Krogman KK (1976) Effect of row spacing and plant population density on grain sorghum production in southern Alberta. *Can J Plant Sci* 56:31–37

Hicks C, Tuinstra MR, Pedersen JF, Dowell FE, Kofoid KD (2002) Genetic analysis of feed quality and seed weight of sorghum inbred lines and hybrids using analytical methods and NIRS. *Euphytica* 127:31–40

Hipp B, Cowley W, Gerard C, Smith B (1970) Influence of solar radiation and date of planting on yield of sweet sorghum. *Crop Sci* 10:91–92

Hopkins AA, Vogel KP, Moore KJ, Johnson KD, Carlson IT (1995) Genotype effects and genotype by environment interactions for traits of elite switchgrass populations. *Crop Sci* 35:125–132

Hopkinson CS, Day JW (1980) Modeling hydrology and eutrophication in a Louisiana swamp forest ecosystem. *Environmental Management*, 4:325–335

Hoseney RC (1994) *Principles of Cereal Science and Technology*. Association of Cereal Chemists, St. Paul, MN, USA.

House LR, Osmanzai M, Gomez MI, Monyo ES, Gupta SC (1995) Agronomic principles. In: Dendy DAV (ed), *Sorghum and Millets: Chemistry and Technology*. American Association of Cereal Chemists, St. Paul, MN, pp 27–68

Hulan HW, Proudfoot FG (1982) Nutritive value of sorghum grain for broiler chickens. *Can J Anim Sci* 62:869–875

Hunter EL, Anderson IC (1997) Sweet sorghum. *Hort Rev* 21:73–104

Janssen G, McClelland C, Metzger W (1930) Superextraction of sorghum and the localization of juice and sugars in internodes of the plants. *J Amer Soc Agron* 22:627:629

Karpenstein-Machan M (2001) Sustainable cultivation concepts for domestic energy production from biomass. *Crit Rev Plant Sci* 20:1–14

Kim JS, Klein PE, Klein RR, Price HJ, Mullet JE, Stelly DM (2005) Chromosome identification and nomenclature of *Sorghum bicolor*. *Genetics* 169:1169–1173

Klein PE, Klein RR, Cartinhour SW, Ulanch PE, Dong J, Obert JA, Morishige DT, Schlueter SD, Childs KL, Ale M, Mullet JE (2000) A high-throughput AFLP-based method for constructing integrated genetic and physical maps: progress toward a sorghum genome map. *Genome Res* 10:789–807

Klein PE, Klein RR, Vrebalov J, Mullet JE (2003) Sequence-based alignment of sorghum chromosome 3 and rice chromosome 1 reveals extensive conservation of gene order and one major chromosomal rearrangement. *Plant J* 34:605–621

Kriegshauser TD, Hancock JD, Tuinstra MR (2006) Variation in nutritional value of sorghum hybrids with contrasting seed weight characteristics and comparisons with maize in broiler chicks. *Crop Sci* 46:695–699

Lemus R, Brummer EC, Moore KJ, Molstad NE, Burras CL, Barker MF (2002) Biomass yield and quality of 20 switchgrass populations in southern Iowa, USA. *Biomass Bioenergy* 23:433–442

Li G, Gu W, Chapman K (2004) *Sweet Sorghum*. Chinese Agriculture Technology and Sciences Publishing House, Beijing

Lin Y, Schertz K, Paterson A (1995) Comparative analysis of QTLs affecting plant height and maturity across the Poaceae, in reference to an interspecific sorghum population. *Genetics* 141:391–411

Lochte-Watson KR, Jackson DS, Weller CL (2000) Fractionation of grain sorghum using abrasive decortication. *J Agric Eng Res* 77:203–208

Maman N, Mason NC, Lyon D, Dhungana P (2004) Yield components of pearl millet and grain sorghum across environments in the Central Great Plains. *Crop Science* 44:2138–2145

Mann IA, Kimber CT, Miller FR (1983) The origin and early cultivation of sorghums in Africa. Report 1454, Texas Agricultural Experiment Bulletin, Texas A&M University, College Station, TX

Mann JA, Kimber, CT Miller FR (1983) The origin of early cultivation of sorghum in Africa. Texas AAT University Bulletin, 1984

Maunder AB (2002) Sorghum worldwide. In: Leslie JF (ed), *Sorghum and Millet Diseases*. Iowa State Press, Ames, IA pp 11–18

Menz MA, Klein RR, Mullet JE, Obert JA, Unruh NC, Klein PE (2002) A high-density genetic map of *Sorghum bicolor* (L.) Moench based on 2926 AFLP(R), RFLP and SSR markers. *Plant Mol Biol* 48:483–499

Menz MA, Klein RR, Unruh NC, Rooney WL, Klein PE, Mullet JE (2004) Genetic diversity of public inbreds of sorghum determined by mapped AFLP and SSR markers. *Crop Science Madison* 44:1236–1244

Miller FR, McBee GG (1993) Genetics and management of physiologic systems of sorghum for biomass production. *Biomass Bioenergy* 5:41–49

Mulkey VR, Owens VN, Lee DK (2006) Management of switchgrass-dominated Conservation Reserve Program lands for biomass production in South Dakota. *Crop Sci.* 46:712–720

Munck L (1995) New milling technologies and products: Whole plant utilization by milling and separation of the botanical and chemical components. In: Dendy DAV (ed), *Sorghum and Millets: Chemistry and Technology*. American Association of Cereal Chemists, St. Paul, MN, pp 223–281

Park S, Bean SR, Wilson JD, Schober TJ (2006) Rapid isolation of sorghum and other cereal starches using sonication. *Cereal Chem* 83:611–616

Paterson AH, Bowers JE, Bruggmann R, Dubchak I, et al. (2009) The *Sorghum bicolor* genome and the diversification of grasses. *Nature* 457:551–556

Peng Y, Schertz KF, Cartinhour S, Hart GE (1999) Comparative genome mapping of *Sorghum bicolor* (L.) Moench using an RFLP map constructed in a population of recombinant inbred lines. *Plant Breed* 118:225–235

Pereira MG, Lee M (1995) Identification of genomic regions affecting plant height in sorghum and maize. *Theor Appl Genet* 90:380–388

Porter K, Axtell J, Lechtenberg V, Colenbrander V (1978) Phenotype, fiber composition, and in vitro dry matter disappearance of chemically induced brown midrib (*bmr*) mutants of sorghum. *Crop Sci* 18:205–208

Porter KB, Jensen ME, Slettern WH (1960) The effect of row spacing, fertilizer, and planting rate on the yield and water use of irrigated grain sorghum. *Agron. J.* 52:431–434

Pratt LH, Liang C, Shah M, Sun F, Wang HM, Reid SP, Gingle AR, Paterson AH, Wing R, Dean R, et al. (2005) Sorghum expressed sequence tags identify signature genes for drought, pathogenesis, and skotomorphogenesis from a milestone set of 16,801 unique transcripts. *Plant Physiol* 139:869–884

Quinby JR (1967) The maturity genes in sorghum. In: Norman AG (ed), *Advance in Agronomy*, Vol. 19. Academic Press, New York, pp 267–305

Ragaee S, Abdel-Aal ES (2006) Pasting properties of starch and protein in selected cereals and quality of their food products. *Food Chem* 95:9–18

Reddy B, Ramesh S, Reddy S, Ramaiah B, Salimath P, Kachapur R (2005) Sweet sorghum—A potential alternate raw material for bio-ethanol and bio-energy. *J SAT Agric Res* 1:1–8

Renewable Energy World (2000) *Bioethanol-Industrial World Perspective*, available at www.jxj.com/magsandj/rew/200_03/bioethanol.html

Ribera LA, Outlaw JL, Richardson JW, Silva JD, Bryant H (2007) *Integrating Ethanol Production into a U.S. Sugarcane Mill: A Risk Based Feasibility Analysis*. Texas AgriLife Research. Texas AgriLife Extension Service, TX, USA

Rooney LW, Awika JM (2004) Specialty sorghums for healthful foods. In: Abdel-Aal ES, Woods P (eds), *Specialty Grains for Food and Feed*. American Association of Cereal Chemists, St. Paul, MN, pp 283–312

Rooney LW, Khan MN, Earp CF (1980) The technology of sorghum products. In: Inglett GE, Munck L (eds), *Cereals for Food and Beverages: Recent Progress in Cereal Chemistry*. Academic Press, New York, NY, pp 513–554

Rooney LW, Serna-Saldivar SO (2000) Sorghum. In: Kulp K, Ponte Jr JG (eds), *Handbook of Cereal Science and Technology*. Marcel Dekker, New York, pp 149–175

Rooney LW, Waniska RD (2000) Sorghum food and industrial utilization, in *Sorghum: Origin, History, Technology, and Production*. C.W. Smith and R.A. Frederiksen (eds.), John Wiley & Sons, Inc., New York, NY, pp 689–729

Salzman RA, Brady JA, Finlayson SA, Buchanan CD, Sun F, Klein PE, Klein RR, Pratt LH, Cordonnier-Pratt M-M, Mullet JE (2005) Transcriptional profiling of sorghum induced by methyl jasmonate, salicylic acid, and aminocyclopropane carboxylic acid reveals cooperative regulation and novel gene responses. *Plant Physiol* 138:352–368

Sang Y, Bean S, Seib PA, Pedersen JF, Shi Y (2008) Structure and functional properties of sorghum starches differing in amylose content. *J Agric Food Chem* 56:6680–6685

Schmer MR, Vogel KP, Mitchell RB, Moser LE, Eskridge KM, Perrin RK (2006) Establishment stand thresholds for switchgrass grown as a bioenergy crop. *Crop Sci* 46:157–161

Schwarte AJ, Gibson LR, Karlen DL, Liebman M, Jannink J-L (2005) Planting date effects on winter triticale dry matter and nitrogen accumulation. *Agron J* 97:1333–1341

Selle P, Cadogan D, Li X, Bryden WL (2010) Implications of sorghum in broiler chicken nutrition. *Anim Feed Sci Technol* 156:57–74

Shelton DR Lee WJ (2000) Cereal carbohydrates. Kulp, K. & Ponte, j. G. (eds). *Handbook of Cereal Science and Technology* (2nd ed.) Marcel Dekker, New York 385–415

Singh V, Moreau RA, Hicks KB (2003) Yield and phytosterol composition of oil extracted from grain sorghum and its wet-milled fractions. *Cereal Chem* 80:126–129

Sivakumar, NV, Seetharama KN, Singh S, Bidinger FR (1979) Water relations, growth, and dry matter accumulation of sorghum under post-rainy season conditions. *Agron. J.* 71:843–847

Snowden, JD (1936) The *Cultivated Races of Sorghum*. Adlard and Son, London

Soltani A, Almodares A (1994) Evaluation of the investments in sugar beet and sweet sorghum production. Presented at the National Convention of Sugar Production from Agricultural Products, Shahid Chamran University, Alwaz, Iran, March 13–16, 1994

Sullivan CY, Blum A (1970) Drought and heat resistances in sorghum and corn. In: *Proc. Of the 25th Annual Corn Sorghum Research Conference*, Wichita, KS, USA pp 555–625

Tao YZ, Jordan DR, Henzell RG, Mcintyre CL (1998) Construction of a genetic map in a sorghum RIL population using probes from different sources and its comparison with other sorghum maps. *Aust J Agric Res* 49:729–736

Undersander DJ, Smith LH, Kaminski AR, Kelling KA, Doll JD (1990) *Sorghum Forage. Alternative Field Crops Manual 1990*. Cooperative Extension Service, University of Wisconsin, Madison, WI

Vanderlip RL (1993) *How a Sorghum Plant Develops*. Contribution No. 1203, Kansas Agriculture Experimental Station, Manhattan, KS

Vavilov NI (1951) The origin, variation, immunity, and breeding of cultivated plants. In *Selected Writings of N.I. Vavilov*. Chronica, Botanica

Vietor D, Miller F (1990) Assimilation, partitioning, and nonstructural carbohydrates in sweet compared with grain sorghum. *Crop Sci* 30:1109–1115

Vogel KP, Brejda JJ, Walters DT, Buxton DR (2002) Switchgrass biomass production in the midwest USA: Harvest and nitrogen management. *Agron J* 94:413–420

Weller CL (2006) *Development of Sorghum Lipids and Nutraceuticals.* Project 2005 Annual Report. University of Nebraska, Lincoln, NE

Wikipedia contributors (2006) Sorghum. Wikipedia, The Free Encyclopedia. Retrieved May 18, 2006 from http://en.wikipedia.org/wiki/Sorghum

Wong J, Lau T, Cai N, Singh J, Pedersen JF, Vensel WH, Hurkman WJ, Wilson JD, Lemaux PG, Buchanan BB (2009) Digestibility of protein and starch from sorghum (*Sorghum bicolor*) is linked to biochemical and structural features of grain endosperm. *J Cereal Sci* 49:73–82

Worley W, Vaughan D, Cundiff J (1992) Energy analysis of ethanol production from sweet sorghum. *Bioresour Technol* 40:263–273

Wortmann C, Ferguson RB, Klein RN, Lyon DJ, Liska AJ, Dweikat I (2010) Dryland performance of sweet sorghum and grain crops for biofuel. *Agron J* 102:319–326

Xin Z, Wang ML, Barkley NL, Burow GB, Franks CD, Pederson GA, Burke JJ (2008) Applying genotyping (TILLING) and phenotyping analyses to elucidate gene function in a chemically induced sorghum mutant population. *BMC Plant Biol* 8:103

20 Soybean

Babu Valliyodan
University of Missouri

Jeong-Dong Lee
Kyungpook National University

Grover J. Shannon and Henry T. Nguyen
University of Missouri

CONTENTS

20.1   Introduction 503
20.2   Brief History of Soybean Production and Oil Improvement 504
20.3   Soybean Oil Quantity 507
20.4   Palmitic Acid 508
20.5   Stearic Acid 511
20.6   Oleic Acid 512
20.7   Linoleic Acid 513
20.8   Linolenic Acid 514
20.9   Environmental Effect on Oil and Fatty Acid Concentration in Soybean 515
20.10 Genetic Engineering for Bioenergy Traits in Soybean 516
20.11 Conclusions and Future Perspectives 517
Acknowledgment 518
References 518

20.1 INTRODUCTION

Bioenergy, the energy derived from biomass, cannot substitute entirely for fossil fuels given present agricultural practices, but it can help to reduce the use of fossil fuels to a certain extent. The main type of bioenergy production is the burning of solid materials or liquid biofuel derived from major bioenergy crops. These diverse and sustainable bioenergy crops will noticeably improve global energy security and environmental benefits. In the United States, most ethanol is produced from corn whereas biodiesel is made from soybean oil or recycled cooking oils. Production and usage of biodiesel, fatty acid alkyl esters, has become more attractive because of its environment friendliness and their origin from the renewable resources (Shieh et al. 2003). Alkali-catalyzed transesterification of oil or fatty acids with the help of an alcohol is the base reaction and produces fatty acid methyl esters and glycerol. Through using the biocatalysts or the enzymatic biodiesel, production has more commercial potential than the complete chemical approach (Hass et al. 2002; Shieh et al. 2003; Tan et al. 2010). Soybean oil has been broadly studied as a raw material for fatty acid methyl ester biodiesel production by transesterification (Abreu et al. 2003; Suppes et al. 2004; Sensoz and

Kaynar 2006). When compared with the other oil seed crops, soybean acreages are greater in the United States and this contributes toward the higher availability of soybean oil for biofuel feedstock. Soybean oil is increasingly used as feedstock for the production of biodiesel.

20.2 BRIEF HISTORY OF SOYBEAN PRODUCTION AND OIL IMPROVEMENT

Soybeans are native to the Far East and were domesticated and grown as early as 5000 years ago by farmers in China. They were first brought to North America in 1804, and late in the century U.S. farmers began to grow soybeans as forage for cattle. In 1904, George Washington Carver began studying the soybean at the Tuskegee Institute and determined that it provided a valuable and versatile source of protein and oil (www.soystats.com). Because of its unique chemical composition, soybean seed has become a very valuable and useful agricultural commodity. Among legumes and cereals it has the highest protein content (40%) and the second-highest oil content (20%). Other valuable components include vitamins, minerals, and isoflavones that are important to human and animal health (Liu 1999).

As the production of and the potential uses for soybeans increased, the need for greater genetic diversity in them was recognized. In the mid-1920s, William J. "Bill" Morse collected more than 10,000 soybean accessions. Several of these germplasm accessions laid the foundation for the development of productive varieties and the rapid ascension of the United States and South America as world leaders in soybean production (www.soystats.com). Now, more than 80% of the world's soybeans are grown by the United States (33%), Brazil (28%), and Argentina (21%), as shown in Figure 20.1. Soybean demand around the world continues to increase because of its usefulness as a renewable source for numerous foodstuffs and industrial products, including bioenergy products from soybean oil (Figure 20.2).

Before World War II, the United States imported 40% of its edible fats and oils. At the advent of the war, this oil supply was cut and processors turned to soybean oil. Soybean oil now provides 56% of the world vegetable oil production, which far exceeds oil produced from any other crop (Figure 20.3). Soybean oil has also been shown to be a renewable source of cleaner-burning biodiesel when used as a substitute for petroleum-based fuels. Nearly 6 L of biodiesel and approximately 22 kg of soybean meal can be produced from 27 kg of soybeans. Recently, high fuel costs prompted the U.S. Congress to create biodiesel tax incentives; biodiesel use has increased from 95 million L in 2004 to 2.6 billion L in 2008 (Figure 20.4).

Although soybeans are used in many products, they are far from perfect. Objectionable characteristics such as a "beany" flavor, indigestible carbohydrates, antinutritional factors, oxidative

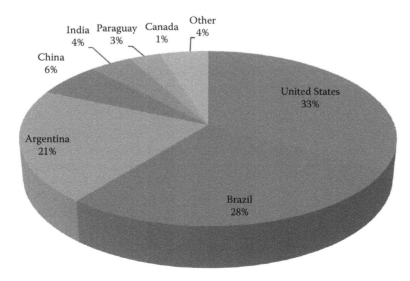

FIGURE 20.1 (See color insert) World oil seed production for major oilseed crops.

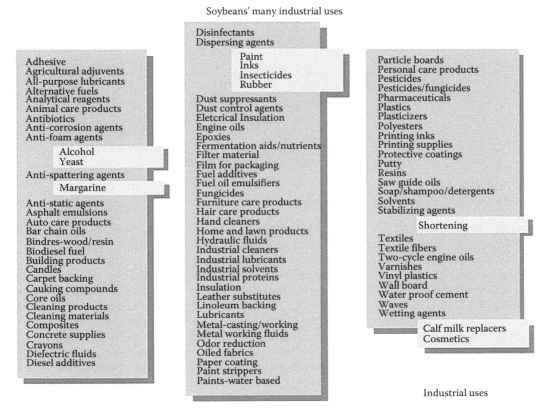

Soybeans' many industrial uses

Industrial uses

FIGURE 20.2 (See color insert) Production of soybeans for major soybean producing countries.

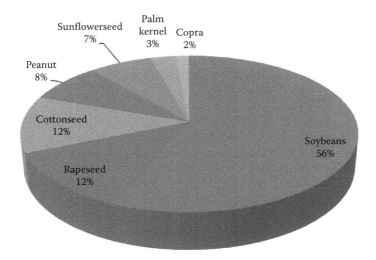

FIGURE 20.3 Industrial use of soybean including use for bioenergy.

stability, poor cold flow of soy biodiesel, and deficiency of sulfur-containing amino acids limit their use for food, feed, and industrial uses (Liu 1999; Wilson 2004).

Five fatty acids make up nearly the entire oil portion of soybean seed. Soybean oil averages 12% palmitic acid (16:0), 4% stearic acid (18:0), 23% oleic acid (18:1), 53% linolenic acid (18:2), and 8% linolenic acid (18:3). The 16:0 and 18:0 fractions are saturated fatty acids and constitute 15% of the

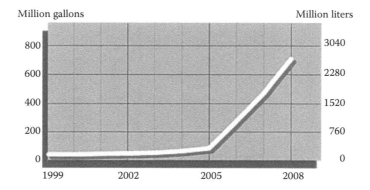

FIGURE 20.4 (See color insert) Biodiesel consumption in the United States.

soybean oil. The remainder of the oil (~85%) is made up of unsaturated fatty acids or 18:1, 18:2, and 18:3. Transgenic, induced, and natural mutations, including those combining two or more genes, have been used in breeding soybeans for enhanced and improved oil content. Several genes have been discovered that affect fatty acid levels in soybeans, enabling breeders to combine genes to tailor novel fatty acid profiles desired for various end uses. Research since the 1970s has led to a greater understanding of how to genetically alter the fatty acid of soybean oil. Because consumer and end-user preferences for soybean oil are changing, breeding is needed to modify the fatty acid composition in soybean oil to fulfill these new demands in industrial, food, and other products (Wilson 2004).

Soybean research priorities have been set, with guidance from consumers and end-users, to initially target fatty acid profiles that have the highest probability to facilitate expanded use of soybean oil in edible and industrial applications in the United States (Wilson 2004). The most visible of these programs was "The Better Bean Initiative" launched in the United States in 2000 by the United Soybean Board. This program's aim is to add value in the oil and protein seed components by genetically changing objectionable characteristics. These deliberations have focused on three different oil phenotypes: saturates, oleic acid, and linolenic acid (Table 20.1). Modification of these oils through breeding and biotechnology is being emphasized to develop the desired fatty acid phenotypes. It is impractical to commercially develop all oil phenotypes. Perhaps the most desired phenotype for soybean oil is less than 7% saturates [palmitic (16:0) and stearic acids (18:0)], more than 55% oleic acid (18:1), and less than 3% linolenic acid (18:3) because of its multiple uses in many food and industrial applications (Wilson 2004).

Human health concerns and improved cold flow of biodiesel have stimulated research to develop soybeans with less saturated fat. The U.S. Food and Drug Administration's requirement for a food to be labeled "low in saturated fat" is less than 1 g of saturated fat per serving. This means that the fatty acid composition of vegetable oil should contain less than 7% total saturated fat to make such a claim. Lowering soybean from 15% saturated fat to 7% or less will make soybean oil more attractive to food manufacturers and health-conscious consumers while additionally improving the cold flow of biofuels made from soybean oil (Wilson 2004). On the other hand, soybean oil high in saturated fats would make hydrogenation less necessary and is in demand for making margarines and shortenings without *trans*-fats for the evolving health-conscious society.

Increased oleic acid (18:1) in soybean oil is important because of its health benefits and increased oxidative stability. A diet in which fat consumption is high in 18:1 is associated with reduced cholesterol, arteriosclerosis, and heart disease (Grundy 1986; Wardlaw and Snook 1990; Chang and Huang 1998). High oleic acid content also increases oxidative stability and extends the utility of soybean oil at high cooking temperatures. It also will significantly increase soybean oil use in pharmaceuticals, cosmetics, and industrial products such as lubricants and biodiesel. To improve acceptance of soydiesel, a high 18:1 level combined with lower saturates (16:0 + 18:0) is needed. This will improve ignition and cold flow in cooler climates (Wilson 2004).

TABLE 20.1

Goals for Redesigning Soybean Oil Composition for Specific Food and Industrial Applications

| | Percent Crude Soybean Oil | | | |
| | | Desired Composition for Specific Use | | |
Fatty Acid	Normal Oil	Frying	Baking	Industrial
Saturated (16:0 + 18:0)	15	7	42	11
Oleic (18:1)	23	60	19	12
Linoleic (18:2)	53	31	37	55
Linolenic (18:3)	9	2	2	22

Source: Wilson, RF., *Soybeans: Improvement, Production and Uses*, 3rd ed, American Society for Agronomy, Madison, WI, 621–677, 2004.

Soybean oil contains a high concentration of polyunsaturated linoleic (18:2) and linolenic (18:3) acids. These fatty acids have a high number of double bonds, which are susceptible to oxidation, resulting in reduced shelf life, low stability at high cooking temperatures, and off-flavors. Oxidation of linolenic acid with three double bonds causes this fatty acid to contribute most to the poor functionality of soybean oil. To improve oxidative stability and undesirable taste, soybean oil is hydrogenated to reduce double bonds, which are sites of oxidative attack and subsequent off-flavor development (Yadav 1996; Liu 1999). Partial hydrogenation of soybean oil increases oxidative stability but leads to the formation of *trans*-fats. Demand from oil seed processors for a lower-cost alternative to catalytic hydrogenation for producing oil products with desired flavor and functionality led to research to breed soybeans containing lower linolenic acid (Liu 1999). More recently, health concerns and labeling laws that require listing the amount of *trans*-fatty acids in foods have prompted food companies to seek alternative oils to replace hydrogenated soybean oil to ensure that their products contain low levels of *trans*-fats. Thus, emphasis on soybean oil with 3% linolenic acid or less has become a high priority (Yadav 1996; Wilson 2004).

In most Asian countries, soybeans are usually consumed directly in products such as tofu, sprouts, soybean paste, and health supplements. As essential fatty acids, linoleic and linolenic acids are desirable components of soybean oil. Linoleic acid ($\omega6$) and linolenic acid ($\omega3$) are potential precursors of eicosapentaenoic acid (EPA; 20:5) and docosahexaenoic acid (DHA) (22:6), which can have multiple positive health benefits in diets including reduction of cardiovascular disease and improved cognitive function (Connor 2000; Brouwer et al. 2004; Gebauer et al. 2006). Studies suggested that adjusting the intake ratio of $\omega6$ to $\omega3$ fatty acids may enhance overall health (Able et al. 2004). Thus, for health benefits, elevated 18:3 genotypes are desirable in food-grade soybeans. Also, oils low in saturates and high in polyunsaturated fatty acids and linoleic and linolenic acids would have applications for replacing drying oils such as tung and linseed oil in oil-based paints and coatings (Wilson 2004).

Because of the demand for healthier, more functional vegetable oils, greater emphasis is being placed on increasing oil content and modifying the fatty acid profile in soybean seed through conventional breeding and promising biotechnology approaches to target genes that affect oil quantity and quality.

20.3 SOYBEAN OIL QUANTITY

Typically, soybean seeds contain approximately 20% oil on a dry weight basis (Wilson 2004). Crude oil contains various glycerolipids, primarily phospholipids, diacylglycerol, and triacylglycerol. Triacylglycerol is the main component of the oil. Phospholipids such as phosphatidylcholine, phosphatidylethanolamine, and phosphatidylinositol have structural functions in cell membranes and may be metabolically involved in triacylglycerol synthesis. Each glycerolipid class is composed of molecular species formed by various combinations of the five fatty acids; 16:0, 18:0, 18:1, 18:2, and 18:3; or polar groups that are esterified at the *sn*-1, *sn*-2, and *sn*-3 stereospecific positions of the glycerol molecule.

The increased demand for renewable fuels such as biodiesel has increased the emphasis on breeding soybeans with higher oil content. Genetic sources are very important for improving soybean oil content. There is a wide range (8–25% on a dry weight basis) in seed oil content among accessions in the U.S. Department of Education (USDA) soybean (*Glycine max*) germplasm collection (USDA, ARS 2007), with most soybean cultivars averaging approximately 20%. However, genes affecting oil content and oil biosynthesis can be affected by the environmental conditions such as temperature and rainfall. Also, oil and protein content in soybean seed are negatively correlated. Thus, it will be difficult to develop soybeans with simultaneously very high oil content and a 40% protein content (dry weight basis) through conventional breeding approaches.

Soybean breeders are interested in increasing soybean oil content using major oil loci for marker-assisted selection (MAS). Soybeans with stable oil content across environments will be developed by combining confirmed quantitative trait loci (QTL). Many genes have been reported to affect oil concentration. Recent genetic maps show 68 QTL affecting soybean oil content have been found on all chromosomes [Chros. (linkage groups (LGs))] except Chro. 3 (LG N). These QTL have been reported in the Breeder's Toolbox in Soybase (http://www.soybase.org) and are shown in Table 20.2.

Diacylglycerol transferase (DGAT) is also a potential candidate gene for increasing oil content in soybean. In developing seeds, DGAT catalyzes the acyl-coenzyme A (acyl-CoA)-dependent acylation of *sn*-1,2-diacylglycerol (DAG) to generate triacylglycerol (TAG) (Weselake 2005). There are two distinct family genes of DGAT: DGAT1 and DGAT2. Since the first identification of the DGAT1 gene from *Arabidopsis* spp. (Routaboul et al. 1999), the influence of DGAT1 in increasing oil and fatty acid production was reported in several crops including soybean (Wang et al. 2006) DGAT1 has resulted in an 11–28% increase in seed oil content in homozygous napin (Jako et al., 2001); 47% in DGAT1-2 transformed maize (Zheng et al. 2008) and 14% in DGAT1-transformed canola (Weselake et al. 2005). In soybean, several reports indicate a role for DGAT in oil accumulation in developing seeds (Kwanyuen and Wilson 1990; Settlage et al. 1998). Sequence and expression levels of a DGAT1 gene were characterized in cultivated and wild soybean (Wang et al. 2006). However, the expression level of this gene was similar in different types of tissues, including mature leaves, flower, and seeds 20 and 30 days after flowering. Therefore, this gene may not be closely involved in oil production in soybean seeds. Two isoforms of DGAT1—DGAT1a (AB257589) with 7575 bp plus DGAT1b (AB257590) with 8164 bp—were reported, and these two proteins differ 4% in amino acid sequence but both have 14 introns and 15 exons. Although DGAT1b has greater activity compared with DGAT1a, TAG biosynthesis activity with DGAT1 from soybeans is 5-fold less than that of a DGAT from *Vernonia galamensis* (Hildebrand et al. 2008). It is not clear which isoform is more critical to oil biosynthesis in soybean.

The DGAT2 gene was discovered after DGAT1, and its influence on oil production in some crops was examined. DGAT2 was first identified in the fungus *Mortierella ramanniana* with two homologs DGAT2A and DGAT2B (Lardizabal et al. 2001). The expression of *M. ramanniana* DGAT2A in soybean led to a 1.5% increase (by weight) in seed oil with no significant impact on yield or protein content (Lardizabal et al. 2008). To date, the role of DGAT2 in oil accumulation in *Arabidopsis* spp. and common oilseed crops such as soybean has not been investigated.

20.4 PALMITIC ACID

Palmitic acid is the predominant saturated fatty acid in soybean oil. Generally, common soybean cultivars contain approximately 12% palmitic acid. Because of health risks associated with the cholesterogenic properties of saturated fatty acids and the poor cold flow properties they cause in biodiesel, reduced levels of palmitic acid are a goal in breeding for improved soybean oil quality.

TABLE 20.2

QTL Locations, Markers, Significance, and Parental Source Associated with Palmitic (16:0), Stearic (18:0), Oleic (18:1), Linoleic (18:2), and Linolenic (18:3) Fatty Acids in Soybean Oil

Fatty Acid	LG	Map Position	Marker	R^2 (%)	Parent 1	Parent 2	References
16:0	A1	3.5	Satt684	33	Cook	N87-2122-4	Li et al. (2002)v
16:0	A2	125.4	Satt133	12	N87-984-16	TN93-99	Panthee et al. (2006)
16:0	B2	38.0	A343_1	24	A81-365022	PI468916	Diers and Shoemaker (1992)
16:0	B2	53.5	A018_1	19	A81-365022	PI468916	Diers and Shoemaker (1992)
16:0	B2	–	UBC122-1500	10	RG10	OX948	Reinprecht et al. (2006)
16:0	D1b	75.7	Satt537	19	N87-984-16	TN93-99	Panthee et al. (2006)
16:0	D2	–	A_537 A_611	–	C1727	PI479750	Nickell et al. (1994)
16:0	D2	57.5	Sat_092	11	RG10	OX948	Reinprecht et al. (2006)
16:0	J	68.3	K375_1	18	A81-365022	PI468916	Diers and Shoemaker (1992)
16:0	L	89.1	–	13	Essex	Williams	Hyten et al. (2004)
16:0	M	67.0	Satt175	9	Cook	N87-2122-4	Li et al. (2002)
16:0	N	–	UBC444-2300	13	RG10	OX948	Reinprecht et al. (2006)
18:0	B2	55.2	Satt168	18	N87-984-16	TN93-99	Panthee et al. (2006)
18:0	B2	72.8–75.3	Satt070 Satt474 Satt556	61	Dare	FAM94-41	Specner et al. (2003)
18:0	C2	121.3	–	13	Essex	Williams	Hyten et al. (2004)
18:0	F	27.1	Sat_090	10	RG10	OX948	Reinprecht et al. (2006)
18:0	G	76.7	Satt288	10–19	RG10	OX948	Reinprecht et al. (2006)
18:0	J	83.3	A233_1	19	A81-365022	PI468916	Diers and Shoemaker (1992)
18:0	J	12.3	Satt249	11	N87-984-16	TN93-99	Panthee et al. (2006)
18:0	L	58.2	–	16	Essex	Williams	Hyten et al. (2004)
18:1	A1	92.3	A104_1	26	A81-365022	PI468916	Diers and Shoemaker (1992)
18:1	A1	92.6	A170_1	23	A81-365022	PI468916	Diers and Shoemaker (1992)
18:1	A1	102.3	A082_1	28	A81-365022	PI468916	Diers and Shoemaker (1992)
18:1	A1	96.0	Satt211	4	G99-G725	N00-3350	Monteros et al. (2008)
18:1	B2	–	A619_1	19	A81-365022	PI468916	Diers and Shoemaker (1992)
18:1	D2	79.2	Satt389	6	G99-G725	N00-3350	Monteros et al. (2008)
18:1	E	11.0	A242_2	20	A81-365022	PI468916	Diers and Shoemaker (1992)
18:1	E	13.6	Pb	21	A81-365022	PI468916	Diers and Shoemaker (1992)
18:1	E	45.4	Satt263	10	N87-984-16	TN93-99	Panthee et al. (2006)
18:1	G	0.0	Satt163	10	RG10	OX948	Reinprecht et al. (2006)
18:1	G	76.7	Satt288	10	RG10	OX948	Reinprecht et al. (2006)
18:1	G	43.4	Satt394	13	G99-G725	N00-3350	Monteros et al. (2008)
18:1	G	96.6	Satt191	7	G99-G725	N00-3350	Monteros et al. (2008)
18:1	L	82.5	–	35	Essex	Williams	Hyten et al. (2004)
18:1	L	30.9	Satt418	9	G99-G725	N00-3350	Monteros et al. (2008)

(Continued)

TABLE 20.2 (Continued)
**QTL Locations, Markers, Significance, and Parental Source Associated with Palmitic
(16:0), Stearic (18:0), Oleic (18:1), Linoleic (18:2), and Linolenic (18:3) Fatty Acids in
Soybean Oil**

Fatty Acid	LG	Map Position	Marker	R^2 (%)	Parent 1	Parent 2	References
18:1	L	71.4	Satt561	.25	G99-G725	N00-3350	Monteros et al. (2008)
18:2	A1	92.3	A104_1	33	A81-365022	PI468916	Diers and Shoemaker (1992)
18:2	A1	92.6	A170_1	30	A81-365022	PI468916	Diers and Shoemaker (1992)
18:2	A1	102.3	A082_1	38	A81-365022	PI468916	Diers and Shoemaker (1992)
18:2	B1	58.9	A118_1	20	A81-365022	PI468916	Diers and Shoemaker (1992)
18:2	B2	–	Fad3i6	70–75	RG10	OX948	Reinprecht et al. (2006)
18:2	E	11.0	A242_2	21	A81-365022	PI468916	Diers and Shoemaker (1992)
18:2	E	13.6	Pb	20	A81-365022	PI468916	Diers and Shoemaker (1992)
18:2	E	44.8	Satt185	13	N87-984-16	TN93-99	Panthee et al. (2006)
18:2	F	93.7	–	10	Essex	Williams	Hyten et al. (2004)
18:2	L	74.5	–	50	Essex	Williams	Hyten et al. (2004)
18:3	B2	–	pB194-1 pB124	85	C1640	PI479750	Brummer et al. (1995)
18:3	B2	87.5	Satt534 Fad3i6	72–78	RG10	OX948	Reinprecht et al. (2006)
18:3	E	6.3	SAC7_1	31	A81-365022	PI468916	Diers and Shoemaker (1992)
18:3	E	11.0	A242_2	23	A81-365022	PI468916	Diers and Shoemaker (1992)
18:3	E	28.3	K229_1	20	A81-365022	PI468916	Diers and Shoemaker (1992)
18:3	E	30.9	A454_1	22	A81-365022	PI468916	Diers and Shoemaker (1992)
18:3	E	34.6	A203_1	22	A81-365022	PI468916	Diers and Shoemaker (1992)
18:3	E	45.4	Satt263	12	N87-984-16	TN93-99	Panthee et al. (2006)
18:3	G	21.9	Satt235	22	N87-984-16	TN93-99	Panthee et al. (2006)
18:3	K	–	A065_3	20	A81-365022	PI468916	Diers and Shoemaker (1992)
18:3	L	36.7	A023_1	26	A81-365022	PI468916	Diers and Shoemaker (1992)
18:3	L	50.6	–	13	Essex	Williams	Hyten et al. (2004)
18:3	L	82.5	–	24	Essex	Williams	Hyten et al. (2004)

Note: The linkage group (LG) and map position (cM) designation of marker is based on the integrated soybean genetic
          linkage map, Gmcomposite2003, reported in SOYBASE (http://soybase.org).

Reduced palmitic acid soybean lines have been developed by chemical mutagenesis, recurrent
selection, and hybridization (Erickson et al. 1988; Bubeck et al. 1989; Wilcox and Cavins 1990;
Burton et al. 1994). At least three recessive genes at the *Fap* locus are responsible for palmitic acid
content in soybean oil (Pantalone et al. 2004; Wilson 2004). Soybean N79-2077-12, with approximately 6% palmitic acid, has a recessive *fap*$_{nc}$ allele and was developed from recurrent selection
(Burton et al. 1994). Low palmitic acid germplasm lines C1726 and A22 were derived from mutagenesis of Century and A1937, respectively. C1726, with approximately 8% palmitic acid, has the
recessive *fap1* allele (Bubeck et al. 1989; Wilcox and Cavins 1990). A22, with 6.8% palmitic acid,
has the recessive *fap3* allele (Fehr et al. 1991a). Palmitic acid content of F_2 progeny from a cross
between A22 × C1726 with the homozygous genotype *fap1 fap1, fap3 fap3,* and line A18 (Fehr
et al. 1991a) had 4.0% 16:0 compared with 7.1 and 8.0%, respectively, for A22 and C1726.

Four recessive alleles have been identified that increased palmitic acid levels. Mutant C1727
(*fap2* allele with 17%), A21 (*fap2b* allele with 20%), and A24 (*fap4* with 18%) were developed by

chemical mutagenesis. Various combinations of these alleles can increase palmitic acid content to 35% of crude oil (Fehr et al. 1991b).

Some agronomic traits were affected in the genotypes with altered palmitic acid genes reported above. Soybean yield was significantly less in genotypes with both the reduced and elevated levels of 16:0 (Wilcox and Cavins 1990; Ndzana et al. 1994; Rebetzke et al. 1998; Hayes et al. 2002). In comparison to related lines with normal 16:0 levels, high-palmitic acid $BC_1F_{2:4}$ lines averaged across each of three different populations typically had a shorter height, smaller seed size, higher protein levels, lower oil levels, and reduced oleic and linoleic acids (Hayes et al. 2002). These associations suggest that pleiotropy and/or linkage drag could hinder efforts to develop competitive cultivars with altered palmitic acid phenotypes (Pantalone et al. 2004).

Molecular markers closely associated with palmitic acid alleles have been discovered (Table 20.2). The *fap2* allele elevated 16:0 in C1727 was mapped to Chro. 17 (LG D2) (Nickell et al. 1994). An elevated palmitic acid soybean line containing the *fap2* mutation was characterized with a candidate gene approach. The underlying mutation in a *KAS II* gene, which is responsible for the elevated palmitic acid phenotype, was discovered and led to the development of a perfect molecular marker assay (Aghoram et al. 2006). The major low palmitic acid allele fap_{nc} ($R^2 = 31$–38%) in NS79-2077-12 was mapped near Satt684 on Chro. 5 (LG A1) (Li et al. 2002). A low 16:0 soybean line with the fap_{nc} allele was characterized and found to contain a deletion in one of the four *FATB* genes catalyzing a thioesterase function in the fatty acid pathway (Cardinal et al. 2007). Again, a molecular marker assay was described that is specific for the mutant allele and could be used in a breeding program.

QTL was associated with reduced palmitic acid in C1726 mapped to Chros. 14 and 12 (LGs B2 and H) using simple sequence repeat (SSR) markers in a BC_1F_2 population from a cross Cook × C1726. The lowest 16:0 contents were observed where the QTL on Chro. 14 (LG B2) was homozygous for the allele from normal soybean Cook, and the QTL on Chro. 12 (LG H) was homozygous for the C1726 allele (Pantalone et al. 2004).

Several QTLs associated with palmitic acid content were mapped in populations in which the parents had normal palmitic acid content. Diers and Shoemaker (1992) used an F_2-derived population of A81-356022 (*G. max*) × PI468916 (*G. soja*) to map the restriction fragment length polymorphism (RFLP) markers associated with 16:0 levels and found three markers: pA-343a ($R^2 = 24\%$) and pA-18 ($R^2 = 19\%$) on Chro. 14 (LG B2) and pK-375 on Chro. 16 (LG J). The low palmitic acid alleles at the Chro. 14 (LG B2) QTL came from *G. soja*, whereas the one at the Chro. 16 (LG J) QTL was from the *G. max* line. A major QTL for reduced 16:0 from an Essex × Williams population was found in an 89.1-cM region on Chro. 19 (LG L) ($R^2 = 13\%$) and was derived from Williams (Hyten et al. 2004). Recently two QTL that reduce palmitic acid—Satt133 on Chro. 8 (LG A2) ($R^2 = 12\%$), located approximately 7 cM upstream from a QTL affecting oil concentration (Brummer et al. 1997), and Satt537 on Chro. 2 (LG D1b) ($R^2 = 20\%$)—were detected from mapping population N87-984-16 × TN93-99, in which both parents had normal fatty acid profiles (Panthee et al. 2006).

20.5 STEARIC ACID

Stearic acid (18:0) content in soybeans averages approximately 3% of the crude oil. Soybean oil higher in the saturated fats 16:0 and 18:0 has improved functional properties for certain food applications, such as solid fats, margarines and shortenings, and confectionery uses. Genotypes high in stearic acid are preferred because, unlike palmitic acid, 18:0 has been shown to be neutral in raising blood serum cholesterol, which is associated with heart disease (Yadav 1996). Soybean genotypes with elevated stearic acid have been developed by chemical or X-ray mutagenesis and from natural mutations. However, high 18:0 genotypes developed from these techniques appear to be associated with poor yield potential (Hayes et al. 2002), so development of productive high stearic cultivars may be difficult.

Six elevated stearic acid germplasm lines were reported to carry homozygous recessive alleles: A6 (*fas^a*) with 30% (Hammond and Fehr 1983b), FA41545 (*fas^b*) with 15% (Graef et al. 1985), A81-606085 (*fas*) with 19% (Graef et al. 1985), KK-2 (*st_1*) with 6% (Rahman et al. 1997), M25 (*st_2*) with 19% (Rahman et al. 1997), and FAM94-41 (*fas_{nc}*) with 9% (Pantalone et al. 2002) stearic acid were reported as genetically altered genotypes. It is known that *fas^a*, *fas^b*, and *fas* are allelic (Graef et al. 1985) and that *fas^a* and *fas_{nc}* are allelic and are different mutations in the same gene (Pantalone et al. 2004). Rahman et al. (1997) reported that the single recessive genes *st_1* and *st_2* for elevated 18:0 are nonallelic. The *st_1st_1st_2st_2* genotype had 30% 18:0 content, but it failed to grow after germination. It is unknown whether *st_1* or *st_2* are allelic to *fas^a*, *fas^b*, or *fas*.

Molecular markers closely associated with stearic acid alleles have been discovered (Table 20.2). A QTL associated with high stearic acid mapped to Chro. 16 (LG J) from a mapping population, A81-356022 × PI468916 (Diers and Shoemaker 1992; Pantalone et al. 2004). A gene derived from FAM94-41 for high 18:0 has been mapped to Chro. 14 (LG B2). Three SSRs—Satt070, Satt474, and Satt556—were highly significant ($P < 0.0001$) with major the QTL near Satt474, having an R^2 value greater than 61% (Spencer et al. 2003). Hyten et al. (2004) also reported on an Essex × Williams population in which two QTL on Chro. 6 (LG C2) and Chro. 19 (LG L) were associated with stearic acid. Panthee et al. (2006) detected two markers, Satt168 [Chro. 14 (LG B2); $R^2 = 18\%$] and Satt249 [Chro. 16 (LG J); $R^2 = 12\%$] associated with stearic acid. They reported that the gene near Satt249 on Chro. 16 (LG J) from N87-984-16 is a novel allele and gave higher concentrations of stearic acid.

20.6   OLEIC ACID

Soybeans typically contain approximately 23% oleic acid (18:1). Increasing 18:1 to a range between 55 and 60% in combination with low 18:3 would have edible and industrial applications with high oxidative stability. This oil would increase stability at high cooking temperatures and reduce the need for hydrogenation reducing *trans*-fats. Oil with a high oleic acid content could be used in the manufacture of soydiesel, lubricants, and hydraulic oils. Genes at the *Fad2* locus have been shown to be responsible for increasing levels of oleic acid (Wilson 2004). Other studies showed that high 18:1 in soybean oil was quantitatively inherited (Burton et al. 1983; Hawkins et al. 1983). The germplasm line N78-2245 was perhaps the first soybean developed with higher levels (51%) of oleic acid by recurrent selection (Wilson et al. 1981). N98-4445A, a mid-oleic soybean accession that has an oleic acid concentration of 40–70% depending on the environmental growing conditions, was developed from combining several oleic acid genes from a three-way cross: N94-2473 × (N93-2007-4 × N92-3907) (Burton et al. 2006).

Molecular markers associated with oleic acid alleles have been disclosed (Table 20.2). Six QTL have been mapped and confirmed for high 18:1 in line N00-3350, a derivative of N98-4445A, on Chros. 5 (LGs A1) ($R^2 = 4\%$) at Satt211, 17 (D2) ($R^2 = 6\%$) at Satt389, 18 (G) ($R^2 = 13\%$) at Satt394, 18 (G) ($R^2 = 7\%$) at Satt191, 19 (L) ($R^2 = 9\%$) at Satt418, and 19 (L) ($R^2 = 25\%$) at Satt561 (Monteros et al. 2008).

Rahman et al. (1994) developed M23 with 46% oleic acid content from irradiating seeds of Bay soybean. The increase in oleic acid content in M23 was controlled by a single partially recessive gene, designated as *ol* (Takagi and Rahman 1996). Another allele *ol^a* at the same *Ol* locus was found in mutant M11 and contains approximately 38% oleic acid (Rahman et al. 1996a).

A recent study revealed that the mutation of the *ol* gene in M23 was the result of a deletion at the *Fad2-1a* locus (Sandhu et al. 2007). A polymerase chain reaction (PCR)-based DNA marker for the high oleic genotype has made it easy to use MAS to select for genotypes with higher oleic acid levels (Alt et al. 2005a). Because oleic acid content is significantly influenced by the environment in which the seed is produced (Oliva et al. 2006), MAS for this trait should increase breeding efficiency.

Combining the M23 gene with genes from other mid-oleic acid sources has resulted in transgressive segregation for higher oleic acid content. Transgressive segregation among lines in the population of N98-4445A × M23 showed oleic acid levels can be increased to more than 70% (Alt et al. 2005b). Although N98-4445A, with more than 50% oleic acid, has been developed, yield potential was less than a cultivar with seed oil containing a normal fatty acid profile (Burton et al. 2006). Like lines high or low in saturates, yield drag may be a significant factor in the development of productive nongenetically modified lines with high oleic acid content.

Low heritability and limited variation have contributed to the slow development of high 18:1 cultivars by conventional breeding (Liu 1999). However, transgenic approaches to increase oleic acid content in soybean oil show great promise. Genes for $\omega 6$ and $\omega 3$ desaturases have been cloned in soybean (Liu 1999). This enabled scientists at E. I. du Pont de Nemours and Company (DuPont) to suppress a $\omega 6$ desaturase gene, which resulted in a transgenic soybean with approximately 80% oleic acid content. Genotypes with this transgene have shown excellent stability for 18:1 content across a range of growing conditions. Knowlton et al. (1996) reported on a transgenic soybean with very high levels of oleic (85.6%), low levels of linoleic (1.6%), and linolenic (2.2%) acids. This transgene has shown no negative effect on grain yield, and levels of 18:1 were very stable across growing environments. On the other hand, high oleic lines derived from nongenetically modified organism sources such as N98-4445A and M23 have yielded less compared with commercial cultivars of similar maturity, and 18:1 levels are influenced by growing environments (Oliva et al. 2006). Recently there has been major progress in elevating oleic acid in soybean oil by conventional breeding with marker-associated selection. Pham et al. (2010) found that when exiting Fad2-1A mutants were combined with the novel mutant Fad2-1B alleles, high oleic acid (80%) was produced in soybean seed.

QTL for oleic acid content have been mapped to several positions in the soybean linkage group. Diers and Shoemaker (1992) detected three QTL associated with variation for 18:1 levels. These QTL were close to RFLP markers pA-82 ($R^2 = 28\%$) on Chro. 5 (LG A1) and pA-619 ($R^2 = 19\%$) on Chro. 14 (LG B2) and were linked to the *pb* locus that determines the sharpness of pubescence ($R^2 = 21\%$) on Chro. 15 (LG E) (Pantalone et al. 2004). A large QTL ($R^2 = 35\%$) was mapped on the position of 82.5 cM on Chro. 19 (LG L) (Hyten et al. 2004), and a single molecular marker related to oleic acid content, Satt263, was mapped on Chro. 15 (LG E) ($R^2 = 10\%$) (Panthee et al. 2006). Bachlava et al. (2009) reported a QTL with a moderate effect on oleic acid concentration that mapped to Chro. 20 (LG I) in close proximity to the *Fad 2-1B* locus.

Plant introductions with higher than average oleic acid concentration (30–50% vs. 24% from most soybeans) are reported in the USDA soybean germplasm collection and are likely to have useful genes for improving oleic acid content (Lee et al. 2009). However, the genetic basis for higher oleic acid in these genotypes has not been reported.

20.7 LINOLEIC ACID

Linoleic acid (18:2) in soybean seeds is produced primarily by desaturation of oleic acid in phosphatidylcholine on the endoplasmic reticulum (Ohlrogge and Browse 1995). Linoleic acid is a predominant fatty acid that is typically approximately 53% in soybean oil (Wilson 2004). The genetic control of linoleic acid has scarcely been studied, perhaps because there is no pressing need to modify its levels in the oil. The accumulation of linoleic acid is the result of a balance of $\omega 6$ and $\omega 3$ fatty acid desaturase enzyme activity.

Molecular markers have been used to map the genes involved in linoleic acid production (Table 20.2). Five RFLP markers—three on Chro. 5 (LG A1), one on Chro. 14 (LG B2), and one on Chro. 15 (LG E)—and one morphological marker (*pb^c*) were found in a mapping population (Diers and Shoemaker 1992). Two large QTLs were mapped on the position of 93.7 cM ($R^2 = 10\%$) on Chro. 13 (LG F) and on 74.5 cM ($R^2 = 51\%$) on Chro. 19 (LG L) (Hyten et al. 2004). A single molecular marker, Satt185, was mapped on Chro. 15 (LG E) ($R^2 = 14\%$). Markers Satt185 and Satt263 (oleic acid marker) map within 0.6 cM of each other, suggesting that the QTL detected there may be influencing an enzyme involved in the process of desaturating oleic acid to linoleic acid (Panthee et al. 2006).

20.8 LINOLENIC ACID

Generally soybean oil has 8–10% linolenic acid (Wilson 2004). Linolenic acid (18:3) has been identified as an unstable component of soybean oil (Liu and White 1992). One of the most important goals of oil quality breeding in soybean has been to reduce its linolenic acid content to improve oxidative stability and flavor and to reduce the need for hydrogenation. Several linolenic acid altered genotypes were developed by recurrent selection, mutagenesis, and germplasm screening.

Through several cycles of recurrent selection, the N79-2245 soybean with 51% oleic acid and 4.2% linolenic acid was developed (Wilson et al. 1981). At least three independent genetic loci are associated with seed linolenic acid levels, with mutant alleles identified at *fan*, *fan2*, *fan3*, and *fanx*. Also, multiple alleles at the *fan* locus have been reported. Mutant C1640 (*fan*, 3.4% 18:3) (Wilcox et al. 1984; Wilcox and Cavins 1985), A5 (*fan*, 4% 18:3) (Hammond and Fehr 1983a; Rennie and Tanner 1991), A23 (*fan2*, 5.6% 18:3) (Fehr et al. 1992), KL-8 (*fanx*) and M-5 (*fan*) (Rahman et al. 1996b), M-24 (*fanx^a*) (Rahman et al. 1998), and RG10 (*fan-b*, < 2.5%) (Stojsin et al. 1998) were developed by mutagenesis. The fan alleles were also found in the USDA Soybean Germplasm Collection. PI123440 and PI361088B contained natural mutations at the *Fan* locus that have been shown to be either allelic or identical to the original *fan* allele in C1640 (Howell et al. 1972; Rennie et al. 1988; Rennie and Tanner 1989a). Evidence of two loci (*fanfan2*) were found in A16 and A17 (Fehr et al. 1992), MOLL (*fanfanx*) (Rahman and Takagi 1997), and LOLL (*fanfanx^a*) (Rahman et al. 2001). A29, a very low (1%) linolenic acid line, was reported (Ross et al. 2000). A29 was developed by combining three independent mutations: *fan* from line A5, *fan2* from A23, and *fan3* from a mutagenized derivative of line A89-144003 (Ross et al. 2000).

The discovery of the molecular genetic basis for the low linolenic acid trait in soybeans was based on using gene information from experiments done in *Arabidopsis* spp. and mutagenized soybean lines identified with the low linolenic acid trait. In *Arabidopsis* spp., it was discovered that a single gene, *FAD3*, controlled the conversion of linoleic acid precursors into linolenic acid precursors in the seed oil (Yadav et al. 1993). The soybean genome was shown to have at least three versions of the *FAD3* gene (Bilyeu et al. 2003; Anai et al. 2005). By sequencing and characterizing the *FAD3* alleles in soybean lines with low linolenic acid, corresponding mutations could be found in three of the genes that affected linolenic acid content (Bilyeu et al. 2003, 2005, 2006; Anai et al. 2005; Chappell and Bilyeu 2006, 2007). Because the sequence changes identified in the alleles defined the mutations, perfect molecular markers were designed to specifically select the desired alleles (Bilyeu et al. 2005, 2006; Chappell and Bilyeu 2006, 2007). Selection by breeder-friendly perfect molecular markers can be an efficient system for soybean improvement, particularly if a backcross strategy is used (Beuselinck et al. 2006; Bilyeu et al. 2006).

Molecular markers closely associated with linolenic acid alleles have been found (Table 20.2). The *Fan* locus has been mapped to Chro. 14 (LG B2) of soybean (Brummer et al. 1995). Four RFLP markers were found using a population from the cross C1640 × PI479750 and were successfully anchored to markers mapping the *Linolen1-2* QTL on Chro. 14 (LG B2) (Diers and Shoemaker 1992; Brummer et al. 1995). In another study, a DNA fragment missing from the ω3 desaturase gene in soybean line A5 was mapped to the same region of Chro. 14 (LG B2) as the *Fan* locus, suggesting that a deletion in this gene was responsible for the reduction in 18:3 in A5 (Byrum et al. 1997; Pantalone et al. 2004).

A study identified and characterized three soybean microsomal ω3 fatty acid desaturase genes—*GmFAD3A*, *GmFAD3B*, and *GmFAD3C*—and determined that the deletion of the *GmFAD3A* gene is responsible for the reduced 18:3 in line A5 (Bilyeu et al. 2003). A point mutation was detected in the *GmFAD3A* gene in mutant C1640 at nucleotide 798 of the coding sequence. A guanine residue was changed to an adenine residue in the C1640 allele, resulting in a change from a tryptophan codon to a premature stop codon. A molecular marker assay based on a PCR reaction was developed to distinguish between wild-type Williams 82 and low 18:3 line C1640 alleles (Chappell and Bilyeu 2006).

Two point mutations, one in *GmFAD3A* and the other in *GmFAD3C*, were detected in the low linolenic acid line CX1512-44. They contributed unequally, but additively to the linolenic acid content (Bilyeu et al. 2005). QTL related to linolenic acid were mapped in two different studies using normal linolenic acid soybean populations. Hyten et al. (2004) found major QTLs positioned at 50.6 cM ($R^2 = 14\%$) and 82.5 cM ($R^2 = 25\%$) on Chro. 19 (LG L), and Panthee et al. (2006) also found molecular markers, Satt263 [Chro. 15 (LG E), $R^2 = 14\%$], and Satt236 [Chro. 18 (LG G), $R^2 = 23\%$]. A major QTL that accounted for 78% of the phenotypic variability for low linolenic acid content derived from RG10 soybean genotypes was found on Chro. 14 (LG B2) (Satt534 and Fad3i6) (Reinprecht et al. 2006).

Mutations were discovered in all three *GmFAD3* ω3 fatty acid desaturase genes in the soybean line A29, which suggested that combinations of mutant alleles at the three *GmFAD3* loci allowed for the development of new germplasm containing 1% linolenic acid in the seed oil along with single nucleotide polymorphism (SNP)-based molecular markers that can be used in a backcross breeding strategy (Bilyeu et al. 2006).

There have been few studies in breeding and mapping for high linolenic acid levels in soybeans. Accessions for this type of soybean are available in the USDA soybean germplasm collection. Eleven *G. max* accessions with over 15% 18:3 and 20 *G. soja* accessions with over 20% 18:3 have been reported in the USDA soybean germplasm collection (USDA, ARS 2007). Genetic regulation of linolenic acid concentrations in wild soybean suggested that the high-linolenic trait in wild soybean genotypes was determined by a set of desaturase alleles that were different from corresponding alleles in *G. max* (Pantalone et al. 1997). Introgression of these alleles from *G. soja* to *G. max* may lead to the production of high linolenic acid soybean oil for various applications such as ω3 fatty acid soy foods and industrial products.

20.9 ENVIRONMENTAL EFFECT ON OIL AND FATTY ACID CONCENTRATION IN SOYBEAN

Genotypes and environmental interactions that influence the oil concentration and fatty acid profile of soybean oil have been addressed in many studies. Studies have indicated that temperature plays an important role in the synthesis of oil and fatty acids. In general, higher temperature increases oil content in soybean seed (Wilson 2004). Soybeans grown under high average temperatures have reduced linoleic acid and linolenic acid and increased oleic acid content; however, contents of saturated fatty acids were changed little by environmental factors (Howell and Collins 1957; Wolf et al. 1982; Rennie and Tanner 1989b; Dornbos and Mullen 1992; Wilson 2004; Hu et al. 2006).

Instability of oleic acid and linolenic acids under various temperature regimes is a concern. Mid-oleic acid and low linolenic acid genotypes with genetically altered fatty acids developed by mutagenesis were more stable across environments than genotypes with altered fatty acid genotypes developed from conventional breeding techniques (Wilcox and Cavins 1992; Schnebly and Fehr 1993; Primono et al. 2002; Oliva et al. 2006). The higher the average linolenic acid content of a genotype, the greater the instability it showed across various growing conditions (Oliva et al. 2006). The 1% linolenic acid genotype IA 3017 was very stable, and the highest linolenic acid genotypes were the least stable for 18:3 across ten growing environments. Thus, selecting for the lowest linolenic acid content should produce the most stable genotypes for 18:3.

Oleic acid was influenced significantly by temperature during the final 30 days of the reproductive period (Oliva et al. 2006). The highest 18:1 genotypes were the least stable across growing environments. In this study, N98-4445, an early group IV, had 55–60% 18:1 when grown in the southern states of North Carolina and Mississippi and 39–45% when grown at Columbia, MO in the central United States. The reduction in 18:1 as growing region moved from south to north was due to lower average temperatures during seed development. On the other hand, M23 was more stable across various environments than N98-4445A in the same study. Thus, genes for elevated oleic acid from different sources may be less influenced by growing conditions. A transgene for high 18:1 was reported to be very stable across growing environments (Knowlton et al. 1996), but to date it has not been approved for use in commercially grown soybeans. Unless nontransgenic sources for 18:1 concentration can be found that are less influenced by temperature, special consideration will have to be given to production in specific regions using cultural practices and earlier maturity groups to produce desired levels of oleic and linolenic acid in soybean oil (Shannon and Sleper 2004). To ensure the desired 18:1 and 18:3 levels, cultivars will need to be produced in warmer regions such as the southern United States or during the warmest periods of the summer in cooler areas. This may require adding high 18:1 and low 18:3 traits to earlier maturing cultivars within a region. Planting early cultivars at early planting dates would increase the chance that temperatures are warmest during reproductive growth, giving the highest probability of producing oil with high 18:1 and low 18:3. On the contrary, high oleic acid is negatively correlated with the polyunsaturated fatty acids 18:2 and 18:3. Thus, if high polyunsaturated fats are desired, cooler temperatures, later planting, and later maturities would appear to favor this phenotype.

20.10   GENETIC ENGINEERING FOR BIOENERGY TRAITS IN SOYBEAN

Progress in the basic understanding of oil biosynthetic pathways and their regulation in plants, coupled with the splicing of genes and the regulatory factors associated with the enzymes of fatty acid modification and oil accumulation, have contributed toward the metabolic engineering of oil seed crops with the agenda of "redesigning oil seeds" with improved total oil accumulation or fatty acid components. Genetic engineering strategies for increased or modified oil content or composition in crop plants have provided alternative routes to the classical plant breeding approaches to attain the above goals. A combination of genetic engineering and classical plant breeding methods helped to re-draw the progress in oil modification or accumulation in crop plants.

Oil storage in seeds constitutes reduced carbon with 1–60% variation between different species. The fatty acid biosynthetic pathway is a primary metabolic pathway, and in plants major fatty acids have a chain length of 16–18 carbons with 1–3 *cis*-double bonds (Ohlrogge and Browse 1995). Biochemical pathways involved in the oil biosynthesis and the major genes associated with these pathways are known in model plants including spp. However, the molecular and biochemical regulatory networks and signaling between sites of biosynthesis such as endoplasmic reticulum and plastids are not completely elucidated. Above all, the environmental effects on these regulatory and signaling pathways are not fully understood.

Although there is significant knowledge on fatty acids biosynthesis, considerably less progress has been made on the improvement of soybeans with increased oil content through genetic engineering. Conjugated double bonds in fatty acids increase their rate of oxidation when compared with polyunsaturated fatty acids with methylene-interrupted double bonds. Cahoon et al. (1999) has reported the production of fatty acid components of high-value drying oils in transgenic soybean embryos. In this study, they have engineered full-length cDNAs for the Momordica (MomoFadX) and Impatiens (ImpFadX) enzymes in somatic soybean embryos, and, as a result, α-eleostearic and α-parinaric acids were accumulated in the transgenic somatic embryos, whereas none of these enzymes were present in the nontransgenic soybean embryos. High oxidative reactivity of soybean oil decreases the shelf life of soybean oil-derived biodiesel (Kinney and Clemente 2005). The oxidative reaction of the fuel will result in the formation of a gel-like substance and solid crystals that will clog the fuel filters and fuel lines (Durett et al. 2008). Seed-specific downregulation of FAD2-1 through genetic engineering has resulted in approximately 80% of 18:1 in seeds (Kinney 1997; Kinney and Clemente 2005), which improved further up to 85% by downregulating the FATB gene in FAD2-1-downregulated transgenic soybean (Buhr et al. 2002). Another important phenotypic observation is the reduced palmitic acid content in the transgenic soybeans with FAD2-1 downregulation. These high oleic transgenic soybeans were used for biodiesel production, and they have shown improved fuel characteristics such as cold temperature flow properties and oxides of nitrogen

(NO$_x$) emissions (Tat et al. 2007). The high 18:1 property in seed oil is a desirable property because of their oxidative stability. Also, engineering of FAD3 in soybean through the seed-specific over-expression helped enhance the 18:3 content up to 50% when compared with the nontransgenic wild type (Cahoon 2003).

Another strategy to improve the oxidative stability of oil content and enhanced fuel characteristics is the increase of stearic acid component in the seeds. Seed-specific downregulation of stearoyal-ACP desaturase in soybean has demonstrated the increased stearic acid content in the seed storage lipids. The desaturation reaction of the enzyme stearoyal-ACP desaturase converts stearic acid to oleic acid. So, the seed-specific downregulation of this protein through genetic engineering will reduce the carbon flow by desaturation and increase the stearic acid content in seeds (Kinney 1998; Kinney and Clemente 2005). It is reported that 20:1 is more oxidatively stable than 18:1 in plants (Isabell et al. 1999). To produce the more oxidatively stable oil component in soybean, Cahoon et al. (2000) has partially reconstructed the biosynthetic pathway producing meadowfoam-type seed oil in transgenic soybean. They have engineered cDNAs for the *Limnanthes douglasii* acyl-CoA desaturase and fatty acid elongase 1 (FAE1) in somatic embryos, and the phenotypic analysis of the transgenic embryos has shown that the 20:1 and 5-docosenoic acid composed up to 12% of the total fatty acids. Genetic engineering of oil biosynthetic pathways in soybean shows the enhanced fuel properties of soybean oil. Also, other biochemical methods such as catalytic cracking of soybean oil clearly demonstrate the acceptable fuel properties when compared with those of petroleum-based fuel (Xu et al. 2010).

20.11 CONCLUSIONS AND FUTURE PERSPECTIVES

Soybean is one of the major legumes rich in protein and oil and is a potential crop for designing for future bioenergy needs. The plant oil commonly called TAG is chemically the most similar to fossil oil and has the greatest potential to be used as one of the major bioenergy resources. Soybean oil has been broadly studied as a raw material for fatty acid methyl esters-based biodiesel production by transesterification. Recent advances in the dissection of oil biosynthesis pathways in model species *Arabidopsis* and the investigation of oil crops such as canola, sunflower, etc. have increased the possibility of deep understanding of oil biosynthetic machinery in soybeans. Also, the availability of genomics technologies and the recent release of soybean genome sequence information offer a unique opportunity to investigate the regulatory networks and signaling pathways associated with oil biosynthesis and modification. Forward and reverse genetic approaches are being used to alter the seed oil content and composition in soybean. Genetic mapping of oil traits in soybean has increased the possibility of identifying the genomic regions associated with fatty acid biosynthesis and has also helped to develop markers (including the SNP markers) for future molecular breeding applications. Also, the classical plant breeding approaches (including mutational breeding) have been successful in generating soybean varieties with improved oil content and composition. On the other hand, metabolic engineering approaches (including the redesigning of the specific steps in the pathway) are also contributing toward enhanced seed oil content and composition in soybean.

A major challenge in redesigning oil crops, especially soybean for enhanced and modified oil content and composition, is the limited knowledge of transcriptional regulatory networks associated with lipid metabolism pathways. Also, the understanding of the tissue-specific or organelle-specific signaling pathways are in the early stages. Utilization of next-generation sequencing technologies coupled with the targeted metabolomic analysis focusing on subcellular compartments will help to advance the construction of these networks and also help to understand the genotypic variation and develop more SNP markers for molecular-assisted breeding purposes.

Another major challenge is the stability of soybean oil content and composition under various environmental conditions. Designing enzymes and specific catalytic reactions controlling the flux of fatty acids between phospholipid and acyl-CoA pools will help improve the production of various fatty acids and storage of total oil content in plants. Several molecular modeling tools can be developed and used to generate hypotheses toward this goal. The enhanced shelf life of soybean oil with high oxidative stability should be the major goals to be addressed in the future.

Another bottleneck is with the difficulty in the application of translational genomics technologies because of the limited genomics information of potential "high oil content and/or modified oil composition organisms" including other eukaryotic and prokaryotic species. Advances in metabolic flux analysis will help to engineer carbon shuttling between source and sink, and more intense investigations are needed in these aspects. In other words, changes in transcript, protein, and metabolite signatures during plant development, reserve mobilization and deposition, and the influence of environmental stresses are not well known, and the efforts toward these objectives will increase the possibility of redesigning soybean for food, feed, and fuel needs. Another viable option is to consider vegetative tissues along with seeds for metabolic engineering toward increased availability of soybean oil feedstock. A comprehensive strategy for increased oil production combined with enhanced soybean yield and stress tolerance will generate cost-effective biofuel feedstock and help in the commercialization of improved soybean varieties.

ACKNOWLEDGMENT

We thank Missouri Soybean Merchandising Council for the funding support toward the improvement of oil content and composition in soybean.

REFERENCES

Abel S, Kock MD, Smuts CM, Villiers C, Swanevelder S, Gelderblom WCA (2004) Dietary modulation of fatty acid profiles and oxidative status of rat hepatocyte nodules: Effect of different n-6/n-3 fatty acid ratios. *Lipids* 39:963–976
Abreu FR, Lima DG, Hamu EH, Einloft S (2003) New metal catalyst for soybean oil transesterification. *J Amer Oil Chem Soc* 80:601–604
Aghoram K, Wilson RF, Burton JW, Dewey RE (2006) A mutation in a 3-Keto-Acyl-ACP synthase II gene is associated with elevated palmitic acid levels in soybean seeds. *Crop Sci* 46:2453–2459
Alt JL, Fehr WR, Welke GA, Sandu D (2005a) Phenotypic and molecular analysis of oleate content in the mutant soybean line M23. *Crop Sci* 45:1997–2000
Alt JL, Fehr WR, Welke GA, Shannon JG (2005b) Transgressive segregation for oleate content in three soybean populations. *Crop Sci* 45:2005–2007
Anai T, Yamada T, Kinoshita T, Rahman SM, Takagi Y (2005) Identification of corresponding genes for three low-[alpha]-linolenic acid mutants and elucidation of their contribution to fatty acid biosynthesis in soybean seed. *Plant Sci* 168:1615–1623
Bachlava E, Dewey RE, Burton JW, Cardinal AJ (2009) Mapping and comparison of quantitative trait loci for oleic acid seed content in two segregating soybean populations. *Crop Sci* 49:433–442
Beuselinck PR, Sleper DA, Bilyeu KD (2006) An assessment of phenotype selection for linolenic acid using genetic markers. *Crop Sci* 46:747–750
Bilyeu KD, Palavalli L, Sleper DA, Beuselinck P (2003) Three microsomal omega-3 fatty-acid desaturase genes contribute to soybean linolenic acid levels. *Crop Sci* 43:1833–1838
Bilyeu K, Palavalli L, Sleper DA, Beuselinck P (2005) Mutations in soybean microsomal omega-3 fatty acid desaturase genes reduce linolenic acid concentration in soybean seeds. *Crop Sci* 45:1830–1836
Bilyeu K, Palavalli L, Sleper DA, Beuselinck P (2006) Molecular genetic resources for development of 1% linolenic acid soybeans. *Crop Sci* 46:1913–1918
Brouwer IA, Katon MB, and Zock PL (2004) Dietary α-linolenic acid is associated with reduced risk of fatal coronary heart disease, but increased prostate cancer risk: A meta-analysis. *J Nutr* 134:919–922
Brummer EC, Graef GL, Orf J, Wilcox JR, Shoemaker RC (1997) Mapping QTL for seed protein and oil content in eight soybean populations. *Crop Sci* 73:370–378
Brummer EC, Nickell AD, Wilcox JR, Shoemaker RC (1995) Mapping the *Fan* locus controlling linolenic acid in soybean oil. *J Hered* 86:245–247
Bubeck DM, Fehr WR, Hammond EG (1989) Inheritance of palmitic and stearic acid mutants of soybean. *Crop Sci* 29:652–656

Buhr T, Sato S, Ebrahim F, Xing AQ, Zhou Y, Mathiesen M, Schweiger B, Kinney A, Staswick P, Clemente T (2002) Ribozyme termination of RNA transcripts down-regulate seed fatty acid genes in transgenic soybean. *Plant J* 30:155–163

Burton JW, Wilson RF, Brim CA (1983) Recurrent selection in soybeans. IV. Selection for increased oleic acid percentage in seed oil. *Crop Sci* 23:744–747

Burton JW, Wilson RF, Brim CA (1994) Registration of N97-2077-12 and N87-2122-4, two soybean germplasm lines with reduced palmitic acid in seed oil. *Crop Sci* 34:313

Burton JW, Wilson RF, Rebetzke GJ, Pantalone VR (2006) Registration of N98-4445A mid-oleic soybean germplasm line. *Crop Sci* 46:1010–1012

Byrum JR, Kinney AJ, Stecca KL, Grace DJ, Diers BW (1997) Alteration of omega-3 fatty-acid desaturase gene is associated with reduced linolenic acid in the A5 soybean genotype. *Theor Appl Genet* 94:356–359

Cahoon EB (2003) Genetic enhancement of soybean oil for industrial uses: Prospects and challenges. *AgBioForum* 6:11–13

Cahoon EB, Carlson TJ, Ripp KG, Schweiger BJ, Cook GA, Hall SE, Kinney AJ (1999) Biosynthetic origin of conjugated double bonds: Production of fatty acid components of high-value drying oils in transgenic soybean embryos. *Proc Natl Acad Sci USA* 96:12935–12940

Cahoon EB, Marillia EF, Stecca KL, Hall SE, Taylor DC, Kinney AJ (2000) Production of fatty acid components of meadowfoam in somatic soybean embryos. *Plant Physiol* 124:243–251

Cardinal AJ, Burton JW, Camacho-Roger AM, Yang JH, Wilson RF, Dewey RE (2007) Molecular analysis of soybean lines with low palmitic acid content in the seed oil. *Crop Sci* 47:304–310

Chang NW, Huang PC (1998) Effects of the ratio of polyunsaturated and monounsaturated fatty acid on rat plasma and liver lipid concentration. *Lipids* 33:481–487

Chappell AS, Bilyeu KD (2006) A *GmFAD3A* mutation in the low linolenic acid soybean mutant C1640. *Plant Breed* 125:535–536

Chappell AS, Bilyeu KD (2007) The low linolenic acid soybean line PI361088B contains a novel *GmFAD3A* mutation. *Crop Sci* 47:1705–1710

Connor WE (2000) Importance of n-3 fatty acid in health and disease. *Amer J Clin Nutr* 71:171S–175S

Diers BW, Shoemaker RC (1992) Restriction fragment length polymorphism of soybean fatty acid content. *J Amer Oil Chem Soc* 69:1242–1244

Dornbos DL Jr, Mullen RE (1992) Soybean seed protein and oil contents and fatty acid composition adjustments by drought and temperature. *J Amer Oil Chem Soc* 69:228–231

Durrett TP, Benning C, Ohlrogge J (2008) Plant triacylglycerols as feedstocks for the production of biofuels. *Plant J* 54:593–607

Erickson EA, Wilcox JR, Cavins JF (1988) Inheritance of altered palmitic acid percentage in two soybean mutants. *J Hered* 79:465–468

Fehr WR, Welke GA, Hammond EG, Duvick DN, Cianzo SR (1991a) Inheritance of reduced palmitic acid content in soybean seed oil. *Crop Sci* 31:88–89

Fehr WR, Welke GA, Hammond EG, Duvick DN, Cianzo SR (1991b) Inheritance of elevated palmitic acid content in soybean seed oil. *Crop Sci* 31:522–1524

Fehr WR, Welke GA, Hammond EG, Duvick DN, Cianzo SR (1992) Inheritance of reduced linolenic acid content in soybean genotypes A16 and A17. *Crop Sci* 32:903–906.

Gebauer SK, Psota TL, Harris WS, Kris-Etherton PM (2006) n-3 Fatty acid dietary recommendations and food sources to achieve essentiality and cardiovascular benefits. *Amer J Clin Nutr* 83:1526S–1536S

Graef GL, Miller LA, Fehr WR, Hammond EG (1985) Fatty acid development in a soybean mutant with high stearic acid. *J Amer Oil Chem Soc* 62:773–775

Grundy SM (1986) Composition of monounsaturated fatty acid and carbohydrates for lowering plasma cholesterol. *New Eng J Med* 314:745–748

Hammond EG, Fehr WR (1983a) Registration of A5 germplasm line of soybean. *Crop Sci* 23:192

Hammond EG, Fehr WR (1983b) Registration of A6 germplasm line of soybean. *Crop Sci* 23:192–193

Haas, MJ, Piazza GJ, Foglia TA (2002) Chapter 29: Enzymatic approaches to production of biodiesel fuels. In: Kuo TM, Gardner HW (eds), *Lipid Biotechnology*. Marcel Dekker, New York, pp 587–598

Hawkins SE, Fehr WR, Hammond EG (1983) Resource allocation in breeding for fatty acid composition of soybean oil. *Crop Sci* 23:900–904

Hayes MF, Fehr WR, Welke GA (2002) Association of elevated palmitate with agronomic and seed traits of soybean. *Crop Sci* 42:1117–1120

Hildebrand DF, Li R, Hatanaka T (2008) Genomics of soybean oil traits. In: Stacey G (ed), *Genetics and Genomics of Soybean*, Vol. 2. Springer, New York, pp 185–209

Howell RW, Brim CA, Rinnes RW (1972) The plant geneticist's contribution toward changing lipid and amino acid composition of soybeans. *J Amer Oil Chem Soc* 49:30–32

Howell RW, Collins FI (1957) Factors affecting linolenic and linoleic acid content of soybean oil. *Agron J* 49:593–597

Hu G, Ablett GR, Pauls KP, Rajcan I (2006) Environmental effects on fatty acid levels in soybean oil. *J Amer Oil Chem Soc* 83:759–763

Hyten DL, Pantalone VR, Saxton AM, Schmidt ME, Sams CE (2004) Molecular mapping and identification of soybean fatty acid modifier quantitative trait loci. *J. Amer Oil Chem Soc* 81:1115–1118

Isabell TA, Abbott TA, Carlson KD (1999) Oxidative stability of vegetable oils in binary mixtures with meadowfoam oil. *Indust Crops Prod* 9:15–123

Jako C, Kumar A, Wei Y, Zou J, Barton DL, Giblin EM, Covello PS, Taylor DC (2001) Seed-specific overexpression of an *Arabidopsis* cDNA encoding a diacylglycerol acyltransferase enhances seed oil content and seed weight. *Plant Physiol* 126:861–874

Kinney AJ (1998) Plants as industrial chemical factories-new oils from genetically engineered soybeans. *Fett/Lipid* 100:173–179

Kinney AJ, Clemente TE (2005) Modifying soybean oil for enhanced performance in biodiesel blends. *Fuel Process Technol* 86:1137–1147

Knowlton S, Ellis SKB, Kelly EF (1996) Performance characteristics of high oleic soybean oil: An alternative to hydrogenated fits. Paper No 29-O, Presented at the 87th American Oil Chemists' Society Annual Meeting and Expo, Indianapolis, IN, April 28 to May 1, 1996

Kwanyuen P, Wilson RF (1990) Subunit and amino acid composition of diacylglycerol acyltransferase from germinating soybean cotyledons. *Protein Struct Mol Enzymol* 1039:67–72

Lardizabal K, Effertz R, Levering C, Mai J, Pedroso MC, Jury T, Aasen E, Gruys K, Bennett K (2008) Expression of *Umbelopsis ramanniana* DGAT2A in seed increases oil in soybean. *Plant Physiol* 148:89–96

Lardizabal KD, Mai JT, Wagner NW, Wyrick A, Voelker T, Hawkins DJ (2001) DGAT2 is a new diacylglycerol acyltransferase gene family. Purification, cloning and expression in insect cells of two polypeptides from *Mortierella ramanniana* with diacylglycerol acyltransferase activity. *J Biol Chem* 276:38862–38869

Lee JD, Woolard M, Slepe DA, Smith JA, Pantalone VR, Nyinyi CN, Cardinal A, Shannon JG (2009) Environmental effects on oleic acid in soybean seed oil of plant introductions with elevated oleic acid concentration. *Crop Sci* 49:1762–1778

Li Z, Wilson RF, Rayford WE, Boerma HR (2002) Molecular mapping genes conditioning reduced palmitic acid content in N87-2122-4 soybean. *Crop Sci* 42:373–378

Liu HR, White PJ (1992) Oxidative stability of soybean oils with altered fatty acid compositions. *J Amer Oil Chem Soc* 69:528–532

Liu K (1999) *Soybeans—Chemistry, Technology and Utilization.* Aspen Publishers, Gaithersburg, MD

Monteros MJ, Burton JH, Boerma HR (2008) Molecular mapping and confirmation of QTL associated with oleic acid content in N00-3350 soybean. *Crop Sci* 48:2223–2234

Ndzana X, Fehr WR, Welke GA, Hammond EG, Duvick DN, Cianzio SR (1994) Influence of reduced palmitate content on agronomic and seed traits of soybean. *Crop Sci* 34:646–649

Nickell AD, Wilcox JR, Lorenzen LL, Cavins JF, Guffy RG, Shoemaker RC (1994) The *Fap$_2$* locus in soybean maps to linkage group D. *J Hered* 85:160–162

Ohlrogge J, Browse J (1995) Lipid biosynthesis. *Plant Cell* 7:957–970

Oliva ML, Shannon JG, Sleper DA, Ellersieck MR, Cardinal AJ, Paris RL, Lee JD (2006) Stability of fatty acid profile in soybean genotypes with modified seed oil composition. *Crop Sci* 46:2069–2075

Pantalone VR, Rebetzke GJ, Burton JW, Wilson RF (1997) Genetic regulation of linolenic acid concentration in wild soybean *Glycine soja* accessions. *J Amer Oil Chem Soc* 74:159–163

Pantalone VR, Walker DR, Dewey RE, Rajcan I (2004) DNA marker-assisted selection for improvement of soybean oil concentration and quality. In: Wilson R, Stalker HT, Brummer EC (eds), *Legume Crop Genomics.* American Oil Chemists Society Press, Champaign, IL, pp 283–311

Pantalone VR, Wilson RF, Novitzky WP, Burton JW (2002) Genetic regulation of elevated stearic acid concentration in soybean oil. *J Amer Oil Chem Soc* 79:549–553

Panthee DR, Pantalone VR, Saxton AM (2006) Modifier QTL for fatty acid composition in soybean oil. *Euphytica* 152:67–73

Pham AT, Lee JD, Shannon JG, Bilyeu KD (2010) Mutant alleles of FAD2-1A and FA D2-1B combine to produce soybeans with the high oleic acid seed oil trait. *BMC Plant Biol* 10:195

Primomo VS, Falk DE, Albert GR, Tanner JW, Rajcan I (2002) Genotype × environment interactions, stability, and agronomic performance of soybean with altered fatty acid profiles. *Crop Sci* 42:37–44

Rahman SM, Kinoshita T, Anai T, Arima S, Takagi Y (1998) Genetic relationships of soybean mutants for different linolenic acid contents. *Crop Sci* 38:702–706

Rahman SM, Kinoshita T, Anai T, Takagi Y (2001) Combining ability in loci for high oleic and low linolenic acids in soybean. *Crop Sci* 41:26–29

Rahman SM, Takagi Y (1997) Inheritance of reduced linolenic acid content in soybean seed oil. *Theor Appl Genet* 94:299–302

Rahman SM, Takagi Y, Kinoshita T (1997) Genetic control of high stearic acid content in seed oil of two soybean mutants. *Theor Appl Genet* 95:772–776

Rahman SM, Takagi Y, Kinoshita T (1996a) Genetic control of high oleic acid content in the seed oil of two soybean mutants. *Crop Sci* 36:1125–1128

Rahman SM, Takagi Y, Kubota K, Miyamoto K, Kawakita T (1994) High oleic mutant in soybean induced x-ray irradiation. *BioSci Biotechnol Biochem* 58:1070–1072

Rahman SM, Takagi Y, Kumamaru T (1996b) Low linolenate sources at the fan locus in soybean lines M-5 and IL-8. *Breed Sci* 46:155–158

Rebetzke GJ, Burton JE, Carter Jr. TE, Wilson RF (1998) Changes in agronomic and seed characteristics with selection for reduced palmitic acid content in soybean. *Crop Sci* 38:297–302

Reinprecht Y, Poysa VW, Yu K, Rajcan II, Ablett GR, Pauls KP (2006) Seed and agronomic QTL in low linolenic acid, lipoxygenase-free soybean (*Glycine max* (L.) Merrill) germplasm. *Genome* 49:1510–1527

Rennie BD, Tanner JW (1989a) Genetic analysis of low linolenic acid levels in the line PI123440. *Soybean Genet Newsl* 16:25–26

Rennie BD, Tanner JW (1989b) Fatty acid composition of oil from soybean seeds grown at extreme temperatures. *J Amer Oil Chem Soc* 66:1622–1624

Rennie BD, Tanner JW (1991) New allele at the fan locus in the soybean line A5. *Crop Sci* 31:297–301

Rennie BD, Zilka J, Cramer MM, Buzzell RI (1988) Genetic analysis of low linolenic acid levels in the soybean line PI361088B. *Crop Sci* 28:655–657

Ross AJ, Fehr WR, Welke GA, Cianzio SR (2000) Agronomic and seed traits of 1%-linolenate soybean genotypes. *Crop Sci* 40:383–386

Routaboul JM, Benning C, Bechtold N, Caboche M, Lepiniec L (1999) The *TAG1* locus of *Arabidopsis* encodes for a diacylglycerol acyltransferase. *Plant Physiol Biochem* 37:831–840

Sandhu D, Alt JL, Scherder CW, Fehr WR, Bhattacharyya MK (2007) Enhanced oleic acid content in the soybean mutant M23 is associated with the deletion in the *Fad2-1a* gene encoding a fatty acid desaturase. *J Amer Oil Chem Soc* 84:229–235

Schnebly SR, Fehr WR (1993) Effect of years and planting dates on fatty acid composition of soybean genotypes. *Crop Sci* 33:716–719

Sensoz S, Kaynar I (2006) Bio-oil production from soybean (*Glycine max* L.); fuel properties of bio-oil, *Indust. Crops Prod* 23:99–105

Settlage S, Kwanyuen P, Wilson R (1998) Relation between diacylglycerol acyltransferase activity and oil concentration in soybean. *J Amer Oil Chem Soc* 75:775–781

Shannon JG, Sleper DA (2004) Breeding soybean for improved functional traits in the U.S. In: *Proceedings of the International Symposium on the Development of Functional Soybean Varieties, New Materials, Medicines, and Foods*. Kyungpook National University, Daegu Metropolitan City, South Korea, pp 81–91

Shieh CJ, Liao HF, Lee CC (2003) Optimization of lipase-catalyzed biodiesel by response surface methodology. *Bioresour Technol* 88:103–106

Spencer MM, Pantalone VR, Meyer EJ, Landau-Ellis D, Hyten Jr DL (2003) Mapping the *Fas* locus controlling stearic acid contents in soybean. *Theor Appl Genet* 106:615–619

Stojsin D, Luzzi BM, Ablett GR, Tanner JW (1998) Inheritance of low linolenic acid level in the soybean line RG10. *Crop Sci* 38:1441–1444

Suppes GJ, Dasari MA, Doskocil EJ, Mankidy PJ, Goff MJ (2004) Transesterification of soybean oil with zeolite and metal catalysts. *Appl Catal A-Gen* 257:213–223

Takagi Y, Rahman SM (1996) Inheritance of high oleic acid content in the seed oil of soybean mutant M23. *Theor Appl Genet* 71:74–78

Tan T, Lu J, Nie K, Deng L, Wang F (2010) Biodiesel production with immobilized lipase: A review. *Biotechnol Adv* 28:628–634

Tat ME, Wang PS, Van Gerpen JH, Clemente TE (2007) Exhaust emissions from an engine fueled with biodiesel from high-oleic soybeans. *J Amer Oil Chem Soc* 84:865–869

USDA, ARS (2007) *National Genetic Resources Program. Germplasm Resources Information Network— (GRIN)*. National Germplasm Resources Laboratory, Beltsville, MD, available at http://www.ars-grin. gov/cgi-bin/npgs/html/desc_find.pl (accessed September 26, 2007)

Wang HW, Zhang JS, Gai JY, Chen SY (2006) Cloning and comparative analysis of the gene encoding diacylg-lycerol acyltransferase from wild type and cultivated soybean. *Theor Appl Genet* 112:1086–1097

Wardlaw GM, Snook JT (1990) Effect of diet high in butter, corn oil, or high-oleic acid sunflower oil on serum lipids and apolipoproteins in men. *Amer J Clin Nutr* 51:815–821

Weselake RJ (2005) Storage lipids. In: Murphy DJ (ed), *Plant Lipids: Biology, Utilization and Manipulation.* Blackwell Publishing, Oxford, UK, pp 162–225

Wilcox JR, Cavins JF (1985) Inheritance of low linolenic acid content of the seed oil of a mutant in *Glycine max. Theor Appl Genet* 71:74–78

Wilcox JR, Cavins JF (1990) Registration of C1726 and C1727 soybean germplasm with altered levels of pal-mitic acid. *Crop Sci* 30:240

Wilcox JR, Cavins JF (1992) Normal and low linolenic acid soybean strains: Response to planting date. *Crop Sci* 32:1248–1251

Wilcox JR, Cavins JF, Nielsen NC (1984) Genetic alteration of soybean oil composition by a chemical muta-gen. *J Amer Oil Chem Soc* 61:97–100

Wilson RF (2004) Seed composition. In: Boerma HR, Specht JE (eds), *Soybeans: Improvement, Production and Uses*, 3rd ed, American Society for Agronomy, Madison, WI, pp 621–677

Wilson RF, Burton JW, Brim CA (1981) Progress in the selection for altered fatty acid composition in soy-beans. *Crop Sci* 21:788–791

Wolf RB, Cavins JF, Kleiman R, Black LT (1982) Effect of temperature on soybean seed constituents: Oil, protein, moisture, fatty acids, amino acids and sugars. *J Amer Oil Chem* Soc 59:230–232

Xu J, Jiang J, Sun Y, Chen J (2010) Production of hydrocarbon fuels from pyrolysis of soybean oils using a basic catalyst. *Bioresour Technol* 101:9803–9806

Yadav NS (1996) Genetic modification of soybean oil quality. In: Verma DPS, Shoemaker RC (eds), *Soybean: Genetics, Molecular Biology and Biotechnology*. CABI Publishers, Wallingford, UK, pp 165–188

Yadav NS, Wierzbicki A, Aegerter M, Caster SC, Perez-Grau L, Kinney AJ, Hitz WD, Booth JR Jr, Schweiger B, Stecca KL (1993) Cloning of higher plant omega–3 fatty acid desaturases. *Plant Physiol* 103:467–476

Zheng P, Allen WB, Roesler K, Williams ME, Zhang S, Li J, Glassman K, Ranch J, Nubel D, Solawetz W, et al. (2008) A phenylalanine in DGAT is a key determinant of oil content and composition in maize. *Nat Genet* 40:367–372

21 Sugarcane

Heitor Cantarella
Agronomic Institute

Marcos Silveira Buckeridge and Marie-Anne Van Sluys
University of São Paulo

Anete Pereira de Souza
State University of Campinas

Antonio Augusto Franco Garcia and
Milton Yutaka Nishiyama, Jr.
University of São Paulo

Rubens Maciel Filho
University of Campinas

Carlos Henrique de Brito Cruz
State University of Campinas and FAPESP

Glaucia Mendes Souza
University of São Paulo

CONTENTS

21.1 INTRODUCTION

Sugarcane is a tropical grass of the Poacea family and *Saccharum* genus and comprises several species although plants presently cultivated are mostly hybrids derived from *S. officinarum*, *S. spontaneum*, *S. robustum*, *S. sinnensis*, and *S. barberi* (Figueiredo 2008). Sugarcane is originated from Asia; the exact center of origin is unknown, but evidence points to Polynesia, New Guinea, India, China, Fiji Islands, and Tahiti. There are reports that *S. officinarum* was known in 6000 BC in India (Daniels et al. 1975). Sugarcane was already cultivated in the Middle East before the Christian era. Later the Arabs introduced the plant in Europe and from there it spread to America and parts of Africa in the early fifteenth century with the start of colonization (Figueiredo 2008).

Sugarcane is highly efficient in biomass accumulation. Its C_4 carbon metabolism allows for increased photosynthesis at high temperatures and efficient carbon assimilation, which leads to the highest yields produced among grasses. In fact, the "C_4" combination of biochemical and morphological specializations was discovered in sugarcane (Kortschak et al. 1965; Hatch and Slack 1966). Nowadays, sugarcane is grown in more than 100 countries (FAOSTAT 2009), mostly between the parallels 35°N and 35°S, covering an area of about 22 million ha and with a yield of approximately 1.6 billion tons of cane (Table 21.1), which represents 0.45% of the world's agricultural area and 1.6% of the arable area, with Brazil, India, and China topping the list. Data of 2007 show that these countries account for almost 63% of the world's cane production; the corresponding figure for Brazil alone is 33% (Table 21.1). In the last 20 years, sugarcane production grew approximately 57% worldwide. In Brazil, the sugarcane production will reach 664 million tons in 2010/2011.

Among the world's four most productive crops—rice, wheat, maize, and sugarcane — sugarcane produces the greatest crop tonnage and provides the fourth highest quantity of plant calories in the human diet (Ross-Ibarra et al. 2007) even though each of the major cereals occupy a severalfold larger fraction of the world's arable land. The stalks of most commercial varieties contain 10–16% fibers and 84–90% juice. The latter contains 75–82% water and 18–25% soluble solids of which the greater part (15–24% of the sugarcane juice) is sucrose; 1–2.5% are nonsugars such as amino acids, fatty acids, waxes, and mineral components (Stupiello 1987). Therefore, 1 t of sugarcane yields approximately 130–170 kg of sucrose.

Sugarcane is nowadays a source of food, feed, biofuels, and bioelectricity. Soon, biopolymers will also be added to the product list. Sugarcane ethanol is produced through the fermentation of

TABLE 21.1

Cane Production of the 20 World Largest Sugarcane Producers in 1987 and 2007

Rank in 2007[a]	Country	Cane Production (Mt)		Variation 1987–2007 (%)	Harvested Area (1000 ha)	Average Yield (t/ha)
		1987	2007			
1 (1)	Brazil	268.5	514.1	91	6712	76.6
2 (2)	India	186.1	355.5	91	4900	72.6
3 (4)	China	52.8	106.3	101	1236	86.1
4 (11)	Thailand	24.4	64.4	163	1010	63.7
5 (6)	Pakistan	43.6	54.8	26	1029	53.2
6 (5)	Mexico	45.9	50.7	10	680	74.5
7 (9)	Colombia	25.0	40.0	60	450	88.9
8 (10)	Australia	24.8	36.0	45	420	85.7
9 (7)	United States	26.5	27.8	5	358	77.6
10 (13)	Philippines	17.2	25.3	47	400	63.3
11 (8)	Indonesia	26.1	25.2	–4	350	72.0
12 (12)	South Africa	21.0	20.5	–2	420	48.8
13 (14)	Argentina	14.5	19.2	33	290	66.2
14 (19)	Guatemala	6.9	18.8	174	225	83.6
15 (16)	Egypt	8.4	16.2	92	136	119.6
16 (23)	Viet Nam	5.5	16.0	193	285	56.1
17 (3)	Cuba	70.8	11.1	–84	400	27.8
18 (17)	Venezuela	8.0	9.3	16	125	74.4
19 (20)	Peru	6.8	8.2	21	68	121.7
20 (24)	Sudan	4.8	7.5	58	72	104.2
	Sum/average	887.6	1426.8	61	19,564	72.9
	World	990.3	1557.7	57	21,977	70.9

Source: Food and Agricultural Organization of the United Nations, 2009. FAO Statitics Division, available at http://faostat.fao.org/ (accessed February 10, 2009).

[a] In parenthesis is rank in 1987. Harvested area and yield data refer to 2007.

sugar. Sugarcane partitions carbon into sucrose that accumulates in the internodes to up to 50% of its dry weight (0.7 M) (Moore 1995). Such high capacity of accumulation in stalks is unique among plants. Most of the ethanol produced in the world derives from plant juices containing sucrose from sugarcane in Brazil and starch from corn in the United States (EIA 2008; UNICA 2009). The United States and Brazil are the top producers of ethanol in the world. The production of bioethanol from sugarcane syrup is well established and nowadays is good enough to provide a product that follows market specification for fuels. The whole process may be considered very robust and does not compete with the food chain because in most of the industrial units the bioethanol is produced with depleted syrup because of simultaneous sugar production. However, in Brazil, where more than 50% of the sugarcane is directed to ethanol production, the whole syrup may be fermented. In addition, there are industrial plants devoted just to ethanol. A close look at the whole production structure allows identification of possible scenarios in which part of the sugarcane bagasse may also be used for bioethanol production when the cellulosic route is sufficiently developed.

In Brazil, bioethanol production from sugarcane is expected to reach 28.5 billion L in 2010/11. Almost 70% of the ethanol produced is hydrated (5.6% water/volume) and 30% anhydrous. Hydrated ethanol is used in automotive vehicles equipped with engines that run exclusively on ethanol or flex-fuel vehicles. Anhydrous ethanol is mixed with gasoline. Several countries are adding ethanol to gasoline to reduce fossil fuel consumption, increase octane rating and reduce pollution. This trend started in Brazil in 1931 and legislation that establishes a mixture of up to 25% of ethanol in gasoline dates to 1966.

Nowadays almost all of the 35,000 gas stations in Brazil have a hydrated ethanol pump (E100). Initially, cars were designed to run on either gasoline or ethanol. Now, flex-fuel cars are available that run on any mixture of ethanol to gasoline (from 0 to 100%). In 2008, flex-fuel cars represented 82% of the new cars sold, totaling around 23% of the automotive fleet in Brazil (around 6 million units). Motorcycle engines are already available that uses ethanol. New progress on combustion engines is underway to allow buses to also run on ethanol. This is greatly advantageous because the production and use of bioethanol as opposed to gasoline and diesel reduces greenhouse gas (GHG) emissions by 90%. This has been calculated on the basis of the whole sugarcane cycle including planting, harvesting, processing, and transporting of the fuel. And, because sugarcane is a renewable source of energy with a rapid growth and up to six annual harvests without the need for replanting, high dry matter yield per unit of fertilizer applied, minimal needs for pesticides—which require great quantities of fossil fuel for their production—and high CO_2 fixing capacity, the use of its ethanol can mitigate global warming.

The success of a biofuel crop is based on its economical and environmental advantages. To achieve sustainability, energy crops should not require extensive use of prime agricultural lands and they should have low-cost energy production from biomass. Basically, the crop energy output must be more than the fossil fuel energy equivalent used for its production. Studies have shown that the output to input ratio of sugarcane first-generation ethanol production is approximately 8–10, compared with 1.6 for maize (Goldemberg 2008). Several crops are being tested for bioethanol production, which can also be produced from starch and sugars from maize, wheat, sugar beet, cassava, and others, but they rarely reach two units of renewable energy produced relative to each unit of fossil fuel energy used. In 2007, the production and use of bioethanol in Brazil reduced GHG in 25.8 million t equivalents of CO_2. This corresponds to 360,000 diesel buses/year (Goldemberg 2008).

After juice extraction, the sugarcane stalk residue (bagasse) can be burned in the sugarcane factories for production of steam and electrical energy. Bioelectricity is the most important new product of the sugarcane business. All mills and distilleries in Brazil are self-sufficient in electric energy through co-generation. Approximately one third of the energy is stored in the cane juice, one third in the bagasse, and one third in the trash. Until recently, most of the sugarcane was burned in the fields to facilitated manual cutting, but increased pollution and decreased the energetic efficiency because the straw and leaves were not used for energy generation. Presently, in the state of São Paulo, responsible for 60% of the sugarcane production in Brazil, almost 50% of harvesting is mechanized without burning. Regulation determines that burning should be prohibited by 2014. This trend is spreading to other regions as well.

Bagasse is burned in highly efficient boilers (over 60 bars) that will allow for surplus energy to be commercialized in the electric grid. In 2008, sugar and ethanol mills produced an average of 1,800 MW. With the more efficient boilers being implemented and new investments in co-generation an estimated 11,500 MW can be produced, which is equivalent to 15% of the electricity demanded by Brazil. Bioelectricity brings several advantages, including low environmental impact, carbon credits, and the need for relatively small and low-risk investments. In Brazil, bioelectricity brings an additional advantage because sugarcane harvest takes place mostly in the dry season when hydroelectric mills are at their lowest production.

Alternatively, ethanol can be produced from bagasse and trash lignocellulose by hydrolysis of the cell wall using enzymes, physical, and chemical treatments (Ragauskas et al. 2006). The processing of lignocellulose and sugar conversion into ethanol is not yet economical but its development is highly desired because it could lead to an increase of 40–50% in ethanol production. Potentially, with the industry of cellulosic ethanol it is expected that the ethanol output might increase from the current 7,500 to 13,000 L/ha. Sugarcane second-generation ethanol has not yet been used commercially but many R&D initiatives are underway.

To improve yield and other traits of interest that will allow for a sustained industry of sugarcane and for the development of an energy-cane, research groups in biotechnology, transgenics, sugarcane genomics, statistical genetics for polyploid genomes, and gene discovery are gathering efforts all over the world to devise the mechanisms involved in the regulation of sucrose content, yield, drought resistance, biomass, and, more recently, cell wall recalcitrance. It is important to note that the whole sugarcane genome sequence is unknown and that sugarcane has a highly polyploid giant genome (~10 Gb). Recent works on expressed sequence tags (EST) added value to this crop's genomics but whole genome sequencing efforts are underway that will make available chromosomal gene structures and allelic variations of sugarcane. The SUCEST (a project that has generated the largest collection of ESTs—http://sucest-fun.org) database consists of 33,620 putative transcripts with a sequence mean size of 864 bp (Vettore et al. 2003) which represents about 30 Mb of sugarcane genome sequence, a small fraction of the complete genome sequence. Only with the recent developments of next-generation sequencing technologies has the identification of genes, alleles, and promoters as well as the definition of the overall structure of the genome been made possible. Also, if sugarcane is to be improved for bioenergy production, a significant number of cultivars and genotypes need to be evaluated at the biochemical and physiological level and molecular biologists have to join efforts with breeders bringing biotechnological tools to the game. Although a lot is known about plant cultivation, the biochemical and genetic characterization of this crop are at early stages. The remainder of this text will focus on the different aspects of bioethanol production using this crop and research developments for the improvement of sugarcane.

21.2 SUGARCANE CULTIVATION

Optimum temperature for sugarcane cultivation is between 30 and 34°C. Plant growth is greatly reduced below 21°C for most varieties. However, in the maturation stage, sucrose accumulation is triggered by dry conditions or low temperature, usually with average temperatures below 20°C. Death of leaves may occur below 2.5°C and apical and lateral buds die at –1 to –3.3°C and –6°C, respectively (Alfonsi et al. 1987; Liu et al. 1999). Usually at least 900–1000 mm of rain are necessary for rain fed production (Inman-Bamber and Smith 2005) but the need of irrigation depends also on rain distribution along the season.

Sugarcane is a semi-perennial bushy plant, in which several long stalks germinate from rhizomes or stools. Long alternated leaves are attached to the stalk nodes. The cylindrical stalks may reach 2–5 m tall and accumulate sugars mainly in the internodes. The world average stalk yields are around 70.9 t/ha but this varies with soil, climate, cycle length, and growing conditions (Table 21.1). Under favorable conditions, yield may reach above 200 t/ha but the theoretical yield according to different authors varies from 285–470 t/ha per year (Landell and Bressiani 2008; Waclawovsky et al. 2010).

In commercial fields sugarcane is planted with stem cuttings (seed cane) instead of seeds. Usually 8–12 t of 8- to 10-month-old stalks containing 12–18 buds per meter of row are planted in furrows spaced at 0.8–2.0 m. Small farmers adopt a narrow spacing whereas wider spacing is used in mechanized fields (Anjos and Figueiredo 2008).

The harvest takes place 10–24 months after planting. After harvest the plant sends up new stalks or ratoons that will be cut again usually within one year. Normally, yields decline in subsequent ratoons but two to ten cuttings can be performed before the crop needs to be planted again, depending on the variety, climate, pest and disease incidence, soil type, and management conditions.

Sugarcane is relatively tolerant to soil acidity but limestone application is recommended when soil pH is below 5.5. Because of the high dry matter yields, fertilizer needs of sugarcane are relatively high. The nutrient content of shoots of a crop yielding 100 t of stalks are around 100–154 kg N, 15–25 kg P_2O_5, 77–232 kg K_2O, and 14–49 kg S (various authors, compiled by Raij et al. 1997; Cantarella et al. 2007; Rossetto et al. 2008). Fertilization of sugarcane varies widely depending on the country, soil type, and yield potential. Sugar crops account for 7.5 million t of the NPK fertilizers used in 2007–2008, representing 4.5% of the world fertilizer consumption (Heffer 2009). It can be assumed that most of this fertilizer goes to sugarcane because sugar beet cropping area corresponds to 0.11% of that of sugarcane worldwide (FAOSTAT 2009). In Brazil 23% of the N, 8.7% of the P, and 21% of the K fertilizer are used in sugarcane (FAOSTAT 2009). Nitrogen and potassium are the nutrients used in largest amounts. Rates of application of N vary from 60 to 200 kg/ha (Raij et al. 1997; Rice et al. 2006). For P, up to 150 kg/ha P_2O_5 may be applied in low fertility soils but rates may be much smaller in many regions (Hartemink 2008). The sugarcane plant extracts large quantities of potassium and fertilization may reach, in kg/ha K_2O, 150 in Brazil and Australia, 175 in Costa Rica, 280 in the United States (Legendre 2001; Rice et al. 2006; Rossetto et al. 2008b).

Despite the high internal requirements of nutrients by sugarcane, the actual chemical fertilizer demand may be low because of recycling of plant and industrial residues because the mills and distilleries export basically carbon, hydrogen, and oxygen in sugar and ethanol. For instance, filter cake, which is generated at a rate of 30–35 kg (18–21 kg dry matter) per tonne of crushed fresh stalks, contains 1–3% P_2O_5 and is returned to the fields in natural form or composted with bagasse. Vinasse, the fluid residue of ethanol fermentation-distillation, is produced at a rate of 10–15 L/L ethanol and contains an average of 2 g K_2O/L. Vinasse is applied at rates varying from 50 to 200 m^3/ha and may supply all of the K needs of the crop that receives the residue. Ashes produced in the furnaces contain several nutrients, including micronutrients, and are also recycled in the fields.

The relatively low amounts of N used in Brazil is taken as evidence that biological nitrogen fixation (BNF) may play a role in nutrition of sugarcane (Urquiaga et al. 1992, 1995; Boddey et al. 2003). Furthermore, Urquiaga et al. (1992) showed that 60–70% of the N accumulated in some sugarcane varieties came from BNF. However, studies carried out in Australia and South Africa failed to show that BNF was a significant source of N to sugarcane (Biggs et al. 2002; Hoefsloot et al. 2005). Recent findings indicate that selected diazotrophic bacteria may effectively increase yield and supply N for sugarcane (Oliveira 2006; Reis et al. 2008). Although BNF offers a promising way to decrease sugarcane dependence of N mineral fertilizers, much work remains to be done.

Nutrient recycling in the field can also be improved if leaf burning that precedes harvesting is avoided. Manual harvesting that prevails in most of the world, especially in developing countries, is made easy by burning the crop but nutrients such as N and S are lost by volatilization and others are spread away with the ashes. When sugarcane is harvested unburned either manually or with a combine machine, a thick mulch of leaves and tops, equivalent to 8–20 t/ha of dry material remains on the soil, recycling nutrients and organic matter to the soil.

21.3 SUGARCANE PESTS AND DISEASES

A number of pests and diseases attack sugarcane. Among the insects the sugarcane borer (*Diatraea saccharalis*) is the most common, and may affect yield and quality of stalk because of the invasion of fungi and bacteria through the holes opened by the borer. The control of sugarcane borer is carried out with integrated pest management (IPM) that involves scouting of the insect population and biological control with one of the various natural enemies including *Cotesia flavipes, Lydella minense,* and *Paratheresia claripalpis.* Insecticides are rarely needed.

Pests that live in the soil, roots or stubble include the spittlebug *Mahanarva fimbriolata* that sucks the plant, the beetle *Migdolus,* and several species of termites. The control involves monitoring of infected areas to assess damage levels, management practices such as mechanical destruction or burning of stubbles, and use of chemical products when necessary. For *M. fimbriolata* biological control with *Metarhizium anisopliae* is also an alternative (Dinardo-Miranda 2008a).

Nematodes of the genera *Meloidogyne* and *Pratylenchus* also attack sugarcane. Resistant varieties usually are not an efficient way to control nematodes but management practices such as crop rotation when sugarcane is replanted, and application of organic matter help to decrease the population of nematodes. In areas where the attack is intense chemical control may be necessary (Dinardo-Miranda 2008b).

Fungi diseases that affect sugarcane include rust (*Puccinia melanocephala*), smut (*Ustilago scitaminea*), eye spot (*Bipolaris sacchari*), red rot (*Colletotrichum falcatum* or *Glomerella tucumanensis*), pokkah-boeng (*Fusarium*), and pineapple disease (*Thielaviopsis paradoxa*). The most important method of controlling fungi diseases is through the use of resistant varieties. Additional measures, which are effective with some diseases, include roguing of diseased stools, burning or plowing out trash or stubbles, and heat treatment of cane seeds (Santos 2008).

Resistant varieties are also the most important method of control of bacteria diseases such as ratoon stunting (*Clavibacter xyli*) and leaf scald disease (*Xanthomonas albilineans*), or viruses such as sugarcane mosaic virus and yellow leaf disease. Periodic surveys of nurseries and heat treatments may also be useful to prevent the spread of some diseases (Almeida 2008; Gonçalves 2008).

Despite the great number of pests and diseases that affect sugarcane, chemical control is applied in less extent than in many other crops because of the effectiveness of biological control methods and especially of resistant varieties. Fungicides and bactericides are seldom used in sugarcane.

21.4 SUGARCANE PHYSIOLOGY

In spite of the fact that sugarcane became one of the main sources of sugar and ethanol in the world, its physiology has been poorly studied in relation to other grass species. Sugarcane plants can be obtained from seeds, germination can be easily obtained in vitro and the structure of the seeds and seedlings of sugarcane is typical of other grasses. Seedlings can be an excellent model to study and understand several aspects of gene expression, biochemistry physiology. For biotechnology purposes though, sugarcane seeds are not useful because what we call sugarcane is not a single species, but a polyploid hybrid. This makes the seed method of reproduction inappropriate for crop cultivation. Crops are planted from clones of designed varieties using stem cuttings. The same plant stays in the field for an average of 5 years. Yields decline with age and after a while it becomes aneconomical so the crop must be replanted. One of the most important physiological events related to sugar production is the drought stress that occurs in wintertime. From the agricultural viewpoint, it is well known that the stress is necessary to induce senescence (by ethylene) of the top shoot of the plant and the consequent storage of sucrose in the stem. Accumulation of sucrose is related to photosynthesis, sink-source relationships, flowering, and water stress. Water stress depends on the level and rate it developed. When applied slowly it leads to developmental changes such as a reduction in leaf expansion and the closing of leaf stomata. Photosynthate translocation is not reduced until the stress becomes severe. Combined, the effect of drought is to lead to accumulation of carbohydrates in the leaves and in storage sinks of the sugarcane plant (Hartt 1936). The accumulation of sucrose in storage parenchyma of sugarcane is called ripening. Ripening is caused by the gradual decrease of tissue moisture, reduction of cell expansion, and the formation of new internodes without much inhibition of photosynthesis. Reduced consumption of sucrose for metabolic energy and new cell formation leads to increased sucrose content (Gosnell and Lonsdale 1974). Sugarcane managing includes the use of drought and growth inhibiting stresses to ripen the crop before harvest.

Sugarcane photosynthesis is based on the so-called Kranz syndrome anatomy. Instead of using the C_3 pathway system, in which CO_2 is converted directly by the enzyme ribulose bisphosphate carboxylase (RUBISCO) into organic acids containing three carbon atoms, sugarcane leaves incorporate CO_2 first into a four-carbon atom compound by a different enzyme (phosphoenol pyruvate carboxylate—PEPc). That is why sugarcane is referred to as a C_4 photosynthesis plant. This four-carbon compound (phosphoenol pyruvate) is subsequently processed to produce CO_2 in another cell type (the cells of the bundle sheath). The process occurs in the leaves in a much higher pressure of CO_2 than in the outside air, and proceeds through the biochemical pathways of C_3 photosynthesis, providing sugarcane with an extremely efficient system that processes light and CO_2 for production of sugars. This photosynthetic system occurs especially in species that live in environments where there is abundance of light and relatively higher temperatures. The C_4 photosynthesis system was discovered in 1966 in Australia in leaves of sugarcane (Hatch and Slack 1966). Later on, other scientists found that plants like maize also display this type of metabolism.

The control of sugar metabolism in sugarcane is associated with the plant hormone ethylene. This hormone cross-talks with several other hormones and also affect the nitrogen metabolism. It is likely that a complex system including multiple genes controlled by environmental factors (mainly water stress), plant hormones that lead to changes in plant metabolism, and ecophysiology (e.g., photosynthesis) are related to the level of production of sucrose and subsequently to biomass and bioethanol.

Physiologically, the accumulation of sucrose in stems of sugarcane appears to be directed for flowering. In 1998 Carlucci et al. reported results of a cultivar (IAC 52–150) growing in Piracicaba, SP, Brazil, that flowers if grown under long days (sugarcane flowers with a 12- to 12.5-h photoperiod). Induction of flowers occurred in March, when humidity was high and flowers started to develop in May, reaching up to 70 cm in June. Flowering initiation is thought to have its optimum between 18 and 31°C. In fact, the difference between maximal and minimal temperatures is crucial. Pereira et al. (1986) and Carlucci et al. 1988 found that flowering was intense when this difference was of 10°C and that flowering did not happen when the maximal–minimal temperature difference was on the order of 14°C. According to these authors, the combination of extreme temperature differences with water stress during the flower induction period negatively affects flowering, retarding or preventing it completely. Thus, flowering seems to be controlled by a combination of factors including temperature and water. Because of that, sugarcane plants hardly flower in the Southeast of Brazil (ca. 24° of latitude), where most of the sugarcane crops are planted. On the other hand, this is probably the reason why a period of water stress is desirable for high accumulation of sugar. The stress is likely to delay flowering initiation, but not induction, which probably leads to a change in the pattern of source-sink relationship, provoking an accumulation of sugar to prepare the plant for flowering. Therefore, avoiding or delaying flowering is very important from the agricultural viewpoint as the sucrose that is stored in the parenchyma cells of the stems is the reserve of carbon that the plant will use to produce the flowers. Thus, by harvesting, farmers intervene in the flowering process, preventing the plant from using the sucrose stored for that purpose and extracting it for sugar or ethanol production. The lack of low temperature or water stress becomes critical when plants are cultivated in the Amazon, in the North of Brazil because it makes it difficult for most varieties to accumulate sucrose; therefore, that region has no favorable conditions to grow sugarcane.

21.5 SUGARCANE GENOME AND GENOMICS INFRASTRUCTURE

Genomics is increasingly recognized as a powerful approach to address scientific questions in biology. It establishes the nucleotide sequence of an organism and as a result enables gene content prediction. The adaptation of an organism to the environment, its performance and its phenotype are a result of multiple gene products interactions. Moreover, knowledge transfer from one model organism to another of yet less information is made possible with comparative analysis. Sugarcane biology will greatly benefit from nucleotide sequence determination, as it will foster a systems biology approach by understanding genome structure and regulatory networks. The challenge in determining the sugarcane genome sequence is the complexity of its genome structure as a polyploid and understanding the balance between alleles. Sugarcane cultivars are known to be a hybrid resulting from crosses between two polyploid genomes *S. spontaneum* ($x = 8$; $2n = 40$–128) and *S. officinarum* ($x = 10$; $2n = 80$). Modern cultivars are polyploid and aneuploid which renders allelic variation/assortment a key aspect in breeding programs (D'Hont 2005; D'Hont et al. 1996).

Collectively, the *Saccharum* complex is diverse in genome content and organization. *Sorghum*, *Miscanthus,* and *Erianthus* are closely related species and represent genetic reservoirs for exploitation of genetic diversity. Molecular phylogenetic studies within the Saccharinae group indicate that *Miscanthus sensu stricto* and *Saccharum* are sister groups*, while *Sorghum* and *Erianthus* share a close relation. Monophyly supports *Miscanthus* and *Saccharum* relation, but a distinct clade named *Miscanthidium* is identified (Hodkinson et al. 2002).

An International Genome Sequencing Initiative has recently been formed to produce a draft sequence from several sugarcane cultivars so that tools are developed for understanding genome ploidy variation, enabling gene discovery and generating a knowledge base molecular infrastructure (http://bioenfapesp.org). Basic research will benefit not only from gene discovery but also from the identification of regulatory sequences involved in sucrose metabolism, carbon partitioning in the plant and responses to restrictive water supply. Breeding programs will have access to the development of new molecular markers. The sugarcane monoploid genome is estimated to be about 1 Gb, comparable in scale to the human and maize genomes. The ground basis to tackle the sugarcane genome are available resources such as the EST collections (see below), array hybridization profiles generated by SUCEST-FUN (described below), a collection of bacterial artificial chromosome (BAC) clones from R570 cultivar, and the recently released *Sorghum* genome (Paterson et al. 2009). Genetic maps are currently being improved by the inclusion of repetitive sequences such as microsatellites and resistance gene analogs (RGAs) (Rossi et al. 2003) thus, increasing resolution. Another class of repetitive sequences is transposable elements (TEs), which are composed in sugarcane by a heterogeneous universe of molecular entities previously described by Araujo et al. (2005). TE selected BAC clones have been sequenced and specific insertion polymorphism studies provide information concerning their association to genetic diversity. The ultimate goal in generating the sugarcane genome sequence is to contribute with a large scientific community effort to improve sugarcane breeding and develop a systems biology-based approach in sugarcane. Initially, shot-gun sequencing and a draft assembly of 1000 BAC clones will provide resources for basic biological processes including access to promoter regions and the possibility of comparative studies among grasses and, specially, cultivars of interest. Furthermore, the sequencing initiative is expected to provide tools for the identification of functional modules of gene variation.

The sugarcane directed genome sequence will provide valuable tools for understanding genome polyploidy variation when compared to *Sorghum* and other Poaceae species (*Miscanthus*, *Erianthus,* and *Oryza*). Sugarcane is a domesticated crop that originated from New Guinea (Asia) about 6000 AD. No more than a hundred years ago modern cultivars have been produced from the cross of two closely related species and have progressively replaced the Noble clones spreading in all of the sugarcane producing areas of the globe. Since then, breeding programs are devoted to addressing the main questions that any crop under a heavy agricultural system is subjected to: growth habit and harvest index; adaptation to photoperiod; resistance to diseases and abiotic stresses (mainly water supply and temperature), flowering and genetic erosion (loss of variability to adapt). Sugarcane varieties are basically maintained from vegetative propagation of selected clones. These clones would in principle keep their agricultural traits. A sugarcane cultivar issued from a breeding program is productive for approximately 15 consecutive years after which cultivar replacement is needed because of loss of quality traits most probably because of genetic erosion and/or instability or changes that occur in pest and disease agents to overcome plant resistance.

The genome of hybrids is highly polyploid and aneuploid (Grivet et al. 1996). Efforts on mapping genes and molecular markers to generate physical maps have been described but because of the genetic complexity of sugarcane, its genome is poorly understood (Ming et al. 1998; Hoarau et al. 2001; Lima et al. 2002; Pinto et al. 2004; Garcia et al. 2006; Raboin et al. 2008). Modern sugarcane cultivars are complex interspecific hybrids with a chromosome number ranging from 100 to 130, of which 15–25% comes from *S. spontaneum*. Considering monoploid genomes, the DNA content is ~930 Mb for *S. officinarum*, ~750 Mb for *S. spontaneum* and approximately 1000 Mb for sugarcane hybirds (D'Hont 2005). At Clemson University Genomic Resources a BAC library prepared from the R570 sugarcane cultivar is available and is represented by 103,296 BAC clones with an average size of 130 kbp (CUGI 2009). This library has been screened for resistance gene analogs (A. D'Hont, personal communication), *adh* locus (Janoo et al. 2007), sorghum euchromatic regions (A. Paterson, personal communication), transposable elements and genes associated to sucrose content and drought responses. These trends render it feasible to undertake a pilot project to sequence the sugarcane genome and address questions related to gene allelic variation and regulatory regions.

A combined approach of new sequencing technologies such as 454 pyrosequencing and Sanger reads will give access to the sugarcane genome sequence uncovering the genetic basis structure sustaining the biological processes. Not only will regulatory regions associated to specific genes of interest be discovered but also gene prediction models compared to sorghum, rice and maize. Breeding strategies can benefit from the comparative genome sequence of homologous regions between R570 and SP80-3280 (the cultivar most represented in the EST collections), thereby allowing for rapid translation of the sequence data into genetic markers. Regulatory sequence variation is also to be uncovered through this comparative approach and, in combination with the expression profile analysis, relevant insights on the evolution of these regions and the contribution of transposable elements will come to light. In a broader view, BAC sequencing will add to the understanding of chromosomal differentiation among Poaceae.

A long-standing goal in polyploid genomes is to understand the relative contribution of each allele to a particular phenotype in a given cultivar. The key problem in achieving this goal is identifying allelic variation and subsidizing breeding programs to quickly select it from among a segregating population. Also, accumulating multiple genes into plant varieties is not yet widely used for polyploid plants. Because of the lack of understanding of the genetic basis of heterozis, genome sequencing of "gene of interest" containing regions will advance knowledge on genome structure creating the molecular basis to explore the genetic diversity among cultivars and breeding populations segregating for characters of interest.

One innovative approach is to understand the relative contribution of TEs to genetic variation in the sugarcane polyploid genome. These mobile elements are ubiquitous among living organisms and constitute intermediate-repeat DNA long considered as selfish (or junk) DNA (Doolittle and Sapienza 1980; Orgell and Crick 1980). Contrary to that, and as previously proposed by McClintock (1984), a new biological concept is arising for transposable elements where, despite their mutagenic capacity, they actively contribute to changes in the gene expression profile and may ultimately result in species divergence (Cordaux et al. 2006; Jordan 2006; Cropley and Martin 2007; Xiao et al. 2008). Their contribution to eukaryotic genome structure is usually associated with gains of nuclear interspersed sequences such as noncoding repetitive DNA between and sometimes within coding units. One means of understanding the contribution of a particular class of TE to the genome is to first identify the gene pool present, its relative amplification across contrasting varieties and its expression pattern. Much of the initial work will be to create a ground basis for identifying TE families associated to particular traits (brix, drought, high CO_2 environment, and regeneration capability). This recognition will impact on balancing selection for (or against) the presence of a given TE family. Clearly, a prerequisite in sequencing polyploid genomes is to be familiarized with its repetitive DNA elements. For sugarcane, 21 families have been identified and further studies were carried out on two retrotransposons (*Hopscotch*-like and SURE) and two transposons (*Mutator*-like and hAT-like). *Hopscotch*-like and *Mutator*-like elements contain lineages that are represented by highly repeated unit spread along the chromosomes with no particular clustering evidenced at telomeric or centromeric regions. Independent of their amplification profiles both high and low copy number elements are expressed in different tissues of sugarcane.

21.6   SUGARCANE BREEDING

There are basically six species within the *Saccharum* genus: *S. officinarum* L. ($2n = 80$), *S. robustum* Brandes and Jeswiet ex Grassl ($2n = 60–205$), *S. barberi* Jeswiet ($2n = 81–124$), *S. sinense* Roxb. ($2n = 111–120$), *S. spontaneum* L. ($2n = 40–128$), and *S. edule* Hassk. ($2n = 60–80$). Because genomes of all of these species may be involved in some form of modern cultivars, sugarcane is considered to have one of the highest genetic complexities among cultivated species.

The genetic breeding of sugarcane in Brazil and worldwide has been explained in great detail in several articles and books (Stevenson 1965; Blackburn 1983; Berding and Roach 1987; Berding and Skinner 1987; Breaux 1987; Heinz and Tew 1987; Hogarth 1987; Tew 1987; Machado Jr et al. 1987; Matsuoka and Arizono 1987; Matsuoka et al. 1999a, 1999b; Landell and Alvarez 1993; Landell and Bressiani 2008). The present text will concentrate on the work of Matsuoka et al. (1999a, b). Matsuoka et al. (1999a) showed in detail how sugarcane breeding began in Brazil in the nineteenth century. The first reports that sugarcane seeds (not the stem) could result in offspring occurred in Barbados in 1858 (Deerr 1921; Stevenson 1965). However, it is presumed that the breeding actually started in 1885 in Java, from the germination of *S. spontaneum*. But a few years earlier in Brazil, Peixoto Lima (1842) had stated in thesis defense that sugarcane would reproduce from seeds obtained by crossing (i.e., by sexual reproduction). This fact, combined with several others, seems to indicate that Brazil was among the pioneers in getting new commercial varieties from seeds.

From the beginning of the twentieth century, there was an increasing concern about the poor sugarcane productivity as well as the increase of pests (Deerr 1921; Aguirre Jr 1936; Edgerton 1955; Dantas 1960; Stevenson 1965; Andrade 1985). As a result, there was an intense exchange of germplasm between different countries, the farmers being primarily responsible for this (Matsuoka et al. 1999a). However, at times when the producers felt pressured by crises in the sector, there were initiatives toward the setting up of experimental stations (Geran 1971).

At present, Brazil has four main sugarcane breeding programs: (1) the sugarcane program from the Agronomic Institute of Campinas (IAC), which started in 1933; (2) the RIDESA Interuniversity Network Sugarcane Genetic Breeding Program, made up of federal universities that started in 1971 as PLANALSUCAR; (3) the Sugarcane Technology Center (CTC), which began work in 1968 as COPERSUCAR; and (4) Canavialis, which began in March 2003 and was recently acquired by Monsanto. Syngenta also started a sugarcane breeding program in Brazil.

Despite there being large differences in the details about how the breeding programs in Brazil (and the world) carried out their activities, there are some points in common which will be highlighted here. In essence, the breeding is based on the selection and cloning of superior genotypes in segregating populations, which are obtained from the sexual crossing between different individuals. The success rate of these processes depends on several factors including the adequate choice of the parents to maximize the chance of response to selection; use of adequate experimental designs; and the correct choice of the traits to be evaluated. Most of the traits considered in the selection process have a quantitative nature and are controlled by quantitative trait loci (QTL), such as soluble solid rate; sucrose content; diameter and number of stalks; fiber content; flowering; precociousness; resistance to pests and diseases, etc.

Each year, breeding programs generate segregating populations formed by thousands of seedlings. The number of seedlings varies according to the program and depends on economic and technical factors. These segregating populations are then submitted to selection in different schemes, which are presented below.

21.6.1   GENERATION OF VARIABILITY

The genetic variability available for selection comes from sexual crossing and can be done in different ways (Matsuoka et al. 1999a, 1999b): (1) biparental crossing, in which crossings are made using two known parents, some of which may be used exclusively as a female; (2) polycrossings, when many genotypes are intercrossed, in which case seeds are gathered from the inflorescence of all parents involved, which prevents the identification of the pollen source; and (3) free pollination, in which seeds are harvested from plants growing freely.

Crossings should be planned in such a way as to maximize the probability of genotype selection that can be released as commercial cultivars. Thus, an alternative that is greatly used is to choose parents with good economically interesting performance traits (Matsuoka et al. 1999a), which naturally occurs for commercially used cultivars. It is noteworthy that this can lead to a narrowing of the genetic basis (Lima et al. 2002).

Because sugarcane is a type of allogamy, the crossings should be planned so as to avoid the occurrence of interbreeding between relatives. This can be achieved based on the genealogies of materials as well as genetic divergence obtained with molecular markers (Lima et al. 2001). Other criteria should also be used such as trait complementarity; ability to combine material during the crossings; and the capacity for each material to produce good populations throughout time.

Some authors mention the use of estimates of the genetic parameter named heritability that in its restricted form takes into account the sexual phase. This can be considered in the prediction of crosses, because heritability indicates how the interest traits are transmitted to the offspring. Some results indicate that the predominant gene action in the Brix content is additive, whereas for the other components the output is not additive (Hogarth 1980; Wu et al. 1980). Hogarth (1977) cites that for stem production, the dominant variance has shown the same magnitude as the additive, where the epistatic variance is predominant for stem weight. Hogarth (1987) showed high heritability values for the fiber content trait. For volume and number of stems, the dominant variance proved to be important (Hogarth et al. 1981). Bressiani (1993) reported high heritability values for length of stem and Brix in Brazilian conditions. For rust, the heritability has been high (Hogarth et al. 1983; Bressiani and Sanguino 1994), which also usually happens with other diseases (Matsuoka et al. 1999a), indicating that the selection of parents may be resistant effective. However, there are examples of transgressive segregation.

21.6.2 INITIAL STAGES OF SELECTION

There are many variations on how the breeding programs develop this stage, especially in terms of selection rates, group size, number of locations, and replications (Simmonds 1979; Skinner et al. 1987; Matsuoka et al. 1999a). In general, a low intensity selection is applied in the early phases, selecting only for traits with high heritability. This is followed by an increase in selection intensity as the experimental precision increases, and cultivation recommendation only when there are many experimental results in different places and years of cultivation.

Broad sense heritability, which should be considered in these phases, has several different values reported in literature. For example, in relation to vigor, some studies indicate low heritability, although several breeding programs have obtained good results with intense selection for this trait (Skinner et al. 1987). These authors report different heritability values for various traits, such as sugarcane yield; Brix yield per hectare; number and diameter of stems; and resistance to rust and smut. These results suggest that in early stages selection is more effective for Brix and for resistance to rust and smut, but these data should be interpreted with caution given the low number of available estimates (Matsuoka et al. 1999a).

Another very important feature to be considered in these phases is the genotype regrowth ability. Despite the small number of available estimates in this respect, many programs consider this aspect in the selection process (Giamalva et al. 1967; Watkins 1967; Giroday 1977; Mariotti 1977; Bond 1978; Skinner et al. 1987; Arizona 1994; Matsuoka et al. 1999a), which can lead to many genotypes in the early phases being discarded.

Several selection strategies for these steps have been used and are briefly described below.

- *Bunch planting:* According to Ladd et al. (1974), this is a recommended method when there is a large number of individuals from a previously untested crossing, because the simultaneous evaluation of many stems in the same area is possible. Normally, 5–10 seedling bundles are planted (Mangelsdorf 1953) and it is expected that natural selection eliminates the inferior ones, based on competition (Urata 1970), although this seems questionable, because it is not known whether this may actually occur in heterogeneous bunches. Matsuoka et al. (1999a) pointed out that the main disadvantages are the inability to select more than one stem per bunch, as the mixture prevents individual identification and makes it difficult to assess tillering, which is positively correlated with budding and yield. Skinner et al. (1987) argue that there are examples of success of the bunch planting, although there are no conclusive data about its superiority.
- *Family selection:* It is based on the fact that for the important traits, the heritability between families is higher than the heritability between individual plants. As a result, several breeding programs prefer to carry out the selection between the families, choosing those with higher phenotypic averages. Note that this procedure also serves to identify superior crosses (Skinner et al. 1987). Matsuoka et al. (1999a) mentioned that this procedure can lead to the disposal of superior individuals in families with a low average, but with high variance.
- *Selection of individual plants:* Many breeding programs carry out this type of selection, using only the traits of high heritability as a criterion in this phase, such as Brix and resistance to flowering and disease. Latter (1964) mentions that in some cases this selection strategy leads to better results than the selection between families and also reduce the risk of discarding superior individuals in families with a low average. According with Matsuoka et al. (1999a) the economic feasibility of assessing each plant individually, rather than evaluating a large number of seedlings should be taken into consideration. In practice, the individual evaluations are usually possible, because in the early phases of selection many genotypes are discarded visually based only on their vigor (Skinner et al. 1987; Matsuoka et al. 1999a).

21.6.3   EVALUATION OF CLONES IN FIELD TRIALS

In this phase, the genotypes selected are compared based on experiments using appropriate statistical designs. Because of the high number of genotypes being evaluated and the small number of stems to be used in experiments, the plots are usually small, e.g., a 5-m-long row per replication, place and time of evaluation. It is very common in this step to use increased complete randomized blocks (Federer 1956). Other traits that were not evaluated in the previous phase are now considered, and assessments of the regrowth capacity are also commonly made. Because of limited experimental precision, the values of heritability are still low and therefore the selection should not be too intense (Matusuoka et al. 1999a). Many sugarcane breeding programs in the world have applied approximately 10–30% intensities (Skinner et al. 1987).

21.6.4   FIELD TRIALS FOR PRECOMMERCIAL GENOTYPES

The clones selected as above are then evaluated in experiments with many replications, in several environments, and throughout various rattons. Such experiments in Brazil are usually installed in randomized block designs. After intensive evaluations through several cuts, environments and years, the new cultivars can be recommended for commercial use. An interesting feature of this phase is that sugarcane growers become greatly involved because most experiments are installed in commercial farms and evaluated under conditions that are very similar to those in actual cultivation.

21.7 SUGARCANE IMPROVEMENT TOOLS: MARKER-ASSISTED BREEDING AND TRANSGENICS

Sugarcane improvement has been achieved through the classical route for over 2 centuries but it is a costly and slow process. Currently, it takes around 12 years for a new cultivar to be released. One strategy to speed up this process is the use of genetic maps and molecular markers through marker-assisted breeding. Another strategy is the use of transgenics.

21.7.1 GENETIC MAPS AND MOLECULAR MARKERS

Most genetic designs used for the construction of genetic linkage maps use populations derived from crosses between inbred lines (e.g., using backcross or selfing). The statistical and genetics methods used in this case are already established and implemented in several softwares, such as Mapmaker/EXP. However, to obtain such strains for sugarcane is impractical, especially because of the large inbreeding depression that occurs when selfing occurs. In this case, the mapping populations are F1 generations obtained from crosses between non-inbred individuals (Lin et al. 2003).

For sugarcane (and several other species for which there are no available inbred lineages, such as fruit and eucalyptus), an alternative that was widely used in the past is called *double pseudo-testcross*. This strategy is the construction of two individual maps (one for each parental), by the polymorphic identification of single dose markers for each parental (Grattapaglia and Sederoff 1994; Porceddu et al. 2002; Shepherd et al. 2003; Carlier et al. 2004). On the basis of this approach, linkage maps for *S. officinarum* ("LA Purple") and *S. robustum* ("Mol 5829") were constructed using random amplified polymorphic DNA (RAPD), amplified fragment length polymorphism (AFLP) and restriction fragment length polymorphism (RFLP) markers in a single dose (Guimarães et al. 1999). Other situations were also considered, but rarely the kind of crossing used involved commercial parents (Al-Janabi et al. 1993; da Silva et al. 1993; Sobral and Honeycutt 1993; da Silva et al. 1995; Grivet et al. 1996; Hoarau et al. 2001). However, from both a biological and statistical point of view, the integration of the information contained in these individual maps into a single integrated map is desirable. This can only be done with the presence of heterozygote markers between parents, which are used to establish linking relationships between the markers individually segregated in each parental (Barreneche et al. 1998; Wu et al. 2002; Garcia et al. 2006; Oliveira et al. 2007).

The construction of integrated genetic maps, using different types of molecular markers with different segregation, has great advantages because it allows the linking map to be saturated and extends the characterization of polymorphic variation throughout the genome. Codominant markers may be useful to bring together cosegregation groups in their respective homology groups, specifically for polyploid species such as sugarcane (da Silva et al. 1993; Grivet et al. 1996). Moreover, the more precise localization of QTL is helped by the availability of an integrated genetic map (Maliepaard et al. 1997). However, there may be different numbers of alleles, in heterozygote parentals for each segregating loci, turning linkage analysis more complex, because the linkage phases in the parentals may be unknown a priori, making it difficult to detect recombination events (Maliepaard et al. 1997; Wu et al. 2002).

Wu et al. (2002) proposed a statistical method on the basis of maximum likelihood analysis that allows the simultaneous estimation of recombination fractions and linking phases between loci in mapping populations derived from crossings between non-inbred individuals (F_1 generation). This method works toward the construction of an integrated genetic map, which is the result of the combination of various pieces of information generated by different types of molecular markers, whose information content varies. Garcia et al. (2006) and Oliveira et al. (2007) constructed integrated genetic maps, consisting of 357 markers distributed throughout 131 co-segregation groups from the crossing between two precommercial sugarcane cultivars (SP80-180 × SP80-4966). The results were better than those obtained when the same data were analyzed using the JoinMap program, which indicates, in this case, better efficiency in the estimation of linking and linking phases of the method proposed by Wu et al. (2002). To do so, a software called OneMap, developed specifically for this purpose (Margarido et al. 2007) was used. It can also be used for other outcrossing species.

Despite the superiority of this new analysis approach in relation to previous methods, in the case of sugarcane, it is still possible to use markers with 1:1 and 3:1 segregations, which are known to be less informative than other types (as for example, those that segregate 1:2:1 and 1:1:1:1 in the case of diploid species). This makes it difficult to get linkage groups and also to order markers within these groups. This leads to less saturated maps with less genome coverage. Furthermore, the integration of maps of the genitors is not always possible.

The first sugarcane genetic maps were built using RAPD and RFLP type molecular markers. Currently, mainly gene or genomic microsatellite molecular markers are used in the setting up of genetic and molecular maps. Microsatellites or SSRs (simple sequence repeats) have become widely used in plant marker studies. These markers are conventionally tandem repeats of small nucleotide sequences of one to six bases in length. The variation in the number of repetitions results in polymorphic loci that are extremely useful, especially in mapping studies. The ability of SSRs to reveal high allelic diversity is particularly useful in the discrimination between genotypes. The successful use of this marker in other species such as barley (Saghai-Maroof et al. 1994; Russell et al. 1997), rice (Wu and Tanksley 1993), wheat (Röder et al. 1995), apple (Szewc-McFadden et al. 1996), and avocado (Lavi et al. 1994), stimulated the application of this technique to more genetically complex species such as sugarcane (Cordeiro et al. 2000).

Nowadays, the search for SSRs is being carried out in expressed sequence tags (ESTs) deposited in public databases, as this alternative is a simpler, faster and more economical strategy for the development of SSRs. EST analysis is a simple strategy to study a portion of the expressed genome, even in organisms with large, complex and highly redundant genomes, such as sugarcane. The basic strategy for obtaining the EST is a fast and efficient method for genome sampling of gene active sequences. As genetic markers, the EST-SSRs have been evaluated in several studies and tend to be considerably less polymorphic than the markers generated from genomic sequences for rice (Cho et al. 2000), sugarcane (Cordeiro et al. 2001; Pinto et al. 2006), wheat (Eujayl et al. 2002), and barley (Thiel et al. 2003).

The analysis in sugarcane of 8678 EST sequences revealed approximately 250 SSRs, the majority made up of perfect trinucleotide repeats where (GCC) n, (CGT) n (CCT) n motifs were the most common (Cordeiro et al. 2001). All selected EST-SSRs were polymorphic in the co-related *Erianthus* and *Sorghum* genera. The lowest value for the polymorphic information content (PIC) was obtained among the varieties of sugarcane (0.23), increasing between the species *S. officinarum* and *S. spontaneum* (0.62) and reaching the highest value (0.80) among the genera *Erianthus* and *Sorghum*. Because of the narrow genetic base of the varieties of sugarcane, the use of EST-SSR can assist in the characterization of the genetic variability available in the germplasm collections of related genera used in introgression programs. Thus, the introgression limitation of the *Erianthus* genome in sugarcane (*Saccharum*) can be overcome by the use of EST-SSR in identifying the portion of the *Erianthus* genome in intergeneric hybrids (Cordeiro et al. 2001).

The application of ESTs was shown to be a successful and efficient means of identifying sugarcane genes. A study by Carson and Botha (2000) showed that of all cDNA clones from the leaf identified in the search for homology, 38% showed significant similarity with known gene sequences. This value can be compared with that observed in the analysis of cDNA libraries from the endosperm and seed corn (39.3%, Shen et al. 1994) and even better than the results obtained using a cDNA library from maize leaves (20%, Keith et al. 1993), tissues of different growth stages of rice (25%, Yamamoto and Sasaki 1997) and portions of RNA from seeds, roots, leaves and inflorescences of *Arabidopsis* (32%, Newman et al. 1994).

21.7.2 Transgenics

Sugarcane biotechnology started in the 1960s with callus induction and rooted callus recovery (Nickell 1964) followed by callus regeneration (Barba and Nickell 1969; Heinz and Mee 1969). The first genetically modified clones were obtained by biolistics (Bower and Birch 1992). Transgenic plants have since then been obtained that incorporate agronomic traits of interest (Gallo-Meagher and Irvine 1996; Arencibia et al. 1997, 1999; Enriquez-Obregon et al. 1998; Ingelbrecht et al.1999; Zhang et al. 1999; Falco et al. 2000; Butterfield et al. 2002; Falco and Silva-Filho 2003; Leibbrandt and Snyman 2003; McQualter et al. 2004; McQualter et al. 2005; Vickers et al. 2005a, b; Wang et al. 2005; Snyman et al. 2006) including drought tolerance (Zhang et al. 2006; Molinari et al. 2007). Several plant tissues can be used to produce callus (Liu 1981 1993; Irvine 1987). Somatic embryogenesis is the most studied (Guiderdoni et al. 1995; Manickavasagam and Ganapathi 1998) and regeneration can be direct (Manickavasagam and Ganapathi 1998) or through the induction of embryogenic callus from immature leaf explants (Guiderdoni 1988). Embryogenic callus can be maintained for months without loss of regeneration capacity (Fitch and Moore 1993). Regeneration efficiency can be optimized for transformation as shown recently (Lakshmanan et al. 2006; Snyman et al. 2006).

Agrobacterium tumefaciens-mediated transformation has also been used in sugarcane (Arencibia 1998; Enriquez-Obregon et al. 1998; Elliott et al. 1998; Manickavasagam et al. 2004). With adequate manipulation of in vitro culture conditions and adequate *A. tumefaciens* cell lines the method can lead to the transference of relatively long DNA fragments, little rearrangements, low copy number and low costs. Sugarcane cultivars differ in their regeneration capacity and the methods must be optimized. Many cultivars that regenerate have already been described (Falco et al. 2000; Lima et al. 2001; Falco and Silva-Filho 2003; Cidade et al. 2006). Although the methods are available, so far there are no genetically modified cultivars released for trading. Transgene expression is largely unstable and many groups are searching for gene promoters that may lead to stable expression in mature plants. Examples of plant promoters useful for transformation but that vary in their efficacy are *CaMV35S, nopaline,* and *octopine* from *A. tumefaciens, Ubi1, Emu, Act1,* but there is no guarantee of tissue specificity (Last 1991; McElroy et al. 1991; Zheng et al. 1993; Green 2002; Neuteboom 2002; Christensen and Quail 1996). *Ubi-1* is the most used in sugarcane for constitutive expression (Lakshmanan 2005) and some studies point to more adequate promoters (Liu 2003; Braithwaite 2004) but still there is little guarantee of targeting the gene expression for specific tissues (Benfey 1989; Neuteboom 2002). It is important to note that constitutive expression of a transgene can lead to phenotypic anomalies and that in the case of drought tolerance a drought inducible promoter can greatly ameliorate the effects (Liu et al. 1998; Huang 2001). Sugarcane promoters active in culms have been described recently (Hansom 1999). Research is also focusing on post-transcriptional silencing events that may render transgene expression unstable.

21.8 SUGARCANE FUNCTIONAL GENOMICS AND BIOINFORMATICS

No matter which route one chooses for sugarcane improvement, target genes for genetic manipulation or use as markers need to be identified. The analysis of the sugarcane transcriptome has been extensively used for many years in a search for genes associated to agronomic traits of interest (Menossi et al. 2008). Transcriptomics complements gene marker identification and either technique can be excellent tools for breeders that wish to improve sugarcane through transgenics or classic breeding. Also, the use of the transcriptome can aid in the identification of markers for QTL mapping and expression-QTL (eQTL), which in sugarcane is difficult until statistical genetic tools are available for highly polyploid genomes. Techniques that have been used for transcriptomics in sugarcane include EST sequencing, nylon-based cDNA macroarrays, cDNA microarrays, long oligonucleotide microarrays (Agilent Technologies®) and short oligonucleotide microarrays (Affymetrix®).

Transcriptome studies in sugarcane were first undertaken by EST sequencing (Carson 2000; Carson and Botha 2002; Casu et al. 2003, 2004; Ma et al. 2004; Vettore et al. 2003; Bowers et al. 2005). The largest collection of ESTs was generated by the SUCEST Project (http://sucest-fun. org), a consortium of over 100 Brazilian laboratories that generated approximately 238,000 ESTs from 26 diverse cDNA libraries which were clustered into 43,000 transcripts or SAS (Sugarcane Assembled Sequences).

The information generated in EST projects has been used in comparative mapping of the family of grasses, using common markers that hybridize with sugarcane, rice, corn, hexaploid wheat, barley and sorghum. However, the molecular information developed to date for sugarcane is minimal when compared to the information that is necessary to identify and characterize the loci that encode the important agronomic characteristics. Table 21.2 lists the genetic maps available for sugarcane. To date, approximately 400 genetic markers have been developed (Cordeiro et al. 2002; Pinto et al. 2006; Oliveira et al. 2009) and used in the construction of the first functional sugarcane genetic map (Oliveira et al. 2008).

All of the EST collections were clustered by the Center for Genomic Research (TIGR) as the Sugarcane Gene Index 2.1, and more recently by the Computational Biology and Functional Genomics Laboratory at the Dana-Farber Cancer Institute as the Sugarcane Gene Index 2.2. In the SUCEST-FUN database (http://sucest-fun.org) a tool is available where clusters generated by the SUCEST collection and the Sugarcane Gene Index collection can be cross-referenced. The SUCEST-FUN has been developed in the concept of the mediator approach that incorporates concepts from *Data Warehouse* and *Federation* approaches. It is a flexible data integration that assembles heterogeneous distributed data sources, experimental data, resources, the application of scientific algorithms and computational analysis. Bioinformatics and the management of scientific data are critical to the support of life sciences discoveries. Nowadays, an explosion of available biological data and research has risen up, most of it compound and stored in dozens of smaller databases. Scientists are not currently able to easily identify and integrate autonomous data sources and exploit this information because of the variety of semantics, interfaces, and data formats used by the underlying data sources. The SUCEST-FUN Database is, therefore, being developed to give access to gene expression studies and make available tools that will allow a Systems Biology approach in sugarcane and the identification of regulatory networks.

The first transcriptomics tools developed made use of existing cDNA clones to produce macroarrays. Macroarrays have been used to define gene expression patterns in immature and mature leaf, immature and mature internodes (Carson and Botha 2002; Carson et al. 2002), sugarcane responses to cold (Nogueira et al. 2003), tissue profiling of transposable element transcripts (Araujo et al. 2005), stem development (Watt et al. 2005), methyl jasmonate responses (De Rosa Jr et al. 2005), the response of sugarcane leaves to ethanol application (Camargo et al. 2007), sink-source activity alterations (McCormick et al. 2006, 2008) and ABA/MeJA-activation of sugarcane transcription factors (Schlogl et al. 2008). Data mining of the SUCEST database led to the identification of 276 sequences homologous to TEs in 21 different families of which 54% correspond to classical transposons and 46% to retrotransposons (Rossi et al. 2001). Retrotransposons mobilize themselves through an RNA intermediate and thus are now considered one of the major forces driving genome expansion in plants (Piegu et al. 2006) while transposons usually move using either a cut/paste or a copy/paste mechanism. Expression profiling of 162 clones (Araujo et al. 2005) showed that callus was the tissue with most expressed TE families. Although it has been proposed several times that tissue culture somaclonal variation could be a result of TE activity, this was the first report that demonstrated that callus is indeed a tissue where different TEs are expressed at the same time, not necessarily in the same cell. One largely unanticipated result was the revelation that within a family there are lineages with varying copy number in the genome (Rossi et al. 2004; Saccaro-Junior et al. 2007). Adding to that, some of the transposon lineages were associated with previously described "domesticated" versions of a transposase (Bundock and Hooykaas 2005; Cowan et al. 2005). Regardless of the mechanism by which a transposable element moves, the field of genetic mobile elements is now flourishing with hypotheses of their impact on genome structure, gene regulation and even function leaving their once considered "junk DNA" status as a secondary role (Casacuberta and Santiago 2003; Kashkush et al. 2003; Bundock and Hooykaas 2005).

TABLE 21.2
Genetic Maps Available for Sugarcane

Mapping population	Progeny	Marker	Marker dosage			Linked markers	N0 CG	*Map coverage	Map Software	Reference
			SD	DD	DM					
ADP068 × SES208	90	RFLP	216	–	–	188	44	1361	Mapmaker	Da Silva et al. 1993
ADP068 × SES208	88	RAPD	208	–	–	176	42	1500	Mapmaker ver 1.0	Al-Janabi et al. 1993
SP701006	32	RFLP	253	–	–	94	25	–	Mapmaker ver 1.0	D'Hont et al. 1994
R570	77	RFLP, Isozyme	505	–	–	408	96	2008	Mapmaker ver 3.0	Grivet et al. 1996
La Purple × Mol 5829	84	RAPD	279	–	–	161	50	1152	Mapmaker ver. 2.0	Mudge et al. 1996
La Purple[a] × Mol 5829[b]	100	RAPD, RFLP, AFLP	341[a] / 301[b]	–	–	283[a] / 208[b]	74[a] / 65[b]	1881[a] / 1189[b]	Mapmaker ver. 2.0	Guimarães et al. 1999
R570	112	AFLP	939	–	–	883	120	5849	Mapmaker ver. 3.0	Hoarau et al. 2001
R570	112	AFLP, EST-RFLP,EST-SSR	939+347	–	–	883 + 282	128	> 5849	Mapmaker ver. 3.0	Rossi et al., 2003
Green German[c] × IND81-145[d] PIN84-1[e] × Muntok Java[f]	85	RFLP	434[c] / 395[d] / 308[e] / 359[f]	132[c] / 54[d] / 86[e] / 159[f]	–	289[c] / 257[d] / 194[e] / 214[f]	75[c] / 70[d] / 71[e] / 73[f]	2466[c] / 2172[d] / 1395[e] / 1472[f]		Ming et al. 2002
IJ76-514 × **Q165**	227	AFLP, RAF, SSR	967	36	123	1074	136	9058.3	Mapmaker ver. 2.0	Aitken et al. 2005
IJ76-514 × Q165	227	AFLP, SSR	240	–	234	534	123	4906.4	JoinMap ver. 3.0	Aitken et al. 2007
Q117 × **MQ77-340**	232	AFLP, SSR	395	58	–	342	101	3582	Mapmaker ver. 3.0	Reffay et al. 2005
SP80-180 × SP80-4966	100	RFLP SSR AFLP	441	677	–	3571 / 2172	1311 / 982	2602.41 / 13402	One Map[1] / JoinMap[2] ver. 3.0	Garcia et al. 2006
R570[g] × MQ76-53[h]	198	AFLP SSR RFLP	1057	–	–	424[a] / 536[b]	86[a] / 105[b]	3144[a] / 4329[b]	Mapmaker Ver. 3.0	Raboin et al. 2006
M134/75 × R570	227	AFLP SSR	557	–	79	474	95	6200	Mapmaker/exp v.3.0	Aljanabi et al. 2007
SP80-180 × SP80-4966	100	AFLP, gSSR, EST-SSR, EST-RFLP	800	869	–	664	192	6261.1	One Map	Oliveira et al. 2007
La Striped[i] × SES 147B[j]	100	AFLP, SRAP, TRAP	247[i] / 221[j]		33[i] / 43[j]	146[i] / 121[j]	49[i] / 45[j]	1732[i] / 1491[j]	JoinMap ver. 3.0	Alwala et al. 2008

a,b,c,d,e,f,g,h,i,j   Refers to the map information of the respective parental. Bold parents correspond to the parental map constructed.

*Cumulative length in centimorgans (cM). SD, single-single markers (simplex marker), marker present only once in the genome segregating in a 1:1 ratio; DD, double-dose marker (double simplex marker), marker present in one copy in both parental genomes, segregating in a 3:1 ratio; DM, duplex marker, marker present twice in one parental genome segregating in a 11:3 ratio ($x = 8$) or in a 7:2 ratio ($x = 10$).

The advent of microarrays increased significantly the number of studies on gene expression profiling. Carbohydrate metabolism has been extensively studied in sugarcane using cDNA microarrays to define gene expression associated with sucrose content. Microarrays were used to profile the developing culm (Casu et al. 2003, 2004) leading to the identification of sugar transporters highly expressed in the maturing stem and the coordinated expression of enzymes involved in sucrose synthesis and cleavage. Transcripts associated with fiber metabolism and defense and stress mechanisms were the most highly expressed transcripts in maturing stem. Stress responses were also defined in roots (Bower et al. 2005) using the same arrays. Customized microarrays containing signal transduction components were used to profile the individual variation of plants cultivated in the field and transcript abundance in six plant organs (flowers, roots, leaves, lateral buds, and 1st and 4th internodes) leading to the identification of genes ubiquitously expressed or tissue-enriched (Papini-Terzi et al. 2005). The same arrays were also used to study signal transduction-related responses to phytohormones and environmental challenges in sugarcane (Rocha et al. 2007) including drought, methyl jasmonate, abscisic acid, insect (*Diatraea saccharalis*), and endophytic bacteria (*Gluconacetobacter* and *Herbaspirillum*) elicited responses. Thirty genotypes contrasting for sucrose content were profiled and over 300 genes associated to sucrose content were discovered (Papini-Terzi et al. 2009). In parallel, sugarcane plantlets were treated with sucrose to define genes directly responsive to this sugar. Interestingly, a large overlap was observed between gene expression responsive to drought and sucrose indicating a common controlling mechanism behind these processes that may rely on protein kinases of the SnRK/SNF1 family. Hormonal regulation associated to sucrose content was also revealed that included ethylene, auxins, jasmonates, absicic, and salycilic acid. Additionally, sugarcane has been expression-profiled under the effect of elevated CO_2 (de Souza et al. 2008). This is an important issue related with sugarcane physiological responses to the global climatic changes (GCC). GCC can be characterized by an elevation of the atmospheric CO_2 concentration (because of fossil fuel usage), which leads to the elevation of temperature and consequently changes in the climate. An experiment performed with sugarcane plants growing in elevated CO_2 showed that they respond by increasing photosynthesis (CO_2 assimilation) and growth (50% more biomass). This culminated with the production of more sucrose and fiber (ca. 29% each). The authors evaluated the gene expression changes associated with the treatment and found carbon metabolism to be affected. Genes related to electron transport in the chloroplasts were more expressed. On the other hand, no effects on CO_2 assimilation were observed, suggesting that sugarcane leaves will respond to the elevation of CO_2 by improving their light harvesting system to use the excess of CO_2 (de Souza et al. 2008).

A large part of the gene expression data is publicly available. Most published sugarcane microarrays were catalogued on public repositories such as Gene Expression Omnibus (GEO-NCBI), Center for Information Biology Gene Expression Database (CIBEX), Microarray Gene Expression Data Society (MGED), and ArrayExpress (Tables 21.3 and 21.4).

TABLE 21.3
Summary of Sugarcane Microarray Data

	Total	SUCEST-FUN	Others
Number of platforms	6	3	3
Number of series	17	12	5
Number of samples	226	182	44

Source: Gene Expression Omnibus database, www.ncbi.nlm.nih.gov/geo/

TABLE 21.4
Summary of All Sugarcane Microarray Experiments

Platform	Samples	Number of Differentially Expressed Genes	Treatment/Conditions
Sugarcane nylon arrays (1536 genes)	12	34	Cold stress-induced
Sugarcane EST nylon arrays (1536 genes)	12	26	MeJA-induced
Affymetrix sugarcane genome array (6024 genes)	12	119	Stem development
	8	74	Leaves sugar accumulation
SUCASTv1 (2208 genes)	16	14	Phosphate starvation
	8	9	Herbivory
	16	13	ABA
	12	42	MeJA
	8	11	N_2-fixing endophytic
	12	80	Drought
	36	172	Mature/Immature Internode high/low brix
	16	126	Internode high/low brix 1,5,9
SUCASTv3 (1920 genes)	26	216	Six different tissues
	4	24	Leaves high/low brix
SUCASTv2 (1920 genes)	28	216	Six different tissues

Source: Gene Expression Omnibus database, http://www.ncbi.nlm.nih.gov/geo (accessed May 5, 2009).

21.9 CONVENTIONAL BIOETHANOL PRODUCTION

Because of rising oil prices and environmental concerns regarding nonrenewable fuels, ethanol has become an alternative to gasoline for auto engines. The Brazilian Government started a program to replace ethanol for gasoline in 1975; in 2008 the use of ethanol surpassed that of gasoline as fuel for cars and light vehicles. Because of the flex-fuel technology that allows engines to operate with any proportion of ethanol and gasoline, since 2008 cars that run on ethanol became dominant in Brazil; in 2009/2010 57% of the sugarcane crushed in Brazil was used for ethanol and 43% for sucrose (Unica 2009) and of the 413 mills operating in Brazil, 398 produce ethanol (MAPA 2009). Except for Brazil, in most countries ethanol comprises a relatively small proportion of the sugarcane industry output and is directed to liquor or chemical purposes; however, it is expected that the use of ethanol as fuel will be the main driver for the expansion of sugarcane production in the world in the near future (Figure 21.1). Presently, approximately 90% of the fuel ethanol production in the world comes from the United States and Brazil.

Sugarcane is used for ethanol production by fermentation followed by distillation of sucrose or molasses. The theoretical yield of ethanol is 617 L/t of sucrose but, at the normal operating conditions of ethanol distilleries, the yield is usually 510–530 L/t sucrose. The average yield of ethanol in Brazil is 82–85 L ethanol/t of fresh cane crushed (Boddey et al. 2008). Anhydrous bioethanol production processed from sugarcane juice in an autonomous distillery is comprised by the following main steps: sugarcane cleaning; extraction of sugars; juice treatment, concentration and sterilization; fermentation; distillation and dehydration; and purification as described in the following items.

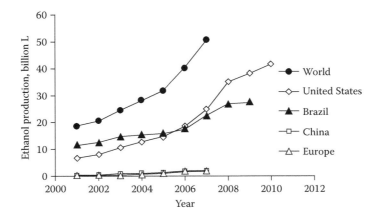

FIGURE 21.1 Fuel ethanol production in the world and in selected regions. Ethanol is produced from corn in the United States and from sugarcane in Brazil. Figures for 2009 and 2010 are projected data. (From Energy Information Administration, Official energy statistics of the U.S. government, 2009, http:www.eia.doe.gov; União da Indústria de Cana-de-Açúcar, Statistics of sugarcane sector—Season 2006/2007, 2009, www.unica. com.br.)

21.9.1 EXTRACTION, JUICE TREATMENT, CONCENTRATION, AND STERILIZATION

Sugarcane is first cleaned in wet or dry cleaning system, which removes 70% of the dirt, before entering the mills, in which sugarcane stems are processed and juice and bagasse are obtained. Mill efficiency is around 96%. Sugarcane juice passes through screens and hydrocyclones that remove dirt, sand and fibers; then phosphoric acid is added to increase phosphate content, and the mixture is heated from 30 to 70°C before addition of lime. The limed juice is heated from approximately 70 to 105°C before the flash tank, where air bubbles are removed from the juice. The degasified juice receives a flocculant polymer and is decanted, aiming at removal of insoluble impurities, including calcium phosphates formed during the liming step. The mud obtained in the decanters is filtered (producing filter cake) and the liquid phase returns to the process just after the liming step.

The clarified juice obtained in the decanters goes through multiple effect evaporators (MEE) to achieve adequate sugars concentration. Only part of the clarified juice must be concentrated. The final juice is made up of clarified and concentrated juice, and contains about 22 wt % sucrose.

To promote sterilization before fermentation, the juice is heated up to 130°C, cooled down to 28°C and fed to the fermentation reactor.

21.9.2 FERMENTATION

The sterilized juice is added to the fermentor along with the yeast media, which is made up of a yeast suspension containing about 28% yeast (vol. basis) and comprises 25% of the reactor volume. Fermentation is carried out at 28°C; ethanol content of the wine can reach 13°GL (approximately 10.5% ethanol on a mass basis, that means around 100 g/L). To achieve this high ethanol content, batch fermentation may have to be carried out for up to 15 h and alternative cooling methods, such as a steam jet system or an absorption machine, are necessary to provide water at temperatures low enough to maintain reactor cooling (Dias et al. 2007), because fermentation is negatively affected by high temperatures. In an integrated process taking into account hydrolyzed lignocellulosic material from sugarcane bagasse or straw, the fermentor could be fed with a blend of molasses and hydrolyzed.

The wine obtained in the fermentor is centrifuged for the recovery of yeast cells. The yeast milk obtained in the second centrifuge contains about 70 vol % yeast, so water is added to this milk to produce a mixture containing 28 vol % yeast that are fed to the reactor for another batch of fermentation. Sulfuric acid is added to the yeast medium to avoid bacterial contamination.

Gases released during fermentation are collected and washed to recover ethanol in an absorption column. The alcoholic solution obtained is mixed with the centrifuged wine and fed to the distillation columns.

21.9.3 DISTILLATION AND DEHYDRATION

Product purification takes place in double-effect and extractive distillation systems. A series of distillation and rectification columns are used for hydrous bioethanol production (93 wt % ethanol). Hydrous ethanol in vapor phase is dehydrated to produce anhydrous ethanol (AE, 99.5 wt % ethanol) in the extractive distillation process with monoethyleneglycol. In the double effect distillation, column reboilers and condensers are integrated, thus diminishing energy consumption on the distillation.

21.9.4 SIMULATIONS FOR INCREASING EFFICIENCY OF HYDRATED BIOETHANOL PRODUCTION

Product purification is a critical step during ethanol production because it is an energy intensive operation. Ordinary distillation is used to produce hydrous bioethanol (approximately 93 wt % ethanol), which is used as fuel in ethanol-based or flex-fuel engines. The conventional configuration of the distillation process used in Brazilian refineries consists of a series of distillation columns (A, A1, and D) and rectification (B and B1) columns. This configuration is depicted in Figure 21.2.

In conventional bioethanol production (first generation), wine obtained from fermentation of sugars containing between 7 and 12 wt % ethanol is the feedstock to obtain ethanol. However, wine produced in the fermentation process with an extractive vacuum flash chamber can achieve 50 °GL (Atala and Maugeri 2006), what may lead to a significant reduction in the separation costs.

21.9.4.1 Ordinary Distillation

Because distillation operations require significant amounts of energy and have a great importance in bioethanol production, the simulation of an operation unit has to be as representative as possible.

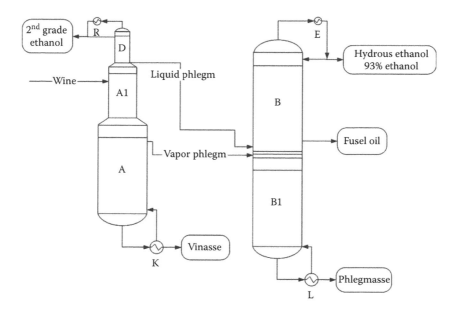

FIGURE 21.2 Configuration of conventional distillation process.

For this reason, simulation of conventional distillation process was carried out in Aspen Plus® considering nonequilibrium stage model. In this model, conservation equations are written for each phase independently and solved together with transport equations that describe mass and energy transfers in multicomponent mixtures; also it is assumed that equilibrium occurs only in the vapor–liquid interface. Besides, in this way, empirical correcting factors, such as efficiencies used in the equilibrium model, are no longer necessary (Pescarini et al. 1996).

Results of the simulation considering nonequilibrium stage model were compared to those obtained using equilibrium stage model with constant plate efficiencies of 55, 70, 85, and 100%. It was observed that, to obtain hydrous ethanol (93%), the idealized equilibrium stage model (efficiency of 100%) predicts an energy requirement that corresponds to 80% of that given by nonequilibrium stage model (~7,000 kJ/kg of hydrous ethanol). In addition, simulations showed that equilibrium stage model with an efficiency of 70% provide results quite similar to nonequilibrium stage model for the conventional distillation process.

21.9.4.2 Multiple Effect Distillation

An alternative to the conventional distillation process is the multiple effect operation of the distillation columns. The operation in different pressure levels gives rise to different temperature levels on condensers and reboilers of the different columns, thus it is possible to integrate the equipment and reduce steam consumption on reboilers.

To optimize bioethanol production, simulations using Aspen Plus were carried out with five different configurations: conventional fermentation and distillation (CFCD), vacuum extractive fermentation and conventional distillation (VFCD), vacuum extractive fermentation and conventional distillation (VFCD), vacuum extractive fermentation and double effect distillation (VFDD), and vacuum extractive fermentation and triple effect distillation (VFTD).

Vacuum extractive fermentation consists of a fermentation reactor coupled to a vacuum flash evaporator, which allows ethanol produced to be simultaneously removed from the reactor. In this study, ethanol concentration in the reactor was kept at low levels (around 8 wt % ethanol) whereas in the flash chamber, wine with 36 wt % ethanol was obtained.

The double effect configuration was similar to the conventional configuration, however, the distillation columns operate under vacuum (19–25 kPa), while rectification columns operate under atmospheric pressure (101–135 kPa). In this way, different temperature levels are observed between columns "A" reboiler and "B" condenser (65 and 78°C, respectively), allowing thermal integration of these equipments and consequently reducing energy consumption on the distillation stage.

In the triple effect configuration, the distillation columns operate under vacuum (19 – 25 kPa), and the liquid phlegm stream produced on column D is split in two: one of them is fed to a rectification column operating under nearly atmospheric pressure (column "B," 70 – 80 kPa) and the other is fed to a rectification column which operates under relatively high pressure (column "B-P," 240 – 250 kPa).

Regarding thermal energy, results showed that the configuration that presents the lowest energy demand is the triple effect configuration (VFTD), providing a reduction in energy consumption of 44 % when compared to the VFCD process and 77 %, when compared to the CFCD process, which is the configuration most commonly used in Brazilian biorefineries.

21.9.5 Simulations for Increasing Efficiency of Anhydrous Ethanol Production

Anhydrous bioethanol, suitable to be used as a gasoline additive or as raw material for production of different renewable materials, must contain at least 99.3 wt % ethanol. Because water and ethanol form an azeotrope with 95.6 wt % ethanol at 1 atm, alternative separation processes are necessary to produce anhydrous bioethanol. The most used processes in Brazilian biorefineries are azeotropic distillation with cyclohexane and extractive distillation with monoethyleneglycol (MEG).

21.9.5.1   Azeotropic Distillation

In azeotropic distillation for bioethanol production, the entrainer (cyclohexane) is added to the binary mixture, producing a heterogeneous azeotrope with water and ethanol. In this process, two distillation columns are used: azeotropic and recovery.

Simulation of the azeotropic distillation process is often complex and extremely sensitive to project parameters and specifications, mainly because of the formation of a second liquid phase inside the azeotropic column. Higler et al. (2004) reported that column efficiencies between 25 and 50% are not uncommon when a second liquid phase is present.

To study the formation of the second liquid phase inside the column, simulations of the azeotropic distillation process with cyclohexane for anhydrous bioethanol production were carried out using the software Aspen Plus with three configurations.

In the configuration that presented the best results, hydrous bioethanol is mixed with the recycle stream of the recovery column and fed to the azeotropic column. Entrainer is comprised of the organic phase obtained in the decanter and a solvent make-up stream. Anhydrous bioethanol is produced on the bottom of the azeotropic column, and the ternary azeotrope in the top. After cooling, the ternary azeotrope splits into two liquid phases in the decanter. The aqueous phase is fed to the recovery column, producing pure water on the bottom.

For all configurations studied, it was observed that the mixture inlet stage in the azeotropic column determines the beginning of a second liquid phase, because this stage is the one that first presents two liquid phases. For this reason, it is important to define the inlet stage near to the bottom of the column; therefore the best feed inlet stage is the last one that allows column convergence.

21.9.5.2   Extractive Distillation

In extractive distillation, also known as homogeneous azeotropic distillation, a separating agent, called solvent or entrainer, is added to the azeotropic mixture to alter the relative volatility of the components in the original mixture.

In the conventional extractive distillation process, solvent is fed to the first column (extractive column), above the azeotropic feed. Anhydrous ethanol is produced on the top of the extractive column, while in the bottom a mixture containing solvent and water is obtained. The solvent is recovered in a second column (recovery column), cooled and recycled to the extractive column (Higler et al. 2004).

Conventional extractive distillation processes used in the industry for the separation of ethanol–water mixtures use monoethyleneglycol (MEG) as a solvent in the distillation column, which is a toxic component.

Energy consumption on column reboilers for the optimized configuration is 1057 kJ/kg of anhydrous ethanol, which is a reasonable value when compared to other distillation-based dehydration processes. Bioglycerol is not harmful to humans or the environment, and its availability has increased in the last years because it is a byproduct of the biodiesel production process.

21.9.6   Simulations for Increasing Efficiency of Extraction Processes

The need to decrease residues generation and the pursuit of cost reduction in bioethanol production has motivated the investigation of more efficient processes to produce this biofuel. One of the proposed options is the use of reaction-separation systems, such as the continuous flash fermentation.

Cardona and Sanchez (2007) point out the reaction–separation integration is a particularly interesting choice for the intensification of ethanol production. The removal of ethanol from the culture broth diminishes the inhibition effect on the growth rate. The continuous extractive fermentation has shown several advantages, such as low vinasse generation and fresh water consumption because of the possibility of feeding molasses at higher concentrations, which reduce costs in waste treatment, and the potentiality of eliminating one distillation column from the process. It is worthwhile

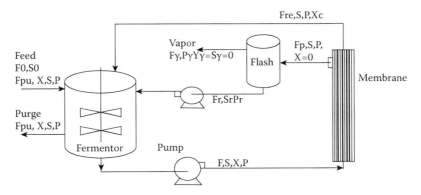

FIGURE 21.3 Schematic diagram of extractive process under vacuum for bioethanol production.

mentioning that the conventional mode of operation produces 13 L of vinasse/L of ethanol (Navarro et al. 2000). Vinasse is one of the main polluting byproducts of alcoholic fermentation because of its low pH, high solids content, etc. Further details of the technical features of extractive processes can be found in Silva et al. (1999) and Atala and Maugeri Filho (2006).

The extractive process was developed by Atala (2004), as cited by Mariano et al. (2008) in an application for the butanol purification. This configuration consists of three interconnected units: fermentor, filter (tangential microfiltration for cell recycling) and vacuum flash vessel (for the continuous removal of ethanol from the broth).

The concentration of sugarcane molasses in the feed stream is from 180 to 330 kg/m³ of total reducing sugar, and the *Saccharomyces cerevisiae* in the steady state reaches 30 kg/m³. The low ethanol concentration is maintained at 40–60 kg/m³. These characteristics of operation guarantee higher yield (10 kg/m³ per hour) than in fed-batch and continuous modes of operation. Figure 21.3 depicts the extractive process formed by a fermentor, pumps, flash vessel, and a membrane system. The kinetic model takes into account the substrate and product inhibitions, the volume occupied by the cells, intracellular ethanol and terms of cell death (considering that continuous processes are operated for long periods). The term ρ is a relation between dry cell mass and the volume of wet cells, and γ, the relation between the inter- and extracellular ethanol concentrations. Having a higher ethanol concentration in the fermentation stage is an interesting alternative, and has a significant impact on the whole process costs because of the reduced effort required in the ethanol purification.

21.10 CELLULOSIC ETHANOL PRODUCTION

Plants are structurally sustained by their walls, which are composites analogous to liquid crystals. Cellulose microfibrils are deposited in large quantities in the vascular tissues and fibers. These microfibrils are covered by hemicellulosic polysaccharides, which in the case of sugarcane are arabinoxylan and β-glucan. The deposition of cellulose is one of the most efficient packing processes in nature. Glucose chains linked by β-1,4-glycosidic linkages are packed together so that very little water is left among the polymers. These interactions and the lack of water prevent the access of enzymes to the glycosidic linkages, avoiding the attack of microorganisms and consequently defending plant tissues against pathogen attack.

As a result of this efficient packing, the plant cell wall is, by far, the most adequate form in which to store carbon and energy in high quantities but at the same time protect against the attack of microorganisms. The potential for production of biomass will therefore increase as a plant makes proportionally more wall. On the other hand, this highly packed form of storage imposes a tremendous barrier to access the energy stored. Enzyme hydrolysis is probably the most efficient way to gain access to the energy stored in glycosidic linkages. What is lacking, though, is the necessary knowledge about how the enzymes of sugarcane or microorganisms attack the wall. It is crucial to find ways to activate these enzymes in vivo so that energy access is gained more easily.

When sugarcane is harvested, sugars are extracted in water by pressing stem tissues. This extracts approximately one third of the energy contained in the plant as soluble sugars. Leaves and other plant parts that are left in the field after stem harvesting correspond to another one third of the sugarcane biomass energy; eventually part or all this material may be collected and used to produce energy. The remaining one third of energy is composed of the bagasse that is already available at the mill facilities and is used to produce vapor, mechanic, and electrical power; in some sugarcane and ethanol plants in Brazil surplus energy is sold and fed to the electric grid. Although part of the bagasse is currently used for production of thermoelectricity, the current efficiency of this process is far beyond the potential of producing energy from the hydrolysis of biomass either with acids or with enzymes (Cortez et al. 2008).

The process of production of free fermentable sugars from hemicelluloses and cellulose is the basis of cellulosic ethanol production. This can be achieved by physical, chemical or biochemical procedures. The great challenge is finding the combination of processes that could produce free fermentable sugars in an economically viable way.

There is great potential for an increase in bioethanol production and process optimization, considering process improvements and the use of sugarcane bagasse as raw material in the hydrolysis process. The evaluation of the energy consumption of the integrated production of ethanol from sugarcane and sugarcane bagasse remains one of the main obstacles for the technical and economical feasibility of the hydrolysis process. Figure 21.4 shows a possible pathway for the integrated process, which can be easily incorporated in existing units as soon as the sugarcane hydrolysis process is sufficiently developed.

A process that may be used to produce bioethanol from lignocellulosic materials, the Organosolv process with dilute acid hydrolysis, is being tested under semi-industrial scale in Brazil for the production of 5000 L/day of ethanol (Rossell et al. 2005). A typical large scale plant, that is, one that produces 1000 m³/day of anhydrous bioethanol crushing approximately 12,000 t sugarcane/day, was considered for the conventional bioethanol production from sugarcane juice. Sugarcane bagasse is one of the main byproducts of conventional bioethanol production from sugarcane juice.

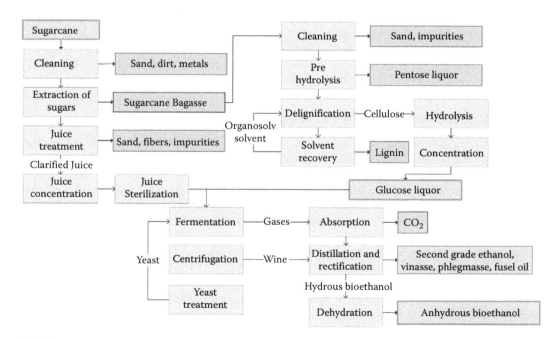

FIGURE 21.4 Integrated bioethanol production from sugarcane and sugarcane bagasse.

In this example 75% of sugarcane bagasse produced in the mills is being used as raw material for bioethanol production in the integrated process. The figure represents a feasible configuration in the years to come, when sugarcane burning is abolished and sugarcane straw is efficiently recovered from the field.

Although polysaccharides can be easily hydrolyzed with acids producing free sugars, when this is done, part of the free monosaccharides produced will be destroyed by the acid and form furfurals, which are toxic to yeast (Cortez et al. 2008). As a result, acid hydrolysis has never become economically viable in industry. An alternative is the use of enzymes, which are much more precise, to break the glycosidic linkages among monosaccharides. However, the problem in this case is to prompt enough water within the microfibrils so that hydrolysis can occur. Hydrolases can be used directly on the fiber residue and a blind search may lead to good extracts from microorganisms. However, if one understands what linkages have to be broken to loosen the wall, much less energy can be used in the process, making it more efficient. The use of polysaccharide hydrolases obtained from microorganisms, insects and the plants themselves, is the strategy that is now being developed in most initiatives to produce ethanol from biomass and sugarcane (Buckeridge et al. 2010). This is a very complex task and will probably be achieved by the combination of the strategies of pretreatment of biomass using physical and chemical methods, together with enzymatic hydrolysis.

21.11   FUTURE PROSPECTS: RESEARCH AND DEVELOPMENT FOR PRODUCTIVITY AND SUSTAINABILITY

Since 1975 R&D has improved the productivity of ethanol from sugarcane more than 80% (from 4200 to 7650 L/ha). The main contributions to these productivity gains came from the development of new varieties through classical breeding, improvements in the fermentation processes and improvements in the agricultural processes. R&D has also been instrumental in obtaining more efficient use of energy in ethanol mills, reducing GHG emissions from the agricultural processes and reducing water, negative environmental impacts and defensive use.

In almost all studies of sugarcane physiology and biochemistry, the ability of the plant to produce sucrose has been the main target of research. This is plausible as this plant is so important to sugar and ethanol world production. As a consequence, several authors have focused their lines of research in sugar biochemistry and source-sink relationships. Moore (2005) has reviewed sugarcane physiology from the viewpoint of systems biology. He proposes that any scientific approach should target the interactions among the several network connections that lead to higher sugar accumulation. These networks include the layers of gene transcription, which leads to protein production and interactions and subsequently leads to metabolic changes in plant tissues. A few relevant questions can be asked such as (1) How is photosynthesis connected to sucrose biosynthesis? (2) How is carbon partitioned between nonstructural and structural carbohydrates? (3) How are the carbon pathways connected with nitrogen metabolism? (4) How is all this integrated or how do hormones perform the cross talk so that communication among biochemical, physiological and ecophysiological stimuli is controlled? and finally (5) Can we alter physiology and metabolism regulation to achieve the attributes of an Energy-Cane? Tools are needed if we want to improve sugarcane either through marker-assisted breeding or transgenic approaches and the sequencing of the sugarcane genome will be an important step toward the development of the biotechnology for this crop. Furthermore, these questions are very complex but interdisciplinary work and integrated databases may solve some of these problems. This is so also because the morphology and physiological performance of the plant is an emergent property of the integration of all of these factors. On the other hand, such an approach can be considerably simplified by using modularity (Wagner et al. 2007). Understanding how plants use the same module in different parts of its body will probably be key not only to understanding how these modules connect, but also to applying what is learnt from how different modules work, by using synthetic biology.

Synthetic biology may also be important to developing alternative uses of ethanol from sugarcane. There are many possible and perhaps suitable ways to use well the available biomass for energy generation in a broader scenario. In fact, many possible pathways to obtain biofuels may be designed based on the available feedstock. The primary use of sucrose for ethanol production is essentially an economical decision and the combination either sequentially or serially of alternative processes is an option to improve feedstock for biofuels and energy production. For first-generation biofuel, the knowledge is in place for the direct use of sugarcane syrup to produce ethanol even when it is depleted through sugar manufacture in integrated bioethanol and sugar production units (in this case it is postulated to be the one and half generation process). The use of lignocellulosic material, in this case either sugarcane bagasse or microalgae from CO_2 as a feedstock, is generally defined as second generation. On the other side, the use of the biomass gasification to further produce liquid biofuels is framed as the third generation pathway. The technology is always improving and in the last years the learning curve has offered already a potential for short term competitive costs for all idealized pathways. It is expected that plastics from sugarcane will be soon available and diesel from the ethylic route will be the next option.

Sugarcane research has been increasing over the years. Figure 21.5 shows the evolution of the number of scientific papers published annually on sugarcane-related subjects, classified according to the country of origin. Brazil is the leader in publications, a trend that started with the public release of the SUCEST EST sequencing data in 2001. In 2008, Brazilian publications accounted for over half of the articles indexed by the ISI/WoS System, 22% of which were originated from the State of São Paulo in Brazil, which also is responsible for 60% of the sugarcane production in Brazil. Recently, Fundação de Amparo à Pesquisa do Estado de São Paulo (FAPESP), one of the largest funding agencies in Brazil started the Bioenergy Program BIOEN (http://bioenfapesp.org) aiming at integrating comprehensive research on sugarcane and other plants that can be used as biofuel sources, stimulating innovative and path changing research in cooperation with research institutions and private companies from other states of Brazil and from other countries.

Research on sugarcane aims at two broad objectives: increasing productivity and increasing sustainability. In Brazil productivity gains have been intense, especially since the establishment of the PROALCOOL Program in (1975) The average yield has increased from 46 to 75 t/ha, mostly through the development of customized varieties, using classical breeding, and through improvements in the agricultural process. As mentioned above, Paul Moore estimated from first principles that the potential yield for sugarcane is expected to be 472 t/ha (Waclawovsky et al. 2010). Considering that yields as high as 212 t/ha have been observed there seems to be room

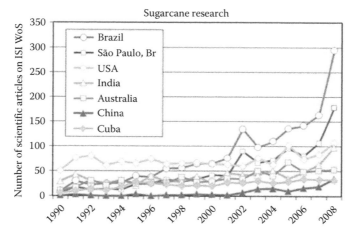

FIGURE 21.5 Evolution of the number of scientific articles published annually on sugarcane, related topics. (Data retrieved from ISI's Web of Science, www.isiknowledge.com).

for improvement. Genomics and a systems biology approach might bring contributions to these developments.

A systemic approach seems to be one of the principal routes to be taken to understand functioning of the sugarcane plants, their integration with the environment and also the production processes that lead from the crop to ethanol. The bioenergy-producing system imposes an immense challenge that includes not only the scientific approaches of agriculture, physics, chemistry, biology, and engineering but also humanities approaches related to economy, health, and sociology. Producing bioethanol in the twenty first century is not an ordinary task as it has to be sustainable from all points of view. For that to happen, top level research will have to be undertaken. This will certainly include the use of synthetic biology strategies that will probably produce much better adapted plants with minimal environmental and social impacts and the development of new methods for converting sugars to biofuels and cellulose to sugars and then to biofuels. While the fermentation of sugar to ethanol is a process that has been well known for centuries, the production of other biofuels remains a scientific and technical challenge. A recent development has been that of engineering microbes to process sugars and secrete certain fuels, including gasoline, diesel, jet fuel, and others. Lee et al. (2008) reviewed several possibilities of metabolic engineering strategies for increasing yields of some biofuels. The technology depends on R&D in metabolic engineering and synthetic biology that might create new means for metabolic engineers to better understand how to adjust the cell pathways to create phenotypes with sufficient efficiency for the production of economically viable biofuels.

Sustainability issues became more prominent as biofuels came to be recognized as serious alternatives to oil, implying the possibility of large scale production. In this theme, land, water, and fertilizer use are relevant topics, as is the precise determination of GHG emissions reduction. In Land Use Change studies, a relevant challenge is to establish the behavior of soil organic carbon (SOC). This knowledge is essential for the correct determination of the GHG emissions balance for sugarcane ethanol. The behavior of SOC depends on the specifics of the land use change action. Table 21.5 shows the expected changes in SOC content related to sugarcane in the Brazilian Atlantic Region (Mello et al. 2006). Large gains in carbon sequestration seem to be obtainable through management modifications and by adequate choice of land. Of course, the contrary is true: for example, planting sugarcane in established forests will cause carbon emissions, as pointed out by Searchinger et al. (2008). It should be mentioned that most of the sugarcane expansion verified in Brazil has been over degraded pastures, and not over forest land (Goldemberg 2008). Much more R&D is necessary, mostly because the behavior of SOC is strongly dependent on specific characteristics of the crop, as shown by Anderson-Teixeira et al. (2009). Water usage is another issue that relates to the sustainability theme. Gerbens-Leenes et al. (2009) reviewed the literature and estimated the Water Footprint (WF) for several biofuels. They pointed out that the WF for

---

TABLE 21.5

Potential for SOC Sequestration Because of Land-Use Change in the Layer from 0 to 20 cm Related to Sugarcane Estimated by Several Authors

Land-Use Change from	Land-Use Change to	Total Area (Mha)	Potential for SOC Sequestration (Tg/year)
Sugarcane, harvested with burning	Machine harvesting	3.3	5.35
Degraded pasture	Sugarcane without burning	1.93	0.19–1.54
Sugarcane harvested with burning	Reforestation	1.93	1.27

Source: Modified from Mello, F.F.C., et al. *Carbon Sequestration in Soils of Latin America*. Haworth Press, New York, pp 349–368, 2006.

sugarcane ethanol is 108 m³/GJ, second only to ethanol from sugar beet (59 m³/GJ) and potato (103 m³/GJ), however it must be remembered that both sugar beet and potatoes have poor GHG and energy balance for ethanol. One difficulty with this kind of estimate is that actual data depend strongly on specific properties of the crop and its management. All of these themes present challenges that require more R&D.

REFERENCES

Aguirre Jr JM (1936) *Creação de novas variedades de canna no Estado de S. Paulo.* Boletim Technico 34. Instituto Agronômico do Estado de São Paulo. Campinas, SP, Brazil

Aitken KS, Jackson PA, McIntyre CL (2005) A combination of AFLP and SSR markers provides extensive map coverage and identification of homo(eo)logous linkage groups in a sugarcane cultivar. *Theor Appl Genet* 110:789–801

Aitken KS, Jackson PA, McIntyre CL (2007) Construction of a genetic linkage map for *Saccharum officinarum* incorporating both simplex and duplex markers to increase genome coverage. *Genome* 50:742–756

Al-Janabi SM, Honeycutt RJ, McClelland M, Sobral BWS (1993) A genetic linkage map of *Saccharum spontaneum* (L.) 'SES 208'. *Genetics* 134:1249–1260

Al-Janabi SM, Parmessur Y, Kross H, Dhayan S, Saumtally S, Ramdoyal K, Autrey LJC, Dookun-Saumtally A (2007) Identification of a major quantitative trait locus (QTL) for yellow spot (*Mycovellosiella koepkei*) disease resistance in sugarcane. *Mol Breed* 19:1–14

Alfonsi RR, Pedro-Jr MJ, Brunini O, Barbieri V (1987) Condições climáticas para a cana-de-açúcar. In: Paranhos SB (coord) *Cana-de-açúcar: cultivo e utilização.* Fundação Cargill, Campinas, SP, Brazil, pp 52–55

Almeida IMG (2008) Doenças causadas por bactérias. In: Dinardo-Miranda LL, Vasconcelos ACM, Landell MGA (eds) *Cana-de-açúcar.* Instituto Agronômico, Campinas, SP, Brazil, pp 437–450

Alwala S, Kimberg CA, Veremis JC, Gravois KA (2008) Linkage mapping and genome analysis in a *Saccharum* interspecific cross using AFLP, SRAP, and TRAP markers. *Euphytica* 164:37–51

Anderson-Teixeira KJ, Davis SD, Masters MD, De Lucia EH (2009) Changes in soil organic carbon under biofuel crops. *Global Change Biol* 1:75–96

Andrade JC (1985) Esforço histórico de antigas variedades de cana-de-açúcar. Associação dos Plantadores de Cana de Alagoas, Maceió, AL, Brazil

Araujo PG, Rossi M, de Jesus EM, et al. (2005) Transcriptionally active transposable elements in recent hybrid sugarcane. *Plant J* 44:707–717

Arencibia A (1998) An efficient protocol for sugarcane (*Saccharum* spp. L.) transformation mediated by *Agrobacterium tumefaciens. Transgen Res* 7:213–222

Arencibia A, Carmona E, Cornide MT, et al. (1999) Somaclonal variation in insect-resistant transgenic sugarcane (*Saccharum* hybrid) plants product by cell electroporation. *Transgen Res* 8:349–360

Arencibia A, Vazquez RI, Pietro D, et al. (1997) Transgenic sugarcane plants resistant to stem borer attack. *Mol Breed* 3:247–255

Arizono H (1994) Métodos e critérios de seleção adotados na obtenção das variedades de cana-de-açúcar (*Saccharum* spp.) RB835089 e RB835486. MS Diss. Escola Superior de Agricultura Luiz de Queiroz, Piracicaba, SP, Brazil

Atala DIP, Maugeri Filho F (2006) Processo fermentativo extrativo a vácuo para produção de etanol, C12P 7/14, C12R 1/865. Brazil, Patent No. PI 0500321-0

Barba R, Nickell LG (1969) Nutrition and organ differentiation in tissue culture of sugarcane—a monocotyledon. *Planta* 89:229–302

Barreneche T, Bodenes C, Lexer C, et al. (1998) A genetic linkage map of *Quercus robur* L. (pendunculate oak) based on RAPD, SCAR, microsatellite, minisatellite, isozyme and 5S rDNA markers. *Theor Appl Genet* 97:1090–1103

Benfey PN, Chua NH (1989) Regulated genes in transgenic plants. *Science* 244:174–181

Berding N, Roach BT (1987) Germplasm collection, maintenance, and use. In: Heinz DJ (ed) *Sugarcane Improvement through Breeding.* Elsevier, Amsterdam, Netherlands, pp 143–210

Berding N, Skinner JC (1987) Traditional breeding methods. In: *Copersucar Int. Sugarcane Breeding Workshop,* São Paulo, SP, Brazil, pp 269–320

Biggs IM, Stewart GR, Wilson JR, Critchley C (2002) N-[15] natural abundance studies in Australian commercial sugarcane. *Plant Soil* 238:21–30

Blackburn FH (1983) *Sugarcane*. Longmans, London

Boddey RM, Soares LHB, Alves BJR, Urquiaga S (2008) Bio-ethanol production in Brazil. In: Pimentel D (ed) *Renewable Energy Systems: Environmental and Energetic Issues*. Springer, New York, pp 321–356

Boddey RM, Urquiaga S, Alves BJR, Reis VM (2003) Endophytic nitrogen fixation in sugarcane: present knowledge and future applications. *Plant Soil* 252:139–149

Bond RS (1978) The mean yield of seedlings as a guide to the selection potential of sugarcane crosses. *Proc Cong Int Soc Sugar Cane Technol* 16:101–110

Bower NI, Casu RE, Maclean DJ, Reverter A, Chapman SC (2005) Transcriptional response of sugarcane roots to methyl jasmonate. *Plant Sci* 168:761–772

Bower R, Birch RG (1992) Transgenic sugarcane plants via microprojectile bombardment. *Plant J* 2:409–416

Bowers JE, Arias MA, Asher R, et al.(2005) Comparative physical mapping links conservation of microsynteny to chromosome structure and recombination in grasses. *Proc Natl Acad Sci USA* 102:13206–13211

Braithwaite KS, Geijskes RJ, Smith GR (2004) A variable region of the SCBV genome can be used to generate a range of promoters for transgene expression in sugarcane. *Plant Cell Rep* 23:319–326

Breaux RD (1987) Breeding for enhance sucrose content of sugarcane in Louisiana. *Field Crops Res* 9:59–67

Bressiani JA (1993) Herdabilidade e repetibilidade dos componentes da produção na cultura da cana-de-açúcar. MS Diss. Escola Superior de Agricultura Luiz de Queiróz, Piracicaba, SP, Brazil

Bressiani JA, Sanguino A (1994) Herança genética da resistência à ferrugem (*Puccinia melanocephala* H&P Sydow) na cana-de-açúcar. Anais Seminário de Tecnologia Agronômica. *Copersucar* 6:151–164

Buckeridge M, Santos WD, de Souza AP (2010) Rotas para o etanol celulósico no Brasil. In: Cortez LAB (coord) *Bioetanol de Cana-de-Açúcar: Pesquisa & Desenvolvimento para Produtividade e Sustentabilidade*. Edgard Blücher, São Paulo, SP, Brazil, pp 365–380

Bundock P, Hooykaas P (2005) An Arabidopsis hAT-like transposase is essential for plant development. *Nature* 436:282–284

Butterfield K, Irvine E, Valdez Garza M, Mirkov E (2002) Inheritance and segregation of virus and herbicide resistance transgenes in sugarcane. *Theor Appl Genet* 104:797–803

Camargo SR, Cançado GM, Ulian EC, Menossi M (2007) Identification of genes responsive to the application of ethanol on sugarcane leaves. *Plant Cell Rep* 26:2119–2128

Cantarella H, Trivelin PCO, Vitti AC (2007) Nitrogênio e enxofre na cultura da cana de açúcar. In: Yamada T, Abdalla SRS, Vitti GC (ed) *Nitrogênio e enxofre na agricultura brasileira*. International Plant Nutrition Institute, Piracicaba, SP, Brazil, pp 355–412

Cardona CA, Sánchez OJ (2007) Fuel ethanol production: Process design trends and integration opportunities. *Bioresour Technol* 98:2415–2457

Carlier JD, Reis A, Duval MF, Coppens D'eecknbrugge D, Leitão M (2004) Genetic maps of RAPD, AFLP and ISSR markers in *Ananas bracteatus* and *A. comosus* using the pseudo test cross strategy. *Plant Breed* 123:186–192

Carlucci MV, Cruz ND, Alvarez R (1988) Fatores ambientes e iniciação floral de cana-de-açúcar no ano agrícola 1984/85, na região de Piracicaba, SP. *Bragantia* 47:15–24

Carson DL, Botha FC (2000) Preliminary analysis of expressed sequence tags for sugarcane. *Crop Sci* 40:1769–1779

Carson DL, Botha FC (2002) Genes expressed in sugarcane maturing internodal tissue. *Plant Cell Rep* 20:1075–1081

Carson DL, Huckett BI, Botha FC (2002) Sugarcane ESTs differentially expressed in immature and maturing internodal tissue. *Plant Sci* 162:289–300

Casacuberta JM, Santiago N (2003) Plant LTR-retrotransposons and MITEs: control of transposition and impact on the evolution of plant genes and genomes. *Gene* 311:1–11

Casu RE, Dimmock CM, Chapman SC, et al. (2004) Identification of differentially expressed transcripts from maturing stem of sugarcane by *in silico* analysis of stem expressed sequence tags and gene expression profiling. *Plant Mol Biol* 54:503–517

Casu RE, Grof CP, Rae AL, McIntyre CL, Dimmock CM, Manners JM (2003) Identification of a novel sugar transporter homologue strongly expressed in maturing stem vascular tissues of sugarcane by expressed sequence tag and microarray analysis. *Plant Mol Biol* 52:371–386

Casu RE, Dimmock CM, Thomas M, et al. (2001) Genetic and expression profiling in sugarcane. *Proc Int Soc Sugarcane Technol* 24:626–627

Casu RE, Manners JM, Bonnett GD, et al. (2005) Genomics approaches for the identification of genes determining important traits in sugarcane. *Field Crops Res* 92:137–147

ChoYG, Ishii T, Temnykh S, et al. (2000) Diversity of microsatellites derived from genomic libraries and GenBank sequences in rice (*Oryza sativa* L.). *Theor Appl Genet* 100:713–722

Christensen AH, Quail PH (1996) Ubiquitin promoter-based vectors for high-level expression of selectable and/or screenable marker genes in monocotyledonous. *Transgen Res* 5:213–218

Cidade DAP, Garcia RO, Duarte AC, Sachetto-Martins G, Mansur E (2006) Morfogênese in vitro de variedades brasileiras de cana-de-açúcar. *Pesq Agropec Bras* 41:385–391

Cordaux R, Udit S, Batzer MA, Feschotte C (2006) Birth of a chimeric primate gene by capture of the trans-posase gene from a mobile element. *Proc Natl Acad Sci USA* 103:8101–8106

Cordeiro GM, Casu RE, McIntyre CL, Manners JM, Henry RJ (2001) Microsatellite markers from sugarcane (*Saccharum* spp) ESTs cross transferable to *Erianthus* and *Sorghum*. *Plant Sci* 160:1115–1123

Cordeiro GM, Taylor GO, Henry RJ (2000) Characterization of microsatellite markers from sugarcane (*Saccharum* spp.), a highly polyploid species. *Plant Sci* 155:161–168

Cortez LAB, Lora EES, Gómez EO (2008) *Biomassa para Bioenergia*. Editora UNICAMP, Campinas, SP, Brazil

Cowan RK, Hoen DR, Schoen DJ, Bureau TE (2005) MUSTANG is a novel family of domesticated transposase genes found in diverse angiosperms. *Mol Biol Evol* 22:2084–2089

Cropley JE, Martin DI (2007) Controlling elements are wild cards in the epigenomic deck. *Proc Natl Acad Sci USA* 104:18879–18880

CUGI (Clemson University Genome Institute) (2009), available at https://www.genome.clemson.edu (accessed April 5, 2009)

D'Hont A, Grivet L, Feldmann P, Rao S, Berding N, Glaszmann JC (1996) Characterisation of the double genome structure of modern sugarcane cultivars (*Saccharum* spp.) by molecular cytogenetics. *Mol Gen Genet* 250:405–413

D'Hont A (2005) Unraveling the genome structure of polyploids using FISH and GISH, examples of sugarcane and banana. *Cytogenet Genome Res* 109:27–33

D'Hont A, Lu YH, Gonzáles de Leon D, et al. (1994) A molecular approach to unravelling the genetics of sugarcane, a complex polyploid of the andropogoneae. *Genome* 37:222–230

Da Silva JAG, Honeycutt RJ, Burnquist WL, et al. (1995) *Saccharum spontaneum* L. 'SES208' genetic linkage map containing RFLP and PCR-based markers. *Mol Breed* 1:165–169

Da Silva JAG, Sorrells ME, Burnquist WL, Tanksley SD (1993) RFLP linkage map of *Saccharum spontaneum*. *Genome* 36:782–791

Daniels J, Roach BT (1987) Taxonomy and evolution. In: Heinz DJ (ed) *Sugarcane Improvement through Breeding*. Elsevier, Amsterdam, Netherlands, pp 7–84

Daniels J, Smith P, Paton N (1975) The origin of sugarcane and centers of genetic diversity in *Saccharum*. *Sugarcane Breed Newsl* 35:4–18

Dantas B, Melo JL (1960) A situação das variedades na zona canavieira de Pernambuco (1954/55 a 1957/58) e uma nota histórica sobre as variedades antigas. Boletim Técnico 11. Instituto Agronômico do Nordeste, Recife, PE, Brazil, pp 29–82

De Rosa-Jr VE, Nogueira FTS, Menossi M, Ulian EC, Arruda P (2005) Identification of methyl jasmonate responsive genes in sucarcane using cDNA arrays. *Braz J Plant* Physiol 17:131–136

de Souza AP, Gaspar M, Silva EA, et al. (2008) Elevated CO_2 induces increases in photosynthesis, biomass, productivity, and modifies gene expression in sugarcane. *Plant Cell Environ* 31:1116–1127

Deerr N (1921) *Cane Sugar*, 2nd ed. Norman Rodger, London

Dias MOS, Maciel Filho R, Rossell CEV (2007) Efficient cooling of fermentation vats in ethanol production—Part 1. *Sugar J* 70:11–17

Dinardo-Miranda LL (2008a) Pragas. In: Dinardo-Miranda LL, Vasconcelos ACM, Landell MGA (ed) *Cana-de-açúcar*. Instituto Agronômico, Campinas, SP, Brazil, pp 349–404

Dinardo-Miranda, LL (2008b) Nematóides. In: Dinardo-Miranda LL, Vasconcelos ACM, Landell MGA (ed) *Cana-de-açúcar*. Instituto Agronômico, Campinas, SP, Brazil, pp 405–422

Doolittle WF, Sapienza C (1980) Selfish genes, the phenotype paradigm and genome evolution. *Nature* 284(5757):601–603

Edgerton CW (1955) *Sugarcane and Its Diseases*. Louisiana State University Press, Baton Rouge, LA

EIA (Energy Information Administration) (2009) Official energy statistics of the U.S. government (2009), available at http://www.eia.doe.gov/ (accessed May 5, 2009)

EIA - (Energy Information Administration) (2008) Oxygenate production, http://tonto.eia.doe.gov/dnav/pet/hist/m_epooxe_yop-nus_1m.htm (accessed 05 May 2009)

Elliott AR, Campbell JA, Bretell RIS, Grof CPL (1998) Agrobacterium-mediated transformation of sugarcane using GFP as a screenable marker. *Aust J Plant Physiol* 25:739–743

Enriquez-Obregon GA, Vazquez PRI, Prieto SDL, Riva-Gustavo ADL, Selman HG (1998) Herbicide resistant sugarcane (*Saccharum officinarum*) plants by Agrobacterium-mediated transformation. *Planta* 206:20–27

Eujayl I, Sorrells ME, Baum M, Wolters P, Powell W (2002) Isolation of EST-derived microsatellite markers for genotyping the A and B genomes of wheat. *Theor Appl Genet* 104:399–407

Eujayl I, Sledge MK, Wang L, et al. (2004) *Medicago truncalata* EST-SSRs reveal cross-species genetic markers for *Medicago* spp. *Theor Appl Genet* 108:414–422

Falco MC, Tulmann-Neto A, Ulian EC (2000) Transformation and expression of a gene for herbicide resistance in a Brazilian sugarcane. *Plant Cell Rep* 19:1188–1194

Falco MC, Silva-Filho MC (2003) Expression of soybean proteinase inhibitors in transgenic sugarcane plants: effects on natural defense against *Diatraea saccharalis*. *Plant Physiol Biochem* 41:761–766

FAOSTAT – Food and Agricultural Organization of the United Nations (2009) FAO Statistics Division: http://faostat.fao.org/ (accessed 10 February 2009)

Federer WT (1956) Augmented (or Hoonuiaku) designs. *Hawaiian Planter's Rec* 55:191–208

Figueiredo P (2008) Breve história da cana-de-açúcar e do papel do Instituto Agronômico no seu estabelecimento no Brasil. In: Dinardo-Miranda LL, Vasconcelos ACM, Landell MGA (ed) *Cana-de-açúcar*. Instituto Agronômico, Campinas, SP, Brazil, pp 31–44

Fitch MMM, Moore PH (1993) Long term culture of embryogenic sugarcane callus. *Plant Cell Tiss Org Cult* 32:335–343

Gallo-Meagher M, Irvine JE (1996) Herbicide resistant transgenic sugarcane plants containing the bar gene. *Crop Sci* 36:1367–1374

Garcia AA, Kido EA, Meza AN, et al. (2006) Development of an integrated genetic map of a sugarcane (*Saccharum* spp.) commercial cross, based on a maximum-likelihood approach for estimation of linkage and linkage phases. *Theor Appl Genet* 112:298–314

Gene Expression Omnibus Database (GEO). http://www.ncbi.nlm.nih.gov/geo (accessed May 5, 2009).

GERAN Brasil (Grupo Especial para Racionalização da Agroindústria Canavieira do Nordeste) (1971) Programa regional de pesquisas canavieiras. Ministério do Interior, Recife, PE, Brazil

Gerbens-Leenes W, Hoekstraa AY, van der Meerb TH (2009) The water footprint of bioenergy. *Proc Natl Acad Sci USA* 106:10219–10223

Giamalva MG, Anzalone L, Chilton SJP, Loupe DT (1967) Evaluation of methods of seedling selection at Louisiana State University. *Proc Cong Int Soc Sugar Cane Technol* 12:916–919

Giroday E (1977) Breeding and selection in Reunion Island. *Sugarcane Breed Newsl* 40:9–14

Goldemberg J (2008) The Brazilian biofuels industry. *Biotechnol Biofuels* 1:6

Gonçalves MC (2008) Doenças causadas por vírus. In: Dinardo-Miranda LL, Vasconcelos ACM, Landell MGA (ed) *Cana-de-açúcar*. Instituto Agronômico, Campinas, SP, Brazil, pp 451–464

Gosnell JM, Lonsdale JE (1974) Some effects of drying off before harvest on cane yield and quality. *Proc Int Soc Sugar Cane Technol* 15:701–714

Grattapaglia D, Sederoff R (1994) Genetic linkage maps of *Eucalyptus grandis* and *Eucalyptus urophylla* using a pseudo-testcross mapping strategy and RAPD markers. *Genetics* 137:1121–1137

Green J, Vain P, Fearnehough MT, Worland B, Snape JW, Atkinson HJ (2002) Analysis of the expression patterns of the *Arabidopsis thaliana* tubulin-1 and *Zea mays* ubiquitin-1 promoters in rice plants in association with nematode infection. *Physiol Mol Plant Pathol* 60:197–205

Grivet L, D'Hont A, Roques D, Feldmann P, Lanaud C, Glaszmann JC (1996) RFLP mapping in cultivated sugarcane (*Saccharum* spp.): genome organization in a highly polyploid and aneuploid interspecific hybrid. *Genetics* 142:987–1000

Grivet L, Glaszmann JC, Arruda P (2001) Sequence polymorphism from EST data in sugarcane: a fine analysis of 6-phosphogluconate dehydrogenase genes. *Genet Mol Biol* 24:161–167

Grivet L, Glaszmann JC, Vincentz M, da Silva F, Arruda P (2003) ESTs as a source for sequence polymorphism discovery in sugarcane: example of the Adh genes. *Theor Appl Genet* 106:190–197

Guiderdoni E, Merot B, Eksomtramage T, Paulet F, Feldman P, Glaszman JC (1995) Somatic embryogenesis in sugarcane (*Saccharum* species). In: Bajaj Y (ed) *Biotechnology in Agriculture and Forestry: Somatic Embryogenesis and Synthetic Seed II*. Springer, Berlin, Germany

Guiderdoni E, Demarly Y (1988) Histology of somatic embryogenesis in cultured leaf segments of sugarcane plantlets. *Plant Cell Tiss Organ Cult* 14:71–88

Guimarães CT, Honeycutt RJ, Sills GR, Sobral BWS (1999) Genetic maps of *Saccharum officinarum* L. and *Saccharum robustum* Brandes and Jew. *Ex Grassl Genet Mol Biol* 22:125–132

Hansom S, Bower R, Zhang L, et al. (1999) Regulation of transgene expression in sugarcane. *Proc Int Soc Sugar Cane Technol* 23:278–289

Hartemink AE (2008) Sugarcane for bioethanol: soil and environmental issues. *Adv Agron* 99:125–182

Hartt CE (1936) Further notes on water and cane ripening. *Hawaiian Planters Rec* 40:355–381

Hatch MD, Slack CR (1966) Photosynthesis by sugar-cane leaves—A new carboxylation reaction and pathway of sugar formation. *Biochem J* 101:103

Heffer P (2009) Assessment of fertilizer use by crop at the global level 2006/07 – 2007/08. (AgCom 09/28, April 2009). International Fertilizer Industry Association (IFA), Paris, France

Heinz DJ, Mee GWP (1969) Plant differentiation from callus tissue of *Saccharum* species. *Crop Sci* 9:346–348

Heinz DJ, Tew TL (1987) Hybridization procedures. In: Heinz DJ (ed) *Sugarcane Improvement through Breeding*. Elsevier, Amsterdam, Netherlands, pp 313–342

Higler AP, Chande R, Baur R, Krishna R, and Taylor R (2004) Nonequilibrium modeling of three-phase distillation. *Comput Chem Eng* 28:2021–2036

Hoarau JY, Grivet L, Offmann B, et al. (2001) Genetic dissection of a modern sugarcane cultivar (*Saccharum* spp.) I. Genome mapping with AFLP markers. *Theor Appl Genet* 103:84–97

Hodkinson TR, Chase MW, Lledó MD, Salamin N, Renvoize SA (2002) Phylogenetics of *Miscanthus*, *Saccharum* and related genera (Saccharinae, Andropogoneae, Poaceae) based on DNA sequences from ITS nuclear ribosomal DNA and plastid trnLintron and trnL-F intergenic spacers. *J Plant Res* 115:381–392

Hoefsloot G, Termorshuizen AJL, Watt DA, Cramer MD (2005) Biological nitrogen fixation is not a major contributor to the nitrogen demand of a commercially grown South African sugarcane cultivar. *Plant Soil* 277:85–96

Hogarth DM (1977) Quantitative inheritance studies in sugarcane. III. The effect of competition and violation of genetic assumptions on estimation of genetic variance components. *Aust J Agric Res* 28:257–268

Hogarth DM (1980) The effect of accidental selfing on the analysis of a diallel cross with sugar cane. *Euphytica* 29:737–746

Hogarth DM (1987) Genetics of sugarcane. In Heinz, DJ (ed) *Sugarcane Improvement through Breeding*. Elsevier, Amsterdam, Netherlands, pp 255–271

Hogarth DM, Ryan CC, Skinner JC (1983) Inheritance of resistance to rust in sugarcane – comments. *Field Crops Res* 7:313–316

Hogarth DM, Wu KK, Heinz DJ (1981) Estimating genetic variance in sugarcane using a factorial cross design. *Crop Sci* 21:21–25

Huang N, Wu L, Nandi S, et al. (2001) The tissue-specific activity of a rice beta-glucanase promoter (Gns9) is used to select rice transformants. *Plant Sci* 161:589–595

Ingelbrecht IL, Irvine JE, Mirkov TE (1999) Posttranscriptional gene silencing in transgenic sugarcane. Dissection of homology-dependent virus resistance in a monocot that has a complex polyploid genome. *Plant Physiol* 119:1187–1198

Inman-Bamber NG, Smith DM (2005) Water relations in sugarcane and response to water deficits. *Field Crop Res* 92:185–202

Irvine JE, Benda GTA (1987) Transmission of sugarcane diseases in plants derived by rapid regeneration from diseased leaf tissue. *Sugar Cane* 6:14–16

Jannoo N, Grivet L, Chantret N, et al. (2007) Orthologous comparison in a gene-rich region among grasses reveals stability in the sugarcane polyploid genome. *Plant J* 50:574–85

Jordan IK (2006) Evolutionary tinkering with transposable elements. *Proc Natl Acad Sci USA* 103:7941–7942

Kashkush K, Feldman M, Levy AA (2003) Transcriptional activation of retrotransposons alters the expression of adjacent genes in wheat. *Nat Genet* 33:102–106

Keith CS, Hoang DO, Barrett BM, et al. (1993) Partial sequence analysis of 130 randomly selected maize cDNA clones. *Plant Physiol* 101:329–332

Kortschak HP, Hartt CE, Burr GO (1965) Carbon dioxide fixation in sugarcane leaves. *Plant Physiol* 40:209–213

Ladd SL, Heinz DJ, Meyer HK, Nishimoto BK (1974) Selection studies in sugarcane (*Saccharum* sp hybrids). I. Repeatability between selection stages. *Proc Cong Int Soc Sug Cane Technol* 15:102–105

Lakshmanan P, Geijskes RJ, Aitken KS, Grof CLP, Bonnett GD, Smith GR (2005) Sugarcane biotechnology: the challenges and opportunities. *In Vitro Cell Dev Biol Plant* 41:345–363

Lakshmanan P, Geijskes RJ, Wang L, et al.(2006) Developmental and hormonal regulation of direct shoot organogenesis and somatic embryogenesis in sugarcane (*Saccharum* spp. interspecific hybrids) leaf culture. *Plant Cell Rep* 25:1007–1510

Landel MGA, Bressiani JA (2008) Melhoramento genético, caracterização e manejo varietal. In: Dinardo-Miranda LL, Vasconcelos ACM, Landell MGA (ed) *Cana-de-açúcar*. Instituto Agronômico, Campinas, SP, Brazil, pp 101–155

Landell MGA, Alvarez R (1993) Cana-de-açúcar. In: Furlani AMC, Viégas GP (ed) *Contribuições do Instituto Agronômico ao Melhoramento Genético Vegetal* v.1. Instituto Agronômico, Campinas, SP, Brazil, pp 77–93

Last KI, Brettell RIS, Chamberlain DA, et al. (1991) pEmu: an improved promoter for gene expression in cereal cells. *Theor Appl Genet* 81:581–588

Latter BDH (1964) Selection methods in the breeding of cross-fertilized pasture species. In: Barnard C (ed) *Grasses and Grasslands*. MacMillan, London, England, pp 168–181

Lavi U, Akkaya M, Bhagwat A, Lahav E, Cregan PB (1994) Methodology of generation and characteristics of simple sequence repeat DNA markers in avocado (*Persea americana* M.). *Euphytica* 80:171–177

Lee SK, Chou H, Ham TS, Lee TS, Keasling JD (2008) Metabolic engineering of microorganisms for biofuels production: from bugs to synthetic biology to fuels. *Curr Opin Biotechnol* 19:556–563

Legendre BL (2001) *Sugarcane Production Handbook* (Pub. 2859). Louisiana State University Agricultural Center, Louisiana Cooperative Extension, Baton Rouge, LA

Leibbrandt NB, Snyman SJ (2003) Stability of gene expression and agronomic performance of a transgenic herbicide-resistant sugarcane line in South Africa. *Crop Sci* 43:671–678

Lima ML, Garcia AA, Oliveira KM, et al. (2002) Analysis of genetic similarity detected by AFLP and coefficient of parentage among genotypes of sugar cane (*Saccharum* spp.). *Theor Appl Genet* 104:30–38

Lima MAC, Garcia RO, Martins GS, Mansur E (2001) Morfogênese *in vitro* e susceptibilidade de calos de variedades nacionais de cana-de-açúcar (*Saccharum officinarum* L.) a agentes seletivos utilizados em sistemas de transformação genética. *Rev Brasil Bot* 24:73–77

Lin M, Lou XY, Chang M, Wu R (2003) A general statistical framework for mapping quantitative trait loci in nonmodel systems: issue for characterizing linkage phases. *Genetics* 165:901–913

Liu DW, Oard SV, Oard JH (2003) High transgene expression levels in sugarcane (*Saccharum officinarum* L.) driven by the rice ubiquitin promoter RUBQ2. *Plant Sci* 165:743–750

Liu DL, Kingston G, Bull TA (1999) A new technique for determining the thermal parameters of phenological development in sugarcane, including sub optimum and supra-optimum temperature regimes. *Agric For Meteorol* 90:119–139

Liu MC (1981) *In vitro* methods applied to sugarcane improvement. In: TA T (ed) *Plant Tissue Culture: Methods and Applications in Agriculture*. Academic Press, New York

Liu MC (1993) Factors affecting induction, somatic embryogenesis and plant regeneration of callus from cultured immature inflorescences of sugarcane. *J Plant Physiol* 141:714–720

Liu Q, Kasuga M, Sakuma Y, et al. (1998) Two transcription factors, DREB1 and DREB2, with an EREBP/AP2 DNA binding domain separate two cellular signal transduction pathways in drought- and low-temperature-responsive gene expression, respectively, in Arabidopsis. *Plant Cell* 10:1391–406

Ma HM, Schulze S, Lee S, et al. (2004) An EST survey of the sugarcane transcriptome. *Theor Appl Genet* 108:851–863

Machado-Jr GR, Silva WM, Irvine JE (1987) Sugarcane breeding in Brazil: the Copersucar program. In: *Copersucar Int. Sugarcane Breeding Workshop*. Copersucar, São Paulo, SP, Brazil, pp 217–232

Maliepaard C, Jansen J, van Ooijen JW (1997) Linkage analysis in a full-sib family of an outbreeding plant species: overview and consequences for applications. *Genet Res* 70:237–250

Mangelsdorf AJ (1953) Sugarcane breeding in Hawaii. Part II – 1921–1952. *Hawaiian Planters Rec* 54:101–137

Manickavasagam M, Ganapathi A, Anbazhagan VR, et al. (2004) *Agrobacterium*-mediated genetic transformation and development of herbicide-resistant sugarcane (*Saccharum* species hybrids) using axillary buds. *Plant Cell Rep* 23:134–143

Manickavasagam M, Ganapathi A (1998) Direct embryogenesis and plant regeneration from leaf explants of sugarcane. *Indian J Exp Biol* 36:832–835.

MAPA—Ministério da Agricultura, Pecuária e Abastecimento. (2009) Brazil, available at http://www.agricultura.gov.br/pls/portal/docs/PAGE/MAPA/SERVICOS/ (accessed June 10, 2009)

Margarido GRA, Souza AP, Garcia AAF (2007) OneMap: software for genetic mapping in outcrossing species. *Hereditas* 144:78–79

Mariano AP, Angelis DF, Maugeri Filho F, P Atala DI, Maciel MRW, Maciel Filho R (2008) An alternative process for butanol production: continuous flash fermentation. *Chem Product Process Model* 3:1–14

Mariotti JA (1977) Sugarcane clonal selection research in Argentina: a review of experimental results. *Proc Congr Int Soc Sugar Cane Technol* 14:121–136

Matsuoka S, Arizono H (1987) Avaliação de variedades pela capacidade de produção de biomassa e pelo valor energético. *Anais Cong Nac STAB* 4:220–225

Matsuoka S, Garcia AAF, Arizono H (1999a) Melhoramento da cana-de-açúcar. In: Borém A (org) *Melhoramento de Espécies Cultivadas* v.1. Universidade Federal de Viçosa, Viçosa, MG, Brazil, pp 205–252

Matsuoka S, Garcia AAF, Calheiros GC (1999b) Hibridação em cana-de-açúcar. In: Borém A (org) *Melhoramento de Espécies Cultivadas* v.2. Universidade Federal de Viçosa, Viçosa, MG, Brazil, pp 221–256

McCormick AJ, Cramer MD, Watt DA (2008) Changes in photosynthetic rates and gene expression of leaves during a source-sink perturbation in sugarcane. *Ann Bot* 101:89–102

McCormick AJ, Cramer M D, Watt DA (2006) Sink strength regulates photosynthesis in sugarcane. *New Phytol* 171:759–770

McElroy D, Blowers AD, Jenes B, Wu R (1991) Construction of expression vectors based on the rice actin 1 (Act1) 5' region for use in monocot transformation. *Mol Gen Genet* 231:150–160

McQualter RB, Chong BF, Meyer K, et al. (2005) Initial evaluation of sugarcane as a production platform for p-hydroxybenzoic acid. *Plant Biotechnol J* 3:29–41

McQualter RB, Burns P, Smith GR, Dale JL, Harding RM (2004) Molecular analysis of Fiji disease virus genome segments 5, 6, 8 and 10. *Arch Virol* 149:713–721

Mello FFC, Cerri CEP, Bernoux M, Volkoff B, Cerri CC (2006) Potential of soil carbon sequestration for the Brazilian Atlantic Region. In: Lal R, Cerri, CC, Bernoux M, Etchevers J, Cerri CEP (ed) *Carbon Sequestration in Soils of Latin America*. Haworth Press, New York, pp 349–368

Menossi M, Silva-Filho MC, Vincentz M, Van-Sluys MA, Souza GM (2008) Sugarcane functional genomics: gene discovery for agronomic trait development. *Int J Plant Genom* 2008:458732 (Epub)

Ming R, Liu SC, Lin YR, et al. (1998) Detailed alignment of *Saccharum* and *Sorghum* chromosomes: comparative organization of closely related diploid and polyploid genomes. *Genetics* 150:1663–1682

Ming R, Wang YW, Draye X, Moore PH, Irvine JE, Paterson AH (2002) Molecular dissection of complex traits in autopolyploids: mapping QTL's affecting sugar yield and related traits in sugarcane. *Theor Appl Genet* 105: 332–345

Molinari HB, Marur CJ, Daros E, et al. (2007) Evaluation of the stress-inducible production of proline in transgenic sugarcane (*Saccharum* spp.): osmotic adjustment, chlorophyll fluorescence and oxidative stress. *Physiol Plant* 130:218–229

Moore G (1995) Cereal genome evolution: pastoral pursuits with 'Lego' genomes. *Curr Opin Genet Dev* 5:717–724

Moore PH (1995) Temporal and spatial regulation of sucrose accumulation in the sugarcane stem. *Aust J Plant Physiol* 22:661–679

Moore PH (2005) Integration of sucrose accumulation processes across hierarchical scales: towards developing an understanding of the gene-to-crop continuum. *Field Crop Res* 92:119–135

Mudge J, Andersen WR, Kehrer R, Fairbanks DJ (1996) A RAPD genetic map of *Saccharum officinarum*. *Crop Sci* 36:1362–1366

Navarro AR, Sepúlveda M del C, Rubio MC (2000) Bio-concentration of vinasse from the alcoholic fermentation of sugar cane molasses. *Waste Manag* 20:581–585

Neuteboom LW, Kunimitsu WY, Webb D, Christopher DA (2002) Characterization and tissue-regulated expression of genes involved in pineapple (*Ananas comosus* L.) root development. *Plant Sci* 163:1021–1035

Newman T, de Bruijn FJ, Green P, et al. (1994) Genes galore: a summary of methods for accessing results from large-scale partial sequencing of anonymous *Arabdopsis* cDNA clones. *Plant Physiol* 106:1241–1255

Nickell LG (1964) Tissue and cell cultures of sugarcane: another research tool. *Hawaiian Planters Rec* 57:223–229.

Nogueira FT, De Rosa-Jr VE, Menossi M, Ulian EC, Arruda P (2003) RNA expression profiles and data mining of sugarcane response to low temperature. *Plant Physiol* 132:1811–1824

Oliveira ALM, Canuto EL, Urquiaga S, Reis VM, Baldani JI (2006) Yield of micropropagated sugarcane varieties in different soil types following inoculation with diazotrophic bacteria. *Plant Soil* 284:230–232

Oliveira KM, Pinto LR, Marconi TG, et al. (2009) Characterization of new polymorphic functional markers for sugarcane. *Genome* 52:191–209

Oliveira KM, Pinto LR, Marconi TG, et al. (2007) Functional integrated genetic linkage map based on EST-markers for a sugarcane (*Saccharum* spp.) commercial cross. *Mol Breed* 21:1–20

Orgel LE, Crick FH (1980) Selfish DNA: the ultimate parasite. *Nature* 284:604–607

Papini-Terzi FS, Rocha FR, Vencio RZ, et al. (2005) Transcription profiling of signal transduction-related genes in sugarcane tissues. *DNA Res* 12:27–38

Papini-Terzi FS, Rocha FR, Vencio RZ, et al. (2009) Sugarcane genes associated with sucrose content. *BMC Genom* 10:120

Paterson AH, Bowers JE, Bruggmann R, et al. (2009) The *Sorghum bicolor* genome and the diversification of grasses. *Nature* 457:551–556

Peixoto Lima GC (1842) *Dissertação a cerca da canna de assucar*. Typographia Universal de Laemmert, Rio de Janeiro, RJ, Brazil

Pereira AR, Barbieri V, Maniero MA (1986) Condicionamento climático da indução ao florescimento em cana--de-açúcar. *Rev Soc Técn Açuc Alcool Bras* 4:56–59

Pescarini MH, Barros AAC, Wolf-Maciel MR (1996) Development of a software for simulating separation processes using a nonequilibrium stage model. *Comput Chem Eng* 20:279–284

Piegu B, Guyot R, Picault N, et al. (2006) Doubling genome size without polyploidization: dynamics of retrotransposition-driven genomic expansions in *Oryza australiensis*, a wild relative of rice. *Genome Res* 16:1262–1269

Pinto LR, Oliveira KM, Ulian EC, Garcia AA, de Souza AP (2004) Survey in the sugarcane expressed sequence tag database (SUCEST) for simple sequence repeats. *Genome* 47:795–804

Porceddu A, Albertini E, Barcaccia G, Falistorco E, Falcinelli M (2002) Linkage mapping in apomictic and sexual kentucky blue grass (*Poa pratensis* L) genotypes using a two way pseudotestcross strategy based on AFLP and SAMPL markers. *Theor Appl Genet* 104:273–280

Raboin LM, Pauquet J, Butterfield M, D'Hont A, Glaszmann JC (2008) Analysis of genome-wide linkage disequilibrium in the highly polyploid sugarcane. *Theor Appl Genet* 116:701–714

Raboin LM, Oliveira KM, Lecunff L, et al. (2006) Genetic mapping in sugarcane, a high polyploid, using biparental progeny: identification of a gene controlling stalk colour and a new rust resistance gene. *Theor Appl Genet* 112:1382–1391

Ragauskas AJ, Williams CK, Davison BH, et al. (2006) The path forward for biofuels and biomaterials. *Science* 311:484–489

Raij B van, Cantarella H, Quaggio JA, Furlani AMC (1997) *Recomendações de Adubação e Calagem Para o Estado de São Paulo* (Boletim Técnico, 100). Instituto Agronômico, Campinas, SP, Brazil

Reffay N, Jackson PA, Aitken KS, et al. (2005) Characterisation of genome regions incorporated from an important wild relative into Australian sugarcane. *Mol Breed* 15:367–381

Reis VM, Urquiaga S, Pereira W, et al. (2008) Resposta de duas variedades de cana-de-açúcar à inoculação com bactérias diazotróficas. *Proc 9th Congr Nac STAB* 9:681–686

Rice RW, Gilber R, Lentini RS (2006) *Nutritional Requirements for Florida Sugarcane* (SS-AGR, 228). University of Florida, Institute of Food and Agricultural Sciences Extension, Gainesville, FL

Rocha FR, Papini-Terzi FS, Nishiyama MY, et al. (2007) Signal transduction-related responses to phytohormones and environmental challenges in sugarcane. *BMC Genom* 8:71

Röder MS, Plaschke J, König SU, et al. (1995) Abundance, variability and chromosomal location of microsatellite in wheat. *Mol Gen Genet* 246:327–333.

Ross-Ibarra J, Morrell PL, Gaut BS (2007) Plant domestication, a unique opportunity to identify the genetic basis of adaptation. *Proc Natl Acad Sci USA* 104(Suppl 1):8641–8648

Rossell CEV, Lahr Filho D, Hilst AGP, Leal MRLV (2005) Saccharification of sugarcane bagasse for ethanol production using the Organosolv process. *Int Sugar J* 107:192–195

Rossetto R, Dias FLF, Vitti AC (2008) Fertilidade do solo, nutrição e adubação In: Dinardo-Miranda LL, Vasconcelos ACM, Landell MGA (ed) *Cana-de-açúcar*. Instituto Agronômico, Campinas, SP, Brazil, pp 221–237

Rossetto R, Dias FLF, Vitti, AC, Tavares S (2008b) Potássio In: Dinardo-Miranda LL, Vasconcelos ACM, Landell MGA (ed) *Cana-de-açúcar*. Instituto Agronômico, Campinas, SP, Brazil, pp 289–312

Rossi M, Araujo PG, Paulet F, et al. (2003) Genomic distribution and characterization of EST-derived resistance gene analogs (RGAs) in sugarcane. *Mol Genet Genom* 269:406–419

Rossi M, Araújo PG, Van Sluys MA (2001) Survey of transposable elements in sugarcane expressed sequence tags (ESTs). *Genet Mol Biol* 24:147–154

Rossi M, Araújo PG, Jesus EM, Varani AM, and Van Sluys MA (2004) Comparative analysis of sugarcane Mutator-like transposases. *Mol Genet Genom* 272:194–203

Rossi M, Araujo PG, Paulet F, et al. (2003) Genomic distribution and characterization of EST-derived resistance gene analogs (RGAs) in sugarcane. *Mol Genet Genom* 269:406–419

Russel JR, Fuller JD, Young G, et al. (1997) Discrimination between barley genotypes using microsatellite markers. *Genome* 40:442–450

Saccaro Jr NL, Van Sluys MA, Varani AM, Rossi M (2007) MudrA-like sequences from rice and sugarcane cluster as two bona fide transposon clades and two domesticated transposases. *Gene* 392:117–125

Saghai Maroof MA, Biyashev RM, Yang GP, Zhang Q, Allard RW (1994) Extraordinary polymorphism microsatellite DNA in barley: species diversity, chromosomal locations, and population dynamics. *Proc Natl Acad Sci USA* 91:466–470

Santos AS (2008) Doenças causadas por fungos. In: Dinardo-Miranda LL, Vasconcelos ACM, Landell MGA (ed) *Cana-de-açúcar*. Instituto Agronômico, Campinas, SP, Brazil, pp 423–435

Schlogl PS, Nogueira FTS, Drummond R, et al. (2008) Identification of new ABA- and MEJA-activated sugarcane bZIP genes by data mining in the SUCEST database. *Plant Cell Rep* 27:335–345

Searchinger T, Heimlich R, Houghton RA, et al. (2008) Use of US croplands for biofuels increases greenhouse gas emissions through land use change. *Science* 319:1238–1240

Shen B, Carneiro N, Torres-Jerez I, et al. (1994) Partial sequencing and mapping of clones from two maize cDNA libraries. *Plant Mol Biol* 26:1085–1101.

Shepherd M, Cross M, Dieters MJ, Henry R (2003) Genetic maps for *Pinus elliottii* var *hondurensis* using AFLP and microsatellite markers. *Theor Appl Genet* 106:1409–1419

Silva FLH, Rodrigues MI, Maugeri Filho F (1999) Dynamic modeling, simulation and optimization of an extractive continuous fermentation process. *J Chem Tech Biotechnol* 74:176–182

Simmonds JC (1979) *Principles of Crop Improvement*. Longmans, London

Skinner JC, Hogarth DM, Wu KK (1987) Selection methods, criteria, and indices In: Heinz DJ (ed) *Sugarcane Improvement through Breeding*. Elsevier, Amsterdam, Netherlands, pp 409–453

Snyman SJ, Meyer GM, Richards JM, Haricharan N, Ramgareeb S, Huckett BI (2006) Refining the application of direct embryogenesis in sugarcane: effect of the developmental phase of leaf disc explants and the timing of DNA transfer on transformation efficiency. *Plant Cell Rep* 25:1016–1023

Sobral BWS, Honeycutt RJ (1993) High output genetic mapping of polyploids using PCR-generated markers. *Theor Appl Genet* 86:105–112

Stevenson GC (1965) *Genetics and Breeding of Sugarcane*. Longmans, London

Stupiello JP (1987) A cana-de-açúcar como matéria prima. In: Paranhos, SB (coord) *Cana-de-açúcar: Cultivo e Utilização*. Fundação Cargill, Campinas, SP, Brazil, pp 761–856

Szewc-McFadden AK, Lamboy WF, Hokanson SC, McFerson JF (1996) Utilization of identified simple sequence repeats (SSRs) in *Malus* x *Domestica* (apple) for germplasm characterization. *Hort Sci* 31:619

Tew TL (1987) New varieties. In: Heinz DJ (ed) *Sugarcane Improvement through Breeding*. Elsevier, Amsterdam, The Netherlands, pp 559–594

Thiel T, Michalek W, Varshney RK, Graner A (2003) Exploiting EST databases for the development and characterization of gene-derived SSR-markers in barley (*Hordeum vulgare* L.). *Theor Appl Genet* 106:411–422

UNICA—União da Indústria de Cana-de-Açúcar. (2009) Statistics of sugarcane sector—season 2006/2007, available at http://www.unica.com.br (accessed May 5, 2009)

Urata R (1970) Seedling propagation and bunch size for field transplanting. In: *Hawaiian Sugar Planters Association Experimental Station, Annual Report 1969*

Urquiaga S, Cruz KHS, Boddey RM (1992) Contribution of nitrogen fixation to sugar cane: Nitrogen-15 and nitrogen-balance estimates. *Soil Sci Soc Amer J* 56:105–114

Vettore AL, da Silva FR, Kemper EL, et al. (2003) Analysis and functional annotation of an expressed sequence tag collection for tropical crop sugarcane. *Genome Res* 13:2725–2735

Vickers JE, Grof CPL, Bonnett GD, et al. (2005a) Overexpression of polyphenol oxidase in transgenic sugarcane results in darker juice and raw sugar. *Crop Sci* 45:354–362

Vickers JE, Grof CPL, Bonnett GD, Jackson PA, Morgante TE (2005b) Effects of tissue culture, biolistic transformation, and introduction of PPO and SPS gene constructs on the performance of sugarcane clones in the field. *Aust J Agric Res* 56:57–68

Waclawovsky AJ, Sato PM, Lembke CG, Moore PH, Souza GM (2010) Improving sugarcane for bioenergy production. *Plant Biotechnol J* 8:1–14

Wagner GP, Pavlicev M, Cheverud JM (2007) The road to modularity. *Nat Rev Genet* 8:921–931

Wang ML, Goldstein C, Su W, Moore PH, Albert HH (2005) Production of biologically active GM-CSF in sugarcane: a secure biofactory. *Transgen Res* 14:167–178

Watkins CD (1967) Some practical aspects of sugar cane selection in British Guiana. *Int Soc Sugar Cane Technol Proc* 12:931–937

Watt D, McCormics A, Govender C, Crame M, Huckett B (2005) Increasing the utility of genomics in unraveling sucrose accumulation. *Field Crop Res* 92:149–158

Wu KK, Heinz DJ, Meyer HK, Ladd SL (1980) Combining ability and parental evaluation in five selected clones of sugarcane (*Saccharum* spp) hybrids. *Theor Appl Genet* 56:241–244

Wu KS, Tanksley SD (1993) Abundance, polymorphism and genetic mapping of microsatellites in rice. *Mol Gen Genet* 241:225–235

Wu R, Ma CX, Painter I, Zeng Z-B (2002) Simultaneous maximum likelihood estimation of linkage and linkage phases in outcrossing species. *Theor Popul Biol* 61:349–363

Xiao H, Jiang N, Schaffner E, Stockinger EJ, van der Knaap E (2008) A retrotransposon-mediated gene duplication underlies morphological variation of tomato fruit. *Science* 319(5869):1527–1530

Yamamoto K, Sasaki T (1997) Large-scale EST sequencing in rice. *Plant Mol Biol* 35:135–144

Zeng Z-B (2001) *Statistical Methods for Mapping Quantitative Trait Loci*. Department of Statistics, North Carolina State University, Raleigh, NC

Zhang L, Xu J, Birch RG (1999) Engineered detoxification confers resistance against a pathogenic bacterium. *Nat Biotechnol* 17:1021–1024

Zhang SZ, Yang BP, Feng CL, et al. (2006) Expression of the *Grifola frondosa* trehalose synthase gene and improvement of drought-tolerance in sugarcane (*Saccharum officinarum* L.). *J Integr Plant Biol* 48:453–459

Zheng Z, Kawagoe Y, Xiao S, et al. (1993) 5' distal and proximal cis-acting regulator elements are required for developmental control of a rice seed storage protein glutelin gene. *Plant J* 4:357–366

22 Switchgrass

Michael D. Casler
U.S. Department of Agriculture

Robert B. Mitchell and Kenneth P. Vogel
University of Nebraska–Lincoln

CONTENTS

22.1 INTRODUCTION

Switchgrass (*Panicum virgatum* L.) is a tall, erect, warm-season perennial native to the tall grass prairie, oak savanna, and associated ecosystems of North America. It can be found in prairies, open woodlands, and brackish marshes east of the Rocky Mountains and generally south of 55° north latitude (Hitchcock 1951; Stubbendieck et al. 1991). Less than 1% of these ecosystems exist today, but these prairie and savanna remnants have served as in situ gene banks, preserving a vast amount of genetic diversity within switchgrass and many other plant species. Switchgrass has a diversity of uses as well, including pasture, hay production, biomass for energy production, soil and water conservation, carbon sequestration, and wildlife habitat.

Switchgrass is adapted to a wide range of habitats, climatic conditions, and management strategies. In North America, switchgrass can be found in hardiness zones 3–9, from southern Canada (Manitoba to Newfoundland) to Baja California and the Gulf Coast (central Mexico to central Florida) (Figure 22.1). Switchgrass tolerates drought, extreme heat and cold, and moderately acid soils, and has relatively few major insect or disease pests. Switchgrass can be defoliated a number of times during the growing season, as in a managed grazing system, or infrequently, as in a bioenergy management system with only one or two harvests per season. Timing of harvests is moderately critical, largely to ensure that the plant is allowed to translocate storage carbohydrates to roots for regrowth after harvest and following overwintering. Grazing systems should incorporate adequate rest periods, although hay or bioenergy management systems should avoid harvesting during the last few weeks of the growing season to allow carbohydrate storage.

Switchgrass is a C_4 species with associated anatomical and physiological characteristics (Waller and Lewis 1979). Switchgrass is very slow to establish, largely because establishment-year development is oriented toward extensive root and crown development, often resulting in intense above-ground competition between switchgrass shoots and annual weeds. Both pre- and postemergence herbicides are valuable tools that aid in the establishment of switchgrass, shortening the time required to reach successful establishment and maximal biomass yields.

Interest in switchgrass as a bioenergy feedstock began when the U.S. Department of Energy (DOE) initiated its Bioenergy Feedstock Development Program (BFDP) by selecting switchgrass as the herbaceous model species. The decision was made largely because of consistently high biomass yields relative to other species across a broad geographic landscape, the relative simplicity of switchgrass propagation by seed, and an existing seed industry (McLaughlin and Kzsos 2005; Parrish and Fike 2005; Sanderson et al. 2007). Accomplishments of this program are credited with the approximately 25% increase in biomass yields of switchgrass because of the improved description and deployment of adapted cultivars, improved harvest and fertility management, and the development of new cultivars with higher biomass yield and expanded adaptation ranges (Sanderson et al. 2007).

FIGURE 22.1   Historical range of switchgrass in North America.

22.2   MORPHOLOGY AND TAXONOMY

Switchgrass grows 0.5–3 m tall and most plants are caespitose in appearance, with an occasional rhizomatous plant. Caespitose plants have short rhizomes and can form a sod over time. The inflorescence is a diffuse panicle 15–55 cm long with spikelets toward the end of long branches (Hitchcock 1951; Gould 1975). Spikelets disarticulate below the glumes and are two-flowered with the upper floret perfect and the lower floret either empty or staminate. Spikelets are 3–5 mm long and florets are glabrous and awnless. The lemma of the fertile floret is smooth and shiny. Leaves have rounded sheaths and firm flat blades that can vary from 10 to 60 cm in length. The number of leaves per culm will vary depending on genotype and environment (Redfearn et al. 1997). The ligule is a fringed membrane 1.5–3.5 mm long and consists mostly of hairs. Switchgrass reproduces by seeds, tillers, and rhizomes. It has the Pancoid type of seedling root (Newman and Moser 1988; Tischler and Voigt 1993). Roots of established plants may reach depths of 3 m (Weaver 1954).

Seed consists of the indurate and smooth lemma and palea, which hold tightly to the caryopsis. The margins of the lemma are enrolled over the margin of the palea. Glumes are almost entirely removed by combining and cleaning. On the average there are approximately 850 seeds/g (Wheeler and Hill 1957). Seed weight differences exist within and among cultivars. As an example, genotypic variation in seed mass within the cultivar Sunburst has been reported with a range of 450–850 seeds/g (Vogel 2002). Switchgrass is easily threshed, cleaned, and planted with commercial planting equipment. A switchgrass seed industry has existed for over 50 years and numerous private companies and public crop improvement associations are involved in seed production, distribution, and marketing.

Switchgrass has two distinct ecotypes—upland and lowland (Brunken and Estes 1975). As the names suggest, the lowland ecotype was originally found on flood plains and riparian zones subject to occasional flooding and/or waterlogging. The upland ecotype was originally found in upland areas that were not subject to flooding and often prone to drought. Plants of the upland and lowland ecotypes are morphologically and genetically distinct from each other. Generally, lowland plants have a later heading date, taller plant height, larger and thicker stems, fewer stems per plant, more upright leaf blades, and a more bluish cast than upland plants. Upland and lowland plants can be easily crossed with each other (Martinez-Reyna et al. 2001) and intermediate types exist in nature, suggesting that upland x lowland crosses have occurred in natural ecosystems, despite the large difference in heading date between upland and lowland ecotypes.

22.3   GENETICS

Switchgrass has a basic chromosome number of $x = 9$ (Gould 1975). A wide range of chromosome numbers has been reported in the literature including somatic counts of 18, 36, 54, 72, 90, and 108 chromosomes (Nielsen 1944; Barnett and Carver 1967). Switchgrass has small chromosomes that are difficult to count. Recent studies aided by the use of flow cytometry indicate that most switchgrass cultivars are either tetraploid ($2n = 4x = 36$) or octaploid ($2n = 8x = 72$) (Hopkins et al. 1996; Lu et al. 1998). The tetraploids and octaploids average 3.1 and 6.1 pg $2C^{-1}$ DNA (Lu et al. 1998). The 2C ("C" stands for "constant") value is the DNA content of a diploid somatic nucleus expressed in pg (picogram or 10^{-12} g) and can be converted to daltons or nucleotide pairs using the formulas: 1 nucleotide pair = 660 Da; 1 pg = 0.965×10^9 nucleotide pairs (Bennett and Smith 1976). To date, all lowland plants appear to be tetraploids on the basis of chromosome counts of mitosis in root tips or flow cytometry analyses, although most upland plants are tetraploids or octaploids. Tetraploid and octaploid plants occur sympatrically in over half of the remnant prairies that were evaluated by Hultquist et al. (1997). They did not report hexaploid plants in remnant prairies. Several researchers have attempted to relate ploidy levels to morphological traits and geographical distribution, but the results were inconclusive (Nielsen 1944, 1947; McMillan and Weiler 1959; Barnett and Carver 1967). There are numerous reports of aneuploidy in switchgrass, particularly at the higher ploidy levels (Costich et al. 2010), but some disagreement as to the frequency and severity of this phenomenon.

Normal bivalent pairing has been reported for tetraploid and octaploid switchgrass plants (Riley and Vogel 1982; Martinez-Reyna et al. 2001). Aneuploid variants and multivalent chromosome associations are more frequent at higher ploidy levels (Barnett and Carver 1967; Brunken and Estes 1975). An analysis of segregation and linkage relationships for random markers distributed across the genome of an upland × lowland tetraploid cross suggested polysomic inheritance (Missaoui et al. 2005a). This conclusion was based on the observed ratios of single- to multiple-dose markers and the observed ratios of loci linked in coupling vs. repulsion phase. Additional analyses, following more complete genomewide marker saturation, clearly identified 18 linkage groups, suggesting an all-tetraploid genome with disomic inheritance (Okada et al. 2010).

The evidence for preferential pairing and disomic inheritance within a polyploid series that ranges from $2n = 2x = 18$ to $2n = 12x = 108$ is surprising. Normally, extensive polyploid series such as this one arise by spontaneous doubling of whole genomes, leading to autopolyploidy (Mable 2003). The evidence for higher frequencies of aneuploid variants and multivalent pairing at higher ploidy levels suggests multiple mechanisms of polyploidization, pairing, and gene inheritance within switchgrass, i.e., apparent allopolyploidy and disomic inheritance at the tetraploid level and possible autopolyploidy and polysomic inheritance at higher ploidy levels. It also raises the intriguing question: If tetraploid switchgrass is an allopolyploid, what are its ancestors and, as a corollary, do upland and lowland switchgrasses have different ancestors? Alternatively, is the polyploid switchgrass genome in the process of diverging into two distinct and duplicate genomes?

Switchgrass is a member of the Paniceae tribe of grasses, which diverged from the Maydeae (maize) tribe approximately 23 Mya. Upland and lowland ecotypes are thought to have diverged from each other sometime between 1 and 2 Mya on the basis of sequences of the nuclear gene encoding plastid acetyl-CoA carboxylase, *Acc-1* (Huang et al. 2003). These authors further speculate that polyploidization events were involved in this divergence, as the *Acc-1* polymorphisms appear to discriminate a "lowland tetraploid" from an "upland octaploid" form. Polymorphisms at the *Acc-1* locus within tetraploid plants of both the upland and lowland ecotypes suggest the possibility of four ancestral diploids (Huang et al. 2003).

Switchgrass has two cytoplasm types, "L" and "U" that are based on chloroplast DNA (cpDNA) polymorphisms that are associated with the lowland and upland ecotypes, respectively, (Hultquist et al. 1996; Missaoui et al. 2006). The "L" cytoplasm types are tetraploids whereas the "U" types can be either tetraploids or octaploids (Hultquist et al. 1996). Martinez-Reyna et al. (2001) used controlled reciprocal crosses between "Kanlow" ("L" tetraploid) and "Summer" ("U" tetraploid) plants and a restriction fragment length polymorphism (RFLP) marker to demonstrate that the chloroplast DNA of switchgrass is maternally inherited. They also determined that the lowland and upland ecotypes and associated cytoplasm types of switchgrass are completely cross-fertile at the tetraploid level and that there is a high degree of similarity among their nuclear genomes as indicated by normal bivalent pairing during meiosis. This is supported by linkage analyses and homology between "Alamo" ("L" tetraploid) and Summer linkage groups (Missaoui et al. 2005a). Despite their homology, upland and lowland ecotypes are genetically distinct, as demonstrated by cpDNA markers (Hultquist et al. 1996; Missaoui et al. 2006) and random amplified polymorphic DNA (RAPD) markers (Gunter et al. 1996). Extreme differential heading and anthesis dates, combined with physical isolation due to habitat differentiation and fragmentation, are the likely causes of relatively recent genetic isolation of the upland and lowland ecotypes. Upland and lowland ecotypes can occur sympatrically, with differential heading and anthesis dates serving to maintain distinct and isolated germplasm pools (Brunken and Estes 1975).

Ecotypic variation in switchgrass derives largely from the broad geographic distribution of the species, which extends from approximately 15 to 55° north latitude in North America. Photoperiod is one of the principal drivers of ecotypic variation, ranging from 13 to 17 h on the summer solstice at these extreme latitudes. Switchgrass is thought to have survived the Pliestocene Glaciation in three refugia located in the southern United States, perhaps including northern Mexico: a western semi-montane (dryland) region, a central humid region with rich, fertile soils, and a southeastern region, perhaps near the Gulf Coast (McMillan 1959). With the retreat of glaciation approximately 11,000 years ago, switchgrass and other members of these prairie remnants rapidly moved north, colonizing a wide range of habitats under increasingly longer day length, but shorter growing seasons. High levels of genetic diversity within these remnant populations allowed natural selection to proceed for traits necessary to survive in northern latitudes, including early flowering and cold tolerance.

As a result of natural selection, switchgrass ecotypes are photoperiod sensitive, requiring short days to induce flowering (Benedict 1941). Photoperiod requirements are based on the latitude-of-origin of individual ecotypes. Flowering is induced by decreases in day length following the summer solstice. In North America, exporting northern ecotypes south exposes them to a shorter-than-normal day length during summer months, which causes early flowering, often drastically reducing their ability to utilize the full growing season because of early senescence (Vogel 2004). The opposite occurs when southern ecotypes are exported north. They remain vegetative for a longer period of time, with a longer photosynthetically active period, often producing more forage than northern ecotypes (Newell 1968a). When grown in the central Great Plains, switchgrasses from the Dakotas (northern ecotypes) flower and mature early and are short in stature whereas those from Texas and Oklahoma (southern ecotypes) flower late and are tall (Cornelius and Johnson 1941; McMillan 1959). This patterned response to photoperiod also occurs for switchgrass cultivars and ecotypes grown in Europe where it is not native (Elbersen et al. 2003).

The photoperiod response also appears to be associated with winter survival and cold tolerance. Southern ecotypes exported too far north will not survive winters because they stay vegetative too late in the fall, lack the ability to store sufficient carbohydrates during winter, and lack sufficient cold/freezing tolerance (Casler et al. 2004, 2007a). As a general rule, switchgrass germplasm should not be exported more than one USDA Plant Hardiness Zone (Cathey 1990) north or south of its area of origin because of these adaptation issues. At this time, the genetic regulation of these latitude-associated traits is unknown.

Switchgrass is a cross-pollinated species with a gametophytic self-compatibility system that is similar to the S-Z incompatibility system found in other Poaceae (Martinez-Reyna and Vogel 2002). Pollen is dispersed by wind. Self-compatibility, as measured by seed set from bagged panicles, is typically less than 1% (Talbert et al. 1983; Martinez-Reyna and Vogel 2002). A postfertilization incompatibility system also exists that inhibits intermatings among octaploid and tetraploid plants (Martinez-Reyna and Vogel 2002). The postfertilization incompatibility system between ploidy levels in switchgrass appears to be similar to the endosperm balance number system found in other species. The postfertilization incompatibility system is probably responsible for the relatively low frequency of hexaploid plants in native prairies. The tetraploid and octaploid plants in native prairies may exist as separate and distinct populations.

As switchgrass migrated north from the glacial refugia, wind pollination and migratory animals that carried seeds, as undigested feed or as hitchhikers in fur or feathers, were likely responsible for considerable genetic mixing along populational boundaries and, perhaps, over large geographic regions. Cross-pollination balanced the effects of natural selection, causing some genetic homogenization across sites and large amounts of genetic variation within sites (Casler et al. 2007b). Even though the tallgrass prairie ecosystem is highly fragmented with only about 1% intact, a vast array of genetic variability has been preserved both within and among prairie remnant sites. Analyses of RAPD markers for plants collected from prairie remnants from the Dakotas to New York suggest that these isolated prairie remnants continue to act as one large remnant population that contains many subpopulations each capable of representing much of the variability present within the population as a whole (Casler et al. 2007b). Although the destruction of the tallgrass prairie and associated ecosystems was nearly complete 100 years ago, self-incompatibility and polyploidy have served to preserve genetic variability within this species.

Because of the effects of photoperiod of flowering and the large temperature differential across the north-south gradient of natural switchgrass populations, latitude is the most important factor determining adaptation of switchgrass ecotypes and cultivars. Differentiation along east-west gradients of switchgrass populations tend to be less obvious than along north-south gradients, most likely because the factors that may cause such differentiation—soil type, historical vegetational succession, and moisture availability—are less important than temperature and day length in generating selection pressures leading to morphological or physiological differentiation (Nixon and McMillan 1964; Casler et al. 2007a). Nevertheless, there is some differentiation between upland switchgrass populations deriving from the tallgrass prairie of the Central Great Plains and the historical Eastern Forest ecosystems (McMillan 1959; Casler et al. 2007a).

The balancing effects of natural selection, creating differentiation among populations and uniformity within populations, and self-incompatibility, which favors homogenization of populations and promotes diversity within populations, suggests that switchgrass germplasm can be classified into relatively few functional gene pools to represent the eastern two-thirds of the United States (Casler et al. 2007a, b). Eight gene pools are proposed to represent four groups of hardiness zones within each of the Great Plains region and Eastern Forest biomes east of the Mississippi River (Figure 22.2). The choice of eight gene pools was somewhat arbitrary and the boundaries between gene pools are fluid to promote flexibility. Some hardiness zones have been combined because it is impractical and unnecessarily reductionist to develop and/or recommend switchgrass germplasm for one hardiness zone. The east-west discrimination of the historical Great Plains and Eastern Forest biomes is based largely on precipitation and humidity, recognizing that eastern germplasm may have reduced drought tolerance relative to western germplasm whereas western germplasm may not have the disease resistance required in the more humid eastern region. For example, biomass yields of Cave-in-Rock tend to be reduced relative to that of other cultivars as this cultivar is moved west of its origin, particularly as moisture becomes limiting (Hopkins et al. 1995a; Casler and Boe 2003; Berdahl et al. 2005). Conversely, Sunburst, originating in the northern Great Plains, tends to have

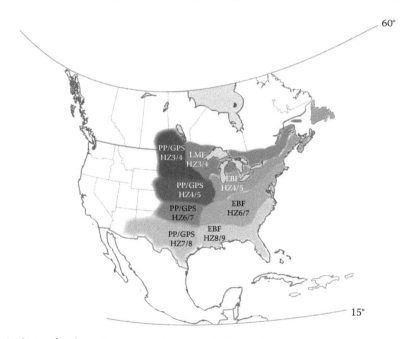

FIGURE 22.2 (See color insert) Proposed gene pools for deployment of regionally adapted switchgrass germplasm and cultivars for use in breeding programs or in conservation and restoration projects. PP, prairie parkland; GPS, Great Plains steppe; LMF, Laurentian mixed forest; EBF, eastern broadleaf forest. (From Bailey, R.G., *Ecoregions: The Ecosystem Geography of the Oceans and Continents,* Springer-Verlag Inc., New York, 1998; Bailey, R. G., *Ecosystem Geography,* 2nd ed, Springer-Verlag Inc., New York, 2009.) HZ, USDA hardiness zone. (From Cathey, H.M., USDA plant hardiness zone map, USDA Misc Pub No 1475. U.S. Department of Agriculture, Washington, DC, 1998; U.S. National Arboretum, 1990. Available at www.usna. usda.gov/Hardzone/ushzmap.html)

reduced relative biomass yields as it is moved east into the historic Eastern Forest biome (Hopkins et al. 1995a; Casler and Boe 2003).

Gene pools are defined so that germplasm originating within any region can be utilized at other sites within that region without concern about lack of adaptation or fear of contaminating local switchgrass populations with exotic genetics (Vogel et al. 2005). The switchgrass gene pool concept is intended to apply to any use of switchgrass germplasm, including development of germplasm pools and breeding populations, parental materials for cultivar development, and deployment of ecotypes and/or natural-track cultivars for use in conservation and restoration projects. The northern gene pools of the United States likely apply to regions of southern Canada that share similar climatic and edaphic conditions, as evidenced by adaptation of numerous switchgrass cultivars in eastern Canada (Madakadze et al. 1998, 1999b).

22.4 PHYSIOLOGY AND GROWTH

The germination and growth of switchgrass seedlings are reduced at soil temperatures less than 20°C (Hsu et al. 1985a, 1985b). Consequently, the recommended seeding dates for switchgrass correspond to those for maize (*Zea mays* L.). Switchgrass seedlings have the panicoid seedling morphology and seedlings emerge by elongation of the mesocotyl or the subcoleoptile internode, which pushes the crown node and the coleoptile, which stays short, to the soil surface (Hoshikawa 1969; Newman and Moser 1988; Tischler and Voigt 1993). When the coleoptile reaches the soil surface, light induces the mesocotyl to stop elongating. Adventitious roots, which are necessary for seedling and plant survival, arise from the crown node at the base of the coleoptile near the soil surface. Planting seed deeper than 1 cm can adversely affect field establishment because more seedling reserves are required for mesocotyl elongation. Dry soil conditions at the soil surface can prevent seedlings from developing adventitious roots to ensure survival, therefore planting dates should be targeted for periods when the probability of rain is high and soil temperatures are sufficiently high for germination (Smart and Moser 1997). Planting too late in the summer will result in stand failures because seedlings will not have adequate time to become established, transition from juvenile to adult phase, and develop the root reserves necessary to become perennial.

Within 6 weeks of emergence several tillers may be produced. Growth of switchgrass in the establishment year depends upon soil moisture, fertility, and competition from weeds and other plants. Switchgrass does not require vernalization to induce flowering. Under optimum conditions, switchgrass will produce seed in the establishment year but flowering occurs several weeks later than in following years. This delay in flowering and seed ripening is likely due to the ineffectiveness of floral induction before the transition from juvenile to adult phase (Poethig 2003). Growth of switchgrass during the establishment year is slow relative to many other grasses, largely because many plant resources are being devoted to development of an extensive root system.

New growth in the spring is initiated from axillary buds on the stem, crown, or rhizomes (Heidemann and Van Riper 1967; Sims et al. 1971; Beaty et al. 1978). The relative amount of new growth from each type of bud varies with ecotype and strain. Bunch types apparently produce new tillers from both crown buds and rhizomes (Heidemann and Van Riper 1967; Sims et al. 1971) but sod-forming plants produce new tillers primarily from rhizomes (Beaty et al. 1978). Depending upon the physiological stage and environmental conditions, new growth may be initiated after harvest from all three types of buds. Plants with short rhizomes produce bunch-type plants, which can be pushed above the soil line by roots, whereas sod-forming plants have longer rhizomes (Beaty et al. 1978). The growth and development of a switchgrass plant depends upon its genotype and the location where it is evaluated. The development of switchgrass is location dependent because flowering depends on photoperiod as discussed previously but also growing-degree-days (GDD) which measure accumulated heat or photosynthesis energy.

The physiological development of switchgrass as determined using a maturity staging system (Moore et al. 1991) is highly correlated to day-of-the-year (DOY) and GDD in temperate climates such as the

Great Plains of the United States (Sanderson and Wolf 1995; Mitchell et al. 1997, 2001). In the Central Great Plains, photoperiod as measured by DOY was more predictive of physiological development than GDD, indicating that photoperiod is the primary determinant of switchgrass development but photosynthesis or heat units can modify the developmental response (Mitchell et al. 1997).

A population of switchgrass plants will have populations of tillers at different stages of development (Mitchell et al. 1997). Genetically broad-based populations will have some plants at anthesis over a 3-week period (Jones and Brown 1951). Florets in an individual panicle will be undergoing anthesis for up to 12 d (Jones and Newell 1946). Peak pollen shedding periods are from 10:00 to 12:00 h or from 12:00 to 15:00 h depending upon environmental conditions (Jones and Newell 1946). Heading dates for cultivars are typically expressed as population means. Because flowering time varies among individual genotypes, the development of ripe seed is also variable within a population or cultivar.

The stem bases, roots, and rhizomes are the primary sites of nonstructural carbohydrate storage. Starch is the primary and most dynamic nonstructural carbohydrate in switchgrass stem bases and rhizomes (Smith 1975). Nonreducing sugars, primarily sucrose, are secondary in importance to starch and fluctuate in a similar manner during the growing season. Total nonstructural carbohydrates (TNC) concentrations in the stem bases of unharvested plants are greatest at the beginning and end of the growing season. Stem base TNC concentrations reach the lowest levels at the time of tiller elongation or when regrowth is initiated following harvest (Smith 1975). A recent fertilization study in which N concentration of biomass was monitored indicates that switchgrass may actively transport N and nonstructural carbohydrates from above ground biomass to stem bases and roots after anthesis but before a killing frost (Vogel et al. 2002a). Harvest time is critical for adequate survival of switchgrass swards, with consistently early harvests reducing switchgrass stands, most likely due to inadequate time for cycling of nitrogen and soluble carbohydrates to storage organs (Casler and Boe 2003).

Switchgrass requires the establishment of a symbiotic relationship with arbuscular mycorrhizal fungi (AMF) in its roots to become established and persist (Brejda et al. 1998). Rhizosphere microflora from numerous native prairies and old seeded stands of switchgrass were effective in enhancing seedling growth of switchgrass in greenhouse trials (Brejda et al. 1998). A field study on two different soils demonstrated that indigenous AMF in cultivated fields of the central Great Plains establish a symbiotic relationship with switchgrass and that inoculation offers little potential to increase switchgrass production unless the soils have been severely degraded (Brejda 1996). Significant interactions between AMF isolates and soil types suggest some differential adaptation of AMF isolates to different soils (Clark et al. 2005).

Switchgrass is relatively drought tolerant with a deep root system likely contributing to its ability to draw water from deep aquifers. Water-use efficiency (WUE) of switchgrass was 1.8–3.6 mg biomass/g water across several soils and nitrogen rates in Pennsylvania (Stout 1992). Switchgrass WUE increased by 30% with an increase in N fertilizer from 0 to 84 kg N/ha. In that study, the WUE of switchgrass was 5.7 times greater than the WUE of orchardgrass (*Dactylis glomerata* L.). In the Great Plains of the United States, the WUE of switchgrass ranged from 3.5 to 5 mg biomass/g water across several locations (Kiniry et al. 2008). The WUE of switchgrass was 1.8–5.0 times greater than the WUE of maize used for grain production, but similar to the WUE of maize biomass production (Kiniry et al. 2008).

22.5 MANAGEMENT

22.5.1 ESTABLISHMENT

Recommended seeding rates are 200–400 pure live seeds (PLS) per m^2 (Vogel 1987). Establishment-year stands with 20 or more plants/m^2 will produce harvestable forage the year of establishment if weeds are controlled and can be in full production the year after establishment (Vogel 1987; Vogel and Masters 2001). Establishment-year stands of 10 plants/m^2 are adequate but will require one or more years to achieve full production yields. Stands of less than 10 plants/m^2 may need to be overseeded or re-seeded. An on-farm study indicated that establishment-year stands of 40% or greater, determined by a frequency grid, can be considered adequate for successful establishment (Schmer et al. 2006).

The minimal germination temperature for switchgrass is 10°C (Hsu et al. 1985a). Temperature-gradient-table studies with several switchgrass cultivars and seedlots demonstrated that near-maximal germination was obtained from 19 to 36°C and optimal germination was between 27 and 30°C (Dierberger 1991). Optimum germination temperatures for switchgrass may be lower than those for seedling development (Panciera and Jung, 1984). Seedling growth of switchgrass at 20°C is much slower than at 25 or 30°C (Hsu et al. 1985b). Although seedlings develop slowly, planting in early spring may be advantageous even though the soil is cold if the seed lot being used has dormant seed. The cold soil may aid in breaking dormancy. Best stands in Iowa were obtained when planted at early to mid-spring (Vassey et al. 1985). In the northeastern United States, a planting window of 3 weeks before and 3 weeks after the recommended maize planting date has been suggested (Panciera and Jung 1984). This general guideline for time of planting would be suitable in most areas where switchgrass is adapted. In some areas "dormant plantings" are made very late in the fall, late enough that the seed will not germinate. The seed remains dormant during winter and the cool moist spring conditions result in a natural cold stratification. Switchgrass should not be planted in late summer because it may not have time to develop sufficient cold tolerance before onset of winter.

Planting seed too deeply often leads to seeding failures with switchgrass and other small seeded warm-season grasses. Switchgrass requires a firm seedbed, allowing the drill to plant the seed approximately 1–2 cm deep. No-till seeding into crop residues or chemically killed sods is often very effective (Samson and Moser 1982). Corrective applications of P or K should be made before seeding but N applications are generally not made until the grass is established because it will stimulate excessive weed growth during the seeding year.

Physiological seed dormancy of some cultivars and seedlots of switchgrass can result in seeding failure. Although alive, dormant seed will not germinate under normally suitable conditions. Simple dormancy will be broken if the seed is aged long enough or if it is given cold treatments or cold stratified to break dormancy (Zheng-Xing et al. 2001). The normal germination test carried out according to Association of Official Seed Analysts procedures (AOSA 1988) includes a period of cold stratification where seed are allowed to imbibe water and are chilled at 4°C for 2–4 weeks to break dormancy. The germination percentage on the seed tag represents the percentage of viable seed but does not represent the actual amount of seed that will germinate upon planting because of dormancy. Producers should conduct a germination test without chilling if they suspect dormant seed and want to determine the percentage of seed that will germinate when planted. With time, much of the dormancy will be broken if seed is stored for 1 year at room temperatures. Seed stored for 3 or more years at room temperature may result in poor stands due to decreased vigor (Vogel 2002). Dormancy of switchgrass seed can be broken by stratification by wet chilling, but drying can cause some of the seed to revert to a dormant condition (Zhang-Xing et al. 2001). Extended stratification (>42 days) significantly reduced the percentage of switchgrass seed that reverted to a dormant condition after drying (Zhang-Xing et al. 2001). It must be emphasized that switchgrass seed should have high germination (>75%) and should not be older than 3 years to ensure successful establishment. Old seed can have good laboratory germination but may have poor seedling vigor and fail to produce acceptable stands under field conditions.

Variation exists among and within cultivars for seed size. Smart and Moser (1999) graded switchgrass seed into lots differing in seed weight and evaluated the seed lots in field plantings. Seedlings from the heavy seed had greater germination, earlier shoot and adventitious root growth than seedlings from light seed but growth and development were similar 8–10 weeks after emergence.

Weed competition is one of the major reasons for stand failure of switchgrass. Seedlings do not develop rapidly until conditions are warm, which is also the same time that annual weeds develop.

Most dicot weeds can be controlled with 2,4-D (2,4-dichlorophenoxyacteic acid). Generally, 2,4-D should be applied after switchgrass seedlings have approximately four to five leaves. Atrazine [6-chloro-N-ethyl-N'-(1-methylethyl)-1,3,5-triazine-2,4-diamine] has been used to improve establishment of switchgrass by controlling broadleaf weeds and C_3 weedy grasses (Martin et al. 1982; Bahler et al. 1984). Switchgrass can metabolize atrazine (Weimer et al. 1988). Acceptable stands of switchgrass could be established at a reduced seeding rate of 107 pure live seed per m^2 when weed interference was reduced following atrazine application at time of planting (Vogel 1987). Imazethapyr {2-[4,5-dihydro-4-methyl-4-(1-methylethyl)-5-oxo-1H-imidazol-2-yl]-5-ethyl-3-pyridine carboxylic acid}, applied at 70 g ai/ha before the grass seedlings emerged, provided excellent weed control and enabled excellent stands of switchgrass to be obtained within 1 year of planting (Masters et al. 1996). The postplant, pre-emergence application of a tank mix of quinclorac (Paramount®; 3,7-dichloro-8-quinolinecarboxylic acid) at 1.1 kg ai/ha plusatrazine at 1.12 kg ai/ha has provided excellent weed control in switchgrass seedlings in Nebraska, North Dakota, and South Dakota (Mitchell, unpublished data). This herbicide treatment controlled broadleaf weeds and weedy grasses and resulted in acceptable stands and high biomass yields. Application of imazapic {2-[4,5-dihydro-4-methyl-4-(1-methylethyl)-5-oxo-1H-imidazol-2-yl]-5-methyl-3-pyridine carboxylic acid} on switchgrass, although effective in some trials, has resulted in significant stand reductions in other tests. The labeled use of imazethapyr and quinclorac on switchgrass as a pre- or postemergent herbicide varies with state or region and year. The efficacy of these herbicides does not change, only the regulations. Maize has been successfully used as a cover crop for switchgrass (Hintz et al. 1998). Atrazine is applied for weed control after both crops are planted. Corn is harvested for grain and is the primary crop the year of establishment. Herbicides should be used only in geographical regions and applications for which they are labeled.

In addition to herbicides that can be used during establishment, other herbicides are available for use on established stands of switchgrass. Switchgrass stands are not affected by metolachlor [2-chloro-N-(2-ethyl-6-methylphenyl)-N-(2-methoxy-1-methylethyl)acetamide] applied at rates needed to control annual weedy grasses (Masters et al. 1996). Commercial products containing both atrazine and metolachlor are labeled for use in seed production in some regions. Metasulfuron (methyl 2-[[[[(4-methoxy-6-methyl-1,3,5-trizin-2-yl)-amino]carbonyl]-amino]sulfonyl]benzoate) and clopyralid (3,6-dichloro-2-pyridinecarboxylic acid) plus 2,4-D can be used for weed control in established stands (Anonymous 2002).

22.5.2 FERTILITY

Switchgrass can tolerate low fertility conditions but it responds to fertilizer (Rehm et al. 1976; Jung et al. 1988). It responds to N fertilization with significant increases in forage and biomass yield (McMurphy et al. 1975; Rehm et al. 1976, 1977; Perry and Baltensperger 1979; Hall et al. 1982; Rehm 1984; Madakadze et al. 1999a; Sanderson et al. 1999; Vogel et al. 2002a). Recommended N fertilization rates vary with location and are primarily dependent upon precipitation, cultivar, and harvest management. In the eastern Great Plains and the Midwest, United States, recommended annual rates of N vary from 90 to 110 kg/ha when switchgrass is managed for hay or pasture whereas further west, where there is less precipitation, rates of 45–70 kg/ha are used. When switchgrass is managed for optimal biomass production in the Midwest, approximately 10–12 kg/ha N needs to be applied for each Mg/ha of biomass yield (Vogel et al. 2002a). At fertility rates above this level, nitrates accumulated in the soil profile. In South Dakota Conservation Reserve Program (CRP) lands dominated by switchgrass, the application of 56 kg N/ha increased total biomass, but there was no benefit to applying more N (Mulkey et al. 2006). Switchgrass may respond to P fertilization if the availability of P in the soil is low (Rehm et al. 1976; Rehm 1984). Switchgrass and other C_4 grasses should be fertilized in late spring when they are initiating growth. Early spring fertilization will stimulate invasion by C_3 grasses and forbs (Rehm et al. 1976). Nitrogen fertilization increases the herbage protein concentration (Rehm et al. 1977; Perry and Baltensperger 1979; Rehm 1984; Vogel et al. 2002a) and in vitro dry matter digestibility (IVDMD) of switchgrass (Perry and Baltsenberger 1979; George et al. 1990).

Fertilizer application rates for switchgrass should be based on the difference between the requirements of the crop and available soil N. However, time of harvest will have a significant impact on the nutrients removed in the harvested biomass. For example, harvesting a switchgrass field at anthesis that produces 11 Mg/ha of DM with a N concentration 1.2% N will remove about 130 kg of N/ha, whereas material harvested after a killing frost may remove only half of that amount of N/ha. Harvesting after a killing frost may reduce N application by 30–40%. Sampling soils to determine available N for switchgrass production must be taken to a depth of 1.5–2 m because of the soil mineralization potential of some soils, atmospheric N deposition, residual soil N from previous crops that may be distributed deep in the soil profile, and the deep rooting capability of switchgrass (Mitchell et al. 2008).

On a strongly acid (pH 4.3–4.9), low P soil, unfertilized switchgrass and big bluestem (*Andropogon gerardii* Vitman) produced 50% as much forage as that receiving a low level of nutrients (Jung et al. 1988). When P declined from 35 to 5 mg/kg, switchgrass yields declined 12% compared to C_3 grasses which declined 35% (Panciera and Jung 1984). On acidic, low water-holding capacity soils, first-cut switchgrass yields were two to three times greater, and four times greater than for tall fescue on sites with N and without N, respectively. Nitrogen-use efficiency was greater for switchgrass than for tall fescue (Staley et al. 1991). The timing of N application is critical in the maintenance of switchgrass stands. If N is applied too early in the spring or in the previous autumn, cool-season plants will utilize it because switchgrass is not active. The stimulated C_3 invaders will increase rapidly and utilize the soil moisture. Later, during the period of switchgrass growth, soil moisture will be depleted and the vigor of switchgrass plants will decline and stands will be invaded by additional C_3 plants which can result in the conversion of a switchgrass pasture into a mixed species cool-season pasture.

22.5.3 Harvest Timing and Frequency

Cellulosic biomass of herbaceous plants can be used as a feedstock for the production of liquid fuels such as ethanol (Lynd et al. 1991) and switchgrass has been identified as a promising species for development into a herbaceous biomass fuel crop in the United States (Vogel 1996; Sanderson et al. 2007). Switchgrass has an array of desirable energy, conservation, environmental, and economic attributes for its use as a bioenergy crop (McLaughlin et al. 2002). These include broad adaptation, high yields on marginal and erosive croplands, harvestability with conventional forage equipment, a very positive energy balance, and relatively easy seed processing.

Several trials have been conducted in the United States and other countries to optimize harvest timing and frequency. In general, a single harvest when switchgrass is fully headed gives the highest yields (Madakadze et al. 1999a, b; Sanderson et al. 1999; Christian et al. 2002; Vogel et al. 2002a). Biomass yield continues to increase up to anthesis, after which biomass yield decreases up to 10–20% before killing frost (Vogel et al. 2002a). There are circumstances, some cultivars at some locations, in which two harvests provide higher biomass yields than one harvest, but the extra fossil fuels required to conduct two harvests may not warrant a two-harvest management system. Harvests after a killing frost usually result in decreased biomass yields but may require lower inputs of N fertilizer, because the plant is able to utilize N mobilized into roots for storage during winter and recovery the following spring. Depending on location and cultivar, biomass yields of the best-adapted cultivars ranged from 10 to more than 20 Mg/ha.

Optimal harvest management for switchgrass use in combustion conversion systems may require delaying harvest until spring when most of the minerals have leached from the plant. Biomass yield reductions during the winter averaged 40% in Pennsylvania (Adler et al. 2006), but this management system may be capable of utilizing some internal N cycling, helping to reduce the N fertilizer requirements (Lowenberg-DeBoer and Cherney 1989). Most of the yield reduction associated with spring harvest was due to harvest losses during baling, suggesting that improvements in harvest machinery could reduce these losses. Losses in plant biomass during late autumn and winter are generally associated with translocation of nitrogenous compounds, soluble carbohydrates, and minerals into underground storage structures and may result in a long-term benefit in a one-harvest biomass production system, promoting a balance between maximal biomass yield and stand longevity while providing a more favorable product for energy from combustion.

22.5.4 SEED PRODUCTION

Management of switchgrass for seed production is based on practices initially recommended by Cornelius (1950) for the Great Plains, subsequent research in other areas of the United States, and on anecdotal results of seed producers. Cornelius (1950) reported that cultivated seed production fields produce more and higher quality seed than from native prairies; row plantings produce more seed than solid stands; fertilization and weed control are necessary for good seed production; and spring burning of seed fields usually improves seed yields. In the central Great Plains where most of the commercially available switchgrass seed is produced, the seed fields are usually planted in rows spaced about 1 m apart, and are fertilized each spring with 50–110 kg/ha N after the fields are burned and cultivated to maintain the grass in rows. In Iowa, Cave-in-Rock had higher seed yields when grown in narrow rows spaced 20 cm apart than in wider rows spaced 1 m apart (Kassel et al. 1985). In contrast, the cultivars Blackwell and Pathfinder had higher seed yields in wide rows. Nitrogen fertilizer significantly increases seed yields in Iowa (George et al. 1990). Phosphorus should be applied when soil tests indicate available soil P is low. Some seed producers irrigate, but many seed fields in the eastern Great Plains are not irrigated. Switchgrass seed, in contrast to seed of many native grasses, is heavy and smooth and is easily combined and cleaned with conventional combines and cleaning equipment (Cornelius, 1950; Wheeler and Hill 1957). Seed is usually harvested by direct combining. Grazing switchgrass seed fields early in the season reduced seed yields in the Midwest United States (George et al. 1990; Brejda et al. 1994). Seed yields in an Iowa study ranged 200–1,000 kg/ha (Kassel et al. 1985). In Missouri, seed yields ranged from 460 to 700 kg/ha (Brejda et al. 1994). The difference in cultivar response was due to differences in lodging.

22.6 CARBON, ENERGY, AND ECONOMIC BUDGETS

The sustainability of switchgrass for bioenergy crops will be determined using carbon, energy, and economic budgets of the feedstock. Feedstocks must be profitable, have high net energy yields, and be either neutral or positive with respect to the environment to be viable and sustainable (Casler et al. 2009).

22.6.1 CARBON SEQUESTRATION

Switchgrass has the potential to extract carbon dioxide from the atmosphere and sequester it in soil. Frank et al. (2004) reported that soil C increased at a rate of 1.01 kg C/m^2 per year, and switchgrass plantings in the northern Great Plains have the potential to store significant quantities of soil organic carbon (SOC). Liebig et al. (2005) reported that switchgrass grown in North Dakota stored 12 Mg/ha more SOC in the 30- to 90-cm depth than a cropland paired field experiment. They concluded that switchgrass effectively stores SOC not just near the soil surface, but at greater depths where C is less susceptible to mineralization and loss. Lee et al. (2007) reported that switchgrass grown in South Dakota CRP lands stored SOC at a rate of 2.4–4 Mg/ha per year at the 0- to 90-cm depth. In a 5-year field study conducted on 10 farms in Nebraska, South Dakota, and North Dakota, SOC increased significantly at 0–30 cm and 0–120 cm soil depths, with an average increase in SOC of 1.1 and 2.9 Mg C/ha per year, respectively (Liebig et al. 2008). However, they noted that change in

SOC varied across sites and ranged from −0.6 to 4.3 Mg C/ha per year for the 0- to 30-cm depth. For example, on four farms in Nebraska, change in SOC for the 0- to 120-cm depth averaged 2.9 Mg C/ha per year. The variation in SOC change reiterated the importance of long-term environmental monitoring sites in major agro-ecoregions.

In addition to increasing soil C, growing switchgrass may increase wildlife habitat, increase landscape and biological diversity, increase farm revenues, and return marginal farmland to production (Sanderson et al. 1996; McLaughlin and Walsh 1998; McLaughlin et al. 2002; Roth et al. 2005). Not harvesting some switchgrass each year would increase the habitat value for grassland bird species that require tall, dense vegetation structure (Roth et al. 2005).

22.6.2 Energy Balance

Energy produced from renewable carbon sources is held to a different standard than energy produced from fossil fuels, in that renewable fuels must have highly-positive energy values. The energy efficiency and sustainability of ethanol produced from grains and cellulosics has been evaluated using net energy value (NEV), net energy yield (NEY), and the ratio of the biofuel output to petroleum input [petroleum energy ratio (PER)] (Schmer et al. 2008) An energy model using estimated agricultural inputs and simulated biomass yields predicted switchgrass could produce greater than 700% more output than input energy (Farrell et al. 2006). These modeled results were validated in a multifarm, field-scale research in the central and northern Great Plains, United States. Switchgrass fields on ten farms in Nebraska, South Dakota, and North Dakota produced 540% more renewable fuel (NEV) than nonrenewable fuel consumed over a 5-year period (Schmer et al. 2008). The estimated on-farm NEY was 60 GJ/ha per year (Schmer et al. 2008), which was 93% greater than human-made prairies and 652% greater than low-input switchgrass grown in small plots in Minnesota (Tilman et al. 2006). The 10 farms and five production years had a PER of 13.1 MJ of ethanol for every MJ of petroleum input, and produced 93% more ethanol per ha than human-made prairies and 471% more ethanol per ha than low-input switchgrass in Minnesota (Schmer et al. 2008). Average greenhouse gas (GHG) emissions from switchgrass-based ethanol in this study were 94% lower than estimated GHG emissions from gasoline (Schmer et al. 2008). In simulated production trials in Wisconsin, switchgrass produced the most net energy, followed by an alfalfa-corn rotation and then continuous corn (Vadas et al. 2008). Producing switchgrass for bioenergy is an energetically positive and environmentally sustainable production system for the central Great Plains and Midwest.

The implementation of switchgrass-based bioenergy production systems will require the conversion of marginal land from annual row crops to switchgrass production and could exceed 10% in some regions depending on the yield potential of the switchgrass strains. In a 5-year study in Nebraska, the potential ethanol yield of switchgrass averaged 3474 L/ha and was equal to or greater than the potential ethanol yield of no-till corn (grain + stover) on a dry-land site with marginal soils (Varvel et al. 2008). Removing an average of 51% of the corn stover each year reduced subsequent corn grain yield, stover yield, and total biomass yield. Growing switchgrass on these marginal sites will likely enhance ecosystem services more rapidly and significantly than on more productive sites.

22.6.3 Economics

Switchgrass is an economically feasible source for cellulosic ethanol. A field-scale study using known farm inputs and actual harvested switchgrass yields conducted on 10 farms over 5 years in Nebraska, South Dakota, and North Dakota determined switchgrass could be delivered at the farm gate for $54/Mg (Perrin et al. 2008). They concluded that the development of new cultivars improved production practices, and an expanded market for switchgrass may reduce the farm-gate cost by as much as 20% (Perrin et al. 2008). Large quantities of switchgrass could be delivered at the farm gate for \$40–45/Mg (Perrin et al. 2008). Assuming a switchgrass farm-gate cost of \$40–54/Mg and conversion of 0.329 L ethanol/kg of switchgrass, the farm-gate feedstock cost would range from \$0.12 to 0.16/L.

22.7 BREEDING AND SELECTION

22.7.1 ECOTYPIC VARIATION

The presence of obvious ecotypic variation has been a strong driver of many switchgrass activities because many of the original collections were made from prairie remnant plants. Although the upland and lowland types form an obvious visual dimorphism, there is a wealth of ecotypic variation within both upland and lowland types. Early plant collectors recognized morphological variations among remnant prairies and sought to collect accessions that would capture much of this variation. Many of the early switchgrass cultivars were simple seed increases of random plants collected from a single source-identified prairie remnant, what are today termed "natural-track" cultivars. A modest amount of selection occurred in determining which of these accessions would be elevated to cultivar status and released to the public. This was accomplished by common-garden experiments in which numerous accessions were evaluated for many years, sometimes followed by more extensive testing at multiple locations. In a few cases, some selection within the strain occurred for vigorous plants and/or traits that would allow for adequate seed production.

These early selection efforts utilized genetic variability among accessions or strains of switchgrass, which is relatively easily observed in replicated experiments (Eberhart and Newell 1959; Hopkins et al. 1995b; Casler 2005). Taken together, these natural-track cultivars provide a fairly effective representation of the morphological and physiological diversity associated with the diverse geography and climates found in the eastern two-thirds of the United States (Table 22.1). Obvious from Table 22.1 and from investigation of switchgrass germplasm resources, such as those present in the USDA National Plant Germplasm System, Germplasm Resources Information Network (GRIN; http://www.ars-grin.gov/npgs/index.html) is the relative difference in frequency of upland vs. lowland accessions, the latter being relatively rare in prairie remnants (Hultquist et al. 1997; Casler 2005).

Lowland ecotypes of switchgrass tend to be more southern adapted than upland ecotypes, as a group. Because of their late heading and ability to maintain photosynthetically active tissue through the longer growing season, lowland ecotypes generally have higher biomass yield than upland ecotypes at southern locations located in USDA hardiness zones 6–9 (Sanderson et al. 1999; Fike et al. 2006b). Lowland ecotypes had 44% higher biomass yield than upland ecotypes under a one-harvest management, whereas this advantage was reduced to 13% under a two-cut management (Fike et al. 2006b). The upland ecotypes used in this study were more closely matched in phenology to the photoperiods of these test locations. Because they were near anthesis at the time of first harvest, there was sufficient time to allow a second period of vegetative development, unlike the lowland ecotypes for which first harvest occurred before their first period of vegetative development was completed (Fike et al. 2006b). Even though the lowland ecotypes had the highest biomass yields, their use in a geographic region considerably north and east of their origin is likely responsible for their more inconsistent responses to harvest management and other increased inputs compared to upland ecotypes (Fike et al. 2006a).

With the recent expansion of breeding and selection activities on switchgrass in the United States, upland-lowland intermediate types have been found with increasing frequency. The exact origin and nature of these accessions is not known. In some cases, they occur sympatrically with obvious upland ecotypes, suggesting their possible hybrid origin (Hultquist et al. 1997). Cave-in-Rock appears to be an intermediate ecotype, sharing traits of both classic upland and lowland phenotypes. Interestingly, Cave-in-Rock is an octaploid, suggesting a fairly complex genetic history if it indeed arose as a hybrid between tetraploid upland and lowland plants. Analysis of DNA from "Miami" switchgrass also provides conflicting data for classification, with cpDNA suggestive of the

TABLE 22.1

Switchgrass Cultivars Representing Various Habitats in the Central and Eastern United States, Largely Representing Local Ecotypes with Minimal Selection for Plant Traits

Cultivar	Ecotype	Ploidy	Year	Origin	Adapted to USDA Hardiness Zones
Alamo	Lowland	4x	1978	Southern Texas	6, 7, 8, 9
Kanlow	Lowland	4x	1963	Northern Oklahoma	6, 7
Miami	U/L[a]	4x	1996	Southern Florida	9, 10
Pangburn	Lowland	4x	NA	Arkansas	6, 7
Dacotah	Upland	4x	1989	Southern North Dakota	2, 3, 4[a]
Falcon	Upland	4x	1963	New Mexico	4, 5, 6
Grenville	Upland	NA	1940	Northeastern New Mexico	4, 5, 6
High Tide	Intermediate[b]	NA	2007	Northeastern Maryland	5, 6, 7
KY1625	Upland	4x	1987	Southern West Virginia	5, 6, 7
Blackwell	Upland	8x	1944	Northern Oklahoma	5, 6, 7
Caddo	Upland	8x	1955	Central Oklahoma	6, 7
Carthage	Upland	NA	2006	North Carolina	5, 6, 7
Cave-in-Rock	Upland	8x	1973	Southern Illinois	5, 6, 7
Forestburg	Upland	8x	1987	Eastern South Dakota	3, 4
Nebraska 28	Upland	8x	1949	Northeast Nebraska	3, 4
Shelter	Upland	8x	1986	Central West Virginia	4, 5, 6

NA, information not available.

[a] Classified as upland on the basis of chloroplast DNA markers and lowland on the basis of nuclear DNA markers. (Gunter, L.E., Tuscan, G.A., and Wullshcleger, S.D., *Crop Sci.*, 36, 1017–1022, 1996.)

[b] Intermediate type on the basis of a mixture of phenotypic traits of upland and lowland ecotypes Unequivocal classification is not possible at this time.

upland type and nuclear DNA suggestive of the lowland type (Gunter et al. 1996). As of this writing, only a tiny fraction of switchgrass germplasm collected in prairie remnants has been characterized for phenotypic traits that distinguish upland from lowland phenotypes, DNA content for ploidy classification, or DNA markers that would help to elucidate origins, genetic nature, and potential breeding value of these putative intermediate types.

In contrast to the cultivars that derive from seed increases of prairie remnant populations or that have undergone very little interplant selection (Table 22.1), there are relatively few cultivars derived from intensive breeding and selection (Table 22.2). Despite their improvement status, bred cultivars of switchgrass are generally no more than three to four cycle or generations of selection removed from wild germplasm. As such, DNA marker profiles of prairie remnant populations and bred cultivars are remarkably similar, showing no separation or discrimination (Figure 22.3). Switchgrass is still a wild and undomesticated plant.

22.7.2 BREEDING METHODS

Switchgrass breeding originates within populations that are direct targets for improvement, relying on genetic variability for target traits within the population of interest. Because switchgrass has two main ploidy levels, tetraploid and octaploid, that are largely cross incompatible (Martinez-Reyna and Vogel 2002), breeding populations should be developed from plants on a uniform ploidy level. Breeding populations can be defined broadly or narrowly, depending on the breeder's objective. Because there is generally so much genetic variability within populations from a single native prairie

TABLE 22.2
Improved Switchgrass Cultivars Representing Significant Breeding and Selection Activities

Cultivar	Ecotype	Ploidy	Year	Principal Traits Selected during Cultivar Development	Adapted to USDA Hardiness Zones
Pathfinder	Upland	8x	1967	Biomass yield and vigor	4, 5
Shawnee	Upland	8x	1996	IVDMD, biomass yield	5, 6, 7
Sunburst	Upland	8x	1998	Heavy seeds	3, 4, 5
Trailblazer	Upland	8x	1984	IVDMD, biomass yield	4, 5
Summer	Upland	4x	1963	Earliness, rust resistance	4, 5
BoMaster	Lowland	4x	2006	IVDMD, biomass yield	6, 7, 8
Performer	Lowland	4x	2006	IVDMD, biomass yield	6, 7, 8

IVDMD, in vitro dry matter digestibility.

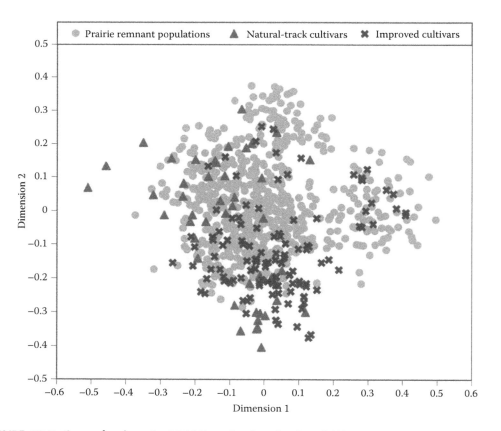

FIGURE 22.3 (See color insert) Multidimensional scale plot of 818 switchgrass plants representing collections made in remnant prairies, public cultivars derived as random seed increases from remnant prairies (natural track), and cultivars improved by selection and breeding, based on random amplified polymorphic DNA (RAPD) markers (From Casler, M.D. et al., *Crop Sci*, 47, 2261–2273, 2007b.)

remnant, populations derived from seed or plants collected at a single site may serve as breeding populations. In this regard, Cave-in-Rock, an accession deriving from one collection site, served as the breeding population for 'Shawnee' (Vogel et al. 1996). Other cultivars, such as Pathfinder, were developed by selection within broader germplasm pools created as composites or synthetics from seed collected at multiple sites that define a target region of interest (Newell 1968b). Multisite composites or synthetics are often thought to have a broader genetic base capable of sustaining genetic gains for more generations of selection and to buffer new cultivars against the possibility of pests or abiotic stresses. DNA marker diversity studies in switchgrass (Casler et al. 2007b; Narasimhamoorthy et al. 2008) and selection experiments in another grass (Burton 1974) suggest that this perception by breeders may not be accurate. Germplasm pools perceived to be either narrow or broad in their genetic diversity are both capable of allowing sustained genetic gains. The polyploid nature of switchgrass serves as a reservoir, storing large amounts of genetic variability.

Genetic variability exists within both broad and narrow gene pools of switchgrass, related to a wide range of traits, including seedling tiller number (Smart et al. 2003a, 2003b, 2004); cell wall composition and forage quality (Vogel et al. 1981; Godshalk et al. 1988); plant height, vigor, and biomass production (Newell and Eberhart 1961; Talbert et al. 1983; Hopkins et al. 1993; Das et al. 2004; Missaoui et al. 2005b; Rose et al. 2007); photoperiod-related traits such as earliness and phytomer number (Van Esbroeck et al. 1998; Boe and Casler 2005; Casler 2005); and biotic or abiotic stress tolerances (Hopkins and Taliaferro 1997; Vogel et al. 2002b; Gustafson et al. 2003). The risk associated with selection for a trait for which little or no genetic variation exists within a population is illustrated by P concentration in Alamo (Missaoui et al. 2005c). P uptake by switchgrass could be improved only by selection for increased biomass yield because of lack of genetic variation for P concentration.

Vogel and Pedersen (1993) and Vogel and Burson (2004) reviewed breeding procedures used to improve switchgrass. Most breeding programs heavily utilize spaced plants as the selection units, in which seeds are germinated in the glasshouse and plants with 3–6 tillers are transplanted to the field on spacings that range from 0.3 to 1.1 m. In most cases, random plants are selected from a bag of seed representing the population to be improved. This is phenotypic selection, sometimes referred to as restricted recurrent phenotypic selection (RRPS) which refers to a number of modifications designed to improve efficiency and the rate of genetic gains (Vogel and Pedersen 1993). Some of these restrictions or modifications include selection for seedling vigor in the glasshouse, removal of some environmental variation in the field using a grid system, and intercrossing selected plants as early and rapidly as possible using excised tillers in the glasshouse (Burton 1974).

One particular aspect of Burton's RRPS, planting extra seedlings in the glasshouse and conducting some form of seedling selection before establishment of field nurseries, may be particularly useful in switchgrass recurrent selection programs. Smart et al. (2003a) demonstrated that seedling tiller number is moderately heritable, creating Cycle-2 populations with mean tiller numbers of 1.2 and 2.0 compared to the base population of 1.6 tillers per plant. Although seedling tiller number did not affect establishment of switchgrass in the field (Smart et al. 2003b), differences in seedling tiller number translated directly to differences in adult-plant tiller numbers (80 vs. 119 tillers per plant). Furthermore, the single-tiller population had a 28% greater leaf elongation rate, 60% more mass per tiller, and 24% higher biomass yield per plant (Smart et al. 2004). Selection of switchgrass seedlings with a single tiller at a defined length of time postgermination appears to be an effective mechanism to improve the efficiency of a field-based recurrent selection program. This is supported by additional studies of adult plants that indicated cultivars with fewer tillers, but more phytomers per tiller and a higher proportion of reproductive tillers, have the highest biomass yield potential (Boe and Casler 2005).

Family selection or genotypic selection methods have also been used for switchgrass improvement. Family selection may involve the use of spaced plants and similar or identical selection protocols as phenotypic selection. In this case, spaced plants are generally arranged in rows where each row represents a family and plants within a row are generally half-sibs of each other, i.e., they each derive from one maternal parent (Vogel and Pedersen 1993). The efficiency of family selection is greatly enhanced by two-stage selection in which the best families are selected based on row means and the best plants are selected from the best rows based on individual-plant observations, i.e., among-and-within-family selection (Vogel and Pedersen 1993; Casler and Brummer 2008). Alternatively, family or genotypic selection may utilize seeded sward plots if family matings allow sufficient seed for plot testing. This breeding method has been rarely utilized in switchgrass breeding at the time of this writing. Implementation of among-and-within-family selection is more challenging when families are seeded into sward plots, because it is impossible to visually distinguish individual genotypes for the second stage of selection. Furthermore, it is not clear that selection of surviving plants from sward plots is advantageous, as it may be relatively inefficient for some species (Casler 2008). As switchgrass swards age, genotypes are lost to interplant competition, stress susceptibilities, and perhaps random mortality. The relative importance of these factors, the rate of genotype loss in switchgrass swards, and the relative fitness of surviving plants (relative to mortal plants) are all important determinants of the efficiency of among-and-within-family selection on sward plots (Casler and Brummer 2008) and all are current unknowns for switchgrass.

Spaced plants are an extremely effective method of collecting data on individual plants and ensuring that the best genotypes can be found and saved for intercrossing once they are identified in the data analysis phase of selection. Spaced plantings are an extremely effective tool for selection based on traits that have moderate to high heritability and are relatively insensitive to interplant competition. Examples include in vitro dry matter digestibility (IVDMD) (Vogel et al. 1981; Casler et al. 2002) and heading date (Van Esbroeck et al. 1998). Conversely, more complex traits that may be highly sensitive to interplant competition, such as biomass yield, may have less predictable selection responses from spaced plants. Selection for increased biomass yield of spaced plants was effective when progeny populations were evaluated as spaced plants (the same trait) (Rose et al. 2007) or as sward plots (a different trait) (Missaoui et al. 2005b). In the case of Rose et al. (2007), selection was more effective under low-input conditions (no supplemental fertilizer or water) compared to high-input conditions. Missaoui et al. (2005b) attributed their gains in biomass yield to a fairly sophisticated approach to controlling interplant competition and spatial variation—use of a honeycomb planting design combined with statistical adjustment for neighbor effects. Selection for increased spaced-plant vigor was also effective for increasing sward-plot biomass yield in WS4U switchgrass, largely by the elimination of unadapted and low-yielding genotypes (Casler 2010). Alternatively, selection for increased biomass yield was ineffective in the EY × FF switchgrass population, due either to lack of genetic variation for forage yield or to a low genetic correlation between forage yield of spaced plants and sward plots (Hopkins et al. 1993).

22.7.3 BREEDING OBJECTIVES

Breeding for improved forage digestibility as measured by IVDMD (Tilley and Terry, 1963) is an effective way to increase switchgrass productivity as measured by beef cattle (*Bos taurus*) production per unit land area (Vogel et al. 1993). Divergent selection was used to develop strains differing in IVDMD from the same base populations. These strains were evaluated in both small plot and grazing trials (Vogel et al. 1981, 1984; Anderson et al. 1988; Ward et al. 1989). On the basis of these trials the high IVDMD strain was released as the cultivar "Trailblazer" (Vogel et al. 1991). In comparison to the control cultivar "Pathfinder" which had similar forage yield and maturity, the single breeding cycle for high IVDMD resulted in the following genetic increases: IVDMD concentration of 40 g/kg, daily live weight gains by beef cattle of 0.15 kg, beef cattle production of 67 kg/ha, and profit of U.S.$59/ha (Casler and Vogel 1999). On the basis of certified seed production, the area seeded to Trailblazer from 1986 to 1997 was over 63,000 ha. The principal area of adaptation for Trailblazer is the central Great Plains of the United States and similar ecoregions. "Shawnee" switchgrass was developed by a single cycle of selection for high IVDMD and high yield from "Cave-in-Rock" (Vogel et al. 1996). It was higher in IVDMD than the parent cultivar with similar biomass yield. Trailblazer and Shawnee are the only switchgrass cultivars developed with improved forage quality.

Three cycles of phenotypic selection for high IVDMD resulted in a linear increase in IVDMD of +1.7%/cycle and an associated linear decrease of –3.3%/cycle in lignin concentration (Casler et al. 2002). Because IVDMD is not a plant trait per se, but defined only by an anaerobic interaction between plant tissue and enzymes secreted by rumen microbes (Tilley and Terry 1963), it is likely that selection acted largely upon lignin concentration and composition. Low-lignin, high-digestibility genotypes tended to also have lower ratios of *p*-coumaric/ferulic acid than high-lignin, low-digestibility genotypes (Sarath et al. 2008). Reductions in lignin concentration appear to have resulted from reductions in cortical fibers and secondary wall thickenings in switchgrass stems (Sarath et al. 2005).

Selection for increased IVDMD has also resulted in a significant decrease in winter survival (Casler et al. 2002). Some families in the high IVDMD populations continue to have high winter survival rates indicating that this apparent genetic correlation may be broken with selection, allowing simultaneous improvements in IVDMD and winter survival (Vogel et al. 2002b). Genetic correlations of forage yield and IVDMD indicate that it should be possible to improve both traits simultaneously (Talbert et al. 1983). The breeding research on improving IVDMD and forage yield demonstrate the need for multiyear evaluation of breeding nurseries in the environments in which the plant materials will be used to ensure that selected plants are exposed to stresses present in normal production environments (Casler et al. 2002; Vogel et al. 2002b).

22.7.4 SWITCHGRASS HYBRIDS

All current cultivars of switchgrass are either open-pollinated seed increases of prairie remnants or synthetic populations of superior plants selected for agronomic performance. In both cases, random intermating occurs among large numbers of plants and there is little opportunity to utilize structural or genomic information that would lead to improved performance, such as hybrid vigor. The partial genetic isolation between upland and lowland ecotypes of switchgrass is reminiscent of complementary heterotic gene pools in maize (*Zea mays* L. ssp. *mays*). Maize breeders created these heterotic gene pools by choosing parental lines simply based on their observed heterotic patterns (Tracy and Chandler 2006). Many generations of selection reinforced and solidified these patterns. Initial crosses of upland × lowland switchgrass have demonstrated an average of 19% midparent heterosis for biomass yield of spaced plants, whereas upland × upland and lowland × lowland hybrids of similar genetic origin showed no heterosis for biomass yield (Martinez-Reyna and Vogel 2008). When evaluated as sward plots, upland × lowland hybrids averaged 30–38% high-parent heterosis (Vogel and Mitchell 2008). There appears to be some natural genetic complementation between upland and lowland ecotypes that results in significant heterosis of these hybrids. These results suggest the need for breeding and selection methods that incorporate efficient methods of evaluating and selecting parents for combining ability. If these results follow the maize model, selection for combining ability could strengthen what already appears to be significant hybrid vigor between upland and lowland ecotypes.

Commercial production of F_1 hybrid switchgrass cultivars will require a mechanism to propagate parental clones in the thousands or tens of thousands. Efficient and repeatable methods for regenerating switchgrass plants from in vitro cultured cells and tissues have been developed (Denchev and Conger 1994; Alexandrova et al. 1996a, b) including a method for regenerating switchgrass plants from cells in suspension culture (Dutta and Conger 1999). Thousands of plants can be transplanted into alternating rows of two parental clones using existing transplanting technologies for horticultural crops. In many cases, this equipment is unused for much of the year and switchgrass hybrid seed production fields can be transplanted at times of the year when vegetable-transplanting equipment is not in use. Use of two heterotic parental clones and physical isolation from other switchgrass will ensure that nearly 100% of the seed harvested from the entire seed production field will be hybrid seed. Because switchgrass is a long-lived perennial and seedling transplants can be planted on spacings that eliminate or minimize interplant competition and the possibility of genetic shifts, hybrid seed production fields can be used for many years, perhaps even the lifetime of a hybrid cultivar.

22.7.5 Transgenic Switchgrass

Transgenic switchgrass plants have been created using bombardment with tungsten-coated particles or Agrobacterium-mediated procedures (Richards et al. 2001; Somleva et al. 2002). Transgenes for herbicide resistance and reporter genes have both been expressed in transgenic switchgrass plants. Controlled crosses between transgenic and nontransgenic plants resulted in the expected expression of both genes in T_1 plants. So far, the greatest limitation to transformation of switchgrass comes from the limited number of genotypes that are capable of plantlet regeneration from tissue culture. Because of this limitation, Alamo is the only switchgrass cultivar that has been successfully transformed and regenerated from culture (Denchev and Conger 1994; Alexandrova et al. 1996a, 1996b; Dutta Gupta and Conger 1999; Richards et al. 2001; Somleva et al. 2002; Mazarei et al. 2008). Because of the climatic limitations of Alamo, deployment of transgenic switchgrass into upland genotypes and to geographic regions to which Alamo is not adapted becomes complicated by the need to transfer transgenes from Alamo to other genotypes. Transfer of transgenes from Alamo to other genotypes will necessarily include large chromosome segments that may be deleterious to northern strains of switchgrass and reduce genetic complementation between upland × lowland hybrids.

Deployment of transgenes in commercial switchgrass cultivars for use in North America will likely require the use of a hybrid system that prevents the introduction of transgenes into wild or natural switchgrass populations. Such a system could be based on the hybrid seed production scheme of Martinez-Reyna and Vogel (2008) in which two parental clones are increased by somatic embryogenesis and transplanted into alternating rows. Using this system, the transgenic parent must be utilized as the female and the nontransgenic parent utilized as the male. In addition, the female transgenic parent must have a sterility system, such as cytoplasmic male sterility, such that the parental clone itself and all of its progeny are male sterile, incapable of releasing transgenic pollen into the wild. Such a system could borrow from the maize hybrid model in which there are fewer rows of male parents relative to female parents, to avoid large increases in seed production costs associated with lack of hybrid seed production on the male parents.

22.7.6 DNA Marker Selection

Development of efficient DNA marker systems and identification of associations between DNA markers and quantitative trait loci (QTL) has been hampered in switchgrass because of the complexity of its genome organization. Existence of diploid ($2n = 2x = 18$) plants would greatly simplify the discovery of efficient and inexpensive markers and the association of those markers with QTL. Only one diploid accession has ever been reported, originating near Chippewa Falls, Wisconsin (Nielsen 1944). In 1996, the site at which this accession was originally collected had been developed into an urban landscape, destroying all native switchgrass, illustrating just how important habitat loss and fragmentation may be to preservation of switchgrass germplasm. More recently, diploid switchgrass plants have been discovered and confirmed, resulting from haploidy in tetraploid seed stocks (Casler and Price 2010, unpublished data; Young et al. 2010).

Although RAPD markers were used for some of the earlier germplasm diversity research, most current efforts are focused on development of expressed sequence tag (EST) markers or EST-SSR (simple sequence repeat) markers (Tobias et al. 2005, 2006; Narasimhamoorthy et al. 2008). Expressed sequence tag markers hold great promise as tools for selection because many EST markers can be traced to functional genes that can be associated with target plant traits and they are amenable to development of single nucleotide polymorphism (SNP) markers.

Strategies for use of DNA markers as selection tools rely on development of predictable and reliable associations between markers and traits that may take on one of several forms. First and simplest would be a SNP marker located within a functional gene or EST combined with phenotypic evaluations of plants containing alternate forms of the polymorphism identifying which allele is favorable. In this case, one nucleotide polymorphism within one gene is focused on one trait, analogous to the brown-midrib genetic mutations involved in the phenylpropanoid pathway and their effect on lignin (Vignols et al. 1995). Second, QTL discovery, using either linkage mapping or association mapping, would identify random DNA markers flanking a QTL of interest. Selection would be based on the flanking markers associated with favorable phenotype, and (hopefully) a small probability of crossover between the flanking markers and the QTL of interest. Third, selection could target several genes for one or more traits, seeking to pyramid multiple favorable genes together into a population using multiple SNP markers. Fourth, marker-assisted recurrent selection (MARS) expands the DNA markers used as selection tools to a broader genomic coverage focused on regions of the genome known to contain QTL of interest, typically based on linkage or association mapping (Johnson 2001, 2004). Least squares and/or maximal likelihood statistical methods are used to develop a marker selection index that is related to plant phenotype in a predictive capacity. Fifth, genome-wide selection goes one step further to saturate the genome with markers, and using best linear unbiased prediction (BLUP) methods to develop a marker index predictive of plant phenotype (Bernardo and Yu 2007).

Any of the above marker selection strategies can be effectively combined with existing selection methodology to implement marker selection for specific plant traits in switchgrass. For phenotypic selection, markers and traits are measured on the same plants to develop the BLUP equations. Plants selected in the first generation are intercrossed to create a new generation for marker evaluation and BLUP selection, eliminating the need for phenotyping in every generation. For among-and-within-family selection, marker scores or indices would be used as the within-family selection criterion on an individual plant basis, whereas biomass yield or other field-based phenotypic traits would be used as the among-family selection criterion.

22.8 SWITCHGRASS AS A BIOENERGY CROP

Designation of switchgrass as an herbaceous model species for bioenergy feedstock development has transformed this species from obscurity to celebrity status. Before the BFDP, switchgrass was used principally for conservation and restoration of tallgrass prairie habitats and livestock production, the latter largely restricted to the Great Plains region of the United States. With the increase in research activity during the BFDP and subsequent USDA-ARS and U.S.-DOE research programs (Bouton 2007; Sanderson et al. 2007), interest in switchgrass research has grown to many public institutions in North America, Europe, and Asia and to some very high-profile commercial research efforts. In 1992, switchgrass breeding was conducted by the USDA-ARS in Lincoln, Nebraska and by South Dakota State University in Brookings, South Dakota. During the BFDP, Oklahoma State University, University of Georgia, and University of Wisconsin (combined with USDA-ARS) initiated new switchgrass breeding programs targeted to different regions and hardiness zones within the United States. During this period, a switchgrass breeding program was also initiated in the private sector near Ottawa, Canada, and a research program focused on molecular genetics was initiated at the University of Tennessee. Since the completion of the BFDP and the rapid expansion of feedstock research in 2005–2008, new breeding and genetics initiatives have been established at Iowa State University, University of Illinois, Texas A&M University, and the Noble Foundation, Ardmore, OK, whereas new private breeding programs have been established at Auburn, AL, and Bryan, TX.

The boom in switchgrass research can be viewed as a double-edged sword. One side of the sword cuts in favor of creating a vast database on basic biology, genetics, and production information to reduce production costs and improve conversion efficiency of switchgrass feedstock to various forms of energy. Vast amounts of funding have brought basic and applied scientists together and formed partnerships between public and private organizations, all working toward the goal of improving production and conversion efficiency. For a plant that, 20 years ago, was little more than an ecofriendly native prairie plant with some applications in livestock agriculture, this represents a social phenomenon on an unprecedented scale. But that brings us to the other side of that sword, which represents all of the other species that are being ignored or have been relegated to "underfunded" status. Meeting societal needs and governmental goals for bioenergy production (Perlack et al. 2005) will require many different crop species and feedstocks, grown on a wide range of soils, habitats, and climatic conditions. Switchgrass cannot, and should not, be the only focus of a feedstock development program for cellulosic bioenergy. There are many other potential candidates for feedstock development that could become as or more important than switchgrass on a regional, habitat-specific, or management-specific basis. Switchgrass is an interesting and amazing plant species that, together with other potential plant species, has huge potential for helping to wean the human race from dependence on fossil fuels.

REFERENCES

Adler PR, Sanderson MA, Boateng AA, Weimer PJ, Jung HG (2006) Biomass yield and biofuel quality of switchgrass harvested in fall and spring. *Agron J* 98:1518–1525

Alexandrova KS, Denchev PD, Conger BV (1996a) In vitro development of inflorescences from switchgrass nodal segments. *Crop Sci* 36:175–178

Alexandrova KS, Denchev PD, Conger BV (1996b) Micropropagation of switchgrass by node culture. *Crop Sci* 36:1709–1711

Anderson B, Ward JK, Vogel KP, Ward MG, Gorz HJ, Haskins FA (1988) Forage quality and performance of yearlings grazing switchgrass strains selected for differing digestibility. *J Anim Sci* 66:2239–2244

Anonymous (2002) *Guide for Weed Management in Nebraska.* University of Nebraska Coop Ext EC-130-D, University of Nebraska, Lincoln, NE

AOSA (1988) Rules for testing seeds. *J Seed Technol* 12:1–122

Bailey, R.G., (1998) *Ecoregions: The Ecosystem Geography of the Oceans and Continents*, Springer-Verlag Inc., New York.

Bailey, R. G., (2009) *Ecosystem Geography*, 2nd ed. Springer-Verlag Inc., New York.

Bahler CC, Vogel KP, Moser LE (1984) Atrazine tolerance in warm-season grass seedlings. *Agron J* 76:891–895

Barnett FL, Carver RF (1967) Meiosis and pollen stainability in switchgrass, *Panicum virgatum* L. *Crop Sci* 7:301–304

Beaty ER, Engel JL, Powell JD (1978) Tiller development and growth in switchgrass. *J Range Manag* 31:361–365

Benedict HM (1941) Effect of day length and temperature on the flowering and growth of four species of grasses. *J Agric Res* 61:661–672

Bennett MD, Smith JB (1976) Nuclear DNA amounts in angiosperms. *Phil Trans Roy Soc Lond B* 334:309–345

Berdahl JD, Frank AB, Krupinsky JM, Carr PM, Hanson JD, Johnson HA (2005) Biomass yield, phenology, and survival of diverse switchgrass cultivars and experimental strains in western North Dakota. *Agron J* 97:549–555

Bernardo R, Yu J (2007) Prospects for genomewide selection for quantitative traits in maize. *Crop Sci* 47:1082–1090

Boe AR, Casler MD (2005) Hierarchical analysis of switchgrass morphology. *Crop Sci* 45:2465–2472

Bouton JH (2007) Molecular breeding of switchgrass for use as a biofuel crop. *Curr Opin Genet Dev* 17:553–558

Brejda JJ, Brown JR, Wyman GW, Schumacher WK (1994) Management of switchgrass for forage and seed production. *J Range Manag* 47:22–27

Brejda JJ, Moser LE, Vogel KP (1998) Evaluation of switchgrass rhizosphere microflora for enhancing yield and nutrient uptake. *Agron J* 90:753–758

Brejda JJ (1996) Evaluation of arbuscular mycorrhiza populations for enhancing switchgrass yield and nutrient uptake. PhD dissertation, University of Nebraska–Lincoln, Lincoln, NE

Brunken JN, Estes JR (1975) Cytological and morphological variation in *Panicum virgatum* L. *Southwest Nat* 19:379–385

Burton GW (1974) Recurrent restricted phenotypic selection increases forage yield of Pensacola bahiagrass. *Crop Sci* 14:831–835

Casler MD (2005) Ecotypic variation among switchgrass populations from the northern USA. *Crop Sci* 45:388–398

Casler MD (2008) Among-and-within-family selection in eight forage grass populations. *Crop Sci* 48:434–442

Casler MD (2010) Changes in mean and genetic variance during two cycles of within-family selection in switchgrass. *BioEnergy Res* 3:47–54

Casler MD, Boe AR (2003) Cultivar x environment interactions in switchgrass. *Crop Sci* 43:2226–2233

Casler MD, Brummer EC (2008) Theoretical expected genetic gains for among-and-within-family selection methods in perennial forage crops. *Crop Sci* 48:890–902

Casler MD, Buxton DR, Vogel KP (2002) Genetic modification of lignin concentration affects fitness of perennial herbaceous plants. *Theor Appl Genet* 104:127–131

Casler MD, Heaton E, Shinners KJ, Jung HG, Weimer PJ, Liebig MA, Mitchell RB (2009) Grasses and legumes for cellulosic bioenergy. In: Wedin WF, Fales SL (eds), *Grassland: Quietness and Strength for a New American Agriculture*. ASA-CSSA-SSSA, Madison, WI, pp 205–219

Casler MD, Stendal CA, Kapich L, Vogel KP (2007b) Genetic diversity, plant adaptation regions, and gene pools for switchgrass. *Crop Sci* 47:2261–2273

Casler MD, Vogel KP (1999) Accomplishments and impact from breeding for increased forage nutritional value. *Crop Sci* 39:12–20

Casler MD, Vogel KP, Taliaferro CM, Ehlke NJ, Berdahl JD, Brummer EC, Kallenbach RL, West CP, Mitchell RB (2007a) Latitudinal and longitudinal adaptation of switchgrass populations. *Crop Sci* 47:2249–2260

Casler MD, Vogel KP, Taliferro CM, Wynia RL (2004) Latitudinal adaptation of switchgrass populations. *Crop Sci* 44:293–303

Cathey HM (1990) USDA plant hardiness zone map, USDA Misc Pub No 1475. U.S. National Arboretum, Agricultural Research Service, U.S. Department of Agriculture, Washington, DC, 1998 U.S. National Arboretum, available at www.usna.usda.gov/Hardzone/ushzmap.html (accessed April 14, 2011)

Christian DG, Riche AB, Yates NE (2002) The yield and composition of switchgrass and coastal panic grass grown as a biofuel in Southern England. *Bioresour Technol* 83:115–124

Clark RB, Baligar VC, Zobel RW (2005) Response of mycorrhizal switchgrass to phosphorus fractions in acidic soil. *Comm Soil Sci Plant Analysis* 36:1337–1359

Cornelius DR (1950) Seed production of native grasses. *Ecol Monogr* 20:1–27

Cornelius DR, Johnston CO (1941) Differences in plant type and reaction to rust among several collections of *Panicum virgatum* L. *J Amer Soc Agron* 33:115–124

Costich DE, Friebe B, Sheehan MJ, Casler MD, Buckler ES (2010) Genome-size variation in switchgrass (*Panicum virgatum*): Flow cytometry and cytology reveal ramp and aneuploidy. *Plant Genome* 3:130–141

Das MK, Fuentes RG, Taliaferro CM (2004) Genetic variability and trait relationships in switchgrass. *Crop Sci* 44:443–448

Denchev PD, Conger BV (1994) Plant regeneration from callus culture of switchgrass. *Crop Sci* 34:1623–1627

Dierberger EM (1991) Switchgrass germination as influenced by temperature, chilling, cultivar, and seed lot. MS Thesis University of Nebraska, Lincoln, NE

Dutta Gupta S, Conger BV (1999) Somatic embryogenesis and plant regeneration from suspension cultures of switchgrass. *Crop Sci* 39:243–247

Eberhardt SA, Newell LC (1959) Variation in domestic collections of switchgrass, *Panicum virgatum*. *Agron J* 51:613–616

Elberson HW, Christian DG, El Bassem N, Bacher W, Sauerbeck G, Alexopoulou E, Sharma N, Piscioneri I, de Visser P, van den Berg D (2003) Switchgrass variety choice in Europe. In: *Switchgrass as An Alternative Energy Crop*, available at http://wwwswitchgrassnlarchivehtm (accessed July 23, 2009), pp 35–41

Farrell AE, Plevin RJ, Turner BT, Jones AD, O'Hare M, Kammen DM (2006) Ethanol can contribute to energy and environmental goals. *Science* 311:506–508

Fike JH, Parrish DJ, Wolf DD, Balasko JA, Green JT Jr, Rasnake M, Reynolds JH (2006a) Long-term yield potential of switchgrass-for-biofuel systems. *Biomass Bioenergy* 30:198–206

Fike JH, Parrish DJ, Wolf DD, Balasko JA, Green JT Jr, Rasnake M, Reynolds JH (2006b) Switchgrass production for the upper southeastern USA: Influence of cultivar and cutting frequency on biomass yields. *Biomass Bioenergy* 30:207–213

Frank AB, Berdahl JD, Hanson JD, Liebig MA, Johnson HA (2004) Biomass and carbon partitioning in switchgrass. *Crop Sci* 44:1391–1396

George JR, Reigh GS, Millen RE, Junczak JJ (1990) Switchgrass herbage and seed yield and quality with partial spring defoliation. *Crop Sci* 30:845–849

Godshalk EB, Timothy DH, Burns JC (1988) Effectiveness of index selection for switchgrass forage yield and quality. *Crop Sci* 28:825–830

Gould FW (1975) *The Grasses of Texas*. Texas A&M University Press, College Station, TX

Gunter LE, Tuscan GA, Wullshcleger SD (1996) Diversity of switchgrass based on RAPD markers. *Crop Sci* 36:1017–1022

Gustafson DM, Boe AR, Jin Y (2003) Genetic variation for Puccinia emaculata infection in switchgrass. *Crop Sci* 43:755–759

Hall KE, George JR, Riedel RR (1982) Herbage dry matter yields of switchgrass, big bluestem, and Indian grass with N fertilization. *Agron J* 74:47–51

Heidemann GS, Van Riper GE (1967) Bud activity in the stem, crown, and rhizome tissue of switchgrass. *J Range Manag* 20:236–241

Hintz RL, Harmony KR, Moore KJ, George RJ, Brummer EC (1998) Establishment of switchgrass and big bluestem in corn with atrazine. *Agron J* 90: 591–596

Hitchcock AS (1951) *Manual of the Grasses of the US*, 2nd ed, USDA Misc Pub 200, U.S. Government Printing Office, Washington DC

Hopkins AA, Taliaferro CM (1997) Genetic variation within switchgrass populations for acid soil tolerance. *Crop Sci* 37:1719–1722

Hopkins AA, Taliaferro CM, Murphy CD, Christian D (1996) Chromosome numbers and nuclear DNA content of several switchgrass populations. *Crop Sci* 36:1192–1195

Hopkins AA, Vogel KP, Moore KJ (1993) Predicted and realized gains from selection for in vitro dry matter digestibility and forage yield in switchgrass. *Crop Sci* 33:253–258

Hopkins AA, Vogel KP, Moore KJ, Johnson KD, Carlson IT (1995a) Genotype effects and genotype by environment interactions for traits of elite switchgrass populations. *Crop Sci* 35:125–132

Hopkins AA, Vogel KP, Moore KJ, Johnson KD, Carlson IT (1995b) Genetic variability and genotype x environment interactions among switchgrass accessions from the Midwestern USA. *Crop Sci* 35:565–571

Hoshikawa K (1969) Underground organs of the seedlings and the systematics of Gramineae. *Bot Gaz (Chicago)* 130:192–203

Hsu FH, Nelson CJ, Matches AG (1985a) Temperature effects on germination of perennial warm-season forage grasses. *Crop Sci* 25:215–220

Hsu FH, Nelson CJ, Matches AG (1985b) Temperature effects on seedling development of perennial warm-season forage grasses. *Crop Sci* 25:249–255

Huang S, Siu X, Haselkorn R, Gornicki P (2003) Evolution of switchgrass (*Panicum virgatum* L) based on sequences of the nuclear gene encoding plastic acetyl-CoA carboxylase. *Plant Sci* 164:43–49

Hultquist SJ, Vogel KP, Lee DJ, Arumuganathan K, Kaeppler SM (1996) Chloroplast DNA and nuclear DNA content variations among cultivars of switchgrass, *Panicum virgatum* L. *Crop Sci* 36:1049–1052

Hultquist SJ, Vogel KP, Lee DJ, Arumuganathan K, Kaeppler SM (1997) DNA content and chloroplast DNA polymorphisms among accessions of switchgrass from remnant midwestern prairies. *Crop Sci* 37:595–598

Johnson L (2001) Marker assisted sweet corn breeding: A model for specialty crops. In: *Proceedings of the 56th Annual Corn Sorghum Industry Research Conference*, December 5–7, 2001, Chicago, IL American Seed Trade Association, Washington DC, pp 25–30

Johnson L (2004) Marker-assisted selection. *Plant Breed Rev* 24:293–309

Jones MD, Brown JG (1951) Pollination cycles of some grasses in Oklahoma. *Agron J* 43:218–222

Jones MD, Newell LC (1946) Pollination cycles and pollen dispersal in relation to grass improvement. *Nebraska Agric Exp Stn Res Bull* 148 Lincoln, NE

Jung GA, Shaffer JA, Stout WL (1988) Switchgrass and big bluestem responses to amendments on strongly acid soil. *Agron J* 80:669–676

Kassel PC, Mullen RE, Bailey TB (1985) Seed yield response of three switchgrass cultivars for different management practices. *Agron J* 77:214–218

Kiniry JR, Lynd L, Greene N, Johnson MVV, Casler MD, Laser MS (2008) Biofuels and water use: Comparison of maize and switchgrass and general perspectives. In: Wright JH, Evans DA (eds), *New Research on Biofuels*. Nova Science Publishers, Hauppauge, NY, pp 1–14

Lee DK, Owens VN, Doolittle JJ (2007) Switchgrass and soil carbon sequestration response to ammonium nitrate, manure, and harvest frequency on Conservation Reserve Program land. *Agron J* 99:462–468

Liebig MA, Johnson HA, Hanson JD, Frank AB (2005) Soil carbon under switchgrass stands and cultivated cropland. *Biomass Bioenergy* 28:347–354

Liebig MA, Schmer MR, Vogel KP, Mitchell RB (2008) Soil carbon storage by switchgrass grown for bioenergy. *BioEnergy Res* 1:215–222

Lowenberg-DeBoer J, Cherney JH (1989) Biophysical simulation for evaluating new crops: the case of switchgrass for biomass energy feedstock. *Agric Syst* 29:233–246

Lu K, Kaeppler SM, Vogel KP, Arumuganathan K, Lee DJ (1998) Nuclear DNA content and chromosome numbers in switchgrass. *Great Plains Res* 8:269–280

Lynd LR, Cushman JH, Nichols RJ, Wyman CF (1991) Fuel ethanol from cellulosic biomass. *Science* 231:1318–1323

Mable BK (2003) Breaking down taxonomic barriers in polyploidy research. *Trends Plant Sci* 8:582–590

Madakadze IC, Coulman BE, Stewart KA, Peterson PR, Samson R, Smith DL (1998) Phenology and tiller characteristics of big bluestem and switchgrass cultivars in a short growing season area. *Crop Sci* 38:827–834

Madakadze IC, Stewart KA, Peterson PR, Coulman BE, Smith DL (1999a) Cutting frequency and nitrogen fertilization effects on yield and nitrogen concentration of switchgrass in a short season. *Crop Sci* 39:552–557

Madakadze IC, Stewart KA, Peterson PR, Coulman BE, Smith DL (1999b) Switchgrass biomass and chemical composition for biofuel in eastern Canada. *Agron J* 91:696–701

Martin AR, Moomaw RS, Vogel KP (1982) Warm-season grass establishment with atrazine. *Agron J* 74:916–920

Martinez-Reyna JM, Vogel KP (2002) Incompatiblity systems in switchgrass. *Crop Sci* 42:1800–1805

Martinez-Reyna JM, Vogel KP (2008) Heterosis in switchgrass: spaced plants. *Crop Sci* 48:1312–1320

Martinez-Reyna JM, Vogel KP, Caha C, Lee DJ (2001) Meiotic stability, chloroplast DNA polymorphisms, and morphological traits of upland x lowland switchgrass reciprocal hybrids. *Crop Sci* 41:1579–1583

Masters RA, Nissen SJ, Gaussoin RE, Beran DD, Stougaard RN (1996) Imidazolinone herbicides improve restoration of Great Plains grasslands. *Weed Technol* 10:392–403

Mazarei M, Al-Ahmad H, Rudis MR, Stewart, Jr. CN (2008) Protoplast isolation and transient gene expression in switchgrass, *Panicum virgatum* L. *Biotechnol J* 3:354–359

McLaughlin SB, De La Torre Ugarte DG, Garten CT Jr, Lynd LR, Sanderson MA, Tolbert VR, Wolf DD (2002) High-value renewable energy from prairie grasses. *Environ Sci Technol* 36:2122–2129

McLaughlin SB, Kszos LA (2005) Development of switchgrass (*Panicum virgatum*) as a bioenergy feedstock in the United States. *Biomass Bioenergy* 28:515–535

McLaughlin SG, Walsh ME (1998) Evaluating the environmental consequences of producing herbaceous crops for bioenergy. *Biomass Bioenergy* 14:317–324

McMillan C (1959) The role of ecotypic variation in the distribution of the central grassland of North America. *Ecol Monogr* 29:285–308

McMillan C, Weiler J (1959) Cytogeography of *Panicum virgatum* in central North America. *Am J Bot* 46:590–593

McMurphy WE, Demman CE, Tucker BB (1975) Fertilization of native grasses and weeping lovegrass. *Agron J* 67:233–236

Missaoui AM, Boerma HR, Bouton JH (2005c) Genetic variation and heritability of phosphorus uptake in Alamo switchgrass grown in high phosphorus soils. *Field Crops Res* 93:186–198

Missaoui AM, Fasoula VA, Bouton JH (2005b) The effect of low plant density on response to selection for biomass production in switchgrass. *Euphytica* 142:1–12

Missaoui AM, Paterson AH, Bouton JH (2005a) Investigation of genomic organization in switchgrass (*Panicum virgatum* L) using DNA markers. *Theor Appl Genet* 110:1372–1383

Missaoui AM, Paterson AH, Bouton JH (2006) Molecular markers for the classification of switchgrass (*Panicum virgatum* L) germplasm and to assess genetic diversity in three synthetic switchgrass populations. *Genet Resour Crop Evol* 53:1291–1302

Mitchell RB, Moore KJ, Moser LE, Fritz JO, Redfearn DD (1997) Predicting developmental morphology in switchgrass and big bluestem. *Agron J* 89:827–832

Mitchell RB, Fritz JO, Moore KJ, Moser LE, Vogel KP, Redfearn DD (2001) Predicting forage quality in switchgrass and big bluestem. *Agron J* 93:118–124

Mitchell RB, Vogel KP, Sarath G (2008) Managing and enhancing switchgrass as a bioenergy feedstock. *Biofuels Bioprod Biorefin* 2:530–539

Moore KJ, Moser LE, Vogel KP, Waller SS, Johnson BE, Pedersen JF (1991) Describing and quantifying growth stages of perennial forage grasses. *Agron J* 83:1073–1077

Mulkey VR, Owens VN, Lee DK (2006) Management of switchgrass-dominated Conservation Reserve Program lands for biomass production in South Dakota. *Crop Sci* 46:712–720

Narasimhamoorthy B, Saha MC, Swaller T, Bouton JH (2008) Genetic diversity in switchgrass collections assessed by EST-SSR markers. *BioEnergy Res* 1:136–146

Newell LC (1968a) Effects of strain source and management practice on forage yields of two warm-season prairie grasses. *Crop Sci* 8:205–210

Newell LC (1968b) Registration of pathfinder switchgrass. *Crop Sci* 8:516

Newell LC, Eberhart SA (1961) Clone and progeny evaluation in the improvement of switchgrass, *Panicum virgatum* L. *Crop Sci* 1:117–121

Newman PR, Moser LE (1988) Grass seedling emergence, morphology, and establishment as affected by planting depth. *Agron J* 80:383–387

Nielsen EL (1944) Analysis of variation in *Panicum virgatum. J Agric Res* 69:327–353

Nielsen EL (1947) Polyploidy and winter survival in *Panicum virgatum. J Amer Soc Agron* 39:822–827

Nixon ES, McMillan C (1964) The role of soil in the distribution of four grass species in Texas. *Amer Midland Nat* 71:114–140

Okada M, Lanzatella C, Saha MC, Bouton JH, Wu R, Tobias CM (2010) Complete switchgrass genetic maps reveal subgenome collinearity, preferential pairing, and multilocus interactions. *Genetics* 185:745–760

Panciera MT, Jung GA (1984) Switchgrass establishment by conservation tillage: Planting date responses of two varieties. *J Soil Water Conserv* 39:68–70

Parrish DJ, Fike JH (2005) The biology and agronomy of switchgrass for biofuels. *Crit Rev Plant Sci* 24:423–459

Perlack RD, Wright LL, Turhollow AF, Graham RL, Stokes BJ, Erbach DC (2005) *Biomass As Feedstock for a Bioenergy and Bioproducts Industry: The Technical Feasibility of a Billion-Ton Annual Supply.* DOE/GO-102005-2135/RONL/TM-2005/66, U.S. Department of Energy, Oak Ridge National Lab, Oak Ridge, TN

Perrin RK, Vogel KP, Schmer MR, Mitchell RB (2008) Farm-scale production cost of switchgrass for biomass. *BioEnergy Res* 1:91–97

Perry LJ, Baltensperger DD (1979) Leaf and stem yields and forage quality of three N-fertilized warm-season grasses. *Agron J* 71:355–358

Poethig RS (2003) Phase change and the regulation of developmental timing in plants. *Science* 301:334–336

Redfearn DD, Moore KJ, Vogel KP, Waller SS, Mitchell RB (1997) Canopy architectural and morphological development traits of switchgrass and the relationships to forage yield. *Agron J* 89:262–269

Rehm GW (1984) Yield and quality of a warm-season grass mixture treated with N, P, and atrazine. *Agron J* 76:731–734

Rehm GW, Sorensen RC, Moline WJ (1976) Time and rate of fertilizer application for seeded warm-season and bluegrass pastures I Yield and botanical composition. *Agron J* 68:759–764

Rehm GW, Sorensen RC, Moline WJ (1977) Time and rate of fertilization on seeded warm-season and bluegrass pastures II Quality and nutrient content. *Agron J* 69:955–961

Richards HA, Rudas VA, Sun H, McDaniel JK, Tomaszewski Z, Conger BV (2001) Construction of a GFP-BAR plasmid and its use for switchgrass transformation. *Plant Cell Rep* 20:48–54

Riley RD, Vogel KP (1982) Chromosome numbers of released cultivars of switchgrass, indiangrass, big bluestem, and sand bluestem. *Crop Sci* 22:1081–1083

Rose IV LW, Das MK, Fuentes RG, Taliaferro CM (2007) Effects of high- vs low-yield environments on selection for increased biomass yield of switchgrass. *Euphytica* 156:407–415

Roth AM, Sample DW, Ribic CA, Paine L, Undersander DJ, Bartelt GA (2005) Grassland bird response to harvesting switchgrass as a biomass energy crop. *Biomass Bioenergy* 28:490–498

Samson JF, Moser LE (1982) Sod-seeding perennial grasses into eastern Nebraska pastures. *Agron J* 74:1055–1060

Sanderson MA, Adler PR, Boateng AA, Casler MD, Sarath G (2007) Switchgrass as a biofuels feedstock in the USA. *Can J Plant Sci* 86:1315–1325

Sanderson MA, Read JC, Roderick RL (1999) Harvest management of switchgrass for biomass feedstock and forage production. *Agron J* 91:5–10

Sanderson MA, Reed R, McLaughlin S, Wullschleger S, Conger B, Parrish D, Wolf D, Taliaferro C, Hopkins A, Ocumpaugh W, Hussey M, Read J, Tischler C (1996) Switchgrass as a sustainable bioenergy crop. *Bioresour Technol* 56:83–93

Sanderson MA, Wolf DD (1995) Switchgrass morphological development in diverse environments. *Agron J* 87:908–915

Sarath G, Akin DE, Mitchell RB, Vogel KP (2008) Cell-wall composition and accessibility to hydrolytic enzymes is differentially altered in divergently bred switchgrass (*Panicum virgatum* L) genotypes. *Appl Biochem Biotechnol* 150:1–14

Sarath G, Vogel KP, Mitchell RB, Baird LM (2005) Stem anatomy of switchgrass plants developed by divergent breeding cycles for tiller digestibility. In: O'Mara FL (ed), *XX International Grassland Congress: Offered Papers.* Wageningen Academic Publishers, Wageningen, The Netherlands, p 115

Schmer MR, Vogel KP, Mitchell RB, Moser LE, Eskridge KM, Perrin RK (2006) Establishment stand thresholds for switchgrass grown as a bioenergy crop. *Crop Sci* 46:157–161

Schmer MR, Vogel KP, Mitchell RB, Moser LE, Perrin RK (2008) Net energy of cellulosic ethanol from switchgrass. *Proc Natl Acad Sci USA* 105:464–469

Sims PL, Ayuko LA, Hyder DN (1971) Developmental morphology of switchgrass and side-oats grama. *J Range Manag* 24:357–360

Smart AJ, Moser LE (1997) Morphological development of switchgrass as affected by planting date. *Agron J* 89:958–962

Smart AJ, Moser LE (1999) Switchgrass seedling development as affected by seed size. *Agron J* 91:335–338

Smart AJ, Moser LE, Vogel KP (2004) Morphological characteristics of big bluestem and switchgrass plants divergently selected for seedling tiller number. *Crop Sci* 44:607–613

Smart AJ, Moser LE, Vogel KP (2003b) Establishment and seedling growth of big bluestem and switchgrass populations divergently selected for seedling tiller number. *Crop Sci* 43:1434–1440

Smart AJ, Vogel KP, Moser LE, Stroup WW (2003a) Divergent selection for seedling tiller number in big bluestem and switchgrass. *Crop Sci* 43:1427–1433

Smith D (1975) Trends of nonstructural carbohydrates in the stem bases of switchgrass. *J Range Manag* 28:389–391

Somleva MN, Tomaszewski Z, Conger BV (2002) Agrobacterium-mediated genetic transformation of switchgrass. *Crop Sci* 42:2080–2087

Staley TE, Stout WL, Jung GA (1991) Nitrogen use by tall fescue and switchgrass on acidic soils of varying water holding capacity. *Agron J* 83:732–738

Stout WL (1992) Water-use efficiency of grasses as affected by soil, nitrogen, and temperature. *Soil Sci Soc Amer J* 56:897–902

Stubbendieck J, Hatch SL, Butterfield CH (1991) *North American Range Plants*. University of Nebraska Press, Lincoln, NE

Talbert LE, Timothy DH, Burns JC, Rawlings JO, Moll RH (1983) Estimates of genetic parameters in switchgrass. *Crop Sci* 23:725–728

Tilley JMA, Terry RA (1963) A two stage technique for in vivo digestion of forage crops. *J Br Grassl Soc* 18:104–111

Tilman D, Hill J, Lehman C (2006) Carbon-negative biofuels from low-input high-diversity grassland biomass. *Science* 314:1598–1600

Tischler CR, Voigt PW (1993) Characterization of crown node elevation in panicoid grasses. *J Range Manag* 46:436–439

Tobias CM, Hayden DM, Twigg P, Sarath G (2006) Genetic microsatellite markers derived from EST sequences of switchgrass (*Panicum virgatum* L). *Mol Ecol Notes* 6:185–187

Tobias CM, Twigg P, Hayden DM, Vogel KP, Mitchell RB, Lazo GR, Chow EK, Sarath G (2005) Analysis of expressed sequence tags and the identification of associated short tandem repeats in switchgrass. *Theor Appl Genet* 111:956–964

Tracy BF, Chandler MA (2006) The historical and biological basis of the concept of heterotic patterns in corn belt dent maize. In: Lamkey KR, Lee M (eds), *Plant Breeding: The Arnel R Hallauer International Symposium*. Blackwell Publishing, Ames, IA, pp 219–233

Vadas PA, Barnett KH, Undersander DJ (2008) Economics and energy of ethanol production from alfalfa, corn, and switchgrass in the upper Midwest, USA. *BioEnergy Res* 1:44–55

Van Esbroeck GA, Hussey MA, Sanderson MA (1998) Selection response and developmental basis for early and late panicle emergence in Alamo switchgrass. *Crop Sci* 38:342–246

Varvel GE, Vogel KP, Mitchell RB, Follett RN, Kimble JM (2008) Comparison of corn and switchgrass on marginal soils for Bioenergy. *Biomass Bioenergy* 32:18–21

Vassey TL, George JR, Mullen RE (1985) Early-, mid-, and late-spring establishment of switchgrass at several seeding rates. *Agron J* 77:253–257

Vignols FJ, Rigau J, Torres MA, Capellades M, Puigdomenech P (1995) The brown midrib3 (*bmr3*) mutation in maize occurs in the gene encoding caffeic acid O-methyltransferase. *Plant Cell* 7:407–416

Vogel KP (1987) Seeding rates for establishing big bluestem and switchgrass with pre-emergence atrazine applications. *Agron J* 79:509–512

Vogel KP (1996) Energy production from forages (or American agriculture—Back to the future). *J Soil Water Conserv* 51:137–139

Vogel KP (2002) The challenge: High quality seed of native plants to ensure establishment. *Seed Technol* 24:9–15

Vogel KP (2004) Switchgrass. In: Moser LE, Burson BL, Sollenberger LE (eds), *Warm-Season (C4) Grasses*. ASA-CSSA-SSSA, Madison, WI, pp 561–588

Vogel KP, Brejda JJ, Walters DT, Buxton DR (2002a) Switchgrass biomass production in the midwest USA: Harvest and nitrogen management. *Agron J* 94:413–420

Vogel KP, Britton R, Gorz HJ, Haskins FA (1984) In vitro and in vivo analyses of hays of switchgrass strains selected for high and low in vitro dry matter digestibility. *Crop Sci* 24:977–980

Vogel KP, Burson BL (2004) Breeding and genetics In: Moser LE, Burson BL, Sollenberger LE (eds), *Warm-Season (C4) Grasses*. ASA-CSSA-SSSA, Madison, WI, p 51–94

Vogel KP, Haskins FA, Gorz HJ (1981) Divergent selection for in vitro dry matter digestibility in switchgrass. *Crop Sci* 21:39–41

Vogel KP, Haskins FA, Gorz HJ, Anderson BA, Ward JK (1991) Registration of 'Trailblazer' switchgrass. *Crop Sci* 31:1388

Vogel KP, Hopkins AA, Moore KJ, Johnson KD, Carlson IT (1996) Registration of 'Shawnee' switchgrass. *Crop Sci* 36:1713

Vogel KP, Hopkins AA, Moore KJ, Johnson KD, Carlson IT (2002b) Winter survival in switchgrass populations bred for high IVDMD. *Crop Sci* 42:1857–1862

Vogel KP, Masters RA (2001) Frequency grid—A simple tool for measuring grassland establishment. *J Range Manag* 54:653–655

Vogel KP, Mitchell RB (2008) Heterosis in switchgrass: Biomass yield in swards. *Crop Sci* 2159–2164

Vogel KP, Moore KJ, Hopkins AA (1993) Breeding switchgrass for improved animal performance. In: Proceedings of the XVII International Grassland Congress, February 17, 1993, North Palmerston, NZ, New Zealand Grassland Society, Dunedin, New Zealand, pp 1734–1735

Vogel KP, Pedersen JF (1993) Breeding systems for cross-pollinated perennial grasses. *Plant Breed Rev* 11:251–274

Vogel KP, Schmer MR, Mitchell RB (2005) Plant Adaptation Regions: Ecological and climatic classification of plant materials. *Range Ecol Manag* 58:315–319

Waller SS, Lewis JK (1979) Occurrence of C_3 and C_4 photosynthetic pathways of North American grasses. *J Range Manag* 32:12–28

Ward MG, Ward JK, Anderson BE, Vogel KP (1989) Grazing selectivity and in vivo digestibility of switchgrass strains selected for differing digestibility. *J Anim Sci* 67:1418–1424

Weaver JE (1954) *North American Prairie*. Jensen Publishers, Lincoln, NE

Weimer MR, Swisher BA, Vogel KP (1988) Metabolism as a basis for inter- and intra-specific atrazine tolerance in warm-season grasses. *Weed Sci* 36:436–440

Wheeler WA, Hill DD (1957) *Grassland Seeds*. D Van Nostrand, Princeton, NJ

Young HA, Hernlem BJ, Anderton AL, Lanzantella CL, Tobias CM (2010) Dihaploid stocks of switchgrass isolated by a screening approach. *BioEnergy Res* 3:305–313

Zheng-Xing S, Parrish DJ, Wolf DD, Welbaum GE (2001) Stratification in switchgrass seed is reversed and hastened by drying. *Crop Sci* 41:1546–1551

Section III

23 Brachypodium

Jennifer N. Bragg
U.S. Department of Agriculture

Ludmila Tyler
University of California, Berkeley, and U.S. Department of Agriculture

John P. Vogel
U.S. Department of Agriculture

CONTENTS

23.1   THE NEED FOR A MODEL GRASS

In a 2005 feasibility study, the U.S. Departments of Energy and Agriculture predicted that within four decades, the United States could sustainably produce over 1 billion dry tons of plant biomass annually for the generation of energy and other products (DOE and USDA 2005). Crop and other residues from grasses, including 256 million tons of corn stover and 52 million tons of wheat straw, represent approximately 348 million tons of this total. An additional 368 million tons is expected to come from the cultivation of perennial grasses and trees as dedicated bioenergy crops (DOE and USDA 2005). Switchgrass (*Panicum virgatum*) and Miscanthus (*Miscanthus × giganteus*) have emerged as particularly attractive potential energy crops (DOE 2007; Dohleman and Long 2009). Given that the grasses being considered as energy crops are essentially undomesticated wild selections, there is considerable potential for improving them. In addition, with the exception of forest trees, many of the traits desirable in an energy crop (e.g., thicker stems, more cell walls) have not been selected for in traditional crops, in which, for the most part, breeding has focused on reproductive organs or digestible leaves, tubers, or stems. Unfortunately, breeding the grasses proposed as energy crops is complicated by their reproductive strategy (self-incompatibility or sterility) which prevents the development of inbred lines, selfing, etc. Basic research on the biology of grasses could be used to design rational approaches to breeding superior energy crops and to accelerate the domestication of these new crops. The most rapid way to gain this basic knowledge is through the use of an appropriate model system.

Arabidopsis thaliana serves as an extremely powerful generalized plant model; however, it is not suitable to study many aspects of grass biology because of the biological differences that have arisen between dicots and monocots in the 150 million years since they last shared a common ancestor. One example that is particularly relevant to the need for a model for bioenergy crops is the dramatic difference between grass and dicot primary cell walls in terms of the major structural polysaccharides present, how those polysaccharides are linked together, and the abundance and importance of pectins, proteins, and phenolic compounds (Carpita 1996; Vogel 2008). A partial list of additional areas in which *Arabidopsis* is not an appropriate model for the study of grasses includes mycorrhizal associations, architecture of the grass plant, grain properties, intercalary meristems, and grass development.

The tremendous importance of grasses as food, feed, and, increasingly, as fuel argues strongly for the development of a truly tractable grass model system. Rice, with its sequenced genome and large research community, at first would seem to fill this need. Upon closer examination, the demanding growing conditions, large size, and long generation time of rice greatly increase the difficulty and expense of conducting high-throughput functional genomic experiments. Furthermore, the semiaquatic and tropical nature of rice limits the applicability of rice as a model for temperate grasses, especially in areas such as freezing tolerance and vernalization. *Brachypodium distachyon* (hereafter referred to as Brachypodium) is well suited to meet the need for an experimentally tractable model for the grasses. In this chapter, we provide a general introduction to Brachypodium, details about using Brachypodium as a model, a summary of genomic resources available for Brachypodium, and examples of how Brachypodium can be applied to the development of grasses as energy crops.

23.2   INTRODUCTION TO *B. DISTACHYON*

The utility of Brachypodium as a model system for the study of grasses was discussed in a 2001 paper that indicated that Brachypodium possesses the biological, physical, and genomic attributes required for use as a model system (Draper et al. 2001). The small size and rapid generation time of Brachypodium enable high-throughput studies. Large numbers of plants (1000 plants/m²) can easily be grown in growth chambers or greenhouses, allowing studies to be conducted under controlled environmental conditions. For comparison, the same space accommodates only 50 wheat plants, 36 rice plants, 13 sorghum plants, 6 maize plants, 6 switchgrass plants, or 2 Miscanthus plants (Rayburn et al. 2009) (Table 23.1 and Figure 23.1, a–c). As a group, the grasses are notorious for very

TABLE 23.1
Comparison of Model and Crop Plants

	Brachypodium distachyon	Arabidopsis thaliana	Oryza sativa	Triticum aestivum	Zea mays	Panicum virgatum	Sorghum bicolor	Miscanthus × giganteus
Common name	Brachypodium	Arabidopsis	Rice	Wheat	Maize	Switchgrass	Sorghum	Miscanthus
Height (cm)	15–20	15–20	100	50	155–215	200	170–320	400
Density (plants/m²)	1000	2000	36	50	6	6	13	2
Growth requirements	Simple	Simple	Demanding	Simple	Simple	Simple	Simple	Simple
Generation time (weeks)	8–12	8–12	30	12	14–20	26	17	N/A
Reproduction	Selfing	Selfing	Selfing	Selfing	Outcrossing: self-compatible	Outcrossing: self-incompatible	Outcrossing: self-compatible	Sterile/rhizome
Genome size (Mbp)	272[a]	119[a]	382[a]	16,000	2500	2400	758[a]	6800
Ploidy	2x	2x	2x	6x	2x	4x–8x	2x	3x
1n chromosome number	5	5	12	7	10	9	10	19
Cell wall type	Type II	Type I	Type II	Type II	Type II	Type II	Type II	Type II

N/A, not applicable.
[a] Assembled genome sizes.

FIGURE 23.1 (See color insert)   Adult plant comparisons and transformation. Flowering plants of maize (a), switchgrass (b), and Brachypodium (c) compared with the same 32-cm ruler, indicated with white arrows. The small size of Brachypodium is an advantage for a model system. (d–f) Comparison of *B. distachyon* and *B. sylvaticum*. (e) The annual species *B. distachyon* (left) next to its perennial relative *B. sylvaticum* (right). Bar is 15 cm. Inflorescences of *B. distachyon* (d) and *B. sylvaticum* (f). Note the exerted anthers of the outcrossing *B. sylvaticum* and the enclosed anthers of the inbreeding *B. distachyon*. (g) Embryogenic Brachypodium callus. The yellow structured regions are competent to regenerate plants and are used for transformation. Bar is 0.5 cm. (h) Plants regenerating from transgenic callus. After transformation and selection, a mixture of dying (brown) and healthy (yellow) callus can be seen. When placed in the light, healthy callus will turn green and regenerate plantlets. Plate is 10 cm in diameter. (i) Close-up of region in regeneration plate designated by the arrow in panel h.

large genomes. Fortunately, the now-sequenced 272-Mbp diploid Brachypodium genome is one of the smallest of any grass (International Brachypodium Initiative 2010). Brachypodium is self-fertile and does not typically outcross (Vogel et al. 2009). This feature is useful for breeding homozygous lines for many applications that require the maintenance of large numbers of independent genotypes (i.e., mapping experiments, mutant analysis, and studies of natural diversity). Furthermore, within the genus *Brachypodium*, there are species that may be useful to study polyploidy and perenniality.

23.2.1 PHYLOGENETIC AND SYNTENIC RELATIONSHIPS OF BRACHYPODIUM TO OTHER GRASSES

The phylogenetic relationship between Brachypodium and the other grasses has been evaluated a number of times with increasing amounts of data. Reports based on internal transcribed spacer (ITS) and 5.8S ribosomal DNA (rDNA) sequence, genomic restriction fragment length polymorphism (RFLP) and random amplified polymorphic DNA (RAPD) markers, and ITS sequence plus the chloroplast *ndfH* gene all placed Brachypodium between rice and a clade containing temperate grains such as wheat, barley, and *Secale* (Hsaio et al. 1994). Additional examinations of a much broader spectrum of grasses used ITS and the *ndfH* sequence, as well as morphological data and chloroplast restriction sites (Kellogg 2001) or the sequence of the *matK* chloroplast gene (Döring et al. 2007). These studies placed Brachypodium in the subfamily Pooideae just below the radiation of the small grains and forage and turf grasses, making Brachypodium a "sister" to this economically important group of grasses. However, phylogenies based on single genes or small sets of genes can produce inconsistent phylogenetic trees (Rokas et al. 2003), and this phenomenon has been observed with rice (Kellogg 1998). Therefore, it was important to examine the phylogenetic relationships of Brachypodium using larger data sets. Analysis of a data set comprising 11 kb of sequence from 20 highly expressed genes verified the relationship between Brachypodium and the small grains (Vogel et al. 2006a). An even larger data set that is based on 335 bacterial artificial chromosome (BAC) end sequences provides further evidence to confirm the placement of Brachypodium within the grass family tree (Huo et al. 2007).

Several genomic regions of *B. distachyon* and *B. sylvaticum* have been compared to rice and wheat, and these have shown general colinearity. One comparison of a 371-kb genomic sequence from *B. sylvaticum* to the orthologous regions of rice and wheat showed perfect macrocolinearity between the three genomes. The order of the shared genes was the same in *B. sylvaticum* and wheat, whereas there was an approximately 220-kb inversion in rice, demonstrating variation in microcolinearity (Bossolini et al. 2007). Markers from a 140-kb region of rice also were compared with *B. sylvaticum* and wheat sequences in efforts to map the *Ph1* locus that controls the pairing of homologous chromosomes in wheat. In this case, the colinearity of *B. sylvaticum* and rice sequences permitted the localization of *Ph1* to a 2.5-Mb interstitial region of wheat chromosome 5B (Griffiths et al. 2006). Furthermore, only 17% of the genes from 55,221 paired *B. distachyon* BAC end sequences were not colinear with the orthologous regions in rice. A Brachypodium physical map covered 88% of the rice sequence and showed conservation of synteny across these genomes (Gu et al. 2009).

23.2.2 RELATED *BRACHYPODIUM* SPECIES

The genus *Brachypodium* contains a relatively small number of grasses and is estimated to have diverged from *Triticeae* and *Poeae* 35–40 million years ago (Bossolini et al. 2007). The genus has been assigned to its own tribe, *Brachypodieae*, within the subfamily Pooideae (Catalan et al. 1995; Catalán and Olmstead 2000). Most of the 12–15 described *Brachypodium* species have been collected from Mediterranean and Eurasian locations, but representatives of this genus have been identified worldwide (Catalan et al. 1995; Catalán and Olmstead 2000). Species originating in the Mediterranean include *B. distachyon* (Figure 23.1 c–e), *B. retusum*, and *B. phoenicoides*. *B. sylvaticum* (Figure 23.1, e and f), *B. glaucovirens*, and *B. pinnatum* are from Eurasian locations, and the European representative is *B. rupestre*. Seven additional taxa are reported from diverse origins: *B. arbuscula* (Canary Islands), *B. boissieri* (southern Spain), *B. kawakamii* Hayata (Taiwan), *B. mexicanum* (Mexico to Bolivia), *B. pringlei* (Central and South America), *B. bolusii* (Africa), and *B. flexum* (Africa).

All members of the genus exhibit a set of common features that include lateral stem development from the coleoptile, small chromosomes, rDNA sequence, repetitive DNA families, and shared nuclear RFLPs. However, variation in morphology, life-cycle, and cytology is sufficient to clearly distinguish the species (Catalán and Olmstead 2000). *B. distachyon* is the only member to have an annual life-cycle, and it is also self-compatible, a trait that is shared with only two perennial species—*B. mexicanum* and *B. sylvaticum* (Khan and Stace 1999). With the exception of *B. mexicanum*, the perennial species are rhizomatous (Catalán and Olmstead 2000).

The phylogenetic relationships between eight *Brachypodium* species (*B. arbuscula*, *B. distachyon*, *B. mexicanum*, *B. phoenicoides*, *B. pinnatum*, *B. retusum*, *B. rupestre*, and *B. sylvaticum*) have been evaluated using multiple data sets: RFLP and RAPD markers, a chloroplast *ndhF* gene sequence, a nuclear rDNA sequence, and rDNA ITS sequence (Shi et al. 1993; Hsaio et al. 1994; Catalan et al. 1995; Catalán and Olmstead 2000). An *Eco*RI site was identified in the rDNA of most perennial species that could be used to distinguish them from *B. distachyon* and *B. mexicanum*, but this approach failed to identify sufficient variation to resolve the relationship between the perennial species. By using RAPD data along with *ndhF* and ITS sequences, *B. distachyon* was identified as the basal lineage of the group, followed by the divergence of *B. mexicanum*, *B. arbuscula*, *B. retusum*, *B. rupestre*, *B. phoenicoides*, *B. pinnatum*, and then *B. sylvaticum* (Catalán and Olmstead 2000).

Polyploidy is common among all taxa, and diploid, tetraploid, hexaploid, and octaploid species have been reported with base chromosome numbers ranging from 5 to 10 (Robertson 1981). In a recent report (Wolny and Hasterok 2009), cytogenetic analyses were performed on six species and two subspecies of *Brachypodium*. The researchers found that fluorescence in situ hybridization (FISH) could help identify the small chromosomes found in *Brachypodium* species that are otherwise difficult to distinguish. Evolutionary relationships between allopolyploid species were also investigated using genomic in situ hybridization (GISH) to assign chromosomes to putative ancestral genomes. Wolny and Hasterok (2009) concluded that *B. pinnatum* ($2n = 28$) is an interspecific hybrid between *B. distachyon* and *B. pinnatum* ($2n = 18$) and suggested that *B. distachyon* is one of the putative ancestral species for both of the allopolyploids *B. phoenicoides* and *B. retusum*. Insight into the organization, phylogeny, and evolution of mechanisms that determine variation in chromosome number will become possible with the development of additional tools such as arm- and region-specific probes for cytogenetic analyses.

Perenniality and self-incompatibility are traits that are present in the wild grasses being developed into bioenergy crops (e.g., Miscanthus and switchgrass) and can also be found in a number of species within the genus *Brachypodium* (Khan and Stace 1999; DOE and USDA 2005). As a result of the close relationship between different *Brachypodium* species, researchers will be able to leverage the resources developed for *B. distachyon* to study these traits in other *Brachypodium* species.

23.3 BRACHYPODIUM AS AN EXPERIMENTAL SYSTEM

Brachypodium displays many traits that make it a tractable and powerful system for research targeted at improving grasses for use as food, feed, and fuel. A large collection of diverse accessions and described inbred lines, simple growth requirements, efficient transformation, and a compact genome make this small grass an attractive choice for an experimental system to understand basic questions in grass biology.

23.3.1 GERMPLASM AND NATURAL DIVERSITY

In contrast to the domesticated cereals which have been subjected to human selection for millennia, Brachypodium is a wild grass. Brachypodium germplasm, for which there are a number of collections, thus provides an excellent resource for investigating natural diversity. The oldest collection consists of approximately 30 population samples dating back to the 1940s. These accessions are available from the U.S. Department of Agriculture (USDA) National Plant Germplasm System (NPGS) (www.ars-grin.gov/npgs/), and relevant passport data can be accessed at www.brachypodium.org. Twenty-seven of the NPGS accessions, 5 diploid and 22 polyploid, were used to generate inbred lines (designated with the prefix "Bd") that are freely available to the research community (Vogel et al. 2006b). The diploid inbred lines include two lines [Bd21, the line used for genome sequencing, and Bd21-3, which was selected for efficient transformation (Vogel and Hill 2008)] that were derived from the same NPGS accession, PI 254867. These lines are genetically distinct and, thus, presumably were derived from different individuals collected at the same location (Vogel et al. 2009). Another collection of diploid and polyploid ecotypes (designated "ABR") is maintained at the University of Wales, Aberystwyth. Some of the ABR ecotypes are unique, whereas others overlap with material in the NPGS collection. Synonymous ABR and NPGS designations are listed at www. brachypodium.org/stocks. A material transfer agreement governs the use of all ABR ecotypes.

Recently, 188 diploid inbred lines and a smaller number of polyploid inbred lines were generated from seeds collected at 53 sites across Turkey; these lines are being freely distributed to the research community (Filiz et al. 2009; Vogel et al. 2009). Inbred lines from seeds collected by M. Tuna were designated with a prefix corresponding to the first three letters of the collection location (e.g., "Tek" for the nearby town of Tekirdag) (Vogel et al. 2009). Lines from seeds collected by H. Budak were grouped based on phenotypic similarity and labeled with the prefix "BdTR" followed by the group number and a letter to designate the specific line (Filiz et al. 2009). To survey the genetic diversity of these newly generated lines, 43 simple sequence repeat (SSR) markers were used to genotype the lines, together with the six previously generated diploid inbred lines (Bd1-1, Bd2-3, Bd3-1, Bd18-1, Bd21, and Bd21-3) (Vogel et al. 2009). The SSR marker profiles were used to create an unrooted phylogenetic tree (Figure 23.2) (Vogel et al. 2009). Interestingly, lines that had been assigned to each BdTR group on the basis of phenotypic similarities clustered together in genetically related groups on the tree despite originating from many different locations. Conversely, lines originating from one location were often genetically distinct. Taken together, these data suggest that there is a significant amount of long-distance seed dispersal. The phylogenetic tree strongly supported a clade containing Bd1-1, the BdTR7 and BdTR8 groups, and the Tek accessions. These lines also shared similar phenotypes, including long vernalization requirement and small, nearly hairless seeds (Figure 23.2) (Vogel et al. 2009).

Diploid Brachypodium accessions have obvious phenotypic differences indicating that they can be exploited to study a number of traits relevant to biomass crop development. The diversity in the flowering times and vernalization requirements of Brachypodium lines is particularly striking. After 2–4 weeks of vernalization, the diploid inbred lines Bd2-3, Bd3-1, Bd21, and Bd21-3 flower relatively rapidly, within 2–3 weeks, under greenhouse conditions, whereas the Bd18-1 and Bd1-1 lines flower much later, even after longer vernalization (Vogel et al. 2006b, 2009). For some Bd lines (Bd2-3, Bd3-1, Bd21, and Bd21-3) extending the day length to 20 h eliminates the need for vernalization (Vogel et al. 2006b, 2009; Vogel and Hill 2008). Very long days do not trigger rapid flowering in any of the new Turkish inbred lines (Vogel et al. 2009). The Tek inbred lines, originating from northern Turkey, are especially late flowering and require 8–16 weeks of vernalization (Vogel et al. 2009). The molecular mechanisms underlying flowering-time regulation in Brachypodium are currently unknown. A report that expression of the floral repressor *Terminal Flower 1* from perennial ryegrass, *Lolium perenne*, delays flowering in Brachypodium provides a starting point for future molecular-genetic studies (Olsen et al. 2006). Additionally, a recent analysis of inbred lines with diverse flowering times suggests that the putative Brachypodium *VERNALIZATION2* and *VERNALIZATION3* genes are involved in controlling flowering time (Schwartz et al. 2010). Other phenotypic differences between lines include the presence or absence of hairs and the number and angle of inflorescence branches (Opanowicz et al. 2008; Filiz et al. 2009; Vogel et al. 2009). Accessions also vary in the degree to which the seed disarticulates from the inflorescence, an agriculturally important trait known as

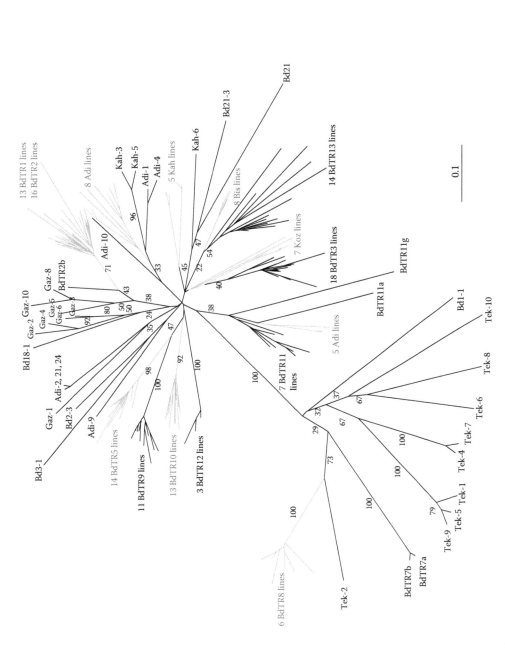

FIGURE 23.2 Neighbor-joining consensus tree for 187 Brachypodium inbred lines on the basis of 43 SSR markers. The unrooted tree was constructed from 100 shared-SSR-allele bootstrap trees. For major branches, bootstrap values >20 are shown. Note the considerable genetic diversity in this collection. (Adapted from Vogel, J. et al., *BMC Plant Biol*, 9, 88, 2009.)

"shattering" (Opanowicz et al. 2008). Diploid seed sizes range from an average of 2.5 mg/seed for the Tek-10 inbred line to 5.9 mg/seed for Kah-7 (Vogel et al. 2009).

Brachypodium also exhibits intraspecific diversity in chromosome number. Most diploid Brachypodium lines have a base chromosome number of 5 ($1n = 5$) (Draper et al. 2001; Vogel et al. 2006b; Filiz et al. 2009). However, accessions with chromosome numbers of $1n = 10$ and $1n = 15$ have also been described (Draper et al. 2001; Hasterok et al. 2004). Studies utilizing FISH and GISH techniques indicate that the $1n = 10$ and $1n = 15$ cytotypes are not merely autopolyploids derived from the $1n = 5$ cytotype (Hasterok et al. 2004). Chromosomes of the $1n = 10$ accession ABR114 are smaller than those of the $1n = 5$ accession ABR1, and the FISH-visualized pattern of rDNA loci in ABR114 is inconsistent with ABR114 being an autotetraploid arising from ABR1 (Hasterok et al. 2004). Thus, ABR114 seems to be a diploid with a base chromosome number of 10. Similar cytogenetic analyses have led to the idea that the $1n = 15$ accession ABR113 is an allotetraploid that arose from the hybridization of ABR1- and ABR114-like parents, with base chromosome numbers of 5 and 10, respectively (Hasterok et al., 2004, 2006). Thus, the different Brachypodium cytotypes should probably be considered different species rather than a simple polyploid series. When examined carefully, the polyploids characterized to date can be easily distinguished from the $1n = 5$ diploid accessions. For example, among the Turkish lines analyzed, one group of polyploids was distinguished by large seeds and thick, hairy stems and a second group of polyploids was characterized by a deep, longitudinal crease in the seed (Vogel et al. 2009). In both groups, anthers exerted more frequently than in the diploid lines. It should be noted that the $1n = 5$ diploid is the cytotype being used for genome sequencing and resource development and that the $1n = 10$ form is known from only one collection to date (Garvin et al. 2008).

23.3.2 GROWTH REQUIREMENTS

The simple requirements for growing Brachypodium make it easy to culture under laboratory conditions. Brachypodium can be grown in growth chambers or greenhouses used for *Arabidopsis*, wheat, barley, switchgrass, or other plants. Our standard conditions for growth chambers are 20-h light:4-h dark photoperiod, 24°C during the day, and 18°C at night with cool-white fluorescent lighting at a level of 150 $\mu Em^{-2}s^{-1}$. Our standard greenhouse conditions are no shading, 24°C in the day and 18°C at night, and supplemental lighting to extend day length to 16 h. Although Brachypodium grows well in a number of different soil types, we have observed that it is highly susceptible to *Pythium* root rot. Plants that are watered excessively or left in standing water often develop disease symptoms under our conditions. We also have observed disease symptoms that correlated with the use of one brand of commercial potting mix and recommend that Brachypodium growers test a few soil formulations before selecting one to grow large numbers of plants (Vogel and Bragg 2009).

Vernalization has been shown to induce flowering in all diploid accessions studied to date, but, as noted above, the time required to induce flowering varies greatly between accessions (Vogel et al. 2006b, 2009; Vogel and Hill 2008). Providing the appropriate conditions to induce flowering is critical to preventing excessive vegetative growth. For a combined stratification and vernalization treatment, we typically sow the seeds and then place them at 4°C for the desired number of weeks (inbred lines Bd21 and Bd21-3 require 2–3 weeks of vernalization to reliably induce flowering unless grown under very long day lengths). After approximately 3 weeks in the cold, seeds begin to germinate. Thus, for vernalization times greater than 4 weeks, we place the pots under fluorescent lighting. Vernalizing seeds/seedlings induces the plants to flower quickly while still small. Alternatively, one can vernalize larger plants if more seed from individual plants is desired. Inbred lines Bd21 and Bd21-3 are particularly responsive to growth under very long day conditions (20 h light:4 h dark) and go from seed to seed in as little as 8 weeks to yield nearly six generations per year. Under these conditions, the plants flower and set seed when they are approximately 15 cm tall—a size that is compatible with high-density planting.

The inbreeding nature of Brachypodium simplifies the maintenance of independent lines under laboratory conditions. The anthers of diploid accessions rarely exert, suggesting a low rate of outcrossing. This was confirmed by measuring pollen flow from transgenic to nontransgenic plants under growth chamber and greenhouse conditions. In a population of more than 2000 progeny, no outcrossing was observed (Vogel et al. 2009). The inbreeding nature of Brachypodium in the wild was confirmed by analyzing SSR profiles of 62 wild individuals. These individuals were overwhelmingly homozygous, despite the presence of multiple SSR alleles in the population, indicating that Brachypodium primarily self-pollinates in the wild (Vogel et al. 2009).

23.3.3 Transformation

The utility of a modern model plant system depends greatly on the development of efficient methods to introduce foreign DNA into its genome. The dicot model *Arabidopsis* benefits from an extremely facile transformation method in which flowers are simply dipped into a solution of *Agrobacterium tumefaciens* for several seconds (Clough and Bent 1998). As a result, the *Arabidopsis* research community has access to invaluable tools such as stable knockout lines for most genes in the genome (Pan et al. 2003). In contrast, grass transformation is a more challenging and labor-intensive endeavor. Routine transformation of grasses requires extensive tissue culture manipulations, and transformation of almost all grasses is very inefficient. Supporting its utility as a model system, Brachypodium has proven to be very responsive to in vitro culture, and current transformation efficiencies are on par with rice, the present gold standard for grass transformation. A major step toward achieving Brachypodium transformation was the development of a method for the induction of embryogenic callus (Figure 23.1g) from Brachypodium seeds and the regeneration of fertile plants (Figure 23.1, h and i) from this callus. In 1995, Bablak et al. established the optimal callus-inducing medium to contain LS salts, 3% sucrose, and 2.5 mg L^{-1} 2,4-Dichlorophenoxyacetic acid (Bablak et al., 1995). Mature seeds from three diploid accessions (B200, B373, and B377) were incubated on callus-inducing media. All were found to produce embryogenic callus, along with several other types of callus, and regeneration was observed on several common media, indicating that Brachypodium had no unusual requirements for regeneration.

Particle bombardment and *A. tumefaciens*-mediated transformation are the two methods most commonly used for plant transformation, and both have been used to successfully transform Brachypodium. Each technique offers unique advantages and disadvantages.

23.3.3.1 Biolistic Transformation

The regeneration of plants from bombarded explants represents the primary determinant of successful biolistic transformation. The first published Brachypodium transformation involved particle bombardment of a polyploid Brachypodium accession (ABR100). In these experiments, the average efficiency was five transformants per gram of starting embryogenic callus (Draper et al. 2001). A subsequent, more detailed, account of biolistic transformation answered the question of whether a diploid accession could be transformed (Christiansen et al. 2005). In this study, the authors successfully transformed the diploid accession BDR018 with an average efficiency of 5.3% of bombarded calluses producing transgenic plants. The authors' failed attempts to transform a second diploid accession (BDR001) demonstrate that, as for other plants, genotype plays a critical role in determining transformation efficiency. These initial Brachypodium studies compare favorably with early reports of biolistic rice transformation that showed an average efficiency of 3.75% (Christou et al. 1991). However, biolistic transformation commonly results in complex transgenic loci. Typically, these loci contain multiple copies of the inserted DNA, including truncated pieces of the target DNA interspersed with genomic DNA (Svitashev and Somers 2002; Kohli et al. 2003). These biolistic insertions often contain many repeats of inserted DNA and can span several megabases of host DNA (Svitashev and Somers 2002). The complexity of these insertions represents a major drawback of biolistic transformation because they can interfere with downstream applications that require relatively simple insertions (e.g., cloning flanking DNA or promoter tagging) and may lead to silencing of transgenes in later generations. Attempts to minimize the complexity of biolistic loci by using linear DNA instead of circular plasmid DNA have produced mixed results (Fu et al. 2000; Loc et al. 2002).

23.3.3.2 Agrobacterium-Mediated Transformation

Compared with biolistic transformation, *Agrobacterium*-mediated transformation has been shown to yield much simpler and lower copy number insertion patterns in rice and *Arabidopsis* (for a direct comparison of methods see Dai et al. 2001 and Travella et al. 2005). Furthermore, transgenic plants contain an average of approximately 1.5 insertions per line, averting the challenges to downstream analyses encountered with the complex loci of the biolistic lines (Feldmann 1991; Jeon et al. 2000). Instead, the difficulties of establishing an efficient *Agrobacterium*-mediated transformation system reside in the host limitations of *Agrobacterium*. Fortunately, Brachypodium has proven amenable to *Agrobacterium*-mediated transformation. In the first report of *Agrobacterium*-mediated transformation, 16 polyploid accessions and 3 diploid accessions were evaluated for transformability (Vogel et al. 2006b). The highest transformation efficiency (14% of the callus pieces cocultivated with *Agrobacterium*-produced transgenic plants) was achieved with the polyploid line Bd17-2. A diploid accession, PI 254867, was transformed at a much lower efficiency (2.5%).

Subsequent studies have made considerable progress at improving the efficiency of the *Agrobacterium*-mediated transformation of Brachypodium. In 2007, three studies reported very high transformation efficiencies for three different Brachypodium lines. The first two papers used the inbred lines Bd21-3 (Vogel and Hill 2008) and Bd21 (Vain et al. 2008), which were separately derived from USDA accession PI 254867 as previously described in Section 23.3.1. The methods described in these two papers share a number of important similarities, including media types, *Agrobacterium* strains, and use of immature embryos as initial explants. Dissecting out immature embryos is the most labor-intensive step in these processes, and both methods take advantage of multiple subculture steps to amplify the callus before transformation so that each dissected embryo gives rise to many transgenic plants. The differences between the methods lie in the following: the use of desiccating conditions to improve transformation of Bd21-3, the formation of a yellow embryogenic callus in Bd21-3 that allows selection of the appropriate callus type without the aid of a microscope, the use of very small embryos and copper sulfate to improve the quality of the Bd21 callus, and visual selection of green fluorescent protein (GFP) and the subculturing callus under a microscope to improve the efficiency of Bd21 selection. In these studies, transformation efficiency was calculated as the percentage of calluses cocultivated with *Agrobacterium* that produced fertile transgenic plants, and the average efficiencies achieved were 37% for Bd21-3 and 17% for Bd21. The third Brachypodium transformation paper reports an extremely high average transformation efficiency of 55% for accession BDR018 (Păcurar et al. 2007). The authors achieve this high efficiency by placing immature embryos on callus-inducing media for 17 days and then cocultivating those embryos with *Agrobacterium*. Efficiency is calculated from the percentage of dissected embryos that form fertile transgenic plants. The limitation of this method is that the embryogenic callus is not subcultured, and therefore no more than one independent transgenic line can arise from each dissected embryo. This increases the labor involved in generating transgenic plants when compared with the methods for Bd21-3 and Bd21 transformation.

The publication of three high-efficiency *Agrobacterium*-mediated transformation methods signals that Brachypodium transformation technology has matured, yet the lessons learned from these three papers inspire continued studies to optimize Brachypodium transformation. For example, adoption of the use of very small embryos (0.3–0.7 mm) and a callus initiation medium that includes 0.6 µg/mL copper sulfate has further increased the average transformation efficiency of Bd21-3 to more than 50% (J. Bragg unpublished; protocol available at http://brachypodium.pw.usda.gov/). Further improvements in Brachypodium transformation will undoubtedly emerge both from investigating the genetic diversity within the growing collections of Brachypodium accessions and from other technical advances. It is noteworthy that the rapid pace of Brachypodium transformation technology development compares very favorably with the development of high-efficiency transformation in rice. In the initial reports, rice transformation efficiencies were less than 1%, and through the efforts of many groups over several years, efficiency was improved to the more than 40% commonly achieved today (Tyagi and Mohanty 2000).

The highly efficient transformation methods developed for Brachypodium lay a strong foundation for Brachypodium as a model system to study fundamental aspects of grass biology. In addition, efficient transformation means that Brachypodium is an excellent test bed for transgenic approaches in the grasses. By using Brachypodium, researchers can much more rapidly test constructs for expression, efficacy, etc., before moving into biomass or other grass crops.

23.3.4   Generation of Brachypodium Mutant Populations

Genome-saturating mutant populations provide a means to explore the relationship between a phenotype and a gene of interest, and both forward and reverse genetic strategies require large populations of mutagenized plants. Diverse approaches for the efficient generation of mutant populations include treatment with mutagenic chemicals, exposure to high-energy radiation, and random DNA insertions. Researchers have had success applying these protocols to Brachypodium, and it appears that there are no limitations in applying common mutagens used with other plants. The development of populations of mutants, similar to those available for *Arabidopsis* and rice, will provide valuable tools for the Brachypodium research community.

23.3.4.1   Ethyl Methanesulfonate Mutagenesis

Ethyl methanesulfonate (EMS) is an efficient chemical mutagen that introduces single base changes and has been used widely to mutagenize plants. In *Arabidopsis*, an extensive review of mutations in 192 genes verified the random nature of EMS mutations and estimated the frequency at one mutation per 170 kb of genomic DNA (Greene et al. 2003). An advantage of the single base changes derived from EMS mutagenesis is that these simple changes can result in partial loss-of-function alleles that may be particularly useful when studying essential genes. We have mutagenized Brachypodium with EMS by adapting a method used to create a population of barley mutants (Caldwell et al. 2004). The frequency of albino plants is often used to measure the success of EMS mutagenesis, and over a mutant population of 2000 M_2 plants, we have observed 2% albinos. This rate is comparable to that typically observed in successful EMS treatments of *Arabidopsis* seeds (Kim et al. 2006). Our initial screens have focused on identifying mutations that alter cell wall and biomass phenotypes, and we have identified more than 25 mutants of interest to date, indicating that the EMS mutagenesis was a success. Visit http://brachypodium.pw.usda.gov/ for our current EMS mutagenesis protocol.

23.3.4.2   Fast Neutron Radiation

Fast neutron radiation (FNR) introduces short deletions into the genomic DNA and is therefore a complementary mutagen to EMS, which induces single base changes. The larger disruptions generated by FNR permit rapid cloning of the mutant loci using a genome tiling array. This is an advantage over EMS mutagenesis, in which the genes must be identified using map-based cloning. However, because the region deleted by FNR typically encompasses several genes, this method is not suitable for the identification of essential genes or genes located adjacent to essential genes. In early experiments, 1–2% of mutagenized plants were albinos, indicating that Brachypodium can be efficiently mutagenized by FNR (D. Laudencia-Chingcuanco and M. Byrne, personal communication).

23.3.4.3   Insertional Mutagenesis

Insertional mutagenesis is a natural complement to chemical and radiation mutagenesis because, although the mutational load is low (only one or a few mutations per line), the ability to sequence DNA flanking the insertion site enables rapid identification of the affected genes, which is required for reverse genetic approaches. The insertion of transferred DNAs (T-DNAs) via *Agrobacterium*-mediated transformation and the movement of transposons are the two most common approaches for transferring known DNA sequences into random sites in the host genome. In the resulting mutant lines, the T-DNA or transposon sequence serves as a tag that can be used to locate the DNA insertion site within the genome. Large, freely available collections of sequence-indexed, tagged lines have been an extremely valuable tool in *Arabidopsis* research, permitting both forward genetic screens for a particular phenotype within the mutant populations and reverse genetic studies in which researchers can identify disruptions of specific genes using a simple BLAST search. The value of a collection of sequence-indexed mutants to facilitate study of candidate genes selected from the enormous amount of sequence and bioinformatic data currently available is tremendous.

Each insertion event has the potential to cause a knockout phenotype; however, vectors can also be designed to achieve various research goals, including the identification of promoters and overexpression of nearby genes. In a gene trap construct, a promoter-less reporter gene (e.g., GUS or GFP) is placed at the end of the T-DNA sequence that is transferred to the plant genome (An et al. 2005). If the vector DNA integrates downstream of a promoter, reporter gene expression could be used to infer the expression pattern of the disrupted gene and provide clues about the role of the disrupted gene. Inclusion of splice acceptor sites adjacent to the reporter genes permits splicing should the vector DNA fall into an intron. Activation tagging constructs place transcriptional enhancers within the vector DNA to increase the transcription of genes close to the insertion site (Weigel et al. 2000; Fits et al. 2001; Nakazawa et al. 2003). Activation tagging is designed to overexpress nearby genes while still maintaining a wild-type expression pattern, and it is particularly well suited to studying genes with redundant functions in which knockouts in one family member fail to produce a phenotype. This strategy can also provide insight into complex processes such as cell wall biosynthesis because activation tagging can activate global control genes.

A large sequence-indexed Brachypodium mutant population would be a powerful research tool, and multiple groups have started assembling such a population. The choice of the most efficient method to produce these collections depends on the efficiency of transformation versus the efficiency of producing transposon-tagged mutants. Although the transposon approach has the potential to rapidly generate a large number of insertional mutants, this technique requires optimization before it will be a productive means of generating Brachypodium mutants. We have transformed Brachypodium with vectors containing the most commonly used transposons (*Ac/Ds* and *En/Spm*) and demonstrated that they were active in the Brachypodium genome. However, most transgenic plants died before setting seed, possibly because the transposons were too active or activated an endogenous transposon (J. Bragg, unpublished). In contrast, the efficiency of T-DNA tagging has increased with the optimization of transformation techniques, and one person can easily produce 100 lines per week. Using this approach, at least two substantial populations are in development. The BrachyTAG project at the John Innes Centre currently lists 4500 T-DNA lines for distribution to the public. Of these, 1005 have flanking sequence tags, and 61 have nearby genes identified. Additionally, we have developed over 8000 T-DNA lines that are available for distribution via the USDA Brachypodium Genome Resources site (see Table 23.2 for links to these resources).

23.3.5 CROSSING BRACHYPODIUM

An efficient method of crossing Brachypodium is required for it to serve as a tractable model genetic system [e.g., to allow positional cloning and mapping of quantitative trait loci (QTLs)]. Brachypodium is primarily an inbreeding species, and flowers rarely open under greenhouse and growth chamber conditions (Figure 23.1e). In observations of rare open flowers, anthers have already dehisced on the stigma. This suggests that even open flowers primarily produce self-pollinations, a notion supported by the highly homozygous nature of wild Brachypodium accessions (Vogel et al. 2009). Two similar methods, one using a microscope and the other a jeweler's loupe, have been developed for crossing Brachypodium. The protocols (available at http://brachypodium.pw.usda. gov/ and http://www.ars.usda.gov/pandp/docs.htm?docid=18531) contain detailed pictures of the flower stages appropriate for crossing and of each step in the procedure. These protocols are based

TABLE 23.2
Internet Resources Available for Brachypodium Research

Resource	Institution	URL	Purpose
Brachypodium.org	Oregon State University	http://www.brachypodium.org/	Overview and links to Brachypodium research and resources
Brachybase	Oregon State University	http://www.brachybase.org/	Genome browser, EST alignments, Illumina-based RNAseq data, and empirical TAU gene model predictions
Brachypodium genome browser	Munich Information Center for Protein Sequence (MIPS)	http://mips.helmholtz-muenchen.de/plant/brachypodium/index.jsp	Genome browser, BLAST server
Phytozome	JGI and Center for Integrative Genomics	http://www.phytozome.net/	Comparative genomic studies of green plants
CoGe	University of California, Berkeley	http://synteny.cnr.berkeley.edu/CoGe/	Comparative genomic studies
Brachypodium Genome Resources	USDA-ARS Western Regional Research Center	http://brachypodium.pw.usda.gov/	Protocols, publications, SSR markers, germplasm, T-DNA insertional mutant lines, maps, and other resources
Garvin Laboratory	USDA-ARS Plant Science Research Unit	http://www.ars.usda.gov/pandp/docs htm?docid=18531	Brachypodium populations and crossing protocol
Modelcrop	JGI	http://www.modelcrop.org/index.html	Genome browser, BLAST server, physical map, and comparative maps
Physical map	University of California, Davis	http://phymap.ucdavis.edu:8080/brachypodium/	Physical map
BrachyTAG	John Innes Centre	http://www.brachytag.org/	T-DNA insertional mutant lines
USDA National Plant Germplasm System	USDA	www.ars-grin.gov/npgs/	Germplasm
GrainGenes	USDA-ARS Western Regional Research Center	http://wheat.pw.usda.gov/GG2/index.shtml	Comparative genomics, cMAP viewer

on the fact that immature anthers will dehisce shortly after removal from the plant. In brief, florets with feathery stigmas and anthers that have not yet dehisced are suitable for crossing. One floret is emasculated by peeling back the lemma and removing the anthers. The anthers that will serve as pollen donors are dissected from florets and permitted to swell for 10–30 min before they release fresh pollen for crossing. Pollen or entire anthers are then introduced to the emasculated floret, and the lemma is closed to prevent desiccation and protect the developing seed. With practice, crosses can be made in 5–10 min or less.

23.4   GENOMIC RESOURCES

An extensive infrastructure of genomic resources has been (and continues to be) assembled for Brachypodium, including cDNA libraries, BAC libraries, a large expressed sequence tag (EST) collection, a high-resolution genetic linkage map, physical maps, SSR markers, bioinformatic resources, and most importantly the recently completed 8× genome sequence.

23.4.1 BIOINFORMATIC RESOURCES

The rapid development of Brachypodium genomic resources presents researchers with an approaching avalanche of genomic data, and it is necessary to develop the appropriate bioinformatic infrastructure to efficiently support and utilize these tools. The available web-based resources are summarized in Table 23.2 and are briefly described below. A Brachypodium-specific web portal, www.brachypodium.org, provides links to numerous sources of Brachypodium information and also houses a newsgroup that links the Brachypodium community. The 8× Brachypodium sequence is housed in three databases that contain or will contain tracks that place genetic markers, BAC clones, ESTs, T-DNA insertion sites, sequences from other species, and other applicable data in the context of the Brachypodium genomic sequence. Three sites designed to support web-based comparative genomics studies—Phytozome, GrainGenes, and CoGe—also contain the Brachypodium 8× genome sequence. Other websites that contain project-specific information include the BrachyTAG and USDA-ARS Genomics and Gene Discovery Unit websites. Where appropriate, the sections below contain references to these resources, and readers are directed to these sites for the most detailed resources currently available.

23.4.2 ESTs AND MICROARRAYS

ESTs generated by randomly sequencing the ends of cDNA clones provide a quick and relatively inexpensive way to learn a great deal about an unknown genome (Adams et al. 1991). As a result, the first significant sequence resource for Brachypodium was the 20,440 ESTs deposited into GenBank in 2005 (Vogel et al. 2006). These ESTs were derived from five cDNA libraries and represent approximately 6000 genes. As part of the genome sequencing project, the U.S. Department of Energy-Joint Genome Institute (JGI) generated approximately 128,000 Sanger ESTs, approximately 2.3 million 454 sequences, and approximately 289 million Illumina ESTs (International Brachypodium Initiative 2010). These newly expanded EST collections were prepared from a diverse set of tissues and treatments (International Brachypodium Initiative 2010). EST sequences are useful for many applications, including microarrays, analyses of gene expression, and manual curation of gene models. The Brachypodium ESTs have already been used to refine the phylogeny of Brachypodium and to identify candidates for all of the genes involved in the biosynthesis of lignin monomers (Vogel et al. 2006a). Soon a Brachypodium whole-genome hybrid exon-scanning/expression/tiling microarray containing 6.5 million features/array will be available on an Affymetrix platform (T. Mockler, personal communication).

23.4.3 BAC LIBRARIES

Sequence data obtained from BAC libraries permit evaluation of genome content and complexity when a full genome sequence is not available and can aid in assembly of genome sequencing data. Furthermore, BAC libraries are useful for comparative genomic analyses that can be exploited to facilitate positional cloning of genes in related species. Eight BAC libraries have been constructed for diploid Brachypodium accessions. The first two libraries were constructed from accessions ABR1 and ABR5 and together contain a total of 9100 clones with an average insert size of 88 kb (Hasterok et al. 2006). These relatively small libraries represent approximately 2.7 haploid genome equivalents. Two BAC libraries made from inbred line Bd21, the line used for genome sequencing, contain a combined total of 110,592 clones with an average insert size of approximately 100 kb. These Bd21 libraries represent 38 haploid genome equivalents and provide greater than 99.99% likelihood a particular gene is included within the library (Huo et al. 2006). These Bd21 libraries were used to generate BAC end sequences from 64,694 clones, and the resulting 38.2 Mbp of sequence covers approximately 11% of the Brachypodium genome (Huo et al. 2007). This sequence was used to anchor the BAC clones to the rice genome and indicated that the Brachypodium genome contains 45.9% GC content, approximately 18% repetitive DNA (11% with homology to known repetitive sequence and 7.3% unique to Brachypodium), and 21.2% coding sequence. In addition, the Arizona Genomics Institute (www.genome.arizona.edu/) has constructed one library from the inbred line Bd3-1 and two libraries from Bd21 (M. Bevan, personal communication).

One BAC library exists for the perennial species *B. sylvaticum*. This library contains 30,228 clones with an average insert size of 102 kb (6.6 genome equivalents, on the basis of a genome size of 470 Mbp) (Foote et al. 2004). From this library, repetitive DNA content was estimated to be approximately 50%, and analyses demonstrated that synteny was maintained among rice, wheat, and *B. sylvaticum* BAC contigs over several regions of chromosome 9. The percentage of repetitive DNA in *B. sylvaticum* is much higher than in *B. distachyon* and largely explains the greater size of the *B. sylvaticum* genome.

23.4.4 Maps and Markers

Mapping resources for Brachypodium are developing rapidly. A physical map has been constructed from two of the Bd21 BAC libraries mentioned above (Gu et al. 2009). This map contains over 67,000 BAC clones assembled into 671 contigs and can be accessed at http://phymap.ucdavis.edu:8080/brachypodium/. In addition, a second physical map using two different Bd21 libraries has been constructed (M. Bevan, personal communication). A high-density linkage map based on 562 single nucleotide polymorphism (SNP) markers has also been constructed (N. Huo, unpublished). The markers fell into five linkage groups corresponding to the five chromosomes of the haploid Brachypodium genome. The resulting map was used to assemble the whole-genome shotgun sequence into chromosome-scale assemblies (International Brachypodium Initiative 2010).

Linking individual BACs contained in physical contigs and ultimately genomic sequences to specific chromosomes can be accomplished through a technique called "BAC landing." In this technique, entire BACs are fluorescently labeled and used for FISH. In this fashion, BACs were assigned to specific chromosomes, and 32 of 39 BACs hybridized to a single locus, underscoring the compact nature of the Brachypodium genome (Hasterok et al. 2006). A more extensive application of the technique will be highly instructive in verifying the whole genome assembly and for comparing the evolutionary relationships among genomes of various grasses (Wolny and Hasterok 2009).

Genetic markers are essential for many experiments, including positional cloning, mapping QTLs, association mapping, ECOTILLING, and analysis of genotypic diversity in populations. Polymerase chain reaction (PCR)-based markers are particularly useful because they are fast, easy to score, and can be used by any laboratory with routine molecular biology tools. A recent publication describes the development of 398 SSR markers for Brachypodium (Vogel et al. 2009). SSRs, also known as microsatellites, are genomic areas with simple, short repeat units. The number of repeats is highly polymorphic, making SSRs powerful markers. As previously discussed in Section 23.3.1, the utility of these SSRs was demonstrated by showing that genetic diversity in a large number of new Brachypodium accessions correlated with significant differences in easily scored phenotypes such as seed size, vernalization requirements, and the presence of hairs (Vogel et al. 2009). Another study used 12 SSR markers to examine introduced populations of polyploid Brachypodium and showed that there have been multiple introductions of Brachypodium into the state of California (Bakker et al. 2009).

23.4.5 Whole-Genome Sequencing

A completely sequenced genome underpins a host of tools, including efficient map-based cloning, sequence-indexed T-DNA populations, gene chips, and reverse genetic approaches such as TILLING and RNA interference (RNAi), and is therefore a requirement for a modern model system. Plans for the development of Brachypodium as a model to accelerate the domestication of grasses for use as biomass crops (e.g., switchgrass and Miscanthus) were spelled out in a U.S. Department of Energy report on the research needed to establish a domestic biofuel industry (DOE 2006). As a result, JGI approved a proposal to sequence the Brachypodium genome with a whole-genome shotgun sequencing strategy through their Community Sequencing Program for 2007. In May 2009, JGI released the final 8× sequence and the version 1.0 annotation of the Brachypodium genome. The assembly incorporates all of the mapping and BAC sequence resources and has been shown to be of very high quality: Gaps in the assembly represent only approximately 0.4% of the genome, and Illumina EST data support more than 92% of the predicted protein-coding genes (International Brachypodium Initiative 2010). Through the Community Sequencing Program for 2009 (http:// www.jgi.doe.gov/sequencing/cspseqplans2009.html), JGI has approved the resequencing of six additional Brachypodium lines using next-generation sequencing platforms. The information gained through this project will permit comparisons of the genetic diversity of phenotypically distinct Brachypodium lines and assist in identifying the genetic basis for phenotypes critical to the improvement of bioenergy and cereal crops as well as many other research interests.

23.5   BRACHYPODIUM AS A MODEL FOR BIOENERGY GRASSES

In the development of biofuel feedstocks, two key considerations are biomass quantity and biomass quality (Carpita and McCann 2008). Biomass quantity refers to the amount of feedstock produced per unit of land. Biomass quality reflects the efficiency with which the feedstock can be converted to the desired biofuel. Biomass quality encompasses feedstock recalcitrance and the feedstock's suitability for the conversion process used. Thus, plant materials that exhibit less resistance to degradation or contain larger relative amounts of easily fermentable sugars are of higher "quality." Brachypodium's typical grass characteristics, together with the many resources available for its study, make Brachypodium a suitable system for investigating the factors that contribute to the quantity and quality of feedstocks.

23.5.1   FACTORS INFLUENCING BIOMASS QUANTITY: PLANT ARCHITECTURE

Genetically influenced traits such as growth habit, stem density, and plant height affect the quantity of biomass that can be harvested. The small size and ease of growth of Brachypodium will facilitate the application of forward and reverse genetic approaches to study these traits. In addition, natural variation in these traits has been demonstrated in recently cataloged collections of Brachypodium accessions, providing researchers with another tool to investigate phenotypes of interest for increasing biomass (Vogel et al. 2009). For example, some accessions have an erect growth habit, whereas others exhibit extensive spreading under the same environmental conditions (Figure 23.3). This trait is relevant to biomass quantity; more erect plants can be planted at higher densities and may be less subject to crop losses due to lodging.

Plant height is another major contributor to biomass yield. Assuming there are no accompanying deleterious characteristics, taller plants will have more useable biomass. For example, a comprehensive study of biomass traits in sorghum recombinant inbred lines found a positive and highly significant correlation between plant height and biomass yield (Murray et al. 2008). Pathways associated with the growth-promoting hormones gibberellins and brassinosteroids are potential targets for increasing plant height (Fernandez et al. 2009). The idea of substantially improving yield by altering phytohormone pathways has a historical precedent: Semi-dwarf rice and wheat varieties were key components of the "Green Revolution" of the 1960s and 1970s, during which the grain yields of cereal crops in developing countries increased dramatically (Evans 1993; Conway 1997; Hedden 2003). These dwarf varieties contained a loss-of-function mutation in a gibberellin (GA) biosynthetic gene in rice (Sasaki et al. 2002; Spielmeyer et al. 2002) and a gain-of-function mutation

FIGURE 23.3   Variation in the growth habit of Brachypodium. Two Turkish inbred lines with erect (left) and spreading (right) growth habits are shown.

in a GA response repressor in wheat (Peng et al. 1999). The resulting disruptions in GA synthesis and signaling enabled the shorter varieties to resist lodging, or falling over, and to channel applied nitrogen fertilizer to grain production rather than stem growth (Evans 1993; Conway 1997; Sasaki et al. 2002; Hedden 2003). The converse goal of increasing vegetative biomass might be achieved through the opposite approach, i.e., upregulating GA biosynthesis and/or de-repressing GA growth responses (Sasaki et al. 2002; Fernandez et al. 2009).

The dwarfing alleles in hexaploid, Green Revolution wheat correspond to the *Rht-B1/Rht-D1* genes, which encode DELLA-motif containing proteins (Peng et al. 1999). A loss-of-function mutation in the orthologous rice gene *SLR1* results in increased plant height (Ikeda et al. 2001). BLAST searches of the Brachypodium genome revealed that Brachypodium, like rice, contains a single DELLA-encoding gene and that the Brachypodium DELLA protein exhibits 86% amino-acid identity to *SLR1*; this finding suggests that the mechanisms governing GA-regulated growth responses are likely to be similar in Brachypodium and the other grasses. Components of the GA pathway and additional candidate genes of interest can be easily identified, cloned, and functionally characterized in Brachypodium thanks to a sequenced genome and an efficient transformation protocol. Brachypodium's short life-cycle and simple growth requirements will further facilitate rapid hypothesis testing for genetic manipulations aimed at improving biofuels feedstocks.

23.5.2   FACTORS INFLUENCING BIOMASS QUANTITY: INTERACTIONS WITH THE ENVIRONMENT

In addition to morphological characteristics, a plant's interactions with the environment also affect biomass yield. Thus, adaptability to a range of soil conditions, efficient water and nutrient use, and resistance to abiotic and biotic stresses are all desirable traits in a bioenergy crop (DOE 2007). For investigating these traits, Brachypodium also provides a wealth of resources. The geographical range of Brachypodium encompasses a diversity of habitats (Garvin et al. 2008; Opanowicz et al. 2008; Vogel et al. 2009), providing the opportunity to investigate adaptations to different environments with a genetically tractable organism. The first proposal of Brachypodium as a model grass included the observation that different Brachypodium ecotypes varied in their responses to the agriculturally important fungal pathogens *Puccinia striformis* f. sp. *triticae* and *Magnaporthe grisea*, the causative agents of wheat yellow stripe rust and rice blast, respectively (Draper et al. 2001). Infections with *Fusarium graminearum* (head blight) (D. Garvin, personal communication) and *Pythium* species (root rot) (Vogel and Bragg 2009) have also been noted. Characterizations of viral infections in Brachypodium are extremely limited—with a report of an unidentified virus on *Brachypodium sylvaticum* (Edwards et al. 1985) and preliminary studies using barley stripe mosaic hordeivirus (A. Jackson, personal communication).

To examine Brachypodium-*M. grisea* interactions in detail, Routledge et al. (2004) challenged 21 Brachypodium ecotypes with four strains of *M. grisea*. The Brachypodium ecotypes exhibited responses ranging from resistance to susceptibility. Significantly, plant cell death was associated with resistance, as seen in rice, and the development of disease in a susceptible ecotype paralleled disease progression in rice (Routledge et al. 2004). Analysis of the progeny obtained by crossing the resistant Brachypodium ecotype ABR5 with the susceptible ecotype ABR1 indicated that a single, dominant locus—likely an *R* gene—conferred resistance (Routledge et al. 2004). After developing a protocol for infecting Brachypodium, rice, and barley with *M. grisea* (Parker et al. 2008), researchers compared metabolite patterns during the early stages of infection and found evidence that *M. grisea* suppresses the production of reactive oxygen species and defensive lignin compounds in all three of the host grasses (Parker et al. 2009). Collectively, this work supports the development of the Brachypodium-*M. grisea* pathosystem as a tool to study the cellular events, gene-for-gene interactions, and other processes underlying plant defense and infection.

A number of observations suggest that Brachypodium can be used to study other aspects of responses to biotic and abiotic stresses. The lack of macroscopic disease symptoms in Brachypodium exposed to the causative agent of powdery mildew on wheat, *Blumeria graminis*, has lead to the proposal that Brachypodium would be a good model for studying non-host resistance to this fungal pathogen (Draper et al. 2001). A Brachypodium proteinase inhibitor gene (*Bdpin1*) that is induced by wounding, methyl jasmonate, and *M. grisea* has been identified (Mur et al. 2004). Because *Bdpin1* belongs to a class of molecules that serve as markers of the wounding response and protect against insect damage, this research suggests that Brachypodium could be suitable for studies not only of plant-microbe interactions, but also of wounding and insect herbivory (Mur et al. 2004). Mycorrhizal fungi can aid plants in the uptake of nutrients, especially phosphate, thereby reducing the cost of agricultural inputs such as fertilizers and facilitating plant growth on marginal land (Morgan et al. 2005). These interactions will be especially important in the context of biomass crops because it is projected that these crops will be grown on marginal lands with minimal fertilizer inputs. Brachypodium, unlike *Arabidopsis*, forms mycorrhizal associations (M. Harrison, personal communication) and thus enables studies of these symbiotic relationships in a model system.

23.5.3 Factors Influencing Biomass Quality: Cell Wall Structure and Composition

Plant-derived biofuel feedstocks primarily consist of plant cell walls. Key cell wall components relevant to fuel production include the carbohydrate polymers cellulose and hemicellulose and the phenylpropanoid polymer lignin. These polymers are cross-linked together to form a composite material that is the cell wall. The production of liquid fuels or specialty chemicals from biomass first requires the deconstruction of the glucan portion of this polymer-composite matrix into its constituent parts, namely monosaccharides. The inherent recalcitrance of the plant cell wall to degradation adds considerably to the expense of this step (DOE 2006). Thus, comprehensive knowledge of the genes that determine grass cell wall structure and composition could be used to design superior feedstocks and accelerate the domestication of the grasses proposed as energy crops (e.g., switchgrass and Miscanthus). The most rapid way to acquire this knowledge is through the use of a model system such as Brachypodium.

Because the compositions of dicot and grass cell walls differ substantially, *Arabidopsis* cannot be used to study all aspects of the grass cell wall (Carpita 1996; Vogel 2008). Commelinoid monocots, including the grasses, have type II primary cell walls. Dicots, such as *Arabidopsis*, and noncommelinoid monocots, including orchids and lilies, have type I primary cell walls. Cell wall composition varies from species to species and even among cell types within a single plant (Knox 2008; Popper 2008). Although the primary and secondary walls of both grasses and dicots contain cellulose, the major hemicellulose is glucuronoarabinoxylan (GAX) in grasses and xyloglucan in dicots. Grass cell walls also contain ferulic acid and ρ-coumaric acid, hydroxycinnamate compounds that are only minor components of most dicot walls. Ferulate-mediated cross-linking of GAX and lignin decreases the digestibility of monocot walls (Grabber 2005) with direct implications for biofuel production. In addition to the differences in hemicellulose and hydroxycinnamates mentioned above, grass primary walls, unlike their dicot counterparts, incorporate mixed linkage (β1,3- and β1,4-linked) glucans and contain few structural proteins and little pectin compared with the large amounts of pectin—up to 35% dry weight—found in type I walls (Vogel 2008). Supporting the use of Brachypodium as a model for grass cell walls is the finding that the monosaccharide profiles of Brachypodium, wheat, barley, and Miscanthus cell walls are similar to each other but different from the profile of *Arabidopsis* (Gomez et al. 2008).

Cell wall composition is critical in the context of the biofuel production process. The conversion of lignocellulosic biomass into liquid fuels generally proceeds via four steps: (1) mechanical and/or thermochemical pretreatment of feedstocks to make the cell wall components more accessible, (2) enzymatic hydrolysis of the pretreated biomass to release sugars from the carbohydrate polymers of the wall, (3) microbial fermentation of the released sugars to produce liquid fuels, and (4) recovery of the biofuels from the fermentation medium (for example, by distillation) (Himmel et al. 2007; Wyman 2007; Kumar et al. 2008). The first two steps—pretreatment and hydrolysis—are necessary to overcome the recalcitrance of plant materials (i.e., their resistance to degradation). Recalcitrance is arguably the largest obstacle to the economical, efficient, and environmentally friendly production of biofuels (Himmel et al. 2007; Wyman 2007). There are various pretreatment options, including treatment with dilute acid, washing with large volumes of hot water, and ammonia fiber expansion (Wyman et al. 2005). Although each pretreatment method has distinct advantages and disadvantages, pretreatments in general require substantial inputs of energy, chemicals, or other resources (Wyman et al. 2005). Also, the production and use of hydrolytic enzymes can be both expensive and limiting (Wyman 2007; Kumar et al. 2008). Additional considerations include the possibility that pretreatments can release or produce inhibitors of fermentation (Himmel et al. 2007) and that the most commonly used fermentative microbes—primarily the budding yeast *Saccharomyces cerevisiae*, but also species such as the anaerobic bacterium *Clostridium thermocellum*—use only six-carbon sugars such as the glucose monomers of cellulose, not five-carbon sugars such as the xylose found in hemicelluloses (Demain et al. 2005; van Maris et al. 2006). From a feedstock development perspective, biomass recalcitrance might be decreased in a number of ways, e.g., by altering the structure or cell-cell-adhesion properties of plant tissues to allow greater access by chemicals and hydrolytic enzymes or by reducing the degree of cross-linking of the cell wall components, the crystallinity of cellulose, or the amount of lignin. Increasing the amount of cellulose or the percentage of hexoses in other cell wall polymers could also improve production efficiency by providing more substrates for microbial fermentation.

The resources available for Brachypodium will be useful for addressing these issues. Because cell walls support and protect plants, there is always the concern that modifying the wall will negatively affect plant fitness. In this respect, the natural variation documented for Brachypodium could prove especially valuable. If an accession with improved digestibility and/or fermentability can be found, its cell wall characteristics could reveal new avenues for altering wall composition and structure without severely decreasing viability. In a complementary approach, the effects of targeted modifications or randomly induced mutations can be analyzed more rapidly in Brachypodium than in grasses with longer generation times. However, to identify naturally occurring or mutant plants with desirable cell wall traits, new screening procedures must be implemented. As a first step, Gomez et al. (2008) showed that treatments with hot, dilute sulfuric acid resulted in limited hydrolysis of Brachypodium cell wall material. This work thus establishes a pretreatment method that can be used to screen for variations—even subtle ones—in the recalcitrance of Brachypodium biomass (Gomez et al. 2008). Further studies applying the tools of cell wall analysis and the techniques of biofuel production to Brachypodium will help to meet the challenge of improving bioenergy feedstocks.

23.6   FUTURE PERSPECTIVES

Brachypodium is rapidly gaining in utility and acceptance as a model grass species. A critical mass of resources and researchers using Brachypodium has been achieved, and this trend will likely accelerate with the completion of the genome sequence. In addition to Brachypodium's relevance to bioenergy, over 30 research papers in the previous 3 years alone have used Brachypodium to investigate topics as diverse as microRNAs (Unver and Budak 2009; Wei et al. 2009), comparative genomics (Bortiri et al. 2008; Huo et al. 2009; Kumar et al. 2009), characteristics of introduced populations (Bakker et al. 2009), and seed storage proteins (Laudencia-Chingcuanco and Vensel 2008). The interest in Brachypodium is also reflected in the large number of people—approximately 75 individuals from more than 20 laboratories—who have recently contributed to the Brachypodium genome annotation effort (International Brachypodium Initiative 2010). In summary, Brachypodium combines the desirable attributes of a model organism with many of the traits of interest for the development and improvement of biofuel feedstocks. Although not itself a bioenergy crop, Brachypodium is an accessible, informative representative of the grasses, a group of plants that are tremendously important to the future of bioenergy.

ACKNOWLEDGMENTS

This word was supported by USDA CRIS project 5325-21000-013-00. "Biotechnological Enhancement of Energy Crops" and by the Office of Science (BER), US Department of Energy, Interagency Agreement No. DE-AI02-07ER64452. Jenniferm N. Bragg and Ludmila Tyler contributed equally to this work as co-first authors.

REFERENCES

Adams M, Kelley J, Dubnick M, Polymeropoulos MH, Xiao H, Merril CR, Wu A, Olde B, Moreno RF, et al. (1991) Complementary DNA sequencing: expressed sequence tags and human genome project. *Science* 252:1651–1656

An G, Jeong D-H, Jung K-H, Lee S (2005) Reverse genetic approaches for functional genomics of rice. *Plant Mol Biol* 59:111–123

Bablak P, Draper J, Davey MR, Lynch PT (1995) Plant regeneration and micropropagation of Brachypodium distachyon. *Plant Cell Tiss Org Cult* 42:97–107

Bakker EG, Montgomery B, Nguyen T, Eide K, Chang J, Mockler TC, Liston A, Seabloom EW, Borer ET (2009) Strong population structure characterizes weediness gene evolution in the invasive grass species Brachypodium distachyon. *Mol Ecol* 18:2588–2601

Bortiri E, Coleman-Derr D, Lazo GR, Anderson OD, Gu YQ (2008) The complete chloroplast genome sequence of *Brachypodium distachyon*: Sequence comparison and phylogenetic analysis of eight grass plastomes. *BMC Res Notes* 1:61

Bossolini E, Wicker T, Knobel PA, Keller B (2007) Comparison of orthologous loci from small grass genomes *Brachypodium* and rice: Implications for wheat genomics and grass genome annotation. *Plant J* 49:704–717

Caldwell DG, McCallum N, Shaw P, Muehlbauer GJ, Marshall DF, Waugh R (2004) A structured mutant population for forward and reverse genetics in Barley (*Hordeum vulgare* L.). *Plant J* 40:143–150

Carpita NC (1996) Structure and biogenesis of the cell walls of grasses. *Annu Rev Plant Physiol Plant Mol Biol* 47:445–476

Carpita NC, McCann MC (2008) Maize and sorghum: Genetic resources for bioenergy grasses. *Trends Plant Sci* 13:415–420

Catalán P, Olmstead RG (2000) Phylogenetic reconstruction of the genus *Brachypodium* P. Beauv. (Poaceae) from combined sequences of chloroplast *ndhF* gene and nuclear ITS. *Plant Syst Evol* 220:1–19

Catalán P, Ying S, Armstrong L, Draper J, Stace CA (1995) Molecular phylogeny of the grass genus *Brachypodium* P.Beauv. based on RFLP and RAPD analysis. *Bot J Linn Soc* 117:263–280

Christiansen P, Didion T, Andersen CH, Folling M, Nielsen KK (2005) A rapid and efficient transformation protocol for the grass *Brachypodium distachyon*. *Plant Cell Rep* 23:751–758

Christou P, Ford TL, Kofron M (1991) Production of transgenic rice (*Oryza sativa* L.) plants from agronomically important indica and japonica varieties via electric discharge particle acceleration of exogenous DNA into immature zygotic embryos. *Nature Biotechnol* 9:957–962

Clough SJ, Bent AF (1998) Floral dip: A simplified method for *Agrobacterium*-mediated transformation of *Arabidopsis thaliana*. *Plant J* 16:735–743

Conway G (1997) *The Doubly Green Revolution: Food for All in the Twenty-First Century*. Cornell University Press, Ithaca, NY

Dai S, Zheng P, Marmey P, Zhang S, Tian W, Chen S, Beachy RN, Fauquet C (2001) Comparative analysis of transgenic rice plants obtained by *Agrobacterium*-mediated transformation and particle bombardment. *Molecular Breeding* 7:25–33

Demain AL, Newcomb M, Wu JHD (2005) Cellulase, clostridia, and ethanol. *Microbiol Mol Biol Rev* 69:124–154

DOE (2006) *Breaking the Biological Barriers to Cellulosic Ethanol: A Joint Research Agenda*. U.S. Department of Energy, Office of Science and Office of Energy Efficiency, Washington, DC

DOE (2007) *Historical Perspective on How and Why Switchgrass Was Selected as a "Model" High-Potential Energy Crop*. U.S. Department of Energy, Oak Ridge National Laboratory, Environmental Sciences Division, Oak Ridge, TN

DOE and USDA (2005) *Biomass as Feedstock for a Bioenergy and Bioproducts Industry: The Technical Feasibility of a Billion-Ton Annual Supply*. U.S. Department of Energy, Oak Ridge National Laboratory, Oak Ridge, TN

Dohleman FG, Long SP (2009) More productive than maize in the midwest: How does *Miscanthus* do it? *Plant Physiol* 150:2104–2115

Döring E, Schneider J, Hilu KW, Röser M (2007) Phylogenetic relationships in the Aveneae/Poeae complex (Pooideae, Poaceae). *Kew Bull* 62:407–424

Draper J, Mur LAJ, Jenkins G, Ghosh-Biswas GC, Bablak P, Hasterok R, Routledge APM (2001) *Brachypodium distachyon*. A new model system for functional genomics in grasses. *Plant Physiol* 127:1539–1555

Edwards M, Cooper J, Massalski P, Green B (1985) Some properties of a virus-like agent found in *Brachypodium sylvaticum* in the United Kingdom. *Plant Pathol* 34:95–104

Evans LT (1993) *Crop Evolution, Adaptation, and Yield*. Cambridge University Press, Cambridge, United Kingdom

Feldmann KA (1991) T-DNA insertion mutagenesis in Arabidopsis: Mutational spectrum. *Plant J* 1:71–82

Fernandez MGS, Becraft PW, Yin Y, Lübberstedt T (2009) From dwarves to giants? Plant height manipulation for biomass yield. *Trends Plant Sci* 14:454–461

Filiz E, Ozdemir B, Budak F, Vogel J, Tuna M, Budak H (2009) Molecular, morphological and cytological analysis of diverse *Brachypodium distachyon* inbred lines. *Genome* 52:876–890

Fits LVD, Hilliou F, Memelink J (2001) T-DNA activation tagging as a tool to isolate regulators of a metabolic pathway from a genetically non-tractable plant species. *Transgen Res* 10:513–521

Foote TN, Griffiths S, Allouis S, Moore G (2004) Construction and analysis of a BAC library in the grass *Brachypodium sylvaticum*: Its use as a tool to bridge the gap between rice and wheat in elucidating gene content. *Funct Integr Genom* 4:26–33

Fu X, Duc LT, Fontana S, Bong BB, Tinjuangjun P, Sudhakar D, Twyman RM, Christou P, Kohli A (2000) Linear transgene constructs lacking vector backbone sequences generate low-copy-number transgenic plants with simple integration patterns. *Transgen Res* 9:11–19

Garvin DF, Gu YQ, Hasterok R, Hazen SP, Jenkins G, Mockler TC, Mur LAJ, Vogel JP (2008) Development of genetic and genomic research resources for *Brachypodium distachyon*, a new model system for grass crop research. *Crop Sci* 48:S69–S84

Gomez LD, Bristow JK, Statham ER, McQueen-Mason SJ (2008) Analysis of saccharification in *Brachypodium distachyon* stems under mild conditions of hydrolysis. *Biotechnol Biofuels* 1:15

Grabber JH (2005) How do lignin composition, structure, and cross-linking affect degradability? A review of cell wall model studies. *Crop Sci* 45:820–831

Greene EA, Codomo CA, Taylor NE, Henikoff JG, Till BJ, Reynolds SH, Enns LC, Burtner C, Johnson JE, Odden AR, Comai L, Henikoff S (2003) Spectrum of chemically induced mutations from a large-scale reverse-genetic screen in Arabidopsis. *Genetics* 164:731–740

Griffiths S, Sharp R, Foote TN, Bertin I, Wanous M, Reader S, Colas I, Moore G (2006) Molecular characterization of *Ph1* as a major chromosome pairing locus in polyploid wheat. *Nature* 439:749–752

Gu YQ, Ma Y, Huo N, Vogel JP, You FM, Lazo GR, Nelson WM, Soderlund C, Dvorak J, Anderson OD, Luo M-C (2009) A BAC-based physical map of *Brachypodium distachyon* and its comparative analysis with rice and wheat. *BMC Genom* 10:1471

Hasterok R, Draper J, Jenkins G (2004) Laying the cytotaxonomic foundations of a new model grass, *Brachypodium distachyon* (L.) beauv. *Chrom Res* 12:397–403

Hasterok R, Marasek A, Donnison IS, Armstead I, Thomas A, King IP, Wolny E, Idziak D, Draper J, Jenkins G (2006) Alignment of the genomes of *Brachypodium distachyon* and temperate cereals and grasses using bacterial artificial chromosome landing with fluorescence in situ hybridization. *Genetics* 173:349–362

Hedden P (2003) The genes of the green revolution. *Trends Genet* 19:5–9

Himmel ME, Ding S-Y, Johnson DK, Adney WS, Nimlos MR, Brady JW, Foust TD (2007) Biomass recalcitrance: Engineering plants and enzymes for biofuels production. *Science* 315:804–807

Hsiao C, Chatterton NJ, Asay KH, Jensen KB (1994) Phylogenetic relationships of 10 grass species: An assessment of phylogenetic utility of the internal transcribed spacer region in nuclear ribosomal DNA in monocots. *Genome* 37:112–120

Huo N, Gu Y, Lazo G, Vogel J, Coleman-Derr D, Luo M, Thilmony R, Garvin D, Anderson O (2006) Construction and characterization of two BAC libraries from *Brachypodium distachyon*, a new model for grass genomics. *Genome* 49:1099–1108

Huo N, Lazo GR, Vogel JP, You FM, Ma Y, Hayden DM, Coleman-Derr D, Hill TA, Dvorak J, Anderson OD, Luo M-C, Gu YQ (2007) The nuclear genome of *Brachypodium distachyon*: Analysis of BAC end sequences. *Funct Integr Genom* 8:135–147

Huo N, Vogel JP, Lazo GR, You FM, Ma Y, McMahon S, Dvorak J, Anderson OD, Luo MC, Gu YQ (2009) Structural characterization of *Brachypodium* genome and its syntenic relationship with rice and wheat. *Plant Mol Biol* 70:47–61

Ikeda A, Ueguchi-Tanaka M, Sonoda Y, Kitano H, Koshioka M, Futsuhara Y, Matsuoka M, Yamaguchi J (2001) slender rice, a constitutive gibberellin response mutant, is caused by a null mutation of the *SLR1* gene, an ortholog of the height-regulating gene GAI/RGA/RHT/D8. *Plant Cell* 13:999–1010

Jeon J-S, Lee S, Jung K-H, Jun S-H, Jeong D-H, Lee J, Kim C, Jang S, Lee S, Yang K, Nam J, An K, Han M-J, Sung R-J, Choi H-S, Yu J-H, Choi J-H, Cho S-Y, Cha S-S, Kim S-I, An G (2000) T-DNA insertional mutagenesis for functional genomics in rice. *Plant J* 22:561–570

Kellogg EA (1998) Relationships of cereal crops and other grasses. *Proc Natl Acad Sci USA* 95:2005–2010

Kellogg EA (2001) Evolutionary history of the grasses. *Plant Physiol* 125:1198–1205

Khan MA, Stace CA (1999) Breeding relationships in the genus *Brachypodium* (Poaceae: Pooideae). *Nord J Bot* 19:257–269

Kim Y, Schumaker KS, Zhu JK (2006) EMS mutagenesis of Arabidopsis. *Meth Mol Biol* (Clifton, NJ) 323:101–103

Knox JP (2008) Revealing the structural and functional diversity of plant cell walls. *Curr Opin Plant Biol* 11:308–313

Kohli A, Twyman RM, Abranches R, Wegel E, Stoger E, Christou P (2003) Transgene integration, organization and interaction in plants. *Plant Mol Biol* 52:247–258

Kumar R, Singh S, Singh OV (2008) Bioconversion of lignocellulosic biomass: Biochemical and molecular perspectives. *J Indust Microbiol Biotechnol* 35:377–391

Kumar S, Mohan A, Balyan HS, Gupta PK (2009) Orthology between genomes of *Brachypodium*, wheat and rice. *BMC Res Notes* 2:93

Laudencia-Chingcuanco DL, Vensel WH (2008) Globulins are the main seed storage proteins in *Brachypodium distachyon*. *Theor Appl Genet* 117:555–563

Loc NT, Tinjuangjun P, Gatehouse AMR, Christou P, Gatehouse JA (2002) Linear transgene constructs lacking vector backbone sequences generate transgenic rice plants which accumulate higher levels of proteins conferring insect resistance. *Mol Breed* 9:231–244

Morgan JA, Bending GD, White PJ (2005) Biological costs and benefits to plant-microbe interactions in the rhizosphere. *J Exp Bot* 56:1729–1739

Mur LAJ, Xu R, Casson SA, Stoddart WM, Routledge APM, Draper J (2004) Characterization of a proteinase inhibitor from *Brachypodium distachyon* suggests the conservation of defence signalling pathways between dicotyledonous plants and grasses. *Mol Plant Pathol* 5:267–280

Murray SC, Rooney WL, Mitchell SE, Sharma A, Klein PE, Mullet JE, Kresovich S (2008) Genetic improvement of sorghum as a biofuel feedstock: II. QTL for stem and leaf structural carbohydrates. *Crop Sci* 48:2180–2193

Nakazawa M, Ichikawa T, Ishikawa A, Kobayashi H, Tsuhara Y, Kawashima M, Suzuki K, Muto S, Matsui M (2003) Activation tagging, a novel tool to dissect the functions of a gene family. *Plant J* 34:741–750

Olsen P, Lenk I, Jensen CS, Petersen K, Andersen CH, Didion T, Nielsen KK (2006) Analysis of two heterologous flowering genes in *Brachypodium distachyon* demonstrates its potential as a grass model plant. *Plant Sci* 170:1020–1025

Opanowicz M, Vain P, Draper J, Parker D, Doonan JH (2008) *Brachypodium distachyon*: Making hay with a wild grass. *Trends Plant Sci* 13:172–177

Păcurar DI, Thordal-Christensen H, Nielsen KK, Lenk I (2007) A high-throughput *Agrobacterium*-mediated transformation system for the grass model species *Brachypodium distachyon* L. *Transgen Res* 17:965–975

Pan X, Liu H, Clarke J, Jones J, Bevan M, Stein L (2003) ATIDB: *Arabidopsis thaliana* insertion database. *Nucl Acids Res* 31:1245–1251

Parker D, Beckmann M, Enot DP, Overy DP, Rios ZC, Gilbert M, Talbot NJ, Draper J (2008) Rice blast infection of *Brachypodium distachyon* as a model system to study dynamic host/pathogen interactions. *Nat Protoc* 3:435–445

Parker D, Beckmann M, Zubair H, Enot DP, Caracuel-Rios Z, Overy DP, Snowdon S, Talbot NJ, Draper J (2009) Metabolomic analysis reveals a common pattern of metabolic re-programming during invasion of three host plant species by *Magnaporthe grisea*. *Plant J* 59:723–737

Peng J, Richards DE, Hartley NM, Murphy GP, Devos KM, Flintham JE, Beales J, Fish LJ, Worland AJ, Pelica F, Sudhakar D, Christou P, Snape JW, Gale MD, Harberd NP (1999) 'Green revolution' genes encode mutant gibberellin response modulators. *Nature* 400:256–261

Popper ZA (2008) Evolution and diversity of green plant cell walls. *Curr Opin Plant Biol* 11:286–292

Rayburn A, Crawford J, Rayburn C, Juvik J (2009) Genome size of three Miscanthus species. *Plant Mol Biol Rep* 27:184–188

Robertson IH (1981) Chromosome numbers in *Brachypodium* Beauv. (Gramineae). *Genetica* 56:55–60

Rokas A, Williams BI, King N, Carroll SB (2003) Genome-scale approaches to resolving incongruence in molecular phylogenies. *Nature* 425:798–804

Routledge APM, Shelley G, Smith JV, Draper J, Mur LAJ, Talbot NJ (2004) *Magnaporthe grisea* interactions with the model grass *Brachypodium distachyon* closely resemble those with rice (*Oryza sativa*). *Mol Plant Pathol* 5:253–265

Sasaki A, Ashikari M, Ueguchi-Tanaka M, Itoh H, Nishimura A, Swapan D, Ishiyama K, Saito T, Kobayashi M, Khush GS, Kitano H, Matsuoka M (2002) Green revolution: A mutant gibberellin-synthesis gene in rice. *Nature* 416:701–702

Schwartz CJ, Doyle MR, Manzaneda AJ, Rey PJ, Mitchell-Olds T, Amasino RM (2010) Natural variation of flowering time and vernalization responsiveness in *Brachypodium distachyon*. *Bioenergy Res* 3:38–46

Shi C, Koch G, Ouzunova M, Wenzel G, Zein I, Lübberstedt T (2006) Comparison of maize brown-midrib isogenic lines by cellular UV-microspectrophotometry and comparative transcript profiling. *Plant Mol Biol* 62:697–714

Spielmeyer W, Ellis MH, Chandler PM (2002) Semidwarf (sd-1), "green revolution" rice, contains a defective gibberellin 20-oxidase gene. *Proc Natl Acad Sci USA* 99:9043–9048

Svitashev SK, Somers DA (2002) Characterization of transgene loci in plants using FISH: A picture is worth a thousand words. *Plant Cell Tiss Organ Cult* 69:205–214

Travella S, Ross SM, Harden J, Everett C, Snape JW, Harwood WA (2005) A comparison of transgenic barley lines produced by particle bombardment and *Agrobacterium*-mediated techniques. *Plant Cell Rep* 23:780–789

Tyagi AK, Mohanty A (2000) Rice transformation for crop improvement and functional genomics. *Plant Sci* 158:1–18

Unver T, Budak H (2009) Conserved microRNAs and their targets in model grass species *Brachypodium distachyon*. *Planta* 230:659–669

Vain P, Worland B, Thole V, McKenzie N, Alves SC, Opanowicz M, Fish LJ, Bevan MW, Snape JW (2008) *Agrobacterium*-mediated transformation of the temperate grass *Brachypodium distachyon* (genotype Bd21) for T-DNA insertional mutagenesis. *Plant Biotechnol J* 6:236–245

van Maris AJ, Abbott DA, Bellissimi E, van den Brink J, Kuyper M, Luttik MA, Wisselink HW, Scheffers WA, van Dijken JP, Pronk JT (2006) Alcoholic fermentation of carbon sources in biomass hydrolysates by *Saccharomyces cerevisiae*: Current status. *Antonie Van Leeuwenhoek* 90:391–418

Vogel J (2008) Unique aspects of the grass cell wall. *Curr Opinion Plant Biol* 11:301–307.

Vogel J, Bragg J (2009) *Brachypodium distachyon*, a new model for the Triticeae. In: Feuillet C, Muehlbauer GJ (eds), *Genetics and Genomics of the Triticeae.* Springer, New York, pp 427–449

Vogel J, Gu Y, Twigg P, Lazo G, Laudencia-Chingcuanco D, Hayden D, Donze T, Vivian, L., Stamova B, Coleman-Derr D (2006a) EST sequencing and phylogenetic analysis of the model grass *Brachypodium distachyon. Theor and Appl Genet* 113:186–195

Vogel J, Hill T (2008) High-efficiency *Agrobacterium*-mediated transformation of *Brachypodium distachyon* inbred line Bd21-3. *Plant Cell Rep* 27:471–478

Vogel JP, Garvin DF, Leong OM, Hayden DM (2006b) *Agrobacterium*-mediated transformation and inbred line development in the model grass *Brachypodium distachyon. Plant Cell Tiss Org Cult* 85:199–211

Vogel JP, Tuna M, Budak H, Huo N, Gu YQ, Steinwand MA (2009) Development of SSR markers and analysis of diversity in Turkish populations of *Brachypodium distachyon. BMC Plant Biol* 9:88

Wei B, Cai T, Zhang R, Li A, Huo N, Li S, Gu YQ, Vogel J, Jia J, Qi Y, Mao L (2009) Novel microRNAs uncovered by deep sequencing of small RNA transcriptomes in bread wheat (*Triticum aestivum* L.) and *Brachypodium distachyon* (L.) Beauv. *Funct Integr Genomics* 4:499–511

Weigel D, JiHoon A, Blázquez MA, Borevitz JO, Christensen SK, Fankhauser C, Ferrándiz C, Kardailsky I, Malancharuvil EJ, Neff MM, Nguyen JT, Sato S, ZhiYong W, YiJi X, Dixon RA, Harrison MJ, Lamb CJ, Yanofsky MF, Chory J (2000) Activation tagging in *Arabidopsis. Plant Physiol* 122:1003–1013

Wolny E, Hasterok R (2009) Comparative cytogenetic analysis of the genomes of the model grass *Brachypodium distachyon* and its close relatives. *Ann Bot* 5:873–881

Wyman CE (2007) What is (and is not) vital to advancing cellulosic ethanol. *Trends Biotechnol* 25:153–157

Wyman CE, Dale BE, Elander RT, Holtzapple M, Ladisch MR, Lee YY (2005) Comparative sugar recovery data from laboratory scale application of leading pretreatment technologies to corn stover. *Bioresour Technol* 96:2026–2032

24 Diesel Trees

Blake Lee Joyce
University of Tennessee

Hani Al-Ahmad
University of Tennessee and
An-Najah National University

Feng Chen and C. Neal Stewart, Jr.
University of Tennessee

CONTENTS

24.1 INTRODUCTION

The natural history of diesel trees has a long interaction with humans in the realm of economic botany. Trees in the genus *Copaifera* belong to the subfamily Caesalpinioideae in the family Fabaceae. In total, there are more than 70 species of *Copaifera* distributed throughout the world with at least 30 species found in South and Central America, primarily in Brazil, four species in Africa, and one in Malaysia and the Pacific Islands (Dwyer 1951, 1954; Hou 1994). The first species in the genus *Copaifera* was described by George Marcgraf and Willem Pies in 1628, but no formal species name was ascribed to the plant, although later it was deemed *Copaifera martii* on the basis of the description by Veiga Junior and Pinto (2002). Oleoresin from a *Copaifera* tree was listed as a drug in the London Pharmacopoeia in 1677 and to the United States Pharmacopoeia in 1820, and Linnaeus first described the genus *Copaifera* in 1762 (Plowden 2004). Later, more descriptions of *Copaifera* species were completed by Hayne in 1825 and Bentham in 1876 (Dwyer 1951). The current taxonomy of the genus has been largely defined by Dwyer and Léonard, who resolved the differences between the genera *Copaifera* and *Guibourtia* and further developed the New World and African species descriptions in the early 1950s (Léonard 1949, 1950; Dwyer 1951, 1954). Some species are still difficult to identify in the field, even to specialists, because of an incomplete taxonomy and esoteric species differences that rely on intricate flower morphology and other transient characteristics than can be difficult to ascertain or collect compared with leaf morphology. To complicate this situation further, *Copaifera* trees have been known to only flower once every 2 or

3 years in Amazônia (Alencar 1982; Pedroni et al. 2002). Furthermore, most up-to-date references on *Copaifera* taxonomy are in Portuguese, which hampers the interchange of information among the mainstream of scientists.

Copaifera species found in Africa are biochemically distinct from those discussed above because they produce resins that harden into a solid copal, which fossilizes into amber, whereas New World species produce a liquid oleoresin because of the higher concentrations of sesquiterpenes (Langenheim 1973). Oleoresin, which results from tapping *Copaifera* trees, was listed as a drug in the London Pharmacopoeia in 1677 and to the United States Pharmacopoeia in 1820. In Brazil, the oleoresin produced by *Copaifera* trees has been used by native people as a local medicine for healing wounds; an antiseptic; to relieve pain; and for a host of skin, respiratory, and urinary ailments (Plowden 2004). They have also been used for more esoteric purposes such as a snake bite remedy, aphrodisiac, removal of intestinal parasites, and as a contraceptive.

More recently, several scientific studies have verified the medicinal properties of various *Copaifera* oleoresin fractions for anti-inflammatory activity (Veiga Junior et al. 2007), stomach ulcers and intestinal damage mitigation (Paiva et al. 1998, 2004), anticancer activity (Ohsaki et al. 1994; Lima et al. 2003; Gomes et al. 2008), reduced pain sensitivity (Gomes et al. 2007), and increased rate of wound healing (Paiva et al. 2002). The oleoresin and oils of *Copaifera* species have also been used in varnishes and lacquers, as lumber, cosmetic products, and tracing paper (Plowden 2004; Lima and Pio 2007).

Additionally, in 1980 the Nobel Prize-winning chemist Melvin Calvin noted that the oleoresin from *Copaifera* trees was being used as diesel fuel directly from the tree with no-to-minimal processing (Calvin 1980). Calvin began his search for plants that could produce liquid fuels to be used directly in engines after the 1973 oil embargo. He later wrote two more papers in 1983 and 1986 on the potential for production of hydrocarbon fuels from living plants, the issue of global warming, and the pressing need to address U.S. foreign oil dependency which now, some 20 years later, seems almost prophetic. Plantations of *Copaifera* trees were established in Manaus, Brazil to test the viability of biofuel production in the 1980s, but they were later shifted to focus on production of timber and the oleoresin for pharmaceutical and industrial purposes (Plowden 2004). The direct reasons for this shift were undoubtedly economic when diesel fuel returned to being relatively cheap.

24.2 CHEMICALS PRESENT IN *COPAIFERA* OLEORESINS

Copaifera oleoresins, in general, are unique because they contain a greater fraction of sesquiterpenes compared with mono- and diterpenes. In *Copaifera multijuga*, approximately 80% of the oleoresin is composed of sesquiterpenes, whereas in *Copaifera guianensis* only approximately 44% of the oleoresin was composed of sesquiterpenes (Cascon and Gilbert 2000). These authors also noted that the majority ratio of diterpene acids and sesquiterpenes oscillated back and forth throughout the growing season in *Copaifera duckei*.

A wealth of original articles and review papers has focused on describing terpene biosynthesis. As such, only a brief description of the major terpene constituent characteristics and their biosynthesis in relation to conifer and *Copaifera* structures will be attempted here.

In short, isoprene units, the building blocks of terpenoids, are derived from either the mevalonic acid (MVA) pathway present in the cytosol of cells, or the 2-C-methylerythritol-4-phosphate (MEP) pathway, also known as the nonmevalonate pathway, which occurs in plastids (Lichtenthaler 1999). Condensation of isopentenyl diphosphate and its isomer dimethylallyl diphosphate, the products of the MVA and MEP pathways, leads to the formation of geranyl pyrophosphate (GPP), farensyl pyrophosphate (FPP), or geranyl geranyl pyrophosphate (GGPP) which are the common precursors for mono-, sesqui-, and diterpenes, respectively. These three intermediates are catalyzed to form mono-, di-, and sesquiterpenes by the action of terpene synthases (TPSs). Individual TPSs can generate either one product or multiple products that, in turn, can be linear or cyclic. Mono- and diterpenes are thought to be derived primarily from isoprenes made in the plastid through the MEP pathway, whereas sesquiterpenes are derived from isoprenes made in the cytosol where the MVA pathway occurs. Movement of intermediates between these two pathways has been demonstrated in plants (Cheng et al. 2007).

The chemical compounds present in *Copaifera* oleoresin vary not only with tissue type (Gramosa and Silveira 2005; Chen et al. 2009) but also seasonally (Cascon and Gilbert 2000; Zoghbi et al. 2007), and among species (Veiga Junior et al. 2007). Therefore, any future genomics-based characterization of *Copaifera* trees must be coupled with close biochemical analysis to correctly match major compounds present in each tissue at the time of sampling. Identification of particular chemicals responsible for the pharmaceutical effects of *Copaifera* oleoresins will be necessary in the future because high chemical variability within samples, seasons, and species will inherently affect the effectiveness, dosage, and safety for patients.

24.3 BIOSYNTHESIS OF *COPAIFERA* OLEORESINS: WHAT CONIFERS CAN TEACH US

Not much is known about the biosynthesis of *Copaifera* oleoresins because most studies have been focused on traditional ecology and forestry of the genus. However, conifer resins have been thoroughly studied for over 40 years. These oleoresins are essentially made of the same basic constituents as *Copaifera* oleoresin: mono-, di-, and sesquiterpenes. Conifer oleoresins usually have an equal part of mono- and diterpene compounds with lower concentrations of sesquiterpenes (Martin et al. 2002).

Monoterpenes are volatile components found in oleoresins. Monoterpene synthases have been extracted from the woody stems of ten conifer species, and their activities have been measured (Lewinsohn et al. 1991). Species with resin ducts showed the highest levels of monoterpene cyclase activity from wood extracts, suggesting that monoterpene synthesis for oleoresins occurs in epithethial cells surrounding the resin ducts. Diterpenoids themselves are not typically found in conifer oleoresins in large quantities. Instead, modifications such as hydroxylation and oxidation occur, so the alcohol, aldehyde, and predominantly acid products are present (Keeling and Bohlmann 2006). These modified diterpene products harden the resin and form rosin after the volatile constituents evaporate. Sesquiterpenes, like monoterpenes, are volatile and are major constituents of *Copaifera* oleoresin. The three major chemical constituents, based on percentage of oleoresin from different species, are presented in Table 24.1. Although the percentages vary, β-caryophyllene is the major sesquiterpene product of oleoresins throughout *Copaifera* species that have been studied to date. Other than these major three sesquiterpenes in each species, there is a great diversity of terpenoids produced in the oleoresin. Nuclear magnetic resonance (NMR) studies have found previously undescribed diterpenes (Monti et al. 1996, 1999) that seem to be unique in biology.

Conifers produce a myriad of specialized tissues to store and secrete oleoresins that range from simple resin blisters to intricate networks of resin ducts (Martin et al. 2002). *Copaifera* trees form resin ducts throughout their xylem tissue that can easily be seen in cross-sections (Calvin 1980). *Copaifera*, *Hymenaea*, and *Daniella* resin ducts display many structural similarities (Langenheim 2003).

Conifer oleoresins accumulate in resin ducts throughout their lifetimes, but a local response can also be induced during mechanical damage, herbivory, or even fungal inoculation. This response activates epithelial cells in resin ducts, signals for formation of special traumatic resin ducts in stem xylem tissue, and induces diterpene biosynthesis gene transcripts (Keeling and Bohlmann 2006). Methyl jasmonate can also induce this response (Martin et al. 2002). Oleoresin production can also be induced in *Copaifera* species. Younger trees that do not produce oleoresin on the first attempt have been known to produce a small amount on a second tapping, putatively through induction by mechanical damage (Plowden 2003; Medeiros and Vieira 2008). Medeiros and Vieira (2008) were also able to draw a weak correlation between trees with termite infestations and production of oleoresin, suggesting that insect damage can induce production of oleoresins.

TABLE 24.1

Three Major Sesquiterpenes Present in the Oleoresins of *Copaifera* Species

Species	Reference	Compound	Percentage in Oleoresin
Copaifera langsdorfii	Gramosa and Silveira 2005	β-Caryophyllene	53.3
		Germacrene B	8.7
		β-Selinene	6.5
Copaifera martii	Zoghbi et al. 2007	β-Caryophyllene	42.6[a]
		δ-Cadinene	15.7[a]
		β-Elemene	5.0[a]
Copaifera multijuga	Veiga Junior et al. 2007	β-Caryophyllene	57.5
		δ-Humulene	8.3
		β-Bergamotene	2.6
Copaifera cearensis	Veiga Junior et al. 2007	β-Caryophyllene	19.7
		δ-Copaene	8.2
		β-Bisabolol	8.2
Copaifera reticulate	Veiga Junior et al. 2007	β-Caryophyllene	40.9
		δ-Humulene	6.0
		β-Bergamotene	4.1
Copaifera trapezifolia	Veiga Junior et al. 2006	β-caryophyllene	33.5
		Germacrene D	11.0
		Spathulenol	7.6

[a] Number represents an average of 11 sampling dates.

The cellular mechanisms involved in transport, storage, and secretion of oleoresin constituents against the concentration gradient present in resin ducts are not well understood (Langenheim 2003; Keeling and Bohlmann 2006). Synthesis of terpenoids present in conifer oleoresins typically involves TPSs and cytochrome P450 oxygenases (P450). A conifer diterpene synthase (PtTPS-LAS) and the first diterpene P450 (PtAO) have been localized to plastids and the endoplasmic reticulum (ER) using a green fluorescent protein in tobacco leaf cells (Ro and Bohlmann 2006). On the basis of the lack of accumulation of diterpenes in cells, these authors suggest that a transport mechanism must be in place to move the diterpenes into the ER or cytosol of cells.

Although *Copaifera* oleoresin exudes from resin ducts during tapping, no experiments have confirmed which tissues are responsible for production of chemical constituents in the oleoresin. Calvin (1980) hypothesized that the constituents in *Copaifera* oleoresin must be synthesized in the canopy of the tree and seep down through the resin ducts. In Norway spruce, diterpene synthases have been localized to epithelial cells surrounding resin ducts using protein-specific antibodies (Keeling and Bohlmann 2006). Recently, we have found that the sesquiterpenes are present in *C. officinalis* oleoresin in leaves and stem tissue of seedlings as well as leaves, stems, and roots of 2-year-old saplings (Chen et al. 2009). The presence of sesquiterpenes in different tissues at different ages could indicate transport or changes in regulation of TPS gene scripts signaled by development. In addition, the terpenes detected in oleoresins also appear in other tissues such as seeds (Gramosa and Silveira 2005). The seeds also have different sesquiterpenes that are not seen in oleoresins such as γ-muurolene, perhaps suggesting that different TPSs function in different tissues.

24.4 BIOLOGICAL FUNCTIONS OF OLEORESIN

The principal chemical constituents of *Copaifera* oleoresin are terpenoids. Therefore, understanding the biological/ecological roles of terpenoids will allow us to understand the roles of *Copaifera* oleoresin. Terpenoids are the largest class of secondary metabolites produced in the plant kingdom. Approximately 50,000 of these have been structurally identified (McCaskill and Croteau 1997). This diverse group of plant metabolites is important for many aspects of plant biology and ecology (Tholl 2006; Yuan et al. 2009). For instance, some terpenoids function in plant defenses against herbivores and microbial pathogens (Gershenzon and Croteau 1991). Other terpenoids produced by flowers as volatiles are involved in attracting insect pollinators for plant cross-pollination (Raguso and Pichersky 1999). Some volatile terpenoids are emitted from herbivore-damaged plants and function as cues to attract natural enemies of the feeding herbivores (Yuan et al. 2008). *Copaifera* oleoresin is generally believed to be involved in plant defenses that can be mainly attributed to terpenoids. Depending on the mechanisms of production, oleoresins may act in constitutive defense, induced defense, or both. *Copaifera* oleoresin could be toxic to herbivorous insects, bacteria, or fungi. Because of high volatility, the terpenoids in *Copaifera* oleoresin may be released from the tree as infochemicals which can deter potential insect pests. Oleoresin may also flow out of the wound to physically push the invading insects out of the entry wound or entomb them, so the insects cannot cause further damage. The wound caused by insect herbivory can be a natural site for invasion of microbial pathogens that would need to be defended against. *Copaifera* oleoresin and its constituents have been documented to have antimicrobial and antifungal activity (Howard et al. 1988; Braga et al. 1998). *Copaifera* oleoresin produced upon insects feeding may therefore prevent further damage caused by pathogens.

Studies on *C. langsdorfii* populations have showed that seedlings have a higher sesquiterpene concentration than their parent trees (Macedo and Langenheim 1989b). Additionally, there was a 48% mortality rate of first-generation oecophorid larvae and pupae when they were reared on seedling leaves, but no mortality was seen on oecophorids reared on parent leaves. The oecophorids that survived feeding on seedling leaves also exhibited a significantly lower weight gain than those feeding on parent leaves. Seedlings had twice as much caryophyllene, the major sesquiterpene present in most species' oleoresin, in leaves when compared with their parents. It is still unknown how tapping *Copaifera* trees for oleoresin affects tree health in the long term. Initial tapping, or even multiple tappings, could harm the tree by removing a source of chemical defense against pathogens and insects and must be considered in future studies.

24.5 OLEORESIN PRODUCTION ECOLOGY

Extractive collection of the oleoresin from wild populations of *Copaifera* trees has long been touted as a means to supplement income for native people in rural and forest areas instead of participating in the destructive practices such as slash-and-burn agriculture and timbering. However, the viability of this practice has been called into question because of the intermittent presence of oleoresin amongst individual trees, low yields of oleoresin per tree, as well as reduced and questionable secondary harvests of trees that produce oleoresins on the first tapping (Plowden 2003; Medeiros and Vieira 2008). Sustainable production of quality oleoresin for medicine and other uses has many problems that must be considered. First, a management system that will maximize production and minimize impact on the forest where harvest is occurring must be described (Rigamonte-Azevedo et al. 2004). This matter is complicated by the fact that the genus *Copaifera* is made up of many species that can produce useful oleoresin, and each of these species will naturally respond differently to each possible management strategy. In addition, anecdotal evidence suggested that each tree could produce between 20 and 30 L of oleoresin from one drill hole every 6 months (Calvin 1980); however, these stories seem to be more myth than fact.

In a study of 43 *C. multijuga* individuals in the Adolpho Ducke Forest Reserve in Manaus, Brazil, about half produced some volume of oleoresin during three tappings (Medeiros and Vieira 2008). Six of these individuals never produced oleoresin at all. On average, trees with a diameter at breast height (DBH) more than 41 cm produced 1.8 L of oleoresin per tree on the first tapping and 0.5 L during the second tapping 1 year later. Trees of 30 and 41 cm DBH produced an average of 0.13 L during the first tapping and 0.16 L during the second tapping. Plowden (2003) studied *Copaifera* oleoresin production from three different species in Pará, Brazil on the Alto Rio Guamá Indigenous Reserve. Trees 55 to 65 cm DBH yielded the most oleoresin, averaging 459 mL after two holes were drilled.

Some of the highest recorded average yields per tree were seen in the southwestern Brazilian Amazon in *C. reticulata* and *C. paupera* trees with 2.92 and 1.33 L, respectively (Rigamonte-Azevedo et al. 2006). However, these numbers were averages among oleoresin-producing individuals only. Only 27% of *C. reticulata* trees and 80% of *C. paupera* trees produced oleoresin. It is not clear whether the lack of uniformity in oleoresin production stems from tapping methodology or whether the oleoresin itself is just not produced constitutively in all trees. Significant variation, both natural and in response to herbivory, in chemical composition of *C. langsdorfii* leaves has been noted (Macedo and Langenheim 1989a, c). This variation, compounded by variation in climate, nutrient availability, and other factors, could also cause sporadic oleoresin production and therefore explain the variation seen in oleoresin collection.

Multiple harvests have also been considered to increase oleoresin yields. Cascon and Gilbert (2000) tapped 300–550 mL of oleoresin from a single *C. duckei* tree ten consecutive times at 4-month intervals, but they never depleted the tree of oleoresin at any point. However, it is impossible to determine how much oleoresin collected at each interval was residual material that had been stored in the tree and how much had been synthesized and replaced between tappings. Most studies suggest that primary tapping accesses oleoresin from accumulations in heartwood that have built up over long periods of time (Plowden 2004) and therefore would not quickly regenerate for a secondary major harvest as Calvin had originally hoped. The density of trees also ranges from 0.1 to 2.0 per hectare depending on location and forest type (Rigamonte-Azevedo et al. 2004).

It is unknown how phenology plays a roll in oleoresin production. As mentioned before, the chemical composition of the *Copaifera* oleoresins changes throughout the year, but no specific cause has been identified as the factor driving this change. Phenology studies of *Copaifera* species are rare and focus more on the flowering, seed set, and leafing patterns (Pedroni et al. 2002). Most of these types of studies have been in *C. langsdorfii*, a species native to the southern parts of Brazil. However, oleoresin collection for commercial products occurs more commonly in the northern half of Brazil and South America. From our experience, the species *C. multijuga* and *C. reticulata* are most commonly available for purchase outside of Brazil, although they are often mislabeled as *C. officinalis.*

In a recent visit to Brazil during July, we were able to observe the oleoresin collection process (Figure 24.1). The trees had to be drilled by hand, and reaching the core of the tree to access the heartwood where the oleoresins are stored was not easy. We observed the tapping of 12 *C. langsdorfii* trees, none of which produced oleoresin. It was suggested that these trees may not produce oleoresin at all, or that they may not be in season because July is during the winter or dry season. This again reinforces the notion that tree species native to the northern parts of Brazil are more suitable for production of oleoresin, or at least traditionally there is a more widespread culture of oleoresin collection in the north.

24.6 COMPARING OLEORESIN TO DIESEL FUEL

Diesel fuel, like gasoline, consists of many different compounds isolated from only one fraction of the greater mixture known as crude oil. Diesel fuel distills from crude oil between the temperatures of 200 and 350°C. Not all diesel fuels come directly from primary distillation; processes such as catalytic cracking, which breaks larger, denser molecules into smaller ones, have been developed to generate more liquid fuels from crude oil barrels (Bacha et al. 2007). In general, diesel fuel is made up of paraffins (alkanes), naphthenes (cycloalkanes), olefins (alkenes), and aromatics. As mentioned before, *Copaifera* oleoresins consist primarily of sesquiterpenes hydrocarbons.

FIGURE 24.1 (See color insert) Identification and tapping of *C. langsdorfii* trees near Nova Odessa, Sao Paulo State, Brazil. (a) A *C. langsdorfii* tree growing near a farm in Nova Odessa, Brazil. The trees grow as single individuals rather than in stands, making it difficult to locate and tap multiple trees. (b) Tapping a *C. langsdorfii* tree with a manual drill. The oils collect in the heartwood and so the hole must be drilled to the very center of the tree, making collection difficult. (c) Botanical characterization of *C. langsdorfii*. Pictures of leaves, seeds with fleshy aril, and seed pods were taken to correctly identify the genus and species of the trees.

The important properties of diesel fuel are the cetane rating, low-temperature operability, and volumetric heating value. Diesel engines produce combustion by compressing air, which, in turn, heats the air; at a designated moment of compression, fuel is injected into the chamber as tiny droplets which vaporize and ignite. The cetane rating measures the ignition quality of fuels, or how readily the fuel burns. A fuel's quality of ignition can have implications in starting engines in cold conditions, as well as emissions, smoothness of operation, noise, and misfires (Bacha et al. 2007).

Low temperatures can cause some constituents in diesel fuels to solidify (such as the paraffins). This, in turn, can clog the fuel filter and stop the flow of fuel to the engine. This effect is measured with "cloud points," the temperature at which the waxes in the mixture begin to solidify, or the "pour point," the temperature when the fuel becomes so thick that it will no longer pour. The volumetric heating value measures how much energy the fuel has per volume. Volumetric heating values influence torque, horsepower, and to some degree, fuel economy.

Monoterpenes and sesquiterpenes are volatile cyclic hydrocarbons. The major sesquiterpene present in most *Copaifera* species, β-caryophyllene, has a chemical structure most similar to a cyclic olefin, or a naphthene, which contains two double bonds. In general, naphthenes have a midrange cetane rating, good low-temperature properties, and an acceptable volumetric heating value. Biofuels from oilseed sources such as soybean and canola have a pour and cloud point of approximately 0°C, making them impractical in areas with cold climates. In addition, fuel additives to improve low temperature properties are not very effective because of the high level of saturated compounds present in the oils (U.S. Department of Energy 2004). Addition of terpenoid components (such as sesquiterpenes) to these types of biofuels could increase their low-temperature properties and complement their high cetane ratings.

Not much is known about the chemical and physical properties of *Copaifera* oleoresin as a diesel fuel. Calvin (1980) submitted a sample of *Copaifera* oleoresin to the Mobil Corporation and obtained a cracking pattern: 50% aromatics, 25% liquid petroleum gas (LPG), 3–4% low-molecular-weight fuel gas, and coke. Later, cracking of *C. officinalis* oleoresin with a zeolite catalyst, ZSM-5, led to production of over 200 compounds from 34 sesquiterpenes present in the original oleoresin (Stashenko et al. 1995). The great variety of resulting products could indicate the utility of these oleoresins in not only fuels but also additional value-added products from a renewable resource. As mentioned before, the seeds of *Copaifera* species not only produce sesquiterpene hydrocarbons but also produce various fatty acids when pressed and extracted (Lima Neto 2008; Stupp et al. 2008). In *C. langsdorfii*, oleic acid (C18:1) made up 33.1% of the fatty acid profile, whereas palmitic acid (C16:0) made up 20.2% of the fatty acid profile. According to Stupp et al. (2008), the major fatty acid that was extracted was linoleic acid (C18:2) which made up 45.3% of the fatty acids and oleic acid making up 30.9%. It would be interesting to test oil pressed from these seeds against other biodiesels and to compare their overall chemical structure to see how the percentages of sesquiterpenes versus fatty acids are present in the seed oil.

24.7 FUTURE SCOPE OF RESEARCH AND DEVELOPMENT

For reasons described earlier, it does not seem economically feasible to create plantations of *Copaifera* trees to produce oleoresin for biodiesel markets. In brief, long generation times, low and sporadic yields per tree, and their tropical nature limit production of oleoresin. Instead, characterization of the unique terpenoid biosynthesis pathway and expressing it in other species already suited for production of biodiesel offers a more reasonable avenue.

Why these oleoresins produce higher amounts of certain terpenoids, sesquiterpenes mostly, is not well understood. Possible mechanisms include differential regulation of sesquiterpene synthesis, or even higher TPS efficiency. Although there has been a lot of work accomplished to characterize the chemical nature of oleoresins, there has been a surprising lack of molecular biology and biochemistry research as to how these oleoresins are created, stored, and transported. Identification, isolation, and characterization of the TPS responsible for the production of the oleoresin constituents will be crucial not only in first determining how these proteins function, but also in localizing them within tissue types to understand production of oleoresin constituents.

Detailed studies on the emission and performance standards of oleoresins as a diesel fuel will also be necessary. The host of traditional diesel classification tests for physical and chemical properties, including density, cloud and pour points, viscosity, heat of combustion, cetane number, etc., should be performed to gauge the usefulness of these compounds in today's markets with current engine technology. This work will also need to include several different possible species because each one has a varying chemical makeup and properties associated with that makeup. These studies will be instrumental in determining whether oleoresin constituents are better suited as a stand-alone biodiesel fuel or as an additive for other petroleum or biofuels.

However, there are barriers to further research on *Copaifera* species. Many publications and historical records about *Copaifera* trees are in Portuguese; this presents a barrier to the larger scientific community and hampers research efforts to assess available information. In addition, obtaining germplasm outside of the native range of the *Copaifera* genus has been exceptionally difficult. Collaborations on an international level will be crucial to establishing successful research initiatives.

We are performing genomics and biochemistry research to characterize and exploit the *Copaifera* terpene biosynthetic pathways. There are at least two end goals of research. First, we need a better understanding of the basic biochemistry of this interesting genus and its oleoresins. Second, genes and gene regulation responsible for hydrocarbon production could be valuable with regards to their use in production of bioproducts and fuels. For example, key genes might be transferred to temperate oilseed species to complement and increase their biofuel production. Taken together, we expect the diesel trees to contribute to new fuels and products beyond diesel.

REFERENCES

Alencar JC (1982) Estudos silviculturais de uma população natural de *Copaifera multijuga* Hayne (Leguminosae) na Amazônia Central IV. Interpretação de dados fenologicos em relação a elementos climáticos. *Acta Amaz* 18:109–209

Bacha J, Freel J, Gibbs A, Gibbs L, Hemighaus G, Hoekman K, Horn J, Ingham M, Jossens L, Kohler D, Lesnini D, McGeehan J, Kikanjam M, Olsen E, Organ R, Scott B, Sztenderowicz M, Tiedemann A, Walker C, Lind J, Jones J, Scott D, and Mills J. (2007) *Diesel Fuels Technical Review. Technical Manual*, Chevron Corporation, available at http://www.chevron.com/products/prodserv/fuels/documents/Diesel_Fuel_Tech_Review.pdf (accessed 02/11/2009)

Braga WF, Rezende CM, Antunes OAC, Pinto ÂC (1998) Terpenoids from *Copaiba cearensis*. *Phytochemistry* 49:263–264

Calvin M (1980) Hydrocarbons from plants: Analytical methods and observations. *Naturwissen* 67:525–533

Cascon V, Gilbert B (2000) Characterization of the chemical composition of oleoresins of *Copaifera guianenesis* Desf., *Copaifera duckei* Dwyer and *Copaifera multijuga* Hayne. *Phytochemistry* 55:773–778

Chen F, Al-Ahmad H, Joyce B, Zhao N, Köllner TG, Degenhardt J, Stewart Jr CN (2009) Within-plant distribution, developmental changes in accumulation and emission of sesquiterpenes from *Copaifera officinalis*. *Plant Phys Biochem* 47:1017–1023

Cheng AX, Lou YG, Mao YB, Lu S, Wang LJ, Chen XY (2007) Plant terpenoids: biosynthesis and ecological functions. *J Integr Plant Biol* 49:179–186

Dwyer JD (1951) The Central American, West Indian, and South American species of *Copaifera* (Caesalpiniaceae). *Brittonia* 7:143–172

Dwyer JD (1954) Further studies on the New World species of *Copaifera*. *B Torrey Bot Club* 81:179–187

Gershenzon J, Croteau R (1991) Terpenoids. In: Rosenthal GA, Berenbaum MR (eds), *Herbivores: Theirs Interactions with Secondary Plant Metabolites, Vol. 1: The Chemical Participants*. Academic Press, New York, 165–219

Gomes NM, Rezende CM, Fontes SP, Matheus ME, Fernandes PD (2007) Antinociceptive activity of Amazonian copaiba oils. *J Ethnopharmacol* 109:486–492

Gomes NM, Rezende CM, Fontes SP, Hovell AMC, Landgraf RG, Matheus ME, Pinto AdC, and Fernandes PD. (2008) Antineoplasic activity of *Copaifera multijuga* oil and fractions against ascetic and solid Ehrlich tumor. *J Ethnopharmacol* 119:179–184. DOI 10.1016/j.jep.2008.06.033.

Gramosa NC, Silveira ER (2005) Volatile constituents of *Copaifera langsdorfii* from the Brazilian northeast. *J Essent Oil Res* 17:130–132

Hou D (1994) Studies in Malesian Caesalpinioideae (Leguminosae). I. The genera *Acrocarpus*, *Afzelia*, *Copaifera*, and *Intsia*. *Blumea* 38:313–330

Howard JJ, Coxin Jr J, Wiemer D F (1988) Toxicity of terpenoid deterrents to the leaf-cutting ant *Atta cephalotes* and its mutualistic fungus. *J Chem Ecol* 14:59–69

Keeling CI, Bohlmann J (2006) Diterpene resin acids in conifers. *Phytochemistry* 67:2415–2423

Langenheim JH (1973) Leguminous resin-producing trees in Africa and South America. In: Meggers BJ, Ayensu ES, and Duckworth WD (eds), *Tropical Forest Ecosystems in Africa and South America*. Smithsonian Institution Press, Washington, DC pp 89–104

Langenheim JH (2003) *Plant Resins: Chemistry, Evolution, Ecology and Ethnobotany*. Timber Press, Portland, OR

Léonard J (1949) Notulae systematicae IV (Caesalpiniaceae-Amherstieae africanae americanaeque). *Bull Jard Bot Brux* 19:384–408

Léonard J (1950) Notulae systematicae IX nouvelles observations sur le genre *Guibourtia* (Caesalpiniaceae). *Bull Jard Bot Brux* 20:270–284

Lewinsohn E, Gijzen M, Savage TJ, Croteau R (1991) Defense mechanisms of conifers. *Plant Physiol* 96:38–43

Lichtenthaler HK (1999) The 1-deoxy-D-xylulose-5-phosphate pathway of isoprenoid biosynthesis in plants. *Annu Rev Plant Biol* 50:47–65

Lima NN, Pio NS (2007) Avaliação da gramatura de cola na propriedade de flexão estática em painéis compensados de *Copaifera duckei* Dwayer e *Eperua oleifera* Ducke. *Acta Amaz* 37:347–352

Lima SRM, Junior VF, Christo HB, Pinto AC, Fernandes P D (2003) *In vivo* and *in vitro* studies on the anticancer activity of *Copaifera multijuga* Hayne and its fractions. *Phytother Res* 17:1048–1053

Lima Neto JS (2008) Constituintes químicos dos frutos de *Copaifera langsdorfii* Desf. *Quim Nova* 31:1078–1080

Macedo CA, Langenheim JH (1989a) A further investigation of leaf sesquiterpene variation in relation to herbivory in two Brazilian populations of *Copaifera langsdorfii*. *Biochem Syst Ecol* 17:207–216

Macedo CA, Langenheim JH (1989b) Microlepidopteran herbivory in relation to leaf sesquiterpenes in *Copaifera langsdorfii* adult trees and their seedling progeny in a Brazilian woodland. *Biochem Syst Ecol* 17:217–224

Macedo CA, Langenheim JH (1989c) Intra- and interplant leaf sesquiterpene variability in *Copaifera langsdorfii*: Relation to microlepidopteran herbivory. *Biochem Syst Ecol* 17:551–557

Martin D, Tholl D, Gershenzon J, Bohlmann J (2002) Methyl jasmonate induces traumatic resin ducts, terpenoid resin biosynthesis, and terpenoid accumulation in developing xylem of Norway spruce stems. *Plant Physiol* 129:1001–1018

McCaskill D, Croteau R (1997) Prospects for the bioengineering of isoprenoid biosynthesis. *Adv Biochem Eng Biotechnol* 55:107–146

Medeiros RS, Vieira G (2008) Sustainability of extraction and production of copaiba (*Copaifera multijuga* Hayne) oleoresin in Manaus, AM, Brazil. *For Ecol Manag* 256:282–288

Monti H, Tiliacos N, Faure R (1996) Two diterpenoids from copaiba oil. *Phytochemistry* 42:1653–1656

Monti H, Tiliacos N, Faure R (1999) Copaiba oil: Isolation and characterization of a new diterpenoid with the dinorlabdane skeleton. *Phytochemistry* 51:1013–1015

Ohsaki A, Yan LT, Ito S, Edatsugi H, Iwata D, Komoda Y (1994) The isolation and *in vivo* potent antitumor activity of clerodane diterpenoid from the oleoresin of the Brazilian medicinal plant, *Copaifera langsdorfii* Desfon. *Bioorg Med Chem Lett* 4:2889–2892

Paiva LAF, Cunha KMA, Santos FA, Gramosa NV, Silveira ER, Rao VSN (2002) Investigation on the wound healing activity of oleo-resin from *Copaifera langsdorfii* in rats. *Phytother Res* 16:737–739

Paiva LAF, Gurgel LA, Campos AR, Silveira ER, Rao VSN (2004) Attenuation of ischemia/reperfusion-induced intestinal injury by oleo-resin from *Copaifera langsdorfii* in rats. *Life Sci* 75:1979–1987

Paiva LAF, Rao VSN, Gramosa NV, Silveira ER (1998) Gastroprotective effect of *Copaifera langsdorfii* oleo–resin on experimental gastric ulcer models in rats. *J Ethnopharmacol* 62:73–78

Pedroni F, Sanchez M, Santos FAM (2002) Fenologia da copaiba (*Copaifera langsdorffii* Desf. – Leguminosae, Caesalpinioideae) emu ma floresta semidecídua no sudeste do Brasil (in Portuguese). *Revista Brasil Bot* 25:183–194

Plowden C (2003) Production ecology of copaiba (*Copaifera* spp.) oleoresin in the eastern *Braz Amaz Econ Bot* 57:491–501

Plowden C (2004) The ethnobotany of copaiba (*Copaifera*) oleoresin in the Amazon. *Econ Bot* 58:729–739

Ro D-K, Bohlmann J (2006) Diterpene resin acid biosynthesis in loblolly pine (*Pinus taeda*): Functional characterization of abietadiene/levopimaradiene synthase (*PtTPS-LAS*) cDNA and subcellular targeting of PtTPS-LAS and abietadienol/abietadienal oxidase (PtAO, CYP720B1). *Phytochemistry* 67:1572–1578

Rigamonte-Azevedo OC, Wadt PGS, Oliveira LHW (2006) Potencial de produção de oleo-resina de copaiba (*Copaifera spp*) de populaces naturais do sudoeste da Amazônia (in Portuguese). *Rev Árvore* 20:583–591

Rigamonte-Azevedo OC, Wadt PGS, Wadt LHO (2004) Copaíba: Ecologia e Produção de Óleoresina. EMBRAPA, Brazil, ISSN: 0104–9046

Stashenko E, Wiame H, Dassy S, Martinez JR, Shibamoto T (1995) Catalytic transformation of copaiba (*Copaifera officinalis*) oil over zeolite ZSM-5. *J High Res Chromatogr* 18:54–58

Stupp T, de Freitas RA, Sierakowski MR, Deschamps FC, Wisniewski A Jr, Biavatti MW (2008) Characterization and potential uses of *Copaifera langsdorfii* seeds and seed oil. *Biores Technol* 99:2659–2663

Tholl D (2006) Terpene synthases and the regulation, diversity and biological roles of terpene metabolism. *Curr Opin Plant Biol* 9:297–304

U.S. Department of Energy (2004) 2004 *Biodiesel Handling and Use Guidelines*. DOE/GO-102004-1999 (Revised 2004), U.S. Department of Energy, Washington, DC

Veiga VF, Pinto AC, de Lima HC (2006) The essential oil composition of *Copaifera trapezifolia* Hayne leaves. *J Essent Oil Res* 18:430–431

Veiga VF, Jr., Pinto AC (2002) O gênero *Copaifera* L. *Quím Nova* 25:273–286

Veiga VF, Jr., Rosas EC, Carvalho MV, Henriques MGMO, Pinto AC (2007) Chemical composition and anti-inflammatory activity of copaiba oils from *Copaifera cearensis* Huber ex Ducke, *Copaifera reticulate* Ducke and *Copaifera multijuga* Hayne—a comparative study. *J Ethnopharmacol* 112:248–254

Yuan JS, Himanen S, Holopainen JK, Chen F, Stewart Jr CN (2009) Smelling climate change: Plant volatile organic compounds in changing environments. *Trends Ecol Evol* 24:323–331

Yuan JS, Köllner TG, Wiggins G, Grant J, Degenhardt J, Chen F (2008) Molecular and genomic basis of volatile-mediated indirect defense against insects in rice. *Plant J* 55:491–503

Zoghbi MGB, Lameira OA, Oliveira ECP (2007) Seasonal variation of oleoresin and volatiles from *Copaifera martii* Hayne growing wild in the state of Pará, Brazil. *J Essent Oil Res* 19:504–506

25 Minor Seed Oils

Sukumar Puhan
GKM College of Engineering & Technology

N. Vedaraman
Central Leather Research Institute

A. Gopinath and V. Edwin Geo
Anna University

K.C. Velappan
Central Leather Research Institute

G. Nagarajan
Anna University

CONTENTS

25.1 INTRODUCTION

World energy consumption is increasing tremendously, but fossil fuel sources are limited and decreasing rapidly. Burning of fossil fuels increases the carbon dioxide (CO_2) in the atmosphere. CO_2 emissions have risen over the last 2 decades, reaching an atmospheric content of 360 ppm, estimating the world CO_2 emissions at approximately 26 billion tons/year, 80% of which comes from the combustion of fossil combustibles such as coal, petroleum, and natural gas (Muezzinoglu et al. 1992; Tuer et al. 1997).

The use of vegetable oils in energy production has almost always concentrated on its use as a substitute for diesel in cars, either directly or in the form of methyl or ethyl esters. The unrefined oils have rarely been used as combustibles in the generation of thermal energy (Lopez Sastre et al. 1998).

Biomass already provides approximately 13% of global energy, but this is largely as domestic firewood used inefficiently for cooking and heating in developing countries (Sims 2001). In developed countries, there is a growing trend toward using modern and efficient bioenergy conversion technologies using a range of biofuels that are becoming commercially competitive with fossil fuels.

25.1.1 POTENTIAL OF NONEDIBLE MINOR VEGETABLE OILS IN INDIA

Depending on the climate and soil conditions, different nations are looking into different vegetable oils for diesel fuel substitutes. Being a tropical country, India is rich in forest resources with a wide range of oil seeds. The potentials of vegetable oils in India are given in Table 25.1. From the table, it is understood that there is a huge market for vegetable oils in India. And in the near future, vegetable oils may play a vital role in the biofuel industry. The production of nonedible oils in India is given in Table 25.2.

25.1.2 DESCRIPTION OF MINOR SEED OILS

25.1.2.1 Neem (*Azadirachta indica* Juss)

Neem grows wild in dry forests and is adapted to all kinds of soils. It is adapted to a wide range of climatic conditions: temperatures of 0–49°C, altitudes up to 1500 m, soil pH up to 8.5, and varying soil depths. Although the tree establishes naturally, tissue culture and stem cuttings can also propagate it. The tree starts producing seeds after 5–6 years. The flowering season spreads over January to April in various parts of the country depending on the climatic conditions. The fruit yield, approximately 37–55 kg/tree, is harvested from June to August. The kernels constitute approximately 45% of the seed and contain 40–45% oil. The oil is dark and bitter with a disagreeable odor. Small-scale laundry shop manufacturers mostly consume it, and the good-quality oil is used for other processes.

25.1.2.2 Mahua (*Mahuca indica*)

Mahua is a large deciduous tree with a short trunk, spreading branches, and large rounded crown. It is found in Maharastra, Bengal, Orissa, in some South Indian forests and Ceylon. The flowering

TABLE 25.1
Potential of Vegetable Oil Seeds in India

Name	Botanical Name	Potential Lakh MT
Neem	*Azadirechta indica*	5
Karanja	*Pongamia pinnata*	2
Kusum	*Scheleichera oleosa*	0.8
Ratanjyot	*Jatropha curcas*	0.5
Sal	*Shorea robusta*	15
Mahua	*Madhuca indica*	5

TABLE 25.2
Potential and Actual Production of Nonedible Oil in India

Oilseed Name	Potential Lakh Tons		Actual Production Lakh Tons	
	Seeds	Oils	Seeds	Oils
Sal	60–70	6–8	2–3	0.25
Mahua	4.30	2.4	0.7–0.85	0.25–0.30
Neem	4.30	1–3	11.5	0.2–0.3
Karanja	1.4	0.3–1	0.25–0.3	0- 0.7–0.1
Kusum	0.9–1.9	0.29	0.20	0.06

season extends from February to April. The mature fruits fall during May to July in North India and August to September in South India. The yield of Mahua seeds depends mainly upon the size of the tree, site condition, age, etc. The yield may vary from 5 kg to approximately 200 kg/tree depending on the size and age of the tree. The average yield of sun-dried Mahua seeds is approximately 1.6 kg/tree. Mahua trees start giving seeds after 10 years, continuing up to 60 years with the yield increasing with age. Some investigations have found that at 60 years of age, the Mahua tree will yield 10 times more than the yield at 10 years of age. The kernel constitutes 70% of the seed and contains 50% oil. The oil has some medicinal value and locally is used on sore, cracked skin. The flowers are used as a vegetable and a source of alcohol. The cake from the oil seed is used as fertilizer. The leaves, flowers, and fruits are eaten by cattle.

25.1.2.3 Karanja (*Pongamia pinnata*)

Karanja is a native of Wetern Ghats, India. It is a hardy tree growing up to 8 m tall and is well adapted to extreme growing conditions such as temperatures of 0–50°C, altitudes up to 1200 m, rainfall of 50–2500 mm, and a soil pH up to 8. Propagation is through seed, stump-cutting, or nursery-raised plants. Seeds germinate within approximately 2 weeks. They flower from April to July and the pods ripen from February to May. Karanja starts producing 4–5 years after planting, and after 7 years it reaches its full productivity of approximately 6–9 t seed/ha per year. The seeds contain 30–40% oil by weight. The oil is nonedible and is used for illumination, lubrication, and synthesis of organic compounds, and its wood is used for firewood, cabinet work, cart wheels, and posts. Leaves of the plant are used as fodder and green manure. The oil cake is used as poultry feed and manure.

25.1.2.4 Linseed (*Linum usitatissimum*)

The crop is harvested when it is fully ripe and the stems are light brown throughout, the leaves shriveled or fallen, the capsules are brittle, and the seeds are brown. Linseed is grown in the latitudes between 10° and 65°, both north and south. Its cultivation is confined to low elevations, but it can be successfully grown up to 770 m. Areas with an annual rainfall ranging from 45 to 75 cm are best suited for its cultivation. The seed crop does well under moderate cold, but the fiber crop grows best in cool moist climates. Linseed can be grown on different kinds of soils, except sandy and badly drained heavy clays or clay loams. It does well on the deep clayey black soils of central and peninsular India and on the alluvium loams of the Indo-Gangetic plains. The main season for sowing linseed is October to November, depending upon the cessation of the rains. The linseed crop starts maturing by mid-February, depending on winter spread and sowing time. Plants turn golden yellow when the crop is mature and ready for harvesting. Harvesting is done when the crop is dead ripe with a sickle or by uprooting the plants. The average yield of a pure crop varies from 210 to 450 kg/ha of seed under rain-fed cultivation. The crop in northern India generally gives a higher yield than in central and peninsular India. The irrigated crop may yield 1200–1500 kg/ha.

Oil content of the seed is 35–45%. Crude linseed oil has a dark amber color and a strong, characteristic odor. Linseed oil is exclusively used as nonedible oil, principally in the manufacture of paints and varnishes, linoleum, oilcloth, and printing ink. India is the third-largest producer of linseed in the world. Indian linseed is mainly cultivated as a rabi seasonal crop with October through November being the main sowing season. February through April is the main harvesting season.

The oil cake left after the oil is pressed out is a most valuable feeding cake, perhaps the most favorite cattle feed. It is good in taste and contains 36% protein, 85% of which is digestible. It is fed to fattening animals. It is also used as organic manure. It contains approximately 5% nitrogen, 1.4% phosphorus, and 1.8% potash. Straw from seed varieties are used in the manufacturer of upholstery, insulating material, rugs, twine, and paper. India accounts for approximately 1.9 million ha, with a seed production of 0.498 million t, and is third among linseed-producing countries. India averages 255 kg/ha.

25.1.2.5 Rubberseed Oil (*Hevea brasiliensis* Muell Arg)

The botanical name of the rubber tree is *Hevea brasiliensis* (Euphporbiaceae). The tree is medium sized, up to 18 m in height, drought-resistant, tolerant of salinity, grows fast, and matures after 4–7 years. In India it is widely grown in the southern states of Kerala, Tamil Nadu, and Karnataka. Rubber seed kernels (50–60% of seed) contain 40–50% of the brown-colored oil. The estimated availability of rubber seeds in India is approximately 30,000 t/year, which can yield approximately 5000 t of rubber seed oil. Rubber trees yield a three-seeded ellipsoidal capsule, each capsule with one seed. Rubber seeds are ellipsoidal, approximately 2.5–3 cm long, mottled brown, lustrous, and weigh 2–4 g each. Fresh oil is pale yellow, but commercial oil is dark in color. It is a semidrying oil used in surface coatings for making alkyl resins. It is a partial substitute for linseed oil in paints and varnishes. It is effective against house flies and lice and is used up to 30% in soap making. Commercial oil contains fatty acids. The rubber seed production in India is approximately 150 kg/ha per year. The estimated availability of rubber seed is approximately 30,000 Mt/year. At present, rubber seed oil has not found any major application and hence the natural production of seeds remains underutilized.

25.2 BIODIESEL PREPARATION

Transesterification is a process of producing a reaction in a triglyceride and alcohol in the presence of a catalyst to produce glycerol and ester. The molecular weight of a typical ester molecule is roughly one-third that of a typical oil molecule and, therefore, it has a much lower viscosity. A catalyst is used to increase the reaction rate and yield. Because the reaction is reversible, excess alcohol is used to shift the equilibrium to the product side. Alcohols are primary and secondary monohydric aliphatic alcohols having one to eight carbon atoms. Among the alcohols that can be used in the transesterification process, methanol and ethanol are more common. Methanol is used extensively because of its low cost and its physicochemical advantages such as polarity and easy solubility for sodium hydroxide (NaOH) (Fanguri and Hanna 1999). To complete a transesterification stoichiometrically, a 3:1 molar ratio of alcohol to triglycerides is needed. The reaction can be catalyzed by alkalis, acids, or enzymes. The alkali include NaOH, potassium hydroxide (KOH), carbonates, and the corresponding sodium methoxide, sodium ethoxide, sodium peroxide, and sodium butoxide Sulfuric acid, sulfonic acids, and hydrochloric acid are usually used as acid catalysis. Lipase can be used as biocatalysts. Alkali-catalyzed transesterification is much faster than acid-catalyzed transesterification and is most often used commercially (Nelson et al. 1996). According to the quality of the parent oil, the selection of a transesterification process can be made as shown in Figure 25.1 and it can be explained as follows:

- If the free fatty acid (FFA) content and moisture content of parent vegetable oils are less than 0.5%, a single-stage transesterification process can be chosen.
- When the FFA content is less than 0.5% but the moisture content is greater than 0.5%, the moisture content can be removed by heating the oil at 110°C for approximately 1 h with stirring. Then, the biodiesel can be produced from single-stage transesterification.

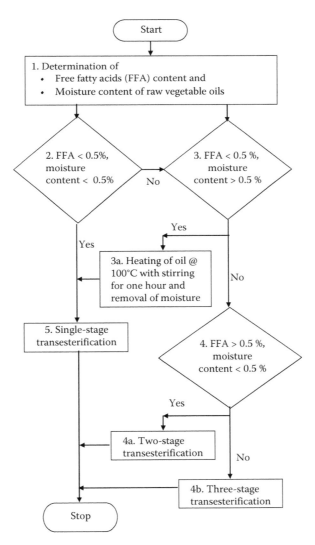

FIGURE 25.1 Flow chart for the selection of transesterification process.

- When the FFA content is greater than 0.5% and the moisture content is less than 0.5%, then a two-stage transesterification process can be selected. If the FFA content and moisture content are greater than 0.5%, then a three-stage transesterification process is selected.

In the work presented here, biodiesel fuels were produced by a single-stage transesterification process.

25.3   STRATEGIES AND METHODOLOGIES

25.3.1   BIODIESEL PROPERTIES AND QUALITY

Biodiesel is a mono-alkyl ester that is derived from vegetable oils or animal fats through transesterification. The purpose of transesterification is to reduce the viscosity of vegetable oils. The transesterification process parameters such as alcohol/oil ratio (6:1 molar ratio), catalyst quantity (NaOH, 0.5 wt %), reaction temperature (65°C), and reaction time (2 h) were optimized for a single oil and then subsequently adjusted for other oils. An alkali catalyst was used because of its low cost and because it is easily miscible with methanol. Methanol was used because of its low cost. The biodiesels produced from different oils were washed and dried. Different biodiesels were then blended to vary the percentage of unsaturation. The biodiesels produced were checked for quality. The important fuel properties were measured as per standard methods and compared with the ASTM limits. The fuel properties were determined following the methods specified in ASTM standards as given in Table 25.3.

25.3.2   COMBUSTION PARAMETERS

25.3.2.1   Ignition Delay

Ignition delay is the time or crank angle between the start of fuel injection into the cylinder and the time of first combustion. The fuel injection is dynamic injection and the first combustion is the heat release. The ignition delay was calculated from the heat release diagram. The dynamic injection timing was calculated based on the negative heat release in the heat release diagram. Because the fuel droplet absorbs heat from the cylinder, there is a negative heat release on the diagram. The point at which the first negative heat release starts is the dynamic injection time and the start of ignition delay period. Similarly, at the point where the first positive heat release is seen on the diagram is the end of the ignition delay period.

25.3.2.2   Heat Release Rate

The heat release rates were determined from the cylinder pressure history. The apparent heat release rate was calculated based on the first law of thermodynamics as given in equation (25.1).

TABLE 25.3
ASTM Methods for Determination of Fuel Properties

Property	Unit	ASTM Standard
Fatty acid composition	wt %	D6584
Density at 15°C	g/cc	D1298
Cetane number	–	D613
Heating value	MJ/kg	D240
Iodine value	g iodine/100 g oil	D1510

$$\frac{dQ_n}{d\theta} = \frac{\gamma}{\gamma - 1} \cdot p \cdot \frac{dv}{d\theta} + \frac{1}{\gamma - 1} \cdot v \cdot \frac{dp}{d\theta}, \tag{25.1}$$

where $g = C_p/C_v$ is the ratio of specific heat, dQ_n/d_θ is the apparent net heat release rate, $dv/d\theta$ is the rate of volume, and $dp/d\theta$ is the rate of pressure.

25.3.3 COMBUSTION PARAMETERS

A single cylinder air-cooled direct injection compression ignition engine developing a power output of 4.4 kW at the rated speed of 1500 rpm was used for experimental studies. The engine was coupled to an electrical dynamometer. The engine was fitted with all accessories to measure the fuel consumption, air consumption, inlet air temperature, and exhaust gas temperature. The engine was started with neat diesel fuel and warmed up. It was allowed to run for 10 min with biodiesel to attain a steady condition at its rated speed of 1500 rpm. The engine was gradually loaded to full load by switching on the load mains. The different biodiesel fuels were tested in a random order. The speed of the engine was maintained at 1500 rpm, and the time taken for 10 cm³ of fuel consumption was measured using a stopwatch. The tests were repeated 5 times, and the average value of the five readings was taken to eliminate uncertainty.

25.3.3.1 Combustion Pressure Measurement

Instrument used

A Kistler piezoelectric transducer with a range of 0–250 bar was used.

Method

- The piezoelectric transducer was mounted on the cylinder head and connected to a charge amplifier for measuring the pressure inside of the engine cylinder.
- By circulating the cooling water through the inlet opening, the transducer was cooled. The central opening was used for sending signal to the amplifier.
- The TDC (top dead center) position signal was fed into the computer through the data acquisition card. In this system, the analogue signal was converted into digital impulses at fixed crank angles using an analogue-to-digital converter (ADC).
- The digital signal was then transmitted to the computer where it was stored. It was immediately processed as soon as all of the data for one or any number of cycles had been completed.
- The processing would be in the form of a pressure-crank angle diagram or a pressure-volume diagram.

The photographic view of the test engine is shown in Figure 25.2 whereas Figure 25.3 depicts the pressure setup. The schematic of the experimental setup is illustrated in Figure 25.4.

The entire experimental work was performed in the laboratory at room temperature (32°C) and an atmospheric pressure of 1.01325 bar.

25.4 OUTPUTS AND IMPACTS

25.4.1 FUEL PROPERTIES

Fuel properties are of foremost importance to the validation of the combustion, performance, and emissions of a fuel in compression ignition engine. Fuel properties affect the engine combustion chemistry in many ways; hence, it is essential to study the indispensable properties to realize the combustion chemistry. Biodiesel fuel combustion chemistry is more complex than that of diesel fuel because of its complex structure and oxygen content. Biodiesel is composed of several fatty acid

FIGURE 25.2 (See color insert)   Photographic view of test engine.

FIGURE 25.3 (See color insert)   Photographic view of pressure measuring setup.

esters that have different physical and chemical properties. Therefore, the fatty acid ester profile of biodiesel affects the fuel properties.

The important fuel properties discussed in this section are density, heating value, cetane number, iodine value, fatty acid ester composition, and percentage of unsaturated fatty acids. The results of fuel tests on different biodiesel fuels are summarized in Table 25.4. The fatty acid methyl ester composition of various test fuels is shown in Table 25.5.

A correlation analysis was made to find out the degree of linear association between different biodiesel properties and percentage of unsaturation. The Pearson product moment correlation coefficient between different properties and the percentage of unsaturation is shown in Table 25.6.

The formula used to calculate the Pearson correlation coefficient (r) is shown in equation (25.1):

$$r = \frac{\Sigma(X-\bar{X})(Y-\bar{Y})}{\sqrt{(X-\bar{X})^2(Y-\bar{Y})^2}}, \tag{25.2}$$

where r is the Pearson correlation coefficient, X is the percentage of unsaturation, and Y represents properties.

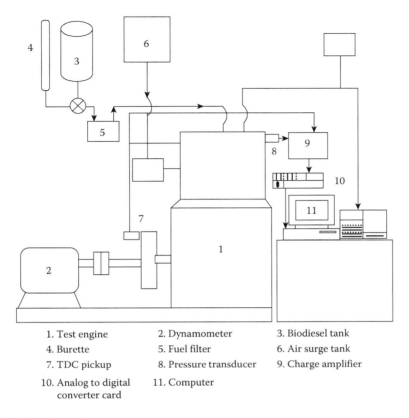

1. Test engine   2. Dynamometer   3. Biodiesel tank
4. Burette   5. Fuel filter   6. Air surge tank
7. TDC pickup   8. Pressure transducer   9. Charge amplifier
10. Analog to digital   11. Computer
     converter card

FIGURE 25.4 Experimental setup.

TABLE 25.4
Properties of Different Biodiesel Fuels

	Density (kg/m³)	Heating Value (MJ/kg)	Cetane Number	Iodine Value (g iodine/100 g oil sample)
ROME	885	39.5	47.8	160
LOME	882	39.3	51.0	140
KOME	878	39.5	52.0	95
NOME	876	39.8	58.7	83
MOME	875	40.5	61.4	65

ROME, rubberseed oil methyl ester; LOME, linseed oil methyl ester; KOME, karanja oil methyl ester; NOME, neem oil methyl ester; MOME, mahua oil methyl ester.

25.4.1.1 Density

The density of a fuel shows how close the molecules are packed in the structure. A higher-density fuel has a more closely packed structure than a lower-density fuel. Fuel density is commonly expressed in kilograms per cubic meter. The greater the fuel density, the greater the mass of fuel that can be stored in a given tank and the greater the mass of fuel than can be pumped for a given pump.

The density of a fuel affects dynamic injection timing, ignition delay, and thereby NOx emissions. However, these properties are uncertatin indications of fuel quality unless correlated with

TABLE 25.5
Fatty Acid Methyl Ester Composition of Biodiesel Fuels

Biodiesel	Percentage of United States	Lauric C12:0	Myristic C14:0	Palmitic C16:0	Stearic C18:0	Oleic C18:1	Linoleic C18:2	Linolenic C18:3
ROME	88.00	Trace	0.10	6.00	5.90	16.00	71.40	0.60
LOME	78.87	Trace	0.24	12.46	8.32	27.78	37.65	13.44
KOME	72.32	Trace	0.05	9.94	7.83	53.19	19.09	0.04
NOME	60.40	Trace	0.47	18.20	20.10	43.70	16.40	0.30
MOME	50.00	Trace	0.00	24.20	25.80	37.20	12.80	0.00

TABLE 25.6
Pearson Correlation Coefficient

x Variable	y Variable	Correlation Coefficient
Percentage of unsaturation	Density	0.955
	Heating value	−0.869
	Cetane number	−0.989
	Iodine value	0.956

other properties. From Table 25.4, it can be observed that the density of different biodiesels is not the same and that the maximum is observed for ROME (885 kg/m^3) and minimum for MOME (875 kg/m^3). It was observed from the literature (Shigley et al. 1995) that the density decreases with an increase in chain length, carbon number, or molecular weight and increases with degree of unsaturation. Density decreases from C8:0 to C18:0 and increases from C18:1 to C18:3. Because ROME is dominated by unsaturated linoleic ester (C18:2), the density of which is higher compared with lower saturated carbon chain esters, it exhibits a higher density compared with other biodiesel fuels. In the case of ROME, the unsaturation (the double bond) and when it deforms the linearity when it is introduced in a structure and forms a bend structure. This is only in the case of the cis-structure in which two hydrogen atoms are on the same side of the carbon atom and they repel each other, thereby giving a bend structure. On the other hand MOME, which is equally dominated by both unsaturated and saturated esters (C12:0), shows a lower density compared with other biodiesel fuels. From Table 25.6, it can be observed that a high positive correlation exists between the unsaturation percentage and density.

The relationship between the fatty acid methyl ester composition and density of biodiesel fuels was investigated, revealing that the higher density of ROME may be more due to the contribution of unsaturated fatty acids in ROME than that of the other biodiesels. Similarly, the lower density of MOME is believed to be due to the lesser contribution of unsaturated fatty acids in MOME. The scatter plot between the density and percentage of unsaturation with the fitted trend line equation is shown in Figure 25.5. The figure provides a clear picture that a highly positive correlation exists between density and percentage of unsaturation. Therefore, it may be stated that the density of biodiesel fuels increases with an increase in the percentage of unsaturation or in the number of double bonds.

By differentiating the fitted line equation $y = 0.268x + 860.47$ with respect to x (i.e., percentage of unsaturation) the gradient between density and percentage of unsaturation can be found as 0.268. This means, for every 1% increase in unsaturation, it may result in 0.268 units (kg/m^3) increase in density, where, r^2 denotes the coefficient of determination.

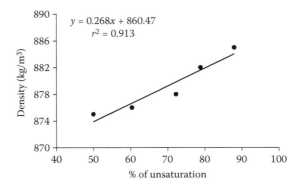

FIGURE 25.5 Variation of density with percentage of unsaturation.

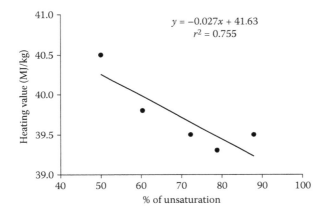

FIGURE 25.6 Variation of heating value with percentage of unsaturation.

25.4.1.2 Heating Value

Heating value, or the energy content or heat of combustion, is a measure of the energy available in a fuel. It is a critical property of fuel intended for use in weight-limited vehicles. The energy content of various biodiesel fuels can be observed from Table 25.5.

From the literature (Demirbas et al. 1998; Knothe 2005), it was observed that the heating value, cetane number, and viscosity increase with an increase in carbon number and decrease with the degree of unsaturation.

- The molecular weight of fatty acids increases with chain length.
- The heating value increases with chain length (molecular weight) and decreases with increase in the number of double bonds.

The increase in heat content may be due to the increase in the number of carbons and hydrogens. The possible cause for decrease in heat content may be the deficiency of hydrogen atoms (greater unsaturation). From Table 25.5, it can be noticed that LOME has the lower energy content and MOME has the higher energy content. Because MOME is dominated by a higher carbon chain (16 to 18), it exhibits a higher energy content, whereas LOME, which is dominated by a lower carbon chain, exhibits lower energy content as compared with other biodiesel fuels. The scatterplot between the heating value and percentage of unsaturation with a fitted trend line equation is illustrated in Figure 25.6.

From the fitted line equation $y = -0.027x + 41.63$, the projected decrease in heating value for every 1% increase in unsaturation is 0.027 units (MJ/kg).

25.4.1.3 Cetane Number

Cetane number (CN) indicates the ignition quality of a fuel, i.e., how fast a fuel takes part in combustion. In other words, it quantifies the activation energy to start the reaction. Higher CN implies lower activation energy and therefore quick reaction. Hence, the time required for a fuel to start the combustion reduces, which is the ignition delay period. Therefore, higher CN indicates a shorter ignition delay period. Delay period is the time interval between the start of fuel injection and the start of combustion. When the delay period is reduced, the fuel accumulated during the delay period also reduces and therefore a small amount of fuel can take part in combustion. In contrast, lower CN fuel has a higher activation energy and longer ignition delay period. A longer ignition delay period accumulates more fuel and higher premixed combustion, which in turn produces more oxides of nitrogen.

CN is a derived property that is influenced by several other physical and chemical properties. In the case of biodiesel, CN is influenced by chain length, alcohol moiety, degree of unsaturation, position of double bond, and cis or trans structure (Knothe 2005). CN increases with chain length and decreases with degree of saturation (Knothe 2005). Gerhard Knothe states that

- CN increases with chain length and decreases with the number of double bonds or unsaturation.
- CN of fatty esters generally increases with the number of methylene groups (CH_2) in the chain of the fatty compound, the number of CH_2 groups in the ester moiety, and the increasing saturation of the fatty compound.
- For the methyl esters, the CN were found to increase in a nonlinear relationship with molecular weight (chain length).

From Table 25.5, it can be observed that MOME has a higher cetane number compared with other biodiesel fuels. The higher CN of MOME can be attributed to the higher carbon number, i.e., stearic ester (C18:0). On the other hand, the ROME has a lower CN than that of other biodiesel fuels. This is because the CN decreases as the number of double bonds (unsaturation) increases. The influence of unsaturation on CN is depicted in Figure 25.7.

From the fitted line equation $y = -0.373x + 80.27$, it can be noticed that a reduction of 0.373 units in CN could be predicted for each percentage increase in unsaturation.

25.4.1.4 Iodine Value

Iodine value (IV) that can be considered as a direct measure of unsaturation may be defined as the amount of iodine (in grams) necessary to saturate a 100-g oil sample. IV is included in the European biodiesel standards. It purportedly addresses the issue of oxidation stability because the IV is a measure of total unsaturation of fatty materials measured in grams of iodine per 100 g of a sample when formally added to the double bonds. Another idea behind the use of the IV is that it indicates the propensity of the oil or fat to polymerize and form engine deposits (Graboski et al. 1998). The engine manufacturers have always been aware of the iodine number which expresses the number of double bonds. The limit was set at 120 in EN14214 and 135 in EN14213 (Prankl 2002). From Table 25.5, it can be noticed that ROME has an IV of 160 and MOME has a value of 65. The influence of fatty acid ester composition on IV for biodiesel samples was studied. From Table 25.6, the correlation analysis exhibits a strong positive correlation between IV and percentage of unsaturation. This was expected because from the definition, IV could be considered as a direct measure of unsaturation. Figure 25.8 depicts the effect of unsaturation on IV.

From the fitted line equation $y = 2.544x - 69.26$, every 1% increase in unsaturation may result in an increase of 2.544 units (g iodine/100 g oil) in IV.

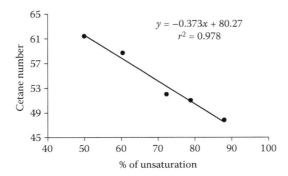

FIGURE 25.7 Variation of CN with percentage of unsaturation.

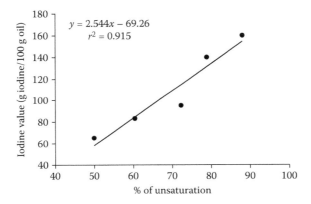

FIGURE 25.8 Variation of IV with percentage of unsaturation.

The gradient between different properties and percentage of unsaturation was based only upon the slope of the fitted lines. It may not be true for all cases. Because a large sample size would be taken for investigation, the gradient of the aforesaid parameters may vary.

25.4.2 Combustion Parameters

The influences of biodiesel properties on different combustion parameters are discussed. The following parameters were considered for the investigation:

- Dynamic injection timing
- Ignition delay
- Peak heat release rate
- Peak cylinder pressure

The experimental values of all of the above parameters at full load for various biodiesel fuels are listed in Table 25.7.

25.4.2.1 Dynamic Injection Timing

Dynamic injection timing is the crank angle point at which the fuel jet actually enters into the engine combustion chamber. Dynamic injection timing is one of the most important parameters because it can have a significant effect on the engine performance and emissions. The advancement of injection

TABLE 25.7

Combustion Parameters

Biodiesel	Dynamic Injection Timing (°CA bTDC)	Ignition Delay (°CA) From–To	Duration	Peak HRR (J/ °CA)	Location of Peak HRR (°CA)	Peak Pressure (bar)	Location of Peak Pressure (°CA)
ROME	14	346–353	7	63	356	63	370
LOME	14	346–353	7	66	357	65	370
KOME	13	347–353	6	67	358	66	371
NOME	13	347–353	6	71	359	67	371
MOME	12	348–354	6	73	360	68	371

timing will increase NO_x emissions (Szybist et al. 2005a). For a given engine at the same operating conditions, the dynamic injection may not be the same for different fuels; that is, dynamic injection timing can greatly be affected by fuel properties. The fuel injection timing can be influenced by the bulk modulus, i.e., the compressibility of the fuel. The bulk modulus or compressibility is a measure of how easily a unit of fluid volume can be decreased when increasing the pressure working on it. A higher bulk modulus indicates that the fluid is relatively incompressible.

The most common type of diesel injector is the pump-line-nozzle injector. In this configuration, the pressure is applied to the fuel upstream of the injector nozzle. The pressure increases until it reaches the nozzle opening pressure whereupon injection occurs. If a fuel is less compressible, the pressure will build more quickly and the fuel will be injected into the combustion chamber earlier in the compression cycle. Fuels with higher compressibility require a longer time to reach the nozzle opening pressure, which results in late injection. The presence of oxygen and double bonds in biodiesel fuels can increase the bulk modulus, thereby advancing the injection timing (McCormick et al. 2005). The actual start of injection (SOI) was not experimentally determined in this investigation. It was calculated based on the dip in the heat release rate diagram.

The bulk modulus of unsaturated biodiesel is higher than that of saturated biodiesel because of the introduction of double bond carbons. A carbon-carbon double bond introduces a bend in the structure and thereby distorts the linearity of a carbon–carbon single bond. This bend configuration may foster intra- or intermolecular interactions in the fuel that reduce compressibility, leading to earlier injection. It was reported (Szybist et al. 2005b) that the bulk modulus of diesel was 2% lower than B20 soybean, yielding a shift in the fuel injection timing by 0.1–0.3 crank angles.

When biodiesel is injected, the pressure rise produced by the pump is quicker as a consequence of its lower compressibility (higher bulk modulus). It also propagates quicker toward the injectors as a consequence of its higher sound velocity. In addition, higher viscosity reduces leakages in the pump, leading to an increase in the injection line pressure. Therefore, a quicker and earlier needle opening is realized with respect to the case of more unsaturated biodiesel fuel. It can therefore be stated that higher density fuels can have a higher value of bulk modulus because of an increase in unsaturation.

From the above statements, it can be concluded that the injection timing is advanced for higher unsaturated and hence higher density fuels. It can be observed from Table 25.7 that high unsaturated content biodiesel (ROME), which has a higher density, advances the injection timing by approximately 2° crank angle (CA) compared with relatively low unsaturated biodiesel (MOME). A correlation analysis was done to find out the relationship between dynamic injection timing, density, and percentage of unsaturation using equation (25.2). The correlation coefficients are shown in Table 25.8.

It can be observed from the table that the dynamic injection timing is highly positively correlated with percentage of unsaturation and the density of biodiesel fuels. Figure 25.9 illustrates the variation of dynamic injection timing with percentage of unsaturation. A good r2 value of 0.879 can be observed from Figure 25.10 between the percentage of unsaturation and dynamic injection timing.

TABLE 25.8
Correlation Coefficient among Dynamic Injection Timing, Percentage of Unsaturation, and Density

x Variable	y Variable	Correlation Coefficient
Percentage of unsaturation	Dynamic injection timing (CA bTDC)	0.935
Density		0.909

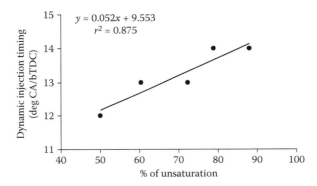

FIGURE 25.9   Variation of dynamic injection timing with percentage of unsaturation.

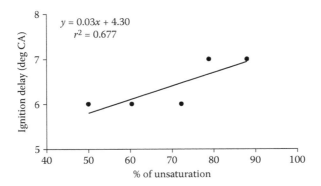

FIGURE 25.10   Variation of ignition delay with percentage of unsaturation.

The positive correlation indicates that the dynamic injection advances with increase in percentage of unsaturation. It can also be proposed that a unit increase in percentage of unsaturation can advance the dynamic injection timing by 0.052° CA.

25.4.2.2   Ignition Delay
The ignition delay can be defined as the time (in seconds or degrees crank angle) interval between the points at which the fuel jet actually enters into the combustion chamber and the start of combustion.

Understanding the physical and chemical processes is important to thoroughly describing the ignition delay.

Hydrocarbon combustion occurs only in the gas phase. Thus, for a liquid fuel, the first step toward ignition involves transitioning from a liquid to a gas phase. The time required for this transition is the "physical delay" in ignition and includes the amount of time required for a droplet of fuel to heat, vaporize, and mix with hot air in the cylinder.

The physical delay is influenced by the density and temperature of air in the cylinder; velocity and turbulence of the air; atomization; penetration; shape of the spray; and the properties of the fuel, including density, viscosity, surface tension, specific heat, enthalpy of vaporization, and vapor pressure.

Combustion is a sequence of chemical reactions in which the gas-phase fuel reacts with oxygen. These reactions proceed stepwise through a mechanism involving free radicals. For ignition to occur, the fuel must be heated to a temperature sufficient for some of the weaker bonds to break and form radicals. The finite rate of these radical-forming oxidation reactions is responsible for the chemical delay in compression ignition. Once a sufficient concentration of free radicals is reached, rapid oxidation occurs (ignition). The heat-release pattern can be considerably influenced by ignition delay, which in turn affects fuel economy and pollutant emissions. The start of injection is usually taken as the time when the injector lifts from its seat (determined by a needle lift indicator).

The correlation coefficients among ignition delay, fuel CN, density, percentage of unsaturation, and start of injection were found and shown in Table 25.9.

From the table, it can be observed that the ignition delay is negatively correlated with CN and positively correlated with fuel density, percentage of unsaturation, and dynamic injection timing. Apart from CN, the reason behind relating ignition delay with density, percentage of unsaturation, and dynamic injection timing can be explained as follows:

- CN has a greater influence on chemical delay whereas fuel density, viscosity, and surface tension can have a significant influence on physical delay (Burman and Deluca 1962).
- The density and CN were found to be highly correlated with one another (McCormick et al. 2005).
- The ignitability of an ester fuel depends not only upon the CN but also upon the fatty acid ester composition (Kinoshita et al. 2006).
- If the injection timing is advanced, the fuel is injected to the combustion chamber at a lower temperature and pressure. If the prevailing temperature and pressure are lower, it is obvious that there would be a longer ignition delay.

From Table 25.9, it can be observed that the ignition delay increases with an increase in percentage of unsaturation. The effect of unsaturation percentage on ignition delay is illustrated in Figure 25.10.

TABLE 25.9

Correlation Coefficient among Ignition Delay, Cetane Number, Density, Percentage of Unsaturation, and Dynamic Injection Timing

x Variable	y Variable	Correlation Coefficient
Cetane number	Ignition delay	−0.771
Density		0.933
Percentage of unsaturation		0.823
Dynamic injection timing		0.873

It can also be proposed from the figure that a unit increase in percentage of unsaturation can increase the ignition delay by 0.03° CA.

25.4.2.3 Heat Release Rate

The details about combustion stages and events can often be determined by analyzing heat release rates (HRRs) as determined from cylinder pressure history. Before further discussion, paying attention to the following points may offer a successful understanding on investigation findings:

- Generally, a fuel that has a longer ignition delay should have a higher value of maximal HRR as compared with fuels that have a shorter ignition delay. However, the maximal HRR not only depends on ignition delay, but also upon heating value and the mass fraction burnt for a given crank angle duration.
- Sauter mean diameter (SMD) has been shown to increase with increasing surface tension, density, and viscosity.
- An increase in droplet size can reduce the fraction of fuel burned in the premixed combustion phase.
- Density increases with increase in unsaturation.

From the aforesaid points it may be concluded that a fuel with more density may lead to an increased droplet size, which in turn reduces the mass fraction burnt in the premixed combustion phase as compared with a lower density fuel. Therefore, a higher density fuel may be expected to have a lower value of maximal HRR. In addition, it was already found that the heating value decreases with increase in unsaturation. Hence, for a given value of mass fraction burnt, fuel with a lower heating value may release less heat energy as compared with the fuel with a higher heating value. From the above discussions it may be concluded that the maximal HRR tended to decrease with an increase in unsaturation. The correlation coefficient among peak HRR, density, unsaturation percentage, and heating value is listed in Table 25.10.

From the table, it can be observed that the maximal HRR is highly negatively correlated with percentage of unsaturation and density. Similarly, the peak HRR is positively correlated with heating value (however, the correlation coefficient is not so significant). From Table 25.7, it can be observed that ROME has a lower value of maximal HRR and MOME has a higher value than that of other biodiesel fuels. The table shows that the order of magnitude of peak HRR for the biodiesel fuels is matched with the reverse order of percentage of unsaturation. This is because the density increases whereas the heating value decreases with increase in percentage of unsaturation.

It can be said that the mass fraction burnt for a given angle decreases with an increase in the percentage of unsaturation and the stoichiometric air-to-fuel ratio increases with an increase in percentage of unsaturation. This can be explained as follows. For the same quantity of supplied air (and hence oxygen), the burning volume for a less unsaturated biodiesel would be more. Because it has a lesser air-to-fuel ratio, the less unsaturated biodiesel could find more oxygen at a given crank angle than that of a more unsaturated biodiesel. From the above discussion, it can be concluded that the peak HRR decreases with an increase in the percentage of unsaturation.

TABLE 25.10

Correlation Coefficient among Peak HRR, Density, Percentage of Unsaturation, and Heating Value

x Variable	y Variable	Correlation Coefficient
Density	Peak HRR	−0.951
Percentage of unsaturation		−0.995
Heating value		0.836

From Table 25.7, it can be seen that the location of peak HRR for MOME occurs at a 360° CA whereas for ROME it occurs at a 356° CA. It may be stated that the location of peak heat release is closer to top dead center (TDC) as the percentage of unsaturation increases (i.e., for more unsaturated biodiesel, the location of peak HRR is before TDC and for less unsaturated, it is away from TDC). As the percentage of unsaturation increases, the peak heat release would shift toward the left side in the HRR diagram. This may be because as the unsaturation composition increases, the dynamic injection timing advances, which in turn can effectively bring the peak heat release before TDC in the HRR diagram. The variation of HRR with percentage of unsaturation is shown in Figure 25.11.

From Figure 25.12, it can be proposed that a reduction of 0.265 units (J) in peak HRR can occur for every 1% increase in unsaturation.

25.4.2.4 Peak Cylinder Pressure

Peak cylinder pressure is the magnitude of maximum pressure developed due to the combustion of fuel by which chemical energy is converted into pressure energy. The magnitude and occurrence of peak pressure affects engine power and emissions. Table 25.11 shows the correlation coefficient between peak cylinder pressure and percentage of unsaturation. From the table, a high positive correlation between peak cylinder pressure and peak HRR can be observed. This is obvious because

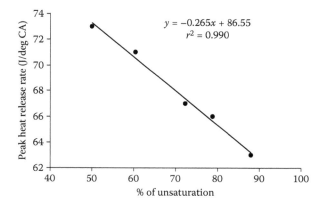

FIGURE 25.11   Variation of peak HRR with percentage of unsaturation.

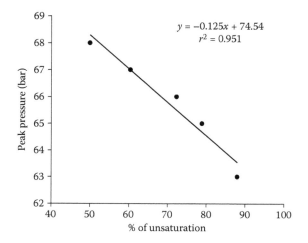

FIGURE 25.12   Variation of peak cylinder pressure with percentage of unsaturation.

TABLE 25.11

Pearson Correlation Coefficient among Peak Cylinder Pressure, Peak HRR, and Percentage of Unsaturation

x Variable	y Variable	Pearson Correlation Coefficient
Peak HRR	Peak cylinder pressure	0.975
Percentage of unsaturation		−0.975

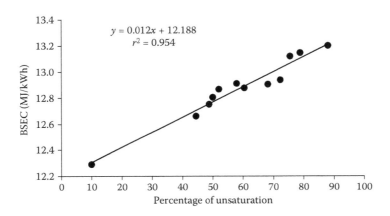

FIGURE 25.13 Variation of BSEC with percentage of unsaturation.

the magnitude of peak cylinder pressure must always be proportionate with peak HRR. It could also be seen from the table that the peak cylinder pressure decreases with an increase in unsaturation percentage.

The influence of unsaturation percentage on peak cylinder pressure is shown in Figure 25.12.

From the fitted line equation $y = -0.125x + 74.54$ shown in Figure 25.13, it can be seen that every 1% increase in unsaturation percentage may cause a reduction of 0.125 bar in peak pressure.

25.4.2.5 Brake Specific Energy Consumption

Brake specific energy consumption (BSEC) is defined as the energy required to develop a unit power in unit time. Brake specific fuel consumption (BSFC) is the quantity of fuel required for developing unit power in unit time. In the case of BSFC, the effect of fuel density alone is considered. But in BSEC, both the density and the heating value of a fuel are taken into account. When comparing different fuels with different densities and heating values, it is essential to discuss BSEC rather than BSFC.

The BSEC can be obtained by multiplying the heating value with BSFC. From Table 25.12, it can be observed that the BSEC is higher for sunflower biodiesel and lower for coconut biodiesel as compared with other biodiesel fuels. From the table, it can also be seen that the order of magnitude of BSEC is exactly matched with the order of magnitude of unsaturation percentage. From the correlation analysis, it was found that the BSEC was highly positively correlated with density and percentage of unsaturation. It can be noted that the BSFC increases with a decrease in heating value. Figure 25.13 depicts the variation of BSEC with percentage of unsaturation.

From the fitted line equation $y = 0.012x + 12.188$, the gradient between BSEC and percentage of unsaturation can be proposed as 0.0116. Every 1% increase in unsaturation may result in an increase of 0.0116 units (MJ/kWh) in BSEC.

TABLE 25.12
Brake Thermal Efficiencies of Biodiesel Fuels at Full Load

Biodiesel	Percentage of United States	BSEC (MJ/kWh)	Brake Thermal Efficiency (%)
SFOME	88.00	13.20	27.27
ROME	78.87	13.15	27.38
JOME	75.50	13.12	27.44
KOME	72.32	12.94	27.82
JT 80:20	68.18	12.91	27.89
NOME	60.40	12.88	27.96
JT 50:50	57.92	12.91	27.88
SFCt 50:50	52.05	12.87	27.98
MOME	50.00	12.81	28.11
POME	48.80	12.75	28.23
JCt 50:50	44.44	12.66	28.43
COME	10.00	12.29	29.28

25.4.2.6 Brake Thermal Efficiency

Thermal efficiency is the ratio between the power output and the energy introduced through fuel injection, the latter being the product of the injected fuel mass flow rate and the lower heating value. Thus, the inverse of thermal efficiency is often referred to as BSEC. Because it is usual to use the brake power for determining thermal efficiency in experimental engine studies, the efficiency obtained is really the brake thermal efficiency. This parameter is more appropriate than fuel consumption to compare the performance of different fuels, besides their heating value.

Brake thermal efficiency can be correlated with fuel burn angle and the generic statement is that the lower the burn angle, the higher the efficiency. But, for the same fuel burn angle, it is difficult to obtain a correlation with burn angle. Rather it can be well correlated to the shape of the heat release diagram. Brake thermal efficiency shown in Table 25.12 is higher for COME and is lower for SFOME. This shows that the order of magnitude of brake thermal efficiency for the biodiesel fuels matches exactly with the reverse order of BSEC. From the correlation analysis, it was found that the brake thermal efficiency decreases with an increase in percentage of unsaturation. This is because the BSEC increases with an increase in percentage of unsaturation. The variation of brake thermal efficiency with percentage of unsaturation is illustrated in Figure 25.14. From the fitted line equation $y = -0.026x + 29.481$, a decrease of 0.026 units (%) could be predicted for every 1% increase in unsaturation.

25.5 CONCLUSIONS

In this study, minor seed oils such as mahua, karanja, neem, rubber seed, and linseed oil methyl esters were prepared and studied in a four-stroke, direct injection diesel engine. The availability of these seeds in India is discussed. Also, the effect of biodiesel fatty ester composition on biodiesel is studied. In addition, the effect of biodiesel composition and properties on combustion parameters is studied. The gradient between biodiesel properties, combustion parameters, and percentage of unsaturation are proposed. It was found that the biodiesel properties that include density and IV are increased with an increase in biodiesel unsaturation, whereas CN and heating value are decreased with an increase in unsaturation percentage. The investigation reveals that the dynamic injection timing advances with increase in unsaturation. The ignition delay increases with an increase in unsaturation whereas the magnitude of the peak heat release rate and the peak pressure decreased

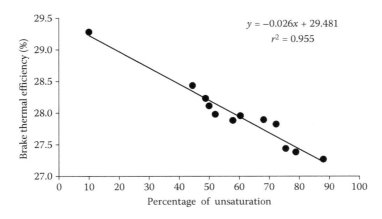

FIGURE 25.14 Variation of brake thermal efficiency with percentage of unsaturation.

with unsaturation. The brake thermal efficiency decreases with increasing unsaturation. On the whole, it is concluded that biodiesel with more unsaturation decreases heating value and CN but increases density and iodine number.

REFERENCES

Burman PG, Deluca F (1962) *Fuel Injection and Controls for Internal Combustion Engines*. Technical Press, London, pp 18–150

Demirbas A (1998) Fuel properties and calculation of higher heating values of vegetable oils. *Fuel* 77:1117–1120

Fangrui M, Hanna MA (1999) Biodiesel production a review. *Biores Technol* 70:1–15

Graboski MS, McCormick RL (1998) Combustion of fat and vegetable oil derived fuels in diesel engines. *Progr Energy Combust Sci* 24:125–164

Kinoshita E, Myo T, Hamasaki K (2006) Diesel combustion characteristics of coconut oil and palm oil biodiesels. Society of Automotive Engineers (SAE) Technical 2006–01–3251, Power Train and Fluid Systems Conference and Exhibition, Toronto, Canada, October 16-19, 2006

Knothe G (2005) Dependence of bio-diesel fuel properties on the structure of fatty acid alkyl esters. *Fuel Process Technol* 86:1059–1070

Lopez Sastre JA, Guijosa L, Sanz M (1998) *Vegetable Oils as Combustibles*. ACREVO (Advanced Combustion Research for Vegetable Oils), Orleans, France

McCormick RL, Tennant CJ, Hayes RR, Black S, Ireland J, McDaniel T, Williams A, Frailey M (2005) *Regulated Emissions from Biodiesel Tested in Heavy-Duty Engines Meeting 2004 Emission Standards*. Society of Automotive Engineers (SAE) Technical Paper Series, 2005-01–2200, SAE, Warrenton, PA

Muezzinoglu NA, Willan ML (eds) (1992) *Industrial Air Pollution*. NATO ASI series, 631. Springer, Berlin, Germany, pp 219–232

Nelson LA, Fogila TA, Marmer WN (1996) Lipase-catalyzed production of biodiesel. *J Amer Oil Chem Soc* 73:1191–1195

Prankl H (2002) High biodiesel quality required by European standards. *Eur J Lipid Sci Technol* 104:371–375

Shigley JW, Bonhorst CW, Liang CC, Althouse PM, Triebold HO (1955) Physical characterization of a) a series of ethyl esters and b) a series of ethanoate esters. *J Amer Oil Chem Soc* 32:213–215

Sims REH (2001) Bioenergy—A renewable carbon sink. *Renew Energy* 22: 31–37

Szybist JP, Boehman AL, Taylor JD, McCormick RL (2005a) Evaluation of formulation strategies to eliminate the biodiesel NO$_x$ effect. *Fuel Process Technol* 86:1109–1126

Szybist JP, Kirby SR, Boehman AL (2005b) NO$_x$ emissions of alternative diesel fuels: Comparative analysis of biodiesel and FT diesel. *Energy Fuels* 19:1484–1492

Tuer S, Uzun D, Ture IE (1997) The potential use of sweet sorghum as a non-polluting source of energy. *Energy* 22:17–19

26 Lower Plants

Michael A. Borowitzka
Murdoch University

CONTENTS

26.1 INTRODUCTION

The algae are an extremely diverse group of (mainly) photosynthetic organisms, ranging in size from approximately 1 μm for some picoplanktonic species to over 30 m in length for some of the large kelps. They are mainly aquatic, growing in fresh to hypersaline waters, but they are also found in soils and on many surfaces such as rocks, trees, and buildings. As photoautotrophs, algae use light energy to fix carbon dioxide (CO_2) into sugars during photosynthesis, and these are then further metabolized. Algae are being explored and developed as a source of a range of renewable, CO_2-friendly energy sources such as biodiesel, bioethanol, hydrogen, and methane (Chynoweth et al. 1987; Sheehan et al. 1998; Miyura 2002; Benemann 2004; Chisti 2007; Brennan and Owende 2010).

This chapter aims to provide a detailed overview of biofuels and energy production from algae, both the seaweeds and the microalgae.

26.2 METHANE

The extremely large biomass of seaweeds available in many parts of the world has the potential to be used as a source of renewable biofuel, either by fermentation to produce methane or through the production of ethanol from the sugars in the biomass. Renewable fuel production by the pyrolysis of macroalgae is also being considered (Ross et al. 2009).

The commercial use of seaweeds, especially the brown algae such as *Laminaria* (*Saccharina*) and *Macrocystis* and red algae such as *Kappaphycus* and *Gracilaria*, is a well established and very large industry. The industry uses wild stocks and cultivated biomass to produce biomass for use as foods and the production of hydrocolloids such as agar, carrageenan, and laminaran (Zemke-White and Ohno 1999; McHugh 2003). The annual harvest of wild seaweeds is estimated at approximately 1,000,000 t wet weight, and the amount produced by aquaculture at approximately 15,000,000 t wet weight (FAO 2006).

Methane production from macroalgal biomass, especially using the kelps *Macrocystis* and *Saccharina* (*Laminaria*), the green alga *Ulva*, and the rhodophytes *Hypnea* and *Gracilaria*, has been studied by several authors (Hansson 1983; Habig et al. 1984; Schramm and Lehnberg 1984; Østergaard et al. 1993) and in the 1970s and 1980s extensive research was carried out on methane production, mainly from *Macrocystis pyrifera*, by the Marine Biomass Program in the United States (Flowers and Bird 1990; Chynoweth 2002).

The best anaerobic digestion systems for macroalgal biomass appear to be vertical flow reactors, especially when operated as an upflow solids reactor in which feed is added to the bottom of the reactor and effluent removed from the top of a nonmixed vessel. This system produces approximately 0.35 m^3 methane/kg volatile solids added at loading rates of 3.2 kg/m^3 per day (Chynoweth et al. 1987). The efficiency of the process can be improved using a two-stage system. In the first stage, digester marine algal hydrolysis and acidification occurs, but not conversion of volatile acids to methane. In the second stage, digester methanogenic bacteria convert the volatile acids to methane (Chynoweth et al. 1987). In the digestion of *Macrocystis* biomass, the methane yields are highly correlated with the mannitol and algin content of the biomass, with mannitol yielding approximately 75% more methane than algin. Similarly, methane yield in the red alga *Gracilaria* is closely correlated with the carbohydrate content or protein and carbohydrate content (Habig et al. 1984). On the other hand, the brown alga *Sargassum* is a poor feedstock, apparently because of the low mannitol content and an unidentified "fiber-like" component (Flowers and Bird 1990). More recently, trials on methane production using *Laminaria digitata* in Europe (Morand et al. 1991) and beach-cast *Laminaria* and *Ulva* in Japan (Koike et al. 2005) have been conducted. In the latter test, the maximum methane yield was 22 m^3/t biomass.

The anaerobic digestion of microalgal biomass to produce methane has also been examined by several workers since the original study of Golueke et al. (1957). Algae harvested from wastewater treatment ponds (Chen 1987; Chen and Oswald 1998; Yen and Brune 2007) and unialgal laboratory cultures of *Chlorella*, *Dunaliella*, *Tetraselmis*, *Scenedesmus*, and *Spirulina* (Asinari Di San Marzano et al. 1982; Samson and Leduy 1982; Sanchez and Travieso 1993; Munoz et al. 2005) have been used as feed biomass. These studies have reported methane yields of 0.09–0.45 L/g volatile solids. High temperatures (>40°C) enhance methane conversion. De Schlamphelaire and Verstaete (2009) have developed a closed-loop system combining an algal growth unit for biomass production, an aerobic digestion unit to convert the biomass to biogas (methane), and a microbial fuel cell to polish the effluent from the digester. This system resulted in a power plant with a potential capacity of 9 kW/ha of solar reactor.

Microalgae generally have a high nitrogen (protein) content and therefore a low carbon-to-nitrogen ratio (C/N). This affects the performance of the anaerobic digester and can result in a significant release of ammonia during anaerobic digestion (Golueke et al. 1957; Samson and Leduy 1986); however, methanogenic bacteria can acclimate to high concentrations of ammonium (Koster and Lettinga 1984). Co-digestion with a high C/N material such as waste paper can result in a significant increase in methane production (Yen and Brune 2007).

The anaerobic digestion of marine microalgae also requires the use of salt-adapted microorganisms, which can tolerate the high salinities (Chen et al. 2008). Methane can also be produced from microalgal biomass by hydrothermal gasification at high temperatures (~350–400°C) and pressure in the presence of a nickel catalyst to produce a synthetic natural gas (Minowa and Sawayama 1999; Haiduc et al. 2009).

Sialve et al. (2009) have suggested that anaerobic digestion of microalgal biomass is the optimal strategy, on an energy-balance basis, for the energetic recovery from microalgal biomass. Furthermore, the nutrient-rich effluent of the digestion potentially can be recycled into new algal growth medium (Phang et al. 2000).

26.3   ETHANOL AND BUTANOL

The sugars and carbohydrates of algae may be fermented to produce ethanol or possibly butanol, both of which can be blended with petrol to produce a renewable transport fuel. For example, the brown kelp *Saccharina latissima* (*Laminaria saccharina*) contains approximately 25% mannitol and 30% laminaran, a linear polysaccharide of (1→3)-β-D-glucopyranose, with the chains terminated by D-mannitol. The bacterium *Zymobacter palmae* has been shown to be able to ferment the mannitol to ethanol (Horn et al. 2000a), and the yeast *Pichia angophorae* can use the mannitol and the laminaran; however, the ethanol yields are still low and the process requires further optimization (Horn et al. 2000b). Laminaran can also be fermented by the yeast *Saccharomyces cerevisiae* when used in combination with the enzyme laminarase (Adams et al. 2008). Recently, a process for ethanol production from seaweeds has been patented (Kim et al. 2008). The sugars produced by microalgae can, of course, also be fermented to produce ethanol (Nakas et al. 1983; Ueda et al. 1996). Microalgae such as *Chlorella* have a high starch content (~30-40% of dry weight) and an up to 65% ethanol conversion efficiency has been reported (Anonymous 1995; Hirano et al. 1997). Ueno et al. (1998) have also produced ethanol from the marine green alga *Chlorococcum littorale* in a dark fermentation process. Ethanol-producing cyanobacteria have also been developed (Fu and Dexter 2007; Lee 2008), and the ethanol can be recovered from the airspace above the medium in which the algae grow (Lee 2008; Woods et al. 2008).

Butanol is produced by anaerobic fermentation using solventogenic clostridia bacteria, such as *Clostridium acetobutylicum* and *C. beijerinckii*, and other bacteria, such as *Butyribacterium methylotrophicum* and *Hyperthermus butylicus* (Dürre 2007). The clostridia secrete a wide range of enzymes that break down polymeric carbohydrates to various monosaccharides that are then taken up by the cells and metabolized. The current state of production of biobutanol has recently been reviewed by Ezeji et al. (2007). Butanol has several advantages over ethanol as a transport fuel in that it relatively less polar than ethanol and more similar to gasoline, making it easier to blend with gasoline. Algae are clearly a potential source of renewable biomass for biobutanol production; however, the only published study so far is on the production of butanol from algae (together with ethanol and 1,3-propanediol) is using glycerol-producing algae, such as *Dunaliella* spp. (Nakas et al. 1983).

26.4   PYROLYSIS

Pyrolysis is a thermochemical decomposition process in the virtual absence of oxygen. The pyrolysis of biomass produces char, a crude "bio-oil" and a noncondensable gas, which contains hydrogen, methane, and higher hydrocarbons. In "fast pyrolysis" the biomass is rapidly heated (in ~5–10 s) to between 400 and 500°C. In "slow pyrolysis," the biomass is heated slower to less than approximately 400°C (Grierson et al. 2009). Fast pyrolysis produces more bio-oil than slow pyrolysis. The application of pyrolysis to produce liquid fuel from microalgae was first proposed by Ginzburg (1993) using *Dunaliella* biomass. Pyrolysis of microalgal biomass from the green algae *Chlorella protothecoides* and *Cladophora fracta* and the cyanobacterium *Microcystis aeruginosa* have given oil yields of up to 57.9% of the biomass dry weight (Peng and Wu 2000; Miao and Wu 2004a; 2004b; Demirbas 2006). Slow pyrolysis trials, which gave good bio-oil yields, have also been carried out with a range of microalgae species (Grierson et al. 2009). However, pyrolysis oils will require upgrading because they are acidic, unstable, viscous, and contain solids and chemically dissolved water (Demirbas 2001; Chiaramonti et al. 2007).

Pyrolysis of dried biomass of the coccolithophores, *Emiliania huxleyi* and *Gephyrocapsa oceanica*, at 300°C produced a high yield of liquid-saturated hydrocarbons, the major components of which were normal alkanes in a series ranging from nC_{11} to nC_{35} (Wu et al. 1999a), and increasing temperature to 400°C resulted a decrease in the liquid saturates and an increase in hydrocarbon gases, mainly methane (Wu et al. 1999b). Coccolithophorid algae seem to be particularly well suited to pyrolysis as they have a high lipid and hydrocarbon content including long-chain (C_{37}-C_{39}) alkenones and alkeonates (Volkman et al. 1995; Bell and Pond 1996; Pond and Harris 1996). There is also good evidence that the yield of hydrocarbon gases is dependent on the lipid content of the biomass (Wu et al. 1996).

26.5   LIPIDS AND BIODIESEL

Microalgae are seen as an important future source of renewable biodiesel. The potential advantages of microalgae as sources of liquid biofuels compared to other oleaginous or sugar producing plants such as canola, oil palms, jatropha, sugarcane, and corn is that they

- Have markedly higher annual productivities than land plants per unit land area when grown in intensive culture,
- Can be grown using saline water (at salinities up to NaCl saturation), thus not competing with food crops for scarce freshwater resources, and
- Can be grown on land unsuitable for agriculture.

Compared with fossil diesel, algae-derived biodiesel has also been shown to have a substantial positive greenhouse gas and energy balance in preliminary modeling (Campbell et al. 2009).

The recognition that algae are potential sources of lipids for the production of biodiesel is not new. Many species of microalgae contain high levels of lipids in the range of 20–60% of dry weight, although contents greater than about 30% are generally only found in nutrient-depleted stationary-phase cultures (Borowitzka 1988; Griffiths and Harrison 2009; Huerliman et al. 2010) (Table 26.1). The lipid composition (i.e., fatty acid composition, saturated/polyunsaturated fatty acid ratio, proportion of phospholipids, etc.) of algae varies between taxa and, to some degree, with growth conditions.

The bulk of microalgal lipids are C_{14} to C_{22} chain length esters of glycerol and fatty acids. Triglycerides are the most common storage lipids, and these may constitute up to approximately 80% of the total lipids in nutrient-starved nongrowing cells (Tornabene et al. 1983). These storage lipids are usually located as droplets in the cytoplasm. The other major algal lipids are sulphoquinovosyl diglyceride, monogalactosyl diglyceride, digalactosyl diglyceride, lecithin, phosphatidyl glycerol, and phosphatidyl inositol as the main membrane lipids (Guschina and Harwood 2006).

The lipids of the different algal taxa vary in the composition of the main fatty acids (Table 26.2), and these differences in the composition of the lipids affect the efficiency of the conversion process to biodiesel and the properties of the biodiesel. Biodiesel is produced from algal lipids by esterifying free fatty acids or transesterifying triacylglycerol fatty acids using an alcohol, usually

TABLE 26.1
Range of Lipid Contents Reported for Microalgae

Algal Class	Total Lipids (% dry weight)	Percent of Total Lipid			Hydrocarbons (% dry weight)
		Neutral Lipids	Glycolipids	Phospholipids	
Cyanophyceae	2–23	11–68	12–41	16–50	0.005–0.6
Rhodophyceae		41–58	42–59		
Cryptophyceae	3–17				0.004
Dinophyceae	5–36				0.004–0.2
Bacillariophyceae	1–39	14–60	13–44	10–47	0.2-0.7
Heterokontophyta	12–72				0.2–70[b]
Chlorophyceae	1–70	21.66	6.62	17–53	0.03–1.0 (39.0)[a]

[a] High value for *B. braunii.*

[b] High value for the coccolithophorid algae (Prymnesiophyceae).

TABLE 26.2
Summary of Principal Fatty Acids for the Major Classes of Microalgae

Class	Major Fatty Acids	Representative Genera
Cyanophyceae	14:0, 16:0, 16:1(n-7), 18:1, 18:2(n-3), 18:3(n-3)	*Oscillatoria, Spirulina*
Rhodophyceae	16:0, 20:4(n-6), 20:5(n-3)	*Porphyridium, Rhodella*
Cryptophyceae	14:0, 16:0, 16:1(n-7), 18:3(n-3), 18:4(n-3), 20:5(n-3)	*Chroomonas, Rhodomonas*
Bacillariophyceae	14:0, 16:0,16:1(n-7),16:3(n-4), 20:5(n-3)	*Chaetoceros, Navicula, Phaeodactylum, Skeletonema*
Raphidophyceae	14:0; 16:0;18:4(n-3) 20:5(n-3)	*Chattonella*
Pavlovophyceae	14:0, 16:0,16:1(n-7), 20:5(n-3)	*Pavlova*
Prymnesiophyceae	14:0, 18:3(n-3), 18:4(n-3), 18:5(n-3), 22:6(n-3)	*Emiliania, Gephyrocapsa, Pleurochrysis*
Eustigmatophyceae	14:0, 16:0, 16:1(n-7), 20:5(n-3)	*Nannochloropsis*
Xanthophyceae	14:0, 16:0, 16:1(n-9), 16:1(n-7), 18:1, 20:5(n-3)	*Monodus*
Dinophyceae	[14:0], 16:0, [18:0], [18:1(n-9)]; 18:5(n-3), [22:2], [22:6(n-3)]	*Alexandrium, Amphidinium, Coolia; Gymnodinium, Heterocapsa*
Prasinophyceae	16:0, 16:4(n-3), 18:1(n-9), 18:3(n-3), 18:3(n-4), 20:5(n-3), [22:6(n-3)]	*Isochrysis, Tetraselmis*
Chlorophyceae	16:0, 16:4(n-3), 18:1, 18:3(n-3), [20:4(n-6)][a]	*Botryococcus, Chlorella, Dunaliella, Oocystis*

Source: Borowitzka, M.A., *Micro-algal Biotechnology*, Cambridge University Press, Cambridge, United Kingdom, pp 257–287, 1998; Volkman, J.K., et al. *J Exp Mar Biol Ecol.*, 128, 219–240, 1989; Volkman, J.K., et al. *Phytochemistry*, 30, 1855–1859, 1991; Volkman, J.K., et al., *J Phycol*, 29, 69–78, 1993; Yongmanitchai, W. and Ward, O.P., *Phytochemistry*, 30, 2963–2967, 1991; Viso, A.C. and Marty, J.C., *Phytochemistry*, 34, 1521–1533, 1993; Zhukova, N.V. and Aizdaicher, N.A., *Phytochemistry*, 39, 351–356, 1995; Pond, D.W. and Harris, R.P., *J Mar Biol Assoc UK*, 76, 579–594, 1996; Cohen, Z., *Spirulina platensis (Arthrospira): Physiology, Cell-Biology and Biochemistry.* Taylor & Francis, London, pp 175–204, 1997; Mostaert, A.S., et al. *Phycol Res.*, 46, 213–220, 1998; Tzovenis I. et al. (2003); Mansour, M.P., et al. *J Appl Phycol.*, 17, 287–300, 2005; Khozin-Goldberg I. and Cohen Z., *Phytochemistry*, 67, 696–701, 2006; Patil, V., et al. *Aquacult Int.*, 15, 1–9, 2007; Petkov, G. and Garcia, G., *Biochem Syst Ecol.*, 35, 281–285, 2007; Usup G., et al. *Phycologia*, 47, 105–111, 2008; Xu Z, et al. *J Appl Phycol.*, 20, 237–243, 2008.

[a] Only reported for *Parietochloris incisa* (Bigogno C., et al. *Phytochemistry* 60, 497–503, 2002.)

Only the major fatty acids making up approximately 75% of the total fatty acids are listed. Note fatty acids in brackets are observed in only some species.

methanol or ethanol (Demirbas 2003). Methanol is more reactive than ethanol, and the fatty acid methyl esters (FAMEs) produced are more volatile than the fatty acid ethyl esters (FAEEs) produced when ethanol is used. Methanol is also cheaper; however, methanol is produced from nonrenewable fossil fuel feedstocks whereas ethanol can be produced from renewable feedstocks (sugars). The transesterification processes are catalyzed with alkalis, such as NaOH, KOH, or sodium metoxide. Alternatively, acid-catalyzed transesterification with simultaneous esterification of free fatty acids can be carried out using sulfuric, hydrochloric, phosphoric, or sulfonic acid (Meher et al. 2006). However, acid-catalyzed transesterification has a slower reaction rate than alkali-catalyzed transesterification, and the acids are more corrosive, thus making the process more expensive. Other methods under development are lipase-enzyme-catalyzed transesterification (Ranganathan et al. 2008; Robles-Medina et al. 2009), noncatalytic conversion by transesterification and esterification under supercritical alcohol conditions (Kudsiana and Saka 2001), and the use of metal-oxide base catalysts at high pressure and temperature (McNeff et al. 2008). A range of solid heterogeneous catalyst processes that are potentially more effective than current methods are also under development (Helwani et al. 2010).

Lipid extracts of many algae have been reported to contain relatively high levels of free fatty acids. These can lead to the formation of soaps in alkali-based transesterification, which reduces yield and increases the level of downstream processing and water use required to remove these soaps. Acid-based catalysis simultaneously esterifies the free fatty acids and transesterifies the triacylglycerols. This is well demonstrated in a study comparing acid- and alkali-catalyzed transesterification of lipids from the diatom *Chaetoceros muelleri* (Nagle and Lemke 1990). Using hydrochloric acid-methanol they achieved a maximum 4% FAME yield, whereas using NaOH as catalyst the yield was only 1.65%. Robles-Medina et al. (2009) have proposed a two-stage process to overcome this problem. In the first stage, acid catalysts are used to convert the free fatty acids to methyl esters, and in the second stage, an alkali-catalyzed process is used to convert the remaining triacylglycerols to methyl esters. However, the high free fatty acid content may be an artifact due to the activity of endogenous lipases during the lipid extraction process. If, when harvested, cells of the diatom *Skeletonema costatum* were immediately treated with boiling water to inactivate the lipases before lipid extraction by the Bligh and Dyer (1959) method; no free fatty acids could be detected (Berge et al. 1995).

Combined extraction and esterification is also possible. Belarbi et al. (2000) used a slurry (82% water by weight) of either the diatom *Phaeodactylum tricornutum* or the green alga *Monodus subterraneus* and transesterified these with methanol and acetyl chloride by heating in a boiling water bath for 120 min at 2.5 atm. They obtained a yield of 77.5% FAMEs.

The properties of the biodiesel are mainly determined by the component fatty acids of the algal lipids used to produce them (Knothe 2005). Of particular interest are the cloud point (the temperature at which the fuel becomes cloudy because of solidification), the pour point (the temperature at which the fuel stops flowing), and the cetane index (related to the ignition delay time and combustion quality of the fuel). Oils with a high content of unsaturated fatty acids result in a biodiesel that is less viscous and has a greater cloud point and pour point, making it more suitable for use in colder climates. However, this biodiesel is more prone to oxidation and has a lower cetane index. Oils with a high proportion of long-chain fatty acids ($>C_{18}$) have a higher cetane index. The oxidative stability of the biodiesel is strongly affected by the position of double bonds in the saturated fatty acids. For example, esters of linoleic acid (double bonds at $\Delta 9$ and $\Delta 12$) oxidize more slowly than esters of linolenic acid (double bonds at $\Delta 9$, $\Delta 12$, and $\Delta 15$) (Frankel 1998). Unsaturation may also decrease the lubricity of the fuel and may contribute to gum formation in the engine.

The European standards for biodiesel for vehicle use (EN14214) and for heating oil (EN 14213) limit the content of FAMEs with four or more double bonds to a maximum of 1 mol % (Knothe 2006). Many oleaginous microalgae, especially the diatoms, cryptomonads, haptophytes, and eustigmatophytes (Brown et al. 1997) have a high content of highly unsaturated fatty acids such as eicosapentaenoic acid (C20:5n-3) and docosahexaenoic acid (C22:6n-3), which means that their lipids are likely to meet the European standards without further treatment such as partial catalytic hydrogenation of the oils (Dijkstra 2006). Much of the research to date has focused on microalgae species with a high content of long-chain polyunsaturated fatty acids for pharmaceutical and nutritional applications (Molina Grima et al. 1999; Kawachi et al. 2002; Tonon et al. 2002), but the great and still largely unexplored diversity of the microalgae does provide the opportunity to seek species and strains with reduced levels of polyunsaturated fatty acids as sources of lipids for the production of biodiesel.

The lipid content of microalgae also varies between species as well as with the growth stage and growth conditions (Borowitzka 1988). Many microalgae increase their lipid content when nutrient limited, especially nitrogen limited (Griffiths and Harrison 2009; Rodolfi et al. 2009). This increased lipid content is mainly due to an increase in triacylglycerols (TAGs), which act as storage lipids (Borowitzka 1988; Roessler 1990). Diatoms also increase their lipid content under silicon limitation (Coombs et al. 1967; Taguchi et al. 1987). In some microalgae species such as *Monodus subterraneus*, *Isochrysis galbana*, *Pavlova lutheri*, *P. tricornutum*, and *Chaetoceros* spp., phosphate limitation also leads to an increase in the TAG content (Khozin-Goldberg and Cohen

2006). On the other hand, other species of algae such as *Dunaliella tertiolecta*, *Tetraselmis* sp., *Nannochloris atomus*, *Biddulphia aurita*, and *Synedra ulna* have a reduced lipid content when nutrient limited (Shifrin and Chisholm 1981; Siron et al. 1989; Reitan et al. 1994). In the green alga *Chlorella vulgaris*, iron supplementation has also been shown to increase the neutral lipid content (Liu et al. 2008). The addition of CO_2 or bicarbonate to the cultures may also increase the lipid content of some (e.g., Muradyan et al. 2004; Chiu et al. 2009; Guihéneuf et al. 2009; Widjaja et al. 2009) but not all species of algae (Raghavan et al. 2008). However, because algal biomass productivity is generally stimulated by CO_2 addition, the lipid productivity may still be enhanced even if the lipid content per cell is not. A more detailed treatment of the effects of nutrition and environmental factors on algal lipid content and the fatty acid profile may be found in Borowitzka (1988), Hu et al. (2008), and Harwood and Guschina (2009).

26.6 HYDROCARBON PRODUCERS

The hydrocarbon content of most microalgae is quite low (Borowitzka 1988), with the marked exception of the colonial green alga *Botryococcus braunii*, which is found in freshwater, brackish lakes, and other water bodies in temperate and tropical zones and which has a very high content of hydrocarbons and ether lipids (Metzger and Largeau 2005). There are three chemical "races" of *Botryococcus*: (1) the A-race that produces essentially n-alkadiene and triene hydrocarbons, odd-carbon-numbered from C_{23} to C_{33}; (2) the B-race, which produces C_{30}-C_{37} triterpenoid hydrocarbons, the botryococcenes, and C_{34} methylated squalenes; and (3) the L-race, which produces a single tetraterpenoid hydrocarbon, lycopadiene. Hydrocarbon contents range from 0.4 to 61% of dry weight for A-race strains, approximately 9–40% for B-race strains, and 0.1–8% for L-race strains (Metzger and Largeau 2005). In the Berkeley strain of *B. braunii*, approximately 7% of the botryococcenes, mainly C_{30} and C_{32} botryococcenes, are located in the cells, whereas the external colonial matrix contains more than 99% of the C_{33} and C_{34} compounds as well as lower-chain-length botryococcenes (Wolf et al. 1985). *B. braunii* also contains "normal" lipids and sterols (Metzger and Largeau 1999).

Hydrocarbon productivity is greatest during the exponential growth phase and does not occur in nitrogen- and phosphorus-deficient media (Largeau et al. 1980; Casadevall et al. 1985; Dayanandra et al. 2007). The hydrocarbon productivity can be enhanced by bubbling the culture with CO_2-enriched air (Casadevall et al. 1985; Ranga Rao et al. 2007). A day/night cycle rather than continuous light also seems to favor hydrocarbon production (Dayanandra et al. 2007). Trials of growing this alga outdoors in tubular photobioreactors up to a volume of 200 L have been carried out (Gudin and Chaumont 1983). *Botryococcus* also grows well on secondarily treated piggery wastewater (An et al. 2003). Recovery of the hydrocarbons can be by solvent extraction (Metzger and Largeau 1999) or by extraction using supercritical CO_2 (Mendes et al. 1994). Frenz et al. (1989a, 1989b) used a novel process in which they extracted the hydrocarbons by a short contact of the wet biomass with a nontoxic solvent such as hexane without reducing cell viability. This process recovered up to 70% of the total hydrocarbons.

The other group of microalgae that produces significant amounts of hydrocarbons and long-chain methyl and ethyl ketones (alkenones) and may have a high lipids content is the coccolithophorids (Fernandez et al. 1994; Bell and Pond 1996). At least one species, *Pleurochrysis carterae*, grows very well in outdoor raceway ponds (Moheimani and Borowitzka 2006, 2007) and is therefore also of interest as a potential source of renewable fuel.

26.7 LARGE-SCALE PRODUCTION OF MICROALGAE

Commercial-scale algae culture has been carried out all over the world for the last 30+ years. The two largest commercial algae production plants are at Hutt Lagoon, Western Australia and Whyalla, South Australia, growing the halophilic algae *Dunaliella salina* for the production of β-carotene for the nutritional supplement industry (Borowitzka and Borowitzka 1989; Borowitzka and Hallegraeff 2007). *Dunaliella* is also commercially grown in Israel and India. The blue-green alga *Spirulina* (*Arthrospira*) is grown in the United States, China, Taiwan, Thailand, and India (Belay 1997; Hu 2004), the green alga *Chlorella* in Taiwan, Indonesia, Japan, and Germany (Lee 1997), and the green alga *Haematococcus* in the United States, Israel, and Sweden (Cysewski and Lorenz 2004). A number of other species such as *Chaetoceros*, *Isochrysis*, *Tetraselmis*, *Nannochloropsis*, and *Nitzschia* are also grown around the world as feed for aquaculture species (Borowitzka 1999b; Muller-Fuega 2004). Although some of these production plants are very large (e.g., the *Dunaliella* plants in Australia have an area >750 ha), they are still very small compared with the production plants needed for biodiesel production. For example, a raceway pond-based production plant producing microalgae with a 40% lipid content at an average daily productivity of 30 g/m² per day would need approximately 75.5 km² of ponds to produce 10,000 barrels (1 barrel = 158.987 L) of oil per day. Therefore, the extremely large scale required for biofuel production using microalgae presents a new and significant challenge (Borowitzka and Moheimani 2011; Fon Sing et al. 2011).

The production of biofuels from algae needs not only to be technically feasible, but also must be commercially viable and environmentally sustainable. The actual production costs of these commercial producers are difficult to obtain; however, the best estimates are given in Table 26.3. How do these costs compare with the production costs required to produce biodiesel? Current prices for comparative feedstock for biodiesel are palm oil (~$U.S. 0.40–0.50/kg) and canola/rape seed oil (~$U.S. 0.60/kg). Assuming that an alga's oil content is approximately 40% of dry weight (achievable average), then to produce oil at $U.S. 0.50/kg, the algal biomass must be produced at a cost of less than approximately $U.S. 0.20/kg (Note: This includes harvesting costs, but it does not include lipid extraction costs). This presents a major challenge for algal oil production for biofuels. To achieve this cost, reliable high productivities all year round of a biomass with a high lipid content (at least 30–40% of dry weight) are essential. Furthermore, culture and harvesting systems have to be constructed and operated at very low cost.

To achieve high biomass productivities, the following key conditions should be met:

- Maximal available sunshine (i.e., as little cloud cover as possible), and
- As little rainfall as possible (because rain means clouds, thus reducing light available, and for better control of salinity).

TABLE 26.3
Estimated Production Costs of Commercially Grown Microalgae

Alga	Culture System	Place	Estimated Production Cost for Dry Biomass in Australian Dollars per Kilogram
Dunaliella salina	Extensive (~270 ha) open ponds	Australia	$6
Chlorella spp.	Open ponds	Taiwan	$17
	Open ponds (shallow cascade system)	Australia (pilot plant operation)	$15
Spirulina	Open raceway ponds	United States	$15–20
Haematococcus pluvialis	Tubular photobioreactor	Israel	>$50
Several species for aquaculture	Mainly closed reactors	Australia, United States, Europe	$50 to >$1000[a]

[a] Some of the high costs can be attributed to the small scale of the operations.

For environmental sustainability, the algae should be grown in saline medium to minimize the need for freshwater.

The lowest cost culture system for microalgae production at this time is the raceway. The best annual average productivity for long-term (>6 months) continuous culture in a raceway reported in the literature is approximately 20 g ash-free dry weight/m^2 per day for the coccolithophorid alga *Pleurochrysuis caterae* in Perth, Western Australia (Moheimani and Borowitzka 2006). In current commercial-scale operations, this is not achieved. In summer, very high productivities of 41–47 g dry weight/m^2 per day have been achieved. The lipid content of this alga is approximately 30–35% of dry weight in log-phase growing cultures. However, it should be noted that the growth conditions or the lipid productivity for this alga have not, as yet, been optimized. To achieve such productivity, the density of the algal culture needs to be controlled to optimize the available light reaching the cells, and for most species CO_2 needs to be added (Borowitzka 1998). The system should also be operated as a continuous culture. This means the algal strain used and the operating regime must be able to exclude, or at least control, the growth of contaminating species and predators such as protozoa. This is possible for quite a number of species such as *D. salina* (because of the very high salinity), *Spirulina* spp. (because of the high alkalinity), *P. tricornutum*, and *Chlorella* spp. (because of high growth rate), *Pleurochrysis caterae* (possibly because of the production of acrylic acid by the alga) and a range of other fast-growing marine algae such as *Nannochloropsis, Tetraselmis, Nannochloris, Pheodactylum*, and *Chlorococcum* (Ben Amotz, personal communication).

For the production of algal lipids to produce biodiesel, it is not the biomass productivity but the lipid productivity that is important (Griffiths and Harrison 2009). Most algae achieve their maximal lipid (oil) content in the stationary phase of growth; however, continuous culture for maximal productivity means that the cells should always be in the log phase of growth. This requires algal strains that have a high lipid content while actively growing so as to have a high lipid productivity. Lipid productivity is a function of the algal productivity and the lipid content. The effect of productivity and cell lipid content on lipid productivity is shown in Figure 26.1. At the potentially achievable scenario of a productivity of 30 g dry weight/m^2 per day with a lipid content of 40% of the biomass, 12 kg oil/ha per day are produced.

The algae also need to have a wide temperature tolerance and grow well outdoors. In the high solar irradiation environments, where high average annual productivities are possible, temperatures in open ponds can reach up to approximately 35°C in summer, and they may cool down to less than 5°C at night in winter. Similarly, closed photobioreactors heat up rapidly during the day on sunny

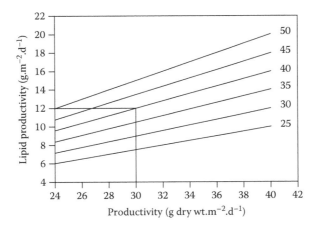

FIGURE 26.1   Effect of productivity and cell lipid content (numbers at end of lines are percent lipid content) on lipid productivity. The fine line shows the potentially achievable target of an annual average productivity of 30 g dry weight/m^2 per day at a lipid content of 40%.

days unless cooled, but they cool down rapidly at night unless heated, and cooling and heating require a significant energy input. The need to reduce the need for freshwater also means that saline water should be used to make up evaporative losses, and this, in turn, means that the algae should be able to grow well over an extended range of salinities.

Harvesting and dewatering also presents a major cost to any microalgae production process (Mohn and Cordero-Contreras 1990; Borowitzka 1999a; Molina Grima et al. 2003) and have been reviewed by Mohn (1988) and Molina Grima et al. (2004). The solids content of microalgal cultures in large-scale systems ranges from 0.1 to approximately 1 g/L, meaning that very large volumes of water have to be processed. Most of the microalgae of interest are also very small (<20 μm dimeter) and have a density very similar to that of the medium they are growing in. Because the water also still contains significant quantities of nutrients, the growth medium must be recycled after harvesting of the algae for economic and environmental reasons. This excludes some types of harvesting. The cheapest harvesting method is filtration; however, most algae, other than filamentous species such as *Spirulina*, are too small for effective filtration. The next best method is settling followed by dewatering, but, where this is not possible, flocculation and flotation (or settling) could be used. The method of harvesting will depend on the species of alga cultured (Shelef et al. 1984; Mohn 1988) and may also be affected by the growth phase (Danquah et al. 2009).

Extraction of the lipids from microalgae will probably need to be by solvent extraction (Molina Grima et al. 1995; Lee et al. 2010).

26.8 OPEN POND CULTURE VERSUS CLOSED PHOTOBIOREACTORS

Closed photobioreactors are often cited as the solution for the production of microalgae for biodiesel (e.g., Chisti 2007; McCall 2008; Rodolfi et al. 2009), and combined closed photobioreactor/open pond systems also have been proposed (Huntley and Redalje 2007). However, despite more than 50 years of work on closed photobioreactors, they have as yet not been shown to be commercially viable for microalgal production, except for the very high-value alga *Haematococcus pluvialis* and *Chlorella* for the health-food market. The various designs of closed photobioreactors recently have been reviewed by Tredici (2004) and Carvalho et al. (2006). Although closed systems appear to be a solution for some of the problems encountered in open systems, such as the potential control of contaminants and greater control of environmental factors, they present other challenges and problems, and it has been our experience that many species of algae cannot be grown in closed systems. Closed reactors also require cooling during periods of high irradiance and are more difficult to scale up (Borowitzka 1996, 1999b). Table 26.4 compares several aspects of open and closed culture systems.

To date, productivities in large-scale closed systems (calculated on a grams-per-liter basis) are only approximately 2–3 times those of open systems. However, the capital costs of closed systems are at least 5–10 times higher and the operating costs, especially energy costs for circulating the algal culture and for temperature control, are also much higher. Furthermore, open raceway culture systems have the advantage of being a well-known, proven, and reliable technology. Details of open pond culture can be found in Borowitzka (2005).

26.9 THE FUTURE

The high activity in research and development on the development of new algae and on processes to produce renewable biofuels from algae will mean that the current economic and technological challenges will be overcome in time (Stephens et al. 2010b). In many laboratories, there is ongoing isolation and screening for better strains suited to large-scale cultivation, and improved lipid productivity will broaden the range of species available. There is also a search for valuable coproducts that may help to improve the economics of microalgal fuel production (Stephens et al. 2010a).

TABLE 26.4

Comparison of the Properties of Open Pond Raceway Culture Systems and Closed Photobioreactors

	Open Ponds	Closed Photobioreactors
Species range	Most species, but control of contamination by other algae and/or predators is a potential problem, but this can be controlled for a number of species (e.g., in all of the ones listed below)	Can only grow species that are shear-tolerant because the need for circulation (air-lifts, pumps) damages many species. "Sticking" of some algae to the walls of the reactor is also a problem
Species that have been shown that they can be cultured reliably for extended (>3 months) periods	*Chlorella, Spirulina, Scenedesmus, Dunaliella, Phaeodactylum, Pleurochrysis, Monodus, Nitzschia, Nannochloropsis*	*Spirulina, Chlorella, Tetraselmis, Isochrysis, Phaeodactylum, Nannochloropsis*
Key Factors limiting productivity		
Light	Some capacity to optimize light environment for cells by controlling pond depth and/or cell density	Can control light environment by controlling diameter of tubing or width of plate reactor as well as cell density. Sticking of the algae to the bioreactor surface also limits light availability and growth in many species. For maximal productivities, the light path (reactor thickness) should not exceed ~400 mm)
Temperature	Limited capacity to control temperature in large ponds; however, maximum pond temperature does not exceed ~30–35°C because of evaporation, so no cooling is required.	In high light the system heats up rapidly and there is a need to cool system. Requires energy (and large amounts of freshwater if evaporative cooling system is used)
CO_2	Can be added	Can be added
O_2 (high O_2 inhibits photosynthesis)	O_2 built up during the day by photosynthesis is lost from system by exchange on pond surface	Requires efficient degassing system. In tubular photobioreactors O_2 limits length of tubes that can be used.
Cost factors		
Capital cost	Relatively high because of the need for pond liners	Very high
Power requirement	Mixing using paddlewheels is very energy-efficient.	Power requirements for circulating the culture are high. Power may also be required for cooling during the day.

See also: Grobbelaar, J.U., *J Appl Phycol.*, 21, 489–492, 2009.

Genetic engineering also presents a new, and as yet little explored option for overcoming some of the limitations of growth and lipid production. One potential strategy that has been proposed is to engineer diatoms, and possibly other algae, to secrete some of their lipids (Ramachandra et al. 2009). For example, in corals the symbiotic dinoflagellates have been shown to apparently secrete lipid-like osmiophilic material (Crossland et al. 1980). The green alga *B. braunii* also accumulates hydrocarbons extracellularly.

Another possibility is to enhance the light utilization efficiency of the algae. For example, Melis et al. (1999) found that mutants of *D. salina* with small chlorophyll antenna size showed higher photosynthetic productivities and photon use efficiencies (see also Mitra and Melis 2008). A mutant of *Chlamydomonas reinhardtii* (*tla1*) with truncated light harvesting antenna size also had enhanced photosynthetic productivity in laboratory and outdoors in a greenhouse (Nakajima and

Ueda 1997, 1999; Polle et al. 2003). Recently, Mussgnug et al. (2007) used RNA interference (RNAi) to downregulate the expression of light harvesting antenna complex proteins of *C. reinhardtii* with the recombinant strain showing higher resistance to photodamage and increased light penetration in the culture.

Alternatively, genetic engineering may be used to enhance key parts of the pathway of lipid synthesis. For example, the acetyl-CoA carboxylase (ACC) gene from *Cyclotella cryptica* has been transformed into the diatoms *C. cryptica* and *Navicula saprophila*, resulting in overexpression of the ACC gene, *acc*1, enhancing enzyme activity 2- to 3-fold. However, there was no significant increase in lipid accumulation in the transgenic diatoms (Roessler et al. 1994; Dunahay et al. 1995, 1996), suggesting that there is a secondary limiting step in the TAG pathway. As an alternative approach, Courchesne et al. (2009) proposed to enhance lipid overproduction by overexpressing transcription factors regulating the metabolic pathways involved in the production of lipids.

These potential improvements in the efficiency of light use and metabolic redirection to lipid synthesis, coupled with essential improvements in culture and harvesting systems, will contribute to the development of commercially viable and environmentally sustainable production of biofuels from algae.

REFERENCES

Adams JM, Gallagher JA, Donnison IS (2008) Fermentation study on *Saccharina latissima* for bioethanol production considering variable pretreatments. *J Appl Phycol* 21:569–574

An JY, Sim SJ, Lee JS, Kim BW (2003) Hydrocarbon production from secondarily treated piggery wastewater by the green alga *Botryococcus braunii*. *J Appl Phycol* 15:185–191

Anonymous (1995) Preparation of ethanol from microalgae Japan Patent 7,087,986

Asinari Di San Marzano,CM, Legros A, Naveau HP, Nyns EJ (1982) Biomethenation of the marine algae *Tetraselmis*. *Int J Sustain Energy* 1:263–272

Belarbi EH, Molina Grima E, Chisti Y (2000) A process for high yield and scalable recovery of high purity eicosapentaenoic acid esters from microalgae and fish oil. *Enz Microb Technol* 26:516–529

Belay A (1997) Mass culture of *Spirulina* outdoors—The Earthrise Farms experience. In: Vonshak A (ed), *Spirulina platensis (Arthrospira): Physiology, Cell Biology and Biochemistry*. Taylor & Francis, London, pp 131–158

Bell MV, Pond D (1996) Lipid composition during growth of motile and coccolith forms of *Emiliania huxleyi*. *Phytochemistry* 41:465–471

Benemann J (2004) Hydrogen and methane production by microalgae. In Richmond A (ed), *Microalgal Culture: Biotechnology and Applied Phycology*. Blackwell Science, Oxford, pp 403–416

Berge JP, Gouygou JP, Dubacq JP, Durand P (1995) Reassessment of lipid composition of the diatom, *Skeletonema costatum*. *Phytochemistry* 39:1017–1021

Bigogno C, Khozin-Goldberg I, Boussiba S, Vonshak A, Cohen Z (2002) Lipid and fatty acid composition of the green oleaginous alga *Parietochloris incisa*, the richest plant source of arachidonic acid. *Phytochemistry* 60:497–503

Bligh EG, Dyer WJ (1959) A rapid method of total lipid extraction and purification. *Can J Biochem Physiol* 37:911–917

Borowitzka MA (1988) Fats, oils and hydrocarbons. In: Borowitzka MA, Borowitzka LJ (eds), *Micro-algal Biotechnology*. Cambridge University Press, Cambridge, United Kingdom, pp 257–287

Borowitzka MA (1996) Closed algal photobioreactors: Design considerations for large-scale systems. *J Mar Biotechnol* 4:185–191

Borowitzka MA (1999a) Economic evaluation of microalgal processes and products. In: Cohen Z (ed), *Chemicals from Microalgae*. Taylor & Francis, London, pp 387–409

Borowitzka MA (1999b) Commercial production of microalgae: Ponds, tanks, tubes and fermenters. *J Biotechnol* 70:313–321

Borowitzka MA (2005) Culturing microalgae in outdoor ponds. In: Anderson RA (ed), *Algal Culturing Techniques*. Elsevier Academic Press, London, pp 205–218

Borowitzka LJ, Borowitzka MA (1989) Industrial production: Methods and economics. In: Cresswell RC, Rees TAV, Shah N (eds), *Algal and Cyanobacterial Biotechnology*. Longman Scientific, London, pp 294–316

Borowitzka MA, Hallegraeff G (2007) Economic importance of algae. In: McCarthy PM, Orchard AE (eds), *Algae of Australia: Introduction*. ABRS, Canberra, Australia, pp 594–622

Borowitzka MA, Moheimani NR (2011) Sustainable biofuels from algae. *Mitigat Adapt Strat Global Change*, doi: 10.1007/s11027-010-9271-9

Brennan L, Owende P (2010) Biofuels from microalgae—A review of technologies for production, processing, and extraction of biofuels and co-products. *Renew Sustain Energy Rev* 14:557–577

Brown MR, Jeffrey SW, Volkman JK, Dunstan GA (1997) Nutritional properties of microalgae for mariculture. *Aquaculture* 151:315–331.

Campbell PK, Beer T, Batten D (2009) *Greenhouse Gas Sequestration by Algae—Energy and Greenhouse Gas Life Cycle Studies*, available at http://www.csiro.au/resources/Greenhouse-Sequestration-Algae.html (accessed July 19, 2009)

Carvalho AP, Meireles LA, Malcata FX (2006) Microalgal reactors: A review of enclosed system designs and performances. *Biotechnol Progr* 22:1490–1506

Casadevall E, Dif D, Largeau C, Gudin C, Chaumont D, Desanti O (1985) Studies on batch and continuous cultures of *Botryococcus braunii*: Hydrocarbon production in relation to physiological state, cell ultrastructure, and phosphate nutrition. *Biotechnol Bioeng* 27:286–295

Chen PH (1987) Factors influencing methane fermentation of microalgae. PhD Thesis, University of California, Berkeley, CA

Chen PH, Oswald WJ (1998) Thermochemical treatment for algal fermentation. *Environ Int* 24:889–897

Chen Y, Cheng JJ, Creamer KS (2008) Inhibition of anaerobic digestion process: A review. *Bioresour Technol* 99:4044–4064

Chiaramonti D, Osamaa A, Solantausta Y (2007) Power generation using fast pyrolysis liquids from biomass. *Renew Sustain Energy Rev* 11:1056–1086

Chisti Y (2007) Biodiesel from microalgae. *Biotechnol Adv* 25:294–306

Chiu S, Kao C, Tsai M, Ong S, Chen C, Lin C (2009) Lipid accumulation and CO_2 utilization of *Nannochloropsis oculata* in response to CO_2 aeration. *Bioresour Technol* 100:833–838

Chynoweth DP (2002) *Review of Biomethane from Marine Biomass*. Department of Agricultural Biological Engineering, University of Florida, Gainesville, FL

Chynoweth DP, Fannin KF, Srivastava VJ (1987) Biological gasification of marine algae. In: Bird KT, Benson PH (eds), *Seaweed Cultivation for Renewable Resources*. Elsevier, New York, pp 285–304

Cohen Z (1997) The chemicals of *Spirulina*. In: Vonshak A (ed), *Spirulina Platensis (Arthrospira): Physiology, Cell-Biology and Biochemistry*. Taylor & Francis, London, pp 175–204

Coombs J. Darley WM, Holm-Hansen O, Volcani BE (1967) Studies on the biochemistry and fine structure of silica shell formation in diatoms. Chemical composition of *Navicula pelliculosa* during silicon-starvation synchrony. *Plant Physiol* 42:1601–1606

Courchesne NMD, Parisien A, Wang B, Lan CQ (2009) Enhancement of lipid production using biochemical, genetic and transcription factor engineering approaches. *J Biotechnol* 141:31–41

Crossland CJ, Barnes DJ, Borowitzka MA (1980) Diurnal lipid and mucus production in the staghorn coral *Acropora acuminata*. *Mar Biol* 60:81–90

Cysewski GR, Lorenz RT (2004) Industrial production of microalgal cell-mass and secondary products—Species of high potential: *Haematococcus*. In: Richmond A (ed), *Microalgal Culture: Biotechnology and Applied Phycology*. Blackwell Science, Oxford, United Kingdom, pp 281–288

Danquah MK, Gladman B, Moheimani N, Forde GM (2009) Microalgal growth characteristics and subsequent influence on dewatering efficiency. *Chem Eng J* 151:73–78

Dayanandra C, Sarada R, Usha Rani M, Shamala TR, Ravishankar GA (2007) Autotrophic cultivation of *Botryococcus braunii* for the production of hydrocarbons and exopolysaccharides in various media. *Biomass Bioenergy* 31:87–93

De Schlamphelaire L, Verstraete W (2009) Revival of the biological sunlight-to-biogas energy conversion system. *Biotechnol Bioeng* 103:296–304

Demirbas A (2001) Biomass resource facilities and biomass conversion processing for fuels and chemicals. *Energy Conver Manag* 42:1357–1378

Demirbas A (2003) Biodiesel fuels from vegetable oils via catalytic and non-catalytic supercritical alcohol transesterifications and other methods: A survey. *Energy Conver Manag* 44:2093–2109

Demirbas A (2006) Oily products from mosses and algae via pyrolysis. *Energy Sources A* 28:933–940.

Dijkstra AJ (2006) Revisiting the formation of trans isomers during partial hydrogenation of triacylglycerol oils. *Eur J Lipid Sci Technol* 108:249–264

Dunahay TG, Jarvis EE, Dais SS, Roessler PG (1996) Manipulation of microalgal lipid production using genetic engineering. *Appl Biochem Biotechnol A* 57–58:223–231

Dunahay TG, Jarvis EE, Roessler PG (1995) Genetic transformation of the diatoms *Cyclotella cryptica* and *Navicula saprophila*. *J Phycol* 31:1004–1012

Dürre P (2007) Biobutanol: An attractive biofuel. *Biotechnol J* 2:1525–1534

Ezeji TC, Qureshi N, Blaschek HP (2007) Bioproduction of butanol from biomass: From genes to bioreactors. *Curr Opin Biotechnol* 18:1–8

FAO (2006) *Yearbooks of Fishery Statistics*. Food and Agricultural Organization, Rome, Italy

Fernandez E. Balch WM, Maranon E, Holligan PM (1994) High rates of lipid biosynthesis in cultured, mesocosm and coastal populations of the coccolithophore *Emiliania huxleyi*. *Mar Ecol Progr Ser* 114:13–22

Flowers A, Bird KT (1990) Methane production from seaweeds. In: Akatsuka I (ed), *Introduction to Applied Phycology*. SPB Academic Publishing, Hague, The Netherlands, pp 575–587

Fon Sing S, Isdepski A, Moheimani NR, Borowitzka MA (2011) Production of biofuels from microalgae. *Mitigat Adapt Strat Global Change*, DOI: 10.1007/s11027-011-9294-x

Frankel EN (1998) *Lipid Oxidation*. The Oily Press, Dundee, United Kingdom

Frenz J, Largeau C, Casadevall E (1989a) Hydrocarbon recovery by extraction with a biocompatible solvent from free and immobilized cultures of *Botryococcus braunii*. *Enz Microb Technol* 11:717–724

Frenz J, Largeau C, Casadevall E, Kollerup F, Daugulis AJ (1989b) Hydrocarbon recovery and biocompatibility of solvents for extraction from cultures of *Botryococcus braunii*. *Biotechnol Bioeng* 34:755–762

Fu PP, Dexter J (2007) Methods and compositions for ethanol producing cyanobacteria. PCT Patent WO2007/084477

Ginzburg B (1993) Liquid fuel (oil) from halophilic algae: A renewable resource of non-polluting energy. *Renew Energy* 3:249–252

Golueke CG, Oswald WJ, Gotaas HB (1957) Anaerobic digestion of algae. *Appl Microbiol* 5:47–55

Grierson S, Strezov V, Ellem G, McGregor R, Herbertson J (2009) Thermal characterisation of microalgae under slow pyrolysis conditions. *J Anal Appl Pyrolysis* 85:118–123

Griffiths MJ, Harrison STL (2009) Lipid productivity as a key characteristic for choosing algal species for biodiesel production. *J Appl Phycol* 21:493–507

Grobbelaar JU (2009) Factors governing algal growth in photobioreactors: The "open" versus "closed" debate. *J Appl Phycol* 21:489–492

Gudin C, Chaumont D (1983) Solar biotechnology study and development of tubular solar receptors for controlled production of photosynthetic cellular biomass. In Palz W, Pirrwitz D (eds), *Proceedings of the Workshop and E.C. Contractor's Meeting*, Capri. Reidel, Dordrecht, The Netherlands, pp 184–193

Guihéneuf F, Mimouni V, Ulmann L, Tremblin T (2009) Combined effects of irradiance level and carbon source on fatty acid and lipid class composition in the microalgae *Pavlova lutheri* commonly used in aquaculture. *J Exp Mar Biol Ecol* 369:136–143

Guschina IA, Harwood JL (2006) Lipids and lipid metabolism in eukaryotic algae. *Prog Lipid Res* 45:160–186

Habig C, DeBusk TA, Ryther JH (1984) The effect of nitrogen content on methane production by the marine algae *Gracilaria tikvahiae* and *Ulva* sp. *Biomass* 4:239–251

Haiduc AC, Brandenberger M, Suquet S, Vogel F, Bernier-Latmani R, Ludwig C (2009) SunChem: An integrated process for the hydrothermal production of methane from microalgae and CO_2 mitigation *J Appl Phycol* 21:529–541

Hansson G (1983) Methane production from marine, green macroalgae. *Resour Conserv* 8:185–194

Harwood JL, Guschina IA (2009) The versatility of algae and their lipid metabolism. *Biochimie* 91:1–6

Helwani Z, Othman MR, Azis N, Fernando WJN, Kim J (2010) Technologies for production of biodiesel focusing on green catalytic techniques: A review. *Fuel Proc Technol* 90:1502–1514

Hirano A, Ueda R, Hirayama S, Ogushi Y (1997) CO_2 fixation and ethanol production with microalgal photosynthesis and intracellular anaerobic fermentation. *Energy* 22:137–142

Horn SJ, Aasen IM, Østgaard K (2000a) Production of ethanol from mannitol by *Zymobacter palmae*. *J Indust Microbiol Biotechnol* 24:51–57

Horn SJ, Aasen IM, Østgaard K (2000b) Ethanol production from seaweed extract. *J Indust Microbiol Biotechnol* 25:249–254

Hu Q (2004) Industrial production of microalgal cell-mass and secondary products—Major industrial species: *Arthrospira (Spirulina) platensis*. In: Richmond A (ed), *Microalgal Culture: Biotechnology and Applied Phycology*. Blackwell Science, Oxford, United Kingdom, pp 264–272

Hu Q, Sommerfeld M, Jarvis E, Ghirardi M, Posewitz M, Seibert M, Darzins A (2008) Microalgal triacylglycerols as feedstocks for biofuel production: Perspectives and advances. *Plant J* 54:621–639

Huerlimann R, de Nys R, Heimann K (2010) Growth, lipid content, productivity, and fatty acid composition of tropical microalgae for scale-up production. *Biotechnol Bioeng* 107: 245–257

Huntley M, Redalje D (2007) CO$_2$ mitigation and renewable oil from photosynthetic microbes: A new appraisal. *Mitigat Adapt Strat Glob Change* 12:573–608

Kawachi, M, Inouye I, Honda D, O'Kelly CJ, Bailey JC, Bidigare RR, Andersen RA (2002) The Pinguiopyceae *classis nova*, a new class of photosynthetic stramenopiles whose members produce large amounts of omega-3 fatty acids. *Phycol Res* 50:31–47

Khozin-Goldberg I, Cohen Z (2006) The effect of phosphate starvation on the lipid and fatty acid composition of the fresh water eustigmatophyte *Monodus subterraneus. Phytochemistry* 67:696–701

Kim GS, Shin MK, Kim YJ, Oh KK, Kim JS, Ryu HJ, Kim KH (2008) Method of producing biofuel using sea algae PCT Patent WO/2008/105618

Knothe G (2005) Dependence of biodiesel fuel properties on the structure of fatty acid alkyl esters. *Fuel Process Technol* 86:1059–1070

Knothe G (2006) Analyzing biodiesel: Standards and other methods. *J Amer Oil Chem Soc* 83:823–833

Koike Y, Matsui T, Saiganji A, Saito H, Amano T (2005) Field test of producing biomass from marine biomass. In: *National Symposium on Power and Energy Systems*, Japan Society of Mechanical Engineers, Tokyo, Japan, pp 217–220

Koster IW, Lettinga G (1984) The influence of ammonium-nitrogen on the specific activity of pelletized methanogenic sludge. *Agric Wastes* 9:205–216

Kudsiana D, Saka S (2001) Methyl esterification of free fatty acids of rapeseed oil as treated in supercritical methanol. *Fuel* 80:225–231

Largeau C, Casadevall E, Berkaloff C, Dhamelincourt P (1980) Studies of accumulation and composition of hydrocarbons in *Botryococcus braunii. Phytochemistry* 19:1043–1051

Lee JW (2008) Designer organisms for photosynthetic production of ethanol from carbon dioxide. PCT Patent WO2008/039450

Lee JY, Jun SY, Ahn CY, Oh MH (2010) Comparison of several methods for effective lipid extraction from microalgae. *Bioresour Technol* 101(Suppl): S75–S77

Lee YK (1997) Commercial production of microalgae in the Asia-Pacific rim. *J Appl Phycol* 9:403–411

Liu ZY, Wang GC, Zhou BC (2008) Effect of iron on growth and lipid accumulation in *Chlorella vulgaris. Bioresour Technol* 99:4717–4722

Mansour MP, Frampton DMF, Nichols PD, Volkman JK, Blackburn SI (2005) Lipid and fatty acid yield of nine stationary-phase microalgae: Applications and unusual C$_{24}$–C$_{28}$ polyunsaturated fatty acids. *J Appl Phycol* 17:287–300

McCall J (2008) Apparatus and method for production of biodiesel. PCT Patent WO 2008/089321

McHugh D J (2003) *A Guide to the Seaweed Industry*. Food and Agricultural Organization Fisheries Technical Paper 441

McNeff CV, McNeff LC, Yan B, Nowlan DT, Rasmussen M, Gyberg AE, Krohn BJ, Fedie RL, Hoye TR (2008) A continuous catalytic system for biodiesel production. *Appl Catalysis A* 343:39–48

Meher LC, Vidya Sagar D, Naik SN (2006) Technical aspects of biodiesel production by transesterification—A review. *Renew Sustain Energy Rev* 10:248–268.

Melis A, Neidhardt J, Benemann J (1999) *Dunaliella salina* (Chlorophyta) with small chlorophyll antenna sizes exhibit higher photosynthetic productivities and photon use efficiencies than normally pigmented cells. *J Appl Phycol* 10:515–525

Mendes RL, Fernandes HL, Coelho JAP, Cabral JMS, Palavra AMF, Novais JM (1994) Supercritical carbon dioxide extraction of hydrocarbons from the microalga *Botryococcus braunii. J Appl Phycol* 6:289–293

Metzger P, Largeau C (1999) Chemicals from *Botryococcus braunii*. In: Cohen Z (ed), *Chemicals from Microalgae*. Taylor & Francis, London, pp 205–260

Metzger P, Largeau C (2005) *Botryococcus braunii*: A rich source for hydrocarbons and related ether lipids. *Appl Microbiol Biotechnol* 66:486–496

Miao XL, Wu QY (2004a) Fast pyrolysis of microalgae to produce renewable fuels. *Anal Appl Pyrol* 71:855–863

Miao XL, Wu QY (2004b) High yield bio-oil production from fast pyrolysis by metabolic controlling of *Chlorella protothecoides. J Biotechnol* 110:85–93

Minowa T, Sawayama S (1999) A novel microalgal system for energy production with nitrogen cycling. *Fuel* 78:1213–1215

Mitra M, Melis A (2008) Optical properties of microalgae for enhanced biofuels production. *Optics Express* 16:21807–21820

Miyura Y. 2002. Microbial process for producing hydrogen. U.S. Patent 6,395,521

Moheimani, NR, Borowitzka MA (2006) The long-term culture of the coccolithophore *Pleurochrysis carterae* (Haptophyta) in outdoor raceway ponds. *J Appl Phycol* 18:703–712

Moheimani NR, Borowitzka MA (2007) Limits to growth of *Pleurochrysis carterae* (Haptophyta) grown in outdoor raceway ponds. *Biotechnol Bioeng* 96:27–36

Mohn FH (1988) Harvesting of micro-algal biomass. In: Borowitzka MA, Borowitzka LJ (eds), *Micro-Algal Biotechnology*. Cambridge University Press, Cambridge, United Kingdom, pp 395–414

Mohn FH, Cordero-Contreras O (1990) Harvesting of the alga *Dunaliella*—Some consideration concerning its cultivation and impact on the production costs of ß-carotene. *Berichte des Forschungszentrums Jülich* 2438:1–50

Molina Grima E, Acién Fernández FG, Robles Medina A (2004) Downstream processing of cell-mass and products. In: Richmond A (ed), *Microalgal Culture: Biotechnology and Applied Phycology*. Blackwell Science, Oxford, United Kingdom, pp 215–251

Molina Grima E, Belarbi EH, Ácién Fernandez FG, Robles Medina A, Chisti Y (2003) Recovery of microalgal biomass and metabolites: Process options and economics. *Biotechnol Adv* 20:491–515

Molina Grima E, Garcia Camacho F, Ácien Fernandez FG (1999) Production of EPA from *Phaeodactylum tricornutum*. In: Cohen Z (ed), *Chemicals from Microalgae*. Taylor & Francis, London, pp 57–92

Molina Grima E, Sanchez Perez JA, Garcia Camacho F, Robles Medina A, Garcia-Jiminez P, Lopez Alonso D (1995) The production of polyunsaturated fatty acids by microalgae—From strain selection to product purification. *Proc Biochem* 30:711–719

Morand P, Carpentier B, Charlier RH, Maze J, Orlandini M, Plunkett BA, Dewaart J (1991) Bioconversion of Seaweeds. In: Guiry MD, Blunden G (eds), *Seaweed Resources of Europe: Uses and Potential*. John Wiley, Hoboken, NJ pp 95–148

Mostaert AS, Karsten U, Hara Y, Watanabe MM (1998) Pigments and fatty acids of marine raphidophytes: A chemotaxonomic re-evaluation. *Phycol Res* 46:213–220

Muller-Fuega A (2004) Microalgae for aquaculture. the current global situation and future trends. In: Richmond A (ed), *Microalgal Culture: Biotechnology and Applied Phycology*. Blackwell Science, Oxford, United Kingdom, pp 352–364

Munoz R, Jacinto M, Guieysse B, Mattiasson B (2005) Combined carbon and nitrogen removal from acetonitrile using algal-bacterial bioreactors. *Appl Microbiol Biotechnol* 67:699–707

Muradyan EA, Klyachko-Gurvich GL, Tsoglin LN, Sergeyenko TV, Pronina NA (2004) Changes in lipid metabolism during adaptation of the *Dunaliella salina* photosynthetic apparatus to high CO_2 concentration. *Russ J Plant Physiol* 51:53–62

Mussgnug JH, Thomas-Hall SR, Ruprecht J, Foo A, Klassen V, McDowall A, Schenk PM, Kruse O, Hankamer B (2007) Engineering photosynthetic light capture: Impacts on improved solar energy to biomass conversion. *Plant Biotechnol J* 5:802–814

Nagle N, Lemke P (1990) Production of methyl ester fuel from microalgae. *Appl Biochem Biotechnol* 24/25:355–361

Nakajima Y, Ueda R (1997) Improvement of photosynthesis in dense microalgal suspensions by reducing the content of light harvesting pigments. *J Appl Phycol* 9:503–510

Nakajima Y, Ueda R (1999) Improvement of microalgal photosynthetic productivity by reducing the content of light harvesting pigments. *J Appl Phycol* 11:151–201.

Nakas JP, Schaedle M, Parkinson CM, Coonley CE, Tanenbaum SW (1983) System development for linked-fermentation production of solvents from algal biomass. *Appl Env Microbiol* 46:1017–1023

Østergaard K, Indergaard M, Markussen S, Knutsen SH, Jensen A (1993) Carbohydrate degradation and methane production during fermentation of *Laminaria saccharina* (Laminariales, Phaeophyceae). *J Appl Phycol* 5: 333–342

Patil V, Källqvust T, Olsen E, Vogt G, Gislerød HR (2007) Fatty acid composition of 12 microalgae for possible use in aquaculture feed. *Aquacult Int* 15:1–9

Peng WM, Wu QY (2000) Effects of temperature and holding time on production of renewable fuels from pyrolysis of *Chlorella protothecoides*. *J Appl Phycol* 12:147–152

Petkov G, Garcia G (2007) Which are fatty acids of the green alga *Chlorella*? *Biochem Syst Ecol* 35:281–285

Phang SM, Miah MS, Yeoh BG, Hashim MA (2000) *Spirulina* cultivation in digested sago starch factory wastewater. *J Appl Phycol* 12:395–400

Polle JEW, Kanakagiri S, Melis A (2003) *tla1*, a DNA insertional transformant of the antenna size. *Planta* 217:49–59

Pond DW, Harris RP (1996) The lipid composition of the coccolithophore *Emiliania huxleyi* and its possible ecophysiological significance. *J Mar Biol Assoc UK* 76:579–594

Raghavan G, Haridevi CK, Gopinathan CP (2008) Growth and proximate composition of the *Chaetoceros calcitrans* f. *pumilus* under different temperature, salinity and carbon dioxide levels. *Aquacult Res* 39:1053–1058

Ramachandra TV, Mahapatra DM, Karthick B (2009) Milking diatoms for sustainable energy: Biochemical engineering versus gasoline-secreting diatom solar panels. *Indust Eng Chem Res* 48: 8769–8788

Ranga Rao A, Sarada R, Ravishankar GA (2007) The influence of CO_2 on growth and hydrocarbon production in *Botryococcus braunii*. *J Microbiol Biotechnol* 17:414–419

Ranganathan SV, Narasimhan SL, Muthukumar K (2008) An overview of enzymatic production of biodiesel. *Bioresour Technol* 99:3975–3981

Reitan KI, Rainuzzo JR, Olsen Y (1994) Effect of nutrient limitation on fatty acid and lipid content of marine microalgae. *J Phycol* 30:972–979

Robles-Medina A, González-Moreno PA, Esteban-Cerdan L, Molina-Grima E (2009) Biocatalysis: Towards even greener biodiesel production. *Biotechnol Adv* 27:398–408

Rodolfi L, Zitelli GC, Bassi N, Padovani G, Biondi N, Bonini G, Tredici MR (2009) Microalgae for oil: Strain selection, induction of lipid synthesis and outdoor mass cultivation in a low-cost photobioreactor. *Biotechnol Bioeng* 102:100–112

Roessler PG (1990) Environmental control of glycerolipid metabolism in microalgae—Commercial implications and future research directions. *J Phycol* 26:393–399

Roessler PG, Bleibaum JL, Thompson GA, Ohlrogge JB (1994) Characteristics of the gene that encodes acetyl-CoA carboxylase in the diatom *Cyclotella cryptica*. *Ann NY Acad Sci* 721:250–256

Ross AB, Anastasakis K, Kubacki M, Jones JM (2009) Investigation of the pyrolysis behavior of brown algae before and after pre-treatment using PY-GC/MS and TGA. *J Anal Appl Pyrolysis* 85:3–10

Samson R, Leduy A (1982) Biogas production from anaerobic digestion of *Spirulina maxima* algal biomass. *Biotechnol Bioeng* 24:1919–1924

Samson R, Leduy A (1986) Detailed study of anaerobic digestion of *Spirulina maxima* algal biomass. *Biotechnol Bioeng* 28:1014–1023

Sanchez EP, Travieso L (1993) Anaerobic digestion of *Chlorella vulgaris* for energy production. *Resour Conserv Recycl* 9:127–132

Schramm W, Lehnberg W (1984) Mass culture of brackish-water adapted seaweeds in sewage-enriched seawater. II. Fermentation for biogas production. *Hydrobiologia* 116/117:282–287

Sheehan J, Dunahay T, Benemann J, Roessler P (1998) *A Look Back at the U.S. Department of Energy's Aquatic Species Program—Biodiesel from Algae*. NREL/TP-580-24190, National Renewable Energy Laboratory, Golden, CO, pp 1–328

Shelef G, Sukenik A, Green M (1984) *Microalgal Harvesting and Processing: A Literature Review*. SERI/STR-231-2396, Solar Energy Research Institute, Golden, CO, pp 1–65

Shifrin NS, Chisholm SW (1981) Phytoplankton lipids: Interspecific differences and effects of nitrate, silicate and light-dark cycles. *J Phycol* 17:374–384

Sialve B, Bernet N, Bernard O (2009) Anaerobic digestion of microalgae as a necessary step to make microalgal biodiesel sustainable. *Biotechnol Adv* 27:409–416

Siron R, Giusti G, Berland B (1989) Changes in the fatty acid composition of *Phaeodactylum tricornutum* and *Dunaliella tertiolecta* during growth and under phosphorous deficiency. *Mar Ecol Prog Ser* 55:95–100

Stephens E, Ross IL, King Z, Mussgnug JH, Kruse O, Posten C, Borowitzka MA, Hankamer B (2010a) An economic and technical evaluation of microalgal biofuels. *Nat Biotechnol* 28:126–128

Stephens E, Ross IL, Mussgnug JH, Wagner LD, Borowitzka MA, Posten C, Kruse O, Hankamer B (2010b) Future prospects of microalgal biofuel production systems. *Trends Plant Sci* 15:554–564

Taguchi S, Hirata JA, Laws EA (1987) Silicate deficiency and lipid synthesis in marine diatoms. *J Phycol* 23:260–267

Tonon T, Harvey D, Larson TR, Graham IA (2002) Long chain polyunsaturated fatty acid production and partitioning to triacylglycerols in four microalgae. *Phytochemistry* 61:15–24

Tornabene TG, Holzer G, Lien S, Burris N (1983) Lipid composition of the nitrogen starved green alga *Neochloris oleoabundans*. *Enz Microb Technol* 5:435–440

Tredici MR (2004) Mass production of microalgae: Photobioreactors. In: Richmond A (ed), *Microalgal Culture: Biotechnology and Applied Phycology*. Blackwell Science, Oxford, United Kingdom, pp 178–214

Tzovenis I, De Pauw N, Sorgeloos P (2003) Optimisation of T-ISO biomass production rich in essential fatty acids II. Effect of different light regimes on the production of fatty acids. *Aquaculture* 216:223–242

Ueda R, Hirayama S, Sugata K, Nakayama H (1996). Process for the production of ethanol from microalgae. U.S. Patent 5,578,472

Ueno Y, Kurano N, Miyachi S (1998) Ethanol production by dark fermentation in the marine green algae, *Chlorococcum littorale*. *J Ferment Bioeng* 86:38–43

Usup G., Hamid SZ,. Chiet PK, Wah CK, Ahmad A (2008) Marked differences in fatty acid profiles of some planktonic and benthic marine dinoflagellates from Malaysian waters. *Phycologia* 47:105–111

Viso AC, Marty JC (1993) Fatty acids from 28 marine microalgae. *Phytochemistry* 34:1521–1533

Volkman JK, Barrett SM, Blackburn SI, Sikes EL (1995) Alkenones in *Gephyrocapsa oceanica*: Implications for studies of paleoclimate. *Geochim Cosmochim Acta* 59:513–520

Volkman JK, Brown MR, Dunstan GA, Jeffrey SW (1993) The biochemical composition of marine microalgae from the Class Eustigmatophyceae. *J Phycol* 29:69–78

Volkman JK, Dunstan GA, Jeffrey SW, Kearney PS (1991) Fatty acids from microalgae of the genus *Pavlova*. *Phytochemistry* 30:1855–1859

Volkman JK, Jeffrey SW, Nichols PD, Rogers GI, Garland CD (1989) Fatty acid and lipid composition of 10 species of microalgae used in mariculture. *J Exp Mar Biol Ecol* 128:219–240

Widjaja A, Chien CC, Ju YH (2009) Study of increasing lipid production from freshwater microalgae *Chlorella vulgaris*. *J Taiwan Inst Chem Eng* 40:13–20

Wolf FR, Nonomura AM, Bassham JA (1985) Growth and branched hydrocarbon production in a strain of *Botryococcus braunii* (Chlorophyta). *J Phycol* 21:388–396

Woods PR, Legere E, Moll B, Unamuzaga C, Mantecon E (2008) Closed photobioreactor system for continuous daily in situ production, separation, collection, and removal of ethanol from genetically enhanced photosynthetic organisms. U.S. Patent 2008/0153080

Wu Q, Dai J, Shiraiwa Y, Sheng G, Fu J (1999a) A renewable energy source—Hydrocarbon gases resulting from pyrolysis of the marine nanoplanktonic alga *Emiliania huxleyi*. *J Appl Phycol* 11:137–142

Wu Q, Shiraiwa Y, Takeda H, Sheng G, Fu J (1999b) Liquid-saturated hydrocarbons resulting from pyrolysis of the marine coccolithophores *Emiliania huxley* and *Gephyrocapsa oceanica*. *Mar Biotechnol* 1:346–352

Wu Q, Zhang B, Grant NG (1996) High yield of hydrocarbon gases resulting from pyrolysis of yellow heterotrophic and bacterially degraded *Chlorella protothecoides*. *J Appl Phycol* 8:181–184

Xu Z, Yan X, Pei L, Luo Q (2008) Changes in fatty acids and sterols during batch growth of *Pavlova viridis* in a photobioreactor. *J Appl Phycol* 20:237–243

Yen HW, Brune DE (2007) Anaerobic digestion of algal sludge and waste paper to produce methane. *Bioresour Technol* 98:130–134

Yongmanitchai W, Ward OP (1991) Screening of algae for potential alternative sources of eicosapentaenoic acid. *Phytochemistry* 30:2963–2967

Zemke-White WL, Ohno M (1999) World seaweed utilisation: An end-of-century summary. *J Appl Phycol* 11:369–376

Zhukova NV, Aizdaicher NA (1995) Fatty acid composition of 15 species of marine microalgae. *Phytochemistry* 39:351–356

27 Paulownia

Nirmal Joshee
Fort Valley State University

CONTENTS

27.1 INTRODUCTION

Rising energy prices and environmental problems have led to increased interest in alcohol as a fuel. Using corn, a human food resource, for ethanol production raises major ethical and moral issues. Today, malnourished people in the world number approximately 3 billion (WHO 2000). The current food shortages throughout the world call attention to the importance of using U.S. exports of corn and other grains for human food and search for alternative sources for biofuel production. For the production of cellulosic ethanol, residue including postharvest corn plants (stover) and timber residues could be used. There is a growing awareness among farmers to establish and use specialized high-biomass "energy crops" such as domesticated poplar trees, switchgrass, bamboo, and Paulownia. The United States uses approximately 140 billion gal of gasoline per year, and there is a growing urgency to replace it with biological resources or biomass (US DOE 2006).

Biomass is a complex mixture of organic materials (such as carbohydrates, fats, and proteins) along with small amounts of minerals (such as sodium, phosphorus, calcium, and iron). The main components of plant biomass are carbohydrates (~75%, dry weight) and lignin (~25%) which can vary with plant type. The carbohydrates are mainly cellulose or hemicellulose fibers which impart strength to the plant structure, and lignin which holds the fibers together. A major advantage of using biomass as a source of fuels or chemicals is its renewability. The net annual production of biomass by photosynthesis has been estimated to be 10 times that of our current annual consumption of fossil fuels (Miyamoto 1997).

It is easier to produce ethanol from sugar feedstocks (e.g., sugarcane, sugarbeets, sweet sorghum, fruits, and other materials known as saccharides) or starchy feedstocks (e.g., corn, cereal grains, sweet potatoes or cassava). However, the cost of production is prohibitive because both of these groups are in the human food chain. For the production of cellulosic ethanol, residue including postharvest corn plants (stover) and timber residues could be used. Therefore, the scientific community is engaged in exploring other systems that are more efficient in breaking down the cellular structures for cellulosic ethanol production.

Paulownia species are highly suitable to revalidate agricultural set-aside areas, to reclaim mining areas, or to restore contaminated sites where major emphasis is on biomass production for chemical or thermophysical processing. Because of the low cost, plentiful supply, and amenability to biotechnology, carbohydrates appear likely to be the dominant source of feedstock for biocommodity processing. In the case of starch, the advantage of enzymatic compared with chemical hydrolysis has already been realized. In the case of cellulose, this has not yet been realized. Cellulose hydrolyzing enzymes can only act effectively after pretreatment to break up the very stable lignin, cellulose, or hemicellulose composites. These treatments are still mostly thermal, thermomechanical, or thermochemical and require a considerable input of energy.

With a deep root system that is fed by underground water at a level below 2 m, Paulownia does not compete with the roots of other crops. Intercropping with Paulownia improves microclimate by reducing the effects of drying winds by 20–50% on average and increasing air moisture. This can considerably increase the yield of some crops, such as ginger, winter wheat, and millet (Zhu et al. 1986). The tree benefits by recovering excess fertilizer that runs deep into the ground and the crops benefit from the nutrients put into the topsoil by fallen leaves. Further, large green leaves are rich in nitrogen and can be used for fodder and green manure. Therefore, the leaves are rich in protein (16.2%), carbohydrates (9.44%), and minerals (Song 1988), making them ideal for animal fodder and green fertilizer (a 10-year-old tree produces 80 kg of dry leaves/year). The fast growth rate of Paulownia may be capitalized upon for agroforestry (Wang and Shogren 1992; Jiang et al. 1994), biomass production (Song 1988), land reclamation (Carpenter 1977), and animal waste remediation systems (Bergmann et al. 1997). Sufficient variation among *P. elongata* clones was revealed for growth parameters and foliar nutrient concentrations to anticipate a benefit from the selection of genotypes that are the most efficient for the remediation of animal waste. The data show that *P. elongata* has potential for use as a swine waste utilization species (Bergmann et al. 1997). This particularly suits the southeastern region of the United States where there is high a concentration of swine and poultry industry. Swine industry effluents and chicken litter are available at comparatively lower prices, replacing the need for synthetic fertilizers.

27.2 HISTORY OF PAULOWNIA IN THE ORIENT AND THE UNITED STATES

Paulownia is known as *kiri* in Japan, specifically referring to *P. tomentosa*. The name *kiri* comes from the word *kiru* (to cut) because it was believed that the tree would grow better and more quickly when it was cut down regularly. It was once customary in Japan to plant a Paulownia tree when a baby girl was born and then make it into a dresser as a wedding present when she gets married.

According to Chinese legend, the mythical phoenix alights only in the branches of the Paulownia tree when it lands on earth (http://www.onmarkproductions.com/html/ho-oo-phoenix.shtml).

China has historically been the largest grower of Paulownia. Chinese people use its wood for making furniture, house construction, toys, plywood, musical instruments, and for packaging. For an ideal multipurpose tree, the Paulownia intercropping models have been applied and extended to 15 million ha in rural areas of the central plains of north China. The total number of Paulownia trees growing in China is approximately 1 billion, including those grown as the shelterbelts at the canal banks and road sides or for ornamental forests around houses and villages. For decades, Japanese craftsmen have used this revered wood in ceremonial furniture, musical instruments, decorative moldings, laminated structural beams, and shipping containers. The tree made its way to the United States in the mid-1800s. Paulownia seeds were used as packaging material for delicate porcelain dishes on their journey across the Pacific. Once unpacked, the tiny wind-blown seeds became naturalized throughout the eastern states.

Paulownia also has a unique biological character (i.e., a root system that grows deep and a crown that develops in loose structure) that makes it suitable for intercropping or in a mixed planting with other shade-enduring trees. The traditional monocropping models in many of these countries have been replaced by Paulownia-crop intercropping fields which have resulted in more reasonable use of sunlight, heat, water, and air resources so that the farmland productivity and the product diversity are raised. The adoption of Paulownia-crop intercropping is also a good solution to the land competition between the development of agriculture, forestry, and animal husbandry (Zhu et al. 1986).

27.3 BOTANY AND DISTRIBUTION

Paulownia species are commonly known as Empress tree, *Kiri* tree, Dragon tree, Royal Paulownia, Princess tree, and many other derivatives. *Paulownia* is a genus of between 6 and 17 species (depending on taxonomic authorities) of plants in the monogeneric family Paulowniaceae, related to and sometimes included in the family Scrophulariaceae. The Latin name *Paulownia* was given by the Swiss botanist Thunberg, and the taxonomic details were published in the "Japanese Flora" in 1781 (Zhao-Hua et al. 1985). They are native to much of China, southern to northern Laos and Vietnam, and have long been cultivated elsewhere in eastern Asia, notably in Japan and Korea. They are deciduous trees 40–50 ft tall, with large leaves 15–40 cm across arranged in opposite phyllotaxy on the stem. Surface studies of structures present on *P. tomentosa* suggest that there are three types of structures that protect young plant parts and/or reproductive organs from herbivores (Kobayashi et al. 2008). These structures have been named as bowl-shaped organs, glandular hairs, and dendritic trichomes. Glandular hairs on leaves, stems, and flowers secrete mucilage containing glycerides and flavonoids and trap small insects. These chemicals seem to play a major role in its antiherbivore property. The flowers are produced in early spring on panicles 10–30 cm long, with a tubular purple corolla resembling a foxglove flower. The fruit is a dry capsule containing thousands of minute winged seeds. The main species in the genus *Paulownia* are *P. tomentosa*, *P. fortunei*, *P. elongata*, *P. albiphloea*, *P. catalpifolia*, *P. australis*, *P. kawakamii*, and *P. fargesii* (Zhu et al. 1986). Molecular marker-based studies including random amplified polymorphic DNA (RAPD) and restriction fragment length polymorphism (RFLP) of chloroplast DNA have been used to establish the hybrid origin of *Paulownia taiwaniana* (Wang et al. 1994). This study provided molecular evidence suggesting that *P. taiwaniana* is the natural hybrid between *P. fortunei* and *P. kawakamii* and that the maternal parent is *P. kawakamii*.

Paulownia species grow on flat or mountainous land, in various types of soil including rich humus soil in temperate areas, dry poor soil, rich forest soil, and light clay soil in the subtropics, laterite soil in the tropics, and dry steppes (semiarid grass covered plains in southeast Europe, Siberia, central Asia, and Central America). It also adapts to various climates, from warm and temperate to tropical, and can even withstand temperatures as low as −20°C. Paulownia can survive between latitudes of 40° north and 40° south and at altitudes of up to 2000 m (El-Showk and El-Showk 2003). *P. elongata* grows faster and is suitable for intercropping with arable crops, whereas *P. catalpifolia* and *P. tomentosa* grow slower but have better wood quality. The annual mean temperature in the Paulownia growing areas varies a great deal, ranging from 10 to 22°C. The optimal temperature range for Paulownia growth is 24–29°C. The lowest temperature that Paulownia can resist is –20°C for *P. tomentosa*, –16°C for *P. elongata*, and –13°C for *P. fortunei*, suggesting that some of these fast-growing species can be grown in large, unused tracts of land in the United States (Zhu et al. 1986). Geographical areas suitable for *P. elongata* cultivation are shown in the Figure 27.1a. Paulownia can adapt to a wide range of precipitation ranging from 500 to 2500 mm. The suitable edaphic conditions for Paulownia growth are fertile sandy loam to heavy loam soils with loose structure and salt content less than 0.05%. *Paulownia* species respond very well to fertilizer application but have little tolerance to water-logging. In a comparative study to assess nutrient requirements and stress response of *Populus simonii* and *P. tomentosa*, it was found that the maximum weight increase was 19% and over 25% per day, respectively. Further, the nitrogen retention in the above-ground parts of *P. tomentosa* was very high in comparison to *P. simonii* [0.26 g dry weight $(gN)^{-1} h^{-1}$ vs. 0.16 g dry weight $(gN)^{-1}h^{-1}$], establishing *Paulownia* sp. as one of the fastest growing tree species for biomass production (Hui-Jun and Ingestad 1984).

The leaf, flower, fruit, and bark of Paulownia have been extensively used in Chinese medicine to treat bronchitis, especially on relieving the cough and reducing phlegm, enteritis, tonsillitis, and dysentery (Jiang et al. 2004). Further, chemical analysis of flowers revealed high quantity of bioactive flavonoid apigenin that has been found to show various pharmacological activities, including anti-inflammatory, antispasmodic, antidiarrheal, vasorelaxant, and an antibacterial activities (Jiang et al, 2004). Antiviral furanoquinone and antimicrobial phenylpropanoid glycosides have been isolated from *P. tomentosa* (Kang et al. 1994, 1999).

Except for *P. tomentosa* (Miller 2004), most *Paulownia* species grown in the United States are noninvasive. Although there is little doubt that it is an exotic genus, the question of its invasiveness is open to conjecture. The prolific small seeds of Paulownia are windblown. However, the seeds do not germinate and survive unless they fall on soils with low pathogen load. Young Paulownia seedlings have a high rate of mortality because of damping-off disease caused by various soil fungi. Generally, Paulownia does not colonize in open areas. Requiring full sunlight for continued development, it is often overtopped by other species and succumbs. Paulownia is usually found on the edge of a forest where sunlight is more available rather than in the interior forest. Because of the strict sunlight and soil requirements, the number of Paulownia plants appears to have declined in the recent years. Seed dispersed from Paulownia plantings does not appear to establish and colonize outside of Paulownia plantations.

27.3.1 REPRODUCTION AND EARLY GROWTH

27.3.1.1 Flowering and Fruiting

The perfect flowers of Paulownia are borne in terminal panicles up to 25 cm (10 in) long in April and May. Their violet, lavender, or blue appearance before the leaves emerge is quite striking (Figure 27.1e). The fruits are ovoid, pointed, woody capsules approximately 30–45 mm long. Immature green fruits from the current year and mature, dehisced, brown to black capsules from the previous year can be spotted on a tree at the same time (Figure 27.1g, h). Capsules turn brown as they mature in September or October and persist on the tree through the winter (Bonner and Burton 1974).

27.3.1.2 Seed Production and Dissemination

Numerous seeds are borne in fruits called "capsules." Each capsule contains up to 2000 seeds, and a large tree may produce as many as 20 million seeds a year. The tiny, flat, winged seeds (Figure 27.1i)

FIGURE 27.1 (See color insert) Various aspects of *P. elongata* biology. (a) Area falling under green line in the U.S. map is suitable for commercial cultivation of *P. elongate*. (b) A 4-year-old, well-managed plantation of *P. elongata* near Lenox, GA. (c) A 20-week-old *P. elongata* tree. (d) Many new shoots sprout from a coppiced tree. (e) Vigorously growing shoots photographed 1 year after coppicing. (f) *P. elongata* in bloom during early spring. Note that there are no leaves at this time of the year. (g) Immature green fruits. (h) Mature and dehisced fruits (capsules). (i) Winged seed. A cluster of seeds attached to the placenta is shown in the inset. (j) Seed germination on tissue culture medium. (k) Six-week-old planting of *P. elongata* at Fort Valley State University farm in Georgia.

are light (1,70,000 seeds/oz). As the capsules break open on the trees throughout the winter and into the spring, wind dissemination occurs easily (Bonner and Burton 1974). Seed surface architecture and RAPD markers have been employed to profile and differentiate *P. fortunei*, *P. tomentosa*, and their hybrid. The surface patterns of winged seeds were examined by scanning electron microscopy. The patterns of reticulation on the wings and seed coat of *P. fortunei* and the hybrid were found to be comparable while that on *P. tomentosa* was different (Kumar et al. 1999).

27.3.1.3 Seedling Development

The seeds of the genus Paulownia germinate quickly and grow rapidly when conditions are favorable (Figure 27.1j). Seed germination is epigeal. Laboratory studies have found that light is required for germination (Bonner and Burton 1974). Cold storage (stratification) reduces the light requirement (Tang et al. 1980). *P. tomentosa* needs bare soil, sufficient moisture, and direct sunlight for good seedling establishment. Seedlings are very intolerant to shade.

27.3.1.4 Vegetative Reproduction

Paulownia roots sprout easily. In fact, lateral root cuttings of 1-year-old seedlings can be used for propagation directly in the field (Tang et al. 1980).

27.3.2 Plant Production

27.3.2.1 Root Cuttings

The propagation of Paulownia is simple and easy; for example, the root, stem, shoot or seed can all be used for propagation with facile methods. Current practice for large-scale production of Paulownia planting stocks is to use the roots of 1- to 2-year old Paulownia seedlings for rooted cutting development. With the combination of advanced propagation techniques and the intensive management in the nursery, a 1-year-old Paulownia tree could have the height of 4 m and root collar diameter of 6 cm.

27.3.2.2 Propagation by Seeds

Seedling and plant development using seed is simple and straightforward. The seeds should be collected from the selected desired pest-free trees that grow well and have not been damaged by artificial hybridization of parent trees. Seed treatment for stimulating the germination is needed by soaking seeds in water until they are fully imbibed. Soaking seeds for 15 min in water (40°C) results in good germination. Maintaining the moisture of seedlings by watering the beds regularly, timely weeding, and thinning are needed. When seedlings grow to 5–10 cm high, they should be transplanted into the nursery field for better development. The related management techniques for seedling development in the field are similar to those for the plants developed through rooted cuttings.

27.3.3 Physical Properties of Paulownia Wood

The wood of Paulownia is strong but light and does not crack or split when nails or screws are used. It is very stable dimensionally and is not subjected to warping, cupping, or splitting, even when exposed to an outdoor environment (http://www.worldpaulownia.com/html/tech.html). Paulownia wood is lightweight, 17–21 lb/ft³ as compared with the Appalachian Red Oak of 39–41 lb/ft³. Paulownia wood takes 30–60 days for air-drying, whereas kiln-drying takes 36–60 h depending on dry kiln configuration, horsepower, and dimension of lumber. The wood is resistant to decay and rotting if it is not in permanent contact with the ground. *Paulownia* species vary in porosity ranging from 75 to 88% in comparison to poplar with 70–72%. The density of Paulownia wood, at 10% moisture content, ranges from 17.8 to 23.2 depending on the species and growing conditions. The thermal conductivity of Paulownia wood is very low, thus giving it excellent heat/cool insulation properties. Further, it has an ignition temperature of 420–430°C, as compared with the average hardwood of 220–225°C. *P. tomentosa* has excellent flame retardancy, and it was found that Paulownia wood contains less lignins in comparison to cedar wood (Li and Oda 2007). Paulownia wood generates very little combustible gas when heated. Because of this property, it has been used to make clothing wardrobes for decades in Japan. In other studies conducted to evaluate *P. elongata* as the raw material for paper production, it was revealed that pulp obtained from Paulownia is not of the highest quality but can be readily used for paper production when mixed with long fibrous material (Ates et al. 2008).

27.3.4 DISEASES OF PAULOWNIA

The main diseases that damage the Paulownia at seedling stage are caused by *Gloeosporium kawkamii* and *Spliacelorna paulowniae,* which attack the juvenile leaves and shoots. No major insect pests are known for royal Paulownia in the United States. Minor damage from several foliage diseases has been reported on the species. Two powdery mildew-causing fungal species, *Phyllactinia guttata* and *Uncinula clintonii,* and another fungal species caused by *Phyllosticta paulowniae* producing small brown spots on the leaves have also been reported (Hepting 1971). No major disease problems have appeared yet in the United States.

Paulownia witch's broom (PWB) is a serious disease affecting Paulownia production. The disease incidence rate of Paulownia at 5 years old was approximately 50–80%, which could reduce 25% of timber production. Primers based on the P1-like adhesion gene sequence of *Mycoplasma pneumoniae* have been successfully used to detect this disease (Zhong and Hiruki 1994). If PWB disease is found, the host seedlings should be immediately removed.

27.4   DEVELOPMENT OF HIGH-YIELDING PAULOWNIA FARMS

On the basis of investigation, within 10 years a well-managed Paulownia plantation can attain a mean height of approximately 16–20 m, a mean diameter at breast height (DBH) of 35–40 cm, and a standing volume of 0.5 m³. However, a species should be chosen depending on the growth characteristics and ecological requirements for planting on specific site conditions. The qualified planting stock for a high-yielding plantation should be 4 m in height and 6 cm in DBH. A well-managed 4-year-old *P. elongata* farm near Lenox, GA that was established for lumber is shown in the Figure 27.1b. Following the guidelines carefully and attending the needs of the growing plants, 15- to 16-ft-tall *P. elongata* trees can be easily obtained in the first season of growth within 4–5 months (Figure 27.1c). Whatever type of planting models is adopted, the intercropping of crops or other shade-enduring economic plants are always beneficial.

Paulownia lumber farming is an environmentally sound alternative to expensive, lightweight hardwoods grown in jungles and rain forests—thriving on marginal or even toxic land (Bergmann et al. 1997). Paulownia roots penetrate down as far as 40 ft, regulating the water table and removing soil salinity. Paulownia trees have been shown to be very effective in absorbing waste pollutants from hog, chicken, and dairy facilities as well as various other pollutants (Bergmann et al. 1997). Paulownia saves forests by producing sawn timber in 6–8 years and growing 2–4 times more lumber than most other commercial trees in the same time period [World Paulownia Institute (WPI) personal communication]. This is vital because the supply of exotic hardwoods rapidly diminishes. After harvesting, a new Paulownia tree grows back from the stump and uses the same well-established root system (Figure 27.1d,e). This saves postharvest clearing costs, land erosion, and runoff. A recently established 6-week-old *P. elongata* demonstration farm (spacing 12 × 12 ft) at the Fort Valley State University is shown in Figure 27.1k.

27.5 A TREE FOR LIGNOCELLULOSIC ETHANOL

Fossil fuels have been the dominant energy resource for the modern world. The major limitations of solid biomass fuels are the difficulty of handling and the lack of portability for mobile engines. To address these issues, research is being conducted to convert solid biomass into liquid and gaseous fuels. Biological (fermentation) and chemical means (pyrolysis and gasification) can be used to produce fluid biomass fuels. Ethanol for automotive fuels is currently produced from starch biomass in a two-step process: starch is enzymatically hydrolyzed into glucose, and then yeast is used to convert the glucose into ethanol. The first four aliphatic alcohols (methanol, ethanol, propanol, and butanol) are of interest as fuels because they can be synthesized biologically and they have characteristics that allow them to be used in current engines. Ethanol is nontoxic, water-soluble, and quickly biodegradable. Blending ethanol in gasoline dramatically reduces carbon monoxide tailpipe emissions. Carbon monoxide emissions are responsible for as much as 20% of smog formation.

27.5.1 BIOMASS PRODUCTION AND MOLECULAR BIOLOGY OF CELLULOSE

Trees constitute most lignocellulosic biomass existing on our planet. Trees also serve as important feedstock materials for various industrial products. Wood from forest trees modified for more cellulose or hemicelluloses could be a major feedstock for fuel ethanol. Xylan and glucomannan are the two major hemicelluloses in wood of angiosperms. However, little is known about the genes and gene products involved in the synthesis of these wood polysaccharides. Further, much research at present is being directed to understand regulatory mechanisms of cellulose synthase (CesA) genes of trees. In the production of cellulose fiber materials, it is highly desirable to engineer trees with more cellulose and controllable cellulose properties such as degree of polymerization and crystallinity. However, little is known about the genes controlling wood cellulose formation. The discovery of differential expression of three secondary cell-wall-related CesA genes in response to tension stress and the identification of an mechanical stress-responsive element (MSRE) containing a DNA fragment in the EgraCesA3 promoter provide an important clue for the future improvement of cellulosic material production in trees (Lu et al. 2008).

Genetic improvement of cellulose biosynthesis in woody trees is one of the major goals of tree biotechnology research. However, progress in this field has been slow because of (1) unavailability of key genes from tree genomes; (2) the inability to isolate active and intact CesA complexes; and (3) the limited understanding of the mechanistic processes involved in the wood cellulose development. Recent advances in molecular genetics of CesA from aspen trees suggest that two different types of CesA are involved in cellulose deposition in primary and secondary walls in xylem (Joshi 2003). The three distinct secondary CesA from aspen—PtrCesA1, PtrCesA2, and PtrCesA3—appear to be aspen homologs of Arabidopsis secondary CesAs, AtCesA8, AtCesA7, and AtCesA4, respectively, on the basis of their high identity/similarity (>80%). These aspen CesA proteins share the transmembrane domain (TMD) structure that is typical of all known "true" CesA proteins: two TMDs toward the N-terminal and six TMDs toward the C-terminal. The putative catalytic domain is present between TMDs 2 and 3. All signature motifs of processive glycosyltransferases are also present in this catalytic domain. In a phylogenetic tree based on various predicted CesA proteins from Arabidopsis and aspen, aspen CesAs fall into families similar to those seen with Arabidopsis CesAs, suggesting their functional similarity. The coordinate expression of three aspen secondary CesAs in xylem and phloem fibers and their simultaneous tension stress-responsive upregulation suggest that these three CesA may play a pivotal role in the biosynthesis of better quality cellulose in the secondary cell walls of plants. These results are likely to have a direct effect on the genetic manipulation of trees in the future.

Further, because lignin limits the use of wood for fiber, chemical, and energy production, strategies for its downregulation are of considerable interest. Transgenic aspen trees, in which expression of a lignin biosynthetic pathway gene Pt4CL1 encoding 4-coumarate:coenzyme A ligase

(4CL) was downregulated by antisense inhibition, exhibited up to a 45% reduction of lignin and it was further compensated for by a 15% increase in cellulose (Hu et al. 1999).

27.5.2 SUSTAINABLE AGRICULTURE AND ENERGY CROPS

Defining agricultural sustainability is subjective, but extractive agricultural practices are always ecologically unsound. The "highest use" for a farm's land must take into account the farm's economics, the environment of which it is a part, and the social value of enabling people to live on the land. The long-term view is important (e.g., for soil productivity and minimization of "boom and bust" economic fluctuations). Woody biomass can give farmers flexibility in fluctuating economic conditions and help to keep farms profitable and intact. Biomass crops can be planted on marginal lands that require vegetative cover for critical periods or that would otherwise provide very little or no income without significant ecological damage. To be profitable, the amount of energy that the biomass fuel provides must exceed that used to produce it. Harvest must be timed to protect important wildlife species. Leaving residues on the field and a root system in the soil will protect against soil erosion and preserve soil structure. A market and cash flow must be guaranteed to reduce the risk for the farmer and provide the profitability needed to think long-term.

Optimizing cellulose processing by refining biomass pretreatment and converting crop residues, first-generation energy crops, and other sources to liquid fuels will be the immediate focus. This will entail reducing cost, enhancing feedstock deconstruction, improving enzyme action and stability, and developing fermentation technologies to more efficiently use sugars resulting from cellulose breakdown. One goal is to decrease industrial risk from a first-of-a-kind technology, allowing more rapid deployment of improved methods. To achieve higher production goals, new energy crops with greater yield per acre and improved processibility are needed.

27.5.3 BIOMASS PRODUCTION POTENTIAL OF PAULOWNIA

The term "biomass" in the present context as a bioenergy feedstock is intended to refer to materials that do not directly go into foods or consumer products but may have alternative industrial uses. Common sources of biomass are (1) agricultural wastes, such as corn stalks, straw, seed hulls, sugarcane leavings, bagasse, nutshells, and manure from cattle, poultry, and hogs; (2) wood materials, such as wood or bark, sawdust, timber slash, and mill scrap; (3) municipal waste, such as waste paper and yard clippings; and (4) energy crops, such as poplars, willows, switchgrass, alfalfa, prairie bluestem, corn (starch), and soybean (oil).

27.5.4 LIGNOCELLULOSIC BIOFUEL AND PAULOWNIA

Paulownia is one of the fastest-growing tree species capable of generating a large amount of biomass in a short period of time (~70–80 lb/tree per year in the first year and ~200 lb/tree per year from the second year onward; unpublished data from WPI), and it can be grown and harvested seasonally or annually in many states of the United States. Research conducted at WPI suggests that up to 68 wet tons of fiber per acre per year can be produced by establishing a Paulownia farm. All of our current micropropagation field trials for biomass yields at different spacing at the Fort Valley State University are being conducted on *P. elongata*.

Environmental concerns regarding the use of fossil fuels and production of carbon dioxide, coupled with increased energy costs, fostered the expansion of the fuel ethanol industry. The industry has become an important partner with U.S. agriculture. The U.S. Department of Agriculture estimates that 17,000 jobs are created for every billion gallons of ethanol produced. The most recent thrust in the United States has been the construction of new corn-based ethanol production facilities. Over the last decade, technology has improved and refined the process of conversion of grains and grasses to alcohol. However, the high cost of feedstock and the cost of production have placed barriers for being widely accepted.

Assuming corn produces 80 gal of ethanol per ton, a 50-million gallon corn-fed facility would require 625,000 tons of feedstock annually, harvested at the rate of 4.29 tons/acre national average yield or the equivalent of 145,690 growing acres (7344 acres corn/day). Paulownia grown specifically for cellulosic ethanol production would be planted on a grid of 8- by 8-ft spacing or 680 plants/acre, and unlike the practices used for forestry (growing for lumber) the plants would be manipulated for multiple sprouts (coppicing). Approximately 8–10 sprouts emerge from each plant. The Paulownia can be harvested each year or left to grow for 2- to 3-year harvest cycles. Once harvested, the Paulownia will regenerate from the stump and the cycle is repeated. Another distinct advantage of tree-type cellulosic feedstock is that one can harvest year-round and supply fiber to the production facility on a daily basis. Corn and row crops must be harvested when the crop is ready because they are seasonal. The latter poses numerous problems—not only the large-scale harvesting, but also the transportation, drying, and storage of hundreds of thousands of tons of fiber, all delivered in one month to be used over the year.

We have evaluated Paulownia wood and found the composition to be 14.0% extractive, 50.55% cellulose, 21.36% lignin, 0.49% ashes, and 13.6% hemicellulose (N. Joshee et al. unpublished research). This material is an attractive candidate for thermochemical conversion to fuels and energy through pyrolysis or gasification. The lignin content may provide the opportunity to extract this lignin for higher-value uses before thermochemical conversion. The resulting syngas will also be lower in tars.

Plant design, bioprocess engineering, and biomass processing strategies are intimately linked. Plants have evolved complex mechanisms for resisting assault on their structural sugars (wall polymers) from the microbial and animal kingdom. Cell-wall polymer organization and interactions are formidable barriers to access by depolymerizing enzymes and must be deconstructed in the pretreatment step to obtain adequate rates of release and sugar yields.

27.5.5 ECONOMIC JUSTIFICATION

A biofuel industry would create jobs and ensure growing energy supplies to support national and global prosperity. In 2004, the ethanol industry created 147,000 jobs in all sectors of the economy and provided more than $2 billion of additional tax revenue (RFA 2005). Growing energy crops and harvesting agricultural residuals are projected to increase the value of farm crops, potentially eliminating the need for some agricultural subsidies. Finally, cellulosic ethanol provides positive environmental benefits in the form of reductions in greenhouse gas emissions and air pollution. Another advantage of tree-type cellulosic feedstock is that one can harvest year-round and supply fiber to the production facility on a daily basis. According to the estimates of our collaborator, WPI (Lenox, GA), Paulownia producing 80 gal of ethanol per dry ton for a 50-million-gallon-per-year facility would require 1712 tens of fiber per day or the equivalent of 50 acres of fiber per day to be harvested. It would require 18,250 acres of Paulownia to sustain the supply to the facility year-round. Excluding the cost of land, Paulownia grown on land contiguous or close to the production facility could be supplied for less than $40.00 per dry ton, one-third of the cost predicted for corn.

The argument in favor of cellulosic ethanol as a replacement for gasoline is compelling. Cellulosic ethanol will reduce dependence on imported oil, increase energy security, and reduce the trade deficit for many nations. The benefit for rural economies will be in the form of increased incomes and jobs. The use of food grains for ethanol production has caused a steep rise in the price of corn- and wheat-based products and the related dairy, meat, and cattle feed industries.

27.6 PAULOWNIA BIOTECHNOLOGY

The wide use of forest-tree products and the progressive deterioration of natural forests mean that foresters can no longer rely on the exploitation of existing forests. Extensive accelerated breeding programs are needed for reforestation and to improve existing forest tree species. Plant genetic transformation techniques and gene isolation and characterization are no longer serious problems; forest-tree species should be a major target for commercial genetic engineering and molecular breeding. The introduction of cloned genes into plant cells and the recovery of stable fertile transgenic plants can be used to make modifications in a plant and has created the potential for genetic engineering of plants for crop improvement.

27.6.1 Tissue Culture

The use of an in vitro propagation technique provides a supply of healthy, homogenous planting material. Micropropagation of tree species offers a rapid means of producing clonal planting stock for afforestation, woody biomass production, and an effective way to capture genetic gains. Generally Paulownia is propagated through seed or by root cuttings. A conventional method of propagation through seed is unreliable because of disease and pest problems, poor germination, altered growth habit, and slower growth than root cuttings (Bergmann 1998; Bergmann and Moon 1997). Most of the tissue culture work conducted on Paulownia has used MS (Murashige and Skoog 1962) and woody plant medium (Lloyd and McCown 1981) with considerable success. For optimal results, it is customary to optimize the composition of growth medium for the variety and the stage of growth.

27.6.1.1 Organogenesis

High-frequency plant regeneration and rapid multiplication are important aspects of plant tissue culture. Plantlet formation from cultured cells and tissues can occur via one of two routes: organogenesis or somatic embryogenesis. Micropropagation research on various species of Paulownia is available. Reproducible shoot and then root induction protocols have been perfected for *P. tomentosa* (Burger et al. 1985; Rao et al. 1996; Rout et al. 2001; Corredoira et al. 2008), *P. elongata* (Bergmann and Whetten 1998), and *P. fortunei* (Rao et al. 1993; Venkateswarlu et al. 2001; Khan et al. 2003). A preliminary study carried out on *P. kawakamii* reported the presence of a differentially expressed cDNA encoding a putative bZIP transcription factor (Low et al. 2001). A 6-fold increased expression of this gene in the shoot apex region suggests involvement of this gene during the adventitious shoot regeneration process in Paulownia. A very interesting research study to investigate the effect of magnetic field was carried out on *P. tomentosa* nodal cultures (Celik et al. 2008). The study revealed that the magnetic field strength and exposure duration are important factors for rapid multiplication, which is probably supported by a higher concentration of chlorophyll a and b and total chlorophyll in treated explants. Most of the micropropagation protocols that have been developed for various species of *Paulownia* to date have predominantly used nodal explants. Figure 27.2a presents a schematic depiction of a nodal explant-based multiple shoot regeneration protocol developed for *P. elongata*. Paulownia plants can be produced commercially and shipped to domestic and international destinations (Figure 27.2b). In general, rooting in Paulownia is fairly easy (Figure 27.2c). Another explant that has been used successfully to a lesser extent is petiole with cut leaf (Figure 27.2d). In our laboratory, we have achieved considerable success in multiple shoot regenerations using shoot tip explants (Figure 27.2e). To assist commercial production by rapid multiplication, we have tried a liquid-culture-based production system using a liquid laboratory rocker system (Caisson Labs, North Logan, UT). Insertion of a filter paper as a substratum in the culture box helps multiplication a great deal. In our experience, plants can be hardened in 7–10 days in a greenhouse under misting conditions to produce a uniform planting stock (Figure 27.2f). There have been reports of vitrification during in vitro propagation of many *Paulownia* species, but reduction of vitrified shoots was possible by adjusting culture medium conditions, especially gelling substance and concentration of sugar (Ho and Jacobs 1995).

27.6.1.2 Somatic Embryogenesis

One of the earliest reports of somatic embryogenesis and plantlets from callus cultures was in *P. tomentosa* (Radojevic 1979). The culture medium used was MS containing 0.7% agar, 200 mg/L

FIGURE 27.2 (See color insert)   Steps in the micropropagation of *P. elongata.* (a) Use of nodal explant for multiple shoot induction. (b) Commercial in vitro production at WPI. Each container has 30 rooted plantlets. (c) Elongated shoots can be easily rooted in the presence of 1 mg/L indole butyric acid supplemented in the medium. (d) Leaf with petiole can also be used for developing multiple shoots through an intervening callus phase. (e) Shoot tips also produced multiple shoots. (f) The "Liquid Lab Rocker" system has been found to be useful in our laboratory for rapid multiplication and reduction in the cost of production. (g) Acclimatized uniform planting stock in the greenhouse.

casein hydrolysate, 100 mg/L myo-inositol, 2 mg/L thiamine, 5 mg/L nicotinic acid, 2 mg/L adenine, and 10 mg/L pantothenic acid. Repeating this media formulation on another four species of Paulownia did not yield any positive results (Yang et al. 1996), suggesting that strong genotypic factors are in play. Callus induced on fertilized ovular explants showed a persistent embryogenic capacity, eventually differentiating into plantlets. Direct and indirect somatic embryogenesis have been reported in *P. elongata* (Ipecki and Gozukirmizi 2003, 2004). Direct somatic embryogenesis was induced on leaf and internodal explants of *P. elongata* and synthetic seeds were also produced (Ipecki and Gozukirmizi 2003). For large-scale production of elite plants, this technology has great relevance because the process can be easily scaled-up using a bioreactor. Advances in somatic embryogenesis have brought mass clonal propagation of the top commercial trees closer to reality, and efficient gene transfer systems have been developed for a number of conifers and hardwoods.

27.6.2 GENETIC TRANSFORMATION

Advances in technology for in vitro propagation and genetic transformation have accelerated the development of genetically engineered trees during the past 15 years. Targeted traits include herbicide tolerance, pest resistance, abiotic stress tolerance, modified fiber quality and quantity, and altered growth and reproductive development. Commercial potential has been demonstrated in the field for a few traits, in particular herbicide tolerance, insect resistance, and altered lignin content. Now that commercial implementation is feasible, at least for the few genotypes that can be efficiently transformed and propagated, environmental concerns have become the main obstacle to public acceptance and regulatory approval. Ecological risks associated with commercial release range from transgene escape and introgression into wild gene pools, to the effect of transgene products on other organisms and ecosystem processes. Evaluation of those risks is confounded by the long life span of trees and by the limitations of extrapolating results from small-scale studies to larger-scale plantations.

Preliminary experiments using in vitro shoots to establish a transformation protocol were carried out using *Agrobacterium tumifaciens* (strains 542, A281, or C58) and *A. rhizogenes* (strain R1601). Opine analyses demonstrated the expression of the introduced gene in proliferating galls or hairy roots (Bergmann et al. 1999). In a separate study, *A. tumifaciens* (LBA4404) harboring binary vector pBI121 (Clontech Laboratories, Inc., Mountain View, CA) was used to transform *P. fortunei* using in vitro grown petiole segments (Mohri et al. 2003). Successful transformation was confirmed by histochemical analysis of β-glucuronidase (GUS) activity in kannamycin-resistant calli; however, the frequency of shoot regeneration was very low. Transformation studies coupled with molecular techniques helped in establishing the role of transcription factor *PkMADS1* isolated from *P. kawakamii* (Prakash and Kumar 2002). In this study, it was seen that the antisense suppression of *PkMADS1* resulted in gross morphological changes such as a change in phyllotaxy. Leaf explants obtained from the antisense *PkMADS1* transgenic plants showed an almost 10-fold decrease in adventitious shoot formation compared with the explants from the sense transgenic lines or the wild-type plants. In a recent study (Castellanos-Hernandez et al 2009), a biolistic protocol for stable genetic transformation was developed using leaf explants. Regenerated plants exhibited the integration of the transgenes as stable expression was demonstrated by GUS assay, determination of NPTII activity, and polymerase chain reaction analysis.

PWB disease, caused by PWB phytoplasma, is one of the most devastating diseases of this genus. PWB seriously slows down tree growth and is even capable of causing seedling death. Introduction of the *shiva-1* gene that encodes an antibacterial peptide using *A. tumifaciens*-mediated gene transfer resulted in plants with fewer phytoplasma and less symptoms in plants (Du et al. 2005). Further analysis of transgenic plants suggested that breeding *shiva-1* Paulownia is an effective strategy to control PWB disease. Developing reproducible transformation systems will be of great help in developing fast-growing lignocellulosic feedstock in the future.

Radical alterations in the quantity and quality of lignin in wood have been shown to be possible in softwoods and hardwoods through identification of naturally occurring mutants and by engineering the lignin biosynthetic pathway with transgenes. The potential environmental and social impacts of the release of transgenic trees have become an increasingly contentious issue that will require more attention if we are to use these technologies to their full advantage.

Current regulations restricting field releases of all transgenes in time and space need to be replaced with regulations that recognize different levels of risk (as determined by the origin of the transgene, its impact on reproductive fitness, and nontarget impacts) and assign a commensurate level of confinement. The next step in determining the acceptability of transgene technology for forest tree improvement is the unconfined release of constructs that pose little risk in terms of gene escape and nontarget impacts (e.g., lignin-altered poplar or pine) to permit evaluation of ecological risks and environmental or agronomic benefits at relevant scales (van Frankenhuyzen and Beardmore 2004).

27.7 SUMMARY

It is clear from the literature that *Paulownia* species have been used as a multipurpose tree in many countries in the east. There has been a growing interest in Paulownia in the western world since the 1970s. Countries that lack large forested areas and must therefore import timber can use Paulownia to help establish a local supply program around it (El-Showk and El-Showk 2003). It is a highly sought after companion tree in many agroforestry models in China. Further, it is well known for its fast growth and high biomass accumulation. Because of this property, it will fit very well in the present scene of lignocellulosic feedstock production.

Genetic modifications by plant transformation allow for stable alterations in biochemical processes that direct traits such as increased yield, disease and pest resistance, increased vegetative biomass, herbicide tolerance, nutritional quality, drought and stress tolerance, and genes to improve the production of ethanol from lignocellulosic biomass. In these methods, foreign constructs are introduced into the plant cell, followed by isolation of cells containing the foreign DNA integrated into the cell's DNA, to produce stably transformed plant cells. To date Paulownia has been selected primarily to enhance its growth rate and wood quality value. Thus, it has been managed primarily as a forest crop for which these traits are important. These targets are quite different from the criteria for biofuel crops for which high biomass yield, high cellulose, and traits involved with bioprocessing to ethanol are essential. Paulownia provides a great opportunity to develop transgenic Paulownia trees with desirable characteristics including higher cellulose and lower lignin content, disease resistance, and tolerance to a wide variety of biotic and abiotic stress conditions. Multipurpose attributes of the Paulownia tree justify thorough research for the deeper understanding of the plant biology related to medicinal properties, biomass production, and positive environmental and economic impact. In addition to this, a resource that has not been tapped to its full potential is marginal farmland, specifically its use for growing tree crops. More than 30 million acres of woodland and idle pasture and cropland exist in the Southeast, and much of this land could be producing valuable tree crops (Clatterbuck and Hodges 2004), with Paulownia being one of them.

ACKNOWLEDGMENTS

Paulownia research at Fort Valley State University (FVSU) is being carried out through an Evans Allen grant (GEOX 5213) funded to the author. Author expresses gratitude to Dr. K. C. Das at the Bioengineering Department (University of Georgia, Athens) for analyzing Paulownia wood, Vicki Owen for maintaining "Paulownia Demonstration Plot" at FVSU and Scot Corbett (World Paulownia Institute, Lenox, GA) for sharing information and pictures, and assisting in setting up "Paulownia Demonstration Plot" Paulownia demonstration farm at FVSU.

REFERENCES

Ates S, Ni Y, Akgul M, Tozluoglu A (2008) Characterization and evaluation of *Paulownia elongata* as a raw material for paper production. *Afr J Biotechnol* 7:4153–4158
Bergmann BA (1998) Propagation method influences first year field survival and growth of Paulownia. *New Forests* 16(3):251–264
Bergmann BA, Moon H-K (1997) Adventitious shoot production in *Paulownia*. *Plant Cell Rep* 16:315–319

Bergmann BA, Rubin AR, Campbell R (1997) Potential of *Paulownia elongata* trees for swine waste remediation. *Trans Amer Soc Agric Biol Eng* 40:1733–1738

Bergmann BA, Lin X, Whetten R (1999) Susceptibility of *Paulownia elongata* to *Agrobacterium* and production of transgenic calli and hairy roots by in vitro inoculation. *Plant Cell Tiss Org Cult* 55:45–51

Bergmann BA, Whetten R (1998) In vitro rooting and early greenhouse growth of micropropagated *Paulownia elongata* shoots. *New For* 15:127–138

Bonner FT, Burton JD (1974) *Paulownia tomentosa* (Thunb.) Sieb. & Zuec. Royal paulownia. In: Schopmeyer CS (tech coord), *Seeds of Woody Plants in the United States.* Agri Handbook 450, U.S. Department of Agriculture, Washington, DC, pp 572–573

Burger DW, Liu L, Wu L (1985) Rapid micropropagation of *Paulownia tomentosa. HortScience* 20:760–761

Carpenter SB. 1977. This "princess" heals disturbed land. *Amer For* 83:22–23

Castellanos-Hernandez OA, Rodriguez-Sahagun A, Acevedo-Hernandez GJ, Rodriguez-Garay B, Cabrera-Ponce JL, Herrera-Estrella LR (2009) Transgenic *Paulownia elongata* S. Y. Hu plants using biolistic-mediated transformation. *Plant Cell Tiss Org Cult* 99:175–181

Celik O, Atak C, Rzakulieva A (2008) Stimulation of rapid regeneration by a magnetic field in *Paulownia* node cultures. *Cent Eur J Agri* 9:297–304

Clatterbuck WK, Hodges DG. 2004. Tree crops for marginal farmland. Paulownia. University of Tennessee Extension PB 1465. pp 1–32, available at http://www.utextension.utk.edu/publications/pbfiles/PB1465.pdf (accessed July 18, 2010)

Corredoira E, Ballester A, Vieitez AM (2008) Thidiazuron-induced high-frequency plant regeneration from leaf explants of *Paulownia tomentosa* mature trees. *Plant Cell Tiss Org Cult* 95:197–208

Du T, Wang Y, Hu Q-X, Chen J, Liu S, Huang W-J, Lin M-L (2005) Transgenic *Paulownia* expressing shiva-1 gene has increased resistance to Paulownia witch's broom disease. *J Integr Plant Biol* 47:1500–1506

El-Showk S, El-Showk N (2003) The *Paulownia* tree. An alternative for sustainable forestry. In: *The Farm*, p 12, available at http://www.cropdevelopment.org (accessed on July 18, 2010)

Hepting GA (1971) *Diseases of Forest and Shade Trees of the United States.* Agri Handbook 386, U.S. Department of Agriculture, Washington, DC

Ho CK, Jacobs G (1995) Occurrence and recovery of vitrification in tissue cultures of *Paulownia* species. *Bull Taiwan For Res Inst New Sr* 10:391–403

Hu W-J, Harding SA, Lung J, Popko JL, Ralph J, Stokke DD, Tsai C-J, Chiang VL (1999) Repression of lignin biosynthesis promotes cellulose accumulation and growth in transgenic trees. *Nat Biotechnol* 17:808–812

Hui-Jun J, Ingestad T (1984) Nutrient requirements and stress response of *Populus simonii* and *Paulownia tomentosa. Physiol Planta* 62:117–124

Ipecki Z, Gozukirmizi N (2003) Direct somatic embryogenesis and synthetic seed production from *Paulownia elongata. Plant Cell Rep* 22:16–24

Ipecki Z, Gozukirmizi N (2004) Indirect somatic embryogenesis and plant regeneration from leaf and internode explants of *Paulownia elongata. Plant Cell Tiss Org Cult* 79:341–345

Jiang T-F, Du X, Shi Y-P (2004) Determination of flavonoids from *Paulownia tomentosa* (Thunb) Steud by micellar electrokinetic capillary electrophoresis. *Chromatographia* 59:255–258

Jiang Z, Gao L, Fang Y, Sun X (1994) Analysis of *Paulownia*—Intercropping types and their benefits in Woyang County of Anhui province. *For Ecol Manag* 67:329–337

Joshi CP (2003) Xylem-specific and tension stress-responsive expression of cellulose synthase genes from aspen trees. *Appl Biochem Biotechnol* 105:17–25

Kang KH, Huh H, Kim B-K, Lee C-K (1999) An antiviral furanoquinone from *Paulownia tomentosa* steud. *Phytotherap Res* 13:624–626

Kang KH, Jang SK, Kim BK, Park MK (1994) Antibacterial phenylpropanoid glycosides from *Paulownia tomentosa* Steud. *Arch Pharm Res* 17:470–475

Khan PSS, Kozai T, Nguyen QT, Kubota C, Dhawan V (2003) Growth and water relations of *Paulownia fortunei* under photomixotrophic and photoautotrophic conditions. *Biol Planta* 46:161–166

Kobayashi S, Asai T, Fujimoto Y, Kohshima S (2008) Anti-herbivore structure of *Paulownia tomentosa*: Morphology, distribution, chemical constituents and changes during shoot and leaf development. *Ann Bot* 101:1035–1047

Kumar PP, Rao CD, Rajaseger G, Rao AN (1999) Seed surface architecture and random amplified polymorphic DNA profiles of *Paulownia fortunei, P. tomentosa* and their hybrid. *Ann Bot* 83:103–107

Li P, Oda J (2007) Flame retardancy of Paulownia wood and its mechanism. *J Mater Sci* 42:8544–8550

Lloyd G, McCown B (1981) Commercially feasible micropropagation of mountain laurel, *Kalmia latifolia*, by use of shoot tip culture. *Proc Int Plant Propag Soc* 30:421–427

Low RK, Prakash AP, Swarup S, Goh C-J, Kumar PP (2001) A differentially expressed bZIP gene is associated with adventitious shoot regeneration in leaf cultures of *Paulownia kawakamii*. *Plant Cell Rep* 20:696–700

Lu S, Li L, Yi X, Joshi CP, Chiang VL (2008) Differential expression of three eucalyptus secondary cell wall-related cellulose synthase genes in response to tension stress. *J Exp Bot* 59:681–695

Miller JH (2004) *Nonnative Invasive Plants of Southern Forests: A Field Guide for Identification and Control*. Gen Tech Rep SRS-62, U.S. Department of Agriculture Forest Service Southern Research Station, Ashville, NC

Miyamoto K (1997) Renewable biological systems for alternative sustainable energy production. *FAO Agri Serv Bull* 128, ISBN 92-5-104059-1

Mohri T, Igasaki T, Shinohara K (2003) *Agrobacterium*–mediated transformation of Paulownia (*Paulownia fortunei*). *Plant Biotechnology* 20:87–91

Murashige T, Skoog F (1962) A revised medium or rapid growth and bio-assays with tobacco tissue cultures. *Physiol Plant* 15:473–497

Prakash A, Kumar PP (2002) PkMADS1 is a novel MADS box gene regulating adventitious shoot induction and vegetative shoot development in *Paulownia kawakamii*. *Plant J* 29:141–151

Radojevic L (1979) Somatic embryos and plantlets from callus cultures of *Paulownia tomentosa* Steud. *Z Pflanzenphysiol* 91:57–62

Rao CD, Goh C-J, Kumar PP (1993) High frequency plant regeneration from excised leaves of *Paulownia fortunei*. *In Vitro Cell Dev Biol-Plant* 29:72–76

Rao CD, Goh C-J, Kumar PP (1996) High frequency adventitious shoot regeneration from excised leaves of *Paulownia* spp. cultured in vitro. *Plant Cell Rep* 16:204–209

RFA (2005) *Homegrown for the Homeland: Ethanol Industry Outlook 2005*. Renewable Fuels Association, available at www.ethanolrfa.org/resource/outlook (accessed on July 18, 2010)

Song Y (1988) Nutritive components of *Paulownia* leaves as fodder. *Chem Indust For Prod* 8:44–49

Tang RC, Carpenter SB, Wittwer RF, Graves DH (1980) Paulownia—A crop tree for wood products and reclamation of surface-mined land. *S J Appl For* 4:19–24

US DOE (2006) *Breaking the Biological Barriers to Cellulosic Ethanol: A Joint Research Agenda*. DOE/SC-0095, U.S. Department of Energy, available at www.doegenomestolife.org/biofuels (accessed July 18, 2010)

van Frankenhuyzen K, Beardmore T (2004) Current status and environmental impact of transgenic forest trees. *Can J For Res* 34:1163–1180

Venkateswarlu B, Mukhopadhyay J, Sreenivasan E, Kumar VM (2001) Micropropagation of *Paulownia fortuneii* through in vitro axillary shoot proliferation. *Indian J Exp Biol* 39:594–599

Wang Q, Shogren JF (1992) Characteristics of the crop Paulownia system in China. *Agric Ecosyst Environ* 39:145–152

Wang WY, Pai RC, Lai CC, Lin TP (1994) Molecular evidence for the hybrid origin of *Paulownia taiwaniana* based on RAPD markers and RFLP of chloroplast DNA. *Theor Appl Genet* 89:271–275

WHO (2000) *Malnutrition Worldwide*, World Health Organization, available at http://whqlibdoc.who.int/hq/2000/WHO_NHD_00.6.pdf (accessed July 18, 2010)

Yang J-C, Ho C-K, Chen Z-Z, Cheng S-H (1996) *Paulownia x taiwaniana* (Taiwan Paulownia). In: Bajaj YPS (ed), *Biotechnology in Agriculture and Forestry. Vol. 35. Trees*. Springer, Berlin, pp 273–289

Zhong Q, Hiruki C (1994) Amplification of *Paulownia* phytoplasma DNA fragments by random primed-PCR. *Proc Jpn Acad* 70:185–189

Zhu ZH, Chao CJ, Lu XY, Xiong YG (1986) *Paulownia in China: Cultivation and Utilization. Singapore*. Asian Network for Biological Sciences and International Development Research Centre, Canada, pp 1–65

28 Shrub Willow

Lawrence B. Smart and Kimberly D. Cameron
Cornell University

CONTENTS

28.1 INTRODUCTION

On a global scale, issues surrounding diminishing fossil fuel supply, increasing demand for energy, and strategic policies for national security, all in the context of rising atmospheric levels of carbon dioxide (CO_2) and increasing evidence of climate change, will contribute to rapid growth in the use of renewable biomass as a source of energy, especially in the United States and Europe. Biomass produced and aggregated in the agricultural and forestry industries can be used in three primary energy sectors: to generate electricity, transportation fuels, or heat. Across multiple scales (state, regional, national, and international), agreed-upon caps or legislated restrictions on carbon emissions are driving up the cost of producing electricity from fossil fuels, especially coal, and are inspiring a shift to the use of biomass to fuel the boilers in power plants. The U.S. Department of Energy and the Biomass Research and Development Board have highlighted the unsustainable rise in demand for foreign petroleum in the United States and have provided leadership in proposing a National Biofuels Action Plan (http://www1.eere.energy.gov/biomass/pdfs/nbap.pdf) to meet a national goal of replacing 15% of gasoline usage with biofuels and to increase biofuel production in the United States to 35 billion gallons per year by 2017. To meet these goals in an environmentally sound and sustainable manner, there will need to be dramatic increases in the total production of biofuel feedstocks. There is great potential for the production, aggregation, and use of plant biomass other than corn grain as a feedstock for the production of biofuels in the United States. The growth of perennial woody crops as a feedstock crop for biofuels and bioproducts offers significant advantages with respect to net energy ratio, soil conservation, nutrient management, biodiversity, and utilization of marginal agricultural land, in addition to diversifying the feedstock commodities available to the biofuels industry (Verwijst 2001; Volk et al. 2004b, 2006).

New York State is the largest user of heating oil in the United States. A vast majority of petroleum used for heating in New York and throughout the Northeast United States is from foreign sources, making the region dependent on economically volatile imports. The combustion of heating oil also releases ancient carbon that contributes to global climate change. Renewable woody biomass holds great potential as a feedstock for heat production because it has numerous environmental benefits and can be produced locally. However, wood chips have not been widely adopted as a fuel for small-scale heating in this region for reasons of delivery and storage logistics, difficulty of use, and cost. Pelletizing woody biomass greatly simplifies the logistics of delivery and makes conversion more reliable because of uniform size, composition, and handling properties. With recent spikes in the cost of heating oil, many consumers in the Northeast have switched over to using wood pellet stoves or boilers, resulting in dramatically greater demand for wood pellets. Wood pellets for heating have, in the past, been made from clean residues collected from the forest products industry, but reduced supply and increased demand for those residues has forced pellet-makers to seek alternative feedstock sources and incorporate wood from forest harvesting operations. With the tightening market for forest product residues, dedicated perennial energy crops could provide a reliable, local source of biomass that can be incorporated into the feedstock mix used to make pellets.

Shrub willow (*Salix* spp.) represents a proven, high-yielding perennial crop that can be grown on underutilized or marginal agricultural land, especially poorly drained sites unsuitable for food crops, and could contribute significantly to the mix of regionally optimized biomass commodity crops. Research conducted or directed over many years by the U.S. Department of Energy at Oak Ridge National Laboratory has concluded that willow has superior properties as a perennial energy crop for the Northeast and Midwest United States (Tolbert and Schiller 1996; Tolbert and Wright 1998). Shrub willow crops have a short harvest cycle, low incidence of pests or diseases in improved varieties, adaptability to a wide range of site conditions, high yield of biomass with low input of fertilizer, efficient recycling of nutrients in leaf litter, and great potential for genetic improvement. For over 20 years, academic research and development in New York State and Canada has demonstrated the potential for shrub willow crops in North America, drawing from a wealth of study and experience in Sweden, the United Kingdom, and Denmark, where willow bioenergy crops are planted on over 20,000 ha (Kuzovkina et al. 2008). Sustained efforts in optimizing agronomic and management techniques, adapting planting and harvesting technology, and demonstrating conversion methods have lead to commercial deployment of shrub willow crops on the rural agricultural landscape in New York since 2005, with over 300 ha in cultivation and dramatic expansion of commercial plantings being proposed to take advantage of nearly 800,000 ha of underutilized agricultural land in New York alone (Volk et al. 2006). According to the National Biofuels Action Plan, willow is characterized as a "third-generation" feedstock crop designed for fuel production. Ongoing breeding and selection of willow as an energy crop has generated a diverse array of varieties that display incremental improvements in yield and are adapted to regional conditions of climate, soils, pests, and diseases. Genomics-assisted breeding of willow will further enhance its sustainable yield potential on vast areas of underutilized agricultural land in the northern United States, northern Europe, and elsewhere.

28.2 DIVERSITY AND ECOLOGY OF *SALIX*

Willows are classified in the genus *Salix* which includes at least 350 species worldwide. Willows are further grouped into three to five subgenera on the basis of morphological characteristics, as described by willow taxonomists for North America, Eurasia, and China (Argus 1997; Skvortsov 1999; Zhenfu et al. 1999; Dickmann and Kuzovkina 2008). The proper identification and classification of willows is notoriously difficult because many of the foliar characteristics are highly variable within a species depending on developmental stage, genotypic variation, and responses to environmental conditions. Considering the potential for variability, these characteristics, such as leaf and stipule size, shape, and pubescence, may be similar across one or more species. To settle on a definitive identification, it is necessary to conduct careful observation of distinguishing floral, stem, bud, and foliar features at many stages of growth. According to the scheme of Argus, species are placed in the subgenera *Protitea* (Ohashi), *Salix*, *Longifoliae* (Andersson), *Chamaetia* (Dumort.), and *Vetrix* (Argus 1997; Dickmann and Kuzovkina 2008). Generally, the tree-form species, including *Salix nigra* and *Salix amygdaloides* are in Protitea Ohashi or *Salix*, whereas *Longifoliae* includes only a few root-suckering shrub species, including *S. interior* Rowlee. The principal species used in the development of bioenergy crops are in the subgenus *Vetrix*, in which species are further classified into 12 sections. Prominent species cultivated for bioenergy include, but are not limited to *S. viminalis* Linnaeus, *S. schwerinii* Wolf, *S. purpurea* Linnaeus, *S. miyabeana* Seemen, *S. dasyclados* Wimmer, *S. sachalinensis* Schmidt, and *S. eriocephala* Michaux (Table 28.1). The remainder of this chapter will focus on those shrub-form species that produce biomass yields sufficient to be considered for the production of bioenergy (Table 28.2).

Willows are native to all of the continents in the Northern Hemisphere, but there is only one species native to the southern hemisphere—the tree *S. humboldtiana* Willdenow—found in Chilé and Argentina (Dickmann and Kuzovkina 2008). Today, willows can be found growing in wide distribution between the Tropic of Cancer and the Arctic Circle—expanded descriptions of the primary habitat, natural ranges, and current locales of commercial acreage will follow. Many species of willow have been favored as ornamental species or for horticultural uses and thus have been widely transported and introduced to many countries through colonial immigration and horticultural propagation. Some of these species have become naturalized, such as *S. purpurea*, *S. fragilis*, *S. alba*, and *S. babylonica* in North America. Through introduction, willows grow well in many countries in temperate regions of the Southern Hemisphere, including Argentina, Chilé,

TABLE 28.1

Taxonomic Organization of *Salix* Species Used in the Commercial Production of Bioenergy, Environmental Engineering, and Other Applications

Subgenus	Section	Species
Vetrix Dumortier	*Cinerella* Dumortier	*S. discolor* Muhlenberg
		S. caprea Linnaeus
	Vimen Dumortier	*S. viminalis* Linnaeus
		S. schwerinii Wolf
		S. sachalinensis Schmidt
		S. dasyclados Wimmer (syn. *S. burjatica* Nasarov)
	Cordatae Barratt	*S. eriocephala* Michaux
	Geyerianae Argus	*S. petiolaris* Smith
	Fulvae Barratt	*S. bebbiana* Sargent
	Helix Dumont	*S. purpurea* Linnaeus
		S. miyabeana Seemen
		S. integra Thunberg
		S. suchowensis Cheng
		S. koriyanagi Kimura
	Daphnella Seringe	*S. daphnoides* Villars
	Hastatae Kerner	*S. cordata* Michaux
Salix	*Humboldtianae* Andersson	*S. amygdaloides* Andersson
		S. nigra Marshall
	Amygdalinae Koch	*S. triandra* Linnaeus
	Subalbae Koidzumi	*S. babylonica* Linnaeus (syn. *S. matsudana* Koidzumi)
	Salix	*S. alba* Linnaeus
	Salicaster Dumortier	*S. lucida* Muhlenberg
Longifoliae Argus	*Longifoliae* Andersson	*S. interior* Rowlee
		S. exigua Nuttall

TABLE 28.2
Willow Varieties Selected for Commercial Bioenergy Production

Variety Epithet	Species or Species Pedigree
Svalöf Weibull/Agrobränsle/Lantmännen Agroenergi (Swedish varieties)	
Jorr, Jorunn, Orm, Astrid	*S. viminalis*
Gudrun, Loden	*S. dasyclados*
Doris, Nora	*S. burjatica* × *S. viminalis*
Inger	*S. triandra* × *S. viminalis*
Tora, Björn	*S. schwerinii* × *S. viminalis*
Olof, Sven, Asgerd	*S. viminalis* × (*S. schwerinii* × *S. viminalis*)
Torhild, Tordis	(*S. schwerinii* × *S. viminalis*) × *S. viminalis*
Karin	[(*S. schwerinii* × *S. viminalis*) × *S. viminalis*] × *S. burjatica*
Klara	[(*S. burjatica* × *S. viminalis*) × *S. burjatica*) × (*S. viminalis* × (*S. schwerinii* × *S. viminalis*)]
Lisa	[(*S. schwerinii* × *S. viminalis*) × *S. viminalis*) × (*S. viminalis* × (*S. schwerinii* × *S. viminalis*)]
Stina	[(*S. schwerinii* × *S. viminalis*) × *S. viminalis*] × [(*S. viminalis* × *S. lanceolata*) × *S. aeygyptiaca*]
Dimitrios	(*S. schwerinii* × *S. viminalis*) × *S. aeygyptiaca*
U.K./European Willow Breeding Program Varieties	
Beagle	*S. viminalis*
Quest	*S. viminalis* × (*S. schwerinii* × *S. viminalis*)
Ashton Stott	*S. viminalis* × *S. burjatica*
Endeavour	*S. schwerinii* × *S. viminalis*
Discovery	*S. schwerinii* × (*S. schwerinii* × *S. viminalis*)
Resolution	[*S. viminalis* × (*S. schwerinii* × *S. viminalis*)] × [*S. viminalis* × (*S. schwerinii* × *S. viminalis*)]
Nimrod	(*S. schwerinii* × *S. viminalis*) × *S. linderstipularis*
Terra Nova	*S. mollisima* × *S. sachalinensis*
Endurance	*S. rehderiana* × *S. dasyclados*
North American Varieties (New York/Ontario Canada)	
Fish Creek, Onondaga, Allegany	*S. purpurea*
Millbrook, Oneida	*S. purpurea* × *S. miyabeana*
Sherburne, Canastota	*S. sachalinensis* × *S. miyabeana*
Otisco, Tully Champion, Owasco	*S. viminalis* × *S. miyabeana*
SV1	*S.* × *dasyclados*
SX61	*S. sachalinensis*
SX64, SX67	*S. miyabeana*
S25	*S. eriocephala*

South Africa, New Zealand, and Australia. A few species have become invasive and detrimental to native ecosystems, such as *S. nigra*, *S. fragilis*, *S. babylonica*, and *S. matsudana* × *S. alba* hybrids in Australia (Stokes and Cunningham 2006; Stokes 2008). Willows are also found in arctic and subarctic regions of Scandinavia, Asia, and North America, indicating that there is potential for extreme cold tolerance in the genus (Jones et al. 1997). In northern Alaska, the land area inhabited by willow shrubs appears to have expanded in recent decades, perhaps as an indicator of climate change (Tape et al. 2006).

Willows are a pioneer species that can tolerate many types of harsh growing conditions and they demand full sunlight. In North America, willows were among the initial colonizers of the wet and gravelly sites left by the recession of glaciers 12,000–13,000 years ago (Hardig et al. 2000). Now they generally occupy lowland and poorly drained sites, including the shorelines of rivers, lakes, and creeks, wetlands, and some upland sites where they may outcompete other shade-intolerant species. Members of the genus *Salix* are deciduous and perennial, most with a remarkable ability to continue growing until a severe frost in the fall because they do not form a terminal vegetative bud. Individual plants have a typical life span of 20–60 years, although some willow shrubs have been grown in nursery stool beds for over 100 years. This longevity often depends on the harvesting or removal of aged stems, which then stimulates the growth of new shoots from buds that emerge from the stool or root crown. Even in situations of extreme defoliation from pests or stem dieback from frost, buds will rapidly break their dormancy and replenish the plant with photosynthetically productive foliage. For most species of willow, when a piece of stem is submerged in water or stuck into moist soil, it can readily produce roots, allowing for the establishment of a new, genetically identical plant and further extending the effective life span of that genet. There is evidence that some populations of *S. purpurea* in New York State have become established strictly through clonal propagation (Lin et al. 2009). Some willow species (e.g., those in the subgenus *Longifoliae*) can produce new plants by root suckering, although this trait has been avoided in the selection of species or varieties to be cultivated for bioenergy production because these plants rapidly spread from the site of the original planting.

All *Salix* are dioecious, meaning that an individual plant produces only male or only female flowers (Figure 28.1). For most species, floral buds are set in late summer or early fall on the distal portions of the shoots and break dormancy early in the spring, maturing to form catkins with several dozen individual flowers each. Although plants established from cuttings may form floral buds by the end of the first growing season, flower production is much more prevalent after the second growing season. Plants established from seed usually do not produce flowers until the second growing season. Cross-pollination is accomplished by both insects—usually bees—and wind, depending on the weather conditions (Argus 1974; Tamura and Kudo 2000; Karrenberg et al. 2002a, 2002b). Once pollinated, the females produce many dozens of seeds for each catkin. The pistil swells to form a capsule, which bursts open when dry and mature to release the seeds to the wind for dispersal—each covered with cottony hairs. These cottony hairs aid the seeds in floating and eventually settling on a suitable substrate for establishment, such as wet sand (Gage and Cooper 2005; Seiwa et al. 2008). Willow seed has strict establishment criteria, and the seeds have a very short life span, resulting in a relatively low rate of successful establishment of the annual seed production for a pioneer species and no seed bank (Karrenberg et al. 2002a, 2002b; Karrenberg and

FIGURE 28.1 Examples of male and female willow flowers. Female catkins produced by *S. viminalis* (left panel) and male catkins of *S. lucida* (right panel).

Suter 2003; Seiwa et al. 2008). However, microbial associations can contribute to the establishment and survival of willows in challenging environments. Species of *Salix* are among the few genera that can form mycorrhizal associations with ectomycorrhizal and endomycorrhizal species (van der Heijden 2001). There is also recent and exciting evidence of willow stem endophytes that may contribute to phytoremediation and/or nitrogen assimilation through nitrogen fixation (Doty 2008; Doty et al. 2009).

Members of the genus *Salix* are important for riparian ecology, providing habitat and forage for numerous species. In North America, willows are a particularly important food source for native beaver, elk, and moose, among others (Peinetti 2002, Peinetti et al. 2001; Baker et al. 2005; Veraart et al. 2006). Shrub habitat tends to be underrepresented in the northern tier of the United States because much shrub land has been developed or converted to agricultural use. Expanded cultivation of shrub willows as a bioenergy crop would increase the proportion of shrub land on the landscape and provide additional habitat for migratory song birds (Dhondt et al. 2007). On occasion, willow plantings have been used to provide fodder for grazing cattle, sheep, and goats, providing some notable nutritional benefits, especially during dry periods when pasture is not available (Moore et al. 2003; McWilliam et al. 2005; Pitta et al. 2007; Lira et al. 2008).

28.3 HISTORICAL CULTIVATION OF WILLOW

Willows have a long history of utility to society in a wide range of applications. Willow bark was the original source of the nonsteroidal anti-inflammatory drug, aspirin (acetylsalicylic acid) (Levesque and Lafont 2000). Historically, willow was used for medicinal purposes by indigenous cultures and immigrants to North American before the development of modern medicines. Even today, among the diversity of the genus *Salix* there is great potential to identify other secondary compounds with pharmaceutical activity (Hostanska et al. 2007; Nahrstedt et al. 2007). In Europe especially, willows have been used extensively for making woven baskets, furniture, and decorative structures. This craft was brought to North America, together with planting stock of European basket willow varieties, by immigrants in the mid- to late-1800s. The industrial production of willow baskets was considerable in the late nineteenth and early twentieth century in upstate New York, centered in Liverpool, where hundreds of thousands of baskets were woven in the early twentieth century. The commercial production and trade in willow baskets declined with the development and distribution of wicker and other cheaper materials, but the craft persists to produce baskets largely for decorative use, particularly in the United Kingdom and in the city of Chimbarongo in Chilé. White willow (*S. alba*) provides the wood used to make cricket bats. Willow has great religious significance as one of the species used in the Jewish holiday of Sukkot and in the Polish and Polish-American celebration of Dyngus Day on the Monday after Easter.

Willows have enjoyed widespread cultivation in landscaping and ornamental settings. Weeping willow (*S. babylonica*) and white willow (*S. alba*) have been particularly popular as an attractive landscaping and shade tree that can tolerate a wide range of climates and especially wet soils (Newsholme 1992). In recent years, a variegated variety of the shrub willow species, *S. integra*, named "Hakuro Nishiki" has become popular in landscaping applications in the northeastern United States. Shrub willow species have been planted and woven to form hedges or living fences, sometimes called fedges. This practice has been further adapted to produce living sound barriers along highways, called "living walls" (Figure 28.2). Many of these have been planted using *S. viminalis* and *S. miyabeana* along highways in and around Montréal in the province of Québec, Canada. A few rows of shrub willow planted adjacent to roads and highways also serve well in trapping blowing and drifting snow, functioning as living snow fences.

More recently, willow has been applied in systems for environmental or ecological engineering. Willow plantations have been successfully implemented for the uptake and control of nutrient effluents, particularly nitrogen and phosphorus, from wastewater treatment systems, from non-point-source agricultural nutrient runoff, or from landfill leachate (Pulford et al. 2002; Adegbidi

FIGURE 28.2   A living sound barrier wall planted with *S. miyabeana* adjacent to a highway.

and Briggs 2003; Dimitriou and Aronsson 2005; Kuzovkina and Quigley 2005; Volk et al. 2006). Planting shrub willow fields adjacent to wastewater treatment facilities captures multiple environmental benefits because the removal of excess nitrogen and phosphorus from wastewater stimulates willow growth and increases yield of biomass that can be used for bioenergy (Aronsson and Perttu 2001). Shrub willow plantations are particularly well suited for the uptake and removal of cadmium from soil (Dimitriou et al. 2006). As a result, sewage sludge may be applied to willow plantings with less chance for long-term accumulation of cadmium in the soil (Labrecque and Teodorescu 2001; Dimitriou et al. 2006). Willow plantations also represent a viable system for the restoration of contaminated urban brownfield sites when they are used in a final polishing phase to remove low levels of organic or heavy metal contaminants and restore the quality of highly disturbed soils (French et al. 2006).

Initially studied as a source of pulpwood, the oil shortages in the late 1970s accelerated the development of perennial woody crops for the domestic production of feedstock for renewable energy. Fast-growing shrub willow was among those perennial crops studied for its fast growth rate and potential for high yields with relatively low inputs on marginal agricultural land (Anderson et al. 1983; McElroy and Dawson 1986; Gullberg 1993). Many of the original willow varieties that were tested in Europe were obtained from the National Willow Collection of the United Kingdom, which was maintained and managed by Kenneth Stott (Newsholme 1992; Stott 1992). Primary among those evaluated by pioneering researchers in northern Europe were *Salix* "Aquatica Gigantea" and *S. dasyclados* Wimm (McElroy and Dawson 1986; Pohjonen 1987; Stott 1991). Varieties selected for basket-making, including genotypes of *S. viminalis* and *S. triandra*, were also included in early trials. Cultivation of willow crops for energy first gained a foothold in commercial markets in Sweden, where it is grown on approximately 13,000 ha and is used as fuel for district heating plants and for combined heat and power (Verwijst 2001). Breeding programs aimed at improving the yield and sustainability of willow bioenergy crops were established in Sweden, England, and Canada (Eriksson et al. 1984; Zsuffa 1990; Lindegaard and Barker 1997; Larsson 1998), which all resulted in the release of improved germplasm that is grown commercially for bioenergy today.

28.4 CULTIVATION OF SHRUB WILLOW FOR BIOMASS PRODUCTION

Although shrub willow research has often been associated with forestry, the agronomic practices are much more akin to conventional agriculture of perennial plants. Proper and thorough site preparation is critical for successful establishment of shrub willow bioenergy crops. A likely scenario for the expansion of perennial bioenergy crop acreage is the conversion of underutilized and abandoned farmland to the production of biomass. Often, these particular fields are underutilized and abandoned for good reason—their characteristics make those sites unsuitable for annual crop cultivation, including but not limited to difficulty in plowing of shallow, rocky, or clay soils; poor drainage leading to periods of flooding; unbalanced pH of soil—usually acidic; and small or irregular parcel size or shape, which complicates harvesting logistics. As agricultural fields fall out of regular cycles of cultivation, they become prone to the establishment of annual and perennial weeds and undesirable perennial plants. Many of these fields may be old pasture, low-value hay fields, or abandoned agricultural fields that have reverted to fallow scrub land. On the basis of these typical characteristics, the conversion of these fields back to cultivation, including the killing and removal of existing vegetation, soil tillage, planting of bioenergy crops, and control of weed competition, can be extremely challenging and prone to failure. The greatest challenge to the establishment of perennial feedstock crops, especially on previously fallow land, is weed management. At its worst, weed competition can cause total crop failure, which is extremely expensive considering that establishment costs for willow are approximately \$1,900–\$2,400/ha (T. Buchholz and T. Volk, personal communication). Poor weed management leads to proportionate decline in establishment and yield per hectare for multiple rotations.

When planting willow bioenergy crops on land that has been in regular annual crop rotation, site preparation in the spring is effective because the field is likely to be largely free of perennial weeds, their seeds, or rhizomes. Proper site conversion from fallow land must be started in the late summer and fall in anticipation of spring planting. This involves application of appropriate broad-spectrum herbicides to kill the existing vegetation. For hay fields, it is best to cut, bale, and remove tall growth a few weeks before herbicide application to ensure maximum effectiveness. Ideally, fields can be cleared, plowed, and planted in mid-summer with a vigorous cover crop such as buckwheat or a legume that can add fixed nitrogen to the soil. Although the standard practice has been to perform moldboard plowing to break up any compaction or plow pan layer, it may be more efficient to use ripper shanks followed by heavy disk harrows or zone tillage equipment, especially on rocky fields. For fields that are prone to erosion, it is best to plant a winter cover crop, such as winter rye (*Secale cereal* L.) (Volk 2002). In the spring, any emerging perennial weeds should be killed with another application of broad spectrum herbicide (such as glyphosate) before disk harrowing and leveling of the field with a cultipacker-type roller.

Planting operations can begin as soon as possible after the threat of frost is over in the spring, which is usually from late April to early June, depending on the local climate. Planting can continue through early summer, as long as there is ample opportunity for rain, which is critical for willow establishment. Obviously, the earlier that plants are established, the greater the opportunity for growth during that first season. Stands of willow energy crops are established by the vegetative propagation of unrooted stem cuttings and, as a result, are populated with clonally-derived germplasm. For small plots, cuttings can be pushed into the soil by hand, but for commercial-scale acreage, mechanical planting systems are used that have been specially designed for this purpose (Figure 28.3). These planters use long pieces of 1-year-old stems, called "whips" or "switches," which are then cut progressively from the base of the whip into approximately 20-cm-long pieces that are pushed into the soil as the planter moves across the field. Because whips are typically 2–3 m in length, a single stem can be used to establish 10–15 new plants. Willow planters have been designed and built in Sweden and Denmark by Salixphere and Egedal, respectively, and attached to a medium-sized (~125–140 hp) tractor. The standard models plant four rows simultaneously,

FIGURE 28.3   Planting shrub willow using whips loaded into a mechanical four-row step planter.

although some six-row willow planters have been designed. Typically each person riding on the back of the planter can load whips into two planting cartridges, thus a crew of two or three plus a driver are needed for planting operations. For a single planter, the average rate of planting is 2–3 acres (1–1.5 ha) per hour or approximately 20 acres (~8 ha) per day with a total seasonal capability of planting approximately 800 acres (~320 ha).

Planting density and row spacing have been optimized to attain maximal yield in a 3- to 4-year harvest cycle while minimizing the cost of planting stock per acre (Kopp et al. 1997; Bergkvist and Ledin 1998; Wilkinson et al. 2007). The usual standard for final planting density is approximately 15,000 plants/ha (~6,100 plants/acre) in a double-row arrangement with 0.6 m (24 in.) between plants in a row and 0.76 m (30 in.) between rows, with 1.52-m (5-ft) alleys between the double rows (Bergkvist and Ledin 1998; Abrahamson et al. 2002). This arrangement allows harvesting equipment to straddle the double row and efficiently harvest two rows in a single pass. However, many growers have noted that the closely spaced double rows make weed control by herbicide application or mechanical cultivation more difficult than wider single-row spacing (F. Allard and D. Rak, personal communication). As larger and larger harvesting machines arrive in the market, it will be necessary to widen the spacing of the alleys so that the harvester tires will fit inside of the rows adjacent to the double row being harvested.

Willow planting stock (whips and cuttings) is grown in dedicated nursery beds that are harvested annually and may be irrigated and fertilized to optimize production. Between the time when the plants have gone dormant and have shed their leaves, but have not yet broken bud in the spring, stems are cut using single-row, whole-stem harvesters that saw the stems just above the ground and collect them in a bunk or wagon. Subsequently, they must be processed by trimming off any branches and removing any portions of stem that do not conform to standards for diameter and straightness. If whips are too large in diameter or if there is curvature, they can jam in the planter, depending on the model. If the diameter is too small, they may bend as they are being pushed into the soil. Depending on the species, a single plant can produce as many as 10–12 stems that are at least 2 m in length and of sufficient diameter such that 50–100 cuttings may be produced per plant. The plants in the nursery beds will resprout the following spring and will continue to produce stems for annual cutting production for decades. Because willow stems exhibit weak dormancy and are capable of initiating growth even at very cool temperatures, whips and cuttings must be stored frozen until just before they are planted. However, exposure of whips or cuttings to temperatures below −12 to −16°C (10 to 3°F) can cause tissue damage and reduce the viability of cuttings (Volk et al. 2004a). Whips and cuttings should also be packaged in bags or in lined boxes so they do not dry out while in the freezer. Within 4–5 days of removal from the freezer to room temperature or above, the stem sections will start to grow roots and the buds will break to form shoots. Refrigerated trucks are critical for the proper transport of whips to the field and short-term on-site storage just before planting.

Immediately after planting, pre-emergence herbicides are applied to prevent the germination of weed seeds. Pesticide regulations and product labeling vary tremendously depending on local authorities, so growers need to be well informed about product applicability. In the United States, some products that have been used in research and development trials include oxyfluorfen [1 lb active ingredient (ai) per acre Goal 2XL, Dow AgroSciences] in combination with simizine (2 lb ai/acre, Princep, Syngenta) or pendimethalin (2 lb ai/acre Prowl 3.3EC, BASF) (Wagner 2000). When pre-emergence herbicides are not used or available, mechanical cultivation should be applied in a timely and frequent manner throughout the season to prevent the establishment and growth of weeds. Mechanical cultivation can also be applied during mid-season in fields where the pre-emergence herbicide protection has begun to fail. Some postemergence herbicides have been used in research trials to control weeds by spraying over the top of willow vegetation, including fluazifop-P-butyl (FusiladeDX, Syngenta) for the control of grass weeds and clopyralid (Stinger, Dow Agrosciences), which can control horseweed, ragweed, thistle, burdock, curly dock, and vetch. Further research is needed to gain an understanding of varietal sensitivity to phytotoxic effects of each product, to optimize application rates for varying soil types, and to gain legal labeling registration for commercial use on willow bioenergy crops.

During the establishment year, willow plants will typically produce two to three stems that will reach heights of approximately 2 m while at the same time building a below-ground diffuse root structure. At the end of this first year when the plants are dormant, current management practices call for the cutback of the stems, also termed "coppicing." This is accomplished using a sharp sickle bar mower attachment on a tractor or in smaller plots by hand using a gas-powered rotary brush saw. Stems are cut within 4–8 cm of the ground, and that biomass is usually left on the field. Coppicing stimulates vigorous sprouting of an increased number of shoots in the spring of the second season. Some varieties can produce more than 40 shoots per plant in the year after coppice. Coppicing also allows the grower the opportunity to mechanically cultivate between rows and/or apply pre-emergence herbicide before willow bud break to accomplish improved weed management. Standard practice is also to apply nitrogen fertilizer to the crop in the spring after coppice at a rate of 80–120 kg N/ha (~100 lb N/acre). Solid fertilizer can be broadcast in the form of ammonium sulfate or sulfur-coated urea, but other environmental benefits can be gained by using material from a waste stream, such as animal manure, wastewater treatment plant biosolids, fermentation waste, composted municipal yard waste, or other organic industrial waste products (Labrecque et al. 1998; Adegbidi et al. 2003; Keoleian and Volk 2005). By the third or fourth growing season after coppice, the stems have grown to heights of 7–9 m (Figure 28.4) with stem diameters averaging 2–7 cm depending on varietal characteristics.

28.5 HARVESTING AND CONVERSION OF WILLOW BIOMASS

Willow is best harvested at the end of the third growing season after coppice which is typically the fourth year of a newly established crop. The product of harvest is woody biomass, but the dimensional specifications of the harvested and processed product will vary for each conversion application. The logistics and energy input required for transport and drying are also critical considerations that need to be matched with downstream conversion processes. Harvesting the stems after the plants have entered dormancy has multiple advantages: the leaves and their significant accumulation of

FIGURE 28.4 Example of a 4-year-old stand of shrub willow.

nutrients have fallen to the ground and will not be removed from the site; the plants have reallocated energy and nutrients to below-ground storage tissues conserving energy to support spring regrowth; the stem biomass holds less water during the winter months than during the growing season; and when the ground is frozen, there is less chance for soil compaction and rutting of the field. However, running harvesting operations during the winter becomes complicated by the presence of deep snow and by the harsh working conditions of extremely cold temperatures. Harvesting can continue over a 4- to 5-month period until just before bud break in the spring. After spring and early summer harvesting, the plants will resprout and grow for the rest of the season, but the harvested biomass will have a higher proportion of moisture and nitrogen.

Commercial harvesting is currently accomplished using either a self-propelled forage harvester with a willow cutting head or using a sugarcane harvester. The willow cutting head on a self-propelled forage harvester has two circular saw blades that cut the stems 6–8 cm above the ground, pulling them into the chopper, which produces chips averaging 2–4 cm in size and blows them into an adjacent forage box, dump wagon, or truck (Figure 28.5). This equipment has been designed and sold commercially by Claas, New Holland, and Coppice Resources Ltd. with the intention that the forage harvester can be used for multiple crops—hay, silage corn, and willow—by exchanging the cutter head. Winter harvest of willow would be scheduled after the fall harvest of silage corn. The sugarcane harvester saws the stems just above the ground and then chops them into 15- to 20-cm-long sections called "billets." These larger pieces of stem are less likely to compost during storage, but they will require further milling before they can be used to generate heat or power. Prototype harvesters that saw, collect, and bundle the stems whole have also been designed, but they are not widely used in commercial production systems.

The chips collected in a dump wagon or forage box are brought to the headlands of the field and are loaded into either an open-top truck, when dump wagons are used, or are unloaded to a blower, which blows the chips into a standard trailer. Because trucking costs are a significant component of the harvesting expense, it is usually only economical to transport wet biomass no more than 75 km (~50 mi) to market, and shorter distances are best (Keoleian and Volk 2005; Tharakan et al. 2005). When used as fuel for a power plant, the harvested willow chips are unloaded from the truck—either using a live bottom truck, dump truck, or by tipping an enclosed trailer on a whole-truck dumper. The chips are usually conveyed to the top of a chip pile, where they are stored—either

FIGURE 28.5 Harvesting shrub willow using a self-propelled forage harvester with a specialized willow cutter head. The forage harvester blows wood chips into a dump wagon being towed alongside.

covered or outside. The moisture content of willow biomass harvested during the winter months is approximately 45–50%, which adds water to any industrial conversion process if used immediately. This may be advantageous for the pretreatment and saccharification steps needed for the biological production of biofuels, but it lowers the net energy content of the biomass for combustion applications. A small amount of microbial respiration occurs in the center of the pile until oxygen is depleted, but the heat generated is sufficient to dry the biomass down to 30–35% moisture content. To prevent the wood chips from further degradation and contamination with microbes, they may be dried using a forced-air dryer, which adds expense, but improves the combustion efficiency and storage life (Gigler et al. 1999).

Willow biomass may be burned directly in a wood-fired boiler or may be co-fired with coal to generate heat and/or power with a net energy ratio of approximately 1:11 (Mann and Spath 1999; Tharakan et al. 2005). In this application, drying to a moisture content of 30–35% is adequate, but the chips may need to be milled to a smaller dimension with a hammer mill. When bulk loads of biomass need to be transported long distances and stored before use, processing to increase the bulk density is justified, including manufacture of wood pellets and briquettes which requires drying the biomass to 11% moisture content. Wood pellets are increasingly being used for home heating in North America; however, much of the equipment is designed to use premium-grade pellets with less than 1% ash content to minimize maintenance concerns (Biomass Energy Resource Center 2007). The ash content of shrub willow wood is 0.5–1.1% and of bark is 4.8–5.9% (Senelwa and Sims 1999; Klasnja et al. 2002), with bark content of mature stems ranging from 14–19% (Adler et al. 2005; Serapiglia et al. 2009). Thus, willow biomass can be mixed with wood residues or debarked tree chips to make premium-grade wood pellets, or they can be used exclusively in the production of standard-grade (1–2% ash) or industrial-grade (>3% ash) wood pellets. There are also several high-efficiency wood chip boilers on the market that can use dried willow chips directly, including models produced by KWB of Austria (www.kwb.at/en/).

Willow biomass has been used as a feedstock for the production of ethanol as a liquid transportation fuel in laboratory research, and commercial deployment of this technology is imminent with near-term advances in pretreatment and saccharification technology (Eklund et al. 1995; Eklund and Zacchi 1995; Hahn-Hagerdal et al. 2006). Because willow is a viable, high-yielding feedstock crop for many locales, it is vital that the bioethanol conversion technologies that emerge will be compatible with willow biomass feedstock characteristics. Recent research has shown that steam pretreatment using 0.5% sulfuric acid at 200°C followed by enzymatic hydrolysis resulted in high sugar recoveries of 92% of glucose and 86% of xylose, which yielded 79% of the theoretical maximal production of ethanol after simultaneous saccharification and fermentation (Sassner et al. 2008). Production of ethanol could be further enhanced by use of a microbe that could ferment all of the available sugars (Sassner et al. 2006). Because there is considerable variation in the chemical composition of the biomass produced from different genotypes of willow, it may well be possible to select varieties that have biomass compositional characteristics that are optimized for particular pretreatment and fermentation technologies (Serapiglia et al. 2008, 2009). Alternatively, willow biomass can be used as a feedstock for thermal conversion processes (Demirbas 2007), including flash pyrolysis or co-pyrolysis for the production of bio-oil (Cornelissen et al. 2008; Lievens et al. 2009) and gasification or hydrogasification for the production of synthetic natural gas (Porada 2009).

28.6 PESTS AND DISEASES OF WILLOW BIOENERGY CROPS

Rust is one of the most damaging diseases of shrub willow. Most rusts that infect willow have been identified as *Melampsora epitea*, which uses larch (*Larix decidua* Mill.) as an alternate host. Rust appears to the naked eye as small (1–2 mm), orange or brown, discrete, slightly raised pustules of teliospores on the surface of the leaf (Figure 28.6). With the onset of severe infections, rust defoliates the plant prematurely and reduces yield significantly (Dawson and McCracken 1994). Furthermore, rust infections can predispose a plant to infection by secondary pathogens, which may lead to further reductions in yield or premature death. Most willow rusts only infect leaves; however, strains of *Melampsora* that infect young leaves and green stems have been identified in some willows, including *S. viminalis* and *S. caprea* (Pei 1997). Species used in bioenergy plantations that display particular susceptibility to rust include *S. burjatica* and *S. viminalis* in Europe and *S. eriocephala* in North America.

Rusts evolve rapidly through hybridization. They usually complete a complex sexual life-cycle that involves five different spore stages, including one that lives on an alternative host, such as European larch. Rust populations multiply through infection and repeated reinfection of the leaf during the summer, finally producing teliospores that overwinter on the dead infected leaves in the litter. More than 12 pathotypes have been identified in England, Scotland, and Northern Ireland (Pei et al. 1993, 1999), some of which have been designated into *formae specialis*; the host range for each *f. sp.* appears to be confined to particular species of willow (Pei et al. 1996). The genetic relationship among *ff. spp.* is complicated (Pei et al. 2005). The pathotypes occurring in rust populations in the United Kingdom and Sweden are similar, but were shown at one point to be different from those occurring in North America (Royle and Ostry 1995). Moist, cool summers and mild, temperate winters are ideal environmental conditions for *Melampsora* growth. Such conditions are often found in Northern Ireland, England, and Sweden where rust has proven to be a major problem in willow plantations (McCracken and Dawson 1992; Hunter et al. 1996).

Resistance to willow rust is under strong genetic control (Rönnberg-Wästljung and Gullberg 1999; Cameron et al. 2008); therefore, selection and breeding programs in Europe and the United States emphasize rust resistance as the primary management for this disease. In Northern Ireland, planting shrub willow in mixed-culture stands has effectively reduced inocula and has proven to be an effective method of nonfungicide control of rust disease (McCracken and Dawson 1997; McCracken et al. 2001; Begley et al. 2009). These stands have been planted by continuously loading the mechanical planter with whips of several different varieties as it proceeds down a row. The physical barrier provided by resistant varieties limits the movement of the inocula and slows the build-up, often delaying the onset of disease. This strategy has successfully slowed the spread of disease and reduced the impact of rust overall on willow plantations (McCracken and Dawson 1997; McCracken et al. 2001, 2005).

Other diseases, such as crown gall disease caused by *Agrobacterium tumefaciens*, anthracnose tip blight caused by *Colletotrichum* spp., black cancer caused by *Glomerella miyabeana*, have not had

FIGURE 28.6 (a) Uredinia on an *S. cordata* × *S. eriocephala* hybrid, (b) giant willow aphid (*Tuberolachnussalignus*) on shrub willow, (c) Japanese beetle (*P. japonica*) damage on *S. dasyclados*, and (d) potato leaf hopper (*Empoascafabae*) damage on an *S. viminalis* × *S. miyabeana* hybrid.

broad effect on bioenergy plantation yield, but may reduce cutting yield in nursery plantations. Fungal pathogens, *Colletotrichum* spp., *Dothiorella* spp., *Botryosphaeriadothidea*, *Cytospora* spp., and *Leucostoma* spp., associated with stem cankers have been identified (S. Kenaley, G. Hudler, Cornell University, unpublished data) in biomass plantations, but these slow-growing diseases localized to the upper portions of stems are regularly removed by harvest on a short rotation. As long as stem canker or tip dieback diseases do not penetrate into the stool, they typically have only a minor impact because buds just below the affected area will rapidly break dormancy and emerge to continue growing.

Leaf-feeding and stem-sucking insects can have a significant negative effect on the yield of short-rotation coppice willow (Kendall et al. 1996; Björkman et al. 2000, 2008; Kreuger and Potter 2001; Bell et al. 2006; Björkman et al. 2008). Partial defoliation, as occurs with beetle feeding, is the most common problem and is usually due to both generalists that can feed on a range of hosts, such as the Japanese beetle (*Popillia japonica* Newman; Figure 28.6) and specialists that are host-specific, such as Chrysomelid beetles, including *Phratora vitellinae* and *Phratora vulgatissima* (brassy and blue willow beetles) in the United Kingdom and *Plagiodera versicolor* (imported willow beetle) in the United States (Peacock et al. 2002, 2004; Nordman et al. 2005). Leaf-sucking insects, such as the potato leaf hopper (*Empoasca fabae*), and stem-sucking insects, such as the giant willow aphid (*Tuberolachnus salignus*; Figure 28.6) and the black willow aphid (*Pterocomma salicis*), can also debilitate fast-growing willows (Collins et al. 2001). Planting fields with a random mixture of multiple, genetically diverse, structurally different varieties has been shown to be effective in limiting the build-up of local populations of insects, which would reduce the need to use pesticides (Peacock et al. 2001; Dalin et al. 2009).

28.7   GENETIC IMPROVEMENT FOR INCREASED YIELD AND RESISTANCE

Increases in yield achieved through genetic improvement will reduce the overall cost of producing willow biomass and will encourage wider adoption and long-term sustainability of the crop. Willow breeding began in North America in the early 1980s at the University of Toronto in Ontario, Canada under the direction of L. Zsuffa (1988) using *S. eriocephala*, *S. exigua*, *S. lucida*, *S. amygdaloides*, *S. bebbiana*, *S. pellita*, *S. petiolaris*, and *S. discolor*. They verified that variations observed in plant biomass, moisture content, and specific gravity were traits associated with species differences (Mosseler et al. 1988). Furthermore, Zsuffa and Mosseler investigated species hybridization, including pollination barriers, crossability relationships, and hybrid performance (Mosseler and Papadopol 1989; Mosseler and Zsuffa 1989; Mosseler 1990).

Breeding programs for willow bioenergy crops also emerged in Sweden in the 1980s, including the industrial breeding program directed by S. Larsson at Svalöf Weibull AB and the academic research programs including U. Gullberg and A. Rönnberg-Wästljung at the Swedish University of Agricultural Science in Uppsala. The original emphasis was placed on producing clones with high biomass and good form using various species. *S. viminalis*, *S. dasyclados*, *S. schwerinii*, *S. triandra*, *S. caprea*, *S. daphnoides*, and *S. eriocephala* were the most common species used (Larsson 1997). The main gene pool for the Swedish breeding program was collected from Europe, central Russia, and Siberia. Experiencing rapid progress and success, they produced varieties with improved yield, erect growth form, and resistance to *Melampsora* rust and insects (Larsson 1998). By 2001, six varieties of shrub willow were available for commercial production with yields averaging 6–14 odt/ ha per year over multiple harvests(Larsson 2001). Since the early 1990s, Svalöf Weibull AB has also been very interested in developing varieties of willow that are frost tolerant with low moisture content. To that end, two varieties, Loden and Gudrun, were developed and commercialized. Most recently, Svalöf Weibull has placed emphasis on developing varieties that display tolerance to high salt concentrations and toxic pollution to be used to remediate polluted soils and leachate water from landfills (Larsson 2001) and has marketed at least 13 varieties for commercial production.

The cultivation and use of willows has a long history in the United Kingdom, and the National Willow Collection, a collection of 1,100 varieties comprising over 120 different species that was managed by Stott for many years, has been a vital resource for ongoing breeding (Stott 1984). Willow breeding by Stott and Lindegaard at the Long Ashton Research Station became focused on bioenergy in the 1980s and 1990s and was formalized by funding that supported the European Willow Breeding Program involving Long Ashton, Svalöf Weibull, and Murray Carter in the mid-1990s (Lindegaard and Barker 1997). The fundamental aim of the program was similar to that of the Swedish program: to produce and commercialize high-yielding, disease- and pest-resistant varieties with a wide and diverse genetic base. Rust, caused by *Melampsora epitea*, was a major problem in the United Kingdom, so special emphasis was placed on rust resistance and developing mixed varietal plantations to prevent rust outbreaks (Lindegaard and Barker 1997). In the early 2000s, the Long Ashton Station was shut down, the National Willow Collection was moved to Rothamsted Research, and willow breeding was reinitiated in the United Kingdom with funding that formed the Biomass for Energy Genetic Improvement Network. Over 300 crosses were attempted in the

5-year period from 2003 to 2008, and the U.K. program has continued to focus on producing and selecting elite genotypes, using molecular techniques to characterize the National Willow Collection and to develop markers for rapid selection of desirable traits (Hanley et al. 2002, 2007; Trybush et al. 2008).

As improved willow germplasm from Canada and Sweden was tested in the New York State in the 1990s, it became apparent that rust was having a significant impact on *S. eriocephala* and potato leaf hopper was debilitating varieties with *S. viminalis* or *S. schwerinii* in their genetic background. Thus, a willow breeding program was initiated by L. Abrahamson, R. Kopp, and L. Smart in New York to develop varieties specifically selected for conditions in North America with improved yield over the varieties selected in the Toronto program. Initial crosses were completed in 1998 at the SUNY College of Environmental Science and Forestry using techniques for controlled pollinations developed by R. Kopp (Kopp et al. 2001, 2002a, 2002b). A large and diverse breeding collection of willows was assembled with major collection efforts in 1995, 2000, and 2001 across portions of the Northeast and Midwest United States (Smart et al. 2005). Most individuals were collected from natural stands of *S. purpurea* and *S. eriocephala* in New York, Pennsylvania, Ohio, and Wisconsin. Molecular marker analyses have been used to demonstrate high genetic diversity and a relatively high degree of heterozygosity among the natural populations of *S. purpurea* and *S. eriocephala* in New York (Lin et al. 2009). By 2007, the willow collection maintained at the Tully Genetics Field Station contained over 700 accessions representing more than 20 species and species hybrids. In addition to the varieties collected, contributions of bred and native collections from U.S. and overseas collaborators and acquisitions of commercial and horticultural varieties from nurseries were also obtained. Since 1998, more than 600 crosses were completed in this program, seven varieties were patented, and elite varieties are being broadly tested in trials across North America (Smart and Cameron 2008).

A major thrust of all of the breeding programs has been to identify heritable traits that are indicative of long-term biomass productivity in field-grown plants and that can be measured in juvenile plants. Cross-sectional stem area, number of stems per plant, and susceptibility to rust have been found to be heritable in *S. eriocephala* (Cameron et al. 2008). Using a multivariate approach, stem area, length of growing time, and insect damage have been identified as traits that can be scored in field-grown juvenile plants and used to evaluate future relative performance (Tharakan et al. 2001). These studies provide willow breeders with the tools they need to quickly and efficiently screen thousands of seedlings for high biomass production.

A second major goal in breeding programs was to identify a parent or a combination of parents that will produce superior offspring. Successfully predicting the performance of an individual parent and/or the performance of a combination of specific parents to generate improved progeny has been shown in a study of 34 full-sib F_1 *S. eriocephala* families (Cameron et al. 2008). Furthermore, parents with low similarity indices, as indexed by amplified fragment length polymorphism (AFLP) fingerprinting, can produce offspring that display large amounts of phenotypic variability, thereby increasing the probability of producing willow clones exhibiting desirable extreme phenotypes (Kopp et al. 2002b). Crossing success rate in a breeding program is often dependent upon the fertility of the parents. Many species of *Salix* are polyploid (Suda and Argus 1968; Thibault 1998), and progeny from interspecific crosses sometime display intermediate levels of ploidy (Thibault 1998; MacAlpine et al. 2008). Flow cytometry has been used to estimate ploidy levels of parents used in crosses and identify triploids and pentaploids that are sterile (MacAlpine et al. 2008). The U.K. program has used this technique to successfully increase their crossing success rate to greater than 58% (MacAlpine et al. 2008).

28.8 FUTURE COMMERCIALIZATION OF WILLOW BIOENERGY CROPS

As a novel crop that requires specialized equipment for planting and harvesting, has relatively high establishment costs, requires a long-term commitment to recover that outlay, and is aimed at the energy sector rather than food or feed markets, the commercial adoption of willow bioenergy crops will be led by entrepreneurial pioneers and forward-thinking businesses and must be supported by government policy. In Sweden, the commercialization of *Salix* has been dominated by Svalöf Weibull and Lantmännen Agroenergie. In the United Kingdom, the willow enterprise has been developed by pioneers, including Coppice Resources Ltd., Strawson's Energy, Murray Carter, and Rural Generation. In North America, the first business engaged in growing willow commercially for bioenergy was Double A Willow in Fredonia, NY (www.doubleawillow.com). This commercial nursery has the capacity to produce 30,000,000 cuttings each year and is active in the development of markets for willow biomass. Projects are developing that will expand the cultivation of willow to provide renewable fuel for combined heat and power projects. Finally, Agro Énergie is a pioneering willow nursery in Québec, Canada. One thing that all of these companies have in common is active and regular engagement with academic and government willow researchers. The two-way communication between industry and academia is crucial to accelerate the development of this novel agricultural enterprise.

The formalization and enforcement of regional, national, and global carbon emission restrictions will drive the expanded use of biomass for energy which will gain further financial advantage as the cost of fossil fuels continues to rise. The community of experts engaged in willow research and extension must grow to meet the needs of a burgeoning industry. Areas that critically need attention include improvements in sustainable land-use conversion, site preparation, and crop establishment; testing of herbicides and development of cover crop systems for weed management; long-term breeding for improved yield, pest and disease resistance, and nutrient and water use efficiency; and harvesting, transport, drying, and storage logistics. As demand for sustainably produced woody biomass rises, so must the intensity of government and industry support for research and development of willow bioenergy crops. One critical example of this is the recent commitment by the U.S. Department of Energy-Joint Genome Initiative to sequence the genome of *S. purpurea*—a project lead by G. Tuskan (Oak Ridge National Laboratory), L. Smart (Cornell University), and C. Town (J. Craig Venter Institute). With a database of the DNA sequences of the entire willow genome in hand, willow breeders and physiologists will command a powerful toolbox that can accelerate the selection of improved varieties and the study of basic woody plant biology. One final research tool that is desperately needed to complement these genomic approaches is the ability to efficiently transform and regenerate willow, which would open new avenues of research into basic aspects of gene function while also stimulating the rapid production of new varieties with specific genetic trait improvements (Smart and Cameron 2008). There is tremendous potential for shrub willow crops to contribute significantly to the expanded utilization of renewable energy, but there is much work to be done before that vision is realized.

REFERENCES

Abrahamson LP, Volk TA, Kopp RF, White EH, Ballard JL (2002) *Willow Biomass Producer's Handbook.* SUNY College of Environmental Science and Forestry, Syracuse, NY

Adegbidi HG, Briggs RD (2003) Nitrogen mineralization of sewage sludge and composted poultry manure applied to willow in a greenhouse experiment. *Biomass Bioenergy* 25:665–673

Adegbidi HG, Briggs RD, Volk TA, White EH, Abrahamson LP (2003) Effect of organic amendments and slow-release nitrogen fertilizer on willow biomass production and soil chemical characteristics. *Biomass Bioenergy* 25:389–398

Adler A, Verwijst T, Aronsson P (2005) Estimation and relevance of bark proportion in a willow stand. *Biomass Bioenergy* 29:102–113

Anderson HW, Papadopol CS, Zsuffa L (1983) Wood energy plantations in temperate climates. *Forest Ecol Manag* 6:281–306

Argus GW (1974) An experimental study of hybridization and pollination in *Salix* (willow). *Can J Bot* 52:1613–1619

Argus GW (1997) Infrageneric classification of *Salix* (Salicaceae) in the New World. In: *Systematic Botany Monographs*, Vol. 52. American Society of Plant Taxonomists, Ann Arbor, MI, pp 1–121

Aronsson P, Perttu K (2001) Willow vegetation filters for wastewater treatment and soil remediation combined with biomass production. *For Chron* 77:293–299

Baker BW, Ducharme HC, Mitchell DCS, Stanley TR, Peinetti HR (2005) Interaction of beaver and elk herbivory reduces standing crop of willow. *Ecol Appl* 15:110–118

Begley D, McCracken AR, Dawson WM, Watson S (2009) Interaction in short rotation coppice willow, Salix viminalis genotype mixtures. *Biomass Bioenergy* 33:163–173

Bell AC, Clawson S, Watson S (2006) The long-term effect of partial defoliation on the yield of short-rotation coppice willow. *Ann Appl Biol* 148:97–103

Bergkvist P, Ledin S (1998) Stem biomass yields at different planting designs and spacings in willow coppice systems. *Biomass Bioenergy* 14:149–156

Biomass Energy Resource Center (2007) *Wood Pellet Heating Guidebook*, available at http://www.mass.gov/ Eoeea/docs/doer/publications/doer_pellet_guidebook.pdf (accessed August 20, 2009)

Björkman C, Bengtsson B, Häggström H (2000) Localized outbreak of a willow leaf beetle: Plant vigor or natural enemies? *Popul Ecol* 42:91–96

Björkman C, Dalin P, Ahrne K (2008) Leaf trichome responses to herbivory in willows: Induction, relaxation and costs. *New Phytol* 179:176–184

Cameron KD, Phillips IS, Kopp RF, Volk TA, Maynard CA, Abrahamson LP, Smart LB (2008) Quantitative genetics of traits indicative of biomass production and heterosis in 34 full-sib F_1 *Salix eriocephala* families. *Bioenergy Res* 1:80–90

Collins CM, Rosado RG, Leather SR (2001) The impact of the aphids *Tuberolachnus salignus* and *Pterocomma salicis* on willow trees. *Ann Appl Biol* 138:133–140

Cornelissen T, Jans M, Yperman J, Reggers G, Schreurs S, Carleer R (2008) Flash co-pyrolysis of biomass with polyhydroxybutyrate: Part 1. Influence on bio-oil yield, water content, heating value and the production of chemicals. *Fuel* 87:2523–2532

Dalin P, Kindvall O, Bjorkman C (2009) Reduced population control of an insect pest in managed willow monocultures. *PLoS One* 4:Article No. e5487

Dawson WM, McCracken AR (1994) Effect of *Melampsora* rust on the growth and development of *Salix burjatica korso* in Northern Ireland. *Eur J For Pathol* 24:32–39

Demirbas A (2007) Progress and recent trends in biofuels. *Prog Energ Combust* 33:1–18

Dhondt AA, Wrege PH, Cerretani J, Sydenstricker KV (2007) Avian species richness and reproduction in short-rotation coppice habitats in central and western New York. *Bird Study* 54:12–22

Dickmann DI, Kuzovkina JA (2008) Poplars and willows of the world, with emphasis on silviculturally important species. In: Isebrands JG, Richardson J (eds), *Poplars and Willows in the World: Meeting the Needs of Society and the Environment*, available at http://www.fao.org/forestry/32608/en/ (accessed August 20, 2009)

Dimitriou I, Aronsson P (2005) Willows for energy and phytoremediation in Sweden. *Unasylva* 56:47–50

Dimitriou I, Eriksson J, Adler A, Aronsson P, Verwijst T (2006) Fate of heavy metals after application of sewage sludge and wood-ash mixtures to short-rotation willow coppice. *Environ Pollut* 142:160–169

Doty SL (2008) Enhancing phytoremediation through the use of transgenics and endophytes. *New Phytol* 179:318–333

Doty SL, Oakley B, Xin G, Kang JW, Singleton G, Khan Z, Vajzovic A, Staley JT (2009) Diazotrophic endophytes of native black cottonwood and willow. *Symbiosis* 47:23–33

Eklund R, Galbe M, Zacchi G (1995) The influence of SO_2 and H_2SO_4 impregnation of willow prior to steam pretreatment. *Bioresour Technol* 52:225–229

Eklund R, Zacchi G (1995) Simultaneous saccharification and fermentation of steam-pretreated willow. *Enzyme Microb Technol* 17:255–259

Eriksson G, Gullberg U, Kang H (1984) Breeding strategy for short rotation woody species. In: Perttu K (ed), *Ecology and Management of Forest Biomass Production Systems*. Swedish University of Agricultural Sciences, Uppsala, Sweden, pp 199–216

French CJ, Dickinson NM, Putwain PD (2006) Woody biomass phytoremediation of contaminated brownfield land. *Environ Pollut* 141:387–395

Gage EA, Cooper DJ (2005) Patterns of willow seed dispersal, seed entrapment, and seedling establishment in a heavily browsed montane riparian ecosystem. *Can J Bot* 83:678–687

Gigler JK, Meerdink G, Hendrix EMT (1999) Willow supply strategies to energy plants. *Biomass Bioenergy* 17:185–198

Gullberg U (1993) Towards making willows pilot species for coppicing production. *For Chron* 69:721–726

Hahn-Hagerdal B, Galbe M, Gorwa-Grauslund MF, Liden G, Zacchi G (2006) Bio-ethanol—The fuel of tomorrow from the residues of today. *Trends Biotechnol* 24:549–556

Hanley S, Barker JHA, Ooijen JWv, Aldam C, Harris SL, Åhman I, Larsson S, Karp A (2002) A genetic linkage map of willow (*Salix viminalis*) based on AFLP and microsatellite markers. *Theor Appl Genet* 105:1087–1096

Hanley SJ, Mallott MD, Karp A (2007) Alignment of a *Salix* linkage map to the *Populus* genomic sequence reveals macrosynteny between willow and poplar genomes. *Tree Genet Genomes* 3:35–48

Hardig TM, Brunsfeld SJ, Fritz RS, Morgan M, Orians CM (2000) Morphological and molecular evidence for hybridization and introgression in a willow (*Salix*) hybrid zone. *Mol Ecol* 9:9–24

Hostanska K, Jurgenliemk G, Abel G, Nahrstedt A, Saller R (2007) Willow bark extract (BNO1455) and its fractions suppress growth and induce apoptosis in human colon and lung cancer cells. *Cancer Detect Prevent* 31:129–139

Hunter T, Royle DJ, Arnold GM (1996) Variation in the occurrence of rust (*Melampsora* spp.) and other diseases and pests, in short-rotation coppice plantations of *Salix* in the British Isles. *Ann Appl Biol* 129:1–12

Jones MH, Bay C, Nordenhall U (1997) Effects of experimental warming on arctic willows (*Salix* spp.): A comparison of responses from the Canadian High Arctic, Alaskan Arctic, and Swedish Subarctic. *Glob Change Biol* 3:55–60

Karrenberg S, Edwards PJ, Kollmann J (2002a) The life history of Salicaceae living in the active zone of floodplains. *Freshwater Biol* 47:733–748

Karrenberg S, Kollmann J, Edwards PJ (2002b) Pollen vectors and inflorescence morphology in four species of *Salix*. *Plant Syst Evol* 235:181–188

Karrenberg S, Suter M (2003) Phenotypic trade-offs in the sexual reproduction of Salicaeae from flood plains *Amer J Bot* 90:749–754

Kendall DA, Hunter T, Arnold GM, Liggitt J, Morris T, Wiltshire CW (1996) Susceptibility of willow clones (*Salix* spp.) to herbivory by *Phyllodecta vulgatissima* (L.) and *Galerucella lineola* (Fab.) (Coleoptera, Chrysomelidae). *Ann Appl Biol* 129:379–390

Keoleian GA, Volk TA (2005) Renewable energy from willow biomass crops: Life cycle energy, environmental and economic performance. *Crit Rev Plant Sci* 24:385–406

Klasnja B, Kopitovic S, Orlovic S (2002) Wood and bark of some poplar and willow clones as fuelwood. *Biomass Bioenergy* 23:427–432

Kopp RF, Abrahamson LP, White EH, Burns KF, Nowak CA (1997) Cutting cycle and spacing effects on biomass production by a willow clone in New York. *Biomass Bioenergy* 12:313–319

Kopp RF, Maynard CA, Rocha de Niella P, Smart LB, Abrahamson LP (2002a) Collection and storage of pollen from *Salix* using organic solvents. *Am J Bot* 89:248–252

Kopp RF, Smart LB, Maynard CA, Isebrands JG, Tuskan GA, Abrahamson LP (2001) The development of improved willow clones for eastern North America. *For Chron* 77:287–292

Kopp RF, Smart LB, Maynard CA, Tuskan GA, Abrahamson LP (2002b) Predicting within-family variability in juvenile height growth of *Salix* based upon similarity among parental AFLP fingerprints. *Theor Appl Genet* 105:106–112

Kreuger B, Potter DA (2001) Diel feeding activity and thermoregulation by Japanese beetles (Coleoptera: Scarabaeidae) within host plant canopies. *Environ Entomol* 30:172–180

Kuzovkina YA, Quigley MF (2005) Willows beyond wetlands: Uses of *Salix* L. species for environmental projects. *Water Air Soil Pollut* 162:183–204

Kuzovkina YA, Weih M, Romero MA, Charles J, Hurst S, McIvor I, Karp A, Trybush S, Labrecque M, Teodorescu TI, Singh NB, Smart LB, Volk TA (2008) Salix: Botany and global horticulture. In: Janick J (ed), *Horticultural Reviews*, Vol. 34. John Wiley, Hoboken, NJ, pp 447–489

Labrecque M, Teodorescu TI (2001) Influence of plantation site and wastewater sludge fertilization on the performance and foliar nutrient status of two willow species grown under SRIC in southern Quebec (Canada). *For Ecol Manag* 150:223–239

Labrecque M, Teodorescu TI, Daigle S (1998) Early performance and nutrition of two willow species in short-rotation intensive culture fertilized with wastewater sludge and impact on the soil characteristics. *Can J For Res* 28:1621–1635

Larsson S (1997) Commercial breeding of willow for short rotation coppice. *Asp Appl Biol* 49:215–218

Larsson S (1998) Genetic improvement of willow for short-rotation coppice. *Biomass Bioenergy* 15:23–26

Larsson S (2001) Commercial varieties from the Swedish willow breeding programme. *Asp Appl Biol* 65:193–198

Levesque H, Lafont O (2000) Aspirin throughout the ages: An historical review. *Rev Med Interne* 21:8S–17S

Lievens C, Carleer R, Cornelissen T, Yperman J (2009) Fast pyrolysis of heavy metal contaminated willow: Influence of the plant part. *Fuel* 88:1417–1425

Lin J, Gibbs JP, Smart LB (2009) Population genetic structure of native versus naturalized sympatric shrub willows (*Salix*; Salicaceae). *Amer J Bot* 96:771–785

Lindegaard KN, Barker JHA (1997) Breeding willows for biomass. *Asp Appl Biol* 49:155–162

Lira CMD, Barry TN, Porlaroy WE, McWilliam EL, Lopez-Villalobos N (2008) Willow (*Salix* spp.) fodder blocks for growth and sustainable management of internal parasites in grazing lambs. *Anim Feed Sci Technol* 141:61–81

MacAlpine WJ, Shield IF, Trybush SO, Hayes CM, Karp A (2008) Overcoming barriers to crossing in willow (*Salix* spp.) breeding. Presented at *Biomass and Energy Crops III*, Sand Hutton, United Kingdom, December 10–12, 2008, pp 173–180

Mann MK, Spath PL (1999) Life cycle comparison of electricity from biomass and coal. Presented at the *1999 ACEEE Summer Study on Energy Efficiency in Industry: Industry and Innovation in the 21st Century*. American Council for and Energy-Efficient Economy, Washington DC, pp 559–569

McCracken AR, Dawson M (1992) Clonal response in *Salix* to *Melampsora* rusts in short rotation coppice plantations. *Eur J For Pathol* 22:19–28

McCracken AR, Dawson WM (1997) Growing clonal mixtures of willow to reduce effect of *Melampsora epitea* var. *epitea*. *Eur J For Pathol* 27:319–329

McCracken AR, Dawson WM, Bowden G (2001) Yield responses of willow (*Salix*) grown in mixtures in short rotation coppice (SRC). *Biomass Bioenergy* 21:311–319

McCracken AR, Dawson WM, Carlisle D (2005) Short-rotation coppice willow mixtures and rust disease development. In: Pei MH, McCracken AR (eds), *Rust Diseases of Willow and Poplar*. CABI Publishing, Cambridge, MA pp 185–194

McElroy GH, Dawson WM (1986) Biomass from short-rotation coppice willow on marginal land. *Biomass* 10:225–240

McWilliam EL, Barry TN, Lopez-Villalobos N, Cameron PN, Kemp PD (2005) Effects of willow (*Salix*) versus poplar (*Populus*) supplementation on the reproductive performance of ewes grazing low quality drought pasture during mating. *Anim Feed Sci Technol* 119:69–86

Moore KM, Barry TN, Cameron PN, Lopez-Villalobos N, Cameron DJ (2003) Willow (*Salix* sp.) as a supplement for grazing cattle under drought conditions. *Anim Feed Sci Technol* 104:1–11

Mosseler A (1990) Hybrid performance and species crossability relationships in willows (*Salix*). *Can J Bot* 68:2329–2338

Mosseler A, Papadopol CS (1989) Seasonal isolation as a reproductive barrier among sympatric *Salix* species. *Can J Bot* 67:2563–2570

Mosseler A, Zsuffa L (1989) Sex expression and sex ratios in intra- and interspecific hybrid families of *Salix* L. *Silvae Genet* 38:12–17

Mosseler A, Zsuffa L, Stoehr MU, Kenney WA (1988) Variation in biomass production, moisture content, and specific gravity in some North American willows (*Salix* L.). *Can J For Res* 18:1535–1540

Nahrstedt A, Schmidt M, Jäggi R, Metz J, Khayyal MT (2007) Willow bark extract: The contribution of polyphenols to the overall effect. *Wien Med Wochenschr* 157:348–351

Newsholme C (1992) *Willows: The Genus* Salix. Timber Press, Portland, OR

Nordman EE, Robison DJ, Abrahamson LP, Volk TA (2005) Relative resistance of willow and poplar biomass production clones across a continuum of herbivorous insect specialization: Univariate and multivariate approaches. *For Ecol Manag* 217:307–318

Peacock L, Harris J, Powers S (2004) Effects of host variety on blue willow beetle *Phratora vulgatissima* performance. *Ann Appl Biol* 144:45–52

Peacock L, Herrick S, Harris J (2002) Interactions between the willow beetle *Phratora vulgatissima* and different genotypes of *Salix viminalis*. *Agric For Entomol* 4:71–79

Peacock L, Hunter T, Turner H, Brain P (2001) Does host genotype diversity affect the distribution of insect and disease damage in willow cropping systems? *J Appl Ecol* 38:1070–1081

Pei MH, Bayon C, Ruiz C (2005) Phylogenetic relationships in some *Melampsora* rusts on Salicaceae assessed using rDNA sequence information. *Mycol Res* 109:401–409

Pei MH, Hunter T, Ruiz C (1999) Occurrence of *Melampsora* rusts in biomass willow plantations for renewable energy in the United Kingdom. *Biomass Bioenergy* 17:153–163

Pei MH, Royle DJ, Hunter T (1993) Identity and host alternation of some willow rusts (*Melampsora* spp.) in England. *Mycol Res* 97:845–851

Pei MH, Royle DJ, Hunter T (1996) Pathogenic specialization in *Melampsora epitea* var *epitea* on *Salix*. *Plant Pathol* 45:679–690

Pei MH, Whelan MJ, Halford NG, Royle DJ (1997) Distinction between stem- and leaf-infecting forms of *Melampsora* rust on *Salix viminalis* using RAPD markers. *Mycol Res* 101:7–10

Peinetti HR, Kalkhan MA, Coughenour MB (2002) Long-term changes in willow spatial distribution on the elk winter range of Rocky Mountain National Park (USA). *Landscape Ecol* 17:341–354

Peinetti HR, Menezes RSC, Coughenour MB (2001) Changes induced by elk browsing in the aboveground biomass production and distribution of willow (*Salix monticola* Bebb): Their relationships with plant water, carbon, and nitrogen dynamics. *Oecologia* 127:334–342

Pitta DW, Barry TN, Lopez-Villalobos N, Kemp PD (2007) Willow fodder blocks—An alternate forage to low quality pasture for mating ewes during drought? *Anim Feed Sci Technol* 133:240–258

Pohjonen V (1987) *Salix* 'Aquatica Gigantea' and *Salix* x *dasyclados* Wimm. in biomass willow research. *Silva Fenn* 21:109–122

Porada S (2009) A comparison of basket willow and coal hydrogasification and pyrolysis. *Fuel Process Technol* 90:717–721

Pulford ID, Riddell-Black D, Stewart C (2002) Heavy metal uptake by willow clones from sewage sludge-treated soil: The potential for phytoremediation. *Int J Phytoremed* 4:59–72

Rönnberg-Wästljung AC, Gullberg U (1999) Genetics of breeding characters with possible effects on biomass production in *Salix viminalis* (L.). *Theor Appl Genet* 98:531–540

Royle DJ, Ostry ME (1995) Disease and pest control in the bioenergy crops poplar and willow. *Biomass Bioenergy* 9:69–79

Sassner P, Galbe M, Zacchi G (2006) Bioethanol production based on simultaneous saccharification and fermentation of steam-pretreated *Salix* at high dry-matter content. *Enzyme Microb Technol* 39:756–762

Sassner P, Martensson C-G, Galbe M, Zacchi G (2008) Steam pretreatment of H_2SO_4-impregnated *Salix* for the production of bioethanol. *Bioresource Technol* 99:137–145

Seiwa K, Tozawa M, Ueno N, Kimura M, Yamasaki M, Maruyama K (2008) Roles of cottony hairs in directed seed dispersal in riparian willows. *Plant Ecol* 198:27–35

Senelwa K, Sims REH (1999) Fuel characteristics of short rotation forest biomass. *Biomass Bioenergy* 17:127–140

Serapiglia MJ, Cameron KD, Stipanovic AJ, Smart LB (2008) High-resolution thermogravimetric analysis for rapid characterization of biomass composition and selection of shrub willow varieties. *Appl Biochem Biotechnol* 145:3–11

Serapiglia MJ, Cameron KD, Stipanovic AJ, Smart LB (2009) Analysis of biomass composition using high-resolution thermogravimetric analysis and percent bark content for the selection of shrub willow bioenergy crop varieties. *Bioenergy Res* 2:1–9

Skvortsov A (1999) *Willows of Russia and Adjacent Countries: Taxonomical and Geographical Review*. University of Joensuu, Joensuu, Finland

Smart LB, Cameron KD (2008) Genetic improvement of willow (*Salix* spp.) as a dedicated bioenergy crop. In: Vermerris WE (ed), *Genetic Improvement of Bioenergy Crops*. Springer Science, New York, pp 347–376

Smart LB, Volk TA, Lin J, Kopp RF, Phillips IS, Cameron KD, White EH, Abrahamson LP (2005) Genetic improvement of shrub willow (*Salix* spp.) crops for bioenergy and environmental applications in the United States. *Unasylva* 221 56:51–55

Stokes KE (2008) Exotic invasive black willow (*Salix nigra*) in Australia: Influence of hydrological regimes on population dynamics. *Plant Ecol* 197:91–105

Stokes KE, Cunningham SA (2006) Predictors of recruitment for willows invading riparian environments in south-east Australia: Implications for weed management. *J Appl Ecol* 43:909–921

Stott KG (1984) Improving the biomass potential of willow by selection and breeding. In: Perttu K (ed), *Ecology and Management of Forest Biomass Production Systems*. Swedish University of Agricultural Science, Uppsala, Sweden, pp 233–260

Stott KG (1991) Nomenclature of the promising biomass coppice willows, *Salix* x *sercians* Tausch ex Kern., *Salix dasyclados* Wimm. and *Salix* 'Aquatica Gigantea'. *Bot J Scot* 46:137–143

Stott KG (1992) Willows in the service of man. *P Roy Soc Edinb B* 98:169–182

Suda Y, Argus GW (1968) Chromosome numbers of some North American *Salix*. *Brittonia* 20:191–197

Tamura S, Kudo G (2000) Wind pollination and insect pollination of two temperate willow species, *Salix miyabeana* and *Salix sachalinensis*. *Plant Ecol* 147:185–192

Tape K, Sturm M, Racine C (2006) The evidence for shrub expansion in Northern Alaska and the Pan-Arctic. *Glob Change Biol* 12:686–702

Tharakan PJ, Robison DJ, Abrahamson LP, Nowak CA (2001) Multivariate approach for integrated evaluation of clonal biomass production potential. *Biomass Bioenergy* 21:237–247

Tharakan PJ, Volk TA, Lindsey CA, Abrahamson LP, White EH (2005) Evaluating the impact of three incentive programs on cofiring willow biomass with coal in New York State. *Energy Policy* 33:337–347

Thibault J (1998) Nuclear DNA amount in pure species and hybrid willows (*Salix*): A flow cytometric investigation. *Can J Bot* 76:157–165

Tolbert VR, Schiller A (1996) Environmental enhancement using short-rotation woody crops and perennial grasses as alternatives to traditional agricultural crops. In: Lockeretz W (ed), *Environmental Enhancement through Agriculture*. Tufts University, Medford, MA, pp 209–216

Tolbert VR, Wright LL (1998) Environmental enhancement of U.S. biomass crop technologies: Research results to date. *Biomass Bioenergy* 15:93–100

Trybush S, Jahodová S, Macalpine W, Karp A (2008) A genetic study of a *Salix* germplasm resource reveals new insights into relationships among subgenera, sections and species. *Bioenergy Res* 1:67–79

van der Heijden EW (2001) Differential benefits of arbuscular mycorrhizal and ectomycorrhizal infection of *Salix repens. Mycorrhiza* 10:185–193

Veraart AJ, Nolet BA, Rosell F, de Vries PP (2006) Simulated winter browsing may lead to induced susceptibility of willows to beavers in spring. *Can J Zool* 84:1733–1742

Verwijst T (2001) Willows: An underestimated resource for environment and society. *For Chron* 77:281–285

Volk TA (2002) *Alternative Methods of Site Preparations and Coppice Management during the Establishment of Short-Rotation Woody Crops*. Forest and Natural Resources Management, SUNY College of Environmental Science and Forestry, Syracuse, NY, p 284

Volk TA, Abrahamson LP, Nowak CA, Smart LB, Tharakan PJ, White EH (2006) The development of short-rotation willow in the northeastern United States for bioenergy and bioproducts, agroforestry and phytoremediation. *Biomass Bioenergy* 30:715–727

Volk TA, Ballard B, Robison DJ, Abrahamson LP (2004a) Effect of cutting storage conditions during planting operations on the survival and biomass production of four willow (*Salix* L.) clones. *New For* 28:63–78

Volk TA, Verwijst T, Tharakan PJ, Abrahamson LP, White EH (2004b) Growing fuel: A sustainability assessment of willow biomass crops. *Front Ecol Environ* 2:411–418

Wagner JE (2000) *Pre-Emergent Herbicide Screening Trials for Willow Biomass Crops*. Faculty of Forestry, State University of New York College of Environmental Science and Forestry, Syracuse, NY, p 187

Wilkinson JM, Evans EJ, Bilsborrow PE, Wright C, Hewison WO, Pilbeam DJ (2007) Yield of willow cultivars at different planting densities in a commercial short rotation coppice in the north of England. *Biomass Bioenergy* 31:469–474

Zhenfu F, Shidong Z, Skvortsov A (1999) Salicaceae. In: Zheng-yi W, Raven P (eds), *Flora of China*. Missouri Botanical Garden Press, St. Louis, MO, pp 139–274

Zsuffa L (1988) A review of the progress in selecting and breeding North American Salix species for energy plantations at the Faculty of Forestry, University of Toronto, Canada. Presented at the International Energy Agency Willow Breeding Symposium. Department of Forest Genetics, Swedish University of Agricultural Sciences, Uppsala, Sweden, pp 41–51

Zsuffa L (1990) Genetic improvement of willows for energy plantations. *Biomass* 22:35–47

29 Sugarbeet

Pawan Kumar, Anjanabha Bhattacharya, and Rippy Singh
University of Georgia

CONTENTS

29.1 INTRODUCTION

The world population is growing at an alarming rate, and with this growth the demand for energy is also increasing at a rapid pace. It is projected that there will be more than a 50% increase in energy demand in 2025, with most coming from rapidly developing nations. This increased demand cannot be met with finite fossil energy sources like petroleum and coal. Therefore, the present-day energy scenario and concerns for global warming have stimulated the search for an alternative to fossil fuel energy which ought to not only be renewable but also ecofriendly with low emission of greenhouse gases (GHGs). Bioenergy from plants is one such source that has the potential to reduce dependence on fossil fuels.

Ethanol produced from plants can be used as an energy source in its pure state or it can be used by blending with petroleum fuel in engines. It offers several advantages over fossil fuel: it has a high octane number (the higher the number, the more efficient the energy usage) (Balat 2009), low toxicity to humans, low volatility, and low evaporation (Hira and de Oliveira 2009). Bioethanol is being produced from sugar-rich crops such as sugarcane and sugarbeet (Wegner and Hagnefelt 2008). Bioethanol production from sugarbeet is of recent origin. It helps to stabilize falling sugar prices in addition to bridging global energy demands. It is more cost-efficient to convert sugars to ethanol than unlocking complex polysaccharides from plant sources because to date there has been limited success in separating lignin from plant material that prevents access of cellulose (polysaccharides) to trap the source molecules needed as raw material in bioethanol production.

Sugarcane is adapted to tropical regions of the world, whereas sugarbeet is predominately grown in temperate regions; therefore, the two sugar-rich crops occupy different geographical niches. Sugarbeet was identified as a crop of choice for sugar production in the temperate regions in the early nineteenth century, when sugar was largely imported from the tropical and subtropical countries and so was considered to be a luxury item that few could afford. Sugar was used largely as a base in traditional medicine such as homeopathy and was a very expensive product that was only derived from cane sugar. Although sugarbeet contains less sugar (17%; Milford and Watson 1971) than sugarcane (18–22%), and it is a poor transformer of sugars from the photoassimilate, it had advantages because it could be grown in temperate climate and the foliage could be used as a cattle feed. This prompted the farmers to start cultivating this crop in the western countries, and now it occupies a sizable acreage. Today it meets almost half of the demand for sugar consumption in the developed world.

29.2 SUGARBEET AS A SUGAR CROP

Sugarbeet (*Beta vulgaris* L.) belongs to family Chenopodiaceae and is an important sugar crop of the world second only next to sugarcane. This is a hardy crop. It is drought and heat tolerant and can grow in saline conditions, although its cultivation is labor-intensive (Streibig et al. 2009). Sugarbeet is a biennial plant predominately grown in temperate regions where beets are planted in spring and harvested in autumn. When grown in warmer regions the beets are planted in autumn and harvested in spring. The recent introduction of tropical sugarbeet has enabled its cultivation in tropical and subtropical regions.

Sugarbeet has been grown for food and fodder (Clarke and Edye 1996) since ancient times, but its value as a sugar crop was realized in the mid-18th century when a method for extracting sugar from the beets was discovered. German chemist Andreas Marggraf discovered that beets contain sucrose, which was similar to that produced from sugarcane. However, an efficient protocol was not developed for another 50 years. In the late eighteenth century, one of Marggraf's students, Franz Karl Achard, succeeded in developing efficient methodology for producing crystalline sugar from sugarbeet on an industrial scale. It was the research effort of Karl Achard that established sugarbeet as an economic source of sucrose in Europe. Karl Archard is now considered to be the father of the sugarbeet industry (Ensminger and Konlande 1993).

29.3 EXTRACTION OF SUGAR

Sugarbeet is harvested mechanically; a series of blades on harvesters cut off the top leaves and the crown. The roots are then cleaned to remove attached soil before transportation. At the processing plant, sugar content of the crop is determined, which in turn determines a grower's payment (Dodic et al. 2009). After thorough cleaning, the beets are chopped into slices called "cossettes." Sugar from beets is removed through a diffusion process. The cossettes are mixed with hot water at approximately 70°C for approximately 90 min. The sugar in the beets passes from the plant cells into the surrounding water by diffusion. This sugar-containing water is referred to as "raw juice," and the left behind cossettes are called "pulp," which is high in moisture but low in sugar content. Remnant sugar from the pulp is squeezed out in a screw press to remove as much juice as possible. This juice is used as part of the water in the diffuser, and the pulp is then sent to a drying plant where it can be processed into other value-added products.

The raw juice needs to be cleaned before crystallization. This is done by a process called "carbonation," where the juice is mixed with milk of lime [calcium hydroxide, $Ca(OH)_2$] and bubbling carbon dioxide through the mixture. Small clumps of chalk start to form in the juice, and as these clumps form, they collect much of the nonsugar from the mixture (Kenter and Hoffmann 2007). After removal of chalk particles, a clean sugar solution is left behind, which is called "thin juice." Thin juice is concentrated into "thick juice" by evaporation in a multistage evaporator. This thick juice contains approximately 60% sucrose by weight. Thick juice is then fed into crystallizers to form sugar crystals, and a centrifugation step separates the sugar crystals from the liquid molasses.

29.4 BIOETHANOL PRODUCTION FROM SUGARS

Biofuels produced from sugar and starch-rich crops (e.g., sugarcane, sugarbeet, and corn) are often referred to as first-generation biofuels because these crops were the first candidates for bioethanol production. The sugar extracted from these crops is fermented anaerobically to produce ethanol. The trend of producing ethanol started from Brazil, where sugar production was much higher than the demand and petroleum was costly. Sugar-based bioethanol allowed Brazil to achieve energy self-sufficiency (Gressel 2008).

Sucrose produced by sugarbeet is a disaccharide molecule made up of glucose and fructose. Monosaccharides like glucose and fructose serve as base material for ethanol production. Therefore, the sucrose molecule first needs to be broken down into its component sugars. This is done by enzymatic hydrolysis of sucrose catalyzed by the enzyme invertase, which converts sucrose into glucose and fructose:

$$C_{12}H_{22}O_{11} \rightarrow C_6H_{12}O_6 + C_6H_{12}O_6 \qquad (29.1)$$

(Sucrose)   (Glucose) (Fructose)

Enzymatic hydrolysis is followed by the fermentation process for the production of bioethanol. Commercial yeast such as *Saccharomyces cerevisiae* converts glucose into ethanol under anaerobic conditions. The recently introduced strain *S. cerevisiae* ATCC 36859 has proven to be an efficient microbe in enhancing ethanol production from sugarbeet molasses (Atiyeh and Duvnjak 2003a).

$$C_6H_{12}O_6 \rightarrow 2C_2H_5OH + 2CO_2 \qquad (29.2)$$

(Glucose)   (Ethanol)

Many other microbes are also capable of ethanol formation. Many bacteria such as *Enterobacteriaceas, Spirochaeta,* and *Bacteroides* spp. and yeasts including *S. cerevisiae, S. uvarum, Schizosaccharomyces pombe,* and *Kluyueromyces* sp. metabolize glucose under anaerobic conditions by the Embden–Meyerhof pathway. In this pathway, one molecule of glucose yields two molecules of pyruvate, which are then decarboxylated to acetaldehyde and then reduced to ethanol. The ethanol thus produced is low in concentration (8–12%). A distillation process is used to increase the concentration of ethanol to the required levels. Modified techniques are used in this process to recover a large portion of ethanol and to prevent ethanol emission into the atmosphere or ethanol losses with water (Krylova et al. 2008).

All of the intermediates produced during sugarbeet processing can be used as raw material for ethanol production. All of these have advantages and limitations. Direct use of beet or pulp as raw material is not very efficient (Turquois et al. 1999) because there is a slow release of sugars from the pulp into fermented solution and the long-term storage of beet leads to a loss of sugar due to enzyme action (Berghall 1997). Solids from raw juice contain 85–90% sugars and 10–15% nonsugar; therefore, raw juice can be used as raw material for fermentation after slight pH adjustments. Easy decomposition by microbes and low storability limits the use of raw juice for direct fermentation. However, thin juice is very suitable for ethanol production (Figure 29.1). Molasses, produced during the crystallization process, is a traditional raw material for distilleries, and more than 90% of ethanol is produced from this raw material.

29.5 SUGARBEET IMPROVEMENT

Sugarbeet continues to be a major crop meeting our daily needs for sugar consumption; thus, major genetic studies have been initiated in recent years to improve the quality and productivity. Other objectives include selection for resistance against insects, pests, and diseases. Crop modeling is another area that should be explored to increase productivity. A specific areawise crop model should be developed that will help us to decide on agronomic practices and harvesting schedules. Sugarbeet can be grown as an intercrop between the main crops.

Thus, increasing the productivity of sugarbeet is of paramount importance and for it to become a successful biofuel crop the need of the hour is to identify and exploit genetic potential of the crop through better selection of traits (Doney and Theurer 1984; Kuhn 1998) that contribute to the

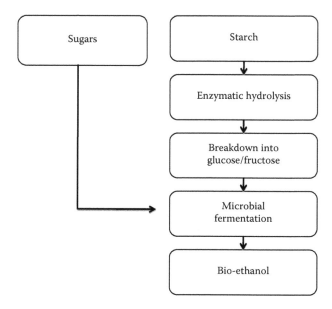

FIGURE 29.1 Flowchart showing stepwise procedure involved in the production of bioethanol from sugar and starch components of sugarbeet.

physiological characters of importance related to crop productivity. Agronomic performance of different sugarbeet varieties and germplasm accession at different locations should also be evaluated to identify varieties adapted to a particular area (Sarwar et al. 2008). Efforts should be given to create a genetic map of the crop species, which will help in advocating better breeding programs focused on crop improvement. Sugarbeet improvement could be aided by identifying quantitative trait loci (QTL) linked to simple sequence repeats (SSRs), random amplified polymorphic DNA (RAPD), expressed sequence tags (ESTs), or amplified fragment length polymorphism (AFLP) markers. Schnedier et al. (2001) identified 21 different QTL associated with sugar content, sugar and beet yield, and amino-nitrogen content in sugarbeet verified at six locations. Further, targeted induced local lesion in genomes (TILLING) should also be undertaken to identify mutants with increased productivity, less canopy area, and decreased days to maturity. Other approaches may include targeted reduction of genes by RNA interference (RNAi) and micro-RNA, which may increase productivity and reduce diversion of photoassimilates toward noneconomic parts of the plants [i.e., vegetative parts particularly genes involved in gibberellin (GA) metabolism]. Also, identification and ectopic expression of transgenes should also be considered. Therefore, cloning genes responsible for rapid conversion of sugars (dextrose, raffinose, and sucrose) to ethanol (Atiyeh and Duvnjak 2003b) and then expression of such genes in a plant system will enhance the production of ethanol.

Smith (2008) reported that manipulation of ADP-glucose pyrophosphorylase, which plays a role in the sugar signaling pathway, to increase the yields of starch have been met with limited success. One of the aspects for biofuel production is synchronous maturity of the crops used for biofuel production. Many chemicals that promote synchronous maturity such as napthalene acetic acid (NAA) could be useful. Also, application of gibberelic acid (GA) may result in high yield. Anti-GA substances such as pacrobutrazol can be used to control plant height, or ectopic expression of GA 2-oxidase will result in dwarf plants (Dijkstra et al. 2008). This will allow for easy harvest of the produce with low cost on resources and saving time. Compact plants are also less prone to damage by biotic and abiotic stresses. Several research groups including researchers from the United Kingdom have concentrated their efforts on controlling stature and flowering in sugarbeet at Boomsbarn, Rothamsted Research, England (Mutasa-Gottgens et al. 2008). Other efforts include transcriptional profiling under stress conditions in sugarbeet, thus increasing our understanding of the crop productivity under different physiological conditions, which could be utilized in crop improvement (Pestsova et al. 2009). Another study by Ozolina et al. (2005) explored the fluctuation of plant hormones and sugar deposition in sugarbeet, thus adding to our existing knowledge base of sugar distribution. Sadaghian et al. (1993) conducted an experiment to identify the genetic basis of number of cells, their length, and GA sensitivity in sugarbeet in relation to bolting, which ultimately determines productivity. Selective photoassimilate partitioning should also be included in the crop improvement regimes. Another strategy could be manipulating the C_3 mechanism of photosynthesis in favor of the C_4 mechanism, which is present in the grass family and at present, can only be grown in tropical regions. Such mechanism should be explored in sugarbeet, thus increasing carbon sequestration. Sévenier et al. (1998) ectopically expressed a gene encoding 1-sucrose::sucrose fructosyl transferase (*1-SST*) from *Helianthus tuberosus* into sugarbeet that increases fructan (forms of simple sugars) content without subsequently altering plant architecture. This could later be used potentially in biofuel production. Further, manipulating genes for enhanced solar energy capture and conversion of photosynthate to sugars (as mentioned in relation to C_3 and C_4 crops), and genes that provide endurance to plants growing under saline conditions or on land not usable for cultivation should be emphasized particularly when soil salinity problems are growing out of proportion in the modern world (Antizar-Ladislao and Turrion-Gomez 2008). Therefore, production of transgenic plants with the above-mentioned traits would help in developing a more robust plant phenotype that was not always possible through conventional or molecular breeding.

Better extension activity should be adopted to disseminate new technologies and crop varieties to the farming community and better postharvest facility should be ensured.

29.6 ALTERNATIVE USES

Sugarbeet is largely used for sugar production and as a root vegetable crop throughout the world. Although it had humble beginnings as a minor crop that was considered a poor man's food or feed for cattle, today it has gained widespread importance. It is in fact the same species as red garden beets. After the sugar has been extracted, the molasses can be used as a material source for biogas production and as a fertilizer (Okiely 1992). The leftover biomass after extraction of the sugar can be used as cattle feed. The green mass left after harvesting the crop can be ploughed back to the soil, restoring soil fertility and organic matter content. Keeping these points in perspective, sugarbeet cultivation may act as a sort of incentive for the farming community in terms of rural employment and socioeconomic development, soil conservation, agricultural sustainabiltity, and as such in the developing world (Venturi and Venturi 2003).

29.7 FURTHER DISCUSSIONS AND CONCLUSIONS

As the demand for alternative biofuel crops intensifies along with the long-standing debate between "food versus fuel" in the modern world, sugarbeet has come a long way to be a part of the solution, being able to provide a green fuel in this era of climate change concerns. What was considered as poor man's crop or as a cattle feed crop species and largely neglected a few centuries ago could now be a lead crop in sugar production for the liquid fuel industry. With falling sugar prices, sugarbeet can be put to another use—to produce bioethanol instead for traditional uses—thus ensuring good returns to the growers. As conversion of sugar to ethanol still remains the most viable and practical option to produce biofuels, the popularity of sugarbeet in biofuel production is bound to increase. This requires a major restructuring of the sugar industry throughout the world backed by reasonable pricing for the producers. Biofuel plants, which use sugarbeet, should be located in the vicinity of the area of production to decrease the price of production and create ample avenues of employment. Further, the production and supply chain need to be straightened, thus eliminating unnecessary middlemen to reduce the price per unit. The need of the hour is to identify or develop superior genotypes and to identify genes responsible for enhancing sugar production in this crop species validated at different climatic regimes. We have much better technologies available on the industrial side for the production of the biofuel; thus, crop productivity to supply the raw material seems to be a bottleneck. Large amounts of subsidy have already been provided in biofuel ventures, and there is much scope of innovation based on local conditions, which will lessen our dependence on gasoline-based industries backed by legislature.

REFERENCES

Antizar-Ladislao B, Turrion-Gomez JL (2008) Second-generation biofuels and local bioenergy systems. *Biofuels Bioprod Biorefin* 2:55–469

Atiyeh H, Duvnjak Z (2003a) Production of fructose and ethanol from cane molasses using *Saccharomyces cerevisiae* ATCC 36858. *Acta Biotechnol* 23:37–48

Atiyeh H, Duvnjak Z (2003b) Utilization of raffinose and melibiose by a mutant of *Saccharomyces cerevisiae*. *J Chem Technol Biotechnol* 78:1068–1074

Balat M (2009) Bioethanol as a vehicular fuel: A critical review. *Energy Source Part A* 31:1242–1255

Berghall S, Briggs S, Elsegood SE, Eronen L, Kuusisto JO, Philip EJ, Theobald TC, Walliander P (1997) The role of sugar beet invertase and related enzymes during growth, storage and processing. *Zuckerindustrie* 122:520–530

Clarke MA, Edye LA (1996) Sugar beet and sugarcane as renewable resources. *Agricultural Materials as Renewable Resources*, ACS Symposium Series. Vol. 647 pp 229–247

Dijkstra C, Adams E, Bhattacharya A, Page A, Anthony P, Kourmpetli S, Power J, Lowe K, Thomas S, Hedden P (2008) Over-expression of a gibberellin 2-oxidase gene from Phaseolus coccineus L. enhances gibberellin inactivation and induces dwarfism in Solanum species. *Plant Cell Rep* 27:463–470

Dodic S, Popov S, Dodic J, Rankovic J, Zavargo Z, Mucibabic RJ (2009) Bioethanol production from thick juice as intermediate of sugar beet processing. *Biomass Bioenergy* 33:822–827

Doney DL, Theurer JC (1984) Potential of breeding for ethanol fuel in sugar beet. *Crop Sci* 24:255–257

Ensminger AH, Konlande JE (1993) *Foods and Nutrition Encyclopedia*. CRC Press, Boca Raton, FL, pp 2411–2416

Gressel J (2008) Transgenics are imperative for biofuel crops. *Plant Sci* 174:246–263

Hira A, de Oliveira LG (2009) No substitute for oil? How Brazil developed its ethanol industry. *Energy Policy* 37:2450–2456

Kenter C, Hoffmann C (2007) Quality and storability of sugar beet at early harvest. *Zuckerindustrie* 132:615–621

Krylova Y, Kozyukov EA, Lapidus AL (2008) Ethanol and diesel fuel from plant raw materials: A review. *Solid Fuel Chem* 42:358–364

Kuhn E (1998) Transgenic sugar beets as industrial plants. *Zuckerindustrie* 123:28–34

Milford GF, Watons DJ (1971) The effect of nitrogen on the growth and sugar content of sugar-beet. *Ann Bot* 35:287–300

Mutasa-Gottgens E, Qi A, Mathews A, Thomas S, Phillips A, Hedden P (2008) Modification of gibberellin signalling (metabolism & signal transduction) in sugar beet: Analysis of potential targets for crop improvement. *Transgen Res* 18:301–308

Okiely P (1992) The effects of ensiling sugar-beet pulp with grass on silage composition, effluent production and animal performance. *Irish J Agric Food Res* 31:115–128

Overstreet LF (2009) Strip tillage for sugar beet production. *Sugar J* 111:292–302

Ozolina NV, Pradedova EV, Salyaev RK (2005) The dynamics of hormonal status of developing red beet root (*Beta vulgaris* L.) in correlation with the dynamics of sugar accumulation. *Biol Bull* 32:22–26

Pestsova E, Meinhard J, Menze A, Fischer U, Windhovel A, Westhoff P (2008) Transcript profiles uncover temporal and stress-induced changes of metabolic pathways in germinating sugar beet seeds. *BMC Plant Biol* 8:122–128

Sadeghian YS, Johansson E (1993) Genetic study of bolting and stem length in sugar beet (*Beta vulgaris* L.) using a factorial cross design. *Euphytica* 65:177–185

Sarwar MA, Hussain F, Chattha AA (2008) After harvest qualitative and quantitative behaviour of some sugar beet varieties. *J Anim Plant Sci* 18:139–141

Sévenier R, Hall RD, van der Meer IM, Hakkert JHC, van Tunen AJ, Koops AJ (1998) High level fructan accumulation in a transgenic sugar beet. *Nat Biotechnol* 16:843–846

Smith AM 2008. Prospects for increasing starch and sucrose yields for bioethanol production. *Plant J* 54:546–558

Streibig JC, Ritz C, Pipper CB, Yndgaard F, Fredlund K, Thomsen JN (2009) Sugar beet, bioethanol, and climate change. *IOP Conference Series on Earth and Environmental Science*, Vol. 6, Session 24, IOP Publishing Limited

Turquois T, Rinaudo M, Taravel FR, Heyraud A (1999) Extraction of highly gelling pectic substances from sugar beet pulp and potato pulp: Influence of extrinsic parameters on their gelling properties. *Food Hydrocolloids* 13:255–262

Venturi P, Venturi G (2003) Analysis of energy comparison for crops in European agricultural systems. *Biomass Bioenergy* 25:235–255

Wegner JP, Hagnefelt T (2008) Danisco Sugar's first beet-based bioethanol plant. *Int Sugar J* 110:697–700

30 Sunflower

Sanjeev K. Sharma
Unichem Laboratories Ltd.

Krishan L. Kalra and Gurvinder S. Kocher
Punjab Agricultural University

CONTENTS

30.1 INTRODUCTION

Limited stocks and exorbitant prices of petroleum products create the need of finding some suitable alternative fuels or fuel supplements (Wyman 1996; Lynd et al. 2005). Ethanol has especially attracted interest as an alternative transportation fuel for two reasons: (1) the oil crisis in the mid-1970s stressed the dependence on the supply of petroleum, which can be reduced by the use of alternative fuels such as ethanol from renewable resources; and (2) if the ethanol production process only uses energy from renewable energy sources, no net carbon dioxide is added to the atmosphere, making ethanol an environmentally safe fuel (Kheshgi et al. 2000).

Ethanol can be blended approximately 10–15% with gasoline, as it is now in the United States, or it can be used as 22–25% blends, as in Brazil. These blends realize three major benefits. First, the gasoline consumption is reduced, thereby lowering oil imports. Second, ethanol increases the octane number of the gasoline, thereby improving the engine performance. Third, ethanol provides oxygen for the fuel, resulting in better combustion. Ethanol can also be used as a nearly pure or "neat" fuel. Hydrous ethanol containing approximately 95% ethanol and 5% water is used as a fuel for ethanol vehicles in Brazil (Wyman and Hinman 1990). Also, low proportions of gasoline can be blended with pure ethanol for use as fuel. Such fuels are environmentally advantageous because of their clean-burning characteristics and reduced emissions of ethanol into the atmosphere. Engines designed for dedicated ethanol use can be tuned to achieve higher efficiencies than conventional gasoline engines (Lynd et al. 1991).

Ethanol can be produced from renewable resources such as raw materials containing sugar (sugarbeet, sugarcane, etc.), starch (potato, com, grain, etc.), and cellulose (lignocellulosic materials). Currently molasses and starch are being used worldwide for ethanol production. These substances are expensive because of their other industrial uses; therefore, they cannot be used for economical alcohol production as an alternative fuel. Molasses is used in refining tobacco, in preparing poultry feed, etc., and starchy materials are hardly sufficient to meet the food requirements of teeming millions. Therefore, research emphasis needs to be directed toward finding a cheap and abundantly available substrate for ethanol production.

World ethanol production is estimated to be 21.6 billion gallons, with approximately 14 billion gallons coming from the United States and Brazil and an Indian share of 0.6 billion gal (Market Research Analyst 2008). Although use of bioethanol as a source of energy would be more for than just complementing solar, wind, and other intermittent renewable energy sources in the long run (Lin and Tanaka 2006), the growing demand of bioethanol can only be met from lignocellulosics (Nguyen and Saddler 1991).

30.2 LIGNOCELLULOSIC BIOMASS

Various lignocellulosic biomass sources, such as agricultural residues (straws, hulls, stems, stalks, bagasse, etc.), fruit and vegetable wastes, deciduous and coniferous woods, municipal solid wastes (paper, cardboard, yard trash, wood products, etc.), waste from the pulp and paper industry, and herbaceous energy crops have the potential to serve as low-cost and abundant feedstocks for production of fuel ethanol. The major component of these materials is cellulose (35–50%), followed by hemicellulose (20–35%) and lignin (10–25%). Proteins, oils, and ash in widely varying ratios make up the remaining fraction of lignocellulosic biomass (Wyman 1994). In lignocellulosic materials, cellulose, a linear polymer of glucose, is associated with hemicellulose and surrounded by a lignin seal. Lignin, a complex three-dimensional polyaromatic matrix, prevents enzymes and acids from accessing some regions of the cellulose polymers.

Crystallinity of the cellulose further impedes acid or enzymatic hydrolysis (Weil et al. 1994; Bothast and Saha 1997). Production of ethanol from lignocellulosic materials involves: (1) detachment of cellulose fibers from the lignocellulosic structure, which is accomplished by physical or chemical means; (2) acid or enzymatic hydrolysis of the cellulose fibers; and (3) fermentation of sugars to ethanol, which is then distilled to yield the final product (von Sivers et al. 1994; Bothast and Saha 1997). During the last 2 decades, advances in technology for ethanol production from biomass have been developed to the point that large-scale production will be a reality in next few years (Yu and Zhang 2004; Moiser et al. 2005).

Although various lignocellulosic crop residues (e.g., wheat straw, rice straw, cotton stalks, corn stalks and cobs, groundnut shells, etc.) have been used for ethanol production, little effort has been made to use sunflower wastes as a substrate for ethanol production. Biomass resources are essential for bioenergy production. There are two major criteria for determining whether a crop is suitable for energy use. The first is the high dry matter yield per land unit and the second is the net gain in energy (the amount of energy produced from the biomass should be higher than the amount of energy required to grow the crop). Being one of the major oil crops cultivated worldwide, sunflower (*Helianthus annuus* L.) has the potential to become a biomass crop. As a member of the Compositae family, sunflower is relatively easy to grow in a wide range of environments, from the equator to a latitude of 55° north. Young sunflower plants withstand mild freezing. Sunflower has a strong root system and shows a considerable level of drought tolerance (Hu 2008). This crop was cultivated in an area of 2.2 million ha with production of 1.50 million t in India in 1998. In the Punjab State of this country, approximately 67,000 ha of land was under sunflower cultivation with a production of 90,000 t in 1999 (Anonymous 1999). Sunflower seed is the third-largest source of vegetable oil worldwide, followed by soybean and palm. Sunflower oil, extracted from the commercially available sunflower varieties containing 39–49% oil in the seed, is used for cooking, as carrier oil, and for production of biodiesel. Among oil crops, sunflower has the highest yield of 33% that has been exploited for biodiesel production as B100 diesel that has been claimed to have numerous advantages over petroleum diesel: it is renewable, nontoxic, biodegradable, and produced by U.S. and Canadian farmers. B100 biodiesel is 100% biodiesel fuel and reduces greenhouse gas emissions by 78.3%, particulate matter by 55.4%, hydrocarbons by 56.3%, mutagenicity by 80–90%, and sulfur by 100% (Biofuel Industries 2002). On December 20, 2008, Nishi-Nippon Railroad Co. Ltd. held successful trial rides of a bus powered by biodiesel fuel (BDF) derived from sunflower seeds (Asia Biomass Office 2008). After extraction of oil, a huge quantity of sunflower stalks (after harvesting of seeds) and sunflower hulls (during the industrial processing of sunflower seeds) are generated that do not find any suitable end use and are generally burnt in the fields, causing environmental pollution (Sharma et al. 2002a; Okur and Saracoglu 2006). Therefore, sunflower stalks and hulls, as lignocellulosics, afford a renewable and low-cost raw material for the production of bioethanol.

30.3 LIGNOCELLULOSIC BIOMASS CONVERSION

The research work on the bioconversion of lignocellulosics to liquid fuels has attained new dimensions today. The structure of these materials is very complex, and native biomass is resistant to enzymatic hydrolysis. The steps for the production of fuel ethanol from lignocellulosic biomass involve feedstock preparation, pretreatment, cellulase production, acid/enzymatic hydrolysis, ethanol fermentation, and ethanol recovery. Although each of the above steps has been extensively studied, the processing techniques required for ethanol production from lignocellulosic materials are presently extensive and costly. Currently, the utilization of cellulosic biomass to produce fuel ethanol presents significant technical and economic challenges, and the success of the process depends largely on the development of environmentally friendly pretreatment procedures, highly effective enzyme systems for conversion of pretreated biomass to fermentable sugars, and efficient microorganisms to convert sugars to ethanol (Gray et al. 2006; Mojovic et al. 2006).

30.4 PRETREATMENT

In general, among lignocellulosics cellulose is a linear polymer of glucose associated with hemicellulose and surrounded by a lignin seal. A lignin seal around cellulose microfibrils and its limited covalent association with hemicellulose prevents enzymes and acid from accessing some regions of the cellulose polymers. The potential formation of six hydrogen bonds (because of the β-1,4 orientation of glucosidic bonds in cellulose) adds to its crystallinity and further impedes acid or enzymatic hydrolysis (Weil et al. 1994). The goal of any pretreatment technology is to alter or remove structural and compositional impediments to hydrolysis to improve the rate of enzyme hydrolysis and increase yields of fermentable sugars from cellulose or hemicellulose. These methods cause physical and/or chemical changes in the plant biomass to achieve this result. Experimental investigation of physical changes and chemical reactions that occur during pretreatment is required for the development of effective and mechanistic models that can be used for the rational design of pretreatment processes. Furthermore, pretreatment processing conditions must be tailored to the specific chemical and structural composition of the various (and variable) sources of lignocellulosic biomass (Mosier et al. 2005). Recent studies indicate that cellulose digestibility is directly correlated with lignin and hemicellulose removal (Kim and Holtzapple 2006).

Several pretreatment techniques have been reported. These include physical methods such as ball milling, hammer milling, boiling, high-pressure steam, electron irradiation, wetting, γ-irradiation, etc. and chemical methods using sodium hydroxide (NaOH), hydrochloric acid, sulfuric acid, sulfur dioxide, alkaline peroxide, phosphoric acid, ammonia, organic solvents (ethylenediamine), supercritical carbon dioxide, etc. (Kosaric et al. 1980, Weil et al. 1994; Zhang et al. 1995). Keller et al. (2003) have even used biological methods for pretreatment of biomass through thermochemical pretreatment methods and have reported these to be promising (Sheehan et al. 2003).

30.4.1 Mechanical Pretreatment/Milling

Milling of lignocellulosic material is a popular method for increasing cellulose digestibility. The material to be treated is subjected to shearing and compressive forces generated by the mill for a specific period of time. Mechanically-based pretreatment technologies aim at reducing the size of biomass to below #20 sieves and to biomass that shows the best mechanical performance (de Sousa et al. 2004). Mechanical pretreatment technologies increase the digestibility of cellulose and hemicellulose in the lignocellulosic biomass, resulting in substantial lignin depolymerization via the cleavage of uncondensed aryl ether linkages (Inoeu et al. 2008). Solubility and fermentation efficiency of these residues is also substantially increased by this pretreatment, leading to value-added utilization of these residues (Qi et al. 2005). The use of mechanical chopping, hammer milling, grind milling, roll milling, vibratory milling, and ball milling are some of the successful low-cost pretreatment strategies (Mtui 2009).

Grinding of lignocellulosic material to a very small particle size makes it susceptible to enzymatic hydrolysis, thus improving digestibility of cellulose and hemicellulose to glycans and xylans, respectively (Chahal 1982). Tewari et al. (1987) studied the effect of grinding on the enzymatic hydrolysis of wheat straw, bagasse, corn cob, and groundnut shell. A mesh size of 40 was found to be the best, except in sugarcane bagasse, in which a mesh size of 20 gave the maximal enzymatic attack (de Sousa et al. 2004). In our laboratory, sunflower stalks ground to a size of 40 mesh have been found to be optimal for biomass digestion (Sharma et al. 2002b). Vaithanomsalt et al. (2009) and Okur and Saracoglu (2006) reported a 40-mesh size for sunflower stalks and a 0.71- to 1.0-mm size for sunflower hulls in their studies, respectively, for optimized pretreatment. Yuldashev et al. (1992) and Alvo and Belkacemi (1997) reported milling as the sole pretreatment sufficient for further optimized saccharification.

30.4.2 Physicochemical Pretreatment

Elevated temperatures and irradiation are the most successful physical treatments in the processing of lignocellulosic biomass. However, combined physical and chemical methods prove more important in the breaking up of the crystalline biomass structure, providing an improved accessibility of cellulose for enzymatic hydrolysis (Hendriks and Zeeman 2009). In addition to steam hydrolysis, liquid hot water, ammonia fiber explosion, and carbon dioxide explosion are other such technologies in which biomass is treated with high pressure, followed by sudden depressurization (Sun and Cheng 2002). The changes in the steam-treated biomass depend on the temperature, pressure, and time of exposure to steam. The organic acids derived from acetylated polysaccharides hydrolyze the hemicellulose to soluble sugars. Secondary reactions that occur under more drastic conditions result in the formation of furfurals, hydroxymethyl furfurals, and their precursors by dehydration of pentoses and hexoses (Chahal 1982; Almeida et al. 2007). It has been suggested that these compounds can reduce enzymatic and biological activity and even cause DNA breakdown (Endo et al. 2008). The addition of sulfuric acid, sulfur dioxide, or carbon dioxide can further improve upon the enzymatic hydrolysis, decrease the production of inhibitory compounds, and lead to more complete liquefaction of hemicellulose, glucan, xylan, etc. (Sun and Cheng 2002). The Stake and Iotech processes that are commercially operating in Canada involve the steam-heating of wood biomass chips to approximately 180–200°C for 5–30 min in a continuous operation and to higher temperatures (245°C) for a shorter time (0.5–2 min) in batch mode, respectively (Wayman 1980).

Dekker and Wallis (1983) reported steam hydrolysis of sunflower seed hulls at 200°C for 5 min followed by explosive defibration that solubilized more than 80% of the total hemicellulose and 85% of the pectic substances. The remaining residue, which consisted of cellulose (38%), lignin (45%), and residual hemicellulose (7%), was highly susceptible to hydrolysis by cellulases. Vaithanomsat et al. (2009) also performed pretreatment of acid-soaked sunflower stalks by steam explosion at 207°C and 21 kg/cm² for 3 min to fractionate the cellulose, hemicellulose, and lignin. Ruiz et al. (2006) used 220°C for 5 min as the pretreatment conditions for sunflower stalks. In our laboratory, we used a low pressure of 1.05 kg/cm² for 30–90 min in an autoclave for NaOH -soaked sunflower stalks and hulls (Sharma 2000; Sharma et al. 2002b). Similarly, Okur and Saracoglu (2006) used relatively mild conditions of 0.7 M acid and 90°C for pretreatment of sunflower hulls.

Acid and alkali hydrolysis have been used for pretreatment of sunflower hulls and stalks. Whereas Jimnez and Bonilla (1993), Vaithanomsat et al. (2009), and Okur and Saracoglu (2006) used acids, Soto et al. (1994) treated the ground sunflower hulls with NaOH (0.5 to 3% w/w) in an autoclave at 120°C for 0.5, 1, and 1.5 h. They reported that the higher the NaOH concentration, the greater the delignification. We also used alkali (0.25–1.5% NaOH; Figure 30.1) and standardized 0.5% NaOH for pretreatment of sunflower stalks and hulls (Sharma et al. 2002b, 2004) that revealed 51 and 53% cellulose, 17 and 17.5% hemicellulose, and 14.6 and 11.40% of lignin in sunflower stalks and hulls, respectively (Sharma 2000; Sharma et al. 2002a). The extraction yield (fraction of sunflower stalks recovered after pretreatment) was 66% (Sharma et al. 2002a).

30.5   CELLULASE PRODUCTION

Even a cursory perusal of current scientific literature shows cellulose hydrolysis by cellulases to be among the most intensively studied topics. Each and every aspect of cellulase production such as isolation and mutation of cellulolytic microorganisms, mode of fermentation, process optimization, genetics and regulation at a molecular level, mode of action, process economics, and cellulase recycling have been investigated comprehensively. Several workers have reviewed the cellulase production and technology aspects (Saddler et al. 1986; Srinivasan and Seetalaxman 1988; Beguin 1990; Wyman 1994; Bothast and Saha 1997; Ward 2002; Juhasz et al. 2004; Immanuel et al. 2006; Kocher et al. 2008).

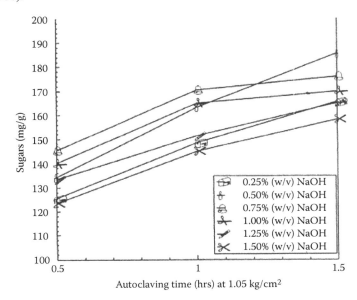

FIGURE 30.1   Effect of NaOH and autoclaving at 1.05 kg/cm³ on the enzymatic hydrolysis of sunflower stalks. Incubation time 48 h.

30.5.1 Cellulolytic Microorganisms

Cellulases are widely distributed in bacteria, actinomycetes, and fungi (Srinivasan and Seetalaxman 1988; Baldrian and Gabriel 2003; Immanuel et al. 2006). Bacteria such as *Cellulomonas*, *Bacteroides*, *Cellvibrio*, *Cytophaga*, *Ruminococcus*, *Pseudomonas*, and *Acetivibrio* are capable of degrading cellulose (Srinivasan and Seetalaxman 1988; Chaudhary et al. 1997). The actinomycetes have also been investigated for their extracellular cellulase, and strains such as *Streptomyces flavogriseus* and *Thermomonospora* have been studied in detail (Srinivasan and Seetalaxman 1988). However, fungi form the most important group for cellulose degradation and formation of cellulolytic enzymes. Among them the fungi imperfecti and basidiomycetes have received the most attention and have yielded promising cultures for cellulose biotechnology. One of the most extensively studied cultures is *Trichoderma reesei* (Mandels 1982; Kuhls et al. 1997). This strain has the advantage of a full complement production of cellulase, stability under enzymatic hydrolysis conditions, and resistance to chemical inhibitors. However, the main disadvantage of *Trichoderma* cellulase is its low activity of β-glucosidase. Other fungi that secrete good levels of extracellular cellulase include *Trichoderma harzianum*, *Penicillium funiculosum*, *Sclerotium rolfsii*, *Aspergillus terreus*, *A. phoenicis*, *Talaromyces emersonii*, *Sporotrichum pulverulentum*, *Humicola grisea*, and *Phanerochaete chrysosporium* (Srinivasan and Seetalaxman 1988; Singh et al. 1995; Falih 1998; De-Paula et al. 1999; Berlin et al. 2006; Kocher et al. 2008). *Aspergillus* spp. are known to produce good levels of β-glucosidases and have been supplemented with *Trichoderma* cellulase to improve the enzyme performance (Wyman 1996; Itoh et al. 2003; Tabka et al. 2006).

30.5.2 The Cellulase Complex

The enzyme system (i.e., cellulase complex) for the conversion of cellulose to glucose has been classified on the basis of the mode of catalytic action and now structural properties (Henrissat et al. 1998) into three enzyme types: (1) endo-l,4-β-glucanase or carboxymethylcellulase (CMC) (EC3.2.1.4); (2) exo-l,4-β-glucanase, (C$_1$) or cellobiohydrolase (CBH) (EC 3.2.1.91); and (3) β-glucosidase or cellobiase (EC 3.2..1.21). The cellulolytic enzymes with β-glucosidase act sequentially and cooperatively to degrade crystalline cellulose to glucose. Endoglucanases act in a random fashion on the regions of low crystallinity of the cellulosic fiber, whereas exoglucanases remove cellobiose (β-1,4-glucose dimer) units from the nonreducing ends of the cellulose chains (Figure 30.2). The presence of carbohydrate binding modules (CBMs) is important for initiation and processivity of exoglucanases (Teeri 1997). Synergism between these two types of enzymes is attributed to the endo-exo form of cooperativity and has been studied extensively between cellulases in the degradation of cellulose by *T. reesei* (Henrissat et al. 1985; Teeri 1997).

In *T. reesei*, two cellobiohydrolases (CBH1 and CBHII), five endoglucanases (EGI, EGII, EGIII, EGIV and EGV), and two β-glucosidases (BGLI and BGL II) have been purified and evaluated on various substrates. The synergism between CBH1 and endoglucanase I and II depends on the structural and ultrastructural features of the cellulosic substrate. Synergism is most marked when highly crystalline substrates were used, low with amorphous cellulose and absent with soluble derivatives. Four types of synergistic phenomenon have been reported: (1) endo-exo synergy between endo- and exoglucanases; (2) exo–exo synergy between exoglucanases processing from the reducing and nonreducing ends of cellulose chains; (3) synergy between exoglucanases and β-glucosidases that remove cellobiose and cellodextrins as end-products of endo- and exoglucanases, and (4) intramolecular synergy between catalytic domains and CBMs (Din et al. 1994; Teeri 1997).

30.5.3 Cellulase Production

As stated earlier, *T. reesei* has been the most extensively studied strain for cellulase production in submerged culture so far. The cellulase production cost represents 40–60% of the overall cost of the process designed to saccharify cellulosic substrates and ferment the products (Ryu and Mandels 1980).

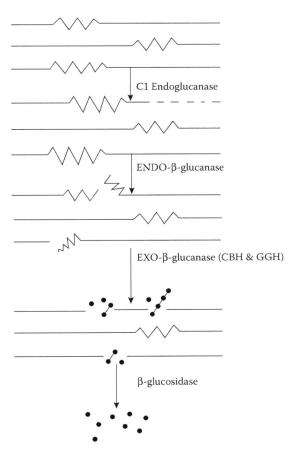

FIGURE 30.2 Model of synergistic action of cellulase complex enzymes. (———, crystalline cellulose; ⁀⋁⋁⋁⌐, amorphous cellulose; − - −, reactive areas; ✐, cellobiose; •, glucose; CBH, cellobiohydrolase; GGH, glucanase glucohydrolase.)

It has been suggested that from a theoretical point of view, 1100 filter paper units (FPU)/L per hour (C_1 activity) are essential for cellulase production to be commercially successful (Bon and Ferrara 2007). Therefore, efforts are needed to economize cellulase production by media optimization and use of supplements/additives. In the literature, many agricultural substrates such as straw, bran, bagasse, etc. have been reported (Deumas et al. 1995; Pei-Jun et al. 2004). In addition, research is being carried out on isolation of potential cellulase-producing microorganisms from diverse habitats (Ray et al. 2007). The best productivities and titers for enzyme production by fungal systems have generally been found for fed-batch operations in which the treated biomass is slowly added to the enzyme production fermentor during the growth of the fungus and the production of cellulase (Watson et al. 1984; Allen and Roche 1989). However, experiments with continuous enzyme production have suffered from lower cellulase productivities (Allen and Roche 1989; Wyman 1994).

In general, cellulase production is not associated with growth and is strongly influenced by the nature and concentration of the cellulosic substrate, the pH of the medium, and nutrients (Acerbal et al. 1986; Kanotra and Mathur 1995; Juhasz et al. 2004). It has been found that high activities are attainable on a crystalline substrate although the growth rates would be slower than with amorphous cellulose (Mes-Hartee et al. 1988; Srinivasan and Seetalaxman 1988; Kollar 1994). In the latter case, the rapid breakdown of the substrate may lead to the formation of free sugars, which may exhibit catabolite repression, leading to poor utilization of the substrate. One of the approaches has been to slow down the rate of growth on a readily metabolizable substrate like Solka Floc by the addition of metal ions such as cobalt and surfactants such as Tween-80 to obtain ready release of the enzyme into the broth (Srinivasan and Seetalaxman 1988; Gashel 1992; Kocher et al. 2008). Cellulosic substrate at a concentration of 1% has been reported to be the best for cellulase production (Wayman and Chen 1992; Steiner et al. 1994; Kanotra and Mathur 1995; Singh et al. 1995; Reczey et al. 1996). Szengyel et al. (1997) reported that a 2% carbon source (Willow) gave the highest cellulase activity, and we found 1% soluble cellulose to be optimal for cellulose production by *T. reesei* Rut C-30 NRRL 11460 (Sharma 2000). In literature, various cellulosic substrates have been evaluated, e.g., crystalline cellulose (Aiello et al. 1996; Geimba et al. 1999), amorphous cellulose (Busto et al. 1996), cotton, wheat straw, barley husk, wheat bran (Gashel 1992; Maheswari et al. 1993; Steiner et al 1994; Awafo et al. 2000), sugarcane bagasse (De-Paula et al. 1999), soyhull (Jha et al. 1995), sunflower (Sharma et al. 1996), corn straw (Yong et al. 1998), corn cobs (Xia and Cen 1999), etc. Yong et al. (1998) observed that batch cultivation of *T. reesei* on 3.75% maize straw gave a cellulase production of 1.76 U/mL (Filter paper (FP) activity) and high cellobiase activity (0.38 U/mL) in 3 days. A pH of 5–6 is optimal for growth of *T. reesei*, but as the pH goes down, the cellulase production increases (Chahal and Wang 1978). However, control of pH is critical because, if left uncontrolled, enzymes will be inactivated by severely acidic conditions (Kosaric et al. 1980; Duff et al. 1987). Abdel Fattah et al. (1995) reported that controlling the pH at 5 in the culture medium during fermentation increased the levels of cellulases, and we observed the same pH for *T. harzianum* (Kocher et al. 2008). Juhasz et al. (2004) reported that nonbuffered media produce high filter paper but lower β-glucosidase activities than buffered media. In addition to pH, temperature also plays an important role in cellulase production. Maximal cellulase production by *T. reesei* has been observed between 25 and 30°C (Maryam et al. 2007) because higher temperatures caused reduced excretion of cellulases. Duff et al. (1987) reported that the optimal temperature for cellulase production by the mixed culture of *T. reesei* Rut-C 30 and *Aspergillus phoenicis* in a 10-L fermenter was 27°C. Janas and Targonski (1995) studied the effect of temperature in the range of 26–38°C on the production of cellulases by four mutant strains of *T. reesei*. We observed that 28°C was the optimum for cellulose production by *T. reesei* in a 15-L fermenter under submerged conditions (Table 30.1; Sharma 2000).

The supply of nutrients in the form of nitrogen as ammonium ions seems to be the most critical factor in regulating the pH of the medium for optimal cellulase yield (Martin et al. 1986). Nitrogen in the form of potassium nitrate (KNO_3) has been reported to be better than ammonium chloride (NH_4Cl) or urea in facilitating the production of cellulase (Gashel 1992). Ammonium sulfate and urea have also been recommended as nitrogen sources for cellulase production (Gutierrez Correa

TABLE 30.1

Batch Production of Cellulase in a 15-L Fermenter by *T. reesei* Rut C-30

Time (days)	Cellulase Activities (IU/g)		
	Filter Paper	CMCase	b-Glucosidase
1	16	49	6
2	29	167	18
3	53	250	30
4	92	364	40
5	112	424	49
6	145	440	54

Fermentation conditions: temperature 28°C, pH 4.6, cellulose 1%, inoculum 3% (v/v), aeration 0.5 L/s, and agitation 250 rpm.

et al. 1999; Kocher et al. 2008). In our studies in shake flasks, maximal filter paper, CMCase, and β-glucosidase activities were optimized as 1.05, 4.62, and 0.42 U/mL, respectively, that were used for saccharification (Sharma et al. 2002b).

30.6 SACCHARIFICATION

A major step in the conversion of cellulose to ethanol or other useful chemicals is the "breakdown" of cellulose to glucose. Two methods are currently suggested as economically feasible: acid or enzymatic hydrolysis. Each method has its advantages and disadvantages, but the overriding factors in the long run must be a low energy requirement and low pollution. Generally, acid hydrolysis procedures give rise to a broad range of compounds in the resulting hydrolysate, some of which might negatively influence the subsequent steps in the process. In addition, expensive corrosion-proof equipment and high temperature and acid concentration are needed for hydrolysis, resulting in high capital cost (Kosaric et al. 1980; Olsson and Hahn-Hägerdal 1996). Enzymatic hydrolysis is not only energy sparing but also avoids the use of toxic and corrosive chemicals. Projected selling prices for ethanol produced from cellulose by acid hydrolysis are currently comparable to those for enzyme-based processes. Enzymatic processes are at a much earlier state of technological maturity; however, in the absence of unforeseen breakthroughs for acid-based processes, research is likely to result in enzyme-based processes that are significantly cheaper than acid-based processes (Wyman 1994; Olsson and Hahn-Hägerdal 1996).

30.6.1 ACID SACCHARIFICATION

Acid hydrolysis of cellulosic biomass is now commercially viable. Two acid-catalyzed processes are known: (1) dilute acid hydrolysis at elevated temperatures and pressures and (2) concentrated acid hydrolysis at low temperature and ambient pressure. In the case of concentrated acids, prehydrolysis and hydrolysis are carried out in one step. However, a weak acid hydrolysis is often combined with a weak acid prehydrolysis (Olsson and Hahn-Hägerdal 1996). The major problems associated with the dilute acid hydrolysis of lignocellulosic biomass are the low sugar yield and the poor fermentability of the produced hydrolysate. The latter is due to the presence of various toxic substances liberated from the structure of lignocellulosics during the hydrolysis process, such as decomposition products of carbohydrates, lignin breakdown products, extraneous materials from biomass, and metal ions from equipment corrosion. Among the identified toxins are furfurals, hydroxymethyl furfural, levulinic acid, acetic acid, formic acid, and various phenolic compounds originating from lignin (Chung and Lee 1985; Wyman 1994; Larsson et al. 1999). Strong acids (e.g., concentrated sulfuric acid or halogen acids) hydrolyze cellulose and hemicellulose at moderate temperatures of 40–50°C with little sugar degradation (Chandel et al. 2007). As a result, concentrated acid processes achieve the high yields of ethanol critical to economic success (Goldstein and Easter 1992; Chandel et al. 2007). However, one of the major problems is the recovery of the expensive acids (Wyman 1994).

The acid saccharification of untreated sunflower stalks in our laboratories under optimized conditions of 5% sulfuric acid at 1 bar for 30 min yielded 30.23 g/100 g of reducing sugars with 33.8% saccharification (Sharma 2000). A two-step sulfuric acid (H_2SO_4) hydrolysis of sunflower seed husks was used by Eklund et al. (1976). The Pentosan fraction was removed using low-temperature and mild acid conditions in the first step and the cellulose fraction was subsequently hydrolyzed under severe reaction conditions. The Pentosan fraction was quantitatively hydrolyzed, but for the cellulose hydrolysis, yield was only 79% of the theoretical yield. Bonillaa et al. (1990) used hydrochloric acid (HCl) between 0.5 and 0.6% at a temperature range of 110–140°C for acid hydrolysis of sunflower stalks. Elsewhere, acid hydrolysis of sunflower stalks has been optimized at 15% solids, 2.5% acid, and 30 min (Tosun 1997). Okur and Saracoglu (2006) degraded sunflower seed hulls using 0.7 M H_2SO_4 at 90°C and achieved 37 g/L of reducing sugars. Iranmahboob et al. (2002) performed concentrated acid hydrolysis (26% H_2SO_4) of mixed wood chips and achieved almost 80% sugar recovery. Agriwastes such as vegetable wastes, pawpaw, etc. have been saccharified by acid or enzymatic hydrolysis yielding 7.8 g/L and 7.6–13.6 g/100 g fermentable sugars (Akin-Osanaiye et al. 2005; Campo et al. 2006).

30.6.2 Enzymatic Saccharification

The pretreated lignocellulosics are vulnerable to enzymatic saccharification. Interest in the enzymatic hydrolysis of lignocellulosic wastes to get ethanol has increased because this involves milder conditions than acidic hydrolysis. Extensive studies on the physiological aspects, kinetics, and economics of enzymatic saccharification have been reviewed (Wyman 1994; Bothast and Saha 1997; Lynd et al. 2002). It has been suggested that for a cellulase preparation to be commercially successful, it should produce 1100 FPU/L per hour, but so far only half of it has been obtained practically (Bon and Ferrara 2007). Hence, many recent studies have been with ready-made enzyme concentrates from different companies such as Dyadic, Genencor, and Novozymes for enzymatic saccharification (Ruiz et al. 2008; Vaithanomsat et al. 2009). On the other hand, research on isolation of new cellulases and hemicellulases from bacterial and fungal sources has continued, and efforts are being made to improve cellulase titers (Aro et al. 2005).

The rate and the extent of saccharification is affected by many factors, such as source of enzyme, nature of the substrate, methods of pretreatment, enzyme and substrate concentration, product inhibition, and enzyme stability, the optimization of which is important for process economics. Tewari et al. (1988) investigated the enzymatic hydrolysis of various agricultural lignocellulosic residues and observed that an increase in enzyme concentration (1–4 IU/mL) and reaction duration (12–72 h) improved saccharification, but an increase in the substrate concentration (>5.0% w/v) had an inhibitory effect. Soto et al. (1994) studied the enzymatic saccharification of alkali-treated sunflower hulls using a commercial cellulase (Celluclast) supplemented with cellobiase (Novozyme) at a cellobiase/cellulase activity ratio of 0.25 and obtained maximal saccharification with an enzyme concentration of 50 FPU/g, with higher concentrations giving a negligible further increase. Overall glucose yield represented conversion of 60% of the cellulose or 39% of the polysaccharides. We optimized 25 FPU/g at 50°C, pH 4.8, and an incubation period of 72 h from *T. reesei* to obtain a saccharification of 49.73% of pretreated sunflower stalks (Sharma et al. 2002b). Ruiz et al. (2006) used a commercial cellulase and obtained 72% saccharification, corresponding to a glucose concentration of 43.7 g/L from the pretreated sunflower stalks. Vaithanomsat et al. (2009) also used a commercial cellulase (Celluclast 1.5 L) and optimized saccharification conditions as 50°C temperature, pH 4.8, and a sunflower stalk pretreated pulp/buffer ratio of 1:6 to obtain 11.97% glucose.

Won Park and Kajiuchi (1995) modified cellulases with alpha-allyl-omega-methoxypolyoxy-alkyline (POA) and maleic acid anhydride (MAA). In comparison with native cellulase, modified cellulase was more stable toward temperature, pH, and organic solvents and gave greater conversion of substrate. Cellulase modification also facilitated strong adsorption of cellulase onto substrate. Kaar and Holtzapple (1998) demonstrated the role of Tween-20 in improving the enzymatic saccharification of corn stover. Addition of Tween-20 improved the conversion of cellulose, xylose, and total polysaccharides by 42, 40, and 42%, respectively. They concluded that Tween-20 acts as an enzyme stabilizer, lignocellulose disrupter, and enzyme effector. A similar effect was also studied by us for saccharification of pretreated rice straw (Kocher et al. 2008). Hang and Woodams (1999) evaluated the efficacy of three commercial fungal enzyme preparations (namely, Celluclast 1.5 L, Rapidase Pomaliq, and Clarex ML) as an enzyme source for the production of soluble sugars from corn husks whereby Rapidase Pomaliq, derived from *Aspergillus niger* and *T. reesei*, produced the highest yield of soluble sugars from corn husks. Nikolov et al. (2000) reported 80% saccharification of pretreated cellulose fibers from the paper industry by the cellulase complex of *T. reesei*.

30.6.3 SIMULTANEOUS SACCHARIFICATION AND FERMENTATION

Cellulose hydrolysis to glucose is a significant component of the total production cost of ethanol from wood (Nguyen and Saddler 1991). An overall economic process must include achieving a high glucose yield (>85% theoretical) at high substrate loading (>10% w/v) over short residence times (<4 days). It has been shown that simultaneous saccharification of cellulose to glucose and fermentation of glucose to ethanol [simultaneous saccharification and Fermentation (SSF)] improves the kinetics and economics of biomass conversion. This SSF reduces the accumulation of hydrolysis products that are inhibitory to cellulase and β-glucosidase, reduces contamination risk because of ethanol, and reduces capital equipment requirements (Sun and Cheng 2002; Karimi et al. 2006; Olofsson et al. 2008). An important drawback of yeast-based SSF is that the reaction has to operate at a compromised temperature of approximately 30°C instead of an optimal enzyme temperature in the range 45–50°C (Taherzadeh and Karimi 2007). Inhibition of cellulose by the produced ethanol is a problem of SSF; Wyman (1996) reported a 5% reduction in cellulose activity by 30 g/L ethanol.

There are no reports on SSF of sunflower wastes except for one by Ruiz et al. (2006), who carried out SSF of steam-pretreated (220°C, 5 min) sunflower stalks and obtained 21 g/L of ethanol while using 10% (w/v) substrate concentration. Elsewhere, through selection of improved cellulase and yeasts better suited to the SSF process, 90–95% yields with 4–5% ethanol concentrations could be achieved in only 3–7 days for various feedstocks (Spindler et al. 1991). Doran et al. (1994) performed simultaneous saccharification and fermentation of pretreated sugarcane bagasse using recombinant ethanol producing *Klebsiella oxytoca* strain P2 and Genencor Spezyme CE. Srinivas et al. (1995) used single-stage bioconversion of cellulosic material to ethanol using intergeneric fusants of *T. reesei* QM9414 and *Saccharomyces cerevisiae* NCIM 3288. Under optimal conditions, the fusant produced 0.17 g ethanol in 30 h. Barron et al. (1995) used a thermotolerant strain *Kluveromyces marxianus* in simultaneous saccharification and ethanol formation from cellulose in which an ethanol yield of 10 g/L at 45°C was obtained, representing 39% of the theoretical yield. Moritz and Duff (1996) described a simultaneous saccharification and extractive fermentation (SSER) process for ethanol production from cellulosic substrates. In batch SSEF reactors with 2–5% aqueous phase, 50% conversion of 25% Solka Floc was achieved in 48 h using 2 FPU cellulase/g. Torget et al. (1996) reported that pretreated poplar saw dust can be converted to ethanol at a yield 91% of theoretical, with an ethanol concentration of up to 4.0% (w/v) in 55 h via a SSF process. Szakacs and Tengerdy (1997) selected *Gliocladium* sp. TUB F-498, a wild strain of a lignocellulolytic fungus, as a potential in situ enzyme source for the bioprocessing of pretreated poplar wood to ethanol in an SSF process.

Krishna et al. (1998) optimized ethanol production by SSF of pretreated sugarcane leaves using a cellulolytic enzyme complex from *T. reesei* QM 9414 and *S. cerevisiae* NRRL-Y-132. The optimal temperature and substrate concentration chosen for SSF were 40°C and 10%, respectively. Meunier Goddik et al. (1999) performed SSF of dilute-acid-pretreated and untreated poplar wood for 5 days using *T. reesei* cellulases and *S. cerevisiae* fermentation. Cellulose conversion varied from 8% for untreated poplar to 78% for the 180°C pretreated poplar.

30.7 ETHANOL FERMENTATION

Efforts directed at ethanol production from biomass at industrial levels have failed because of economic constraints. The main problems encountered in the efficient conversion of the lignocellulosic hydrolysates to ethanol are twofold. Firstly, after pretreatment, the hydrolysate contains not only fermentable sugars but also a broad range of compounds having inhibitory effects on the microorganisms used for fermentation (Endo et al. 2008). The composition of these compounds depends on the type of lignocellulosic material used and the chemistry and nature of the pretreatment process. Secondly, the hemicellulose hydrolysates contain not only hexoses but also pentoses. Hexoses can easily be fermented by *S. cerevisiae* with well-known processes, but this is not so for the pentoses. The latter which may constitute upto 30% as in corn stover, if fermented to ethanol can improve the overall economy of ethanol production from lignocellolosics (Sakai et al. 2007). Ethanol production from lignocellulosic materials has been reviewed (Sun and Cheng 2002; Hahn-Hägerdal et al. 2007; Sakai et al. 2007).

S. cerevisiae is the most widely used yeast for hexose fermentation to ethanol that provides high yields and productivities in addition to remarkable ethanol tolerance, but pentoses have to be converted to xylulose before fermentation to ethanol (Deng and Ho 1990; Sakai et al. 2007). The pentose fraction in lignocellulosic hydrolysates consists mainly of D-xylose, which can be converted to ethanol by a number yeasts like *Pachysolan tannophilus*, *Pichia stipitis*, *Candida shehatae*, etc. The first step in xylose degradation is conversion to xylulose through xylitol, involving a xylose reductase that reduces xylose to xylitol and a xylitol dehydrogenase to convert xylitol to xylulose (Bothast and Saha 1997). However, these yeasts have a relatively low ethanol yield and inhibitor tolerance. *P. stipitis* is able to ferment glucose, xylose, mannose, galactose, and cellobiose and has the ability to produce cell mass from L-arabinose, but not ethanol (Agbogbo and Coward-Kelly 2008). Moreover, the genome sequence of *P. stipitis* has recently been published. The sequence showed numerous genes encoding xylanase, endo-1,4-β-glucanase, exo-1,3-β-glucosidase, β-mannosidase, and α-glucosidase (Jeffries et al. 2007). The presence of these genes in *P. stipitis* suggests the presence of useful traits for the SSF of cellulose and hemicellulose (Berson et al. 2005). Even *S. cerevisiae* strains have been genetically engineered to use xylose (Hahn-Hägerdal et al. 2007; Chu and Lee 2007).

30.7.1 Ethanol Production

The production of ethanol has been studied in a number of pretreated and saccharified lignocellulosics. In our laboratories, we standardized fermentation conditions for ethanol production using sunflower-stalk- and hull-saccharified worts as substrates, in which a fermentation period of 24 h, temperature of 30°C, pH of 5, and inoculum size of 3% (v/v) were found to be optimal. These conditions when scaled-up in 1- and 15-L fermenters revealed ethanol yields of 0.439 and 0.437 g/g from sunflower-stalk-saccharified worts, respectively, and 0.449 and 0.446 g/g, from sunflower hulls, respectively (Sharma et al. 2002a, 2004; Table 30.2). Okur and Saracoglu (2006) reported 0.41 g/g ethanol from 35 g of reducing sugars obtained from acid hydrolysis of sunflower hulls. Vaithanomsat et al. (2009) reported a maximal ethanol yield of 0.028 g/100 g sunflower stalks. Jargalsaikhan and Saracoglu (2009) optimized sunflower-hull hemicellulosic hydrolysate using *P. stipitis* and reported an ethanol yield of 0.32 g/g.

TABLE 30.2
Ethanol Production from Enzymatic Hydrolysate of Sunflower Stalks and Hulls in 1- and 15-L Fermenters by *S. cerevisae* var. *ellipsoideus*

| Fermentation Time (h) | Sunflower Stalks | | | | Sunflower Hulls | | | |
| | Ethanol Yield (g/g) | | Fermentation Efficiency (%) | | Ethanol Yield (g/g) | | Fermentation Efficiency (%) | |
	1 L	15 L	1 L	15 L	1 L	15 L	1 L	15 L
6	0.89	0.090	17.45	17.64	0.113	0.111	22.16	21.76
12	0.264	0.266	51.76	52.16	0.272	0.274	53.33	53.73
18	0.439	0.437	86.08	85.68	0.449	0.446	88.04	87.45
24	0.425	0.426	83.33	83.53	0.436	0.437	85.49	85.68

Fermentation conditions: sugars 40 g/L, temperature 30°C, pH 5.0, aeration 1 L/min for first 10 h, and agitation 150 rpm.

Elsewhere, reports on ethanol production from other lignocellulosics have been reported. Singh et al. (1984) reported ethanol production from the acid hydrolysate of bagasse using *S. cerevisiae* with an efficiency of 82.6% (low because of the presence of 0.3 g/dm^3 of furfural) and a yield factor Yp/s of 0.422. Tewari et al. (1985) carried out the fermentation of acid and enzymatic hydrolysate of saw dust by *S. cerevisiae* var. *ellipsoideus*. The enzymatically hydrolyzed samples supported better ethanol production on the basis of reducing sugars than the acid-treated samples. Similar observations have been observed in our laboratories with sunflower stalks and hulls in which ethanol yields reduced from 0.444 and 0.454 to 0.439 and 0.419 g/g, respectively (Sharma 2000). Dhillon et al. (1988) fermented the rice straw hydrolysate containing 7.68% reducing sugars by *S. cerevisiae* with ethanol production of 2.89%. Laplace et al. (1991) investigated the combined fermentation of glucose and xylose to ethanol by separate or co-culture processes using *P. stipitis, C. shehatae, S. cerevisiae,* and *Z. mobilis.*

Roberto et al. (1994) studied the influence of aeration and pH on xylose fermentation to ethanol by *P. stipitis.* The best ethanol yields (0.35 g/g) were obtained in flasks agitated at 100 and 150 rpm for respective VF/VM (volume of flasks/volume of the medium) ratios of 5.0 and 2.5. Furlan et al. (1994) studied the effect of oxygen on ethanol and xylitol production by xylose fermenting yeasts. *P. stipitis* and *Candida parapsilosis* were the most effective in the production of ethanol and xylitol, respectively. The optimal oxygen transfer coefficients were 4.8 and 16.3 per hour in the two cases. The highest ethanol productivity was obtained under microaerobic conditions (Kruse and Schugerl 1996). Kastner et al. (1996) investigated the effect of pH on cell viability and ethanol yields in D-xylose fermentation by *C. shehatae*. Ethanol yield increased from 0.25 to 0.37 g/g as the pH was increased from 2.5 to 6. A pH of 6 also extended the cell viability during anaerobic conditions.

Larsson et al. (1999) studied the effect of acetic acid, formic acid, levulinic acid, furfural, and 5-hydroxymethyfurfural (5-HMF) (compounds generated during dilute acid hydrolysis of softwood) on fermentability of dilute acid hydrolysate of softwood by *S. cerevisiae*. Ethanol yield and volumetric productivity decreased with increasing concentrations of acetic acid, formic acid, and levulinic acid. Furfural and 5-HMF decreased the volumetric productivity but did not influence the final yield of ethanol. Saha and Cotta (2006) reported ethanol production with a yield of 0.23 g/g from enzymatically saccharified wheat straw.

30.7.2 Ethanol Recovery

Recovery of ethanol from fermentation broth is at least a three-step process: (1) distillation of dilute aqueous alcohol to its azeotrope (95.57% ethanol by weight); (2) distillation using a third component—either an organic solvent or a strong salt solution to break up the azeotrope and remove the remaining water; and (3) distillation to separate water from the third component so that it can be recycled. Most of the energy consumption occurs in distilling above 85% ethanol. The ethanol produced after fermentation in our studies was distilled from the fermentation broth and dehydrated with calcium chloride ($CaCl_2$) (4–20%), and a maximal dehydration of aqueous ethanol (95.5%, v/v) was observed with 18% $CaCl_2$.

30.8 CONCLUSIONS

Sunflower is a known crop for edible oil and biodiesel production. It has been suggested that using sunflower-based biodiesel in combination with bioethanol can counter the problems of the high sulfur and flash point of potato-based biodiesel (Ghobadian et al. 2008). Moreover, sunflower processing also generates significant quantities of lignocellulosics in the form of stalks and seed hulls that can be used for bioethanol production. Our experiments on utilization of these lignocellulosic substrates have also revealed their potential for bioethanol production. In this endeavor, the pretreatment of sunflower stalks and seed hulls was standardized by using them as 40-mesh ground forms that were treated with 0.5% NaOH followed by steaming in an autoclave at 15 psi for 1.5 h, resulting in an extraction yield of 66 (stalks) and 57.6% (seed hulls), respectively. The equivalent hydrolysis of cellulose, hemicellulose, and lignin was 12.6 and 11.5%, 66.3 and 62.9%, and 44.9 and 71.9%, for sunflower stalks and hulls, respectively. However, more efforts are needed to improve upon the pretreatment strategies. In this study, we compared the acidic saccharification of untreated and enzymatic saccharification of pretreated sunflower stalks and hulls, which revealed a significantly better saccharification with our indigenous cellulase (25 FPU and incubation conditions of 50°C, pH 4.8, and 72 h) from *T. reesei* RUT C-30. The ethanol fermentation of saccharified substrates by cofermentation with *S. cerevisiae* and *P. tannophilus* revealed an ethanol yield of 0.4 and 0.45 g/g from sunflower stalks and hulls, respectively, in a 15-L fermenter that may be scaled-up.

REFERENCES

Abdel-Fattah AF, Ismail AMS, Abdel Naby MA (1995) Utilization of water hyacinth cellulose for production of cellulase by *Trichoderma viridae* 100. *Cytobios* 82: 151–157

Acerbal C, Ferrer A, Ledesma A (1996) Enhanced cellulase production from *T. reesei* QM 9414 on physically treated wheat straw. *Appl Microbiol Biotechnol* 24:218–223

Agbogbo FK, Coward-Kelly G (2008) Cellulosic ethanol production using the naturally occurring xylose-fermenting yeasts, *Pichia stipitis. Biotechnol Lett* 30:1515–1524

Aiello C, Ferrer A, Ledesma A (1996) Effect of alkaline treatment at various temperatures on cellulase and biomass production using submerged sugarcane bagasse fermentation with *Trichoderma reesei* QM 9414. *Bioresour Technol* 57:13–18

Akin-Osanaiye BC, Nzelibe HC, Agbaji AS (2005) Production of ethanol from *Carica papaya* (pawpaw) agro waste: Effect of saccharification and different treatments on ethanol yield. *Afr J Biotechnol* 4: 657–659

Allen AN, Roche CD (1989) Effects of strain and fermentation conditions on production of cellulase by *Trichoderma reesei. Biotechnol Bioeng* 33:650–656

Almeida JR, Modig T, Peterson A et al. (2007) Increased tolerance and conversion of inhibitors of lignocellulosic hydrolysates by *Saccharomyces cerevisae. J Chem Technol Biotechnol* 82:340–349

Alvo AN, Belkacemi K (1997) Enzymatic saccharification of milled timothy (*Phleum pratense* L.) and alfalfa (*Medicago sativa* L.). *Bioresour Technol* 61:185–198

Anonymous (1999) *Statistical Abstracts of Punjab, 1999*, Economic and Statistical Organization, Punjab, India, pp178–179

Aro N, Pakula T, Penttillar M (2005) Transcriptional regulation of plant cell wall degradation by filamentous fungi. *FEMS Microbiol Rev* 29:719–739

Asia Biomass Office (2008) *Diversification in Biofuel Produced from Sunflower Seeds*, available at www.asiabiomass.jp (accessed June 30, 2009)

Awafo VA, Chahal DS, Simpson BK (2000) Evaluation of combination treatments of NaOH and steam explosion for the production of cellulase systems by two *T. reesei* mutants under solid state conditions. *Bioresour Technol* 73:235–245

Baldrian T, Gabriel J (2003) Lignocellulose degradation by *Pleurotus osttreatus* in the presence of cadmium. *FEMS Microbiol Lett* 220:235–240

Barron N, Marchant R, Mc Hale L, Mc Hale AP (1995) Studies on the use of a thermotolerant strain of *Kluyveromyces marxianus* in simultaneous saccharification and ethanol formation from lignocellulose. *Appl Microbiol Technol* 43:518–520

Beguin P (1990) Molecular biology of cellulase degradation. *Annu Rev Microbiol* 44:219–248

Berlin A, Balakshin M, Gilkes M, Kadla J, Maximenko V, Kubo S, Saddler J (2006) Inhibition of cellulase, xylanase and beta- glucosidase activities by soft wood lignin preparations. *J Biotechnol* 125:198–209

Berson RE, Young JS, Kamer SN, Hanley TR (2005) Detoxification of actual pretreated corn stover hydrolysate using activated carbon powder. *Appl Biochem Biotechnol* 121:923–934

Biofuel Industries (2002) *Sunflower Biodiesel*, available at www.sunflowerbiodiesel.com (accessed July 5, 2009)

Bon PS, Ferrara MA (2007) Bioethanol production via enzymatic hydrolysis of cellulosic biomass. Food and Agricultural Organization Seminar, Rome, Italy, October 12, 2007, available at http://www.fao.org/biotech/seminaroct2007.htm (accessed May 25, 2009)

Bonilla JL, Chica A, Ferrer JL, Jimenez L, Martin A (1990) Sunflower stalks as a possible fuel source. *Fuel* 69:792–794

Bothast, RJ, Saha BC (1997) Ethanol production from agricultural biomass substrates. *Adv Appl Microbiol* 44:261–286

Busto MD, Ortega N, Perez Mateos M (1996) Location, kinetics and stability of cellulases induced in *Trichoderma reesei* cultures. *Bioresour Technol* 57:187–192

Campo D, Alegría I, Zazpe M, Echeverría M, Echeverría I (2006) Diluted acid hydrolysis pretreatment of agri-food wastes for bioethanol production. *Indust Crop Prod* 24:214–221

Chahal DS (1982) *Enzymatic Hydrolysis of Cellulose, State of the Art*. Report submitted to Division of Energy, National Research Council of Canada, Ottawa, Ontario, Canada, pp 1–73

Chahal DS, Wang DIC (1978) *Chaetomium cellulolyticum* growth behaviour and protein production. *Mycologia* 70:160–170

Chandel AK, Chan ES, Rudravaram R, Lakshmi Narasu, Venkateswar Rao L, Ravindra P (2007) Economics and environmental impact of bioethanol production technologies: An appraisal. *Biotechnol Mol Biol Rev* 2:14–32

Chaudhary P, Kumar NN, Deobagkar DN (1997) The glucanases of *Cellulomonas*. *Biotechnol Adv* 15:315–331

Chu BCH, Lee H (2007) Genetic improvement of *Saccharomyces cerevisiae* for xylose fermentation. *Biotechnol Adv* 25:425–441

Chung IS, Lee YY (1985) Ethanol fermentation of crude acid hydrolysate of cellulose using high level yeast inocula. *Biotechnol Bioeng* 27:308–315

De-Paula EH., Ramos LP, de Oliveira M (1999) The potential of *Humicola grisea* var. *thermoidea* for bioconversion of sugarcane bagasse. *Bioresour Technol* 68:35–41

de Sousa MV, Monteiro SN, d'Almeda JRM (2004) Evaluation of pretreatment, size and molding pressure on flexural mechanical behaviour of chopped bagasse-polyester composites. *Polymer Testing* 23:253–258

Dekker RFH, Wallis AFA (1983) Autohydrolysis explosion as pretreatment for the enzymatic saccharification of sunflower seed hulls. *Biotechnol Lett* 5:311–316

Deng XX, Ho NWY (1990) Xylulokinase activity in various yeasts including *S. cerevisae* containing the cloned xylulokinase gene. *Appl Biochem Biotechnol* 24–25:193–199

Deumas R, Tengerdy RP, Qutierrez-Correa M (1995) Cellulase production by mixed fungi in solid substrate fermentation of bagasse. *World J Microbiol Biotechnol* 11:333–337

Dhillon GS, Grewal SK, Singh A, Kalra MS (1988) Production of sugars from rice straw. *Acta Microbiol Polonica* 37:167–173

Din N, Damude HG, Gilkes NR, Miller RC, Warren RA, Kilburn DG (1994) C_1-C_x revisited: Intramolecular synergism in a cellulase. *Proc Natl Acad Sci USA* 91:11383–11387

Doran JB, Aldrich HC, Ingram LO (1994) Saccharification and fermentation of sugarcane bagasse by *Klebsiella oxytoca* P2 containing chromosomally integrated genes encoding the *Zymomanas mobilis* ethanol pathway. *Biotechnol Bioeng* 44:240–247

Duff SJB, Cooper DG, Fuller OM (1987) Effect of media composition and growth conditions on production of cellulase and beta glucosidase by a mixed fungal fermentation. *Enz Microb Technol* 9:47–52

Eklund E, Hatakka A, Mustranta A, Nybergh P (1976) Acid hydrolysis of sunflower seed husks for production of single cell protein. *Eur J Appl Microbiol* 2:143–152

Endo A, Nakamura T, Ando A, Tokuyasu K, Shima J (2008) Genome-wide screening of the genes required for tolerance to vanillin, which is a potential inhibitor of bioethanol fermentation, in *Saccharomyces cerevisiae*. *Biotechnol Biofuels* 1:3

Falih AM (1998) Impact of heavy metals on cellulolytic activity of some soil fungi. *Kuw J Sci Eng* 25:397–407

Furlan SA, Bouilloud P, Castro HF, de Castro HF (1994) Influence of oxygen on ethanol and xylitol production by xylose fermenting yeasts. *Process Biochem* 29:657–662

Gashel BA (1992) Cellulase production and activity by *Trichoderma* sp. A-001. *J Appl Microbiol* 73:79–82

Geimba MP, Riffel A, Agostini V, Brandelli A (1999). Characterization of cellulose hydrolyzing enzymes from the fungus *Bipolaris sorokiniana*. *J Sci Food Agric* 79:1849–1854

Ghobadian B, Rahimi H, Tavakkoli Hashjin T, Khatamifar M (2008) Production of bioethanol and sunflower methyl ester and investigation of fuel blend properties. *J Agric Sci Technol* 10:225–232

Goldstein IS, Easter JM (1992) An improved process for converting cellulose to ethanol. *TAPPI J* 75:135–143

Gray KA, Zhao L, Emptage M (2006) Bioethanol. *Curr Opin Chem Biol* 10:141–146

Gutierrez CM, Portral L, Moreno P, Tengerdy RP (1999) Mixed culture solid substrate fermentation of *Trichoderma reesei* with *Aspergillus niger* on sugarcane bagasse. *Bioresour Technol* 68:173–178

Hahn-Hägerdal B, Karhumaa K, Fonseca C, Spencer-Martins I, Gorwa-Grauslund MF (2007) Towards industrial pentose-fermenting yeast strains. *Appl Microbiol Biotechnol* 74:937–953

Hang YD, Woodams EE (1999) Enzymatic production of soluble sugars from corn husks. *Lebensmttel Wissenehaft Technol* 32:208–210

Hendriks ATMW, Zeeman G (2009) Pretreatments to enhance the digestibility of lignocellulosic biomass. *Bioresour Technol* 100:10–18

Henrissat B, Dsiguez H, Viet C, Schulen M (1985) Synergism of cellulases from *Trichoderma reesei* in the degradation of cellulose. *Biotechnology* 3:722–726

Henrissat B, Teri TT, Warren RA (1998) A scheme for designating enzymes that hydrolyse the polysaccharides in the cell walls of plants. *FEBS Lett* 425:352–354

Hu J (2008) Sunflower as a potential biomass crop. Presented at Huazhong Agricultural University, International Symposium on BioEnergy and Biotechnology, Wuhan, China, March 16–20, 2008, p 14

Immanuel G, Dhanusha R, Prema P, Palavesam A (2006) Effect of different growth parameters on endoglucanase enzyme activity by bacteria isolated from coir retting effluents of estuarine environment. *Int J Environ Sci Technol* 3:25–34

Inoue H, Yano S, Endo T, Sakaki T, Sawayama S (2008) Combining hot-compressed water and ball milling retreatments to improve the efficiency of the enzymatic hydrolysis of eucalyptus. *Biotechnol Biofuels* 1:1–9

Iranmahboob J, Nadim F, Monemi S (2002) Optimizing acid hydrolysis: A critical step for production of ethanol from mixed wood chips. *Biomass Bioenergy* 22:401–404

Itoh H, Wada M, Honda Y, Kuwahara M, Watanabe T (2003) Biorganosolve pretreatments for simultaneous saccharification and fermentation of beech wood by ethanolysis and white rot fungi. *J Biotechnol* 103:273–280

Janas P, Targonski Z (1995) Effect of temperature on the production of cellulases, xylanases and lytic enzymes by selected *Trichoderma reesei* mutants. *Acta Mycol* 30:255–264

Jargalsaikhan O, Saracoglu N (2009) Application of experimental design method for ethanol production by fermentation of sunflower seed hull hydrolysate using *Pichia stipitis* NRRL 124. *Chem Eng Commun* 196:93–103

Jeffries TW, Grigoriev IV, Grimwood J, Laplaza JM, Aerts A, Salamov A, Schmutz J, Lindquist E, Dehal P, Shapiro H, et al. (2007) Genome sequence of the lignocellulose-bioconverting and xylose-fermenting yeast *Pichia stipitis. Nat Biotechnol* 25:319–326

Jha K, Khare SK, Gandhi AP (1995) Solid state fermentation of soyhull for the production of cellulase. *Bioresour Technol* 54:321–322

Jimenez L, Bonilla JL (1993) Acid hydrolysis of sunflower residue biomass. *Process Biochem* 28:243–247

Juhasz T, Szengyel Z, Szijarto N, Reczey K (2004) Effect of pH on cellulase production of *Trichoderma reesei* RUT C30. *Biotechnology* 113:201–212

Kaar WE, Hotzapple MT (1998) Benefits from Tween during enzymatic hydrolysis of corn stover. *Biotechnol Bioeng* 59:419–427

Kanotra S, Mathur RS (1995) Isolation and partial characterization of a mutant of *T. reesei* and its application in solid state fermentation of paddy straw alone or in combination with *Pleurotus sajor-caju. J Environ Sci Health A* 30:1339–1360

Karimi K, Kheradmandinia S, Taherzadeh MJ (2006). Conversion of rice straw to sugars by dilute-acid hydrolysis. *Biomass Bioenergy* 30:247–253

Kastner JR, Roberts RS, Jones WJ (1996) Effect of pH on cell viability and product yields in D-xylose fermentation by *Candia shehatae. Appl Microbiol Biotechnol* 45:224–228

Keller FA, Hamilton JE, Nguyen QA (2003) Microbial pretreatment of biomass: Potential for reducing severity of thermochemical biomass pretreatment. *Appl Biochem Biotechnol* Spring:102–108

Kheshgi HS, Prince RC, Marland G (2000) The potential of biomass fuels in the context of global climatic changes; focus on transportation fuels. *Annu Rev Energy Emiss* 25:199–244

Kim HT, Holtzapple MT (2006) Lime pretreatment and enzymatic hydrolysis of corn stover. *Bioresour Technol* 96:1994–2006

Kocher GS, Kalra KL, Banta G (2008) Optimization of cellulase production by submerged fermentation of rice straw by *Trichoderma harzianum* Rut–C 8230. *Int J Microbiol* 5, available at http://www.ispub.com/ journal/the_internet_journal_of_microbiology/volume_5_number_2_18/article/optimization_of_ cellulase_production_by_submerged_fermentation_of_rice_straw_by_trichoderma_harzianum_ rut_c_8230.html (accessed on May 25, 2009)

Kollar A (1994) Characterization of specific induction, activity and isozyme polymorphism of extracellular cellulases from *Venturia inaequalis* detected in vitro and on the host plant. *Mol Plant-Micr Interact* 7: 603–611

Kosaric N, Ng DCM, Russell I, Stewart GS (1980) Ethanol production by fermentation: An alternative liquid fuel. *Adv Appl Microbiol* 26:147–227

Krishna SH, Prasanthi K, Chowadry GV, Ayyanna C (1998) Simultaneous saccharification and fermentation of pretreated sugarcane leaves to ethanol. *Process Biochem* 33:825–830.

Kruse B, Schugerl K (1996) Investigation of ethanol formation by *Pachysolan tannophilus* from xylose and glucose/xylose co-substrates. *Process Biochem* 31:389–408

Kuhls K, Lieckfeldt E, Samuels GJ, Meyer M, Kubicek CP, Borner T (1997) Revision of *Trichoderma* sect *Longibrachiatus* including related teleomorphs based on analysis of ribosomal DNA internal transcribed spacer sequences. *Mycologia* 89:442–460

Laplace JM, Delgenes JP, Moletta R, Navarro JM (1991) Combined alcoholic fermentation of D-xylose and D-glucose by 4 selected microbial strains, process considerations in relation to ethanol tolerance. *Biotechnol Lett* 13:445–450

Larsson S, Palmqvist E, Stenberg K, Tenborg C, Stenberg K, Zacchi G, Nilvebrant N-O (1999) The generation of fermentation inhibitors during dilute acid hydrolysis of softwood. *Enz Microb Technol* 24:151–159

Lin Y, Tanaka S (2006) Ethanol fermentation from biomass resources: Current state and prospects. *Appl Microbiol Biotechnol* 69:627–642

Lynd LR, Cushman JH, Nichols RJ, Wyman CE (1991) Fuel ethanol from cellulosic biomass. *Science* 251:1318–1323

Lynd LR, van Zyl WH, McBride JE, Laser M (2005) Consolidated bioprocessing of cellulosic biomass: An update. *Curr Opin Biotechnol* 16:577–583

Lynd LR, Weimer PJ, van Zyl WH, Pretorius JS (2002) Microbial cellulose utilization: Fundamentals and biotechnology. *Microbiol Mol Rev* 66:506–577

Maheshwari DK, Jahan H, Paul J, Verma A (1993) Wheat straw, a potential substrate for cellulase production using *Trichoderma reesei*. *World J Microbiol Biotechnol* 9:120–121

Mandels M (1982) Cellulases. *Ann Rep Ferment Process* 5:35–77

Market Research Analyst (2008) *World's Ethanol Forecast 2008–12. Jan 26, 2008*, available at www.market researchanalyst.com (accessed June 29, 2009)

Martin RS, Blanch HW, Wilke CR, Sciamanna (1986) Production of cellulase enzymes and hydrolysis of steam exploded wood. *Biotechnol Bioeng* 28:564–569

Maryam L, Esfahani ZH, Barzegara M (2007) Evaluation of culture conditions for cellulase production by two *Trichoderma reesei* mutants under solid state fermentation conditions. *Bioresour Technol* 98:3634–3637

Mes-Hartree M, Hogan CM, Saddler JN (1988) Influence of growth substrate on production of cellulase enzymes by *Trichoderma harzianum* E58. *Biotechnol Bioeng* 31:725–729

Meunier Goddik L, Bothwell M, Sangseethong K, Piyachomkwan K, Chung Y-C, Thammasouk K, Tanjo D, Penner MH (1999) Physicochemical properties of pretreated poplar feedstocks during simultaneous saccharification and fermentation. *Enz Microb Technol* 24:667–674

Moiser N, Wyman C, Dale B, Elander R, Lee YY, Holtzapple M, Ladisch M (2005) Features of promising technologies for pretreatment of lignocellulosic biomass. *Bioresour Technol* 96:673–686

Mojovic L, Nikolic S, Rakin M, Vukasinovic M (2006) Production of bioethanol from corn meal hydrolysates. *Fuel* 85:1750–1755

Moritz JW, Duff SJB (1996) Simultaneous saccharification and extractive fermentation of cellulosic substrates. *Biotechnol Bioeng* 49:504–511

Mtui GYS (2009) Recent advances in pretreatment of lignocellulosic wastes and production of value added products. *Afr J Biotechnol* 8:1398–1415

Nguyen QA, Saddler JN (1991) An integrated model for technical and economic evaluation of an enzymatic biomass conversion process. *Bioresour Technol* 35:275–282

Nikolov T, Bakalova N, Petrova S, Benadova R, Spasov S, Kolev D (2000) An effective method for bioconversion of delignified waste cellulose fibers from the paper industry with a cellulase complex. *Bioresour Technol* 71:1–4

Okur M, Saracoglu NE (2006) Ethanol production from sunflower seed hull hydrolysate by *Pichia stipitis* under uncontrolled pH conditions in a bioreactor. *Turk J Eng Env Sci* 30:317–322

Olofsson K, Bertilsson M, Lidén G (2008) A short review on SSF, an interesting process option for ethanol production from lignocellulosic feedstocks. *Biotechnol Biofuels* 1:1–14

Olsson L, Hahn-Hagerdal B (1996) Fermentation of lignocellulosic hydrolysates for ethanol production. *Enz Microb Technol* 18:312–331

Pei-Jun LI, De-bing J, Qi-xing Z, Chun-gui Z (2004) Optimization of solid fermentation of cellulase from *Trichoderma koningii*. *J Environ Sci* 6:816–820

Qi BC, Aldrich C, Lorenzen L, Wolfaardt GW (2005) Acidogenic fermentation of lignocellulosic substrate with activated sludge. *Chem Eng Commun* 192:1221–1242

Ray AK, Bairagi A, Ghosh KS, Sen SK (2007) Optimization of fermentation conditions for cellulase production by *Bacillus subtilis* CY5 and *Bacillus circulans* TP3 isolated from fish gut. *Acat Icht Pist* 37:47–53

Reczey K, Szengyel Z, Eklund R, Zacchi G (1996) Cellulase production by *Trichoderma reesei*. *Bioresour Technol* 57:25–30

Roberto IC, Mancilha IM, Felipe MGA, Silva SS, Sato S (1994) Influence of aeration and pH on xylose fermentation to ethanol by *Pichia stipitis*. *Arquivos Biol Technol* 37:55–63

Ruiz E, Cara C, Ballesteros M, Ballesteros M, Manzanares P, Ballesteros I, Castro E (2006) Ethanol production from pretreated olive tree wood and sunflower stalks by an SSF process. *Appl Biochem Biotechnol. Part A: Enz Eng Biotechnol* 130:631–643

Ryu DDY, Mandels M (1980) Cellulases: Biosynthesis and applications. *Enz Microb Technol* 2:91–102

Saddler JN, Hogan C, Louis-Seize G, Yu EKC (1986) Factors affecting cellulase production and efficiency of cellulose hydrolysis. In: Young MM, Hasnain S, Lampley J (eds), *Biotechnology and Renewable Energy*. Elsevier Science Publishing, New York, pp 83–92

Saha BC, Cotta MA (2006) Ethanol production from alkaline peroxide pretreated enzymatically saccharified wheat strains. *Biotechnol Progr* 22:449–453

Sakai S, Tsuchida Y, Okino S, Ichihashi O, Kawaguchi H, Watanabe T, Inui M, Yukawa H (2007) Effect of lignocellulose-derived inhibitors on growth of and ethanol production by growth arrested *Corynebacterium glutamicum* R. *Appl Environ Microbiol* 73:2349–2353

Sharma SK (2000) Saccharification and bioethanol production from sunflower stalks and hulls. PhD Dissertation, Punjab Agricultural University, Ludhiana, India

Sharma N, Aggarwal HO, Bhatt AK, Bhalla TC (1996) Saccharification of physico-chemically pretreated lignocellulosics by partially purified cellulase of *Trichoderma viridae*. *Natl Acad Sci Lett* 19:141–144

Sharma SK, Kalra KL, Grewal HS (2002a) Fermentation of enzymatically saccharified sunflower stalks for ethanol production and its scale up. *Bioresour Technol* 85:31–33

Sharma SK, Kalra KL, Grewal HS (2002b) Enzymatic saccharification of pretreated sunflower stalks. *Biomass Bioenergy* 23:237–243

Sharma SK, Kalra KL, Kocher GS (2004) Fermentation of enzymatic hydrolysate of sunflower hulls for ethanol production and its scale up. *Biomass Bioenergy* 27:399–402

Sheehan J, Aden A, Paustian K, Killian K, Brenner J, Walsh M, Nelson R (2003) Energy and environmental aspects of using corn stover for fuel ethanol. *J Indust Ecol* 7:117–146

Singh A, Das K, Sharma DK (1984) Production of xylose, furfural, fermentable sugars and ethanol from agricultural residues. *J Chem Technol Biotechnol* 34A:51–61

Singh S, Sandhu DK, Brar JK, Kaur A (1995) Optimization and natural variability of cellulase production in *Aspergillus terreus*. *J Ecobiol* 7:249–255

Soto ML, Dominguez H, Nunez MJ, Lema JM (1994) Enzymatic saccharification of alkali treated sun flower hulls. *Bioresour Technol* 49:53–59

Spindler DD, Wyman CE, Grohmann K (1991) The simultaneous saccharification and fermentation of pretreated woody crops to ethanol. *Appl Biochem Biotechnol* 28–29:773–781

Srinivas D, Rao KJ, Theodore K, Panda T (1995) Direct conversion of cellulosic material to ethanol by the intergeneric fusant *Trichoderma reesei* QM 9414/ *Saccharomyces cerevisiae* NCIM 3288. *Enz Microb Technol* 17:418–423

Srinivasan MC, Seetalaxman R (1988) Microbial cellulases: A status report on enzyme production and technology aspects. *Indust J Microbiol* 28:266–275

Steiner J, Socha C, Eyzaguirre J (1994) Culture conditions for enhanced cellulase production by a native strain of *Penicillium purporogenum*. *World J Microbiol Biotechnol* 10:280–284

Sun Ye, Cheng J (2002) Hydrolysis of lignocellulosic materials for ethanol production: A review. *Bioresour Technol* 83:1–11

Szakacs G, Tengerdy RP (1997) Lignocellulosic enzyme production on pretreated poplar wood by filamentous fungi. *World J Microbiol Biotechnol* 13:487–490

Szengyel Z, Zaachi G, Récezy K (1997) Cellulase production based on hemicellulose hydrolysate from steam pretreated willow. *Appl Biochem Biotechnol* Spring:351–362

Tabka MG, Herpoël-Gimbert I, Monod F, Asther M, Sigoillot JC (2006) Enzymatic saccharification of wheat straw for bioethanol production by a combined cellulase xylanase and feruloyl esterase treatment. *Enz Microbiol Technol* 39:897–902

Taherzadeh MJ, Karimi K (2007) Enzyme-based ethanol: A review. *Bioresour* 2:707–738

Teeri TT (1997) Crystalline cellulose degradation: New insight into the function of cellobiohydrolases. *Trends Biotechnol* 15:160–167

Tewari HK, Marwaha SS, Kennedy JF, Singh L (1988) Evaluation of acids and cellulase enzyme for the effective hydrolysis of agricultural lignocellulosic residues. *J Chem Tech Biotechnol* 41:261–275

Tewari HK, Marwaha SS, Singh L (1985) Ethanol production from acid and enzymatic hydrolysate of saw dust. *Ann Biol* 1:261–275

Tewari HK, Singh L, Marwaha SS, Kennedy JF (1987) Role of pretreatments on enzymatic hydrolysis of agricultural residues for reducing sugar production. *J Chem Tech Biotechnol* 38:153–155

Torget R, Hatzis C, Hayward TK, Hsu TA, Philippidis GP (1996) Optimization of reverse flow, two temperature, dilute acid pretreatment to enhance biomass conversion to ethanol. *Appl Biochem Biotechnol* 57–58:85–101

Tosun A (1997) Dilute acid hydrolysis of sunflower residue (cellulosic wastes) prior to enzymatic hydrolysis. *Fresenius Environ Bull* 6:296–301

Vaithanomsat P, Chuichulcherm S, Apiwatanpiwat W (2009) Bioethanol production from enzymatically saccharified sunflower stalks using steam explosion as pretreatment. *Proc World Acad Sci Eng Technol* 37:140–143

von Sivers M, Zacchi G, Olsson L, Hahn-Hagerdal B (1994) Cost analysis of ethanol production from willow using recombinant *E. coli*. *Biotechnol Progr* 10:555–560

Ward OP (2002) Bioethanol technology, development and perspective. *Adv Appl Microbiol* 51:53–80

Watson TG, Nelligan I, Lessing L (1984) Cellulase production by *Trichoderma reesei* Rut C-30 in fed batch culture. *Biotechnol Lett* 6:667–672

Wayman M (1980) Alcohol from cellulosics: The autohydrolysis extraction process. In: *Proceedings of the 4th International Alcohols Fuels Technol Symposium*, Sao Paolo, Brazil, pp 79–84

Wayman M, Chen S (1992) Cellulase production by *Trichoderma reesei* using whole wheat flour as a carbon source. *Enz Microb Technol* 4:825–831

Weil J, Westgate P, Kohlmamm K, Ladisch MR (1994) Cellulose pretreatments of lignocellulosic substrates. *Enz Microb Technol* 16:1002–1004

Won Park J, Kajiuchi T (1995) Development of effective modified cellulase for cellulose hydrolysis process. *Biotechnol Bioeng* 45:366–373

Wyman CE (1994) Ethanol from lignocellulosic biomass: Technology, economics and opportunities. *Bioresour Technol* 50:3–16

Wyman CE (1996) Ethanol production from lignocellulosic biomass: Overview. In: Wyman E (ed), *Handbook on Bioethanol: Production and Utilization*. Taylor and Francis, Bristol, PA, pp 1–18

Wyman CE, Hinman ND (1990) Ethanol: Fundamentals of production from renewable feedstocks and use as a transportation fuel. *Appl Biochem Biotechnol* 24/25:735–753

Xia LM, Cen PL (1999) Cellulase production by solid state fermentation on lignocellulosic waste from the xylose industry. *Process Biochem* 34:909–912

Yong Q, Cheng YP, Yu SY (1998) Effects of cellobiose and starch hydrolysate on cellulase production on steam exploded corn straw. *J Nanjing For Univ* 22:53–56

Yu ZS, Zhang HX (2004) Ethanol fermentation of acid hydrolysed cellulosic pyrolysate with *Saccharomyces cerevisiae*. *Bioresour Technol* 93:199–204

Yuldashev BT, Rakimov MM, Rabinovich ML (1992) Effect of pretreatment on the enzymatic hydrolysis of cellulose containing wastes of cotton processing. *Prikladnaia Biokhim Mikrobiol* 28:443–448

Zhang Y, Eddy C, Deanda K, Finkelstein M, Picataggio S (1995) Metabolic engineering of pentose metabolism pathway in ethanologenic *Zymomonas mobilis*. *Science* 267:240–243

31 Sweetpotato

Karine Zinkeng Nyiawung, Desmond Mortley, Marceline Egnin, Conrad Bonsi, and Barrett Vaughan
Tuskegee University

CONTENTS

31.1 INTRODUCTION

Sweetpotato, *Ipomoea batatas* (L.) Lam., is a root crop that belongs to the family Convolvulaceae, the morning glory family. It is a creeping dicotyledonous plant and an important crop, widely grown in tropical, subtropical, and warm temperate regions. It ranks as the world's seventh most important crop, with an estimated annual production of approximately 122 million metric tons (Collado et al. 1999; FAO 2006). On the basis of analysis of morphological characters of sweetpotato and the wild *Ipomea* species, the center of origin of *I. batatas* was thought to be somewhere between the Yucatan Peninsula of Mexico and the mouth of the Orinoco River in Venezuela (Austin 1987). Recent evidence revealed by the use of molecular markers suggests that Central America is the primary center of diversity and most likely the center of origin, considering the richness of the wild varieties of sweetpotato (Haung and Sun 2000; CIP 2006). Sweetpotato is especially valued because it is highly adaptable and tolerates high temperatures, low fertility soil, and drought (Yencho et al. 2002).

The growth cycle of sweetpotato is generally from 3.5 to 7 months and takes place in three phases:

1. From planting to formation of storage roots (40–60 days),
2. From storage root formation to the time of maximal leaf development (60–120 days), and
3. From maximum leaf development to the total development of storage roots (45–90 days).

Normally the cycle is completed within 100–150 days for the short-term varieties, at which time the plant can be harvested (Ramirez 1992). A temperature range between 15 and 33°C is required during the vegetative cycle, the optimal temperature being 20–25°C. The highest yields are obtained when temperatures are high during the day (25–30°C) and low at night (15–20°C); low night temperatures favor the development of storage roots, and high day temperatures favor vegetative development. Storage root development only occurs within a temperature range of 20–30°C; 25°C is optimal, and growth stops below 10°C (Ramirez 1992).

Sweetpotato is a short-day plant that needs light for maximum development, although the growth of storage roots appears not to be influenced by photoperiod alone. It is most likely that temperature and fluctuations in temperature, together with short days, favor the development of storage roots and limit the growth of foliage (Yong 1961). Moisture has a vital influence on its growth and production. The water content of the leaf is 86%, stem 88.4%, and storage root 70.6% depending on the cultivar. Moist soils and 80% relative humidity favor the development of the vegetative part of the plant. It is important to have moist soils at planting to achieve good establishment. The soil must be kept moist during the 60-to 120-day growth period, although the humidity must be low at harvesting to prevent rotting of the storage root (Carballo 1979). Sweetpotato can be cultivated in a wide range of soils. Soil that is friable with a depth of more than 25 cm and has a good superficial and internal drainage system is ideal. It also prefers lightly acid or neutral soils, with an optimal pH between 5.5 and 6.5. Excessively acid or alkaline soils often encourage bacterial infections and negatively influence yields (Cairo 1980). Sweetpotato is multiplied by sexual and asexual means, although the former is only of interest to geneticists and plant breeders. Asexual reproduction using tubers and stems is the quickest and most economic and therefore the form of production most commonly used (Ramirez 1992).

The chemical composition (Table 31.1) of sweetpotato varies greatly according to genetic and environmental factors (FAO 1998). Storage roots may have a smooth or irregular surface, and the skin and the flesh may range from almost pure white through cream, yellow, orange, and pink to a very deep purple (Onwueme 1978).

31.2 SWEETPOTATO AS FEEDSTOCK FOR ETHANOL

Most agricultural biomass containing starch can be used as a potential substrate for the production of gaseous or liquid fuels (Nigam and Singh 1995). Two characteristics of a crop suitable for energy use include high yield per unit area and high dry matter content (Bonelli et al. 2007). A high yield reduces land requirements and lowers the cost of producing energy from biomass (Demirbas 2007). Sweetpotato satisfies these requirements, making it very suitable for bioethanol production. Ziska et al. (2009) evaluated cassava, sweetpotato, and field corn as potential carbohydrate sources for bioethanol production in Alabama and Maryland. They found that sweetpotato yielded 1.5 times as much carbohydrate as corn in Alabama and 2.3 times as much in Maryland. Sweetpotato also yielded the highest concentration of root carbohydrate: 9.4 mg/ha in Alabama and 12.7 mg/ha in Maryland. Corn is the primary feedstock for bioethanol production in the United States, by a production process that consumes 75–90% as much energy as is available from the fuel (Marris

---

TABLE 31.1
General Composition of Sweetpotato Storage Roots

Constituent	Percent or (mg/100 g)
Moisture	50–81%
Protein	1.0–2.4%
Fat	1.8–6.4%
Starch	8.0–29%
Nonstarch carbohydrates	0.5–7.5%
Reducing sugar	0.5–7.5%
Ash	0.9–1.4%
Carotene (average)	4 mg/100 g
Thiamine	0.10 mg/100 g
Ascorbic acid	25 g/100 g
Riboflavin	0.06 mg/100 g

Source: Adapted from FAO, *Storage and Processing of Roots and Tubers in the Tropics*, Food and Agriculture Organization of the United Nations, 1998, available at http://www.fao.org/DOCREP/X5415E/x5415e01.htm (accessed June 2010).

2006). Because corn is a staple food not only in the United States but worldwide, this has led to controversy and concerns related to the large amounts of arable land required for the crops and the impact on grain supply. Studies have shown that sweetpotato storage root, carbohydrate, and ethanol yields are approximately 3 times that of corn (Rodgers et al. 2007; ARS 2008). To improve the potential of sweetpotato as a feedstock for bioethanol production, varieties with high yields and dry matter content termed industrial sweetpotatoes (ISPs) have been developed at North Carolina State University (NCSU), which have been shown to have ethanol yields as high as 67.8 g/L of ethanol for flour-based fermentation and 34.9 g/L for fresh sweetpotato sugar fermentation (Santa-Maria et al. 2006; Duvernay 2008). A limitation of using sweetpotato for bioethanol production is its economical conversion to simple sugars before fermentation. Researchers are evaluating the potential of transforming sweetpotato biotechnologically with a hyperthermophilic amylolitic enzyme under control of a root-specific promoter. The genetically engineered sweetpotatoes so produced when heated are expected to bioprocess their own starch into simple sugars (Santa-Maria et al. 2006).

Sweetpotato is a nutritious and generous food source for humans and animals as well as a raw material for manufacturing starch, sugar, and alcohol (Kozai et al. 1997; Saiful Islam et al. 2002; Zhang et al. 2002). Only orange flesh sweetpotato varieties are recommended for marketing in the United States and they include Beauregard, Garnet, Hernandez, and Jewel, with Beauregard being the no. 1 industry standard (USDA 2005). Orange flesh sweetpotatoes are especially preferred for food because they contain significant amounts of dietary fiber, minerals, vitamins (especially vitamins C, B_6, and folate), and antioxidants, such as phenolic acids, anthocyanins, tocopherol, and β-carotene (Woolfe 1992). Teow et al. (2007) reported significant variations in respect to β-carotene content among sweetpotato genotypes, with orange flesh varieties having higher β-carotene content than white flesh. That white flesh sweetpotatoes lack these nutritional qualities makes them more suitable for ethanol production because they reduce the food fuel competition particularly in the United States (Eggen 2006; Service 2007). Moreover, orange flesh clones generally have lower dry matter and hence extractable starch content (Barbet et al. 1999).

31.2.1 STARCH

Starch offers a high yielding ethanol resource (Nigam and Singh 1995). Starch is the main storage carbohydrate of plants. It is deposited as semicrystalline granules in storage tissues (grains, tubers, and roots) and it also occurs to a lesser extent in most vegetative tissues of plants (Copeland et al. 2009). It is a polysaccharide of α-D-glucose and consists of the polymers amylose and amylopectin. Amylose is essentially linear, and the glucose units are bound together with α-(1,4)-linkages and very few α-(1,6)-bonds. Amylopectin is larger than amylose, is highly branched, and has α-(1,6)-bonds in the branching points in addition to the α-(1,4)-linkages in the linear chains (Stevneb et al. 2006). Amylose has a molecular weight range of approximately 10^5–10^6, corresponding to a degree of polymerization (DP) of 1000–10,000 glucose units. The molecular weight of amylopectin is approximately 10^8 and it has a DP that may exceed 1 million. Most starches contain 60–90% amylopectin, although high-amylose starches with as little as 30% amylopectin and waxy starches with 100% amylopectin are known. Approximately 5% of amylopectin glucoses are in α-(1,6) linkages, giving it a highly branched, tree-like structure and a complex molecular structure architecture that can vary substantially between different starches with regard to placement and length of branches (Copeland et al. 2009). The moisture content of native starch granules is usually approximately 10%. Amylose and amylopectin make up 98–99% of the dry weight of native granules, with the remainder including small amounts of lipids, minerals, and phosphorus in the form of phosphates esterified to glucose hydroxyls. The proportion of amylose and amylopectin depends on the source of the starch (Allen et al. 1997).

Starch granules range in size (from 1 to 100 μm diameter) and shape (polygonal, spherical, and lenticular) and can vary greatly with regard to content, structure, and organization of the amylose and amylopectin molecules; the branching architecture of amylopectin; and the degree of crystallinity (Lindeboom et al. 2004) (Figure 31.1).

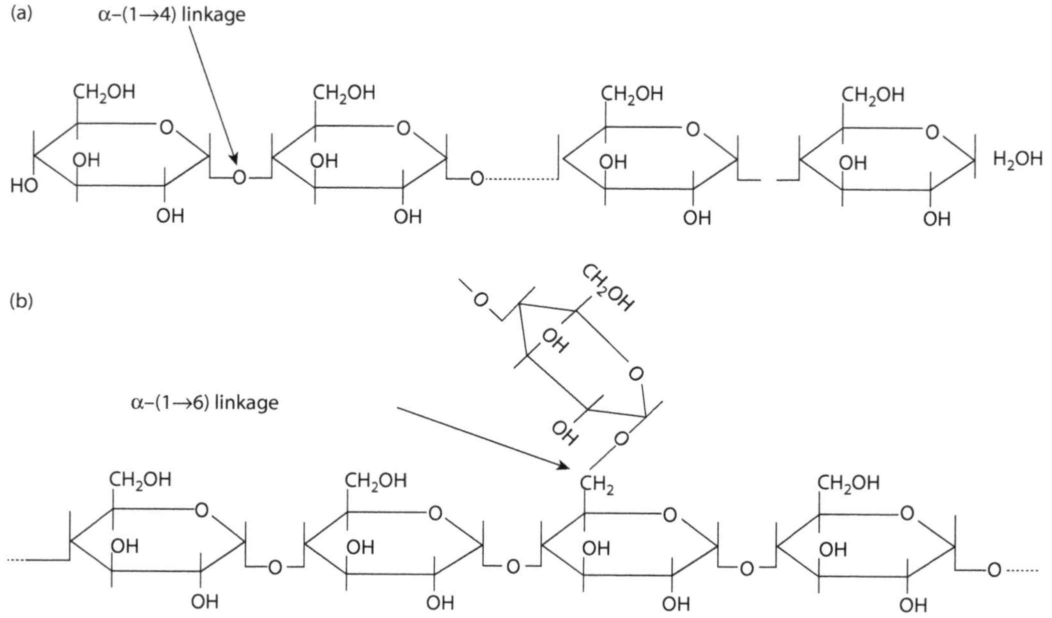

FIGURE 31.1 Structure of amylose (a) and amylopectin (b) showing α-(1,4) and β-(1,6) linkages. (Adapted from Nowjee N.C., *Melt Processing and Foaming of Starch*, University of Cambridge, 2004. available at http://www.cheng.cam.ac.uk/research/ groups/polymer/RMP/nitin/Foaming.html (accessed June 2010).)

The extent of crystallinity of native starch granules ranges from approximately 15% for high amylose starches to approximately 45–50% for waxy starches (Copeland et al. 2009). Amylopectin chains with more than ten glucose units are organized into double helices, which are arranged into either A- or B- crystalline forms that may be identified by characteristic x-ray diffraction spectral patterns. The double helical structures within the A- and B-type crystalline forms are essentially the same (Gidley 1987; Imberty et al. 1991). However, the packing of the helices in the A-type crystalline structure is more compact than in the B-type structure, which has a more compact structure with a hydrated core. The A-type crystal pattern has amylopectin molecules with shorter chains (Jane 2006). Tuber starches and amylose-rich starches yield the B-type pattern, although both types could occur (Buléon et al. 1998). Starch varies greatly in form and functionality between and within botanical species and even from the same cultivar grown under different conditions, hence providing starches of diverse properties (Copeland et al. 2009).

The susceptibility of starch to hydrolysis by α-amylase has been shown to vary with botanical origin (Srichuwong et al. 2005). Starch digestibility is not necessarily related to total starch content because the amylose/amylopectin ratio and processing affect the extent of starch digestion (Reynolds et al. 1997). Although a minor component by weight, lipids can have a significant role in determining the properties of starch. The lipid of native starches is highly correlated with amylose content: the higher the amylose content, the higher the lipid present. Inclusion complexes with lipids form mainly with the amylose component of the starch; hence, the amylose/amylopectin ratio is an important factor that produces variability in the ability of natural starches to bind lipids (Copeland et al. 2009). Because of its high degree of branching, the lipid binding capability of amylopectin is considered to be much weaker than amylose. There is little direct evidence to suggest that amylopectin forms true inclusion complexes with lipids, although some studies have led to proposals that favor the interaction of some lipids and surfactants with outer amylopectin branches (Hahn and Hood 1987; Eliasson 1994; Villwock et al. 1999). Complexes between amylose and lipids can significantly modify the properties and functionality of starch. For example, these complexes reduce the solubility of starch in water, alter the rheological properties of pastes, decrease swelling capacity, increase gelatinization temperature, reduce gel rigidity, retard retrogradation, and reduce the susceptibility to enzyme hydrolysis (Holm et al. 1983; Karkalas and Raphaelides 1986; Billaderis and Seneviratne 1990; Nierle and El Bayâ 1990; Guraya et al. 1997; Crowe et al. 2000; Kaur and Singh 2000; Oczan and Jackson 2002; Tufvesson et al. 2003a, b). According to Stevneb et al. (2006), barley cultivars with low levels of amylose have a higher degree of starch hydrolysis than normal and high amylose content for all time intervals ($P < 0.05$). Their study also showed that the degree of starch hydrolysis is higher for B-granules than for A-granules despite an increased amylose content and higher amount of amylose-lipid complexes in B-granules; hence, granule size has a profound effect on the degree of starch hydrolysis, consistent with the low level of crystallinity and a higher surface area in B-granules compared with A-granules. Studies have also showed that the hydrolysis rate of maize starches is proportional to the surface area of the granules, which may be closely related to the adsorption of enzyme onto the granule surface (Li et al. 2004). External chains of amylopectin that construct the crystalline structures of starch granules likely affect the rate of hydrolysis. Srichuwong et al. (2005) showed that starch hydrolysis by α-amylase is positively correlated with the proportions of unit chains with a DP of 8–12 and negatively correlated with a DP of 16–26 ($n = 15$, $P < 0.01$). Longer chains would make long helices and strengthen hydrogen bonds between chains, spanning the entire crystalline region. On the other hand, the existence of shorter chains form short or weak double helices, producing inferior crystalline structures (Jane et al. 1999).

Crystalline polymorphic form, fraction of crystalline structures in starch, molecular associations between starch components, amylose content, granule size, granule shape, and surface pores are all factors that have been mentioned in enzyme digestibility. Of these factors, granular structure is considered to be the most important in defining the rate and extent of enzymatic hydrolysis (Zhang and Oates 1999). The natural variability in amylose and amylopectin molecules is due to the complexity of starch biosynthesis (Copeland et al. 2009).

31.2.2 STARCH BIOSYNTHESIS

The biosynthetic pathway of starch synthesis involves several types of enzymes. An overview of starch biosynthesis as shown in Figure 31.2.

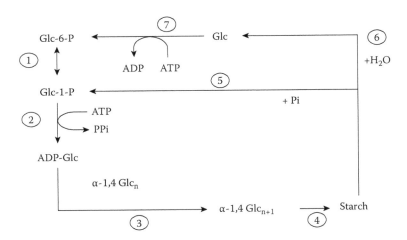

FIGURE 31.2 General scheme for starch biosynthesis. (1) phosphoglucomutase; (2) ADP-glucose pyrophosphorylase; (3) GBSS and soluble SS; (4) branching enzymes; (5) starch phosphorylase; (6) amylases, debranching enzymes, and maltases; (7) hexokinase.

Adenosine-5'-diphosphate glucose (ADPGlc) pyrophosphorylase (AGPase, EC 2.7.7.27) is the enzyme responsible for the production of ADPGlc, the soluble precursor and substrate for starch synthases in all plant tissues capable of starch biosynthesis. The AGPase reaction is the first committed step in the biosynthesis of starch (Tetlow 2006). AGPase catalyzes the rate-limiting step in starch biosynthesis and exerts a high degree of control on the flux of carbon into this pathway. In 1966, Ghosh and Preiss established that plant AGPases were regulated through allosteric activation by 3-phosphoglyceric acid (3-PGA) and inhibition by orthophosphate. A review by Preiss and Levi (1980) reported in vivo concentrations for the effectors and substrates of AGPase that fully support the hypothesis that the 3-PGA/inorganic phosphate ratio regulates the rate of starch biosynthesis in plant leaves.

Starch synthases (SS, EC 2.4.1.21) catalyze the transfer the transfer of the glucosyl moiety of the soluble precursor ADP glucose to the reducing end of a pre-existing α-(1,4)-linked glucan primer to synthesize the insoluble glucan polymers amylose and amylopectin (Tetlow 2006). Plants posses multiple isoforms of SS, categorized according to conserved sequence relationships. The isoforms within each of the major classes of SS genes are highly conserved in higher plants (Ball and Morell 2003). The major classes of SS genes are broadly split into two groups: the first group is primarily involved in amylose synthesis, and the second group primarily confined to amylopectic biosynthesis (Tetlow 2006). The first group of SS genes contains granule-bound starch synthase (GBSS) and includes GBSSI and GBSSII, which function in the elongation of amylose (De Fekete et al. 1960; Nelson and Rines 1962; Fujita and Taira 1998; Nakamura et al. 1998; Vrinten and Nakamura 2000). The second group of SS genes, designated SSI, SSII, SSIII, and SSIV, are exclusively involved in amylopectin biosynthesis (Ball and Morell 2003).

According to Tetlow (2006), starch-branching enzymes (SBEs, EC 2.4.1.18) generate α-(1,6)-linkages by cleaving internal α-(1,4) bonds and transferring the released reducing ends to C_6 hydroxyls to form the branched structure of the amylopectin molecule. SBE activity is also a

$$
\begin{array}{ccc}
\text{Pi + G-G-G-G-G - ....} & \xrightarrow{\text{degradative}} & \text{G-P} \quad \text{G-P + G-G-G-G- ....} \\
\text{starch chain} & \text{synthetic} & \\
& & \textit{α-Glc-1-P} \quad \text{degraded starch chain} \\
& \text{Phosphorylase} & \text{(so-called primer)}
\end{array}
$$

function of multiple isoforms (Yamanouchi and Nakmura 1992; Gao et al. 1997; Morell et al. 1997; Sun et al. 1998; Regina et al. 2005).

Starch phosphorylase (SP, EC 2.4.1.1) catalyzes the reversible transfer of glucosyl units from glucose 1-phosphate to the nonreducing end of α-1,4-linked glucan chains and may be driven in a synthetic or degradative direction by relative concentrations of the soluble substrates (Tetlow 2006).

Because a degraded starch chain is the product of the degradation reaction of SP, it is thus a required substrate in the reverse synthetic reaction as shown in the reaction above. This was the origin of the primer requirement for starch biosynthesis from the nonreducing end and has pretty much been retained for 65 years (Bocca et al. 1997; Ball et al. 1998; Tomlinson and Denyer 2003). However, Mukerjea and Robyt (2005) showed that three kinds of starch granules are capable of incorporating D-glucose from ADPGlc into their starches in the absence of any added maltodextrin primers to suggest that starch synthesis does not require a primer. They showed that biosynthesis of starch chains occurs de novo with the addition of two glucose units from ADPGlc to the active site of starch synthase in a nonprimer, reducing-end, two-site insertion mechanism.

Murata and Akazawa (1968) showed that starch synthesis in sweetpotato indispensably requires K^+ for the starch synthesizing reaction catalyzed by the granule-bound enzyme of sweetpotato roots—SS. It has long been known that root growth (starch accumulation) of sweetpotato is most effectively stimulated by the K^+ fertilizer compared with some other plants. Later in 1969, they demonstrated that the activity of starch synthetase in sweetpotato roots was also enhanced by ^{14}C-maltooligosaccharides. In addition, they also found that the stimulatory effect of K[+] and maltooligosaccharides were nearly additive if ADPGlc was used as a glucose donor. In a study conducted by Aighewi (1984), the total carbohydrate contents of sweetpotato did not appear to be significantly influenced by drip irrigation and fertilizer nitrogen regardless of the method of application. He found a significant increase in root yield in response to the drip irrigation treatments. Water availability is critical in the first 8 weeks.

REFERENCES

Aighewi IT (1984) Nitrogen use efficiency, carbohydrate and protein contents of sweetpotato in response to drip irrigation/nitrogen fertilizer methods. MS Thesis, Tuskegee University, Tuskegee Institute, AL

Allen MS, Oba M, Choi BR (1997) Nutritionist's perspective on corn hybrids for silage. In: *Silage: Field to Feedbank*, NRAES-99. Northeast Regional Agricultural Engineering Service, Ithaca, NY pp 25–36

ARS (2008) Sweetpotato out-yields corn in ethanol production study, Report, U.S. Department of Agriculture, Agricultural Research Service, available at http://www.ars.usda.gov/is/pr/2008/080820.htm (accessed June 2010)

Austin DF (1987) The taxonomy, evolution and genetic diversity of sweetpotato and related wild species. In: *Exploration, Maintenance and Utilization of Sweetpotato Resources, Proceedings of the 1st Planning Conference*, International Potato Center (CIP), Lima, Peru, pp 27–59

Ball SG, Morell MK (2003) From bacterial glycogen to starch: Understanding the biogenesis if the plant starch granule. *Annu Rev Plant Biol* 54:207–233

Ball SG, Van de Wal HBJM, Visser RGF (1998) Progress in understanding the biosynthesis of amylase. *Trends Plant Sci* 3:1360–1385

Barbet C, Reynoso D, Dufour D, Mestres C, Arredondo J, Scott G (1999) Starch content and properties of 106 sweetpotato clones from the World Germplasm Collection held at CIP, Peru. In: *CIP Program Report 1997-98*, International Potato Center (CIP), Lima, Peru, pp 279–286

Billaderis CG, Seneviratne HD (1990) On the supermolecular structure and metastability of glycerol monostereate-amylose complex. *Carbohydr Polym* 13:185–206

Bocca SN, Rothschild A, Tandecarz JS (1997) Initiation of starch biosynthesis: Purification and characterization of UDP-glucose: Protein transglucosylase from potato tubers. *Plant Physiol Biochnol* 35:205–212

Bonelli PR, Buonomo EL, Cukierman AL (2007) Pyrolysis of sugarcane bagasse and pyrolysis with an Argentinean subbitumous coal. *EnergyEnergy Sour Part A* 29:731–740

Buléon A, Colonna P, Planchot V, Ball S (1998) Mini review. Starch granules: Structure and biosynthesis. *Int J Biol Macromol* 23:85–112

Cairo P (1980) Soil Editorial. Puebloy Educacion, Habana, Cuba

Carballo N (1979) *Effect of Soil Humidity on the Sub-Period of Growing in the Cultivation of Sweetpotato (Ipomea batats L.)*. I Forum Cienific Technic, Santa Clara, Cuba

CIP (2006) *About Sweetpotato*, International Potato Center (CIP), Lima, Peru, available at http://www.cipotato.org/sweetpotato (accessed June 2010)

Collado LS, Mabesa RC, Corke H (1999) Genetic variation in the physical properties of sweetpotato starch. J Agric Food Chem 47:4195–4201

Copeland L, Blazek J, Salman H, Tang MC (2009) Form and functionality of starch. *Food Hydrocolloid* 23:1572–1534

Crowe TC, Seligman SA, Copeland L (2000) Inhibition of enzymic digestion of amylose by free fatty acids in vitro contributes to resistant starch formation. *J Nutr* 10:2006–2008

De Fekete MAR, Leloir LF, Cardini CE (1960) Mechanism of starch biosynthesis. *Nature* 187:918–919

Demirbas A (2007) Combustion of biomass. *Energy Sour Part A* 29:549–561

Duvernay WH (2008) Conversion of Industrial Sweetpotatoes for the Production of Ethanol. MS Thesis, North Carolina State University, Raleigh, NC

Eggen D (2006) Corn farmers smile as ethanol prices rice, but experts on food supplies worry. *New York Times*, January 16

Eliasson AC (1994) Interactions between starch and lipids studied by DSC. *Thermochim Acta* 246:343–356

FAO (1998) *Storage and Processing of Roots and Tubers in the Tropics*, Food and Agriculture Organization of the United Nations, available at http://www.fao.org/DOCREP/X5415E/x5415e01.htm (accessed June 2010)

FAO (2006), Food and Agriculture Organization of the United Nations, available at http://faostat.fao.org/site/339/default.aspx (accessed June 2010)

Fujita N, Taira T (1998) A 56-kDa protein is a novel granule-bound starch synthase existing in the pericarps, aleurone layers, and embryos of immature seed in diploid wheat (*Triticum monococcum* L.). *Planta* 207:125–132

Gao M, Fisher DK, Kim KN, Shannon JC, Guiltinan MJ (1997) Independent genetic control of maize starch-branching enzymes IIa and IIb. *Plant Physiol* 114:69–78

Ghosh HP, Preiss J (1966) Adenosine diphospate glucose pyrophosphorylase: A regulatory enzyme in the biosynthesis of starch in spinach leaf chloroplasts. *J Biol Chem* 241:4491

Gidley MJ (1987) Factors affecting the crystalline type (A-C) of native starches and model compounds: A rationalization of observed effects in terms of polymorphic structures. *Carbohydr Res* 161:301–304

Guraya HS, Kadan RS, Champagne ET (1997) Effect of rice starch-lipid complexes on in vitro digestibility, complexing index, and viscosity. *Cereal Chem* 74:561–565

Hahn DE, Hood LF (1987) Development of an equilibrium dialysis technique for quantifying starch starch-lipid complexes. *Cereal Chem* 64:77–80

Haung JC, Sun M (2000) Genetic diversity and relationship of sweetpotato and its wild relatives in Ipomea series Batatas (Convolvulaceae) as revealed by inter-simple sequence repeat (ISSR) and restriction analysis of chloroplast DNA. *Theor Appl Genet* 100:1050–1060

Holm J, Bjorck I, Ostrouska S (1983) Digestibility of amylose-lipid complexes in vitro and in vivo. *Stärke [Starch]* 35:294–297

Imberty A, Buléon A, Tran V, Perez S (1991) Recent advances in knowledge of starch structure. *Stärke [Starch]* 53:205–213

Jane JL (2006) Current understanding on starch granule structures. *J Appl Glycosci* 53:205–213

Karkalas J, Raphaelides S (1986) Quantitative aspects of amylose-lipid interactions. *Carbohydr Res* 157:215–234

Kaur K, Singh N (2000) Amylose-lipid complex formation during cooking of rice flour. *Food Chem* 71:511–517

Kozai T, Yoshinaga K, Kubota C (1997) Rapid production of sweetpotato cuttings under controlled environment: CO_2 enrichment and cutting preparation affected the growth and production rates. *HortScience* 32(3):481

Li JH, Vasanthan T, Hoover R, Rossnagel BB (2004) Starch from hull-less barley. V. In vitro susceptibility of waxy, normal, and high amylose starches towards hydrolysis by alpha-amylases and amyloglucosidase. *Food Chem* 84:621–632

Lindeboom N, Chang PR, Tyler RT (2004) Analytical, biochemical and physiochemical aspects of starch granule size, with emphasis on small granule starches: A review. *Stärke [Starch]* 56:89–99

Marris E (2006) Drink the best and drive the rest. *Nature* 444:670–672

Morell MK, Blennow A, Kosar-Hashemi B, Samuel MS (1997) Differential expression and properties of starch branching enzyme isoforms in developing wheat endosperm. *Plant Physiol* 113:201–208

Mukerjea R, Robyt JF (2005) Starch biosynthesis: The primer nonreducing-end mechanism versus the nonprimer reducing-end two-site insertion mechanism. *Carbohydr Res* 340:245–255

Murata T, Akazawa T (1968) Enzymic mechanism of starch synthesis in sweetpotato roots I. Requirement of potassium ions for starch synthetase. *Arch Biochem Biophys* 126: 873–879

Nakamura T, Vrinten P, Hayakawa K, Ikeda J (1998) Characterization of a granule-bound starch synthase isoform found in the pericarp of wheat. *Plant Physiol* 118:451–459

Nelson OE, Rines HW (1962) The enzymatic deficiency in the waxy mutant of maize. Biochem *Biophys Res Commun* 9:297–300

Nierle W, El Bayâ AW (1990) Lipids and rheological properties of starch 1. The effects of fatty acids monoglycerides and monoglyceride ethers on pasting temperature and viscosity of wheat starch. *Stärke [Starch]* 42:268–270

Nigam P, Singh D (1995) Enzyme and microbial systems involved in starch processing. *Enz Microb Technol* 17:770–778

Nowjee NC (2004) *Melt Processing and Foaming of Starch*. University of Cambridge, available at http://www.cheng.cam.ac.uk/research/ groups/polymer/RMP/nitin/Foaming.html (accessed June 2010)

Oczan S, Jackson SD (2002) The impact of thermal events on amylose-fatty acid complexes. *Stärke [Starch]* 54:593–602

Onwueme IC (1978) *The Tropical Tuber Crops: Yams, Cassava, Sweet Potato, and Cocoyams*. John Wiley, New York, USA, 234 p

Preiss J, Levi C (1980) Starch biosynthesis and degradation. In: Preiss J, Stumpf PK (eds), *The Biochemistry of Plants: A Comprehensive Treatise. Carbohydrates Structure and Function*, Vol. 3 Academic Press, New York pp 371–423

Ramirez GP (1992) Cultivation, harvesting and storage of sweetpotato products. In: Machin D, Nyvold S (eds), *Roots, Tubers, Plantains and Bananas for Animal Feeding*. Proceedings of the FAO Expert Consultation held in CIAT, Cali, Colombia. Food and Agriculture Organization of the United Nations, Rome, Italy, pp 203–215

Regina A, Kosar-Hashemi B, Li Z, Pedler A, Mukai Y, Yamamoto M, Gale K, Sharp PJ, Morell MK, Rahman S (2005) Starch branching enzyme IIb in wheat is expressed at low levels in the endosperm compared to other cereals and encoded at a non-syntenic locus. *Planta* 222:899–909

Reynolds CK, Sutton JD, Beever DE (1997) Effects of feeding starch to dairy cattle on nutrient availability and production. In: Garnsworthy PC, Wiseman J (eds), *Recent Advances in Animal Nutrition*. Butterworths, London, pp 105–134

Rodgers JA, Bomford MK, Geier BA, Silvernail AF (2007) *Evaluation of Alternative Bioethanol Feedstock Crops*. Kentucky Academy of Sciences Annual Meeting, Louisville, KY

Saiful Islam AFM, Kubota C, Takagaki M, Kozai T (2002) Sweetpotato growth and yield from plug transplants of different volumes, planted intact or without roots. *Crop Sci* 42:822–826

Santa-Maria MC, Yencho CG, Thompson WF, Kelly R, Sosinski B (2006) Genetic engineering of sweetpotato for endogenous starch bio-processing. American Society of Plant Biologists, available http://abstracts. aspb.org/pb2006/public/P46/P46020.html (accessed June 2010)

Service R (2007) Biofuel researchers prepare to reap a new harvest. *Science* 315:1488–1489, 1491

Srichuwong S, Sunarti TC, Mishima T, Isono N, Hisamatsu M (2005) Starches from different botanical sources I: Contribution of amylopectin fine structure to thermal properties and enzyme digestibility. *Carbohyd Polym* 60:529–538

Stevneb A, Sahström S, Svihus B (2006) Starch structure and degree if starch hydrolysis of small and large starch granules from barley varieties with varying amylose content. *Anim Feed Sci Technol* 130:23–28

Sun C, Sathish P, Ahlandsberg S, Jansson C (1998) The two genes encoding starch-branching enzymes IIa and IIb are differentially expressed in barley. *Plant Physiol* 118:37–49

Teow CC, Truong V, McFeeters RF, Thompson RL, Pecota KV, Yencho GC (2007) Antioxidants activities, phenolic and β-carotene contents of sweetpotato genotypes with varying flesh colors. *Food Chem* 103:829–838

Tetlow IJ (2006) Understanding storage and starch biosynthesis in plants: A means to quality improvement. *Can J Bot* 84:1167–1185

Tomlinson K, Denyer K (2003) Starch synthesis in cereal grain. *Adv Bot Res* 40:1–61

Tufvesson F, Wahlgren M, Eliasson AC (2003a) Formation of amylose-lipid complexes and effects on temperature treatment. Part 1. Monoglycerides. *Stärke [Starch]* 55:61–71

Tufvesson F, Wahlgren M, Eliasson AC (2003b) Formation of amylose-lipid complexes and effects on temperature treatment. Part 2. Fatty acids. *Stärke [Starch]* 55:138–149

USDA (2005) *Quality of Food Vegetables*, U.S. Department of Agriculture, available at http://www.fns.usda. gov/tn/Resources/quality_veg.pdf (accessed June 2010)

Villwock VK, Eliasson AC, Silverio J, BeMiller JN (1999) Starch-lipid interactions in common, waxy, ae du and ae su2 maize starches examined by differential scanning calorimetry. *Cereal Chem* 76:292–298

Vrinten PL, Nakamura T (2000) Wheat granule-bound starch synthase I and II are encoded by separate genes that are expressed in different tissues. *Plant Physiol* 122:255–263

Woolfe JA (1992) *Sweetpotato. An Untapped Food Resource*. Cambridge University Press, Cambridge, United Kingdom

Yamanouchi H, Nakamura Y (1992) Organ specificity of isoforms of starch branching enzyme (Q-enzyme) in rice. *Plant Cell Physiol* 33:958–991

Yencho GC, Pecota KV, Schultheis JR, Sosinski BR (2002) Grower-participatory sweet potato breeding efforts in North Carolina. *Acta Hort* 583:69–76

Yong CK (1961) Effects of thermoperiodism on tuber formation in *Ipomoea batatas* under controlled conditions. *Plant Physiol* 36(5):680–684

Zhang T, Oates CG (1999) Relationship between α-amylase degradation and physic-chemical properties of sweetpotato starches. *Food Chem* 65:157–163

Zhang Z, Wheatley CC, Cooke H (2002) Biochemical changes during storage of sweetpotato roots differing in dry matter content. *Post-Harvest Biol Technol* 24:317–325

Ziska LH, Runion GB, Tomecek M, Prior SA, Torbet HA, Sicher A (2009) An evaluation of cassava, sweetpotato and field corn as potential carbohydrate sources for bioethanol production in Alabama and Maryland. *Biomass Bioenerg* 33:1503–1508

32 Organic Farm Waste and Municipal Sludge

Milenko Roš
Institute for Environmental Protection and Sensors

Gregor D. Zupančič
Institute for Environmental Protection and Sensors

CONTENTS

32.1   INTRODUCTION

Like settlements, towns and industries, organic farms also produce polluted wastes and wastewaters. In some countries, about half of all serious water pollution incidents are due to manure runoff from farms. Poultry, cows, and pigs are the farm animals most responsible for the pollution. Livestock production occupies 70% of all land used for agriculture and 30% of the planet's land surface. It is responsible for 18% of the world's greenhouse gas emissions as measured in CO_2 equivalents. It also generates 64% of the ammonia, which contributes to acid rain and acidification of ecosystems (FAO 2006; Action 2008; Watson 2008).

When large numbers of animals are farmed intensively in industrial units in an attempt to maximize financial gain, it is known as "factory farming." During the second half of the twentieth century agricultural practices went through massive changes, particularly mechanization, the use of chemicals, and large-scale intensive farming.

As a result of increasing the density of domestic farm animals, reported farm pollution incidents have remained high and farm waste remains a major problem (Ritter et al. 1984; FAO 2006).

Another serious problem is sludge from municipal and industrial biological wastewater treatment plants that can be transformed into biogas at proper conditions because such sludge contains mainly degradable organic compounds.

Farm waste is a mixture of animal feces and urine, plus milk and chemicals such as antiparasitic drugs or antibiotics (Boxall et al. 2003; Boxall and Long 2005; Reinoso and Becares 2008). A large quantity of animal waste is generated by concentrated animal feeding operations, and disposal of the waste has been a major problem. Factory farms (Wikipedia 2008b) collect the animal waste and mix it with water to form slurry. Slurry is a type of liquid manure that can be used on fields as fertilizer (Oleszkiewicz and Koziarski 1981; Schröder 2005; Westerman and Bicudo 2005; Chong et al. 2008). If the soil or plants are unable to absorb the slurry or if the slurry is spread in too high a concentration, the run-off can get into water systems.

Slurry is generally more polluting than raw sewage. When slurry tanks are accidentally or deliberately breached, large amounts of slurry can spill into rivers, streams, or lakes, including wetlands, causing severe environmental problems. Animal waste is found in soil, surface water, groundwater, and seawater (Ritter et al. 1984).

Slurry disturbs aquatic ecosystems by increasing nitrogen and phosphorus levels, leading to the growth of toxic algae, which poison the fish and decrease oxygen levels, causing fish to suffocate (Smith et al. 1998).

32.2 WASTE AND WASTEWATER POLLUTION

32.2.1 TYPES OF WASTE POLLUTION

Increased use of agrochemicals, farm machinery, and irrigation in recent years has made the pollution problem worse.

Milk spills (Nguyen 2000; Elmquist 2005; Filipy et al. 2006; Havlkova et al. 2008) are another major environmental hazard for aquatic ecosystems. Milk is a highly polluting substance, and when it gets into waterways, it is a threat to fish and other animals living within the waterway. This is because the bacteria feeds on the milk and uses up oxygen that fish and other animals need to survive. The dairy industry sometimes accidentally loses huge volumes of milk from its tankers. Sometimes milk that cannot be sold is deliberately dumped and gets into waterways.

Chemicals used on farms, such as pesticides, fungicides, or fertilizers, are found in waterways (Oturan et al. 2001; Guida et al. 2008; van den Berg and van Lamoen 2008). Common farm chemicals include aluminum phosphide, cresol, organophosphorus pesticides, pyrethroids, methyl bromide, strychnine, and tryquat. These chemicals are sprayed on farmland using tractors and boom sprayers, or aerial sprays from light planes. Droplets are produced that can linger in air and may be carried by wind away from the intended area. This is known as "spray drift." Chemical spray drift cannot always be contained and might still occur despite correct application. These chemical sprays often drift over neighboring properties or waterways and can affect human health, animals, and the environment (Muller et al. 1983; Chung et al. 1985; Fawcett 1991; Bicudo and Svoboda 1995; Schierhout et al. 1997; Brewer et al. 1999; Cumby et al. 1999).

Factory-farmed animals are given antibiotics in their food to prevent infection in their overcrowded conditions. Animal waste contains substantial amounts of bacteria, and because only about a quarter of the antibiotic is digested by an animal, the waste may also contain antibiotics. This combination is a perfect opportunity for the development of antibiotic-resistant bacteria which may pass to humans when water or soil is contaminated with the bacteria from farm waste. Vegetables and fruit can also become contaminated with antibiotic-resistant bacteria if animal waste that contains the bacteria is spread onto agricultural fields of fruit or vegetables.

Many farmers in developing countries want to maintain traditional farming methods, which are often better for the environment, but competition from corporations contributes to making these unviable, forcing small farmers into large-scale farming or condemning them to poverty.

Permaculture (Wikipedia 2008c) and small-scale organic farming are better alternatives to monocultures and factory farming. But these systems are still only models for farming and are not being used. There is still a long way to go before agriculture as a whole is environmentally sustainable.

We need more government controls on farming-related pollution and waste. There are measures that can be taken to ensure that manure from farms is treated and does not reach waterways and pollute them. There should be improvements in farm practices through an appropriate combination of educating land owners and land managers, providing funds for improvement programs, offering incentives and subsidies, improving regulations, and better enforcement of them.

Farm waste has huge potential as a source of energy (Cantrell et al. 2008; Lansing et al. 2008). It is a major source of methane, a greenhouse gas contributing to climate change. When possible the methane should not be allowed to escape into the atmosphere. It should be captured and the energy potential should be harnessed. Typical values of biogas potential are presented in Table 32.1.

32.2.2 Types of Water Pollution

There are two main types of water pollution from agriculture—point source pollution and diffuse pollution.

32.2.2.1 Point Source Pollution

Pollution of a waterway from a particular source such as a tank, building, or field is referred to as point source pollution. This can result when slurry, silage effluent, or uncollected dirty water reaches a waterway. It can also be caused by fuel oil, pesticides, or sheep dip, which can poison or damage river life or make groundwater unfit for use.

When farm wastes get into a waterway, they are broken down by microorganisms. This process uses oxygen needed by river life, including plants and fish, to survive. Biochemical oxygen demand (BOD) is a measure of the amount of oxygen needed by these microorganisms to break down the organic material. BOD is, therefore, a measure of the polluting strength of organic wastes. Farm wastes have a high BOD, some values are shown in Table 32.2. In serious pollution cases, all river life can be killed for considerable distances downstream.

32.2.2.2 Diffuse Pollution

Diffuse pollution cannot be attributed to a single event or action and is generally caused by high concentration of organic compounds, nitrates, phosphates, or pesticides.

TABLE 32.1
Biogas Potential from Farm Wastes

Substrate	Dry Matter (%)	Organic Matter (% DM)	Biogas yield (m³/kg⁻ OM)
Liquid cattle manure	6–11	68–85	0.1–0.8
Liquid pig manure	3–10	77–85	0.3–0.8
Excreta from sheep	18–25	80–85	0.3–0.4
Excreta from horses	25–28	72–75	0.4–0.6
Excreta from poultry	10–29	67–77	0.3–0.8
Low fat milk	6–8	92–95	0.6–0.7
Whey	4–6	80–92	0.5–0.9

DM, dry matter; OM, organic matter

TABLE 32.2
Typical BOD Values for Farm Wastes

Waste	Biochemical Oxygen Demand (mg/L of oxygen)
Whole milk	100,000
Silage effluent	65,000
Pig slurry	25,000
Cattle slurry	17,000
Dirty yard water	1500
Raw vegetable washings	500–3000
Dilute dairy and parlor washings	1000–2000
Raw domestic sewage	300

Diffuse pollution is originated through agriculture (fertilizing, sprinkling, watering, and irrigation). Pollutants such as nitrates, phosphates, pesticides, and heavy metals are mainly spread out over larger surfaces and have mainly two ways to reach water bodies (Núñez-Delgado et al. 2002; Centner et al. 2008; Drolc and Zagorc Koncan 2008; Ribbe et al. 2008). First, during the rain they rinse into surface water and, second, they trickle through land into groundwater.

32.3 WASTE TREATMENT

32.3.1 WASTE TREATMENT

When we talk about waste treatment we have generally in mind detoxification of pollutants and organic mineralization. Most agriculture waste contains organic compounds that are mainly well biodegradable. That is why we can mineralize them either through wastewater treatment, composting, or through digestion in special reactors called digesters. All systems either run under aerobic or anaerobic conditions.

32.3.1.1 Composting

Composting (Wikipedia 2008a) is the aerobic decomposition of biodegradable organic matter, producing compost. Or in a simpler form: composting is the decaying of food, mostly vegetables or manure. The decomposition is performed primarily by facultative and obligate aerobic bacteria, yeasts and fungi, helped in the cooler initial and ending phases by a number of larger organisms, such as springtails, ants, nematodes, and oligochaete worms.

Composting can be divided into home composting and industrial composting. Essentially, the same biological processes are involved in both scales of composting; however, techniques and different factors must be taken into account.

Compost also known as brown manure, is the aerobically decomposed remnants of organic matter. It is used in landscaping, horticulture, and agriculture as a soil conditioner and fertilizer. It is also useful for erosion control, land and stream reclamation, wetland construction, and as landfill cover.

Composting recycles or "downcycles" organic household and yard waste and manures into an extremely useful humus-like, soil end product called compost. Examples are fruits, vegetables, and yard clippings. Ultimately this permits the return of needed organic matter and nutrients into the food chain and reduces the amount of "green" waste going into landfills. Composting is widely believed to considerably speed up the natural process of decomposition as a result of the higher temperatures generated. The elevated heat results from exothermic processes, and the heat in turn reduces the generational time of microorganisms, and thereby speeds up the energy and nutrient exchanges taking place. It is a popular misconception that composting is a "controlled" process; if the right environmental circumstances are present the process virtually runs itself. Hence, the popular expression: "compost happens." It is, however, important to engineer the best possible circumstances for large amounts of organic waste to break down properly. This is especially so when it is accompanied by heating, because at elevated temperatures oxygen within the piles is consumed more rapidly, and if not controlled, will lead to malodor.

Decomposition similar to composting occurs throughout nature as garbage dissolves in the absence of all of the conditions and weather patterns, that modern composters talk about; however, the process can be slow. For example, in the forest, bark, wood and leaves break down into humus over 3–7 years. In restricted environments, for example, vegetables in a plastic trash container, decomposition with a lack of air encourages growth of anaerobic microbes, which produce disagreeable odors. Another form of degradation practiced deliberately in the absence of oxygen is called anaerobic digestion—an increasingly popular companion to composting as it enables capture of residual energy in the form of biogas, whereas composting releases the majority of bound carbon-energy as excess heat (which helps sanitize the material) as well as copious amounts of biogenic CO_2 to the atmosphere.

It is important to distinguish among terms such as *biodegradable*, *compostable*, and *compost-compatible*. Another term for composting is "yeasting;" it is literally translated into changing garden while making gestures.

- A biodegradable material is capable of being broken down completely under the action of microorganisms into carbon dioxide, water, and biomass. It may take a very long time for a material to biodegrade depending on its environment (e.g., hardwood in an arid area), but it ultimately breaks down completely.
- A compostable material biodegrades substantially under composting conditions, into carbon dioxide, methane, water and compost biomass. Compost biomass refers to the portion of the material that is metabolized by the microorganisms and which is incorporated into the cellular structure of the organisms or converted into humic acids, etc. Compost biomass residues from a compostable material are fully biodegradable. "Compostable" is thus a subset of "biodegradable." The size of the material is a factor in determining compostability, because it affects the rate of degradation. Large pieces of hardwood may not be compostable under a specific set of composting conditions, whereas sawdust of the same type of wood may be.
- A compost-compatible material does not have to be compostable or even biodegradable. It may biodegrade too slowly to be itself compostable, or it may not biodegrade at all. However, it is not readily distinguishable from the compost on a macroscopic scale and does not have a deleterious effect on the compost (e.g., it is not a biocide). Compost-compatible materials are generally inert and are present in compost at relatively low levels. Examples of compost-compatible materials include sand particles and inert particles of plastic.

Although composting has historically focused on creating garden-ready soil, it is becoming more important as a tool for reducing solid waste. More than 60% of household waste is recyclable or compostable. The decomposition of material sent to landfills is a principal cause of methane, an important greenhouse gas, so reducing the amount of waste sent to landfills is a key element of the fight against climate change. Surveys have shown that the first reason municipalities don't compost their waste is because they feel the process is complicated, time-consuming, or requires special equipment. However, especially in rural areas, much of the solid waste could be removed from the waste stream by promoting "extremely passive composting" where consumers simply discard their yard waste and kitchen scraps on their own land, regardless of whether the material is ever re-used as "compost."

Many different materials are suitable for composting organisms. Composters often refer to "C:N" requirements; some materials contain high amounts of carbon in the form of cellulose, which the bacteria need for their energy. Air spaces are left in the compost because decomposing the organisms requires oxygen. Other materials contain nitrogen in the form of protein, which provides nutrients for the energy exchanges. It would, however, be an oversimplification to describe composting as about carbon and nitrogen, as is often portrayed in popular literature. Elemental carbon—such as charcoal — is not compostable nor is a pure form of nitrogen, even in combination with carbon. Not only this, but also a great variety of man-made, carbon-containing products, including many textiles and polyethylene, are not compostable—hence the push for biodegradable plastics. Suitable ingredients with relatively high carbon content include

- Dry, straw-type material, such as cereal straws
- Autumn leaves
- Sawdust and wood chips
- Paper and cardboard (such as corrugated cardboard or newsprint with soy-based inks).

Ingredients with relatively high nitrogen content include

- Green plant material (fresh or wilted) such as crop residues, hay, grass clippings, and weeds
- Manure of poultry and herbivorous animals such as horses, cows, and llamas
- Fruit and vegetable trimmings.

The most efficient composting occurs by seeking to obtain an initial C:N mix of 25–30 to 1 by dry chemical weight. Grass clippings have an average ratio of 10–19 to 1 and dry autumn leaves from 55–100 to 1. Mixing equal parts by volume approximates the ideal range.

Poultry manure provides much nitrogen but with a ratio to carbon that is imbalanced. If composted alone, this results in excessive N-loss in the form of ammonia—and some odor. Horse manure provides a good mix of both, although in modern stables, so much bedding may be used as to make the mix too carbonaceous.

For home-scale composting, mixing the materials as they are added increases the rate of decomposition, but it can be easier to place the materials in alternating layers, approximately 15 cm thick, to help estimate the quantities. Keeping carbon and nitrogen sources separated in the pile can slow down the process, but decomposition will still occur.

Some people put special materials and activators into their compost. A light dusting of agricultural lime (not on animal manure layers) can curb excessive acidity, especially with food waste. Seaweed meal provides a ready source of trace elements. Finely pulverized rock (rock flour or rock dust) can also provide minerals, whereas clay and leached rock dust are poor in trace minerals.

Some materials are best left to a high-rate thermophilic composting system, as they decompose slower, attract vermin and require higher temperatures to kill pathogens than backyard composting provides. These materials include meat, dairy products, eggs, restaurant grease, cooking oil, manure and bedding of nonherbivores, and residuals from the treatment of wastewater and drinking water. Meat and dairy products can be recycled using bokashi, a fermentation method, which uses bokashi bran (Wikipedia 2009), wheat bran inoculated with effective microorganisms (EMs).

There are two major approaches to composting: active and passive. These terms are somewhat of a misnomer because both active and passive composts can attain high heating, which increases the rate of biochemical processes. But the terms active and passive are appropriate descriptions for the nature of human intervention used.

An active compost heap, steaming on a cold winter morning, is kept warm by the exothermic action of the bacteria as they decompose the organic matter.

Active (hot) composting is composting at close to ideal conditions, allowing aerobic bacteria to thrive. Aerobic bacteria break down material faster and produce less odor and fewer pathogens and destructive greenhouse gases than anaerobic bacteria. Commercial-grade composting operations actively control the composting conditions such as the C:N ratio. For backyard composters, the charts of carbon and nitrogen ratios in various ingredients and the calculations required to get the ideal mixture can be intimidating, so many rules of thumb exist for approximating it.

Pasteurization is a misnomer in composting, as no compost will become truly sterilized by high temperatures alone. Rather, in very hot compost where the temperature exceeds 55°C for several days, the ability of organisms to survive is greatly compromised. Nevertheless, there are many organisms in nature that can survive extreme temperatures, including the group of pathogenic Clostridium, and so no compost is completely safe. To achieve the elevated temperatures, the compost bin must be kept warm, insulated and damp.

Aerated composting is an efficient form of composting from the chemical point of view as it produces ultimately only energy in the form of waste heat and CO_2 and H_2O. With aerated composting, fresh air (i.e., oxygen) is introduced throughout the mix of materials using any appropriate mechanism. The air stimulates the microorganisms that are already in the mix, and their byproduct is heat. In a properly operated compost system, pile temperatures are sufficient to stabilize the raw material, and the oxygen-rich conditions within the core of the pile eliminate offensive odors. High temperatures also destroy fly larvae and weed seeds, yielding a safe, high-quality finished product.

Finally, aeration expedites the composting process through the mechanism of heating insofar as the elevated heat will drive biochemical processes faster, so that a finished product can be rendered in 60 to 120 days. Aerated compost is an excellent source of macro- and micronutrients as well as stable organic matter, all of which support healthy plant growth. In addition, the microorganisms in compost aid in the suppression of plant pathogens. Finally, compost retains water extremely well resulting in improved drought resistance, a longer growing season, and reduced soil erosion.

In Thailand, this system has been used by farmer groups for more than 445 sites (May 2008). The process needs only 30 days to finish without the need for turning and 10 metric tons of compost is obtained each time. A blower (6 cm squirrel-caged blower with 2.2 kW motor) is needed to force the air through 10 static piles of compost 15 minutes for twice a day. The raw materials consist of agricultural wastes and animal manure in the ratio of 3:1 by volume.

32.3.1.2 Passive Composting

Passive composting is composting in which the level of physical intervention is kept to a minimum, and often as a result, the temperatures never reach much above 30°C. It is slower but is the more common type of composting in most domestic garden compost bins. Such composting systems may be either enclosed (home container composting, industrial in-vessel composting) or in exposed piles (industrial windrow composting). Kitchen scraps are put in the garden compost bin and left untended. This scrap bin can have a very high water content, which reduces aeration and therefore it becomes odorous. To improve drainage and airflow, a gardener can mix in wood chips, small pieces of bark, leaves or twigs, or make physical holes through the pile.

32.3.1.3 Natural Composting

An unusual form of natural composting in nature is seen in the case of the mound-builders (megapodes) of eastern Indonesia, New Guinea, and Australia as well as in the case of bowerbirds of New Guinea and Australia. These megapodes are fowl-sized birds famous for building nests in the form of huge compost heaps containing leaf litter, in which they incubate their eggs. The birds work constantly to maintain the correct, almost exact, incubation temperatures, by adding and removing leaves from the compost pile. In effect, this teaches us that thermophilic high-temperature composting is not man-made.

32.3.1.4 Moisture and Heat

An effective compost pile is about as damp as a well wrung-out sponge. This provides the moisture that all life requires. Microorganisms vary by their ideal temperature and the heat they generate as they digest. Mesophilic bacteria survive best at temperatures of 20°C to 44°C. Thermophilic (heat-surviving) bacteria grow optimally at around 55°C, and can attain the fastest decomposition, because metabolic processes proceed more rapidly under higher temperatures. Elevated temperature is also preferred because it causes the most rapid pathogen reduction, and is more destructive of weed seeds. To minimally achieve it, the heap should be about 1 m wide, 1 m tall, and as long as is practicable. This provides enough insulating mass to build up heat but also allows aeration. The centre of the pile heats up the most.

If the pile does not heat up, common reasons include that

- The heap is too wet, limiting the oxygen which bacteria require
- The heap is too dry for the bacteria to survive and reproduce
- There is insufficient protein (nitrogen-rich material)

The necessary material should be added, or the pile should be turned to aerate it and bring the outer layers inside and vice versa. You should add water at this time to help keep the pile damp. One guideline is to turn the pile when the high temperature has begun to drop, indicating that the food source for the fastest-acting bacteria (in the center of the pile) has been largely consumed. When turning the pile does not cause a temperature rise, it brings no further advantage. When all of the material has turned into dark brown crumbly matter, it is ready to use.

32.3.1.5 Industrial and Agricultural Composting

Industrial and agricultural composting systems are increasingly being installed as a waste management alternative to landfills, along with other advanced waste processing systems. Industrial and agricultural composting or anaerobic digestion combined with mechanical sorting of mixed waste streams is called mechanical biological treatment increasingly used in Europe because of stringent new regulations controlling the amount of organic matter allowed in landfills. Treating biodegradable waste before it enters a landfill reduces global warming from fugitive methane; untreated waste breaks down anaerobically in a landfill, producing landfill gas that contains methane, a greenhouse gas even more potent than carbon dioxide.

Most commercial, industrial, and agriculture composting operations use active composting techniques (Yadav et al. 1982; Baeten and Verstraete 1988; Georgacakis et al. 1996; Imbeah 1998; Mohaibes and Heinonen-Tanski 2004; Zhu et al. 2004; Wakase et al. 2008). These ensure that the process does not get out of control, especially with the high throughput demand imposed by contracted, incoming waste. This means that as short as possible a processing time must be maintained to keep the facility properly functioning. Partly for this reason composters have declined to support compost maturity standards if it would increase the required holding time. The greatest amount of technological control of composting is seen in systems that use an enclosed vessel and control their temperature, airflow, moisture, and other parameters.

32.3.2 Anaerobic Digestion

Anaerobic digestion (Wikipedia 2008a) is a series of processes in which microorganisms break down biodegradable material in the absence of oxygen. It is widely used to treat wastewater sludges and organic wastes because it provides volume and mass reduction of the input material with biogas as byproduct. As part of an integrated waste management system, anaerobic digestion reduces the emission of landfill gas into the atmosphere. Anaerobic digestion is a renewable energy source because the process produces a methane and carbon dioxide rich biogas suitable for energy production helping replace fossil fuels. Also, the nutrient-rich solids left after digestion can be used as fertilizer.

32.3.2.1 Biochemical Reactions in Anaerobic Digestion

There are four key biological and chemical stages of anaerobic digestion (Figure 32.1):

1. Hydrolysis
2. Acidogenesis
3. Acetogenesis
4. Methanogenesis

In most cases, biomass is made up of large organic compounds. For the bacteria in anaerobic digesters to access the chemical energy potential of the organic material, the organic matter molecular chains must first be broken down into their smaller constituent parts. These constituent parts or monomers such as sugars are readily available to bacteria to process. The process of breaking these chains and dissolving the smaller molecules into solution is called hydrolysis. Therefore, hydrolysis of these high molecular weight molecules is the necessary first step in anaerobic digestion. Through hydrolysis the complex organic molecules are broken down into simple sugars, amino acids, and fatty acids. Hydrolysis can be biological (using hydrolytic microorganisms), biochemical (using extracellular enzymes), chemical (using catalytic reactions), as well as physical (using thermal energy and pressure) in nature.

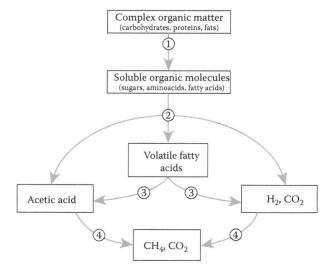

FIGURE 32.1 Anaerobic pathway of complex organic matter degradation.

Acetate and hydrogen produced in the first stages can be used directly by methanogens. Other molecules such as volatile fatty acids (VFAs) with a chain length that is greater than acetate must first be catabolized into compounds that can be directly utilized by methanogens. The biological process of acidogenesis is where there is further breakdown of the remaining components by acidogenic (fermentative) bacteria. Here VFAs are created along with ammonia, carbon dioxide, and hydrogen sulfide as well as other byproducts. The process of acidogenesis is similar to the way that milk sours.

The third-stage anaerobic digestion is acetogenesis. Here simple molecules created through the acidogenesis phase are further digested by acetogens to produce largely acetic acid as well as carbon dioxide and hydrogen.

The terminal stage of anaerobic digestion is the biological process of methanogenesis. Here methanogens utilize the intermediate products of the preceding stages and convert them into methane, carbon dioxide and water. It is these components that make up the majority of the biogas emitted from the system. Methanogenesis is sensitive to both high and low pH values and occurs between pH 6.5 and pH 8. The remaining nondigestible material that the microbes cannot feed upon, along with any dead bacteria, constitutes the digestate.

A simplified generic chemical equation for the overall process outlined next is as follows:

$$C_aH_bO_cN_dS_e + \left(a - \frac{b}{4} - \frac{c}{2} + \frac{3d}{4} + \frac{e}{2}\right)H_2O \rightarrow$$

$$\rightarrow \left(\frac{a}{2} - \frac{b}{8} + \frac{c}{4} + \frac{3d}{8} + \frac{e}{4}\right)CO_2 + \left(\frac{a}{2} + \frac{b}{8} - \frac{c}{4} - \frac{3d}{8} - \frac{e}{4}\right)CH_4 + dNH_3 + eH_2S$$

If this equation is applied to basic components of organic materials—carbohydrates, fats, and proteins—the following results will be obtained:

1. Carbohydrates, example glucoses

$$C_6H_{12}O_6 \rightarrow 3\,CO_2 + 3\,CH_4$$

$$50 \qquad 50 \qquad \text{(gas composition, vol. \%)}$$

2. Fats, example palmitine—fatty acid

$$C_{16}H_{32}O_2 + 7H_2O \quad \rightarrow \quad 4.5\ CO_2 + 11.5\ CH_4$$

$$\qquad\qquad\qquad\qquad\qquad 28 \qquad 72 \qquad \text{(gas composition, vol. \%)}$$

3. Proteins. For proteins there is no general formula. An approximate empirical formula for amino acids is $C_{13}H_{25}O_7N_3S$ and would yield:

$$C_{13}H_{25}O_7N_3S + 6\ H_2O \rightarrow 6.5\ CO_2 + 6.5\ CH_4 + 3\ NH_3 + H_2S$$

$$\qquad\qquad\qquad\qquad 46.5 \qquad 46.5 \qquad 0 \qquad 7 \quad \text{(gas composition, vol.\%)}$$

In this case NH_3 shows 0% in gas composition because the reactor conditions are usually such that NH_3 stays completely dissolved (as NH_4^+ ion).

If we summarize the examples above and presume that equal parts of groups (proteins, carbohydrates and lipids) are present in the substrate, biogas has following composition:

Methane (CH_4)	56%
Carbon dioxide (CO_2)	42%
Hydrogen sulfide (H_2S)	2%

On the basic relations described above, we can conclude that lipids (fat-soluble molecules) give the highest production of biogas with highest percentage of methane, proteins give little less biogas production with relatively low percentage of methane, and carbohydrates give the lowest amounts of biogas also with relatively low percentage of methane.

In general, most authors state the amount of methane between 55 and 70% vol. percent, rarely more. A second component is carbon dioxide with 27–44%. There are also other trace gases usually present, such as hydrogen H_2 (1–2%), hydrogen sulfide (up to 3%) and gases usually below the limit of detection (NH_3—ammonia, CO—carbon monoxide, and N_2—nitrogen).

32.3.2.2   Factors That Affect Anaerobic Digestion

As with all biological processes the optimal environmental conditions are essential for successful operation of anaerobic digestion (Table 32.3). The microbial metabolism processes depend on many parameters; therefore these parameters must be considered and carefully controlled. Furthermore, the environmental requirements of acidogenic bacteria differ from environmental requirements of methanogenic bacteria. In aspiration for providing optimum conditions for each group of microorganisms a two-stage process of waste degradation was developed, containing a separate reactor for each stage. The first stage is for hydrolysis/acidification and the second for acetogenesis/methanogenesis. Provided that all of the degradation process has to take place in one single reactor (one-stage process), usually methanogenic bacteria requirements must be fulfilled with priority. Namely, methanogenic bacteria have much longer regeneration time, much slower growth, and are more sensitive to environmental conditions then other bacteria present in the mixed culture (Table 32.4). However, there are some exceptions to the case:

- With cellulose containing substrates (which are slowly degradable), the hydrolysis stage is the limiting one and needs prior attention.
- With protein-rich substrates, the pH optimum is equal in all anaerobic process stages. Therefore a single digester is sufficient for good performance.
- With fat-rich substrates, the hydrolysis rate is increasing with better emulsification, so that acetogenesis is limiting. Therefore, a thermophilic process is advised.

TABLE 32.3
Environmental Requirements

Parameter	Hydrolysis/Acidogenesis	Methanogenesis
Temperature	25–35°C	Mesophilic: 30–40°C
		Thermophilic: 50–60°C
pH Value	5.2–6.3	6.7–7.5
C:N ratio	10–45	20–30
Redox potential	+ 400 to –300 mV	Less than –250 mV
C:N:P:S ratio	500:15:5:3	600:15:5:3
Trace elements	No special requirements	Essential: Ni, Co, Mo, Se

Source: Deublein, D. and Steinhauser, A., *Biogas from Waste and Renewable Resources*. Willey-VCH Verlag GmbH & Co. KGaA, Weinheim, Germany, 2008.

TABLE 32.4
Regeneration Time of Microorganisms

Microorganisms	Time of Regeneration
Acidogenic bacteria	Less than 36 hours
Acetogenic bacteria	80–90 hours
Methanogenic bacteria	5–16 days
Aerobic microorganisms	1–5 hours

32.3.2.2.1 Temperature

Anaerobic digestion can operate in a wide range of temperature, between 5°C and 65°C. Generally there are three widely known and established temperature ranges of operation: psychrophilic (15–20°C), mesophilic (30–40°C), and thermophilic (50–60°C). With increasing temperature the speed of anaerobic digestion increases. For instance, with ideal substrate thermophilic digestion can be approx. four times faster than mesophilic. However, using real waste substrates, there are other inhibitory factors that influence digestion that make thermophilic digestion only approximately two times faster than mesophilic.

The important thing is, when choosing a certain temperature it should be kept constant as much as possible. In the thermophilic range (50–60°C) fluctuations as low as ±2°C can result in 30% less biogas production (Zupancic and Jemec 2010). Therefore, it is advised that temperature fluctuations in the thermophilic range should be no more than ±1°C. In the mesophilic range, the microorganisms are less sensitive; therefore fluctuations of ±3°C can be tolerated.

For each range of digestion temperature there are certain groups of microorganisms present that can flourish. In the temperature ranges between the three established temperature ranges the conditions for each of the microorganisms are less favorable. In these ranges anaerobic digestion can operate, however much less efficiently. For example, mesophilic microorganisms can operate up to 47°C, thermophilic microorganisms can already operate as low as 45°C. However, the speed of reaction is low and it may happen that the two groups of microorganisms may exclude each other and compete in the overlapping range. All this results in poor efficiency of the process, and therefore, these temperatures are rarely applied.

32.3.2.2.2 Redox Potential

In the anaerobic digester, low redox potential is necessary. Methanogenic microorganisms need redox potential to be between –300 and –330 mV for optimum performance. Redox potential can increase up to 0 mV in the digester; however, it should be kept in the optimum range. To achieve that, no oxidizing agents, such as oxygen, nitrate, nitrite, or sulfate, should be added to the digester.

32.3.2.2.3 C:N Ratio and Ammonium Inhibition

In biomass of microorganisms, the ratio of C:N:P:S is approx. 100:10:1:1. The ideal substrate C:N ratio is then 20–30:1 and C:P ratio 150–200:1. A C:N ratio higher than 30 causes slower multiplication of microorganisms because of low protein formation and thus low energy and structural material metabolism of microorganisms. Consequently, lower substrate degradation efficiency occurs. On the other hand, a C:N ratio as low as 3:1 can result in successful digestion. However, when such low C:N ratios and nitrogen rich substrates are applied (which is often the case using farm waste) a possible ammonium inhibition must be considered. Ammonium, although it represents an ideal form of nitrogen for cells growth of microorganisms, is toxic to mesophilic methanogenic microorganisms at concentrations over 3000 mg/L and pH over 7.4. With increasing pH the toxicity of ammonium increases (Figure 32.2).

Thermophilic methanogenic microorganisms are generally more sensitive to ammonium concentration. Inhibition can occur already at 2200 mg/L of ammonium nitrogen. However, the ammonium inhibition can very much depend on the substrate type. A study of ammonium inhibition in thermophilic digestion done by Sung and Liu (2003) shows an inhibiting concentration to be over 4900 mg/L when using nonfat waste milk as substrate.

Ammonium inhibition can likely occur when digester leachate (or water from dewatering the digested substrate) is recirculated to dilute the solid substrate for anaerobic digestion. Such recirculation must be handled with care and examined for potential traps such as ammonium or other inhibitory ions build up.

To resolve ammonia inhibition when using farm waste in anaerobic digestion, several methods can be used:

- The first possibility is carefully combining different substrates to create a mixture with lower nitrogen content. Usually some plant biomass (such as silage) is added to liquid farm waste in such case.
- The second possibility is diluting the substrate to such an extent that concentration in the anaerobic digester does not exceed the toxicity concentration. This method must be handled with care. Only in some cases dilution may be a solution. If the substrate requires too much dilution, a microorganism washout may occur, which results in process failure. Usually, there is only a narrow margin of operation. The original substrate causes

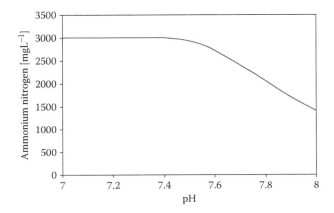

FIGURE 32.2 Ammonium nitrogen toxicity concentration to methanogenic microorganisms.

ammonium inhibition when diluted to the extent necessary to stop ammonia inhibition, and already a washout because of dilution occurs.

- It is also possible to remove ammonium from the digester liquid. This method is usually most cost effective and rarely used. One such process is stripping ammonia from the liquid. It is also commercially available (GNS 2009).

32.3.2.2.4 pH

In anaerobic digestion the pH affects the most methanogenic stage of the process. The optimum pH for the methanogenic microorganisms is between 6.5 and 7.5. If the pH decreases below 6.5, more acids are produced and that leads to imminent process failure. In real digester systems with suspended biomass and substrate containing suspended solids, normal pH of operation is between 7.3 and 7.5. When pH decreases to 6.9, serious actions to stop process failure must be taken. When using UASB systems (or other systems with granule-like microorganisms), which utilize liquid substrates with a low suspended solids concentration, the normal pH of operation is 6.9–7.1. In such cases, the pH limit of successful operation is 6.7.

In normally operated digesters, there are two buffering systems that ensure that pH stays in a desirable range:

- Carbon dioxide–hydrogen carbonate–carbonate buffering system. During digestion CO_2 is continuously produced and released into gaseous phase. When pH value decreases, CO_2 is dissolved in the reactor solution as uncharged molecules. With increasing pH value, dissolved CO_2 forms carbonic acid which ionizes and releases hydrogen ions. At pH 4 all CO_2 is in form of molecules and at pH 13 all CO_2 is dissolved as carbonate. The center point around which pH value swings with this system is at pH 6.5. With concentrations between 2500 and 5000 mg/L, hydrogen carbonate gives strong buffering.
- Ammonia–ammonium buffering system. With decreasing pH value, ammonium ions are formed with releasing of hydroxyl ions. With increasing pH value, more free ammonia molecules are formed. The center point around which pH value swings with this system is at pH 10.

Both buffering systems can be overloaded by the feed of rapidly acidifying (quickly degradable) organic matter, toxic substances, decrease of temperature, or a too high loading rate to the reactor. In such case, a pH decrease is observable, combined with CO_2 increase in the biogas. Measures to correct the excessive acidification and prevent a process failure are as follows:

- Stopping the reactor substrate supply for the time to methanogenic bacteria can process the acids. When the pH decreases to the limit of successful operation, no substrate supply should be added until pH is in the normal range of operation or preferably in the upper portion of normal range of operation. In suspended biomass reactors, this pH value is 7.4 and in granule microorganism systems this pH value is 7.0.
- If procedure from the point above has to be repeated many times, the system is obviously overloaded and the substrate supply has to be diminished (increasing the residence time of the substrate).
- Increasing the buffering potential of the substrate. By the mixture of certain substrates that contain some alkaline substances, the buffering capacity of the system can be increased.
- Addition of the neutralizing substances. Typical are lime (CaO or Ca(OH)$_2$), sodium carbonate (Na$_2$CO$_3$), or sodium hydrogen carbonate (NaHCO$_3$), and in some cases sodium hydroxide (NaOH). However, with sodium substances, precaution must be practiced because sodium inhibition can occur with excessive use.

32.3.2.2.5 *Inhibitory Substances*

In anaerobic digestion systems, a characteristic phenomena can be observed. Some substances, which are necessary for microbial growth in small concentrations, inhibit the digestion in higher concentrations. A similar effect can be had from a high concentration of total volatile fatty acids (tVFAs). Although they represent the very substrate that methanogenic bacteria feed upon, the concentrations over 10,000 mg/L may have an inhibitory effect on digestion (Mrafkova et al. 2003; Ye et al. 2008).

Inorganic salts can significantly affect anaerobic digestion. Table 32.5 shows the optimal and inhibitory concentrations of metal ions from inorganic salts.

In real operating systems, it is unlikely that inhibitory concentrations of inorganic salts of metals would occur, mostly because in such high concentrations ions would precipitate in nonsoluble salts in alkaline conditions, especially if H_2S is present. The most real threat in this case is sodium inhibition of anaerobic digestion. This can occur in cases where substrates are wastes with extremely high salt contents (some food wastes, tannery wastes, etc.) or when excessive use of sodium substances were used in neutralization of the substrate or the digester liquid. A study done by Feijoo et al. (1995) show that concentrations of 3000 mg/L already show occurrence of sodium inhibition. However, anaerobic digestion can operate up to concentrations as high as 16,000 mg/L of sodium, which is close to the saline concentration of seawater. Measures to correct the sodium inhibition are simple. The high salt substrates must be pretreated to remove the salts (mostly washing). The use of sodium substances as neutralizing agents can be substituted with other alkaline substances (such as lime).

Heavy metals also have stimulating effects on anaerobic digestion in low concentrations; however, higher concentrations can be toxic. In particular lead, cadmium, copper, zinc, nickel and chromium can cause disturbances in anaerobic digestion process. For farm waste especially, zinc is present (in pig slurry) and it usually originates from pig fodder that contains zinc additive as an antibiotic. Inhibitory and toxic concentrations are shown in Table 32.6.

Other organic substances, such as disinfectants, herbicides, pesticides, surfactants, and antibiotics can often flow with the substrate and also cause nonspecific inhibition. All of these substances have a specific chemical formula and it is hard to determine what the behavior of inhibition will be. Therefore, when such substances do occur in the treated substrate, specific research is strongly advised to determine the concentration of inhibition and possible ways of microorganism adaptation.

32.3.3 ANAEROBIC DIGESTION TECHNOLOGY

Basic block schematic of anaerobic digestion (Figure 32.3) shows that technological process of anaerobic digestion is relatively simple. It consists of three basic phases: substrate preparation and pretreatment, anaerobic digestion itself, and posttreatment of digested material with biogas use. In this section all of the processes will be elaborated in detail.

32.3.3.1 Substrates and Pretreatment

In general, all types of biomass can be used as substrates as long as they contain carbohydrates, proteins, fats, cellulose, and hemicellulose as main components. It is, however, important to

TABLE 32.5
Optimal and Inhibitory Concentrations of Ions from Inorganic Salts

	Optimal Concentration [mg/L]	Moderate Inhibition [mg/L]	Inhibition [mg/L]
Sodium	100–200	3500–5500	16,000
Potassium	200–400	2500–4500	12,000
Calcium	100–200	2500–4500	8000
Magnesium	75–150	1000–1500	3000

TABLE 32.6
Inhibitory and Toxic Concentrations of Heavy Metals

Metal Ion	Inhibition Start[a] [mg/L]	Toxicity to Adopted Microorganisms[a] [mg/L]
Cr^{3+}	130	260
Cr^{6+}	110	420
Cu	40	170
Ni	10	30
Cd	70	600
Pb	340	340
Zn	400	600

[a] As inhibitory concentration, it is considered the first value that shows diminished biogas production and as toxic concentration it is considered the concentration at which biogas production is diminished by 70%.

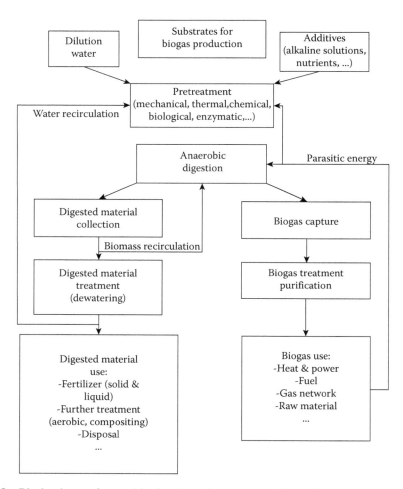

FIGURE 32.3 Block scheme of anaerobic digestion, biogas use, and digested material use.

consider several points before considering the process and biomass pretreatment. The contents and concentration of substrate should match the selected digestion process. For anaerobic treatment of liquid farm waste, the most appropriate concentration is between 2 and 8% of dry solids by mass. In this case, conventional single-stage digestion or two-stage digestion is used. If the treatment of solid farm waste using a solid digestion process is considered, the concentration is between 10 and 20% by mass. The concentration of dry matter in farm waste depends on the feces/urine ratio, whether straw or other material is used for animal litter, and the quantity of water used to flush the material away. The empirical value of used water should not exceed 100 L per LSU (livestock unit) to achieve desired liquid manure concentration suitable for anaerobic digestion. The liquid manure can also contain other foreign matter; fodder residue can for instance contribute to better biogas yield, others usually impair the process of digestion. Such materials are

- Sand from mineral materials in animal feed
- Sawdust from scattering
- Large pieces of straw from litter
- Soil from roughage
- Soil which is carried from meadows
- Skin and tail hair, bristles, and feathers
- Cords, wires, plastics, stones, and other similar material.

The presence of foreign matter in the substrate can lead to increased complexity in the operating expenditure of the process. During the process of digestion of liquid manure from pigs and cattle, the formation of a scum layer on the top of the digester liquid can be formed, caused by straw and muck. The addition of rumen content and cut grass (larger particles than silage) can contribute to its formation. If the substrate consists of undigested parts of corn and grain combined with sand and lime, the solid aggregates can be formed at the bottom of the digester and can cause severe clogging problems.

In all such cases, the most likely solution is pretreatment to reduce the size of solids. Naturally, all of the nondigestible solids (soil, plastic, metal, stones, etc.) should be separated from the substrate flow. On the other hand, grass, straw and fodder residue can contribute to the biogas yield. When properly pretreated they are accessible to the digestion microorganisms. Such pretreatments are generally physical, chemical, or combined in nature.

As physical pretreatment, disintegration is the most common (the most known are grinding and mincing). In grinding and mincing, the energy required for operation is inversely proportional to the particle size. Because such energy contributes to the parasitic energy, it should be kept in the limits of positive margin (the biogas yield increased by pretreatment is more than energy required for it). In the case of farm waste, the empirical value for such particle size is between 1 and 4 mm.

Chemical pretreatment can be used when treating lignocellulosic material, such as spent grains or even silage. Very often chemical treatment is used combined with heat, pressure, or both. It is common to use acid (hydrochloric, sulfuric, or others) or an alkaline solution of sodium hydroxide (in some cases potassium hydroxide). Such a solution is added to the substrate in quantities that surpass the titration equilibrium point, and then it is heated to the desired temperature and possibly pressurized. Retention times are generally short (up to several hours) compared to retention times of the anaerobic digesters. The pretreated substrate is then much more degradable. The downside of this pretreatment is an excessive demand for parasitic energy and the cost of chemicals required. It rarely outweighs the costs of building a bigger digester. Therefore, it is used mostly in treating industrial waste (such as brewery) where there is plenty of waste lye or acid present and waste heat can be regenerated from the industrial processes as well.

Thermal pretreatment can yield up to 30% more biogas if properly applied. This process occurs at temperature range of 135–220°C and pressures above 10 bar. Retention times are short (up to several hours) and hygienization is automatically included. Pathogenic microorganisms are completely inhibited. The process runs economically only with heat regeneration. When heat is regenerated from outflow to inflow of the pretreatment process, it takes only slightly more heat than conventional anaerobic digestion. Such a process is very appropriate for cell like material such as waste sludge.

It is also possible to use biological processes as pretreatment. They are less expensive but emerging in the world. During the process of silage, disintegration takes place by the lactic acids, which decompose complex components of certain substrates (Hendriks and Zeeman 2009). Therefore, for example, clover should be harvested and processed before blooming whereas corn is best cut at the end of maturation. Recently also disintegration with enzymes has been quite successful, especially using cellulase, protease, or carbohydrases at a pH of 4.5 to 6.5 and a retention time of at least 12 days, preferably more (Hendriks and Zeeman 2009).

32.3.3.2 Anaerobic Digestion Process

For anaerobic digestion, several different types of anaerobic processes and several different types of digesters are applicable. It is hard to say, which digester type will serve best treating the selected substrate. For treating farm waste, it is suggested that anaerobic digesters are decentralized to serve each farm separately, to make it an economic and technological unit combined with the farm. It is important to study the waste of each such unit carefully to be able to determine optimal conditions for substrate digestion. Farm waste can differ greatly even in the same geographical areas; therefore, it is strongly recommended to conduct laboratory- and pilot-scale experiments before constructing the full-scale digester. Considering the costs of the full-scale digester, conducting pilot-scale experiments is a minor item, especially if you can get good data and construct an optimal digester. The biggest economic setback is if a digester is constructed and it does not perform optimally and therefore requires a rebuilding.

There are several processes available to conduct anaerobic digestion. Roughly, the digestion process can be divided into solid digestion and wet digestion processes. Solid digestion processes are in fact anaerobic composters. In this process, substrate and biomass are in solid form (although having approx. 20% of dry matter and 80% water). Such processes have several advantages. The main advantage is reducing the reactor volume because of less water in the transport. Four times more concentrated substrate equals approximately four times less reactor volume. It is also possible that some inhibitors (such as ammonium) can have less inhibitory effects in solid digestion process. The biggest disadvantage of the solid digestion process is the substrate transport. Substrate in solid form requires more energy for transport in and out of the digesters. There is also a stronger possibility of air intrusion into the digesters, which poses a great risk to process stability and safety. It has been only recently that such processes have gained ground for a wider use. A fine example is the commercially available Kompogas® process (Kompogas 2009).

A much larger variety of digestion processes are wet digestion processes. In these processes, the conventional concentration of the digester suspension rarely exceeds 5% of dry solids by mass. There are several reactor technologies available to successfully conduct anaerobic digestion. Roughly, they can be divided into batch-wise (Figures 32.4 and 32.5) and continuous processes. Furthermore, continuous processes can be divided into single stage (Figure 32.6) or two-stage processes (Figure 32.7). In most of the wet digestion processes, microorganisms are completely mixed and suspended with substrate in the digester (suspended solids of substrate and microorganisms are impossible to separate). If the substrate contains few solids and is mostly dissolved organic liquid, we can apply flow-through processes. In these processes, microorganisms are in granules and granules are suspended in liquid that contains dissolved organic material. In such anaerobic processes, microorganisms are easily separated from the substrate. A typical representative of such a process is the UASB (upflow anaerobic sludge blanket) process (Figure 32.8).

32.3.3.3 Batch Processes

In the batch process, all of the different stages of substrate treatment happen in one tank. Typically the reaction cycle of the anaerobic sequencing batch reactor (ASBR) is divided into four phases: fill-load, react-digestion, settle, and decant-unload (Figure 32.4). The reactor is filled with substrate at

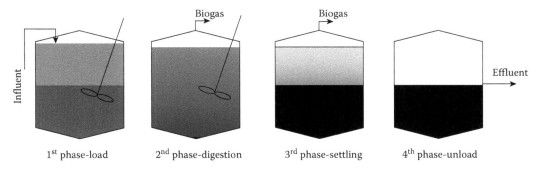

FIGURE 32.4 Batch ASBR process schematic.

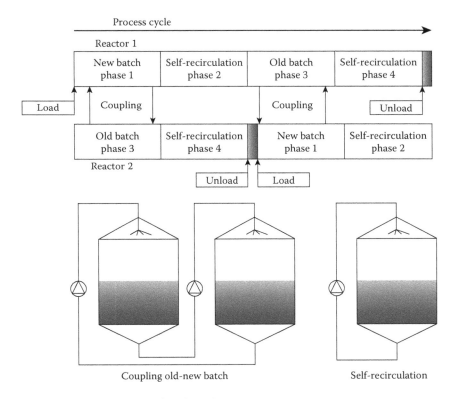

FIGURE 32.5 Batch solid anaerobic digestion scheme.

once and degrades without any interference until the end of the cycle phase. This leads to temporal variation in microbial community and biogas production. Therefore, batch processes require more precise measurement and monitoring equipment to function optimally. Usually, these reactors are at least built in pairs, sometimes even in higher numbers. This achieves a more steady flow of biogas for instant use. The tank is usually emptied between the cycles to a certain exchange volume, which in case of anaerobic digestion is rarely more than 50% of total reactor volume. The residue in the tank is microbial inoculum for the next cycle. This makes batch reactors more voluminous than conventional continuous reactors; however, they do not require equalization tanks and the total reactor volume is usually less than in conventional processes. They can be coupled directly to the waste discharge; however this limits the use to more industrial processes (for example food industry) and less to farm waste production. The typical cycle time is 1 week.

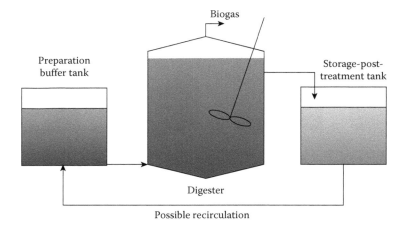

FIGURE 32.6 Single-stage conventional anaerobic digestion.

Better used in farm waste treatment would be batch processes that treat solid substrate (semi-solid manure). In this case, the cycle is also divided into four phases that are somehow different than in an ASBR process. This process requires always digesters to be in pairs. The reactor is almost completely emptied between cycles; therefore, it requires inoculation through leachate exchange between the two digesters (from the one in the peak biogas production to the one at the start of the process). In the other phases, leachate is self-circulated (Figure 32.5). Typical cycle time is between 30 and 60 days. Although solid substrate reduces the reactor volume, the volume is still rather large because of long cycle times compared to conventional digesters that process wet substrates. The advantage of these digesters is that they are applicable for smaller scale farms and require less complicated monitoring equipment. For biogas use a buffer gas tank is required.

32.3.3.4 Continuous Processes

Most of the commercial biogas plants use a conventional continuous process as the technology for anaerobic digestion. By conventional it is meant fully mixed, semi-continuous, or continuous load and unload reactor at the mesophilic temperature range (35–40°C; Figure 32.6). In the majority of cases, the substrate is loaded to the reactor once to several times a day; rarely is it loaded continuously. Continuous load can lead to shot circuit, which means that fresh load can directly flow out of the reactor if mixing is too intense or inflow and outflow are improperly positioned. The digester is usually single stage. Although they are built in pairs, they do not function as a stage separated process. Usually digesters are equipped with a preparation tank, where various substrates are mixed and prepared for loading, which also serves as a buffer tank. In many cases also a post treatment tank is added (it is also called a postfermentor), where treated substrate is completely stabilized and prepared for further treatment. The posttreatment tank can also serve as a buffer to further treatment of the substrate. Generally postfermentors do not add much to overall biogas yield (up to 5%) if the digester operates optimally. The size of the preparation and posttreatment tank are determined according to the necessary buffer capacity for continuous operation. The size of the digester is determined with hydraulic retention time (HRT) and with organic loading rate (OLR). HRT is defined as digester volume divided by substrate flow and it tells us how many days it takes on average for a certain portion of the substrate to pass through the reactor. For mesophilic digesters, the usual values are between 20 and 40 days, depending on the substrate degradability. In thermophilic digesters, HRT between 10 and 20 days can achieve the same treatment efficiency.

OLR (sometimes also called volume load) is defined as mass of organic material fed to the digester per volume per day. Typical values for mesophilic digesters is 2.0–3.0 kg m^{-3}/day and for thermophilic digesters, 5.0 kg m^{-3}/day. Maximum OLR depends very much on the substrate biodegradability; a mesophilic process can rarely achieve higher loads than 5.0 kg m^{-3}/day and thermophilic 8.0 kg m^{-3}/day, respectively. Locally in the digester for a short period of time, higher loads can be achieved; however, it is not advisable to run continuously on such high loads.

To achieve better efficiency and higher loads stage separated process can be applied (Figure 32.7). In this case, the whole substrate or just portions of the substrate that are not easily degradable are treated first in hydrolysis-acidogenic stage reactor and after that in the methanogenic reactor. By separating the biological processes in two separate tanks, each can be optimized to achieve higher efficiency as if in just one tank, where all biological processes are occurring simultaneously. Much research has been published giving considerable attention to this kind of process (Dinsdale et al. 2000; Song et al. 2004; De Gioannis et al. 2008; Ponsá et al. 2008). Both stages can be either mesophilic or thermophilic; however, it is preferred that the hydrolysis-acidogenic reactor is thermophilic; and methanogenic is mesophilic. Typical HRT for hydrolysis-acidogenic reactor is 1–4 days (thermophilic), depending on the substrate biodegradability. Typical HRT for the methanogenic reactor is 10 to 15 days (mesophilic) and 10–12 days (thermophilic). Advantages of this process beside shorter HRT are higher overall volume loads (20% or more). Many authors also reported slightly better biogas yields (Messenger et al. 1993; Han et al. 1997; Roberts et al. 1999; Tapana and Krishna 2004). The only disadvantage is that this process requires more sophisticated process control and equipment, but the construction is more cost effective.

32.3.3.5 Flow-Through Processes

Flow-through processes, such as the UASB process (Figure 32.8), are used only for substrates where organic material is in dissolved form with solids content at maximum 1–5 g/L. In this substrate category are mostly wastewaters of industrial origin (beverage industry). A detailed depiction of the process is presented therefore in the wastewater treatment section.

32.3.3.6 Posttreatment and Substrate Use

After the substrate has been digested, it needs additional treatment. There are several possibilities of digested substrate use. The most common and most used possibility in farm waste treatment is using the digested substrate as a fertilizer. It can be used wet or dewatered. Wet substrate (total solids concentration 1–5% by mass) is taken from the post-treatment tank and used. However, it must be considered that fertilizing is possible only twice (sometimes once) annually. The post-treatment container, therefore, must be designed accordingly. A possible solution is a lagoon, where digested substrate is stored and additionally stabilized and mineralized during the storing time. When using

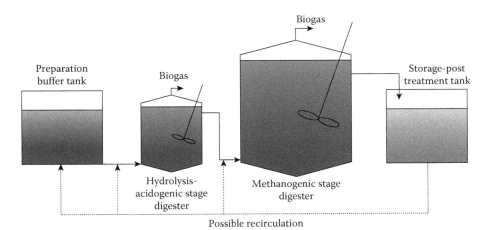

FIGURE 32.7 Two-stage anaerobic digestion.

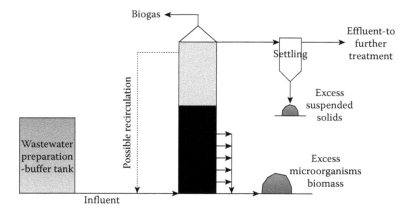

FIGURE 32.8   UASB process.

solid substrate (total solids concentration 20–30% by mass), the digested substrate is mechanically dewatered first (by belt press or centrifuge) and then liquid and solid parts are used separately. Solid remains after dewatering can be used as a fertilizer directly, or they can be material for composting (see further section). The liquid part can be used in the substrate preparation as dilution water; however, great caution must be given to nutrient build-up and consequently possible inhibition in the anaerobic digestion. Usually only a portion of that liquid is used in the substrate preparation. The rest must be further aerobically treated as a wastewater.

32.3.3.7   Biogas Production, Storage, Treatment, and Use

When operating a biogas plant, biogas is the main product and considerable attention must be given to its production, storage treatment, and use. Biogas production completely depends on the efficiency of the anaerobic digestion and its microorganisms. Previous sections have shown what conditions must be met to successfully operate anaerobic digestion. There are two distinct parameters that describe the biogas production:

1. *Specific biogas productivity—SBP* (also called biogas yield) is defined as volume of biogas produced per mass of substrate inserted into digester (m^3/kg). There are variations; SBP can be expressed in m^3 per kg of substrate mass, total solids, volatile-organic solids or COD. SBP tells us how much biogas we have gained from the substrate. Maximum possible SBP for a certain substrate is called biogas potential. Biogas potential can be determined by the standard method (ISO 1998).
2. *Biogas production rate—BPR* is defined as volume of biogas produced per volume of the digester per day ($m^3 \ m^{-3}$/day). BPR tells us how much biogas we can gain from the active volume of a digester in a day.

Typically, SBP values of an optimally operating digester are 80–90% of the biogas potential. Typical values of SBP for farm waste and liquid manure are shown in Table 32.7 and Table 32.8. Typical values of BPR for mesophilic digesters are from 0.9 to 1.3 $m^3 \ m^{-3}$/day. Lower values indicate the digester is oversized; higher values are rare or impossible, because of anaerobic process failure. For thermophilic or two-stage digesters, the typical BPR values are from 1.3 to 2.1 $m^3 \ m^{-3}$/day, respectively. UASB reactors are much less volume demanding and can achieve a BPR of up to 10 $m^3 \ m^{-3}$/day.

Biogas production is rarely constant and fluctuates (because of different loading rates, possible inhibitions, etc.). Therefore, a buffer is required for the biogas storage. This enables the biogas user

TABLE 32.7
Farm Waste Substrates and Biogas Yields

Substrate for Biogas Production	TS (%)	VS in TS (%)	Biogas Yield (m³/kg of Organic TS)
Spent grain	20–26	80–95	0.5–1.1
Yeast	10–18	90–95	0.72
Stomach content of pigs	12–15	80–84	0.3–0.4
Rumen content (untreated)	12–16	85–88	0.3–0.6
Vegetable wastes	5–20	76–90	0.3–0.4
Fresh greens	12–42	90–97	0.4–0.8
Grass cuttings (from lawns)	20–37	86–93	0.7–0.8
Grass silage	21–40	87–93	0.6–0.8
Corn silage	20–40	94–97	0.6–0.7
Straw from cereals	86	89–94	0.2–0.5
Cattle manure (liquid)	6–11	68–85	0.1–0.8
Cattle excreta	25–30	75–85	0.6–0.8
Pig manure (liquid)	2–13	77–85	0.3–0.8
Pig excreta	20–25	75–80	0.2–0.5
Chicken excreta	10–29	67–77	0.3–0.8
Sheep excreta	18–25	80–85	0.3–0.4
Horse excreta	25–30	–	0.4–0.6
Waste milk	8	90–92	0.7
Whey	4–6	80–92	0.5–0.9

TABLE 32.8
Liquid Manure and Biogas Yield per LSU

	LSU	Liquid Manure [m³ per animal] per Day	per Month	per Year	TS (%)	Biogas Yield (m³ LSU/Day)
Cattle						
Feeder cattle, cow	1	0.05	1.5	18.0		
Dairy cow, stock bull, trek ox	1.2	0.055	1.65	19.8		
Feeder bull	0.7	0.023	0.69	8.3	7–17	0.56–1.5
Young cattle (1–2 years)	0.6	0.025	0.75	9.0		
Calf breeding (up to 1 year)	0.2	0.008	0.24	2.9		
Feeder calf	0.3	0.004	0.12	1.4		
Pigs						
Feeder pigs	0.12	0.0045	0.14	1.62		
Sow	0.34	0.0045	0.14	1.62		
Young pigs (up to 12 kg)	0.01	0.0005	0.015	0.18	2–13	0.60–1.25
Young pigs (12–20 kg)	0.02	0.001	0.03	0.36		
Young pigs (over 20 kg)	0.06	0.003	0.09	1.08		
Young pigs (45–60 kg) Young sow (up to 90 kg)	0.16	0.0045	0.14	1.62		
Poultry						
Young feeder, poultry, young hens (up to 1200 g)	0.0023	0.0001	0.006	0.07		
Young feeder, poultry, young hens (up to 800 g)	0.0016	0.0001	0.006	0.07	20–34	3.5–4.0
Lying hen (up to 1600 g)	0.0030	0.0002	0.006	0.07		

LSU, Livestock unit (500 kg of live weight)

to get a constant biogas flow. Most modern biogas plants are equipped with co-generation units, also called combined heat and power units (CHP), which require constant gas flow for steady and efficient operation. There are several possibilities of biogas storage: they can roughly be divided into low-pressure (10–50 mbar) and high-pressure storage (over 5 bar). Low-pressure storage is used in on-site installations and for gas grid delivery; high pressure storage is used for long tem storage, for distant transport in high pressure tanks, and in installations with scarce space for volume extensive low pressure holders.

Low pressure biogas holders come in many variations. It is possible to include a biogas holder in the design of the digester. The most known is the digester with a movable cover. These digesters are less common because a movable cover requires increased investment and operating expenditure. More common are external biogas holders that are widely commercially available. An example of a modern biogas holder is presented in Figure 32.9.

Low pressure biogas holders require an extensive volume of 30 to 2000 m³ (Deublin and Steinhauser 2008). Usually, the pressure is kept constant and the volume of the bag is varied. High pressure biogas holders are made of steel, are of constant volume, and are subject to special safety requirements. They do require more complex equipment for compression and expansion of the gas and are harder and more cost effective to operate and maintain.

Biogas is not pure methane and carbon dioxide, but contains many components such as water droplets, dust, substrate microparticles and trace gases. Therefore, biogas treatment is necessary to preserve equipment. Solid particles can be filtered out, and sludge and foam are separated in cyclones. For removal of trace gases, where hydrogen sulfide (H_2S) is the most disturbing because of its corrosive properties, processes like scrubbing, adsorption, and absorption are used. In some cases, drying is also required (usually to the relative humidity of less than 80%).

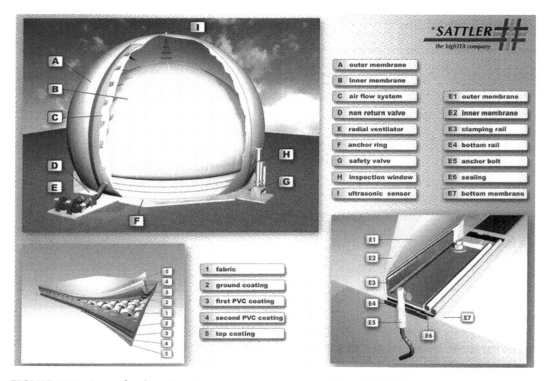

FIGURE 32.9 (See color insert) An example of a commercially available biogas holder. (From Sattler, Biogas storage systems, 2009. http://www.sattler-ag.com/sattler-web/en/products/190.htm)

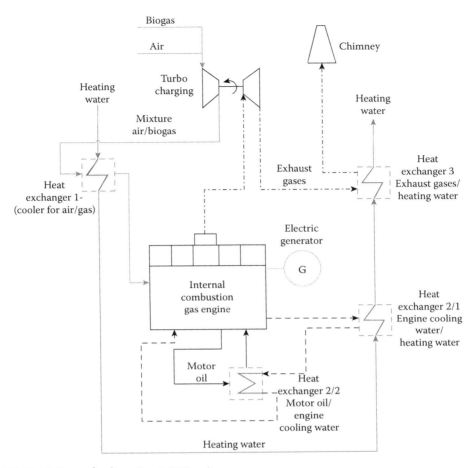

FIGURE 32.10 (See color insert) A CHP unit.

After treatment, biogas is used to produce energy. The most common way is to use all biogas in a cogeneration plant (CHP unit) to produce power and heat simultaneously (Figure 32.10). In this case, we can achieve maximum power production and enough excess heat to run the digesters. The energy required for operation of the digester is also called parasitic energy. The anaerobic digesters require heat for heating the substrate to operating temperature and compensating the digester heat losses. The digester also requires energy for mixing, substrate pumping, and pretreatment. The largest portion of heating demands in the digester operation is substrate heating. It requires over 90% of all heating demands, and only up to 10% is required for heat loss compensation (Zupancic and Ros 2003). In mesophilic digestion, a CHP unit delivers enough heat for operation, whereas in thermophilic digestion, additional heat is required. This additional heat demand can be covered with heat regeneration form substrate outflow to substrate inflow (Figure 32.11). Usually, a conventional countercurrent heat exchanger is sufficient; however, a heat pump can be applied as well.

Electric energy is also required for digester operation. It is required for pumping, mixing, and control and regulation equipment. In practice, no more than 10–15% of electric energy produced should be used for parasitic demands. Naturally, the pretreatment process may require also electric or heat energy. Pretreatment in the rule enhances anaerobic digestion, producing higher biogas production rates. However implications of pretreatment methods must be carefully considered. The golden rule is that pretreatment should not spend more energy than the energy of the biogas it is producing. If the energy use and production balances out, pretreatment may have benefits such as

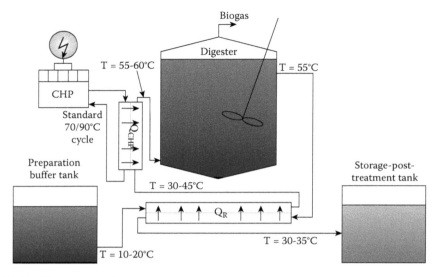

FIGURE 32.11   Heat regeneration.

more stable digested substrate, smaller digesters, pathogen removal, etc. There are substrates that require extensive pretreatment; especially this is the case for lignocellulosic material (like spent brewery grains) that require energy intensive pretreatment to be successfully digested at all. In such cases, the energy loss for pretreatment must be accounted for in the energy production. In many cases it cannot outweigh the economy of the process; it may well happen that the parasitic energy demand is too high.

32.4   WASTEWATER TREATMENT

32.4.1   AGRICULTURE WASTEWATER TREATMENT

The different types of contamination of wastewater require a variety of strategies to remove the contamination.

32.4.1.1   Solids Removal

Most solids can be removed using simple sedimentation techniques with the solids recovered as slurry or sludge. Very fine solids and solids with densities close to the density of water pose special problems. In such case, filtration or ultrafiltration may be required. Alternatively, flocculation may be used using alum salts or the addition of polyelectrolytes

32.4.1.2   Removal of Biodegradable Organics

Biodegradable organic material of plant or animal origin is usually possible to treat using extended conventional wastewater treatment processes such as activated sludge or trickling filter.

Activated sludge is a biochemical process for treating sewage and industrial wastewater that uses air (or oxygen) and microorganisms to biologically oxidize organic pollutants, producing a waste sludge (or flock) containing the oxidized material. In general, an activated sludge process includes:

- An aeration tank where air (or oxygen) is injected and thoroughly mixed into the wastewater.
- A settling tank (usually referred to as a "clarifier" or "settler") to allow the waste sludge to settle. Part of the waste sludge is recycled to the aeration tank and the remaining waste sludge is removed for further treatment and ultimate disposal.

32.4.1.3 Treatment of Other Organics

Synthetic organic materials including solvents, paints, pharmaceuticals, pesticides, coking products, and so forth can be very difficult to treat. Treatment methods are often specific to the material being treated. Methods include advanced oxidation processing, distillation, adsorption, vitrification, incineration, chemical immobilization, and landfill disposal. Some materials such as some detergents may be capable of biological degradation and in such cases, a modified form of wastewater treatment can be used.

32.4.1.4 Treatment of Toxic Materials

Toxic materials, including many organic materials, metals (such as zinc, silver, cadmium, thallium etc.) acids, alkalis, and nonmetallic elements (such as arsenic or selenium), are generally resistant to biological processes unless very dilute. Metals can often be precipitated out by changing the pH or by treatment with other chemicals. Many, however, are resistant to treatment or mitigation and may require concentration followed by landfilling or recycling. Dissolved organics can be incinerated within the wastewater by advanced oxidation processes.

Wastewater originated at farms can be treated with different systems. Simple systems are anaerobic lagoons (Ritter et al. 1984; Oleszkiewicz 1985; Boiran et al. 1996), where organic compounds could be effectively treated by a pretreatment system. The problem with lagoons is bad construction, especially the lagoon bottom. Such lagoons may influence groundwater because of percolation of wastewater through soil.

Another very frequently used wastewater system is the UASB reactor (Kalyuzhnyi et al. 1998; Correa et al. 2003; Garcia et al. 2008; Mahmoud 2008; Yetilmezsoy and Sakar 2008, which is useful for strong wastewaters. During the treatment at anaerobic conditions, we can obtain biogas that contains up to 70% methane gas.

Many wastewaters, especially from pig farms, are treated in SBR systems (Bernet 2000; Ra 2000; Tilche 2001; Kishida et al. 2003; Obaja et al. 2003, 2005; Ndegwa 2004; Zhang et al. 2006).

32.4.2 Upflow Anaerobic Sludge Blanket

UASB technology (Figure 32.12), normally referred to as the UASB reactor, is a form of anaerobic digester that is used in the treatment of highly polluted wastewater.

The UASB reactor is a methanogenic (methane-producing) digester that evolved from the anaerobic digester. A similar but variant technology to UASB is the expanded granular sludge bed (EGSB) digester.

UASB uses an anaerobic process while forming a blanket of granular sludge that suspends in the tank. Wastewater flows upwards through the blanket and is processed (degraded) by the anaerobic microorganisms. The upward flow combined with the settling action of gravity suspends the blanket with the aid of flocculants. The blanket begins to reach maturity at around three months. Small sludge granules begin to form, whose surface area is covered in aggregations of bacteria. In the absence of any support matrix, the flow condition creates a selective environment in which only those microorganisms, capable of attaching to each other, survive and proliferate. Eventually the aggregates form into dense compact biofilms referred to as "granules."

Biogas with a high concentration of methane is produced as a byproduct, and this may be captured and used as an energy source to generate electricity for export and to cover its own running power. The technology needs constant monitoring when put into use to ensure that the sludge blanket is maintained and not washed out (thereby losing the effect). The heat produced as a byproduct of electricity generation can be reused to heat the digestion tanks.

The blanketing of the sludge enables a dual solid and hydraulic (liquid) retention time in the digesters. Solids requiring a high degree of digestion can remain in the reactors for periods up to

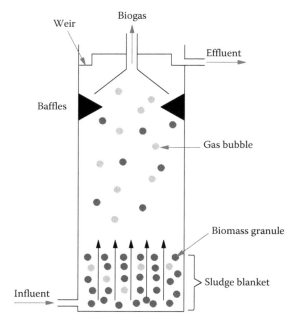

FIGURE 32.12 (See color insert)   Scheme of UASB.

90 days. Sugars dissolved in the liquid waste stream can be converted into gas quickly in the liquid phase, which can exit the system in less than a day.

32.4.2.1   Wastewater Characteristics

Wastewater that contains substances that can adversely affect sludge granulation, cause foaming, or cause scum formation is of concern. Wastewaters with higher concentration of proteins and/or fats tend to create more and more such problems. The fraction of particulate versus soluble COD is important in determining the design loadings for UASB reactors and the applicability of the process. As the fraction of solids in the wastewater increases, the ability to form a dense granulated sludge decreases. At a certain solids concentration (greater than 6 g TSS/L) anaerobic digestion and anaerobic contact processes may be more appropriate.

32.4.2.2   Volumetric Loading

Typical COD loadings as a function of the wastewater strength, fraction of particulate COD in the wastewater, and TSS concentration are summarized in Table 32.9. Removal efficiencies of up to 95% for COD have been achieved at COD loadings ranging from 12 to 90 kg_{COD} m^{-3}/day on a variety of wastes at 30° to 38°C with UASB reactors. Values for hydraulic retention time (HRT) for high-strength wastewater have been as low as 4 to 8 h at these loadings. Where less than 90% COD removal and higher-effluent TSS concentrations are acceptable, higher upflow velocities can be used that will develop a more dense granulated sludge by flushing out other solids. Thus, the higher volumetric COD loadings are shown for these conditions.

Recommended loadings as a function of temperature for wastewaters with mainly soluble COD are presented in Table 32.10. These loadings apply to the sludge blanket volume, and a reactor effectiveness factor of 8.8 to 0.9 is used to determine the reactor liquid below the gas collector. The higher loading recommendation for the wastewater containing mainly volatile fatty acids (VFA) is based on the potential of obtaining a more dense granulated sludge.

A typical HRT is shown in Table 32.11.

TABLE 32.9

Recommended Volumetric COD Loading for UASB Reactors at 30°C to Achieve 85–95% COD Removal

	Volumetric Loading (g COD m⁻¹/day)			
Wastewater COD (mg/L)	Fraction as Particulate COD	Flocculent Sludge	Granular Sludge with High TSS Removal	Granular Sludge with Little TSS Removal
1000–2000	0.10–0.30	2–4	2–4	8–12
	0.30–0.60	2–4	2–4	8–14
2000–6000	0.10–0.30	3–5	3–5	12–18
	0.30–0.60	4–8	2–6	12–24
	0.60–1.00	4–8	2–6	NA
6000–9000	0.10–0.30	4–6	4–6	15–20
	0.30–0.60	5–7	3–7	15–24
	0.60–1.00	6–8	3–8	NA
9000–18,000	0.10–0.30	5–8	4–6	15–24
	0.30–0.60	NA	3–7	NA
	0.60–1.00	NA	3–7	NA

NA, not applicable.

TABLE 32.10

Recommended Volumetric Organic Loadings as a Function of Temperature for Soluble COD Substrates for 85 to 95% COD Removal (Average Sludge Concentration Is 25 g/L)

	Volumetric Loading (kg sCOD m⁻³/day)			
	VFA Wastewater		Non-VFA Wastewater	
Temperature (°C)	Range	Typical	Range	Typical
15	2–4	3	2–3	2
20	4–6	5	2–4	3
25	6–12	6	4–8	4
30	10–18	12	8–12	10
35	15–24	18	12–18	14
40	20–32	25	15–24	18

Source: Lettinga, G. and Hulshoff, L.W., UASB-process design for various types of wastewaters. *Water Sci Technol*, 24, 87–107, 1991.

32.4.2.3 Upflow Velocity

The upflow velocity, based on the flow rate and reactor area, is a critical design parameter. Recommended design velocities are shown in Table 32.12. Temporary peak superficial velocities of 6 m·s⁻¹ and 2 m·s⁻¹ can be allowed for soluble and partly soluble wastewater, respectively.

For weaker wastewaters, the allowable velocity and reactor height will determine the UASB reactor volume, and for stronger wastewater it will be determined by the volumetric COD loading.

TABLE 32.11

Applicable HRT for Treatment of Raw Domestic Wastewater in a UASB Reactor

Temperature (°C)	Average HRT (h)	Maximum HRT for 4- to 6-h Peak (h)
16–19	10–14	7–9
22–26	7–9	5–7
>26	6–8	4–5

Source: Lettinga, G. and Hulshoff, L.W., UASB-process design for various types of wastewaters. *Water Sci Technol*, 24, 87–107, 1991.

TABLE 32.12

Upflow Velocities and Reactor Heights Recommended for UASB Reactors

Wastewater Type	Upflow Velocity [m/h] Range	Typical	Reactor Height [m] Range	Typical
COD nearly 100% soluble	1.0–3.0	1.5	6–10	8
COD partly soluble	1.0–1.25	1.0	3–7	6
Domestic wastewater	0.8–1.0	0.7	3–5	5

32.4.3 SEQUENCING BATCH REACTOR

Sequencing or sequential batch reactors (SBR) are industrial processing tanks for the treatment of wastewater. SBR reactors treat wastewater such as sewage or output from anaerobic digesters or mechanical biological treatment facilities in batches. Oxygen is bubbled through the wastewater to reduce biochemical oxygen demand (BOD) and chemical oxygen demand (COD) to make it suitable for discharge into sewers or for use on land.

A sequencing batch reactor in the system with activated sludge (Orhon 1994) operates on the basis fill-and-draw. In recent years, the modification of the fill-and-draw process is intensifying as an SBR system. SBR offers various advantages in comparison with a conventional activated sludge system.

All wastewater treatment plants that were in operation between 1914 and 1920 were designed as fill-and-draw systems. When continuous flow activated sludge systems were developed, interest for sequencing batch reactors extremely declined. In the early 1960s, SBR systems began to reappear with the development of new technology and equipment (Dennis 1979; Hoepker et al. 1979; Irvine et al. 1979; Ketchum 1979; Irvine and Moe 2001).

The SBR process is composed of one reactor or by a series of parallel reactors where complete treatment procedures occur: wastewater treatment and separation of sludge from treated wastewater. The operating principles of the SBR are characterized by five discrete periods: fill, react, settle, decant, and idle. When the SBR is subjected to sequential redox environments (anaerobic/anoxic/aerobic conditions) during the react period, it provides the removal of organic substrate and nutrients simultaneously. Ammonium is oxidized to nitrite and nitrate (nitrification) in the aerobic phase and nitrate is reduced to N_2 (denitrification) in the anoxic phase of the react period. Organic substrate from the wastewater is oxidized in the anoxic phase in the denitrification process.

Although the SBR process looks like the classical *fill-and-draw* process with activated sludge, the development of SBR is most recent (Irvine et al. 1979; Chambers 1993; Wilderer 2001).

The purpose of recent research was to stress advantages of SBR in comparison with conventional flow systems. The principal investigations are publications of Dennis and Irvine (Dennis 1979; Irvine 1979; Ketchum 1979), in which they studied the effects of *fill/react* ratios. Hoepker et al. (1979) found out that smaller loading has given better quality of effluent. Ketchum et al. (1979) studied the possibilities of SBR for tertiary treatment, especially for phosphorus removal. Obaja and Mace (1985) presented the possibilities of nitrification and denitrification of a given plan and control of the process.

The majority of advantages of the SBR can be attributed to the flexible nature of operation parameters (Hvala 2001; Kazmi 2001; Miklos 2001; Morling 2001; Ng 2001; Ruiz 2001; Tilche 2001; Yalmaz 2001; Yoong 2001). A large choice of system parameters can be the consequence of constant volume, where we can change *fill/react* ratios and the time of aeration. Flexibility of operation parameters also enables understanding of basic mechanisms of the process and critical phases that are very important for further application (Wareham 1993; Orhon 1994; Zec 1997; Paul 1998; Wu 2001). Recently, a unified basis of design for SBRs was prepared mainly covering practical aspects of SBR technology and emphasizing the need for appropriate design guidelines (Artan 2001; Artan et al. 2006).

Although there are several configurations of SBRs, the basic process is similar. The installation consists of at least two identically equipped tanks with a common inlet that can be switched between them. The tanks have a "flow through" system, with raw wastewater (influent) coming in at one end and treated water (effluent) flowing out the other. While one tank is in settle/decant mode, the other is aerating and filling. At the inlet is a section of the tank known as the bioselector. This consists of a series of walls or baffles that direct the flow either from side to side of the tank or under and over consecutive baffles. This helps to mix the incoming influent and the returned activated sludge, beginning the biological digestion process before the liquor enters the main part of the tank.

There are four stages to treatment: fill, aeration, settling and decanting. The aeration stage involves adding air to the mixed solids and liquid either by the use of fixed or floating mechanical pumps or by blowing it into finely perforated membranes fixed to the floor of the tank. During this period, the inlet valve of the tank is open and a returned activated sludge pump takes mixed liquid and solids (mixed liquor) from the outlet end of the tank to the inlet. This "seeds" the incoming sewage with live bacteria.

Aeration times vary according to the plant size and the composition/quantity of the incoming liquor, but are typically 60–90 minutes. The addition of oxygen to the liquor encourages the multiplication of aerobic bacteria and they consume the nutrients. This process encourages the production of nitrogen compounds as the bacteria increase their number, a process known as nitrification.

To remove phosphorus compounds from the liquor, aluminum sulfate (alum) is often added during this period. It reacts to form nonsoluble compounds that settle into the sludge in the next stage.

The settling stage is usually the same length of time as the aeration. During this stage the sludge formed by the bacteria is allowed to settle to the bottom of the tank. The aerobic bacteria continue to multiply until the dissolved oxygen is all but used up. Conditions in the tank, especially near the bottom, are now more suitable for the anaerobic bacteria to flourish. Many of these, and some of the bacteria that would prefer an oxygen environment, now start to use nitrogen as a base element and extract it from the compounds in the liquid, using up the nitrogen compounds created in the aeration stage. This is known as denitrification.

As the bacteria multiply and die, the sludge within the tank increases over time and a waste activated sludge pump removes some of the sludge during the settle stage to a digester for further treatment. The quantity or "age" of sludge within the tank is closely monitored, as this can have a marked effect on the treatment process.

The sludge is allowed to settle until clear water is on the top 20–30% of the tank contents.

The decanting stage most commonly involves the slow lowering of a scoop or "trough" into the basin. This has a piped connection to a lagoon where the final effluent is stored for disposal to a wetland, tree growing lot, ocean outfall, or to be further treated for use on parks, golf courses, etc.

The operating principles of a batch activated sludge process, or SBR, are characterized in six discrete periods:

1. Filling
2. Reacting
3. Settling
4. Decanting
5. Idling
6. Sludge wasting

32.4.3.1 Fill

During this stage the SBR tank is filled with the influent wastewater. To maintain suitable F/M (food to microorganism) ratios, the wastewater should be admitted into the tank in a rapid, controlled manner. This method functions similarly to a selector that encourages the growth of certain microorganisms with better settling characteristics.

32.4.3.2 React

This stage involves the utilization of biochemical oxygen demand (BOD) and ammonia nitrogen, where applicable, by microorganisms. The length of the aeration period and the sludge mass determines the degree of treatment. The length of the aeration period depends on the strength of the wastewater and the degree of nitrification (conversion of the ammonia to a less toxic form of nitrate or nitrite) provided for in the treatment.

32.4.3.3 Settle

During this stage, aeration is stopped and the sludge settles, leaving clear, treated effluent above the sludge blanket. Duration of settling varies from 5 to 60 minutes depending on the number of cycles per day and sludge settling characteristics.

32.4.3.4 Decant

At this stage of the process, effluent is removed from the tank through the decanter without disturbing the settled sludge.

32.4.3.5 Idle

The SBR tank waits idle until it is time to commence a new cycle with the filling stage.

32.4.3.6 Sludge Wasting

Excess activated sludge is wasted periodically during the SBR operation. As with any activated sludge treatment process, sludge wasting is the main control of the effluent quality and microorganism population size. This is how the operator exerts control over the effluent quality by adjusting the mixed liquor suspended solids (MLSS) concentration and the mean cell residence time (MCRT).

32.5 MUNICIPAL SLUDGE TREATMENT

Sludge (solids) stabilization processes are very important for reliable performance of any wastewater treatment plant. For successful operation of solids treatment facilities, the sources, characteristics, and quantities of the treated solids have to be carefully evaluated, for the design of these facilities to offer optimum performance for minimum investment and operational costs.

32.5.1 Sludge Sources, Characteristics, and Properties

There are several sources of sludge or solids. Sludge can originate from chemical processes such as precipitation or coagulation; physical processes such as sedimentation; or biological processes such as in the wastewater treatment plants. Solids are constituted of organic and inorganic solids. Municipal sludge is mainly of biological origin and is produced in wastewater treatment plants. Such solids are mostly organic (from 50 to 90% is organic component). The sources of solids vary according to the type of the plant and its method of operation. The principal sources of solids and the types are presented in Table 32.13.

To treat and dispose of the solids originated in wastewater treatment plants effectively, it is important to know the characteristics of the solids to be processed. The characteristics vary depending on the origin of the solids, the amount of aging that has taken place, and the type of processing to which they have been subjected. Solids characteristics are gathered in Table 32.14.

32.5.2 General Composition

Typical data on the chemical composition of untreated solids are reported in Table 32.15. Many chemical constituents, including nutrients, are important in considering the ultimate disposal of the processed solids and liquid removed during processing. The measurement of pH, alkalinity, and organic acid content is important in process control of sludge digestion. The content of heavy metals, pesticides and hydrocarbons has to be determined when land application methods are considered. The thermal value of solids is important where incineration is considered.

Waste sludge is generally composed of solids and water. Solids are only up to 10% of sludge (often referred to as total dry solids or just total solids); the rest is water. Solids are then generally divided in organic solids (often referred to as volatile solids) and inorganic solids.

One of the most important factors in sludge digestion is the organic component in solids (volatile solids). From this parameter depends how much solids can be removed. The higher the organic content in solids, the better the removal rate will be, and in case of anaerobic digestion, this means higher biogas production. Also, the thermal value of sludge increases with the content of organic solids in sludge. The inorganic components of sludge (inorganic solids) mainly stay inactive throughout the sludge digestion process.

TABLE 32.13
Sources of Solids from Conventional Wastewater Treatment Plants

Unit operation or process	Types of solids	Remarks
Screening	Coarse solids	Coarse solids are removed by mechanical and hand cleaned bar screens. In small plants solids are often comminuted for removal in subsequent treatment units
Grit removal	Grit and scum	Scum removal facilities are often omitted in grit-removal facilities.
Preaeration	Grit and scum	In some plants, scum removal facilities are not provided in preaeration tanks. If the preaeration tanks are not preceded by grit removal facilities, grit deposition may occur in preaeration tanks.
Primary sedimentation	Primary solids and scum	Quantities of solids and scum depend upon the nature of the collection system and whether industrial wastes are discharged to the system.
Biological treatment	Suspended solids	Suspended solids are produced by the biological conversion of BOD. Some thickening may be required to concentrate the waste sludge stream from the biological treatment system.
Secondary sedimentation	Secondary biosolids and scum	For good wastewater treatment it is necessary to remove solids from secondary settling tanks.

TABLE 32.14
Characteristics of Solids Produced during Wastewater Treatment

Solids or Sludge	Description
Screenings	Screenings include all types of organic and inorganic materials large enough to be removed on bar racks. The organic content varies, depending on the nature of the system and the season of the year. This sludge is usually not the subject of digestion processes, because of the large size of particles/pieces. It is generally disposed in a landfill.
Grit	Grit is usually made up of the heavier inorganic solids that settle with relatively high velocities. Depending on the conditions, grit may also contain significant amounts of organic matter, especially fats and grease.
Scum/grease	Scum consists of the floatable materials skimmed from the surface of primary and secondary settling tanks and from grit chambers and chlorine contact tanks, if so equipped. Scum may contain grease, vegetable and mineral oils, animal fats, waxes, soaps, food wastes, vegetable and fruit skins, hair, paper and cotton, cigarette tips, plastic materials, condoms, grit particles, and similar material. The specific gravity of scum is less than 1.0 (usually around 0.95).
Primary sludge	Sludge from primary tanks is usually gray in color, slimy and in most cases, has an extremely offensive odor. If sludge is left in storage, it can readily be partially digested.
Sludge from chemical precipitation	Sludge from chemical precipitation with metal salts is usually dark in color, though its surface may be red if it contains much iron. Lime sludge is gray-brown. The odor of chemical sludge may be unpleasant, but not as offensive as of primary sludge. The hydrate of iron or aluminium in it makes the chemical sludge gelatinous. If the sludge is left in the tank it undergoes similar decomposition as primary sludge, only at slower rate.
Activated sludge	Activated sludge generally has a brown flocculent appearance. If the color is dark, the sludge may be approaching a septic condition. If the color is lighter than usual, there may have been under aeration with a tendency for the solids to settle slowly. Sludge in good condition has an inoffensive "earthy" odor. The sludge tends to become septic rapidly and then has an odor of putrefaction Activated sludge will digest alone or when mixed with primary sludge.
Trickling filter sludge	Humus sludge from trickling filters is brown in color, flocculent and relatively inoffensive when fresh. It generally undergoes decomposition more slowly then other undigested sludges. When trickling filter sludge contains many worms, it may become inoffensive quickly. Trickling filter sludge digests readily.

32.5.2.1 Specific Constituents: Trace Compounds, Nitrogen, Phosphorous, Potassium, and Heavy Metals

The consideration of these constituents is mainly important when sludge is applied to land. Sludge characteristics that affect the suitability for land application and for beneficial use include organic content (usually measured as volatile solids), nutrients, pathogens, metals, and toxic organics. In most cases when sludge is applied to land, there are sufficient nutrients present for good plant growth. However, trace elements and heavy metals may limit the sludge application to land. Trace elements are those inorganic chemical elements that in very small quantities can be essential or detrimental for plants and animals. Heavy metals are some of these trace elements in the sludge and mostly are the limiting factor for sludge land application. The typical heavy metal content in wastewater sludge is presented in Table 32.16.

32.5.2.2 Sludge Quantities

Data on the sludge quantities produced from various processes and operations are presented in Table 32.17. Although data in Table 32.8 can be useful as presented, it must be noted that quantity of sludge can vary widely. The contribution of raw municipal wastewater (in the U.S.) is about 400 L PE^{-1}/day (Novotny et al. 1989; Hammer 1996) with total solids of less than 0.1%, 100–350 mg/L suspended solids and 110–400 mg/L BOD$_5$. These numbers add up to sludge quantity of 54

TABLE 32.15
Typical Chemical Composition of Untreated Sludge

Item	Untreated Primary Sludge Range	Untreated Activated Sludge Range
Total dry solids (TS), %	5–9	0.8–1.2
Volatile solids (% of TS)	60–80	60–90
Grease and fats (% of TS)	13–35	5–12
Protein (% of TS)	20–30	32–41
Nitrogen (N, % of TS)	1.5–4.0	2.4–5.0
Phosphorous (P_2O_5, % of TS)	0.8–2.8	2.8–11
Potassium (K_2O, % of TS)	0–1	0.5–0.7
Cellulose (% of TS)	8–15	–
Iron (not as sulfide)	2.0–4.0	–
Silica (SiO_2, % of TS)	15–20	5–20
pH	5.0–8.0	6.5–8.5
Alkalinity (mg/l as $CaCO_3$)	500–1500	580–1100
Organic acids (mg/l as HAc)	200–2000	1100–1700
Energy content (kJ/kg TS)	23000–29000	19000–23000

TABLE 32.16
Heavy Metal Content in Wastewater Sludge

Metal	Dry Solids (mg/kg) Range	Median
Arsenic	1.1–230	10
Cadmium	1–3410	10
Chromium	10–99,000	500
Cobalt	11.3–2490	30
Copper	84–17,000	800
Iron	1000–154,000	17,000
Lead	13–26,000	500
Manganese	32–9870	260
Mercury	0.6–56	6
Molybdenum	0.1–214	4
Nickel	2–5300	80
Selenium	1.7–17.2	5
Tin	2.6–329	14
Zinc	101–49,000	1700

g PE^{-1}/day for primary sludge (thickened with solids content of 5 mass %) and secondary sludge of 25 g PE^{-1}/day at trickling filters (solids content of 4 mass %) or 35 g PE^{-1}/day at activated sludge process (solids content of 0.7 mass %) (Novotny et al. 1989). In Europe, the contribution of raw municipal wastewater is 200 L PE^{-1}/day with BOD$_5$ value of 60 g BOD$_5$ PE^{-1}/day. These numbers add up to sludge quantity of 45 g PE^{-1}/day for primary sludge (thickened with solids content of 5 mass %) and secondary sludge of 25 g PE^{-1}/day at trickling filters (solids content of 4 mass %) or 35 g PE^{-1}/day at activated sludge process (solids content of 0.7 mass %) (Novotny et al. 1989). Thickened mixed primary and secondary sludge quantity (in both U.S and Europe) amounts 80 g PE^{-1}/day (solids content of 4 mass %). The typical sludge concentrations are presented in Table 32.18.

TABLE 32.17

Typical Data for the Physical Characteristics and Quantities of Sludge Produced from Various Wastewater Treatment Operations and Processes

Treatment Operation or Process	Specific Gravity of Solids	Specific Gravity of Sludge	Dry Solids (kg 10^{-3}/m^3)	
			Range	Typical
Primary sedimentation	1.4	1.02	110–170	150
Activated sludge (waste sludge)	1.25	1.005	70–100	80
Trickling filter (waste sludge)	1.45	1.025	60–100	70
Extended aeration (waste sludge)	1.30	1.015	80–120	100[a]
Aerated lagoon (waste sludge)	1.30	1.01	80–120	100[a]
Filtration	1.20	1.005	12–24	20
Algae removal	1.20	1.005	12–24	20
Chemical addition to primary tanks for phosphorous removal				
Low lime (350–500 mg/L)	1.9	1.04	240–400	300[b]
High lime (800–1600 mg/L)	2.2	1.05	600–1300	800[b]
Suspended growth nitrification	Negligible			
Suspended growth denitrification	1.20	1.005	12–30	18
Roughing filters	1.28	1.02	–	–[c]

[a] Assuming no primary treatment.
[b] Solids in addition to that normally removed by primary sedimentation.
[c] Included in solids production from secondary treatment processes.

32.5.2.3 Mass–Volume–Concentration Relationships

The volume of sludge depends mainly on its water content and only slightly on the character of the solid matter. A conventional 4% sludge, for example, contains 96% water by weight, has a concentration of 40.2901 g/L and a density of 1007.25 kg/m^3. Examples for the sludge relationships between concentration, density and percent of solids are gathered in Table 32.19.

32.5.2.4 Anaerobic Sludge Digestion, Biogas Production, and Use

Waste municipal sludge can be easily digested. Anaerobic digestion is the worldwide most common treatment for municipal sludge before disposal or further treatment. The properties and principles of anaerobic digestion have been thoroughly discussed in the previous section; in this section we will focus on specifics that concern municipal sludge itself. Generally municipal sludge does not cause any problems in anaerobic digestion if no inhibitors and toxic substances are present in the wastewater. Primary sludge (or any organic sludge that comes from sedimentation) requires no special pretreatment for anaerobic digestion. Secondary sludge (biological sludge from wastewater treatment processes) like activated sludge or trickling filter sludge are in fact cellular bacterial matter and take usually more time to be digested. The cells of microorganisms need to be ruptured for anaerobic microorganisms to process the material. Conventionally no special pretreatment methods are applied for primary sludge, except homogenization if large chunks are present. For secondary sludge, being cellular material some pretreatment methods can enhance the digestion and therefore lower the retention time in the digester (Pham et al. 2008; Salsabil et al. 2008). There are several methods present and commercially available; some of the most common are ultrasonic pretreatment, pressure drop pretreatment, and different kinds of thermal pretreatment. Ultrasonic pretreatment uses vibrating forks of 0.5 m in size at the frequency of 40,000–50,000 kHz, to generate ultrasound and rupture the cell walls. Pressure drop pretreatment uses pressures of 200 bar and subsequently a quick release to normal conditions. In this case, rapid expansion causes the cell to rupture. Thermal

TABLE 32.18

Expected Solids Concentrations from Various Treatment Operations and Processes

Operation or Process Application	Solids Concentration (% Dry Solids)	
	Range	Typical
Primary settling tank		
Primary sludge	5–9	6
Primary sludge to a cyclone degritter	0.5–3	1.5
Primary sludge and waste activated sludge	3–8	4
Primary sludge and trickling filter sludge	4–10	5
Primary sludge with iron salt addition for phosphorous removal	0.5–3	2
Primary sludge with low lime addition for phosphorous removal	2–8	4
Primary sludge with high lime addition for phosphorous removal	4–16	10
Scum	3–10	5
Secondary settling tank		
Waste activated sludge with primary settling	0.5–1.5	0.8
Waste activated sludge without primary settling	0.8–2.5	1.3
High purity oxygen with primary settling	1.3–3	2
High purity oxygen without primary settling	1.4–4	2.5
Trickling filter sludge	1–3	1.5
Rotating biological contactor sludge	1–3	1.5
Gravity thickener		
Primary sludge	5–10	8
Primary sludge and waste activated sludge	2–8	4
Primary sludge and trickling filter sludge	4–9	5
Dissolved air flotation thickener:		
Waste activated sludge with polymer addition	4–6	5
Waste activated sludge without polymer addition	3–5	4
Centrifuge thickener (waste activated sludge only)	4–8	5
Gravity belt thickener (waste activated sludge with polymer addition)	4–8	5
Anaerobic digester		
Primary sludge	2–5	4
Primary sludge and waste activated sludge	1.5–4	2.5
Primary sludge and trickling filter sludge	2–4	3
Aerobic digester		
Primary sludge	2.5–7	3.5
Primary sludge and waste activated sludge	1.5–4	2.5
Primary sludge and trickling filter sludge	0.8–2.5	1.3

pretreatment causes solubilization and thermal decomposition of lignocellulosic material to more simple molecules and therefore makes the sludge more easily digestible. In some cases sludge can be so solubilized that it can be treated in UASB reactors (under certain conditions) massively reducing digester size (D'abbieri et al. 2008). Pretreatments of sludge usually do not require more than 20% of parasitic energy.

The technology used for anaerobic digestion is usually a simple, single-stage process; rarely in practice are two-stage processes used, although they do offer a reduction in digester size. However, the control of the process is more demanding; therefore WWTP operators usually choose simpler technology. The vast majority of digesters are mesophilic, although in recent years thermophilic digesters are surfacing more often. It was believed for quite some time that mesophilic digesters use too much parasitic energy for operation and that was the reason for much larger application

TABLE 32.19

Sludge Concentration–Density–Percent of Solids Relationships (at 20°C)

Percent Sludge (w_s)	Concentration (g/L)	Sludge Density (kg/m³)	Concentration (g/L)	Percent Sludge (w_s)	Sludge Density (kg/m³)
0.5	5.0045	1000.90	5	0.4996	1000.90
1	10.0180	1001.80	10	0.9982	1001.80
2	20.0723	1003.61	20	1.9928	1003.60
3	30.1629	1005.43	30	2.9839	1005.40
4	40.2901	1007.25	40	3.9714	1007.20
5	50.4541	1009.08	50	4.9554	1009.00
6	60.6551	1010.92	60	5.9359	1010.80
7	70.8933	1012.76	70	6.9129	1012.60
8	81.1688	1014.61	80	7.8864	1014.40
9	91.4820	1016.47	90	8.8565	1016.20
10	101.8330	1018.33	100	9.8232	1018.00
15	154.1624	1027.75	150	14.6056	1027.00
20	207.4689	1037.34	200	19.3050	1036.00

of mesophilic technology. However, when heat regeneration is used (between sludge outflow and inflow) the parasitic energy demand of thermophilic and mesophilic processes are equal (Zupancic and Ros 2003). The mesophilic sludge digestion process can sustain OLR of 3.0–3.5 kg of volatile solids per m³ of digester per day (kg m^{-3}/day), whereas a thermophilic process can sustain OLR of up to 8.0 kg m^{-3}/day. This offers a digester size reduction of at least 50%, which can be beneficial in construction costs.

The biogas yield (SBP) of anaerobic sludge digestion (mesophilic or thermophilic) is between 400 and 600 L per kg of volatile solids inserted (L/kg), depending on sludge composition. This yields 24–36 L per PE per day of biogas; 7–10 W per PE of power potential and 2–3 W per PE of electric power potential. Biogas production rate varies in mesophilic digesters from 0.7 to 1.1 cubic meters per volume of digester per day (m^3·m^{-3}·d^{-1}); in thermophilic from 1.3 to 1.7 m^3 m^{-3}/day. Biogas composition is at 60–80% methane, the rest is mostly CO_2. H_2S is present in relatively low concentrations (up to 0.4%). A specific problem in sludge biogas may be the presence of siloxanes (Dewil et al. 2006). They originate from municipal wastewater, especially from cosmetic and detergent products containing silicon based compounds and which are washed out to wastewater from human use. At high temperatures of combustion, siloxanes are oxidized to SiO_2 which remains on the surface of the machine parts. There it causes abrasion of the pistons and that results in CHP breakdown. How much this phenomena affects the power generation fuel cells has not yet been investigated. To avoid this problem biogas must be treated and siloxanes removed. There are several treatments available (Schweigkofer and Niesser 2001); one of the modern treatment possibilities is removal with membranes (Ajhar and Melin 2006). Resent research is also focused from removal on siloxanes from the sludge before anaerobic digestion (Appels et al. 2008).

32.5.2.5 Further Treatment and Use of Digested Sludge (Biosolids) and Supernatant Water

After anaerobic digestion sludge is usually mechanically dewatered (25–40% of dry solids), the solid and liquid fraction are then further treated. Landfilling of biosolids is prohibited in most EU countries (EEP 1986), so other options must be accounted for. The most environment-friendly option would be fertilizer use. Contrary to farm waste, which is usually suitable for fertilizer use, biosolids from municipal sludge may contain trace elements that may prohibit fertilizer use (Singh and Agrawal 2008). Especially high concentrations of heavy metals such as Cu, Mn, Pb and Zn and some organic pollutants may restrict the use of biosolids for fertilizer (Aparicio et al.). If the pollutants are below the limits, biosolids can be beneficially used for fertilizers and soil enhancers. Biosolids can be used directly (some unpleasant odor problems may arise) or additionally composted to produce odorless soil like material. Anaerobically digested sludge still has enough organic material present to be composted (autothermally at 60°C) (Cukjati 2008). Subsequent sections present detailed information about composting.

Polluted biosolids are mostly incinerated or co-incinerated with other fuels. It is possible to recover some energy from the sludge incineration process. The main problem with incineration is low combustion value of dewatered wet sludge (Figure 32.13). Although dry sludge has the combustion value (CV) of 15–22 MJ/kg (fresh) or 8–15 MJ/kg (digested), it contains more than 50% water, which reduces the combustion value significantly. Because of sludge composition (cellular material) it is extremely hard and energy demanding to dewater sludge mechanically to values more than 40%. Therefore, sludge must be thermally dried, and this significantly reduces the efficiency of sludge incineration process. To improve the incineration process, some of the heat necessary for sludge drying may be regenerated within the incineration process.

Better efficiency of sludge treatment can also be achieved with modern thermal processes, either low temperature (Vieira et al. 2009) or high temperature method, like pyrolysis (Kim and Parker 2008; Hossain et al. 2009), producing many useful fuels.

The liquid fraction of sludge dewatering (sludge supernatant) must also be treated. It is rich with nutrients (total nitrogen over 1500 mg/L) and poor in COD (200-300 mg/L), and therefore a direct use is not advisable. Sludge supernatant is usually returned to the WWTP inflow and mixed with raw wastewater. Sludge supernatant does not present a significant portion in the volume flow or the COD/BOD load of the raw wastewater. However, it does present approximately. 50% of the WWTP ammonium load (30–40% total nitrogen load) (Table 32.20). The consequences of this fact are

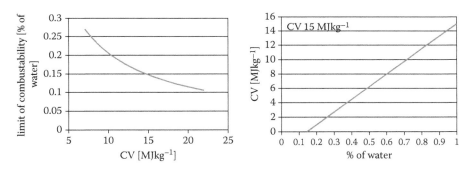

FIGURE 32.13 Combustion value (CV) and limit of combustibility of dewatered municipal sludge.

TABLE 32.20

Ammonium in the WWTP (Case of 200,000 PE WWTP)

	Flow (m³/year)	Average Ammonium Concentration (mg/L)	Ammonium (g)	Percent of Flow	Percent of Ammonium Load
Raw wastewater influent	6,744,312	10.6	71,489,707	99.3	49.9
Sludge supernatant	48,112	1487	71,542,544	0.7	50.1

Source: Zupancic, G.D. and Ros, M., Ammonia removal in sludge digestion utilizing nitrification with pure oxygen aeration. *Nutrient Management in Wastewater Treatment Processes and Recycle Streams*: IWA specialized conference, Krakow, Poland, IWA Publishing, London, 2005.

that the volumes of aerobic wastewater treatment tank for nitrogen removal significantly increase. Recent research pointed out some solutions for this problem. The addition of an aerobic stage to anaerobic digestion may transform 80% of the ammonia to nitrate (Zupancic and Ros 2008), which has a beneficial effect of raw wastewater. It provides oxygen in form of nitrate for anoxic treatment and consequently decreases BOD for the first stage of wastewater treatment.

32.5.2.6 Biosolids Composting

Activated sludge contains a lot of nutritive compounds and also many microorganisms. During the aerobic composting (the consequence of microorganisms), the energy is released in the form of heat.

The time courses for the important factors in the composting process for activated sludge (the amount of CO_2 generated, the temperature, weight change and water content of the composting materials), were simulated by means of regression analysis. Furthermore, we could accurately estimate the temperature change of the composting materials in the early phase of composting from the heat balance, including microbial heat generation estimated by regression analysis. At the optimum aeration regime, we can minimize the time required for the temperature to increase to 65°C, at which most pathogens in compost materials are killed (Kishimoto et al. 1987).

32.5.2.6.1 The Composition

The most important elements for microbe decomposition during the composting are carbon (C) and nitrogen (N). Carbon represents 50% of cell biomass and is bounded to all energetic processes in the cell. Nitrogen is a crucial element in proteins, amino acids, nucleic acids, and enzymes, one of the most important components of cell. The ratio C:N in compost substrate is between 25:1 to 35:1 (Tchobanoglous et al. 2003). When the ratio C:N is smaller than 25:1, larger losses of nitrogen in shape of free ammonia are occurring, which causes unpleasant smell (Larsen and McKartney 2000).

32.5.2.6.2 Microorganisms

Composting is a process caused by different microorganisms that decompose organic material to mineralized form. They need for their own development and activity suitable conditions in the environment: aeration (ventilation), moisture, and ration of nutrients. We can distribute the course of composting in three temperature ranges:

1. Mesophilic range, where the temperature is between 20 and 45°C. These fungi and bacteria prevail in the compost: *Alcaligenaceae, Alteromonadaceae, Bacillaceae, Burkholderiaceae, Bradyrhizobiaceae, Caryophanaceae, Caulobacteraceae, Cellulomonadaceae, Flavobacteriaceae, Flexibacteraceae, Hyphomicrobiaceae, Intrasporangiaceae, Methylobacteriaceae, Microbacteriaceae, Micrococcaceae, Moraxellaceae, Neisseriaceae, Nitrosomonadaceae, Nocardiaceae, Nocardiopsaceae, Paenibacillaceae, Phyllobacteriaceae, Propionibacteriaceae, Pseudomonadaceae, Pseudonocardiaceae, Rhodobacteraceae, Sphingobacteriaceae, Staphylococcaceae, Xanthomonadaceae* (Haruta et al. 2005).
2. Thermophilic range, where the temperature is between 45°C and 65°C. These fungi and bacteria prevail in the compost (Schloss et al. 2003): *Micromonasporaceae, Streptomycetaceae, Thermoactinomycetaceae, Thermomonosporaceae, Streptosporangiaceae, Hydrogenobacte, Thermus* spp.
3. Final mesophilic range, where the number of thermophilic organisms decreases, and in the same time increases the number of mesophilic bacteria and fungi. In this phase also higher organisms appear: protozoa, arthopods, and worms (Schurard 2005).

32.5.2.6.3 Oxygen Content

Maximal biodegradation in the compost runs in the range of over 15% of oxygen content. The concentration from 10% to 15% is still acceptable; values of oxygen under 5% decrease the process of biodegradation (Day and Shaw 2001).

Oxygen can be added into the compost mixture in different ways, natural ventilation being the simplest. At natural ventilation movement of air occurs upwards from the bottom, because of temperature difference between material and air. Efficiency of this process depends also upon structure and finenesses of material (Barrington et al. 2003).

The second way is artificial aeration from the bottom. With this manner we accelerate the procedure of composting and take care of moisture in the compost heap. The most effective manner is the usage of closed bioreactors where we can control all process parameters.

32.5.2.6.4 Temperature

Temperature is one of essential and most obvious indicators of occurrence in a compost heap. Increased temperature in the compost heap is the consequence of release of heat, because of microbe oxidative decomposition of organic substances (Liang et al. 2002). The optimal thermophilic range for the phase of composting is between 55 and 65°C (Roš 2001). Also removing of pathogenetic organisms is in this range maximum.

32.5.2.6.5 Moisture

Moisture is, in addition to oxygen content, one of the most essential factors of composting. The content of moisture in a compost mixture should be in the 50–70% range (Richard et al. 2002). Moisture is in compost in two forms: as free liquid and as metabolic moisture that occurs with the decomposition of organic substances.

When the material is more humid, as is for optimal condition in compost heap, the limit of moisture is getting off as a free liquid. The lowest content of moisture that still allows bacterial activity is 12–15%.

32.5.2.6.6 pH Value

A large number of different microorganisms takes part in a process of biological decomposition of compost mixture. Microorganisms are differently adapted to changes of value pH. The optimal value in the compost heap is in the pH range of 5.5 to 8.5. Humic and fulvic acids occur during decomposition of organic substances in compost. That is why at the beginning of composting the pH value decreases, and with additional aeration of compost mixture the pH value increases again (Tang et al. 2003).

32.5.2.6.7 Ash

Compost mixture contains organic part, inorganic part, and water. When we incinerate the compost mixture, organic part of compost evaporates and inorganic part remains as an ash. Raw materials for composting contain a larger or smaller share of organic substance, depending upon the type of waste materials (Forster-Carneiro et al. 2007).

REFERENCES

Ajhar M, Melin T (2006) Siloxane removal with gas permeation membranes. *Desalination* 200(1–3): 234–235.

Aparicio I, Santos JL, Alonso E (2009) Limitation of the concentration of organic pollutants in sewage sludge for agricultural purposes: A case study in South Spain. *Waste Manag* 29(5):1747–1753.

Appels L Baeyens J et al. (2008) Siloxane removal from biosolids by peroxidation. *Energy Conver Manag* 49(10): 2859–2864.

Artan N, Tasli R, Orhon D (2006) Rational basis for optimal design of sequencing batch reactors with multiple anoxic filling for nitrogen removal. *Process Biochem* 41: 901–908.

Artan N, Wilderer P (2001) The mechanism and design of sequencing batch reactor system for nutrient removal—The state of art. *Water Sci Technol* 43(3): 53–60.

Baeten D, Verstraete W (1988) Manure and municipal solid waste fermentation in flanders: An appraisal. *Biol Wastes* 26(4): 297–314.

Barrington S, Choiniere D, Trigui M, Knight W (2003) Compost convective airflow under passive aeration. *Bioresour Technol* 86(3): 259–266.

Bernet N, Delgenes N, Akunna JC, Delgenes JP, Moletta R (2000). Combined anaerobic-aerobic SBR for the treatment of piggery wastewater. *Water Res* 34(2): 611–619.

Bicudo J, Svoboda IV (1995) Effects of intermittent-cycle extended-aeration treatment on the fate of nutrients, metals and bacterial indicators in pig slurry. *Bioresour Technol* 54(1): 63–72.

Boiran B, Couton Y, Germon JC (1996) Nitrification and denitrification of liquid lagoon piggery waste in a biofilm infiltration-percolation aerated system (BIPAS) reactor. *Bioresour Technol* 55: 63–77.

Boxall A, Long C (2005) Veterinary medicines and the environment. *Environ Toxicol Chem* 24(4): 759–760.

Boxall ABA, Kolpin DW, Halling-Sørensen B, Tolls J (2003). Are veterinary medicines causing environmental risks? *Environ Sci Technol* 37(15): 286A-294A.

Brewer AJ, Cumby TR, Dimmock SJ (1999) Dirty water from dairy farms, II: treatment and disposal options. *Bioresour Technol* 67(2): 161–169.

Cantrell KB, Ducey T, Ro, KS, Hunt, PG (2008) Livestock waste-to-bioenergy generation opportunities. *Bioresour Technol* 99(17): 7941–7953.

Centner TJ, Wetzstein ME, Mullen JD (2008) Small livestock producers with diffuse water pollutants: adopting a disincentive for unacceptable manure application practices. *Desalination* 226(1–3): 66–71.

Chambers B (1993) Batch operated activated sludge plant for production of high effluent quality at small works. *Water Sci Technol* 28(10): 251–258.

Chong C, Purvis P, Lumis G et al. (2008) Using mushroom farm and anaerobic digestion wastewaters as supplemental fertilizer sources for growing container nursery stock in a closed system. *Bioresour Technol* 99(6): 2050–2060.

Chung K-T, Tseng H-C, Lai Y-F et al. (1985) Microbiological, physical and chemical studies of Livestock-farm water in Taiwan. *Agri Wastes* 14(1): 1–18.

Correa SMBB, Ruiz E, Romero F (2003) Evolution of operational parameters in a UASB wastewater plant. *Water SA* 29(3): 345–352.

Cukjati N (2008) Comparative composting of mesophilic and thermophilic stabilized sludge from municipal wastewater treatment plant. PhD Thesis, Faculty of Chemistry and Chemical Technology, University of Ljubljana, Ljubljana, Slovenia, 104 p.

Cumby TR, Brewer AJ, Dimmock SJ (1999) Dirty water from dairy farms, I: biochemical characteristics. *Bioresour Technol* 67(2): 155–160.

D'abbieri M, Massetti F, Molinari M, Reversi D (2008) A new process for the reduction and energy recovery of excess biological sludge generated by wastewater treatment plants. In: *2nd Int Symp on Energy from Biomass and Waste*, IWWG, Venice, Italy, 9 p [CD-ROM].

Day M, Shaw K (2001) Biological, chemical, and physical processes of composting. In: Stoffella PJ, Kahn BA (eds) *Compost Utilization in Horticutural Cropping Systems*. CRC Press, Boca Raton, FL, pp 17–50.

De Gioannis G, Diaz LF, Muntoni A, Pisanu A (2008) Two-phase anaerobic digestion within a solid waste/ wastewater integrated management system. *Waste Manag* 28(10): 1801–1808.

Dennis RW, Irvine RL (1979) Effect of fill:react ratio on sequencing batch biological reactors. *J Water Pollut Control Fed* 51(2): 255–263.

Deublein D, Steinhauser A (2008) *Biogas from Waste and Renewable Resources*. Wiley-VCH Verlag GmbH & Co. KGaA, Weinheim, Germany

Dewil R, Appels L et al. (2006) Energy use of biogas hampered by the presence of siloxanes. *Energy Conver Manag* 47(13–14): 1711–1722.

Dinsdale RM, Premier FC, Hawkes FR, Hawkes DL (2000) Two-stage anaerobic co-digestion of waste activated sludge and fruit/vegetable waste using inclined tubular digesters. *Bioresour Technol* 72(2): 159–168.

Drolc A, Zagorc Koncan J (2008) Diffuse sources of nitrogen compounds in the Sava river basin, Slovenia. *Desalination* 226(1–3): 256–261

EEP (1986) Council directive 86/278/EEC on the protection of the environment, and in particular of the soil, when sewage sludge is used in agriculture. Official J Brussels, *Eur Environ Protec* L181: 6–12.

Elmquist H (2005) Environmental systems analysis of arable, meat and milk production. Doctoral thesis. Department of Biometry and Engineering. Faculty of Natural Resources and Agricultural Sciences, University of Upsala, Upsala, Sweden, 69 P.

FAO (2006) Livestock impacts on the environment. http://www.fao.org/ag/magazine/0612sp1.htm (Accessed 2 Dec 2008).

Fawcett HH (1991) What you need to know to live with chemicals. By RI and SL Freudenthal, *J Hazard Mater* 26(3): 371.

Feijoo G, Soto M, Mendez R, Lema JM (1995). Sodium inhibition in the anaerobic digestion process: antagonism and adaptation phenomena. *Enzym Microb Technol* 17(2): 180–188.

Filipy J, Rumburg B, Mount G et al. (2006) Identification and quantification of volatile organic compounds from a dairy. *Atmos Environ* 40: 1480–1494.

Forster-Carneiro T, Perez M, Romero LI (2007) Composting potential of different inoculum sources in the modified SEBAC system treatment of municipal solid wastes. *Bioresour Technol* 98: 3354–3366.

Garcia L, Rico C, Garcia PA, Rico JL (2008) Flocculants effect in biomass retention in a UASB reactor treating dairy manure. *Bioresour Technol* 99: 6028–6036.

Georgacakis D, Tsavdaris A, Bakouli J, Symeonidis S (1996) Composting solid swine manure and lignite mixture with selected plant residues. *Bioresour Technol* 56: 195–200.

GNS (2009) Nitrogen removal from manure and organic residues by ANAStrip - process (System GNS): http://www.gns-halle.de/english/site_1_6.htm (Accessed 28 Sept 2009).

Guida M, Inglese M, Meriç S (2008) A multi-battery toxicity investigation on fungicides. *Desalination* 226(1–3): 262–270.

Hammer MJ (1996) *Water and Wastewater Technology*. Prentice Hall, New York, 519 p.

Han Y, Sung S, Dague RR (1997) Temperature-phased anaerobic digestion of wastewater sludges. *Water Sci Technol* 36(6–7): 367–374.

Haruta S, Nakajama T, Nakamura K (2005) Microbial diversity in biodegradation and reutilization processes of garbage. *J Biosci Boeng* 99: 1–11.

Havlkova M, Kroeze C, Huijbregts MAJ (2008) Environmental and health impact by dairy cattle livestock and manure management in the Czech Republic. *Sci Total Environ* 396: 121–131.

Hendriks ATWM and Zeeman G (2009) Pretreatments to enhance the digestibility of lignocellulosic biomass. *Bioresour Technol* 100(1): 10–18.

Hoepker EC, Schroeder ED (1979) The effect of loading rate on batch-activated sludge effluent quality. *J Water Pollut Control Fed* 51(2): 264–273.

Hossain MM, Lopez D, Herrera J, de Lasa HI (2009) Nickel on lanthanum-modified [gamma]-Al2O3 oxygen carrier for CLC: Reactivity and stability. *Catalysis Today* 143 (1–2): 179–186.

Hvala N, Zec M, Ros M, Strmcnik S (2001) Design of a sequencing batch reactor sequence with an input load partition in a simulation-based experimental environment. *Water Environ Res* 73(2): 146–153.

Imbeah M (1998) Composting piggery waste: A review. *Bioresour Technol* 63: 197–203.

Information for Action Farm Waste (2008): http://www.informaction.org/cgi-bin/gPage.pl?menu=menua.txt&main=farmwaste_gen.txt&s=Farm%20Waste (Accessed 28 Sept 2009).

Irvine RL (1979) Sequencing batch biological reactors - an overview. *J Water Pollut Control Fed* 51(2): 235–243.

Irvine RL, Miller G, Bhamra AS (1979) Sequencing batch treatment of wastewaters in rural areas. *J Water Pollut Control Fed* 51(2): 244–254.

Irvine RL and Moe WM (2001) Period biofilter operation for enhanced performance during unsteady-state loading conditions. *Water Sci Technol* 43(3): 231–239.

ISO (1998) EN ISO 11734 (1998) Water quality—Evaluation of the ultimate anaerobic biodegradability of organic compounds in digested sludge—Method by measurement of the biogas production, International Standard Organization.

Kalyuzhnyi S, Fedorovich V, Nozhevnikova A (1998) Anaerobic treatment of liquid fraction of hen manure in UASB reactor. *Bioresour Technol* 65: 221–225.

Kazmi AA, Fujita M, Furumai H (2001) Modelling effect of remaining nitrate on phosphorus removal in SBR. *Water Sci Technol* 43(3): 175–182.

Ketchum LH, Irvine RL, Liao PC (1979) First cost analysis of sequencing batch reactors. *J Water Pollut Control Fed* 51(2): 288–297.

Ketchum LH, Liao PC (1979) Tertiary chemical treatment for phosphorus reduction using sequencing batch reactor. *J Water Pollut Control Fed* 51(2): 298–304.

Kim Y, Parker W (2008). A technical and economic evaluation of the pyrolysis of sewage sludge for the production of bio-oil. *Bioresour Technol* 99(5): 1409–1416.

Kishida N, Kim J-H, Chen M et al. (2003) Effectiveness of oxidation-reduction potential and pH as monitoring and control parameters for nitrogen removal in swine wastewater treatment by sequencing batch reactors. *J Biosci Bioeng* 96(3): 285–290.

Kishimoto M, Preechaphan C, Yoshida T, Taguchi H (1987) Simulation of an aerobic composting of activated sludge using a statistical procedure. *MIRCEN* J 3: 113–124.

Kompogas (2009) Energy production with optimised ecobalance. http://www.compogas.com/ (Accessed 20 Jan 2009).

Lansing S, Botero RB, Martin JF (2008) Waste treatment and biogas quality in small-scale agricultural digesters. *Bioresour Technol* 99: 5881–5890.

Larsen KL, McKartney DM (2000) Effect of C:N ratio on microbial activity and N retention bench-scale study using pulp and paper biosolids. *Compost Sci Utiliz* 51: 157–166.

Lettinga G, Hulshoff LW (1991) UASB-process design for various types of wastewaters. *Water Sci Technol* 24(8): 87–107.

Liang C, Das KC, McClendon RW (2002) The influence of temperature and moisture contents regimes on the aerobic microbial activity of a biosolids composting blend. *Bioresour Technol* 86(2): 131–137.

Mahmoud N (2008) High strength sewage treatment in a UASB reactor and an integrated UASB-digester system. *Bioresour Technol* 99: 7531–7538.

Messenger J, de Villers HA, Ekama GA (1993). Evaluation of the dual digestion system: Part 1: Overview of the Milnerton experience. *Water SA* 19(3): 185–192.

Miklos J, Plaza E, Kurbiel J (2001) Use of computer simulation for cycle length adjustment in sequencing batch reactor. *Water Sci Technol* 43(3): 61–68.

Mohaibes M, Heinonen-Tanski H (2004) Aerobic thermophilic treatment of farm slurry and food wastes. *Bioresour Technol* 95(3): 245–254.

Morling S, Person T, Johanson B (2001) Performance of an SBR-plant for advanced nutrient removal, using septic sludge as a carbon source. *Water Sci Technol* 43(3): 131–138.

Mrafkova L, Goi D, Gallo V, Colussi I (2003) Preliminary evaluation of inhibitory effects of some substances on aerobic and anaerobic treatment plant biomasses. *Chem Biochem Eng* Q17(3): 243–247.

Muller BW, Brodd AR, Leo JP (1983) Hazardous waste remedial action—Picillo farm, Coventry, Rhode Island: An overview. *J Hazardous Materials* 7(2): 113–129.

Ndegwa PM (2004) Limitation of Orthophosphate Removal during Aerobic Batch Treatment of Piggery Slurry. *Biosyst Eng* 87(2): 201–208.

Ng WJ, Ong SL, Hu JY (2001) Denitrifying phosphorus removal by anaerobic/anoxic sequencing batch reactor. *Water Sci Technol* 43(3): 139–146.

Nguyen LM (2000) Organic matter composition, microbial biomass and microbial activity in gravel-bed constructed wetlands treating farm dairy wastewaters. *Ecol Eng* 16: 199–221.

Novotny V, Imhoff KR, Olthof M, Krenkel PA (1989) *Karl Imhoff's Handbook of Urban Drainage and Wastewater Disposal*. John Wiley, New York, 390 p.

Núñez-Delgado A, López-Períago E, Diaz-Fierros-Viqueira F (2002) Pollution attenuation by soils receiving cattle slurry after passage of a slurry-like feed solution: Column experiments. *Bioresour Technol* 84(3): 229–236.

Obaja D, Mac S, Mata-Alvarez J (2005) Biological nutrient removal by a sequencing batch reactor (SBR) using an internal organic carbon source in digested piggery wastewater. *Bioresor Technol* 96: 7–14.

Obaja D, Mace S, Costa J et al. (2003) Nitrification, denitrification and biological phosphorus removal in piggery wastewater using a sequencing batch reactor. *Bioresor Technol* 87: 103–111.

Oleszkiewicz A (1985) Cost-effective treatment of piggery wastewater. *Agric Wastes* 12(3): 185–206.

Oleszkiewicz JA, S Koziarski (1981) Management and treatment of wastes from large piggeries. *Agric Wastes* 3(2): 123–144.

Orhon D, Artan N (1994) *Modelling of Activated Sludge Systems*. Technomic Publ, Lancaster, Basel, Switzerland, 589 p.

Oturan MA, Oturan N, Lahitte C, Trevin S (2001) Production of hydroxyl radicals by electrochemically assisted Fenton's reagent: Application to the mineralization of an organic micropollutant, pentachlorophenol. *J Electroanalyt Chem* 507(1–2): 96–102.

Paul E, Plisson-Saune S, Mauret M, Canet J (1998) Process state evaluation of the alternating oxic-anoxic activated sludge using ORP, pH and DO. In: *IAWQ 19th Int Conf*, IAWQ. Vancouver, Canada, pp 299–306.

Pham TTH, Brar SK, Tyagi RD, Surampalli RY (2009) Ultrasonication of wastewater sludge—Consequences on biodegradability and flowability. *J Hazard Materials* 163 (2–3), pp. 891–898.

Ponsá S, Ferrer I, Vázquez F, Font X (2008) Optimization of the hydrolytic-acidogenic anaerobic digestion stage (55 °C) of sewage sludge: Influence of pH and solid content. *Water Res* 42(14): 3972–3980.

Ra CS, Lo KV, Shin JS, Oh JS, Hong BJ (2000) Biological nutrient removal with an internal organic carbon source in piggery wastewater treatment. *Water Res* 34(3): 965–973.

Reinoso R and Becares E (2008) The occurrence of intestinal parasites in swine slurry and their removal in activated sludge plants. *Bioresour Technol* 99: 6661–6665.

Ribbe L, Delgado P, Salgado E, Flügel WA (2008) Nitrate pollution of surface water induced by agricultural non-point pollution in the Pocochay watershed, Chile. *Desalination* 226(1–3): 13–20.

Richard TM, Hamelers HVM, Veeken A, Silva T (2002) Moisture relationships in composting processes. *Compost Sci Utiliz* 10(4): 286–302.

Ritter WF, Walpole EW, Eastburn RP (1984) Effect of an anaerobic swine lagoon on groundwater quality in Sussex county, Delaware. *Agric Wastes* 10(4): 267–284.

Roberts R, Davies WJ, Forster CF (1999) Two-stage, thermophilic-mesophilic anaerobic digestion of sewage sludge. *Process Saf Environ Protec* 77(2): 93–97.

Roš M (2001) Biološko čiščenje odpadne vode (Biological wastewater treatment, in Slovenian). GV Založba. Ljubljana, Slovenia, 243 p.

Ruiz C, Torrijos M, Sousbie P, Lebrato Martinez J, Moletta R (2001) The anaerobic SBR process: Basic principles for design and automation. *Water Sci Technol* 43(3): 201–208.

Salsabil MR, Prorot A, Casellas M, Dagot C (2008) Pre-treatment of activated sludge: Effect of sonication on aerobic and anaerobic digestibility. *Chem Eng J* 148(2–3): 327–335.

Sattler (2009). Biogas storage systems: http://www.sattler-ag.com/sattler-web/en/products/190.htm (Accessed 9 Jan 2009).

Schierhout GH, Midgley A, Myers JE (1997) Occupational fatality under-reporting in rural areas of the Western Cape Province, South Africa. *Saf Sci* 25(1–3): 113–122.

Schloss PD, Hay AG, Wilson DB, Walker LP (2003) Tracking temporal changes of bacterial community fingerprints during the initial stages of composting. *FEMS Microbol Ecol* 46: 1–9.

Schröder J (2005) Revisiting the agronomic benefits of manure: a correct assessment and exploitation of its fertilizer value spares the environment. *Bioresour Technol* 96(2): 253–261.

Schurard F (2005) Composting of organic waste. In: Joedering HJ, Winter J (eds) *Environmental Biotechnology Concepts and Applications*. Wiley-VCH, Weinheim, Germany, pp 333–354.

Schweigkofer M, Niesser R (2001) Removal of siloxanes in biogases. *J Hazard Materials* 83(3): 183–196.

Singh RP, Agrawal M (2008) Potential benefits and risks of land application of sewage sludge. *Waste Manag* 28(2): 347–358.

Smith SR, Woods V, Evans TD (1998) Nitrate dynamics in biosolids-treated solids. III. Significance of the organic nitrogen, a twin-pool exponential model for nitrogen management and comparison with the nitrate production from animal wastes. *Bioresour Technol* 66: 161–174.

Song Y-C, Kwon S-J, Woo J-H (2004) Mesophilic and thermophilic temperature co-phase anaerobic digestion compared with single-stage mesophilic- and thermophilic digestion of sewage sludge. *Water Res* 38(7): 1653–1662.

Sung S, Liu T (2003) Ammonia inhibition on thermophilic anaerobic digestion. *Chemosphere* 53: 43–52.

Tang JC, Inoue Y, Yasata T et al. (2003) The chemical and microbial properties of various compost products. *Soil Sci Plant Nutr* 49: 273–280.

Tapana C, Krishna PR (2004) Anaerobic thermophilic/mesophilic dual-stage sludge treatment. *J Environ Eng* 126(9): 796–801.

Tchobanoglous G, Burton FL, Stensel HD (2003) *Wastewater Engineering: Treatment and Reuse*. Metcraft & Eddy, Inc., Boston, MA.

Tilche A, Bortone B, Malaspona F, Piccinini S, Stante L (2001) Biological nutrient removal in a full-scale SBR treating piggery wastewater: results and modelling. *Water Sci Technol* 43(3): 363–371.

Van den Berg VS, van Lamoen F (2008) An integrated approach on pollution abatement in rural areas; regional pilot projects in the Province of North-Brabant. *Desalination* 226(1–3): 183–189.

Vieira GEG, Romeiro GA, Sella SM et al. (2009) Low temperature conversion (LTC)—An alternative method to treat sludge generated in an industrial wastewater treatment station—Batch and continuous process comparison. *Bioresour Technol* 100(4): 1544–1547.

Wakase S, Sasaki H, Itoh K et al. (2008) Investigation of the microbial community in a microbiological additive used in a manure composting process. *Bioresour Technol* 99: 2687–2693.

Wareham DG, Hall KJ, Mavinic DS (1993) Real-time control wastewater treatment systems using ORP. *Water Sci Technol* 28(11–12): 273–282.

Watson B (2008) Climate change: An environmental, development and security issue: http://www.bsas.org.uk/Animal_Bytes/Global_Climate_Change/ (Accessed 29 Nov 2008).

Westerman PW, Bicudo JR (2005) Management considerations for organic waste use in agriculture. *Bioresour Technol* 96(2): 215–221.

Wikipedia (2008a) Composting: http://en.wikipedia.org/wiki/Composting (Accessed 29 Nov 2008).

Wikipedia (2008b) Factory farming. from http://en.wikipedia.org/wiki/Factory_farming (Accessed 29 Nov 2008).

Wikipedia (2008c) Permaculture: http://en.wikipedia.org/wiki/Permaculture (Accessed 29 Nov 2008).

Wikipedia (2009) Bokashi composting: http://en.wikipedia.org/wiki/Bokashi_composting (Accessed 29 Nov 2008).

Wilderer PA, Irvine RL, Goroncy MC (2001) *Sequencing Batch Reactor Technology*. IWA Publishing, London, p 76.

Wu W, Timpany P, Dawson B (2001) Simulation and application of a novel modified SBR system for biological nutrient removal. *Water Sci Technol* 43(3): 215–222.

Yadav KS, Mishra MM, Kapoor KK (1982) The effect of fungal inoculation on composting. *Agric Wastes* 4(5): 329–333.

Yalmaz G, Ozturk I (2001) Biological ammonia removal from anaerobically pre-treated landfill leachate in sequencing batch reactors (SBR). *Water Sci Technol* 43(3): 307–314.

Ye C, Cheng JJ, et al. (2008) Inhibition of anaerobic digestion process : A review. *Bioresour Technol* 99(10): 4044–4064.

Yetilmezsoy K, Sakar S (2008) Development of empirical models for performance evaluation of UASB reactors treating poultry manure wastewater under different operational conditions. *J Hazard Materials* 153: 532–543.

Yoong ET, Lant PA (2001) Biodegradation of high strength phenolic wastewater using SBR. *Water Sci Technol* 43(3): 299–306.

Zec M, Hvala N, Roš M, Vrtovšek J (1997) Operation and control of sequencing batch reactor for wastewater treatment. In: *1st International Conference on Environmental Restoration*, Ljubljana, Slovenian Association on Water Pollution Control.

Zhang Z, Zhu J, King J, Li WH (2006) A two-step fed SBR for treating swine manure. *Process Biochem* 41: 892–900.

Zhu N, Deng C, Xiong Y, Qian H (2004) Performance characteristics of three aeration systems in the swine manure composting. *Bioresour Technol* 95(3): 319–326.

Zupancic GD, Jemec A (2010) Anaerobic digestion of tannery waste: Semi-continuous and anaerobic sequencing batch reactor processes. *Bioresour Technol* 101: 26–33.

Zupancic GD, Ros M (2003) Heat and energy requirements in thermophilic anaerobic sludge digestion. *Renew Energy* 28(14): 2255–2267.

Zupancic GD, Ros M (2005) *Ammonia Removal in Sludge Digestion Utilizing Nitrification with Pure Oxygen Aeration*. Nutrient management in wastewater treatment processes and recycle streams: IWA specialized conference, Krakow, Poland, IWA Publishing, London.

Zupancic GD, Ros M (2008) Aerobic and two-stage anaerobic-aerobic sludge digestion with pure oxygen and air aeration. *Bioresour Technol* 99(1): 100–109.

33 Vegetable Oils

Gerhard Knothe
U.S. Department of Agriculture

CONTENTS

33.1 INTRODUCTION

Numerous vegetable oils and their derivatives, most commonly biodiesel, the mono-alkyl esters of oils or fats, have been investigated for their suitability as fuel, mainly for transportation purposes, although some reports exist on the use of vegetable oils or biodiesel as heating fuel. Indeed, in addition to the commonly applied biodiesel standards ASTM D6751 (United States) and EN 14214 (Europe), a separate standard when using biodiesel as heating oil has been developed, EN 14213 (Europe). The standard ASTM D396 (fuel oils) now covers blends of up to 5% biodiesel with fuel oil if the biodiesel used satisfies the specifications in D6751.

Classical commodity oils such as soybean, canola (rapeseed), palm, and some other oils used as energy sources are covered in other chapters of this book. This chapter deals with less common oils that have been studied for their potential to provide a source of energy. A few other oils will also be briefly discussed if, although commercialized for reasons such as economics, they are not usually considered to be biodiesel feedstocks. Some of the feedstocks discussed in this chapter have also been used untransesterified or transformed to what is known as "renewable diesel."

Several driving forces are causing the search for additional feedstocks for biodiesel (or biofuels in general). Although it is beyond the scope of this chapter to discuss these driving forces in detail, they can be summarized as the needs to increase the potential supply of energy feedstocks in light of diminishing petroleum reserves, to provide domestic sources of renewable energy to decrease dependence on imported petroleum, and to enhance economic development by providing a source of jobs and income, an aspect of particular interest in more indigent sections of developing countries. Issues that have also affected this search are the food-versus-fuel discussion, i.e., the standpoint that sources of food should not be used as sources of fuel [although this discussion does not seem to affect other nonfood uses such as lubricants, polymers, etc. (FAO 2009) and the carbon footprint of some feedstocks]. The food-versus-fuel issue has caused some shifting toward emphasizing feedstocks that yield inedible oils. On the other hand, soybeans have historically been grown for their protein content so that if production is increased for the sake of the oil, the protein, used as animal feed, should become more plentiful and therefore less expensive, reducing the cost of feeding animals. The carbon footprint issue, incorporating the issue of land-use change, currently affects largely commodity oils. For palm oil, concern has been voiced over the clearing of tropical rainforest for the sake of new palm plantations, but greenhouse gas mitigation by using palm-based biodiesel has also been discussed (May 2005). Although similar concerns have been raised over the increased cultivation of soybeans, especially in tropical areas, biodiesel from soybean oil possesses a positive energy balance (Hill et al. 2006). For rapeseed/canola (but other feedstocks as well), concerns exist over the use of fertilizers and nitrous oxide (N_2O) emissions (Crutzen et al. 2007), runoff, and the carbon footprint of producing fertilizers. It is not clear how some of these issues will in the future affect feedstocks, which are discussed in this chapter because of their current nascent status because of which many issues have not yet been addressed.

In some cases, other distinct aspects play a role. For example, the major background for the investigation of coffee oil as a biodiesel feedstock was to possibly find a use for damaged coffee beans to minimize the resulting economic losses (Oliveira et al. 2008). No significant differences in the fatty acid profiles of healthy and damaged coffee beans were observed (Oliveira et al. 2006). Another report (Mariod et al. 2006) discusses oils obtained from bugs, one of them a pest of watermelons, and both oils are used for cooking and medicinal applications. Although formally not vegetable oils, for the sake of interest they are included in this chapter.

In any case, with the increasing interest in these issues, the number of publications concerned with feedstock variety has grown considerably. A significant number, if not the majority, of the oils briefly discussed in this chapter can be obtained from plants found most commonly in tropical or subtropical climates because many of these oils are of limited physiological use and the issue of energy supply is becoming more pressing in the developing countries located in these regions of the world.

33.1.1 FATTY ACID PROFILE/FUEL PROPERTIES

An important issue that does not always find the interest it deserves when the above issues are discussed is that of the composition of the fuels being discussed. Ultimately, it is the composition of the fuels and the properties resulting from this composition that determine if a feedstock is viable as a source of fuel. Therefore, a brief discussion of fuel properties imparted by the fatty acid profile of a vegetable oil is presented here with reference to the corresponding specifications in biodiesel standards. Furthermore, most classical commodity oils and most oils discussed in this chapter largely contain the same five major fatty acids in their profiles: palmitic (hexadecanoic; C16:0), stearic (octadecanoic; C18:0), oleic [9(Z)-octadecenoic; C18:1], linoleic [9(Z),12(Z)-octadecadienoic; C18:2], and linolenic [9(Z),12(Z),15(Z)-octadecadienoic; C18:3].

Cetane number, kinematic viscosity, oxidative stability, and cold-flow specifications in biodiesel standards are those most directly affected by the fatty acid profile. Minor components of biodiesel (e.g., mono- and di-acylglycerols formed during the transesterification reaction, sterol glucosides, and antioxidants) can also influence these properties, especially oxidative stability and cold flow, but they will not be discussed here.

The cetane number is a dimensionless descriptor of the ignition quality of a diesel fuel. It is related to the ignition delay time that a fuel experiences upon injection into the combustion chamber of a diesel engine. Hexadecane (trivial name: cetane) is the high-quality reference compound on the cetane scale and has been assigned a cetane number of 100. The shorter the ignition delay time, the higher the cetane number and vice versa. Generally, higher cetane numbers are more desirable.

The cetane number of a fatty acid chain increases with increasing chain length and increasing saturation. Thus, methyl stearate has a high cetane number (> 90) and methyl linolenate has a low cetane number (~25). The cetane number of methyl stearate is also higher than that of methyl laurate (~65). The cetane number of a mixture (e.g., biodiesel) approximately correlates with the cetane numbers of the individual components proportionally taking their amounts into consideration. Minimum cetane numbers in biodiesel standards are 47 in ASTM D6751 (United States) and 51 in EN 14214 (Europe). Most biodiesel fuels possess cetane numbers in the range of the high 40s to lower 60s.

The high viscosity of vegetable oils, approximately an order of magnitude greater than that of petroleum-based diesel fuel (petrodiesel), is the major reason why these feedstocks are transesterified to biodiesel. The high viscosity of vegetable oils, influencing penetration and atomization of the fuel in the combustion chamber, leads to operational problems such as engine deposits. Biodiesel fuels possess viscosity values closer to those of petrodiesel. Thus, kinematic viscosity is prescribed in biodiesel standards with the ranges being 1.9–6 mm^2/s (ASTM D6751) and 3.5–5 mm^2/s (EN 14214). Most biodiesel fuels exhibit kinematic viscosity in the range of 4.0–5 mm^2/s. Again, compound structure significantly influences viscosity. Viscosity increases with chain length and decreasing *cis*-unsaturation. The kinematic viscosity of methyl laurate is 2.43 mm^2/s, that of methyl palmitate is 4.38 mm^2/s, methyl stearate is 5.85 mm^2/s, methyl oleate is 4.51 mm^2/s, methyl linoleate is 3.65 mm^2/s, and methyl linolenate is 3.14 mm^2/s (Knothe and Steidley 2005). Thus, the kinematic viscosity of a biodiesel fuel depends on its fatty acid profile.

Oxidative stability is one of the major technical issues affecting the commercial use of biodiesel. Oxidation of fatty acid chains is a complex reaction, consisting initially of the formation of hydroperoxides followed by secondary reactions during which products such as acids, aldehydes, ketones, hydrocarbons, etc., can be formed. Unsaturated fatty acid chains, especially the polyunsaturated species (i.e., esters of linoleic and linolenic acids) are susceptible to oxidation. Relative rates of oxidation given in the literature (Frankel 2005) are 1 for oleates, 41 for linoleates, and 98 for linolenates. Small amounts of unsaturated fatty esters probably affect oxidative stability more than their small amounts indicate. Oxidative stability is addressed in biodiesel standards primarily by the corresponding specification which prescribes the use of a Rancimat instrument. This instrument permits an accelerated test to be conducted with the goal of judging the oxidative stability of a sample. The lower the induction time by this method, the less oxidatively stable the sample. Minimum induction times prescribed in biodiesel standards by this test are 3 h (ASTM D6751) and 6 h (EN 14214). However, an antioxidant will almost always be needed to achieve these specifications because the induction time of methyl oleate is 2.79 h, that of methyl linoleate is 0.94 h, and that of methyl linolenate is 0 h. Methyl esters of saturated fatty acids possess induction times greater than 24 h (Knothe 2008). No typical oxidative stability times for specific biodiesel fuels can be given because oxidative stability is strongly influenced by factors such as storage conditions and the presence of extraneous materials, including antioxidants. However, to achieve the mentioned minimal times in standards, the use of antioxidants is almost always necessary as mentioned above. It may be noted that the European biodiesel standard EN 14214 contains some specifications that can also be related to the phenomenon of oxidative stability. These specifications are the iodine value, a crude measure of total unsaturation of a sample, a maximum of 12% for linolenic acid methyl esters and a maximum of 1% for esters of fatty acids with more than three double bonds.

In addition to oxidative stability, cold flow is another major technical issue that affects the commercial use of biodiesel. Almost all biodiesel fuels have relatively poor cold-flow properties, as demonstrated by relatively high cloud points. The cloud point (i.e., the temperature at which the first solids are visible in a sample upon cooling) for the methyl esters of soybean oil is approximately 0°C and only slightly lower, approximately –3°C, for the methyl esters of rapeseed (canola) oil. The esters of biodiesel fuels with high amounts of longer-chain saturated fatty esters may have even higher cloud points. For example, the cloud point of the methyl esters of palm oil which contains approximately 40% methyl palmitate, is 15°C or even higher. The melting points of fatty esters can serve to illustrate the potential effect on cloud point and thus cold flow. The melting point of methyl palmitate is approximately 30°C, and that of methyl stearate is 38°C, but that of methyl oleate is –20°C. Again, chain length and degree of unsaturation have a significant influence on this property. Because of the different requirements on cold flow caused by time of year and geographic location, cold flow is a "soft" specification in biodiesel standards, with a report for cloud point being required by ASTM D6751 and limits varying by time of year and location using a method termed the "cold filter plugging point" are prescribed in EN 14214. It is important to note that improvement of biodiesel properties is rendered difficult because improvement of cold flow (e.g., less saturates, more unsaturates) negatively affects oxidative stability and cetane number and vice versa.

33.1.2 HISTORICAL PERSPECTIVES

The first vegetable oil to be tested in a diesel engine was peanut oil (Diesel 1912, 1913). This event occurred at the 1900 World Exhibition in Paris at the request of the French government as the inventor of the diesel engine, Rudolf Diesel (1858–1913), himself states (Diesel 1912). Diesel experimented later with vegetable oils and expressed support for the concept (Diesel 1912). There is abundant literature dating from the 1920s to the late 1940s describing the use of vegetable oils as diesel fuel (Knothe 2005). Interestingly, a theme in many of these reports was to provide the tropical colonies of European countries with a domestic source of fuel to ensure some degree of energy independence. Accordingly, the first description of what is today known as biodiesel, the mono-alkyl esters of vegetable oils or animal fats (or other lipid feedstocks), can be found in the Belgian patent 422,877 issued in 1937 (Chavanne 1937). A later report provides considerably more details on this project, mainly concerned with the ethyl esters of palm oil, including the first commercial bus running on this fuel (van den Abeele 1942). Several reports from this time describe or summarize the use of various vegetable oils as fuel (Walton 1938; Chowhury 1942; Pacheco Borges 1944). Vegetable oils in addition to the aforementioned peanut and palm oils that were investigated in "historic times" include castor, cottonseed, olive, rapeseed, soybean, sunflower, and others (Knothe 2005 and references therein); however, no fuel meeting the current definition of biodiesel was prepared from them.

Table 33.1 lists vegetable oils that are emerging energy feedstocks by their common names and/or their scientific names and provides relevant literature references. Table 33.2 provides the fatty acid profiles of the oils listed in Table 33.1 as far as they have been reported. Tables for analytical data and fuel property data for these oils are not provided here, although the original literature often reports such data. However, many literature data do not appear to stand up to close scrutiny of their correctness. A major issue is kinematic viscosity which in many cases in the literature appears to be too high (often by ~0.5–1 mm²/s) to be correct and may be an indication of incomplete conversion of the investigated oil to biodiesel. Other data, such as cetane number and properties related to cold flow, also do not always appear correct. For example, some papers may mention the "melting point" of an oil or fat or the corresponding biodiesel fuels, but, because of their multicomponent nature, this is not correct; instead, these materials have cloud and pour points (i.e., melting ranges). Therefore, analytical or fuel property data will only be mentioned in the text for some oils, which are discussed in more detail. However, considering that most biodiesel fuels obtained from feedstocks discussed in this chapter largely contain the five major fatty acids previously mentioned in their fatty acid profiles, most of these fuels will, if properly prepared, meet the specifications in biodiesel standards. Note also that various methods have been used for preparing the biodiesel fuels referenced here, also taking into consideration the various degrees of refining of the available feedstocks. Many of these oils exhibit high contents of free fatty acids which make acid pretreatment necessary before the usual alkaline transesterification to biodiesel. Furthermore, exhaust emissions tests have been performed on the biodiesel fuels obtained from some of these oils, with the results generally agreeing with the exhaust emissions tests conducted with biodiesel fuels derived from commodity oils.

TABLE 33.1

Noncommodity or Minor Vegetable Oils of Which Mono-Alkyl Ester (Biodiesel) Derivatives or Other Energy Uses Have Been Reported

Oil	References
Andiroba (*Carapa guianensis*)	Abreu et al. 2004
Azadirachta indica	Munavu and Odhiambo 1984; Azam et al. 2005
Babassu	Abreu et al. 2004; Lima et al. 2007; da Silva et al. 2008; Urioste et al. 2008
Coffee	Oliveira et al. 2006, 2008
Cumaru (*Dipteryx odorata*)	Abreu et al. 2004
Cuphea	Geller et al. 1999; Knothe et al. 2009
Cynara cardunculus	Encinar et al. 1999; Fernández et al. 2006
Desert date (*Balanites aegyptiaca*)	Chapagain et al. 2009
Grape seed	Fernández et al. 2010
Hazelnut	Koçak et al. 2007; Xu and Hanna 2009
Idesia polycarpa var. vestita	Yang et al. 2009
Jojoba (*Simmondsia chinensis*)	El Kinawy 2004; Huzayyin et al. 2004; Canoira et al. 2006; Radwan et al. 2007; Saleh 2009; Selim 2009
Kusum (Schleichera triguga)	Sharma and Singh 2010
Macadamia nut	Knothe 2010
Maclura pomifera (Rafin.) Schneider	Saloua et al. 2010
Mahua [Hippe; *Madhuca longifolia (indica)*]	Chandraju and Prathima 2004b; Ghadge and Raheman 2005; Puhan et al. 2005a, 2005b; Bhatnagar et al. 2006; Ghadge and Raheman 2006; Kaul et al. 2007; Kumari et al. 2007; Raheman and Ghadge 2007; Kapilan and Reddy 2008; Bhale et al. 2009; Godiganur et al. 2009
Meadowfoam	Moser et al. 2010
Melon bug (*Aspongubus viduatus*)	Mariod et al. 2006
Milkweed	Holser and Harry-O'Kuru 2006
Moringa oleifera	Rashid et al. 2008
Nahor (*Mesua ferra* L., Guttiferae)	De and Bhattacharyya 1999
Piqui (*Caryocar* sp.)	Abreu et al. 2004
Pistacia chinensis	Su et al. 2007
Polanga (nyamplung, *Calophyllum inophyllum*)	Warnigati et al. 1992; Azam et al. 2005; Sahoo et al. 2007; Sahoo et al. 2009; Sahoo and Das 2009
Pongamia pinnata (karanja, koroch, honge, *Pongamia glabra*)	De and Bhattacharyya 1999; Chandraju and Prathima 2004a; Karmee et al. 2004; Azam et al. 2005; Karmee and Chadha 2005; Sarma et al. 2005; Sumar et al. 2005; Bhatnagar et al. 2006; Meher et al. 2006a, 2006b; Kaul et al. 2007; Rathore and Madras 2007; Banamurpath et al. 2008; Kalbande et al. 2008; Naik et al. 2008; Scott et al. 2008; Sharma and Singh 2008; Srivastava and Varma 2008; Baiju et al. 2009; Mukta et al. 2009; Sahoo and Das 2009; Sahoo et al. 2009
Poon (*Sterculia foetida*)	Devan and Mahalakshmi 2009a, 2009b
Pumpkin (*Cucurbita pepo* L.)	Schinas et al. 2009
Rice bran	Özgül and Türkay 1993; Özgül-Yücel and Türkay 2002, 2003a, 2003b; Ryu and Oh 2003; Lai et al. 2005; Zullaikah et al. 2005; Einloft et al. 2008; Kasim et al. 2009; Sinha et al. 2008; Lin et al. 2009; Saravanan et al. 2009; Subramani et al. 2009
Roselle (*Hibiscus sabdariffa* L.)	Nakpong and Woortthikanokkhan 2010
Rubber seed	Ikwuagwu et al. 2000; Ramadhas et al. 2005a, 2005b, 2005c

(*Continued*)

TABLE 33.1 (Continued)
Noncommodity or Minor Vegetable Oils of Which Mono-Alkyl Ester (Biodiesel)
Derivatives or Other Energy Uses Have Been Reported

Oil	References
Sclerocarya birrea	Mariod et al. 2006
Sesame	Chandarju and Prathima 2004b; Banamurpath et al. 2008
	Saydut et al. 2008
Sea mango (*Cerbera odollam*)	Kansedo et al. 2009
Sorghum bug (*Agonoscelis pubescens*)	Mariod et al. 2006
Tall	Liu et al. 1998; Coll et al. 2001; Altparmak et al. 2007
Terminalia catappa (Castanhola)	dos Santos et al. 2008
Tomato seed	Giannelos et al. 2005
Tobacco seed	Giannelos et al. 2002; Usta 2005a, 2005b; Veljkovi et al. 2006
Tucum	Lima et al. 2008
Tung	Shang et al. 2010
Yellow horn	Zhang et al. 2010
Zanthoxylum bungeanum	Yang et al. 2008

Some of the more common vegetable oils of the emerging oils are discussed here in separate brief sections. For example, pongamia oil has probably received the most attention in the literature of the oils discussed here. Others that are briefly mentioned here include moringa, rice bran, olive oil, and a few others. Although olive oil is a classical commodity oil, it is usually not considered a typical biodiesel feedstock. It may be noted that *Jatropha curcas* is probably the most prominent of the emerging energy crop plants. However, it is dealt with in Chapter 14 and therefore will not be discussed here. Numerous other noncommodity vegetable oils have been investigated as fuel sources, and those not discussed in other chapters (e.g., Brassicas are discussed in Chapter 2.8) are briefly discussed here. The fatty acid profiles of 75 plant oils and the cetane numbers of their methyl esters have been compiled (Azam et al. 2005), a few of which are listed here.

33.2 EXAMPLES OF EMERGING VEGETABLE OIL FEEDSTOCKS

Many of the emerging feedstocks that are potential energy sources and that have been recently studied occur commonly in tropical or subtropical regions of the world. Often they are obtained from trees and or bushes in arid or semiarid regions. Although some oils have nutritional or other physiological applications, many are not considered edible. Details on some individual oils follow.

33.2.1 PONGAMIA

Of the emerging vegetable oil feedstocks discussed in this chapter, pongamia oil has probably received the most attention in the technical literature. Different names have been used for this oil in the literature. In addition to pongamia (*Pongamia pinnata*, *Pongamia glabra*), the terms "karanja" and "honge" also refer to pongamia in biodiesel-related literature. A review on this subject was published recently (Scott et al. 2008), and the brief discussion of this oil presented here draws heavily on it. Another recent article (Kesari et al. 2008) deals with the selection of candidate plus trees for breeding and also with oil extraction.

    P. pinnata is a fast-growing leguminous tree (Leguminosae family, subfamily Papilionodeae, tribe Millettieae) native to the Indian subcontinent and Southeast Asia (Malaysia, Indonesia, and Myanmar), but it is now also grown in other tropical areas as well as parts of Australia, China, New, Zealand, and the United States (Scott et al. 2008). The seeds contain 30–40% oil, and it and can

TABLE 33.2

Fatty Acid Profile of Noncommodity or Minor Vegetable Oils Studied as Energy Source

Oil	8:0	10:0	12:0	14:0	16:0	16:1	18:0	18:1	18:2	18:3	20:0	20:1	22:0	22:1	Other
								Fatty Acid Profile							
Andiroba (Carapa guianensis)					27	1	7	49	16						
Azadirachta indica					14.9		14.4	61.9	7.5		1.3				
Babassu	5.0	6.0	44.0	17.0	8.0		4.5	14.0	2.0						
Calophyllum inophyllum					17.9	2.5	18.5	42.7	13.7	2.1					2.6 (24:0)
Coffee					34		7	9	44	1.5	3		0.7		
Cuphea PSR 23	0.3	64.7	3.0	4.5	7.0		0.9	12.2	6.7						4.2 (6:0)
C. viscosissima VS-320	40.2	36.9	4.8	6.8	3.3		0.15	1.4	2.05	0.1					
Cumaru (Dipteryx odorata)					23		7	37	29						
Cynara cardunculus					11		4	25	60						
Desert date (Balanites aegyptiaca)					12.7–16.0		10.5–12.1	23.5–43.7	31.5–51.6						
Grape seed					6.9		4.7	18.7	68.8	0.5	0.2	0.2			
Hazelnut					4.5–5.9		0.5–2.8	68.8–78.6	14.2–23.3	0.1–0.2		0.3			
Idesia polycarpa			0.31		15.06	6.5	1.18	5.5	70.6	1.1		6.2			
Kusum (Scheichera trijuga)				15.5	10.35		11.1	27.1	6.1		15.8		0.01		
Macadamia				0.6–0.8	8.5–8.9	15.9–20.2	3.3–3.6	55.3–58.7	1.8–2.1		2.3–2.7	2.6	0.7–0.9		3.8–4.0 (18:1 Δ11)

(Continued)

TABLE 33.2 (Continued)
Fatty Acid Profile of Noncommodity or Minor Vegetable Oils Studied as Energy Source

Oil	8:0	10:0	12:0	14:0	16:0	16:1	18:0	18:1	18:2	18:3	20:0	20:1	22:0	22:1	Other
Maclura pomifera (Rafin.) Schneider					6.7		2.3	15.3	75.4						
Mahua (*Madhuca indica*)					24.5		22.7	37.0	14.3		1.5				
Meadowfoam					0.6		0.2	1.0	0.9		0.8	64.2 (Δ5)	0.2	10.2	18.9 (22:2)
Melon bug (*Aspongubus viduattus*)					30.9	10.7	3.5	46.6	3.9						2.4 (17:0)
Milkweed					5.9	6.8	2.3	34.8	48.7	1.2	0.2				
Moringa oleifera					6.5		6.0	72.2	1.0		4.0	2.0	7.1		
Nahor (*Mesua ferra*)					15.3		12.0	53.2	17.0		2.0		0.5		
Piqui (*Caryocar sp.*)					40		2	47	4						
Polanga					12.01		12.95	34.09	38.26	0.3					
Pongamia (Honge, karanja)					11.3–14.1		7.5–12.9	41.4–56.0	15.0–26.7		1.75–2.1		1.9		
Poon					22.4		7.3	16.4	45.9		6.47				
Pumpkin (*Cucurbita pepo L.*)					12.5	0.2	5.4	37.1	43.7	0.2	0.4				
Rice bran				tr	20	2–4	42	32	0.5–1.8						
Roselle (*Hibiscus sabdariffa L*)					18.15		4.1	33.3	38.2	2.1					
Rubber seed					10.2		8.7	24.6	39.6	16.3					
Sclerocarya birrea					14.2	0.2	8.8	67.3	5.9						
Sea mango (*Cerbera odollam*)					30.3		3.8	48.1	17.8						

Sesame (*Sesasum indicum*)						41.3	43.7						
Sorghum bug (*Agonoscelis pubescens*)			12.2	1.0	7.3	40.9	3.9						
Terminalia catappa L.			35.0		5.0	32.0	28.0						
Tomato seed	0.10		12.26		5.15	22.17	56.12	2.77	0.41				
Tobacco seed	0.09		10.96	0.2	3.34 3.49	14.54	69.49	0.69	0.25	0.12	0.09		
	0.17		8.87	0.0		12.4	67.75	4.20					
Tucum	1.94	2.12	52.51	25.04	7.46	0.45	8.39	2.11					
Yellow horn			7.1		3.25	25.3	38.6		0.4	9.6	0.7	10.8	
Zanthoxylum bungeanum			10.3	10.8	1.1	35.9	24.8	15.9					3.7 (24:1)

Fatty acid profiles in this table do not necessarily add to 100%. The reasons for this can be rounding, unidentified species, or incomplete fatty acid profiles given in the literature. tr, traces. Data are from the references in this chapter.

grow on marginal land. It grows best at latitudes from sea level to 1200 m with an optimal annual rainfall of 500–2500 mm; it is drought and saline tolerant (Scott et al. 2008 and references therein). Products derived from pongamia have included not only fuel but also traditional medicines, animal fodder, green manure timber, and fish poison (Scott et al. 2008). As is the case with many of the emerging energy crops mentioned here, successful establishment of pongamia as a fuel source would require work in the area of agronomy, plant propagation, genetics, and molecular biology (Scott et al. 2008). Pongamia has been stated to compare favorably to other emerging energy sources for biodiesel in terms of land-use requirements (Azam 2005; Scott et al. 2008). Relevant predictions for biodiesel production from pongamia include 20,000 seeds/year from a 10-year-old tree with 1.8 g oil/seed (36 kg oil/tree per year) at 350 trees/ha (leading to 12,600 kg oil/ha per year) (Scott 2008).

The fatty acid profile of pongamia consists of oleic acid (40–55%), palmitic acid (5–15%), stearic acid (5–10%), and linoleic acid (15–20%) with smaller amounts of the longer-chain saturated fatty acids arachidic acid (C20:0), behenic acid (C22:0), and lignoceric acid (C24:0) (Scott et al. 2008). However, wide variability in the oil content and fatty acid profile was observed in germplasm accessions collected in a region of central India (Mukta et al. 2009). The cetane number has been reported as 55.84 (Scott et al. 2008), which appears to agree with the given fatty acid profile. However, the cloud point is reported as 8.3°C (Scott et al. 2008), which would coincide with the saturated fatty acid content reported but would also render this feedstock of reduced interest for use in more moderate climates.

33.2.2 MAHUA

Another oil that has been reported as an energy source relatively frequently in the literature is mahua (*Madhuca indica*) oil. Mahua is a forest tree growing in many parts of India (Kumari et al. 2007). The oil (fat) derived from it has a very high content of saturated fatty acids (C16:0 and C18:0), leading to poor cold-flow properties of the resulting biodiesel, probably even problematic in warmer climates.

33.2.3 MORINGA OIL

A comprehensive review (Morton 1991) provides detailed information on moringa (*Moringa oleifera*). It is also called horseradish tree and various other names. Similar to pongamia, moringa is a single-genus family of oilseed tree species that grows in tropical and subtropical regions and tolerates drought and poor soil. The kernels contain 30–49% oil (Sengupta and Gupta 1970). Moringa, a single-genus oilseed tree is truly wild only in the western Himalayas and Punjab was cultivated in India and distributed to other parts of tropical Asia and Oceania and is now also grown in the Caribbean, parts of Africa and Latin America, and in the United States (Morton 1991 and references therein). Almost all parts of the tree have been used in folk medicine (Morton 1991). Various parts have been used for nutritional purposes and nonphysiological applications (Morton 1991). Moringa oil is also known as "ben oil" or "behen oil" because of its content of behenic (docosanoic) acid. The absence of linolenic acid and small amounts of linoleic acid are partially responsible for the relatively high oxidative stability of the oil (Sengupta and Gupta 1970). Biodiesel has been prepared from moringa oil (Rashid et al. 2008). This biodiesel fuel is discussed in Section 33.2.6.

33.2.4 RICE BRAN OIL

Rice bran oil (*Oryza sativa*) is an edible oil of nutritional value with a nutlike taste that finds use in cooking and nutritional applications. It appears that this could be a biodiesel feedstock that is significantly affected by the food-versus-fuel issue. Rice bran oil is ultimately derived from rice—one of the basic foodstuffs for a significant portion of humankind. The annual production of rice in 1999 was approximately 600 million t (Gunstone and Harwood 2007 and references therein); however, the production of rice bran oil is approximately 0.7–1.0 million t (Gunstone and Harwood 2007) with India, China, and Japan being the major producers of this oil. Rice bran, obtained after removing the hull and abrading the bran layer, contains 18–24% oil and 4–6% free fatty acids, fibers, protein, and carbohydrates. The major fatty acids in rice bran oil are palmitic, oleic, and linoleic acids with smaller amounts of stearic and linolenic acids and traces of other fatty acids such as 14:0, 16:1, 20:0, 20:1, and 22:1. Rice bran oil contains a number of other constituents, including waxes (2–4%; esters of saturated fatty acids with saturated alcohols), monomethylsterols, dimethylsterols, and tocotrienols, the latter together with oryzanols imparting high oxidative stability to rice bran oil (Gunstone and Harwood 2007). Numerous publications have dealt with biodiesel derived from rice bran oil (Table 33.1).

33.2.5   Olive Oil

Olive oil is a commercialized oil of recognized nutritional value and is expensive compared with other oils. It is therefore not listed in Tables 33.1 and 33.2. It can only be seen as an experimental source of fuel, if at all, and it is only briefly mentioned in this section. It also appears that this oil would be potentially strongly affected by the food-versus-fuel issue. A study has dealt with Turkish sulfur olive oil as fuel source (Aksoy et al. 1988), and other reports have examined olive oil as a supplement to petrodiesel fuel (Rakopoulos et al. 1992a, 1992b; Rakopoulos et al. 2006). Other works have focused on used olive oil, which would be less affected by the aforementioned issues, as a source of biodiesel (Dorado et al. 2003a, 2003b, 2004; Yuste and Dorado 2006).

33.2.6   Oils with Varying Fatty Acid Profiles

Most commodity oils such as soybean and rapeseed (canola) oils as well as most lower-volume and "emerging" oils such as those discussed in this chapter possess fatty acid profiles consisting largely of the five major species—C16:0, C18:0, C18:1, C18:2, and C18:3. As a result, the biodiesels derived from these feedstocks exhibit the common problems of poor cold flow and oxidative stability with varying severity. Feedstocks with high amounts of C16:0 and/or C18:0 will demonstrate poor cold flow. The presence of even longer-chain unsaturated fatty acids such C20:0 and C22:0 in some feedstocks will exacerbate the cold-flow problem. For example, biodiesel derived from moringa oil, which contains approximately 23.5% saturated fatty acids, 4% being C20:0 and 7% being C22:0, together with approximately 72% C18:1, has a cloud point of 18°C (Rashid et al. 2008). Conversely, feedstocks with high amounts of polyunsaturated fatty acids will possess insufficient oxidative stability. For example, camelina oil (Chapter 2.8) used for biodiesel production had around 38.4% C18:3 (Bernardo et al. 2003; Froehlich and Rice 2005). However, it must be considered that small amounts of polyunsaturated fatty acid chains may exert greater influence on oxidative stability than the small amounts indicate.

Some of the oils listed in a compilation of 75 nontraditional seed oils for potential biodiesel use in India contain major amounts of fatty acids beyond the five most common ones (Azam et al. 2005). The biodiesel fuels from oils with significant amounts of eleostearic acid displayed low cetane numbers, with the cetane number of biodiesel from *Aleuritis montana* (64.6%—eleostearic acid = 9*t*,11*t*,13*t*-octadecatrienoic acid) reported as 20.56, that of biodiesel from *Aleuritis fordii* (tung oil; 81.5%—eleostearic acid = 9*c*,11*t*,13*t*-octadecatrienoic acid) given as 36.25, and that of biodiesel from *Mallotus phillippinensis* (72% kamlolenic acid = 18-hydroxyeleostearic acid) listed as 36.34. Biodiesel from tung oil displayed a relatively high viscosity of approximately 7 mm^2/s (Shang et al. 2010), a possible reason being its low stability which may lead to polymer formation. Other authors reported some fuel-related properties of 17 untransesterified Kenyan vegetable oils (Munavu and Odhiambo 1984) and of six other seed oils from tropical Africa (Eromosele and Paschal 2003).

To address these technical issues, feedstocks with inherently different fatty acid profiles or that have been altered to provide such profiles are of interest. Esters of palmitoleic acid [9(Z)-hexadecenoic acid; C16:1] would be especially desirable to be enriched in a biodiesel fuel (Knothe 2008) because the cetane number and low melting point (melting point of methyl palmitoleate is approximately −34°C) are of interest. Macadamia nut oil is a model feedstock that is moderately enriched in palmitoleic acid, but its content of saturated fatty acids precludes improvement of the cold-flow properties of the corresponding biodiesel fuel (Knothe 2010). When considering saturated fatty acid chains, esters of capric acid (decanoic acid; C10:0) offer a compromise of properties (cetane number of methyl decanoate 51.6, melting point of −13°C) because esters of octanoic acid possess cetane numbers that are too low. Esters of dodecanoic (lauric) acid possess melting points that are too high (melting point of methyl laurate is ~4.5°C) (Knothe 2008). Accordingly, methyl esters of a variety of cuphea oil (PSR 23, a cross of *Cuphea viscosissima* and *Cuphea lanceolata*) containing approximately 65% methyl decanoate, 12.2% methyl oleate, 7% methyl palmitate, and 6.7% methyl linoleate with the rest consisting of other saturated esters showed a cloud point of −9 to −10°C and a cetane number of 55–56 (Knothe et al. 2009). Other work dealt with fuel properties of the genetically altered *C. viscosissima* strain VS-320, which contains C6:0 and C8:0, in the untransesterified form (Geller et al. 1999). Although cuphea oil, of which approximately 260 species with greatly varying fatty acid profiles are known (Isbell 2002), has not been commercialized because of issues such as seed shattering, the fatty acid profile of the C_{10}-enriched oil used here can serve as model for other oils that are potentially enriched in this fatty acid.

An oil containing high amounts of unsaturated fatty acid chains in the C_{20}–C_{22} range is meadowfoam; however, the corresponding biodiesel fuel exhibits relatively high oxidative stability, which may be due to the position of the double bonds in the chain (Moser et al. 2010).

33.2.7 JOJOBA

Jojoba (*Simmondsia chinensis*) is a woody perennial bush shrub native to arid regions such as the deserts of the southwestern United States and neighboring areas of Mexico (Miwa 1971), but it is now also commercially grown in other countries such as Australia, Argentina, Chile, Egypt, India, Peru, and South Africa (International Jojoba Export Council 2009). It has a seed containing approximately 45–55% of a wax-like material known as "jojoba oil." Jojoba oil is not a typical oil consisting mainly of triacylglycerols; rather, it is a mixture of long-chain esters containing smaller amounts of triacylglycerols and other materials such as phospholipids and tocopherols. The esters that largely comprise jojoba oil are those of monounsaturated fatty acids with alcohols of C_{16}–C_{24} chain length. The major fatty acids and alcohols are C20:1 and C22:1, with double bond positions at C_{11} and C_{13}, respectively, but minor amounts of acids and alcohols with double bonds in other positions are also present (Miwa 1971; Hamilton et al. 1975). Because of its properties, it has largely found application in skin care products and cosmetics. Work on producing a biodiesel fuel from jojoba oil was reported (Canoira et al. 2006), but the resulting methyl esters still contained approximately 21% jojoba alcohols and therefore the viscosity was high (9–10 mm²/s), well above the prescribed maximal values in the ASTM D6751 and EN 14214 biodiesel standards.

33.2.8 TALL OIL

Tall oil fatty acids are obtained from the wood pulp industry from the digestion of pine wood chips with an alkaline solution of sodium sulfate or an acidic solution of sodium sulfite (Gunstone and Harwood 2007). Upon acidification with sulfuric acid of the aqueous alkaline solution, a mixture of resin acids and fatty acids results. Tall oil is produced mainly in North America and Scandinavia. However, the compositions vary because of the nature of the pulped wood.

33.2.9 RENEWABLE DIESEL

In addition to biodiesel (mono-alkyl esters of vegetable oils and animal fats), through a hydrotreating process similar to the processing of petroleum mixtures compositionally resembling petroleum-based diesel fuel can be obtained from vegetable oils (for example, see Rantanen et al. 2005). This kind of process has been described for various vegetable oils and could also apply to the feedstocks discussed in this chapter.

33.3 CONCLUSIONS AND FUTURE PROSPECTS

The fact that all biodiesel feedstocks combined do not have the potential to provide sufficient amounts of biodiesel for replacing all petroleum-based diesel fuel has been a major factor in the search for additional feedstocks. This search is reflected in the use of vegetable oils as energy sources summarized in this and other chapters. It is likely that this search will continue in the future. However, issues such as yield improvement, sustainability, and acceptable fuel properties will influence the use of any potential feedstock. For many of the emerging feedstocks, these issues have not been sufficiently determined. Therefore, it remains to be seen which feedstocks will be able to partially meet the need of yielding fuels that can replace fossil fuels.

REFERENCES

Abreu FR, Lima DG, Hamú EH, Wolf C, Suarez PAZ (2004) Utilization of metal complexes as catalysts in the transesterification of Brazilian vegetable oils with different alcohols. *J Mol Cat A Chem* 209:29–33

Aksoy HA, Kahraman I, Karaosmanoglu F, Civelekoglu H (1988) Evaluation of Turkish sulfur olive oil as an alternative diesel fuel. *J Amer Oil Chem Soc* 65:936–938 Erratum (1989): *J Amer Oil Chem* Soc 66:837

Altiparmak D, Keskin A, Koca A, Gürü M (2007) Alternative fuel properties of tall oil fatty acid methyl ester—Diesel fuel blends. *Bioresour Technol* 98:241–246

Azam MM, Waris A, Nahar NM (2005) Prospects and potential of fatty acid methyl esters of some non-traditional seed oils for use as biodiesel in India. *Biomass Bioenergy* 29:293–302

Bernardo A, Howard-Hildige R, O'Connell A, Nichol R, Ryan J, Rice B, Roche E, Leahy JJ (2003) Camelina oil as a fuel for diesel transport engines. *Indust Crops Prod* 17:191–197

Bhale PV, Deshpande NV, Thombre SB (2009) Improving the low temperature properties of biodiesel fuel. *Renew Energy* 34:794–800

Bhatnagar AK, Kaul S, Chhibber VK, Gupta AK (2006) HFRR studies on methyl esters of nonedible vegetable oils. *Energy Fuel* 20:1341–1344

Canoira L, Alcántara R, Garcia-Martínez MJ, Carrasco J (2006) Biodiesel from jojoba-oil wax: Transesterification with methanol and properties as fuel. *Biomass Bioenergy* 30:76–81

Chandraju S, Prathima BK (2004a) Ethyl ester of honge oil and palm oil blends with diesel as an ecofriendly fuel in heavy vehicles; An investigation. *Asian J Chem* 16:147–155

Chandraju S, Prathima BK (2004b) Ethanol derivatives of hippe oil (*Madhuca indica*) and sesame oil blends with diesel as an ecofriendly fuel in heavy duty vehicles: An investigation. *Indian Sugar* 54:253–264

Chapagain B.P, Yehoshua Y, Wiesman Z (2009) Desert date (*Balanites aegyptiaca*) as an arid lands sustainable bioresource for biodiesel. *Bioresour Technol* 100:1221–1226

Chavanne CG (1937) Procédé de transformation d'huiles végétales en vue de leur utilisation comme carburants (Procedure for the transformation of vegetable oils for their uses as fuels). Belgian Patent 421,877, August 31, 1937

Chowhury DH, Mukerji SN, Aggarwal JS, Verman LC (1942) Indian vegetable fuel oils for diesel engines. *Gas Oil Power* 37:80–85, *Chem Abstr* 36:5330

Coll R, Udas S, Jacoby WA (2001) Conversion of the rosin acid fraction of crude tall oil into fuels and chemicals. *Energy Fuels* 15:1166–1172 da Silva TB, Neto AFL, dos Santos LSS, Lima JRdO, Chaves MH, dos Santos JR Jr, de Lima GM, de Moura EM, de Moura CVR (2008) Catalysts of Cu(II) and Co(II) ions adsorbed in chitosan used in transesterification of soybean and babassu oils—A new route for biodiesel synthesis. *Bioresour Technol* 99:6793–6798

De BK, Bhattacharyya BK (1999) Biodiesel from minor vegetable oils like karanja oil and nahor oil. *Fett/Lipid* 101:404–406

Devan PK, Mahalakshmi NV (2009a) Performance, emission and combustion characteristics of poon oil and its diesel blends in a DI diesel engine. *Fuel* 88:861–867

Devan PK, Mahalakshmi NV (2009b) Study of the performance, emission and combustion characteristics of a diesel engine using poon oil-based fuels. *Fuel Process Technol* 90:513–519

Diesel R (1913) *Die Entstehung des Dieselmotors.* Verlag von Julius. Springer, Berlin, Germany

Diesel R (1912) The Diesel oil-engine and its industrial importance particularly for Great Britain. *Proc Inst Mech Eng* 179–280, *Chem Abstr* 7:1605 (1913)

Dorado MP, Ballesteros E, Arnal JM, Gómez J, López Giménez FJ (2003a) Testing waste olive oil methyl ester as a fuel in a diesel engine. *Energy Fuels* 17:1560–1565

Dorado MP, Ballesteros E, Arnal JM, Gómez J, López FJ (2003b) Exhaust emissions from a diesel engine fueled with transesterified waste olive oil. *Fuel* 82:1311–1315

Dorado MP, Ballesteros E, Mittelbach M, Lopez FJ (2004) Kinetic parameters affecting the alkali-catalyzed transesterification process of used olive oil. *Energy Fuels* 18:1457–1462

dos Santos ICF, de Carvalho SHV, Solleti JI, de La Salles WF, de La Salles KT, Meneghetti SMP (2008) Studies of *Terminalia catappa* L. oil: Characterization and biodiesel production. *Bioresour Technol* 99:6545–6549

Einloft S, Magalhães TO, Donato A, Dullius J, Ligabue R (2008) Biodiesel from rice bran oil: Transesterification by tin compounds. *Energy Fuels* 22:671–674

El Kinawy OS (2004) Comparison between jojoba oil and other vegetable oils as a substitute to petroleum. *Energy Sour* 26:639–645

Encinar JM, Gonzalez JF, Sabio E, Ramiro MJ (1999) Preparation and properties of biodiesel from *Cynara cardunculus* L. oil. *Indust Eng Chem Res* 38:2927–2931

Eromosele CO, Paschal NH (2003) Characterization and viscosity parameters of seed oils from wild plants. *Bioresour Technol* 86:203–205

FAO (2009) *The Market and Food Security Implications of the Development of Biofuel Production,* Food and Agriculture Organization of the United Nations, available at ftp://ftp.fao.org/docrep/fao/meeting/016/k4477e.pdf (accessed June 15, 2010)

Fernández J, Curt MD, Aguado PL (2006) Industrial applications of *Cynara cardunculus* L. for energy and other uses. *Indust Crops Prod* 24:222–229

Fernández CM, Ramos MJ, Pérez A, Rodríguez JF (2010) Production of biodiesel from winery waste: Extraction, refining and transesterification of grape seed oil. *Bioresour Technol* 101:7019–7024

Frankel EN (2005) *Lipid Oxidation,* 2nd ed. The Oily Press, Bridgwater, United Kingdom

Fröhlich A, Rice B (2005) Evaluation of *Camelina sativa* oil as a feedstock for biodiesel production. *Indust Crops Prod* 21:25–31

Geller DP, Goodrum JW, Knapp SJ (1999) Fuel properties of oil from genetically altered *Cuphea viscosissima.* *Indust Crops Prod* 9:85–91

Ghadge SV, Raheman H (2005) Biodiesel production from mahua (*Madhuca indica*) oil having high free fatty acids. *Biomass Bioenergy* 28:601–605

Ghadge SV, Raheman H (2006) Process optimization for biodiesel production from mahua (*Madhuca indica*) oil using response surface methodology. *Bioresour Technol* 97:379–384

Giannelos PN, Sxizas S, Lois E, Zannikos F, Anastopoulos G (2005) Physical, chemical and fuel related properties of tomato seed oil for evaluating its direct use in diesel engines. *Indust Crops Prod* 22:193-199

Giannelos PN, Zannikos F, Stournas S, Lois E, Anastopoulos G (2002) Tobacco seed oil as alternative diesel fuel: Physical and chemical properties. *Indust Crops Prod* 16:1–9

Godiganur S, Murthy CHS, Reddy RP (2009) 6BTA 5.9 G2-1 Cummins engine performance and emissions tests using methyl ester mahua (*Madhuca indica*) oil/diesel blends. *Renew Energy* 34:2172–2177

Gunstone FD, Harwood JL (2007) Occurrence and characterisation of oils and fats. In: Gunstone FD, Harwood JL, Dijkstra AJ (eds), *The Lipid Handbook,* 3rd ed. CRC Press, Boca Raton, FL, pp 37–142

Hamilton RJ, Raie MY, Miwa TK (1975) Structure of the alcohols derived from wax esters in jojoba oil. *Chem Phys Lipids* 14:92–96

Hill J, Nelson E, Tilman D, Polasky S, Tiffany D (2006) Environmental, economic, and energetic costs and benefits of biodiesel and ethanol biofuels. *Proc Natl Acad Sci USA* 103:11206–11210

Holser RA, Harry-O'Kuru R (2006) Transesterified milkweed (*Asclepias*) seed oil as a biodiesel fuel. *Fuel* 85:2106–2110

Huzayyin AS, Bawady AH, Rady MA, Dawood A (2004) Experimental evaluation of diesel engine performance and emission using blends of jojoba oil and diesel fuel. *Energy Conver Manag* 45:2093–2112

Ikwuagwu OE, Ononogbu IC, Njoku OU (2000) Production of biodiesel using rubber [*Hevea brasiliensis* (Kunth Muell.)] seed oil. *Indust Crops Prod* 12:57–62

International Jojoba Export Council, available http://www.jojoba-oil.org (accessed June 15, 2010)

Isbell TA (2002) Current progress in the development of cuphea. *Lipid Technol* 14:77–80

Kalayasiri P, Jeyashoke N, Krisnangkura K (1996) Survey of seed oils for use as diesel fuels. *J Amer Oil Chem Soc* 73:471–474

Kalbande SR, More GR, Nadre RG (2008) Biodiesel production from non-edible oils of jatropha and karanj for utilization in electrical generator. *Bioenergy Res* 1:170–178

Kansedo J, Lee KT, Bhatia S (2009) *Cerbera odollam* (sea mango) oil as a promising non-edible feedstock for biodiesel production. *Fuel* 88:1148–1150

Kapilan N, Reddy RP (2008) Evaluation of methyl esters of mahua oil (*Madhuca Indica*) as diesel fuel. *J Amer Oil Chem* Soc 85:185–188

Karmee SK, Mahesh P, Ravi R, Chadha A (2004) Kinetic study of the base-catalyzed transesterification of monoglycerides from *Pongamia* oil. *J Amer Oil Chem Soc* 81:425–431

Karmee SK, Chadha A (2005) Preparation of biodiesel from crude oil of *Pongamia pinnata*. *Bioresour Technol* 96:1425–1429

Kasim NS, Tsai T-H, Gunawan S, Ju Y-H (2009) Biodiesel production from rice bran oil and supercritical methanol. *Bioresour Technol* 100:2399–2403

Kaul S, Saxena RC, Kumar A, Negi MS, Bhatnagar AK, Goyal HB, Gupta AK (2007) Corrosion behavior of biodiesel from seed oils of Indian origin on diesel engine parts. *Fuel Process Technol* 88:303–307

Kesari V, Krishnamachari A, Rangan L (2008) Systematic characterisation and seed oil analysis in candidate plus trees of biodiesel plant, *Pongamia pinnata*. *Ann Appl Biol* 152:397–404

Knothe G (2005) The history of vegetable oil-based diesel fuels. In: Knothe G, Krahl J, Van Gerpen J (eds), *The Biodiesel Handbook*. America Oil Chemists Society Press, Champaign, IL, pp 1–16

Knothe G (2008) "Designer" biodiesel: Optimizing fatty ester composition to improve fuel properties. *Energy Fuels* 22:1358–1364

Knothe G (2010) Biodiesel derived from a model oil Enriched in palmitoleic acid, macadamia nut oil. *Energy Fuels* 24:2098–2103

Knothe G, Cermak SC, Evangelista RL (2009) Cuphea oil as source of biodiesel with improved fuel properties caused by high content of methyl decanoate. *Energy Fuels* 23:1743–1747

Knothe G, Steidley KR (2005) Kinematic viscosity of biodiesel fuel components and related compounds. Influence of compound structure and comparison to petrodiesel fuel components. *Fuel* 84:1059–1065

Koçak MS, Ileri E, Utlu Z (2007) Experimental study of emission parameters of biodiesel fuels obtained from canola, hazelnut, and waste cooking oils. *Energy Fuels* 21:3622–3626

Köse Ö, Tüter M, Aksoy HA (2002) Immobilized *Candida antarctica* lipase-catalyzed alcoholysis of cotton seed oil in a solvent-free medium. *Bioresour Technol* 83:125–129

Kumari V, Shah S, Gupta MN (2007) Preparation of biodiesel by lipase-catalyzed transesterification of high free fatty acid containing oil from *Madhuca indica*. *Energy Fuels* 21:368–372

Lai CC, Zullaikah S, Vali SR, Ju Y-H (2005) Lipase-catalyzed production of biodiesel from rice bran oil. *J Chem Technol Biotechnol* 80:331–337

Lima JRdO, da Silva RB, da Silva CCM, dos Santos LSS, dos Santos JR Jr, Moura EM, de Moura CVR (2007) Biodiesel from babassu (*Orbignya* sp.) synthesized via ethanolyic route. *Quim Nova* 30:600-603.

Lima JRdO, da Silva RB, de Moura EM, de Moura CVR (2008) Biodiesel of tucum oil, synthesized by methanolic and ethanolic routes. *Fuel* 87:1718–1723

Lin L, Ying D, Chaitep S, Vittayapadung S (2009) Biodiesel production from crude rice bran oil and properties as fuel. *Appl Energy* 86:681–688

Liu, DDS, Monnier J, Tourigny G, Kriz J, Hogan E, Wong A (1998) Production of high quality cetane enhancer from depitched tall oil. Petrol Sci Technol 16:597–609.

Mariod A, Klupsch S, Hussein OH, Ondruschka B (2006) Synthesis of alkyl esters from three unconventional sudanese oils for their use as biodiesel. *Energy Fuels* 20:2249–2252

May CY, Ngan MA, Weng CK, Basiron Y (2005) Palm diesel. An option for greenhouse gas mitigation in the energy sector. *J Oil Palm Res* 17:47–52.

Meher LC, Dharmagadda VSS, Naik SN (2006a) Optimization of alkali-catalyzed transesterification of *Pongamia pinnata* oil for production of biodiesel. *Bioresour Technol* 97:1392-1397.

Meher LC, Kulkarni MG, Dalai AK, Naik SN (2006b) Transesterification of karanja (*Pongamia pinnata*) oil by solid basic catalysts. *Eur J Lipid Sci Technol* 108:389–397

Miwa TK (1971) Jojoba oil wax esters and derived fatty acids and alcohols: Gas chromatographic analyses. *J Amer Oil Chem Soc* 48:259–264

Morton JF (1991) The horseradish tree, *Moringa pterigosperma* (Moringaceae). A boon to arid lands. *Econ Bot* 45:318–333

Moser BR, Knothe G, Cermak SC (2010) Biodiesel from meadowfoam (*Limnanthes alba* L.) seed oil: Oxidative stability and unusual fatty acid composition. *Energy Environ Sci* 3:318–327

Mukta N, Murthy IYLN, Sripal P (2009) Variability assessment in *Pongamia pinnata* (L.) pierre germplasm for biodiesel traits. *Indust Crops Prod* 29:536–540

Munavu RM, Odhiambo D (1984) Physicochemical characterization of nonconventional vegetable oils for fuel in Kenya. *Ken J Sci Technol* 5:45–52

Nakpong P, Woortthikanokkhan S (2010) Roselle (*Hibiscus sabdariffa* L.) oil as an alternative feedstock for biodiesel production in Thailand. *Fuel* 89:1806–1811

Oliveira LS, Franca AS, Mendonça JCF, Barros-Júnior MC (2006) Proximate composition and fatty acids profile of green and roasted defective coffee beans. *Bioresour Technol* 99:3244–3250

Oliveira LS, Franca AS, Camargos RRS, Ferraz VP (2008) Coffee oil as a potential feedstock for biodiesel production. *Bioresour Technol* 99:3244–3250

Özgül S, Türkay S (1993) *In situ* esterification of rice bran oil with methanol and ethanol. *J Amer Oil Chem Soc* 70:145–147

Özgül-Yücel S, Türkay S (2002) Variables affecting the yields of methyl esters derived from *in situ* esterification of rice bran oil. *J Amer Oil Chem Soc* 79:611–614

Özgül-Yücel S, Türkay S (2003a) Purification of FAME by rice hull ash adsorption. *J Amer Oil Chem Soc* 80:373–376

Özgül-Yücel S, Türkay S (2003b) FA monoalkylesters from rice bran oil by *in situ* esterification. *J Amer Oil Chem Soc* 80:81–84

Pacheco Borges G (1944) Use of Brazilian vegetable oils as fuel. *Anais Assoc Quim Brasil* 3:206–209, *Chem Abstr* 39:5067 (1945)

Puhan S, Vedaraman N, Ram BVB, Sankaranarayanan G, Jeychandran K (2005a) Mahua oil (*Madhuca Indica* seed oil) methyl ester as biodiesel—Preparation and emission characteristics. *Biomass Bioenergy* 28:87–93

Puhan S, Vedaraman N, Sankaranarayanan G, Ram BVB (2005b) Performance and emission study of mahua oil (Madhuca Indica seed oil) ethyl ester in a 4-stroke natural aspirated direct injection diesel engine. *Renew Energy* 30:1269–1278

Radwan MS, Ismail MA, Elfeky SMS, Abu-Elzayeed OSM (2007) Jojoba methyl ester as a diesel fuel substitute: Preparation and characterization. *Appl Therm Eng* 27:314–322

Raheman H, Ghadge SV (2007) Performance of compression ignition engine with mahua (*Madhuca indica*) biodiesel. *Fuel* 86:2568–2573

Rakopoulos CD (1992a) Olive oil as a fuel supplement in DI and IDI diesel engines. *Energy (Oxford)* 17:787–790

Rakopoulos CD (1992b) Comparative performance and emission studies when using olive oil as a fuel supplement in DI and IDI diesel engines. *Renew Energy* 2:327–331

Rakopoulos CD, Antonopoulos KA, Rakopoulos DC (2006) Comparative performance and emissions study of a direct injection diesel engine using blends of diesel fuel with vegetable oils or bio-diesels of various origins. *Energy Conver Manag* 47:3272–3287

Ramadhas AS, Jayaraj S, Muraleedharan C (2005a) Biodiesel production from high FFA rubber seed oil. *Fuel* 84:335–340

Ramadhas AS, Jayaraj S, Muraleedharan C (2005b) Characterization and effect of using rubber seed oil as fuel in the compression ignition engines. *Renew Energy* 30:795–803

Ramadhas AS, Muraleedharan C, Jayaraj S (2005c) Performance and emission evaluation of a diesel engine fueled with methyl esters of rubber seed oil. *Renew Energy* 30:1789–1800

Rantanen L, Linnaila R, Aakko P, Harju T (2005) NexBTL—Biodiesel Fuel of the Second Generation. Society of Automotive Engineers (SAE) Technical Paper Series 2005-01-3771, SAE, Warrendale, PA

Rashid U, Anwar F, Moser BR, Knothe G (2008) *Moringa oleifera* oil: A possible source of biodiesel. *Bioresour Technol* 99:8175–8179

Rathore V, Madras G (2007) Synthesis of biodiesel from edible and non-edible oils in supercritical alcohols and enzymatic synthesis in supercritical carbon dioxide. *Fuel* 86:2650–2659

Ryu K, Oh Y (2003) A study on the usability of biodiesel fuel derived from rice bran oil as alternative fuel for IDI diesel engine. *Kor Soc Mech Eng Int J* 17:310–317

Sahoo PK, Das LM, Babu MKG, Naik SN (2007) Biodiesel development from high acid value polanga seed oil and performance evaluation in a CI engine. *Fuel* 86:448–454

Sahoo PK, Das LM (2009) Process optimization for biodiesel production from jatropha, karanja and polanga oils. *Fuel* 88:1588–1594

Sahoo PK, Das LM, Babu MKG, Arora P, Singh VP, Kumar NR, Varyani TS (2009) Comparative evaluation of performance and emission characteristics of jatropha, karanja and polanga based biodiesel as fuel in a tractor engine. *Fuel* 88:1698–1707

Saleh HE (2009) Experimental study on diesel engine nitrogen oxide reduction running with jojoba methyl ester by exhaust gas recirculation. *Fuel* 88:1357–1364

Saloua F, Saber C, Hedi Z (2010) Methyl ester of [*Maclura pomifera* (Rafin.) Schneider] seed oil: Biodiesel production and characterization. *Bioresour Technol* 101:3091–3096

Saravanan S, Nagarajan G, Rao GLN (2009) Feasibility analysis of crude rice bran oil methyl ester blend as a stationary and automotive diesel engine fuel. *Energy Sustain Dev* 13:52–55

Saydut A, Duz MZ, Kaya C, Kafada AB, Hamamci C (2008) Transesterified sesame (*Sesamum indicum* L.) seed oil as a biodiesel fuel. *Bioresour Technol* 99:6656–6660

Schinas P, Karavalakis G, Davaris C, Anastopoulos G, Karonis D, Zannikos F, Stournas S, Lois E (2009) Pumpkin (*Cucurbita pepo* L.) seed oil as an alternative feedstock for the production of biodiesel in Greece. *Biomass Bioenergy* 33:44–49

Scott PT, Pregelj L, Chen N, Hadler JS, Djordjevic MA, Grasshoff PM (2008) *Pongamia pinnata*: An untapped resource for the biofuels industry of the future. *Bioenergy Res* 1:2–11

Selim MYE (2009) Reducing the viscosity of jojoba methyl ester diesel fuel and effects on diesel engine performance and roughness. *Energy Conver Manag* 50:1781–1788

Sengupta A, Gupta MP (1970) Studies on the seed fat composition of Moringaceae family. *Fette Seifen Anstrichm* 72:6–10

Shang Q, Jiang W, Lu H, Liang B (2010) Properties of tung oil biodiesel and blends with diesel. *Bioresour Technol* 101:826–828

Sinha S, Agarwal AK, Garg S (2008) Biodiesel development from rice bran oil: Transesterification process optimization and fuel characterization. *Energy Conver Manag* 49:1248–1257

Srivastava PK, Verma M (2008) Methyl ester of karanja oil as an alternative renewable source energy. *Fuel* 87:1673–1677

Su EZ, Xu W-Q, Gao K-L, Zheng Y, Wei DZ (2007) Lipase-catalyzed in situ reactive extraction of oilseeds with short-chained alkyl acetates for fatty acid esters production. *J Mol Catal B* 48:28–32

Subramani S, Nagarajan G, Rao GLN (2008) High free fatty acid crude rice bran oil—A renewable feedstock for sustainable energy and environment. *Clean Soil Air Water* 36:835–839

Sumar AK, Konwer D, Bordoloi PK (2005) A comprehensive analysis of fuel properties of biodiesel from *Koroch* seed oil. *Energy Fuels* 19:656–657

Urioste D., Castro MBA, Biaggio FC, de Castro HF (2008) Synthesis of chromatographic standards and establishment of a method for the quantification of the fatty ester composition of biodiesel from babassu oil. *Quim Nova* 31:407–412

Usta N (2005a) Use of tobacco seed oil methyl ester in a turbocharged indirect injection diesel engine. *Biomass Bioenergy* 28:77–86

Usta N (2005b) An experimental study on performance and exhaust emissions of a diesel engine fueled with tobacco seed oil methyl ester. *Energy Conver Manag* 46:2373–2386

van den Abeele M (1942) L'Huile de palme: Matière première pour la préparation d'un carburant lourd utilisable dans les moteurs à combustion interne (Palm oil as raw material for the production of a heavy motor fuel). *Bull Agric Congo Belge* 33:3–90, *Chem Abstr* 38:2805 (1944)

Veljković VB, Lakievic SH, Stamenkovic OS, Todorovic ZB, Lazic ML (2006) Biodiesel production from tobacco (*Nicotiana tabacum* L.) seed oil with high content of free fatty acids. *Fuel* 85:2671–2675

Walton J (1938) The fuel possibilities of vegetable oils. *Gas Oil Power* 33:167–168, *Chem Abstr* 33:8336 (1939)

Warnigati S, Agra IB, Prasetyaningsih E (1992) Alcoholysis of nyamplung (*Callophyllum inophyllum* Linn) seeds oil to diesel oil-like fuel. *Renew Energy Technology Environment, Proceedings of the 2nd World Renewable Energy Congress* 3:1479–1483

Xu YX, Hanna MA (2009) Synthesis and characterization of hazelnut oil-based biodiesel. *Indust Crops Prod* 29:473–479

Yang FX, Su Y-Q, Li X-H, Zheng Q, Shi RC (2008) Studies on the preparation of biodiesel from *Zanthoxylum bungeanum* seed oil. *J Agric Food Chem* 56:7891–7896

Yang, F-X, SuY-Q, Li X-H, Zhang Q, Sun T-C (2009) Preparation of biodiesel from *Idesia polycarpa* var. vestita fruit oil. *Indust Crops Prod* 29:622–628

Yuste AJ, Dorado MP (2006) A neural network approach to simulate biodiesel production from waste olive oil. *Energy Fuels* 20:399–402

Zhang S, Zu Y-G, Fie Y-J, Luo M, Zhang D-Y, Efferth T (2010) Rapid microwave-assisted transesterification of yellow horn oil to biodiesel using a heteropolyacid solid catalyst. *Bioresour Technol* 101:931–936

Zullaikah S, Lai CC, Vali SR, Ju YH (2005) A two-step catalyzed process for the production of biodiesel from rice bran oil. *Bioresour Technol* 96:1889–1896

Index

Milton Keynes UK
Ingram Content Group UK Ltd.
UKHW050458071024
449327UK00015B/434